Machine Learning and Artificial Intelligence: Concepts, Algorithms, and Models

ISBN: 979-8-9921621-1-0

To my mother for her unconditional love and respect for science.

To my father for his energy, love, hope, and confidence.

To my lovely wife for her love, life, freedom, and happiness.

To my daughter for the love, happiness, and hope she brings to our life.

Acknowledgment

First of all, I would like to express my deepest gratitude to all those who, in one way or another, discouraged my pursuit of computer science, algorithms, mathematics, and other passions due to perceived limitations—whether it be my English skills, mathematical abilities, nationality, race, religious beliefs, or the schools I attended. To those who made decisions based on personal interests, university rankings, or biases, your actions have been a powerful motivator in my journey. This book is my answer to you, and it's damn satisfying.

I also wish to dedicate this book to those who defy limitations imposed by chance, man-made borders, talent, intelligence, institutional privilege, or societal norms. These are the individuals who follow their passions regardless of external voices. Many of my role models are individuals I've found through motivational videos and books, and I owe them a great deal of gratitude for their inspiration.

Having worked and lived in eight countries and four different U.S. states, I have learned that *"Science does not belong to anybody, and nobody can prevent us from learning or building new things"*.

I am especially grateful to those who have a personal connection with me: to God, for giving me the strength to undertake this endeavor; to my wonderful parents, Dr. Maryam Chitsaz and Dr. Akbar Rawassizadeh, for their unwavering support; to my loving and energetic wife, Dr. Tahereh (Zohre) Javaheri, whose curiosity and passion for knowledge drive me forward. She can read about everything except this book and our delightful daughter, Raha, who fills our lives with joy (even though she made sleep a rare commodity during her early years). Special thanks to my spiritual guide, Seyed Morteza Tababayi, who encouraged me to embark on this journey.

I am also grateful for the enriching academic environments I've experienced, from the Iranian culture that values education to the German academic system, which promotes fairness and does not rank or discriminate based on qualifications. Austria, with a similar system (at least on paper), shares these positive aspects. I am lucky to have worked with amazing colleagues and students at Boston University. Their intellectual challenges and insightful feedback have significantly shaped this work. I would especially like to thank my students for pointing out errors in my slides and writing during my courses in "Foundations of Machine Learning," "Web Mining and Data Analytics," "Advanced Programming," "Advanced Machine Learning and Neural Networks," "Generative AI," and "Deep Reinforcement Learning".

Additionally, I owe much to those who assisted with technical edits and ongoing translations. Nassir Zarrinpanah and Zohre (my wife) for their last-minute but critical corrections; Dr. Shahin Gheytanchi for reviewing early chapters; my former students who volunteered for the translation of this book into Chinese and Russian; and finally, Dr. Khatoon Khedri for her detailed technical edits on most of the chapters.

Lastly, I apologize in advance if any of the examples or comics in this book offend readers; they were not intended to do so.

Illustrations and Comics:

    Somayeh Yazdan-Panah,

    Shakiba Sadeghi,

    Beatrice Bassolli,

Cover page Illustration:

    Ali Khalkhali (comic), Shakiba Sadeghi (doodles)

About this book

Three revolutions shift our lives, mostly in a positive way: the agricultural revolution, the industrial revolution, and the digital revolution. Despite the countless benefits of these revolutions, some problems arise as well. For example, the Industrial Revolution led to the colonization of the world by a few European countries, and there were many wars around Europe, originating in Europeans' need to access land resources.

If you are about 40 and over years old, you are lucky to have observed life before and after the Internet. You observed before and after the third biggest human revolution, the digital revolution. The digital revolution introduced the "information age" with the existence of the Internet, and it is slowly reducing the gap between nations caused by the industrial revolution. The Internet seems to be the most successful product of the digital revolution. It has some drawbacks, such as wide access to porn content and addictive social media life, but its advantages are countless. One of its salient advantages is constructing a vast amount of digital data. On the other hand, technological advancements in hardware enable the analysis of a vast amount of data. For example, Genome sequencing is possible in a short amount of time, and related devices are getting cheaper and more affordable.

These changes lead us to live in the era of data science. Some scientists even call it the fourth paradigm [Hey '09]. Some state an industrial shift has emerged, i.e., Industry 4.0 [Kagermann '11], and this shift originated from several advances such as data science, which caused automation and removed the need for human labor in many industries. Despite its risks, its benefit seems more significant for humankind. The proliferation of digital data is emerging in several different disciplines, ranging from political science and medicine to architecture and art. Therefore, in the near future, learning to work with data might gain similar importance as computer literacy.

Since you are reading these lines, we assume you are already familiar with the importance of working with data, and we only wasted your time. Due to the availability of data and artificial intelligence systems, we are living in a fast pace of scientific growth; there are recent trends that try to provide education with simpler and faster content for learning. In this book, we aim for "simplicity" and "fast learning" to be two core elements. The foundation of data science is mathematics, but many of us have not studied mathematics in depth, or our traditional educational systems failed to make it engaging. Therefore, this book starts without diving into complex math. Instead, we gradually introduce more mathematical concepts as the reader progresses.

We believe the best way to learn data science is by starting with algorithms rather than focusing on mathematics. We simplify the use of each algorithm with simple examples. Then, slowly, as you progress through this book, we will add more mathematics. We also use humor and informal language, occasionally incorporating comics to make the narrative more engaging. Any resemblance between the characters in our comics and real people, organizations, or events is purely coincidental.

Our hope is that this book will make the knowledge of data science accessible to people around the world, particularly those who might not have access to quality education due to injustice, corruption, or dysfunctional educational systems.

Who should read or not read this book?

We assume you learned the basics of algebra and probability in high school and used simple data analysis tools such as spreadsheets, Microsoft Excel, Apple Numbers, etc. While programming experience might be helpful, it is not required. This book is primarily focused on concepts rather than coding, and we recommend using resources such as large language models to help with coding implementations. However, there is a GitHub page[1] that we built to maintain codes written for each chapter. There, you can see examples of algorithms and models being used, though we cannot guarantee their continuous maintenance.

This book is about concepts and algorithms and has a high-level view of a wide variety of concepts related to statistics, data science, machine learning, and artificial intelligence. It does not delve deep into theories, and in the earlier chapters, we avoid providing mathematical formalization and equations. As we progress to the book, we assume the reader becomes more knowledgeable and thus can understand mathematics better.

Here are a few questions to help you decide if this book is for you:

* I am a mathematician, and I want deep mathematical understanding and theoretical proofs. Is this book for me?

Unfortunately, no. This book explains concepts, algorithms, and models. In the early chapters, we do not delve deeply into mathematics, but we gradually increase the mathematical depth as the book progresses. We skip mathematical proofs unless it is necessary. For example, we explain the gradient and derivative in Chapter 8, and many readers learn them in high school.

* I want to be a good data scientist with hands-on experiments. At the end of reading this book, theory, and coding are both important to me. Is this book good for me?

Yes, while we do not provide extensive code samples in the book, you can easily find implementations using large language models or online resources to make a hello world code for most models and all algorithms. We have implemented lots of code to write this book, and we have shared them in the described Github link.

There are too many concepts we must learn, and experts in this area often complain that they feel that they fall behind and don't know what exactly is worth learning. We tried to cover most of the useful things that are required to be learned by a data scientist, artificial intelligence practitioner, and machine learning expert.

* You expect me to read about 1200 pages in US letter format without a single line of code? Why do I need to waste time learning concepts when I do not learn how to implement them?

---

[1] https://github.com/rezar/mlbook

This book focuses on science and not technology (implementation). We believe that a strong conceptual understanding is crucial for becoming a proficient data scientist or AI practitioner. With this foundation, you'll be better equipped to choose the right algorithms and innovate when faced with new challenges. As for implementation, nowadays, with Large Language Models and a vast amount of libraries available in Python, R, etc., there is no need to put the code inside the book. You can easily use these resources to generate example code or find implementations online. This approach allows us to focus on timeless concepts rather than specific implementations that may become outdated. Nevertheless, you can find code examples for the concepts covered in this book in the GitHub link provided.

* My background is in art, humanitarian sciences, political sciences, biology, carpeting, health sciences, construction work, plumbing, etc. I have zero software, mathematics, or computer science skills, but I like to learn about data science and artificial intelligence. Is this book good for me?

Yes, it is useful, and while writing this book, we have also considered a large audience with backgrounds like yours. There is a wide gap between mathematicians and other people using data science and artificial intelligence. This book tries to bridge this gap.

I am an experienced programmer/software developer, and I want to move into data science, machine learning, and artificial intelligence, is this book good for me?

Yes, this is the right choice because you do not have a problem searching online or using large language models for implementations of the algorithms and concepts and using them to learn coding it as well.

Nowadays, Who has time to read >1000 pages of text? I can watch online videos and learn everything from the best lecturers. Besides, I can use these AI tools and ask them to write the code. Why do I need to spend my time reading this book?

We agree that most of the concepts that we covered in this book are explained online with plenty of examples and implementations. However, this book offers an advantage: it provides a comprehensive, structured overview of the entire field, carefully curated. While working on data, you do not work with one single algorithm; you will try many different possibilities until you get your desired answer. If you get a task to perform clustering, perhaps ten years ago, you experimented only with k-means, but now you need to report the result in many different clustering algorithms and see which one works better, plus a detailed hyperparameter tuning. This book covers a vast number of algorithms and concepts that help you make the right decision.

Furthermore, any problem could be resolved in different ways. Some information retrieval methods and data structures seem unrelated to machine learning, such as sliding windows, Bloom Filter, SkipList, and MinHashing, but they are extremely useful methods for developing your own algorithm or using it in your machine learning pipeline. There are very few resources available to list and describe them in a single place. We have tried our best to list everything that we believe is important in one single book.

This book builds structural thinking for you. This structural thinking approach enables you to see connections between different areas of data science and AI, fostering creativity and innovation in your problem-solving. We have published many good scientific papers, consulted several small companies, experienced working in the enterprise software industry, and built a few real-world applications. Therefore, we are very confident about what we recommend here.

I do not have enough time or resources, and I want to learn only one topic. Is this book good for learning this topic?

Yes, but check the "Good News Potato" messages at the beginning of each section. It specifies minimum pre-requirements for reading a chapter. Then, you need to read pre-required chapters and the chapter you think has the information you need.

Earlier, you claimed this book is suitable for any specific area that needs to learn data science. I am an alienologist, and I have not found any practical examples of UFOs. Why?

Algorithms and concepts are the same. If you like to use the data in your field, you need to have domain knowledge to adapt the existing algorithms to your needs. There are many good books, especially written about data science in health, finance, biology, etc. However, they use the same algorithms and not different algorithms. For example, linear regression does not change between different fields; we need the domain knowledge to prepare our data for the linear regression.

Table of Content

xiv

Part ii: Unsupervised Learning

xvi

Part iii: Supervised Learning

xxii

xxvi

xxxii

Part i: Preliminary Concepts

We start this book with basic concepts that we need to be familiar with for learning machine learning, artificial intelligence, data mining, and anything related to data science.

In the first part, we focus on describing basic concepts, different types of data, and how to evaluate machine learning algorithms.

Next, we discuss visualization methods briefly. We recommend never undermining the impact visualization; if possible, start by visualizing the collected data at the beginning; it gives us very useful first insight into the data.

This book is not intended to go deep into mathematical proofs, but we must know basic statistical and algebra concepts. Therefore, in the third chapter of this part, we explain the statistics required for machine learning. In later chapters, we also introduce the mathematics required to understand that chapter. You might find some concepts boring or too much to learn. However, our experience studying many models and algorithms shows that it is crucial to get a very good understanding of statistics, visualizations, and other basic concepts required to understand complex models.

Chapter 1: Basic Concepts

This chapter focuses on concepts we need to learn to start the exciting journey of artificial intelligence, machine learning, and data science. We aim to keep all of these chapters separate and independent from each other. Unless you are an experienced data scientist, do not skip this chapter. By the end of this chapter, you should have gained a good basic understanding of different data science branches, including machine learning, data mining, and artificial intelligence.

Ultra Short History of AI and Machine Learning

It is estimated that the first mechanical machines that operated intelligently were introduced by Heron of Alexandria in ancient Greece, specifically in Hellenistic Egypt, which was a wind power. Later, the Banu Musa brothers, based in Baghdad (which retains the same name today), ruled by Abbasid Caliphate, proposed devices that performed computing in the 9th century, which are the most basic machines that store programs and perform tasks automatically [Hill '79]. In the same century, another Persian mathematician, Muhammad ibn Musa al-Khwarizm, known as al-Khwarizmi made significant contributions to algebra and algorithms, which led to the introduction of the concept of Algorithm. His birthplace, Khwarazm, is in present-day Uzbekistan and Turkmenistan.

Intelligent device development continued from that point and was applied in both warfare and the arts. However, significant advancements were not made until the 18th century, during the Industrial Revolution. The modern history of theoretical computer science can be traced back to German scientist Leibnitz, who made significant contributions to the theoretical foundations of computing, including his work on the binary system in 1686 [Wiener '48]. About 100 years later, an English scientist, Thomas Bayes, introduced the Bayes theorem [Bayes '63], which was revealed two years after his death in 1763. Later, in 1837, Charles Babbage, another English scientist, introduced a mechanical computer, named an "Analytical Engine," that could do more than calculation and execute algorithms. In particular, The Analytical Engine was a mechanical machine that was designed to solve mathematical equations. Ada Lovelace wrote notes on Babbage's paper, which included what many consider to be the first algorithm intended for processing by a machine [Fuegi '03]. However, the construction of this machine was not completed during their lifetime.

Afterward, in 1931, Kurt Gödel, an Austro-Hungarian mathematician, proposed foundational principles in mathematical logic, including his incompleteness theorems. Because of World War II, the development of electronic computers marked a significant milestone in the history of intelligent devices. The ENIAC, developed in the 1940s, was one of the first fully operational electronic digital computers. In 1950, Alan Turing proposed a "learning machine" that could learn and eventually became the definition of artificial intelligence.

This period also saw the development of programming languages. In 1956, the term "artificial intelligence" was coined at a Conference at Dartmouth College (located in New Hampshire, US) to discuss simulating human intelligence.

The initial enthusiasm for AI in the 1950s and 1960s led to high expectations, but progress in AI did not meet these expectations. In the 1970s, funding for AI research decreased in developed countries, and many researchers moved away from the field. This is known as the first AI winter. On the other hand, in the 1970s, the rise of personal computers (PCs) like the Apple and IBM PC makes computing accessible to the masses in the world. The ARPANET evolves into the Internet, fostering global communication and information sharing. The 1980s saw the development of knowledge representation and expert systems for capturing human knowledge. This led to an increase in attention to AI.

The second AI winter occurred in the late 1980s and early 1990s. Similar to the first AI winter, again, overblown expectations, unfulfilled promises, and a lack of significant breakthroughs led to a loss of confidence in AI as a field of study.

Later, in the 1990s and early 2000s, AI progressed until the breakthrough of neural networks in 2012, which leveraged advances in computing powers and led to the introduction of deep learning.

What is Data Science, Machine Learning, Data Mining, and Artificial Intelligence?

Data mining is the process of extracting (a.k.a mining) knowledge from the data. For instance, given a sequence of words in the following order: "{run, eat, smoke, happy, heart-attack, run, drink, run, smoke, happy, eat, heart-attack}", through data mining, we can extract a piece of knowledge as follows: the word "smoke" is followed by the word "happy". However, in the real-world, knowledge associations are usually more complex and not as easily identifiable by predefined human rules. For instance, we encounter "heart-attack" after "smoke" too, but that is not as easily recognizable as the previous example. This is where computers could assist us in recognizing these patterns. In real-world problems, data is much more complex and changing rapidly. Therefore, instead of defining "rules" to extract knowledge from data, we use machine learning.

Machine learning is a technique or method in which patterns and knowledge are extracted from data. Mitchell [Mitchell '97] states that machine learning is the study of *"how to construct computer programs that automatically improve with experience"*. Géron [Géron '19] defines machine learning as a *"science of programming computers so they learn from data"*.

Therefore, we can say that data mining is more focused on the applications and applying machine learning methods to the data [Provost '13].

Systems, algorithms, and models that mimic a human cognitive capability, such as seeing or answering questions, are known as *Artificial Intelligence (AI)*. Another definition of AI was proposed by Russell and Norvig [Russel '09], and they categorize existing definitions of AI into one or more of the following categories: *Thinking Rationally, Acting Rationally, Thinking Humanly, Acting Humanly*.

Data Science is the umbrella term that covers anything associated with data analysis.

We can use data for three purposes: (i) to describe or diagnose a phenomenon, (ii) to predict events or changes based on the available data, and (iii) to create a system that leverages the data to mimic the cognitive capability of humans.

Describe or diagnose a phenomenon: We use machine learning methods to find a characteristic in the data that describes or diagnoses a phenomenon, e.g., *"smoking leads to happiness"*. Or discovering patterns that exist but we are unaware of it, e.g., after "eat" comes "heart attack". The algorithms and machine learning models that are used for describing or diagnosing a phenomenon are usually "classification," "regression," or "clustering" algorithms. Another example is using the model to analyze pathological images and identify cancerous cells in microscopic images.

Predict events or changes: The second use of data is usually referred to as "prediction", i.e., the use of the existing data to describe what is happening in the future. Prediction means what is happening in the future, e.g., if (smoke), then (heart-attack) is going to happen. Recommendation systems are a type of machine learning application that uses prediction as well, e.g., after "run", recommend "eat" happens. Therefore, immediately after the user is finished running, a recommender system of a food corporation starts to recommend eating junk food, and the recommendation system makes more money for the food corporation, not the poor user who ruins their workout.

Mimicking Human Cognitive Capabilities: This is the most recent use of data due to advances in deep neural network models. For example, we ask a large language model questions about our headaches, which mimic a physician, or finding a cat in a picture, understanding a handwritten text. These systems are referred to as *artificial intelligence (AI)*.

These terms can be overlapped and used in the mix. For example, when we started this book back in 2017, the terms data mining and machine learning were popular; when we ended the first version of this book in 2024, the terms data science and artificial intelligence were more popular.

Why is data science important?

Unlike living creatures, traditional computers were not inherently capable of learning and adapting; they operated based on fixed instructions (algorithms). They operated based on a given set of instructions (a.k.a. algorithm) to do a specific job. Machine learning is the science of enabling computers to *learn* and *adapt* their reaction [Chapmann '17]. In simple terms, instead of manually describing instructions for a machine, these algorithms or models enable the computer code to identify instructions based on the data we feed to it. Therefore, computers are able to learn by themselves.

Any scientific phenomenon should be repeatable and reproducible. For example, if a fire burns a piece of paper, if we doubt it, we can experiment with it by putting a piece of paper in the fire.

This is a scientific phenomenon. If we told you that we saw in our dream that an angel came from heaven and told you to pay us, this is not a scientific phenomenon because it is not repeatable and reproducible. Data science and all its fields, similar to any scientific discipline, operate based on the repeatability and reproducibility of data.

What is an algorithm?

We start with the definition of the algorithm. The naming algorithm is a Latinized name of a Persian mathematician, "Abdullah Mohammad Ibn Musa Al-Khawrazmi". He is one of the pioneers who created algebraic methods and a systematic method to study the linear or quadratic equation [Knuth '97] and ruined our lives with algebra and mathematics. In simple terms, an algorithm is a *series of instructions* that perform a *task*.

All computer-related techniques are operated as algorithms. We give one or more *inputs* into the algorithm, system, or machine and it produces one or more *outputs*.

Check Figure 1-1 and assume we have a bunch of lovely chickens, they go into the imaginary machine, Happiness. The machine receives chickens as input and produces bowls of crispy and tender chicken nuggets as output. Mathematically, a function can be written as: $f(x) = y$. Here, x is the input of function f and y is the output. In our example, we can do the same and write: $HAPPINESS(chickens) = nuggets$. Clearly, a system is composed of more than one algorithm. For instance, several different algorithms could be running inside a Happiness machine. Some could work in a sequence (the output of an algorithm will be the input of the following algorithm), or some can work in parallel. As we progress through the book, we learn more about these possibilities.

Figure 1-1: Happiness machine receives chickens as input and delivers chicken nugget as output.

Evaluating a Machine Learning Algorithm

Let's assume we want to buy a Happiness machine to feed it with chickens and get back chicken nuggets. The Happiness machine is an algorithm that receives chickens as input and delivers chicken nuggets as output. Different vendors are creating these types of machines as well. It is important that the machine we purchase does not consume too much electricity (efficiency) does not give out anything other than chicken nuggets (accuracy). How do we evaluate which machine learning algorithm is better than others that do the same thing?

Each machine learning algorithm has two core aspects: *efficiency* and *accuracy*. Accuracy refers to the quality of the machine learning algorithm or model output. Efficiency is the cost with several elements to be analyzed, including training time, model size, scalability, energy usage, memory use, GPU use, CPU use, etc. However, computational complexity is usually a good indicator of the algorithm's efficiency unless the application and its features also require a detailed analysis of the other costs. For example, an algorithm run on a mobile device must be resource-efficient and avoid draining the battery quickly.

Computational Complexity and Efficiency

The efficiency of an algorithm is strongly associated with *computational complexity*. Before discussing computational complexity, let's review some simple algebraic relationships. X and Y are in a relationship and living happily together. Assume n is a number to represent their relationship. If $X = nY$, then we can say X is growing "linearly". If $X = Y^n$, then we can say X is growing "exponentially". See Figure 1-2, where X has been plotted with respect to Y in two different scenarios.

Each algorithm is associated with two types of complexities (*space complexity* and *time complexity*), which are known as computational complexities. Referring to the "efficiency"

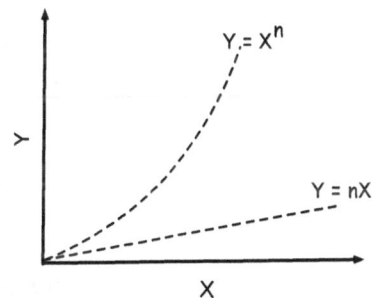

Figure 1-2: Linear and exponential growth examples.

of the algorithm traditionally refers to reporting the computational complexity. However, after the introduction of neural networks, due to their intensive complexity, these factors are not common to report.

The time complexity affects the latency (response time), and space complexity affects the memory, network utilization, and storage (e.g., disk space). It is common to report only time complexity, but in applications, especially those involving large datasets or limited memory environments, space complexity is also a critical factor as well.

The time complexity of an algorithm is calculated based on the number of operations the algorithm performs. Assume we have an algorithm; as input, it receives an ant or an elephant. Since the elephant's size is 1,000,000 times larger than the ant, if an algorithm takes one operation and one second to produce output for the ant and 1000,000 seconds but again only one operation for the elephant (see below Figure 1-3), we can say the algorithm has linear computational complexity. In other words, the number of operations increases linearly with the input size. We can have another algorithm that also receives an ant or elephant as input. If we fed one ant as input, and the output costs one operation, but it takes 10,000,000,000 ($100,000^2$) operations for the elephant (as input), we say the algorithm has *quadratic complexity*. This means that the number of operations grows proportionally to the square of the size of the input. In other words, for an input of size n, the number of operations is proportional to 2^n. On the other hand, if an algorithm requires approximately $2^{100,000}$ operations for an input of size 100,000, we say the algorithm has *exponential complexity*. This means that the number of operations grows exponentially with respect to the input size.

The complexity of an algorithm is shown with a big O, and the "number of operations" it requires to perform, such as $O(n), O(n \log n), O(n^2)$, etc. This is also called "asymptotic notation" or "Big-Oh notation".

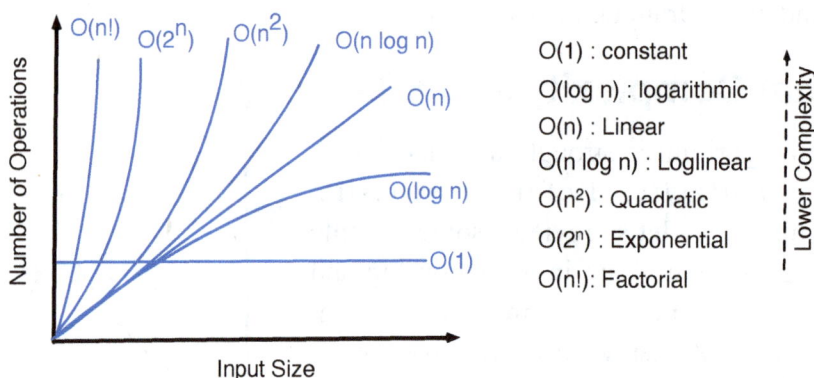

Figure 1-3: Common computational complexities and their approximate growth pace.

Figure 1-3 shows the different growth rates of the algorithm based on the number of operations. Note that depending on the size of the n these shapes could change, but we present an approximation to give the reader a sense of complexity. If an algorithm has $O(1)$ computational complexity, it is called constant complexity. If it has $O(n)$, $O(n . \log (n))$, or $O(\log n)$ it has a linear complexity and the rest of complexities, including $O(n^2)$ is referred to as polynomial complexity, $O(2^n)$ is referred to as exponential complexity, and $O(n!)$ as exponential complexity[1].

Let's get back to our Happiness machine to understand better how the complexity is calculated and assigned. Assume we have n chickens; if, with only n number of operations, our n number of chickens have been converted into chicken nuggets, the complexity will be linear $O(n)$. If each chicken requires m different operations to convert into chicken nuggets, then we will have $m \times n$ operations. Therefore, the computational complexity will be $O(m \times n)$. If it is not clear whether

---

[1] If you do not recall what is factorial, $n! = (n - 1) \times (n - 2) \times \ldots \times 2 \times 1$

8

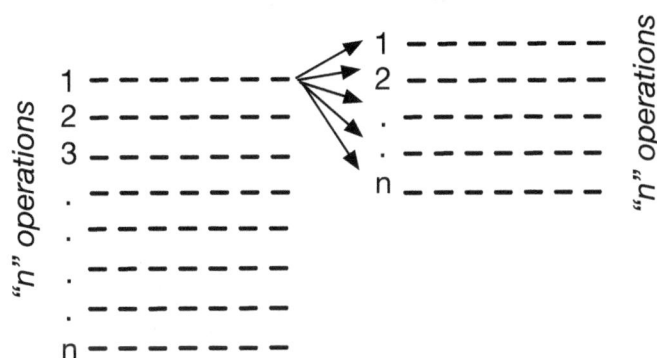

Figure 1-4: An example of n^2 number of operations.

m is larger or smaller than n, usually we assume the complexity is $O(n \cdot log(n))$. In other words, when the complexity is not linear but smaller than quadratic, we call it Log-linear.

Figure 1-4 illustrates n^2 operations. In this figure, each line is one operation that an algorithm performs, and it demonstrates how n^2 operations will be implemented.

In addition to the computational complexity, to report the efficiency of the algorithm, we can report on hardware resource utilization, including CPU usage, memory usage, response time (or latency), energy use, scalability of the algorithm, etc. If the algorithm runs remotely on a network, we could report about network bandwidth utilization as well. An efficient algorithm should use the least amount of hardware resources wisely and favor lower computational complexity. Nevertheless, many algorithms, especially deep neural network algorithms, are computationally complex and very resource-intensive, and reporting their computational complexity is too complex. Instead, their number of parameters, training time, and other information will be reported.

You might think that understanding resource utilization is not essential for your work and that they are too technical, but they are very important. Assume your boss gives you a day to prepare a plan for him so that he can convert his underperforming crappy company into a revolutionary AI enterprise. The algorithm you are planning to use could take several days to run. They gave you a stone-age laptop at work and expect you to run an algorithm and predict the market changes by tomorrow morning. You should take into account the computational complexity and resources available when deciding about an algorithm. This example shows that any data scientist should take into account algorithm cost (computational complexity and resource efficiency) to decide wisely about the algorithm used for an application.

Algorithms Runtime

In terms of running the algorithm, there are two types of algorithms: the ones that process the data in *batch* mode and the ones that process the data *online* (or *incremental*). Batch algorithms take longer to run and use the entire dataset to learn from data. Incremental or online algorithms use the most recent data that has been arrived at to update the existing model that has been previously constructed. For example, credit card fraud detection should operate in real-time, but predicting the market in the next month could be done in a batch mode. More about models will be explained shortly. Online algorithms do not run precisely in real-time, but they are faster than batch algorithms.

Accuracy

All machine learning algorithms make decisions based on patterns they learn from data, such as predicting what will happen when new data arrives in a system. Even when an algorithm focuses on diagnosing or simulating human behavior, it is associated with a decision that this decision is associated with a type of quality. In other words, we expect a machine learning algorithm to provide precise and accurate output. Machines' predictions are prone to wrong predictions but are significantly more accurate than those of human fortune tellers, politicians, and bankers. Usually, (not always) algorithms do not lie as much as humans do. However, in the recent decade, humans have been successful in sharing this attitude with algorithms and also building algorithms that can produce fake data. We will see algorithms that generate fake data (e.g., synthetic human faces) in Chapter 11.

The most important element to assess the quality of a machine learning algorithm output is the *accuracy* of the predictions. *Accuracy means how "precise" an algorithm can solve our problem.* Accuracy is not about efficiency; it is about the *quality* of the algorithm's output. In other words, it measures the *correctness of the output*.

We can say that accuracy is a representation of error in the context of machine learning. The error can be referred to as the differences between actual (what is happening in reality) and predicted (what the machine estimates about reality) values:

$$Error = Actual - Predicted$$

Looking back at our Happiness algorithm, we should be sure that we receive the chicken nuggets as output and that anything else, instead of a chicken nugget, is an error. For instance, if a Happiness algorithm returns 90% chicken nuggets and 10% cats (which is very odd), then the *precision* of our Happiness algorithm is 90%.

Usually, an algorithm receives different parameters from the user along with the input dataset. These parameters that are configured by the user are referred to as *hyperparameters*. One challenge we usually encounter is identifying the best values of hyperparameters to get the highest possible accuracy. For instance, we test different combinations of buttons on each Happiness machine to identify which one works the best for the target dataset of chickens to produce the least number of cats and the highest number of chicken nuggets.

For example, the Happiness algorithm receives chickens and delivers chicken nuggets. You know that this world is not perfect. The input dataset (chickens) we give to the algorithm (Happiness machine) and the output (chicken nuggets) are not perfect. There will be a level of error, i.e., chicken not delivered as a chicken nugget, and we use accuracy-related metrics to measure errors.

There are types of information used to estimate the accuracy of a machine learning algorithm: *True Positive, True Negative, False Positive, and False Negative.*

True Positive (TP) is the number of input chickens that are correctly converted into chicken nuggets (output). True Negative (TN) is the number of other input objects (not chickens), but they are correctly not converted into a chicken nugget. False Positive (FP) is the number of other input items (not chickens) incorrectly converted into chicken nuggets. False Negative (FN) is the number of chickens that are incorrectly converted into something else instead of a chicken nugget. Figure 1-5 presents the results of these four important mapping of input to output. Try to memorize these four states; for measuring the accuracy of the machine learning algorithm, we need to understand these four states perfectly fine.

A *confusion matrix* is used to present the described variables for the accuracy of an algorithm. Imagine we have 111 inputs (should be chickens but are not necessarily 100% chicken) that will be fed into the Happiness machine. Table 1-1 presents the confusion matrix of these inputs. The top horizontal line is being used to present output or predicted values. The vertical line on the left side is being used to present input that is given to the system. By wrong cases, we mean any other input rather than chicken, such as cat, elephant, human, etc. By wrong output, we mean an output that is not a chicken nugget.

True-Positive:

True-Negative:

False-Positive:

False-Negative:

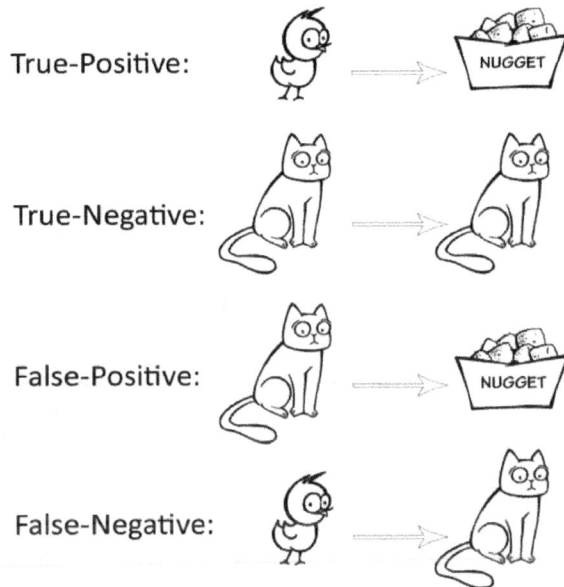

Figure 1-5: Different mappings of the input to output in the Happiness algorithm.

		Output	
n = 111		Wrong Output	Correct Output
Input	Wrong Input	TN = 12	FP = 10
	Correct Input	FN = 9	TP = 80

Table 1-1: Confusion matrix of feeding 111 inputs into the Happiness machine.

Different formulas are being used to measure accuracy based on the four described definitions (TP, TN, FP, and FN). Usually, the important ones are *precision* (or *positive predicted values*), *recall* (or *sensitivity*), *true negative rate* (or *specificity*), *F-score* (or *F-Measure*), and *Accuracy*. Their equation is written as follows:

$$Precision = \frac{TP}{TP + FP}$$

$$Recall = \frac{TP}{TP + FN}$$

$$True\ Negative\ Rate = \frac{TN}{TN + FP}$$

$$F-score = 2 \times \frac{Precision.Recall}{Precision + Recall}$$

$$Accuracy = \frac{TP + TN}{TP + TN + FP + FN}$$

These equations are extremely important to memorize them … kidding … whenever you need to analyze the accuracy of an algorithm, open your browser and search them online, or purchase a paper version of this book and open this page. Also, there is an angel in heaven who is praying for you if you pay the money for this book and help the author to get rich.

The Balance between Accuracy and Efficiency

Based on the knowledge we intend to extract from the data or the type of application. sometimes efficiency is important, and accuracy is less critical; sometimes accuracy is essential, and efficiency is not that important.

For instance, imagine you own a factory that produces different products. You would like to predict market sales for each of your products. You are using a machine learning algorithm to increase your revenue. The market fluctuation is high, and you need to analyze the market every hour. Thus, you need a fast algorithm to run every hour and produce results in a reasonable time. In this scenario, accuracy might not be the most important thing because people love your products, and you will sell all of them in the end. Your current focus is to sell more in a shorter amount of time.

In another scenario, assume you are a biologist and are one step away from curing cancer with your revolutionary medication. You develop a pill that changes something in the human genome and cures obesity. However, this drug is only used for patients who are obese. Otherwise, if a patient does not have obesity and consumes your drug, he or she might be converted into a hamster. In this scenario, the algorithm's accuracy in detecting whether a patient is obese is very important to avoid converting non-obese humans into hamsters.

Now, we understand how to measure an algorithm's (i) efficiency and (ii) accuracy. Likewise, we know that any algorithm has a computational complexity. No algorithm is perfect in all aspects. It is our job to find a balance between resource efficiency and accuracy.

NOTES:

* While discussing computational complexity, some report it as the algorithm runtime, which is not the execution or response time but the number of operations. Nevertheless, the number of operations usually correlates with the execution or response time.

* Usually, space complexity is not as important as time complexity. In most scientific literature, it is important to report time complexity. Still, unless the algorithm deals with a large amount of data (big data), it is expected that the authors of an algorithm neglect to report its space complexity.

* The notion of big-Oh and its number of operations is being used to report the worst-case scenario. In real-world settings, the algorithm is not necessarily performed that badly, but we report the worst-case scenario.

* $O(n!)$ algorithm exists, and some of the biggest challenges in computer science are very slow algorithms with $O(n!)$, e.g., travel salesman problem [Bhargava '16], has n factorial (!) operation.

Types of Input Dataset

Obviously, we should feed digital data and not chickens into an algorithm. In this section, we discuss what an algorithm gets as input. We feed a *dataset* into a machine learning algorithm. A dataset is a set of *data points* (a.k.a *data objects*, *observations*, *samples*), and it could have different formats, such as a single number (scalar), vector, images, sound, audio, different textual formats, or textual data with a notion of time, i.e., sequential, temporal, and timestamped data [Han '11]. Furthermore, along with the dataset, an algorithm receives some configurational values known as hyperparameters. We have explained earlier that hyperparameters configure how to execute the algorithm, e.g., the maximum number of iterations allowed by the algorithm.

Usually, the input dataset should get converted into numbers (even text, images, and sounds) because algorithms are designed to work with numbers, and computers can deal with numbers more easily than other data types. For example, an image could be represented as a series of pixels; each pixel can define a set of numbers, including the X and Y coordinates of the pixel and a number for its color.

Different types of datasets can be fed into a machine learning algorithm; we list some of the common ones in the following;

Tabular Dataset

In traditional machine learning tools, the input format is usually provided in a table format. Most of the time, tabular data are presented as Comma Separated Values (CSV) or Tab Separated Values (TSV).

Table 1-2: A sample table and name of its properties.

Table 1-2 presents a sample table (tabular data) and its components. In a tabular dataset, a table row is called a *record* or *feature vector*, a column is called an *attribute* or *feature*, and each value is referred to as a *feature value* or *field*.

In addition to CSV format, there are other data formats available, but rarely for machine learning algorithms. For instance, XML or JSON will be used when the number of fields is different in each record, and we need to have a nested hierarchical data structure. Besides, many other types of databases exist, such as spatio-temporal datasets, relational databases, and document databases. but we will skip describing them here, and they are out of the context of machine learning concepts.

Temporal Datasets

A dataset with a notion of time (either an implicit notion of time is there or explicitly a timestamp is presented) is called a *temporal dataset*. There are common categories of temporal datasets, including *sequential*, *timestamped*, and *time series*. Sequential datasets include an implicit notion of time, and data points are stored in the order, The following shows two examples of sequential records:

- *fried meat, eat, butter, more fried meat, eat, heart attack*

- *wake up, breakfast, commute, work, lunch, work, coffee, work*

The term '*heart attack*' occurs after '*more fried meat*' and '*eat*', but there is no information about the exact time of each event. We call this dataset sequential because its data has an implicit notion of time. Nevertheless, it does not have a precise time stamp, and the time notion of each data point is defined based on its position along with other data points. If you are interested in these data, James Allen [Allen '83] has introduced reasoning algebra, which can be used to deal with sequential and event occurrences.

Another type of temporal data is *timestamped* data, which has an explicit notion of time. An example of two records that have timestamped data is presented as follows:

- Monday 10th Nov 2017, 12:42, start eating the lunch, body_temperature= 37°c

- Monday 10th Nov 2017, 13:42, stop eating the lunch, body_temperature= 39°c

In a timestamped dataset, each record has an explicit notion of time. Usually, the number of columns in a dataset is equal, or if a record does not have a value for each column, we fill the dataset with null or zero values. This process is known as imputation and will be described in Chapter 16.

Time series refers to a dataset in which we have one variable, which is collected on a specific interval of time, such as a heart rate signal collected every minute. The interval does not change and is constant, e.g., every minute, every second, etc. Besides, the time interval does not have a timestamp. The following, presents an example of cross-entropy

{time:1, heart-rate: 120}, {time:2, heart-rate: 124}, {time:3, heart-rate: 118}

Any data that collects one single variable in the context of time could be considered as time series. Even a system could collect multiple time series. For example, fitness trackers use an accelerometer, and an accelerometer creates a time series in each X, Y, and Z dimension.

Data Streams

Many data sources continuously produce data, and it is not cost-effective or feasible for a system to store the data. Examples include sensor networks, financial markets, and social media platforms. The algorithm can use a *window* to access the data in these cases. Besides, it is impossible to know the dataset's begin and end. By using a window, we can gain access to a small segment of data. This is different from a database (a database is the same as a dataset). In a database, the data points are all static and accessible. A window allows systems to work with a manageable subset of data at a time. This can be a time-based window (e.g., the last 10 minutes of data) or a window based on the number of data points (e.g., the last 1000 records). We will learn more about windowing in Chapter 5.

Figure 1-6 shows that there are chickens that are running, assume them as data points, and the cat (algorithm) can access them only from the window in a short amount of time. Usually, in the software engineering world, a database is represented as a cylinder because it refers to the old notion of a magnetic disk used to store data. Sometimes, we also use a cylinder to present a dataset in this book. Note the difference between a dataset, which is a collection of data, often viewed as a single entity for analysis. A database, on the other hand, is a structured system for storing, managing, and retrieving data, and it can contain multiple datasets. In this book, we deal with the dataset and not databases.

Graph Datasets

A graph is a set of objects (nodes or vertices) and their relation, defined by edges. From a mathematical perspective, a graph is a collection of *edges* and *nodes*, as shown in Figure 1-7.

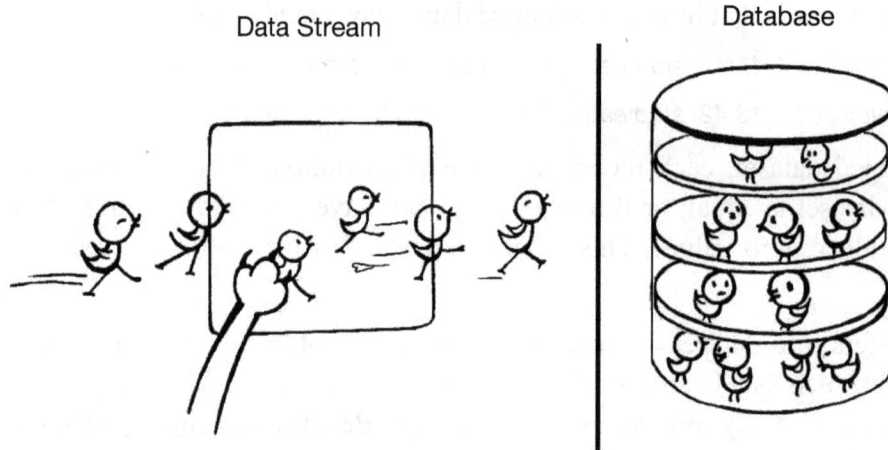

Figure 1-6: Data stream on the left and traditional database on the right. A window that is being used to access the stream data (running chickens).

Vertices in this figure are human, chicken, cat, and dog. Edges represent a relationship between vertices. In this figure, the edges describe the relationship, i.e., pet, eat, etc.

Graph databases were used when relationships between data objects were *many-to-many*. If you are familiar with databases, we use relational databases and convert relations between data entities into one-to-one or many-to-one. If we intend to keep the many-to-many relationships between data, we could use a graph.

Figure 1-7 shows a graph that includes a human, cat, dog, chicken and their relations. Graph databases are rarely used in machine learning, but graph algorithms have lots of algorithms, and we dedicate Chapter 15 to graph algorithms.

Due to the importance of the relationship between data points, some scientific disciplines, such as genomics or molecular structure, require data to be presented in graphs. Mining knowledge from graphs is a valuable solution to answer some of the most important questions of humankind, such as analyzing a social media network to understand "who is the rich person in my connections that posts a new picture of their expensive travels" to make their

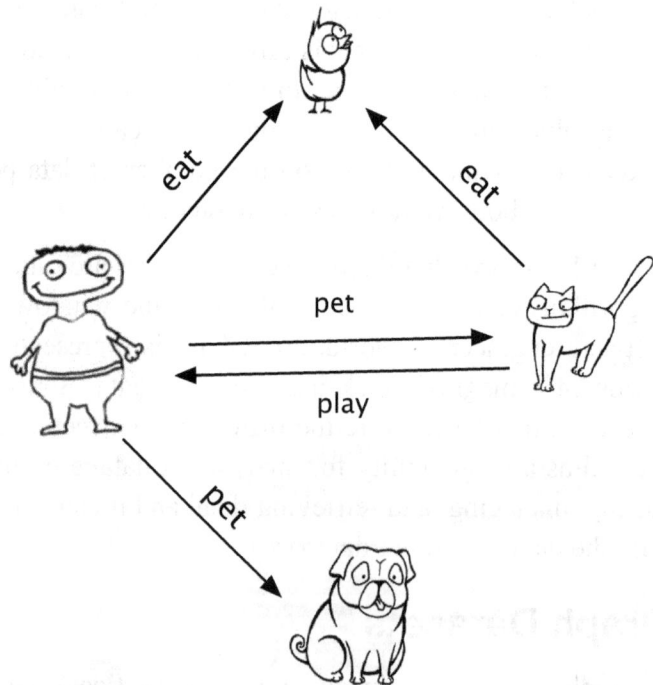

Figure 1-7: A simple graph example.

social media connections jealous of them. Also, it has some other less important applications, such as protein structure, biological network analysis, and chemical structure modeling, that are used to cure diseases. Some other applications of graphs are analyzing the relations between web links and road network modeling.

Text

A large part of the machine learning and artificial intelligence community focuses on working with unstructured data in text format, including Natural Language Processing (NLP), text mining, textual information retrieval, and Language Models.

All these disciplines focus on analyzing textual data in an unstructured format, such as HTML files collected from the web, social media posts, and digital text documents. Structure formats are treated as tabular data.

Text data could be used in structured, tabular, or unstructured formats. In a structured format, we can store text data in tabular format. While structured formats are often treated as tabular data, it's worth noting that structured text data can also exist in other forms, such as XML or JSON, which are structured but not tabular.

In an unstructured form, we can use HTML page content. The large language models, such as ChatGPT[2], are trained on the textual data available on the Internet.

There are many applications for text data analysis, such as opinion mining and sentiment analysis, conversational agent construction, recommendation systems, knowledge extraction from medical transcripts, etc. We will describe text analysis in several chapters, including Chapter 6, Chapter 10, and Chapter 12.

Image and Video

Another very popular application of machine learning and artificial intelligence is image and video data. Both are considered unstructured data, and advances in deep neural networks greatly extend the variety of their applications, from medical image analysis to social media filtering and text-to-image conversion models.

Images are converted into a tensor (we will describe tensor shortly). We describe several concepts, algorithms, and models related to image analysis in Chapter 6, Chapter 10, Chapter 11, and Chapter 12. Video is also another medium that could be fed into a machine learning algorithm. Some algorithms and models are particularly designed to deal with videos, but since working with images is easier, video data frames are usually extracted and treated as a series of images for analysis. Video data is a type of unstructured data that contains a sequence of images or frames, each of which has a width, a height, and a number of channels (such as RGB). If audio is included, it adds another complex layer to the data. Video analysis applications range from surveillance and sports analysis to video content recommendation and autonomous vehicles.

---

[2] https://chat.openai.com

Audio

Another type of data that can benefit from machine learning is audio. Several audio-related applications employ machine learning and artificial intelligence, such as synthetic voice construction, denoising audio, music generation, sound classification, etc.

One key focus in audio data analysis is speech recognition, where machine learning algorithms are trained to understand and transcribe human speech, a technology that is fundamental to the development of voice-activated assistants and automated transcription services. Another significant area is sound classification and event detection, where algorithms are used to identify specific sounds or events within an audio stream. This has applications in areas ranging from wildlife monitoring to human coughing sound classification. Synthetic voice construction and voice cloning are other applications of artificial intelligence and machine learning in audio.

Usually, audio is treated as a signal, and we will learn more about the signal process in Chapter 6 and Chapter 7.

Numerical Data Types

If, similar to the author, you have a passive-aggressive relationship with mathematics, we have bad news. The core of machine learning and many other scientific disciplines is mathematics. Nevertheless, the good news is that we tried our best to describe them in a way that even readers without a mathematical background understand.

We have explained that computers are adept at working with numbers, which can be represented in various formats and dimensions. In terms of formats, numbers are commonly represented as integers (whole numbers) or floating-point numbers (which include decimals). As for dimensions, numbers can be represented as scalars (single values), vectors (arrays of numbers), and even more complex forms like matrices and tensors for multidimensional data.

We have four different numerical dimensions that can be fed into a machine learning algorithm: *scalar*, *vector*, *matrix*, and *tensor* (see Figure 1-8).

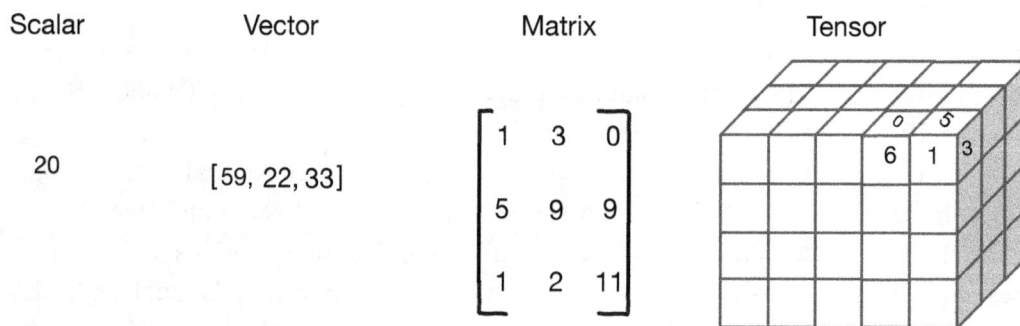

Figure 1-8: Numerical data structures that we feed into a machine learning algorithm.

Scalar is a single number representing the simplest form of data. For example, our Happiness algorithm gets a scalar value, which indicates the crunchiness level. It means it will get a number to understand how crunchy a chicken nugget should be. A scalar is usually presented with a Greek lowercase letter, e.g., $\lambda, \theta, \beta, \ldots$

Vector is a series (or array) of numbers (or scalars). In computer science, a vector is a one-dimensional array of numbers, and It can represent a series of values, like a list of numbers.

If you are familiar with coding, you might think of an array. An array can have a text input, but a vector is always numbers. For instance, the Happiness machine receives two information objects, an array that includes chicken names, i.e., *{Mr. Foo, Mrs. Bar, Dr. Null}* and a scalar value of '3', which indicates chicken should be fed with food from box number 3.

A Vector could be presented with large or small letters, e.g., a vector *v*, which has *n* members, is written as: $v = (x_1, x_2, \ldots x_n)$. Its members (scalar values) are usually presented with small English letters with an index as a subscript. In most of the text in this book, we write vectors with small letters, but there might be places where we use capital letters for vectors to be sure our formalization follows the original work formalization.

Matrix is a two-dimensional array of numbers (vector of numbers). For instance, the Happiness machine requires the ratio of weight to height of each chicken, which is a number. Therefore, we can use a matrix to feed this input into the Happiness algorithm. A matrix has height and width. If the height of a matrix is three and its width is two, we say it has three rows and two columns. We present a Matrix as $M_{3\times2}$ (3 rows, 2 columns). A matrix is usually presented with a capital letter, e.g., matrix *A,* which has *"n"* rows and *"m"* column, will be written as follows:

$$A_{n\times m} = \begin{bmatrix} x_{11} & x_{12} & \cdots & x_{1m} \\ \cdot & \cdot & \cdot & \cdot \\ x_{n1} & x_{n2} & \cdots & x_{nm} \end{bmatrix}$$

A **Tensor** can be considered an n-dimensional array, extending the concept of a matrix to higher dimensions. In other words, it is a generalization of a matrix in higher dimensions. You can imagine a tensor as a sequence of matrices with equal height and width. For instance, the tensor we show in Figure 1-8 comprises three matrices. The tensor that is shown on the right side is presented as $\mathcal{A}_{4\times5\times3}$ (height × width × depth). Each of the indices presents one dimension of the $\mathcal{A}$ tensor. A tensor is usually presented with a calligraphic capital letter or just a capital letter.

A basic example of Tensor is the image. Another good example of a more than three-dimensional tensor is a video file. A video file can be presented as a sequence of images. Therefore, each image has a pixel that has a color. Each pixel of a picture itself could be presented with a number that specifies red, green, and blue ratio and X, Y coordinates. Therefore, even a single picture is presented as three-dimensional tensor.

Additionally, considering the changes in each pixel based on time (because it is a video) could present the 4th dimension of data. In other words, we will have color, coordinates, and time. We will describe tensors in more detail in Chapter 7.

NOTES:

* An algorithm normally has more than one input parameter; we usually feed a dataset and a set of parameters into an algorithm. Parameters are usually in the form of numbers (scalar or vectors) but could be text, such as the name of the distance metrics as well.

* While working with a neural network, even vector and matrix are considered tensors; a vector is called a one-dimensional; a matrix is two-dimensional, and more than two-dimensional data is referred to as a tensor. If you encounter a person who says vector is a rank-1 tensor, try not to get angry; those are people who are involved in working with tensors. We will explain these names later in Chapter 7.

* All machine learning algorithms (and all scientific disciplines) operate based on one important principle: *the data will be repeated,* and there is a "finite" set of values for each data point. In mathematical calculations, numbers do not have limits and move toward infinity. However, in the context of machine learning, the dataset is limited to a set of possibilities, even a stream of datasets.

* There are algorithms that receive different file formats, but tabular files such as CSV and TSV are the most common ones. Some machine learning tools that can be used on different data formats, such as XML or JSON. Also, there are tools that connect the algorithm directly to a relational database and get data from there. Nevertheless, you do not need to worry about them. At the time of writing this book, they are used very rarely. Recent products are moving toward separating software engineering details from machine learning.

Tasks that Machine Learning Perform

A machine learning algorithm or model receives input data, processes it, and provides an output. The Happiness machine (algorithm) gets chickens as input and outputs chicken nuggets. Each category of machine learning algorithms receives a dataset as an input and outputs either the same dataset with some additional information (annotated dataset) or some knowledge related to the input dataset. The output format could be in the form of the original dataset with additional information or vector, tensor, matrix, image, audio, text, etc.

Traditionally, machine learning algorithms are categorized as *supervised learning, unsupervised learning,* and *reinforcement learning.* In the recent decade, a new category of neural network algorithms, called *self-supervised learning,* has been introduced. They are still unsupervised learning algorithms, but due to their wide variety, they are classified as another category of algorithms. Figure 1-9 summarizes different directions in machine learning, for now, this is enough, but the rest of this book expands and explains this figure.

In the following, we briefly describe detailed categories of machine learning algorithms to understand which algorithm provides which type of output.

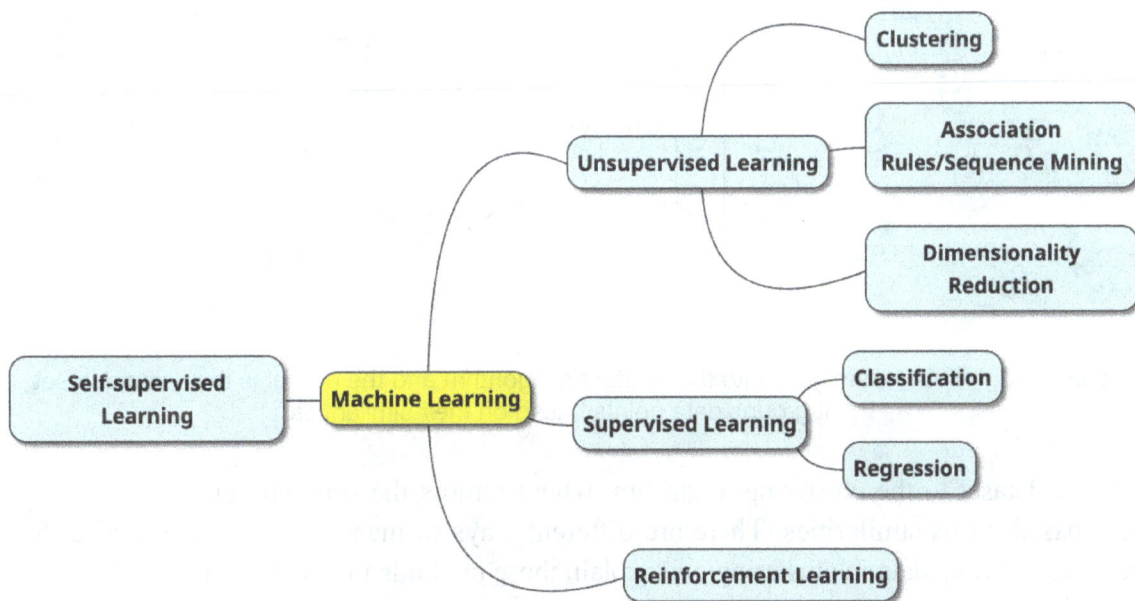

Figure 1-9: An overall taxonomy of machine learning algorithms.

Unsupervised Learning

Unsupervised learning algorithms are used to analyze a given dataset with no human intervention. In particular, we give the dataset to the algorithm, and it extracts patterns or makes smaller groups of the given dataset without the need for human intervention. Due to their lack of need for human intervention, they are the easiest types of machine learning algorithms. Unsupervised algorithms are categorized into clustering, association rules and sequence mining, dimensionality reduction, and anomaly detection. Each item is briefly described here to be sure we will get familiar with them, but later in the book, each will be explored in detail.

Clustering

Clustering refers to a group of algorithms in which a user gives metrics for measuring distances between the data points, and based on using those metrics, the clustering groups similar data points together. Assume we feed a dataset of animals into a clustering algorithm (Figure 1-10). The clustering algorithm identifies three groups of similar animals and groups them together.

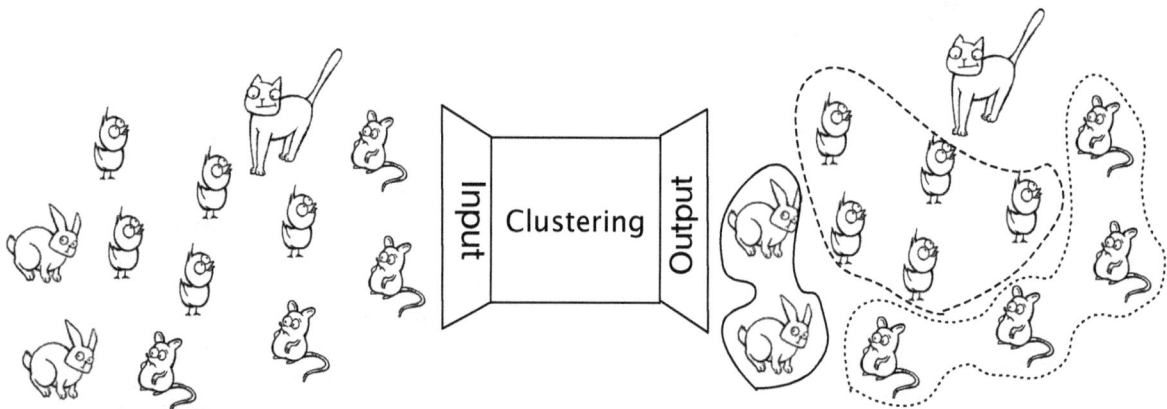

Figure 1-10: Input dataset goes into the clustering algorithm and the output is the same dataset, but it groups data points based on their similarities.

We give a dataset to the clustering algorithm, which returns the original dataset but groups its content based on its similarities. There are different ways to measure similarities and different approaches to group data in clustering; we explain these methods in detail in Chapter 4.

Association Rule and Sequence Mining

In some applications, we must identify events that occur together (co-occurrence identification). For example, check if event X happens Y also will happen or not. In other words, we intend to find if there is an association between X and Y. Association rule mining algorithms are to discover interesting relations between variables in the given dataset. They aim to identify patterns, associations, correlations, or causal structures among sets of items. For instance, it can determine if the occurrence of event X in a dataset implies the occurrence of event Y. This is often expressed as 'if-then' rules, like '*if X, then Y*'.

To understand these algorithms, assume we try to concentrate on a task from 10:00 to 10:40 and put away our smartphone; instead of doing our task every 10 minutes, we go to the window and take a look outside. Figure 1-11 shows the animals we see outside at different times. We are curious to know which animals usually appear together. There could be exceptions, but we are looking for a generalized rule. As Figure 1-11 shows, chickens usually (but not always) appear with cats. We can say that when we encounter a chicken, then with high probability, a cat will appear as well. The output of these algorithms is sets that indicate co-occurrences. For example, if we feed the data of Figure 1-11 into an association rule mining algorithm, the output will be something such as: *{chicken, cat, confidence=60%}*. This means that 60% of the time, a chicken and a cat appear together.

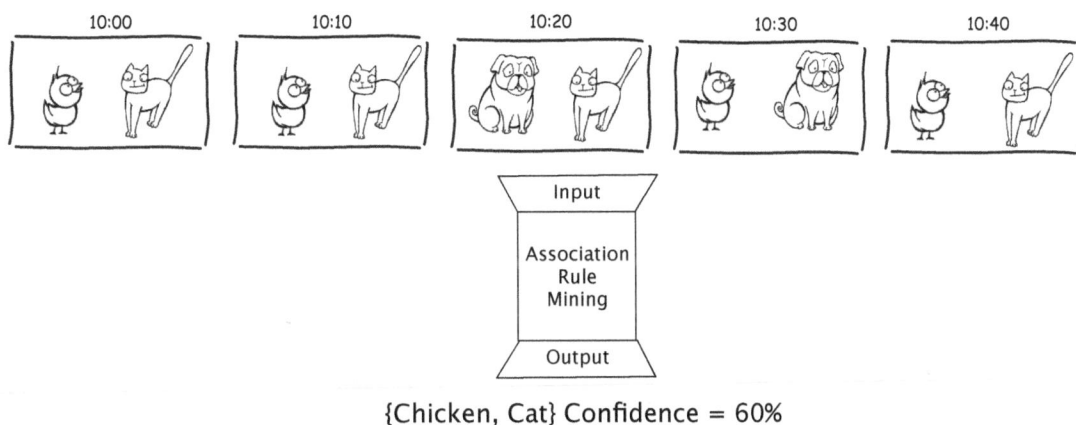

Figure 1-11: Co-occurrences of different animals at different time.

When working with datasets that have an implicit notion of time or order, traditional association rule mining may be insufficient because it does not take the order of the data into account. Association rule mining is typically concerned with the co-occurrence of items within a dataset, regardless of their order. However, in many scenarios, the sequence in which events or items occur is crucial for understanding patterns and relationships. Sequential mining algorithms are used to analyze co-occurrences with respect to the order of data. We explain classic algorithms for association rule and sequence mining in Chapter 5 and the neural network approaches (RNNs) in Chapter 10.

Dimensionality Reduction

While using machine learning algorithms, dealing with high-dimensional data often leads to complexities and inefficiencies, a problem known as the 'curse of dimensionality'. This can result in overfitting (we will explain it in Chapter 8), increased computational costs, and difficulties in visualizing data. Dimensionality reduction methods address these challenges by simplifying the data structure. They reduce the number of variables under consideration by projecting them in a lower dimensional space. This results in enhancing model efficiency, simplifying data visualization, and helping to reveal hidden patterns within the data, with less computational cost and without significant loss of important information

Deviation and Anomaly Detection

One important challenge while working with real data is to identify changes that affect the dataset quality or identify data points that are significantly different from other data objects. These data points are called *anomalies* or *outliers*.

Outlier or anomaly refers to one or more data points in a dataset that are significantly different or distant from other data objects. Usually, we discard those data points, but in some cases, there is a reason behind an anomaly, and it is interesting for us to identify that reason. In these cases, an anomaly could be called a *surprise*. For example, Figure 1-12 has one mouse as an anomaly, among other data objects (chickens). It could be discarded while analyzing the data, but on the other hand reveals a mouse existence among chickens as well.

Figure 1-12: An anomaly inside a dataset, which is marked in a red circle.

In some cases, the deviated data points (e.g., outliers) do not include anything meaningful; in that case, we call them noise and usually identify them to remove them from the original dataset. Anomaly (outlier) detection algorithms, change point detection algorithms, noise reduction, and many other algorithms have been developed for these use cases. However, they do not fall under the category of unsupervised learning, and some of them belong to supervised learning methods. We explain more about these types of algorithms in Chapter 16.

Supervised Learning

Supervised learning relies on humans to label the data. Then, the machine learning algorithm uses the human-assigned labels to build a model of the data. It is the most popular group of machine learning algorithms.

To understand how supervised learning works, look at Figure 1-13. There, we assign some labels to the data objects (three chickens and two mice). We give the dataset and labels to the machine learning algorithm, which assigns the rest of the labels, as shown in this figure. However, due to the lack of labels, we can see mistakes in machine generated labels as well. For example, the only cat is labeled as a mouse.

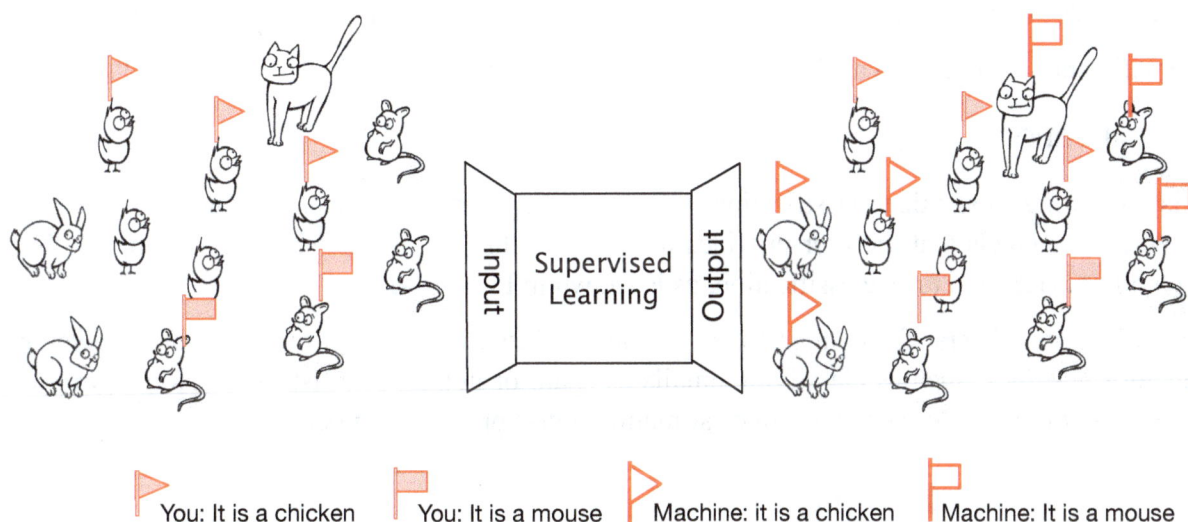

You: It is a chicken   You: It is a mouse   Machine: it is a chicken   Machine: It is a mouse

Figure 1-13: A classification example. The user provides two labels and the machine labels the rest of data which are not labeled.

While there is a diverse range of unsupervised learning methods available, it is observed that a significant portion of machine learning applications and algorithms utilize supervised learning. This is because supervised learning, relying on labeled datasets, often provides a more straightforward pathway to train models for specific tasks, including classification or regression.

Regression and Correlation

Regression, a type of supervised learning algorithm, is utilized for predicting continuous outcomes by establishing a mathematical relationship between a dependent variable (the target) and one or more independent variables (the predictors). The goal of regression is to create a model (mathematical equation) that can take input data and accurately predict the output, often in the form of a numeric value. For example, a regression can let us know how much playing video games affects the mood. It is not just a number instead it is a mathematical equation, e.g. $video - game = 1.3 \times mood - score + 0.4$.

In simpler terms, regression is applied to predict the values of a dependent variable (output) based on the values of independent variables (inputs). This contrasts with *correlation*, which quantifies the strength and direction of the relationship between two variables. *While regression results in an equation defining a relationship, correlation computes a single scalar*, known as the correlation coefficient, that indicates how closely two variables are related. This coefficient, calculated using methods like Pearson, Kendall, or Spearman (See Chapter 3), is a measure of association rather than a predictive model.

For example, we use correlation analysis to understand how much body weight is related to exercise sessions. Correlation takes two vectors as input (representing the two variables) and outputs a scalar indicating their relationship's strength, unlike regression, which produces an equation. Thus, regression's output is an equation describing a predictive relationship, whereas correlation's output is a scalar value indicating the degree of association between variables.

We will explain correlation analysis in Chapter 3 and regression algorithms in Chapter 8.

Classification

It is not wrong to say that classification is the most popular use of machine learning. Learning based on the labels that the user provides to the algorithm to distinguish different objects from each other is referred to as classification, as is shown in Figure 1-13.

Classification differs from regression because classification algorithms output discrete, categorical values, such as classifying emails as spam or not. In contrast, regression algorithms output *continuous numeric* values, like estimating house prices or temperatures.

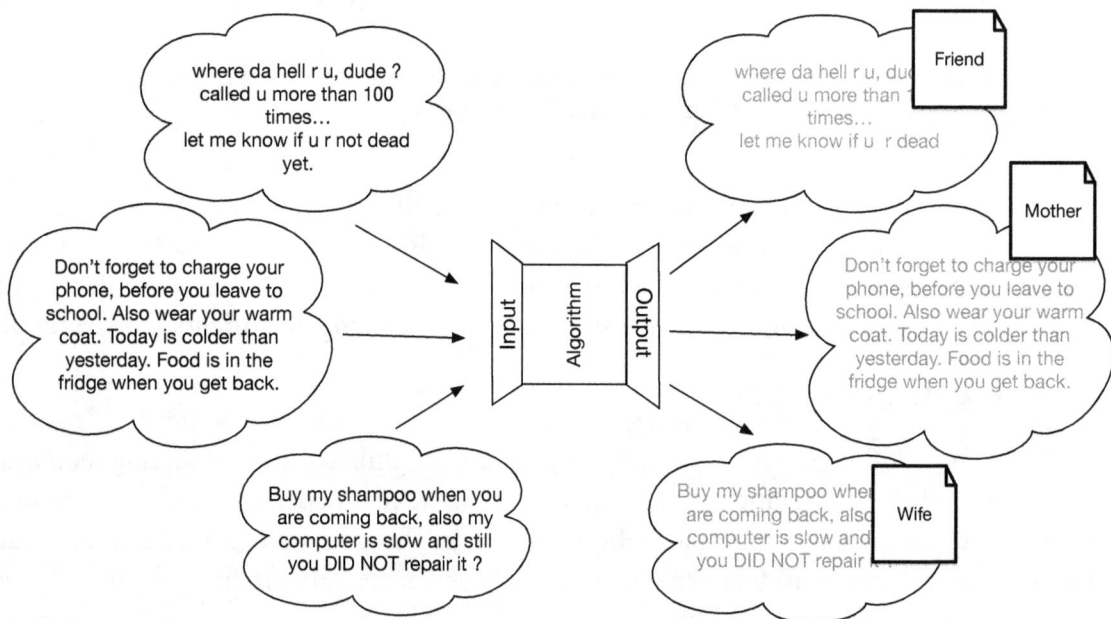

Figure 1-14: A classification example of identifying who is talking.

An analogy to understand classification is how a child learns to differentiate between a mouse and a rabbit. A child will learn the difference between a mouse and a rabbit by observing several examples and learn to distinguish them without explicit rule-based instructions. We show examples of mice and rabbits to the child and do not define rules for the child to learn, e.g., we don't say if an ear is longer than 5 centimeters, then it belongs to the rabbit. The child's brain itself identifies rules.

As another example, assume we have an algorithm that can feed different voices from different conversations between individuals into this algorithm. Based on the style and content of the conversation, it recognizes who the person is (see Figure 1-14). This task seems uncommon, but it could be done with machine learning if we provide enough audio data about each person. This is an example of classification.

In general, the classification algorithm requires some *prior knowledge* to decide about labeling, and this prior knowledge will be given to the algorithm by labeling. For instance, we gave an algorithm 100 other text and labeled the subject of the text. Then, the algorithm identifies the subject of the newly given by leveraging previous data. We explain classification algorithms in Chapter 9 and continue using them in later chapters.

Self-Supervised Learning

Self-supervised learning is another group of machine learning algorithms where the model learns to understand data by using a part of the data itself as the label. Essentially, it creates its own supervised learning problem from unlabeled data. For example, a self-supervised learning model might be given a sentence with a missing word and learn to predict the missing word based on the context provided by the rest of the sentence.

This approach enables the model to learn representations of the data without the need for explicitly labeled datasets, making it particularly useful for tasks where labeled data is scarce or expensive to obtain. It's commonly used in areas like natural language processing and computer vision. We will learn about these methods in Chapter 11.

Generative AI

While not exclusively a form of self-supervised learning, Generative AI often leverages principles akin to self-supervised techniques. Recent advancements in neural networks have led to the development of models capable of generating synthetic data, such as text, images, music, and even videos, that closely resemble human-created content. At the time of writing, these neural network models were so sophisticated that they could produce artificial data, like human faces, that were indistinguishable from real ones[3]. Look at Figure 1-15. These faces, while convincingly realistic, do not correspond to any real individuals; they are entirely machine-generated creations.

---

[3] Images are generated using https://thispersondoesnotexist.com web page, which uses StyleGan 2 and we will explain it in Chapter 11.

Figure 1-15: Artificial faces generated by a generative AI model.

Beyond images, these generative AI models are also adept at creating text, audio, and even complex structures like molecular compositions [Topol '19], demonstrating remarkable versatility and realism in their outputs. We explain more about generative AI models in Chapter 11.

Reinforcement Learning

Reinforcement learning represents a branch of machine learning algorithms centered around the concept of an *agent* interacting with its *environment*. In this setup, the agent makes decisions and

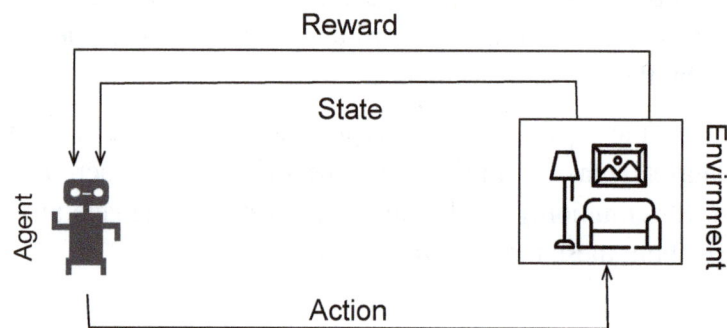

Figure 1-16: Basic concept of reinforcement Learning

takes *actions*, which may be based on prior knowledge or exploration of the environment. Following each action, the agent receives feedback in the form of *rewards* (a negative reward is a resemblance of a penalty) based on the outcome of its actions.

This feedback guides the agent in learning and adjusting its future actions to maximize rewards over time. Reinforcement learning is characterized by this cycle of action, feedback, and learning, enabling the agent to improve its performance in complex, often dynamic environments. The agent learns based on experimenting (trial and error).

Figures 1-16 visualize how a very basic reinforcement learning algorithm operates. For example, a robot (agent) orders equipment inside a living room (environment). Reinforcement learning

28

algorithms were used in robot control, computer playing games against humans, and recommendations, but less in other fields until the emergence of ChatGPT in November 2022, which uses reinforcement learning in a part of its architecture to provide the best possible answers. We will explain reinforcement algorithms in Chapter 13 of this book and more about the use of reinforcement learning in large language models in Chapter 12.

NOTES:

* Most machine learning algorithms are complex, and the algorithm user should not care about the underlying logic of the algorithm. However, there are some specific algorithms that the algorithm should be explainable to the user, such as medical data analysis. These algorithms are referred to as *explainable* algorithms. Some other algorithms, such as deep neural networks, are too complex and impossible for us as humans to see how the algorithm makes inferences from the data. These are known as *black box* algorithms.

* While classifying data, we can take two approaches: *inductive learning* and *transduction learning*. In inductive learning, *the resulting model is trained on a labeled training dataset and then applied to make predictions on unseen test data*. Its goal is to generalize from the training data to any data from the same distribution. Transductive learning *focuses on labeling specifically the given, observed (unlabeled) data points, but it's not necessarily about training on both labeled and unlabeled data*. In transductive learning, the model is trained using both labeled and unlabeled data, and the goal is to predict the labels for the unlabeled data points in the same dataset.

Training and Evaluation in Supervised Learning

As we have explained, a large part of machine learning works are focused on supervised learning models. To evaluate the output of a supervised learning model, we should separate a dataset into *test* and *train,* or *test*, *train*, and *validation*. In this section, we describe this style of evaluation. When we get the labeled data, we separate it into two (train/test) or three sections (train/test/validation), as it is shown in the upper part of Figure 1-17.

Then, we run the supervised machine learning algorithm on the train data. The result will be a model. We use the test data to evaluate the accuracy of the model. The train data is used for building the model,

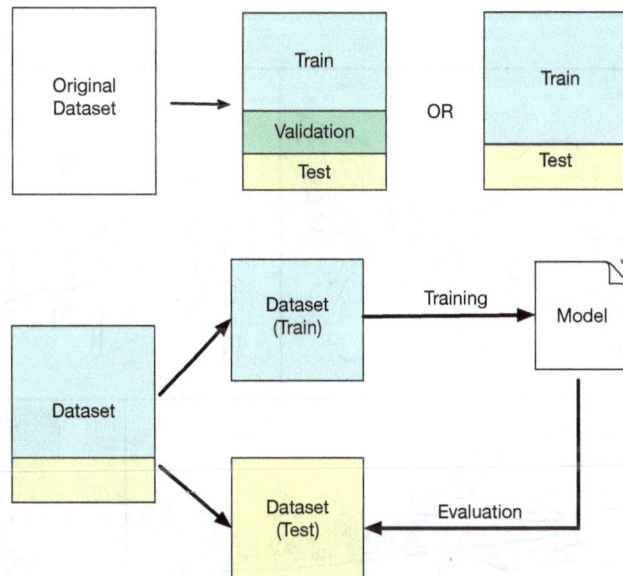

Figure 1-17: (Top) Different separations of the dataset for experimenting with a supervised machine learning algorithm. (Bottom) Building a model in a supervised learning algorithm.

as shown in Figure 1-17, and the test data is not used to construct the model.

Sometimes, the model is built with several iterations of training and testing. In this scenario, we separate the training data into Train and Validation. Then, we test the model of trained data on the Validation dataset and improve the model. Afterward, when we believe the model is ready, we will test it on the Test dataset. In other words, the Validation dataset acts as a proxy for the test set, allowing us to make improvements without "touching" the test data, which should be used only for the final evaluation.

There are three other concepts that are important to remember while dealing with the training and evaluation of supervised learning, including *ground truth dataset*, *cross-fold validation,* and *hold-out validation.*

Ground Truth Dataset

While reading about machine learning, we often encounter the *term ground truth or gold standard* dataset. The ground truth dataset is a dataset that is annotated manually by expert humans, and we assume the annotations are ideally correct. To measure the accuracy of an algorithm we will compare its output to the ground truth dataset.

For example, assume you are a rich kid and don't have anything to do except enjoy your lavish life and post pictures on social media. Nevertheless, each fun activity gets boring quickly, and you are suffering from a dilemma of not having continuous happiness in your life. You hire an event manager to plan your fun activities in advance. His job is to plan your fun packages. He is also a good data scientist, and he is trying to create an algorithm to create the best fun package for you. To have a ground truth dataset for evaluating his algorithm, he asks you about your previous activities, and you assign each one a score based on how much fun you had in them.

Let's say the list of your fun activities is as follows:

1. Going to your dad's private island with your personal yacht and enjoying the privacy there (fun level: 6/10).

2. Flying to a luxury hotel in the middle of Sin City with a private pool and spa in your suite, take as many pictures as possible, and share them on your social media (fun level: 5/10).

3. Gathering your friends at your home, partying, drinking, and smoking until you pass away (fun level: 2/10).

4. Driving to meet up with other rich kids and show off your latest cars and clothes (fun level: 3/10).

5. Going to a charity and helping prepare food for people who are in need of healthy food (fun level: 8/10).

This is a ground truth dataset because the data is manually labeled by an expert. Next time, when your event manager is willing to develop an algorithm to plan a package of fun activities for you, he can use this ground-truth dataset to evaluate the algorithm result. The algorithm can mix these activities to make a fun package, e.g., one month: 1,3,5, another month: 2,4,5, etc.

k-fold Cross Validation

k-fold cross validation is used to ensure that the accuracy of applying a machine learning algorithm on a dataset is generalizable. It divides the dataset into k numbers of equal-sized subsets (or folds). In each iteration of cross validation, one of the k subsets is used as the test set, and the remaining $k - 1$ subsets are combined to form the train set. The model is trained on the train sets and evaluated on the single test set. This process is repeated k times, with each of the k subsets used exactly once as the test set. The results from the k iterations are then averaged to measure the accuracy of the final model. Figure 1-18 visualizes a 5-fold cross validation.

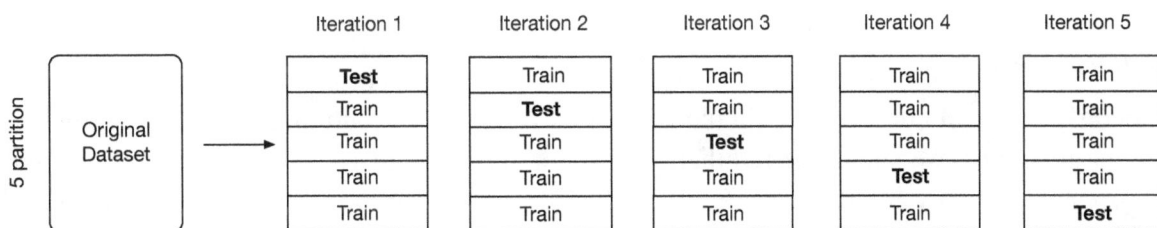

Figure 1-18: 5-fold cross validation example. We can see that the dataset is segmented into 5 partitions and each time one partition will be used as test set, and the rest will be evaluated with respect to the test set.

The test set is removed from the dataset, and we use the training dataset as input for our algorithm. Then, we acquire a result. We test the validity of this result on the test dataset and see whether it is in line with the result we got from the training dataset.

k-fold cross-validation partitions the dataset into k sub-dataset and repeats this process *k* times. Figure 1-18 illustrates how k-fold cross validation works. You can observe from this figure that each time, a different partition will be the test dataset, and other partitions will be the training partitions. Results acquired from this method ensure that it is accurate because it frequently changes the test dataset, and thus, there will be no impact from outliers.

What is the rationale for doing this and not just doing it once? The answer is that using different subsets of the data for validation provides a better insight into how well the model performs on unseen data, hence offering a more realistic picture of the model's ability to generalize. Also, it ensures every data point is used for both training and testing, thus maximizing the use of available data.

If we partition data only once into test and train sets and perform the validation, this is called *hold-out validation* or *1-fold cross validation*. Usually, two-thirds of a dataset is used for training, and one-third is used for testing.

Working with Data Step by Step

Sometimes, the data is available online, and we use it for our application. In some cases, the data is unavailable to collect, which is the most time consuming task in the context of data science. Here, we describe our recommendation for performing data analysis and executing machine learning. Others might recommend different steps, but the process flow is usually similar among different recommendations.

(1) Defining the question and collecting data

In real-world projects, before starting to feed our data into machine learning algorithms or creating our model, the first step is to define the question. We should clarify the objectives of our studies and what we need to know. What are we looking for? For example, do we need to analyze the data to increase profits? Do we need to identify bottlenecks in a system? Do we need to find the hidden patterns that impact the overall performance of a system? do we want to train our robot to walk? etc.

To identify the right question, we should be familiar with the domain of study. For example, if it is about genomics, we should be familiar with genome data analysis or have an expert available with domain expertise in our team. Based on the question(s) we intend to answer, either we will conduct experiments and collect new data, or we will use an existing dataset available online and use it completely or as part of our needs.

(2) Explore the data with visualization and statistics

The second step involves exploring the dataset we receive to gain insights. At this stage, we might realize the need for more data, and we need to either collect different types of data or increase the size of our dataset for the study by using synthetic data. Therefore, we will go back to the previous step to collect more data, perhaps data with higher quality or even synthetic data.

Two approaches can be helpful to gain an overall understanding of the dataset: (i) visualization and (ii) statistical analysis. Chapter 2 describes visualization, and Chapter 3 describes statistics in a bit more detail.

Visualization offers an easy approach to starting with data and describing the characteristics of data on a very high level, especially when we have a large amount of raw data with a small number of dimensions. While doing visualization, we can identify potential correlations among the data with visualizations. Besides, we can identify and remove outliers. In Chapter 16, we describe outlier removal in detail.

In addition to visualization, statistical analysis is also important to conduct at this stage. As an example, assume you are a data scientist and work with a third-party data provider. The provider who has a dataset believes that their data is top secret. Thus, they only give you 100 records from one million to extract knowledge from them.

You think we could use mathematical arguments to demonstrate the need for more data. In this case, descriptive statistics can help. You plot the distribution of your 100 sample data points. They probably do not follow any standard distribution (we will learn them in Chapter 3). You use these pieces of evidence to argue that you can not make any valid judgments without sufficient data. This might convince them to give you more data until, statistically, you can justify the results from analyzing the dataset. In summary, you should have enough data to be able to generalize the results of your statistical analysis to the new upcoming data as well. We say the sample data is *biased* if the sampled data are not generalizable.

One major challenge in data science is to collect enough data. Data often originates from heterogeneous sources like web pages or relational databases, requiring software programming to collect and transform it into a homogeneous format for analysis. Software engineers are equally essential as machine learning experts in a data science or AI modeling project.

(3) Data cleaning and wrangling

Assume you read all of this book, and now you are one of the best data scientists or AI engineers in the world. Now, you go and get hired in a big AI enterprise. Your employer asks you to analyze a dataset. That seems easy; you will get a cleaned CSV file, feed it into an algorithm, and by the end of the day, the report is ready. Unfortunately, this is an illusion. The most time-consuming part of working with data involves cleaning the dataset, removing noise and outliers, and transforming data format into a format that you can use for the machine learning algorithm. This is another part that requires some software engineering skills.

The process of removing outliers, missing data, and noise is referred to as *data cleaning*. Data cleaning, however, is not limited to noise and outlier removal; we might have an abundant amount of missing data that cannot be simply removed. We need to reconstruct the missing data, which is called *imputation,* and keep them in the dataset. We will discuss imputations and outlier detections more in Chapter 16.

The process of transforming data from its raw format into a format appropriate for the target algorithm is called *data wrangling* or *data munging*. There are standard textual data formats described to mitigate the effort of data transformation, such as XML, JSON, RDBMS (Relational

Database Management System), and CSV. The conditions are required for images, audio, numerical data, and other data types as well. Nevertheless, different applications provide different output formats, and thus, we (or software engineers) should implement a lot of data wrangling and cleaning to get data into an appropriate format understandable by the target algorithm. Usually, de-facto standard machine learning tools such as R and Python handle these issues very well. Again, software development is an essential need at this stage that we can not neglect.

Most of the time, we need to convert the data type to prepare them for the machine learning algorithm. Most machine learning algorithms are designed to work with *numerical data*, and some can handle *categorical data* (numbers and alphabets) as well. We often need to normalize numerical values into a specific range (e.g., 0 to 1) or convert the dataset's data type to match the input type accepted by the target algorithm. More about normalization will be described in Chapter 3 and Chapter 6.

There is bad news: about 80% of a project's time is spent on data collection and cleaning, while only 20% is used to apply an existing machine learning algorithm to the data.

(4) Preparing Data for the Algorithm

Now, we have clean data that is ready for the algorithm. Usually (not always), other steps must be taken to prepare the data for the machine learning algorithm, including feature engineering and changing data dimensions.

Note that this step and the next step (running the machine learning algorithm) will be done iteratively. It means that we select some features, create the input data, and experiment with the algorithm. Then, either manually or by using another algorithm (automatically), we can check the quality of results (e.g., the accuracy); if the results are not satisfactory or optimal, we return and test again with a different selection of data. This process continues until we find the best possible combination of inputs for the target algorithm.

Feature Engineering: After we have enough data prepared in the correct format for the algorithm, we usually need to select the information (features) to be used. Furthermore, sometimes, we need to alter the data representation as well, such as transforming high-dimensional data into a lower-dimensional space.

We might be able to feed the entire dataset to the algorithm and check the result. However, feeding all information is associated with several challenges, including resource usage or distorting the algorithm's accuracy by keeping unrelated features in the dataset. Therefore, feature engineering is used to select a related set of features that provide the best possible result.

Feature engineering involves *feature selection, feature generation,* and *feature extraction.* Feature selection is the process of selecting a subset of the features that provide the most relevant results. For example, while using an algorithm to predict weight changes of chickens in aviculture, we do not need to feed the names of chickens into the algorithm. Feature generation combines features, usually with mathematical methods, and creates new features. When we filter useful features from many possible features, this process is called feature extraction. More about both tasks will be explained later in Chapter 6.

Dimensionality Reduction: Many machine learning algorithms are designed to work optimally with one (vector) or two (matrix) dimensional data. Therefore, converting a dataset with multiple variables (e.g., a table with numerous numerical fields) into lower-dimensional data can be challenging. There are methods to reduce the dimensionality of the data by transforming them into lower dimensions, i.e., *dimensionality reduction*. In other words, the process of dimensionality reduction reduces the complexities of data by reducing its dimension while ensuring the new data includes the same characteristics as previous data. Dimensionality reduction methods will be further explained in Chapter 7.

(5) Running the Machine Learning Algorithm

We have explained that machine learning is the process of enabling a computer program to learn from data. From previous descriptions, we recognized that there are different types of machine learning algorithms.

Some algorithms, such as clustering and association rule mining, automatically classify the given input dataset. These algorithms are called unsupervised learning. In contrast, some other algorithms rely on manual users' labels and annotation as a pattern to automatically label the rest of the data objects. For instance, we can recognize chickens and cats, but we do not know exactly their differences, or we are too lazy to define their differences for the algorithm. Instead, we label some instances in the dataset and then feed them into the algorithm. Then, the algorithm tries to label the rest of the data objects. These algorithms are called "supervised learning". "Regression" and "classification" algorithms fall into the category of supervised algorithms.

Typically, a process of supervised machine learning is performed by (i) recognizing "features" that the dataset has and preparing the input for the algorithm. (ii) Then, we separate the dataset into two sections, a "train" and a "test" dataset (or train, validation, and test). (iii) Then we run the algorithm on the train data, and its result is the model. (iv) Next, the model will be tested on the "test" dataset.

A model is either a set of *if-then-else*, rules (if we use decision trees), e.g., *if "ear is long" then "it is a rabbit"; if "color is yellow", then it is "chicken",* or a mathematical equation, which will see about them in Chapter 8 and chapters after that.

After training is finalized and the model is ready, we use the existing model to make inferences about the new data that has arrived in the system. This stage is also known as *inference time.* In other words, the model is the piece of knowledge which is used by the algorithm to create the label for the new input data. Zheng and Cassari [Zheng '18] stated that a model describes the relationship between different aspects of data. Deisenroth et al. [Deisenroth '19] state that the goal of the machine learning algorithm is to find good models that generalize well to yet unseen data. For instance, we trained a model that can recognize animals; when a new animal arrives in the machine, we do not compare it with all previous animals to identify the newly given animal. The system compares it with a model, which holds meta-data of each animal, e.g., if the ears are too long, then it is a rabbit; if it is yellow, then it is a chick, etc.

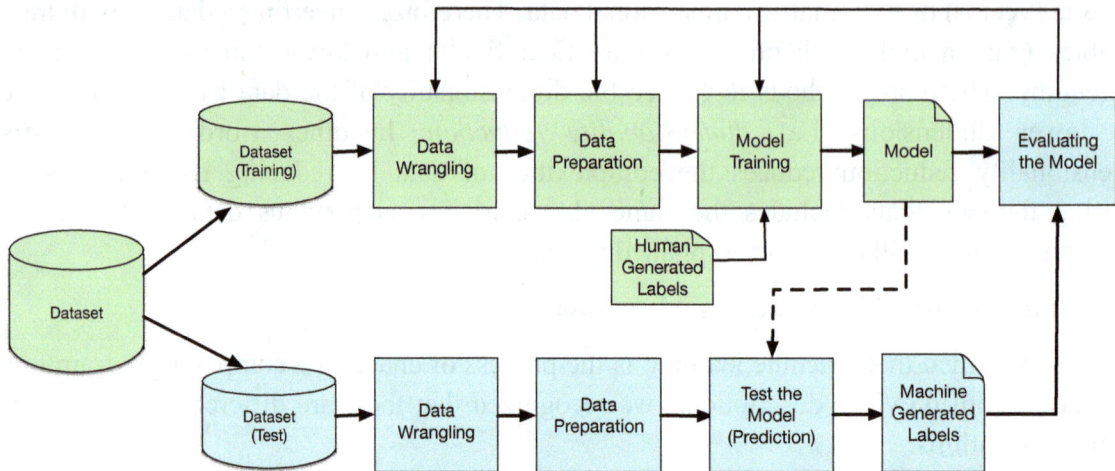

Figure 1-19: Supervised machine learning pipeline. Training is shown in green and testing in blue.

Figure 1-19 illustrates the use of the process of supervised learning in detail. If you do not understand some parts of it, do not worry. We will return to them later in this book, just try to get some overview of the process.

There are challenges associated with modeling, such as changing the data and making the model that is built on old data invalid, i.e., *data drift* (Chapter 16). As we progress through the book, more challenges will be explained and described.

To summarize, we have the following flow: Data → Supervised Algorithm → Model. We use the model for prediction or deciding about the newly encountered data.

(6) Evaluation

After the algorithm runs and produces the result, the next step is to evaluate the result it has produced. We have described how we can evaluate the resource utilization of an algorithm or its accuracy. For example, we do not need to care about resource utilization if we can access a supercomputer to run a machine learning algorithm in batch mode. On the other hand, if we are developing software that needs to run the model on a smartphone with limited resources compared to desktops, we should care about its resource use.

Note that steps 4 to 6 (data preparation to evaluation) will be executed iteratively. This means that based on evaluation results, we go back and change input parameters or algorithms, tune parameters, and make another run until we get a satisfactory result or the results do not improve anymore. Besides, we cannot only go from steps 1 to 6 once and solve all the world's problems. These six steps are iterative. As we have explained before, in addition to a dataset, a machine learning algorithm gets some parameters as input.

Usually, in each iteration, we need to play with the algorithm's input parameters (hyperparameters), until we identify the optimal values for these parameters. A parameter that will be given by the user to the algorithm to configure something in the algorithm while running the algorithm is called a *hyperparameter*. The optimal value of a hyperparameter can be estimated (i) by running the algorithm with different hyperparameter values and comparing their results, or (ii) it can be identified by solving through hyperparameter optimization algorithms (Chapter 14).

Another type of parameter is called the "model parameter". Model parameters will be identified by the machine learning algorithm via the cost or objective function. We have explained that most models are mathematical equations. Model parameters are the coefficients of the variables in those equations. More about them will be described in Chapter 8. Usually, when we read about neural networks, the weight and biases of a neural network are referred to as parameters of the neural network, which are model parameters.

Don't worry if recent paragraphs are not easy to understand; we will learn them as we progress through this book.

NOTES:

* Usually, a project includes a "pipeline" of algorithms and is not limited to one single algorithm. For instance, first, we classify a dataset with a supervised learning algorithm. Then, we focus on one set of data labels and apply clustering methods to those labels. Then, we focus on one data cluster and apply an association rule mining algorithm on that particular cluster, etc.

* We explained that we should separate datasets to test and train sub-datasets. If the training data is being used as the test data, this phenomenon is called *data leakage,* and we should avoid it.

* When there are a small number of labels available, and we employ a supervised learning approach for classifying the data, it is called "semi-supervised learning". We do not use this term in this book, but we list it here to be sure you will be familiar with it.

* There is a theory called "No Free Lunch Theory" [Wolpert '97], and it states that for every classification and regression algorithm, any two algorithms are equivalent when their performance (e.g., accuracy) is averaged across all possible problems. In other words, the computing cost of finding a solution is the same for any algorithm when we average them across all problems in the class. In other words, the No Free Lunch Theory highlights the importance of understanding the specific problem at hand and choosing or designing algorithms that are well-suited to that problem rather than expecting a single algorithm to be universally superior across all tasks.

Summary

We have described the basics of machine learning algorithms. As input, it usually receives a number. A number could be scalar, vector, matrix, or tensor, but it could also be in other formats, such as images or text documents. Then, it delivers the output based on the type of the algorithm, such as clustering, regression, generative AI, etc.

Until now, we have learned that there are four main categories of machine learning algorithms: unsupervised learning, supervised learning, reinforcement learning, and semi-supervised learning. In unsupervised learning, we do not need to label any data; we feed the data in, and the algorithm performs the desired task without labels, which could be clustering or association rule mining. In supervised learning, we feed labels for each data point into an algorithm, and the algorithm uses the data along the label to build a model and then labels the new incoming data. In reinforcement learning, the agent performs a task and receives a reward or punishment that it then uses to adapt and improve its performance. Self-supervised learning is trained to predict part of the input data from other parts of the same input data. The goal is to learn useful features or representations of the data in an unsupervised manner. For example, in the context of image processing, a self-supervised learning task might involve predicting one part of an image given another part or predicting the next frame in a video sequence. In Natural Language Processing (NLP), it could predict a missing word in a sentence.

Chapter 2: Visualization

In the previous chapter, we described visualization as an essential step in getting an overview of data. Visualizations enable us to gain intuitive insight into the data. Even when we are analyzing the output of a machine learning algorithm, we should still deal with visualizations. Therefore, we dedicate this short chapter to it.

Cleaning the data is an essential step toward preparing it for the algorithm. Before and after cleaning the data, we can benefit from visualizations of the dataset we intend to study. Visualization gives us a good insight into the data. It is a helpful approach, especially while working with tabular data. However, it is not easy to build a visualization from unstructured data, but at the end of this chapter, we learn many methods to address them.

Many underestimate the importance of visualization; it is one of the most under-utilized aspects of data science. Based on our experience, you might get a fat promotion in your job just by doing some appealing visualizations, which identify the system bottleneck that several senior software developers and data scientists have not identified before. Besides, visualization can be used to detect anomalies, group data, and identify latent patterns in the dataset.

This book will not delve deep into visualizations, and there are plenty of fantastic books about this topic, which we introduce at the end of this chapter. Nevertheless, we describe major visualization methods to familiarize you with them. By learning these descriptions, you can decide on your visualization plan correctly.

Since the data-cleaning process is time consuming, we might do both visualizations and statistical analysis in parallel.

Background and History

Visualizations allow us to represent large, complex datasets in more straightforward and more interpretable forms. They provide *abstractions* that can reveal *insights*, *patterns*, and *relationships* in data. Visualizations for understanding data have a long history dating back millennia.

One of the oldest known visualizations is the Babylonian Map of the World presented in Figure 2-1, discovered in the late 1800s in the ruins of Sippar, south of Baghdad (today belongs to Iraq). This artifact dates back to around 600 BCE during the Babylonian civilization. The Babylonian map depicts the world as the Babylonians conceptualized it at the time. As one of the first known visualizations of data, it demonstrates how visual abstraction allows a large amount of knowledge to be presented in a simplified pictorial form for communication. The approach of using visualizations for understanding data continues to be relevant today.

Another historical visualization example used to present the importance of visualization is the London Cholera map, which is the first modern use of this information presentation technique. In

1854, there was a cholera outbreak in London. At that time, people used to pump their water

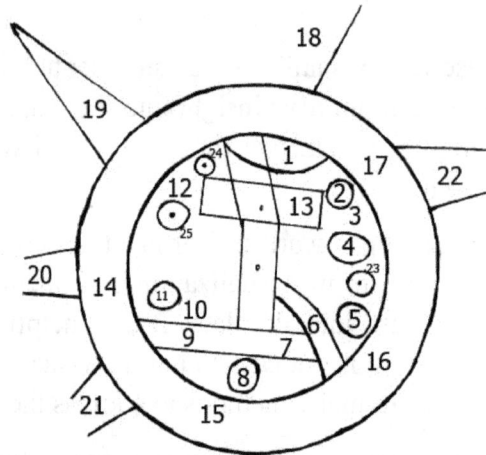

Figure 2-1: Babylonian world map, the first abstract visualization known.

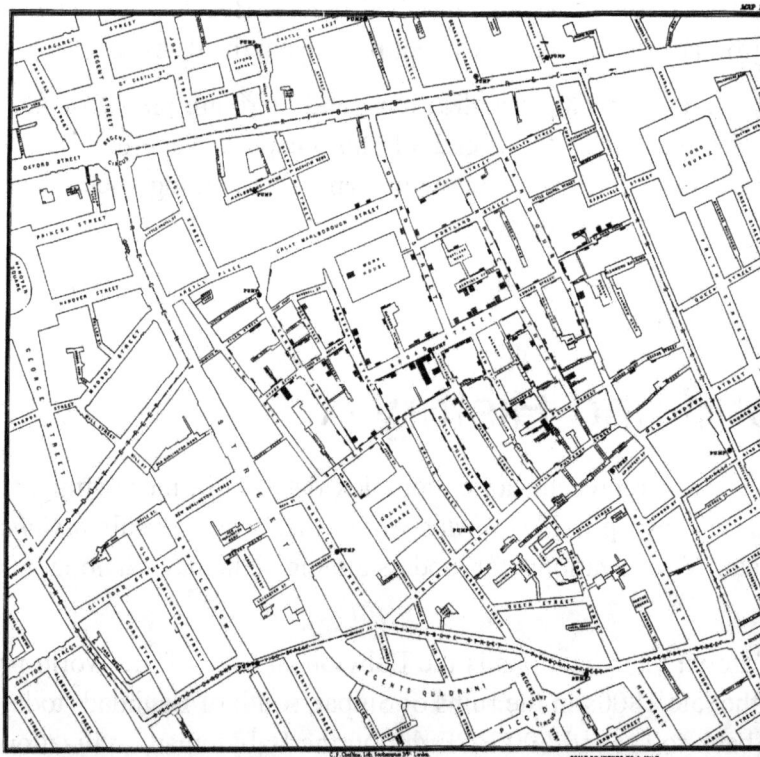

Figure 2-2: John Snow's map of cholera death distributions in London, based on well placement and reported deaths (Image source: https://en.wikipedia.org/wiki/1854_Broad_Street_cholera_outbreak).

from wells in the city, but there was no proper way of managing sewage. Therefore, the waters were contaminated with sewage, rotten garbage, and the feces of animals and humans, which led to the cholera outbreak. A physician, John Snow, sampled water from different wells and realized there were water supply issues in specific locations. Much like right now, authorities were careless about the scientific community. Snow published a map of death locations and well placements in that region. By looking at his map, authorizes quickly realizes that the outbreak's origin was the wells in particular locations. Figure 2-2 presents the map he designed. Then, local authorities closed those wells, and the cholera problem was resolved.

In both examples, we can recognize that visualizations provide concise and abstract explanations of the data.

Basic Visualizations

Figures 2-3 present four basic visualization forms, including; a pie chart, a bar chart/histogram, a line chart, and a scatter plot.

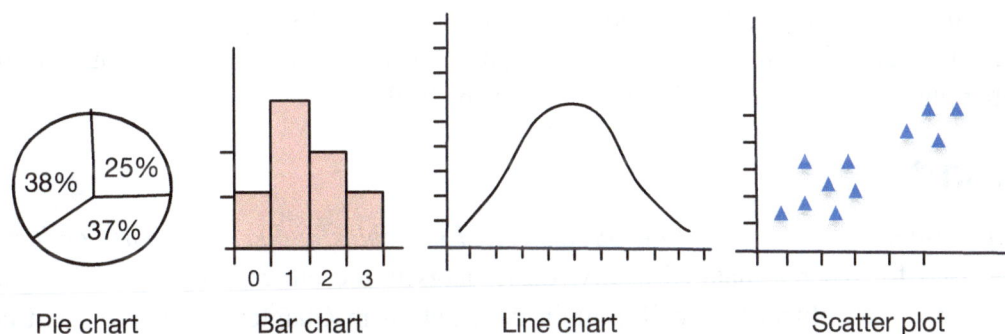

Pie chart	Bar chart	Line chart	Scatter plot

Figure 2-3: From left to right: Pie chart, Histogram/bar chart, line chart and scatter plot.

Pie Chart

A pie chart presents data objects of a dataset in *distinct groups*. It is useful when we intend to compare the proportional size of other groups of data. If group sizes are similar, using a pie chart may not be the best choice. Sometimes, a pie chart is also called a donut chart, which has an empty circle in the middle, and thus, it looks like a donut. In Figure 2-3, the circle filled with percentages is a pie chart.

Bar Chart and Histogram

Bar chart and Histogram is used when we intend to compare the size of different distinct groups of data, and it compares the relative size of data. In other words, if the data groups have similar sizes, bar charts are more readable and reflect small differences among the data groups of different sizes.

When a bar chart hosts more than one data object, it is called a *segmented bar chart* or *split-category bar chart*. For example, assume we have two or more variables, and we intend to compare them in different conditions. In this scenario, as shown in Figure 2-4, we have used

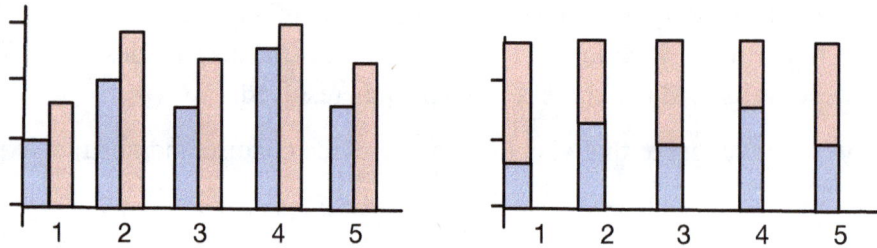

Figure 2-4: (left) An splitted bar chart example. (right) Another splitted bar chart example. Both present which presents two different data objects.

different colors for each feature (column if we have tabular data).

A *histogram* is the same as a bar chart but is used to show "distributions of data points", where bars represent the frequency of data points in numerical intervals. If you are interested in more details of bar charts, Dawn Griffiths provides more detailed explanations about the different styles of bar charts in Chapter 1 of her book [Griffiths '09].

Line Chart

Line chart presents a *series of continuous numerical variables, unlike bar charts or histograms, which are* used for discrete data[1]. Usually, these datasets are either "time series" or statistical distribution of a data object (we will explain distribution in Chapter 3). A line chart connects sequences of data points by a straight line. It is used when a variable continuously changes, and we are interested in understanding the changes. Using more than one line, we can present multiple data series.

Sometimes, if we have many data points, even for discrete data points, we use a line chart instead of a histogram to visualize them.

Scatter Plot

A scatter plot is very similar to a line chart, but it is used when we have a large number of disconnected data points. Similar to a bar chart and a histogram, it is also useful when we deal with two variables in the dataset. We can use a scatter plot to show relationships (correlation) between two or more variables. In Figure 2-3, we use blue triangles to show each data point; we can show another variable with another shape and color, such as a red circle.

Scatter plots can also be used to show the distribution of data. For example, a scatter plot of the heights of a group of people will show the distribution of heights in that group. A scatter plot with a wide distribution will show that there is a wide range of heights in the group. A scatter

---

[1] It is fairly easy to understand the differences between continuous and discrete variables. We post pone its explanation to the next Chapter.

plot with a narrow distribution will show that the heights in the group are more similar. If the term distribution is not clear, wait; we will learn it in the next chapter.

Broken Axis Chart

Broken axis charts are used when the ratio of information we would like to present is very different, and we can not present them in the basic form of a bar chart or line chart. One approach is to use a broken bar chart, as shown in Figure 2-5; the same could be done with a broken line chart as well. In Figure 2-5, we can see that Column4 has a significantly larger value than other columns. If we vertically stretch the bar chart, Column1, Column2, and Column3 information is unreadable. To solve this issue, we can use a broken chart for this column.

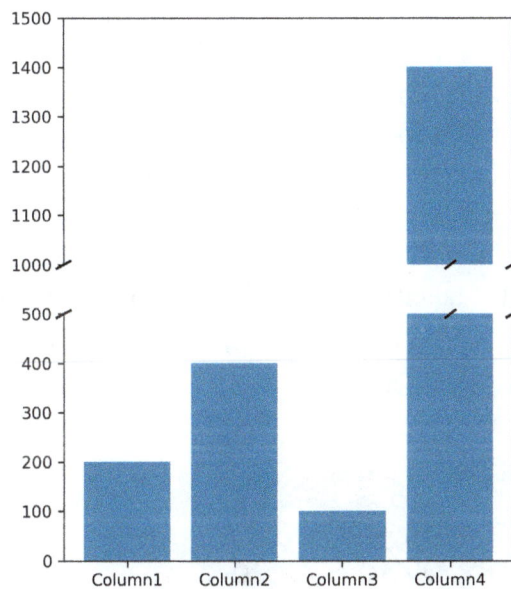

Figure 2-5: A broken barchart. We can see that Column 4 has a very high value and the visualization uses broken barchart to skip the data between 500 and 1000.

More than One Dimensional Data Visualizations

What if our data has more than two elements, not just X and Y coordinates, that must be plotted? Usually, our brain can handle up to three-dimensional data visualization, and our drawing surfaces, such as paper and digital monitors, are limited to two-dimensional data presentation.

In the following, we list some common visualization techniques used for more than two-dimensional data presentation.

Surface Plot

It is used to present data points with three dimensions, as shown in Figure 2-6. We use surface plots when we are interested to see data changes based on their different characteristics. The surface plot is also called a *3D scatter plot*.

Let us introduce you to Mr. Nerd, who is the celebrity of this chapter. He is a decent software developer who intends to learn artificial intelligence as well. He is keen on using new technologies and keeping his house up-to-date with the newest gadgets. Recently, he has started to gain some weight, and he hesitate to do outdoor activities, including physical activities, or have interactions with other people. Therefore, he decided to reduce his weight by reducing his food consumption. He bought a new AI refrigerator equipped with sensors and a camera that can track how many times he has taken food out, the weight, and the calories of the food taken. The refrigerator has an extensive display that provides him with visualizations of his food consumption. In this example, we have the three-dimensional data, including (i) the time of the day Mr. Nerd opens his refrigerator door, (ii) the calories of the food that is taken out, and (iii) the weight of food that has been taken out.

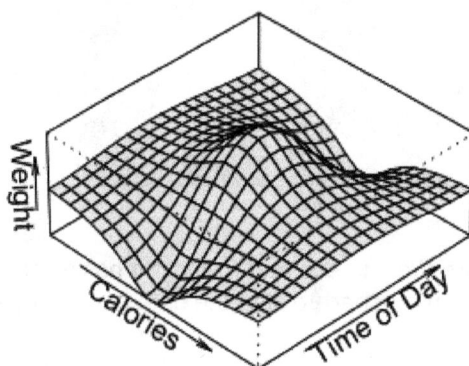

Figure 2.6: Example of three dimensional data presentation on surface plot.

Figure 2-6 is a surface plot and presents an example of the visualization the refrigerator provides to Mr. Nerd at the end of every day. As we can see, a surface plot can visualize three-dimensional numerical data, such as time of day, weight, and calories in this example.

Contour Plot

The surface plot is very attractive and enticing, but it cannot visualize the data when the structure of the data is not symmetrical. For instance, we have a dataset that, when we visualize in the surface plot, has too many mountains and valleys, making it hard to understand. By using a surface plot, we will get something like Figure 2-7 (a). We can play with the color of the surface plots and get Figure 2-7 (b), which is still not easy to understand.

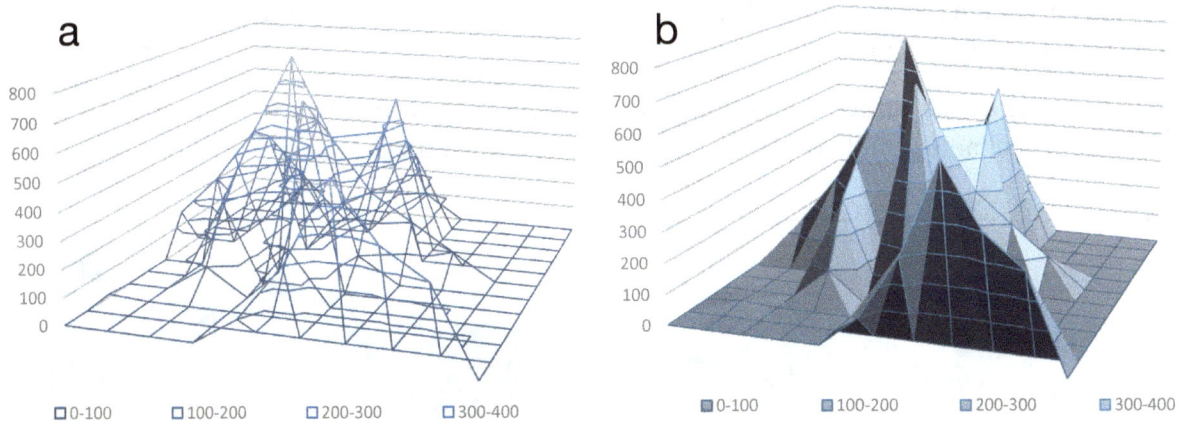

Figure 2-7: (a) a surface plot of dataset in Table 2-1 (b) a colored surface plot of the same dataset.

In these cases, we use a contour plot (level plot) to visualize the 3D data on a 2D surface (see Figure 2-8). This ugly plot looks like an infection on the skin, but it is very practical, especially when we need to visualize data distribution. It is often used in cartographical data to visualize the height of a geographical region. A contour plot and its underlying dataset are shown in Figure 2-8.

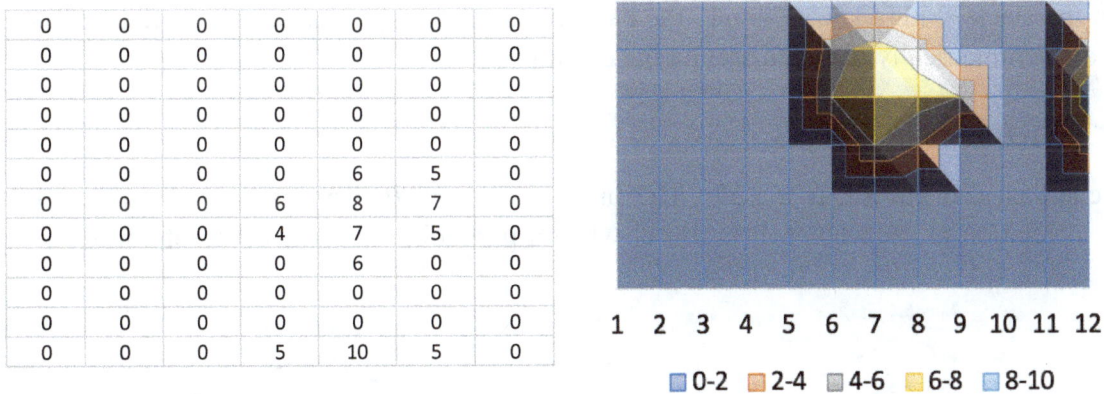

Figure 2-8: (left) a table of data, (right) the contour plot visualization of the left side table.

Area Plot

There are cases in which we intend to analyze different time series together to see how they correlate, most commonly when reporting trends or changes during the time. For instance, since 2010, Mr. Nerd has been purchasing a technology magazine that reports the currently trending technologies. The magazine writes a lot about the hype in the market. He is continuously learning to keep himself a knowledgeable developer. Figure 2-9 presents the number of articles per year and their topic in the magazine. By looking at Figure 2-9, we can see that the Semantic Web was dead around 2012, and the magazine's interests have shifted from web services and big data to deep learning and machine learning. However, until 2021, there are still articles about Web Services and Big Data.

In this example, we have three-dimensional data: frequency of technology written about (Y or vertical axis), technology field (colored streams), and time (vertical Axis or X).

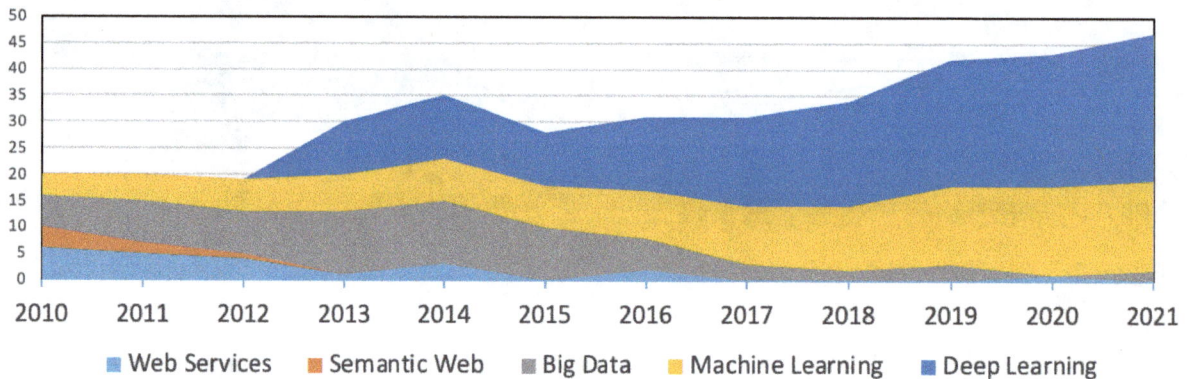

Figure 2-9: Area plot example.

Another form of Area plot is known as a stream graph, which we didn't show due to our laziness. In an area chart, all data changes are stacked on top of each other over a straight baseline at the bottom of the stack. With a stream graph, the baseline is set through the center of the chart, and the areas are symmetrically gathered around the central line.

Radar Chart

A radar chart is also known as a *web chart* or *spider chart*. It displays several quantitative variables in relation to each other. It is useful when a particular data has different variables to

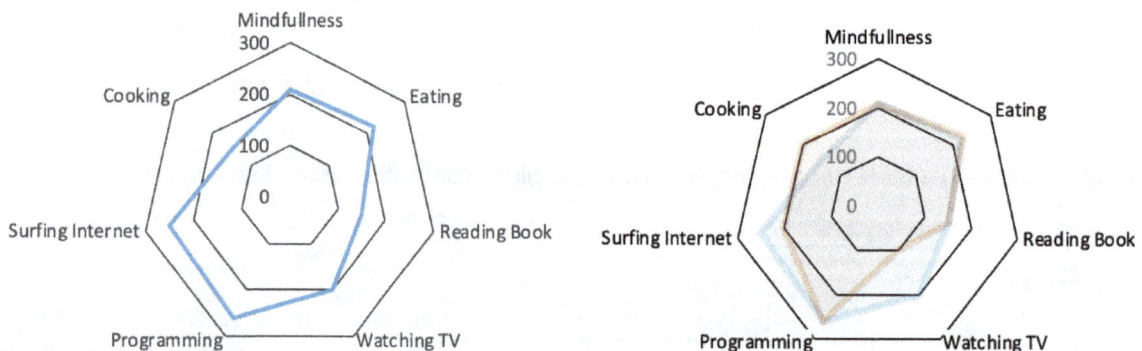

Figure 2-10: (left) a radar chart example. (right) a radar chart example that compares two different entities with the same variables.

46

describe it, and these variables stay in the same numeric range. In other words, radar charts are particularly useful for comparing multiple variables across the same scale. We can also use it to compare different datasets with the same variables together (right side of Figure 2-10).

Radar charts display multivariate data, where each variable is represented on a separate axis that starts from the same point. Each axis represents a different variable, and the scale of each axis must be consistent to make an accurate comparison.

For instance, Mr. Nerd writes down his time on his tasks at home and has a report like the radar chart on the left side of Figure 2-10. Later, after he had used his new refrigerator, he measured the time he had spent on his tasks. By comparing his behavior before purchasing the refrigerator and after purchasing it on the chart (the right side of Figure 2-10), he observed that TV watching and Internet browsing decreased. Instead, he spent more time on cooking at home. Then, he stays happy in his nerdy life until the world's end.

Heatmap

Heatmap is used to provide an overview of three-dimensional data. The heatmap visualizes the intensity of a data distribution over a 2D space, and it is useful when there is a large variety of data ranges in each dimension. It uses color to convey information about the magnitude or frequency of values in a matrix or map, enabling users to quickly identify patterns, trends, and outliers in a large dataset. Figure 2-11 presents a sample heatmap.

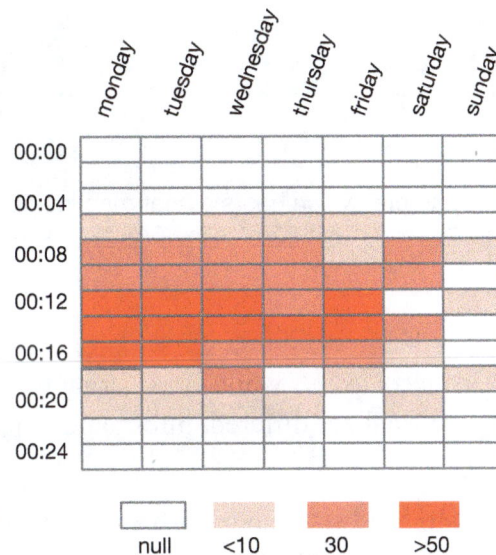

Figure 2-11: Heatmap example.

For example, assume Mr. Nerd falls in love with his new smart refrigerator. He intends to measure how many times he talks with his refrigerator during the day. He made an application that records his communications. He realizes that he usually talks more to the refrigerator during weekdays (not weekends) around lunchtime. Figure 2-11 presents his communications in minutes while talking to his refrigerator. By studying Figure 2-11, we can see that between 14:00 and 18:00 on weekdays, there is a higher likelihood that he talks to his awesome refrigerator.

Calendar Plot

Another form of a heatmap that is presented inside a calendar is a calendar plot. A calendar plot is used to display the frequency of events that occurred over a period of time, typically days or months. Each cell of the calendar grid corresponds to a specific day or month, and the color intensity or shading of each cell indicates the frequency or magnitude of events on that particular day or month. This allows us to grasp patterns and trends in event occurrences over time quickly.

It is easy to understand and interpret as it shows a day as a single unit. There are also varieties that make a month or week a single unit. Figure 2-12 presents a calendar plot example.

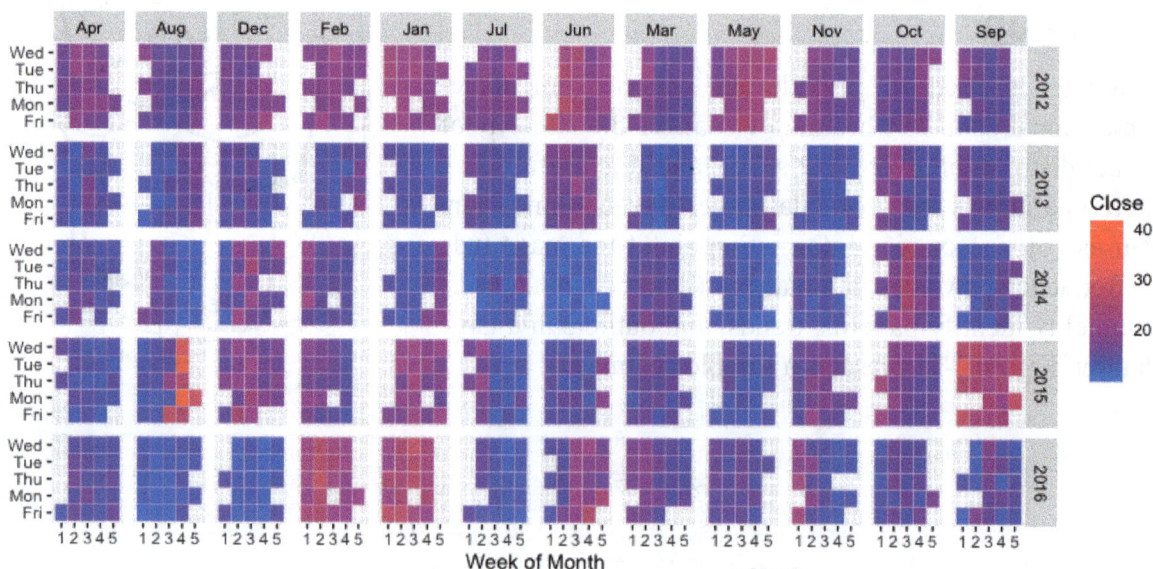

Figure 2-12: Calendar plot example, which hosts a heat map inside. The dataset is taken from: https://raw.githubusercontent.com/selva86/datasets/master/yahoo.csv

Timeline Plot

A timeline plot visualizes a *sequence of events* while considering their correlations, such as having overlap or discontinuity. It visualizes different information objects in chronological order (ordered by time).

For example, an obese friend of Mr. Nerd starts to track his important life events. He has also started to study whether there is an association between his overeating and health. Therefore, he uses the timeline shown in Figure 2-13.

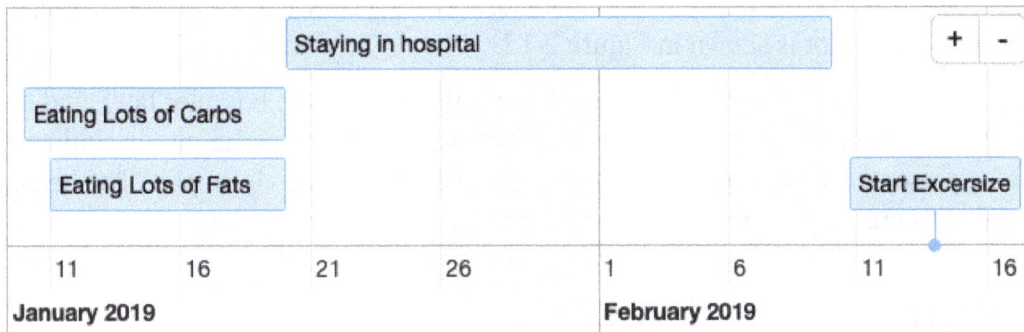

Figure 2-13: Example of timeline visualization.

Box plot and Violin plot

A *Box plot* or *Whisker plot* is a widely used plot to present different variables with some statistical details. A box plot is useful for understanding the central tendency, variability, and skewness of the data, as well as identifying outliers. It's particularly helpful for comparing distributions across different groups. We will learn more about the box plots in Chapter 3 after we become familiar with some statistical concepts. A simple box plot is shown in Figure 2-14.

A more detailed box plot is a violin plot, which includes information about the distribution of data as well. A violin plot combines the features of a histogram and a box plot, displaying the distribution of a single data object through a combination of density traces and box plot elements. It provides a detailed visualization of the shape and spread of the data.

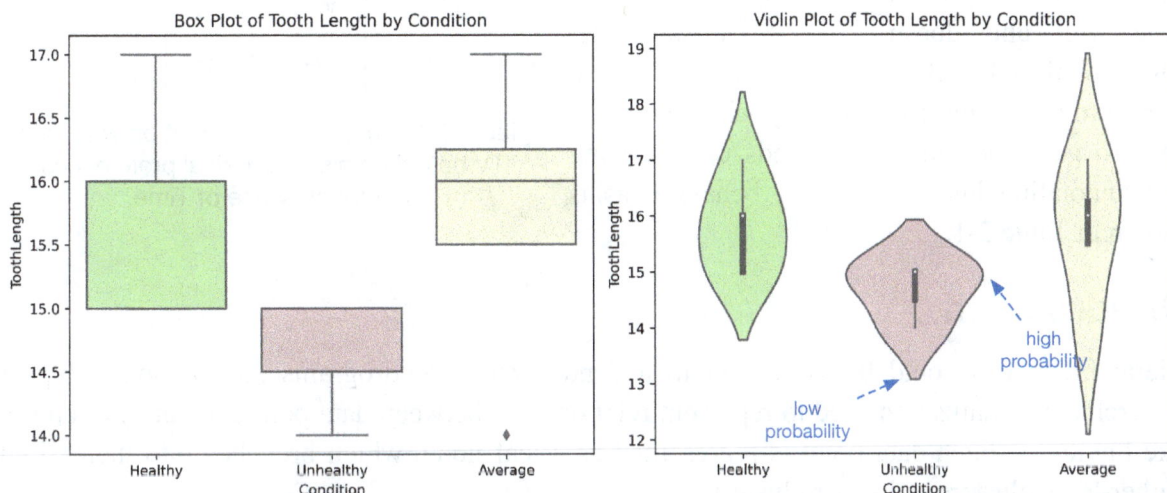

Figure 2-14: (left) box plot example (right) violin plot example.

In other words, a violin plot visualizes the numerical data distribution and probability density by showing the shape of the distribution, which enables quick visual assessment and comparison of data objects. A violin plot is shown in Figure 2-15.

Don't worry if you don't understand it, we will learn more about the box plot later, and then you can come back and check the violin plot. For now, remember that a violin plot is similar to a box plot but also presents the density of the data. Figure 2-15 presents a violin plot. The horizontal stretch on this figure presents the density at each value.

Hierarchical Visualizations

Many datasets have a hierarchical structure in which data objects are grouped or related in a tree-like structure. In simple words, inside some information, there is other information, e.g., inside a job element of an applicant's CV, there are several companies that s/he had worked for. Examples include organizational hierarchies, any dataset with a tree structure, geographic data, etc.

Hierarchical visualizations are designed to represent multi-level relationships visually. These visualizations allow users to observe and understand patterns at both local levels and globally across the dataset structure.

For instance, Mr. Nerd wants to use his spare time well and doesn't like to miss some educational and useful programs on the TV. He started to record his TV-watching behavior every day. To gain such insight, he ends up recording and annotating his TV-watching behavior, as is shown in Table 2-1.

Program	Category	Time Spent
Animal & Nature	Useful	2
Documentaries	Useful	3.2
Reality Shows	Waste of Time	7.8
Political News	Waste of Time	1
Indie Movies	Useful	2.9
Blockbuster Movies	Waste of Time	4.5
Technology News	Useful	2
Ads	Waste of Time	5.1
Celebrity News	Waste of Time	1

Table 2-1: Time Mr. Nerd spent on watching TV based on his categorical preferences: Useful or Waste of Time.

Dendrogram

Hierarchical data could be visualized in a dendrogram. Dendrograms are a popular type of hierarchical visualization used to represent relationships between data points organized in a tree-like hierarchy. Each data point is represented as a leaf node, which branches into increasingly higher-level clusters and super-clusters.

Dendrograms are common visualizations that present the results of topic modeling (we will explain them in Chapter 7) and hierarchical clustering results (we will explain them in Chapter 4). Figure 2-15 presents a dendrogram of Mr. Nerd's TV-watching behavior. We can see in this structure that data are categorized into two groups: "useful" and "wasting time". The next level is

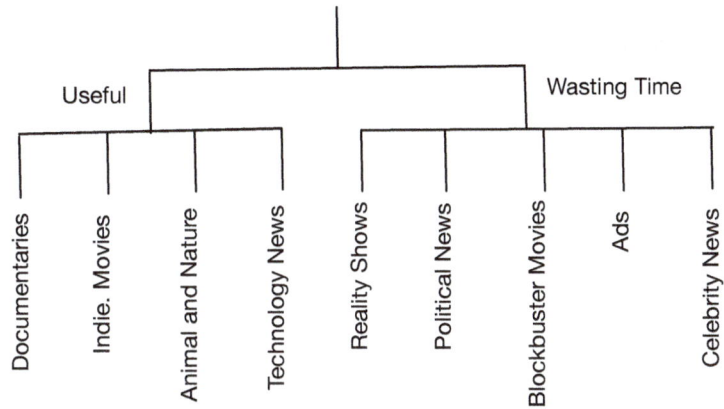

Figure 2-15: Dendrogram example with only two levels of hierarchy.

types of TV programs, and it does not go deeper into hierarchies, but dendrograms can go deep into several levels.

TreeMap

A treemap is another hierarchical visualization technique used to represent the dataset in a hierarchical manner. It uses nested rectangular shapes to display the relative sizes of different categories or subcategories within the hierarchy. Dendrograms mainly show the hierarchical topology, but treemaps can visualize additional attributes like the quantity of each data object via the rectangle sizes.

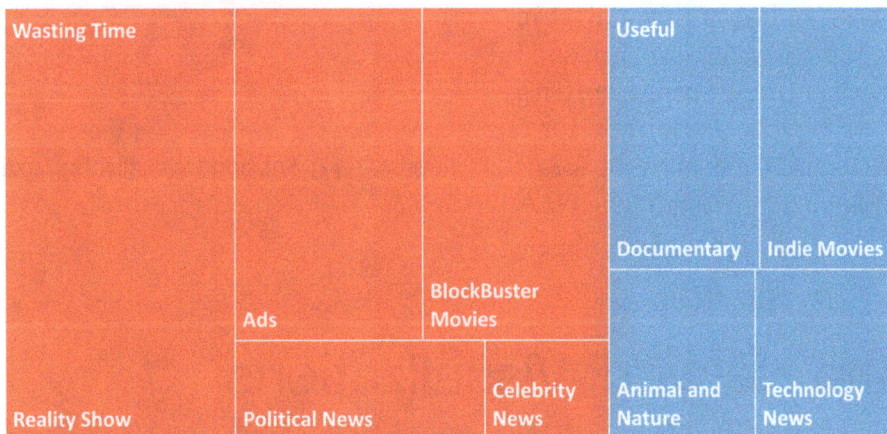

Figure 2-16: Treemap of watched TV programs based on being useful or wasting time.

Figure 2-16 presents an example of Treemap visualization. In this example, we have two higher levels; thus, we use two colors to separate them in the Treemap.

Treemaps are good at visualizing the relative sizes of different parts of a hierarchy, but they can be challenging to understand for large hierarchies. Additionally, treemaps cannot show the relationships between different levels of a hierarchy, which is something that dendrograms can do. In other words, Treemaps don't illustrate the *relationships between different levels of the hierarchy*. They only show the hierarchy within a single level.

An alternative to standard treemaps is the *Voronoi Treemap*; these use *Voronoi tessellation* to subdivide space, giving a more organic layout that some find visually appealing. Usually, they are shown in a circular fashion, and we didn't present them here. We didn't present it here; you can search online and find them.

Sunburst

Sunburst charts are a radial, space-filling visualization that represents hierarchical data in a circular layout (similar to pie charts). Like treemaps, sunbursts use nested shapes and divisions to show hierarchy levels and categories. However, rather than rectangles, sunburst charts use nested pie slices emanating from the center outward. Sunburst is a common visualization for genomics data exploration, file system analysis, and social network analysis.

Figure 2-17 presents a sunburst visualization of the same data presented in Figure 2-16. but here, we can add more levels of hierarchies as well.

The *Icicle plot* resembles a sunburst chart but uses a stacked, bar-like layout. Each level in the hierarchy is represented as a horizontal bar, with subdivisions representing subcategories. Icicle plots work well for deep hierarchies and can be more space-efficient than circular representations. We didn't present it here; you can search online and find them.

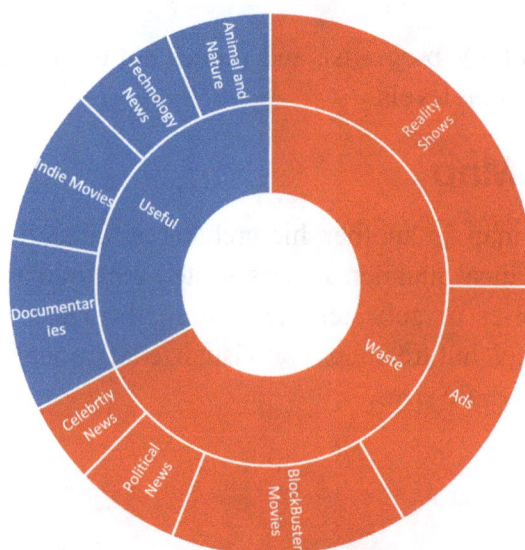

Figure 2-17: Subburst visualization example.

Graph and Network Visualizations

Some data science and machine learning problems can be solved with graphs. In this book, we dedicate one chapter to graph algorithms (Chapter 15). Graphs are used to present and model data in a network fashion. Here, we list techniques used to present networked or graph data.

Node-link Diagram

A traditional method to present graphs is using a node-link diagram. The nodes represent the entities in the network, and the links/edges represent the relationships between the entities.

Nodes will be presented as a single dot or circle, and edges will be presented as lines. Figure 2-18 visualizes a graph in node-link diagram style, the most common approach to visualizing graphs.

When you read the term "graph," an image of a node-link diagram will likely come to your mind. Those images are node-link diagram visualization. All graph visualizations we describe in Chapter 15, use a node-link diagram visualization.

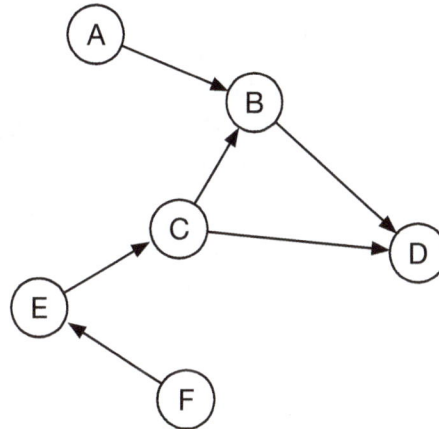

Figure 2-18: Node-link diagram visualization of a graph

Arc Diagram

An arc diagram is a visual representation of a network where nodes are arranged in a horizontal line, and the connections between them are displayed as arcs or curves above or below the line. In addition to graph visualization, it is useful to visualize co-occurrences between specific data objects.

In some arc diagram visualizations, arcs have different thicknesses to demonstrate the frequency or density of the connection. However, the method is not extensively employed due to potential cluttering, mainly when dealing with large graphs, which makes it difficult to understand. Figure 2-19 presents a sample arc diagram plot.

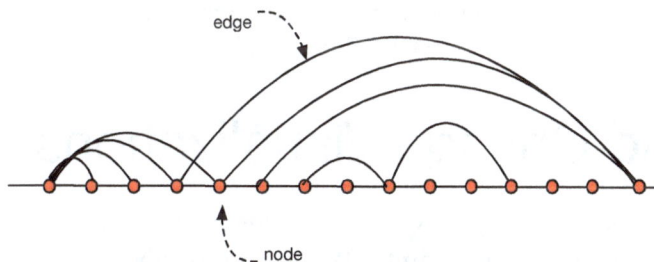

Figure 2-19: Arc diagram example.

Chord Diagram (Circular Link)

The chord diagram or circular link visualization is very similar to the arc diagram, presenting connections among different nodes of a network. The input is a matrix that represents the degree of relation between nodes. In other words, this visualization utilizes a matrix to display the degree of relation between nodes, and the connections are presented as chords within a circular layout.

For example, assume Mr. Nerd has four colleagues working with them on the same project. Since Mr. Nerd's office policy is close to zero, like the author of this book, he has decided to use visualization and observe how many times other colleges are saying good things about other colleges. Based on the fact that "an apple polisher will get the promotion in the end", he tries to predict who will get the promotion at the end of the year.

Let's call his coworkers *'A'*, *'B'*, *'C'* and *'D'*. His observation results are presented in a matrix (left side of Figure 2-20). Read this matrix as follows: College 'A' said two times positive things about college 'B'. College 'B' said nine times positive things about college 'D'. Now, based on this matrix shown on the left side of Figure 2-20, the chord diagram that presents college relationships is shown on the right side of Figure 2-20. In this example, reading the matrix seems very easy, but when the number of data objects to compare gets large, we cannot read them, and a chord diagram can help to understand the relations between data points.

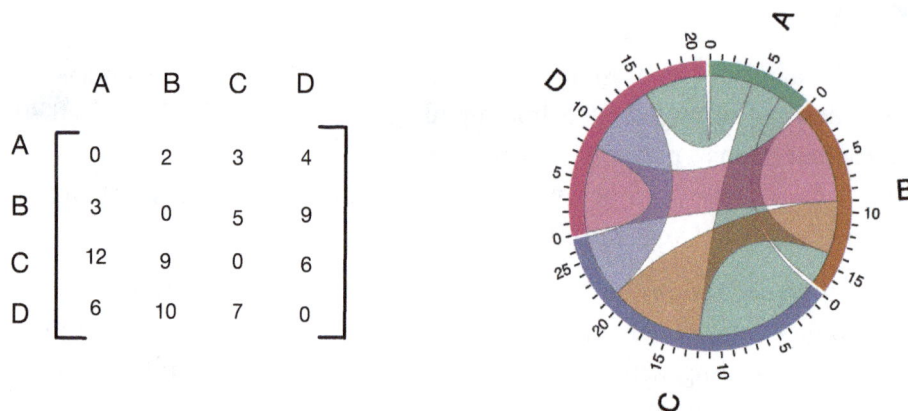

Figure 2-20: Chord diagram of the matrix on the left. Note that usually the number of data points is too large that reading and interpreting them from matrix is too complicated and chord diagram is very helpful to resolve that complexity.

Deviation and Change Visualizations

In specific applications, it is necessary to visualize changes or deviations of specific data points compared to a threshold or other data points. We have described in Chapter 1 that deviation analysis, such as anomaly detection, is a large domain of data analysis applications. We can use deviation visualizations in those cases, including diverging bars (or dots) and dumbbell plots.

Divergent Bar Plot

It is a deviation visualization type that presents binary changes of a bar plot. In other words, it is a type of bar plot used to visualize metric values that can be positive, negative, or zero.

Divergent bar plots show bars extending from either side of an axis. The bars can be arranged vertically or horizontally. The side of the axis that the bars extend from is the value of changes, such as when there are positive and negative values, or the category, such as when there are two types of things we want to show.

For example, Mr. Nerd decided to observe if his professional relationships with his colleagues improved. He scores his relationships with each colleague. If a relationship has improved, it receives a positive score and is shown with the green bar. If a relationship has worsened, it receives a negative score and is shown with the red bar. The result is shown in Figure 2-21 as a divergent bar plot. By looking at Figure 2-21, Mr. Nerd can see his relationships ranked from best to worst. Thus, he can invest more effort in those with negative scores to improve those relationships.

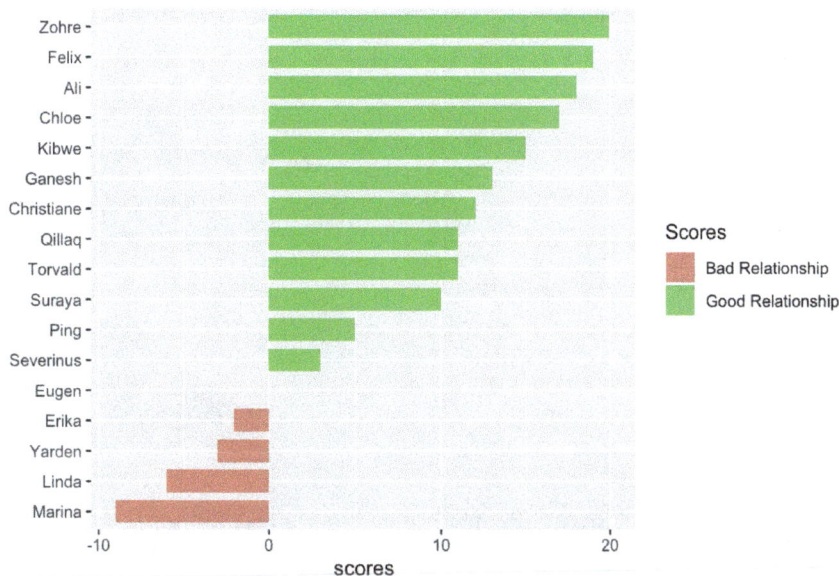

Figure 2-21: A divergent bar visualization by using diverging bar. Names (except Zohre) were generated random by using online random name generators, and there is no intention for name selection.

Dumbbell Plot

A dumbbell plot is used to visualize changes for a specific data point at two different times or conditions. This reveals that the dumbbell plot is another form of deviation visualization, but it also incorporates the magnitude of change. The dumbbell plot consists of two points and a bar connecting them: one point for the value at the first measurement and another for the value at the second measurement. The distance between the two points represents the change in the value.

For example, Mr. Nerd intends to reduce his high-calorie food consumption; with the help of his smart refrigerator, he intends to measure the amount of food he consumes during the two-month period of monitoring his diet.

He starts to count the number of food items he consumes and uses the dumbbell plot to visualize them, as shown in Figure 2-22.

The first month of consumption is marked with green circles, and the second month with orange circles. We can see that if the green is located on the right side of the orange, it indicates a decrease in consumption.

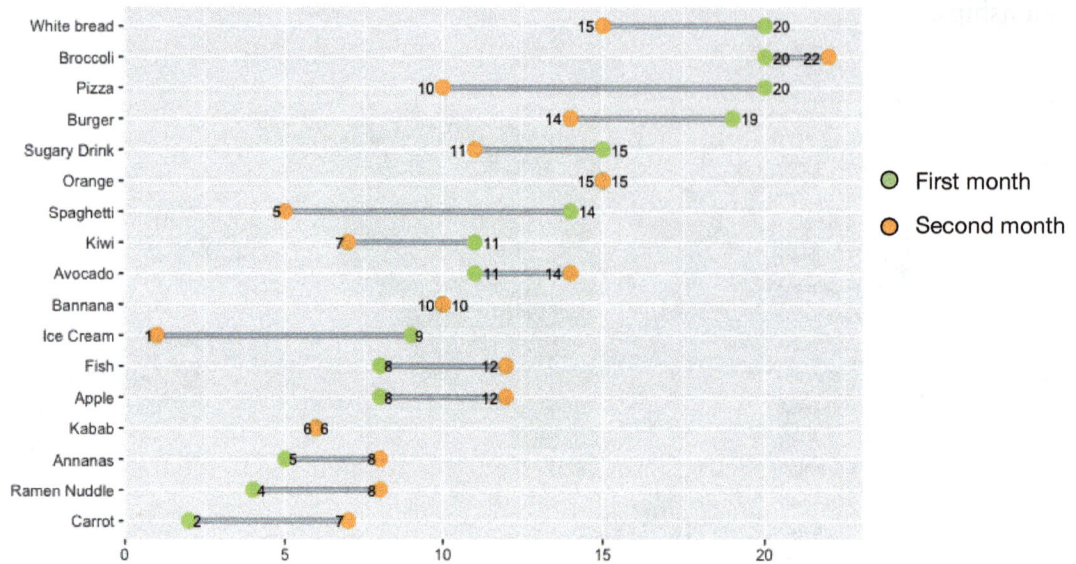

Figure 2-22: Dumbbell bar visualization.

Waterfall

Waterfall visualization is another form of visualizing changes in a variable. It shows how an initial value of a variable is increased and decreased by a series of intermediate values. In simple words, it shows the cumulative effect of a series of positive and negative changes. It could start from left to right and use up and down bars. It could also be vertical, using left-moving bars to indicate a decrease and right-moving bars to indicate an increase. In the Waterfall charts presented in Figure 2-23, we use orange to represent negative changes (costs) and dark blue to represent positive changes (gains).

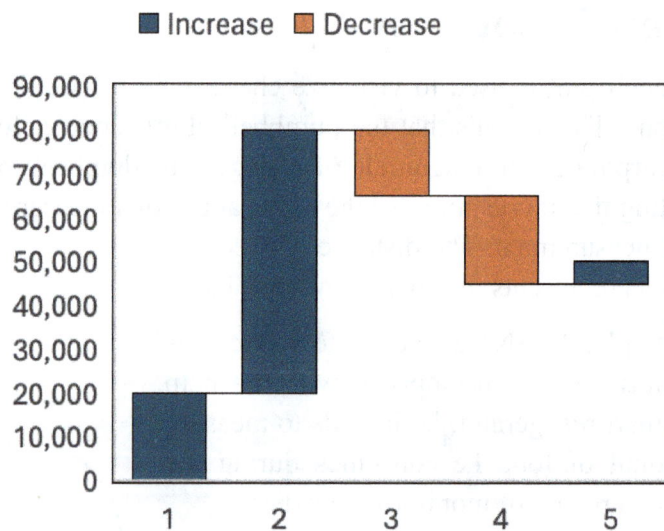

Figure 2-23: An example of Waterfall chart, negative changes presented in orange and positive changes in dark blue.

Waterfall visualizations and divergent bar plots share some similarities but are not the same. Waterfall plots are great for showing cumulative effects and sequences of positive and negative changes, while divergent bar plots are ideal for comparing magnitudes and directions of differences between two groups or categories. Later in Chapter 16, we will see some more Waterfall examples.

Beeswarm

A beeswarm plot is a type of data visualization to show distributions of data points across few different categories. They are useful for displaying datasets where there are many data points that may overlap if plotted as a simple scatter plot. The Beeswarm plot spreads out data points while minimizing their overlap, making it easier to see each individual data point. This feature is useful when dealing with large datasets where a scatter plot might result in many points plotted directly on top of one another. Besides, they can show how a particular measurement varies across different categories. For example, Figure 2-24 presents an

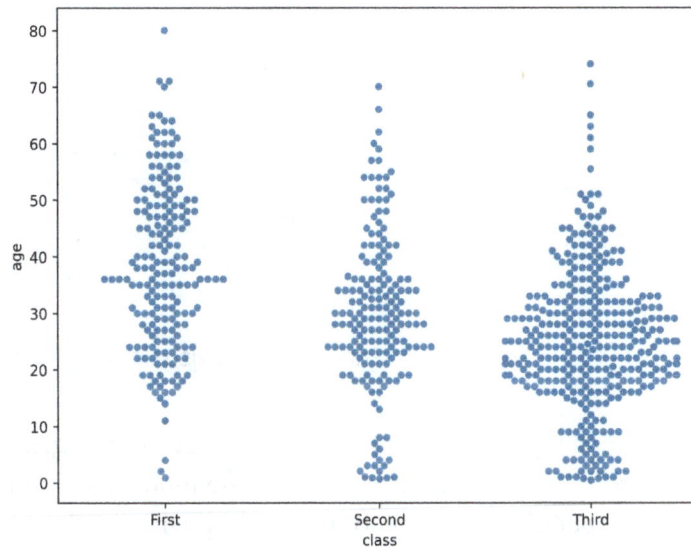

Figure 2-24: An example of Beeswarm visualization. The dataset isTitanic passenger ages and their distribution across different passenger classes.

example of Beeswarm visualization for the ages and distribution of Titanic ship passengers in different passenger classes.

This example uses a simple model; sometimes, there are two more different categories included as well. To handle that, we could use colors on each data point to present another data dimension. Later in Chapter 16, we will see some more Beeswarm examples.

3D Data Visualizations

There are different uses for three-dimensional (3D) data visualization. One common application is simulating geometric shapes that exist in three dimensions in the real world. Another common use is to include more variables in 3D visualization.

When dealing with a geometric object, its 3D coordinates are crucial. Point cloud, mesh, or voxel visualizations can be employed. Figure 2-25 illustrates these three 3D visualization approaches for presenting and analyzing a geometric shape in 3D.

If a set of points in 3D space is used to represent an object or a scene, it is known as *point cloud* representation. Point clouds are often used when 3D scanning real-world objects since the scanner captures a discrete set of points on the surface. Point clouds allow efficient display and rendering of complex 3D shapes. Point clouds are commonly used in applications like 3D scanning and LiDAR-based terrain mapping.

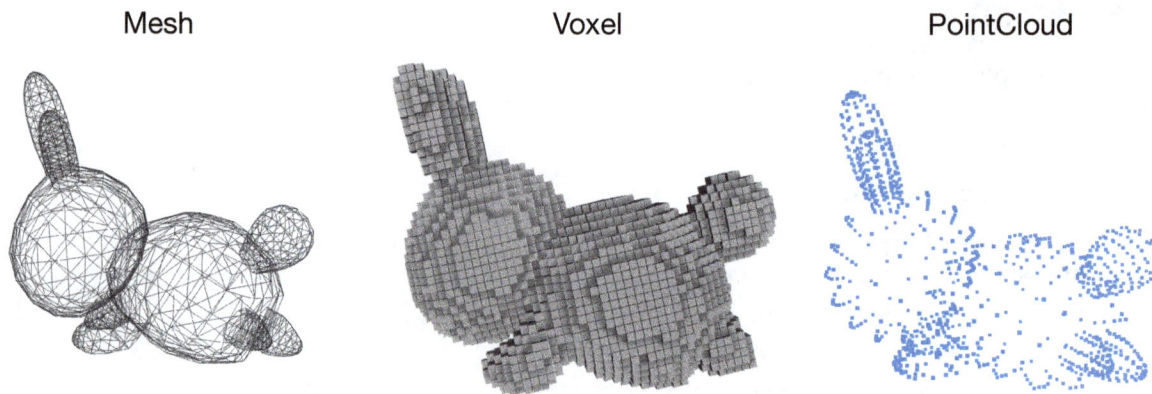

Figure 2-25: Three different visualization form of a 3D object.

If the shape is represented with a set of connected triangles, it is known as a *Mesh representation*. Meshes are a common representation of 3D models in computer graphics. The mesh connectivity allows for operations like smoothing, subdivision, and animation. Mesh representations are used in computer graphics, simulations, and video games due to their ability to represent complex geometries accurately.

If the shape is represented by 3D cubes (voxel), it is known as *a voxel representation*. Voxels represent 3D space as a grid of cubic cells, like pixels for 2D images. They provide a straightforward way to integrate 3D geometry with grid-based physical simulations or volumetric data. Memory requirements scale with volume rather than surface area. Voxel representations are

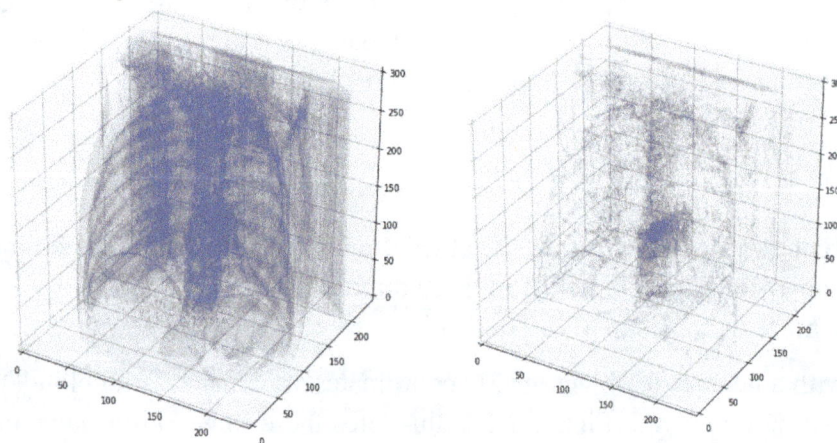

Figure 2-26: Point cloud visualization examples, (right) CT image of a lung, (left) the Covid-19 infection inside the lung has been highlighted and the algorithm removes the rest of points.

commonly used in medical imaging, such as CT scans, as well as in fields like architectural modeling and scientific visualization.

As a practical example of point cloud, we can refer to the CovidCTNet [Javaheri '21], a neural network model to extract Covid-19 infections from CT images of lungs, as shown in Figure 2-26. We can see on the right side the CT image of a patient, which is converted into a point cloud, and on the left side, the CovidCTNet extracts the infected area of the lung.

In addition to visualizing these geometric shapes, we can use 3D visualization to incorporate more than two or three dimensions of data.

For example, Figures 2-27 present *multidimensional heatmaps*[2]. In this example, eight heatmaps have been presented, which means if we have a feature that has eight different features, we can present this feature as follows.

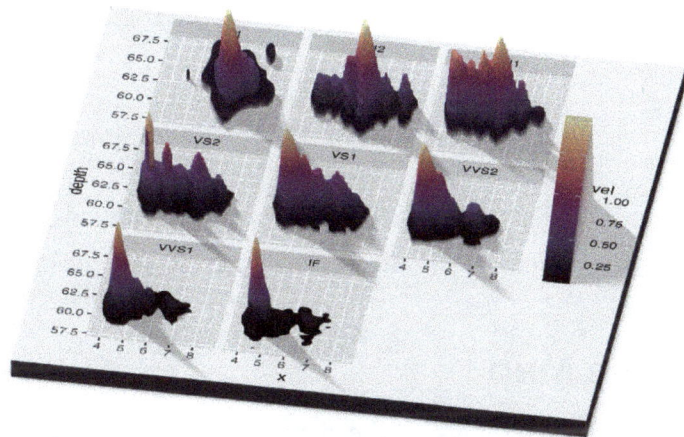

Figure 2-27: Multidimensional heatmap visualization.

Geographical Map Visualizations

As we can see from Figure 2-28, visualizing information on the map is used to incorporate geographical location with information. There are several approaches that can employ maps for visualizations, including Choropleth, Bubble map, Connection map, etc. There is no need to explain them in detail, but we briefly present two more popular, i.e., the Choropleth map and the Bubble map.

Choropleth

It is very similar to a heat map but utilizes geographical locations to represent the density of information within a specific region. In a Choropleth map, distinct regions are shaded or colored to indicate the magnitude of a particular variable. This visualization effectively communicates spatial patterns and variations, presenting data with geographical significance. Choropleth maps are commonly used to display election results, economic data for each region, population health metrics, etc. Figure 2-28 illustrates a Choropleth map visualizing the GDP of world countries in 2014. The darker the color, the higher the GDP of the target country.

---

[2] This visualization is built with Rayshader package of R https://www.rayshader.com

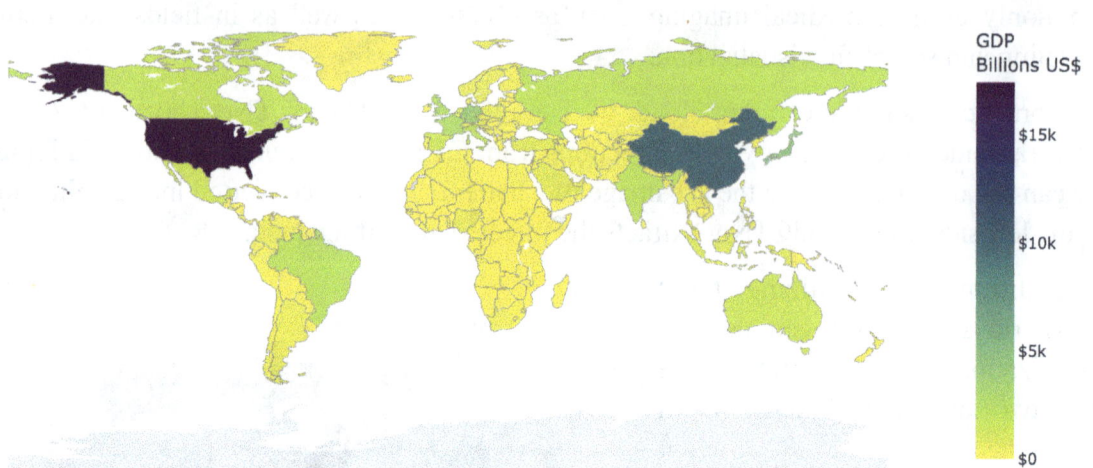

Figure 2-28: A choropleth map visualization of countries based on their GDP.

Source: the dataset file is downloaded from https://www.kaggle.com/wsaqaf/
2014_world_gdp_with_codes.csv

Bubble Map

Another way to represent the density of a variable on a geographical map, allowing users to compare the variable across different geographic locations, is by depicting the variable's magnitude using circles of varying sizes on the map. It is very similar to the Choropleth map but uses a different style of visualization.

The size of each circle on a Bubble Map corresponds to the value of the variable being represented. Larger circles indicate higher values, and smaller circles correspond to lower values. This visual representation immediately conveys the spatial distribution and intensity of the variable, enabling viewers to identify patterns, trends, and areas of interest.

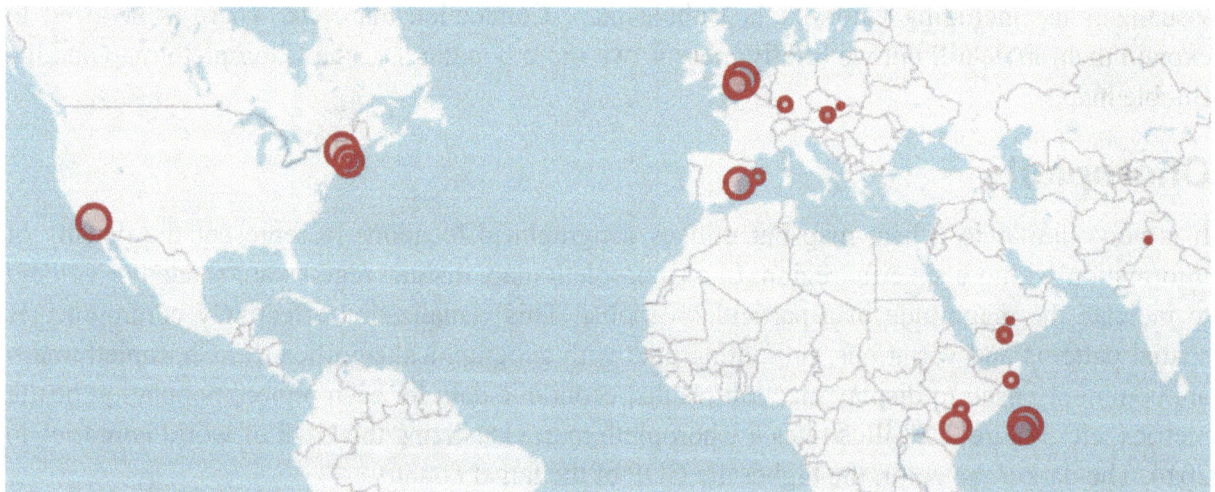

Figure 2-29: Sample bubble map.

60

Figure 2-29 presents an example of a bubble map. You can see from this figure that the population is very much distributed in the west, east, and southeast sides of the United States.

There are also some aesthetically appealing maps that combine, for example, 3D histograms with 2D maps, but we skip explaining them, and building them at the time of revising this chapter requires Javascript knowledge.

Customized and Other Multidimensional Visualizations

For datasets with more than three variables, we have explained that there is no standard method. However, it is common to repeat a visualization if one dimension has low cardinality (the variety of data is small). Sometimes, when there are more than three variables, we should build our own customized and unconventional visualizations. Most data scientists use a straightforward visualization but add them on top of each other (such as Figure 2-27) to present multi-dimensional data. Another example is Figure 2-30 which shows the impact of three parameters for a daily behavior tracking algorithm [Rawassizadeh '16]. These parameters include θ, λ, and a time, which will be one of the following: {5',15',30',60',90',120'}. You can see that the set of {5',15',30',60',90',120'} has a cardinality degree of six, which we can present this dimension as six separate plots.

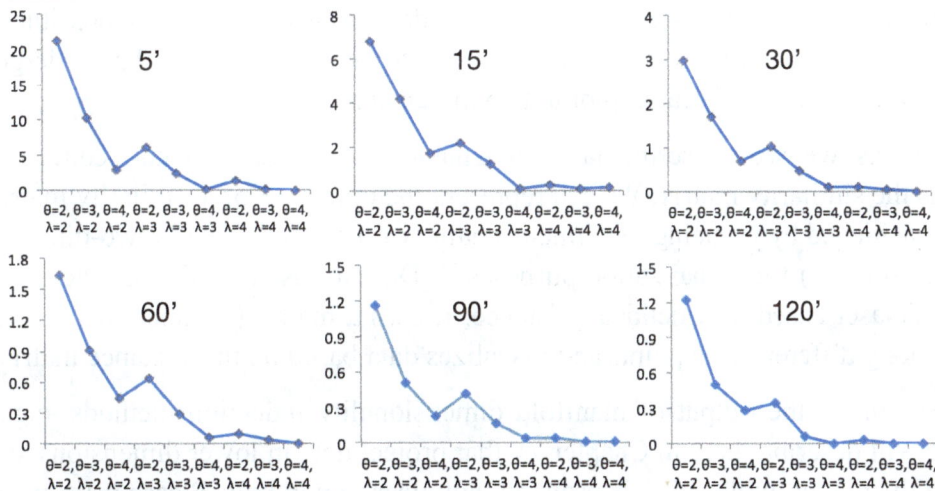

Figure 2-30: A combination of line charts to visualize four variables, including three parameters (θ, λ, and time segments), and the output will be presented in Y Axis.

Sometimes, combining existing visualizations does not resolve our needs, and we must design a customized visualization. For example, we designed Figure 2-31 to visualize three days of smartphone use by a single user [Rawassizadeh '19]. It is not a scatter plot, but it has some

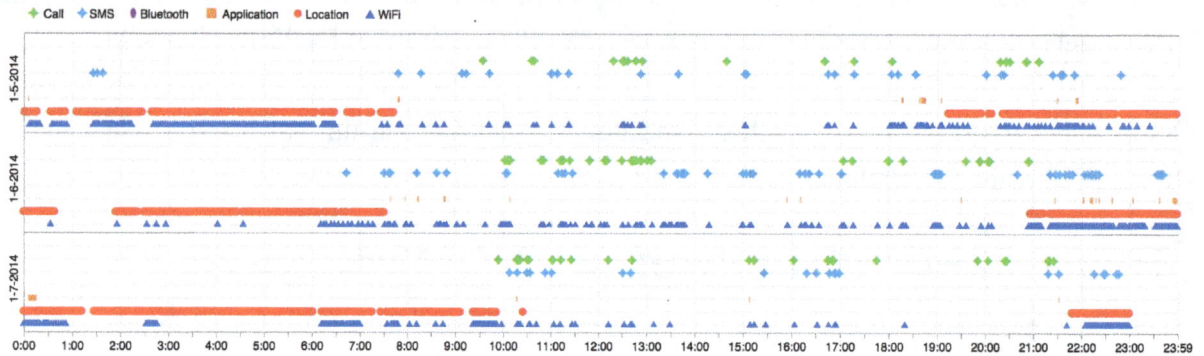

Figure 2-31: An example of three days visualization of user activities. The data is collected by the smartphone.

similarities with a scatter plot. This figure is a multivariate visualization, because it visualizes the call, SMS, Bluetooth devices in proximity, application use, geographical location, and WiFi devices in proximity, all based on their time of the day, collected by a lifelogging smartphone application [Rawassizadeh '13].

Multi-Dimensional Scaling

Multi-Dimensional Scaling (MDS) encompasses statistical techniques designed to project high-dimensional data into lower dimensions, often 2D or 3D, for visualization purposes. The goal is to maintain the distances or similarities between data points as much as possible during this projection process. A commonly employed approach to visualize data that has undergone MDS is to transform it into a colored scatter plot in two dimensions.

For instance, as we will describe later in Chapter 4, the fundamental concept underlying clustering is the similarity matrix. We can represent the outcomes of a similarity matrix through a scatter plot, achieved by reducing the dimensionality of the results data to two-dimension (2D) or three-dimension (3D) for visualization purposes. MDS can visualize the distances between data points of a dataset based on Euclidean distance; it uses a matrix (distance matrix) that includes distances among different data points and visualizes data based on this distance matrix.

Another example is the output of manifold dimensionality reduction methods, e.g., tSNE and UMAP (we will describe them in Chapter 7), that project data in lower dimensions and use MDS to visualize them. Our explanation is probably not clear to you now. Please wait; when we delve into dimensionality reduction methods in Chapter 7, we will elaborate on the MDS visualization. However, because it belongs to visualization, we must list it here.

NOTES:

* Visualization is also used for identifying trends and outliers. It is a common and easy approach to identify that information.

* We can use an area plot to show trends of changes during the time between different time series. We will describe the time series in Chapter 6.

62

* Many visualization literature recommend using the minimum amount of features for visualizing data and removing all other unnecessary information from the shape, including background, grid lines, etc. There is a term used in visualization called *data-ink ratio*. It is a concept introduced by Edward Tufte [Tufte '01]. It quantifies the proportion of ink used to present the data, and if it is removed, the visualization will lose its content. He recommends removing non-data-Ink information from visualization to make it simpler and easier to understand. For example, the right plot in Figure 2-32 has a high data-ink ratio, but by removing unnecessary elements, we end up having a minimal Data-Ink ratio, as presented in the left plot of Figure 2-32.

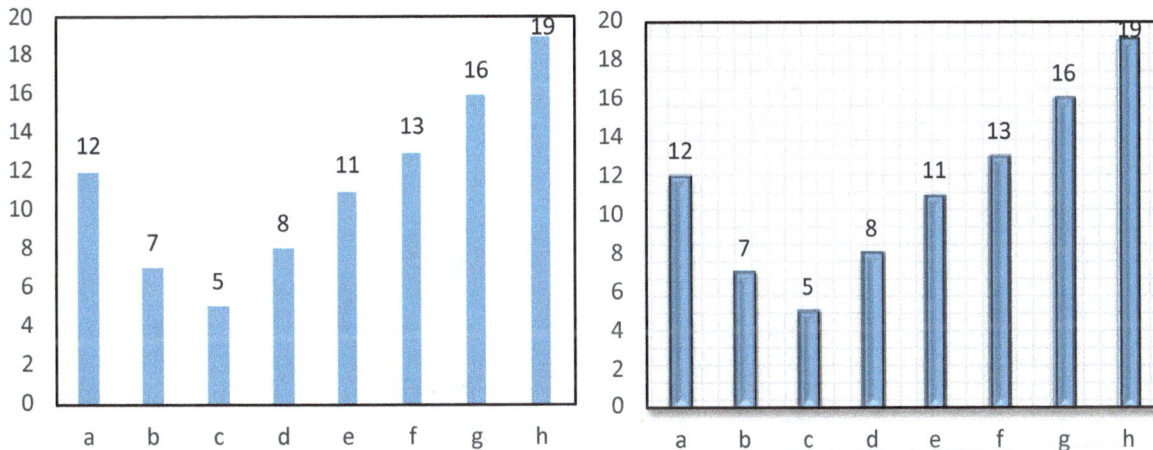

Figure 2-32: (a) low data-ink ratio, (b) high data-ink ratio.

* While visualizing data, consider that your audience might have a visual impairment, so don't rely too much on color to convey your message, e.g., use color for information distinction. Besides, remember the black and white printers while deciding the colors of your visualization if it goes to get printed as black and white.

Summary

Usually, when we start to work with data, we need to perform lots of data wrangling and cleaning before we apply any algorithm to the data. Visualization is helpful because it gives the first impression of the data we intend to work with.

In real-world scenarios, we will get a large dataset, and we will cry for a couple of days to realize what we should do with this dataset.

We recommend starting with visualization, which is one of the best techniques to start working with a new dataset, even if it is not expected and does not stay in your deliverable list. However, this is not always possible; when we are dealing with massive data or unstructured data, visualization does not make much sense.

Table 2-2 summarizes what we have learned in this chapter about visualizations and when to use them. A large category of visualizations that are not listed in Table 2-2 are the ones that we should design and implement on our own for our specific task.

Visualization Category	Visualization Methods
Simple one dimensional data	Piechart, Histogram, line chart, scatter plot
More than Two Dimensional Data	Surface plot, Contour plot, Area plot, Radar chart, Heatmap, Calendar plot, Timeline, Box plot, Violin plot
Hierarchical Data	Dendrogram, Treemap, Sunburst
Graph and Network Data	Node-link Diagram, Arc Diagram, Chord Diagram
Deviated and Changed Data	Divergent barplot, Dumbbell chart, Waterfall, Beeswarm
3D Shapes	Point cloud, Mesh, Voxel
Geographical Map	Choropleth, Bubble Map

Table 2-2: Summary of described visualizations (not customized ones).

Further Reading and Watching

* Chapter 7 of the Abela book [Abela '13] includes a guideline about how to design charts.

* If you are using visualization as an auxiliary tool, we do not believe it is worth investing your time and learning about visualization unless your study subject is something related to design, art, or visualization. This doesn't mean that visualization is not essential, but we often end up using one of the described existing solutions for our data visualization.

* Cole Nussbauer Knaflic has written a book [Knaflic '15] about how visualization can resolve communication issues between engineers and managers by telling a story using data. You can find many more explanations about telling a story with your data. Visualizations are helpful when it comes to communicating your data with both technical and non-technical people. However, that book does not provide much technical content.

* Many plots in this section were designed using a super-secret high-tech software called Microsoft Excel. Also, the *ggplot2* package of R is good. Take a look at the *ggplot2* package of R. It has a number of fantastic examples. At the time of writing this chapter (in 2017), the R programming language had ggplot2 (https://ggplot2.tidyverse.org/index.html), which is a very flexible visualization package but not easy to learn and use. Python also does not fall behind and has good visualization packages, including Matplotlib (https://matplotlib.org) and Seaborn (https://seaborn.pydata.org).

* If you want to build your customized interactive visualization, the best choice is to use Javascript or HTML5. Especially after the introduction of LLMs that can code, they are fairly easy to implement, and there is no need to go into detail.

* Recent advances in data science have led to many new visualizations. Claus O' Wilke did amazing work in making unique guidelines for visualization [Wilke '19]. If you are interested in learning visualization for scientific purposes, we believe this is a good book.

* To draw 3D models, we have used Blender (https://www.blender.org), and to make point clouds and meshes, we have used Mesh Lab (https://www.meshlab.net) and for the voxel visualization, we have used Voxelizer (https://drububu.com/miscellaneous/voxelizer).

Chapter 3: Statistics & Probability

In the previous two chapters, we explained that the first step of working with a dataset is to clean it and draw some visualizations to gain some insight into the structure of the dataset. As the next step after or in parallel to visualization, we could perform some statistical analysis on the dataset.

We use statistics to simplify the process of understanding the complexities of the data. Statistics in the context of data science goes into two branches: *descriptive statistics* and *inferential statistics*. Descriptive statistics involve using data to summarize the narrative within a dataset. In other words, descriptive statistics are employed to characterize the data's features. Identifying narratives in the dataset enables us to draw conclusions about its characteristics. Inferential statistics involve making inferences from data and extracting knowledge, aligning with the goals of machine learning.

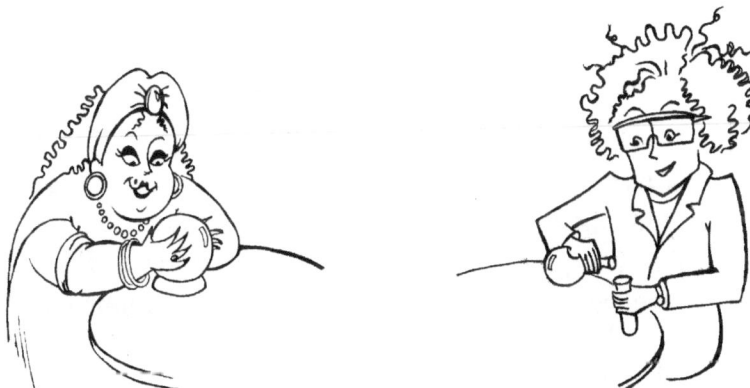

Some old-school scientists who hesitate to learn deep learning believe statisticians are about inferences (like a real scientist) and machine learning makes predictions without evidence (like a fortune teller).

This chapter starts by defining extremely basic but very important concepts that need to be familiar with to learn machine learning. Then, it describes probability, including too many different distributions, but they are all necessary to learn. Next, it describes normalization, followed by a hypothesis test. Afterward, it explains effect size and its related tests. Next, it describes the concept of entropy and information gain. Finally, probability estimation methods will be explored. Note that the focus of this book is not on statistics. The software you use for machine learning, e.g., Python or R, handles the statistical analysis. Therefore, the mathematical details of some methods will be skipped. However, do not skip this chapter. It might be boring, but it is crucial for the rest of your machine learning journey.

Concepts and Definitions

In this section, we define some general terms that we should memorize, and we will refer to them later a lot, even until the end of this book. If you get confused while reading a statistical description in other chapters, please return to this chapter.

Variable and Value

A *variable* is a mathematical object unit that can have multiple *values*. In other words, value(s) is fixed and assigned to a variable, which can vary. A widely used notion to present a variable is to use the x character. For instance, $x = 3$ means that x is variable and 3 is the value. $y = \{1,2\}$ means that y is a variable with two values, 1 and 2.

x	y	z
a_1	b_1	c_1
a_2	b_2	c_2
a_3	b_3	c_3

Table 3-1: Three variables (x,y,z), and each has three values.

The statistical analysis focuses either on a single variable (univariate) or a combination of more than one variable (multivariate). This means that while working with statistics, our focus is usually on a variable or set of variables and not the entire dataset. Therefore, in this section, we are talking mostly about variables and not datasets.

A variable in the dataset refers to one column of data; the variable name is the column name, and its values are column content.

In summary, the column's name identifies the variable, and its value(s) are represented in the column content. When talking about more than one variable, we are usually referring to multiple columns of a dataset. Table 3-1 presents three variables: x, y, and z. Each of them has three different values. It means, we match column name to variable and column data to values.

Random or Stochastic Variable

In statistics, we will encounter the term random variable a lot. A *random variable* or *stochastic variable* is a variable whose possible values are numerical outcomes of a random phenomenon. These values are not fixed but are determined by the outcomes of the random process.

Continuous and Discrete Variable

We call a variable a *continuous variable* if, between two continuous variables, there are an infinite number of other values, such as the numbers between 3.1 and 3.3. There will be infinite numbers between these two numbers, e.g., 3.297824, 3.1435345834, and so forth.

In contrast, a *discrete variable* can have countable and finite values only, such as days of a week that are only seven possible choices, Monday, Tuesday, etc.

Different Data Types

In the previous chapter, we explained that a data object or an event will be repeated in a dataset, which is the nature of any scientific phenomenon. If a value of a data object is not a number but is derived from a set of known objects, these values are called *categorical (nominal) data*. If a value is a number, we call it *numerical data*.

Ordinal data are categorical data that are ordered based on a condition(s), such as "Monday, Tuesday, .., Sunday" or "high, medium, low".

Interval data is another type of variable with a meaningful order, and the intervals between values are equal, but there is no true zero point. A classic example of interval data is temperature (in Celsius or Fahrenheit). The difference between 10°C and 20°C is the same as between 20°C and 30°C, but 0°C does not mean the absence of temperature. Another example is calendar dates.

Ratio data is a specific type of variable and has all the properties of interval data but with an important addition: a true zero point. This zero point indicates the absence of the quantity being measured. Common examples of ratio data include weight, height, age, and income.

To summarize these three data types, Ordinal data is categorical data with an order. Interval data is numerical data with equal intervals between values but no true zero point. Ratio data is numerical data with equal intervals and a zero value. Ordinal, interval, and ratio data are not very common in a machine learning context, but we should remember this definition, and when we encounter it in a text, do not freak out.

Dependent, Independent, and Control Variables

Other important definitions are characteristics of a variable in statistical inferences or mathematical equations. We have three types of variables: *independent variable, dependent variable,* and *control variable*.

Dependent variables (output) represent the outcome of an analysis or study.

Independent variables (input) influence a dependent variable's value (output). In simple words, consider the independent variable (input) as something that causes changes to the dependent variable (output).

Control variables may influence the dependent variable(s), but we are not interested in studying them. Sometimes, we intend to keep the control variables unchanged.

Figure 3-1 shows an example of these variables. We are interested in getting more fruit from our tree. We give water (independent variable) to a tree to get its fruits (dependent variable).

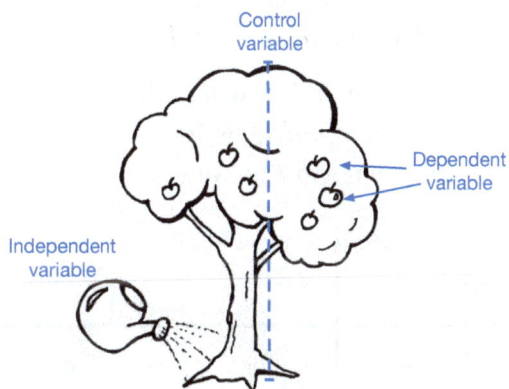

Figure 3-1: An example of different variable types.

Meanwhile, the tree is a certain height, but the height is not important in our study (control variable). Note that variables' roles are dependent on a study; for instance, in one study, a variable could be independent, but in another study, the same variable could be a dependent or control variable. In another example, we might be interested in the height of the tree, based on the soil, and thus, it will be considered as the dependent variable.

Independent versus Dependent Trials

While sampling for statistical analysis, the collected data are called *trials* or *samples*. They could be dependent or independent. If we flip a coin 10 times, it gets head or tail each time. However, each trial (coin flipping event) does "not" have any connections to the previous trials, and thus we call each coin flipping event an independent trial.

In some trials, the outcome of a trial is *dependent* on the previous trial. For instance, consider the scenario in which our friend starts to learn machine learning and gets depressed by not learning it. Now, our friend should use Prozac daily to cope with his depression. If he gets one Prozac daily, its impact is good on his mood; if he gets two Prozacs, its impact is better than one pill; the third Prozac makes him sleepy, and the fourth Prozac pill makes him very depressive. In this case, we have dependent trials of using Prozac pills per day.

A classic example is to take a ball from a box that contains red and blue balls. We randomly took one ball from that box, and it was blue in color. Since one ball is removed from inside the box and it is blue, the chance of taking out a blue ball again is reduced, and the chance of having the next ball red increases. This is also a dependent trial. To summarize; an independent trail means *the previous trails are not affecting the current trail.*

First Insight on the Data & Basic Statistical Concepts

In this section, we describe basic statistical concepts, including different types of *mean*, definitions of the *median, variance, standard deviation, covariance, quartile, and whisker plot (box plot)* with examples. If you are familiar with these concepts, you can skip this section, but there are some concepts, such as *multimodal data*, that you might encounter while working with real-world data and miss them if you do not read this section.

Assume we are working with a dataset of chickens in aviculture. Table 3-2 presents the weight and the number of chickens for each weight in one room of the aviculture. This table could be called a *frequency table* as well.

Weight	0.2kg	0.4kg	0.5kg	0.6kg	0.7kg	0.8kg	0.9kg	1kg	1.2kg
Number of chickens	1	2	6	8	7	4	3	1	1

Table 3-2: Frequency table that presents of chickens for each weight.

If we plot the frequency or distribution of these numbers in a histogram or line plot, we get Figure 3-2. We can use Figure 3-2 to see that the most frequent weight of chickens is between 0.5 and 0.7. This is a very simple inference, but we can use statistics to understand the characteristics of this dataset in more detail.

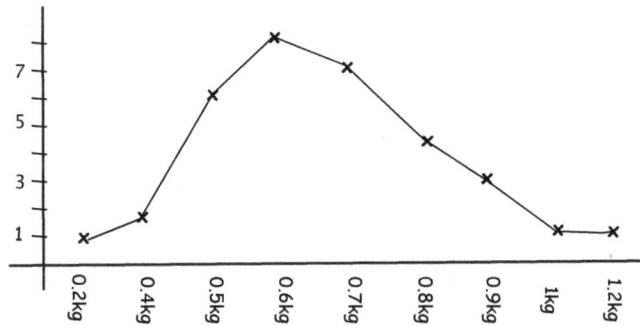

Figure 3-2: Distribution of chickens based on their weight.

Mean

Arithmetic Mean

Mean (average), or *arithmetic mean*, is shown as μ, and it is calculated by summing all values of a variable (weights are variable, and the number of chickens is the value), then dividing by the number of variables (chickens, the total number of chickens = 34). The mean for this example is computed as follows:

$$\frac{0.2 \times 1 + 0.4 \times 2 + 0.5 \times 6 + 0.7 \times 7 + 0.8 \times 4 + 0.9 \times 4 + 1 \times 1 + 1.2 \times 1}{34} = 0.53$$

The *mean* is used when the data is approximately *symmetric*. If the data distribution is skewed, then it is better to use the *median* and *mode* along with the mean to understand the average in the dataset. Symmetric in this context refers to drawing a vertical line from the peak of the data crossing the x-axis, i.e., 8 on the y-axis and 0.6 kg on the x-axis in Figure 3-2. The result will be two shapes, and if they have an equal size, they are called symmetric. If you draw such a line, you can realize that Figure 3-2 is not symmetric (it is asymmetric). It is something usual when we work with a real-world dataset, and if we need to have a symmetric shape, we can sometimes get closer to symmetry by increasing the sample size (dataset size). The reason for favoring having a symmetric shape will be described later in detail. Note that many statistical and machine learning methods can handle asymmetry.

Geometric Mean

The geometric mean is the nth root of the product (multiplication) of n values is computed as follows.

$$Geometric\ mean = \sqrt[n]{x_1, x_2, \ldots x_n}$$

The geometric mean is very sensitive to zero and outliers. By sensitive, we mean zero or outliers that have a significant impact on the result. The geometric mean is used when we want an average in *multiplicative conditions* (subject to multiplication) or when we want to know *the product of values*.

For instance, the capacity of each chicken room in our aviculture is measured with three dimensions: width, length, and height. Therefore, to compare different chicken rooms with a number, we use geometric mean, i.e., $\sqrt[3]{width \cdot length \cdot height}$. As another example, let's analyze a 'Chicken Entertainment Inc.' stock. In the first year, their stock grows 50%, next year, it grows 20%, and in the third year, it grows 10%. Therefore, in the first year, the stock is 1.5 times higher than the previous year, then the second year is 1.5×1.2 higher than two years ago, and the third year is $1.5 \times 1.2 \times 1.1$ higher than three years ago. Since the growth in each year affects the value in future years as well, the geometric mean is effective at measuring the average annual growth. To calculate the average growth in stock, we can also use the geometric mean.

Harmonic Mean

The *harmonic mean* is useful when numbers are defined in a relationship to something. It is capable of dealing with outliers in the dataset that cannot be removed in the preprocessing stage. For example, we are studying a new medication's impact on participants, and outliers are important, so we can't remove them.

The outlier affects the data, but in some cases, we can not remove them. If we have *n* numbers of variables, the equation to compute harmonic is written as follows.

$$Harmonic\ mean = \frac{n}{\frac{1}{x_1} + \frac{1}{x_2} + \ldots + \frac{1}{x_n}}$$

For instance, assume a chicken running from a cat. The running chicken runs the first 8 meters at a 2 km/hour speed. The following 12 meters at 5 km/hour and the last 30 meters at 25 km/hour speed. The arithmetic mean of the running chicken is 10.6 km/hour, which is not precise. However, by using a harmonic mean as follows, we get 16.6 km/hour, which is a more realistic estimation.

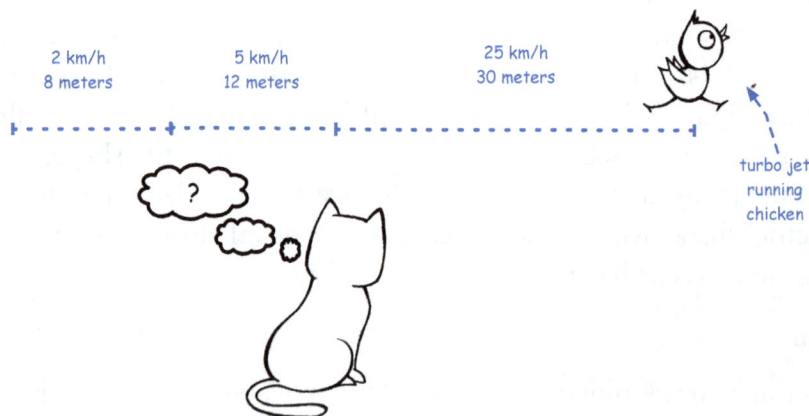

When we encounter a text referring to mean without saying its type (arithmetic, geometric, harmonic), we should assume it is an arithmetic mean. Here, we do the same; when we say mean or average, we refer to the arithmetic mean. Otherwise, we explicitly mention the type of mean.

Median

The median is calculated by ordering the discrete data and identifying the number in the middle of the ordered set. For example, assume we have a dataset of chicken weights, ordered, as follows {1,1,1,2,4,4,6,7,8}, and the median in this dataset is 4, which is the fifth element. If the set of numbers is odd, the middle one is the median, but if the set of numbers is even, there will be two medians, the two middle numbers. The median is useful to generalize data when the peak of the data (Figure 3-2) is skewed toward the right or left, and the curve is not symmetric. In these cases, the mean is not the best representative of the average from the data. For instance, assume we have a dataset as follows: {1,2,3,4,5,6,7,8,85,88}. In this dataset, we have many small numbers and few big numbers. The mean is 20.9, and the median is 6. However, the number 6 seems to capture better where the middle of the data is, and it is more descriptive for this particular dataset.

Mode

The mode is the value in the dataset that occurs with the highest frequency in the dataset. For instance, the peak of the curve in Figure 3-2 belongs to 0.6 kg. This means that the mode is 0.6, or, in other words, most of the chickens (eight of them) weigh 0.6 kg.

Sometimes, a dataset has more than one peak, such as Figure 3-3, which has three different peaks. These datasets are known to have *multimodal* distribution. Many mix multimodal with multivariate, but they are not the same concept. Recall that multivariate means we have multiple variables, such as Figure 2-12 in Chapter 2, in which variables are not necessarily related.

In statistics, multimodal means having multiple peaks in one single dataset, but in fields outside statistics, such as computer science or psychology, multimodal can refer to information from different sources, but for one specific action as well, e.g., facial state, smile, and voice are different information, which can be used to express the human emotion (i.e., the specific action).

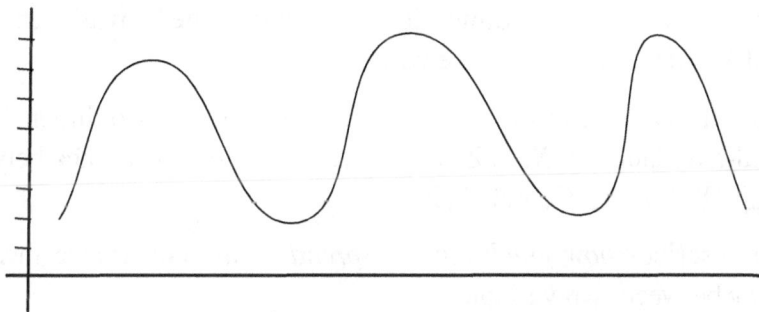

Figure 3-3: Example of a multimodal dataset that has three modes.

These methods are used for discrete data. For continuous data, the mode and median are calculated differently. We don't present these calculations at this stage.

Variance, Standard Deviation, and Covariance

$$cov(X, Y) = \frac{\sum_{i=1}^{n} (x_i - \mu_x)(y_i - \mu_y)}{n}$$

Variance measures the variability, spread, or inconsistency in a dataset. The variance of the discrete finite-sized dataset is the following equation (σ^2):

A low variance in the dataset means more consistency among data points. A high variance among data points means they are less consistent.

Standard deviation is the root square of variance and is shown as *SD* or $\sigma = \sqrt{variance}$. It is used to measure the distance from the mean. In other words, standard deviation describes how far the data is from the mean or how far they deviate.

The variance equation shown above is used for the census and not sampled data. A *census* in the context of statistics refers to a process of obtaining data from every member of a dataset. In contrast, *sampling* refers to the use of a policy to select some members of a large dataset and not all of them. The variance of the sampled dataset has *n-1* instead of *n* in its denominator. More details about the sampling will be explained in this Chapter and later in Chapter 16.

The standard deviation or variance belongs to one-dimensional data. However, if our dataset is multiple-dimensional data, we use covariance instead of variance. Note that the covariance calculation is always used for two dimensions of data. The covariance of the vector $X = \{x_1, x_2, \dots\}$, and $Y = \{y_1, y_2, \dots\}$ will be written as follows:

The variable names are the same, except we add X and Y, which are two dimensions. If we have more than dimensional-two data, e.g. X, Y, Z. We can calculate the covariance between every two pairs of data, including $(X, Y), (X, Z)$, and (Y, Z).

We can say variance describes *how much data is spread*, while covariance measures both the *spread* and *dependency* between two variables.

When multiple dimensions are considered, covariance can be shown as a matrix. For example, if we have three variables in our dataset as the following matrix, the covariance between column 1 and column 3 is presented in the covariance matrix at cells [3(row),1(column)], and cell [1,3], i.e., 2.66. Respectively, the covariance between column 1 and column 2 is presented in the covariance matrix at cells [1,2] and cell [2,1], i.e., 0.66. With the same approach, we can find the covariance between column 2 and column 3, which is 0.88.

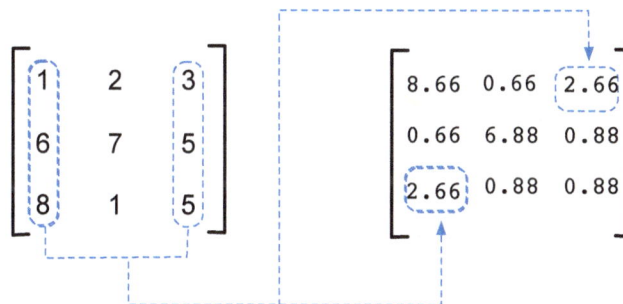

$$\begin{bmatrix} 1 & 2 & 3 \\ 6 & 7 & 5 \\ 8 & 1 & 5 \end{bmatrix} \qquad \begin{bmatrix} 8.66 & 0.66 & 2.66 \\ 0.66 & 6.88 & 0.88 \\ 2.66 & 0.88 & 0.88 \end{bmatrix}$$

Range and Quartile

The range can describe *how data is spread*, and unlike mean and variance, it doesn't say anything about the distribution of the data. In other words, range refers to the difference between maximum and minimum values in a dataset.

To calculate the range of a variable, the values of this variable should be "ordered". Then, we subtract the largest number, i.e., *upper bound*, in the dataset from the smallest number, i.e., *lower bound*. For example, the range for the dataset given in Table 3-2 is calculated as follows: $1.2kg - 0.2kg = 1kg$. The range has a considerable weakness. In real-world datasets, we always have outliers. If there is one weird, alien, undocumented immigrant number (outlier) in the dataset, it affects the range. For instance, if the dataset has an outlier and instead of chickens, it is a sheep, which weighs 30 kg, then the range of Table 3-2 will be $30kg - 0.2kg = 29.8kg$. The weight of *29 kg* is very odd in the dataset of chickens' weight.

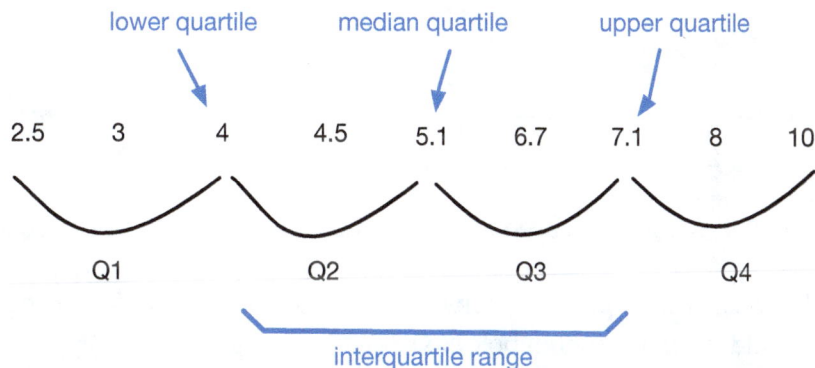

Figure 3-4: A dataset is ordered in ascending order to identify its quartiles.

To mitigate errors caused by outliers, first, we order the dataset in ascending order. Then, we should split the dataset into four segments of equal size. Each of these segments is called a *quartile*. For instance, assume we have a dataset of nine chickens with different weights. The chicken weights are ordered as shown in Figure 3-4. Here, 4 is the minimum quartile, which is called the *lower quartile*. 7.1 is the *upper quartile*. The middle number is 5.1, and it is called the *median quartile* (see Figure 3-4). After we defined four quartiles, instead of calculating the range by subtracting the upper bound from the lower bound (10-2.5=7.5), we subtract the upper quartile from the lower quartile, 7.1 – 4 = 3.1. It is called the *interquartile range*. In other words, *Interquartile range = upper quartile – lower quartile*.

Interquartile is handling outliers by neglecting the 25% of the data from both the beginning and end sides of a dataset; it focuses on the 50% of data in the middle, and thus, *it can resolve the negative impact of outliers*.

Quartiles are dividing a dataset into four segments. What about dividing it into 100 pieces? In this case, we can say our data is transformed into *Percentile*. Quartile, percentile, and median are used to make a narrative for the dataset.

Whisker plot or Box plot

There is a specific visualization to plot the result of quartiles and ranges. It is called the *Whisker* plot, *box plot*, or *box and whisker diagram*. Figure 3-5 presents a single box plot, but usually, they are different objects used together, as shown in Figure 3-6, and each data object is described in a box plot.

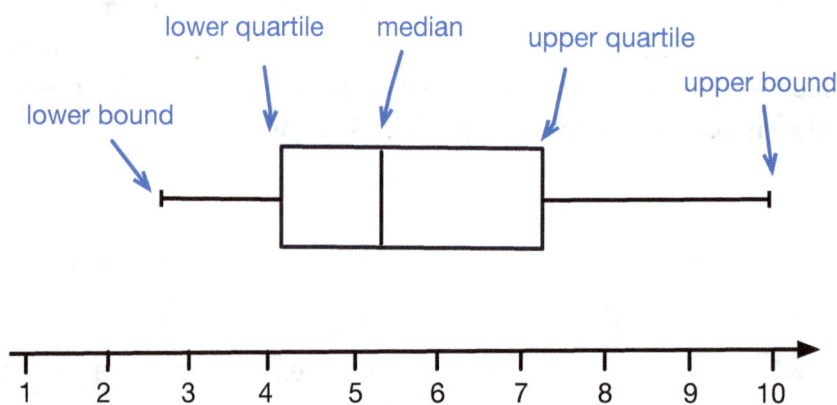

Figure 3-5: Whisker plot used to visualize quartiles, and upper/lower bounds.

A question might arise: what type of narrative could we say about the data using the box plot? Or what inferences could we make within box plots?

To answer this question, we use an imaginary scenario: you are a lovely animal lover who tries to provide shelter for feral and stray cats in the city but can't provide for all of them. Three cats came to your street: Cat no.1, Cat no.2, and Cat no.3.

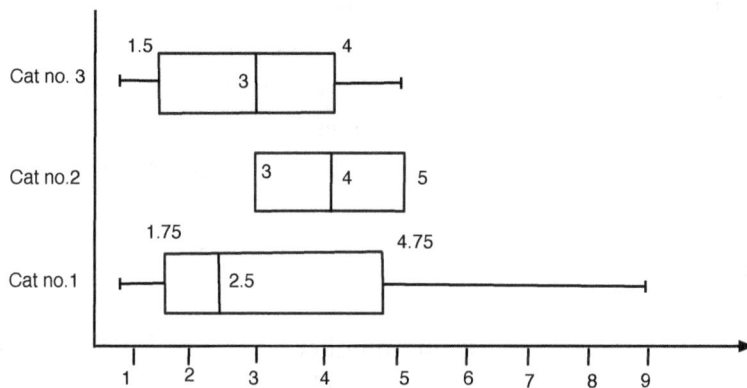

Figure 3-6: Whisker-plot of our street cats and their injuries in fight.

Cat no.1   Cat no.2   Cat no.3

If you live in areas that have feral cats, you hear their fighting voices often, especially in mating seasons. These three cats also fight a lot and get many injuries after each fight. You like to help them, but you have space to rescue only one. You decide to save the weakest ones who get the most of the injuries. You collect some data about their injuries from their previous fights. Table 3-3 presents the injuries of each cat in the last 10 fights they had.

Cat no.1	1	2	3	4	7	9	4	2	2	1
Cat no.2	4	3	4	5	5	3	5	3	3	4
Cat no.3	3	4	5	3	5	1	4	2	2	1

Table 3-3: Number of injuries each cat receives in the last 10 fights

However, by looking at this table, we still can't understand which one is the weakest one. At first glance, it might seem that Cat no.1 is the poorest cat because it has once 7 and another time 9 injuries. Also, look how tiny and dumb he is. Nevertheless, box plots in Figure 3-6, drawn from Table 3-3, show that Cat no.2 is the weakest one, and has received the most amount of injuries. Hence, we should don't trust our eyes when math is available, and as a result, it is better to adopt cat no.2 instead of others. Box plots are similar to bar charts and can be drawn horizontally as well.

Degree of Freedom

In statistics, the *degree of freedom* refers to the number of values in the final calculation of a statistical method that is free to vary. Degree of Freedom is calculated as the total number of observations (N) minus the number of parameters estimated (P), i.e., N-P. In some tests, we use N-1 instead of N or P-1 instead of P.

Let's consider a situation where you're tracking your calorie intake over a month while allowing yourself a weekly sweet treat. Although you are on a diet, you allow yourself to take one sweet treat per week. Over a month, you have four weeks to eat the sweet, and you've selected four different sweets: a chocolate bar, ice cream, a piece of cake, and a small candy box. Each week, you will choose one of these sweets, but the total calories from these treats must not exceed a certain limit for the month. Let's say your calorie limit for all these sweets together is 1000 calories. If the chocolate bar is 300 calories, the ice cream 200 calories, the cake 250 calories, and the candy box 250 calories, you have some flexibility (degrees of freedom) in how you consume these treats across the weeks, but the total must always sum up to 1000 calories or less.

In a statistical sense, the degree of freedom here is the number of choices you can make before your options are constrained by the calorie limit. For the first week, you can choose any of the four sweets (the degree of freedom is 3, because once you choose one, you're left with three items whose total calories must sum to a certain number). In the second week, your choice is slightly more limited because you need to keep the total calorie count under the limit, and so on. By the time you reach the last week, if you have adhered to the calorie limit, you will likely have only one sweet left that fits into the remaining calorie allowance. Therefore, your degree of freedom is reduced to 0 since your choice is entirely determined by your previous choices. This example illustrates how degrees of freedom in statistics represent the number of independent choices left before constraints (like total calorie count) limit your options.

Probability

Probability is one of the core needs in machine learning, and the poor author of this book spent a huge amount of his time learning it. Here, we start with basic probability concepts and then we move to explain distributions that are used in machine learning and deep neural network models.

Basic Probability Concepts

Probability is the extent to which an event is likely to occur. In other words, how likely an event is probable is specified by a probability. For instance, The probability of rolling a dice and getting six is 1/6. We use $P(x)$ to demonstrate the probability of a variable x.

Tossing a coin has two possible events, including getting head (H) or tail (T), and their probabilities are equal, so we can say that $P(H) = 0.5$ and $P(T) = 0.5$. The probability function is formalized as follows:

$$P(desired\ event\ happen) = \frac{number\ of\ desired\ event}{total\ number\ of\ events}$$

A probability can have a value range from 0 = "will never happen", to 1 = "always happen". $P(x')$ means the probability of not having x is: $P(x') = 1 - P(x)$. This is called the *marginal probability* rule. The marginal probability rule, also known as the sum rule, is used to find the probability of an event occurring without considering any other events or conditions.

Probability has its own rules and definitions. We describe some of the important probability rules.

Joint Probability

Imagine you are at a party, and there is a bowl of nuts. The host brings it to you and asks you to take some nuts. Your favorite nuts are pistachio and cashew. The pot includes both of them, but it also involve lots of hazelnuts and almonds, which were not your most desired nuts.

Assume that the pot includes 10% pistachio, 20% cashew, 40% hazelnut, and 30% almonds. To be polite, you intend only to take two pieces of nuts from the pot without choosing them. What is the probability of having only one pistachio and one cashew in your hand? In this case, we should use the joint probability.

A *union* between two events is written as follows: $P(A \ or \ B) = P(A \cup B) = P$ (*A* occurs, or *B* occurs, or both occur). In some cases, when events overlap, we compute the joint probability as follows: $P(A \cup B) = P(A) + P(B) - P(A \cap B)$.

The union of events A and B, denoted as $A \cup B$, means that on a particular trial of the experiment, either A or B occurred (or both did). The following Additive Rule of Probability is a useful formula for calculating the probability of $A \cup B$.

$$P(A \cup B) = P(A) + P(B) - P(A \cap B)$$

Respectively, a *joint probability of intersection* between the two events happening is written as follows: $P(A \ and \ B) = P(A \cap B) = P$(*A* occurs and *B* both occur). In other words, in this case, we have $P(A \cap B) = P(A).P(B)$. This applies only when events A and B are independent. Independence means that the occurrence of one event does not affect the probability of the occurrence of the other event. In cases where events A and B are not independent, the general formula for the joint probability is: $P(A \cap B) = P(A).P(B|A) \ or \ P(B).P(A|B)$.

A *conditional probability* is shown as $P(A|B)$ and it means the probability of A given B has occurred. The sign '|' reads as "given" and the conditional probability is computed as follows:

$$P(A|B) = \frac{P(A \cap B)}{P(B)}$$

This form of conditional probability is also called *Kolmogorov probability*.

For our example $P(Pistachio \cap Cashew) = (10/100).(20/100) = 0.02$. It is very unlikely, and you have only a 2% chance of being successful. However, some statisticians might say that we should consider removing one item and thus have: $(10/100).(20/99) \approx 0.02$ or $(10/99).(20/100) \approx 0.02$ Still, with this approach, the chance is very low.

These numbers show that you should reduce your expectations. What about having at least one pistachio or at least one cashew from your selection of two nuts? In this case, we can use the union probability to find this.

Therefore, we have:

$P(Pistachio \cup Cashew) = P(Pistachio) + P(Cashew) - P(Pistachio \cap Cashew) =$

$10/100 + 20/100 - 0.02 = 0.28$

Based on the result, it is more likely that you will get at least one of your two preferred nuts compared to the probability of having both pistachios and cashews.

Bayes Rule

Bayes Rule is an alternative approach to calculate P(A|B). It can be derived from the conditional probability above and written as follows:

$$P(A|B) = \frac{P(B|A).P(A)}{P(B)}$$

We skip its mathematical proof because this book is not about mathematics. To better understand the use of the Bayes rule, consider a scenario where eating too many nuts increases the person's triglycerides. Suppose 20% of guests have high triglycerides, we have $P(Trig) = 20\%$. Also, 40% guests love nuts and eat lots of nuts, i.e., $P(NutLove) = 40\%$.

You know that *60%* of those who love nuts have high triglycerides $P(Trig|NutLove) = 60\%$. Now, we can find if loving nuts might cause high triglycerides or find $P(NutLove|Trig)$.

$$P(NutLove|Trig) = \frac{P(Trig|NutLove).P(NutLove)}{P(Trig)} = \frac{0.6 \times 0.4}{0.35} = 0.68$$

The result says it is 68% of people with high triglycerides love nuts. There is a direction in machine learning that uses probabilistic reasoning on data, but we skip it in this book and later in Chapter 9 explain Naive Bayes rule classification.

Probability Density/Mass Functions and Cumulative Distribution Functions

The *Probability Density/Mass Function (PDF/PMF)* shows how dense the probability is at each data point. In other words, a variable could have different values, and PDF or PMF shows how likely a value can be assigned to the target variable. For discrete data, we use PMF, and for continuous data, we use PDF. PDF doesn't show the probability of a value (since it's zero) but rather helps in calculating the probability over a range of values.

Imagine you are living in the future; corporation greed and corrupt governments have ruined the planet's resources. There are new businesses that make good profits. Edible insect cultivation is one of those businesses. You are a world-known AI expert and recently got a new consulting project.

An insect farmer came to you and asked you to use data science and other scientific methods to improve the taste of his bugs. He told you that he guessed that rain positively impacts his bugs' moods and a good mood will result in tastier bugs.

We should measure the amount of rain to better understand insects' moods. Figure 3-7 (a) shows the amount of rain in centimeters and the number of days for each rainfall, which is a discrete variable. In this figure, 6 *cm* of rain is the most frequent amount of rain we had because it occurred in eight days.

Note that the sum of probabilities in PMF is equal to 1, and the area under the curve in PDF is equal to 1. Why do we use PMF and not PDF? Because rain per day is a discrete value, we don't have any other dates between yesterday and today.

Cumulative Distribution Function (CDF) is a function that describes a *distribution* of a variable (either discrete or continuous variable). In other words, CDF is a function that describes the probability that a variable (discrete or continuous) will take a value less than or equal to a specific value.

To plot the CDF of our example, first, we need to plot the PMF because we are dealing with a discrete variable, i.e., the amount of rain. Figure 3-7 (b) presents the PMF of Figure 3-7 (a). Figure 3-7 (c) presents the CDF plot, which looks like steps, and each step is the size of the PMF. The *Y*-axis of the PMF (or PDF) diagram shows the density of a value, which is a number between 0 and 1, and the *X*-axis represents the values as in PMF.

By using CDF, we can answer the probability that rain is going to be less or larger than a particular value. For example, what is the probability of having less than *6 cm* or $P(x < 6cm)$ rain? In other words, CDF is the probability of being less or greater than x (x is a value). To answer this question, by plotting CDF, we can add up all probabilities until that particular point

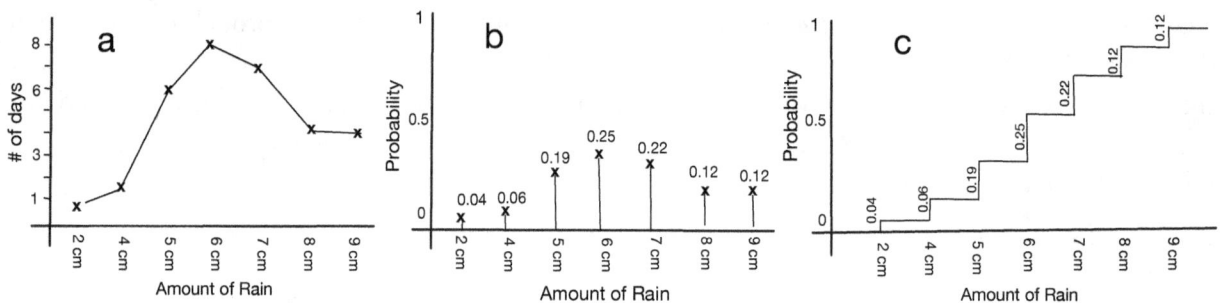

Figure 3-7: (a) The amount of rain based on the frequency of rain, (b) the PMF of rain, (c) CDF of the Rain.

from the CDF plot. Therefore, it will be $0.04 + 0.06 + 0.19 = 0.29$.

Once again, let us remind you that in this example, we consider rain as discrete data and use a histogram to plot its PMF. PDF uses a line chart because it is used for continuous data.

Figure 3-7 (b) shows the PMF plot, which is the sum of all probabilities, and it should be equal to 1. Then, by using the PMF values, the CDF will be designed as shown in Figure 3-7 (c). You can see from Figure 3-7 (c) and Figure 3-7 (b) that the size of the steps in CDF is taken from values in PMF.

Statistical Distribution

A dataset has a characteristic like ours. For instance, an unknown person (assuming we are not talking about the first author of this book) is overeating, but he complains that he has a slow metabolism and easily gains weight. It is his characteristic. A dataset has characteristics, too, but instead of using plain language to describe it, we use statistical distribution to describe its characteristics, i.e., descriptive statistics.

Some models and algorithms, such as Gaussian Mixture Model clustering or generative AI models, operate based on the data distribution. Therefore, we use inferential statistics to make a decision (e.g., group the data) about the data. You can check the earlier sections in this chapter to recall the differences between inferential and descriptive statistics.

Again, we remind ourselves that any variable must include repetitive values in a dataset because any scientific phenomena should be reproducible, and the entire notion of machine learning and data science operates based on the notion of reproducibility and repeatability. Usually, the more repeated values we have in the dataset, the better the accuracy of our machine learning algorithm.

> If you skip learning so many distributions, you will suffer in your entire Machine Learning Life.

> Oh, I forgot to introduce myself: I am a good-news-potato, and I always scare the hell out of you by bad news.

Sometimes, we smooth the dataset to increase its repetitiveness. More about generalization and smoothing will be explained later in other chapters.

We can use distribution to describe a narrative about a dataset or for the prediction and sample it for generative AI applications. Distribution usually focuses on one single variable, such as a column in a CSV file, which is called a *univariate distribution*. In other words, it presents *how often all possible values of a single variable occur* (probability or frequency of their occurrences). We can plot the distribution of a variable in a two-dimensional space, i.e., the X-axis presents values, and the Y-axis presents the *volumes*. A multivariate distribution is a distribution that can hold more than one variable (see Figure 3-8).

There are specific known distributions, but a real-world dataset might not always fit into any known distributions [Skiena '17]. However, being familiar with the known distribution and their use. Whatever you do with data, there will be a nerd with statistical knowledge to say: *"I can't believe this until you show me the data distribution."* Therefore, plot your data distribution before using any machine learning model or algorithm.

In the following, we describe distributions and when to use each distribution. Be patient if they sound boring; you need to understand them, and they are very crucial for the rest of this book, especially when we discuss neural networks. Keep in mind that any distribution can be mathematically formalized (described) using *means, variances, covariances*, and perhaps other additional parameters.

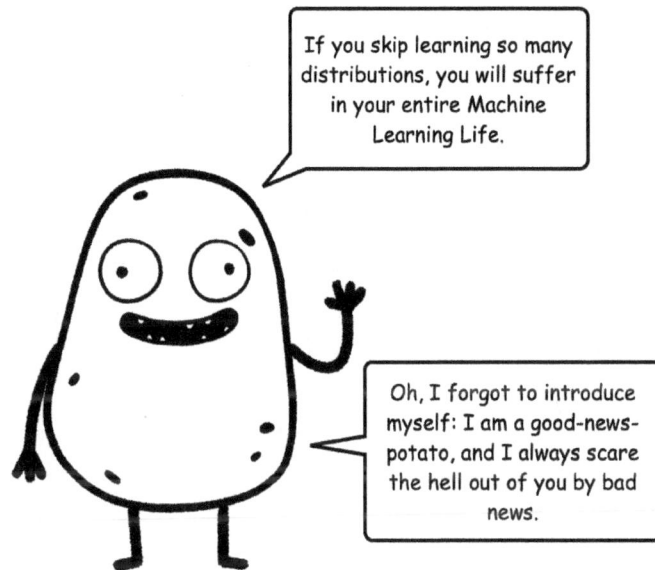

Some distributions are discrete, and some are continuous. Besides, we describe their PDF or PMF as well, but to help us keep our brain unexploded by the end of this chapter, we skip explaining their CDF, and you can check them online.

Normal (Gaussian) Distribution

The most common distribution that exists in nature is *normal distribution,* also known as *Gaussian* distribution or *Bell-shaped* distribution. It is typically used for a continuous variable. This distribution has a symmetric bell shape. As you see in Figure 3-8, the peak of the curve in the normal distribution is the mean, and most of the data in this distribution are located near the mean. Having a symmetric shape is an ideal case for studying a distribution.

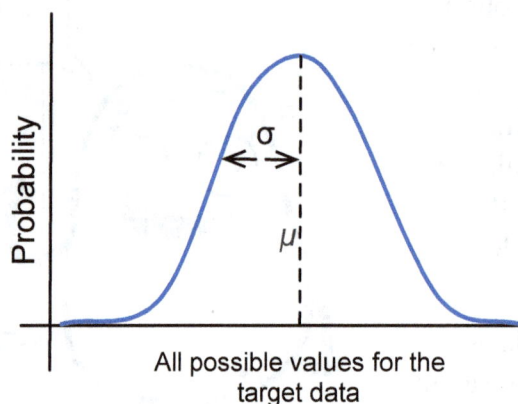

Figure 3-8: Normal distribution with its bell shape curve

The Gaussian distribution is very prevalent because often, in social and natural sciences, a random variable that has been appropriately collected should follow a normal distribution. Also, it has several sub-types of distribution, such as *z-distribution* or *t-distribution*, which we will explain very briefly later. Gaussian distribution is always defined by mean μ and standard deviation σ, as shown in Figure 3-8. The following equation presents the PDF of Gaussian distribution for any given variable of x. Here μ is the mean and σ is the standard deviation.

$$f(x) = \frac{1}{\sigma\sqrt{2\pi}} e^{-\frac{(x-\mu)^2}{2\sigma^2}}$$

If you can't recall from school both Pi (π) and Euler (e) numbers ($\pi = 3.14$, and $e = 2.71$) are constant numbers in mathematics.

A multivariate normal (Gaussian) distribution is using a covariance matrix instead of variance. For example, a two-dimensional normal distribution uses the following vector for the mean and covariance matrix (Σ), assuming ρ is a correlation between two dimensions.

$$\mu = \begin{bmatrix} \mu_1 \\ \mu_2 \end{bmatrix}, \quad \Sigma = \begin{bmatrix} \sigma_1^2 & \rho\sigma_1\sigma_2 \\ \rho\sigma_1\sigma_2 & \sigma_2^2 \end{bmatrix}$$

To understand ρ lets assume we use bivariate Gaussian distribution (the simplest form of multivariate Gaussian distribution). Assuming we have X and Y variables with mean of μ_X and μ_Y and standard deviation of σ_X and σ_Y, their correlation coefficient is computed as follows:

$$\rho = \frac{Cov(X, Y)}{\sigma_X \sigma_Y}$$

Usually, a normal distribution is written with calligraphic N and its mean and variance, as follows: $\mathcal{N}(\mu, \sigma^2)$.

Uniform Distribution

A uniform distribution (rectangular distribution) is a distribution in which all its *outcomes are equally likely* (they have the same probability). For example, while throwing dice, all chances have equal probability; we have 6 options, and the chance to get a particular number is 1/6. Another good example is flipping a coin, we can get either head or tail, and both have an equal probability of 1/2. These two examples are discrete uniform distributions.

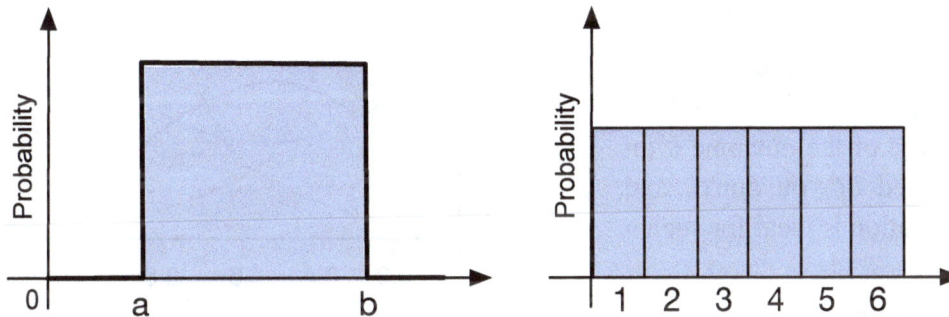

Figure 3-9: (left) Continuous uniform distribution (right) dice tossing distribution which is discrete uniform distribution.

Uniform distribution could also be from continuous data. For example, an algorithm that generates random data has a continuous uniform distribution. The result of trials lay between certain bounds in this distribution, and these bounds are defined as parameters a (minimum) and b (maximum), and the Uniform distribution is written as $U(a, b)$.

The left part of Figure 3-9 presents a shape of continuous Uniform distribution, and the right part presents the Uniform distribution of tossing a dice, which can have one of the six equal outcomes. The PDF of continuous Uniform distribution is written as follows:

$$f(x) = \begin{cases} \frac{1}{a-b} & for \quad a \leq x \leq b \\ 0 & for \quad x < a \text{ or } x > b \end{cases}$$

Beta Distribution

Another distribution that is often used in specific algorithms, such as Thomson Sampling (we will describe it in Chapter 13), is *Beta distribution*. It is particularly effective for modeling variables that represent probabilities, which are naturally constrained between 0 and 1. This distribution is useful in scenarios where the exact probability of a binary outcome (such as success/failure or yes/no) is uncertain. The Beta distribution allows for a flexible way to express this uncertainty. It's especially useful in contexts where we have some prior knowledge or data about the likelihood of these binary outcomes and wish to incorporate this information to understand better or predict future occurrences.

The beta distribution represents all possible values of a probability when (i) we don't know what the probability distribution of the target dataset is, and (ii) we know that the probabilities are not moving toward infinity and have a range between [0,1].

The Beta distribution is defined by two parameters α and β. Figure 3-10 presents the beta distribution based on different values for α and β. As we can see, Beta distribution can plot many different behaviors just by changing its α and β values.

For instance, assume you are guessing the probability of being elected as the president of your company. To model this probability, we can use a Beta distribution. This choice is based on the binary nature of the outcome (you either get elected or you don't), and the Beta distribution is ideal for such a scenario.

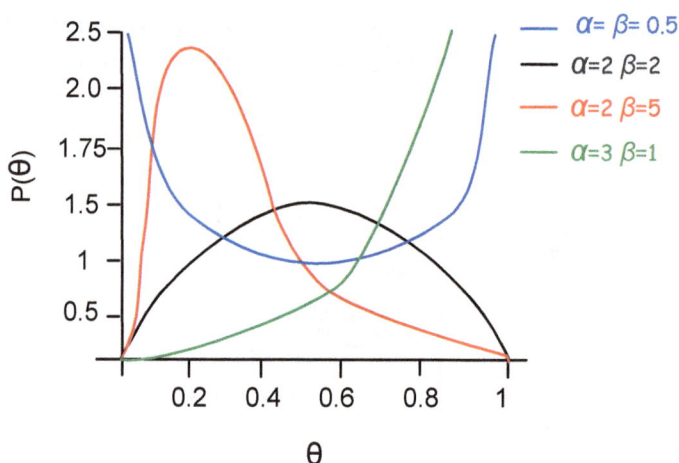

Figure 3-10: Beta distribution with different values for alpha and beta.

Let's use another example to remember the beta distribution. We are living in a world full of plastic pollution now; we would like to know how probable it is that in the future, our chickens hatch eggs in plastics. The author's aunt said it is true; a decent scientist who read this book, like you, said it is impossible. To model the range of beliefs or estimates about this probability, we can use the Beta distribution. Why Beta distribution? Because we are dealing with a binary outcome (the event either happens or not), we want to capture the range of uncertainty or variability in people's beliefs about this probability. To summarize, Beta distribution is useful to represent *all possible values of a probability*.

To formalize, a Beta distribution is a distribution that is parameterized by θ given α and β parameters, and its PDF is written as follows:

$$P(\theta \mid \alpha, \beta) = \frac{\theta^{\alpha-1}(1-\theta)^{\beta-1}}{B(\alpha,\beta)} \propto \theta^{\alpha-1}(1-\theta)^{\beta-1}$$

θ is in the range between 0 and 1, $\theta \in [0,1]$. B is the normalization constant that ensures the total probability is 1. Therefore, we can take out $B(\alpha,\beta)$ and say that the $P(\theta \mid \alpha, \beta)$ is proportional[1] to $\theta^{\alpha-1}(1-\theta)^{\beta-1}$. By substituting values for α and β in $\theta^{\alpha-1}(1-\theta)^{\beta-1}$, we can get distribution shapes shown in Figure 3-10.

Dirichlet Distribution

A multivariate generalization of Beta distribution (beta distribution for more than one variable) is called *Dirichlet distribution*. In other words, Dirichlet distribution is over vectors of variables and not a single variable. It could be assumed as a *distribution over distributions*.

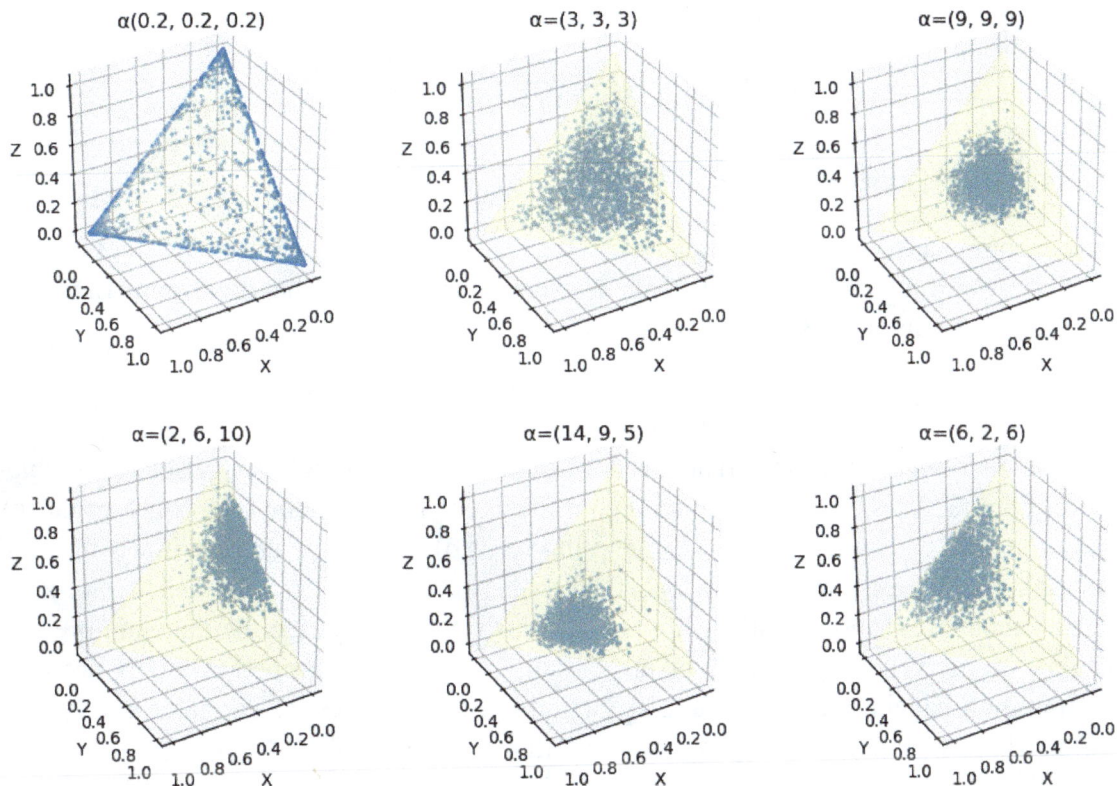

Figure 3-11: Example of multivariate Dirichlet distribution with six sample vectors, each of them has alpha, a vector of size three.

---

[1] $\propto$ sign is used in mathematics to show something is proportional to something.

Similar to the Beta distribution, this distribution is parameterized by a vector of $\alpha = \{\alpha_1, \alpha_2, \ldots, \alpha_k\}$ parameters. Each α_i influences the shapes of the distribution at the dimension i. In other words, for the k-dimensional Dirichlet distribution, the parameter vector α has k elements. Look at Figure 3-11, which presents a ternary counterplot (a counterplot inside an equilateral triangle) to visualize the Dirichlet distribution. Assuming we have $k=3$ outcome as vectors ranging from 0 to 1. Figure 3-11 presents shapes that are created based on different parameters for α. For $\alpha < 1$, we get concentrations at the corners of the triangle. For $\alpha > 1$, the distribution tends toward the center of the triangle. As α increases, the distribution becomes more tightly concentrated around the center of the triangle. This distribution is used in a *Latent Dirichlet Allocation*, a topic modeling approach, which we explain in Chapter 7.

Before explaining the PDF of Dirichlet distribution, we should describe the *Gamma function* (Euler Gamma Function). The Gamma function behaves like a factorial for natural numbers but generalizes to positive real numbers (a continuous set). This is useful for modeling situations involving continuous change and it is written as Γ, along with improper integral[2] as follows:

$$\Gamma(n) = \int_0^\infty x^{n-1} e^{-x} dx \quad , \quad n > 0$$

The PDF of the Dirichlet distribution is as follows[3]:

$$f(\theta_1, \theta_2, \ldots, \theta_m) = \frac{\Gamma(A)}{\prod_{i=1}^m \Gamma(a_i)} \prod_{i=1}^m \theta_i^{a_i - 1}$$

Here $A = \sum_{i=1}^m a_i$ and $a_1, \ldots a_m$ are the parameters for $i = 0, \ldots m$

If it is too complicated to learn, do not worry. We rarely need to recall its PDF unless you have one of those weird job interview questions. We should only know what algorithm (LDA, which we will learn in Chapter 7) uses this distribution.

Binomial Distribution

Most of the distributions we explain are used for discrete variables. A binary discrete variable has only two discrete states, such as flipping a coin (head or tail), the door state (open or close), or whether you find this book fantastically helpful (yes, no). When we are dealing with a binary variable that has a fixed number of independent trials, we can use *Binomial distribution* to make inferences about the dataset. Statisticians use this distribution to judge the data and predict the probability of an event, e.g., buying or not buying a stock, using a medication or not using a medication on a patient, etc. Binomial distribution deals with states or variables that have only two values (binary states), "bi" stays for two in Latin.

To understand the Binomial distribution, consider an example: we have three fat chickens and feed them junk food to make them heavier. Any of these poor chickens either get heart attacks (h) because of overeating or do not get heart attacks (n). Therefore, one of the eight following

---

[2] We will briefly review Integral in Chapter 8.

[3] The sign Π means the multiplication of variables in front of it. We use Σ the sign for the summation.

situations could happen for these three chickens: {h,h,h}, {h,h,n}, {h,n,n}, {n,n,n}, {n,h,h}, {n,h,n}, {n,n,h}, {h,n,h}. We show each probability with a $P(.)$ function. Therefore, we can have the following inferences:

$P(no\ heart\ attack) = P(\{n, n, n\}) = 1/8$
$P(\{one\ gets\ heart\ attack\}) = P(\{h, n, n\}) + P(\{n, h, n\}) + P(\{n, n, h\}) = 1/8 + 1/8 + 1/8 = 3/8$
$P(\{two\ get\ heart\ attack\}) = P(\{h, h, n\}) + P(\{n, h, h\}) + P(\{h, n, h\}) = 1/8 + 1/8 + 1/8 = 3/8$
$P(\{all\ three\ get\ heart\ attack\}) = P(\{h, h, h\}) = 1/8$

By using x as a variable that specifies the number of chickens getting heart-attack, we can rewrite the above probabilities as follows:

$P(zero\ heart\ attack) = P(x = 0)$
$P(one\ heart\ attack) = P(x = 1)$
$P(two\ heart\ attack) = P(x = 2)$
$P(three\ heart\ attack) = P(x = 3)$

Since Binomial distribution deals with discrete values, we use histograms to present the distribution. We can plot the $P(x)$ with a histogram as shown in Figure 3-12.

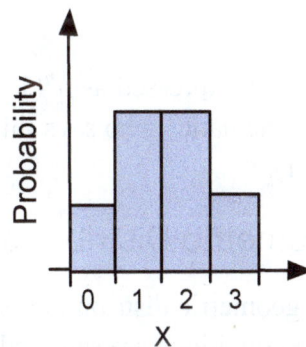

Figure 3-12: Binomial distribution of three chickens getting heat-attack from eating too much junk food.

The binomial distribution is appropriate if the following three conditions are all true. (i) We have a series of independent trials. It means that, in our example, a chicken's heart attack does not impact the heart attack of another chicken. (ii) Trial output is binary and denoted as success or failure, but it could be other information as well, such as yes/no, true/false, etc. (iii) The number of trials is not infinite, and there is a finite number of trials.

The binomial PDF enables us to get the probability of observing x successes in n trials, with the probability p of success on a single trial. The binomial PDF for the given value x and given pair of parameters n and p are written as follows[4]:

$$f(x \mid n, p) = \binom{n}{x} p^x q^{(n-x)}$$

$f(x \mid n, p)$ presents the probability of observing exactly x successes in n independent trials, where the probability of success in any given trial is p, and the probability of failure in any given trial is q.

Here, we learn Binomial distribution, which is for one-dimensional data. If we have more than one-dimensional data, the Binomial distribution will be referred to as the *multinomial* distribution.

Bernoulli Distribution

Bernoulli distribution is a specific case of Binomial distribution; it is discrete and *has only one trial.* An experiment with random results and only having one binary outcome is known as a Bernoulli trial. For instance, we have one chicken, and this chicken has overeaten; either it gets a heart attack or does not get a heart attack. In this case, the probability of getting a heart attack is 1/2, and the probability of not getting a heart attack is $1 - (1/2) = 1/2$.

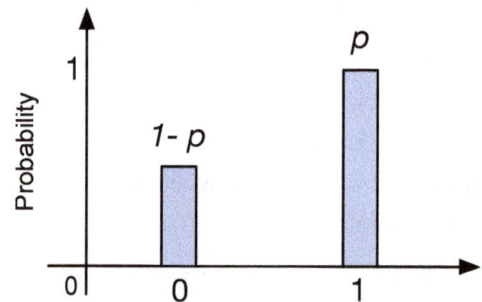

Figure 3-13: Bernoulli distribution.

Figure 3-13 visualizes a Bernoulli Distribution. The Bernoulli distribution is discrete, and its PMF is written as follows:

$$P(n) = \begin{cases} 1-p & for \quad n = 0 \\ p & for \quad n = 1 \end{cases}$$

It can be expressed as $p^n(1-p)^{1-n}$, where $n \in \{0,1\}$. In the future, when you encounter a nerd of mathematics who says this is a Bernoulli vector, she or he means the values are binary (either 0 or 1).

Geometric Distribution

The geometric distribution is similar to the Binomial distribution. It is used if all of the following three conditions are satisfied: (i) there is a series of 'independent' trials. (ii) The trial's output is binary, e.g., success/failure, yes/no, true/false, etc. (iii) As the desired binary state is acquired, the trial stops immediately. The first two conditions are similar to the Binomial distribution, but the third condition makes it different from the Binomial distribution.

---

[4] $\binom{n}{k}$ is referred as the binomial coefficient and it is calculated as: $\binom{n}{k} = \dfrac{n!}{k!(n-k)!}$

This distribution, similar to the Binomial distribution, is also used to make inferences about the data. We use geometric distribution when a statistician wants to answer the question, *how many trials do we need to get a successful output?*

For instance, assume there is a pigeon in the street that relieves himself on the clean window of a car. Whenever he encounters a clean car window, his diarrhea starts immediately. He is a gentle pigeon and tries to control himself, but it is hard. This should happen at least once a day, and after the first relief, the other cars stay clean (the trials stop after the first success). If your car window received its pigeon defecate share yesterday, you have cleaned it. It is also possible to receive it today as well. It means there is no relation between yesterday's today's events (independent trials).

Based on your past observation, you have found that the probability of any car parked in his territory and getting dirty is 0.6. Today, you bring your car fresh out of the carwash, but unfortunately, you should park it in his territory again. There are three other clean cars in the street, too (four cars in total). So, you would like to calculate the probability that our pigeon leaves the other three cars clean, and relieves himself at your car window?

The probability that he makes a clean window dirty is $P = 0.6$, so not making it dirty is $1 - 0.6 = 0.4$. Assuming X denotes the number of cars he passes without making them dirty. Assuming *"not making dirty"* = *"success"* and *"making dirty"* = *"failure"*, the probability of a car staying clean is as follows:
$P(X = 1) = P(success\ in\ the\ 1^{st}\ trial) = 0.4$
$P(X = 2) = P(failure\ in\ the\ 1^{st}\ trial) \times P(success\ i\ nthe\ 2^{nd}\ trial) = 0.6 x 0.4 = 0.24$
$P(X = 3) = P(failure\ in\ the\ 1^{st}\ trial) \times P(failure\ in\ the\ 2^{nd}\ trial) \times$
$P(success\ in\ the\ 3^{rd}\ trial) = 0.6 x 0.6 x 0.4 = 0.144$
$P(X = 4) = P(failure\ in\ the\ 1^{st}\ trial) \times P(failure\ in\ the\ 2^{nd}\ trial) \times$
$\times P(failure\ in\ the\ 3^{rd}\ trial) \times P(success\ in\ the\ 4^{th}\ trial) = 0.6 \times 0.6 \times 0.6 \times 0.4$
$= 0.086$

Figure 3-14 visualizes the result, and it shows that as soon as the pigeon encounters a clean car, the chance is higher to stay clean, but if it passes three cars, the chance of keeping his stomach clean for the fourth car is very low at 0.086. Therefore, we can conclude that as much as there are clean cars on the street, your car's chance of receiving our lovely pigeon stomach residuals is reduced. This means as soon as it encounters a clean car video, its stomach gets bad. In other words, if he encounters your car as the first car and tries to control himself, the chance is 0.4 that he keeps your window clean, but if he passes three other clean cars and yours is fourth, your chance of keeping your window clean is only 0.086. Figure 3-15 shows the plot of this geometric distribution.

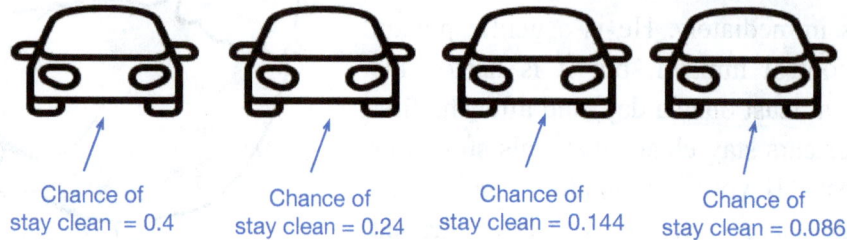

| Chance of stay clean = 0.4 | Chance of stay clean = 0.24 | Chance of stay clean = 0.144 | Chance of stay clean = 0.086 |

Figure 3-14: Our lovely pigeon fly on clean cars and when he encounters a clean window his stomach starts to relief.

As you can see from Figure 3-15, geometric distributions always have a right-skewed shape (the concentration is on the left side of the X-axis), and the density of data reduces as we move right

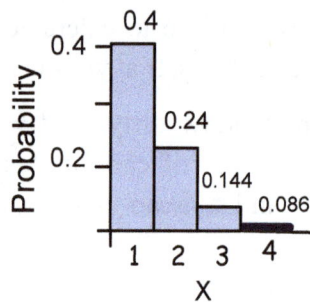

Figure 3-15: Geometric distribution of the probability of not getting a dirty window by the pigeon.

in the histogram, i.e., there is a long tail on the right side. This distribution is written as $X \sim Geo(p)$. X is a geometric distribution where the probability of success is p. Since p provides the probability of success, x is the number of trials ($x = 1,2,...$), The PMF of Geometric distribution is written as: $f(x) = p(1-p)^x$

Once again, let's emphasize the differences between geometric and Binomial distribution. The geometric distribution is similar to the Binomial distribution, but the experiment with geometric distribution stops after the first desired state (e.g., the pigeon relives itself). In contrast, in Binomial distribution, the experiment could continue (e.g., the pigeon continues to relive itself after the first relive).

Poisson Distribution

Poisson distribution is being used to model the intervals of rare events. In other words, we can use Poisson distribution to model this dataset when dealing with a dataset that includes rare system events, such as machine malfunctions. We should know the average occurrences of these rare events (in time), and their time is not changing.

In Poisson distribution, a rare event occurs *randomly* and *independently*. It means the probability, or rate, of this event happening does not change through time, and we can guess how often this could happen. For instance, a rare event occurs two times per year on average. We would like to know the probability that in one year, this event occurs exactly five times. In this case, Poisson distribution will help.

Geometric and Binomial distributions involve a series of trials, while the Poisson distribution models the number of rare event occurrences in a particular interval. While formalizing this distribution, the mean and variance are equal, and written as λ (lambda). Remember that we said the average (mean) is not changing. The following equation is used to calculate the PMF of the Poisson distribution:

$$P(X = r) = \frac{e^{-\lambda} \cdot \lambda^r}{r!}$$

'e' is the Euler number and it is 2.718

Mean

The number of rare event occurrences

This distribution can model the number of times the rare event, which is discrete, occurs in an interval such as time, space, volume, etc.

Let's discuss some examples to understand its use. By having data from pandemics, we can use Poisson distribution to predict when will the next pandemic, like COVID-19, affect the lives of millions of people?

As another example, assume we are running a machine that produces chicken nuggets. It receives chickens as input and provides chicken nuggets as output. Sometimes, instead of a chicken nugget, it magically converts chickens into a cat (a rare event), and the output is a cat instead of a chicken nugget. We have a small room to keep a few cats temporarily, and every month, an animal shelter vehicle collects the cats from our room. We need to estimate cat production in a time interval to assign food and temporary shelter to these cats. Cat production is a rare event in this system. Nonetheless, it happens with our machine, and we know it occurs about twice a month (λ=2). We would need to estimate what is the probability of getting exactly 3 cats out of this machine in four months because our temporary room has space for only three cats, and the animal shelter told us they can't come earlier than four months. We want to present statistics to the animal shelter and convince them to send their pickup earlier.

Going back to the definition we described, we have a rare event (cat as output), and the mean occurrences of this event per month are 2, $\lambda = 2 \times 4$ (month) = 8, in other words, the mean in four months is 8. Therefore, X presents the number of cats a system can produce.

we have $P(X = 3) = \dfrac{e^{-8}.8^3}{3!} = 0.286$.

Consequently, we can answer other questions as well. What is the probability that we get zero cats in a month (cat per month presented as λ=2)?

$$P(X = 0) = \frac{e^{-2}.2^0}{0!} = \frac{e^{-2}.1}{1} = 0.135$$

What is the probability that we get 1 cat in a month?

$$P(X = 1) = \frac{e^{-2}.2^1}{1!} = 0.270.$$

We have followings probability of the number of cats getting produced in a single month:
$P(getting\ 0\ cat\ in\ a\ month) : P(X = 0) = 0.135$
$P(getting\ 1\ cat\ in\ a\ month) : P(X = 1) = 0.270$
$P(getting\ 2\ cat\ in\ a\ month) : P(X = 2) = 0.270$
$P(getting\ 3\ cat\ in\ a\ month) : P(X = 3) = 0.180$
$P(getting\ 4\ cat\ in\ a\ month) : P(X = 4) = 0.090$
$P(getting\ 5\ cat\ in\ a\ month) : P(X = 5) = 0.036$

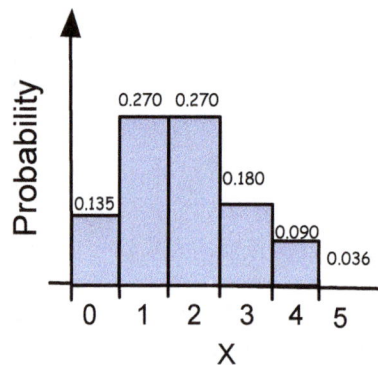

Figure 3-16: An example of a Poisson distribution.

The plotting of this distribution is shown in Figure 3-16. Poisson distribution is right skewed. If the λ is large, it is getting more symmetrical, and if the λ is near zero, it is getting right-skewed. Figure 3-17 present different lambda in this distribution. Remember that discrete distributions use histograms; Poisson distribution is also a discrete distribution, but for the sake of readability, a line chart presented as connected dots is usually used.

Weibull Distribution

The Poisson distribution is appropriate if the rare events we mentioned occur constantly. Nevertheless, if they do not occur at a constant rate and time, we cannot identify their rate; we can use the Weibull distribution. It models a distribution where the rate of rare events is not constant and may vary over time. The shape of a Weibull distribution depends on a parameter k, which is known as the *shape factor*.

The Weibull distribution is a continuous probability distribution. Thus, it is characterized by a Probability Density Function (PDF) rather than a Probability Mass Function (PMF). Its PDF is written as follows:

$$f(x; \lambda, k) = \begin{cases} \frac{k}{\lambda} \left(\frac{x}{\lambda} \right)^{k-1} e^{-(x/\lambda)^k} & \text{for } x \geq 0, \\ 0 & \text{for } x < 0. \end{cases}$$

Here, x is the variable (the value for which we are calculating the PDF), λ is the scale parameter and k is the shape factor. The sign ';' means that the right side of the probability are parameters that changing them affect the left side of the probability.

The right side of Figure 3-17 shows the Weibull distribution example with different combinations of λ and k.

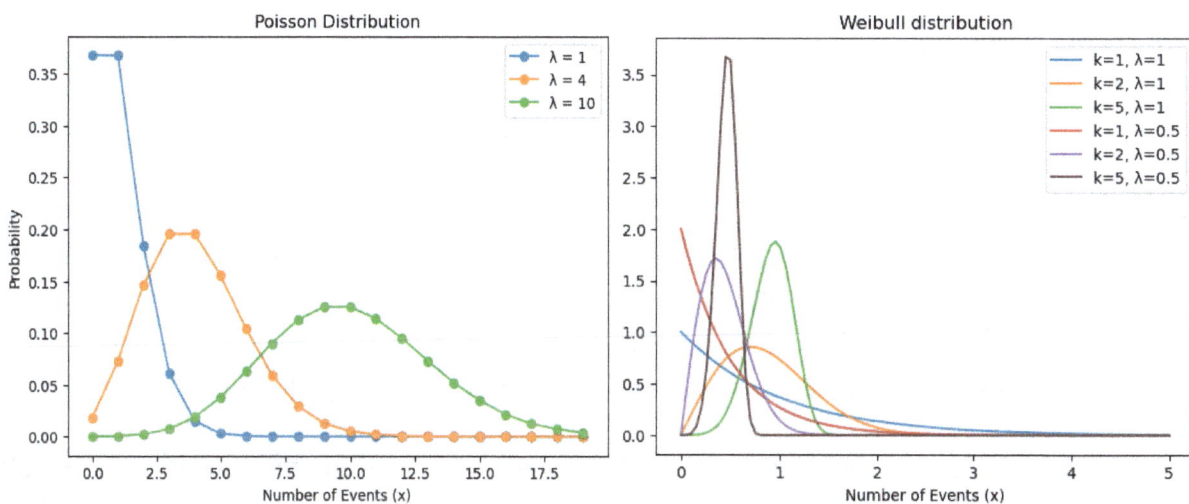

Figure 3-17: (left) The direction of the skewness based on the lambda in Poisson distributions. (right) Different Weibull distributions for different values of λ and k.

Power Law (Long Tail), Exponential, and ZipfLaw Distributions

Do you remember those nerds at school who have nothing to do except do their homework and compete to get the top mark? Then, the teacher gives the most challenging possible quiz, and when you nag about your grade, the teacher always points you to them as an example to prove the test was not hard. Imagine you are at the school and have a class of 11 students; here are the marks of a quiz ranging from 0 to 100: {99, 90, 35, 34, 30, 27, 25, 20, 15, 12, 8, 7}. Figure 3-18 plots the marks on the left side. You see two nerds on top and the other student on the bottom. If you have finished school and have a job, probably one of those nerds is your boss now, and their salary is similar to their grades. Therefore, we could say salary has a power law distribution.

On the right side of Figure 3-18, you see the common shape of the Power law distribution. If you recall, in Chapter 1, we described exponential growth. The Power law has exponential growth as well. In particular, this distribution presents exponential changes in a dataset.

The PDF of the power law is $f(x) = x^{\alpha}$. Here, α is called the power law exponent and causes these exponential changes; it is constant.

There is another distribution similar to this one called *exponential distribution*. The PDF of the exponential distribution is written as $f(x) = \alpha^x$. It means the exponent is a variable. Note that

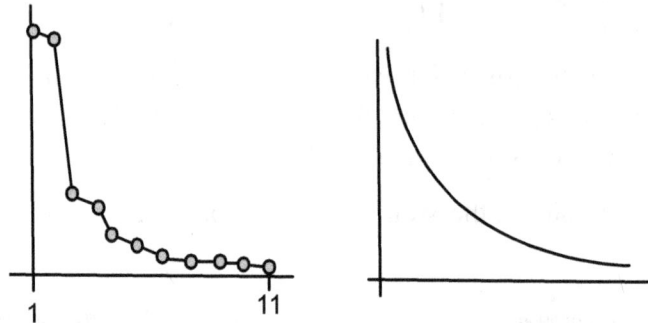

Figure 3-18: (left) Distribution of grades in your class that has two nerds on the top. (right) Abstract shape of power law distributions.

mean and standard deviation in the power of law distributions do not make sense because, due to extreme values, which are rare (on the right side of the distribution plot), we can not characterize this distribution by mean and standard deviation.

These two distributions are often more observed in the real-world. For instance, consider the size of cities in big countries, except a few countries like Germany; usually, the population density is concentrated in a few cities, and the rest of the cities are not dense. Or consider the wealth distribution, which is very fair around the world! At the time of writing this chapter (2017), half of the world's wealth is in the hands of 1% of billionaires, and we should pray to god that writing this book has some financial benefit to pay our debts.

There are many other real-world examples in addition to wealth distribution and sizes of cities in a country, such as the magnitude of earthquakes and word frequencies in a text.

Zipf law

Another specific case of power law includes Zipf law. To understand Zipf's law, we use a simple typical example of word frequency in an English book. The most used word is "The", then the second most used word will be "of", which will be half (1/2) of the most used word (the). The third most used word will be "and", which will be a third (1/3) of the most used word. The word after that is "to", which will be a quarter (1/4) of the most used word, and this ratio continues. Such a ratio distribution is called Zipf's law, a type of power law distribution. Figure 3-19 shows an example of Zipf's law, which is a distribution of English words in a book.

Pareto principle

Pareto principle is a type of power of law distribution. It is also known as the 80/20 rule. It says 80% of effects come from 20% of causes, and the 20% remaining effects come from 80% of

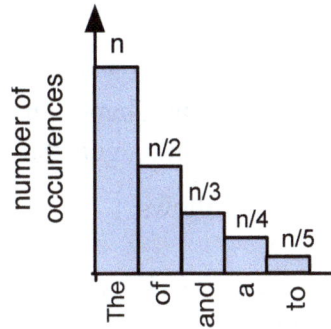

Figure 3-19: Zipf law shows the approximate distribution of words in an english text corpus.

remaining causes. For instance, Pareto, back in 1896, shows that 80% of the land in Italy is owned by 20% of the population.

Chi-Square Distribution

Another continuous distribution that we should know is the Chi-squared (χ^2) distribution. The chi-square distribution is non-symmetrical, skewed to the right side of the X-axis, in which the shape of the distribution is very much dependent on the degree of freedom k. The mean in the chi-square distribution is equal to the degree of freedom; as the degree of freedom increases, this distribution is skewed toward a normal distribution.

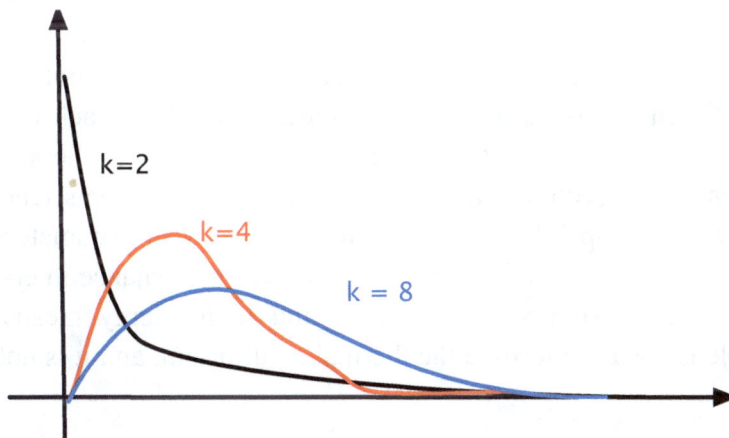

Figure 3-20: Chi-square distribution with three different degree of freedom.

Figure 3-20 shows this distribution for three different degrees of freedom, 2,4, and 8. This distribution is being used for the chi-square test, including the test for Goodness-of-Fit and dependence between two categorical variables; we explain both of these uses later. Besides, it is used to test whether a sample dataset variance equals a hypothesized population variance and estimate confidence intervals for given variance and standard deviation.

For now, just remember that the total area under the curve equals one, and this distribution depends on the degree of freedom.

Setting k to 1 or 2 makes the shape of Chi-distribution a smooth curve, starting high and getting low. Setting k to a higher value than 2 makes the shape curved and right-skewed. The right skewness decreases as k increases until this distribution gets the shape of a normal distribution.

PDF of Chi-square distribution is computed as follows:

$$f(x, k) = \frac{x^{k-1}e^{-x^2/2}}{2^{(k/2)-1}\Gamma(\frac{k}{2})} \quad for \quad x \geq 0 \text{ for } x \geq 0$$

Here, Γ is the Gamma function (we have explained it earlier), k refers to the degree of freedom or independent (input) variables with normal distributions.

We emphasize that we do not need to memorize the details of the PDF of this distribution, by plugging numbers into this distribution and playing with parameters, we can get different shapes of χ^2 distribution.

Boltzmann Distribution

Boltzmann (Maxwell-Boltzmann or Gibbs) distribution [Gibbs '02] is used to model the following statement: the energy distributed from high density to places that have lower density until there is a balance between densities (thermal equilibrium). In school, we learned that by increasing the heat (energy), gas molecules start moving faster, and their kinetic energy increases. The temperature is proportional to the average kinetic energy.

As an example, to understand the described equilibrium, think if we spray perfume in a room; at the beginning, that region has the smell of spray, but then the sprayed molecules move into the room until their kinetic energy is equal to other molecules or they reach thermal equilibrium. Therefore, after a few seconds, we can't smell the perfume sprayed in the air. At this stage, we can say they have reached an equilibrium. In simple terms, all systems tend to move toward thermal equilibrium. *Thermal equilibrium* refers to a condition when parameters do not exchange any energy (the high energy moves to low energy until there is a balance in energy everywhere). Low energy means high probability in that state, and high energy means low probability. However, this example is used to illustrate the thermal equilibrium, and it is not about the energy equilibrium.

To understand energy equilibrium, consider that we have three containers (A, B, and C) of a gas molecule. Container A's temperature is 300 Kelvin, container B's temperature is 200 Kelvin, and container C's temperature is 100 Kelvin. The average kinetic energy of molecules in container A is higher than the average kinetic energy of molecules in container B, and container B's average kinetic energy of molecules is higher than container C. This means molecules in this container are moving faster (higher velocity). If we plot their PDF, we will have a shape similar to Figure 3-21. These distributions are Boltzmann, also known as Boltzmann (Maxwell-Boltzmann or Gibbs) distribution.

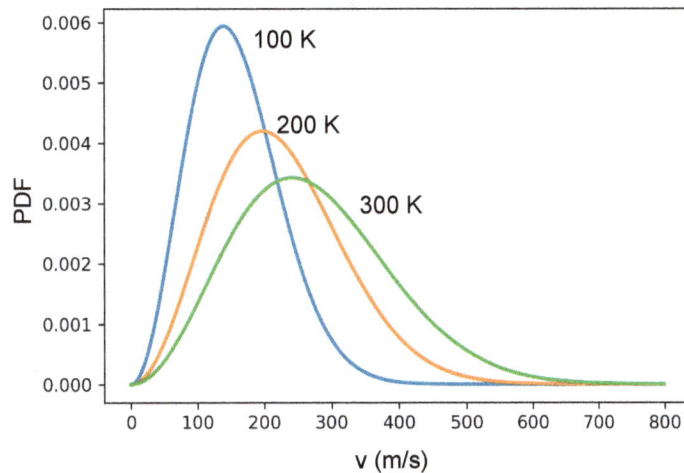

Figure 3-21: Boltzmann distribution PDF three different temperatures.

In the context of machine learning, the Boltzmann (Maxwell-Boltzmann or Gibbs) distribution represents the probability for the distribution of the states in a system based on the different energy levels in the system.

The PDF of the Boltzmann distribution is computed as follows:

$$p_i = \frac{e^{-\epsilon_i/kT}}{\sum_{j=1}^{M} e^{-\epsilon_j/kT}}$$

In this equation, p_i is the probability of state i, ϵ_i presents the energy at state i, and k is the Boltzmann constant (1.380649×10^{-23} joule per kelvin), T is the thermodynamic temperature or temperature of the system. Remember that Boltzmann distribution is based on the probability of a state in the system, and it is inversely related to the energy of the system at that state. If we connect all containers A, B, and C together. Their temperature will be changed to something like the average temperature.

This distribution does exist in some natural phenomena, such as gas distribution, where there is no dense energy; the gas will be distributed equally; if the energy increases at some point, the density of gas decreases and changes in that place. Later in Chapter 11, we will revisit this distribution.

We smell a bit of smoke coming out of your brain now. It is ok; after learning all these distributions, your brain is about to explode. However, the bad news is that there are more types of distribution that we do not explain here, but the good news is that in the context of machine learning, they are not used very often. For example, a Laplace distribution is useful to enforce scarcity and a Dirac distribution is useful to enforce domain knowledge [Goodfellow '16]. The distributions we have explained here are very popular in statistical problem solving, and we will encounter them in the next chapters. For example, we will encounter Dirchilet distribution in Chapter 4, Boltzmann Distribution in Chapter 11, and Beta distribution in Chapter 13.

Distribution check with P-P Plot and Q-Q Plot

We can use a Probability-Probaility plot (P-P plot) or Quantile-Quantile plot (Q-Q plot) to visually assess whether a dataset follows a particular distribution.. The result of the probability plot is a scatter plot or four scatter plots with a diagonal in the middle of each scatter plot. If the result of the probability plot approximately draws a straight line in the diagonal of the rectangle, then we can claim that our dataset follows the assumed distribution. The P-P plot draws CDF of the data against the CDF of another dataset or theoretical distribution. Q-Q plots the quantiles from the data against the quantiles of a theoretical distribution or another dataset. Q-Q plot visualizes four different plots.

Figure 3-22 shows two different sub-plots of Q-Q plots. We can check if the data had a normal distribution, one follows the normal distribution because data points are roughly distributed around the diagonal, and the other one does not follow the normal distribution because data points are deviated from the diagonal line of the plot.

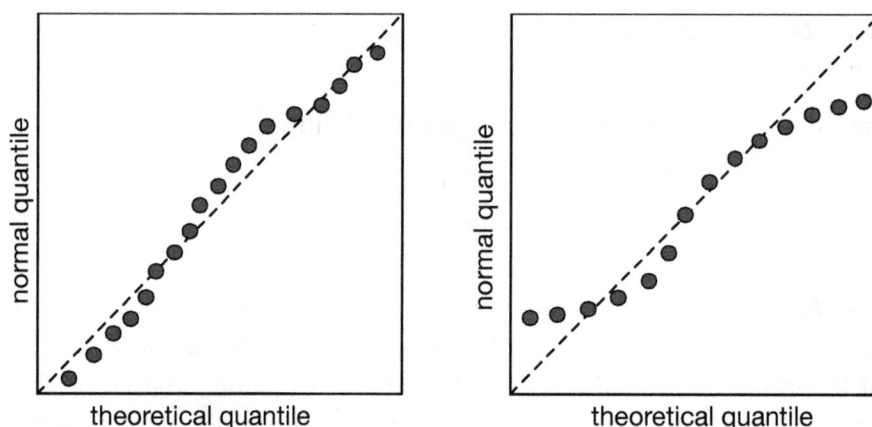

Figure 3-22: Two sample diagrams from Q-Q plots, the one on the left is close to a straight line, which means our dataset is following the desired distribution. The one on the right does not follow the desired distribution because it does not create a straight line.

Your software tool will perform the Q-Q plot, and you don't need to learn how it works in detail. In short, the Q-Q plot algorithms take our input dataset, sort it in ascending order, and then plot our sample data versus the quantiles calculated from the theoretical distribution. The number of data points will match the number of our sample dataset.

Since most of the statistical assumptions were based on the fact that the dataset follows a normal distribution, some tests are used to identify whether the dataset is normally distributed. By plotting the density of the data, we can visually inspect the normal distribution of the data. However, they can be used for other distributions as well.

There are statistical tests available for that as well, such as the Shapiro-Wilk [Shapiro '65] test for the test of normal distribution, the KS-Test and the Chi-Square Goodness of Fit test to check the distribution of data.

NOTES:

* If we are sampling a part of a dataset and calculating its mean, we usually use $\bar{X}$ instead of μ.

* We use Binomial distribution when we like to know the "probability of getting a certain number of successes". We use geometric distribution when we like to know "how many trials we need before the first success".

* For both Binomial and geometric distribution, the probability of success in each trial should be equal. Otherwise, none of those distributions could be used.

* Geometric distribution and Binomial distribution are very similar. However, geometric distribution "stops" as the first failure or success or any other target boolean variable it may encounter. For example, the Pigeon empties its stomach once it sees a clean car window and will continue doing it. In Binomial distribution, we are interested in the number of successes in the independent trials.

* In Poisson distribution, λ (lambda) is being used to present the mean and not μ, because, in Poisson distribution, variance is equal to the mean. Therefore, using μ or σ^2 might be confusing.

* Use Poisson distribution if the events are independent. For instance, malfunction events occur in a given interval, and we know the value of λ in that interval.

* Binomial, geometric, and Poisson distributions are for discrete data, and for discrete data, we use the histogram. Nevertheless, since the number of data points is usually large, a line chart is being used to demonstrate distribution.

* When the number of samples is too large, it is better to use the Poisson distribution rather than the Binomial distribution. Because when n is large, the system must calculate $n!$ and it will eat lots of computer memory, and it can come out of the monitor and eat the person who put this input data into the machine as well. However, the choice between these two distributions is based on the nature of the data and the underlying processes, not primarily on computational considerations.

* The Power Law distributions can be used for both continuous and discrete variables as well. Gaussian are Chi-square are used for continuous distribution only.

* We can use the histogram to draw "discrete" data distributions.[5] While working with continuous data, we have many different numbers to present. Therefore, the range will be used to present these numbers, and a line chart will be used. Usually, the line chart is used for continuous data, and a histogram is used for discrete data. When there are too many data points

---

[5] There is a good link in Wikipedia that list all distributions https://en.wikipedia.org/wiki/ List_of_probability_distributions#Continuous_distributions

to plot, and for the sake of readability and simplicity, most of the time, instead of a histogram, a line chart is used to present a distribution of discrete data as well.

* Statistical skewness is a measure that describes the asymmetry of a distribution. It helps to quantify the extent to which a dataset's values are concentrated on one side or the other of the distribution's mean (average).

Expected Value and Expectation of a Function

The *expected value* is a type of arithmetic mean but with the *weight* or *probability* for each value. It could be called weighted average as well. It is presented with $\mathbb{E}[\,.\,]$ in the context of artificial intelligence machine learning. For instance, imagine you are an expert AI engineer, and you are doing technical consultation and offering three different AI courses. Introduction to AI, which costs \$300 per person; intermediate AI, which costs \$700 per person; and Advanced AI for experts, which costs \$1000 per person. Then 100 people subscribed for your course. 2% have subscribed for the expert level, 8% for the intermediate level, and %90 for the beginner level. The expected value or weighted average of the earnings is as computed as follows:

$EV = 0.02 \times 1000 + 0.08 \times 700 + 0.90 \times 300 = 20 + 56 + 270 = 346,$

In other words, \$346 is the expected value of your earnings by teaching AI.

By working with a function, usually, the average of the values we inject into the function is known as the *expectation of a function*, and it is presented as $\mathbb{E}[f(x)]$, which presents the expected value of the function f where x is a random variable. In reality, it is nothing than plugging n numbers into the function and computing their average; thus, we can have the following:

$$\mathbb{E}[f(x)] = \frac{1}{n} \sum_{n} f(x)$$

Later in Chapter 11 and many times in Chapter 13, we will encounter $\mathbb{E}_{x \sim A, y \sim B}[\,\ldots\,]$. We should read it as the excepted value for x and y which x is sampled from A distribution and y is sampled from B distribution. See, it is very simple; don't freak out while encountering these weird, super long mathematical terms. To summarize, *when we say we sample data x from the distribution of* $p(x)$, *we use this notation:* $x \sim p(x)$.

Normalization

While working with data in different ranges, we should make them comparable by adjusting values on different scales to a relevant scale. Besides, if the data are on large scales, it is better to scale them down for the machine learning algorithm while maintaining their characteristics, e.g., bring them into a smaller number range. These processes are referred to as normalization.

To better understand the need for normalization, consider the following scenario that describes the need to bring data into a common scale. Mr. Foo is a wise CEO of a giant corporation. Recently, his corporation faced new challenges in sales that require lots of new resources to solve. It means he should consider hiring more data scientists and machine learning engineers for his data science division. However, similar to other super-riches, he loves to keep his money for himself and not spend it on new hiring. On the other hand, it is essential to have a robust solution for these new issues.

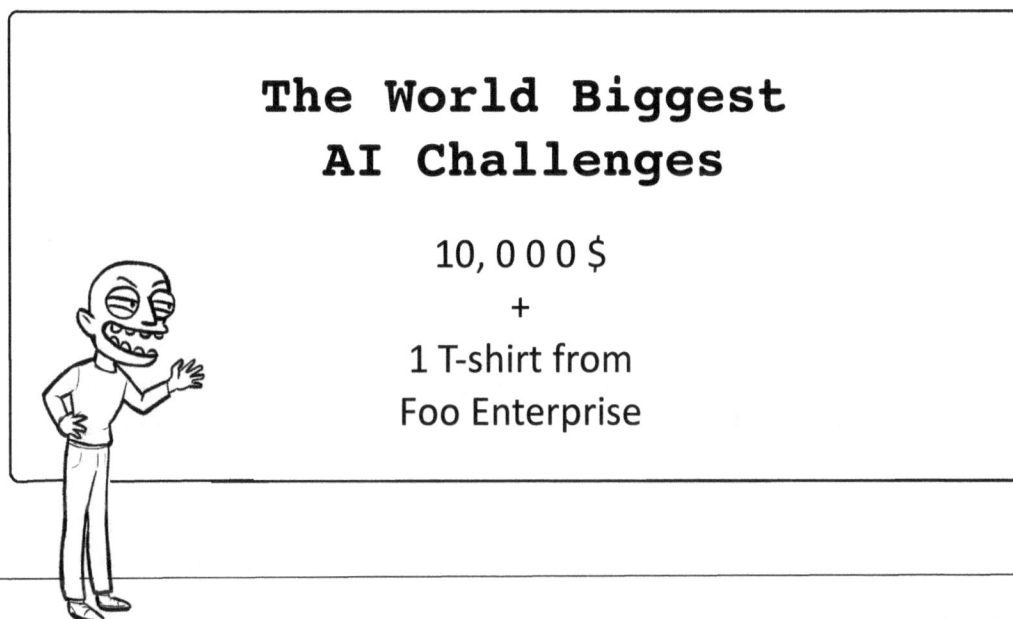

The World Biggest AI Challenges

$10,000\$$

+

1 T-shirt from
Foo Enterprise

He looks at other businesses and finds a wise way to get a cheap workforce to solve his problems. He plans to launch a competition in his corporation, and data scientists worldwide can compete to see which group can solve his problem most efficiently. As a reward, they will receive a certificate of success from the Foo Enterprise, which is very prestigious for their CV, a few thousand dollars, and a T-shirt. With these over-generous gifts, Mr. Foo saves millions, and instead of hiring new staff, he launches a data science competition every year.

One secret of Mr. Foo's success is to show off himself as a very fair entrepreneur and respect diversity. Therefore, he mandates competitors to build teams, including international data scientists. There is a problem in evaluating participants, who are primarily juniors. The only way to judge their qualifications is through their university grades. The grading system is different around the world. In China and the US, it is based on alphabets, e.g., A, B, C. In India, it is ranked from 100 (best) to 0 (worst); in Germany, it is ranked from 1 (best) to 5 (failed); in Iran, it is ranked from 20 (best) to 0 (worst), and so forth.

There should be a way to show all these grades into a unique score so that participants' qualifications can be easily compared together. The process of such a grade scale transformation to a common score is normalization.

z-score

One of the known standard approaches to normalizing numerical values is using the *z-score* or *standard score*. The z-score is calculated for each data point, and the difference from the mean is divided by the standard deviation. Note that each data point in the dataset has a single z-score, and there is no z-score for the entire dataset.

Mr. Foo's competition administrator can use the z-score to transform the participants' grades into a number that enables them to compare them with different measurement systems.

When we transform all data points of a dataset to their z-score, the mean of the new z-score is always 0, and the standard deviation is always 1. Often, a normalization (z-score or other normalization) tries to transform the data between the range 0 and 1 or -1 and 1.

$$z = \frac{x_i - \mu}{\sigma}$$

A single data point — Mean — z-score — Standard Deviation

z-score is useful when we must compare two different distributions (e.g., one is a normal distribution, but the other is not). We can transform them both into z-scores to be able to compare them.

If we plot the z-score normalized data points, we will encounter a specific kind of normal distribution called *z-distribution*[6]. A z-distribution is a normal distribution with a mean of zero and a standard deviation of one (Recall that distributions are shown with mean, variance, and other parameters).

There is another distribution similar to the z-distribution with similar characteristics called the *t-distribution*. It is bell-shaped and symmetric, similar to z-distribution. However, it is shorter than the z-distribution, and its curve is flattened, as shown in Figure 3-23. The t-distribution is used to study the mean of a population if the dataset is normally distributed.

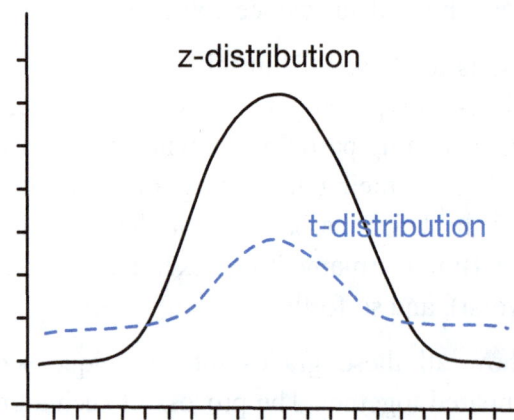

Figure 3-23: t-distribution and standard z-distribution, t-distribution is always flatter than the normal distribution.

---

[6] You might ask why we explain normalization so late, because we would be sure that you understand the concept of distribution, then we are able to describe normalizations. Otherwise, you didn't know what z-distribution or t-distribution means.

How Much Data is Enough?

We might dream of getting hired as a data scientist or AI engineer, starting our data analytical company, or starting to learn something at the university and working with these sexy machine learning and AI algorithms. Well, that is a nice dream, but most of the time, we are responsible from A to Z for working with the data, and sometimes we are even responsible for collecting the data, preparing coffee for ourselves, cleaning the microwave after use, wiping our desks, cleaning the data, etc.

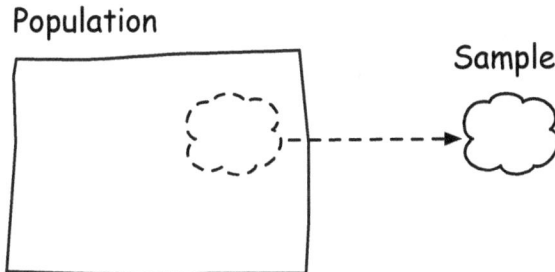

One of the most cumbersome tasks in data science is collecting data to run the experiment. Once we start experimenting with the collected data, we should ask ourselves whether we have enough data to generalize our findings and make any inferences about the data. In other words, the important question is: "How much data is enough?".

Unfortunately, there is no precise answer, but two approaches are being used to provide insight into the dataset size. Obviously, the more data we collect, the better analysis we can perform. Nevertheless, our resources (time, money, energy, CPU, project deadline) are limited, and we cannot continue collecting data infinitely.

There is a term called *sampling*, which means selecting some data points from an entire *population* (the statistical name for the entire dataset). The small dataset that is selected from the population is called the *sample dataset*. If we use the entire population and not sample the data, this is called *census data*. More about sampling approaches will be described in Chapter 16.

For the sake of brevity, we refer to the sample dataset as a sample and the population dataset as a population. Please remember the definitions of population versus sample, and we are going to refer to them a lot. If a sample describes the same characteristics in the data, we say the sample is representative of the population.

To understand if the sample data is representative of the population, an easy step is to plot the sample distribution and compare it with the population distribution. If their distribution shapes are similar (clearly, the sample dataset is smaller), it means that the sample dataset is the correct representation of the population. Figure 3-24 shows an example of correct and incorrect sample datasets. Nevertheless, it is usually impossible to access all data of the population (the entire dataset), and even if it is possible, it is not cost-effective. Therefore, we should select a wise number of samples from the entire dataset and create a sample that is representative of the population.

If the population follows a distribution, the sample population must follow that distribution as well. To distinguish between the sample mean and the original one, we use different signs. The sample mean is shown as $\bar{x}$ (read as x bar), but the original mean is shown as μ. An ideal sample has a mean equal to the population, and a good sample has a mean close to the population mean.

There are methods used to sample data, such as clustering, random sampling, etc., which we will explain later in Chapter 16.

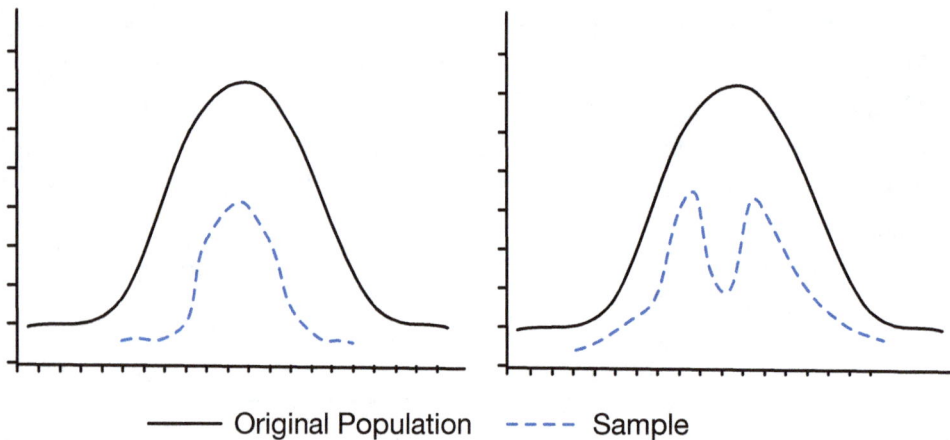

Figure 3-24: Example of correct or unbiased (left) and incorrect or biased (right) samples of the population.

Central Limit Theorem (CLT)

The *Central Limit Theorem* describes that the distribution of the sample means (or sums) approximates a normal distribution as the sample size becomes large enough. It is significant because the normal distribution is often used in statistical modeling due to its ability to encompass a wide range of uncertainties with minimal prior assumptions. If the concept of uncertainty in this context isn't clear now, don't worry; as you delve deeper into statistics and data science, these ideas will become more understandable.

Law of Large Numbers (LLN)

The Law of Large Numbers is a theorem that explains that experimenting several times on the sample dataset brings the sample parameters (e.g., mean, standard deviation) closer to the population parameters. In short, as the number of experiments increases, the sample characteristics should get closer to the population.

Note that The LLN applies to repeated trials or observations drawn from the population itself, not from a fixed sample dataset. Each trial adds more information about the population's characteristics.

Bias in Sampling

An incorrect sampling causes an unforgivable sin, which is called *bias*. *Bias in statistics refers to a systematic error that skews results and leads to inaccurate conclusions*. This can arise from various sources, including how data is collected, processed, or analyzed. In the sampling context, bias occurs when the sampling method systematically favors certain outcomes, leading to samples that do not accurately represent the whole population.

Bias can be unintentional due to methodological errors and intentional, where data is manipulated to achieve a desired outcome. Ensuring random and representative sampling is one of the key ways to minimize sampling bias. More about bias and methods used to mitigate bias will be explained in Chapter 16.

Confidence Interval

The analysis we perform on a sample is an estimate of the population (entire dataset). In other words, we work with an estimate of the dataset and not the entire dataset. We are not sure about the accuracy of this estimate and how close the sample is to the population. In short, we should find a way to check: *How good is our estimate?*

Two approaches are used to measure the correctness of sample data: *confidence interval* and *significant test*. To gain a better understanding of the accuracy of our sample, we use *Confidence Interval (CI)* [Neyman '37]. A CI provides a range of values within which the correct population parameter (like a mean or standard deviation) is likely to fall. It gives an estimate of the uncertainty surrounding a sample statistic. In other words, CI is presented as a range, and it is used to identify the *interval* or a *range of values* from the sample to estimate the chance of whether our sample reflects the data in the population.

CI is operated based on a *confidence level*. The confidence level is the probability (in percentage) that the population's mean falls within the given interval. In simple words, if we make a sample dataset, a certain percentage of the sampled data points (confidence level) will contain the mean equal to the original dataset (population) mean.

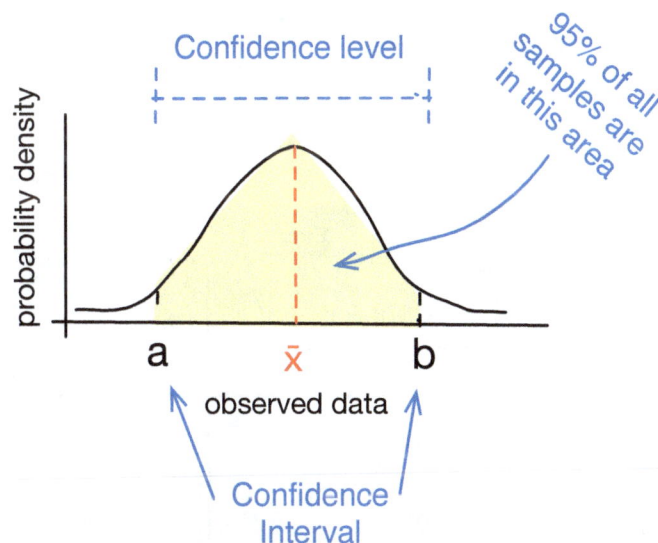

Figure 3-25: Confidence Interval, which shows a range marked with a and b. In this figure 95% of data located between a and b.

The confidence level usually is 90%, 95%, or 99%. There are other levels as well, but 95% is the most commonly used value for confidence level in computer science. In biology and other fields, it might be different. In the rest of this book, if we talk about confidence level, we mean 95% .

Figure 3-25 presents the distribution of the sample dataset. The area between points 'a' and 'b' includes the data from the population with 95% percentage of accuracy. Here, accuracy is defined as the mean of data in the interval is equal to the mean of the population. Therefore, we can say 95% of the data from the population are in a range between 'a' and 'b'. Since we don't know the mean of the entire population, we can use our sample dataset and estimate the range for the mean of the population, i.e., between 'a' and 'b'. In other words, between 'a' and 'b' is a range that is called a confidence interval. 'a' is referred to as the lower bound, and 'b' is referred to as the upper bound.

The larger the size of the sample dataset, the smaller the width of the confidence interval, which means the sample dataset provides a more precise estimate of the population parameter. In the ideal case, assuming the sample size is equal to the population size, the confidence interval becomes extremely small, approaching zero, and both 'a' and 'b' overlap on the $\bar{X}$.

Let's repeat again: it is hard to identify the mean of the entire population. Otherwise, we don't perform sampling from the dataset, and we use the entire population (original dataset). If the sample size has a normal distribution, CI is calculated based on the mean of the "sample" with the following equation:

$$CI = \bar{X} \pm Z \frac{\sigma}{\sqrt{n}}$$

Desired Confidence Interval (Confidence Level)	Z-score
90%	1.65
95%	1.96
99%	2.58

Table 3-4: Conventional z-score for confidence interval calculation.

Z is a constant value or z-score associated with the CI, and we get its value from Table 3-4. We do not describe how it gets calculated, and you don't need to learn it. Remember, the confidence interval expresses the range of population mean; it means it quantifies the margin of error. The $Z(\sigma/\sqrt{n})$ is called the *margin of error*. In other words, CI is nothing more than a statistic variable that is mean plus/minus margin of error.

It is essential to understand the concept of a confidence interval. Most programming languages include math libraries that can easily calculate it, just as they do for other statistical formulas. Here, we present the equation to give you insight into how it's derived. We believe there's no need to memorize equations—save your valuable brainpower for understanding concepts and algorithms, which are numerous enough on their own.

To understand the usage of the confidence interval in a real-world scenario, consider the following example. The insect farmer we introduced earlier would need you to quantify the quality of his edible bugs, but he can't identify the average weight of his bugs.

They have different sizes and weights. Can you help him? Yes, of course, you can do this excellent humanitarian work for social good and save our planet while leaving corporations to continue their money-making businesses.

As the first step, you go to his farm and collect 30 of those lovely edible bugs. Then, you calculate the mean weight of these 30 bugs, i.e., 3.2 grams, and the standard deviation, i.e., 2.7 grams of the 30 insects. Next, you calculate the CI by the described formula. Based on a 95% confidence level, we can say the mean of his insect is between 2.234 and 4.166. Then, you show this to the farmer, and he answers you: *"It is too vague; the difference between 2.2 and 4.1 highly deviates."*

How can you resolve this by increasing the sample size and thus reducing the confidence interval until it is convincing for him? You go to the farm and collect more bugs. Now, you collect 50 bugs, and based on a 95% confidence level, the mean of their weight will be between 2.9 and 3.4. Then, the farmer is convinced and happy because he knows the average weight of his bugs is in this range.

Based on the confidence interval, we can estimate the number of samples we require. The equation to estimate n is as follows:

$$n \geq \left(\frac{Z.\sigma}{Margin \ of \ Error}\right)^2$$

For instance, if we decide on confidence, e.g., 95% with a z-score of 1.96 (see Table 3-3). Then, we use a small sample set to identify the standard deviation. Yes, we are sampling to estimate the correct sample size, and it seems weird. Also, we should provide our desired margin of error. The

result of this equation will be rounded up, and the correct number of samples will be shown for the given margin of error, standard deviation, and confidence level.

For instance, we would like to know the optimal weight of a bug at the farm, with a CI of 90% (Z=1.65), and we don't want the margin of error to be more than 0.5 grams. We sample 10 bugs and compute the standard deviation as 1.8. Therefore, using the above equation, we can say that the number of samples should be as follows:

$$(\frac{1.6 \times 1.8}{0.5})^2 = 33.17$$

This means at least 33 bugs should be sampled.

Usually, the more we sample, the larger the confidence interval until, at a point, it does not get larger. This is due to the *Law of Large Numbers (LLN)*.

Hypothesis and Significance Tests

Previously, we described using a confidence interval and margin of error to understand whether we have collected enough samples. Assume you analyzed the data statistically and made some novel discoveries. For instance, after several years of hard work, your company discovers that *'all cucumbers are green'*, *'corporations do not give a damn about the earth and pollution. Instead, they asked you to eat bugs.'*, *'Mass media are promoting hate among different nations, religions, and races.'* etc.

Now, you need to show that these findings are generalizable and that your experiment results are not biased or discovered by accident (random). These findings are called *hypotheses*, and to demonstrate the generalizability of a finding, we use *significance tests*, which we will explain in detail with an example.

We start with our previous example. The insect farmer is happy about your previous work, and now he has asked you to help him identify which type of bugs are tastier and worth further breeding. To answer him, you start eating some sample bugs and write down the taste and weight of the bug legs (we are really sorry for your job now). Then, you find that bugs with a better taste have an average leg weight of about 0.2 grams. The rest of the bugs taste either too greasy (fat bugs) or too crunchy (thin bugs).

We make a hypothesis as follows: *"The tastiest bugs have a leg weight of 0.2 grams"*. How can we claim this finding is true?

Statistical significance tests are used to check the correctness of our claim (hypothesis). For instance, *"if you accuse some of our media corporations of promoting hate among different nations and religions"*, you should use statistical significance to prove it. If you find that bugs with 0.2 grams of leg weight are the tastiest bugs, you should use a statistical significance test to prove it.

To conduct a significance test, we deal with two hypotheses: the *null hypothesis*, i.e., H_0, and the *alternate hypothesis*, i.e., H_1 or H_A. H_0 states that our finding, which is driven by analyzing the data (or hypothesis we make about the sample), is NOT true. H_1 states what we think should be correct about the data, but H_0 says our hypothesis is wrong (H_1 is the hypothesis that says H_0 is false). Instead of proving H_1 in a statistical significance test, we should reject H_0. In other words, to claim H_1 is true, we must reject H_0.

In the bug example, we can say $H_0 =$ *"bugs with an average weight of 0.2 g. leg DO NOT taste better than other bugs"*, $H_1 =$ *"bugs with average leg weight of 0.2 g. DO taste better than other bugs"*. So we have:

H_0: $\mu \neq 0.2$ gram
H_1: $\mu = 0.2$ gram

As we described, we should conduct a significance test that rejects H_0, and then we can say our lovely H_1 is correct. If H_0 is correct, we should use a larger sample size or change our hypothesis, or if we are a politician, just change the problem and put it under the rug.

How can we test H_0 and determine whether to accept or reject it?

The result of the significance test is presented as a *p-value* (probability value). In technical terms, the p-value is the probability that H_0 is correct, and in turn, there is no conclusion to be inferred from the data. In our example, if the p-value is large enough, it means there is no relationship between the leg's average weight of 0.2 grams and the tastiness of bugs. P-value will be a variable between 0 and 1, determining whether we can reject H_0.

If the p-value is less than a value called *significance level (α),* then we can reject the H_0, and therefore our claim (H_1) is correct.

Therefore, we should remember:
p-value $< \alpha \rightarrow$ reject H_0 (Good)
p-value $\geq \alpha \rightarrow$ reject H_1 (Bad)

Let's define the statistical significance test in another way. Put this definition under your pillow to read it every night before sleep:

The purpose of the significance test is to identify whether the differences between the two groups of data we are comparing are by chance or if there is a significant difference.

Usually, the convention is to set $\alpha = 0.05$. An alpha level of 0.05 means that there exists a 5% risk of concluding that there is an effect (or a difference) when there is none (the grey area in Figure 3-26). Therefore, we can say that H_1 is true, and $1 - 0.05$ the data is covered by H_1. Based on Figure 3-25 and Figure 3-26, H_1 is the white area inside the curve, and H_0 is the sides in grey color. Therefore, if the p-value is less than or equal to 0.05, it falls into the grey area; H1 is acceptable, and we can reject H_0.

Note that the probability of H_0 is either smaller, more significant, or not equal to the α value. In our example, we say that $\mu \neq 0.2$ is H_0 (legs with 0.2 grams are not the tastiest).

The critical values are used to distinguish the white area in Figure 3-26. They will be calculated based on the given α, and the statistical software you use will do it for you. This means that the

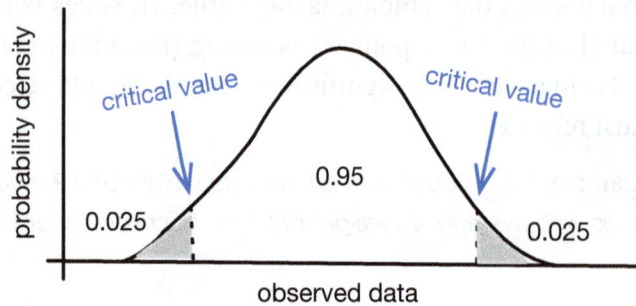

Figure 3-26: Normalized sample data where α = 0.05 for our bug leg example.
The *p-value* on the left and right sides (inside the two grey regions) together occupies 0.05 of the distribution.

software will calculate the position of the critical value on the X-axis. We should distribute the %5 on the left and right sides of the X-axis. Therefore, on each side, we have %2.5.

Keep in mind that the sum of both grey sides of the distribution in Figure 3-26 should be smaller than the significance level, and the p-value should fall into the grey area to accept H_1. Otherwise, if the p-value is inside the white area, we should accept H_0.

After experimenting and selecting our hypothesis, there are four steps we should perform to test our hypothesis. These steps are listed as follows:

(1) Choose the significance level α. For example, the farmer chooses 0.05, the most common significance level choice. α sometimes is called type I error rate, and we will explain more about it shortly.

(2) Collect sample data. This means you will collect bugs, measure their average leg weight, eat them, and write down the taste of each bug. We are sorry for you; life is hard, and we all suffer.

(3) Calculate the test statistics p-value[7] using a significance test (we will explain Significance tests later). A test statistic is a standardized value calculated from sample data for the hypothesis test. The test statistics measure the degree of agreement between sample data and H_0. We will briefly describe the use of different test statistics later. In short, we calculate a value, i.e., test statistics, to measure whether we can reject the H_0 and thus accept our lovely H_1.

(4) The result of a test statistic is the p-value. Now, we can compare the obtained p-value with α (significance level), e.g., 0.05. If the p-value is smaller than α, p-value <0.05, then we can happily reject the H_0. This means our H_1 is valid, and the tastiest bugs are bugs with an average leg weight of 0.2 g. If the p-value is equal to or larger than α, then the H_0 is true, and thus, our H_1 is not acceptable.

---

[7] The term "statistic" in "test statistic", refers to a quantity derived from sample dataset.

Hypothesis Errors

Earlier in this chapter, we explained that the process of making inferences about the data is called *inferential statistics*. In inferential statistics, no hypothesis test is certain or absolutely correct. We always deal with errors.

There are two types of errors in inferential statistics as shown in Table 3-5, *type I error* and *type II error*. When a null hypothesis is true, but we reject it by mistake, we make a Type I error, i.e., false positive error. We can reduce the chance of type I errors by increasing the value of α. When the null hypothesis is false, and we fail to reject it, we make type II errors, i.e., false-negative errors (Check Chapter 1). We can reduce the chance of type II errors by increasing the sample size.

	Reality	
	H0 = True	H0 = False
Accept H0	Correct Decision	Type II error
Reject H0	Type I error	Correct Decision

Table 3-5: Hypothesis test outcomes.

To summarize, assume we have collected some sample data and made some fantastic novel inferences from the data, like *'most cucumbers are green'* (more than 95%). We need to prove it with inferential statistics. To prove it, we can use the p-value to reject our null hypothesis, i.e., *'95% of cucumbers are not green'*.

NOTES:

* As the sample size increases, the margin of error decreases. However, increasing the sample size after a certain point does not have any effect on the margin and error, and keep in mind that sampling is an expensive process.

* There is no optimal test to report the exact sample size we require. However, by comparing two sample datasets, we can say if the sample size is big enough or not. Therefore, if the significance test fails, we should either use a larger sample size or give up on our alternate hypothesis.

* A significance test, in this context, helps us understand whether the result we have gained from an experiment is random (false) or it is valid and has a cause (actual). If our results were random, this means that we could not generalize our findings from our sample to the population.

A/B Testing and Significance Test

Most scientific experiments we conduct disregarding its field (e.g., human-computer interaction, biology, psychology, etc.) require A/B testing to compare the results of experiments. A/B testing refers to grouping data into two (could be more) groups, one group control, and the other treatment or experiment. Then, apply something to these groups and compare the results.

For example, we are building a computer game that brings the user joy while playing. We claim it has a more positive emotional impact on the player than other games. We can prove our claim by using A/B testing. In particular, we experimented with two groups of users. One group uses our game; they are called the treatment or experiment group. The other groups do not use our game; we call them the control group. Then, we measure their joy level, and we can conclude whether our computer game brings more joy to the treatment group than the control group.

Now, let's study an example. We are giving weight loss medication (A and B) to a group of individuals: group A (control and do not receive the medication) and group B (treatment and get the medication).

The weight loss in two groups is reported as negative, and weight gain as positive. The result is as follows:
Group A: -4kg, 0kg, 3kg, 1kg, 1kg, => mean: 1kg
Group B: -1kg, -1kg, -1kg, 4kg, -1kg => mean: 1kg

We can see that everyone in group B lost weight except for one person who gained 4kg. However, the mean weight change is equal to that of group A. It shows that just comparing the mean is usually not enough. To solve this issue, we can use a significance test. The significance test compares two groups of data reports if there is a statistically significant difference between the two groups of data. The goal of a significance test in this context is to determine whether the observed differences in outcomes (in this case, weight changes) between the two groups are likely due to the treatment (medication) rather than occurring by chance.

To summarize, make a slot in your brain and write the following, "A significance test is comparing if two groups of data are different in their statistical characteristics". Nonetheless, significance test results only tell us if there is a significant difference between those two groups, but they cannot measure the magnitude of the difference.

There are two categories of significance tests: parametric significance tests and non-parametric significance tests. Parametric significance tests assume that all samples have a normal distribution. Non-parametric significance tests do not rely on the normal distribution of samples.

Whenever the term non-parametric is encountered, it means that there is no information about the distribution (distributions are characterized by their parameters).

Parametric Significance Tests

In this section, we will describe useful tests for a given condition, but we skip computation details. Our focus is on learning statistics for machine learning. Plenty of fantastic statistical books or online resources are available if you are interested in the details of each test. At the end of this chapter, we introduce some of them.

t-test

We use a t-test to compare the mean of two groups of data where the sample size is small, i.e., less than 30 sample data points are available, and we do not know the standard deviation of the population distribution (or the other group distribution). We could assume that the population or other group dataset is *approximately normally distributed,* and we use a t-distribution to test the null hypothesis. This test operates on a *small number of data points* without knowing the variance of the other group. In fact, the t-test uses the sample (one dataset) variances to estimate the population (another dataset) variances.

The t-test checks whether the *means* of the two groups are significantly different. As shown in Figure 3-23, the t-distribution was another form of normal distribution but more flattened than the z-distribution with a fatter tail[8]. Therefore, since the sample size is small, we can say that increasing the sample makes the distribution similar to normal distribution.

There are three types of t-tests: the *one-sample t-test*, the *independent (or unpaired) t-test,* and the *paired (or dependent) t-test*.

One-sample t-test

It is used when we have *a single group of sample data*, the sample size is small, and we would like to compare it with a known population mean. However, we do not know the standard deviation of a population. In other words, we only have the sample dataset and the mean of the population dataset. The purpose of a one-sample t-test is to compare the mean of this sample against a known mean of the population.

For example, assume your second job is selling ice cream on the street. Every week, you bring 100 ice creams and go to sell them (the population size is 100). Your average daily sales are 50 ice-creams (population mean is 50), and we assume α=0.05, with a standard deviation of 12.

The insect farmer told you to sell his bug-infused ice cream and boost your sales. You accept his proposal to sell his bug-infused ice-creams as well. After a couple of weeks, you sample 20 days and study your average daily ice cream sales (all types of ice creams). It is 60, with a standard deviation of 15. Does the bug-infused ice cream change your sales at all (either positive or negative)?

To answer this question, you can use a significance test because you can compare two groups of data together. The one-sample t-test is applicable here because we have a small sample (n = 20) with a known standard deviation (σ = 15) and mean ($\bar{x}$ = 60) and a population with a known

---

[8] Distribution tail is also referred to as Kurtosis.

mean ($\mu = 50$). However, we do not know the standard deviation of the population. In summary, we have the following:

Before selling bug-infused ice cream: *Group 1: sample size=?, SD= ?, mean = 50*

After selling bug-infused ice cream:

Group 2: sample size= 20, SD= 15, mean = 60

Using software to calculate the t-test, the result of the one-sample t-test shows that *p*-value < *0.05*. Therefore, we find that adding bug-infused ice cream has some impact on our ice cream sales. We can only say there is a significant difference between the two groups, and we can not provide any more justification.

Independent t-test

Independent t-test, also known as *two-sample t-test*, is the most commonly used t-test. We use this t-test to compare the mean of *two groups* that are *independent* and report whether there are significant differences among them. In addition to the normal distribution of data, the mean of both datasets should be different as well. The independent t-test assumes that variances for the two datasets are equal. However, there is a variation of the independent t-test, known as *Welch's t-test*, that does not assume equal variances.

To understand the independent t-test, we use an example. Assume you decided to do something very important for humanity and changed your job from a data scientist and ice cream seller to a biologist. You discovered a medication that can cure obesity. To prove if your drug is successful, you test it on two groups of users (groups A and B) whose members have the same diet and the same amount of physical activity.

Group A receives the medication (treatment group), and group B does not receive the drug (control group). Group A has 15 members (n_1=15), and group B has 20 members (n_2=20). The mean weight of group A members after using the drug is 72kg, and the mean weight of group B members is 73kg. Group A members' weight standard deviation is 4.2 kg, and group B members' weight standard deviation is 1.1kg. To summarize, we have the following information:

Using your obesity medication: *Group A: sample size = 15, SD= 4.2, mean = 72*
Not using your obesity medication: *Group B: sample size = 20, SD= 1.1, mean = 73*

By using an independent t-test, we compute the *p*-value, and we regret to inform you that the *p*-value of the unpaired t-test shows the result of 0.031, which is < 0.05. Therefore, your obesity medication is effective.

Paired t-test

The Paired t-test is used when we have *one group* of data measured at *two different times*. It is another form of a one-sample t-test. Usually, this test is employed to check if the new treatment, method, etc., is effective on the same data points and works better than the previous method or not.

For instance, your biological startup applies some gene modification to the weight loss medication. Then, measure human weights before you give them genetically modified medication, and then again, you measure their weights afterward. A paired t-test will be used to identify the statistical significance between these two measurements. This means that the data points are the same, but we measure them at two different times.

ANOVA, MANOVA and ANCOVA

The t-test is limited to comparing only two groups of data (or one group in two different conditions). However, Analysis of Variance (ANOVA) is a statistical method used to analyze differences among statistical characteristics of *two or more groups of data*. The simplest form of ANOVA generalizes the significance test for more than two groups. The H_0 in ANOVA assumes that all groups' means are equal, and H_1 assumes that at least two of the group means are different.

H_0: $\mu_1 = \mu_2 = \mu_3$
H_1: Means are not all equal.

ANOVA also assumes homogeneity of variances (the variances are similar across groups). Similar to the t-test, ANOVA assumes all samples follow a normal distribution and that the variances of samples are not significantly different.

ANOVA works with factors (variables) and levels (values). Factors are variables such as gender, which male, female, and transgender are levels for the gender factor. The result of the ANOVA test will be presented as an F-ratio. F-ratio is a ratio of two variances[9]. ANOVA operates based on a hypothesis test called the F-test. It compares the variability between groups to the variability within groups.

There are three well-known types of ANOVA, One-Way ANOVA, Repeated-Measures ANOVA, and Factorial ANOVA. Also, there are two known extensions of ANOVA, including MANOVA and ANCOVA. We explain them briefly in the following.

One-Way ANOVA

In this test, we have *only one variable (factors)* with *at least two values (levels)*, and levels are *independent*. For instance, assume you are now a successful biologist, and you get a new contract from the insect farmer to use some drugs on his bugs and make their size optimal for a better taste. When you were a data scientist, you got to know that muscular insects have less fat and taste better. Now, you start experimenting with feeding insects testosterone to increase their musculature tissues. You use testosterone in three different dosages: 0 mcg, 5 mcg, and 20 mcg.

In other words, the factor is a testosterone dosage, and levels are 0, 5, and 20 mcg. Analyzing the differences between these three groups of testosterone treatment could be done with One-Way ANOVA.

---

[9] If you are looking for a resource to understand ANOVA deeply, we recommend you check Chapter 9 of Statistics II for Dummies [Rumsey '09]. Here, you only need to learn when to use it.

Repeated Measure (Dependent) ANOVA

In Dependent ANOVA, we have *only one variable* with at *least two values*, but the values are *dependent*.

For instance, assume, we measure the weight of bugs who have received 20 mcg of testosterone. If we measure their weight on 3 different days, e.g., Day 1, Day 3, and Day 6, we need to use repeated measure ANOVA to see if there is any statistical difference between them. Because bugs are the same, their weights each day have changed. Analysis of the impact of testosterone will be done on one variable with dependent values; in this case, we use dependent ANOVA.

It is very similar to a paired t-test, but we use the t-test when our dataset is small.

Factorial ANOVA

When we have *more than one variable (factor)*, we use this test. Note that variables must be *independent*. A well-known type of Factorial ANOVA is *Two-Way ANOVA*. Two-way ANOVA is used when we have one dependent variable and *two independent variables*, and there might be an interaction between independent variables.

For instance, assume we are measuring the weight changes (dependent value) in different days (variable) for male and female bugs separately (two independent variables). Or we are measuring mood changes (assume bugs have mood, e.g., happy, sad, etc.), i.e., dependent values, of male and female bugs (two independent variables) to different dosages of testosterone (independent values). For this type of analysis, we deal with two independent variables and different values (dependent or independent); we use Factorial ANOVA[10].

MANOVA

Multivariate Analysis of Variance (MANOVA) is a significance test for sample datasets with *more than one dependent variable* and *one independent variable*. ANOVA is limited to one dependent variable, but MANOVA can handle more than one dependent variable. In the previous example, we give testosterone to bugs and measure their weight in one day. However, the farmer is obliged to adhere to ethical codes and would like to be sure that testosterone does not have any adverse effect on bugs' moods. Therefore, in addition to measuring weight, every day, a therapist (who can talk in bug languages) talks with bugs and measures their level of happiness.

Similar to ANOVA, if there is one independent variable and more than one dependent variable, we use *One-Way MANOVA*. If there is more than one independent variable and more than one dependent variable, we use *Two-Way MANOVA*.

If you think it is not easy to remember them, we do agree with you that you should preserve your brain cells for the next chapters. Just try to identify the characteristics of your dataset and read again these descriptions to decide about the best possible test, you can use Table 3-6 to decide about your test as well.

---

[10] A very good and brief description of ANOVA exists here: http://statisticslectures.com/topics/introanova, and we adopt our example from this link.

ANCOVA

Another variation of factorial ANOVA is ANCOVA, which stands for Analysis of Covariance (ANCOVA). There are additional variables that are not independent or dependent variables, but they affect the dependent variable.

These variables are called *covariates*. A covariate is a type of *control variable that is measured but not manipulated in the experiment*. Covariates are extraneous variables that are not of primary interest to our analysis but can influence the dependent variable.

To understand what covariate is, let's use an example, While we are measuring bugs' mood, we might not consider that the weather has an impact on their mood. On rainy days, bugs are unhappy, and on sunny days, they are happy. Weather is a covariate in this example. The goal of a scientific process is to establish a relationship between the independent variable and dependent variable without any external influence, but covariates can have an influence.

Dependent Variables

Independent Variables		1	>1
	1	One-Way ANOVA	One-Way MANOVA
	>1	Factorial ANOVA	Two-Way MANOVA

Table 3-6: Deciding about the best ANOVA test based on number of independent and dependents variables.

Since ANCOVA is the analysis of covariance, it decomposes the variance of the dependent variable into variance explained by the covariates and variance explained by the independent variable, plus residual variance. In simple terms, think of ANCOVA as adjusting the dependent variable by the group mean of the covariates. We are very sure you understand the previous sentences are perfectly fine, but don't worry. We will not need to know how it works.

Remember, we should use ANCOVA instead of ANOVA when we have covariates. The same is applicable to MANOVA and MANCOVA. When we need MANOVA but we have covariates, we go for MANCOVA[11].

NOTES:

* Some argue that probability plots are visual tools, so subjective judgment is involved in interpreting them. They complement, but don't replace, formal statistical tests for distribution fit.

---

[11] We do not describe the statistical analysis with covariate in detail if you are interested in learning more, check this fantastic tutorial: http://www.statsmakemecry.com/smmctheblog/stats-soup-anova-ancova-manova-mancova

* When there is more than "one group" of normally distributed datasets to compare, we go for ANOVA and its derivations for statistical significance tests.

* Where we have a small (< 30 samples) sample size and we intend to find if there are any significant differences between the population mean and hypothesized value, we can use the t-test.

* To conduct a t-test, we should know the mean for both datasets that will be compared together. t-test H_0 states that two population means are the same, $H_0: \mu_1 = \mu_2$, while the alternative hypothesis H_1 says they are not the same, $H_1: \mu_1 \neq \mu_2$.

* The null hypothesis in ANOVA assumes that all samples' means are equal $H_0: \mu_1 = \mu_2 = \mu_3$. The alternate hypothesis, H_1, says that at least two means are different.

* None of the statistical tests are ideal, and all of them make mistakes. Usually, it is better to test your data with different tests and see which one provides a good answer. However, we should have a good justification for selecting a specific test and rejecting others.

* Both t-test and ANOVA operate based on the assumption that the population and samples follow a normal distribution (according to the Central Limit Theorem). If the data does not follow the normal distribution or we do not have this assumption, we should go for nonparametric tests.

* In addition to covariates, there is another term called *confounding variables*, which are variables that distort the relationship between the independent and dependent variables.

Non-Parametric Significance Tests

Back in the early decade of 2000, very few computer scientists knew statistics, and a common point to criticize statistical analysis was using a parametric significance test for the dataset, which we didn't know whether had a normal distribution. Both t-test and ANOVA (plus its variances) compare samples presumably normally distributed. There are significant tests that do not need the assumption of normal distribution for the underlying dataset. They are called *Non-Parametric* tests. We can call them distribution-free tests because they do not require much prior knowledge about the distribution. In the rest of this section, we list some of the common ones.

Chi-Square Test

Chi-Square (χ^2) is one of the common non-parametric tests; it is used for two different purposes: (i) testing the *independence of two categorical variables* or (ii) testing the *goodness-of-fit*. In simpler terms, by the test of independence, we mean this test is used to examine whether two categorical variables are independent. By goodness-of-fit, we mean testing if two distributions fit each other, for example, to see if a dataset has a normal distribution, and we use goodness-of-fit to compare a theoretical normal distribution with our dataset.

Test of independence

A usual practice in data analysis is checking if there is a relationship between two variables. Remember, each variable can have different values; thus, we are comparing two sets of their values together. If both variables are numerical, we use correlation to analyze their relation, which will be explained later in this chapter. If both variables are categorical, we use a *Chi-Square* test (χ^2-test) to determine whether there is a relationship between those two variables.

This test operates based on a contingency table. *Contingency*, *Crosstab*, or *R×C table* (read as R by C table, R stayed for row and C for column) is a table that presents the frequency of different variables in a dataset and their relations together. For instance, look at Table 3-7, which presents the relationship between bug tastes and the leg weights of 161 bugs. We would like to use this table to see whether there is any relationship between bugs' leg weight and taste.

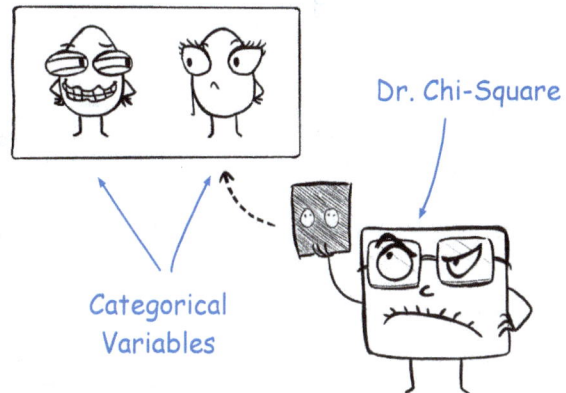

Dr. Chi-Square

Categorical Variables

Let's review our hypothesis:

H_0: Bugs with leg weight of not 0.2 gr. (heavier or lighter than 0.2 gr.) taste better.
H_1: Bugs with leg weight of 0.2 gr. tastes better than other bugs.

The chi-square test calculates the p-value based on the differences between *observed* and *expected values* from the contingency table. Assuming E presents the expected value, it is easy to calculate the expected value as follows:

$$E = \frac{total\ row \times total\ column}{sample\ size}.$$

To calculate the expected values for the observed values presented in the contingency table, we multiply the row total by the column total and divide by the grand total. Table 3-7 presents observed values, and Table 3-8 computes the expected values of Table 3-7.

After we have calculated the expected values, we can use the following equation to calculate the chi-square score (χ^2):

$$\chi^2 = \sum_{i=1}^{r} \sum_{j=1}^{c} \frac{(O_{ij} - E_{ij})^2}{E_{ij}}$$

In this equation, O_{ij} is the observed value on row i and column j. Respectively, E_{ij} is the expected value on row i and column j. In our example, this number will be:

$(47 - 29.9)^2/29.9 + (14 - 25.6)^2/26.6 + \ldots$

Taste

Leg Weight		Good	Too Oily	Too Crunchy	Total
	=0.2	47	14	22	83
	>0.2	8	34	12	54
	<0.2	3	1	20	24
	Total	58	49	52	161

Table 3-7: A contingency table that reports bug leg weight and taste. This is the observed dataset.

Taste

Leg Weight		Good	Too Oily	Too Crunchy	Total
	=0.2	(83x58)/161 = 29.9	(83x49)/161 = 25.6	(83x52)/161 = 26.8	83
	>0.2	(54x58)/161 = 19.4	(54x49)/161 = 16.4	(54x52)/161 = 17.44	54
	<0.2	(24x58)/161 = 8.6	(24x49)/161 = 7.3	(24x52)/161 = 7.7	24
	Total	58	49	52	161

Table 3-8: Expected values contingency table calculated from observed values from Table 3-6.

Also, the degree of freedom should be calculated as well, i.e. $(R - 1) \times (C - 1)$, in our example, the degree of freedom will be $(3 - 1)(3 - 1) = 4$ because we have three columns and three rows. The software package will use the chi-square table[12], contingency table, degree of freedom, and a given α (significance level, which is the threshold for deciding whether to reject or fail to reject the null hypothesis) to perform a chi-square test and provide us with a p-value.

Although we don't need to learn these steps, it is important to know the contingency table. It is used in many data analyses, e.g., Odds-Ratio calculation, which we will explain later.

To summarize, to use the chi-square test for the relationship between two categorical variables, we should give the contingency table and significance level as inputs. The software package will do the rest, and based on the output p-value, we can see whether the two categorical variables are independent or not.

---

[12] We do not explain these tables in this book, but just keep in mind they are tables with constant information.

Goodness-of-Fit

The second use of the chi-square test is the Goodness-of-Fit test. Goodness-of-Fit is used to check how well a *sample (observed)* dataset fits an *expected (hypothesized) distribution*. In particular, this test evaluates whether the observed dataset fits the expected (or predicted) dataset. It could also be used to see if a dataset fits a particular distribution, e.g., normal distribution, or not.

Usually, by looking at the distribution of data, we are able to make some predictions about the data. Nevertheless, in real-world settings, there is no guarantee that what we observe is similar to what we expect, and thus, we use the chi-square test (Goodness-of-Fit test).

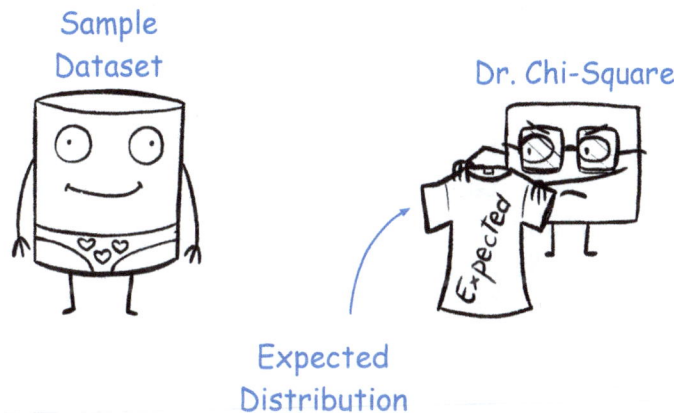

We explained how an expected variable is calculated in each cell of the RxC table. The Goodness-of-Fit is calculated by the following equation, which is a very similar equation of Chi-square to check the relationship between two categorical variables:

$$Goodness - of - Fit : \chi^2 = \sum_{i=1}^{n} \frac{(O_i - E_i)^2}{E_i}$$

Similar to the previous equation, O presents the *observed value* in each data point, and E presents the *expected value* for that particular data point.

The Goodness-of-fit test has the following hypothesis:
H_0: *The sampled (observed) dataset is not significantly different from the expected dataset.*
H_1: *The sampled (observed) dataset significantly differs from the expected dataset.*

If the result of the Goodness-of-Fit test is small, we can reject the alternate hypothesis. Otherwise, the observation does not fit the expectation, and we reject the null hypothesis. Similar to the test of independence, the output of the Goodness-of-fit will present a justification in the p-value based on the given α. The p-value indicates the probability of observing a Chi-square statistic as extreme as, or more extreme than, the one calculated from our dataset under the assumption that the null hypothesis is true. If the p-value is less than the significance level, we reject the null hypothesis.

Kolmogorov-Smirnov

Kolmogorov-Smirnov test (KS-test) is another non-parametric test. Remembering the name of this test for non-Russian speakers is not easy, but try to memorize it and pronounce the names correctly. It is very helpful to show off your statistical knowledge in any meeting, even at parties. We did it many times, and we were successful in entertaining others on how knowledgeable we are.

The KS-Test requires fewer technical assumptions compared to the t-test and ANOVA. It doesn't make any assumption about the distribution of two different samples, and this makes it a very popular significant test.

KS-Test measures the differences between the Cumulative Distribution Function (CDF) of two datasets, as shown in Figure 3-27. In particular, it reports the maximum differences between the two CDFs. We hope you remember what CDF is; otherwise, please check the earlier sections of this chapter.

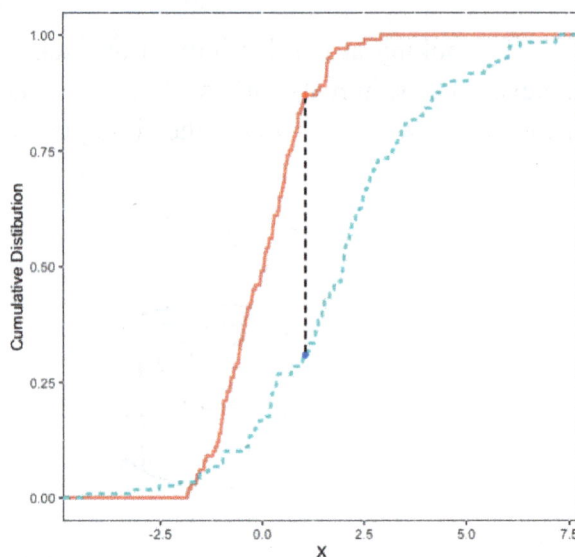

Figure 3-27: KS-Test based on comparison of the distance between CDFs of two samples.

Following are the null and alternate hypotheses for the test:

H_0: the two datasets being compared are drawn from the same continuous distribution.

H_1: the two datasets being compared are not drawn from the same continuous distribution.

The KS-Test can also be used for Goodness-of-fit.

Kruskal-Wallis Test

Kruskal-Wallis Test (KW-Test) is a non-parametric test analogous to one-way ANOVA, and it is used to compare more than one sample.

As the first step, it combines data from all samples and then ranks them in ascending order (from smallest to largest). Then, it searches for patterns of how these rankings are distributed among our samples. If two samples have an equal mix of ranks, they are assumed to be similar. Otherwise, if two samples do not have similar ranks, they are assumed to accept the null hypothesis. In particular, it compares medians of more than two samples and reports if they are equal (H_0) or are not equal (H_1), similar to ANOVA.

To conduct this test, the following conditions must be met: all samples should follow the same distribution, their variance should be the same, and they should be independent samples.

	Taste (1 is worst, 10 is best)									
Room no.1	1	2	2	1	1	3	3	2	3	2
Room no.2	1	3	2	1	5	1	1	2	3	1
Room no.3	3	1	2	4	2	1	1	3	1	6

Table 3-9: Bug tastes score for each room.

Let's go back to our attractive example, the insect farm. The insect farm has three bug rooms, and bugs in each room receive a unique diet, and thus, they have different taste.

Since each room is treated with a unique diet, the farmer would like to know which diet makes the tastiest bugs. The first question is to check, whether there is any difference among different diets. Unfortunately, again, you should go for sampling and choose to select 10 random bugs from each room. Then, you eat them and rank their taste from 1 to 10. 1 is the worst taste, and 10 is the best taste. You have already eaten all 30 bugs (again, we are very sorry for you) and use Table 3-9 to report their taste.

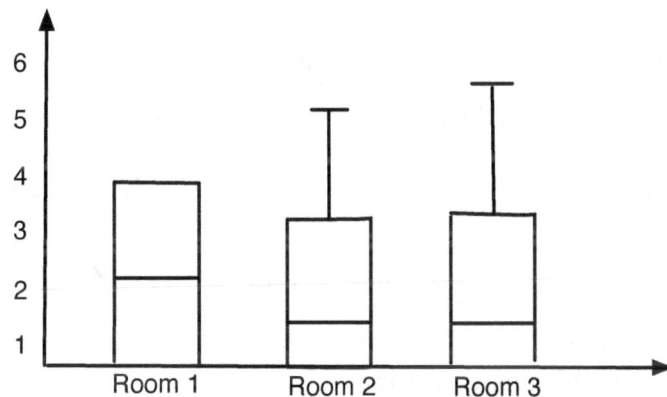

Figure 3-28: Boxplot of sample data to check if they are from a similar distribution or not.

To plot the distribution, we use a box plot. The box plot shown in Figure 3-28 illustrates that all three samples of Table 3-9 follow a similar distribution. We use a box plot, because in a non-parametric significance test, we deal with the median rather than the mean. Now, we can order and rank all data, as shown in Table 3-10. Consider that we have 12 times rank 1, we should sum them all and assign them a unique rank, i.e.:

$$\frac{1+2+3+4+5+6+7+9+10+11+12+13}{13} = 6.4$$

Therefore, the rank assigned to all 1 will be 7. Respectively we have 8 times 2, and following the same path for 2s, will lead to the following.

$$\frac{14+15+16+17+18+19+20+21}{8} = 17.5$$

and for 7 times 3, we have:

$$\frac{22+23+24+25+26+27+28}{7} = 25$$

4, 5, and 6 will be 29, 30, and 31.

Room no.1	6.4	6.4	6.4	17.5	17.5	17.5	17.5	25	25	25
	1	1	1	2	2	2	2	8	8	8
Room no.2	6.4	6.4	6.4	6.4	6.4	17.5	17.5	25	25	30
	1	1	1	1	1	2	2	8	8	8
Room no.3	6.4	6.4	6.4	6.4	6.4	17.5	17.5	25	29	31
	1	1	1	1	1	2	2	8	4	8

Table 3-10: All sample data ordered and ranked.

After all, ranks have been assigned, the test calculates the "Kruskal-Wallis test statistic", T, using the sum of the ranks for each room, i.e., Room no.1=16.5, Room no.2=14.8 and

Room no.3= 15.3, by using the following equation:

$$\frac{12}{n(n+1)} \sum_{i=1}^{g} \frac{T_i^2}{n_i} - 3(n+1)$$

In this equation, n is equal to all sample data, 30.

After the Kruskal-Wallis test statistic is calculated, the next step is to compare this test statistic with a chi-square distribution to determine the p-value. The degrees of freedom for this comparison are equal to the number of groups minus one, i.e., $k - 1$.

The resulting p-value is 0.9094. Therefore, the null hypothesis is not rejected (because p-value > 0.05), and you can report to the farmer that a different diet does not impact his bug taste. Even if the result of this test was p-value < 0.05, it still does not give any more detail about the differences; it just rejects the null hypothesis.

Of course, you can eat more bugs to make a larger dataset and repeat this experiment. We sincerely feel sorry for you getting trapped in this project. Hopefully, the next project will be in a chocolate factory.

Mann-Whitney-U Test

Mann-Whitney (Mann-Whitney-Wilcoxon, MWM-Test. Wilcoxon rank-sum test) is another non-parametric used for hypothesis tests to identify precisely which sample differs from other samples. It tests two related samples or samples from repeated measurements. This test could be used as a substitute for the *Paired (Two-Sample) T-test*, while there is no guarantee about the normal distribution of the data.

KW-Test only identifies if they are similar or different, but it can not identify exactly which sample dataset differs from others. Therefore, after the KW-Test rejects the H_0, we can use Mann-Whitney-U Test (U-Test).

For U-Test we have the following assumptions.
H_0: *The distributions of both samples are equal.*
H_1: *The distributions of both samples are not equal.*

To conduct the comparison, we should run a test on each pair of the samples until we find which one is different from the other ones. This process is called *pairwise comparisons* or *multiple comparisons*. Rejecting H_0 means the two samples we are comparing have a different median. To understand the process of this test, assume that in the previous example (Table 3-8), we would like to answer the following question: Is there a significant difference between bug tastes in Room no. 3 and Room no. 2?

The Mann-Whitney test compares every data point from Room no. 2 to every data point from Room no. 3. The test starts by ordering all numbers from both rooms. By ordering them, we have 20 elements in a set as follows. The number before ":" is just presenting the order, so we have "order: score".

{1:1, 2:1, 3:1, 4:1, 5:1, 6:1, 7:1, 8:1, 9:1, 10:2, 11:2, 12:2, 13:2, 14:3, 15:3 ,16:3, 17:3, 18:4, 19:5, 20:6}.

Now, as the second step, the algorithm ranks the equal ones and assigns them a number. For the equal ones, it calculates the averages of ranks and assigns the new score to them. In our example, for all 1s, we will have: $\frac{1+2+3+4+5+6+7+8+9}{9} = \frac{45}{9} = 5$, for all 2s, we will have $\frac{10+11+12+13}{4} = 11.5$, and for all 3s, we have $\frac{14+15+16+17}{4} = \textbf{15.5}$. The rank for 4 is 18, for 5 is 19 and for 6 is 20. The algorithm assigns these ranks instead of the original data, and we have Table 3-11 as a result.

Room no.2	5	5	5	5	5	11.5	11.5	15.5	15.5	19	Sum 98
Room no.3	5	5	5	5	5	11.5	11.5	15.5	18	20	101.5

Table 3-11: All sample data ordered and ranked and scored based on their rank

Afterward, the test sums all numbers, and this number is called the *Sum of Ranks (SR)*. It uses the following equation to calculate the U score:

$$U = SR - \frac{n(n+1)}{2}.$$

$U_{Room-no.2} = 43$ and $U_{Room-no3} = 46.5$. Then, it looks up in a constant table (which is predefined, and we don't need to learn it) and uses our given α (let's assume 0.05) and U to calculate the p-value. In this example, p-value = 0.46812, and therefore, it is not < 0.05. It means we can not reject H_0. The same test could be done between Room no. 1 and Room no. 2, and Room no. 1 and Room no. 3, to identify bugs in which rooms are different from bugs in other rooms.

While describing the KW-test, we find that there are no differences among those rooms, but again, we repeat the experiment with the U-Test. In real-world settings, one test might fail to reject H_0, and the other one could be able to reject H_0. Then, we do not know which one to select, and there is no ultimate solution for this phenomenon.

Significance Test Error Correction

One of the challenges we face when using multiple significance tests for a single dataset is rejecting null hypotheses by mistake. This phenomenon is known as Type I error. It means that when we conduct multiple hypothesis tests, the risk of increasing false positive rates increases (check Chapter 1 to recall false positives). Some tests, such as Bonferroni correction [Bonferroni '36] or Tukey correction [Tukey '49], can be used to reduce the likelihood of Type I error by adjusting the *p*-value.

Bonferroni Correction

Bonferroni correction adjusts the p-value by dividing the original α value by the number of analyses on the dependent variable when two or more statistical significance tests are applied to the same dataset. To calculate Bonferroni correction, we should create a new α, let's call it α', which is used instead of the original α. It is calculated easily by the following equation

$$\alpha' = \frac{\alpha}{number\ of\ comparisons},$$ the sign in the denominator is called a combination.

For example, if we choose $\alpha = 0.05$ (which is very common to 95% confidence), and we have four comparisons, then we have $\alpha' = \dfrac{0.05}{4} = 0.0125$. Therefore, instead of having a p-value < 0.05 to reject the null hypothesis, we should have p-value < 0.0125 to reject the null hypothesis.

We should be aware that Bonferroni correction increases the chance of Type II error. Type II error occurs when we do not reject the null hypothesis, but we must reject it. Besides, Bonferroni is not good for more than a couple of comparisons. If we have more than five comparisons, it reduces the new p-value, which makes it harder to reject the null hypothesis. Therefore, another method, such as Tukey, will be used.

Tukey Correction

Tukey correction, a.k.a, Tukey method or Tukey's Honestly Significant Difference (HSD) test, is used in conjunction with ANOVA, and its purpose is to determine which mean amongst a set of means differs significantly from other means.

It is used after ANOVA has rejected the null hypothesis. Then, a Tuckey correction can be used to make pairwise comparisons between all possible pairs of means divided by standard error to identify exactly where the differences lie.

Its null and alternative hypotheses between the two means are as follows:

H_0: $\mu_i = \mu_j$

H_1: $\mu_i \neq \mu_j$

First, for each pair of mean, it computes the q value by using the following equation:

$$q_{i,j} = \frac{(\mu_i - \mu_j)}{\sqrt{MSE/n}}$$

Here, MSE refers to the mean square error read from ANOVA table results (we didn't explain its details), n is the size of the sample dataset for each group.

Next, it determines the critical value for q (q_{crit}) based on the total number of comparisons, confidence level, and degrees of freedom. This value is read from a statistical table specifically designed for the Studentized range distribution.

Afterwards, it computes the HSD score as follows:

$$HSD = q \times \sqrt{\frac{MSE}{n}}$$

We can reject H_0 reject if $|\mu_i - \mu_j| > HSD$. Tukey's HSD test has assumptions, which make its applicability limited. One of these assumptions is that the variability (spread) of the data should be roughly the same across all the groups being compared. This is called the homogeneity of variances. Another assumption is that any errors or differences within each group should follow a normal distribution.

NOTES:

129

* If there is no relation between those two variables, they are independent variables, and if there is a relationship, they are dependent variables.

* Rejecting the H_0 in a Chi-square means that the two target variables have a relation, but we can not identify their type of relationship with the Chi-square test.

* The Goodness-of-Fit test compares the differences between the frequencies we "observe" (model) and frequencies we "expect".

* Sometimes, there is a normal distribution among the dataset we have, and thus, we must use parametric tests. The test result might lead to a wrong conclusion. Therefore, if we are not sure about the distribution we could also perform a non-parametric test on the sample data as well.

* Parametric significance tests usually rely on the mean, but non-parametric significance tests usually rely on the median.

* While dealing with non-parametric tests, we always first need to identify if there is a difference between samples, using tests such as KW-Test or Chi-Square), then we can use another test, such as MWM-Test, to identify exactly which one of these samples is different from others.

* When we are conducting many analyses on a dependent variable, by chance, we might make a Type I error (H_0 is true, but we reject H_0, which is a false positive). To reduce the likelihood of making Type I errors, we can use Bonferroni Correction [Dunn '61].

* Significant tests will be implemented on more than one group of data, and based on the comparison between those groups we can conclude if our groups are significantly different or not.

Effect size

If your brain is still alive after reading statistical significance and all the methods we have described, let's repeat our motivation: why do we use the statistical significance test? A statistical significance test gives us a p-value and lets us know if there is a significant difference between the two groups we are comparing. The answer will be yes (p-value <0.05, H_0 is rejected) or no (p-value >0.05, H_1 is rejected).

However, the significance test does not tell anything about the *magnitude* or *size of the difference*. The effect size tells us about the size of differences between the two groups [Sullivan '12]. Let's go back to the example we have used to describe p-Value. There, we have the following.
H_0: "Bugs with leg weight of 0.2 g. DO NOT taste better than other bugs".
H_1: "Bugs with leg weight of 0.2 g. DO taste better than other bugs".

Now, assume our significance test returns that there is a significant difference between the two groups, "*bugs with an average weight of 0.2 grams*" versus *"bugs without the average weight of 0.2"*.

How much are they tastier? Ten times, a bit more, etc.? To answer this question and quantify the amount of the difference, we use the effect size test.

In this section, we list three approaches for identifying the effect size, including *mean differences*, *categorical differences*, and *correlation-based differences*. For each category, we describe one method.

Cohen's d Test

Cohen's d Test [Cohen '92] is the most straightforward test used widely for mean differences. It computes θ as follows:

$$\theta = \frac{\mu_1 - \mu_2}{\sigma}$$

In this equation, σ is a standard deviation based on both datasets, i.e. $\sigma = (\sqrt{\sigma_1^2 + \sigma_2^2})/2$.

The Cohen's d output could be interpreted as large, medium, or small. $\theta = 0.2$ is considered as a 'small' effect size, $\theta = 0.5$ a medium effect size, and $\theta = 0.8$ a large effect size. For instance, we measure the chickens' height, and we notice that the roosters (male chickens) are taller than female chickens. The mean height of male chickens is 0.6m, the mean weight of female chickens is 0.5m, the standard deviation of male chicken height is 0.2m, and for female chickens is 0.3. Therefore, the Cohen's d index will be:

$$\theta = \frac{0.6 + 0.5}{(\sqrt{0.2^2 + 0.3^2})/2} = 6.11. \text{ Since } \theta = 6.11 \text{ the effect size is large.}$$

Cohen's d test is parametric, and when we deal with non-parametric data, we can use Cliff's d test [Cliff '93].

Odds Ratio

The Odds Ratio (OR) is a measure of categorical differences between two groups of data. It measures the association between two properties and is called a relative measure of effect. Odds-Ratio uses a 2×2 contingency table. Assume we conduct an experiment and give a medication to bugs; our medication name is called 'taste booster'. We would like to know if it affects their taste or not. We conducted an A/B testing experiment, and we grouped bugs into two groups; one group received our taste-boosting medication, we call the experiment (or treatment) group, and the other group did not receive our taste-boosting medication (or control group). If we call boosted taste an 'Event' and not boosted taste 'None Event', we can use Table 3-12 to present the contingency table.

	Experiment	Control
Event	a	b
None Event	c	d

Table 3-12: Contingency table example.

The Odds-Ratio is calculated as follows:

$$OR = (a \times d)/(b \times c)$$

If the $OR = 1$, this means there is no effect identified between the two groups. if $OR > 1$ it means the experiment performs better than the control (positive association), if $OR < 1$ it means the control performs better than the experiment (negative association).

Correlation Coefficients

A correlation coefficient is a score between -1 and +1, presented as r. This score indicates the linear relations between two variables. There are three well-known correlation coefficients in use: *Pearson*, *Spearman*, and *Kendall*. Unlike previous methods, which are used once and done, correlation coefficients are usually used in the software development process. It means we might need to create a software application that frequently computes correlation coefficients. Therefore, we also need to be careful about their computational complexity.

Why did we explain something about software here? To brag about the author's software development skills.

The most popular correlation coefficient is the *Pearson coefficient* [Stigler '92]. The outputs of the Pearson coefficient are two objects, ρ[13] for a population and the letter r for the result. They are very easy to calculate and skip describing them here in detail.

Let's go for a correlation test and as soon as we receive the positive results then we will marry.

---

[13] Read it as "rho", which is a Greek letter.

If the r value is positive, two variables have a positive correlation. A positive correlation means increasing one variable results in increasing the other variable. If the r value is negative, they have a negative correlation, i.e., decreasing one variable results in increasing another variable and vice versa. If $r = 0$, they don't have any correlation. Pearson's computational complexity is $O(n)$.

Pearson is limited to measuring the *linear relationship* between two variables, and it does not support correlation when there is a *monotonic* or *non-monotonic* relation (consistently increasing or decreasing, but not necessarily at a constant rate) existing in the shape of correlation. Figure 3-29 illustrates a simple comparison between linear, monotonic, and non-monotonic relationships. For example, if we have 1 unit of sugar, we will get 2 negative points in our diet; if we have 2 units of sugar, we get 4 negative points in our diet. Such a relationship can be supported perfectly by Pearson.

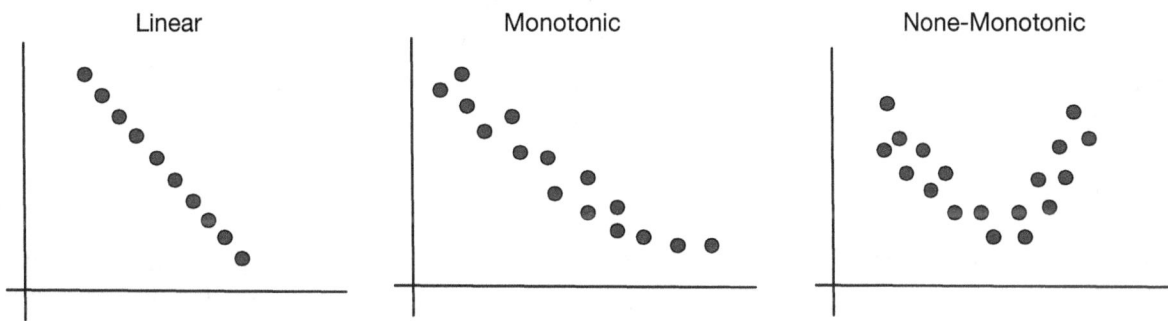

Figure 3-29: Monotonic vs None-Monotonic relationship between X and Y.

To mitigate this issue in cases where there might be a non-constant slope in linear correlation, we can use *Spearman Rank* correlation [Spearman '04], shown with ρ. Spearman correlation is very similar to Pearson, but it is able to measure the correlation between two variables, and it is non-parametric (distribution-free). Figure 3-30, which compares scores for Pearson and Spearman correlations. Spearman Rank's computational complexity is $O(n \log n)$.

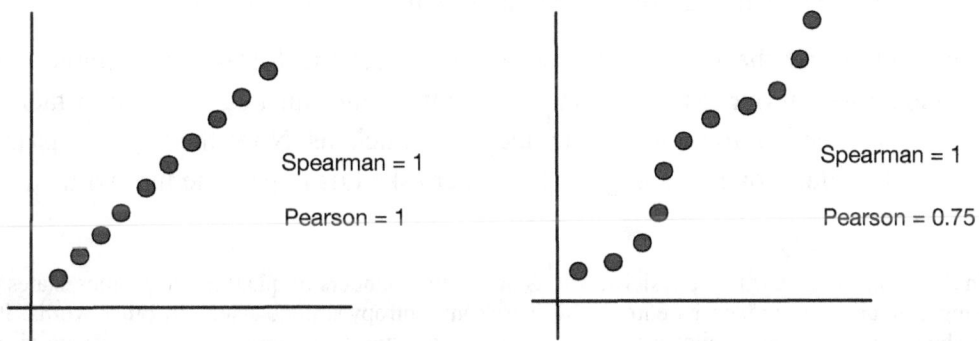

Figure 3-30: Spearman vs Pearson coefficient score for linear (left) and monotonic (right) between X and Y axis.

Another method is called *Kendall's Tau* correlation [Kendall '38], which is similar to Spearman correlation non-parametric, but usually, it returns smaller values than Spearman.

Kendall's Tau, comparing all possible pairs of observations. While it typically produces smaller values than Spearman, it has certain statistical advantages, particularly for smaller sample sizes However, its computational complexity is $O(n^2)$. We can also use covariance to measure whether two variables and their values are correlated together.

When implementing these methods in software, computational complexity becomes important: Pearson has $O(n)$, Spearman has $O(n \ log \ n)$ due to ranking, and Kendall's Tau has $O(n^2)$ complexity, due to pairwise comparisons.

Entropy & Information Gain

If you have started to read this chapter from the beginning, which we highly suggest, we know you are too tired. However, the good news is that most of what you need for applied statistics has been covered. The bad news is that two extremely important topics about machine learning remain to be learned. Brace yourself; another brain-eating concept is about to come.

Entropy

Entropy is a measurement of disorder or measurement of impurity[14]. In simple words, entropy is measuring the *uncertainty* of a variable [Shanon '49]. Bishop [Bishop '06] defines entropy as *"the average amount of information needed to specify the state of a random variable"*.

The higher the entropy, the more information we need to be able to make a valid justification. Higher entropy is statistically always more likely to occur. We can refer to entropy as a number, which explains how unpredictable our probability distribution is. The entropy of random variable X (e.g. flipping a coin), with n number of outcomes, can be defined as follows[15].

$$H(X) = - \sum_{i=1}^{n} p(x_i) log_2 P(x_i)$$

Here, H(X) represents the entropy of variable X, n represents the number of outcomes or events, and $p(x_i)$ represents the probability of the variable having the value of x_i.

To better understand the concept of entropy, consider the following example. Assume some yellow chickens from our aviculture sneak their way into our bug cultivation facility and enjoy our tasty bugs. Each bug room might include some chickens. Now, the bug rooms are mixed with chickens, and the farm owner is angry. The owner asked us to provide him with an estimate from

---

[14] Erwin Schrödinger, an Austrian physicist, who is one of the founders of quantum mechanics, states that a hallmark of a living creature is to reduce its entropy by increasing entropy around itself. In other words, it says the total entropy should increase, but it allows it to decrease in some places as long as it is increasing elsewhere. Such a weird phenomenon, isn't it? Some make war in a distant place to keep their economy balanced (our disorder decreases) and profit from the other nations' disorder.

[15] If you are not familiar with the concept of the logarithm; if we have $log_x a = y$, it means $x^y = a$. Also, remember $log_{10} P(x_i) = log P(x_i)$.

each room, including the "impurity" of each room is (100% bugs in the room mean a very pure room, and chickens inside bug rooms are impurity).

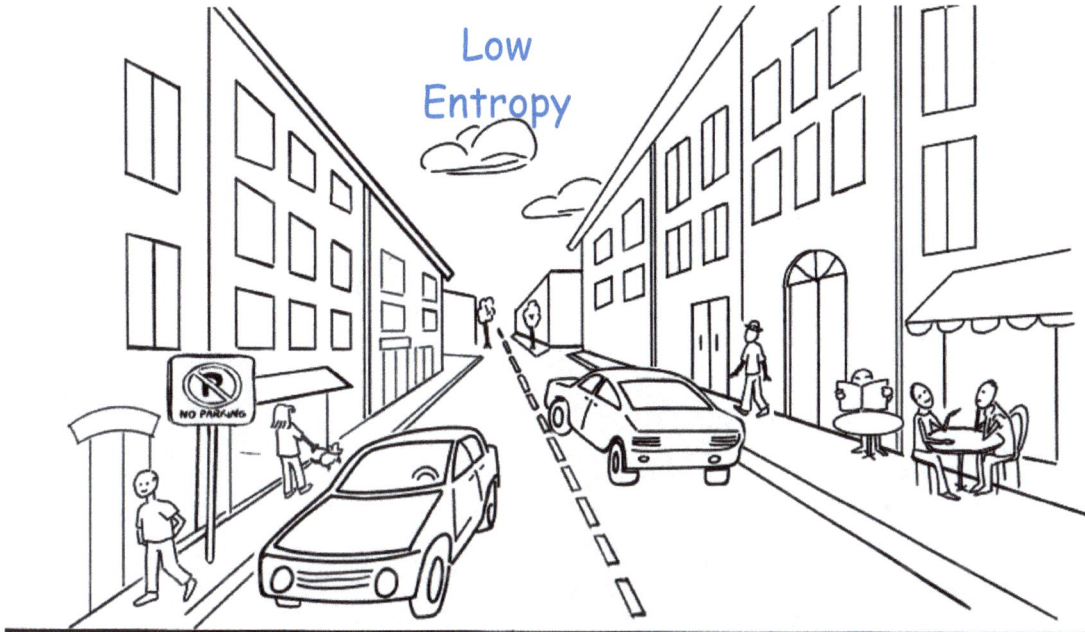

We sample some data from each room, and we have the following probabilities:

Room 1: P(bug)=0.5, p(chicken)=0.5 —> Entropy = - [0.5 log₂(0.5) + 0.5 log₂(0.5)] = 1

Room 2: P(bug)=0.65, p(chicken)=0.35 —> Entropy = - [0.65 log₂(0.65) + 0.35 log₂(0.35)] = 0.93

Room 3: P(bug)=0.75, p(chicken)=0.25 —> Entropy = - [0.75 log₂(0.75) + 0.25 log₂(0.25)] = 0.81

Room 4: P(bug)=0.95, p(chicken)=0.05 —> Entropy = - [0.95 log₂(0.95) + 0.05 log₂(0.05)] = 0.29

It means Room 1 is the most impure because entropy is at its highest level. Room 4 has the purest boxes because its entropy is 0.29 and the lowest. The value of entropy is not necessarily between 0 and 1, it could get larger than 1 as well, especially when we have more than one variable to analyze in the dataset, and the value of entropy gets larger than one.

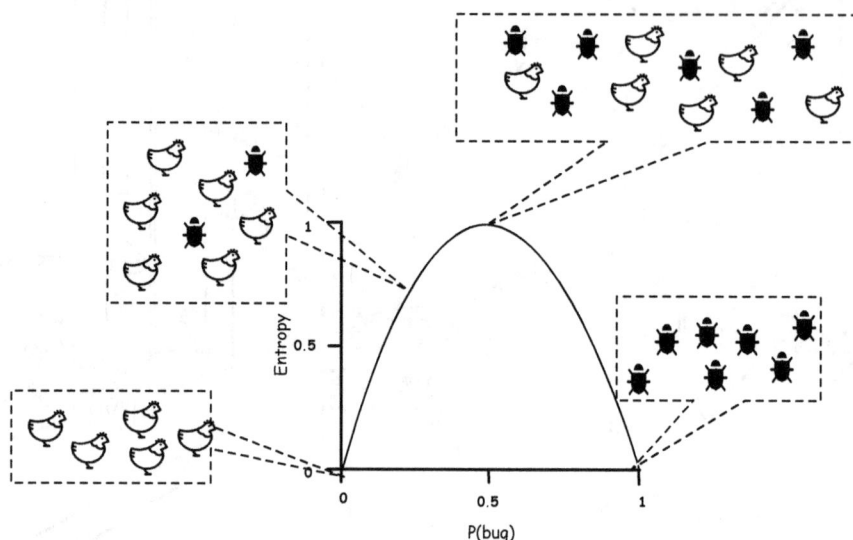

Figure 3-31: Relation between entropy and bug probability. The highest entropy =1 means the number of bugs and chicken are equal. In the lowest entropy we either have no bug at all, P(bug)=0, or all of them are bugs, i.e. P(bug)=1.

Figure 3-31 presents the relation between entropy and the probability of having bugs. Based on this Figure, we can see that high entropy means low certainty and low entropy means high certainty.

Entropy is used to compute *Information Gain (IG)*, which *measures changes in entropy* based on the amount of new information that is added to the dataset.

Usually, while dealing with real-world datasets, we need to apply a pipeline of machine learning algorithms on the dataset to filter the data for the next level, and in the next level, we gain better predictive results. This means that the amount of information we gain from the data increases.

For instance, Figure 3-32 presents the entropy of a dataset including chicken and bugs in two different stages. At first, we had high entropy because the mixture of chicken and bugs was equal. Then, we apply some algorithm (Magic), which acts like a filtering mechanism, and this algorithm divides the dataset into two datasets, each having lower entropy than the original one. We can say the magic algorithm increases the information gained by dividing our dataset into two sub-datasets with more relevant data. Later, we will learn about algorithms that reduce entropy, such as Clustering algorithms.

In Chapter 9, we will learn more details about the information gain. For now, just understanding its application is enough.

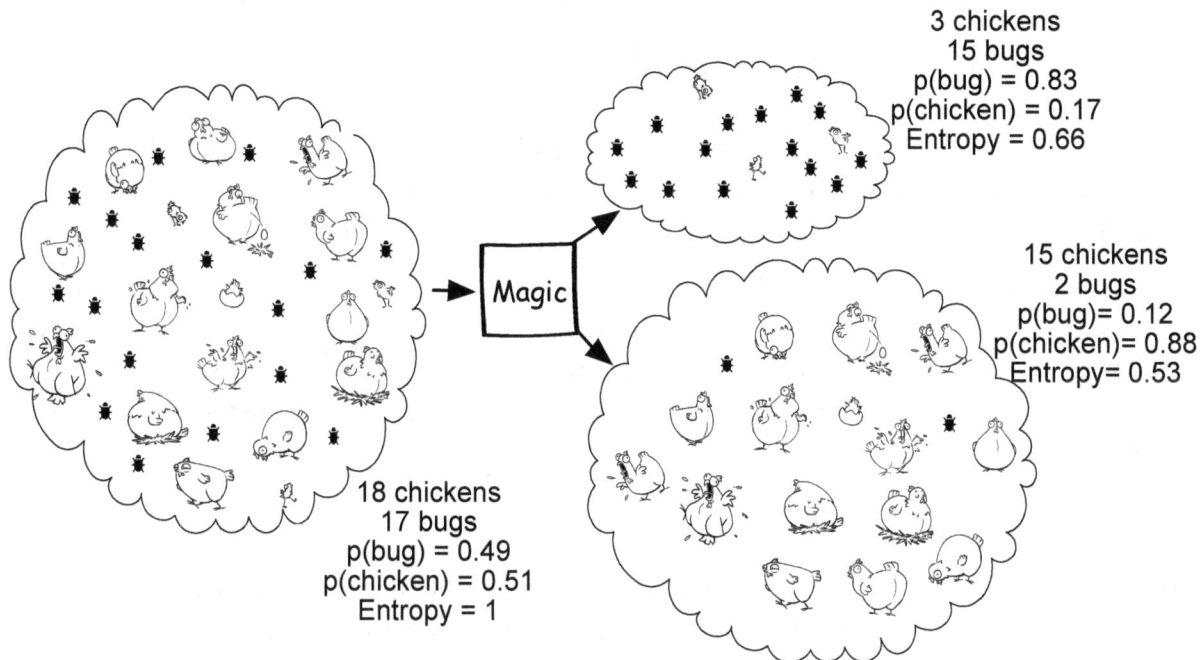

Figure 3-32: High entropy dataset fed into the Magic algorithm and the Magic outputs two datasets with lower entropy.

Measuring Distances Between Distributions

There are methods to measure the differences between data distributions based on Entropy, including *Cross-entropy*, *Kullback-Leibler Divergence* (*KL-Divergence*) [Kullback '51], a.k.a. *relative cross-entropy*, and *Jensen Shanon Divergence* (*JS-Divergence*).

KL-Divergence

KL-Divergence is used to compare two distributions or measure the similarities between two distributions. For example, compare the result of predicted distribution constructed by approximation method to actual distribution. We will encounter KL-Divergence a lot in this book, especially when dealing with generative models and neural networks.

For example, assume we are using algorithm A and algorithm B to predict the bugs eaten by chickens in room #4. Assuming the actual probability of bugs in room #4 is $p = 0.29$. We can compare the results of $D_{KL}(p||A) = 0.18$ and $D_{KL}(p||B) = 0.65$, and the results show that the algorithm A performs better because 0.18 is closer to 0.29.

KL-Divergence is the difference between the cross-entropy of two distributions, i.e., $H(p,q)$, and one of their entropy, i.e., $H(p)$. Soon, we will explain the cross-entropy. The KL-Divergence between two discrete distributions p and q is written as $D_{KL}(p||q)$ and computed as follows:

$$D_{KL}(p||q) = H(p,q) - H(p) = \sum_i p_i \, log(\frac{p_i}{q_i})$$

137

If we need to deal with continuous distributions, their KL-Divergence is computed as follows:

$$D_{KL}(p||q) = \int_i p_i \, log(\frac{p_i}{q_i})$$

Therefore, if $p = q$, then $D_{KL}(p||q) = 0$. The result of a KL-Divergence will be a number between 0 and ∞, when two distributions have no overlap, their KL-Divergence will be ∞ (see Figure 3-33).

Keep in mind that $D_{KL}(p||q) \neq D_{KL}(q||p)$. As soon as somebody asks you something that includes a comparison between two distributions, you should interrupt him or her and throw your knowledge on the table by saying: "I recommend using KL-Divergence". KL-Divergence is often used in information theory, topic modeling, feature selection, and comparing the output of generative models.

Cross-entropy

Cross entropy measures the difference between two probability distributions over the same set of events. Given two probability distributions, p and q, over a set of I events, the cross-entropy between them is defined as $H(p, q)$ and computed as follows:

$$H(p,q) = -\sum_{i \in I} p_i \log_2(q_i)$$

The cross-entropy can be computed as the sum of the KL-Divergence and the entropy of the first distribution: $H(p, q) = D_{KL}(p||q) + Entropy(p)$. Unlike KL-Divergence, cross-entropy is symmetric, and thus, we have $H(p, q) = H(q, p)$.

Cross-entropy is used as a cost function in classification tasks in neural networks by measuring the differences between predicted and actual distributions. We will use it later in Chapter 10 and other chapters discussing Neural Networks.

Jensen Shanon Divergence (JS-Divergence)

We have explained that KL-Divergence can be used to measure distances between two distributions, but it is not symmetric, i.e., $D_{KL}(p||q) \neq D_{KL}(q||p)$. Besides, the KL-Divergence range is unbounded. It means that it varies between 0 to ∞. The ∞ result occurs when two distributions do not have any overlap. To handle these limitations, we can use *Jensen Shanon Divergence (JS-Divergence)*. It is symmetric, bounded, and computed based on KL-Divergence with the following equation:

$$JS(p,q) = \frac{1}{2}KL(p||m) + \frac{1}{2}KL(q||m)$$

where m is the average of p and q, $m_i = (p_i + q_i)/2$.

The range of JS is between 0 ($p = q$) to 1 (p and q do not have any overlap at all).

To understand the differences between KL-Divergence and JS-Divergence, check Figure 3-33. In this figure we have two normal distributions in different locations from each other, and their JS-

Divergence and KL-Divergence scores are written on their side. As it is shown, when two distributions stay apart, the KL-Divergence goes toward infinity, but JS-Divergence still provides some numeric data.

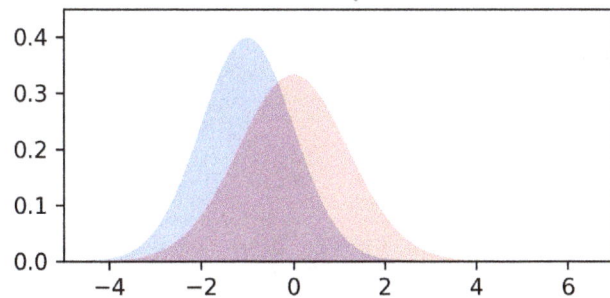

KL-Divergence (p,q) = 0.38
KL-Divergence (q,p) = 0.38
JS-Divergence (p,q) = 0.09
JS-Divergence (q,p) = 0.09

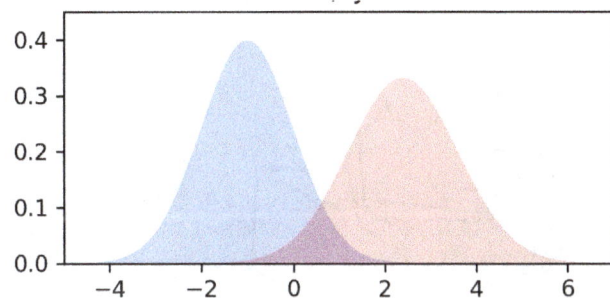

KL-Divergence (p,q) = 4.04
KL-Divergence (q,p) = 5.83
JS-Divergence (p,q) = 0.54
JS-Divergence (q,p) = 0.54

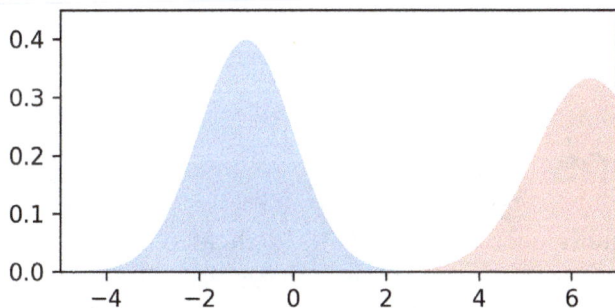

KL-Divergence (p,q) = ∞
KL-Divergence (q,p) = ∞
JS-Divergence (p,q) = 0.70
JS-Divergence (q,p) = 0.70

Figure 3-33: KL-Divergence and JS-Divergence of two normal distributions, based on their overlap.

NOTES:

* Sometimes, we need to quantify how predictable our target dataset is. In those cases, we can use entropy to quantify the predictability of the dataset. Higher entropy could indicate less predictability, and low entropy in a dataset is a sign of high predictability.

* Here, we have learned that JS-Divergence, Cross Entropy, and KL-Divergence can be used to compare two distributions. In addition to these three methods, we could also use Kolmogorov-Smirnoff as well. We have explained it as KS-Test, but some also use this method to compare two distributions.

* Comparing distributions is the core of neural networks, especially generative AI models. We will explain them in Chapter 11 and Chapter 13. Therefore, please be sure that you have learned everything we explained here perfectly fine.

Probability Estimations

A large part of the modern artificial intelligence community is dedicated to self-supervised learning algorithms. Many generative models use self-supervised learning. We describe generative models in Chapter 11 briefly; generative models aim to learn the distribution of the data. They can generate new data instances that resemble the training data. Many generative models start with estimating (guessing) the probability of data. Then, compute the error of their guess and later refine their guess based on the error.

There is a baseline method to estimate the probability of data, i.e., *Maximum Likelihood Estimation (MLE)*, and a very common approach to implement it is *Expectation Maximization (EM)*.

Before we explain the details of the MLE approach, we describe two concepts of MLE: *probability* and *likelihood*.

Probability means what is the chance of observing X in the given sample dataset.

Likelihood means, given the observed subset X of a dataset, what are the best distribution parameters (e.g., mean, variance, ...) that fit the given dataset?

In simple words, probability is a measure of how likely is (likelihood) an event will occur. On the other hand, the likelihood measures how well a particular set of data points fits a specific model or hypothesis.

You might not understand this part, but you can skip this section and come back later after you are done with Chapter 8 and Chapter 11, and understand optimizations, parameter estimation, etc.

Maximum Likelihood Estimation (MLE) Approach

Earlier in this chapter, we said a dataset has a characteristic, and this characteristic is presented as a distribution. Any distribution is specified by its parameters. The MLE approach uses a sample dataset to determine the *best distribution parameters*, such as mean and variance for normal distribution, of the original dataset [Edgeworth '08].

As we said several times, in real-world scenarios, we do not have access to the entire dataset; we have only a part (sample observation) of the entire dataset. Therefore, by using the available sample dataset, MLE estimates the distribution parameters of the original dataset.

For example, we have a small part of a dataset, and we assume the original dataset has a Gaussian distribution (of course, it is just an assumption, but it is a common practice). We don't know what the parameters of that Gaussian distribution are because there could be infinite numbers of Gaussian distributions. The MLE tries to approximate the mean and standard deviation by using the sample dataset and constructs a distribution that is the closest match to the original dataset distribution.

Please make some space in your brain for one sentence to memorize: *the goal of MLE is to find the best distribution parameters that fit the original dataset (population)*.

The population dataset distribution is presented with an unknown set of parameters is denoted as θ, and the distribution of the dataset depends on these parameters. The function that quantifies the likelihood of observing the given data X and unknown θ parameters is the likelihood function, denoted as $L(X; \theta)$. We use a method called the MLE function to estimate θ, and it is presented as $\hat{\theta}$. Formally, the MLE for the observed dataset X can be written as:

$$\hat{\theta} = \arg\max_{\theta} \hat{L}(X; \theta)$$

We use the semicolon ';' instead of the '|', ';' is a sign of joint probability, which means θ it is specified and occur along with X. Nevertheless, you might encounter some literature that uses '|'

instead of ';'. When we encounter the '^' sign, it means it is something that the output of the algorithm will be specified (e.g., predict it).

For computational efficiency, instead of maximizing the likelihood directly, we often use the *logarithm*[16] of likelihood (log-likelihood). Therefore, log-likelihood can be written as:

$$\ell(\theta) = ln[L_n(X; \theta)] \text{ or } \hat{\theta} = \arg \max_{\theta} ln[L_n(X; \theta)]$$

Even we can use negative log-likelihood:

$$\hat{\theta} = - ln[L_n(X; \theta)] \text{ or } \hat{\theta} = \arg \min_{\theta} - ln[L_n(X; \theta)]$$

The logarithm of numbers smaller than one is negative, and the negative log brings them back to positive. It means we calculate the inverse of minimization, which is equal to maximization[17]. In this context, minimizing the error is equivalent to maximizing the log-likelihood.

How does MLE get implemented? There are different approaches to implement it, such as regression algorithms, i.e., Ordinary Least Squares (OLS), or numerical approaches, such as gradient descent. After Chapter 8, we get a good understanding and can learn algorithms to resolve MLE. One popular algorithm for implementing MLE is Expectation Maximization, which we describe here.

You might find this not a very useful concept or hard to digest at this point; you can skip it and return to this part after you read Chapter 6 and gain a more solid understanding of the use of distributions in machine learning.

Expectation Maximization (EM)

Although MLE does not have the entire dataset, it assumes the dataset is complete or fully observed. Nevertheless, part of the dataset could be missing in the observation subset that is used by MLE. Those missing parts could construct parameters that are known as latent variables. Here, latent variables refer to parameters that do not exist in the observed dataset.

The Expectation Maximization (EM) algorithm [Dempster '77] is an algorithm that implements the MLE approach and can handle the latent variable in the sample dataset as well. Therefore, if there are missing data in our observed dataset, the EM algorithm is recommended.

The objective of EM, similar to the MLE objective, is to find unknown parameters θ that find the best distribution fit to the original dataset, i.e., approximate maximum likelihood. To approximate the maximum likelihood, this algorithm operates iteratively in two steps. The first step is the estimation step (E-step), and the second step is maximization (M-step). We can summarize it as follows:

(i) *E-step* makes an initial guess of parameters for the expected distribution.

---

[16] if you can not recall logarithm: $log_a b = x$ means that $a^x = b$. Also, natural logarithm is written as ln, and $ln\ a = x$ and it means $a = e^x$.

[17] We should express our gratitude to uncle Kia (Prof. Kia Teymourian), who clarified the rationale of this for us at work and challenged the author a lot for the explanation of this part.

142

(ii) *M-step* starts after the E-step, and when newly observed data is fed into the model. In this step, the EM algorithm tweaks the estimated parameters (from E-Step) to cover newly observed data as well.

(iii) From M-step, the process will be repeated until the created distribution does not change in E-step or M-step and it reaches a stable state (converged) or a maximum threshold of iteration.

The random assignment of the initial parameter is a bit tricky because sometimes the EM algorithm might be stuck in the local maximum/minimum (optimum) and, by mistake, assumes it as a global optimum. Check Figure 3-34 to understand the concept of global and local optimum. More about this will be explained in Chapter 8.

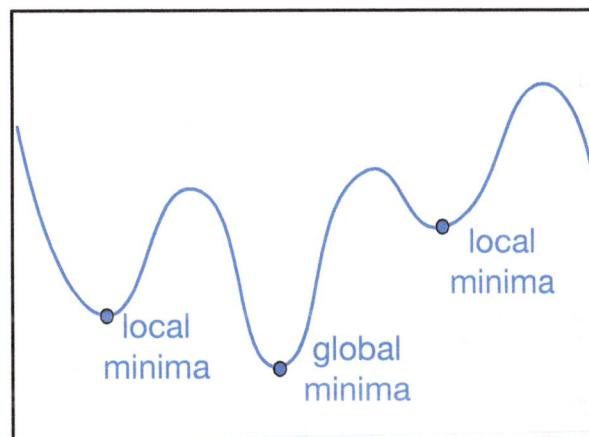

Figure 3-34: Local versus global minima on a function.

The Gaussian Mixture Model (GMM) and Hidden Markov Model (HMM) are two examples of using the EM algorithm approach for MLE. We will explain these two algorithms in Chapter 4 and Chapter 5.

The computational complexity of MLE and EM depends on many factors, including the shape of the likelihood, the condition of the algorithm, etc. Therefore, it is not something generalizable that we can report here.

Summary

We started this section by describing statistical concepts required for machine learning algorithms and AI models, including variable vs value, types of variables, and some basic statistical operations on the data. Next, we have explained probabilities and terms used in probability, including PMF, PDF, CDF, and expected values.

Then, we switch to distributions and explain some of the common ones within their needs. Table 3-13 summarizes the distributions we have described and explains which one is used in which scenario.

Normal	Uniform	Beta	Dirchilet
- bell curved - used to present wether we have collected enough data or not. - two of its well-known subtypes are z-distribution and t-distribution	- used when all outcomes of sample has equal probability. - Its discrete version has finite outcome and its a continuous version has an infinite outcome.	- used when there is an uncertainty in binary trail and we don't have information about their underlying probabilities. - defined by two parameters α and β	- multivariate generalization of Beta distribution - defined by two parameters α and β - use in topic modeling, i.e. latent dirchilet analysis

Binomial	Bernoulli	Geometric	Power Law / Exponential
- used when there is a series of binary independent trail. - it can be used for making inferences about the binary trails in terms of probability.	- A special type of Binomial distribution that has only one trial.	- very similar to Binomial, except that after the first encounter the trail stops. - it can be used for making inferences about the binary trails in terms of probability.	- observed in many real-world phenomena including physics, biology, literature,... - it is characterized by a parameter called α. - Two subtype of this distribution are Zipf law and Pareto distribution.

Poisson	Weibull	Chi-Square	Boltzmann
- used to model a rare event that is happening in the system, in a particular interval. - it can be used for making inferences about the binary trails in terms of probability.	- used to model a rare event that is happening in the The system, in the different intervals. - it can be used for making inferences about the binary trails in terms of probability.	- used to test Goodness-of-Fit - used to test dependence between two categorical variables	- Describes the probability of a system being in a certain state as a function of that state's energy and the temperature of the system. - The energy gets distributed from high density to places with lower density until there is a balance between energy distribution density (thermal equilibrium).

Table 3-13: Summary of described distributions.

After distributions, we briefly describe the normalization, followed by answering how much data is enough for sampling. The next question we answered here was about the significance tests and which test was appropriate in which condition. In particular, we use a significance test to observe whether there is a significant difference between the two groups of data. If the data is normally distributed, a parametric significant test can be used; if not, a non-parametric significant test will be used. The significance tests we have explained are summarized in Figure 3-35.

Significant tests provide us with a binary result. It states whether or not two group differences are significant. There are methods used to quantify the magnitude of differences between two groups of data, including Odds Ratio, Cohen's d (Cliff's d for non-parametric data), and correlation coefficients.

Then, we introduced uncertainty and its related concepts, including entropy, information gain, and methods to measure the distances between two distributions. The more uncertainty we have, the lower the chance of predicting a data behavior. Therefore, it is useful to increase the information gain and mitigate uncertainty before feeding our data into a machine learning algorithm.

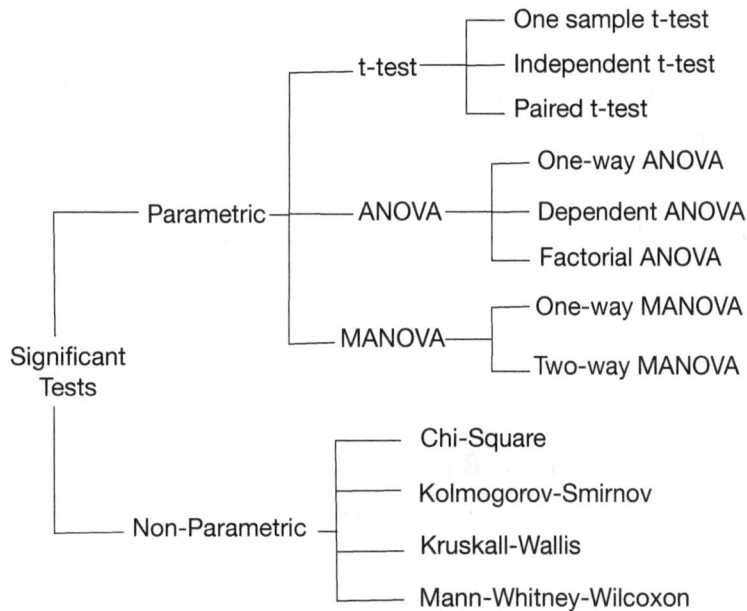

Figure 3-35: Summary of described significance tests.

At the end of this chapter, we discussed the MLE approach and why it is used in generative models. Then, the EM algorithm has been introduced. EM is an iterative algorithm. First, it computes an initial guess on distribution parameters, then it evaluates the estimates, and again, it uses the previous estimates to compute better estimates and continues this process iteratively until a threshold of iteration is reached or the distribution can not be improved.

Further Readings and Watching

* There are fantastic books that we can recommend you to read to understand the basics of probabilities, such as "Head First Statistics" [Griffithis '09], "U Can: Statistics for Dummies" [Rumsey '15], or "Statistics in Nutshell" [Boslaugh '12].

* Penn State Stat-500 and MiniTab: https://onlinecourses.science.psu.edu/stat500 Another good source that I would say is a free online book for statistics. They also provide minitab (statistical software) examples for their code. The minitab weblog, http://blog.minitab.com/, has fantastic descriptions as well, and we have used it a lot for writing this section. If you can not afford to pay the license fee, there is a wide availability of free R and Python statistical packages, and you can use them.

* Stats How To (http://www.statisticshowto.com): This is a web page with a short and clear description of statistical information. We have also used this page to check the validity of our example. The good thing about this page is that you can learn your required content in a short amount of time.

* Vassar Stats (http://vassarstats.net) This is another helpful webpage where you can add your numbers and make the calculations online. Of course, these pages are mostly for teaching

purposes, and when you have a large amount of data, you should use a software package such as R, Python, etc.

* There is an excellent explanation of entropy in the 'Data Science for Business' book [Provost '13]; if you are willing to learn more about entropy and its related concepts, this is a good book.

* "Practical Statistics for Data Science" [Bruce '17] is one of the best books we can refer to for learning statistics. It is concise and directly goes onto the concept without any time wasted on mathematical explanation.

* Icons that have been used in this section are from https://visualpharm.com and www.iconfinder.com

* The statistics book written by Rumsey [Rumsey '15] also has lots of good explanations for beginners, and it is worth taking a look if you would like to focus on the statistical aspects of machine learning and artificial intelligence.

* Maximum likelihood estimation has been described in detail in Duda's book [Duda '73].

Part ii: Unsupervised Learning

In Chapter 1, we explained that machine learning is the process of enabling a computer program to learn from the input data we provide to the machine. There, we explained that computers use different ways to learn from the data, including supervised, unsupervised, self-supervised, and reinforcement learning.

Supervised learning builds a model that learns based on labels that are manually assigned to the data. For instance, from a set of mixed animals, we can label and separate chickens in one group and cats in another group from other animals. However, we don't need to know their differences, or we do not intend to formalize their differences to a computer because the formalization process is expensive. Therefore, we label some instances in the dataset and then feed them into the algorithm. The algorithm tries to understand the rest of the data (classify) based on the given labels. Usually, we label data based on what we need from the data, and thus, we have prior knowledge or domain knowledge about the dataset.

There are two main challenges with labeling. First, our expectations from the data are not always well-defined. The data structure might be very complex, and at the initial stage, we may not know what to extract from it. Second, manually annotating data with labels can be difficult or even unfeasible. This could be due to the high cost of labeling or the lack of domain expertise. In such cases, it's important to first explore the data. By exploring, we can gain a better understanding of its characteristics. Similar to visualization and statistical analysis, *unsupervised learning* provides a way to explore data without needing labels. Unsupervised learning, however, focuses on uncovering hidden patterns or structures within unlabeled data, such as grouping similar datapoints or identifying events that are co-occurring.

Chapter 4: Clustering

One of the most common forms of unsupervised learning is *clustering*. Clustering is the process of grouping *similar* data points of a dataset together, a.k.a. *similarity*. It means that members of one group are similar together and dissimilar to the members of another group (recall Figure 1-10 from Chapter 1).

Unlike supervised learning, clustering assumes we do not have the knowledge or workforce to label the data. However, we can use some sort of similarity measurements to group similar data together. In addition to exploring data, clustering has many other applications in other fields, especially information retrieval. For instance, to search a large amount of data, we can first cluster them. Next, focus on one target cluster and then search only the related groups and not all data [Rawassizadeh '18]. In this case, clustering acts as a sort of data filtering mechanism. Furthermore, we can use clustering for anomaly detection [Chandola '09]. Many other tasks benefit from clustering.

The process of applying several machine learning algorithms on a dataset until we get our desired output is called a *machine learning pipeline*. It is common that while dealing with a complex dataset, as the first step in a machine learning pipeline, we cluster data into smaller groups and then focus on one or more groups in more detail. For instance, first, we cluster the data. Then, in the second step, we focus on one of the clusters and label only the data points of that cluster; next, we perform a classification on those labeled data points.

Similarity and Dissimilarity

We have explained that clustering operates based on the concept of similarity. As the first step to applying clustering, we should build a similarity matrix, which is used to compare data points of a dataset. Then, the clustering algorithm uses this similarity matrix to construct clusters.

Usually, but not always, a clustering algorithm receives a tabular dataset as input. Categorical and nominal data are converted into numbers that, in the end, will represent all data points in numerical format for the matrix. However, converting them into a number is not always a straightforward process and requires some additional analysis.

For instance, we have tabular data of chickens, and each chicken has the following data: *name, weight, height, body shape,* and *date of birth*. Therefore, the chicken dataset could be converted into a matrix, as is shown in Table 4-1. We want to cluster them based on their body style; thus, not all data are useful for clustering, and we do not include information that is not useful, including the chicken's *'name'* and *'date of birth'*. This process of selecting important data and neglecting unimportant ones is feature selection, which we will explain in Chapter 6.

Furthermore, since we cannot add textual data, such as body shape, into the matrix, we should convert them into numerical numbers or encode them by assigning 0 to thin, 1 to fat, and 2 to

Chicken Dataset

Name	weight	heigh	body shape	date of birth
Chicken 1	200gr	12cm	thin	1st Feb 2018
Chicken 2	250gr	11cm	thin	12th Feb 2018
Chicken 3	700gr	18cm	fat	28th Jan 2018
Chicken 4	600gr	19cm	muscular	25th Jan 2018

$$
\begin{array}{c}
 & \text{Weight} & \text{Height} & \text{Body shape} \\
\text{Chicken 1} & 200 & 12 & 0 \\
\text{Chicken 2} & 250 & 11 & 0 \\
\text{Chicken 3} & 700 & 18 & 1 \\
\text{Chicken 4} & 600 & 19 & 2
\end{array}
$$

Table 4-1: Converting some columns of a data table into a matrix for a clustering algorithm.

muscular. Note that any encoding we use for an interval (e.g., body shape) or other categorical variables should not change the underlying relation of the data.

When we look at the matrix at the bottom of Table 4-1, all matrix members are numbers, but they still do not seem rational. The first column is in grams, the second in centimeters, and the third in a range from 0 to 2.

By using a normalization, such as z-score (we have described in Chapter 3) we standardize the

data into a unit-less format. Han et al. [Han '11] recommend standardizing the data for a cluster. First, we should calculate Means Absolute Deviation (MAD) because it is more robust toward

A single data point → x_i Mean → μ

z-score → $z = \dfrac{x_i - \mu}{MAD}$

Mean Absolute Deviation →

A single data point → x_i Mean → $\bar{x}$

$$MAD = \dfrac{\sum_{i=1}^{n} |x_i - \bar{x}|}{n}$$

Total number of all data points →

outliers in comparison to standard deviation. The following two equations[1] show how to calculate MAD and z-score by using MAD instead of standard deviation.

Then, we can calculate the z-score by using MAD. It is helpful to learn the underlying math, but you do not need to learn it; your software supports normalizing the data for you. After converting all data into z-scores, we can

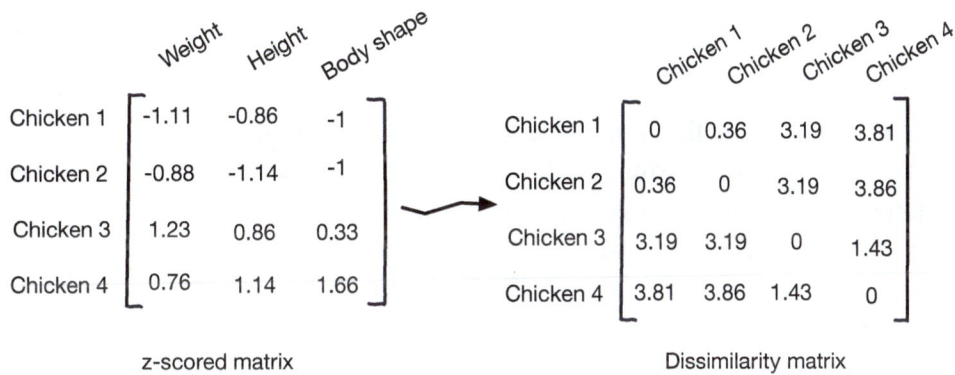

	Weight	Height	Body shape
Chicken 1	-1.11	-0.86	-1
Chicken 2	-0.88	-1.14	-1
Chicken 3	1.23	0.86	0.33
Chicken 4	0.76	1.14	1.66

z-scored matrix

→

	Chicken 1	Chicken 2	Chicken 3	Chicken 4
Chicken 1	0	0.36	3.19	3.81
Chicken 2	0.36	0	3.19	3.86
Chicken 3	3.19	3.19	0	1.43
Chicken 4	3.81	3.86	1.43	0

Dissimilarity matrix

	Chicken 1	Chicken 2	Chicken 3	Chicken 4
Chicken 1	0			
Chicken 2	d_{21}	0		
Chicken 3	d_{31}	d_{32}	0	
Chicken 4	d_{41}	d_{42}	d_{43}	0

Figure 4-1: (top) Computing a dissimilarity matrix from a z-scored normalized matrix. (bottom) A typical dissimilarity matrix structure.

compare distances between each data point and present them in a dissimilarity or similarity matrix.

---

[1] Note that in these equations, sometimes we write mean as μ and sometimes as $\bar{x}$. There are no differences between them, and we can write in both forms. You might also encounter both styles in machine learning text.

The matrix on the left of Figure 4-1 shows the chicken data in *z*-score and their dissimilarity matrix based on Euclidean distance in the right. The dissimilarity matrix is a *symmetric* matrix, and all its main diagonal numbers are obviously zero because the distance between a data point and itself is zero. "chicken 1" and "chicken 1" (self) have 0 distance. The lower the number, the more similar the two objects are. The bottom part of Figure 4-1 shows the structure of the dissimilarity matrix.

Similarity Measurement Methods

As we have described, clustering algorithms operate based on computing distances between data points of a dataset. Different distance functions are used to calculate distances, called *similarity measures*, *similarity metrics*, or *similarity functions*.

The number of methods for similarity measurement is vast, and here, we describe the commonly used ones. For some specific cases, the algorithm developer also creates its similarity measurements. For instance, a face recognition clustering algorithm could use its similarity measures, which is different from a vehicle clustering algorithm used by traffic cameras.

First, we describe methods used to measure similarities between numerical (quantitative) data points. Then, we describe methods that are used to measure similarities between categorical (none-quantitative) data points, textual data, or a mixture of different data types. Nevertheless, by transferring none-quantitative data into a numerical range, we can also use numerical similarity measurements for none-quantitative data.

Euclidean Distance

The most widely used distance measurement for numerical data is Euclidean distance. Assume we want to measure the distance between data objects A and B, which each have n numbers of attributes[2]. The distance between A and B will be presented as d by using the following equation:

$$d(A, B) = \sqrt{(a_1 - b_1)^2 + (a_2 - b_2)^2 + \ldots + (a_n - b_n)^2}$$

Here, a_i represents the *i*th attribute of data object A and b_j represents the *j*th attribute of data object B. For instance, for our chicken dataset, we have three attributes useful for comparison: weight, height, and body shape, and each has equal importance. The following equation presents the Euclidean distance between two chickens, *C1* and *C2*:

$$d(C1, C2) = \sqrt{(C1_{weight} - C2_{weight})^2 + (C1_{height} - C2_{height})^2 + (C1_{bodyshape} - C2_{bodyshape})^2}$$

When we are summing the same categories of data together, we can use Σ sign in mathematics, and thus, we can write the Euclidean distance as follows:

$$d(A, B) = \sqrt{\sum_{i=1}^{n} (a_i - b_i)^2}$$

---

[2] As a reminder, a record in tabular format is composed of different attributes or features. Each column could be interpreted as a feature.

In some applications, one attribute is more important than other attributes. For instance, in an Aviculture, the height of a chicken might not be as important as its weight. Therefore, we can assign a weight w_i to each attribute and thus have the following equation for the Euclidian distance:

$$d(A, B) = \sqrt{\sum_{i=1}^{n} w_i(a_i - b_i)^2}$$

Euclidean distance is useful when we are dealing with a low-dimensional dataset (each data object has a few attributes) and data objects are not sparse (there are not too many unknown or null values in the dataset). We can use Manhattan distance for a high-dimensional dataset [Aggarwal '01].

The time complexity of Euclidean distance is usually linear, i.e., $O(n)$. If you can't recall time complexity, check Chapter 1.

Manhattan Distance

If you have seen the Manhattan area of New York City (NYC), it is similar to a grid. The Manhattan distance assumes data points are located on a grid, and their distances are based on a grid, similar to poor drivers in the Manhattan area of NYC with its brutal traffic and angry drivers. In other words, between two points, we are only allowed to move horizontally or vertically (only along the x-axis or y-axis).

Figure 4-2 illustrates Manhattan distance vs Euclidean distance. We aim to move from the start point (A) to the destination point (B). A car can move in the streets with a grid style (Manhattan distance). On the other hand, if we can fly with a drone, we can use Euclidean distance to calculate the distance between A and B.

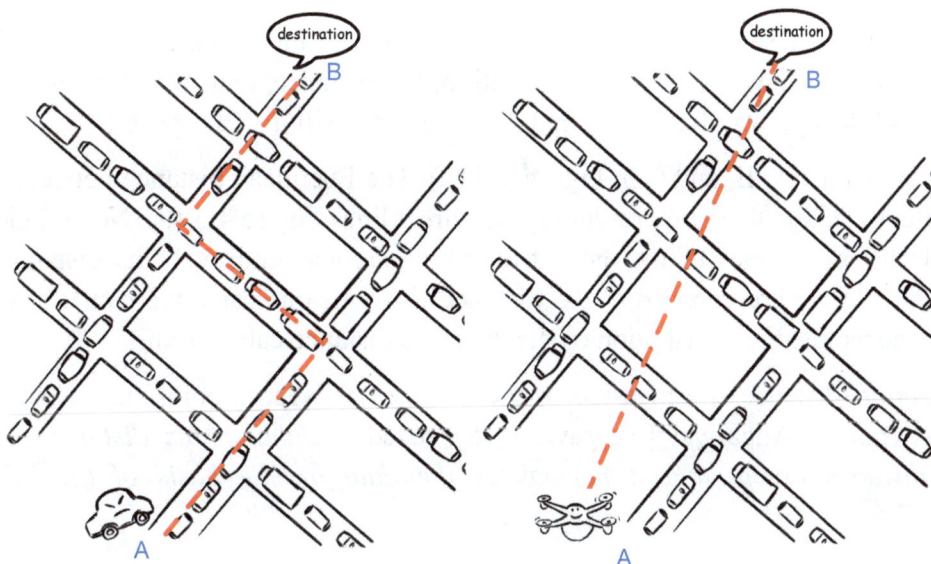

Figure 4-2: (left) Manhattan distance, (right) Euclidean distance.

The Manhattan distance between data object A and point B can be calculated as follows:

$$d(A, B) = |a_1 - b_1| + |a_2 - b_2| + \ldots + |a_n - b_n|$$

Mathematically written, the Manhattan distance can be formalized as follows:

$$d(A, B) = \sum_{i=1}^{n} |a_i - b_i|$$

Generalizing it to our car and drone movement example, a_1 and b_1 are moving along the x-axis and a_2 and b_2 are moving along the y-axis. The time complexity of Manhattan distance is linear, $O(n)$, similar to Euclidean distance.

We can see that Euclidean and Manhattan distance equations are very similar. Using a Manhattan distance to summarize a dataset content is called L_1-norm, and using the Euclidean distance for the same purpose is called L_2-norm.

We can generalize them to L_p norm (Minkowski distance) and formalize them as follows:

$$d(A, B) = (\sum_{i=1}^{n} |a_i - b_i|^p)^{1/p}$$

If we intend to incorporate the weight of features into the L_p-norm, then a weight of α could be multiplied by $|a_i - b_i|$; thus, we will have the following *Minkowski* distance function:

$$d(A, B) = (\sum_{i=1}^{n} \alpha |a_i - b_i|^p)^{1/p}$$

As we have explained earlier, by weight, we mean some features for comparison are more important than others; hence, we increase their weight.

Mahalanobis Distance

In some cases, the dataset distribution affects the similarity measurement, and we also want to incorporate that factor into the distance calculation. For instance, look at Figure 4-3 (a), it shows a distribution of data points of a dataset in a 2D space. We will use 2D space[3].

We have three points: A, B, and C in Figure 4-3 (a). The Euclidean distance between $A - B$, and $A - C$ are identical, and they are presented with direct lines Figure 4-3 (b). Nevertheless, B stays a bit outside the dense part of the dataset distributions, and intuitively its distance should be larger to A, while we are comparing B's distance to C. In this case, we can use Manalonobis distance to incorporate the distribution of the data into distance calculation.

The Mahalanobis distance is a measure of the distance between a point and a distribution, not between two points. Aggarwal [Aggrawal '15] stated: *"Mahalanobis distance is similar to Euclidean distance except that it normalizes the data on the basis of the inter-attribute correlations."*

---

[3] For the sake of readability in this chapter we use 2D datasets, but usually, we deal with more than two dimensional datasets, but we can't visualize them.

Mahalanobis distance operates based on the distribution of data points and their correlations.

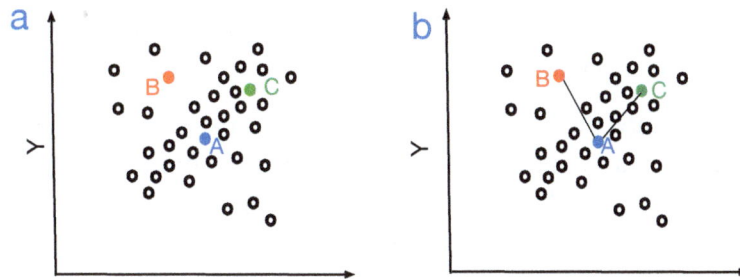

Figure 4-3: Visualization of distances that have equal Euclidean distance, but the distribution of the dataset is not taken into account.

The distances between data points of a dataset are computed in four steps, as follows:

- *Finding Centroid point:* First, it identifies the center (mean), a.k.a, centroid, of the data points. Let's call this point the average point.

- *Covariance Matrix Calculation:* Then, it tries to figure out how all data points are spread on the dataset. To do so, it calculates the covariance matrix. The covariance matrix captures how each dimension (or variable) in the dataset varies and how these dimensions are correlated with each other.

- *Measuring and Scaling the Distance:* Next, it measures the distance between each data point and the average point. In particular, it uses the inverse of the covariance matrix to scale the distance of each data point to the average point. The distance of a point from the average point is calculated for each dimension, and then these distances are adjusted (scaled) by the corresponding variances and covariances. This step is where the Mahalanobis distance differs from the Euclidean distance. In simple words, instead of measuring this distance in a straight line (like Euclidean distance), it measures it in a unique way that considers how the points are spread out.

- *Adjusting for Spread and Correlation:* Finally, the distances are adjusted based on the spread of data points. If the points are spread out in one direction, then the distance in that direction counts for less. To perform this adjustment, it takes the scaled distances for each dimension, squares them, and adds them up. This sum is then square-rooted to get the final Mahalanobis distance.

Given that the dataset has a distribution of Q, the distance of the data point, $\overrightarrow{A}$ with n number of dimensions $\overrightarrow{A} = \{a_1, a_2, \ldots a_n\}$, is computed as follows:

$$d(\overrightarrow{A}, Q) = \sqrt{(\overrightarrow{A} - \overrightarrow{\mu})^T S^{-1} (\overrightarrow{A} - \overrightarrow{\mu})}$$

155

Where $\vec{\mu}$ presents the the average point, T is the transposed of the vector, and S is the covariance matrix. Since transposing a matrix has exponential complexity, Mahalanobis distance will have exponential complexity as well.

When do we use *Mahalanobis distance?* We use it in scenarios where incorporating the distribution is important for distance measurement.

For instance, assume you are the hiring manager of a company. You are willing to enroll good employees who are good and get paid well. Your company is not creating a slavery paradise by giving them free food, coupons, and colorful bicycles. Best talents usually end up in those corporations, and you would like to avoid losing your staff to them. There are two factors that are important for you when hiring a new team member: personality and technical qualifications. Based on your previous hires, you plot your good staff based on their personality (x-axis) and technical qualification (y-axis) (Figure 4-4 (a)). Now you have one open position and two new candidates A and B gets to the final round. Their level of personality and technical qualifications are illustrated in Figure 4-4 (b). We know the best possible candidate is closer to the center. By using Mahalanobis distance, we recognize that candidate A is better than candidate B because she stays in the second circle around the average point but B stays in the third circle close to the average point.

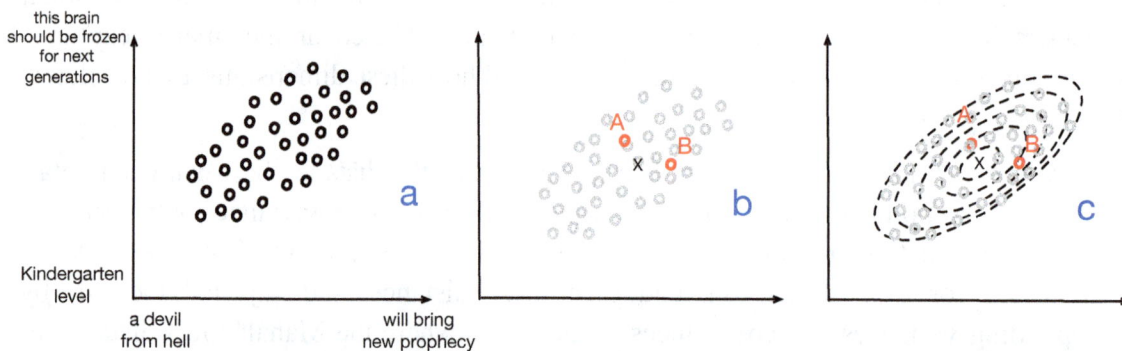

Figure 4-4: Calculating Mahalanobis distances of two new candidates among centroid previous good candidates.

Hamming Distance

It is a distance metric similar to the Manhattan distance but between two vectors of bits[4]. This distance measures the number of bits that need to be changed to convert x, a string of bits, into y, another string of bits. For instance, assuming we have $x = 01010011$ and $y = 01010100$, the hamming distance to convert x to y is going to be 3 because from 01010**011**, (011=highlighted bits) should be changed to 01010**100**. Hamming distance can be used for string transformation as well, where instead of bits, we count character changes between two strings, e.g., the distance between "hello" and "hallo" is one character.

---

[4] A bit is the smallest unit of data in a computer, and it can be either 0 or 1.

Note that hamming distance measures the number of symbol changes in both strings. Similar to the Euclidean and Manhattan distances, the time complexity of the Hamming distance is linear, too.

Levenshtein (Edit) Distance

The Levenshtein distance is another typical distance to measure the similarity between two strings. This distance function counts the number of *minimum* edits (delete, insert, or substitute) required to transform the source string data into the target data. For instance, the Levenshtein distance from "TGGCC**A**A" to "TGGCC**GA**C" is two because we need one substitution and one insertion. It is common to use this distance for similarity metrics in bio-informatics.

As another example consider, we would like to measure the distance between xyxyxyxy and yxyxyxyx. For the strings "xyxyxyxy" and "yxyxyxyx", using only insertions, deletions, and substitutions, the minimum number of edits to transform one string into the other would indeed be 8 (all 'x's to 'y's and all 'y's to 'x's). The operation of removing one character and adding it to the end, while intuitively simpler, is not a single edit operation in the context of the Levenshtein distance. Therefore, in the standard Levenshtein distance calculation, the distance between these two strings would be 8, not 2.

The time complexity of edit distance is usually not easy to calculate [Pighizzini '01], but we could assume $O(n \cdot m)$, where n and m are the lengths of the two strings.

Longest Common Subsequence (LCS) Distance

Another method for text matching and similarity calculation in sequential data such as genome sequences is LCS. It is very useful to deal with genome data. This method measures the number of similar sequences between two-character sequences (not string sequences) of data. For instance, assume we have a sequence of "a, x, b, c, g, w, s" and "x, a, x, y, b, c". The longest common subsequence or LCS distance between these two strings is {a,x,b,c}. It's important to note that a subsequence doesn't have to consist of consecutive elements; the elements need to be in order but not necessarily contiguous.

The larger the LCS value, the higher the common subsequence. It uses dynamic programming, we do not explain here[5], but remember the time complexity of this algorithm is $O(m.n)$ for two strings with sizes of m and n.

Cosine Distance

The cosine similarity metric is used to calculate the distance between two vectors of data, and it is computed as follows:

$$d(A, B) = \frac{A \cdot B}{||A||_2 \cdot ||B||_2} = \frac{\sum_{i=1}^{n} a_i b_i}{\sqrt{\sum_{i=1}^{n} a_i^2} \sqrt{\sum_{i=1}^{n} b_i^2}}$$

---

[5] While describing Reinforcement Learning in Chapter 13, very briefly we explain dynamic programming.

Note that the $||.||_2$ signs around a vector means L_2 of that vector. Norms in machine learning describe a magnitude. of a vector, and later, we will encounter these terms a lot (especially in Chapter 8 and Chapter 14).

Assume we have assigned an organization to analyze chicken emails to the aviculture. We will analyze their report to be sure they are all happy with their life condition and keep their living standards high. Nevertheless, there are too many emails, and one chicken staff cannot read them all. She uses an approach called *topic modeling* to group similar reports based on the words that have been used in their text. Topic modeling is a clustering technique used for textual documents, and we will explain this later in Chapter 7. Now, just knowing that it is a text clustering is enough.

The appearances and frequency of the words in four sample emails are presented in Table 4-2. This word frequency is used for clustering similar emails together.

In other words, to measure similarities between textual documents, we can assume the text is a "bag of words", and by counting the frequency of similar words between two documents, we can measure their similarities. For example, to measure the similarity between "email 1" and "email 2", we use cosine similarity as follows:

$A = (8,0,2,14), B = (9,1,4,11)$

$A.B = 8 \times 9 + 0 \times 1 + 2 \times 4 + 14 \times 11 = 234$

$||A|| = \sqrt{8^2 + 0^2 + 2^2 + 14^2} = 16.248$

$||B|| = \sqrt{9^2 + 1^2 + 4^2 + 11^2} = 14.79$

$d(A, B) = \dfrac{218}{18.11 \times 14.79} = 0.97$

	food	sleep	temperature	crowdedness
email 1	8	0	2	14
email 2	9	1	4	11
email 3	1	12	19	0
email 4	0	7	12	0

Table 4-2: Term(word) frequency among each email.

The similarity between "email 1" and "email 2" is 0.97. Therefore, they are highly similar (the number is close to one).

The time complexity of cosine similarity is usually quadratic.

Same as the Euclidean norm, the *Lp norm* is presented as follows:

$$||\bar{x}||_p = \sqrt[p]{|x_1|^p + |x_2|^p + \ldots}$$

For instance, we described L_1-norm *(Manhattan distance)* as the sum of all elements in the vector as follows:

$$||\bar{x}||_1 = \sqrt[1]{|x_1|^1 + |x_2|^1 + \ldots} = x_1 + x_2 + \ldots$$

Jaccard Distance or Jaccard Index

The Jaccard similarity is operated based on the union and intersection of two sets, and it is written as follows:

$$d(A, B) = \frac{A \cap B}{A \cup B}$$

If you are unfamiliar with set algebra, the intersection sign is $\cap$, and it is presented as a dotted area on the right side of Figure 4-5, and the union sign is presented with $\cup$ on the left side of Figure 4-5.

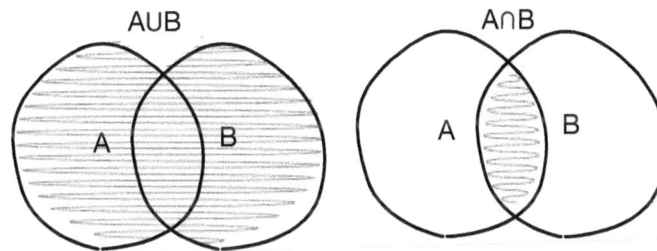

Figure 4-5: (left) union of two sets A and B, (right) the intersection of these two sets on the left.

The Jaccard index is useful when we are dealing with non-numerical data points, especially when it is useful for *binary* comparisons. For instance, we collect the restaurants a user has visited to identify his taste for a food recommendation. In this case, we do not intend to compare the names of visited places. We intend to have a binary comparison of food menus for each restaurant. Some visited restaurants have overlaps in their menus, such as seafood restaurants and Japanese sushi. Therefore, we go for the Jaccard Index, which uses a binary comparison of the food items on their menu.

Furthermore, Jaccard is useful when two sets do not have equal sizes, and they are significantly different in size.

The Jaccard computational complexity depends on the implementation. For instance, if sets are represented as lists, the complexity might be higher due to the need to check for membership. However, if sets are implemented as hash sets (we will learn about it in Chapter 5), the intersection and union operations can indeed be near-linear in many cases.

Dynamic Time Warping (DTW)

DTW is an effective distance measurement to measure *time series* similarities. DTW is used for time series data, and we will encounter lots of time series in the real-world. Therefore, we describe it here. While reading it, please be patient; it is a bit challenging to understand.

Time series is a numerical vector of data in a fixed time duration, such as a heart rate per minute, accelerometer data per second, etc.

For instance, assume a chicken uses a device to measure its heart rate continuously. The heart-rate data is collected every minute; it is a time series, and Figure 4-6 shows a time series of our chicken heart rate during the day.

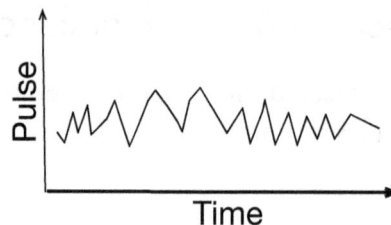

Figure 4-6: A time series example.

Assuming we have two similar time series if we use Euclidean distance instead of DTW, we can observe that the two time series will get a high dissimilarity and low similarity score. However, DTW handles its distance calculation significantly better. See Figure 4-7, it compares two time series with two different distance measurements, Euclidean and DTW. By looking at matching between peaks and valleys, the DTW provides more appropriate distance mapping between two time series that are similar to Euclidean distance.

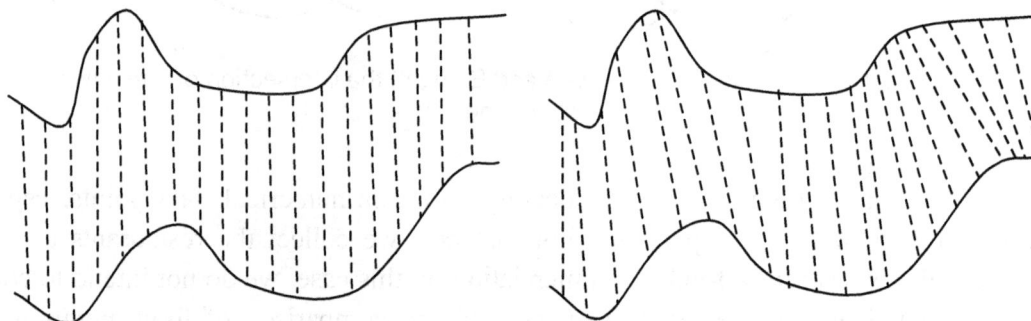

Figure 4-7: Distance comparison between two time series. (left) Euclidean distance and (right) DTW distance.

In a more technical sense, Minkowski distances (e.g., Euclidean and Manhattan) measure distances based on a one-to-one mapping between data points in a sequential order. DTW, however, measures the distance based on the mapping of one to many data points. The optimal choice of warping is determined via dynamic programming.

Assuming each time series can be presented as a vector, e.g., $\vec{A} = \{a_1, a_2, \ldots a_n\}$ and $\vec{B} = \{b_1, b_2, \ldots b_n\}$. the equation to compute DTW between two time series $\vec{A}$ and $\vec{B}$, is written as follows:

$$DTW(\vec{A}, \vec{B}) = \left| A_i - B_j \right| + min(DTW(i-1,j), DTW(i,j-1), DTW(i-1,j-1))$$

160

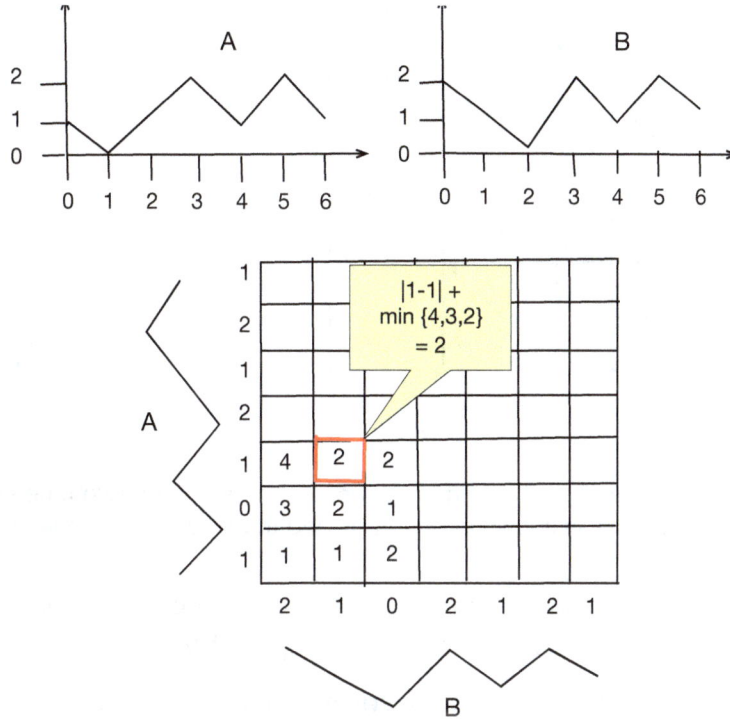

Figure 4-8: Cumulative matrix creation for dynamic time warping. Each cell content is calculated by differences from A_i and B_i, plus a minimum number from its neighbor cells.

It is better to understand this equation with examples. Assume we have two time series $\vec{A} = 1,0,1,2,1,2,1$ and $\vec{B} = 2,1,0,2,1,2,1$. First, a *cumulative matrix* will be constructed using the described equation, as shown in Figure 4-8. Take a look Figure 4-8 and the described equation to see how the value of cells i, j is calculated based on subtracting A_i from B_j and adding the minimum identified value from cells $(i - 1, j)$, $(i, j - 1)$ and $(i - 1, j - 1)$ to the result. The content of the red cell in Figure 4-8, which is 0, will be calculated as shown on the yellow area on top of this cell.

After the cumulative matrix is populated with the data, the algorithm starts from the main diagonal[6] of the matrix and selects the minimum values in the surrounding space with the range of w, until it reaches the end of the matrix diagonal. Figure 4-9 presents this process, but note that it did not use the numbers in Figure 4-8. It shows the main diagonal in dark grey dots and the path that can be used to connect time series data points in grey. This path is called the *warping path*.

We should also specify a hyperparameter, w, which is called a *warping constraint*. This parameter presents the *maximum* amount of deviation allowed from the diagonal. In Figure 4-9, we set $w = 1$ (the light grey area around the main diagonal). Then, between each cell on the main diagonal and the w cells around it (green rectangle in Figure 4-9), the cell with the

---

[6] The line in the middle of the matrix that starts from the bottom-left top-right cell is called diagonal.

161

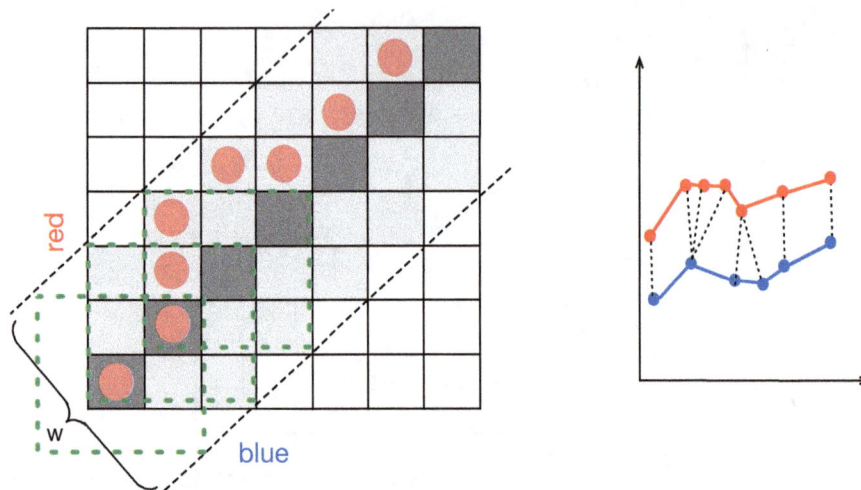

Figure 4-9: Warping path based on the warping constraint of w=1 for a two time series shown on the right. The main diagonal is shown in dark grey and the area that fits in w=1 is light grey.

minimum value is selected (we show it in red color). These red cells will be the mapping points between two time series, as shown on the right side of Figure 4-9.

In Figure 4-9, we presented the first three comparisons with dotted green boxes. The rest will be done with the same method, and we will not describe them here.

The selection of value for w is important because it affects the accuracy of the DTW calculation. Standard DTW (the one we explained here) has $O(n^2)$ time complexity, which is not efficient for a distance measurement function because a distance measurement function will compare all distances between data points together. Nevertheless, there are good implementations of DTW that have $O(n)$ complexity.

Graph Similarities

There are various methods used to measure similarities between two graphs, and four major approaches to measure graph similarities are recommended by Aggrawal [Aggrawal '15]. We briefly describe them here. Later in Chapter 15, we will describe graph algorithms in detail.

Common Subgraph Distance: the similar subgraph (a smaller graph inside the original graph) of two graphs is a representation of graph similarity.

Substructure-based Similarity: When finding identical subgraphs is challenging, we can look for structural similarities. This approach is akin to subsequence similarity in the Longest Common Subsequence (LCS) distance.

Graph Edit distance: Similar to Levenshtein distance, this method counts the number of edits required to transform one graph into the other.

Graph Kernel distance: We can use kernel functions to measure the similarity between two graphs. Kernel functions will be described later in Chapter 9.

Graph clustering algorithms are referred to as community detection, and the ones that we explain do not rely much on the described graph similarity metrics.

NOTES:

* While converting interval, ordinal, or categorical variables into a number for a clustering algorithm, remember not to harm the relationship between data points and keep them after the conversion. For instance, if we have Big, Medium, and Small. It could be converted into 1,0.5,0, and converting it to some other structure, such as 0.5,1,0, will harm the clustering process.

* Since Euclidean distance uses a squared root, it can not tolerate noise better than Manhattan distance, and thus, Manhattan distance is better at dealing with noise and outliers.

* Euclidian distance, Manhattan distance, and, in general, L_p-norm distances are not good for sparse and high dimensional datasets because those datasets are sparse and contain too many features, noise, and outliers.

* To measure similarities between ordinal data points, we usually convert them into a range of values. It is recommended [Han '11] to map the ordinal data between 0 and 1. For instance, chickens weighted from 1-300 grams could be assigned 0, from 300-500 grams could assigned 0.5, and more than 500 grams could be assigned 1. Therefore, we can use these numbers, which are a realistic mapping to the original data, and use a numerical distance.

* A real-world dataset is usually multivariate; thus, it includes different data types, including numerical, binary, categorical (nominal), etc. We can focus on one attribute/feature/field for clustering. Nevertheless, it is not useful to resolve a problem only for one attribute. A better method is to convert all data points into a single data type, such as numerical values between 0 and 1, and then apply the clustering. However, a conversion to another format is associated with smoothing the data, and smoothing can negatively affect the precision of the algorithm. Furthermore, we should be able to select a specific number of columns from the dataset, which is called feature selection, and we will explain it later in Chapter 6.

* Equally important to the selection of clustering algorithm is the selection of similarity measurement method. Therefore, while we are planning to cluster our data, we should dedicate some effort to selecting the similarity metrics as well.

* Cosine similarity is useful for textual information clustering. Cosine is not very flexible when comparing dissimilarities.

* Jaccard is useful when we are comparing two data points based on their concept and dissimilarity.

* When we are dealing with strings of different sizes, we should use Levenshtein distance. Hamming distance can not handle strings of different sizes.

* To choose the right similarity method, consider the distribution of the dataset, dimensions, and data types. For example, for binary comparison, we can use Jaccard, but for time series, it is better to use DTW.

* DTW cannot measure the distance between two time series with different lengths [Ratanamahatana '05], you might see this by mistake in some textbooks. It can tolerate small size differences but not large ones. It has been recommended that they be interpolated to get equal sizes and then compared. Interpolation will be explained in Chapter 16.

Clustering Algorithms

Clustering Algorithms

Please accept our congratulations if you have finished the similarity metrics, and if you have not and you are not familiar with similarity methods, go back, read the previous section, and then come to this section.

There are many different clustering algorithms and many different categorizations of clustering algorithms [Xu '05]. Here, we focus only on the ones that are widely in use and worth learning. There are five major categories of clustering algorithms, including *partition based clustering, density based clustering, hierarchical clustering,* and *probabilistic clustering.*
Furthermore, there is another category of clustering, which is used for text mining, i.e., topic modeling, but due to their flexibility, they can be used for non-textual documents as well. We will explain topic modeling later in Chapter 7.

Partitioning (k-representative) Methods

k-mean

It is the most used clustering algorithm. To use k-mean we should specify the number of expected clusters, k, before running the algorithm. It means we are dealing with one hyperparameter (k) here, which is the number of desired clusters. Then, the k-mean algorithm partitions the dataset into the k numbers of distinct clusters. The hyperparameter k of k-mean is the number of clusters we intend to have.

To understand how k-mean works, check the visualized example presented in Figure 4-10 (a). There, we have a 2D plane (two-dimensional data)[7], and data points were distributed among x-axis and y-axis. Assume we give $k=2$ to the k-mean algorithm.

Initialization: As the first step, based on the number of expected clusters (k), the algorithm chooses k random points on the plane that includes the dataset. △ signs in Figure 4-10 (b) present those two random data points, also called *seed points*.

Distance Calculation: Second, the algorithm finds the distances of all other data points to these two △ points. Figure 4-10 (c). shows some distances from △ signs to other data points, but we do not show all of them. Otherwise, there will be too many lines, and you cannot read them. Now, each data point has its distance computed with each of those △ points.

Cluster Assignments: In the third step, the algorithm assigns each data point to its nearest △, See the result in Figure 4-10 (d). This means that for each △ we have one cluster, and every data point's distance to one △ will be computed, and the data point is assigned to the △ cluster that has the shortest distance. Figure 4-10 (d). shows two different clusters; one is blue, and the other cluster is red.

Updating Centroids: In the fourth step, the algorithm calculates the mean (center) in each cluster and moves each △ to the center of its cluster. See Figure 4-10 (e), the grey △ was their previous placement, and after calculating the mean, they have moved into the center.

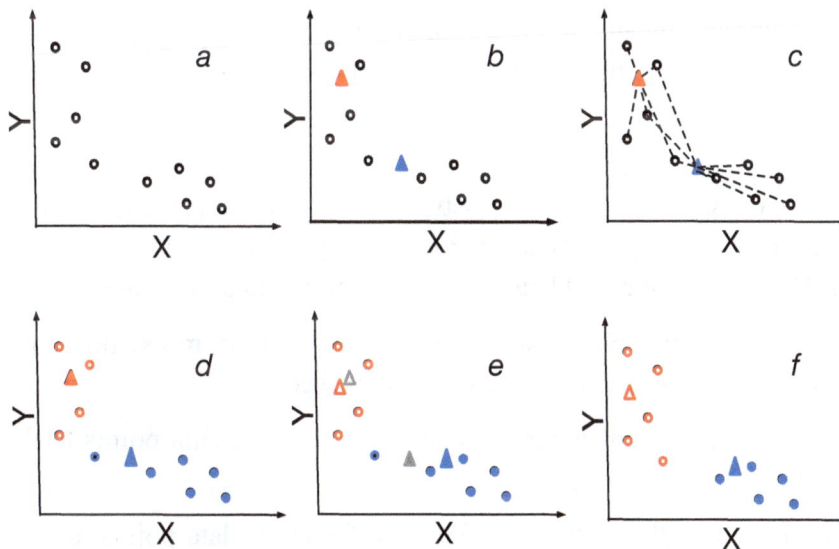

Figure 4-10: k-mean algorithm step by step.

Iteration: Then, the algorithm repeats processes from Figure 4-10 (c) to Figure 4-10 (e), until all △ signs stay in the center of their cluster and they do not move any further, a.k.a., cluster centers

---

[7] Of course, we could have more than two-dimensional data, but for the sake of simplicity, as we said before, we will explain most clustering algorithms in a two-dimensional space.

converge. If they don't stop, there could be a maximum number of iterations given by the user as a hyperparater, and then after it reaches this maximum, it will stop.

These △ points are called the *centroid* of a cluster. You can see that from Figure 4-10 (e) to Figure 4-10 (f), one data point changes from the blue cluster into the red cluster. This is due to the movement of △ signs into the center of their clusters.

Assuming n presents the number of data points, k the number of clusters (centroids), i is the number of iterations, and the computational complexity of the k-mean algorithm is $O(n \times k \times i)$, which is linear.

k-median

k-median is similar to k-means and k-medoids, but it's distinct in its approach. It uses the median of the points in a cluster to define the center and typically uses Manhattan distance for clustering. Recall that k-mean uses Euclidean distance.

k-medoid

It is another data partitioning, and similar to k-mean, it tries to keep the data within the minimum distance of their cluster and creates clusters with the maximum distance from each other.

k-medoid can handle noise better than k-mean because instead of using Euclidean distance between data points, it uses pairwise dissimilarities between data points. The seed points are selected randomly, similar to k-mean, but not a random point in the dataset range. Instead, it uses random data points from the existing data points. Figure 4-10 (b) shows that seed points (△) were two points calculated randomly in the dataset range (and not selected from existing data points).

To understand how it works, we study each illustration in Figure 4-11. Figure 4-11 visualizes the process of k-medoid clustering approach for $k = 2$, by using a PAM (Partitioning Around Medoid) algorithm [Han '11]. Figure 4-11 (a) shows data points in a 2D space.

Initialization: First, the algorithm selects two random data points from existing data points. The dotted circles in Figure 4-11 (b) present those two random seeds.

Distance Calculation: Then, it calculates the distances of all other data points to these two data points Figure 4-11 (c).

Cluster Assignments: Based on the distances calculated for those data points, each data point is assigned to a cluster, as shown in Figure 4-11 (d).

Updating Medoid: Next, for each cluster, it finds the Medoid data point, which is the center data point and not the mean between data points. Then, it moves the medoid center to those points.

Iteration: Again, it calculates the clusters' Medoid and continues from step Figure 4-11 (c) to Figure 4-11 (e) until there will be no changes in the medoid member of each cluster or a maximum number of iterations reaches (hyperparameter).

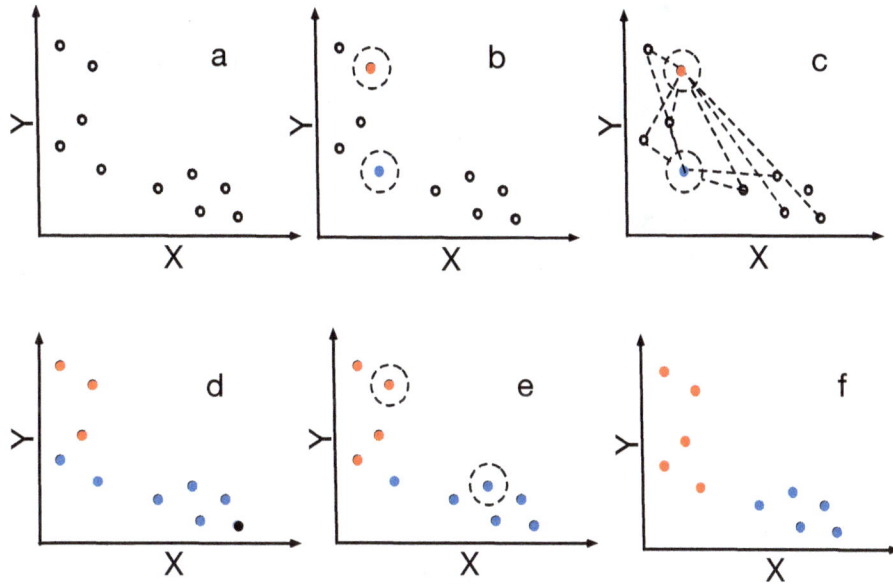

Figure 4-11: k-medoid algorithm step by step.

You can see it is very similar to k-mean, but it is more tolerant to outliers because when there is a significantly different variable, it can affect the mean, but choosing a medoid resolves the outlier issue. For instance, consider we have a vector (one dimensional data) of *{1,3,6,1000}*. The mean of this vector is *252.5*, which does not really represent a cluster. However, if we calculate the medoid, it will be *6*, i.e., the closest value to the median). Using *6* as the center of the cluster is more realistic than *252.5*.

Assuming k is the number of desired clusters, and n is the number of data points, the computational complexity of traditional k-medoid (PAM algorithm) is $O(k(n-k)^2)$, which is quadratic, and thus k-mean (linear) is more efficient. The complexity arises because, for each pair of a non-medoid point and a medoid point, the algorithm must compute the total cost of swapping them and then potentially perform the swap, which is an expensive operation.

Density Based Methods

When the data distribution has a non-convex shape in 2D space, partitioning clustering methods do not provide an accurate result. Convex shapes are curvatures that have rounded exteriors, and non-convex shapes have rounded interiors, as shown in Figure 4-12. A convex is defined as a shape in that all its interior angles are less than 180°; if at least one angle is larger than 180°, then it is not a convex shape. In other words, every straight line between two points on the shape is inside the shape completely.

Another category of clustering algorithms is performing clustering based on the density of data, and they can cluster datasets that have non-convex shapes.

Before going into the technical details of density based clustering, we should understand why partitioning clusterings do not work on non-convex shapes. Imagine there is a jungle that grows a very rare plant called fruit-for-privileged-AI-scientist (FPAS). By eating FPAS from this plant, you will learn everything in data science. There is a very big jungle, and you have heard that few people have found this plant there. You are planning to go and search the jungle. Nevertheless, the jungle is too big, and you can't finish searching before the deadline of your project. Therefore, you should customize your search and focus on places with a higher chance of finding the FPAS. You get a map showing where fruits grow in the jungle. Take a look at Figure 4-13 (a). It shows the placement of fruits in the jungle. You would like to use a clustering algorithm to efficiently search the jungle in three days and separate search locations into three places, one for each day.

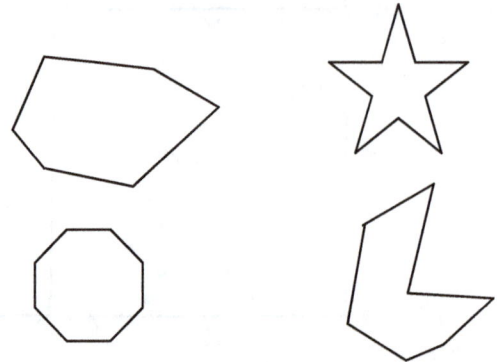

Figure 4-12: (left) Convex shapes, (right) non-convex shapes.

the rare plant

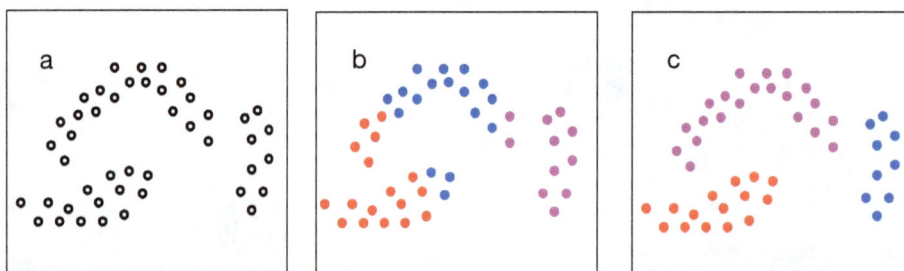

Figure 4-13: (a) Plants distribution in jungle. (b) The results of clustering plants by a partitioning algorithm. (c) The results of clustering plants by a density based algorithm.

If you use a partitioning algorithm such as *k*-mean, your result might end up in three clusters like Figure 4-13 (b), which is clearly not efficient for you to search. However, a density based clustering provides a result similar to Figure 4-13 (c). which is the most efficient clustering that you can use for this type of search. In the following, we describe two well-known density based clustering, DBSCAN and OPTICS.

Density-Based Spatial Clustering Applications with Noise (DBSCAN)

The term DBSCAN [Ester '96] stayed for Density-Based Spatial Clustering Applications with Noise. DBSCAN automatically finds the number of clusters, unlike k-means and k-medoid, which we should give the number of desired clusters as input (hyperparameter). Nevertheless, it requires us to specify two other input parameters: Epsilon 'ε' (eps, radius, or search distance) and minimum points (*MinPts*). Epsilon specifies the radius size around the selected data point, and MinPts is the minimum number of neighbor data points in the circle, which has a radius of eps size, required to consider them as a member of the current cluster (See Figure 4-14).

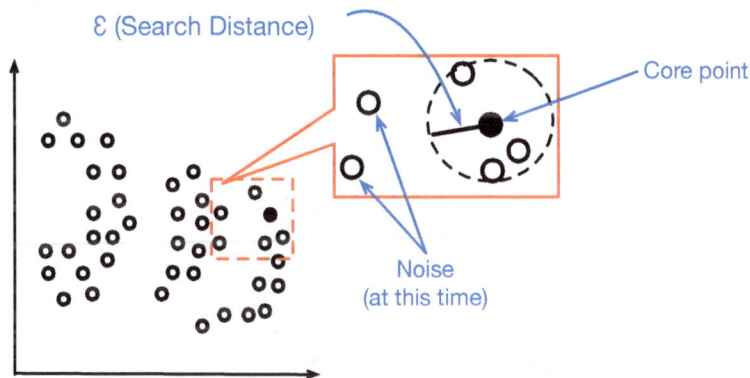

Figure 4-14: The element of DBSCAN clustering algorithm.

The DBSCAN algorithm starts with selecting a random data point, i.e., *core point,* and draws a circle with the size of *eps* around the core point. Then, it counts the data points inside the circle. If the number of data points with a radius of eps is larger than MinPts, they will be considered to be in the same cluster as the data point. DBSCAN assumes that data points are either *core, border,* or *noise.* It draws a circle with the radius on eps around the core point (one data point is randomly selected at the beginning of the algorithm). The data points that stay on the border of the circle are called *border points*. The data points that were not clustered at the end are considered *noise.* In other words, they do not stay inside any circle with *MinPts* number of points.

If the number of data points inside a circle is less or equal to *MinPts*, the algorithm does not assign them to the same cluster. If data points are assigned to a cluster, the same process continues from one of the new data points that has been assigned to the cluster. This process will be repeated until no other data points can get into the current cluster.

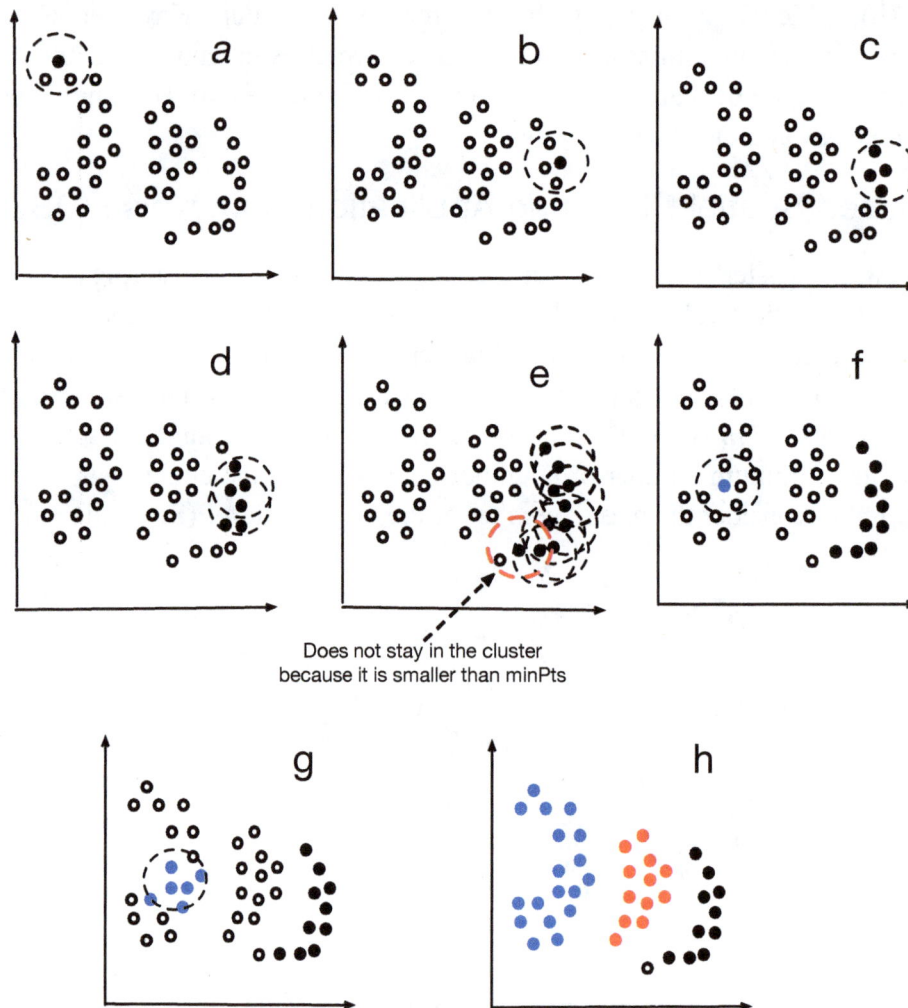

Figure 4-15: DBSCAN clustering algorithm, step by step. dataset with non-convex shape on the left.

Next, the algorithm selects another random data point and repeats this process until it clusters all possible points around that new core point. The algorithm continues this repetitive process until all points have been processed and assigned either to a cluster or as noise.

To better understand the DBSCAN process, check Figure 4-15, which presents a non-convex 2D data distribution. In this Figure, we can see there are three clusters of data, all similar to a banana shape. Using a partitioning algorithm is not going to provide a good clustering result. To identify these three clusters, we should use another method that can deal with non-convex shape of data clusters, such as DBSCAN.

Figure 4-15 (b) to (h) present DBSCAN steps. Assuming the *minPTS* is set to two, and we don't discuss *eps* for the sake of simplicity. The algorithm starts by choosing a random data point in the dataset, e.g., the black dot in Figure 4-15 (a). Then, it uses this point as a core point and draws a circle with a radius of "eps" around it. Some neighbor points could stay in this circle. If the number of neighbor points is not fewer than *MinPts*, then these points were assumed to be in the same cluster. If there are fewer than MinPts points around the core point, the core point is not a good selection, and the algorithm should select another core point. As shown in Figure 4-15 (a), because only two data points stayed in the circle of the core point, it selects another core point, see Figure 4-15 (b).

After the algorithm selects another random point (Figure 4-15 (b)) as a new core point, it checks its *MinPts*. Now, we have more points than *MinPts* inside the circle. Therefore, all these points will stay in the same cluster. Then, another point from these points is selected as a new core point to repeat the process. Note that we do not select any random point this time because the MinPts condition is true; another point is selected from points inside the circle around the current core point.

As shown in Figure 4-15 (c) to 4-15 (d), this process recursively continues, and we assign data points to the same cluster until we encounter a set of data points that includes less than *MinPts* points in the circle (Figure 4-15 (e)). Then, the points inside that circle are not considered as cluster members, and the algorithm assumes them to be noise; it stops assigning points to that cluster and goes for another random point. For instance, in Figure 4-15 (e) (see the red circle), one data point failed the *MinPts* condition at the end of this iteration, and this data point was assumed to be noise.

After it finishes calculating all adjacent data points into the same cluster, then it selects another random point (the blue points in Figure 4-15 (f)) and checks the *MinPts* condition. If the condition is true (Figure 4-15 (g)), it takes the neighbor points in the same cluster and continues this process until all data points inside circles are visited (with the condition of more than *MinPts*) and no data point remains unvisited at the end (Figure 4-15 (h)). The points that are not clustered were considered as noise.

You understand how DBSCAN works, and now we can realize how important the selection of *MinPts* and *eps* is. The complexity of DBSCAN is quadratic or $O(n^2)$, but if a spatial index is used (which we do not describe) it will be $O(n \ log \ n)$.

OPTICS

Ordering points to identify clustering structure (OPTICS) [Ankerst '99] was released after DBSCAN to address some of its shortcomings. DBSCAN is very sensitive to its parameter *MinPts* and *eps*. The algorithm user should set these two parameters, which requires many trials and errors until she can find the best combination of those two parameters. There is no way to get rid of *MinPts*, but OPTICS tries to mitigate the need for optimal *eps* value. Still, OPTICS did not completely eliminate *eps*, but it can reduce its impact on the final result.

Besides, DBSCAN cannot handle clusters within clusters and clusters with different densities. For instance, in Figure 4-16, we can see three clusters. Cluster C3 is embedded inside cluster C2. DBSCAN uses a fixed size *eps*. If we set *eps* to a very small value, it cannot identify C1 and C2. If we make it big enough to identify C1 and C2, it will be too big to identify C3. Therefore, a fixed size *eps* can find clusters. In other words, one global *eps* cannot identify all clusters.

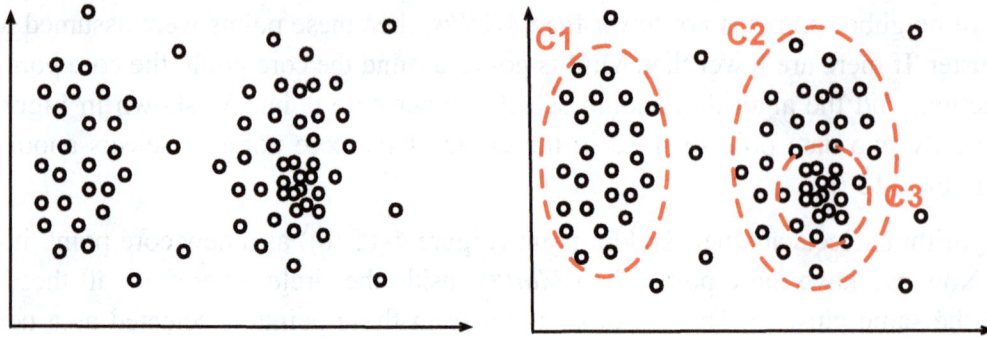

Figure 4-16: Dataset on the left, and clustered dataset on the right. Note that cluster C3 is inside C2.

Interestingly by using OPTICS, a data point member could be a subset of two clusters, as is shown in Figure 4-16; all cluster C3 members are cluster C2 members, too.

OPTICS can mitigate this challenge by not producing clusters explicitly. This algorithm outputs an ordered list of data points based on their minimum *reachability distances* and *core distances*. The output can be graphically represented in a histogram. The peaks and valleys in the histogram could be used to find clusters. With the histogram, we can also identify clusters easily by looking at the plot.

Technically speaking, OPTICS is an extended version of DBSCAN and it uses ε' instead of ε and $0 < \varepsilon' < \varepsilon$. DBSCAN has a fixed value for ε, but OPTICS uses ε' which changes.

Instead of identifying clusters, it creates and stores the *Ordered list* or *Seed list*. This list is composed of the data points which have been processed. Plotting this list enables us to easily identify clusters by looking at the hills and valleys in the histogram.

How does the OPTICS algorithm create that sexy ordered list? Before describing the list creation process, we first define some concepts. Please read them carefully because we refer to them a lot during the algorithm explanation.

Core distance: It is the minimum distance required from the core point to reach *MinPts* (minimum number of points). The core distance is the same as the search distance (ε) in DBSCAN, and it is fixed. However, OPTICS's core distance is flexible and could be changed until it finds all *MinPts*. In other words, the algorithm inflates the circle of core distance (pink circle in Figure 4-17) until it finds *MinPts* number of points. Then, the radius of this circle will be the core distance.

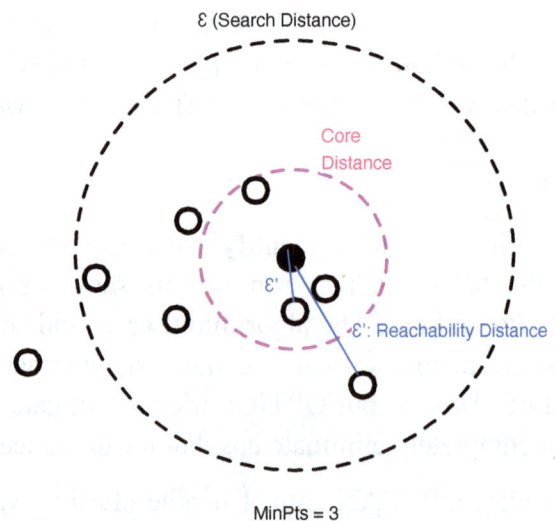

Figure 4-17: OPTICS algorithm parameters for MinPts = 3.

172

Search Distance (ε): It is the search radius we explained in DBSCAN. It is fixed and given by the algorithm user as a hyperparameter. The core distance can inflate until it reaches the search distance, but not more. We still need to provide this hyperparameter, but its impact is not as severe as that of DBSCAN.

Reachability Distance: It is presented as ε' the reachability distance of a data point p with respect to another data point o is defined as:

$$reachability - distance\ (p, o) = max\ [core - distance(o), distance\ (p, o)]$$

This distance is always smaller or equal to the search distance, i.e., $\varepsilon' <= \varepsilon$. The reachability distance exists only if the distance to the target data point is within the ε neighborhood of the core point and there are enough MinPts within the ε neighborhood of the core point to make it a core point. Figure 4-17 shows the differences between ε, ε' and the core distance. We present the reachability distance in blue between two points, one inside the core distance and the other one outside the core distance.

To understand OPTICS, we should memorize these definitions, so repeat after me: ε' *(reachability distance)* $< \varepsilon$ *(search distance)* also *core distance* $<$ *search distance*. Good, now repeat after me: *"Reachability distance and core distance are both smaller than search distance"*, now louder, … excellent.

The OPTICS algorithm calculates core distance and reachability distance for each data point and stores them in the ordered list. The name of this list is *OrderSeeds* or *Seed List*. Elements of this list are ordered by the *shortest reachability distance* from their closest neighbor data point (smallest *reachability distance*). Unlike DBSCAN, which explicitly produces clusters, OPTICS only calculates all data points' reachability distances and core distances within the given ε radius. The *OrderSeeds* is an ordered list by the smallest *reachability distance*. Therefore, the denser clusters stayed on top of this list. Figure 4-18 (j) presents the cluster order of data points based on their reachability-distance. Such an ordering guarantees that higher density clusters are created first. The deeper the valley, the denser the cluster will be.

Figure 4-18 presents an example of the OPTICS algorithm, we assume $MinPts = 2$. We present points in the core distance with red, points accessible within the search distance as blue, and visited data points or core points as black. The first circle around the core point is a core distance, and the second one, which is dotted, is the search distance ε.

First, the algorithm chooses a random data point from the dataset, Figure 4-18 (a), assuming it selects point D. The algorithm assumes that the first point has an undefined reaching distance, and the plot in Figure 4-18 (a) shows a long vertical bar for that. Data point D has three other data points in its search distance, H, C, and G. Since *MinPts* is 2, H and C are in the search distance because two points are enough and C and H are closer than G. Therefore, we add them within their reachability distance into the Seed List. The next step is to find the shortest distance from the Seed List and choose that point as a core point, which is H. Then, as shown in Figure 4-18 (b), now, H is the core point. At this stage, G and D will be within the reachability distance, and J is also available within search distance. H, G, and D are enough to meet the *MinPts* condition, and we should add G and D to the *Seed List*.

Nevertheless, D has been processed, and we only add G to the *Seed list*. A comparison between

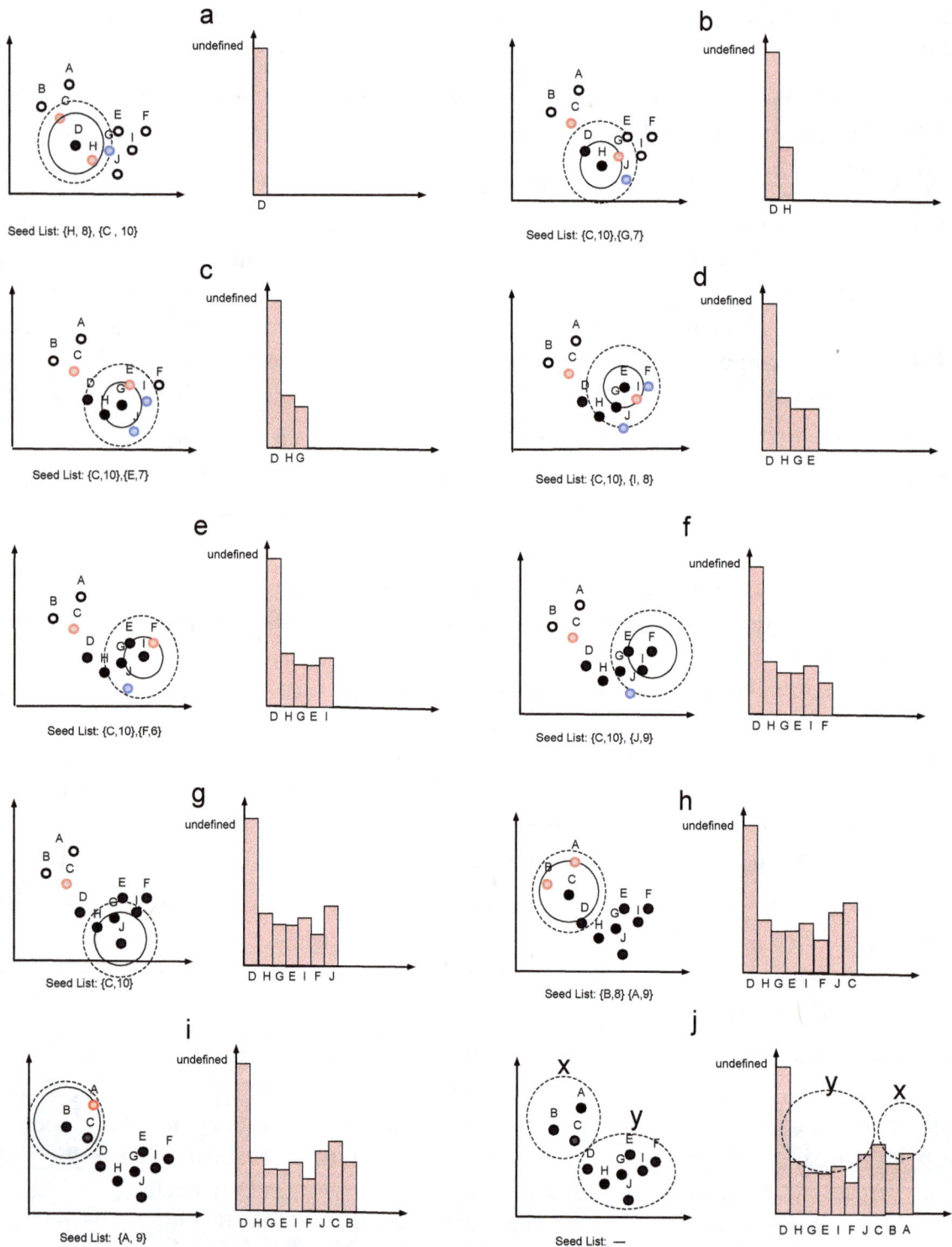

Figure 4-18: Step-by-step implementation of the OPTICS algorithm. The search distance is represented by a dotted circle, and the core distance by a solid circle. Blue dots are within the search distance, red dots are within the reachability distance, and black dots have already been processed.

G and C in the *Seed List* reveals that G has a shorter distance, and thus, it will be added to the reachability plot and selected as the next core point, as shown in Figure 4-18 (c). The new core point is G now, and E stays in its shortest reachability distance. Therefore, we add E and its distance into the *Seed List* and consider E as a core point now, Figure 4-18 (d). This process continues until all data points are visited and the reachability plot has been drawn completely, as shown in Figure 4-18 (i). By looking at the valleys and hills in the *Seed list*, we can identify clusters. Figure 4-18 (j) shows two clusters, X and Y, from the *Seed list* and on the 2D plane.

The x-axis represents the points in the order generated by the OPTICS algorithm. The y-axis represents the reachability distance, which is a measure of how close points are to their neighbors. Valleys in this plot generally represent clusters, and mountains (or high reachability distances) represent the transitions between clusters. We can imagine a "horizontal line" for deciding a threshold for cluster separation. In particular, lowering the line (choosing a higher reachability distance threshold) will merge some of the valleys, resulting in fewer, larger clusters. Raising the line (choosing a lower reachability distance threshold) will separate more valleys, leading to smaller clusters. Figure 4-19 is another example that shows a reachability plot of a two-dimensional dataset with four clusters.

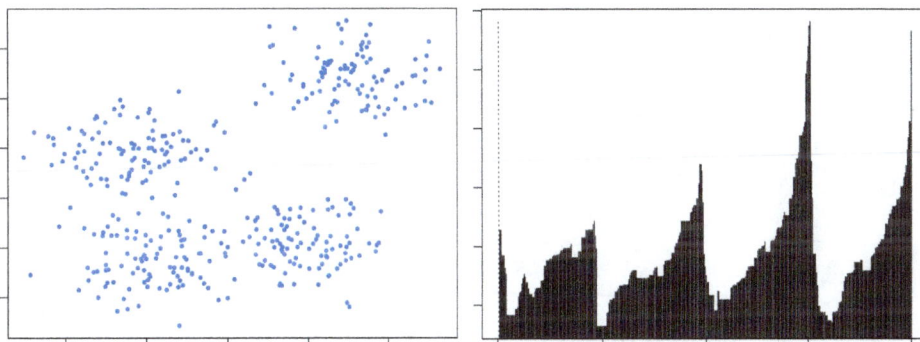

Figure 4-19: (left) Original data points and we can see there are four clusters. (right) Reachability plot that is the result of feeding these data points into the OPTICS algorithm.

Similar to DBSCAN, the computational complexity of OPTICS is also $O(n^2)$, but using a spatial index enables us to have a $O(n\ log\ n)$ complexity.

Hierarchical Methods

Density based clustering methods resolve the problem of manually defining the number of clusters (unlike partition based clustering). Nevertheless, density based clustering methods depend on different hyperparameters, i.e., ε and *MinPts*. Hierarchical clustering mitigates the dependency on the number of clusters we expect from the algorithm. The result of hierarchical clustering is presented in a *dendrogram* (see Figure 4-20), and by looking at the dendrogram, we can get an overview of the clustering result. It assists us in deciding on the optimal number of clusters. These clustering methods create a hierarchy of data points, and based on the hierarchy, the user can decide on the number of clusters. For instance, Figure 4-20 presents a dendrogram of hierarchical clustering output. We can draw different lines at different levels, such as the right side of Figure 4-20, to have different numbers of clusters. In this example, line X results in two clusters, and line Y results in seven clusters.

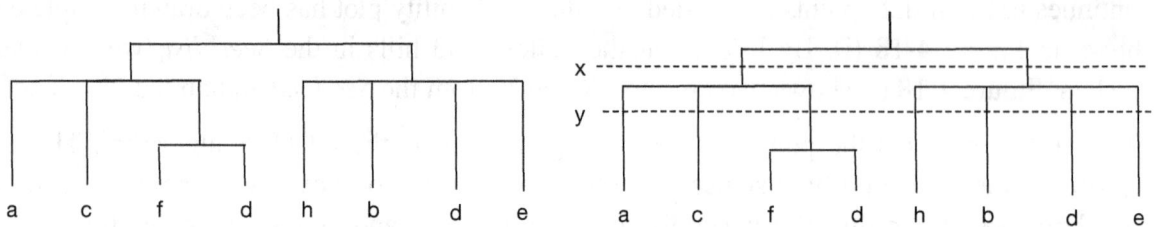

Figure 4-20: (Left) A sample dendrogram of a dataset, (right) based on the line we draw on the dendrogram we can get different number of clusters. Line X results in two clusters and line Y results in seven clusters.

Representing data in hierarchical form has some practical applications. For instance, assume an aviculture that grows chickens and has different customers. There is a customer who is looking for muscular chickens and is preparing them for a chicken weightlifting competition. There is another customer who is managing a chicken running competition, and he is looking for speed runner chickens. Another customer is looking for chickens who can work as politicker, so he is looking for lazy chickens with no talent and too much self-confidence.

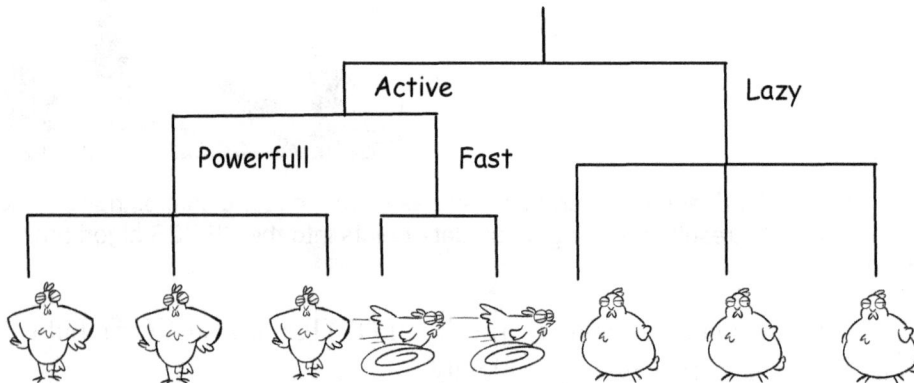

By clustering them hierarchically, the aviculture owner can categorize chickens into different categories and sell them based on customer demand. At a higher level, it could have a cluster of active and lazy chickens, i.e., two clusters of data. Then, going one step deeper inside the active cluster, we can have a cluster of "fast" and a cluster of "powerful" chickens. The fast ones are good for sprinting (running a short distance very fast), and the powerful ones are good for weight lifting. We can say that hierarchical clustering methods create a *taxonomy* of the data. Taxonomy is a sexy substitute for grouping; after you finish reading this book, always say taxonomy instead of grouping, which shows off your literacy level. We are doing this, and we are enjoying it.

There are two methods to build hierarchical clusters: agglomerative (bottom-up) and divisive (top-down). Agglomerative clustering starts by considering each data point as a single cluster.

Then, it groups similar data points until reaching a point where all data points are gathered in one single group. Divisive clustering acts in the opposite. It starts by considering all data as a single group and then separates data points into smaller groups in each step until there is no possibility of creating a new branch or a specified threshold is reached. In simple words, all data belong to one big group at the beginning, and then they are separated into smaller clusters. For instance, Figure 4-21 visualizes how both types of algorithms start to cluster data. We can see in this figure that the agglomerative method has a bottom-up approach, and the divisive method has a top-down approach.

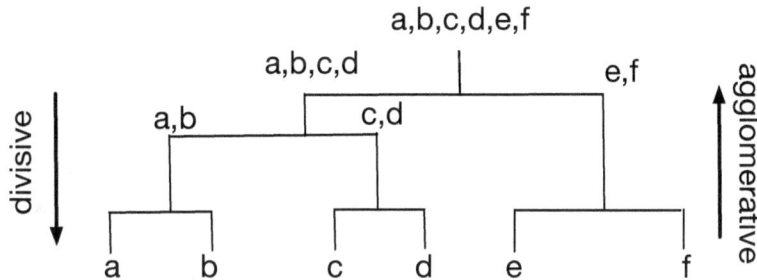

Figure 4-21: Agglomerative and divisive clustering approaches.

Assuming n is the number of data points, agglomerative hierarchical clustering methods have a time complexity of $O(n^3)$, and divisive ones have a time complexity of $O(2^n)$. Both are not very efficient and thus, there are improved agglomerative hierarchical clustering methods that reduce this complexity to $O(n \log n)$, such as Single-linkage (SLINK) [Sibson '73] and Complete-Linkage (CLINK) [Defays '77] methods. Here, we describe DIANA [Kaufman '09] as divisive clustering and SLINK as an agglomerative model.

Single-Linkage (SLINK)

SLINK [Sibson '73] operates based on the shortest distance between elements of a similarity (distance) matrix. We hope you remember the similarity matrix, which is the core of clustering; if you forget it, let us know to kill ourselves; we cannot carry this amount of pain you give us.

To understand it better, we explain SLINK with examples. Assume we have a dataset of six data points: a,b,c,d,e,f. Their coordinate values are as follows: {(2,6),(5,7),(4,9),(13,12),(15,14),(17,13)}. The result of the similarity matrix, based on Euclidean distance, is presented in the matrix of Figure 4-22 (a).

The first step in the SLINK algorithm is to identify the smallest distance in the matrix, which is the distance between b and c, i.e., $dist(b,c) = 2.23$ as shown in Figure 4-22 (a). This means they are the nearest data points in a dendrogram, and we can put them in the nearest neighborhood. In other words, in this matrix, the shortest distance is the distance between 'b' and 'c'.

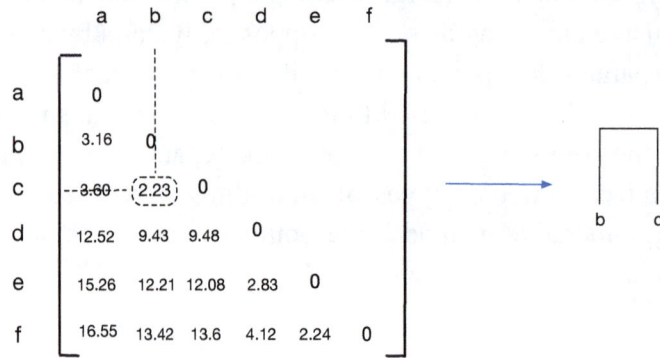

(a)

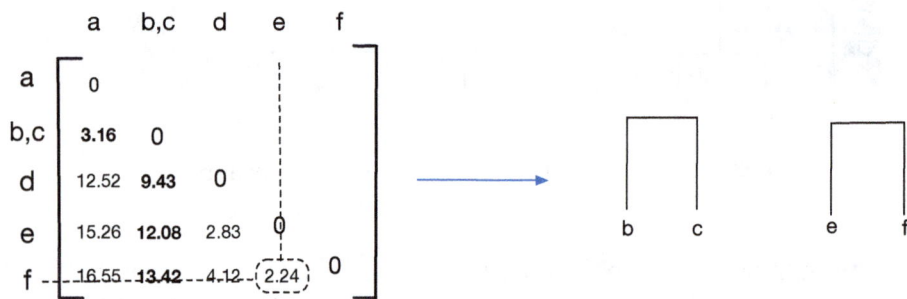

(b)

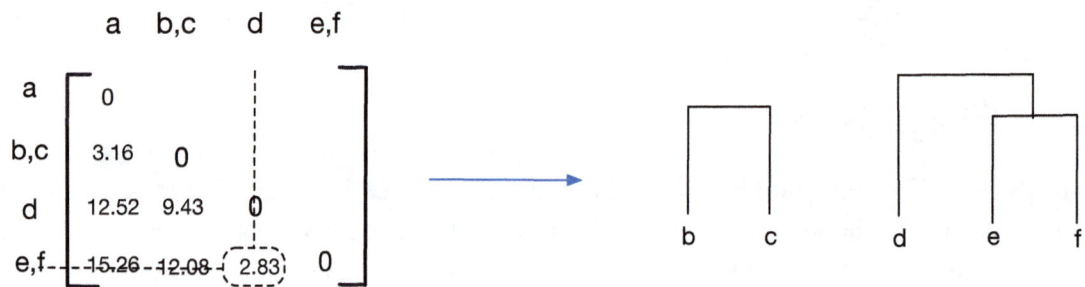

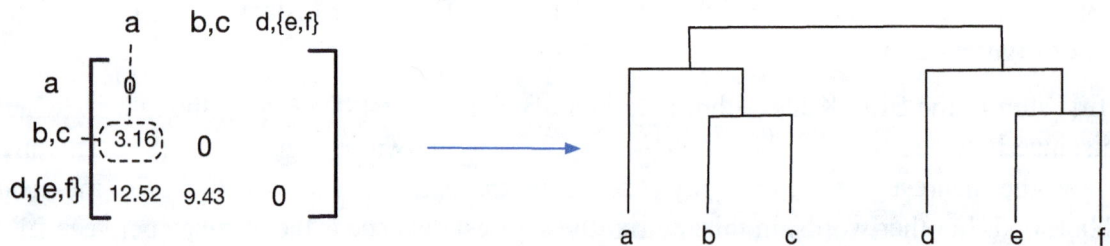

(d)

Figure 4-22: SLINK algorithm clustering steps.

Therefore, they are adjacent in the dendrogram, and a dendrogram is presented on the right side of Figure 4-22 (a). Next, the algorithm compares all other data points with these two identified data points (b, c), and chooses the minimum distances from the comparison. The following lists comparisons with these two identified data points:

$a : min[dist(a,b), dist(a,c)] = min[3.16,3,60] = 3.16$
$d : min[dist(d,b), dist(d,c)] = min[9.43,9.48] = 9.43$
$e : min[dist(e,b), dist(e,c)] = min[12.21,12.08] = 12.08$
$f : min[dist(f,b), dist(f,c)] = min[13.42,13.60] = 13.42$

Then, the matrix is updated by merging b and c into one data point. They are substituted into the new matrix, as shown in Figure 4-22 (b). Then again, the algorithm finds the minimum distance, i.e., between e and f, which is 2.24. Then, it draws them in a dendrogram, as shown in the dendrogram of Figure 4-22 (b). This process will be repeated, and again, it calculates the distances of all points to e and f:

$a : min[dist(a,e), dist(a,f)] = min[15.26,16.55] = 15.26$
$b,c : min[dist((b,c),e), dist((b,c),f)] = min[12.08,13.42] = 12.08$
$d : min[dist(d,e), dist(d,f)] = min[2.83,4.12] = 2.83$

Again, it chooses the shortest distance, which is 2.83, the distance between d and {e,f}. Therefore, the dendrogram will look like the dendrogram in Figure 4-22 (c). The new matrix will be created based on the most recent results.

The described process will be repeated, and from the matrix, we can realize the shortest distance is between {b,c} and a.

$a : min[dist(a,b,c), dist(a,d,(e,f))] = min[3.16,12.52] = 3.16$

Therefore, as is shown in the dendrogram of Figure 4-22 (d), they stay in the same cluster. Now, nothing is left except connecting the two sub-clusters with a line together, and the final dendrogram is ready.

Assuming we have n number of data points, the computational complexity of SLINK is $O(n^2)$. There are two other similar algorithms called ALINK (Average Linkage) and CLINK (completer linkage), we skip explaining them due to their similarities with SLINK.

Divisive Analysis (DIANA)

Divisive analysis (DIANA) [Kaufman '09] is a divisive clustering method. Usually, divisive methods are computationally more expensive than agglomerative approaches, and they have not been widely covered in the literature.

Divisive clustering starts with a single cluster and continues to create sub-clusters by splitting the top cluster until each datapoint is a cluster, and we cannot split them any further. This algorithm reminds us very much of the approach some countries took or are taking to colonize a nation and the concept of divide and conquer in computer science.

Assume we are living in the future, our resources on the earth are about to finish, and our lovely politicians are planning to contact aliens on a remote planet and get natural resources. We were contacted by a planet. However, after they came to earth, seeing the amount of garbage we produce and our greed for unsustainable reproduction, they decided not to collaborate with us. Now, our politicians have decided to follow the 18th and 19th century patterns of Europe and colonize their planet. However, attacking them with the military is infeasible because they are largely populated, and military operations are very expensive and infeasible. Aliens on those

planets are living together in peace, while they have different shapes and faces, which means different races. Based on the 18th and 19th centuries' experience, the best way to conquer that planet would be to divide their aliens. We assign them a,b,c,d,e,f, and the following matrix presents dissimilarities between those ethnicities.

	a	b	c	d	e	f
a	0					
b	3	0				
c	6	2	0			
d	4	7	6	0		
e	11	11	5	3	0	
f	12	9	10	11	1	0

Our politicians use *divisive clustering* to break their alliance. They use DIANA algorithm, which starts by considering all data points as a single cluster and splits this cluster into two clusters. Based on the given dissimilarities, we can see which ethnicities are best to invest in and get them out of the alliance. In other words, this step focused on understanding which data point has the highest value of dissimilarity compared to other data points. To calculate that, we simply measure the average dissimilarity score. We name this function $avg.dis(.)$.

$avg.dis(a) = (3 + 6 + 4 + 11 + 12) \div 5 = 7.2$
$avg.dis(b) = (3 + 2 + 7 + 11 + 9) \div 5 = 6.4$
$avg.dis(c) = (6 + 2 + 6 + 5 + 10) \div 5 = 5.8$

$$avg.dis(d) = (4 + 7 + 6 + 3 + 11) \div 5 = 6.2$$
$$avg.dis(e) = (11 + 11 + 5 + 3 + 1) \div 5 = 6.2$$
$$avg.dis(f) = (12 + 9 + 10 + 11 + 1) \div 5 = 8.6$$

We can see that f is the most dissimilar ethnicity, thus f separates from the other data points. Now we have two clusters; one includes {a,b,c,d,e}, and the other includes {f}, called the *splinter group*.

	Avg. dis. to other data	Avg. dis. to splinter data	Differences between two columns
a	(3+6+4+11)÷4= 6	12	6 - 12= -6
b	(3+2+7+11)÷4=5.75	9	5.75 - 9 = -3.25
c	(6+2+6+5)÷4=4.75	10	4.75 - 10 = -5.25
d	(4+7+6+3)÷4=5	11	5 - 11 = -6
e	(11+11+5+3)÷4=7.5	1	7.5 - 1 = 6.5

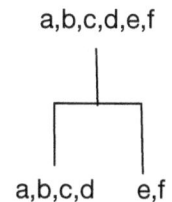

Table 4-3: Comparison between splinter data and remaining data for DIANA algorithm.

	Avg. dis. to other data	Avg. dis. to splinter data	Differences between two columns
a	(3+6+4)÷3 = 4.33	(11+12)÷2 = 11.5	4.33 - 11.5 = -7.17
b	(3+2+7)÷3 = 4	(11+9)÷2 = 9	4 - 9 = -5
c	(6+2+6)÷3 = 4.67	(5+10)÷2 = 7.5	4.67 - 7.5 = -2.83
d	(4+7+6)÷3 = 5.67	(3+11)÷2 = 7	5.67 - 7 = -1.33

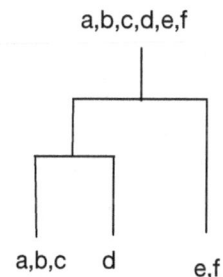

Table 4-4: Comparison between splinter data and remaining data for DIANA algorithm (2).

	Avg. dis. to other data	Avg. dis. to splinter data (d)	Differences between two columns
a	(3+4)÷2 = 3.5	4	3.5 - 4 = -1.5
b	(3+7)÷2 = 5	7	5 - 7 = -2
c	(6+2)÷2 = 4	6	4 - 6 = -2

Table 4-5: Comparison between splinter data and remaining data for DIANA algorithm (3).

The next step is to use the larger group (not the splinter group) and again compare their dissimilarity together, but in addition to this step, calculate their dissimilarity to the splinter group as well. This process is presented in Table 4-3. It has four columns: data point name, average distance of each data point to other data points, average distance to splinter data, and differences between both columns.

By looking at Table 4-3, we can recognize that the only difference that is not negative is e. Therefore, e changes its side to the splinter cluster, and we get {a,b,c,d} in one cluster and {e,f} in the other cluster (see right side of Table 4-3). In other words, they stay in their group if the difference is negative. Otherwise, they join the splinter group.

This computation will be repeated with the new non-splinter cluster, as shown in Table 4-4. You can see in Table 4-4 that all differences are negative, which means no element will be moved between those two clusters. It means the process stops at this level, and now it is the time to go and perform the process between elements inside each cluster. In other words, now we try to divide {a,b,c,d}, and thus we consider following the dissimilarity matrix.

$$
\begin{array}{c} & \begin{matrix} a & b & c & d \end{matrix} \\ \begin{matrix} a \\ b \\ c \\ d \end{matrix} & \begin{bmatrix} 0 & & & \\ 3 & 0 & & \\ 6 & 2 & 0 & \\ 4 & 7 & 6 & 0 \end{bmatrix} \end{array}
$$

We redo the process with this dissimilarity matrix.
$avg.dis(a) = (3 + 6 + 4) \div 3 = 4.4$
$avg.dis(b) = (3 + 2 + 7) \div 3 = 4.4$
$avg.dis(c) = (6 + 2 + 6) \div 3 = 4.67$
$avg.dis(c) = (4 + 7 + 6) \div 3 = 5.67$

Now, d is the most dissimilar data, and it separates d from the other data, and the result will be a new dendrogram. At this stage, 'd' is a splinter cluster and separate, as shown on the right side of Table 4-4. Then, chooses the large group and compares their dissimilarities with the splinter cluster, as shown in Table 4-5. In Table 4-5, all distances are negative. Therefore, there will be no more splinter cells, and elements inside each branch should be analyzed separately (once for {a,b,c}, and once for {e,f}).

We do not write the rest, but the same process, but the algorithm performs these steps iteratively until all data points are analyzed as a separate point (each data point is a cluster itself) or a threshold has been reached, which means the algorithm will stop.

The computational complexity of DIANA is typically $O(n^2 \, log \, n)$, where n is the the number of data points to be clustered.

Large Scale Hierarchical Clustering

Usually, in the real world, we deal with a large dataset, which is why we choose clustering. Nevertheless, there is always a problem of dealing with limited resources and how to load the dataset into memory for clustering. The baseline hierarchical clustering methods we have described previously are not working well with large data. Therefore, we need some methods to cope with the resource limitations of the computer, especially with memory.

At the time of writing this section (in 2018), there were some arguments that memory doesn't matter and that memory will get big enough that there will be no limitation on memory size in the near future. As computer capacities grow, the miniaturization trend of computers grows as well [Kurzweil '06].

Therefore, memory and other resources are always a problem. Do you remember when you were a child, your parents or your teacher told you if you studied hard in the future, you would have an easy life.

A resolution to resource access is another lie in the same level of studying hard to have an easy life. Therefore, it is important to be familiar with clustering algorithms that are designed to mitigate resource limitations, especially memory limitations. We introduce BIRCH and CURE two examples of these memory efficient hierarchical clustering algorithms.

BIRCH

Balanced Iterative Reducing and Clustering using Hierarchies (BIRCH) [Zhang '96] is a resource efficient integration of agglomerative clustering with other clustering methods. BIRCH performs clustering in two phases, which are briefly described as follows. Next, we describe it with examples.

Phase 1: This phase focuses on producing micro-clusters of the original data. First, it scans the data and creates a CF-Tree (Cluster Feature Tree) structure. We will describe CF-Tree with an example. In a more technical sense, this phase scans all data and builds a CF-Tree in the memory. Then, it condenses the CF-Tree into the desirable length and compresses it into a smaller CF-Tree. It is used to reduce the complexity and increase the scalability. The first scan (pass) results in lots of small clusters (see Figure 4-23).

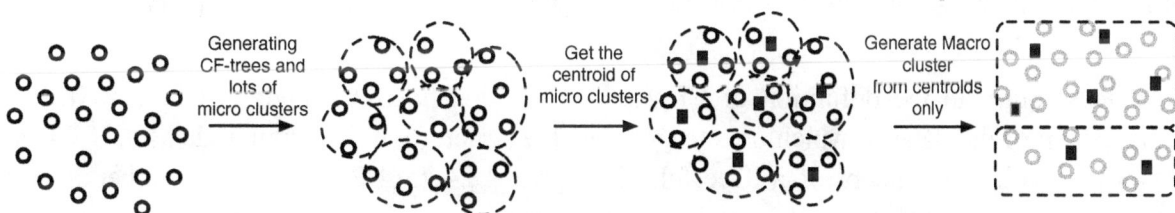

Figure 4-23: A general overview on BIRCH clustering algorithm.

Phase 2: In this phase, the algorithm uses an arbitrary clustering algorithm (e.g., k-mean) to perform the global clustering. This pass improves the clustering results of the first pass and creates macro-clustering results. This pass applies a clustering algorithm (any arbitrary clustering) on CF-Tree leaf nodes and not the original dataset. In simple words, it takes the centroid of each cluster in the previous pass and performs a clustering only on centroid data. Remember that each centroid is the representative of a cluster that has been created in the first pass. Therefore, instead of loading all data into memory, it loads only centroids of the first phase (micro-clusters) into memory for clustering. Figure 4-23 provides a visualization of BIRCH's two phases.

In the following, we describe the CF-Tree construction process and the rest in detail; we warn that it might be boring to read. However, the policy is very interesting and you can use the same approach while dealing with large data.

Phase1 Description: Before describing CF-Tree we should define CF. Cluster Feature (CF) is a *summary statistics* of data, and it is defined as a three-tuple data $CF = <N, LS, SS>$. N stays for the number of data points. LS stands for the linear sum of all data points, and SS stands for square sum of data. They are computed as follows:

$$LS = \sum_{i=1}^{N} x_i \text{ and } SS = \sum_{i=1}^{N} x_i^2.$$

For example, we have the following 2D dataset in a micro-cluster: $\{(1,2),(2,1),(3,2)\}$. Then, $CF = <3,(1+2+3,2+1+2),(1^2+2^2+3^2,2^2+1^2+2^2)> = <3,(6,5),(14,9)>$.

There are other important parameters used to build CF-Trees, including the *centroid* of a cluster $\overline{X}_0$, *radius* of a cluster, i.e., R, and *diameter* of a cluster, i.e., D. Assuming that n is the number of data points, these parameters can be calculated as follows:

$$\overline{X}_0 = \frac{\sum_{i=1}^{n} \bar{x}_i}{n} = \sqrt{\frac{LS}{n}}$$

$$R = \sqrt{\frac{\sum_{i=1}^{n} (\bar{x}_i - \bar{x}_0)^2}{n}} = \sqrt{\frac{SS - (LS)^2/n}{n}}$$

$$D = \sqrt{\sum\sum (\bar{x}_i - \bar{x}_j)} = \sqrt{\frac{\sum_{i=1}^{cluster-size} (2.n.SS_i - 2.LS_i^2)}{n(n-1)}}$$

The centroid ($\overline{X}_0$) is the middle point inside the cluster. Radius (R) is the average distance from all members of the cluster to the centroid point. Diameter (D) is the average pairwise distance within a cluster.

Now, let's return to the definition of CF-Tree, a tree that hosts CF-nodes. It is a multi-level compression of the dataset, which tries to preserve the structure of the original dataset. If you are unfamiliar with the concept of a Tree, check Figure 4-24 to 4-28 or just remember it is an abstract representation of the dataset in the tree style. We will explain more about trees in Chapter 5.

CF-Tree requires three input parameters: *branching factor B, threshold T,* and *number of entries in a leaf node L. B* is the maximum number of children allowed for each non-leaf node. *T* is the maximum radius of a cluster in a leaf node. *L* is the maximum number of data allowed in a leaf node. Don't forget that you need to specify *B, T,* and *L* as input parameters of the BIRCH algorithm.

For example, we have six data points, and we intend to cluster them with BIRCH. Our data points are $x1 = 1.5, x2 = 0.9, x3 = 0.5, x4 = 0.4, x5 = 1.3$. We choose *T=0.2, L=2,* and *B=2* as values for these hyperparameters. We present the root nodes with rectangles and leaf nodes with ovals. While reading the rest of the explanation, we recommend writing *LS, SS,* and *R* on a paper and calculating them in each step yourself because it helps you understand the algorithm.

The algorithm starts by $x1 = 1.5$. It initializes the *root node* for CF-Tree with the CF values of *x1.* Also, a leaf node is created to store the $x2$ (See Figure 4-24).

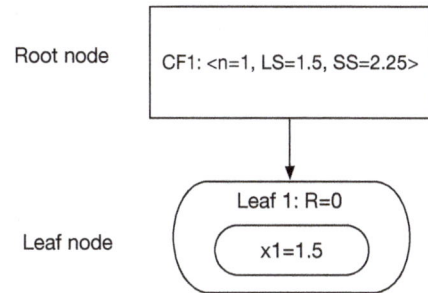

Figure 4-24: CF-Tree after the algorithm encounters X1.

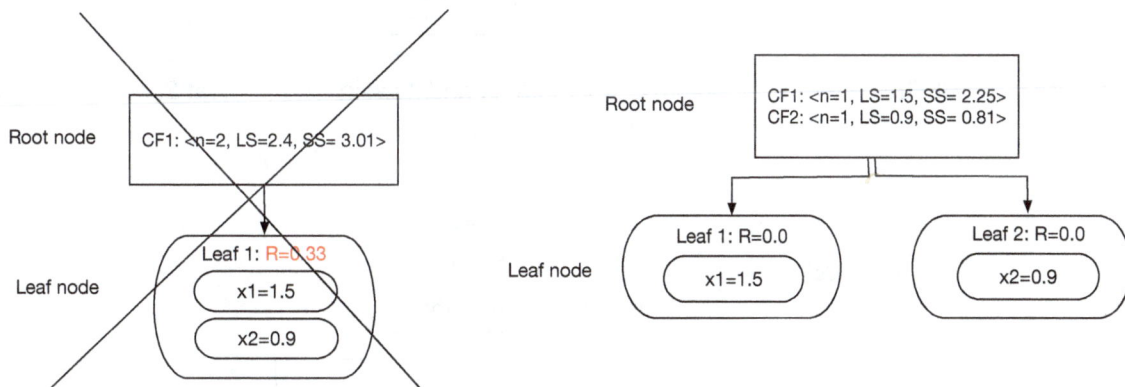

Figure 4-25: CF-Tree after the algorithm encounters X2. It cannot keep X2 and X1 in the same leaf node, because R > T.

Then, it reads the second value, i.e., $x2 = 0.9$. First, it passes this node into Leaf 1, but temporarily. Now, $R = 0.33 > T = 0.2$, which means $x2$ cannot remain in Leaf 1, and it splits the Root node into Node 1 and Node 2, and another leaf node for *x2* will be created. The new CF-Tree is shown on the right side of Figure 4-25.

Next, the algorithm encounters the third data point, $x3 = 0.5$. It compares $x3$ to the location of Leaf 1 (i.e., LS/n) and location of Leaf 2

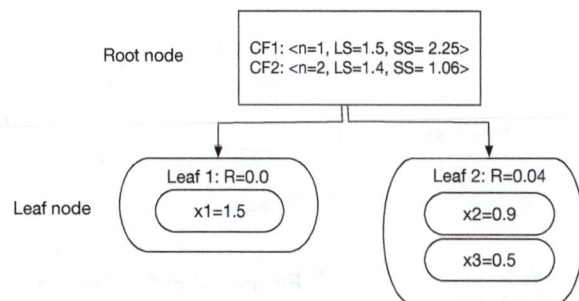

Figure 4-26: CF-Tree after the algorithm encounters X3.

(i.e., LS/n). The distance between $x3$ to Leaf 1 is $(1.5 - 0.5) = 1$, and $x3$ to Leaf 2 is $(0.9 - 0.5) = 0.4$.

Therefore, it is closer to Leaf 2, and it will be assigned to Leaf 2, as is shown in Figure 4-26. Note that R, LS, and SS have been changed because of the new node $x3$.

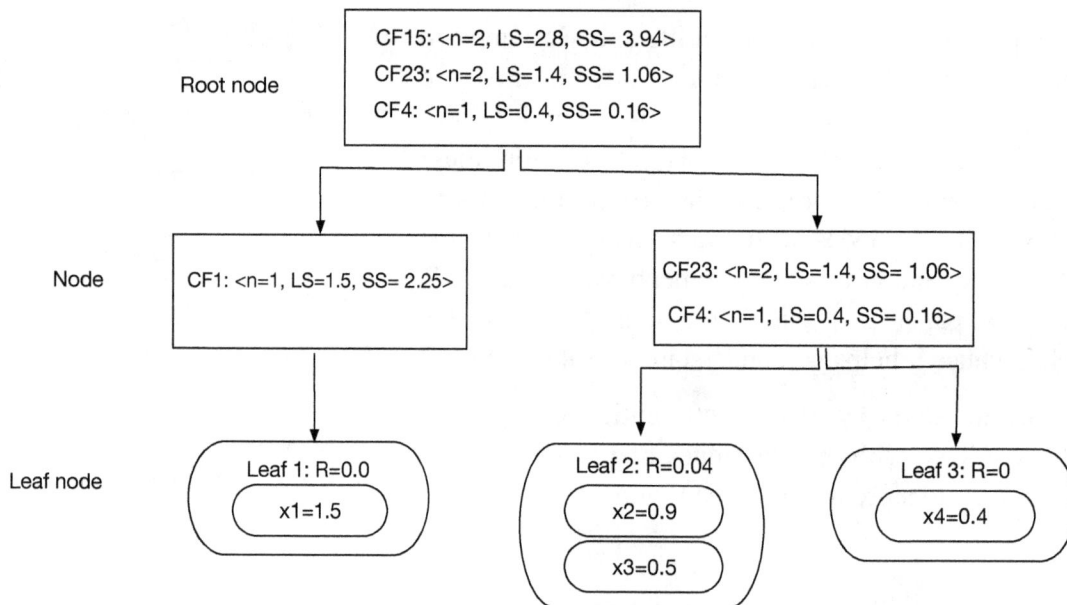

Figure 4-27: CF-Tree after the algorithm encounters X4 and it assigned it to Leaf 2, because it has a shorted distance to leaf 2.

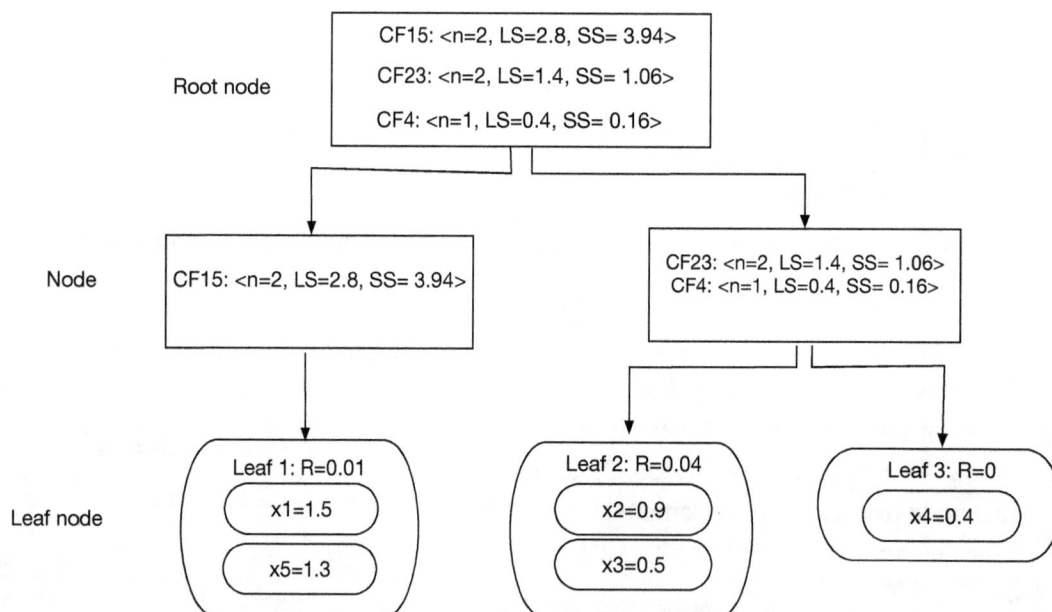

Figure 4-28: CF-Tree after the algorithm processes X5.

186

Next, the algorithm encounters $x4 = 0.4$. Again, it compares $x4$ to the location of Leaf 1 (LS/n) and Leaf 2 (LS/n). The distance between $x4$ to Leaf 1 is $(1.5 - 0.4) = 1.1$ and $x4$ to Leaf 2 is $(1.4/2 - 0.4) = 0.3$. Since the distance to Leaf 2 is shorter than Leaf 1, it is assigned $x4$ to Leaf 2. However, it is not possible because $L = 2$, this means we cannot have more than two data in a single leaf. Therefore, again, it makes another branch, i.e., Leaf 3, and the CF-Tree will look like Figure 4-27.

The last data point the algorithm encounters is $x5 = 1.3$. At the root node, the algorithm compares it with CF1 and CF23. There, we have the following distances: $(x5, CF1) = (1.5/1 - 1.3) = 0.2$ and $(x5, CF23) = (1.8/2 - 1.3) = -0.4$. Therefore, $x5$ is closer to Leaf 1, and it assigns it to Leaf 1. The CF-Tree result of constructing CF-Tree with these five numbers is shown in Figure 4-28.

As we can see in Phase 1, the algorithm scans the dataset record by record and assigns data points into either an existing CF-node or creates a new CF-node. This example helps us understand how the algorithm constructs the CF-Tree.

Phase 2 Description: Now we have constructed the CF-Tree in the memory, which is a sort of compressed version of the dataset. We kept the previous example short to help you understand it rapidly, but it has only two CF-nodes. Assume we have a larger CF-Tree including CF1, CF2,..., CF5, and their centroid are as follows: $\bar{x}_{CF1} = 0, \bar{x}_{CF2} = 1, \bar{x}_{CF3} = 1.1, \bar{x}_{CF4} = 3, \bar{x}_{CF5} = 3.2$. We can use any clustering algorithm and cluster these centroid data. For instance, if we use k-mean with $k = 2$, we will end up having two clusters. One cluster includes $CF1, CF2, CF3$, and the other cluster includes $CF4, CF5$. In general, the second pass performs macro clustering; it uses a clustering method, and because the original data is presented to CF elements, any clustering algorithm can operate very efficiently.

BIRCH scales linearly, and this feature makes it capable of clustering data streams; when new data points arrive, it can dynamically calculate its CF and assign it to its related cluster.

Nevertheless, BIRCH has some limitations as well. Each CF-node in the CF-Tree can hold a limited number of entries. Furthermore, a CF-Tree node is not always a representation of a cluster. Besides, since it uses the notion of radius or diameter to control the boundary of a cluster, it might not perform well if the dataset does not have a spherical shape.

The computational complexity of generating a CF-Tree (Clustering Feature Tree) in BIRCH can vary depending on several factors, such as the structure of the data, the branching factor (the number of children each node in the tree can have), and the threshold settings for the clustering features. However, the complexity of CF-Tree construction is often considered to be approximately linear, i.e., $O(n)$, while assuming n is the number of data points. However, in worst case scenarios, the complexity could increase to $O(n^2)$.

CURE

In Clustering Using Representative (CURE) algorithm [Guha '98], each data cluster is presented by a collection of data points called *representative data points*. This algorithm also performs clustering in two phases. The first phase includes four steps, listed as follows.

Memory

Data points

(i) The algorithm selects some *random data points from* the dataset and loads them into the memory. Remember that the entire dataset is too large to fit into memory, and this is the reason CURE uses only some random data points.

(ii) Then, it performs a clustering (any arbitrary clustering algorithm) on those random data points that reside in memory.

(iii) After it constructs clusters, in each of these clusters, it selects *k representative* points, which are usually the points that have the largest distance to the center of clusters.

(iv) It identifies the centroid of each cluster and moves these k points (representative points) toward the cluster's centroid. A hyperparameter will determine the amount of movement, and it is called α. The algorithm shrinks the clusters by their representative points because it increases the accuracy of data point assignment to their closest cluster.

In the second phase, the algorithm goes through the entire dataset, which is on disk and not memory, and checks the distance of each data point to the clusters created in the last step of the first phase, then assigns them to the *closest cluster*.

To understand CURE, we study the example presented in Figure 4-29 to understand better how CURE works. We have a large dataset as shown in Figure 4-29 (a). The algorithm randomly selects some data points, as is shown in Figure 4-29 (b). The black circles are selected points, and the grey ones are not selected points. In other words, the black data points are in the memory, and now the algorithm focuses only on these data points. Then, it clusters them, as shown in Figure 4-29 (c). Next, it starts by selecting k number of data points from each cluster. Those data points are representative points that present their cluster. In this example, we set $k=3$ (k is something we should give as input), as is shown in Figure 4-29 (d). It selects the first data point with the maximum distance to the cluster's centroid. The next data point will be the data point with the largest distance to the previously selected data point. This process will continue until all k points are selected.

In other words, the algorithm chose those points based on their maximum distance to provide the best representation of the cluster. After representative data points have been selected, the algorithm moves them toward the centroid of the cluster with α size — see Figure 4-29 (e). As a

188

188

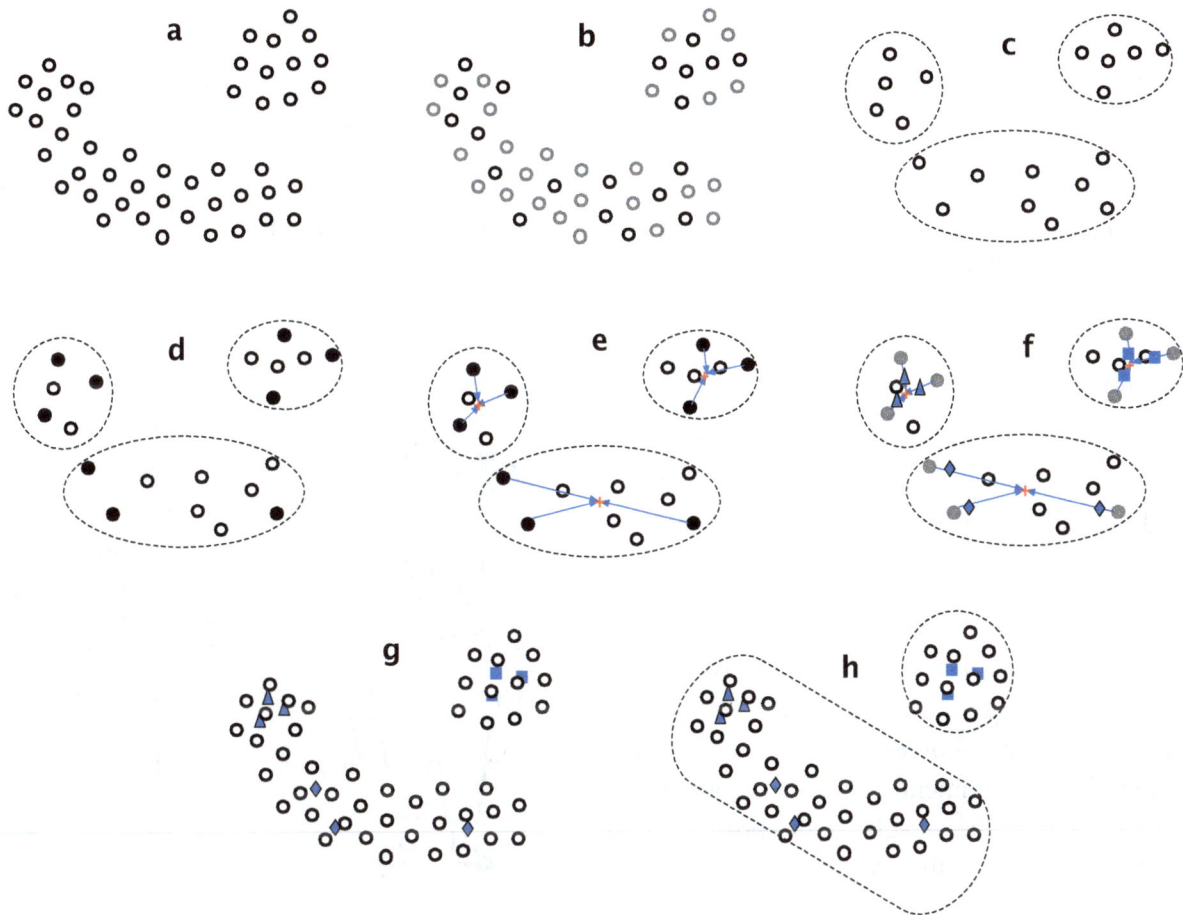

Figure 4-29: CURE clustering algorithm.

result, we will have new synthetic data points substituted by original representative data points. We use a triangle, square and diamond in blue color to present them — see Figure 4-29 (f)[8].

Now, the algorithm keeps these synthetic points and scans all the data points of the dataset . Check Figure 4-29 (g) to compare their distance with these synthetic points. Based on the distance of the real points to these synthetic points, it assigns them to their related cluster. The result will be something similar to Figure 4-29 (h).

CURE is useful for dealing with large datasets and reduces the impact of outliers. The time complexity of CURE is also not easy to compute, but in the worst case scenario, it has $O(n^2 \log n)$ complexity.

Probabilistic and Fuzzy Clustering

All previously described clustering methods assume each data point will be assigned to one cluster, and a single data point cannot participate in more than one cluster. However, there is

---

[8] For the sake of readability, we choose different shapes for each cluster.

another group of clustering algorithms that do not make such an assumption valid, and they are known as *probabilistic clustering* or *model-based clustering*.

Probabilistic clustering enables a data point to be a member of more than one cluster, and because of this characteristic, they are also referred to as *soft clustering*.

To better understand the concept of soft clustering, assume you are a gorgeous, intelligent lady with high intellectual capabilities, and many single men would love to marry you. You have three conditions for your dream partner: (i) he should be humble and family loving, not selfish; (ii) he should care about the environment, e.g., separate his recyclable garbage, be a nature lover, don't change his properties too often, and (iii) he should have a good look.

x_1 x_2

We have three clusters of those behaviors *C1 (nature lover)*, *C2 (humble)*, and *C3 (good looking)*, and we have *n* different data points (*n* men are in love with you), i.e., $x_1, x_2, \ldots x_n$.

By using *k*-mean clustering, each man will be assigned only to one cluster. Nevertheless, each of them has some adjectives from other clusters as well. For example, he could care about the environment but not care about his looks. Therefore, each data point could participate in more than one cluster.

Assume we are analyzing two candidates, x_1 and x_2. By using a probabilistic clustering, we can have the following assignment, which is more realistic.

x_1 *to* $C1(0.499)$, *to* $C2(0.499)$, *to* $C3(0.002)$
x_2 *to* $C1(0.199)$, *to* $C2(0.001)$, *to* $C3(0.8)$

Commonly, the sum of probabilities for all clusters membership of any data point should be equal to 1.

This type of clustering assumes the dataset is a *mixture of two or more probability distributions*. Each distribution presents a cluster, and a distribution is described by a *density function* with a *weight (mixing coefficient)*.

In other words, mixture model clustering methods assume that a dataset can be divided into *probability distributions (clusters)* and clusters following a known distribution, e.g., Gaussian distribution. Therefore, we can say that a dataset is composed of a combination of different distributions. A *mixture model is a weighted sum of distributions*.

Now, a question might arise: how is a dataset converted into a combination of different distributions? Before describing this, ensure you are familiar with the Expectation Maximization

algorithms and Maximum Likelihood described in Chapter 3. If you are unfamiliar with them, please read them and be sure you know them well.

Gaussian Mixture Model (GMM)

If we assume all result clusters should have a Gaussian distribution, we can use Gaussian Mixture Model (GMM) to cluster the dataset. GMM can extract each Gaussian distribution from a multimodal Gaussian distribution.

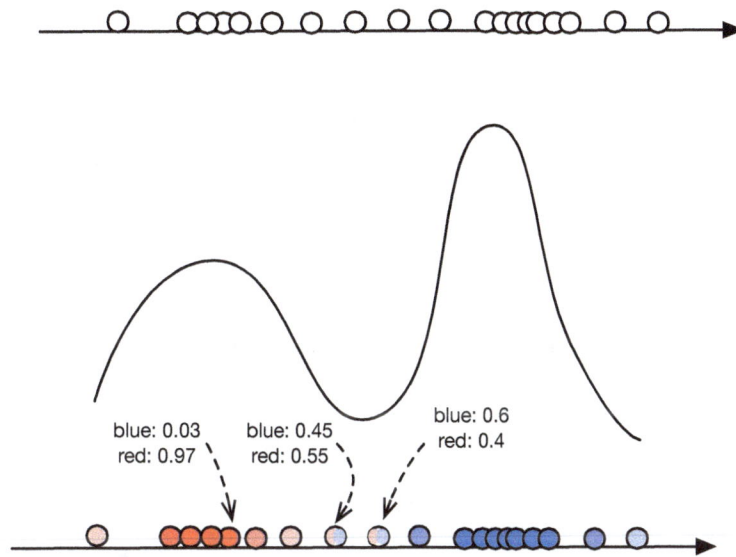

Figure 4-30: On dimensional data distribution on top and its multimodal distribution on the bottom.

As it has been described earlier, each cluster is modeled by a distribution. The Gaussian distribution is defined by the *mean* (corresponding to the center of the distribution) and *standard deviation* (how the spread of the data is around the center).

The distributions of a dataset constitute the clustering, which is formalized as a parameter estimation problem. We need to identify the parameter, and GMM uses the EM algorithm to accomplish this task.

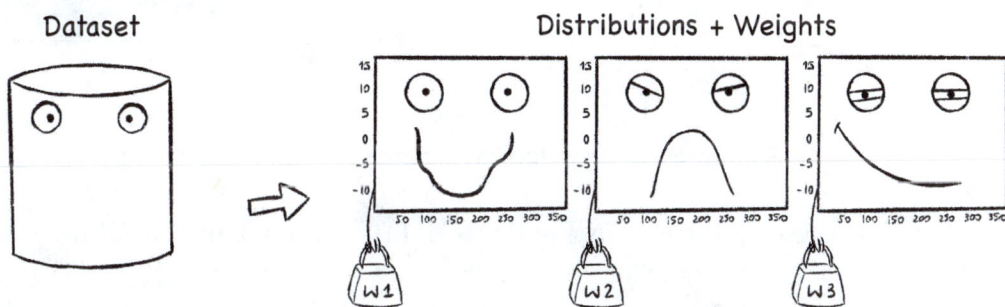

We need to identify the θ parameter, and GMM uses the EM algorithm to accomplish this task [Dempster '77].

Each cluster in GMM is characterized by the mean vector, covariance matrix, and associated probability. The EM algorithm assigns each data point a score, which indicates their cluster contribution. In other words, the task of GMM is to find distribution parameters for each data cluster. The entire dataset could be represented as a mixture of n Gaussian distributions, and thus, it has n number of clusters. Note that in soft clustering methods, each data point is presented with a probability of belonging to a cluster. Therefore, a data point could belong to more than one cluster.

To better understand how GMM works, check Figure 4-30. This figure shows a distribution of a one dimensional dataset. The dataset of Figure 4-30 can be represented as a combination of two distributions. Based on the distribution of this figure, we can estimate there are two clusters of data; we used blue and red colors to present them.

We can see that some data points contribute in both blue and red clusters, which is presented as a probability. We would like to extract two clusters (we should give the number of clusters to GMM as a hyperparameter). We call these clusters blue (b) and red (r).

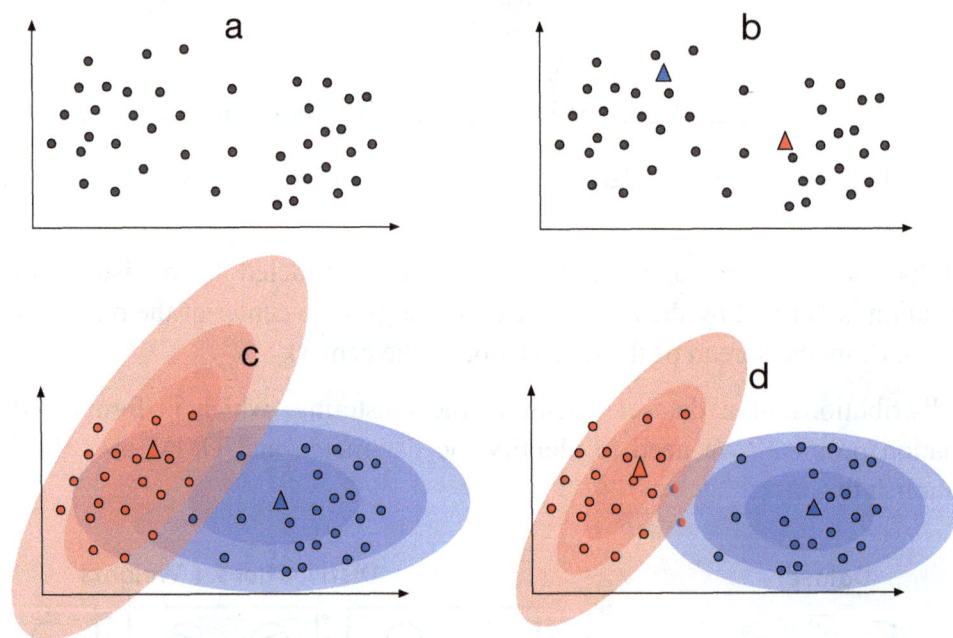

Figure 4-31: Step by Step use of the EM algorithm for GMM clustering.

Assume we have a dataset presented in Figure 4-31 (a) and want to run GMM on it. The algorithm starts by choosing two random points, shown as a triangle in Figure 4-31 (b). GMM typically starts by initializing the parameters of the Gaussian distributions. Then, it defines a random Gaussian distribution with some random mean and variance (recall that the Gaussian distribution is characterized by mean and variance parameters).

192

The first step of the EM algorithm is the Estimation step (E step). This step compares the distances (any distance metrics can be used) from all other points to the two selected points. Then, based on the calculated distances, each data point receives a probability score, which specifies its contribution to one or more clusters, see Figure 4-31 (c). For each data point x_i in each cluster, the algorithm calculates $P(b|x_i)$ and $P(r|x_i)$. $P(b|x_i)$ identifies the likelihood score that x_i is coming from cluster b, and $P(r|x_i)$ identifies the likelihood score that x_i is coming from cluster r.

The next step is the Maximization step (M step), and it is similar to k-means. This step uses $P(b|x_i)$, and $P(r|x_i)$ for every data point, moving the mean points, i.e., red and blue triangles, into each cluster's center. In other words, the M step involves updating the parameters of the Gaussian distributions (means, variances, and mixture weights) based on the responsibilities estimated in the E-step.

After moving them, we will get something such as Figure 4-31 (d). Again, the algorithm reassigns probabilities for each data point based on the new locations of means (E step). This process iteratively continues until there are no changes in mean point; they stay at their location, and then both clusters are finalized. In other words, the EM algorithm stops the iteration either by observing that the mean is not moving or by reaching a predefined threshold; this threshold is a hyperparameter.

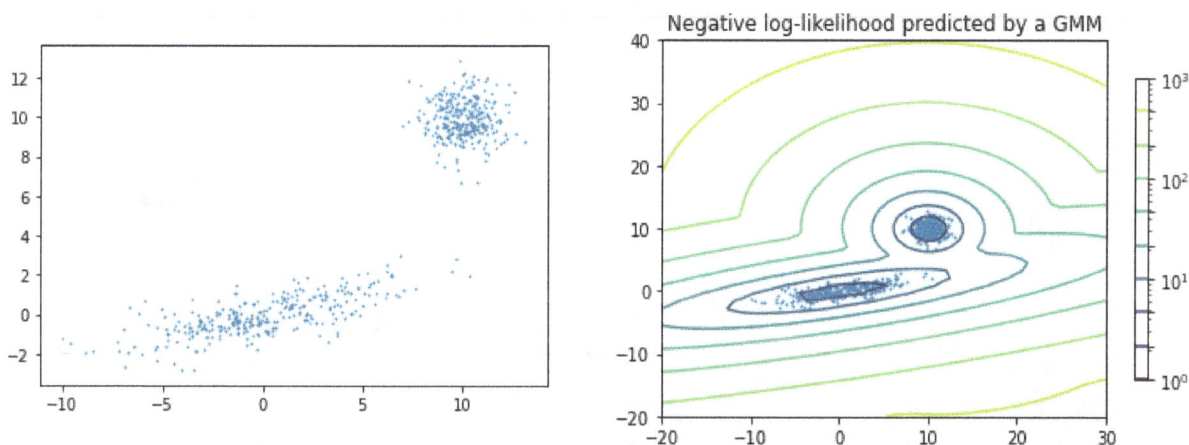

Figure 4-32: (left) Original dataset, (right) the GMM is executed, and the distributions of each cluster are plotted.

GMM is similar to the k-mean, but GMM does soft clustering, and some data points have a probability of contributing to both clusters, unlike the k-mean. Besides, GMM uses the EM algorithm to change the center of clusters and assign data points to a cluster.

Figure 3-32 shows another example visualization, which could help us understand the iteration of the GMM algorithm. The original dataset is shown on the left side, and the clusters have been identified based on the distributions on the right side.

Fuzzy C-Mean Clustering (FCM)

FCM is another soft clustering algorithm in that data points could be a member of more than one cluster. This clustering approach relies on the concept of fuzziness [Zadeh '96], which defines uncertainty in mathematics. In the context of clustering, we can say dataset members could have a range of probabilities and belong to more than one cluster.

It is very similar to k-mean and starts by defining some random points in the space. This algorithm operates as follows:

(i) Similar to *k*-means, it starts by randomly selecting *c* data points as the centroid of clusters, and *c* is a hyperparameter that defines the number of clusters.

(ii) It calculates the fuzzy membership of each member (data point) in a cluster and assigns the probabilities of membership to each data point. In FCM, the membership degree of each data point to each cluster is inversely related to the distance (e.g., Euclidean distance) to the cluster's centroid, adjusted by a fuzziness parameter.

(iii) Next, it computes the centroid of each cluster and moves the centroid points toward the center of each cluster. The new centroid of each cluster is calculated as a weighted average of all points in the dataset, where the weights are the membership degrees of the points in that cluster. The mathematical equation for updating C_j the centroid of the *jth* cluster is as follows:

$$C_j = \frac{\sum_{i=1}^{N} (u_{ij})^m \cdot x_i}{\sum_{i=1}^{N} (u_{ij})^m}$$

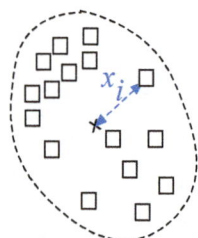

number of cluster members

distance of i-th member to the mean of the cluster

mean of all cluster members

$$SSE = \sum_{i=1}^{n} (x_i - \bar{x})^2$$

Here, N is the total number of data points, u_{ij} is the membership degree of the *i*th data point in the *j*th cluster, x_i is the current data point and *m* is the fuzziness parameter (a higher m results in fuzzier clusters).

(iv) Steps (ii) and (iii) repeat iteratively until centroid points do not change or the maximum number of iterations (given by the user) is reached.

Other Types of Clustering

There are other categories of clusterings, such as topic modeling for unstructured text or graph community detection (clustering for graphs). We will explain topic modeling in Chapter 7 and graph clustering in Chapter 15. Although topic modeling methods have been designed for text, we can use them to cluster many different data types. Also, note that we don't need to limit ourselves to the described clustering methods. Based on our domain knowledge, we can design a customized method for our target dataset.

For example, when we were young and idealists, we thought about enabling users to search and talk to their devices about their activities collected by their smartphone lifelogging app [Rawassizadeh '13]. One approach was to develop clustering for the daily behavioral activities of users based on their routine location changes [Rawassizadeh '19], which were read by their smartphones. Figure 4-33 presents an abstract representation of two consecutive days, and there, we can see the routine behaviors of the user based on the time of the day. We use the user's location changes (read from GPS or WiFi) to identify if they are moving or stationary. Then, based on the location state changes, we create a clustering algorithm that clusters similar spatio-temporal events together. In other words, it uses the time of the day and the user's location to cluster her activities. For example, the red dotted area presents the user's gym cluster.

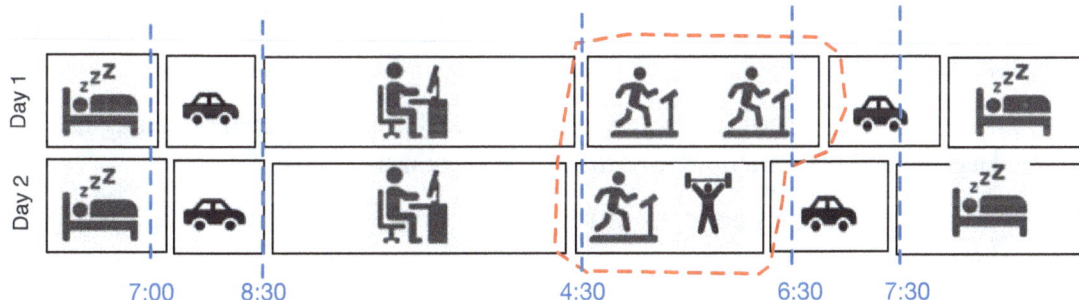

Figure 4-33: Spatio-Temporal clustering of daily activities [Rawassizadeh '19].

Keeping advertising our work aside, you can see that we are not limited to de-facto clustering techniques, and it is possible to build a custom and application specific clustering method as well.

NOTE:

* In contrast to soft clustering methods, both in partitioning methods and density based methods, clusters are *mutually exclusive* (it is a sexy word for not being a member of two sets or two events cannot occur together). This means that a member of a cluster cannot be in any other cluster.

* It is common to run a clustering algorithm several times with different hyperparameters until we get the desired result. This could be done via hyperparameter tuning, which we will explain in Chapter 16.

* DBSCAN is useful when we have a non-convex shape. It can remove outliers as well. Nevertheless, if the cluster is too small, it considers them an outlier, and if a cluster is too large, it considers it as two or more than two clusters.

* While using DBSCAN, setting MinPts to a large size may lead to merging the smaller cluster into a larger cluster. Therefore, setting MinPts depends on your application.

* OPTICS clustering algorithm is less sensitive to the ε value, which is its main advantage over DBSCAN. Nevertheless, we must provide something for ε. However, there is an optimized version of DBSCAN, called HDBSCAN [McInnes '17], and it does not need to get ε. We did not describe HDBSCAN in this book.

* OPTICS is usually more accurate than DBSCAN because it can identify clusters with different densities, but DBSCAN cluster identification is limited to one single ε value. However, the best approach to decide about them is to experiment.

* Usually, the hierarchical clustering method provides better quality than partitioning (k-representative) methods. However, they are computationally complex.

* There are two challenges with agglomerative hierarchical clustering. The first challenge is when to stop combing clusters, and the second challenge is how to determine the closeness of clusters to each other (as the distance increases, the clustering result improves).

* Due to the computational complexity of the hierarchical clustering methods, they are not used for large datasets, and they are mostly used for small datasets.

* Agglomerative clustering usually identifies small clusters, and divisive clustering usually identifies large clusters.

* BIRCH is a scalable clustering algorithm because it can handle large datasets and can undo what was done in the previous step. Most clustering methods are stated forward, and they cannot revise their previous steps.

* While using BIRCH, we might run out of memory, and the CF-tree cannot fit into the memory. To handle this issue, we can increase the value of T.

* The CURE algorithm can handle non-convex clusters of data and has no assumption about the distribution of data [Leskovec '14] (Chapter 7). Other hierarchical and partitioning methods usually have limitations.

* We explained probability before, and as a reminder, probability describes what the chance of observing X in the given sample dataset is. In other words, if the normal dataset has a normal distribution (described by parameters $\theta = \mu, \sigma$) what is the chance of observing X. *Likelihood* means given the observed data X, what are the best distribution parameters that describe the distribution of X (e.g., find μ, σ for normal distribution that describes the distribution of X).

* Mixture model clustering methods, such as GMM, usually suffer from an overfitting problem.

* Clustering and other methods that are operating by using the distribution of the data are called *generative models.*

* Model based clusterings (probabilistic clustering) are computationally complex and also require the user to assume the underlying model, which is usually not known in advance.

Clustering Result Evaluation

We have learned well-known clustering methods in the previous pages, and it is our greatest pleasure to announce that you have received an Honorary Ph.D. in clustering from the prestigious Institute of Nowhere University of Technology.

We ran a clustering algorithm on our dataset and got some clusters as a result. We also need to evaluate the quality of generated clusters and see whether our result is valid. Determining the number of clusters is crucial. We have learned that most of the clustering algorithms require us to provide a number of clusters we would expect to have at the end, or if they don't, such as hierarchical clustering, they need another parameter to be given by the user (e.g., the branch level to decide in hierarchical clustering). Moreover, knowing the correct number of clusters is essential for the granularity of clustering results. The second challenge is to determine which clustering algorithms work better for our dataset. We can resolve or mitigate these challenges by using cluster evaluation methods and experimenting with different clustering algorithms.

Quantitative or extrinsic (not subjective and not related to human annotations) and some qualitative or intrinsic (subjective, your or other human annotations will be involved) methods used for the evaluation of clustering results. Intrinsic methods focus on reporting how well clusters are separated. Extrinsic methods use a ground truth dataset (human-labeled dataset) to evaluate the correctness of clusters. However, we usually do not use extrinsic methods because building a ground truth dataset needs manual labels, and we use clustering (unsupervised

Quantitaitve Evaluation

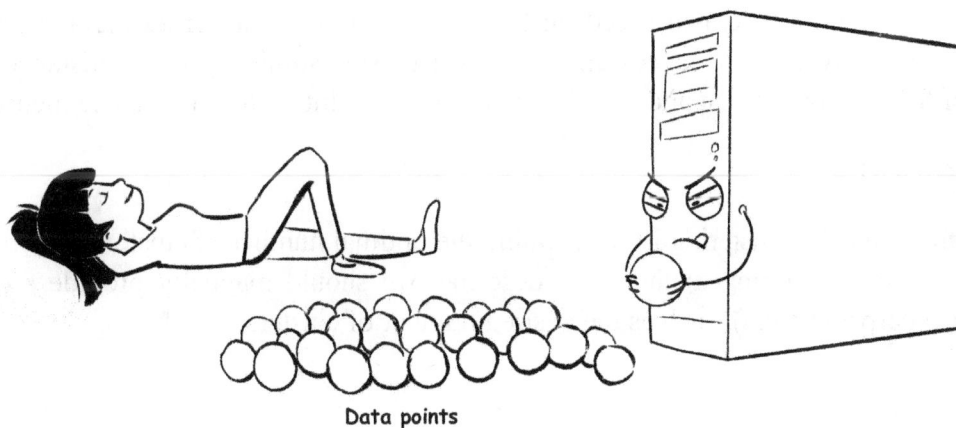

Data points

learning) to eliminate the process of manually labeling data. Note that there are no optimal methods, and all these evaluation methods have their shortcomings. We describe some of the common ones in this section.

Intrinsic Cluster Evaluation Methods

Intrinsic methods are used to measure how tight clusters are. The optimal case has the smallest distance between cluster members, i.e., internal distance, and the largest distance between clusters, i.e., the external distance.

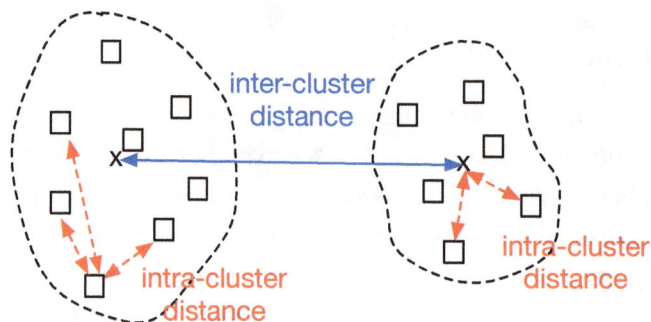

Figure 4-34: Inter-cluster distance between two clusters and two different forms of intra-cluster calculation. The red arrows on the left are the center of the cluster, calculating all distances of all data points from each other and averaging them. The red arrows on the right calculate the distances of all data points to the centroid and average them. This means we use two different methods to calculate inter-cluster distance.

Before describing Intrinsic methods, check Figure 4-34, which visualizes and describes the inter-cluster and intra-cluster distances. These terms are used to define cluster evaluation metrics.

Intra-cluster distance is the average distance of each node inside the cluster to the center of the cluster. It could also be calculated as the average distance between each node. *Inter-cluster distance* is the distance between the center point of two clusters. This center point could also be calculated with different methods, such as mean, median, etc.

All intrinsic methods are operated based on the fact that a good clustering method has a high intra-cluster similarity (low distance) and low inter-cluster similarity (high distance). In the remainder of this section, we list and describe some common intra-cluster distance metrics.

Elbow Method

One of the challenges of clustering is determining the optimal number of clusters. You remember that in most of the clustering methods we describe, we should manually provide some input parameters (hyperparameters), such as the desired number of clusters.

198

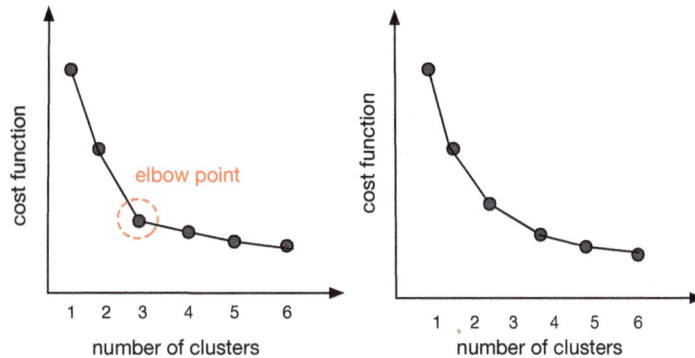

Figure 4-35: Elbow method example, (left) there is an elbow and can conclude three is the number of optimal clusters. (Right) There is no elbow, and thus, we cannot use the elbow method to determine the optimal number of clusters.

There is no ultimate solution for identifying the optimal number of clusters. The Elbow can be used to determine the optimal number of clusters. It is mainly used for partition-based clustering methods, e.g., k-means.

The elbow method presents the relation between a "cost function" and a "number of clusters", as shown in Figure 4-35. The point that constructs the elbow in the shape of a function presents the optimal number of clusters; see the left side of Figure 4-35. However, sometimes there could be no elbow, such as on the right side of Figure 4-35. In these cases, we cannot use the elbow method, and we recommend checking for another method.

The Elbow method often uses the *F-test* as a cost function, which is the percentage of variance. You don't need to remember the F-test, and we will explain later about the cost function. Usually, F-test is explained as the ratio of the between-group variance to the total variance. In simple words, first, the sum of centroids of each cluster will be calculated. Then, the sum of all squared distances will be used to present our cost function.

Within Cluster Sum of Squares Errors (WSS)

In Chapter 1, we have said *Error = Actual - Predicted*. The Sum of Squares Error (SSE) is a sum of the squared differences between each data point and its cluster's mean, as it has been shown in the following equation.

The SSE gets near zero as elements of a cluster get closer to the center of the cluster. Therefore, we can test a clustering algorithm with different parameters. The one that produces the lowest SSE is the best one. Assuming we have three clusters, we are looking for the following result: $SSE_{total} = SSE_{c1} + SSE_{c2} + SSE_{c3}$. We can use WSS as a cost function for the elbow method and use the elbow method to identify the optimal number of clusters, i.e., where the elbow happens is the optimal number.

Silhouette Index

Silhouette coefficient [Rousseeuw '87] is a score range from -1 to +1. A value of +1 indicates that the data point is very well matched to its own cluster and poorly matched to neighboring clusters. A value of 0 indicates that the data point is on or very close to the decision boundary between two neighboring clusters. It measures how similar a data point (member of a cluster) is to its own cluster, i.e., *cohesion*, compared to other clusters, i.e., *separation*.

Note that we can judge silhouettes by observing all data points and not a single data point. To calculate the Silhouette index, we first need to calculate the Silhouette coefficient. The silhouette coefficient for data point i is calculated based on two things $a(i)$ and $b(i)$.

$a(i)$ is the *average distance between member i and other members of the cluster* (the cluster which i belongs to it). In simple words, $a(i)$ it means how well the data point i has been related to its cluster. $b(i)$ is the *average distance of i to all data points in the nearest cluster.* It computes the average distance to all points in cluster j and the average distance to all points in cluster k and other clusters. Then it selects the smallest of these averages. Silhouette coefficient is presented with the following equation:

$$s(i) = \frac{b(i) - a(i)}{max(b(i), a(i))}$$

After identifying the Silhouette coefficient for each data point, we should calculate the average of the Silhouette score in each cluster.

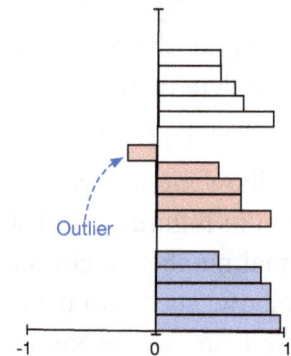

Figure 4-36: Silhouette coefficient visualization.

Next, when the Silhouette coefficient for each cluster is ready, assuming k is the number of clusters, we can easily calculate the Silhouette index for the entire dataset as follows:

$$S_n = \frac{S_1 + S_2 + \ldots + S_k}{k}$$

Usually, we can plot the output of the Silhouette coefficient with a bar chart and see which data points are properly clustered and which are not. For instance, Figure 4-36 visualizes the result of applying the Silhouette coefficient on a dataset with three clusters. Each cluster is presented with one color, and the red cluster has one outlier because its silhouette coefficient is negative.

Dunn Index

Dunn Index [Dunn '74] is another method that reports the intra-class distance between two clusters divided by the maximum value of the inter-class distance from one of those clusters. Assuming c_1, c_2 are clusters, the DI is computed as follows:

$$DI = \frac{min\ (inter\_disnace(c_1, c_2))}{max\ (intra\_distnace(c_1\ or\ c_2))}$$

A higher Dunn index reveals that members of clusters are compact and clusters are well separated from each other.

200

Davies-Bouldin Index

Davies-Bouldin index (DBI) [Davies '79] is another intrinsic cluster evaluation method, and based on our descriptions, it can be written as follows.

There are more methods of intrinsic cluster evaluation, such as Calinski-Harabasz [Caliński '74], but we do not explain all of them. Also, we briefly explained both DBI and DI, which we believe is unnecessary to memorize how they work. Moreover, we do not report the time complexity of these evaluation methods because we should do them once to decide about parameters and algorithms, and they will not be used in the target application that hosts a clustering algorithm.

Extrinsic Cluster Evaluation Methods

Extrinsic evaluation will be done when a ground truth dataset is available. We use unsupervised learning when we do not have labeled data. However, this evaluation style requires a small dataset (ground truth dataset). In the remainder of this section, we explain two common extrinsic evaluation methods here, *Purity* and *RandIndex*.

Ground truth Dataset

Data points

Purity (Entropy)

Purity is the result of comparing our algorithm result with the ground truth dataset. Assume we have n data points, and we select m data points ($m<n$) to manually label them. By labeling, we get a labeled dataset g as a ground truth dataset. It means that we have a set of clusters, e.g., $\{g_1, g_2, g_3\}$, which were created by our labels. We also manually assign each data point d_x to a cluster g_x, e.g.: $\{ <d_1, g_1> , <d_2, g_1> , <d_2, g_1> \ldots <d_m, g_3> \}$.

We have another dataset, which is the result of running our clustering algorithm on the dataset, and it has labeled them as well,   e.g.,
$\{ <d1,c1>, <d2,c2>, <d2,c2> \dots <dm,c3> \}$, c_x presents the cluster that a data point belongs. Therefore, we can have a set of clusters created by our algorithm, e.g., $\{c_1, c_2, c_3\}$. Now we can create a confusion matrix (check chapter Chapter 1, if you can't recall it) for ground truth and c (algorithm result) datasets.

		Actual		
		g_1	g_2	g_3
Clustering Results	c_1	16	8	0
	c_2	4	8	2
	c_3	10	12	6

Table 4-6: Confusion matrix example, used to compute the purity of clustering result.

Note that the number of clusters in the ground truth dataset and the clustering algorithm should be exactly the same. Assume we have 66 data points to cluster $d = 66$, and we set the number of clusters to three (both by the algorithm and ground truth dataset). Therefore, the confusion matrix is something like Table 4-6.

Now that we have the confusion matrix, we can use the following equation to calculate the purity:

$$Purity = \frac{1}{n} \sum_{i=1}^{k} max_j |c_i \cap g_j|.$$

The purity uses the confusion matrix, selects the maximum of each cluster (every row), and uses it to calculate the purity score as follows:

$$DBI = mean\ (max\ \{\frac{intra\_distance(c1) + intra\_distance(c2)}{inter\_distance(c1,c2)}\})$$

$$Purity = \frac{16 + 8 + 12}{66} = 0.54.$$

The Purity of our clustering algorithm is 0.54, which is not close to 1, and it is not good. We can compare different parameters of a clustering algorithm or different clustering algorithms based on their purity to identify the most accurate settings.

Rand Index

Rand index is another extrinsic clustering evaluation method that requires a ground truth dataset. Recall from Chapter 1 that we explained true positive, true negative, false positive, and false negative. The rand index will be calculated as follows:

$$RI = \frac{TP + TN}{TP + FP + FN + TN}$$

As you can see from its equation, the Rand Index operates based on giving weight to false positives and false negatives. However, in some clustering algorithms, it is not easy to distinguish between false positives and false negatives. Therefore, an easier approach to calculating the Rand Index is written as follows:

$$RI = \frac{a + b}{a + c + b + d}$$

In this equation, Consider C as clustering results of algorithms and G are clustering results of ground truth.

a: the number of data objects that are in the *same* clusters of C and in the *same* clusters of G

b: the number of data objects that are in *different* clusters of C and in *different* clusters of G

c: the number of data objects that are in the *same* clusters of C and in *different* clusters of G

d: the data objects that are in *different* clusters of C and in the *same* clusters of G

We could say that $a + b$ specifies the number of agreements and $c + d$ specifies the number of disagreements.

Summary

We start this chapter by introducing some similarity metrics used for clustering techniques. Some of these methods are good for numerical data, some for text, etc. Table 4-7 summarizes similarity metrics.

Then, we describe some common clustering methods and their categorizations, which are summarized in Table 4-8. In particular, there are four types of clustering algorithms, and for each category, we try to use two examples. Table 4-8 also briefly describes when to use which algorithm.

At the end, we finalize this section with a description of cluster evaluation methods, which is summarized in Figure 4-37.

Distance Method	Description
Eculidean	Lp2 norm, a type of Minkowski distance. Useful for sparse data.
Manhattan	Lp1 norm, a type of Minkowski distance. Handles outlier better than Euclidean distance.
Mahanolobois	A numerical distance measurement, which takes into account the distribution of the data as well.
Hamming	A type of edit distance, similar to Manhattan distance but for bit vectors.
Levenshtein	A type of edit distance counts the number of minimum changes required to convert a string. It's widely used in applications like spell-checking and DNA analysis.
Longest Common Sequence Distance	A type of edit distance useful for genome sequences. It counts the longest common sequence among two sequences.
Cosine	A numerical similarity measurement is very useful for comparing vectors that have different sizes, e.g., textual data.
Jaccard	Measures similarity in a binary comparison, and it is very useful for any type of information that is discretely tokenized.
Dynamic Time Warping	Good for measuring the similar between time series. Nevertheless, its original version is computationally complex, and because of its complexity, it is not easy to integrate it into a real-world product.

Table 4-7: A summary of similarity methods used for clustering.

Category	Algorithm Name	Input Parameters	Descriptions
Partition based	k-means	1- number of clusters 2- threshold for iteration (optional)	the most used clustering algorithm due to its simplicity
	k-medoids	1- number of clusters 2- threshold for iteration (optional)	very similar to k-mean but more tolerant to outliers
Density based	DBSCAN	1- epsilon 2- minimum points	can clusters data points with non-convex shape
	OPTICS	1- minimum points 2- epsilon (search distance)	OPTICS reduces the sensitivity of the epsilon parameter (in comparison to DBSCAN). It can also handle clusters with different densities.
Hierarchical	SLINK (agglomerative)	No parameter required	both of these methods are computationally inefficient, but they are baseline algorithms for hierarchical clusterings.
	DIANA (divisive)	No parameter required	For efficient hierarchical clustering, we use BIRCH and CURE.
Hierarchical for Large Dataset	BIRCH	1- CF-Tree branching factor 2- CF-Tree threshold 3- CF-Tree max. number of entries in a leaf node	very scalable and able to undo what has been done in its previous step. Sometimes CF-Tree can not fit into the memory, and we get of memory error
	CURE	1- number of clusters 2- number of representative points 3- shrink factor	useful to handle large datasets with non-convex shapes.
Probabilistic	GMM	1- number of clusters	enables soft clustering, operates based on the two steps of expectation and maximization
	Fuzzy C-Mean	1- number of clusters	enables soft clustering, operates very similar k-mean

Table 4-8: Summary of clustering methods and their characteristics.

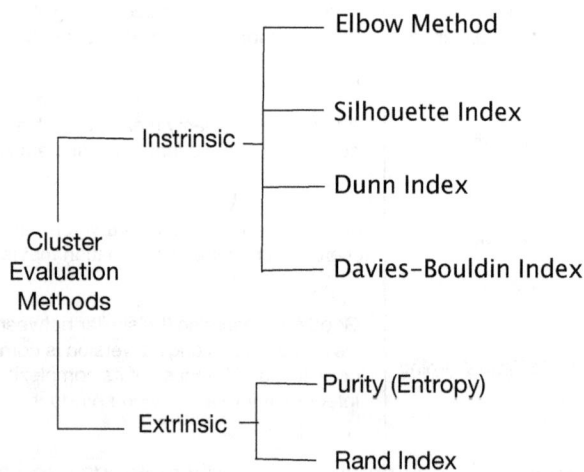

Figure 4-37: Intrinsic and extrinsic cluster evaluation methods.

We use a combination of all these described methods with visualization to identify the best combination of clustering. They are easy to learn, and we don't need to memorize their equation; knowing their name is enough and how to call them in the related programming platform.

Further Reading and Watching

* A tutorial about DTW is provided by Mueen from the University of New Mexico: http://www.cs.unm.edu/~mueen/DTW.pdf

* If you are interested in delving very deep into similarity metrics and measurements, Deza and Deza provide a good resource [Deza '09].

* A very good online interactive visualization for DBSCAN is provided by Naftali Harris, and it is available under this link: https://www.naftaliharris.com/blog/visualizing-DBSCAN-clustering

* The example of DIANA is inspired by the example described in "Finding Groups in Data" book [Kaufman '09].

* The example we used for BIRCH is inspired by the *Data Mining and Predictive Analytics* book [Larose '15]. Also, there are online videos explaining BIRCH available by Balaraman Ravindran and Jiawei Han.

* A good explanation of the Mixture Model and how we use it for clustering is provided by Fong Chun Chan under this link: http://tinyheero.github.io/2015/10/13/mixture-model.html

* The Manning et al. Introduction to Information Retrieval [Manning '10] is a very useful book about entropy and clustering accuracy comparison.

Chapter 5: Frequent Itemset, Sequence Mining, and Information Retrieval

Do you remember that in the first chapter of this book, we talked about events that occur in a sequence? No worries if you forget it; we don't remember this example, either. Therefore, we repeat the example again. Assume we have the following sequence of events for a person who has suffered from two heart attacks.

exercise, read, over-eat, smoke, stress, drink, heart-attack, stress, over-eat, read, sleep, smoke, drink, heart-attack, …

Now, imagine a friend of yours starting overeating, smoking, and drinking. Can you predict what will happen to him from the given sequence of events? Yes, unfortunately, he could potentially get a heart attack. Now, you want to create a set of devices that monitor user's habits and behavior, such as a lifelogging application that collects data to predict what might happen next.

This application requires the identification of frequent patterns that occur in the data, and capability is supported by a group of algorithms called *Frequent Itemset* or *Association Rule Mining* algorithms. To summarize, *frequent pattern mining is the process of identifying events, tasks, entities, etc., that frequently occur together*.

In simple terms, these algorithms are looking for associations between items in the dataset. We use these algorithms to answer questions such as: *"What entities are occurring frequently and appearing together?"*.

A classic example of frequent itemset mining is the supermarket purchase data. It states that if a person bought bread and butter, the next most probable item would be milk because historical records show that most people bought milk, butter, and bread together.

> You might think this chapter has nothing to do with machine learning. You should learn them all; otherwise, you will suffer in your work.

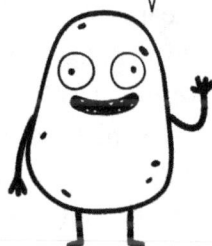

Frequent itemset mining algorithms are common approaches to extracting knowledge from cooccurring events due to their simplicity and ease of implementation. Sequence mining algorithms are useful for extracting patterns from sequential data or identifying frequently occurring subsequences in sequential data.

Despite its simplicity, searching plays an important role in these algorithms. Search space refers to the scope of the data the algorithm should search. A substantial part of this chapter is dedicated to search methods and how to improve the search space. If your main interest is in computer science, or you are willing to develop your machine learning algorithm or artificial intelligence application and can not use the existing ones, we strongly recommend that you study and learn the search methods in detail; they are more useful than you can imagine.

First, start with some concepts that are common in frequent itemset mining and sequence mining. Next, we describe frequent itemset mining followed by sequence mining. Finally, we conclude this chapter with a discussion on Hidden Markov Model (HMM).

Basic Concepts

All frequent itemset and sequence mining algorithms have *Support* and *Confidence*. To understand them, let us give another deeply philosophical example so that these two concepts can be better understood and remembered. Suppose you exercise and regularly go to a luxurious gym. More important than going to the gym for your physical health is your outfit and style inside the gym. You should feel perfect about your outfit before going to the gym. Otherwise, you can't concentrate on your exercise, and your exercise will not be pleasant.

Apart from the importance of your outfit, the following list of events frequently occurs in your gym sessions. We use an abbreviation from the alphabet to refer to them easily.

a - Wear sloppy clothes or a dirty shirt.
b - Encountering your crush in the gym training close to you.
c - Forget your headphones at home, which their colors are set with the color of your training shoes.
d - Skip cardio and go to weight training, or skip weight training and just do cardio
e - Take a selfie in front of the gym mirror

session	events
1	a, b, c, d
2	b, d
3	b, e
4	a, c, d
5	a, d
6	a, c, e
7	b, c
8	c, e
9	d, e
10	a, c

Table 5-1: Gym sessions and events

Table 5.1 presents events that have occurred in your last 10 gym sessions. Since you need to devote a significant amount of your time reading this book to learning machine learning and artificial intelligence, you do not have enough time to match your clothes colors and go to the gym always in perfect style. Therefore, sometimes you just go to your closet, grab a shirt/shorts, and go to the gym. At these times, you hope that nobody is in the gym and will be alone there. It seems that every time you do not care about your outfit (event 'a' occurs), your crush is also in the gym and watching you (event 'b' occurs). You start to get superstitious and think some events are occurring together through magic, and you have begun to doubt whether they happen by chance or whether there is a connection between the co-occurrences of these events.

To solve this issue, you start gathering data about your gym experience at each session you've led to going to the gym. Table 5-1 presents the result of your data collection and the frequency of events that have been described. To be able to analyze the dataset presented in Table 5-1, we should be familiar with the following concepts.

Transaction here means a unit that has entities or events. For instance, in the example we describe in Table 5-1. each gym session is a transaction. In this table, each record or each line is a transaction.

Itemset is a collection of one or more events, entities, data objects, or elements that appear in the dataset. For instance, in session #2 we have three different itemsets: {b}, {d}, {b,d}. In session #6, we have seven different itemsets: {a}, {c}, {e}, {a,c}, {a,e}, {c,e}, {a,c,e}.

Support specifies how frequently an item occurs in the dataset. In other words, support is the frequency of an itemset occurring in all transactions. For example, itemset {a,c,d} occurs two times in the dataset, once in session #1 and the second time in session #4. Therefore, we can write: support {a,c,d} = 2/10. Support is computed by the following equation:

$$support = \frac{number\ of\ time\ entities\ occur\ together}{number\ of\ transactions}$$

The support of A and B could be formalized as a union of A and B probabilities: $Support(A, B) = P(A \cup B)$

Confidence is a probability that presents that if an entity occurs, then another entity occurs as well. The confidence can be computed as follows:

$$confidence(A \implies B) = \frac{Number\ of\ times\ A\ and\ B\ occurs\ together}{Number\ of\ times\ A\ occurs}$$

For instance, based on Table 5-1, we can say that the confidence of event (c => e) happening is computed as follows: $confidence(c => e) = 2/6$. Confidence can be computed with respect to support as follows:

$$confidence(A \implies B) = P(A|B) = \frac{support(A \cup B)}{support(A)}$$

In other words, this equation quantifies the association between events (confidence) by counting their frequencies (support). Both confidence and support values are numbers between 0 and 1. However, sometimes, frequent-itemset mining algorithms transform the confidence and support values to percentages instead of using a value between 0 and 1.

Minimum support is a value given by the user as a hyperparameter for the algorithm. It defines a minimum number of supports that should be in a dataset to make our assumption acceptable. For instance, if we set minimum support of event {a,c,d} is equal to 3/10, then our assumption fails, but if we set it to 2/10 or 1/10, the rule or assumption is valid. In other words, it represents the minimum level of support that an itemset should have to be considered frequent. From now on, when you encounter the term min_support, we mean minimum support.

Minimum confidence defines a minimum number of confidence that a set of events should occur that the frequent itemset mining algorithm considers as frequent. This value is also configured by the user as a parameter for the algorithm. For instance, if we set the minimum of confidence to 1/6, then (c=>e) is considered a frequent pattern because:

minimum confidence = 1/6 < confidence (c => e) = 2/6.

Let's get back to our initial question: is there any magic correlation between wearing sloppy clothes (a) and encountering your crush in the gym (b)?

We choose minimum support to 20% and confidence to 20% to say if there are equal or more than 20% of co-occurrences. Based on Table 5.1. The support of events 'a' and 'b' occurring together will be as follows:

$support(a, b) = 1 \div 10 = \% 10$

$confidence(a|b) = 0.1 \div 0.5 = \% 20$

This means support does meet our assumption. Therefore, there will be no co-occurrences. You care too much about your outfit, take it easy, and enjoy your exercise in the gym, disregarding other gym members, especially your crush.

Two good books on this topic by Han et al. [Han '11], and Zaki and Meira [Zaki '14] state that association rule mining includes two major steps. The first step is finding frequent itemsets (or candidate generation), and the second step is generating associate rules from the frequent itemsets (or calculating supports and confidence values for each candidate).

Most of the search optimization algorithms will be conducted at the first step, which is the high-cost part of these algorithms. By not using a search improvement method, assuming n is the number of possible itemsets, the computational complexity of the itemset mining algorithm will be $O(2^n)$. This means it is not feasible to implement such an algorithm even on small datasets.

Temporarily, we suspend our discussion about frequent pattern mining and delve deep into search and information retrieval methods that are widely used in association rule and sequence mining algorithms.

Information Retrieval Methods

You might ask what is the relation between information retrieval and why we need to learn it for machine learning and artificial intelligence. To answer you, we recommend waiting until the end of this chapter, and you will realize why.

Suppose that we have a dataset composed of 1,000 transactions, and each transaction has 100 members (itemset). something like the data presented in Table 5-2. Assume we set minimum support to a low value, such as two percent, and we like to identify the frequent itemset.

For the first transaction, the algorithm should compare the following items together:

$\{i_1, i_2\}, \{i_1, i_3\}...\{i_1, i_{100}\}, \{i_2, i_3\}, ...$

Therefore, for a single transaction, we need $100 \times 100 = 10,000$ comparisons. Assuming we have 1000 transactions, then this number of comparisons will be extremely large. This method is called *brute force*. If we run a brute force comparison for this reasonably small dataset on an average computer, it puts it under extreme pressure, and we probably get an out-of-memory error.

In this section, we describe methods that search the data in a more subtle approach than brute force. According to Norvig and Russel [Russell '09], a good search algorithm puts aside less relevant data and focuses on a smaller set of data, which is more relevant to the question.

After Brute force search execution.

From a technical perspective, search optimization's main objective is to reduce time and space complexities. Usually, time complexity refers to the number of data objects that the algorithm must search, and space complexity refers to the amount of memory consumed as the algorithm searches the data.

First, we describe some general concepts. Then, we briefly describe three approaches that are widely used for improving the performance of an algorithm, especially in frequent itemset mining. These methods include the use of *hash data structures*, *tree structures*, *bloom filters*, *sliding windows*, and *skip list*. If you are only using machine learning algorithms to resolve your problem and are not developing an algorithm, you can skip the following text, but be sure that god will save a room in hell for you because you did not bother yourself and read our explanations.

Concepts

Similar to other sections, first, we describe some concepts, and then we move to algorithms.

Key-Value: Many search algorithms rely on the concept of *key* and *value*. *Value* is the information we seek, and the *key* is the object we use to access the saved information.

Dictionary or Catalog: Keys and values are saved in a storage called a dictionary, catalog, or *symbol table*. Two operations are important while using dictionaries: (i) searching the content of the dictionary and (ii) inserting new entries.

Inverted Index: If you look at the back of a book, you see a section called *index*. This section states which keyword appears on which pages. By looking at the index first, we can reduce our search scope to the related page instead of the entire book. Similarly, an

transaction	items
1	$i_1, i_2, \ldots, i_{100}$
2	.
3	.
⋮	.
1000	.

Table 5-2: A dataset with 1000 transactions and each transaction has 100 items.

inverted index in the context of computer science is a data structure used to store a mapping from content, such as words or numbers, to its locations in a database, or a document, or a set of documents. Table 5-3 presents a sample inverted index, but it includes more than one book.

Hash Function: It is a computationally efficient mathematical algorithm that takes input data (of any size) and converts it into a fixed-size value, typically known as a *hash value* or *hash code*. In other words, a hash function is a function that maps the original data (value) to data of a smaller size (key).

A common problem in a hash function is *collision*, which means two different keys get the same hash value. There are reverse hash functions, which decode the hash value and give back the original value of the data, but it is not always possible with all hash algorithms.

A good hash function should reduce the chance of collision as much as possible. It is a very fast and

Keyword	Text Location
weight lose	Book 1, pages 34-37 Book 1, pages 123-124 Book 3, pages 53-59
cardio	Book 1, pages 13-18 Book 2, pages 9-19 Book 6, pages 90-92
bench press	Book 3, pages 101-102
. . .	. . .

Table 5-3: An inverted index based on keyword and related paragraphs inside all books.

memory efficient way to transform data with $O(1)$ or $O(n)$ computational complexity.

Hash Structures

In the following, we describe some hash data structures that are useful to design a search algorithm.

Hash Table

In the context of information retrieval, the hash value is often used in data structures like *hash tables* to index data, enabling the algorithm to store or retrieve the data efficiently. The hash table is a key-value storage (hash key and hash value) to facilitate access to the table values. In other words, a hash table is a data structure that efficiently maps keys to values using a hash function. It offers fast search, insertion, and deletion operations.

We explain the inverted index with examples. Assume you have created a robot with a language model that can coach you in the gym. There was an odd object before the smartphone era called "book", and people read that instead of playing with their phones. You purchase many books about health and fitness and ask your robot to learn them. When you ask your robot some related question, it should search the content of those books and retrieve information to answer you. Assume there are 100 books, and when you ask a question, the robot should find the answer from those textbooks in its memory. Nevertheless, existing large language models do not have all the information you need, and using a brute force method to search the content of these books will take a huge amount of time.

To mitigate this issue, we use an *index*. By indexing the data, we can retrieve information efficiently. An index is a data structure that improves the speed of data retrieval operations by organizing data in a searchable format.

We can build an inverted index for our robot's memory to store books' content to enable fast access to the related information. For example, we could build an inverted index like the one presented in Table 3-5. Of course, we will use algorithms to build this index (not manually). We will explain some keyword extraction algorithms in Chapter 6.

To better understand the impact of the index, let's assume you ask your robot the following questions: *"I feel I need to lose some weight; what type of exercise do you recommend me to perform? Is Cardio good for me?"*. The robot understands that you have asked about two keywords: (i) weight loss and (ii) cardio exercise. Therefore, it searches for information about both of these two keywords. Assume the robot constructs its knowledge based on the information from Table 5.3. Using this table, the robot can find the information in books and read them as the answer to the user's query. In this example, the index is the keyword and the data are locations of the text in the book.

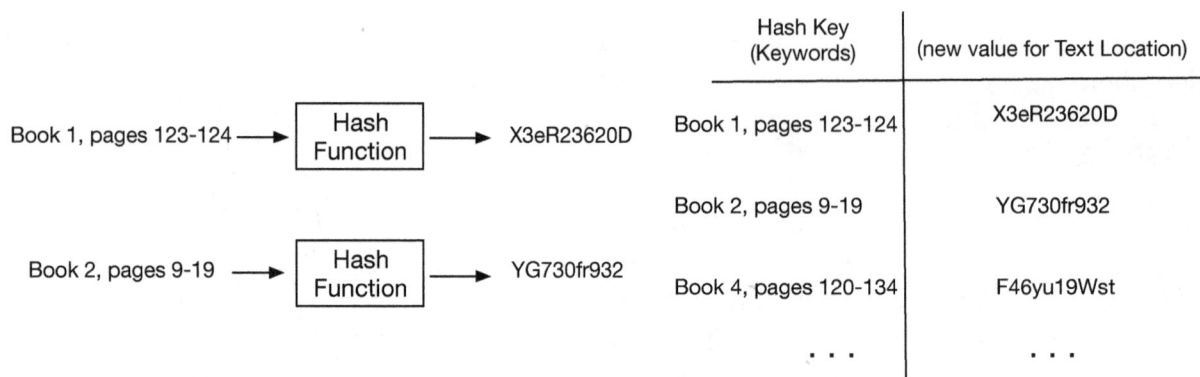

Hash Key (Keywords)	(new value for Text Location)
Book 1, pages 123-124	X3eR23620D
Book 2, pages 9-19	YG730fr932
Book 4, pages 120-134	F46yu19Wst
. . .	. . .

Figure 5-1: Hash function converting the given strings into fixed size string.

Table 5-4: An index table based on keyword and hash values.

Nevertheless, our robot cannot access this information very fast in its small brain, and we can use a hash function to convert some text information into a smaller text. For instance, we will use the hash function (See Figure 5-1) on the text locations and get a set of single strings with identical size. These strings are smaller than the original text, and thus, they can be stored in the robot's memory (See Table 5-4).

Hash tables are composed of keys and values. In our example, text locations have been converted to hash keys, and keywords have been converted to hash values. We use this example to show that the hash key is not necessarily always the index. It could be used for encrypting information as well.

In this example, a hash table is used to compress the data, but a hash table is more than just compression. It also enables the system to access the data easily. McDowell [McDowell '16] defines a hash table as follows: *"a data structure that associates a key to a value to enable the algorithm very quickly lookup for the original data"*.

Reading a record from the hash table, adding a new record into a hash table, or updating an existing record in the table all have $O(1)$ complexity; in worst cases, when our hash function is not good, it has $O(n)$ complexity.

Minwise Independent Permutations Hashing (MinHash)

Some forms of hash functions are popular while working with machine learning and information retrieval problems. Minwise Independent Permutations Hashing (MinHash or MinHashing) [Broder '97] is a popular algorithm for efficiently estimating similar sets of information objects. An information object could be an image, audio, video, time series (audio signal, heart rate), or anything.

Minhashing converts any information object into a set or vector. Then, it uses a Jaccard index to identify their similarities. We do hope you recall from Chapter 4 that Jaccard similarity is used to quantify the similarity between two sets of information, and it is written as:

$$J(S_1, S_2) = \frac{S_1 \cap S_2}{S_1 \cup S_2}$$

The process of changing the information from its original format to another format is called *representation,* and we will use it a lot in machine learning. The Minhashing algorithm changes the representation of information objects into vectors (signature vectors).

Minhash is using Jaccard efficiently. It maps each set of information into *signature vectors,* which are vectors of fixed length. We will explain these vectors shortly. Using signature vectors enables the Jaccard to identify how similar the two sets are without enumerating each of their elements.

Gym Member	Exercises
Member 1	bench press, leg press, squat
Member 2	bicep curls, over head press
Member 3	bicep curls, tricep curls, over head press
Member 4	bench press, squat

	Member 1	Member 2	Member 3	Member 4
bench press	1	0	0	1
leg press	1	0	0	0
squat	1	0	0	1
bicep curls	0	1	1	0
tricep curls	0	0	1	0
over head press	0	1	1	0

Table 5-5: (Left) Sample gym members with their training routine, (Right) the same data but presented as a sparse matrix.

To implement MinHash, we explain it step by step, to makes it easy to understand.

Step 0: The step is preprocessing, which converts our dataset into a boolean (0 or 1) sparse matrix. A matrix that includes lots of zeros is called a *sparse matrix.*

For example, assume our robot is willing to find us a training partner who will always come to the gym on time and train with us. Such a fictional character has never been encountered in the author's life, but let's live in our imagination and assume the robot will find it for us. There are three gym members who have the same interest, and their training plans for each session are listed on the left table in Table 5-5. To build the sparse matrix, the right side table is reconstructed from the data in the left side table. This example is very simple: instead of users, we can have documents, and instead of fitness activities, we could have words or n-grams (wait until Chapter 6). Instead of users, we can have patients, and instead of fitness activities, we can have side effects. There are many ways to apply MinHashing and benefit from its performance.

215

	C_1	C_2	C_3	C_4
a	1	0	0	1
b	1	0	0	0
c	1	0	0	1
d	0	1	1	0
e	0	0	1	0
f	0	1	1	0

permutation 1 →

	C_1	C_2	C_3	C_4
f	0	1	1	0
d	0	1	1	0
c	1	0	0	1
e	0	0	1	0
b	1	0	0	0
a	1	0	0	1

Hash result for each set
$H_1(C_1) = \{c\}$
$H_1(C_2) = \{f\}$
$H_1(C_3) = \{f\}$
$H_1(C_4) = \{c\}$

permutation 2 →

	C_1	C_2	C_3	C_4
c	1	0	0	1
d	0	1	1	0
a	1	0	0	1
e	0	0	1	0
b	1	0	0	0
f	0	1	1	0

Hash result for each set
$H_2(C_1) = \{c\}$
$H_2(C_2) = \{d\}$
$H_2(C_3) = \{d\}$
$H_2(C_4) = \{c\}$

permutation 3 →

	C_1	C_2	C_3	C_4
a	1	0	0	1
c	1	0	0	1
d	0	1	1	0
f	0	1	1	0
b	1	0	0	0
e	0	0	1	0

Hash result for each set
$H_3(C_1) = \{a\}$
$H_3(C_2) = \{d\}$
$H_3(C_3) = \{d\}$
$H_3(C_4) = \{a\}$

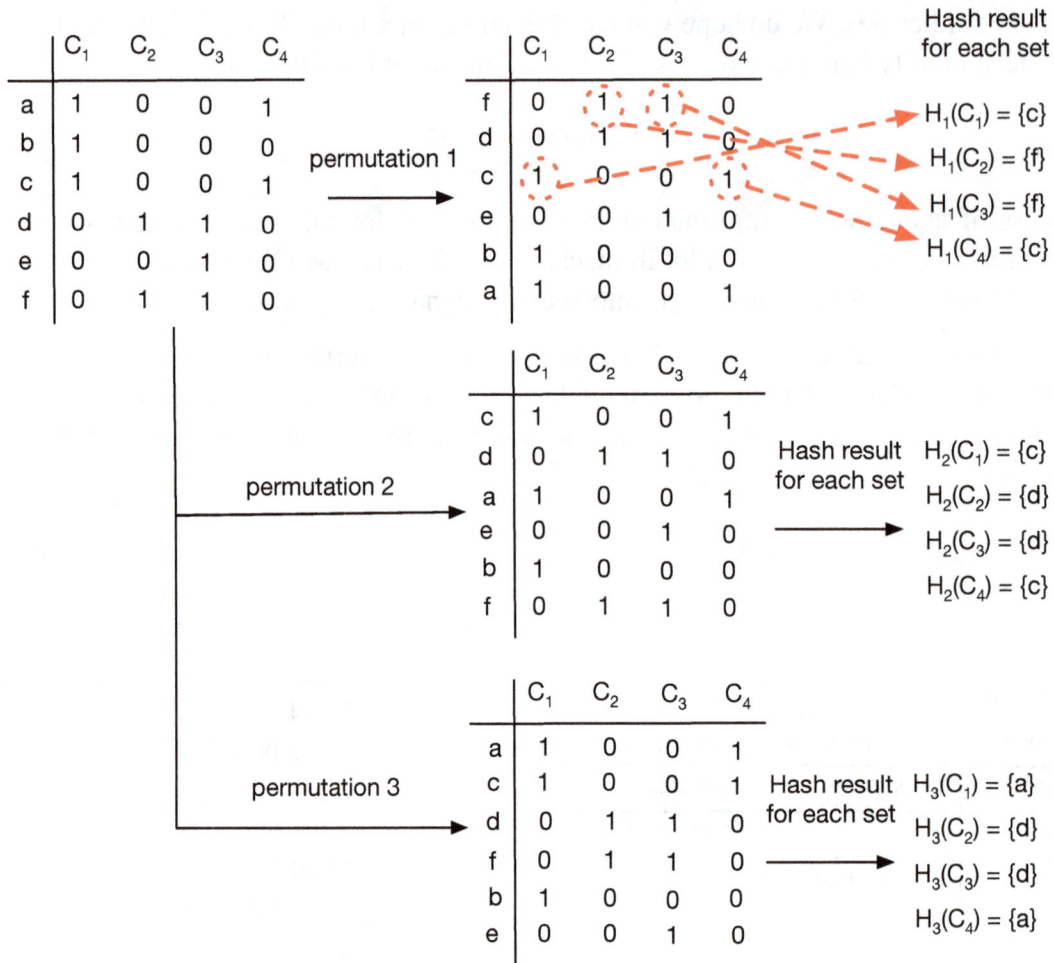

Figure 5-2: (Left) Simple representation of the right side of Table 5-5, (Right) three different permutations of the left side table rows. You can observe the minHash (H function) for each set. Then their hash value will be calculated as the row name of the first non-zero value that they encounter.

Step 1: For the sake of simplicity, we rename columns as C with an index and rows as alphabets ($a,b,c,...$) in the matrix presented on the right side of Table 5-5. Then, the MinHash algorithm randomly permutes (shuffles) rows and generates several matrices, as they are shown on the right side of Figure 5-2. In this example, we permuted the table three times (this number has been chosen arbitrarily), but based on our signature vector size and dataset size, we can create more permutations.

Now, the MinHash function for permutation i is presented as $H_i(C_j)$. J is the column name (gym member). The Hash value is the row name (fitness activity) which has the first non-zero value in the table. Take a look at the red arrows in Figure 5-2. In the table 'permutation 1', for the column C_1, the first non-zero row name is 'c'. Therefore, we will have $H_1(C_1) = c$. In the table 'permutation 2', for the column C_2 the first non-zero row name is 'd'. Therefore, we will have $H_2(C_2) = d$, and so forth.

Step 2: Based on the hash values for three permutations in Figure 5-2, it computes the signature vector $S(.)$, as a set of its hash values. In this example, we have the following signature vectors:

$$S(C_1) = \{H_1(C_1), H_2(C_1), H_3(C_1)\} = \{c, c, a\}$$
$$S(C_2) = \{H_1(C_2), H_2(C_2), H_3(C_2)\} = \{f, d, d\}$$
$$S(C_3) = \{H_1(C_3), H_2(C_3), H_3(C_3)\} = \{f, d, d\}$$
$$S(C_4) = \{H_1(C_4), H_2(C_4), H_3(C_4)\} = \{c, c, a\}$$

In other words, we have used a hash function $H(.)$ that converts each set gym member into a signature vector (here is a single row name or fitness activity) in each permutation. The signature vector is the set of hash values for each permutation, but it is smaller than the original data, and thus, it occupies less memory.

Step 3: The Jaccard function can compare two signature vectors, instead of comparing the original data. For example, for the similarity between C_1 and C_2, or the similarity between C_2 and C_3, we have the following:

$$J(C_1, C_2) = \frac{\{c, c, a\} \cap \{f, d, d\}}{\{c, c, a\} \cup \{f, d, d\}} = Null$$

$$J(C_2, C_3) = \frac{\{f, d, d\} \cap \{f, d, d\}}{\{f, d, d\} \cup \{f, d, d\}} = 1$$

We can conclude that member 1 (C_1) and member 2 (C_2) have no similarity, and member 2 (C_2) and member 3 (C_3) are very similar.

In this example, the number of data is too small, but in real-world scenarios, each dataset has many members, and if we intend to compare the contents instead of the signature, it does not fit into the memory. Therefore, MinHashing is very useful for large information searches.

Tree Data Structure

A tree is a hierarchical data structure composed of nodes connected by edges. It features a single root node at the top, and each node may have zero or more child nodes, forming a branching structure without cycles (see Figure 5-3). A tree has depth, and it refers to the number of edges in the longest path from the root node to the furthest leaf node.

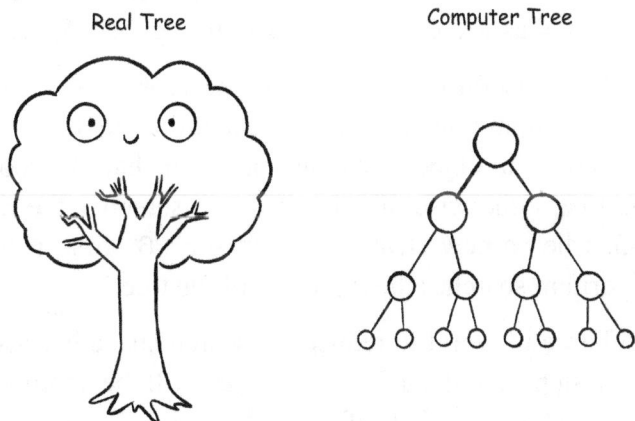

Real Tree   Computer Tree

Unlike general graphs, which can contain loops and multiple paths between nodes, a tree is a type of acyclic graph with a unique path between any two nodes. This property makes trees especially useful for efficiently organizing, indexing, and searching data. They are widely employed in various algorithms, including those in machine learning.

Figure 5-3 presents an example that demonstrates how trees are useful in reducing the search space. Assume we intend to search for the data object #8. Without an index, the algorithm should start scanning from #1, #2, #3... until it reaches #8. Nevertheless, tree-indexing can be used by scanning 'a', 'c', and 'f' branches. From there, it scans #7 and #8, which, in total, only four scans are required to get there.

Indexes are widely used to facilitate data access in databases. Most indices were implemented through trees, but there are other forms as well. We will explain tree indexing algorithms here to give you a sense of comparison between different types of trees and their use in search space reduction.

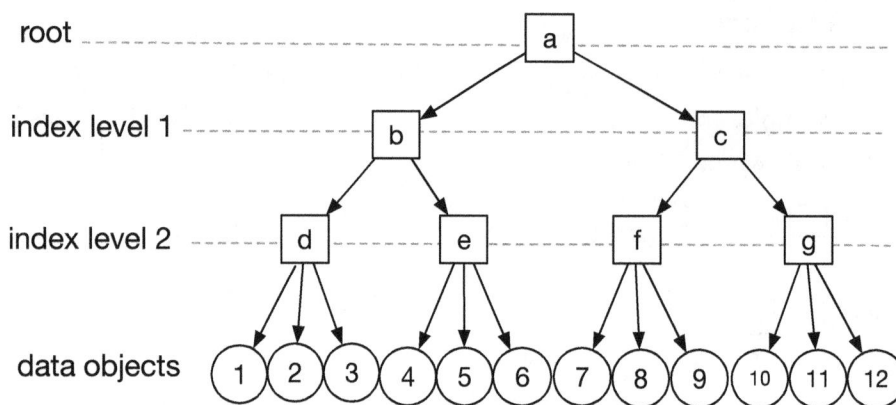

Figure 5-3: Example of indexing 12 data points with tree.

Binary Tree

When each root node in the tree has only two child nodes, it is called a binary tree. The example provided in Figure 5-4 is not a binary tree because the first root node includes three child nodes. The Binary tree is used for indexing and facilitating search. On the other hand, examples provided in Figure 5-5 and Figure 5-6 are both binary trees.

Binary search tree (BST): It is a type of binary tree in which all node values on its left should be smaller than the nodes on the right side, *all left nodes' values < all right nodes' values.* The left and right sides of the BST are BST as well. All nodes are stored as key/value pairs, and keys are presented as values on nodes, e.g., in Figure 5-5, the root node key is 4.

BSTs are useful when the data objects (or keys) have a sort of order inside the tree because when the algorithm intends to compare a new data object with the content of the tree, it checks whether the new data object is smaller or larger than the root. If it is smaller, it searches the left side of the tree. Otherwise, if it is greater, it searches the right side of the tree. For example, in Figure 5-5, when a new element arrives, e.g., '6', first, it is compared to '4', and since 6 > 4, then the algorithm searches the right side of the tree.

BST tree is useful to reduce the search space because as we go down deep into tree branches in each step, about half of the nodes will be removed from the scanning listing (if the tree is balanced, exactly half of the nodes will be removed). For example, if we intend to search for a

218

position of '6' and in the following set: {4,3,2,6,1,5,9}, with a brute force search, we should pass 4,3,2 and then reach to '6', but by using a tree, we pass '4', and then we reach to the right side of the tree, and the first item is '6' (see Figure 5-5).

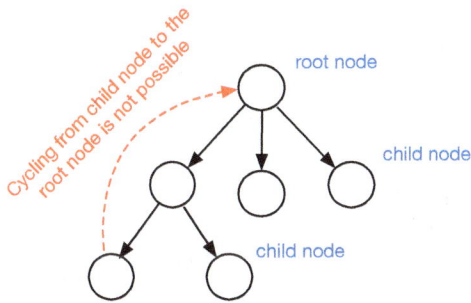

Figure 5-4: Tree structure and its elements. Note that, unlike graph, cycling is not possible in tree.

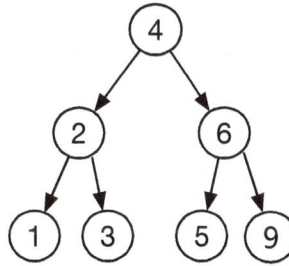

Figure 5-5: An example of balanced binary tree, which is also balanced.

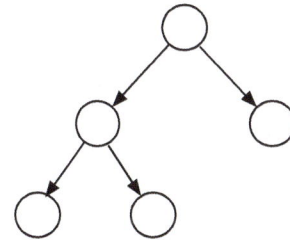

Figure 5-6: Unbalanced binary tree.

Searching information in a BST usually takes *O (log n)*. Nevertheless, inserting a new key (data object) inside a BST that is not balanced is $O(n)$, and when the tree is empty, it is $O(1)$, because the algorithm can navigate through a correct branch to fit the new key into its correct position. Therefore, it is recommended to use *BST when the data structure is static and not changing*. For instance, data streams usually have a dynamic structure (they are rapidly changing) and are not static. They allow for fast lookup, addition, and removal of items and can maintain order when the tree is balanced. On the other hand, BSTs can become inefficient if the data is inserted in a sorted manner, which can result in an unbalanced tree.

Balanced and Full Binary Tree: A full binary tree is a binary tree in which each node of the tree has either zero or two children (both Figure 5-5 and Figure 5-6 are full binary trees).

A balanced binary tree is typically defined as a binary tree in which the height (or depth) of the two subtrees of every node differ by no more than one. Figure 5-5 is a perfectly balanced binary tree; here, we have two branch depths on the left and two branch depths on the right. Figure 5-6 is an example of an unbalanced binary tree because the depth on the left and right sides are not equal.

All balanced binary trees are BSTs, but not all BSTs are balanced. Balanced binary trees are rebalanced after each insertion or deletion to the tree. The advantage of a balanced binary tree over BST is that their search and insert costs are guaranteed that, in the worst case, they will be $O(log\ n)$.

2-3 Search Tree

It is not trivial to construct a balanced tree. A BST can easily become unbalanced with some skewed distribution insertions.

By enabling each node to hold more than one key (see Figure 5-7), we can create a *2-3 search tree.* When there are *n* keys in a node, it always has *n* + 1 children. In Figure 5-7, {B, C} node has three children, and if we continue drawing the children for {J, K, L} node, it should have four leaf nodes.

This tree still keeps tree nodes at a balanced depth. Besides, keys are ordered. Similar to BST, a 2-3 search tree is guaranteed to have search, insert, and even delete costs all to be *O(n log n)*.

It is more efficient than BST, because by storing multiple keys in a single node, the tree preserves more space compared to BSTs. Besides, with multiple keys per node, fewer comparisons might be needed during the search, potentially leading to faster lookups.

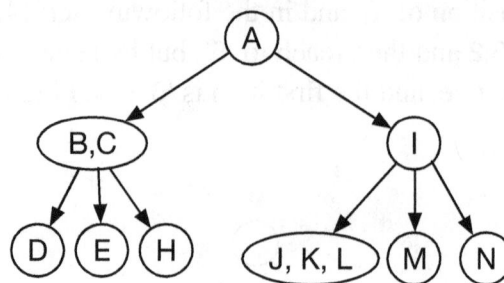

Figure 5-7: 2-3 search tree, which some node hosts more than one key.

B-Tree and B+Tree

B-Tree and B+Tree are generalizations of 2-3 search trees. There are two of the most common indexing approaches used by relational databases (SQL scripts) for accessing the data and reducing the search space in relational databases and file systems [Comer '79]. There is no explanation of the term B in the original paper [Bayer '70], but some speculation assumes it stays for Balanced.

B-Tree and B+Tree have a *B* (*branching factor* or *minimum degree*) parameter that follows rules four rules:

(i) *B* $\leq$ children per non-root node *<2B*

(ii) *B-1* $\leq$ keys per node *< 2B -1*

(iii) All node keys in B-tree are sorted in increasing order

(iv) All leaves appear at the same level, which is the characteristic of balanced trees.

B+Tree is an improvement of B-Tree. While B-Trees store keys and data in every node, B+Trees store keys in internal nodes and the actual data only in leaf nodes. Also, in B+Trees, leaf nodes are often linked to each other in a linked list, facilitating efficient range queries. If you are not familiar with Linked List, there is no need to learn it unless you give a job interview on the algorithm design.

The computational complexity of B-Tree and B+Tree for search, insert, and delete operations is *O(n log)*, where n is the total number of elements in the tree. The main difference between B+Tree and B-Tree is the range query. The computational complexity of the range query in B+Tree is *O(log n + m)*, where *n* is the number of entries in the tree and *m* is the number of elements in the range query.

Red-black Tree

A red-black tree is an approximately balanced tree. They are not absolutely balanced, but they are approximately balanced. Figure 5-8 shows an example of a red-black tree. Red-black trees have four properties:

(i) Tree nodes have a color, either black or red (a bit can be used to set the color of the node).

(ii) The root of the tree is always black.

(iii) It is not possible to have a parent node and its child node both in red (in other words, every red node must have two or null black children).

(iv) Every path from the root to a leaf node must contain the same number of black nodes (in the example presented in Figure 5-8, you can see every path has two black nodes and no more or less than two nodes). The colors of this tree can be quantified by a boolean flag.

The insert, search, and delete in the red-black tree is $O(log\ n)$, but it usually requires less memory than the balance tree. The red-black tree is useful when we intend to insert and remove data into the tree frequently.

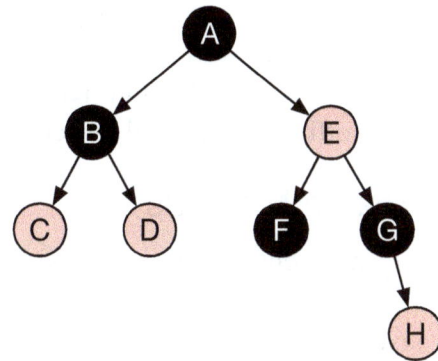

Figure 5-8: Red-Black tree example.

Trie and Radix Tree

Another widely used tree is the *trie* or *pre-fix tree*, which originates from the information retrieval community. Nodes in the trie, unlike other trees, do not hold a key. Instead, they hold a partial key, which facilitates the navigation and information access in the tree.

Trie is used to store words; the path from the root to the last child presents a word, and each node stores a character. Together, those characters build a word. It is very common for basic styles of auto-complete and spell-checking.

Trie nodes are assumed to be ordered from the root, and the root node of the trie is always empty. Figure 5-9 shows a simple example of a trie. As we add more words (or information) into a trie, its size grows, but it requires less time to find the right place for the newly inserted information because it holds more words as the tree gets bigger.

Trie can find the word in a dictionary in $O(n)$ time, assuming n is the length of words (characters of words). Besides, it can insert a word in $O(n)$ time. Building a trie at the beginning takes a significant time, but usually, it is worth the time.

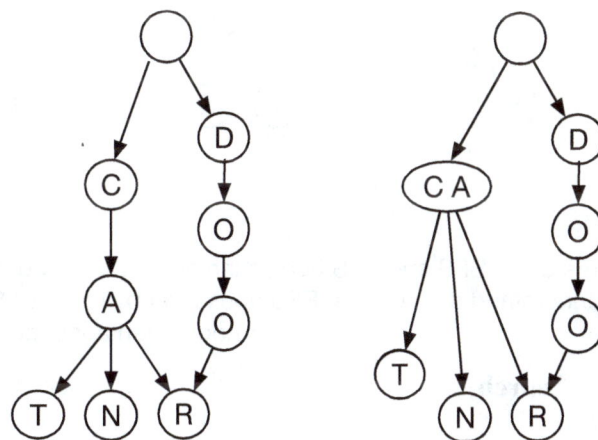

Figure 5-9: (left) Trie example, (Right) Radix tree of the same trie example.

Trie structure is not limited to words and alphabets, and we can use it for different types of data objects as well. For instance, it could be used to store addresses based on hierarchical level, e.g., starting from city name, region name, street name, house number, and apartment number (leaf nodes are apartment numbers).

There is another style of Trie, which is known as the *Radix tree*. A Radix Tree, also known as a Compact Prefix Tree, is a space-optimized version of a Trie. In a Radix Tree, any node that is the only child is merged with its parent. It means it collapses the edges and stores multiple characters in a single node. This structure reduces the number of nodes and edges in the tree, making it more space-efficient, especially when the stored strings have fewer common prefixes.

Tree Search Methods

There are different approaches to scanning the tree. Two baseline methods are Breadth-First-Search (BFS) and Depth-First-Search (DFS), which we explain here. Next, we will explain two more advanced ones, Beam Search and Monte Carlo Tree Search.

BFS and DFS

Breadth-First-Search (BFS) starts from the root and explores all children of the root first, then it goes one step further and scans all children of each child. DFS starts at the root (or another chosen node) and explores each branch as far as possible before backtracking. It means that it goes down one path (child and its descendants) as far as it can until it hits a node with no unvisited children. Then, it backtracks and continues the process with the next available path. Figure 5-10 present their differences in a simple visualization.

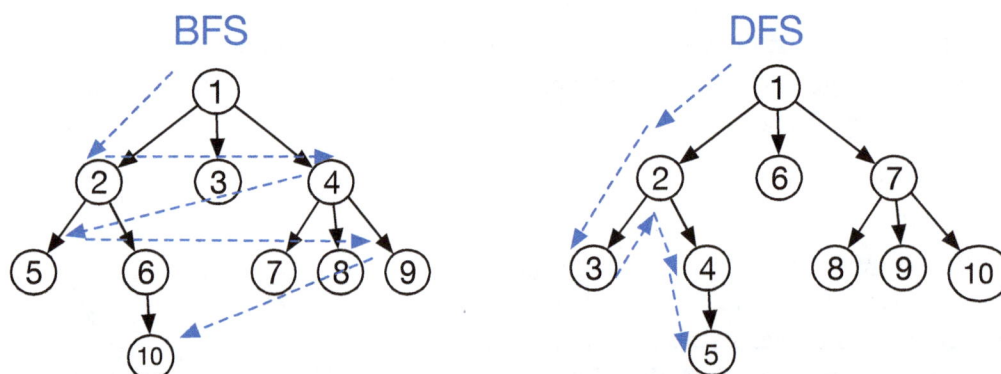

Figure 5-10: BFS and DFS tree scanning methods. Follow the numbers on the nodes and blue arrows to understand their orders. BFS goes level by level, DFS goes on node deep down and backtracks to an upper node, then again does it for another node.

Beam Search

Beam search is a tree search algorithm that is used in Natural Language Processing (NLP) models, such as sequence-to-sequence models (we will learn them in Chapter 12), which have applications in machine translation, text generation, and speech recognition. The primary function of beam search in the context of NLP applications is to generate a sequence of output tokens, which could be words or characters.

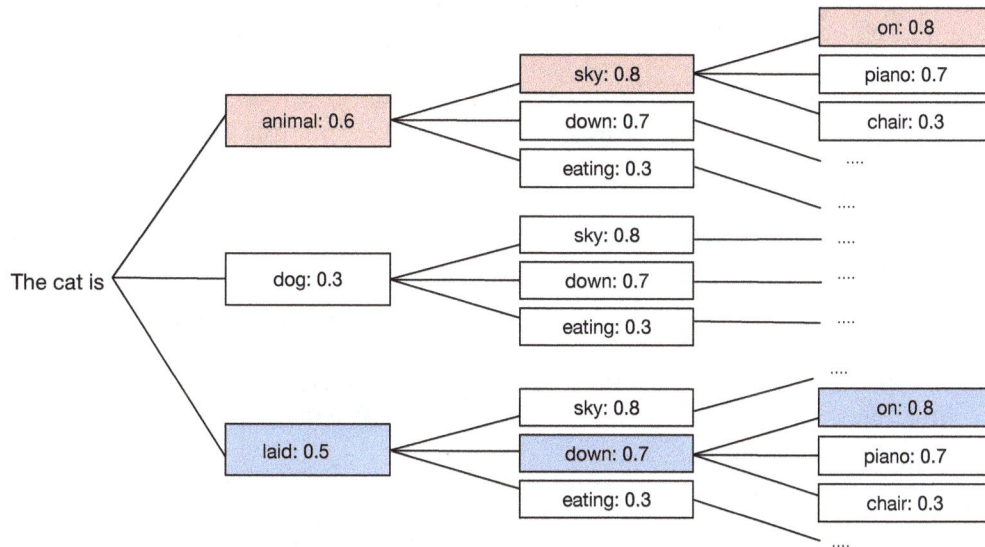

Figure 5-11: The difference between greedy search (red) and beam search (blue).

One important task of the neural network decoder (we will learn the decoder in Chapter 11) is to predict the next token. To predict the next token, the algorithm can use a greedy wise[1] approach and select the word with the highest probability. For example, let's assume the decoder tries to predict the next token for "The cat is". There are possible choices for the next toke with the following probabilities: {animal:0.6, dog:0.3, laid:0.5}. The highest probability belongs to the "animal", and the greedy wise search selects it. Now, we have the sentence: "The cat is animal...". Then, the next possible token is one of the following words, along with the probabilities: {sky:0.8, down:0.7, eating:0.3}. The greedy wise search selects the one with the highest probability (sky:0.8), and as a result, we have "the cat is animal sky...". This sentence doesn't make sense. However, "the cat laid down on..." is correct and makes more sense. The question is, how can we resolve it?

The beam search algorithm decides on the next token based on the joint probabilities of a specific number of next tokens. The beam size is the hyperparameter that specifies the number of tokens to consider together for their joint probability calculation. In other words, it broadens the width of searching for the next tokens. For example, let's assume we set the beam size to three by using beam search instead of selecting only the next highest probability token; beam search selects the three tokens with the next highest probability.

In Figure 5-11 example, the sum of the blue path (laid + down + on) probability is 2, but the sum of the red path (animal + sky+on), which is a result of the greedy wise approach, is 2.1. Therefore, the beam search selects the blue path.

---

[1] A greedy algorithm in the context of computer science always refers to short-sightedness. It means the algorithm favors the immediate highest value than checking the value after that. Think about the politician who is ruling your country; their approach is exactly a greedy and wise approach; I will fix things that are working in my term and don't give a damn about the long-term future.

With a beam search, we can have multiple candidate sequences, and they are scored based on their probabilities. The candidates with the highest probabilities will be selected.

However, this is the simplest form of beam search; in reality, beam search computes all probabilities for the top k combinations of words, and it maintains multiple candidate sequences.

Monte Carlo Tree Search

Before we start explaining this search tree, note that it includes some terms that might be unfamiliar to you, and we will explain them in Chapter 13. Therefore, feel free to skip it now and return after reading Chapter 13. In this chapter, we explain trees, and thus, we describe the Monte Carlo Search tree here.

Monte Carlo Tree Search (MCTS) [Coulom '07] is a *heuristic²* tree construction and expansion method used in decision making process. It is a method used to search the game tree to explore potential moves in a computer game. Each node in the game tree presents states in the game, and edges are actions that lead to a change in the game state. Figure 5-12 shows a small tic-tac-toe game tree. MCTS is used to teach an AI agent to play multiplayer games that are played turn by turn, such as Chess and Go. Some prominent deep reinforcement learning algorithms that play board games, including Chess, Shogi, and Go [Silver '18], use MCTS. We will explain reinforcement learning algorithms and models in Chapter 13.

MCTS constructs and expands the game tree in four iterative phases: selection, expansion, simulation, and backpropagation.

Selection: First, there is nothing more than a root node, which represents the current state of the game. Starting from the root node, the algorithm selects child nodes down the tree until it reaches a node that presents an unexplored move or until it reaches a terminal state in the game. The selection policy is often based on a balance between exploration and exploitation (we will explain them later in Chapter 13), such as the Upper Confidence Bounds (UCB), which again we will explain in Chapter 13.

Expansion: If a leaf node with unexplored moves is reached, the algorithm expands it by generating some of the possible child nodes corresponding to possible moves in the game. These

---

² A heuristic search refers to a technique that may not always find the perfect solution but is nonetheless useful for making decisions or solving problems efficiently when perfect solutions are impractical.

newly created nodes are added to the tree, and one is selected for further exploration based on the selection policy (e.g., the one with the highest UCB).

In other words, the selection process continues until a leaf node is reached. The term leaf node in the context of MCTS typically refers to a node at the edge of the current tree that has not yet been expanded. Meaning it doesn't have all possible child nodes. Then, it expands the game tree

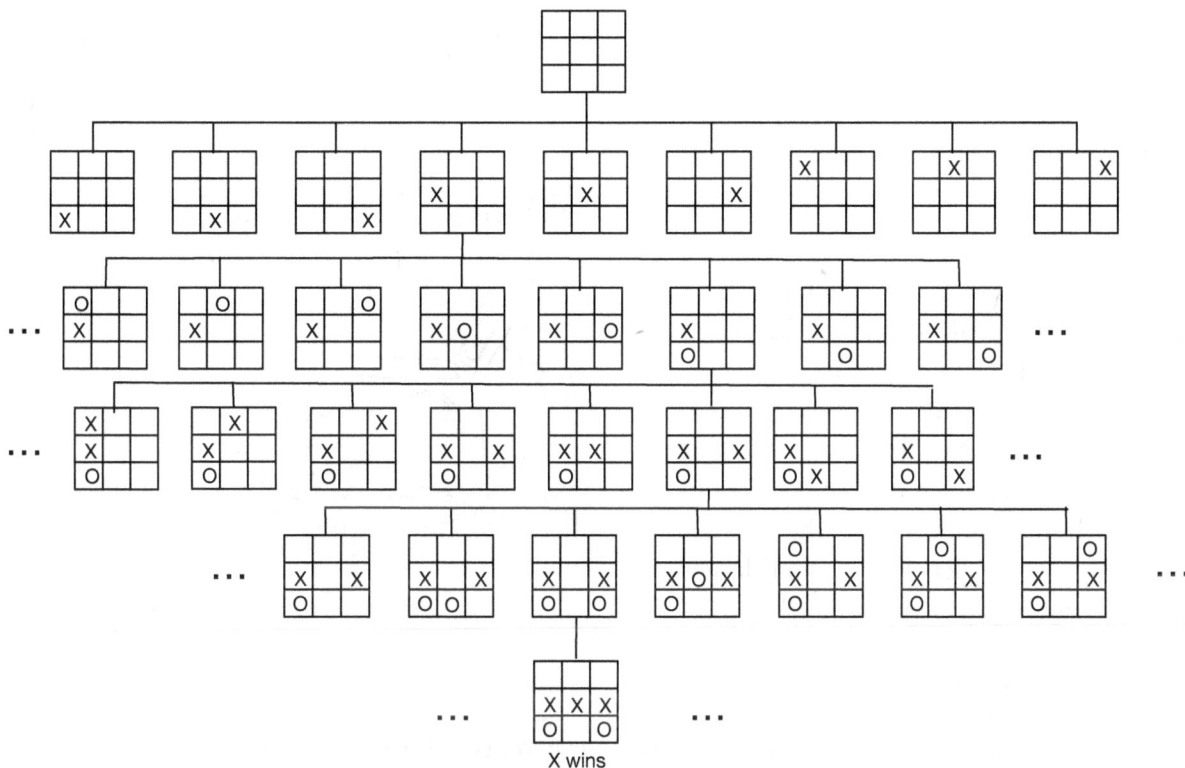

Figure 5-12: A small portion of decision tree for tic-tac-toe game is visualized. MCTS does the same, it does not construct a game tree that covers all possible states. Instead its game tree includes a limited number of possible states.

at that node by performing random sampling from the search space.

Note that the MCTS does not expand all possible leaves; it performs random sampling because it is impossible to study all possible choices. For example, even for a simple game like tic-tac-toe, we have $9! = 362,880$ possible states. Figure 5-12 does not draw the game tree completely, and MCTS does the same. It does not cover all possible game states in the tree.

Simulation (Playout or Rollout): After expansion, the algorithm performs a simulation or playout starting from the selected leaf node. The playout involves making random moves until the game or the scenario reaches a terminal state. The game's terminal state is typically a *winning, losing, or drawn state*. In other problem-solving scenarios, it could be a state where a solution is found, or a maximum number of steps is reached. The simulation step in MCTS is used to estimate the potential outcomes of different moves or decisions, which is essential for the search algorithm to identify the best options and make informed choices.

Backpropagation: Once we reach a terminal state (the playout is completed), such as losing or winning, the algorithm updates the statistics of the nodes along the path from the root to the currently selected leaf node. The statistics usually include the number of visits to the node and the accumulated values or scores of the playouts. These statistics are backpropagated up the root node of the tree, incrementing the visit count and updating the value estimates of the parent nodes.

The process of selection, expansion, playout, and backpropagation is repeated for a certain number of iterations (given as a hyperparameter) or until a time limit is reached.

There are already too many trees to learn. Why does the author explain something about reinforcement learning here?
My brain is exploding.

The result of MCTS is a tree that the algorithm can use to play a game against a human opponent or another AI agent. Increasing the number of playouts increases the quality of game playing. Therefore, a game that has different levels of difficulty can use MCTS to build different game trees. The smaller ones are for easier levels, and larger game trees belong to harder levels.

Bloom Filters

We hope you were lucky enough to live with your parents before becoming independent. Assume you live with them and have lost a pair of socks and searched the entire house to find it, but you didn't find it. One of the most known capabilities of mothers is to show us the placement of the things lost in the house. You ask your mother, who tells you where the socks could be located. We don't know the scientific explanation of how they can find it, but we speculate they use something like a bloom filter in their brain.

Bloom filter [Bloom '70] is a memory-efficient data structure that efficiently checks whether data might exist in a set or does not exist at all. However, the bloom filter can not specify the absolute position of the information we are looking for; it can only check its existence. Just checking the existence is also very helpful. Consider that your mother filters different places for your missing socks and tells you where to look for them. She told you to look inside the washing machine, bathroom, and bedroom and not inside the kitchen or living room. This reduction of search scope facilitates finding the information.

In a more technical sense, the bloom filter can be used to reduce our search space by not searching all the possible locations for the data; instead, at the beginning, it can check the existence of the search term by looking at the content of the bloom filter's bit vector.

It contains a *bit vector*, which is nothing more than a vector of bits (0 or 1). This bit vector holds information about information objects inside a set. When an information object is added to the set, this information will be converted into a bit stream, and those bits are added to the bit vector. A hash function converts the recently added information object into bits and fills its bits into the bit vector. That hash function could be any standard hash function.

Figure 5-13 shows the process of adding two keywords to our set. We can see in Figure 5-13 (a) the bit vector is empty, and no information is available in the set. Then we add the "bedroom" into the set, and the hash function sets bit numbers 3 and 14 to 1, as shown in Figure 5-13 (b). Then, we add another keyword, i.e., "kitchen", and it sets bit numbers 7 and 12 to one as well; see Figure 5-13 (c).

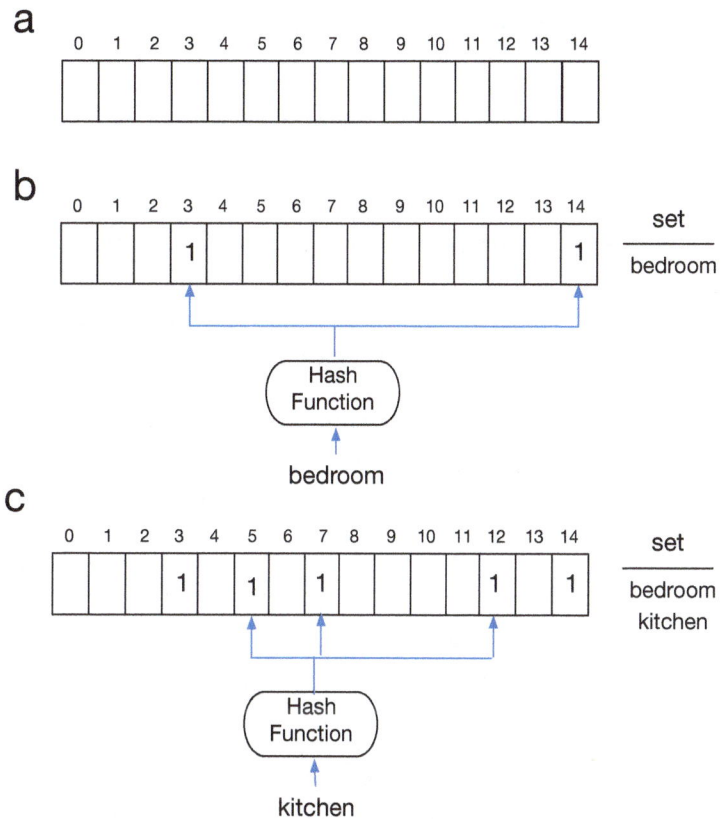

Figure 5-13: A Bloom filter example, which we add two strings to the set and it sets their related bit offset to 1.

If we want to search whether information exists in the set, we can leverage the bloom filter instead of searching inside the set, which could be a resource-consuming process. In particular, we give the search subject to the hash function; it will be converted into a bit vector; then, we can check if those bits are set to one inside a bit vector. The result of the Bloom filter check tells us that either the information objects *might* exist in the set or it does not exist in the set. For example, we intend to search for "apple", as shown in Figure 5-14. The hash function sets three bits to one for the word "apple", including bits 2, 8, and 12. Since bits 2 and 8 were not set before (they have zero value and not one), the word "apple" does not exist in the set. Then, the algorithm answers that it does not exist in the set.

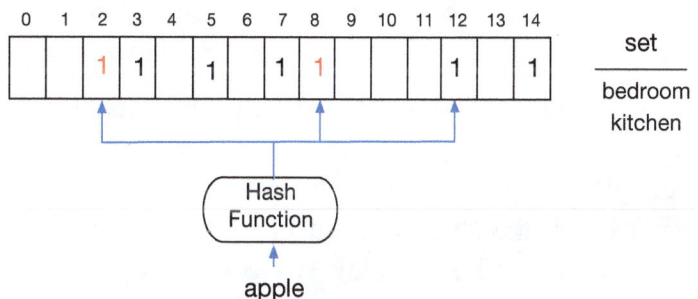

Figure 5-14: A new entry, "apple" is searched and since two of its bits does not change to one in the bloom filter, the result will be no, "apple" does not existed in the set.

Note, if all bits of the word "apple" were previously set to one, the algorithm answer is that *maybe* the word apple exists, and then it is worth going and searching inside the set with a brute force search.

The Bloom filter has many usages. For example, we intend to build a search engine for the documents stored in several different places on the company's server. Searching each particular server's disk is resource consuming, or we can use a bloom filter and first search the bloom filter. If the answer to the bloom filter is not no, then our search algorithm can search that server disk's content. We have used a combination of Bloom filter and Huffman encoding (check Chapter 14) to implement a fast natural language search on smartwatches and smartphones [Rawassizadeh '23] that resembles the Q&A of a language model.

The result of a Bloom filter is prone to false positives because it could claim an element belongs to a set while it is not in the set. This risk can be reduced by using a larger size bit vector.

Sliding Windows

The sliding window is not an algorithm but a data sampling policy to collect data from data streams, such as time series, audio, video, or other fast-producing data sources. While it is mainly used for data streams, it is not limited to data streams and could be used for other datasets as well.

We explain the importance of sliding windows with an example. Assume we have 999 records, with a long sequence of events inside each record; see the top part of Figure 5-15. We need to identify what events occur together inside transactions. In this example, we can see to compare transaction contents together, we should compare the content of txn #1 to txn #2, txn #3, ... and txn #999. Then, we should do the same for the content of txn #2 to all other transactions, etc. This is known as brute force comparison and it would be impossible without a supercomputer. One way to handle that is to compare the content of a few (let's say n number of transactions) of sequential transactions together (see Figure 5-15) and find the ones with common events. It means that we compare only a few transactions together and not all of them.

Sliding Window Types

There are three types of sliding windows: *disjoint, overlapping*, and *landmark* sliding windows. These three styles of sliding windows are shown in Figure 5-16.

Disjoint sliding windows do not share any data between consecutive windows, and windows do not overlap. Overlapping sliding windows share some data between consecutive windows. For example, the first window includes #1, #2 and #3. The second window includes #3, #4, and #5; we can see that the #3 is repeated in both windows. The landmark sliding window has a flexible window size, and the window size is dependent on a landmark data point or event. For instance, in Figure 5-16, we define encountering number 5 or 9 as a landmark, and as soon as the sliding window encounters them, it closes the landmark sliding window and opens another one. Figure 5-16 visualizes these types of sliding windows.

Brute force comparison

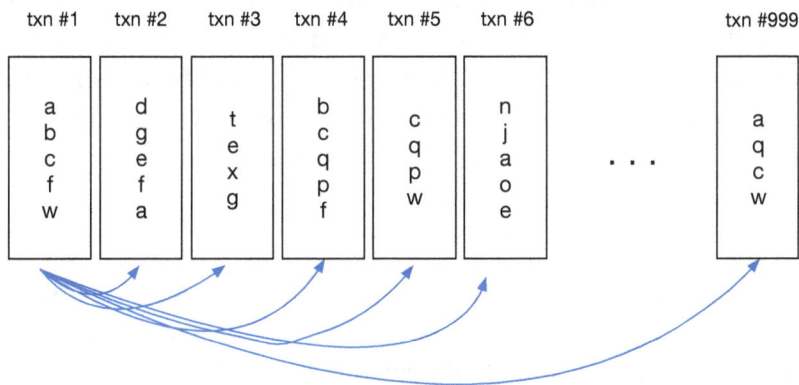

txn #1   txn #2   txn #3   txn #4   txn #5   txn #6   txn #999

a	d	t	b	c	n	. . .	a
b	g		c	q	j		q
c	e	e	q	p	a		c
f	f	x	p	w	o		w
w	a	g	f		e		

Sliding window comparison

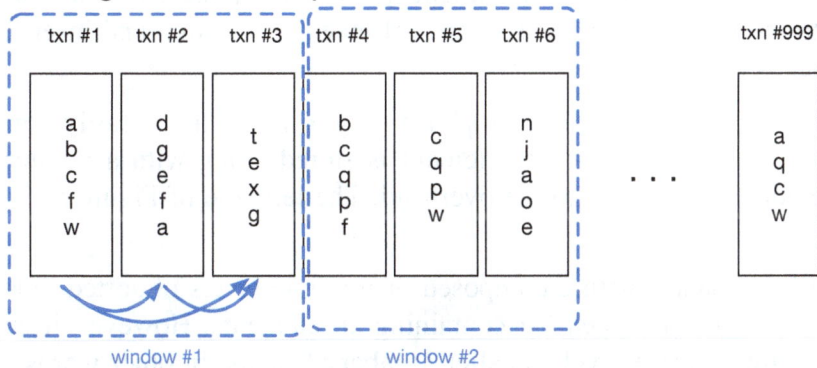

txn #1   txn #2   txn #3   txn #4   txn #5   txn #6   txn #999

a	d	t	b	c	n	. . .	a
b	g		c	q	j		q
c	e	e	q	p	a		c
f	f	x	p	w	o		w
w	a	g	f		e		

window #1   window #2

Figure 5-15: Brute force comparison versus sliding window comparison.

disjoint window

1 2 3 4 5 6 7 8 9 10

overlapping window

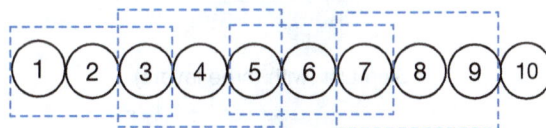

1 2 3 4 5 6 7 8 9 10

landmark window

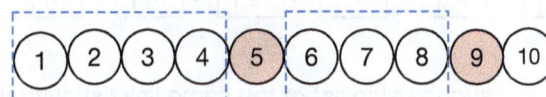

1 2 3 4 5 6 7 8 9 10

Figure 5-16: Three types of sliding windows. The landmark window stops when it encounters a particular data points.

The sliding window has one limitation: it can not be searched outside the window content. In particular, events that cooccur but stay far away from each other (do not fit into the window) will be neglected. In a research work [Rawassizadeh '16], we have created a frequent itemset mining algorithm, which is very energy efficient and lightweight and could be run on smartwatches as well. We use this algorithm to learn daily human behavioral routines, like going to work, to the gym, etc. We have used a sliding window to run the algorithm on a small device.

Skip List

A skip list is a data structure that allows for fast search, insertion, and deletion operations, similar to a balanced tree, but with simpler implementation and performance guarantees based on probabilistic balancing rather than strict structural constraints.

Before explaining skip list, we should learn the concept of a linked list. A *linked list* is a data structure in which the data are stored in a sequence, with each node containing a value and a reference (or link) to the next node in the sequence. Linked List search and insertion time both take $O(n)$ time, which is linear.

Despite its efficiency in certain operations, implementing a linked list is fairly complex, and it has a memory overhead because each data element is stored along with a pointer to the next element, and these pointers contribute to the overhead. The left side of Figure 5-17 represents a linked list.

A *skip list* [Pugh '90] is a data structure composed of multiple levels of sorted linked lists. The bottom-most layer is an ordered linked list containing all elements. However, in higher layers, we have linked lists with a progressively smaller number of items. In other words, a skip list is essentially a hierarchical linked list designed to improve the efficiency of operations compared to a standard linked list. The advantage of the skip list is that it can do a search and insertion in $O(log\ n)$ time. The right side of Figure 5-17 represents the skip list in comparison to the linked list. How do some elements stay at the bottom of the layer of the skip list, and some appear on a higher level? Elements move to top layers by random selections.

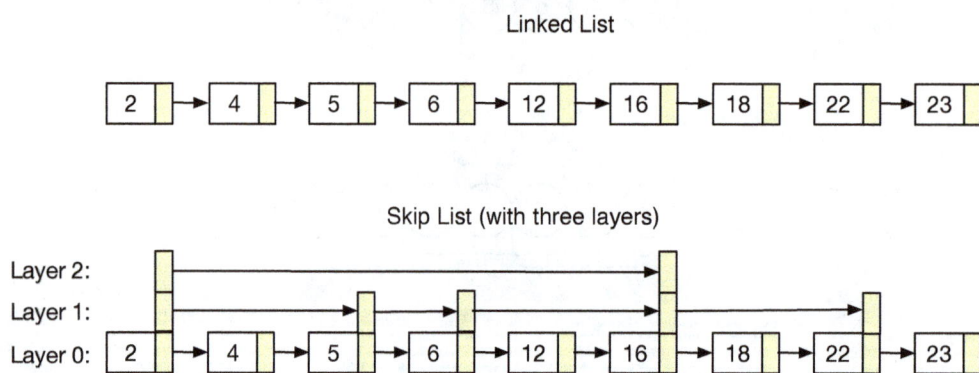

Figure 5-17: A link list on top and the skip list of the same link list, with three layers in the bottom.

To understand how a skip list makes the search for information faster, let's analyze it with an example, as shown in Figure 5-17. We have a skip list, as shown in this Figure, and we want to add the number 19 to this list. If we add 19 to the standard linked list at the top of the Figure, we need to traverse through elements 2, 4, 5, 6, 12, 16, and 18 before adding 19. This means performing seven comparisons.

However, using a skip list makes this process more efficient. We start at the topmost layer, which contains the highest-level pointers. In this example, it has two elements: 2 and 16. We first compare 19 with 2 and then with 16, making two comparisons. Since 19 is greater than 16, we move one layer down. In the second layer, we compare 19 with the next node, 22. Since 19 is less than 22, we focus on the left side and move down another layer. At the bottom layer, we compare 19 with 18 and then with 22, making two more comparisons and finally inserting 19 before 22. In total, we perform only four comparisons instead of seven, highlighting the efficiency of skip lists in reducing the number of comparisons needed for search and insertion operations by leveraging multiple levels of linked lists to "skip" over large portions of the list.

Skiplist is widely used in databases, including Casandra[3], Redis[4], and different database technologies. Casandra is a distributed database system that uses skip lists as part of its in-memory representation of data before it is written to disk. Redis is a key-value database and uses skip lists for its MemTable.

NOTES:

* Frequent itemset mining only identifies the cooccurrences of items, and it can not be used to identify causality (cause and effect). In other words, correlation is not causality.

* We wonder why many recent machine learning literature neglect these important topics. Association rule mining or frequent itemset mining has many applications in genomics and recommendation systems. Due to their simplicity, they are one of the easiest unsupervised methods we can start to test on your dataset.

* Another similar concept to the hash table is the hash map. In some programming languages, such as Java, there are small differences between hash maps and hash tables. The hash map allows us to have null values for a key, but in the hash table, we are not allowed to have null values. Nevertheless, independent of programming language, you can assume they are the same.

* Usually, for the sake of brevity, a binary search tree will be referred to as a search tree, and the term binary is dropped.

* A balanced tree assumes everything will be available in memory. Nevertheless, when the size of the dataset grows, it does not fit into the memory, and the algorithm should read data from a disk. Memory is faster and more expensive than disk but has lower storage than disk.

* Hash tables excel at exact match searches where the key must be identical to any key stored within the table to retrieve its corresponding value. Tries, on the other hand, support prefix-

---

[3] https://cassandra.apache.org

[4] https://redis.io

based searches, making them ideal for applications like autocomplete systems. Besides, trie does not need a hash function, but hash table creation usually needs one. However, the hash table is more space-efficient than a trie.

* Hash functions used for bloom filter should be fast and independent, such as MurmurHash[5] or fnv-Hash[6].

Frequent Pattern Mining Algorithms

Earlier, we briefly discussed different methods to reduce search space and time. Now, we get back to the primary motivation of this chapter and describe frequent itemset mining algorithms. Here, we list three of the baseline frequent item mining algorithms.

Apriori

It is the first itemset mining algorithm introduced by Agrawal and Srikant [Agrawal '94] and operates on transactional datasets. It uses a bottom-up approach to search for frequent itemset. The more advanced version of this algorithm creates a trie or other trees to optimize the search in the dataset to identify frequent itemsets. Nevertheless, we explain its first version, which iteratively scans the dataset. In each iteration, the algorithm builds two lists, a candidate list and a frequent list, which have a size of k (k-itemset). Then, in the next iteration, the algorithm increases the size of k and again creates both frequent and candidate lists. This process continues until none of the items in the frequent list is greater or equal to minimum support. Then, frequent itemsets that meet the minimum support condition will be separated to calculate their confidence. Next, itemsets that accept both minimum support and confidence will be returned as frequent itemsets.

We stick to our holy tradition and use an example to better understand this algorithm in detail. Assume we have a dataset with five transactions, as shown in Figure 5-18. We would like to set *min-support* to 60% (3/5) and *confidence* to 80% (4/5). Check the beginning of this chapter if you don't recall min-support and confidence.

The first step is to create the *candidate list* by counting the support (frequency) of each item in the dataset. Assuming k stays for the number of the current iteration, the result will be shown in table C_k. Then, based on minimum support, the algorithm creates a list of frequent itemset, which will be presented as table F_k. In other words, as the first step, the algorithm calculates the support (by counting their frequencies) of each single item. The result will be presented in Table C_1 of Figure 5-18. Then, each result in the candidate list (Table C_1) will be compared to minimum support. If they are equal to or greater than minimum support, we consider them in the frequent itemset list (Table F_1).

In the second step, the algorithm uses items in Table F_1, and increases itemize size by one more item, and creates Table C_2, i.e., we have {a} and {b} in F_1 and it *joins them*, i.e., {a,b} or it can be

---

5 https://en.wikipedia.org/wiki/MurmurHash
6 https://en.wikipedia.org/wiki/Fowler%E2%80%93Noll%E2%80%93Vo_hash_function

min_support = 0.6 or 3/5, min_confidence = 0.8 or 4/5

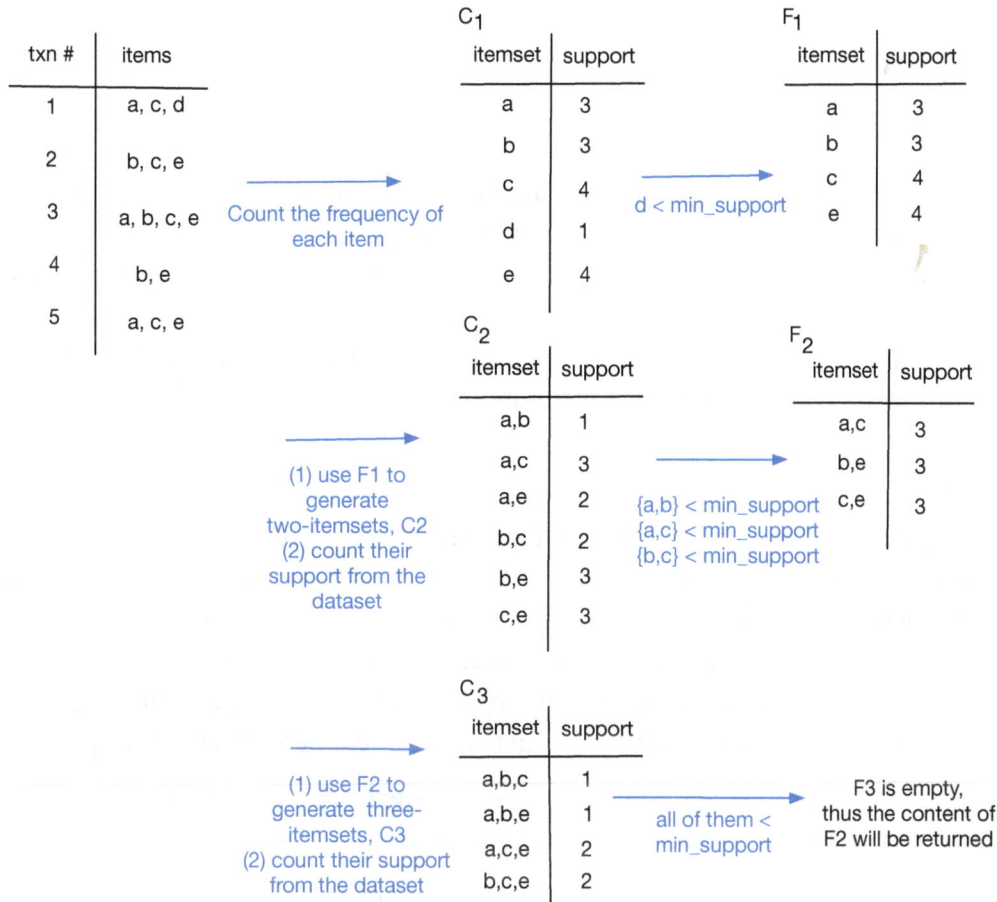

txn #	items
1	a, c, d
2	b, c, e
3	a, b, c, e
4	b, e
5	a, c, e

Count the frequency of each item →

C_1

itemset	support
a	3
b	3
c	4
d	1
e	4

d < min_support →

F_1

itemset	support
a	3
b	3
c	4
e	4

(1) use F1 to generate two-itemsets, C2
(2) count their support from the dataset →

C_2

itemset	support
a,b	1
a,c	3
a,e	2
b,c	2
b,e	3
c,e	3

{a,b} < min_support
{a,c} < min_support
{b,c} < min_support →

F_2

itemset	support
a,c	3
b,e	3
c,e	3

(1) use F2 to generate three-itemsets, C3
(2) count their support from the dataset →

C_3

itemset	support
a,b,c	1
a,b,e	1
a,c,e	2
b,c,e	2

all of them < min_support →

F3 is empty, thus the content of F2 will be returned

Figure 5-18: Apriori algorithm step by step.

written as $a \bowtie b = \{a, b\}$. Note that Apriori assumed items are stored in lexicographic order and thus it does not make {b,a}. Therefore, instead of writing $a \times b$, we should write $a \bowtie b$. Lexicographic order is just another term for alphabetical order. To prove you that we are elite and know mathematics, we show off and write it in lexicographic order.

Then, again, by comparing the content of C_2 with the minimum support the algorithm creates F_2. In the third step, the algorithm creates C_3, but all of its items are smaller than minimum support. Therefore, it can not create F_3, and thus, it assumes the content of F_2 is the final version of its frequent itemsets. In some literature, the process of creating the C tables is called the *join step,* and the process of creating F tables is called the *prune step.*

After it finds the final version of frequent item sets, it starts to calculate the confidence for these frequent itemsets, which were identified in table F_2. Recall that the confidence of (A=>B) is equal to the number of times A and B occur together, divided by the number of times A occurs. Therefore, we will have the followings:

confidence (a,c) = 3/3 = 1
confidence (c,a) = 3/4 = 0.75
confidence (b,e) = 3/3 = 1
confidence (e,b) = 3/4 = 0.75
confidence (c,e) = 3/4 = 0.75
confidence (e,c) = 3/4 = 0.75

Based on the given minimum confidence and minimum support, we get two frequent itemsets, {a,c} and {b,e}, because all other values are less than 80% (or 4/5).

The original Apriori operates without any trees, but its optimized version [Zaki '14] uses a trie to create C_k lists efficiently, then the tree will be scanned in BFS style (we have explained BFS earlier in this chapter). Another version of Apriori uses a hash set [Park '95]. We skipped describing them, but you can check related literature for it.

FP-Growth

Apriori is among the first algorithms of its kind and useful for small datasets. However, it faces two major issues. First, when the dataset size is large, it creates a very large set of a candidate list (C_1, C_2, …). Second, in every iteration, it scans the dataset to generate the frequent itemset list (F_1, F_2 , …). Therefore, it consumes too much memory, and we can't use it for large datasets. Keep in mind that one of the major challenges of data mining, machine learning, algorithm design, and everything in computer science is to manage limited memory and disk space.

A popular successor of Apriori is the *Frequent Pattern Growth (FP-Growth)* algorithm [Han '00]. It employs the imperialistic approach of divide and conquer.

Apriori

Your Computer Memory

FP-Growth algorithm operates in two major steps. The first step is to create an FP-Tree (frequent pattern tree), which is a compressed version of the dataset in a tree. This tree holds itemset association information. The second step divides the compressed dataset (FP-Tree) into a set of

(a)

#txn	items
1	a, c, d, e
2	a
3	c, d, e
4	a, b, c, e
5	b, d, e
6	a, d, e
7	a, c, d
8	d, e

(b)

item	support
a	5
~~b~~	~~2~~
c	4
d	6
e	6

(c)

item	support
e	6
d	6
a	5
c	4

Table 5-6: (a) Sample dataset with eight transactions, (b) supports of each item has been counted, (c) items which has less than minimum support has been removed.

smaller datasets. These small datasets are called *conditional base patterns*, and the FP-Growth algorithm extracts frequent itemsets from these conditional base patterns.

To learn it with an example, assume we have Table 5-6 (a), which includes eight transactions and different items inside each transaction. The minimum confidence is 50%, and the minimum support is three items (we round numbers to their nearest lowest numbers). As the first step, the algorithm scans the dataset and counts each item, see Table 5-6 (b). based on the given minimum support. Next, it removes items whose count is less than minimum support, here 'b' is removed. Then, in the third step, it orders items from Table 5-6 (b), and we will get the result in Table 5-6 (c). Afterward, it scans each transaction in the original dataset and orders their items based on the frequency of support. As a result, we will get the ordered items based on the min-support column in Table 5-7. Note that the content is identical to the "items" columns but is ordered based on Table 5-6 (c).

#txn	items	ordered items based on min support
1	a, d, e, c	e, d, a ,c
2	a	a
3	c, d, e	e, d, c
4	a, c, e	e, a, c
5	d, e	e, d
6	a, d, e	e, d, a
7	a, c, d	d, a, c
8	d, e	e, d

Table 5-7: Original Data with additional column which orders items based on the result of Table 5-6 c.

Next, the algorithm uses the "ordered items based on min support" column of Table 5-7 and builds the FP-Tree. The FP-Tree is a type of a trie, and similar to the other trie, it starts with a null root node.

To build the FP-Tree, it starts with the first transaction, txn#1, and it creates a node for each new item it encounters and writes the number of times it encounters that item in front of the node. For example: 'e:2' means it encounters the item 'e' twice. Based on the third right column of Table 5-7, i.e., 'ordered items based on min support', the ordered items for txn#1 are {e,d,a,c}, thus it creates the first branch nodes and writes the number of time it encounters each node in front of

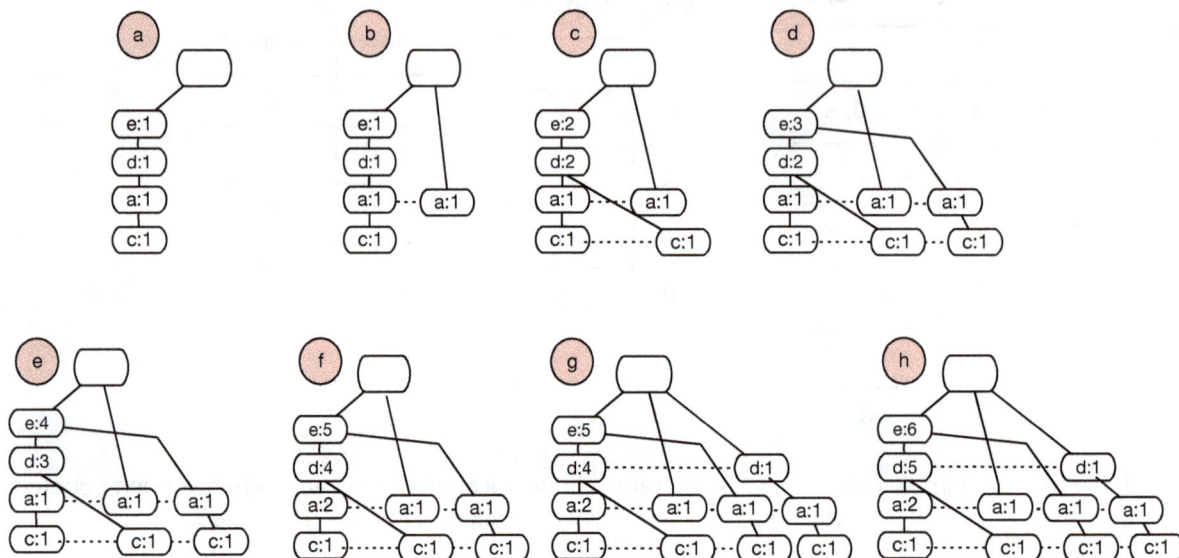

Figure 5-19: FP-Tree creation step by step based on the right most column of Table 5-7.

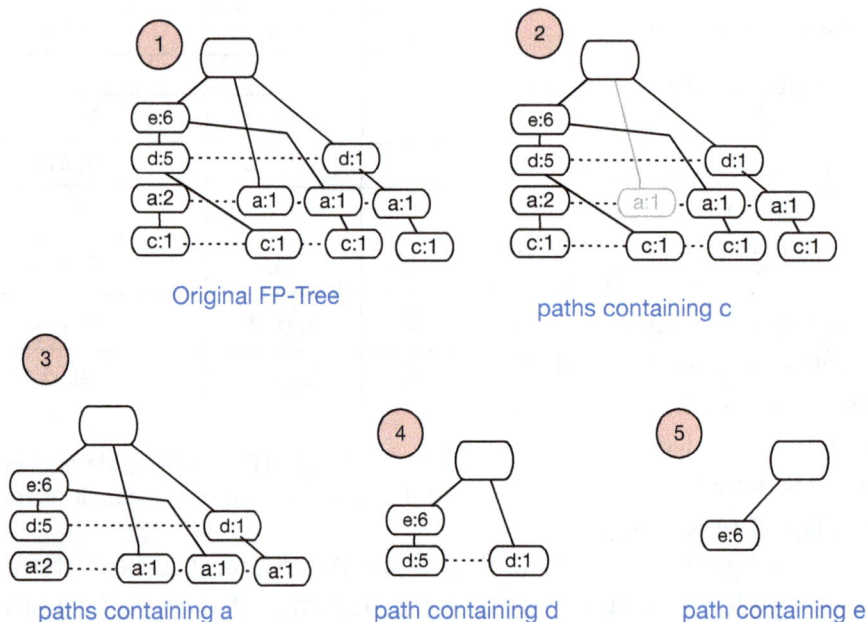

Figure 5-20: Subtree creation for each node, from the original FP Tree.

the item inside its node, as shown in Figure 5-19 (a). Then, it moves to the next transaction, txn#2, and its items, {a}. The tree does not start with 'a', thus the algorithm creates a new branch to handle the item 'a', and a new branch with 'a' will be created. The newly created branch of the FP-tree is presented in Figure 5-19 (b). Next, it encounters #txn 3, which we have {e,d,c} and thus both of their values will be increased by 1, but after 'd' there is a 'c', which is not in the tree path, and thus the algorithm should create a new branch, i.e., Figure 5-19 (c).

The next transaction is txn#4, which includes {e,a,c}. The algorithm encounters the 'e' node as the start and increases its value (e:3), but the next node is 'a', and it does not exist after 'e', thus, it creates another branch, as is shown in Figure 5-19 (d). This process continues until all transactions are parsed, and we end up having Figure 5-19 (h). Try to continue the rest of the figure on your own. We will ask you about them in your dream, and in the middle of the night, we come into your nightmare with a flaming sword if you forget it.

Once the FP-Tree has been constructed, its size is smaller in comparison to the original dataset, because some itemsets share the same path in FP-Tree. FP-Tree contains everything we need to know from the frequent itemsets inside the dataset. The next step is to extract the frequent item sets from the FP-Tree. To perform this step, the algorithm takes a bottom-up approach, which means starting from a leaf node to the root node and identifying all the paths that lead to frequent itemsets.

To start this step, first, it finds conditional trees. In particular, it starts from the bottom of the tree (lowest leaf node) to the root node, which will be in the following order {c, a, d, e}. Then, it extracts subtrees ending for each node. Figure 5-20 presents subtrees extracted from the original tree. Figure 5-20 (1), presents the original FP-Tree, Figure 5-20 (2) presents the subtree for node 'c', Figure 5-20 (3) presents the subtree for node 'a', etc.

Each subtree will be processed from bottom to top to extract frequent itemsets. For example, we check whether 'c' is a frequent itemset based on the given minimum support (min-support=3). Subtree no. 2 in Figure 5-20 is the subtree containing all paths from 'c'. The algorithm counts 'c's in this subtree (counting all values along the dotted lines for 'c'), and there are 3 'c' s that are not smaller than minimum support. The next subtree is 'a' (Subtree no. 3 in Figure 5-20). We have $2 + 1 + 1 + 1 = 5$ 'a', which meets the condition of minimum support. Therefore, it lists frequent itemsets that have 'a' in their endings as follows: {e,d,a}, {a}, {d,a}, {e,a}. The next subtree is 'd'. We have $5 + 1 = 6$ that is greater than minimum support. Therefore, it lists frequent item sets which have "d" in their endings as follows: {d,e}, {d}. The same process will be done for 'e', and it has {e} frequent itemset, i.e. {e}.

These listed itemsets are called *conditional base patterns*. The result of this process is shown in Table 5-8. There, we can see a column for each item and a column called "conditional pattern base". This column is nothing more than just removing the leaf node, which is shown on the left-most column, 'item', gets its support value from the leaf node, and writes its path to the root of the FP-Tree. Since 'e' is directly connected to the root node and does not have a path, we do not write it in this table. Figure 5-20 presents how the data of the "conditional pattern bases" column has been constructed.

After conditional base patterns have been created, the next step is to create conditional FP-Trees. To create conditional FP-Trees the algorithm goes through a "conditional base pattern", adds the values of each item together, and removes items whose values are less than minimum support. The result constructs the conditional FP-Tree column. In other words, "conditional FP-Tree"

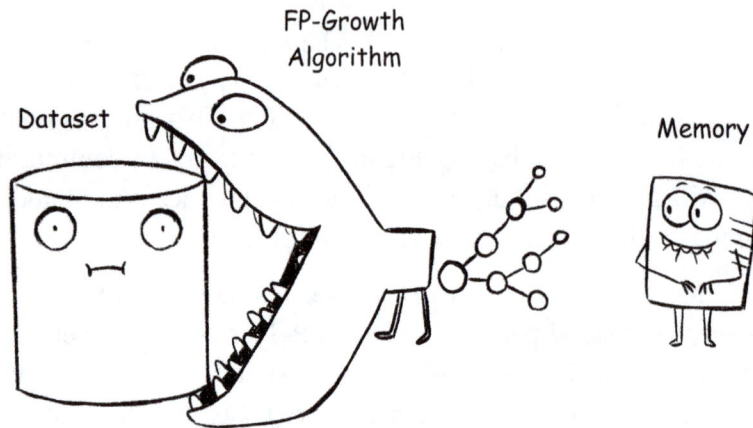

FP-Growth
Algorithm

Dataset

Memory

holds a set of single items that meet the minimum support condition (see the third column in Table 5-8).

After the "conditional FP-Tree" has been created, the next step is to create *frequent patterns*. It is easily created by combining the "item" column with "conditional FP-Tree" in the same row. The

item	conditional base pattern	conditional FP-Tree	frequent patterns
c	{a,d,e:1}, {a,e:1}, {d,e:1}, {a,d:1}	{a,:3}, {e:3}, {d:3}	{c,d:3}, {c,a:3}, {c,e:3}
a	{d,e:2}, {d:1}, {e:1}	{d:3}, {e:3}	{a,d:3}, {a,e:3}
d	{e:5}	{e:5}	{d,e:5}
e	–	–	–

Table 5-8: Frequent pattern identification based on the FP-Trees (subtrees).

result is presented in the "frequent pattern" column.

Check the dotted lines in Table 5-8 for examples of how frequent patterns are constructed by adding items to conditional FP-trees.

For example, in Table 5-8, by combining item "d" to its "conditional FP-Tree" we will get a frequent pattern of {d,e:5}. By looking at the "frequent patterns" column in Table 5-8, we can see all the occurrences are equal to or larger than minimum support (min support = 3). As a result, we will have {c,d},{c,a}, {c,e}, {a,d},… as frequent patterns.

Finally, we are done with identifying frequent patterns. Prepare a tea or coffee for yourself because we have bad news, and the algorithm still does not identify confidence. The confidence check is straightforward and will be done similarly to the Apriori.

We should calculate confidence for {c,d}, {c,a}, {c,e}, {a,e}, {a,d} and {d,e}. Confidence of (X=>Y) is equal to the number of times X and Y occur together, divided by the number of times X occurs.

We use Table 5-6 (a) and calculate confidences as follows:
confidence (d,e) = 5/6 = 0.83
confidence (e,d) = 5/6 = 0.83
confidence (a,e) = 3/5 = 0.6
confidence (e,a) = 3/6 = 0.5
confidence (a,d) = 3/7 = 0.42
confidence (d,a) = 3/5 = 0.6

Other confidence values, such as the one for (c,d), (d,c), (c,a), (a,c), (c,e), and (e,c) will be calculated in the same manner, which we do not write them here. As a result, {d,e}, {a,e} or {e,a} and {d,a} have occurred more than minimum confidence (50%), and thus will be returned by the algorithm as frequent itemsets.

By using a trie, the FP-Growth algorithm significantly mitigates the memory issue of Apriori, and thus, it is widely used.

ECLAT

Equivalence Class Transformation (ECLAT) [Zaki '97], is another efficient frequent itemset mining algorithm. The Apriori algorithm scans the dataset k times to find the support of $k+1$ itemsets. Frequent scans of the entire dataset make the Apriori algorithm computationally inefficient. This algorithm only has minimum support, and it does not require confidence; this makes it faster than Apriori.

Eclat assumes items are sorted in lexicographical order. It identifies the support in a different way than Apriori or FP-Growth. In particular, it searches the tree of items in DFS mode. We have explained DFS earlier in this chapter. Using DFS allows faster discovery of long frequent patterns, which their discovery is usually not memory efficient.

In the first step, Eclat converts the dataset into a vertical format, known as a *transaction identification list (tidlist)*, unlike FP-Growth and Apriori, which scan data horizontally. Figure 5-21 (a) shows the original dataset, and Figure 5-21 (b) shows the *vertical format* of the same dataset. A tidlist is a list of transaction IDs, in which that particular item has appeared. After tidlists have been created, then the algorithm iterates through newly generated tidlists, and compares the intersection of each item tidlist with all other tidlists. In other words, after the kth frequent itemset has been identified, the algorithm increases from k to $k+1$ and iterates again on the latest item list to identify frequent itemsets (based on the intersection between tidlists of items).

(a)   (b)

frequent 1-itemset

#txn	items
1	a, c, d
2	a, b
3	c, d
4	a, d
5	c, d

item	tid list
a	1,2,4
b	2
c	1,3,5
d	1,3,4,5

(c)   (d)

frequent 2-itemset

item	tid list
a, b	a ∩ b = 2
a, c	a ∩ c = 1
a, d	a ∩ d = 1,4
b, c	b ∩ c = null
b, d	b ∩ d = null
c, d	c ∩ d = 1,3,5

frequent 3-itemset

item	tid list
a, b, c	null
a, b, d	null
b, c, d	null
a, d, c	1

There will be no more
tidlist larger than
min_support

2 = min_support   {a, d}
                       {c, d}

3 > min_support

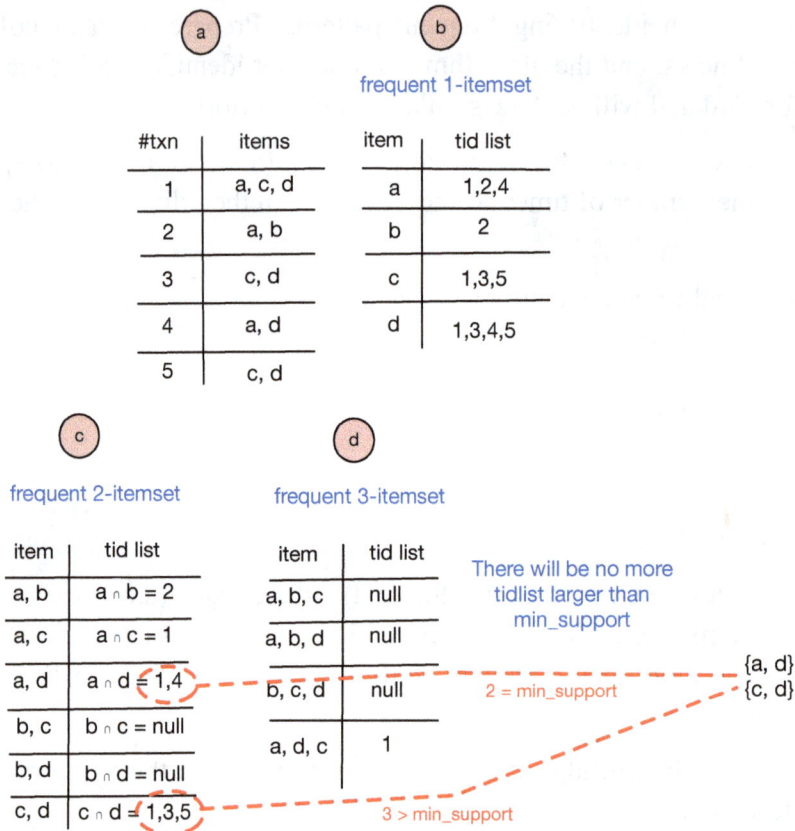

Figure 5-21: Step by step process of ECLAT algorithm.

This process continues until no more intersections are left, and all intersection results will be null or less than minimum support. Then, based on the minimum support, the final list of frequent itemsets will be returned as identified frequent itemsets.

To better understand this algorithm, check the example provided in Figure 5-21 and assume min_support is two. Here, we have a dataset of transactions in Figure 5-21 (a), and as the first step, the ECLAT algorithm transforms the dataset into Figure 5-21 (b). The algorithm performs this step by scanning the original dataset (Figure 5-21 (a)) and sorting all transaction IDs (tids) based on items to create Figure 5-21 (b). This table is called the "vertical transformation" of the Figure 5-21 (a). Next, the algorithm increases the k (number of frequent item counters) to one more value and lists all possible combinations of items and their intersections. The result is shown in Figure 5-21 (c) (left column). The right column of Figure 5-21 (c) presents the intersection of tids for each item on the left column. Again, in the next iteration, the algorithm increases the k and calculates the intersections between items. The result has been shown in Figure 5-21 (d). By increasing k (k=3), all intersections will be either null or their number of transactions is smaller than min-support. Therefore, what we get in Figure 5-21 (d) is a finalized frequent itemset. The algorithm compares them with minimum support and returns the ones that are not smaller than minimum support as frequent itemsets.

In addition to the simplicity of this algorithm, we don't need to calculate confidence for this algorithm; the only input parameter is minimum support. The computational complexity of the ECLAT algorithm is not straightforward to express in terms of simple Big O notation. However, in the worst-case scenario, assuming m is the number of unique items, the time complexity of the ECLAT algorithm is $O(2^m)$.

NOTES:

* While dealing with frequent items, the order of items appearing inside the transaction is unimportant. For instance, in Figure 5-18, it does not matter 'a' is before or after 'b'. What matters is co-occurrences inside transactions.

* The execution time of FP-Growth is better than Apriori because it uses a compact version of the dataset, i.e., FP-Tree. The size of the FP-Tree is smaller than the original dataset.

* In some literature [Aggrawal '15], the trie that is used in association rule mining is called an *enumeration tree*, which is a lexicographically ordered trie. This lexicographic ordering imposes a hierarchical structure on the transactional dataset.

* The computational complexity of association rule mining algorithms is not straightforward. Different factors come into play, such as the location of frequent items in the tree, the number of frequent items, the depth of the tree, etc.

Sequence Mining Algorithms

Association rule mining algorithms do not consider the order of items inside transactions. They expect users to order items and then feed them into the algorithm lexicographically. What if the order of data is important? For instance, remember that earlier in this chapter, we introduced a robot coach to help us exercise and stay fit while coping with the stress of learning machine learning. Our lovely robot can program our daily exercise routines as a sequence of events, e.g., first, 30 minutes jogging on a treadmill (cardio training), then 20 minutes of large muscle training (strength training), and spend 10 minutes for abdominal muscle training (light strength training). If there is no order in performing these exercises, they could harm our health instead of improving it. As another example, you can ask your robot to do some household work in parallel, and some should be done in a sequence.

There are many more examples of these kinds of datasets, such as time series, daily currency or stock exchanges, weather, sequence of nucleotide pairs on the strand of DNA, word sequences in a sentence, etc.

Datasets, whose order of events is important, can not be fed into association rule mining algorithms. Another category of algorithms is *sequence mining* algorithms, which can resolve time and order issues. The main difference between a sequence mining algorithm and an association rule mining algorithm is that we take into account the orders in a transaction. Here, we describe some well-known sequence mining algorithms, including GSP, SPADE, FreeSpan, and PrefixSpan.

Before describing algorithms, we should describe some brief notations. The transaction content will be shown as something like <a,b,(c,d), f>. This means that the first event is 'a', then 'b', next 'c' and 'd' (occurs together), and the last 'f' has occurred. Parentheses mean that the content inside parentheses occurred together.

Generalized Sequential Pattern (GSP)

The Generalized Sequential Pattern (GSP) algorithm [Srikant '96] is one of the first sequential pattern mining algorithms. The first step is *candidate generation*, and the algorithm calculates the frequency for each item in the dataset. The second step is *pruning*, and it removes items with a frequency is smaller than minimum support. Next, as the third step, it merges items and creates $k+1$ size sequence of elements (k is the number of iterations). Again, it calculates the frequency of items, and if they are smaller than minimum support, it removes them. Then, it scans the dataset again, and this iterative process continues until no new itemset is left to be created. In summary, the algorithms operate the following in a loop (iteratively): (i) scan the dataset for k frequent sequences. (ii) generate $(k+1)$ size candidate sequences, (iii) prune the newly generated sequence by comparing their frequency with *min_support*, and (iv) increase the value of k until no sequence is left.

As an example to understand this algorithm, consider the table that includes sequential data on the left side of Figure 5-22, and the minimum support is equal to one. In the first step, the algorithms list all items and compute their support. If they are smaller than the minimum support, then it removes them. In the right table of Figure 5-22, we can see 'g' and 'h' have been removed because their frequency is smaller or equal to the minimum support.

Seq. ID	Sequence Data
1	<(b,c),c,d,(a,c)>
2	<(b,f),(c,e),b,(f,g)>
3	<(a,h),(b,f),a,b,f>
4	<(b,e),(c,e),d>
5	<a,(b,d),b,c,b,(a,d,e)>

item	Support
a	3
b	5
c	4
d	3
e	3
f	2
~~g~~	~~1~~
~~h~~	~~1~~

Figure 5-22: Dataset with sequential data on the left and its items based on their support.

Next, the algorithm computes the 2-length products of all item combinations, as presented in Figure 5-23. Since some items might occur together, we should design two tables, one for sequential occurrences and the other for parallel occurrences. Note that the right side of Figure 5-23 does not require the other half to be filled because the same combination existed in the table, e.g., (a,b) = (b,a). In other words, here, the order doesn't matter.

Again, using Figure 5-22, the algorithm generates a list of candidates with k+1 size and removes the ones that are smaller than minimum support. By checking all elements in both tables in Figure 5-23, we will have the following sequences with the length of two:

<a,a>, <a,b>, <a,c>, <a,d>, <a,e>, <a,f>, ... <f,a>, <(f,b)> ...<(f,f)>

Then, the algorithms count these sequences in the dataset and remove the ones that are less than minimum support. For example, by checking Figure 5-22, <a,a> existed in Seq ID. 3 and Seq ID. 5, thus it keeps it, but <a,f> exists only in Seq ID. 3. Since its occurrences are equal or smaller than minimum support, it will be removed.

	<a>	<b>	<c>	<d>	<e>	<f>
<a>	<a,a>	<a,b>	<a,c>	<a,d>	<a,e>	<a,f>
<b>	<b,a>	<b,b>	<b,c>	<b,d>	<b,e>	<b,f>
<c>	<c,a>	<c,b>	<c,c>	<c,d>	<c,e>	<c,f>
<d>	<d,a>	<d,b>	<d,c>	<d,d>	<d,e>	<d,f>
<e>	<e,a>	<e,b>	<e,c>	<e,d>	<e,e>	<e,f>
<f>	<f,a>	<f,b>	<f,c>	<f,d>	<f,e>	<f,f>

	<a>	<b>	<c>	<d>	<e>	<f>
<a>		<(a,b)>	<(a,c)>	<(a,d)>	<(a,e)>	<(a,f)>
<b>			<(b,c)>	<(b,d)>	<(b,e)>	<(b,f)>
<c>				<(c,d)>	<(c,e)>	<(c,f)>
<d>					<(d,e)>	<(d,f)>
<e>						<(e,f)>
<f>						

Figure 5-23: k+1 size sequence generation process. Right table presents parallel occurrences of items, and left table presents their sequential occurrences.

Again, the algorithm increases the iteration value (k=2) and performs another iteration, which results in the following sequences:

<a,a,a>, <a,a,b>, <a,a,c>, <a,a,d>, … <f,f,f>, <(a,b),a> … , <(e,f),f>

This process continues until there is no sequence left that has support equal to or greater than minimum support, then the algorithm ends and returns frequent sequences.

SPADE

Sequential Pattern Discovery using Equivalent class (SPADE) [Zaki '01] is a sequential pattern mining algorithm that operates in vertical data format. We will explain the vertical data format shortly.

①

Seq. ID	Sequence Data
1	<(cd),(ab),(abf),(acdf)>
2	<(abf),e>
3	<(abf)>
4	<(dgh),(bf),(agh)>

②

SID	EID	Element
1	1	cd
1	2	ab
1	3	abf
1	4	acdf
2	1	abf
2	2	e
3	1	abf
4	1	dgh
4	2	bf
4	3	agh

a:

SID	EID
1	2, 3, 4
2	1
3	1
4	3

b:

SID	EID
1	2,3
2	1
3	1
4	2

c:

SID	EID
1	1,4

d:

SID	EID
1	1, 4
4	1

e:

SID	EID
2	2

f:

SID	EID
1	3,4
2	1
3	1
4	2

g:

SID	EID
4	1, 3

h:

SID	EID
4	1, 3

Figure 5-24: Horizontal dataset to vertical dataset conversion and then separating each element based on their SID and EID.

First, it performs one scan on the dataset and transforms it into a vertical format. Then, it identifies k-sequences by joining the ID of (k-1) sequences. For example, to find sequences with size two, all single items are joined together if they share the same sequence IDs, and their frequency is larger than minimum support. Then, it checks if those items' element IDs are ascending; then, they will be considered frequent sequence patterns.

We know it is impossible to understand the explanation, and weirdly, many references just provide this description and the algorithm pseudocode. Let's use our lovely style of explanation with examples to describe it.

Assuming the minimum support is two, we accept two as well, and we have a dataset of sequences shown in Figure 5-24 (1). Recall that elements inside parenthesis are occurring together; thus, there is no order for them; the algorithm considers them unique elements. The first step is to create Figure 5-24 (2) from Figure 5-24 (1), which lists all elements inside each transaction. This table has two columns, SID, which is Seq. ID, and EID, which specifies the element's position inside a sequence.

In the next step, the algorithm separates each data item (a,b,c,d,e,f,g, and h) within its SID and EID, as we can see eight small tables in Figure 5-24, tables (a) to (h). In other words, for every item, it constructs a table. The next step is to check which of these tables has fewer records than minimum support, and we can see tables 'c', 'g', 'e', and 'h' will be removed.

Next, the algorithm creates a combination of elements based on their table. We have 'a', 'b','d', and 'f' remained.

Then, the algorithm should construct the second-order combination of them, i.e., 'aa', 'ab', 'ad', 'af', etc. To build the second-order combination, these small tables are added together and combined. They will be joined based on common SID, and the prior EID should be smaller than the posterior EID. In other words, the combination is built based on (i) equal SID, and (ii) numerical order of EID. For example, consider the process of creating a table for 'aa'. Look at Figure 5-25, the first EID is '2'. Then, it is compared to '2', and since both are equal, the

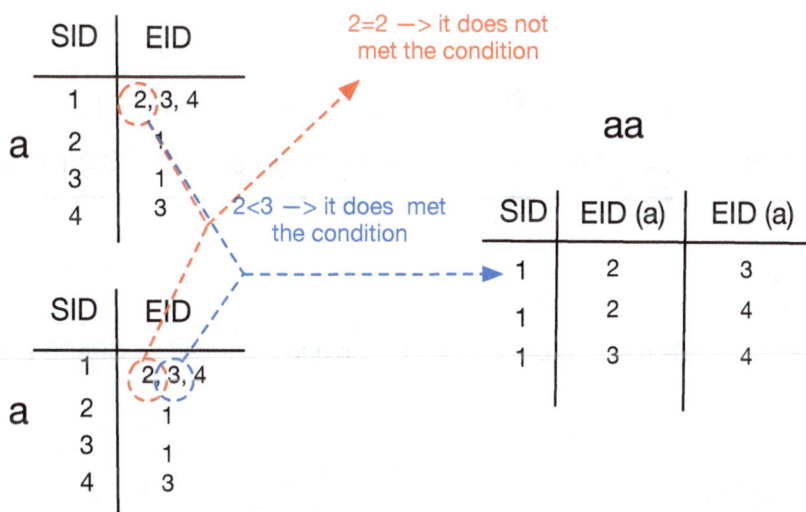

Figure 5-25: SPADE combining two elements based on their EID and SID to create their combination table.

algorithm does not write them in the result table (see the red dotted lines in Figure 5-25). Next, the algorithm moves to the next element. Now, it intends to compare 2 with 3, and 2<3. Therefore, it can add them to the 'aa' table (see the blue dotted lines in Figure 5-25). Once again, we repeat that the condition to add them to the combined table is *that the first EID should be smaller than the second EID.*

Figure 5-26 visualizes combinations of two elements together. To understand this figure, look at the data of 'a','b','d', and 'f' in Figure 5-25. The next step is to combine three elements, e.g., 'aab', 'aad', 'aaf', 'abb', etc. The algorithm joins the content of three related tables to construct combinations of three elements. For example, 'aab' will be constructed by the join between 'a', 'b', and 'a'. 'abf' will be construed by the join between 'a','b', and 'f'.

aa

SID	EID (a)	EID (a)
1	2	3
1	2	4
1	3	4

ad

SID	EID (a)	EID (d)
1	2	4
1	3	4

ba

SID	EID (b)	EID (a)
1	2	3
1	3	4
1	2	4

bd

SID	EID (b)	EID (d)
1	2	4
1	3	4

ab

SID	EID (a)	EID (b)
1	2	3

af

SID	EID (a)	EID (f)
1	2	3
1	3	4
1	2	4

bb

SID	EID (b)	EID (b)
1	2	3

bf

SID	EID (b)	EID (f)
1	2	3
1	3	4
1	2	4

da

SID	EID (d)	EID (a)
1	1	2
1	1	3
1	1	4
4	1	3

dd

SID	EID (d)	EID (d)
1	1	4

fa

SID	EID (f)	EID (a)
1	3	4
4	2	3

fd

SID	EID (f)	EID (d)
1	3	4

db

SID	EID (d)	EID (b)
1	1	2
1	1	3
4	1	2

df

SID	EID (d)	EID (f)
1	1	3
1	1	4
4	1	2

fb

SID	EID (f)	EID (b)
-	-	-

ff

SID	EID (f)	EID (f)
1	3	4

Figure 5-26: All combination of two EIDs for the a, d, b and f, created by SPADE algorithm.

Figure 5-27 shows one example of the combination of three elements. We can see from this example that only two tables on the top have records that are more than the minimum support. That is correct because we should consider events inside parenthesis that occur not in a sequence but in parallel; thus, the algorithm correctly does not consider the content of parenthesis as a sequence.

a,b,f

SID	EID(a)	EID(b)	EID(f)
1	2	3	4

a,b,d

SID	EID(a)	EID(b)	EID(f)
1	2	3	4

a,d,f

SID	EID(a)	EID(b)	EID(f)
-	-	-	-

b,d,f

SID	EID(a)	EID(b)	EID(f)
-	-	-	-

Figure 5-27: Some (not all) combinations of EIDs for the a, d, b and f, created by SPADE.

FreeSpan

Both SPADE and GSP are very useful, but they have three drawbacks. They perform many scans on the dataset. Although SPADE uses a vertical database format, it could generate large candidate lists, and neither SPADE nor GSP are optimal in mining long sequences of patterns (very similar to Apriori issues). More efficient algorithms were introduced later that employ the concept of *pattern-growth* for finding sequential patterns, such as FreeSpan [Han '00] and PrefixSpan [Pei '04]. The pattern growth concept involves efficiently discovering frequent itemsets by progressively extending existing patterns rather than generating candidates, thereby reducing the search space and improving computational efficiency.

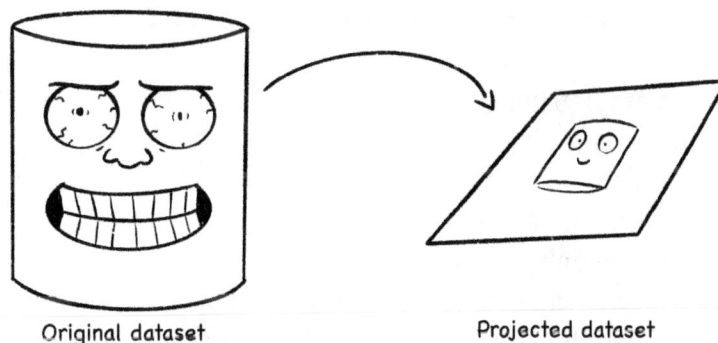

Original dataset   Projected dataset

Frequent Pattern-Projected Sequential Pattern Mining (FreeSpan) [Han '00] is a sequence mining algorithm that operates based on this concept. FreeSpan recursively creates "projected datasets" to reduce the search space into smaller partitions. In other words, a projected dataset means a smaller size dataset derived recursively[7] from the original dataset.

#txn	items
1	<a,(abc),(ac),d,(cf)>
2	<(ad),c,(bc),(ae)>
3	<(ef),(ab),(df),c,b>
4	<e,g,(af),c,b,c>

count each item support
and order them based
on their support dataset

item	support
a	4
b	4
c	4
d	3
e	3
f	3
~~g~~	~~1~~

Figure 5-28: Original dataset on the left and its ordered item based on support on the right.

When deriving a projected dataset from the original dataset, a projected dataset only includes sequential patterns (ordered sets of frequent items). In other words, this algorithm recursively creates projected datasets from the previous dataset (the first one is the original dataset), which those sub-datasets (projected datasets) do not include non-frequent sequential patterns. As a result, the algorithm will end up with a list of sequential patterns in each projection dataset and eliminate non-frequent items.

To understand it, we explain an example provided in Chapter 11 of Frequent Pattern Mining Book [Aggrawal '14], adapted from Han et al. [Han '05] in more detail. Assuming minimum support is two, we have a sample dataset as has been shown on the left side of Figure 5-28, and the convention to write this dataset is as before (parenthesis used to state those elements occur in parallel is not in a sequence). The first step, similar to GSP, is to scan the dataset and order all of its items based on their support count. Therefore, we end up having the right side of Figure 5-28. Since the support of 'g' is less than min-support, we removed it and did not consider it in our calculation. According to Figure 5-28, which lists the supports of each item, we will have six subsets. Then, the algorithm constructs their projected datasets, which will be obtained by one scan of the dataset. These sub-datasets do not overlap, and they are as follows:

(1) subset of items containing only 'a'.
(2) subset of items containing 'b' but not items after 'b', i.e., 'a' and 'b' only.
(3) subset of items containing 'c', but not items after 'c', i.e., 'a', 'b' and 'c'.
(4) subset of items containing 'd', but not items after 'd', i.e., 'a', 'b', 'c' and 'd'.
(5) subset of items containing 'e', but not items after 'e', i.e., 'a', 'b', 'c', 'd' and 'e'.
(6) subset of items containing only 'f'.

Note that all these six projected datasets will be constructed once during one dataset scan.

From these six projected datasets, patterns will be extracted as follows:

---

[7] If you do not know what is recursive algorithm it is worth spending some time and reading about it; in short, it is a function that calls itself until a condition is met.

#txn	items
1	<a,(abc),(ac),d,(cf)>
2	<(ad),c,(bc),(ae)>
3	<(ef),(ab),(df),c,b>
4	<e,g,(af),c,b,c>

→ create <a> projected dataset →

#txn	items
1	<a,a,a>
2	<a,a>
3	<a>
4	<a>

→ identify sequential pattern →

<a,a>:2
<a>:2

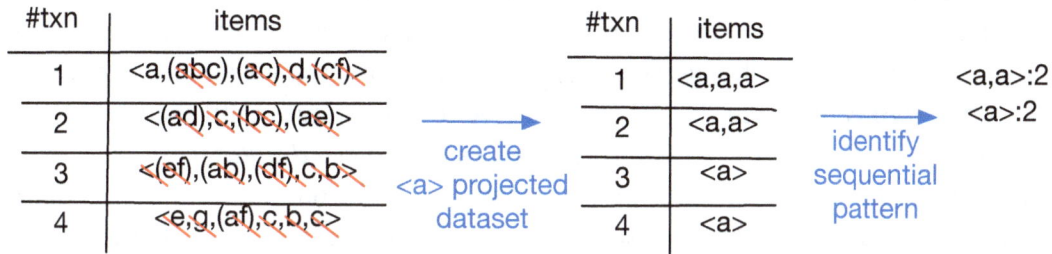

Figure 5-29: Constructing 'a' projected dataset and extracting patterns of items containing item 'a', but not after 'a'.

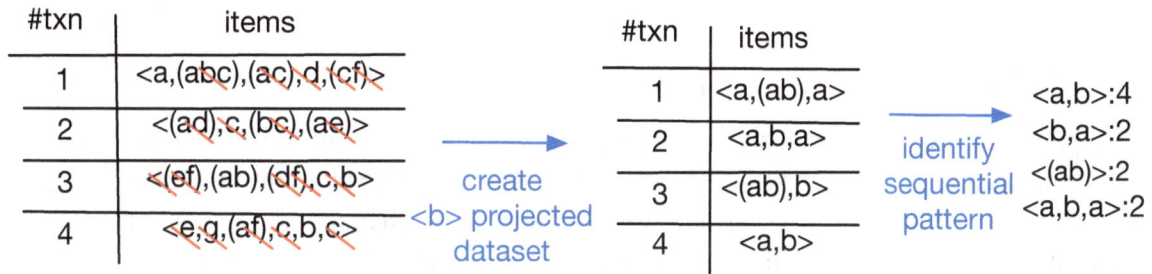

#txn	items
1	<a,(abc),(ac),d,(cf)>
2	<(ad),c,(bc),(ae)>
3	<(ef),(ab),(df),c,b>
4	<e,g,(af),c,b,c>

→ create <b> projected dataset →

#txn	items
1	<a,(ab),a>
2	<a,b,a>
3	<(ab),b>
4	<a,b>

→ identify sequential pattern →

<a,b>:4
<b,a>:2
<(ab)>:2
<a,b,a>:2

Figure 5-30: Constructing 'b' projected dataset and extracting patterns of items containing item 'b', but not after 'b'.

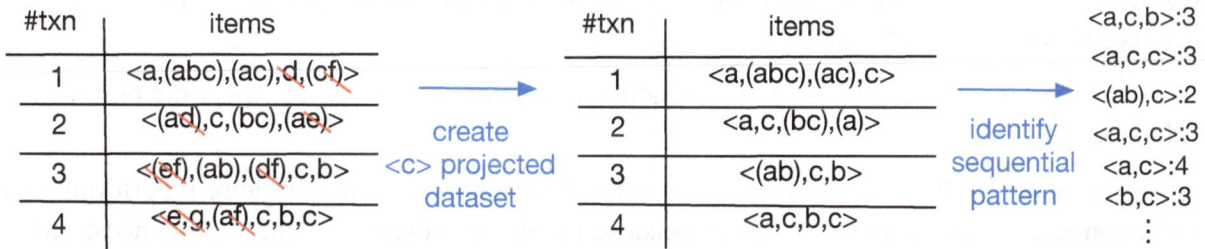

#txn	items
1	<a,(abc),(ac),d,(cf)>
2	<(ad),c,(bc),(ae)>
3	<(ef),(ab),(df),c,b>
4	<e,g,(af),c,b,c>

→ create <c> projected dataset →

#txn	items
1	<a,(abc),(ac),c>
2	<a,c,(bc),(a)>
3	<(ab),c,b>
4	<a,c,b,c>

→ identify sequential pattern →

<a,c,b>:3
<a,c,c>:3
<(ab),c>:2
<a,c,c>:3
<a,c>:4
<b,c>:3
⋮

Figure 5-31: Constructing 'c' projected dataset and identifying sequential patterns of items containing item 'c', but not after 'c'. We do not list all possible frequent patterns for the sake of space.

(i) The 'a' projected dataset is {<a,a,a>, <a,a>, <a>, <a>}. To understand how we get it take a look at Figure 5-29. The algorithm extracts sequential patterns from the projected dataset, and they are "<a,a>:2, <a>:2", here ':2' means the frequency number. Therefore, we can conclude from the given dataset {<a,a,a>, <a,a>, <a>, <a>}, following two sequential patterns <aa> and <a> are extracted.

(ii) The 'b' *projected dataset* is {<a,(ba),a>, <a,b,a>, <(ab),b>, <a,b>}. Check Figure 5-30 to understand how the projected dataset is created. By searching this dataset, we get four new sequential patterns: {<a,b>:4, <b,a>:4, <(ab)>:2, <aba>:2}.

(iii) The 'c' *projected dataset* is {<a,(abc),(ac),c>, <a,c,(bc),a>, <(ab),c,b>, <a,c,b,c>}. The algorithm identifies the frequent sequential patterns, as shown in Figure 5-31. In particular, first, the algorithm identifies sequential patterns with length 2, for <c> *projected dataset,* and they are:

{<a,c>:4, <(bc)>:2, <b,c>:3 ,<c,c>:3, <c,a>:2, <c,b>:3}. Then, to identify length 3 sequential patterns, for each of the identified patterns, it creates another projected dataset. For example, <a,c>-*projected dataset*={<a,(abc),(ac),c>, <a,c,(bc),a>, <(ab),c,b>, <a,c,b,c>}, <b,c>-*projected dataset*=..., <c,c>-*projected dataset*= .... This recursive loop continues until there is no sequential pattern existed to be identified. As a result, from <a,c>-*projected dataset* we have {<a,c,b>:3, <a,c,c>:3, <(ab),c>:2, <a,c,a>:2} sequential patterns identified from <a,c>-*projected dataset*, but if we look for <a,c,b>-*projected dataset*, <a,c,c>-*projected dataset*, etc. there will be no 4 lengths sequential pattern and the recursive-loop breaks. Nevertheless, we did not present this amount of detail in our figures. This process continues for all subsets (4), (5), and (6) until no subset remains with sequential patterns and sequential pattern results will be returned by the algorithm.

PrefixSpan

PrefixSpan [Han '01] uses projected sequences of the dataset (partitioning the search space into smaller datasets) to reduce the search space. It is similar to FreeSpan, but instead of creating the projected sequence datasets from all possible frequent subsequences, the projection dataset will be created based on *prefixes*. A prefix is an item at the beginning of the sequence that is frequent in the dataset. Note that a prefix is not just a single item but can be any prefix subsequence that leads to the generation of a projected database. The *suffix* is the remaining part of the sequence without a prefix. Assuming we have the sequence <a,(abc),(ac),d,(cf)>, assuming prefix as: <a>, then its suffix is: <(abc)(ac),d,(cf)>

'_' is called a *placeholder* that is used to denote the position in a sequence where the prefix ends and the suffix begins.

Let us repeat again: PrefixSpan is usually more efficient than FreeSpan because it performs less projection and shrinks sequence datasets because FreeSpan projects the entire sequences; rather PrefixSpan projects only suffixes. However, we recommend you to experiment with both, then decide. Similar to FreeSpan, this algorithm operates in three steps. The first step is to identify and order length 1 item in the sequential dataset based on their support and remove items that have less than minimum support (Figure 5-28) (assuming min_support=2).

The second step is to divide the search space into smaller size datasets and create a projected dataset for each prefix. We could have six subsets (projected dataset):
(1) subset of items containing only 'a',
(2) subset of items containing 'b' but not items after 'b', i.e. 'a' and 'b' only
(3) subset of items containing 'c', but not items after 'c', i.e. 'a', 'b' and 'c'
(4) subset of items containing 'd', but not items after 'd'
(5) subset of items containing 'e', but not items after 'e'
(6) subset of items containing only 'f'.

Note that all these six projected datasets will be constructed once during one scan of the dataset. Besides, while projecting the dataset into the suffix dataset, every other item (disregard their order) before the prefix item will be removed. For example, for prefix <b>, the original <ef,(ab),

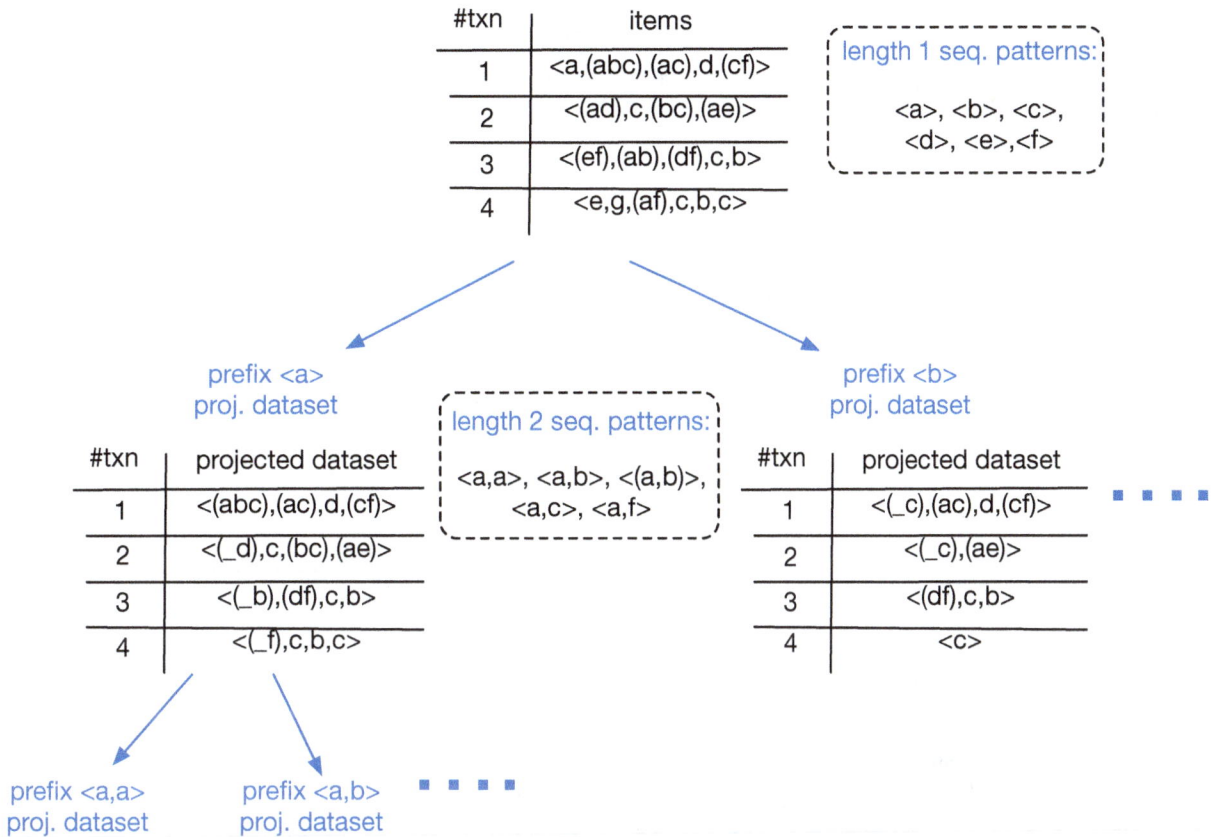

Figure 5-32: Recursive increase in the prefix size and sequence length.

#txn	items
1	<a,(abc),(ac),d,(cf)>
2	<(ad),c,(bc),(ae)>
3	<(ef),(ab),(df),c,b>
4	<e,g,(af),c,b,c>

length 1 seq. patterns:

<a>, <b>, <c>, <d>, <e>,<f>

prefix <a> proj. dataset

#txn	projected dataset
1	<(abc),(ac),d,(cf)>
2	<(_d),c,(bc),(ae)>
3	<(_b),(df),c,b>
4	<(_f),c,b,c>

length 2 seq. patterns:

<a,a>, <a,b>, <(a,b)>, <a,c>, <a,f>

prefix <b> proj. dataset

#txn	projected dataset
1	<(_c),(ac),d,(cf)>
2	<(_c),(ae)>
3	<(df),c,b>
4	<c>

prefix <a,a> proj. dataset

prefix <a,b> proj. dataset

(df),c,b> will be converted to <(df),c,b>, and everything before 'b' will be removed, disregard their order.

The third step is to find sequential patterns inside those datasets. Then, increase the length of the prefix (e.g., prefix <a> will be prefix <af> and prefix <aa>.) and find sequential patterns with a longer length inside new datasets. This process of length increase will continue until no candidate frequent sequence remains to be generated. Figure 5-32 shows prefix size and sequence length increases[8].

Similar to the FreeSpan algorithm for each prefix, the algorithm creates a projected dataset, as shown in Figure 5-33. There, we have six projected datasets, and recursively, the algorithm searches and identifies frequent sequences of suffix datasets and their prefix.

The right side of Figure 5-33 presents frequent sequences for each projected dataset. We do not present the details in Figure 5-33, but each prefix will include a tree of the projected dataset as it has been shown Figure 5-33. However, Figure 5-33 only shows the final results from each single length prefix.

---

[8] We made this figure by hand, so you might find some errors in our calculation.

Original dataset

Projected dataset

Sequential patterns (min_support = 2)

#txn	items
1	<a,(abc),(ac),d,(cf)>
2	<(ad),c,(bc),(ac)>
3	<(ef),(ab),(df),c,b>
4	<e,g,(af),c,b,c>

prefix: <a> →

#txn	projected dataset
1	<(abc),(ac),d,(cf)>
2	<(_d),c,(bc),(ac)>
3	<(_b),(df),c,b>
4	<(_f),c,b,c>

→ <a> <a,a> <a,b> <a(bc)>
<a,(bc),a> <a,b,a> <a,b,c> <(ab)>
<(ab),c> <(ab),d> <(ab),f>
<(ab),d,c> <a,c> <a,c,a> <a,c,b>
<a,c,c> <a,d> <a,d,c> <a,f>

#txn	items
1	<a,(abc),(ac),d,(cf)>
2	<(ad),c,(bc),(ac)>
3	<(ef),(ab),(df),c,b>
4	<e,g,(af),c,b,c>

prefix: <b> →

#txn	projected dataset
1	<(_c),(ac),d,(cf)>
2	<(_c),(ac)>
3	<(df),c,b>
4	<c>

→ <b> <b,a> <b,c> <(bc)> <b,(ac)>
<(bc),a> <b,d> <b,d,c> <b,f>

#txn	items
1	<a,(abc),(ac),d,(cf)>
2	<(ad),c,(bc),(ac)>
3	<(ef),(ab),(df),c,b>
4	<e,g,(af),c,b,c>

prefix: <c> →

#txn	items
1	<(ac)d(af)>
2	<(bc)(ac)>
3	<b>
4	<bc>

→ <c> <c,a> <c,b> <c,c> <c,(ac)>

#txn	items
1	<a,(abc),(ac),d,(cf)>
2	<(ad),c,(bc),(ac)>
3	<(ef),(ab),(df),c,b>
4	<e,g,(af),c,b,c>

prefix: <d> →

#txn	items
1	<(cf)>
2	<c,(bc)(ac)>
3	<(_f),c,b>
4	-

→ <d> <d,b> <d,c>
<d,c,b> <d,f>

#txn	items
1	<a,(abc),(ac),d,(cf)>
2	<(ad),c,(bc),(ac)>
3	<(ef),(ab),(df),c,b>
4	<e,g,(af),c,b,c>

prefix: <e> →

#txn	items
1	-
2	-
3	<(_f),(ab),(df),c,b>
4	<(af),c,b,c>

→ <e> <e,a> <e,a,b> <e,a,c> <e,a,c,b>
<e,b> <e,b,c> <e,c> <e,c,b> <e,f>
<e,f,b> <e,f,c> <e,f,c,b>

#txn	items
1	<a,(abc),(ac),d,(cf)>
2	<(ad),c,(bc),(ac)>
3	<(ef),(ab),(df),c,b>
4	<e,g,(af),c,b,c>

prefix: <f> →

#txn	items
1	-
2	-
3	<(ab),(df),c,b>
4	<c,b,c>

→ <f> <f,b> <f,b,c> <f,c> <f,c,b>

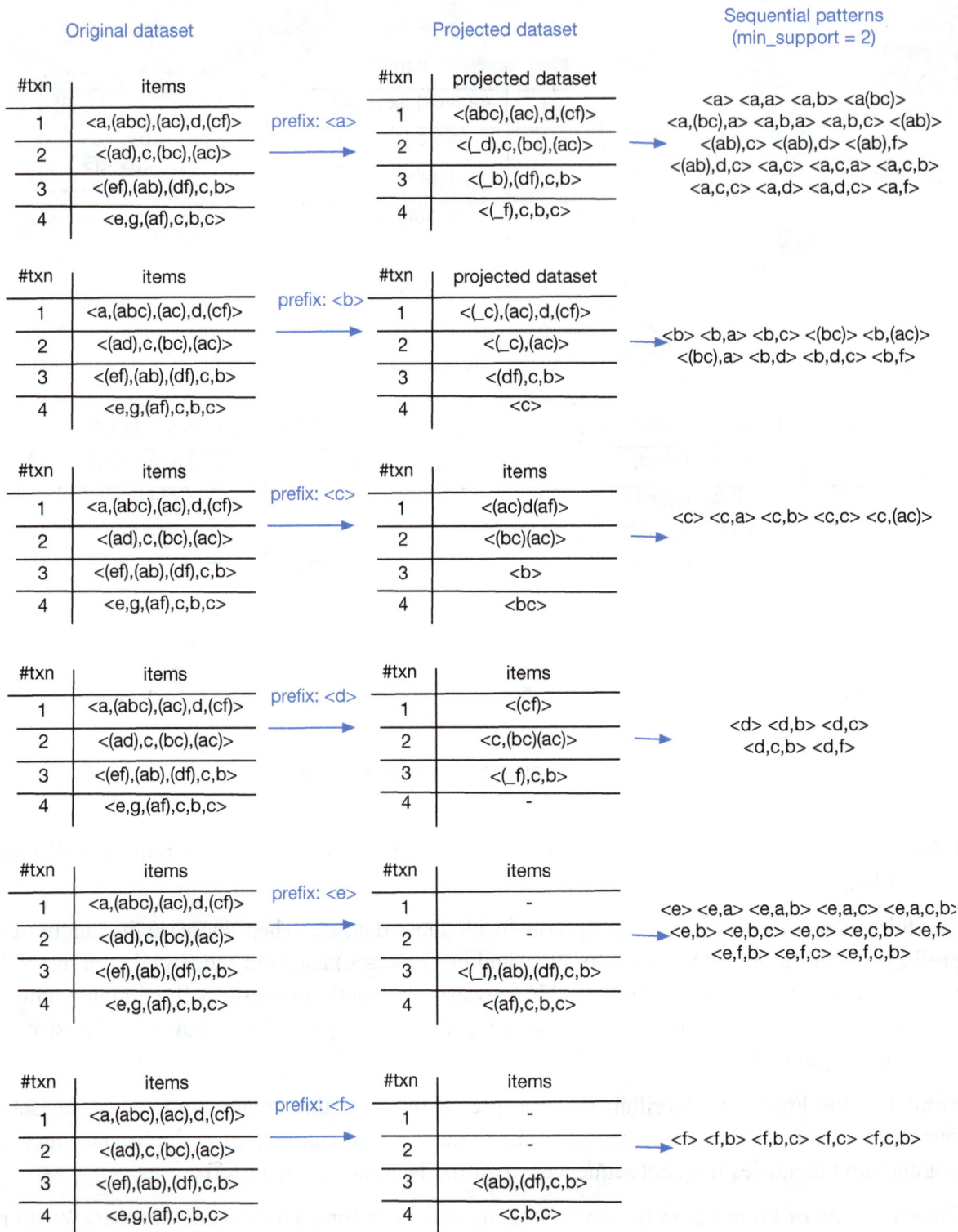

Figure 5-33: Original dataset on the left, projected datasets in the middle based on length 1 sequential patterns (prefix length is 1). On the left you can see identified sequential patterns.

The major cost of PrefixSpan is projecting many datasets into the memory. In other words, suffixes of sequences often appear repeatedly in recursive projected datasets. The authors of this algorithm introduce a concept of pseudo-projection. It is used to hold the pointer of the suffix in the memory and does not repeat the entire suffix.

Congratulations, now, you stay in front of the mirror and give yourself an honorary professorship

Students after learning itemset mining
and sequence mining algorithms

in unsupervised learning, but only if you read the previous chapter as well.

NOTES:

* A major advantage that makes GSP dissimilar to naive brute force is its pruning step. The pruning process in each iteration reduces the search space.

* SPADE and GSP operate based on an Apriori principle, which states that if a sequence is not frequent, none of its super-sequences is also frequent. For example, if x,y is infrequent, the x,y,z and (x,y),z will be infrequent too.

* Candidate generation methods (SPADE and GSP) are not effective for finding long, frequent patterns because they need many passes of scanning and generate lots of candidates.

* One might argue why not using FP-Tree to improve the Apriori style of sequence generation (GSP and SPADE) in sequence mining. Items in a sequence can not be re-ordered or collapsed, and thus, the FP-Tree gets too large and inefficient.

* Both PrefixSpan and FreeSpan use a type of projection-guided depth-first search, which we did not explain in detail.

* If a sequential pattern appears in all projected datasets, the size of the projected datasets does not shrink in FreeSpan, which is a major drawback in comparison to PrefixSpan. The size of the projected dataset reduces only when there is an item that appears less than minimum support.

* There are many frequent pattern mining algorithms, such as constraint-based patterns, negative patterns, compressed patterns, etc. We do describe the important ones in this section. You can find the rest either online or by checking their papers. Besides, due to the advances in neural networks, nowadays these are not very popular algorithms. However, we still believe they are important to learn.

Sequential Data Prediction with Hidden Markov Model

Hidden Markov Model (HMM) is a set of sequence mining algorithms traditionally used for speech tagging, weather forecasting, stock market prediction, etc. For instance, read the following passage from a person with narcissistic mental issues: *"I am perfect; I can achieve whatever I want. I am doing always perfect."* Assume another human asks you what the next sentence could be. With high confidence, you can say it is another narcissistic term. A computer can use HMM to find this annotation or tag (narcissistic is a tag) based on the previous annotations of the sentence, e.g., *"narcissist, narcissist, ?"* Then the '?' will probably also be a 'narcissist' tag. In particular, the Markov model tries to predict a future state, but we need information about the current state. In simple language, from things we can observe, we infer things we can not observe.

Before explaining it, we should describe a few terms, including *probabilistic graphical models* and *Bayesian networks*. Brace yourself because there are many terms we need to learn before learning about HMM.

Probabilistic Graphical Model

A *probabilistic model* is a set of random variables and their probability of occurrence. Probabilistic models use the distribution of events to describe phenomena or events. *A probabilistic graphical model* is a more specific probabilistic model. It uses graph structure to visualize the structure of probabilistic models. Probabilistic graphical models present the Bayesian network, which is a directed graph. The nodes of the probabilistic graphical model are "variables," and edges are the probabilistic relation between variables. A probabilistic graphical model forms a *Bayesian network,* which represents a set of variables and their conditional dependencies via a Directed Acyclic Graph (DAG).

For instance, Figure 5-34 presents a probabilistic graphical model or Bayesian network. This Bayesian network presents a joint distribution among three variables x, y, and z. In other words, we can decompose $P(x, y, z)$ into $P(z \mid x, y)P(x)P(y \mid x)$. A *joint distribution* can be generalized with the following formalization:

$$P(x_1, x_2 \ldots, x_n) = P(x_n \mid x_1, x_2 \ldots, x_{n-1}) \ldots P(x_2 \mid x_1)P(x_1)$$

Let's use a more complex example and check Figure 5-35, which presents a more complex graph. The Bayesian network of Figure 5-35 will be written as follows:

$$P(x_1, x_2, x_3, x_4, x_5) = P(x_1)P(x_2 \mid x_1)P(x_3 \mid x_1, x_2)P(x_4 \mid x_2, x_3)P(x_5 \mid x_1)$$

We can write the relationships of the graph nodes as a joint probability distribution via the following equation[9]:

$$P(X) = \prod_{n=2}^{n} P(x_n \mid x_{n-1})$$

A probabilistic graphical model is defined based on two types of variables: *observed* and *hidden variables*. Observed variables are what we see or what we can directly measure. Hidden variables or latent variables are what we infer from observed variables.

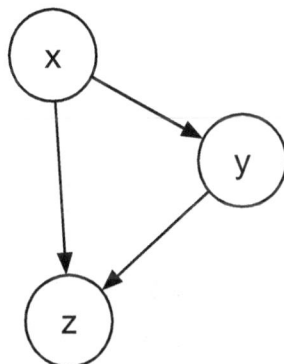

Figure 5-34: Simple example of graphical model probability distribution.

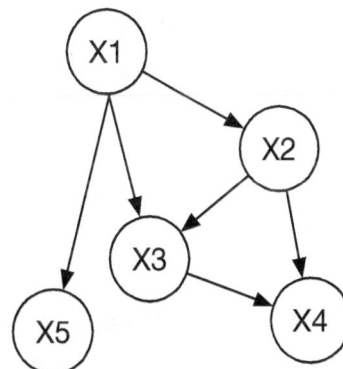

Figure 5-35: A bit more complex example of the Probabilistic Graphical model.

Hidden Markov Model Concept

A *Markov Process or Markov Chain* is a process for which predictions can be made regarding future outcomes based solely on its current state, without needing to consider the entire history of states. We can think of it as a memory-less process.Stating it formally, the Markov process is a stochastic (random) process that has the Markov property. The *Markov property* means that the *conditional probability distribution of future states depends only on the current state*, not on the sequence of events that preceded it. Let's use an example to understand the application of the Markov process.

---

[9] If you are not familiar with $\prod$, it is similar to $\sum$ but used for multiplications. $\prod x = x_1.x_2 \ldots$

Assume we have a sequence of events (s stands for the event) happening together for Luckless Larry on a day that he should attend an important meeting:

* s1: His phone's clock rang in the morning, and he accidentally pressed the silent button instead of the snooze button. As a result, he woke up an hour later.
* s2: He starts running around the house to find his clothes and wear them.
* s3: As he ran toward the restroom, his toe hit the bed and began to inflate.
* s4: He takes his phone to the bathroom to send an e-mail because he's going to be late, but... his phone falls into the toilet bowl.
* s5: He spent an extra hour washing and disinfecting his phone and tying up his bloated toe.
* s6: He left the house without breakfast, but when he ran to get the bus, he slipped and fell into a hole full of mud.
* s7: He was late in the meeting, and a colleague made a sarcastic joke about a guy who is late and wear muddy clothes.

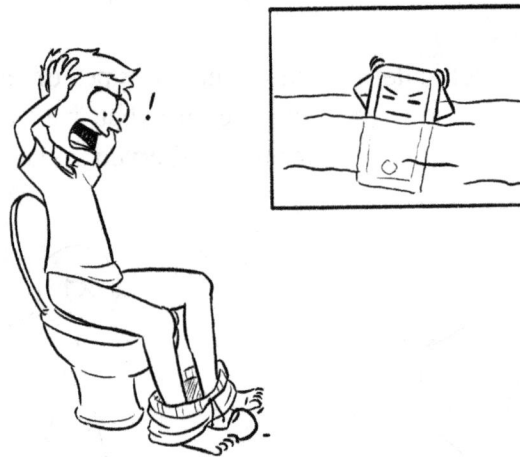

Figure 5-36 is one way of interpreting the probabilistic graphical models of events that happened to Larry. Note that events are nodes, and the transition from one event to the next is a link between two nodes. If the probability of each event depends on its previous event only and not other events, it is called a *first-order Markov chain*, and Figure 5-36 is an example of it. For example, 's4' only depends on 's3' and not 's2' or 's1', which in reality might not be correct. In Larry's example, we know 's3' (hitting his toe on a bed base) happened because of 's1', and thus we can say 's3' depends on 's1'.

If the probability of each event depends on its two previous events, it is called *the second-order Markov chain*. For example, to predict a state ('s3'), we should look back into two previous states ('s2' and 's1'), it is second-order Markov because we looked back into 2nd previous states.

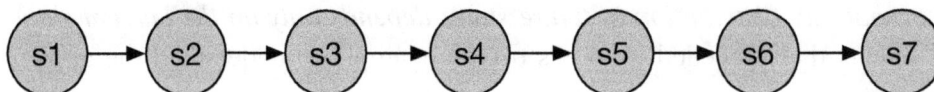

Figure 5-36: Sequence of events (observed variables) which has happened to Larry Luckless.

The next day, he purchased an AI assistant to manage his lifetime better. The AI assistant starts to analyze his previous day to learn how it can help him. It should trace back all events (observation variables) which has resulted in event 's7' and analyze them to see which ones lead to 's7'. In particular, the assistant should study the impact of 's1' on 's2', 's1' on 's3', both 's1' and 's2' on 's3', 's2' on 's3', … this continues until the end of the world. Hence, such an approach is computationally inefficient for an algorithm. Figure 5-37 shows the spaghetti of comparisons we need for such a simple understanding of event 's7'. This figure shows that it is impractical to analyze the dependency of an event on all its previous events.

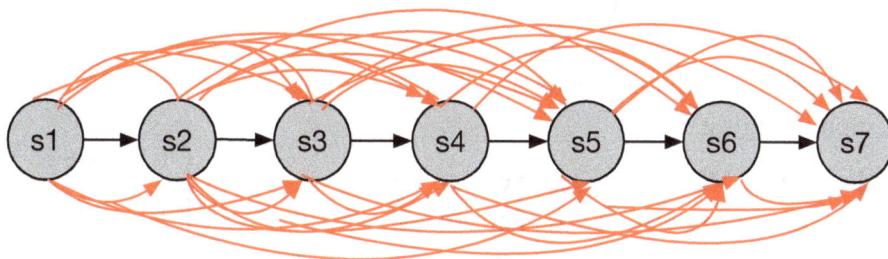

Figure 5-37: A jungle of sequential impacts of events on s7, which should be studied by our robot.

To resolve this jungle of events and relations, we can assign each observation state a *hidden state*, which controls the observation state. In particular, for each observable variable 's_m', we introduce the hidden variable 'h_m'. Therefore, instead of transitioning from observable variables, hidden variables can handle the transition between variables. This is called the "Hidden Markov Model" and it is shown in Figure 5-38.

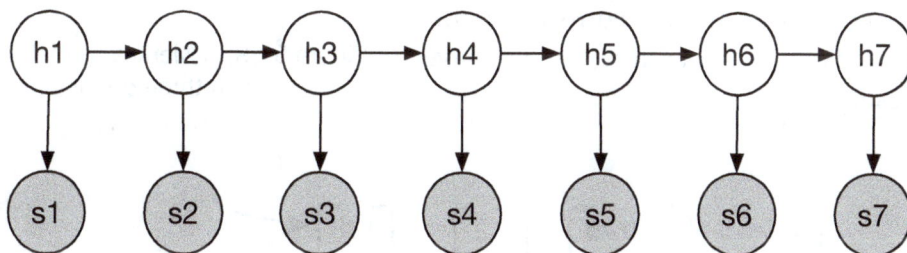

Figure 5-38: A Markov chain that presents observational variables and hidden variables. It is common to have observational variables filled and hidden variables not color filled.

Markov Model for Prediction

Suppose we intend to create a model to predict the weather, one of the most popular uses of the Markov model. We want to use the Markov Model for prediction, and the model has three states: Sunny, Rainy, and Cloudy.

When the weather is sunny, the chance of having it sunny tomorrow is 70%, the chance of having it rainy the next day is 10%, and the chance of having it cloudy is 20% (blue lines and texts in Figure 5-39). When the weather is rainy, the chance of having it rainy the next day is 10%, cloudy 45%, and sunny 45% (red lines and texts in Figure 5-39). When the weather is cloudy, the chance of having rain is 70%, having clouds is 20%, and sunny is 10% (See black lines and text in Figure 5-39).

How did we choose these numbers? We watch the previous data, and by leveraging the historical data, we identify these numbers. Each of these three observed variables (Rainy, Cloudy, and Sunny) are called *states*. Another definition we need to learn is the *state transition probability*, which presents the probability of moving from one state into another state, e.g., from the state "Sunny" to the state "Cloudy" is 25%. We can present all state transition probabilities as a matrix A on top of the graph in Figure 5-39.

$$A = \begin{array}{c} \\ rainy \\ cloudy \\ sunny \end{array} \begin{array}{ccc} rainy & cloudy & sunny \\ \begin{bmatrix} 0.1 & 0.45 & 0.45 \\ 0.7 & 0.2 & 0.1 \\ 0.05 & 0.25 & 0.7 \end{bmatrix} \end{array}$$

Besides, to predict with a Markov model, we need an *initial distribution* or *starting probability*, and π sign is used to present it. The initial distribution is a vector of numbers between 0 and 1. It is used to present the distribution of each state at time 0. In this example, let's assume that since the first day God created that region, it was rainy with a probability of 0.2, cloudy with a probability of 0.1, and sunny with a probability of 0.7. This is what we know (a priori knowledge), and we present it as a vector, i.e., $\pi = (0.2, 0.1, 0.7)$.

Now, if we multiply the transition probability with the initial distribution, we can get a vector that predicts what is going to happen tomorrow in terms of weather and probabilities. Therefore, we will have the following results:

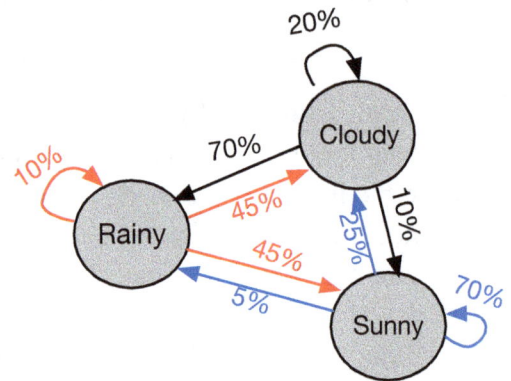

Figure 5-39: A Markov model of weather with three conditions.

$$\begin{bmatrix} 0.1 & 0.45 & 0.45 \\ 0.7 & 0.2 & 0.1 \\ 0.05 & 0.25 & 0.7 \end{bmatrix} \times \begin{bmatrix} 0.2 \\ 0.1 \\ 0.7 \end{bmatrix} = \begin{bmatrix} 0.38 \\ 0.23 \\ 0.19 \end{bmatrix} \begin{array}{l} \leftarrow rainy \\ \leftarrow cloudy \\ \leftarrow sunny \end{array}$$

As another example, let's predict the probability of having four sequences of days, such as "cloudy, rainy, rainy, and sunny" sequences. By using the Markov model, we can write it as joint conditional probability as follows:

$$P(S_{cloudy}) \cdot P(S_{rainy}|S_{cloudy}) \cdot P(S_{rainy}|S_{rainy}) \cdot P(S_{sunny}|S_{rainy}) = (0.2) \cdot (0.45) \cdot (0.1) \cdot (0.05) = 0.00045$$

It is clear that it will be a small value because as much as the size of the sequence increases, the probability or predictability power reduces.

Let's summarize the information we need to create a Markov model for prediction as follows:

* $S = s_1, s_2, \ldots s_n$ is the set of n *states*.

* $A = a_{11}, a_{12}, \ldots, a_{mn}$ is a *transition probability matrix*, a_{ij} represents the probability of moving from state i to state j.

* $\pi = \pi_1, \pi_2, \pi_3, \ldots, \pi_n$ is the *initial probability distribution*, and $\sum_{i=1}^{n} \pi_i = 1$

Hidden Markov Model for Prediction

Now that we have gained enough understanding about how a Markov model performs sequence prediction, we can start the HMM prediction. Before proceeding further, please be sure you remember all previously described terms and concepts.

Similar to the Markov Model, the HMM is also used to assess the current state by leveraging the history of past events. However, we use HMM when we can not observe some states. This means that the events we are interested in are *hidden*, unlike states inside the Markov model, and we cannot observe them.

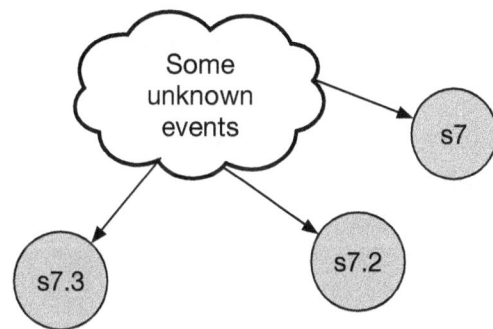

Figure 5-40: What Patty sees from Larry's events.

The difference between the Markov model and the Hidden Markov model is that all states are observable in the Markov model, but states themselves are not directly observable. Instead, each state produces an observation with a certain probability (observation probabilities), and these observations are visible.

Let's get back to our tradition and explain it with an example. Going back to Larry Luckless's example, let's define two more states.

s7.2: Mr. Luckless arrives on time, and everything works fine, with no conflict with his colleagues.
s7.3: Mr. Luckless didn't show up in the office at all.

Another person is working in Larry's office; she is Patty Peacemaker. She is a manager who tries to resolve issues between staff members, and she is experienced in managing conflicts. She doesn't know (can not observe) what has happened with Larry that led to events 's7', 's7.2', or 's7.3'. She can only see Larry arrived wet, muddy, and late (it could be because of a personal issue,

Patty Peacemaker

falling into a hole in the street, or something else). In other words, she doesn't know what events occurred before Larry arrived at the office. In the context of HMM, they are hidden events or states.

Figure 5-40 is a presentation of Patty's observation. She only observes events happening inside the office. States inside the cloud are called *hidden states*[10], which are not observable states. For example, the events that have happened inside Larry's home or on the route to work are hidden states for Patty, who sits inside the office from early morning.

In addition to S, A, and π to model and predict this with HMM, we should introduce two more definitions, *observable variables* and *observation probability matrix (emission probabilities matrix)*. Observable variables (or states or sequence) $O = O_1, O_2, \ldots O_T$, present a sequence of T number of observations. The observation probability matrix or emission probabilities matrix expresses the observation O_t being generated from the state s_i, i.e., $B = b_i(O_t)$. Here, b_i is a function that gives the probability of observing a particular output from the state i. O_t denotes the observation at time t.

In the given example, *s7, s7.2,* and *s7.3* are observable variables for Patty. Emission probabilities are probabilities of their occurrences. HMM is presented as a graphical Figure 5-41. h1, ...,h7 is a Markov chain, which we have described before, but s1 to s7 are observation variables $(o_1, o_2, \ldots)$. Patty decides to solve Larry's latency and absence issues and reduce his conflicts in the work environment. She has observed Larry's arrival in the office and could use the observation information to gain insight into the events ($S_1 \ldots S_6$ are events that she cannot observe).

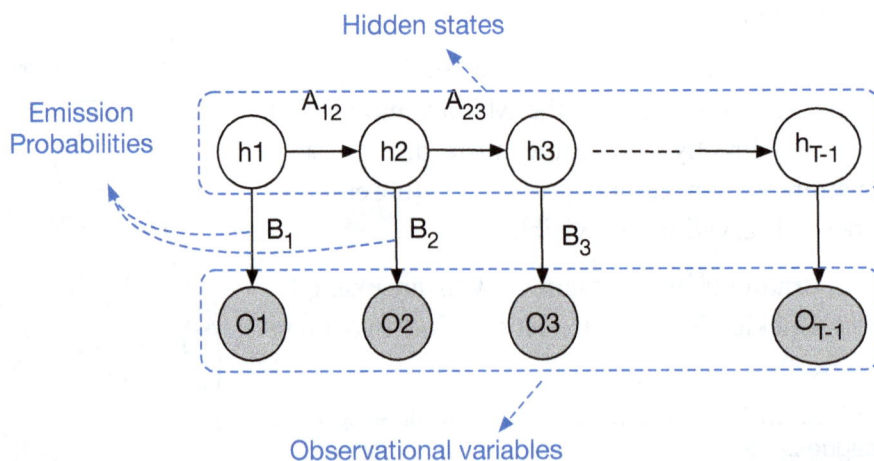

Figure 5-41: Probabilistic Graphical Model HMM, along with its variables and notations.

---

[10] In the probabilistic graphical model, we don't use the cloud; here, we use it to present hidden states.

For the sake of simplicity, to continue this example, let's use only three of those events S_1, S_2, and S_3. Based on the previous discussion with Larry, she understands that the "transition probability matrix", A, and the "emission probability matrix", B, will be as follows:

$$
A = \begin{array}{c} \\ S1 \\ S2 \\ S3 \end{array}
\begin{array}{ccc} S1 & S2 & S3 \\ \left[\begin{array}{ccc} 0.1 & 0.5 & 0.4 \\ 0.4 & 0.5 & 0.1 \\ 0 & 0.7 & 0.3 \end{array}\right] \end{array}
\qquad
B = \begin{array}{c} \\ S1 \\ S2 \\ S3 \end{array}
\begin{array}{ccc} S7 & S7.2 & S7.3 \\ \left[\begin{array}{ccc} 0.7 & 0.05 & 0.25 \\ 0.4 & 0.5 & 0.1 \\ 0 & 0.7 & 0.3 \end{array}\right] \end{array}
$$

Note that both A and B have been constructed based on previous observations. A combination of A, B, and π is called the *Hidden Markov Model,* and it is formalized as follows:

$$\lambda = (A, B, \pi)$$

These algorithms rely on dynamic programming. Dynamic programming uses a recursive algorithm to find the overlapping subproblems. If you are not familiar with dynamic programming, let's make a super simple, non-scientific definition. Dynamic programming is a way to make an algorithm efficient by storing some of the results, and it works for algorithms that have repetitive computations. We will encounter Dynamic programming later in this book as well, in Chapter 13 (Reinforcement Learning).

Likelihood (Forward Algorithm)

When we need to identify the *probability of a sequence of observations O*, we use the Forward algorithm. In other words, the objective of this algorithm is to find the *likelihood P(O| λ)* given $\lambda = (A, B$ and $\pi)$ as *input* to the algorithm.

To better understand the use of this algorithm, read the following example. Next week, Larry should go to an expensive course on behalf of the company. The course is five sessions long, and he should attend at least three out of five sessions absolutely on time (and one session could be tolerated as a late arrival). In total, he should be there four sessions out of five in general. Patty knows that the first and second sessions might not be very important, but the last three sessions are very important for him to learn the material needed by the company. Assuming O represents the set of observations, she needs to calculate the probability of the following series for Larry's course attendance as follows: $O = \{absent, late, ontime, ontime, ontime\}$ or $O = \{S7.3, S7, S7.2, S7.2, S7.2\}$.

Note that she can not simply use the average on O events and find the probability of her desired sequence because those events are associated with probabilities, so she should use HMM. For this scenario, she should use the *forward algorithm* of the HMM. When she has a good estimate of Larry's attendance, she may be able to recommend whether to send Mr. Luckless to the course or look for another candidate.

S1	S2	S3
T	T	T
T	T	F
T	F	T
T	F	F
F	T	T
F	T	F
F	F	T
F	F	F

Table 5-9: All possible combination of events

Since we do not know whether *S1, S2, and S3 have occurred,* we consider them binary variables. It means either Larry shows up on time (true) or does not show up (false). To model the probability with the Forward algorithm, the algorithm needs all possible sequences of these three events. There will be 2^3 different sequences, as shown in Table 5-9. This means that it should compare the probability of the desired observations (O) over the *sum of all possible state sequences, weighted by their probability.* Assuming hidden states are presented $H = \{h_1, h_2, \dots\}$, the HMM uses the following equation to compute the Likelihood, which could be generalized for the Forward algorithm as well.

$$P(O) = \sum_{i=1}^{n} P(O, H_i) = \sum_{i=1}^{n} P(O|H_i)P(H_i)$$

For our example (based on conditions in Table 5-9) we will have:

P(*absent, late, ontime, ontime, ontime*) =

        P(*absent, late, ontime, ontime, ontime | S1=T, S2=T, S3=T*). *P(S1=T, S2=T, S3=T)* +

        P(*absent, late, ontime, ontime, ontime | S1=T, S2=T, S3=F*). *P(S1=T, S2=T, S3=F)* +

        P(*absent, late, ontime, ontime, ontime | S1=T, S2=F, S3=T*). *P(S1=T, S2=F, S3=T)* +

        P(*absent, late, ontime, ontime, ontime | S1=T, S2=F, S3=F*) ....

To calculate each of those hidden variable probabilities, we can use matrix *A*. You can see it is very inefficient, and as much as the number of hidden states grows, the computational complexity of the algorithm exponentially increases, and we have *(#hidden-states)#observations* sequences.

To handle this, we should use the Forward algorithm, which is efficient, and its computational complexity is O(*#hidden-states²* × *#observations*). The Forward algorithm uses dynamic programming. It uses the following equation to compute the likelihood:

$$P(O|\lambda) = \sum_{i=1}^{N} \alpha_T(i)$$

Here, *O* is the observation sequence, λ presents the parameters of HMM, and N specifies the number of states in HMM model. $\alpha_T(i)$ is the *forward probability* at the final time *T* for state *I*, indicating the probability that ends in state *i*. $\alpha_T(i)$ is computed with dynamic programming as:

$$\alpha_T(i) = [\sum_{j=1}^{N} \alpha_{T-1}(j)a_{ji}]b_i(O_T)$$

a_{ji} is transition probability read from transition probability matrix, $b_i(O_T)$ is the probability of observing O_T at state *i*.

Once again, let's emphasize that dynamic programming is a sort of memory management, and when something happens once, it will be kept in memory, so the next time the code does not redo the process, it instead gets it from memory. In particular, the Forward algorithm stores intermediate values as it calculates the observation sequence probabilities.

To summarize, we use the likelihood algorithm to understand: given observation sequences, what are the probabilities that are being produced for each observation by a particular HMM model?

Decoding (Viterbi Algorithm)

When we need to discover the most probable hidden state sequence H from available observations, we use the Viterbi algorithm. In other words, the objective of this algorithm is as follows: given λ $(A, B,$ and $\pi)$ as input to the algorithm, find a sequence with the highest of probability. In simple terms, we have a set of observation sequences and HMM model available as inputs. We would like to know the single optimal hidden sequence for the given HMM models and observation sequences.

Assume Patty needs to schedule a morning meeting between Larry and the Big Boss in the next few months. She should know what is the best single possible sequence of Larry's appearances in the office. In other words, based on the available knowledge about Larry (given λ), she will need to find one single sequence that presents the best estimation for Larry's office arrival probabilities. To find this, she observes Larry's office arrival patterns weekly and uses the Viterbi algorithm to identify the most likely sequence.

Note that, unlike the forward algorithm, the output of the decoding algorithm is one single sequence, which has the maximum probability. Also, the Viterbi algorithm uses dynamic programming, and it does not re-calculate an already calculated probability.

The Viterbi algorithm operates in four steps. First, for each state i, it computes the initial probability as follows: $V_t(i) = \pi_i . b_i(O_t)$

Here $V_t(i)$ presents the highest probability of any state sequence ending in state i, at time t. π_i is the initial probability of state i, and $b_i(O_t)$ is the probability of observing O_t given state i. At the initialization step, $b_i(O_i) = b_i(O_1)$. The next step is the recursion step for each next observation O_t, and for each state I, it updates the probability by using $V_t(i) = max_j[V_{t-1}(i) . a_{ji}] . b_i(O_t)$

equation. In this equation, a_{ji} is the transition probability from state j to state i. The last step ends by identifying the $V_t(i)$, which has the highest probability.

To summarize, we use the decoding algorithm to understand the most probable hidden sequence from a given sequence of observations and the HMM model.

Training (Baum-Welch) Algorithm

Patty's model sounds attractive, and she intends to make an HMM model for each of her employees based on her observations. To do so, she needs HMM parameters $\lambda = (A, B, \pi)$ for each employee based on her observations about that employee.

We use the training algorithms when we intend to find *transition probabilities (A)* and *emission probabilities (B)*, and initial distribution, π, assuming our inputs are *observations sequence O*. This is known as learning or training the model $\lambda = (A, B, \pi)$ to fit the observation (O). In simple words, the input is A, B, π, and the output is (A, B, π), but improved to model the observations sequence O.

The most common training algorithm is the Baum-Welch algorithm [Baum '70]. The Baum-Welch algorithm applies the expectation maximization (EM) algorithm (Chapter 3) to find the unknown parameters (A, B, and π) of the HMM. In particular, it uses the expectation maximization iterative procedure to train the parameters (A, B, π) of an HMM model that best fits a set of observations O. This means that we start with some random possible transition probabilities (A), emission probabilities (B) and initial distribution (π), but the goal is to find accurate A, B, and π that fit the observations sequence O.

Baum-Welch uses the *Forward-Backward* algorithm [Baum '67] for the expectation step (E step) to re-estimate the model parameters. The Forward-Backward algorithm computes the probabilities of being in a particular state at a particular time, given the observation sequence. These probabilities are then used to update the model's parameters in the Maximization (M) step, improving the model's ability to represent the observation sequence O. In particular, the M step update A, B, and π that fit better the observations sequence O.

In the remainder of this section, we describe the Forward-Backward algorithm. The Forward-backward performs reasoning based on (i) *Forward pass*, looking into past observations; (ii) *Backward Pass*, estimating future observations; and (iii) *Combination*, combining them to perform accurate reasoning, a.k.a, posterior probability calculation.

Forward pass: The objective of the forward pass is to calculate the probability of ending up in each state after observing the first t observations in the sequence. Assuming $\alpha_t(i)$ is the forward probability from the previous step, to compute the forward pass for each time step t and state i, it computes the forward probability $\alpha_t(i)$ as follows:

$$\alpha_t(i) = [\sum_{j=1}^{N} \alpha_{t-1}(j)a_{ji}]b_i(O_t)$$

We explained the rest of the variables in this equation earlier, and due to laziness, we did not explain it again.

Backward pass: The objective of the backward pass is to calculate the probability of observing the future observations from time t+1 to the end, given that it is in state i and time t. Assuming $\beta_t(i)$ is the probability of the future observations given state i at time t, to compute backward pass for each time step t (going backward from the last observation) and each state i, it computes the backward probability as follows:

$$\beta_t(i) = \sum_{j=1}^{N} a_{ij} b_j(o_{t+1}) \beta_{t+1}(j)$$

Combination: To compute the probability of being in state i at time t, i.e., $\gamma_t(i)$, given the entire sequence of observations, it combines forward and backward passes by the following equation:

$$\gamma_t(i) = \frac{\alpha_t(i)\beta_t(i)}{\Sigma_j = 1^N \alpha_t(j)\beta_t j}$$

NOTES:

* An HMM is considered an unsupervised learning model. This means it learns from observation sequences only, without the need for explicitly labeled data. It uses algorithms like the Baum-Welch algorithm to discover hidden states and their transition probabilities based on the data itself. However, HMMs can also be trained in a supervised fashion if you have access to labeled data. This means each observation point has a corresponding label indicating the hidden state. This can be helpful for improving accuracy or learning specific classifications.

* In the era of RNN and Transformer (which we will explain them in the neural network chapters of this book), who needs to learn and use HMM? Big data or strong machines for training are not always available. Therefore, HMM models are still in use in real-world applications.

Summary

If you read this chapter from beginning to end, you might find it extremely boring, but if you intend to design your algorithm and not use the existing one, this is one of the most useful chapters of this book. Knowing algorithms that improve the performance of search space gives you an upper hand as an algorithm developer or data scientist to deal with resource limitations.

We start with model concepts required for association rule mining algorithms, including support, confidence, itemsets, and transactions. Then, we suspend discussion about association rule mining algorithms and describe search improvement methods. We have explained Hash Tables, different Tree structures, sliding windows, Bloom filter, and Minhashing as algorithms used to reduce the search space. Table 5-10 summarizes the tree structures we have described, and they can be used to improve search methods.

Search Algorithm	Computational Complexity (Search and Insert)	Applications
Binary Search Tree (BST)	Balanced: O(log n) Unbalanced: O(n)	When the structure of data is static and not changing
2-3 Tree	O(log n)	Try to mitigate the data size limitation of binary search tree.
B-Tree	O(log n)	One of the most popular index that is used in relational databases.
Red Black Tree	O(log n)	When we need frequent insert and remove into/from the tree.
Trie	O(m), m: the maximum length of string	Used to store words. e.g, "auto complete" features in browsers or "spell checkers" in word processing applications.
Radix Tree	O(m), m: the maximum length of string	Used to store words. e.g, "auto complete", indexing, string search.

Table 5-10: Some common tree structures that are used to reduce the search space.

Next, we described associate rule mining algorithms and then sequence mining algorithms. The difference between sequence mining algorithms and associate rule mining algorithms is the order of the data. Sequence mining algorithms are used when the order of the data is important and should be taken into account while searching for frequent patterns. Association rule mining and sequence mining algorithms are summarized in Table 5-11.

At the end of this chapter, we described HMM and the rationale for using HMM for prediction. In summary, HMM can resolve three types of problems:

1- The model parameters and observed data will be given as input. The HMM estimates the optimal sequence of hidden states.

2- The model parameters and observed data will be given as input. The HMM calculates the model likelihood.

3- The observed data will be given as input. The HMM estimates the model parameters.

	Ordered	Characteristics	Complexity	use cases
Apriori	No	(i) Create a candidate list (ii) Search in BFS model (iii) Scans the dataset k times to find the support of $k+1$ itemsets	$O(2^n)$	Baseline algorithm, good for small dataset.
FP-Growth	No	(i) Create FP-Tree and divide it into smaller datasets (ii) Search in BFS model (iii) Only scan the dataset twice	$O(n^2)$	Large dataset, it it more memory efficient than Apriori
ECLAT	No	(i) Use item as key (ii) Search in DFS model (iii) Only scan the dataset twice	$O(2^m)$ m= number of unique items	Large dataset, it it more memory efficient than Apriori
GSP	Yes	(i) Calculate the frequency of each item and then prune the dataset (ii) Scan the dataset k+1 times (k=number of iteration in the dataset) (iii) brute force comparison.	$O(n^2.m)$ m= average length of sequence, n= is the number of sequences	Baseline algorithm, good for small dataset. Not good for long sequential pattern.
SPADE	Yes	(i) convert dataset from vertical into horizontal structure (ii) creates the combination of sequences (subsequence generation) and scans the dataset for each new subsequence (iii) brute force comparison and make a large candidate list, similar to Apriori	$O(n.m^2)$ m= average length of sequence, n= is the number of sequences	Baseline algorithm, good for small dataset. Not good for long sequential pattern.
FreeSpan	Yes	(i) find a sequential pattern with length k (at first k=1) (ii) divide the search space based on prefixes and construct new subsequence (projected dataset) (iii) repeat step (i) and add increase k until no frequent patten left.	$O(m.n)$ m= average length of sequence, n= is the number of sequences	Rely on pattern growth, good for handling along sequences
PrefixSpan	Yes	Similar to FreeSpan, but slightly more efficient than FreeSpan, because projected dataset were created from "prefixes", and not all of the possible frequent subsequences,	$O(m.n)$ m= average length of sequence, n= is the number of sequences	Rely on pattern growth, good for handling along sequences. No need for candidate list generation and projected datasets shrinking in each iteration

Table 5-11: An overview on association rule mining and sequence mining algorithms.

Further Reading and Watching

* Chapter (i) of *Artificial Intelligence, the Modern Approach* from Russel and Norving [Russell '09] has some explanation of search and problem solving by searching. It was a classic reference book, but its content seems a bit outdated in the current state of AI.

* IFor a more modern take on algorithms for search, we recommend Chapter 3 of *Algorithms* book by Robert Sedgewick [Sedgewick '11]. That section has a well-written section on trees and how they search or insert their content. If you are interested in learning more about trees, *Cracking the Interview Code* [McDowell '16] has a detailed easy to understand description of

inserting and searching in trees. Cormen et al. provide a very detailed description of tree structures. *Introduction to Algorithm* book [Cormen '22].

* Bill Mill provides an interactive tutorial about the Bloom filter: https://llimllib.github.io/ bloomfilter-tutorial. It has a good demo for the Bloom filter.

* Vaidehi Joshi proposed a good online tutorial for both B-Tree and Trie. https://medium.com/ basecs. Also, her other tutorials are good.

* If you are interested in learning more sequence mining and frequent itemset mining algorithms, check Aggarwal's book [Aggarwal '14]. It covers many different algorithms and provides a good introduction to those algorithms. Some of them are not easy to understand, but worth taking a look at them. Also, Zaki et al. [Zaki '14] book covers some of these algorithms in detail.

* It is really hard to construct an example for the Prefixspan algorithm without human error; we are thankful for Wen-Chih Peng slides that are available online and enable us to check if we have correctly constructed our example.

* Bishop's *Pattern Recognition* book [Bishop ' '06] has very good explanations of graphical models and Markov models; it is among the fewest parts of this book that we understand not by re-reading every paragraph more than five times.

* There are many online tutorials for HMM. A good one (https://www.cs.sjsu.edu/~stamp/rua/ hmm.pdf) is from Mark Stamp, "A Revealing Introduction to Hidden Markov Models". Also, "Speech and Language Processing" book by Jurafsky and Martin [Jurafsky '18] Appendix A (https://web.stanford.edu/~jurafsky/slp3/A.pdf), has a good explanation of HMM models with examples.

Chapter 6: Feature Engineering

If you read previous chapters, you can imagine that you are a pioneer in both unsupervised machine learning and strong in statistical analysis. However, let's step outside for a minute out of our imagination; in real-world settings, we apply an algorithm on a dataset, and we can easily realize that feeding a dataset into the algorithm and then immediately getting back a good result is impossible. There is blood and sweat involved, as we have described in Chapter 1; most of our time will be spent on preparing the dataset, such as cleaning, extracting related information from the raw data, convincing the boss for better hardware, getting more time, not passing breakfast and having fast food lunch to save time, and many other issues related to data cleaning and preprocessing.

This chapter focuses on one of the important middle steps in the pipeline of machine learning, i.e., feature engineering. In simple words, feature engineering is preparing the data for the machine learning algorithm. Real-world data includes lots of noise, missing data points, and sometimes, the data comes from many different sources. Therefore, iterative preprocessing is required to prepare the data for the algorithm. Why iterative? Because we will repeat this process several times until we get an acceptable result from our machine learning algorithm. Therefore, we can say that feature engineering is a trial-and-error process.

More about data cleaning and preparing the data for the machine learning algorithm will be explained in Chapter 16. In this chapter, our focus is on feature engineering.

what is the reality of a dataset

what we think about a dataset

Basic Concepts of Feature Engineering

First, we clarify some definitions required for this chapter. Then, we describe feature engineering methods and algorithms for different types of data.

A *feature* is a representation of raw data in an understandable machine format. Assuming the dataset as a table, each column of the table presents an attribute or feature, which can or can not contribute to the execution of the machine learning algorithm. We can also convert features from one format to another format. For example, we can convert {null, low, medium, high, all} to a numerical feature as follows: {0, 0.25, 0.5, 0.75, 1}.

Feature engineering is the process of identifying, filtering, and/or reconstructing the most relevant data (from the original raw data) that we intend to use and removing the irrelevant data that might affect the quality of the machine learning method's output. Neural network models, which are a big part of machine learning and recent advances in artificial intelligence, do not require traditional feature engineering but still require data preprocessing, which overlaps with the feature engineering methods we learn in this chapter.

Feature selection is the process of selecting a set of attributes that are useful for the machine learning algorithm and removing the attributes that are unimportant for the algorithm. Chanrashekar and Sahin [Chanrashekar '14], described feature selection as selecting inputs that reduce noise and the effect of invariant variables.

Feature generation is the process of making new features from a combination of existing features, which are useful for the machine learning algorithm. For example, assuming we have X, Y, Z, and an arithmetic mean of three attributes together, we can generate a new feature, which might help the algorithm better discriminate the data for classification rather than each of these attributes alone.

We have explained hyperparameters in Chapter 1. To review it, assume that we have a machine that it has lots of knobs and buttons on it. These buttons and knobs are called hyperparameters. By playing with these knobs and buttons, we will manipulate the behavior of the machine. Our machine receives input and produces output, but to produce an optimal output, we need to tune its hyperparameters. Tuning these hyperparameters could be a subset of feature engineering as well.

Feature Selection

As we have described, most of the time, the raw data is not good enough to feed them directly into a machine learning algorithm. One of the initial tasks we need to do with the data is to select what data (features) we need and filter the data[1] we don't need. Now, the question is: how do we identify useful features, and how do we select them?

As an example, consider that we have a database of chicken information that presents our chickens' attributes. Since the chicken cultivation business is good, we launched a new company that produces outfits and clothes for chickens.

We are planning on launching a fashion show for chickens, and we should select some fit chickens for the show. Nevertheless, preparing chickens for the show costs a lot, and when they are chicks, we should select the best talents among them. We should work on enabling them to participate in entertainment shows when they get old enough. Therefore, we are dealing with a prediction. The goal of our predictive model is to identify attractive chickens who will have the best body shapes when they grow up.

We have the following information in our chicken table: 'Chicken id', 'name', 'weight', 'height', 'daily mood score', 'eye shape', 'singing voice score', 'peak size', 'feet size', 'last vaccine received date', ' the number of vaccines received', 'flu-shot received', ' the number of eggs hatching per week', 'feather color'.

Looking at this dataset, we observe that there is too much irrelevant information. By feeding all this information into a machine learning algorithm, we might not be able to get a good prediction about the fittest chickens for entertainment shows. Therefore, we need to remove unnecessary features (attributes or column names) and concentrate on selecting features that present whether a chicken is fit for the show. For instance, *the* 'number of eggs hatching per week' has nothing to do with the chicken's appearance and style, and thus, we could remove this feature.

This example is simple, and we can easily remove features, but in some examples, we could not realize which feature is important to keep and which feature is not. Therefore, we should find some more automatic methods to remove the unnecessary features. There are three types of feature selection methods: Filtering, Wrapper, and Embedded.

Filtering methods

Filtering methods process features and remove the ones that are not relevant to the machine learning algorithm goal, e.g., prediction. While we are dealing with numerical data, *correlation analysis* is one of the most convenient filtering methods to remove redundant features. By removing features that are highly correlated, we can make the dataset smaller and thus enable the machine algorithm to operate while using less amount of resources.

---

[1] Here by referring to the data, we mean fields or columns in tabular data and not records or rows.

Contrary to what we have explained, sometimes, we focus on one important feature and remove features that are not correlated with that particular important feature. To check the correlation between numerical data, we can use correlation coefficient tests, including Pearson, Kendall, and Spearman (Check Chapter 3).

Another method to identify correlative features among numerical variables is to use *covariance*. In Chapter 3, we explained how we can identify the relation between two vectors by using covariance. If the covariance between two variables is positive, they tend to change together. In other words, positive covariance suggests that the two variables tend to increase or decrease together. If the two variables are independent (they don't have a correlation), their covariance is zero. Nevertheless, a covariance of zero does not necessarily mean that they are necessarily independent.

We could use the *chi-square test* (check Chapter 3) for correlation analysis between categorical data. The initial assumption of the chi-square test is that two given vectors of data points are independent (not correlated), and the null hypothesis is that both input vectors are dependent (correlated).

Wrapper methods

These methods use a black box algorithm (we do not know what is happening inside the algorithm), and the algorithm decides which features to use. Ideally, the algorithm should combine n number of features in n different methods to identify the best features. However, combining n number of features in n different methods will burn the machine into ashes because of the high computational overload. Therefore, heuristic algorithms[2] or search optimization algorithms will be used to improve the search process.

There are two categories of wrapper methods: *sequential search* methods and *heuristic search* methods. We start by describing two of the well-known sequential search methods. Then, we describe the basics of the genetic algorithm as a heuristic search method.

Sequential Feature Selection and Sequential Backward Selection

The *Sequential Feature Selection (SFS)* algorithm starts with an empty set, and features will be used by the machine learning algorithm one by one, and then the result will be stored. This process continues until the highest accuracy (or lowest error) is achieved. Figure 6-1 presents how the SFS wrapper method operates. First, it gets one feature, as shown in Figure 6-1 (a), then it gets another feature, as shown in Figure 6-1 (b), until Figure 6-1 (c), in which the accuracy will be reduced. SFS realizes the accuracy degradation, and thus, it does not use a feature added in Figure 6-1 (c), and puts it back into the feature pool, see Figure 6-1 (d). The feature pool is some artificial concept we use for you to understand it, and we assume the dataset (cylinder shape in Figure 6-1) on the left is a feature pool.

---

[2] In many real-world applications, it is either computationally very expensive or infeasible to get to the best answer. Instead, we try to approximate the best answer. Algorithms that approximate the best answer and do not get the best answers are usually referred to as heuristic algorithms.

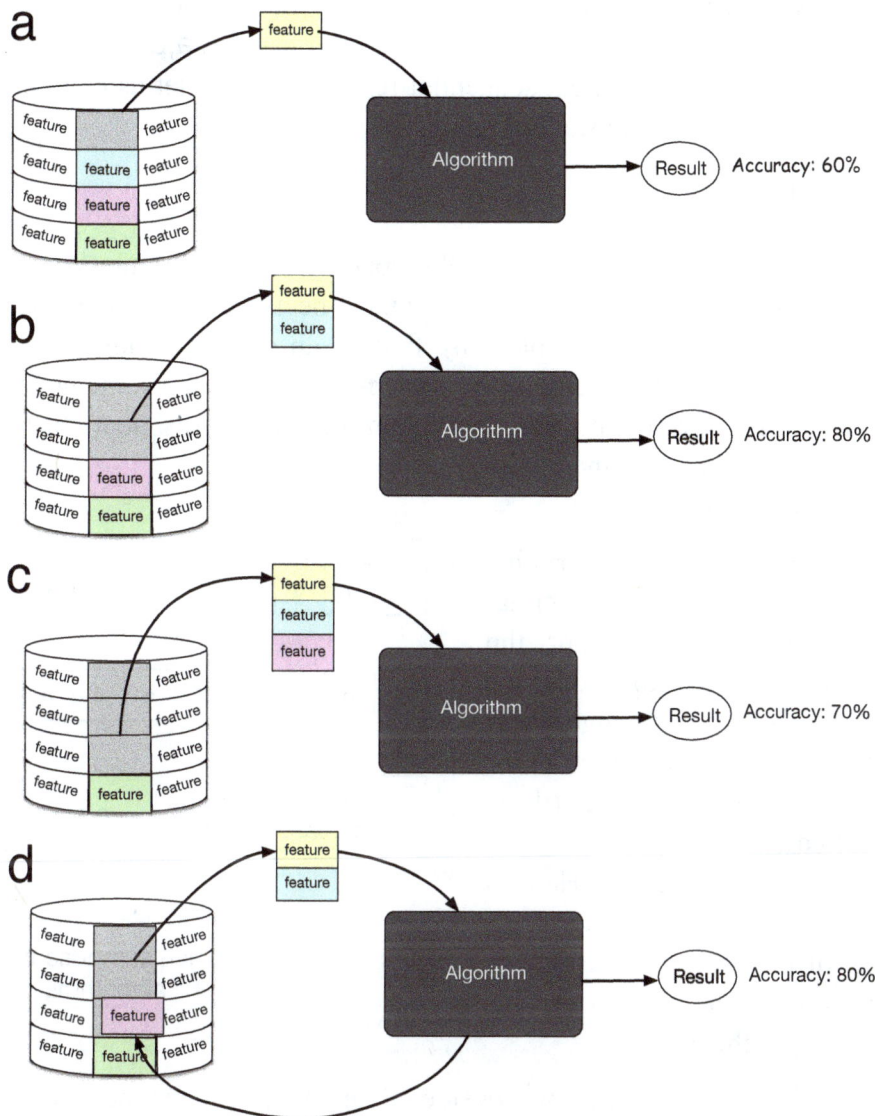

Figure 6-1: A toy example to present how sequential feature selection algorithms work.

Another baseline algorithm is *Sequential Backward Selection (SBS)*, which starts with all sets of features, then removes features one by one, and checks the accuracy (or other objective function) until a desired accuracy has been achieved. It operates in contrast to SFS because first, it feeds all features into the algorithm, then it removes them one by one to study changes in the error or accuracy. In other words, it operates in contrast to SFS. It is easy to visualize in your brain how SBS works as well. If you can't, send us a nice fat check, and we will do the visualization for you.

Note, that these are basic methods and rarely used because they are prone to stuck in local minima or maxima. Usually, your machine learning tool or library uses a more advanced version of these algorithms for wrapper methods. To understand sticking in local minima, assume a scenario in SBS, we add all features a,b,c,d, and the accuracy is 80% now we remove feature d, and the accuracy is 85%, then we remove feature **c** and the accuracy get 87%, and we think feature **a,b** are the best combinations. However, if we used features **a,b,d** we get an accuracy of 90%, which is the highest accuracy, but sequential removal doesn't let us experiment with this. In other words, we are stuck in the local optimum.

Genetic Algorithm

Another fairly known but uncommon group of Wrapper methods for feature selections is *genetic algorithms*. Genetic algorithms are well known for meta-heuristic search approaches. Heuristics are techniques that "try to solve a problem by finding an approximation instead of the best answer". Heuristic methods are specific to a problem. Meta-heuristics methods are similar to heuristic methods but not specific to a single problem. In other words, heuristics could fall into the trap of local optima, but meta-heuristics can come outside the local optima trap[3].

Genetic algorithms are meta-heuristics algorithms that try to find an optimal solution. The approximation is not the optimal solution but sufficient to resolve the problem. In other words, Genetic algorithms are used to find a 'near optimal' solution for a problem that is hard to solve (these problems are called NP-Hard problems, and we will explain them in Chapter 16). Hard to solve, in this context, refers to the operational cost of the problem that is too high, which makes it infeasible to get it solved with existing methods.

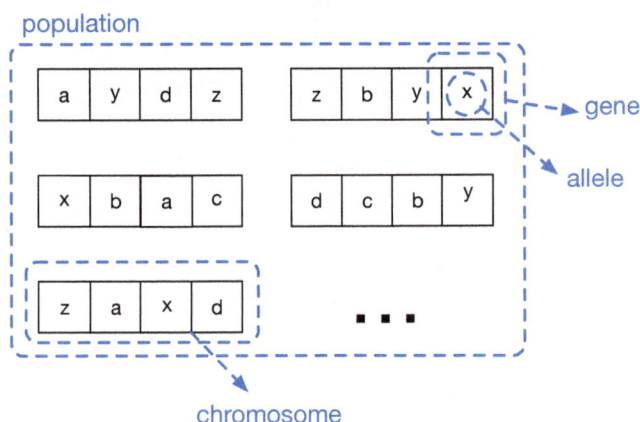

Figure 6-2: Entities which compose the population in genetic algorithms.

Furthermore, genetic algorithms can be considered optimization methods operated based on search. They operate based on Darwin's theory of "Survival of the Fittest". The remainder of this section explains an overview of genetic algorithms and their components. Genetic algorithms try to search for a solution by somehow a random search, but they perform better than purely random or brute force search. Brute force search could provide the best result, but they are computationally expensive.

Genetic algorithms have a pool or *population* of solutions (or answers). Each solution is called a *chromosome,* and each chromosome has a set of variables or parameters, which is called a *gene*. The value of each gene is called an *allele*. Think of a solution as a set of parameters whose combination yields a result, and this result will be represented as a *fitness score*. In the following we lists the concepts we should learn to understand the genetic algorithm.

---

[3] If you can't understand the local optimum and the local minimum, don't worry now; we will come back to this discussion in Chapter 8.

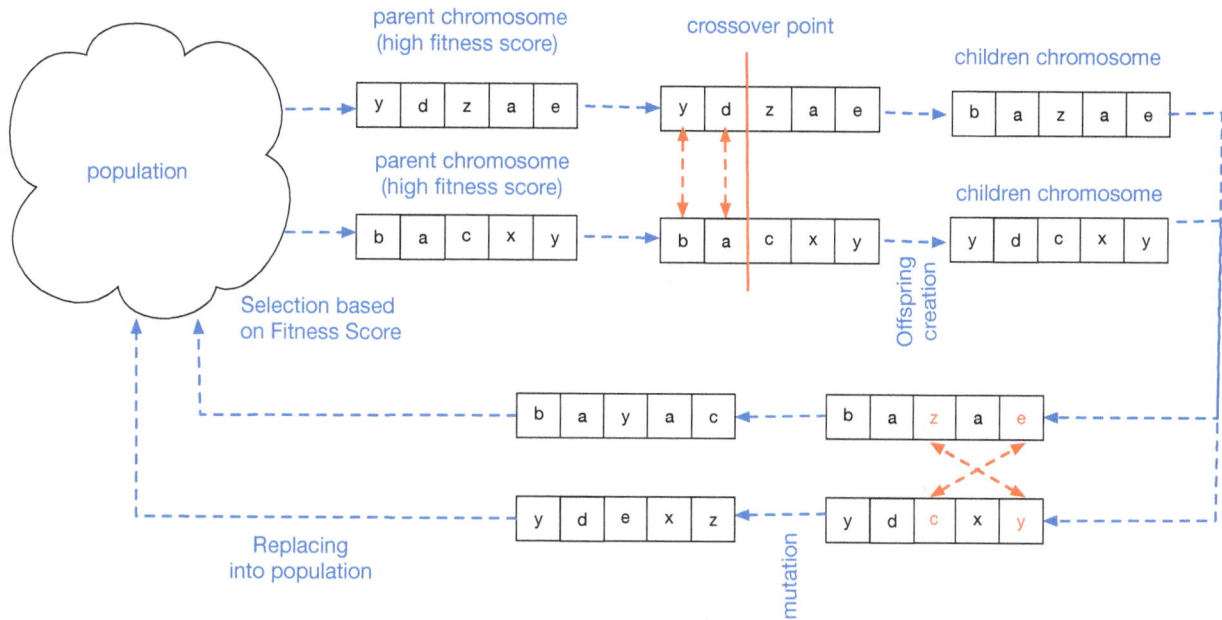

Figure 6-3: Two chromosomes with high fitness scores will be selected. Then, based on the given crossover point, two children's chromosomes will be created. Next, a mutation will be implemented to increase their diversity, and two newly generated chromosomes will replace two chromosomes in the population with a lower fitness score.

Population: A set of potential solutions for the given problem, but not necessarily all possible solutions.

Chromosome: A single solution that is composed of a set of genes.

Gene: Variables or parameters that characterize the chromosome.

Allele: The value inside a gene. In other words, the value of the variable (gene).

Figure 6-2 presents an abstract form of entities in a genetic algorithm. Besides, genetic algorithms have some functions and operations, which are listed below. Please memorize them because we need them to understand how genetic algorithms operate.

Fitness score & Fitness function: A cost of a solution (chromosome) is acquired by a function called fitness function, and this cost is presented in a value called fitness score.

Selection: it is a process in which it tries to find chromosomes with random fitness scores for mating. The mating process here means they pass their genes together and make new children.

Offspring: It is a new chromosome, a.k.a. *children's chromosome*, which has been created as a result of mating among two parents' genes. In other words, offspring creation is a process in which two chromosomes (parents) are combined together and pass over their genes to create new children (offspring). For example, Figure 6-3 shows a crossover point, and this point specifies from which element the parents' gene will be flipped. This particular offspring creation method is known as *crossover*. There are many other methods available for offspring creation, and we

275

only show the crossover here due to its simplicity. The upper part of Figure 6-3 presents how two parents' genes make two offspring with the crossover of two genes.

Mutation: To ensure the diversity of offspring's chromosomes and that they are significantly different from their parent chromosomes, we use mutation. In particular, if the parents are very similar, their offspring could be similar as well (premature convergence). In this case, to maintain diversity, a process called mutation is helpful. In Figure 6-3, the mutation swaps two genes at different positions. There are different methods for mutation, and we chose to show flipping for the sake of simplicity.

Note that chromosomes in Figure 6-3 are not very similar and thus mutation is not necessary. Still, mutation is performed to cover all important concepts in the genetic algorithm. After the mutation process, newly generated chromosomes will go back into the population and replace older chromosomes that have lower fitness scores.

This process (shown in Figure 6-3) continues iteratively until the algorithm can not generate any (the population has converged) new offspring that have significantly higher genetic scores or a certain threshold has been reached.

Embedded methods

If you remember, from Chapter 1, we explained that supervised learning includes training. The embedded methods perform feature selection as a part of the training process. Wrapper methods can be more flexible and might provide better performance since they evaluate feature subsets based on the model performance, but they are usually computationally more expensive than embedded methods. However, since the feature selection process is embedded in the training phase, their computational costs are significantly lower.

There are two good examples of embedded methods, decision trees, and regularizations, such as LASSO, Ridge, and ElasticNet [Tibshirani '96]. We will explain them in Chapter 8 and Chapter 9 in more detail.

Feature Generation

Feature generation is the process of combining, mixing, merging, etc., two or more raw data objects and creating a new data object from them. Why do we do that? Because a newly generated feature improves the accuracy or reduces the computational cost while maintaining the accuracy of the data.

For instance, assume we are building a fitness tracker for our chickens to monitor their physical activities. The fitness tracker uses an accelerometer sensor, which produces three different streams of data, i.e., the X, Y, and Z axes. Nevertheless, in comparison to our hand, chicken's feet (hand) are too small, and it is not possible to put many hardware parts in their fitness tracker and store the data. The size of hardware has a direct correlation with hardware capacity [Rawassizadeh '15]. Therefore, we need a way to reduce the size of the data we collect. We could multiply X, Y, and Z of three accelerometer data and store one result instead of storing three different axes. In other words, from three existing data, i.e., X, Y, and Z, we create one new data object, e.g., multiplication of them $X \times Y \times Z$. This process is referred to as feature generation.

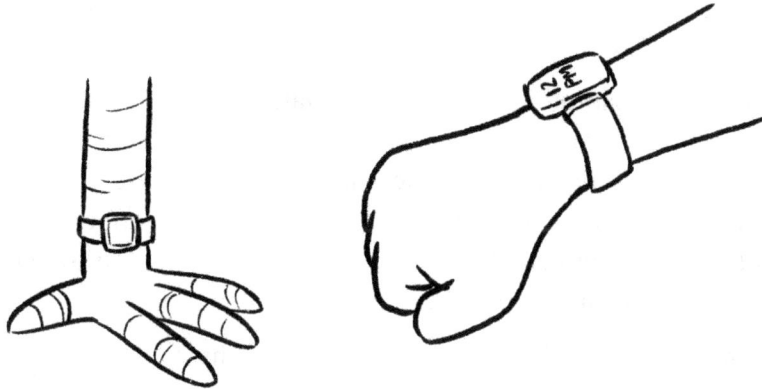

A very simple feature generation method multiplies all features together. This is called *interaction* features or *polynomial feature* creation. For instance, if we have the following features: x_1, x_2, x_3, x_4. their interaction feature with a polynomial degree of two could be as follows: $x_1^2, x_1 . x_2, x_1 . x_3, x_1 . x_4, x_2^2, x_2 . x_3, x_2 . x_4, x_3^2, x_3 . x_4, x_4^2$.

While this method can significantly increase the feature space, potentially leading to computational complexity, it is valuable for capturing complex relationships within the data. Besides, it could be extended to more features. For example, we did not consider $x_1 . x_2 . x_3, x_2 . x_3 . x_4, \ldots$ as features. In other words, we can say that for *n* number of features, there could be many combinations of features, which is too much and, thus, inefficient from a computational perspective. Nevertheless, due to the superior accuracy these methods or similar methods provide, they are being used.

Another well-known, widely used feature generation method is the dimensionality reduction method, which we will explain in the next chapter.

In summary, the main objective of feature engineering is usually to reduce the number of attributes of the dataset while keeping the accuracy (or other objective function) of the algorithm high.

Feature Engineering for Numerical Data

The most common type of data we encounter in machine learning is numerical data. As we have stated before, even when data is not in numerical format, we convert it to numerical format so that the machine learning algorithm can process it. This is due to the fact that computers are designed to operate with numerical data, and conversion to numbers makes it less complex to process.

The first step in handling numerical data should involve performing a *sanity check*. This process checks whether the data is in proper form. For example, it checks whether there is a negative value, missing values, validating range and scales, etc.

Normalization typically refers to the process of scaling numerical data to fall within a certain range, often between 0 and 1, or to have a specific distribution, such as a normal distribution with a mean of 0 and a standard deviation of 1.

Binning the data means substituting the original values of the data with discrete, more generalized values. In other words, binning or quantization divides the entire range of values into a series of intervals and then converts numeric values into categorical counterparts. The term quantization is widely used in neural networks to convert float-32 numbers into smaller numbers such as int-8. Therefore, we recommend using the term binning instead to avoid any confusion. We will explain more about this in Chapter 14.

We have explained more than a trillion times that the machine learning process is based on repeatability. Therefore, we might need to change data to make them repeatable. In this case, we could bin the data, which is especially good for numerical data. For example, we can bin the time of the day into five bins as follows:

* original data: {00:00, 01:00 , 02:00, … 23:59 }

* binned data: {mid-night, morning, afternoon, evening, night}

As we see in this example, to bin the data, we should identify the size of each bin. We could quantize the time of day into two bins as {before 12:00', after 12:00'} too, which two bins are more generalized than having five bins. The more generalization we perform, the easier it gets for the algorithm to work with the data, but too much generalization disrupts the semantics of the data.

Assume you have a biomedicine enterprise and are developing a medication to reverse aging. The drug is more effective for the elderly and not younger people. Therefore, you need a way to categorize your test participants based on their age. Instead of using exact ages, you can group them into categories such as young, middle-aged, and older participants.

Your enterprise realizes that you should inject participants with their medication after lunchtime. Sometimes, they finished lunch at 12:30', sometimes at 12:35', sometimes at 12:37', and so forth. We can generalize after lunchtime between 12:30 and 13:00. In other words, we discretize the continuous data (time of day, e.g., 12:37') into discrete units, i.e., 12:30-13:00.

Scaling Data

Several machine learning models, particularly those involving matrix operations like linear regression and neural networks, are affected by the scale of features within the data. Processes like matrix multiplication become computationally expensive with large numbers, which can strain memory resources. Scaling addresses this issue by transforming the numerical values of features to a smaller range.

In addition to the computational complexity, we should be sure that all features contribute equally to the machine learning decision and that the magnitude of the features does not increase or decrease their role in the final mode, and again, scaling resolves this issue.

Scaling refers to the process of normalizing or standardizing the numerical features of a dataset. It is typically done to ensure that the features are on a similar scale.

Some common scaling methods include *min-max scaling, standardization (variance scaling)*, L_2 *norm,* and L_1 *norm.*

Min-Max Scaling

It is used to compress (squeeze) or stretch feature values in the range of 0 to 1. Assume we have x, a vector of variable its Min-Max scaling factorial be shown as $\tilde{x}$.

It is written as $\tilde{x} = \dfrac{x - min(x)}{max(x) - min(x)}$

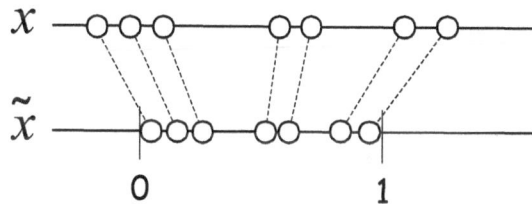

Figure 6-4: Min-Max scaling example from a vector of data.

Figure 6-4 presents this scaling method on a vector of data.

Standardization and z-score Standardization

Another widely used scaling method is called standardization, variance scaling, or z-score standardization. Assuming x is the individual data points that we intend to standardize, $\bar{x}$ is the dataset mean and σ is its standard deviation, the z-score of x is $\tilde{x}$, and it will be calculated as follows:

$$\tilde{x} = \frac{x - \bar{x}}{\sigma}$$

We have explained the z-score in Chapter 3 as well. However, again, we refer to it here to remind you of its importance.

L_1 and L_2 norms

Two common scaling methods are L_1 norm and L_2 norm, which are usually used for the vector of data, and thus, they are referred to as vector norms as well.

L_2 norm or Euclidean norm normalizes the data by using the following equation:

$$\tilde{x} = \frac{x_i}{||x||_2} \quad \text{and} \quad ||x||_2 = \sqrt{x_1^2 + x_2^2 + \ldots x_n^2}$$

For example, assume we have data points in dimensional space as follows: {(2, 3),(-4, 5),(4, 6),(-5, -3),(-3, -4), (4, 4), (2, 5)}. By applying L_2 norm, we get the following scaled data: {(0.21, 0.25), (-0.42, 0.43),(0.42, 0.51),(-0.52, -0.26),(-0.31, -0.34), (0.42, 0.34), (0.21, 0.43)}. We can see that the L_2 norm brings all original data between zero and one.

Also, a less common norm is L_1 norm (Manhattan norm), which normalizes the data by using the following equation:

$$||x||_1 = |x_1| + |x_2| + |x_3| + \ldots = \Sigma_{r=1}^{n}|x_i|$$

Then, similar to the L_2 norm, each data point will be normalized as follows:

$$\tilde{x} = \frac{x_i}{||x||_1}$$

Note that the sign $||.||_x$ is used to present L_x norm.

To our knowledge, there are no specific guidelines on when to use each approach; the choice of normalization depends on the application. Experimentation is often the best way to determine the most suitable normalization method.

The Magic Power of Transformation

Let's assume you are an owner of aviculture and you would like to create a special program that improves chickens' physical performance. You schedule a running competition among chickens. They ran some distance at different times. Most chickens are around 0-30 seconds, and few of them ran that distance more than 50 seconds. Figure 6-5 plots the distribution of their performance. Here, we can see most chickens reach the end line in about 0-30 seconds, and few end their competition around 60-90 seconds.

Figure 6-5 shows power law distribution (check Chapter 3). Chickens who reach the finish line longer than 30 seconds might require some extra fitness programs. However, suppose we use a statistical method or machine learning algorithms to make a model to predict chickens' performance. In this scenario, the large part of the dataset (chickens who ran between 0-30 seconds) might hide the minorities (chickens who ran slowly and longer than 30 seconds). In other words, there will be no place for the minority group in the machine learning model.

To summarize, in some regions of the dataset, there is a large density of data, and in other regions, there is a very low density of data. However, the data for both regions are important to

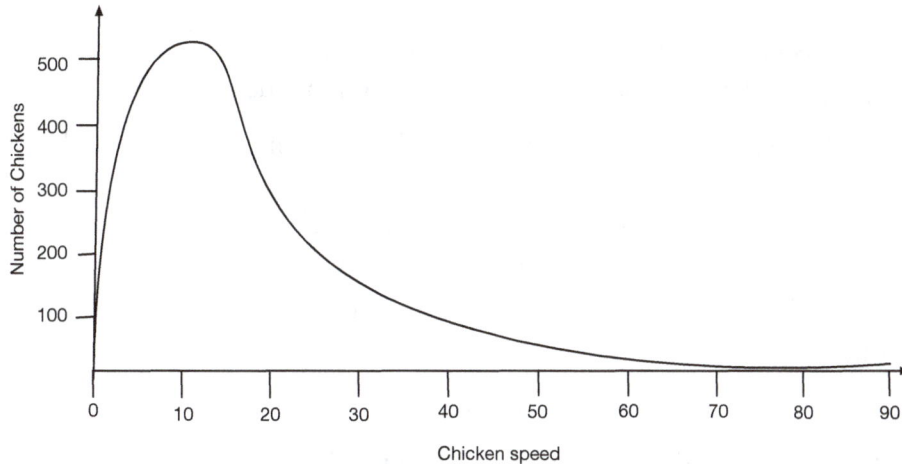

Figure 6-5: Number of chickens distribution based on their running speed.

analyze, and we should not let the algorithm neglect the low-density data because of their small numbers. We use data transformation to change the representation of our dataset, such as making its distribution look more like a Gaussian distribution.

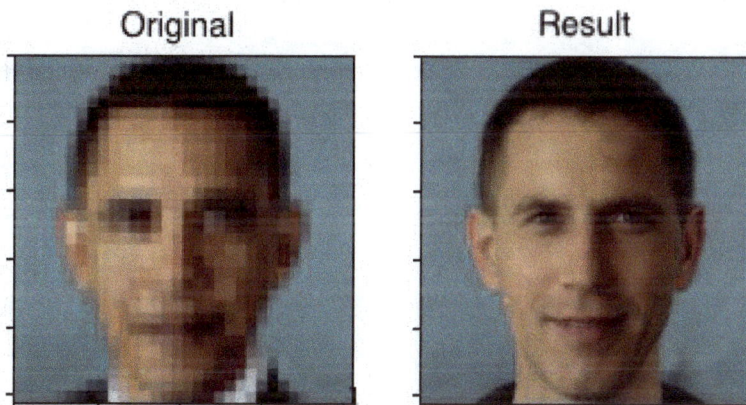

Figure 6-6: A mistake caused by depixelation of the give face, due to lack of diversity in the training data.

Not applying transformation might result in biased justification by the machine learning algorithm, such as the example shown in Figure 6-6. That machine learning model trained on lots of images to de-pixelize an image [Menon '20]. A Twitter user[4] gave an image which was Barak Obama, the former president of the United States (another awful politician), and gets a result as a white person from the model. This bias came from the lack of data on minorities, i.e., Afro-American people, in the training sct. We will learn more about bias in Chapter 16.

Applying transformation could mitigate this issue. There are two well-known transformation models: *logarithmic transformation* and *Box-Cox transformation*.

---

[4] https://twitter.com/Chicken3gg/status/1274314622447820801

Logarithmic Transformation

Logarithm transform is one of the most basic but useful transformations. It is used to reduce the emphasis on high-density data and increase the emphasis on low-density data.

If, similar to the author, you studied in a pure theoretical schooling system, you might encounter a *logarithm* when you are in high school. You probably did not understand its use and might hate it at school, like the author of this book. Nevertheless, it has amazing capabilities in mathematics and data science. The logarithm is written as $log_a(y) = x$, identical to $a^x = y$. $log_a(1) = 0$, because $a^0 = 1$. Note that while writing the logarithm function, both a and x are positive values. Also, keep the following in mind:

$$log\ a = x \leftrightarrow a = 10^x \text{ and } ln\ a = x \leftrightarrow a = e^x$$

To better understand the use of the transformation function, let's go back to our chicken example. There are a few chickens that do not perform well. We should respect diversity and consider that slow chickens could use the same algorithm, i.e., the algorithm is not customized only for fast chickens. To make such a balance, we should decrease the emphasis on fast chickens and increase the emphasis on slow chickens. The log transform will do this for us. Check the differences between Figure 6-7 (a). and Figure 6-7 (b). Figure 6-7 (a) presents the original dataset, and Figure 6-7 (b) presents the same dataset, but a log transformation is applied.

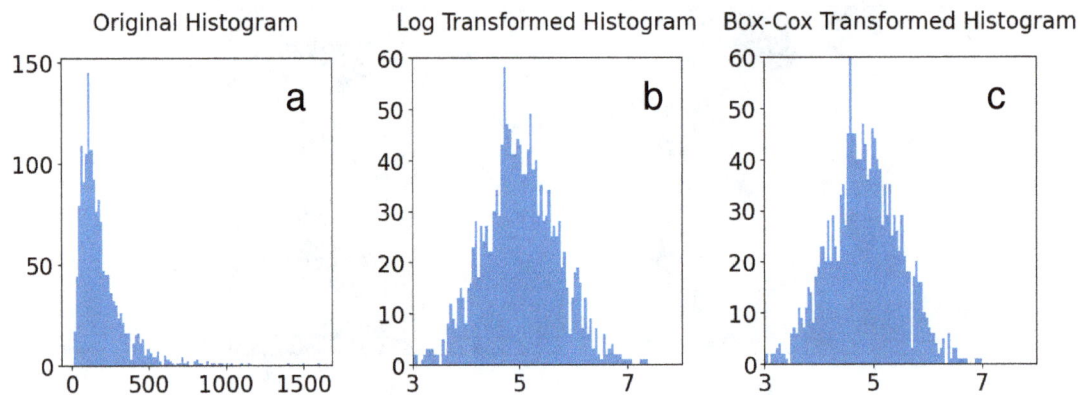

Figure 6-7: Different transformation of the data (left) original data, which a small part of the has a high concentration and the large part is neglected (center) logarithmic transformation (right) Box-cox transformation which some times performs a bit better than logarithmic transformation.

We can observe that the distribution is shifted toward being normally distributed. In summary, by using logarithmic transformation, we increase the emphasis on small parts of the dataset (slow chickens) and decrease the emphasis on the large part of the dataset (fast chickens).

Box-Cox Transformation

Box-Cox transform [Box '64] provides more flexibility than the logarithm transform because it introduces a parameter (λ) that allows the different shapes of skewness. This transformation also tries to convert the shape of a non-normal distribution into a normal distribution, but note that it is not always successful. Nevertheless, it is a widely used transformation function for feature engineering/generation. Assuming we transform the variable y, it operates based on the following equation.

$$y(\lambda) = \begin{cases} \frac{y^\lambda - 1}{\lambda} & if \quad \lambda \neq 0 \\ log(y) & if \quad \lambda = 0 \end{cases}$$

Although the equation might seem scary, it is very simple. It operates based on a parameter called lambda (λ). The value of λ varies between -5 and +5. If the λ is zero, we only use log transformation. Other values of y will be transformed to $(y^\lambda - 1)/\lambda$. The software package you use usually provides a default value for λ.

Remember that this equation works only for positive numbers; if our data have negative values, there is another variation of this equation that all numbers will be converted to positive numbers by adding a constant variable to all data. Figure 6-7 (c) presents the Box-cox transformation of the raw information presented in Figure 6-7 (a). You can see from Figure 6-7 (c). that Box-cox transformation usually performs better than log transformation.

NOTES:

* Feature selection can increase the bias as well. Usually, the process of feature selection in some deep learning models, such as autoencoders (Chapter 11) and convolutional neural networks (Chapter 10), is done by the model itself, and we give all possible features to the model. Then, the algorithm sorts out the useful features itself. However, we do not advocate generalizing a method that works for deep learning algorithms to all other machine learning models. Especially for battery-powered devices, which are associated with energy limitations, checking too many features is a major challenge.

* Remember that both Box-cox and log transform operate when the data is positive. If we deal with a dataset that includes negative values as well, we can add a fixed number (constant) to all numbers (the number could be equal to the positive value of the lowest number) and to all data points and make all our data points positive. For instance, we have {-5,-2,1,3,...} we can add 5 (which is the positive value of -5) and convert the vector to {0,3,4,7,...}.

* The logarithm transform and box-cox transform are members of a group known as the *power-transform* group. These transformations operate based on "stabilizing the variance" of the dataset. In other words, power transformation changes the distribution in a way that variance is no longer dependent on the mean.

* Remember that scaling is a process used to transform data into a consistent range, such as scaling it to a range between 0 and 1. On the other hand, normalization (sometimes referred to as standardization) involves transforming the data to follow a normal distribution, typically with a mean of 0 and a standard deviation of 1.

* Sometimes, none of these methods works, and still, the data quality is unsatisfactory. In those cases, we might need to increase the frequency of sampling from an existing data, which is

called *upsampling*. Also, sometimes, we need to reduce the frequency of sampling from an existing dataset that we have collected due to some limitations, such as hardware capabilities, which is called *downsampling*. A good example is a traffic sensor reading. Assume we collect data every three hours to model congestion, but it is not satisfactory, and the congestion peak, which only lasted for two hours, was neglected. In this case, we need upsampling. However, now we collect data every minute, and the computing machine should run the algorithm on a very large dataset, which can not handle that. Therefore, we change the data collection frequency to half an hour, i.e., downsampling. We will encounter these two terms several other times in this book.

Feature Engineering for Categorical Data

We are very sure that you remember the categorical data from Chapter 3. Let us repeat the thing we have mentioned a billion times in this book: machine learning algorithms rely on the repeatability of the data, and therefore, the range and variety of variables in categorical data are always finite. Categorical variables are usually non-numeric. For instance, the number of cities in any country is categorical data, the days of the week are categorical data, the races of chickens or humans are categorical data, etc.

There are some known methods used for feature extraction among categorical variables, including One-Hot encoding, Dummy coding, Effect coding, and Target encoding. Since these methods were developed by scientists from different disciplines, there are no differences between the term encoding versus coding in these methods. Therefore, don't bother yourself wondering why one method is called coding and another is called encoding.

One-hot Encoding

One-hot encoding converts any variable values of a dataset into a bit (0 or 1). For instance, assume we have a set of chickens with four different body shapes, i.e., 'ultra fat', 'fat', 'fit' and 'thin'. By using one-hot encoding, we can present each body shape as follows:

Ultra fat Chicken

Ultra fat: 1, 0, 0, 0

fat: 0, 1, 0, 0

fit: 0, 0, 1, 0

thin: 0, 0, 0, 1

We should memorize the names of these simple methods. One might ask why we need to convert data into one-hot encoding or other methods. The answer is that computers can work better with numbers rather than text and even better with bit vectors than numbers.

Dummy Coding

It is very similar to One-hot Encoding with one difference: instead of using k bits to present the k numbers of data points, it uses $k-1$ bits to present data. The Dummy coding of the same example can be presented as follows:

Ultra fat:	1, 0, 0
fat:	0, 1, 0
fit:	0, 0, 1
thin:	0, 0, 0

The feature ('thin' in our example) in which all its bits are zero is called "reference category". Since one reference category exists, there is no need to have an extra bit. In other words, dummy coding occupies three bits instead of four bits (One-hot encoding).

Effect Coding

It is similar to dummy coding, but the reference category (all 0 bits) is presented with -1. The Effect coding of the above example will be:

Ultra fat:	1, 0, 0
fat:	0, 1, 0
fit:	0, 0, 1
thin:	-1, -1, -1

Effect coding makes the results of linear regression easier to interpret. Don't worry if you are not familiar with linear regression; we will explain linear regression in Chapter 8.

All these three methods are very similar; by one-hot encoding, we can use all zeros for missing data, but dummy coding can not handle missing data. Dealing with missing data is an important issue, which we will explain in Chapter 16. The computational complexity of all these three methods is linear. These are basic encoding functions, and when the number of features gets very large, it is not easy to use these methods.

Feature Hashing

There are methods used to deal with a large number of features, including feature hashing, which focuses on compressing features. Feature hashing uses a hash function and maps each feature into a hash value [Moody '88]. We have explained the hash function in Chapter 5. The original feature presents the hash key, but instead, the algorithm will use its values. For example, assume we have many long textual features as follows, and by converting them to hash key, their size will get reduced.

"the cat on the roof"	→ 00a9
"the cat in the bin"	→ 01xb
"the chicken who eats the cat"	→ 82n4
"the chicken who knew how to deal with cats"	→ z12d

It reduces the dimensionality of feature spaces, which is particularly useful when dealing with high-dimensional data like text. By mapping features to a lower-dimensional space using a hash function, feature hashing helps in handling large sets of features with potentially reduced memory usage and computational cost. The challenge here is to work with hashed features (hash values), and it is not easy to get back to the original value of each feature (hash keys).

Generally, the computational complexity of applying the hash function is constant per feature but linear in the total number of features being processed because each feature is hashed independently.

Bin Counting

Another method to deal with large features is bin counting. Bin counting also compresses feature space. It converts the target data points into a set of probabilities. Therefore, instead of dealing with many different features, the algorithm deals with a set of probabilities. These probabilities were extracted by analyzing historical data. In other words, the attribute or feature will be converted into a likelihood. In many machine learning tasks, the informative data is associated with historical data values, such as click probability on online advertisements.

machine ID	input chicken	#chicken nugget	#cat	#elephant	#burgers
1	3200	3000	200	0	0
2	5631	5000	-	31	-
3	2000	1500	200	-	-
4	1500	1000	100	200	100

Table 6-1: Performance of four machines which receive chickens as input and should deliver nuggets as output. Other outputs are assumed to be incorrect.

machine ID	Correct output	Incorrect output
1	3000÷3200=0.93	200÷3200=0.06
2	5000÷5631=0.89	31÷5631=0.005
3	1500÷2000=0.75	200÷2000=0.1
4	1000÷1500=0.66	(100+200+100) ÷1500=0.267

Table 6-2: Performance of four machines based on the likelihood of correct or incorrect output.

To better understand this method, assume we have Table 6.1, which reports the performance of happiness machines, which receive chickens as input and produce either chicken nuggets or errors (cat, elephant, burger, etc.) as output. Just note that in this example, the output of the machine should not necessarily be equal to the input of the machine.

We can calculate the number of correct outputs and change the presentation of the table to deal with probabilities instead of original data. Table 6-2 presents the result of this calculation. We can substitute original features with their likelihoods, as it has been shown in Table 6-2. However, converting original feature values into a probability might lead to some data loss as well. In this example, we do not consider them for the sake of simplicity.

Target Encoding

While dealing with categorical data, we can compute an average for each distinct value of a feature and then replace the data with its mean. This sounds odd, but since machine learning algorithms can work with numbers better, replacing each variable's value with its average makes it easier and probably computationally efficient for the target algorithm to process it. This process is referred to as target encoding.

To better understand the target encoding method, look at Table 6-3; on the right table, we have a dataset with three columns. The second and third columns include categorical data, and they will be changed by Target encoding to their average, but the first column is numerical, and it stays intact.

Column 1	Column 2	Column 3		Column 1	Column 2	Column 3
1	x1	a		1	0.375	0.75
2	x2	a		2	0.625	0.75
0	x2	a		0	0.625	0.75
1	x1	a	Target encoding	1	0.375	0.75
1	x2	b		1	0.625	0.25
2	x1	b		2	0.375	0.25
0	x2	a		0	0.625	0.75
0	x2	a		0	0.625	0.75

Table 6-3: Converting categorical data into numbers by using target encoding.

Target encoding is useful in one of the most accurate classification algorithms, CatBoost (we will explain them in Chapter 9). Nevertheless, relying on an average value when the number of values used in the average is low might cause overfitting (we will explain this problem in Chapter 8).

Another common issue when using target encoding is *Target/Data Leakage*. In short, *data leakage refers to a process in which the test dataset uses the data from the training dataset as well.*

By using the probability of the target to change the categorical features, we are feeding them with the information we are trying to model. In other words, the model will learn from a variable that contains the target in itself. If it is now hard to digest this concept, get back to this section after reading Chapter 8, then you will understand this issue.

To mitigate these issues, cross-fold validation (we will learn it in Chapter 8) is recommended, and the result data should be ensured that it is not overfitted and that there is no target leakage.

Feature Engineering for Textual Data

A large segment of artificial intelligence and data science is dedicated to working with textual data and extracting knowledge from textual corpus, such as getting data from online news media, books and articles, transcripts of movies, etc, and using them to build Large Language Models. This leads to the introduction of different disciplines in computer science, such as natural language processing and text information retrieval.

Information retrieval is defined as activities used by machines to search data (mostly textual data), and it is different from extracting patterns from the textual data, i.e., text mining. At the time of preparing this material, there were five main types of tasks performed with text: classification, clustering, sentiment analysis, keyword and theme extraction, conversational agent and question answering (Chatbot), and text generation (generative AI).

Classification has been described in Chapter 1, and hopefully, you will remember it. More about it will be described in Chapter 9. Sentiment analysis is the process of identifying subjective information about an object from the text. In simple words, it is the process of identifying the *valence of emotion* (happy, sad, …) from textual data. We are sure you are familiar with chatbots and generative AI, such as ChatGPT. Thus, there is no need for further explanation.

We explain the text feature extraction with a love story about chickens. There is a very attractive social media influencer, a female chicken, and many male chickens fall in love with her and write her love letters. Unlike our corporate-funded celebrities who only care about humankind and global warming (while riding their private jets) and not themselves, she only cares about money and lavish life.

She received lots of emails from her followers, and she organized them based on the topic. The one that asks her for a date in an expensive place or sends her a gift will be read, and the rest will be discarded. This process of filtering email is the classification problem.

After filtering many emails, she still has many offers. The next step she does is to remove anybody who is not praising her from her life. Therefore, by using an opinion-mining (sentiment analysis) tool, her assistant will filter emails that do not have a strong positive tone and not praise her. Besides, she is very much interested in luxury hotels. Thus, her assistant uses a theme extraction to analyze every email and choose the one that is about luxury travel.

To perform classification or clustering on raw texts, we explain a de facto standard procedure, which is usually taken to extract meaningful features from the text. The first step with the text is preprocessing, which includes three to four major steps. Assume the following text has been extracted from an email:

"I am not rich, but I am a true lover, and I wish I had enough money to bring you to super-expensive resorts. Every night, I dream about you; together, we stand on the deck of a ship like those two humans in that American movie and go to Richland Island. There, we can book a luxury hotel and eat our breakfast in our private swimming pool every day while having an overview of the mountains of the island."

The first step, which will be done after data collection, is *tokenization*. *Tokenization* is the process of converting a text corpus into a set of tokens (segments). The token is the unit of processing text, e.g., sentences and words. A token should be a useful unit for linguistic processing, e.g., a single punctuation such as a comma is not useful for identifying a sentiment of an opinion, but a whole sentence will be useful. For instance, the above text could be tokenized as follows:

- I am not rich, but I am a true lover, and I wish I had enough money to bring you to super expensive resorts.
- Every night, I dream about you; together, we stand on the deck of a ship like those two humans in that American movie and go to Richland Island.
- There, we can book a luxury hotel and eat our breakfast in our private swimming pool every day while having an overview of the mountains of the island.

In this example, we have converted each sentence into a token using the "." as the token identifier.

The next step is to remove *stop words*. Stop words include, 'and', 'you', 'is', 'at', 'the', 'in', 'a', ... In particular, "articles", "prepositions", "conjunctions" and sometimes "pronouns" are considered stop words.

After stop word removal, we will have the following tokens:

- not, rich, but, true, lover, wish, enough, money, bring, super, expensive, resorts.
- Every, night, dream, about, together, stand, deck, ship, like, those, two, humans, movie, going, rich-land, island.
- There, book, luxury, hotel, eat, breakfast, private, swimming, pool, every, day, while, having, overview, mountains, island.

The step after that is *stemming*. Stemming is cutting the end or beginning of the word to convert it into its original root. For example, 'having', 'swimming', 'humans', and 'going' from the above examples will be converted into 'have', 'swim', 'human', and 'go'. These three steps usually occur in most text-processing approaches.

The step after these steps is usually text *normalization*. A widely used text normalization is Term Frequency/Inverse Document Frequency (TF/IDF), a vector space representations-based normalization. TF/IDF quantifies how important the word is in the document. TF measures the frequency of a term (word) in a document, or, in other words, the number of target terms divided by all terms in that document.

$$TF = \frac{\#\ term\ appearance\ in\ the\ document}{\#\ total\ terms\ in\ the\ document}$$

Nevertheless, the size of the document could be large or small, and this affects the importance of a term. IDF measures how rare the term is in the target document. It will be computed as follows:

$$IDF = log_e(\frac{\#\ total\ documents}{\#documents\ which\ includes\ the\ term})$$

Here, e is the Euler constant, 2.718. TF/IDF is computed as a product of TF and IDF, i.e. $TF \times IDF$.

Figure 6-8 presents these basic preprocessing steps in their usual order, but they might be implemented in different order.

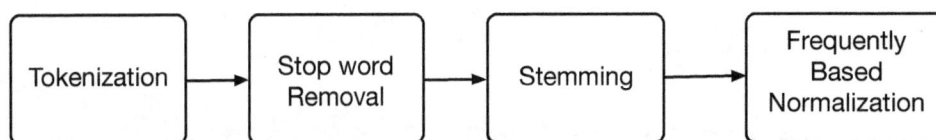

Figure 6-8: A common processing flow of a text corpus.

After preprocessing, there are some common methods to process text for a machine learning task, including Bag-of-words, N-Grams, Part-of-Speech tagging, and word embeddings. These methods could be used in combination together and not alone. We describe each separately, but if you intend to build your own Natural Language Processing (NLP) or text mining algorithm, keep in mind that you will benefit a lot from a combination of these methods.

Bag-of-Words

Bag-of-words (BoW) method treats the tokenized document (e.g., a sentence, a web page, a paragraph, or...) as a list of words, disregarding their order of appearance. This means that it does not care about their position in the sentence. For example, the bag-of-words method does not distinguish between the following two lists:

- there, book, luxury, hotel, eat, breakfast, private, swimming, pool, every, day, while, having, overview, mountains, island.

- while, having, overview, mountains, island, swimming, pool, private, every, day, there, book, luxury, hotel, eat, break fast.

After the text had been tokenized, all tokens were transferred into a list along with their frequency. For example, 'I am working hard, I am not a loser, I will change my life', will be fed into a bag-of-words algorithm, and the following is the algorithm output:

I = 4, am = 3, will = 1, loser = 1 ...

Similar sentences could be compared together based on the similarity between their words and the frequency of each word. Therefore, bag-of-words algorithms are very useful for clustering algorithms or topic modeling, which we will learn in Chapter 7.

Subword Tokenization

By looking at the space between words in a text, we can tokenize a text easily into words, but some languages, such as Chinese, Japanese, and Korean, do not have space between their words, and therefore, we need another approach to performing tokenization. *Subword tokenization* mitigates this challenge. For example, the word 'education' will be broken down into 'edu', 'cat', 'ion'.

Subword tokenization can also handle out-of-vocabulary (OOV) words in languages that use spaces. By breaking down words into smaller units, models can better handle rare or unknown words by understanding their sub-components.

Subword tokenization can be used even on the character level, and despite its extreme resource utilization, it works with acceptable accuracy for NLP tasks with neural networks. However, character level tokenization is too inefficient from a resource usage perspective.

A well-known approach for subword tokenization is Byte Pair Encoding (BPE), which we will describe in Chapter 14. BPE is a compression algorithm, but it is common for subword tokenization as well.

N-gram

Bag-of-words is a very useful method, but it doesn't care about the sequence of words at all. For example, 'I am not a loser.', is a sentence that says something positive about the subject, but if we treat it as a bag-of-words, it does consider 'loser' as a single term and 'not' as another term. Most probably, the algorithm interprets it as a negative sentence, which is incorrect. N-grams will help in handling these cases. N-gram is similar to bag-of-words, but it separates words based on their adjacent words. For example, a 2-gram (bi-gram) of 'I am not losing'. will be 'I am', 'am not' and 'not losing', respectively its 3-gram (tri-gram) will be 'I am not' and 'am not losing'.

The advantage of N-gram over Bag-of-Word is that it captures the language structure to some extent. The optimal size of the N depends on the application; we usually need several trials and errors to identify it.

Part of Speech Tagging

Another method to treat the text data is to explain and annotate what words have a role in the sentence. For example, 'I am very happy chicken and I love myself' sentence has pronoun, verb, adjective, etc. The Part of Speech tagging (PoS) will annotate this sentence as follows:

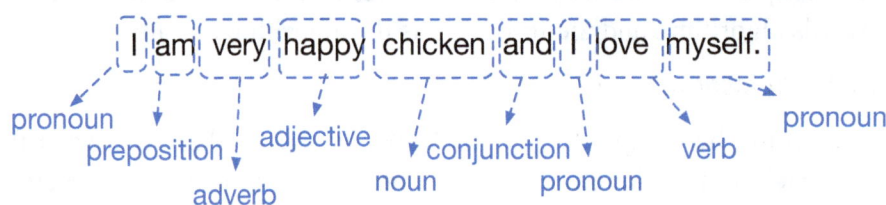

There the algorithm will output a list of two tuples, the word in the sentence, and its role. such as <I, pronoun>, <am, verb>, <very, adverb>, …

Algorithms that analyze text, such as chatbots, text mining systems, or NLP translators, can function more accurately when they understand the grammatical roles of words in a sentence. For instance, when analyzing text for geographical information, PoS tagging helps in identifying nouns and proper nouns, providing a foundation for more advanced processing.

However, to accurately recognize "Chicken Land" as the name of a city, the algorithm would rely on *Named Entity Recognition (NER)*, which uses context, PoS tags, and additional linguistic resources. NER is designed to identify named entities such as places, people, and organizations within the text. In this case, after PoS tagging identifies "Chicken" and "Land" as nouns (potentially with proper noun tags if capitalized), NER would classify "Chicken Land" as a single entity, recognizing it as a place name rather than two unrelated nouns.

Word Embeddings and Language Models

The representations we have explained previously (BoW, N-Gram, PoS) are all discrete representations of text, and they have some drawbacks. One drawback is that the machine learning algorithm can not recognize synonyms without having a dictionary where 'Good' and 'Nice' belong to the same category of words. For example, consider the following scenario; once in a document, a person could be presented as "Salvador Moustachio" and once 'ID2345'. Both of these terms refer to the same person, and previous methods are not capable of identifying that

'ID2345' refers to 'Salvador Moustachio'. Another drawback is that there is no accurate association between words, e.g., "chicken" is associated with "bird" or "boy" is associated with "man".

Word embedding approaches try to identify the role of the word based on its context in the document. Algorithms work with numbers, and word embedding approaches such as Word2Vec [Mikolov '13], GloVe [Pennington '14], or FastText [Bojanowski '17] convert each word (textual data) into a *numerical vector*, and because of these characteristics, they are referred as *vector space models*.

For example:

chicken = [0.9, 0.8, 0.01, 0.02, 0.01]

bird = [0.7, 0.9, 0.02, 0.0, 0.01]

boy = [0.0, 0.01, 0.0, 0.01, 0.2]

Man = [0.01, 0.05, 0.01, 0.02, 0.4]

We could say that the objective of word embedding methods is to map similar words with similar vectors in the vector space model. For example, 'water', 'steam', and 'ice' have similar vector representations, also, 'orange' and 'apple' have similar vector representations. Converting a word into a vector helps the algorithm to have similar vectors for synonyms.

Additionally, by converting a text into a numerical vector, we could also perform a numerical comparison between words. For example, we can also compare chicken ID2345 with another chicken in the text. We can add or subtract vectors using Euclidean, cosine, or other distances. For example, 'king - man + woman = queen', 'Syria - Damascus = Iraq - Baghdad', or 'Syria - Damascus + Baghdad = Iraq'.

Figure 6-9 presents words of a text corpus in a 2D space. There, we feed a corpus of textual data (Wikipedia) using GloVe and train a *language model* on Wikipedia. Nevertheless, the word is translated to a long vector, and we use PCA dimensionality reduction (we will learn it in Chapter 7) to reduce each word into 2D space.

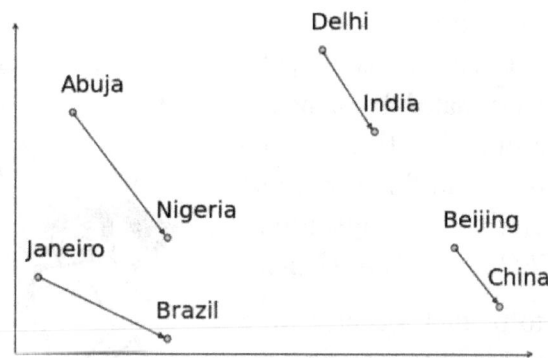

Figure 6-9: Using word embedding and dimensionality reduction to represents some words in 2D space.

Word embedding approaches use a neural network to convert a given text corpus into a vector of words, and they are the backbone of State-of-the-Art methods in text analysis. We will explain neural networks in Chapter 10, 11 and 12. Because of this characteristic, in some literature, they are referred to as 'Neural Word Embeddings'.

Training a Word Embedding

A word embedding is trained on a large corpus of text. For example, we can give a book, Wikipedia, or any other large corpus of textual data into a neural network to train the language model. As a result, each word has a set of numerical coordinates (vector). We will explain more about the neural network in Chapter 10 and language models in Chapter 12.

A simple approach to constructing a word embedding or language model is moving a sliding window over the input text. The algorithm starts using a sliding window and moves it along the text. The content of each window is the training instances used to construct the model. The next word will be the label of the instance. For example, assuming the window size is three words, 'The wound is the place where the Light enters you' will be converted into a language model as:

Instance: 'the wound is', label: 'the'

Instance: 'is the place', label: 'where'

...

After a word embedding is trained, each word in the given corpus will be associated with a vector of numbers.

A refined version of word embedding is called a language model. When a language model is trained on a very large corpus of text, it is called a Large Language Model (LLM). Language models are widely used in NLP and even other fields of machine learning. We will explain the latter in detail in Chapter 13. To summarize, training a word embedding or language model refers to *feeding a large text to a neural network, and the neural network builds a vector representation for the given text.*

Word Embedding Models

The basic idea of word embedding is to identify the context (c) by a vector of input words, i.e. $\{w_1, w_2, \ldots w_t\}$, which can be formalized as $p(c \mid w_t)$, and this probability will be estimated by a neural network. In other words, the first objective of word embedding neural network is to transfer words into vectors. Then, it can perform two operations: *Continuous Bag of Words (CBOW)* and *Skip-Gram.*

The objective of CBOW is to predict a center word based on the given input words that describe the context. For example, consider this sentence: 'The white cow is heavy'. Using the 'cow' as the input, the algorithm can predict the neighbor words, including 'heavy' and 'white'.

Skip-Gram uses a set of words to predict the context. In other words, the Skip-gram predicts surrounding words (based on the window size) of the given target word. For example, we use neighbor words such as 'white' and 'heavy' in the word embedding algorithm to predict the context, i.e., 'cow'. Some said Skip-Gram is the reverse of CBOW.

Let's analyze another example to understand the difference between CBOW and Skip-gram. We give the input (context) *'Today is ... and the weather is also very good'*, CBOW can predict the missing word is 'amazing'. The missing word is selected from a set of words (e.g. {amazing, probability =0.8}, {snowy, probability = 0.03}, ...), and each of them has a probability. The word *'amazing'* has the highest probability, and thus it is selected by CBOW to be returned as the predicted word. Look at Figure 6-10 to see the differences between CBOW and Skip-gram in the word embedding architectures.

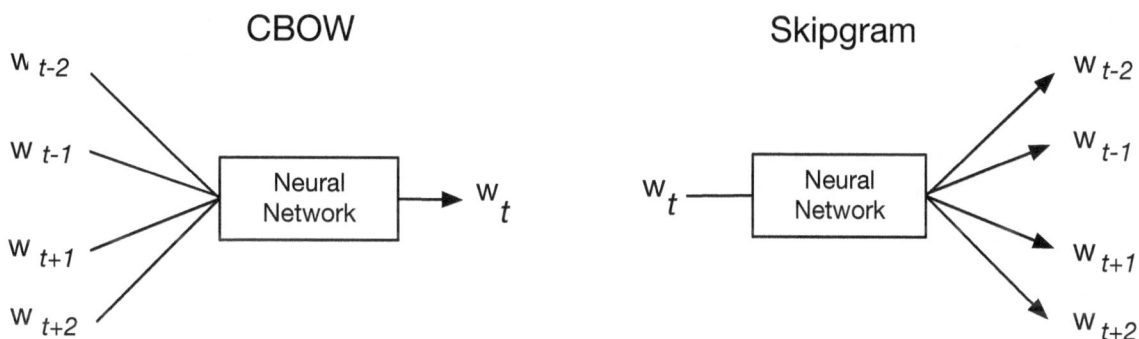

Figure 6-10: CBOW and Skip-gram of word embedding models.

In the following, we briefly describe three popular word embedding models without going into the details of their architecture.

Word2Vec

One of the most impactful word-embedding approaches still in use was introduced in 2013, word2Vec [Mikolov '13]. It can perform two tasks, CBOW and Skip-gram. It uses a neural network to learn the distribution of each word in the text, and this distribution is presented as a vector. Authors of Word2Vec use a neural network. We will explain neural networks in Chapter 10; you don't need to worry about learning them now. Therefore, feel free to skip the following paragraph. The input layer converts each word into a one-hot encoded vector, where only one element (corresponding to the index of the word in the vocabulary) is 1, and all other elements are 0. The projection layer, also known as the embedding layer, indeed maps the one-hot encoded input vectors into dense vectors of lower dimensionality. These dense vectors represent the word embeddings and are typically much smaller in dimensionality compared to the one-hot encoded vectors. In the original Word2Vec architectures (e.g., Continuous Bag of Words (CBOW) and Skip-gram), there's typically no hidden layer.

GloVe

Global Vector for Word Representation (GloVe) was introduced in 2014 [Pennington '14], one year after Word2Vec, and it addresses one of Word2Vec limitations, known as word co-occurrence limitation. Word2Vec goes through each word in the entire corpus and predicts the surrounding words in the corpus once for every word. In particular, Word2Vec's skip-gram does not efficiently incorporate the frequency of word co-occurrences. Besides, a frequent co-occurrence of words creates more instances for the training set; they increase the train dataset size.

GloVe resolves this challenge by scanning the entire corpus once and counting the co-occurrences once for all words. The result will be a matrix that involves all co-occurrences, i.e., a window-based *global co-occurrence matrix*. It is a dense representation that presents the overall (global) statistics of word co-occurrences. GloVe uses this matrix to learn word embeddings by considering the global statistical properties of word co-occurrences, not just the *local context* (Word2Vec focuses on local context only).

For example, consider the two sentences and their simplified symmetric *global co-occurrence matrix*, in which we only check one neighbor word (window size=1). Based on the existence of a window word, it adds one point to the matrix. Since 'I' is a neighbor to 'like' two times, their value in the matrix will be 2.

I like NLP.
I like Word-Embedding methods.

	I	like	NLP	Word-Embedding	methods
I	0	2	0	0	0
like	2	0	1	1	0
NLP	0	1	0	0	0
Word-Embedding	0	1	0	0	1
methods	0	0	0	1	0

By looking at this matrix, we can see 'NLP' and 'Word-Embedding', which are two similar words that have the same vector representation. This is a very simple example, and usually, there are simple approaches to eliminate the too frequent words, such as 'prepositions', like stop word removal, that we have explained earlier. Another simple approach is to define a threshold and ignore all words that are too frequent.

To summarize the differences between GloVe and Word2Vec; Skip-gram of Word2Vec models tries to capture co-occurrence one window at a time, but GloVe tries to capture the overall counts of how often the co-occurrence appears.

FastText

Despite the success of Word2Vec and GloVe, a challenge they have is the generalization to unknown words. FastText [Bojanowski '17], which was introduced in 2016, focuses on addressing this issue. Skip-gram of word2Vec uses uni-gram (one complete word) and takes every single word as a single unit. FastText uses n-gram and considers n number of characters together as a single unit. In simple words, while using FastText, instead of having a vector for a single word, each word is represented by a combination of its character n-gram vectors. The model aggregates these n-gram vectors, possibly weighted, to form the word's final vector representation.

For example, assuming an n-gram character size of three, in the training set, the word 'aquarium' can have one vector in word2Vec, but in FastText, we will have 'aqu', 'qua', 'uar', 'ari', and 'ium'. Now, if in the test set, we encounter the word 'aqua' for the first time, FastText can realize it has some similarities to 'aquarium', because the vector values of 'aqu' and 'qua' are highly similar.

In summary, using n-gram FastText solves a major issue of other word embeddings, which is out of vocabulary words. Previous methods can not produce a vector for a word that does not exist in their training set, but by using n-gram, FastText resolves this issue.

We have learned briefly three approaches to word embeddings without delving into the details of neural network architecture. Nevertheless, you can check them on your own after you learn Chapter 8 and Chapter 10. All current state-of-the-art NLP models use word embeddings at the beginning.

Theme and Keyword Extraction

Another common feature while using textual data is theme and keyword extraction. We will explain in Chapter 7 topic modeling approaches, which are one style of clustering text documents, and we can manually assign them keywords and themes. Other approaches are algorithms that, in particular, identify keywords, such as TF/IDF, which we have explained earlier, TextRank [Mihalcea '04], Rake [Rose '10], Yake [Campos '18], LLM based methods.

TF/IDF

TF-IDF can be used to evaluate the importance of a word to a document in a collection or corpus. By looking at the described equation, we can see the importance increases with the number of times a word appears in the document but is balanced by the word's frequency across the entire corpus. This balancing process helps to highlight words that are significant in individual documents, distinguishing them from those that are common throughout the corpus.

TextRank

TextRank [Mihalcea '04] is a graph-based ranking model for text processing that is particularly useful for keyword and sentence extraction. It is based on the PageRank algorithm (we will explain it in Chapter 15). In TextRank for keyword extraction, words are treated as nodes, and the connections between them (based on co-occurrence within a certain window of words) form the edges of the graph. Words are then ranked according to their importance within the graph, with higher-ranked words selected as keywords. It doesn't make much sense to explain how it works without knowing PageRank, which is explained in Chapter 15. After learning Page rank, we can come back and check this method.

Rapid Automatic Keyword Extraction (RAKE)

Rake [Rose '10] identifies key phrases in a text by analyzing the co-occurrence of words and their frequencies. It focuses on phrases containing words that frequently appear together and are relatively rare in the document as a whole. RAKE is particularly good at extracting multi-word phrases as keywords. For example, assume this sentence: *"Climate change is a global challenge that has no borders and to combat it requires coordinated work by all countries, thus corporations head, and politicians fly their private jets to gather together and discuss this issue frequently"*.

We want to extract keywords from this sentence using RAKE. First, the text is preprocessed to remove punctuation and stop words. The remaining words are analyzed to identify phrases by looking at co-occurrences within the text. In this case, phrases might be identified based on the remaining words after stop words are removed. RAKE calculates a score for each word within these phrases, often based on word frequency and degree (how often it co-occurs with other words). For simplicity, let's say the scoring is just based on frequency in this example. Next, Phrases are ranked based on the scores of their words, and the top-ranking phrases are selected as keywords. We might have phrases like "Climate change", "global challenge", and "private jets".

Yet Another Keyword Extractor (YAKE)

Yake [Campos '18] is an unsupervised method to extract keywords from documents. It starts by preprocessing the document, including converting text to lowercase letters, removing stop words, punctuations, etc.

Next, it uses five features to identify the characteristics of each word in the text; these features include (i) Casing (ii) Word Positions (values more those words occurring at the beginning of a document based on the assumption that relevant keywords often tend to concentrate more at the beginning of a document), (iii) Word Frequency, (iv) Word Relatedness to Context, and (v) Word DifSentence. Word Relatedness to Context computes the number of different terms that occur on the left side of the candidate word. The higher the number of different terms that co-occur with the candidate word, the more meaningless is the candidate word. This feature is more about the diversity of the context in which a word appears. A high diversity suggests the word is less focused and potentially less relevant as a keyword. Word DifSentence specifies how often a candidate word appears within different sentences.

Then, it combines these features into a single measurement that calculates a keyword score for each word. This score tries to balance between phrase coverage (how much of the document the phrase covers) and phrase significance (if the phrase is salient compared to other phrases). Afterward, it ranks candidates by the keyword score and returns the top-scored keywords as extracted keywords.

LLM Keyword Extraction

With the advent of LLM and transformers, models such as BERT or Llama have become increasingly popular. These methods can involve fine-tuning LLM on a keyword extraction task or using embeddings to represent text segments and then applying clustering or ranking algorithms to identify keywords. These methods can capture the semantic relationships between words more effectively than the other three methods we explained. After reading Chapter 12, we get familiar with LLM models and learn how they are fine-tuned for a specific task such as keyword extraction.

NOTES:

* Sometimes, stemming and normalization of text might have an adverse effect on the accuracy of the target algorithm. Hence, it might be better to learn the target application needs before applying them. Besides, test the system in both conditions, with or without normalization and stemming.

* While dealing with textual data, most of the time, Cosine similarity and Jaccard index will be used, and rarely, we use Euclidean similarity.

* Bag of words failed to capture the structure of the sentence. Therefore, we could use N-Gram, which is slightly more efficient than bag of words (in terms of resource use). Part of Speech (PoS) tagging is usually more accurate than N-Gram.

* Usually, algorithms use a combination of Bag of Words, PoS, and N-Grams to provide superior accuracy. Therefore, we need to learn all of them and how they process text documents.

* While performing stemming, there is another process called lemmatization. Lemmatization is similar to stemming, but it also considers the role of the word in the sentence as well, by using a dictionary. For example, in a document, the term "cars" will be "car" after stemming, but with lemmatization, the term "automobile" will be converted to the "car" too.

* All described word embedding libraries (word2Vec, GloVe, and Fasttext) can be used for both CBoW and Skip-Gram.

Feature Engineering for Image and Video Data

After the availability of neural network models, image (or video) advanced a lot, and neural networks revolutionized the computer vision field. We will explain neural networks in Chapter 10 and Computer Vision models in Chapter 12. However, we list basic models for image feature detection here. Please do not skip learning them; similar to other concepts, they are very useful if you intend to build your framework.

Features in image data are regions that we can label, e.g., the face of an individual in a picture, cars in the street, legs of chickens, etc. The image processing algorithms transform an image into a matrix or tensor and then process it as numerical data. Before describing image features, we briefly describe some concepts for digital image processing.

Image Processing Concepts and Components

An atomic information unit that presents an image is a *pixel*. A pixel holds a color intensity as a scalar number and has X and Y coordinates inside an image, representing its position. We can present a black and white image with a matrix. Color images are typically represented using the RGB color model, where a pixel's color is determined by three values corresponding to its Red, Green, and Blue intensity levels. These can each range from 0 to 255 in an 8-bit image. Therefore, each pixel in a color image is commonly represented by three data points: one for Red, one for Green, and one for Blue. Image data for color images are stored in a 3D tensor, where the dimensions represent the height, width, and color channels (RGB) of the image.

Figure 6-11 presents a toy example of how a black and white image has been converted into a matrix of data. For the sake of simplicity, we choose to have a very simple black-and-white image in low-pixel format and present its matrix in a binary format (zero is white, and one is black).

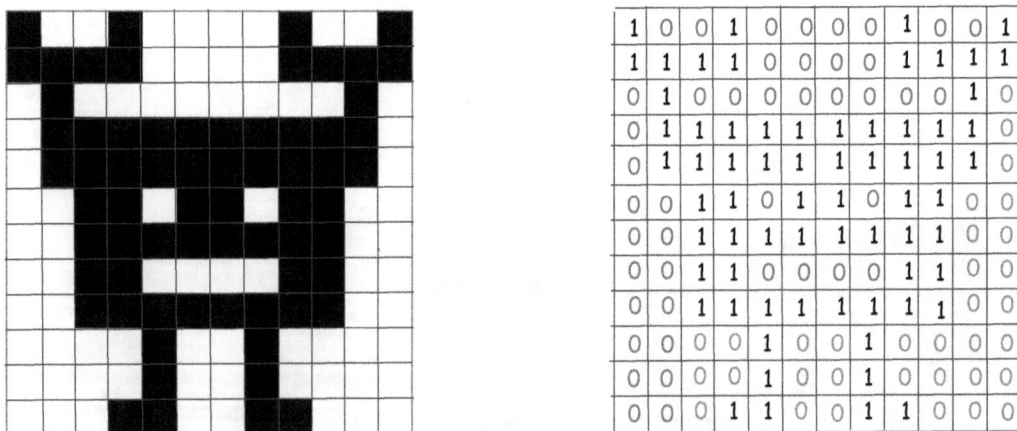

1	0	0	1	0	0	0	0	1	0	0	1
1	1	1	1	0	0	0	0	1	1	1	1
0	1	0	0	0	0	0	0	0	0	1	0
0	1	1	1	1	1	1	1	1	1	1	0
0	1	1	1	1	1	1	1	1	1	1	0
0	0	1	1	0	1	1	0	1	1	0	0
0	0	1	1	1	1	1	1	1	1	0	0
0	0	1	1	0	0	0	0	1	1	0	0
0	0	1	1	1	1	1	1	1	1	0	0
0	0	0	0	1	0	0	1	0	0	0	0
0	0	0	0	1	0	0	1	0	0	0	0
0	0	0	1	1	0	0	1	1	0	0	0

Figure 6-11: Simple black and white image which will be read as a matrix by the computer. Black pixels are presented with one and white pixels with zero.

Each pixel with coordinates (x, y) has eight neighbor pixels, as it has been shown in Figure 6-12. Neighbor pixels of pixel (x, y) are located at following positions: $(x + 1, y)$, $(x - 1, y)$, $(x, y + 1)$, $(x, y - 1)$, $(x + 1, y + 1)$, $(x + 1, y - 1)$, $(x - 1, y + 1)$, $(x - 1, y - 1)$. Depending on the field and application, "neighbor" might refer to the 8-neighborhood (as described) or the 4-neighborhood, which only includes horizontal and vertical neighbors.

If the image is colored, we will have Red, Green, and Blue (RGB) colors as well, or the color could be written in HSV format (hue, saturation, and value). Therefore, we will use a tensor to present an image instead of a matrix.

Path (curve): A digital is a path from one pixel to another pixel, with distinct pixels in between, which will be written as a set of coordinates.

Figure 6-12: Sample pixel with its eight neighbors.

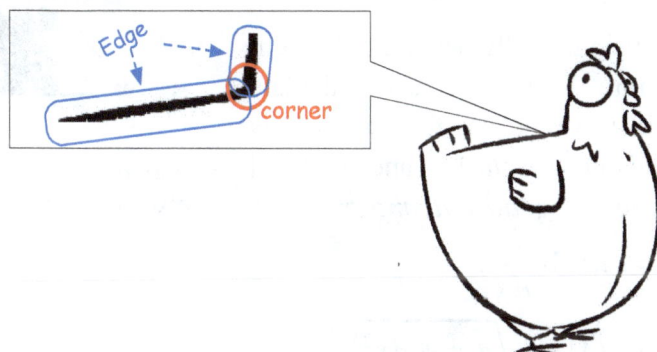

Region: A set of adjacent pixels is called a region, and an image is composed of a set of regions.

Segmentation: Partitioning an image into different regions (based on color similarity and edge discontinuity) can enable the machine to process the image more easily. This process is called image segmentation.

Edge and Corner: A region in which its pixel's color intensity changed significantly (more than a specific threshold) is called an *edge* or *boundary*. An intersection of two edges creates a *corner*. In other words, significant changes in two directions of the curve make a corner. Check Figure 6-13 to better understand the differences between corners and edges.

The feature selection process means finding the corner in the image and assigning labels or descriptions to that region. Corners are widely used in machine learning algorithms for tracking an object, searching for an image, etc.

Skeleton: The skeleton is the thinning of the boundary of an image to get a one-pixel width line that presents the image shape. Skeleton construction from an image has many different applications, from pose estimation to medical analysis. Figure 6-14 presents an image with its skeleton, identified by a pose estimation algorithm.

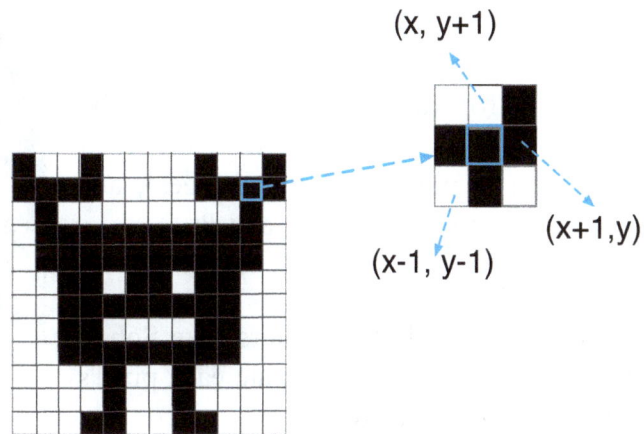

Figure 6-13: Differences between edge and corner.

Gradient: The gradient is a *directional change* in the *intensity* or *colors* in the neighbor pixels or region. In other words, it is a difference in changes between neighboring pixels or regions.

To calculate the gradient of an image, we need to convert the image to grayscale (using the luminance formula, for example). The luminance formula is defined as *Y* as a scalar that presents luminosity (color changes), and each color has a value for *R, G,* and *B*:

$$Y = 0.2126R + 0.7152G + 0.0722B$$

Next, we apply gradient operators (like Sobel) to the grayscale image to compute each pixel's gradient magnitude and direction.

The gradient is a vector that includes magnitude and orientation. For example, in Figure 6-15 from the pixel (x,y) to (x+1,y), the amount of changes is zero because both pixels are black. Figure 6-16 shows an example of the

Figure 6-14: Original image and its skeleton shown on the image with colored lines.

gradient of a pixel among its neighboring pixels. The algorithm usually calculates the gradient *magnitude* and *orientation*. To compute the magnitude and orientation for each pixel, it defines a vertical distance (dx) and a horizontal distance (dy). Typically, dx and dy are computed based on intensity values (e.g., brightness) of the pixels, by following equations:

$$dx = (x - 1, y) - (x + 1, y)$$
$$dy = (x, y - 1) - (x, y + 1).$$

By using the Pythagorean theorem (Figure 6-16), we can calculate both values as follows (θ is *gradient orientation* and $||\nabla f||$ presents *gradient magnitude*):

$$\theta = tan^{-}1(\frac{dy}{dx})$$

$$||\nabla f|| = \sqrt{dx^2 + dy^2}.$$

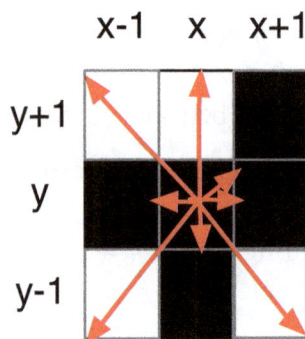

Figure 6-15: Gradient of a pixel to the neighbor pixels.

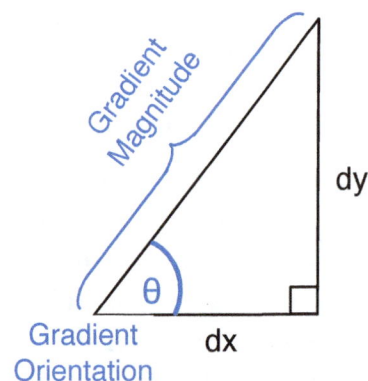

Figure 6-16: Gradient magnitude and orientation will be calculated by using pythagorean theorem.

By having a gradient magnitude and orientation, we can draw a vector that presents the gradient of that specific pixel.

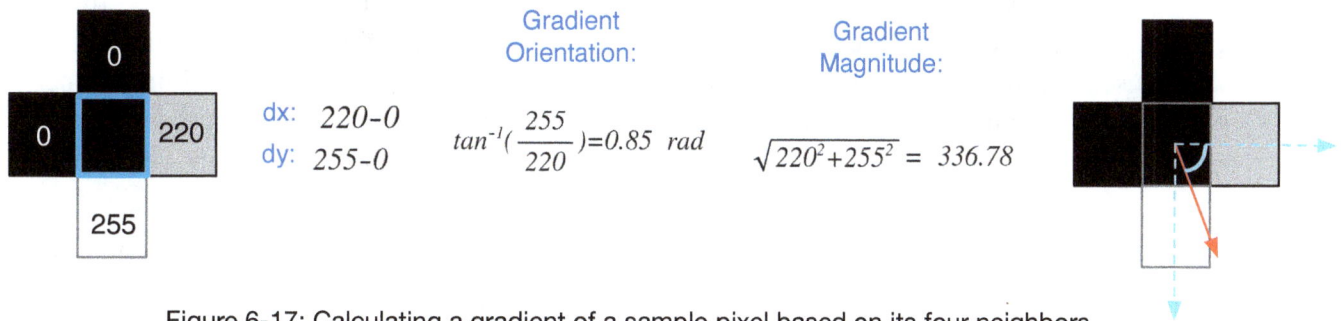

Figure 6-17: Calculating a gradient of a sample pixel based on its four neighbors.

Check Figure 6-17 as an example in which the gradient of the blue bordered pixel based on its four neighbors has been calculated.

Features Extraction from Image Data

In this section, we first explain a few concepts used for advanced feature extraction from image data. Then, we explain popular feature extraction algorithms. Since 2012, deep learning has revolutionized image and video processing and its methods for extracting features. Later, in Chapters 11 and 12, we explain how deep learning extracts image features. However, we do not need to learn the details of every algorithm; most of the time, we use a trained model for our tasks. There are scenarios in which we can not use neural networks, such as lack of transparency, and we should use traditional features engineering using computer vision algorithms. This section briefly lists those features.

Objection Detection: The process of identifying an object inside an image is referred to as object detection. For example, Figure 6-18 presents an image of a cat, and the algorithm can detect a Persian cat on a Persian carpet under a table with some chair legs. The object detection algorithm detects the object in the middle picture of Figure 6-18. In particular, it detects the cat object in this figure and puts a bounding box around the cat.

Object Localization: Object localization is indeed about specifying the location of an object within an image, often through a bounding box, but it's typically described as part of the object detection process. For example, in the middle picture of Figure 6-18, the cat's location is specified with the bounding box. The distinction is that localization refers to finding the position of one object in an image, whereas detection might involve identifying and locating multiple objects.

Semantic Segmentation: Localization is not accurately specifying the boundaries of an object inside the image. One of the major tasks in computer vision is extracting objects from an image and separating and segmenting the image into meaningful components. For example, a self-driving car segments a street into different components, including pedestrians, signs, other cars, etc. Semantic segmentation is the process of separating objects inside an image and identifying the shape of each object. In a more technical sense, image segmentation partitions an image on the pixel level to accurately identify the boundaries of each object inside the image, see the right side of Figure 6-18.

303

The result of image segmentation is a set of contours extracted from the image. Image segmentation provides an accurate interpretation of the image data for the algorithm, and it is widely used for self-driving vehicle algorithms and medical image analysis algorithms. Since image segmentation is a popular approach, there are several neural network advancements in image segmentation, such as U-Net (which will be described in Chapter 11), which we describe later.

Instance segmentation: Semantic segmentation does not differentiate between individual instances of the same object class; it labels each pixel for a category without distinguishing between different objects of the same type. When there is more than one object of the same type existing in the image, the instance segmentation enables the semantic segmentation process to recognize every pixel belonging to which segment (of the same type). We can say instance segmentation is an extension of semantic segmentation.

For example, in an image that has three humans in it, the instance segmentation recognizes each person from the other. In our example presented in Figure 6-18, the segmentation process recognizes two different chairs, one marked in blue and one marked in green. Therefore, this semantic segmentation includes instance segmentation as well. Later in Chapter 12 we will learn more about segmentation and instance segmentation.

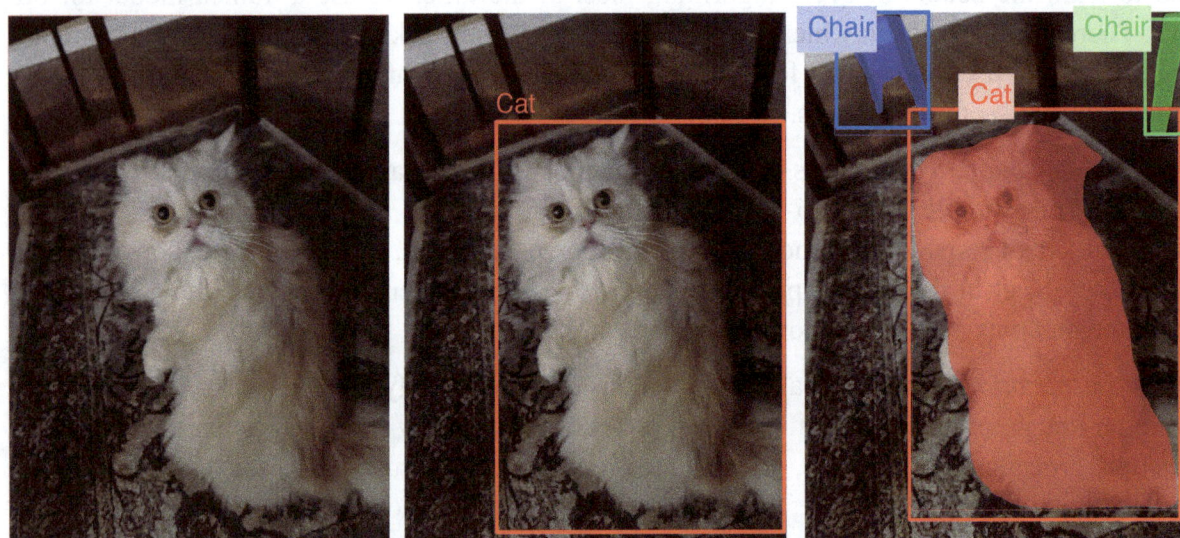

Figure 6-18: (Left) An image of a cat under a table on a carpet. (Middle) an object detection algorithm specifies the cat in the image and draws a boundary around its position in the picture, i.e., localization. (Right) A segmentation algorithm identifies objects in the picture and specifies them. It identifies two chairs and the cat.

304

Image Feature Extraction Algorithms

Feature extraction from image data involves two steps: feature detection and feature description. Feature detection is the process of finding a region or boundary in the image data, and feature description is the process of assigning labels to the detected features.

An image has some basic features, including compactness, circularity, eccentricity, and Euler number. You can read more about them in Chapter 11 of Gonzalez and Woods's book [Gonzalez '16]. These are too basic to be used for feature extraction. Thus, we do not describe them and dedicate our focus to feature extraction algorithms, which are used more in real-world algorithms. Before we proceed further, note that image features should be independent of location, scale (zoom in/out), and rotation.

Harris-Stephens (HS) Corner Detection

Harris-Stephens (HS) corner detector [Harris '88] is a fairly simple algorithm for detecting corners. As we have stated before, the corner is defined as an intense change in the direction of the edge or curve. This algorithm moves a sliding window on the image to detect its corners. This window, called the detector window, is used to calculate the intensity of changes. We don't go into its mathematical detail, but if it senses significant intensity changes in both the x and y directions, it assumes a corner is identified ('a' in Figure 6-19). If it senses significant intensity changes on one axis ('b' in Figure 6-19), then it assumes that the region includes an edge. If there are no changes, it is a flat region ('c' in Figure 6-19).

This algorithm receives two parameters, "threshold" T and "sensitivity factor" k. A corner will be identified if $k > T$ inside the patch at that location. Otherwise, the window moves to the next location, and no corner will be identified.

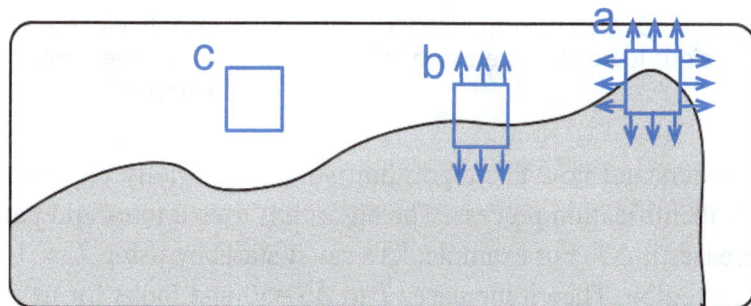

Figure 6-19: The HS algorithm square moves along the image to identify corners.

Assuming the image has a size of $m \times n$, and l is the size of the square window ($l \times l$), the computational complexity of HS corner detection is quadratic, i.e., $O(mnl^2)$.

Maximally Stable Extremal Regions (MSER)

MSER is another algorithm [Matas '04], which focuses on *blob detection* instead of corner detection. Blobs are regions that have different properties in comparison to other regions in the image, such as different colors or different brightness. In the MSER algorithm, we call them the *extremal region*. In other words, MSER regions are areas with uniform intensity surrounded by a background with different intensities. MSER is trying to identify those regions by trying different thresholds.

For example, assume Figure 6-20 (a), which includes three blob regions (two white and one grey) on a black plate. This means we have three significant changes in color intensity: white, grey, and black. According to the RGB code, the white color code is 255, grey is 128, and black is 0. If we plot each color code in 3D, we get Figure 6-20 (b).

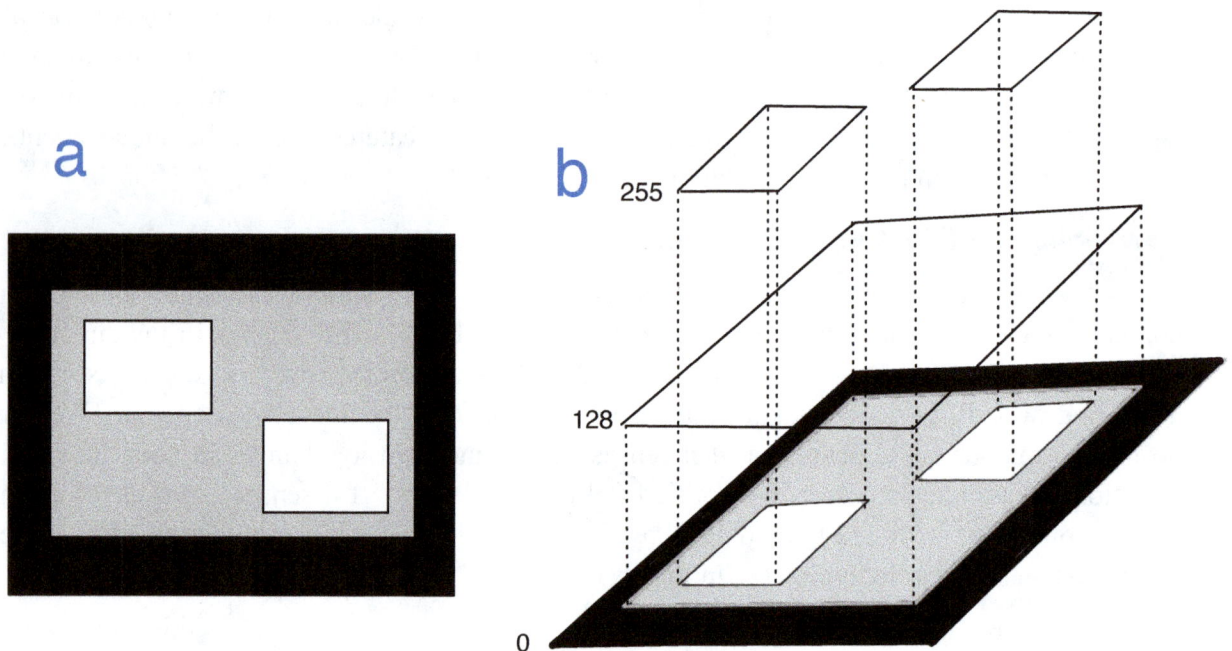

Figure 6-20: A sample figure on the left, with its density plot based on color intensity, is plotted on the right side. To better understand MSER, assume it uses the threshold parameter to cut this density plot horizontally.

To understand how this algorithm works, we briefly explain its iterative threshold increase and blob identification process. The algorithm uses a threshold parameter T, and in each iteration, it increases it ΔT. For example, let's say it starts by using $T = 10$ and the only blob it identifies is a black region. Then it increases T to $\Delta T=50$ and looks for new blobs until $T + \Delta T \geq 128$, which identifies the grey region as well. Again, this process continues until we reach the maximum boundary of $T + \Delta T$ (which is 255 for RGB colors), and in the last iteration, it identifies two white square blobs.

MSER operates based on four parameters: (1) a Δ parameter (stability score), which controls the stability of a blob. In other words, presents a maximum variation a blob can hold to stay stable as a blob. (2) Another parameter is the similarity of a blob to its lower-level blob (parent blob). This is used to prune regions that are too similar. Two other parameters are (3) maximum and (4) minimum sizes (minimum and maximum number of pixels they can hold) a blob can hold, and if it passes those thresholds, then it will not be considered a blob.

Assuming n is the total number of pixels in the image, and t is the threshold, the computational complexity of MSER is $O(n \ log \ n + nt)$.

Histogram of Oriented Gradients (HOG)

Histogram of Oriented Gradients (HOG) is a feature detection algorithm based on the image's gradients [Dalal '05]. It is a classic approach for object detection in an image, e.g., a vehicle in the street, a pedestrian walking in the street, etc.

This algorithm operates based on the assumption that an image could often be described as a set (distribution) of gradients. In other words, we can convert an image into a set of gradients, such as Figure 6-21. Another example is presented in Figure 6-22, in which the gradient of a real image is calculated and presented. Figures 6-22 presents the gradient in Sobel X and Sobel Y. In particular, the Sobel X filter detects edges with a change in intensity along the X-axis (horizontal edges). It calculates the gradient by moving from left to right and emphasizes the differences in intensity between adjacent pixels in the left-right direction. The Sobel Y filter detects edges with a change in intensity along the Y-axis (vertical edges). It calculates the gradient by moving from top to bottom and emphasizes the differences in intensity between adjacent pixels in the top-bottom direction.

The algorithm operation is straightforward. First, it divides the image into a grid of sub-images (cells), as you can see from Figure 6-21. Second, it calculates a histogram gradient of each cell. As a result, it will get something like the right side of Figure 6-21. The set of histograms is used as a "feature vector" that describes the image or the object. Then, we fed these histograms into a classification algorithm and classified objects in our image. Classification algorithms will be described in more detail in Chapter 9.

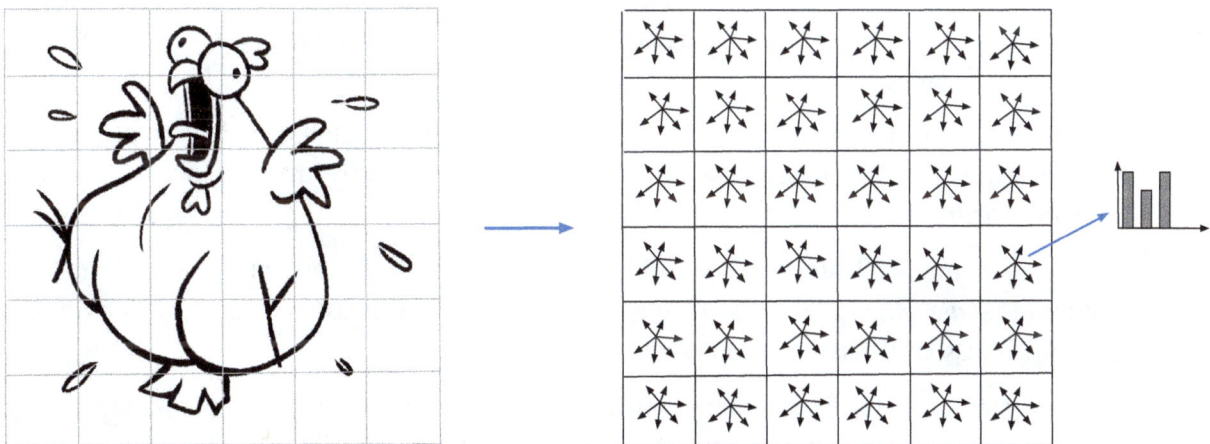

Figure 6-21: (Left) A sample Figure on the left is binned into 6x6. (Right) each bin is presented with its gradient histogram (vectors are not real; they are just to show you how the image gets converted). Also, instead of presenting them on the X or Y axis, its vectors are presented among a central dot.

One preliminary step is required at the beginning. Cells' illumination invariance (e.g., lights and shadows) should be normalized or scaled, e.g., Min-Max scaling. Also, the authors of the HOG paper identified that L_2 normalization performs better than other forms of normalization.

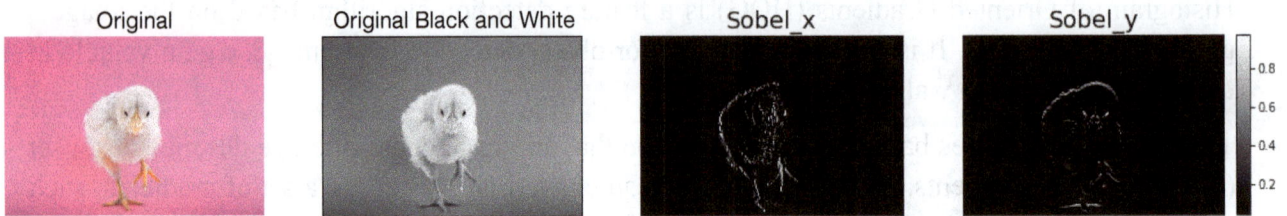

Figure 6-22: (Left) A sample picture, which is then turned into black and white. Afterward, the X gradient and Y gradient of the image are calculated.

Since the HOG algorithm focuses on the gradient of a cell (local) rotation, scaling does not affect HOG values. Converting an image into gradients also causes the HIG feature descriptors to get compressed, which is very practical, especially on small devices, because image processing applications are usually resource-intensive.

Assuming the image has $m \times n$ size, the computational complexity of HOG is $O(mn)$, which is linear.

Scale Invariant Feature Transform (SIFT)

SIFT [Lowe '99] is one of the most powerful image feature extraction algorithms, which can tolerate different types of noises, such as changes in illumination, viewpoints, and scaling. When all images have the same size, orientation, and illumination, a simple corner detection such as HS or MSER works well. However, if images have different scales, viewpoints, rotations, or illuminations, SIFT can help.

Nevertheless, SIFT can not provide a unique best possible answer, and thus it uses a heuristic method (it identifies a solution near the best one but not necessarily the best one). The SIFT algorithm works as follows.

(i) The first step of the SIFT algorithm is to identify image locations invariant to scale changes. This stage builds *scale space* to identify *stable features*. The scale space is created by convolving the image with Gaussian filters at various scales.

(ii) The second step involves finding keypoints in the scale space created in step (i) and refining them based on their stability. In particular, SIFT finds keypoints by using the "Gaussian filtered images". To find keypoints the algorithm blurs the image and shrinks it into several different sizes[5]. Then, it employs blurred images and creates a new set of images by computing the Difference of Gaussians (DoG).

---

[5] if you want to learn how this process is performed, you can read about the Laplacian of the Gaussian method.

(iii) In the third step, *orientation* is assigned to keypoints, and any further calculations are done based on their orientation.

(iv) In the last step, *descriptors (features)* will be generated to identify SIFT features. In particular, a descriptor for each keypoint is generated based on the gradients, composed of magnitude and orientation, around the keypoint. These descriptors actually allow features to be matched between different images. The descriptors are highly distinctive, which enables robust matching across different views of the same scene or object.

Let's use an example with some visualizations to better learn each of these steps in the SIFT algorithm.

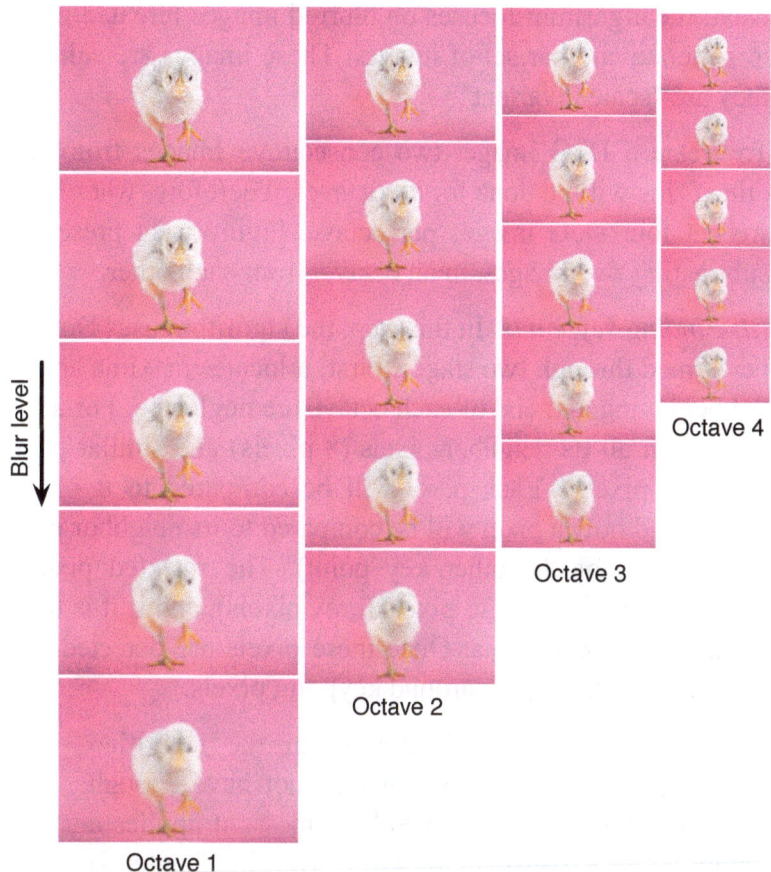

Figure 6-23: Scale space construction for SIFT algorithm.

(i) Construct scale space: In this step, the algorithm takes the original image, uses Gaussian Filtered to blur the image, and then it scales it to a smaller size. The authors of the SIFT paper recommend having four octaves (octaves present each time the scaling is performed) and five blur levels. Each octave is half of its previous octave. It includes a specific number of images, and each image is progressively blurred with the Gaussian Blur operator. Figure 6-23 presents this process.

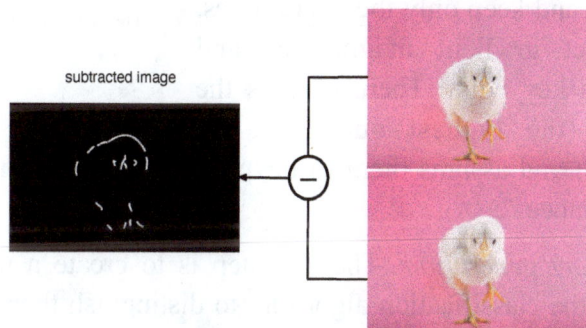

Figure 6-24: Subtraction of two images from each other. Note that in the result, we highlight changes to make them eyes-friendly. Otherwise, changes are so insignificant that it is not easy for our eyes to recognize them.

Next, the algorithm focuses on blurred images (created in the previous step) and combines them to generate another set of images. These images are called Difference of Gaussians (DoG), and they are "scale invariant".

To create a DoG image, two consecutive images from each octave are subtracted from each other. This will be done for all octaves. Therefore, when we compare five images in each octave, we get four DoG images per octave. Figure 6-24 presents a simplified image of the result of subtracting two images, in an octave, from each other.

(ii) Finding keypoints: In this step, the algorithm uses DoG images to extract keypoints. It will be performed through two stages. First, it locates maxima and minima in images. The two images at the beginning and end of each octave are neglected. For all images in between, we compare each pixel with all its neighbor pixels (8 pixels) and similar pixels in the two other neighbor images (twice 9 pixels). Each pixel will be compared to $8 + 9 + 9 = 26$ pixels. See Figure 6-25 to understand how a pixel will be compared to its neighbor pixels (grey pixels for one key point and blue pixels for the other key point). The selected pixel (red pixel in Figure 6-25) will be compared to all other 26 pixels (grey pixels), and if it is the largest or smallest, it will be marked as maxima and minima. Once these pixels are detected, sub-pixel values are generated by the expansion of the image around keypoint pixels.

Refining keypoints: Since not all of the identified keypoints are useful. Those that do not have enough contrast are considered useless. Besides, it applies a Taylor series expansion (we will learn it in Chapter 8) for more accurate keypoint localization and the elimination of keypoints based on the edge response. In particular, it removes the ones that are edges or do not have enough contrast and keeps the corner ones. In this stage, the algorithm checks the intensity of each keypoints, and removes keypoints with low contrast features such as intensity.

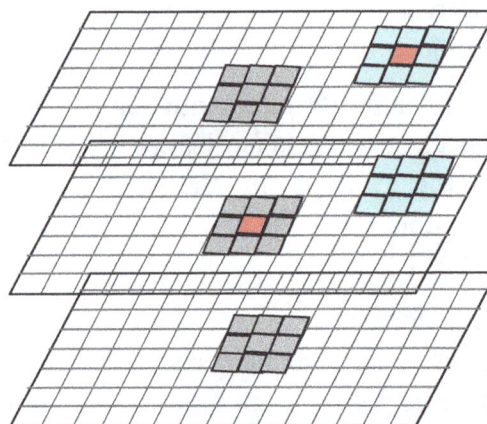

Figure 6-25: The red pixel is the selected keypoint, and the grey ones are the neighbor pixels, which will be compared to this one.

(iii) *Assigning orientation to key points:* After we get rid of weak key points and keep only the useful ones, the algorithms collect gradient magnitudes and directions around each key point. Then, it finds the *largest orientations* (the highest one from the histogram of gradients) and assigns them to keypoints. Assigning orientation to each keypoint ensures "rotation invariances".

(iv) Feature (Descriptor) generation: The last step is to create a unique fingerprint for each keypoint, that enables the classification algorithm to distinguish them from other keypoints. For example, the eyes inside an image of a face should have a unique fingerprint easily distinguishable from other keypoints.

310

To perform this step, first, the algorithm creates a 16 x 16 window around each keypoint. Each pixel is associated with its gradient orientation. Then, these 16 x 16 windows will be segmented into sixteen 4 x 4 windows. For each of these new windows, a histogram of directions will be created. Therefore, we have 4 x 4 x 8 = 128 directions values for each keypoint. The last step is to normalize these 128 values. Figure 6-26 visualizes this process for one sample keypoint.

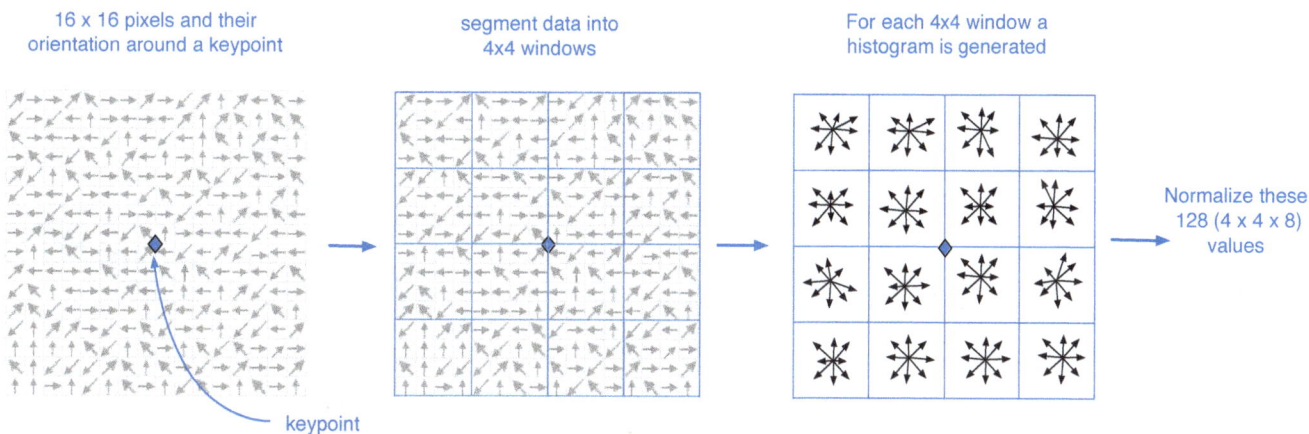

16 x 16 pixels and their orientation around a keypoint

segment data into 4x4 windows

For each 4x4 window a histogram is generated

Normalize these 128 (4 x 4 x 8) values

keypoint

Figure 6-26: Visualizing the feature generation process for each keypoint.

SIFT has many steps, and it is computationally very complex. Nevertheless, we can not neglect using it because of its superior accuracy compared to other image-based feature recognition methods.

The computational complexity of the SIFT algorithm involves multiple factors, primarily dependent on the size of the input image, denoted as $m \times n$, and the number of keypoints identified, k. The process can be broken down into several key stages, each contributing differently to the overall complexity, including scale-space construction, keypoint detection, orientation assignment, and descriptor creation.

The computational complexity of the SIFT can be computed as $O((m \times n) \cdot s \cdot l + k \cdot (d + o))$, where $m \times n$ is the size of the input image, s denotes the number of scales (octaves) considered in the scale-space construction. l represents the number of blur levels per octave. k is the number of keypoints detected. d accounts for the complexity of generating descriptors for each keypoint. o represents the operations involved in assigning orientations and other per-keypoint computations.

Now, take a bit of rest. SIFT is not easy to understand, and we didn't explain lots of its details, but it is time to open your CV and make it shine as a person who knows feature extraction algorithms, even for images.

I am enriched with feature engineering

Your CV

Watershed Transformation

The watershed algorithm is a segmentation algorithm that takes a grey-scaled image as input and views it as a topographic surface (the left side of Figure 6-27). The valleys could have a black color, and the maximum peak will be white (you can think of something like MSER).

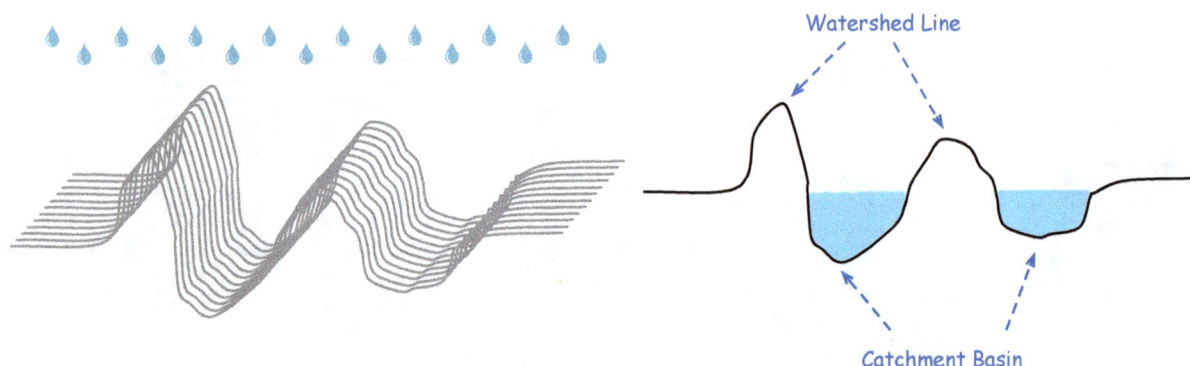

Figure 6-27: (Left) A topographical map that rain drops on it. (Right) A 2D view of the map shows that the first catchment basin is the first place that goes underwater. Then, at the end, there is a watershed line. Watershed lines specify the border lines to separate segments.

There are different implementations of a watershed algorithm [Kornilov '18], but we describe a simple one here. Assuming we have a topographic surface, the watershed algorithm floods the topographic surface. As rain increases, valleys or *catchment basins*, i.e., local minima, get full of water, and the last parts that go under the water are peaks. In a technical sense, first, an image should be converted into a greyscale, and then, based on the color intensity of the image, it will be converted into the topographic surface. In this context, darker areas (closer to 0 intensity) are typically considered lower elevation (basins or valleys) and lighter areas (closer to 255 intensity) as higher elevation (peaks).

Then, the local minima (valleys) will be identified, and pixels around them will be filled with water if there are no significant changes in the elevation (intensity) of adjacent pixels. This process continues until the water reaches a maximum peak (local maxima). In other words, the second step involves identifying catchment basins and watershed lines by flooding the image. Flooding with water in this context refers to identifying areas of the image that are connected by low intensity (or elevation) and determining the boundaries where different regions meet.

These peaks or barriers that stay outside water are boundaries of segmentation, and they are called *Watershed lines*. These watershed lines prevent water coming from different areas from merging together. By using watershed lines, the image can be segmented.

In summary, the process of flooding and identifying watershed lines means segmenting the image into regions associated with different local minima.

312

The actual computational complexity of the watershed algorithm can vary significantly depending on the implementation details, such as the use of hierarchical queues, efficient data structures for managing the watershed process, and the method of handling markers/seeds. However, assuming n is the number of pixels and t presents the time complexity of the sorting algorithm (the number of markers or seeds used to guide the segmentation process). The simplified computational complexity of the watershed algorithm could be $O(n \log n + nt)$.

There are several more image segmentation algorithms [Gonzalez '17]. Still, we only focus on Watershed and MSER algorithms because they are among the few image segmentation algorithms that could perform with accuracy comparable with neural network segmentation algorithms, especially when the dataset size is small. Later, in Chapters 11 and 12, we explain more neural network segmentations, such as U-Net and SAM series of models, which are state-of-the-art segmentation models.

NOTES:

* Usually, we do not calculate the gradient for border pixels because they don't have a neighbor. We could extend them with the value of null changes and then calculate gradients for them as well.

* Any region detection algorithm should consider handling changes, transformation, rotation, and scaling the target image.

* Although MSER is more accurate than HS, it is sensitive to natural lighting effects, such as the light of the day or the movement of the shadow.

* When images are similar (scale and orientation), we can use MSER. However, if their scale changes, or rotated or light has been changed, we should use SIFT instead.

* How many histogram bins do we need when we are quantizing the image data using a gradient? HOG authors [Dalal '05] recommended that nine bins spanning from 0° to 180° are enough. SIFT authors [Lowe '99] recommended that eight bins spanning from 0° to 360° is enough.

* Usually, color normalization should be done at the beginning of image processing to remove illumination and contrast variance, especially since this preprocessing is important for SIFT and HOG, which are sensitive to color changes.

Video Feature Extraction

There are a few approaches known to extract features from videos, and we briefly list them here. Most video feature extractions use convolutional neural networks, which we will learn about in Chapter 10. A video is presented as a sequence of images, and each of these images is referred to as a frame.

Motion Vector

A Motion vector is used to represent the motion of a block or a group of pixels between two or more frames in a video sequence. It is used to capture the movement of objects and camera motion between video frames. It can be used for action recognition and video classification.

Motion vectors are usually computed for blocks of pixels (e.g., 8×8 or 16×16 blocks) rather than every pixel, providing a more sparse representation of motion compared to optical flow. The main use of motion vectors is in video compression standards like MPEG and H.264/AVC, where they reduce temporal redundancy between frames and thus reduce the video size. Figure 6-28 visualizes the motion vector of a circle movement between two frames.

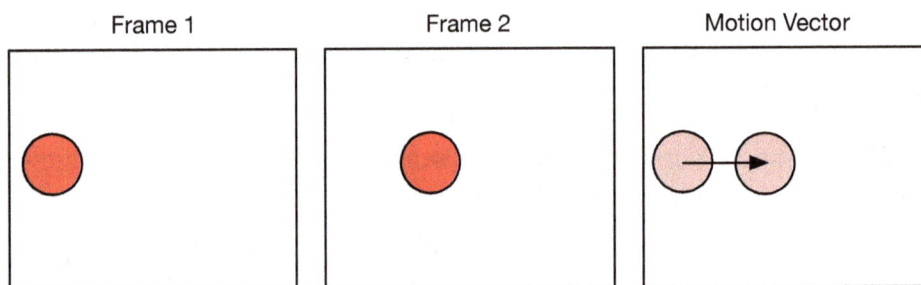

Figure 6-28: Motion vector presents the movement between *Frame 1* and *Frame 2*.

Optical Flow

Optical flow is another approach to quantifying the motion of objects by measuring changes in brightness patterns between consecutive frames. Specifically, it measures how pixels' brightness values shift from one frame to the next. By examining these changes, we can infer the movement of objects or the observer's motion. However, we should keep in mind that changes in lighting can result in apparent optical flow even in the absence of actual object or observer motion.

There are two types of optical flows: *sparse optical flow* and *dense optical flow*. Sparse optical flow identifies the movement vectors of select "interesting features" (such as specific pixels s edges or corners'representing an object) within a frame. On the other hand, Dense optical flow calculates the movement vectors for every pixel in the frame, essentially providing one flow vector per pixel. Dense optical flow has higher precision, but it is slower and utilizes more computational resources.

For example, Figure 6-29 presents optical flow changes, and we can see that in sparse optical flow, some dumbbells show green lines, which are the paths of changes. However, since his hand crosses on these dumbbells by mistake, the sparse optical flow assumes them to be moving. The red dots present the edges that the implementation library detects[6].

There are more video features that exist, such as differentiating between two frames to detect the moving parts, background modeling to subtract the background from a video, etc. We do not explain them in more detail because most works in this area benefit from Optical flow or Motion vectors.

Provides useful motion-based features. Optical flow methods estimate the motion between two video frames based on the apparent movement of objects, surfaces, and edges. This information can be used directly as features or to improve the performance of other feature extraction methods by providing motion cues.

3D Convolutional Neural Networks (3D CNNs)

Similar to images, video feature extractions are enriched a lot by neural networks. We explain convolutional neural networks in Chapter 10. If you have difficulties understanding them now, come after you about CNN in Chapter 10.

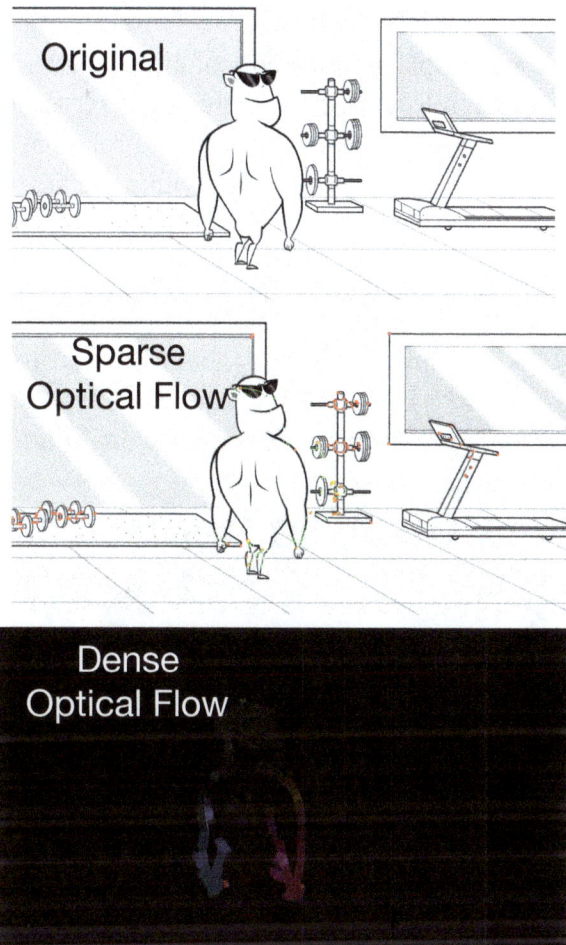

Figure 6-29: In the original video on top, the character moves, the background is steady, and its sparse and dense optical flows are presented.

We can see that in sparse optical flow, even some dumbbells have green lines, which present the flow of changes, because the character's hands cross them when he is moving.

Unlike traditional CNNs that operate on 2D data, 3D CNNs extend the convolution operation to the temporal dimension, allowing for the extraction of features that capture both spatial and temporal information simultaneously [Ji '12, Karpathy '14]. In other words, 3D CNNs apply convolutional filters not only across the height and width of the video frames (as in 2D CNNs) but also across consecutive frames or time steps. This is particularly useful for action recognition and event detection in videos.

Some recent approaches incorporate 3D CNN with RNN (we will learn them in Chapter 10) or attention mechanism (we will learn Attention in Chapter 12) [Wang '16].

---

[6] We have used open CV2 and Python to build these images.

Graph Convolutional Networks

We will explain Graph convolutional networks (GCNs) in Chapter 15. They can treat video data as a graph [Wang '18], where nodes represent significant points or regions in video frames, and edges represent spatial or temporal relationships. This approach is useful for capturing complex interactions and relationships within the video data.

GCNs can learn hierarchical representations of graph data, which is very useful for video analysis as it allows for capturing both low-level details (such as individual pixel movements) and high-level patterns (such as the overall motion of an object across frames). Besides, by using GCN, the graph can be dynamically constructed and updated across frames to reflect changes over time. This dynamic nature allows GCNs to adapt to the evolving content of a video, providing a flexible framework for analysis.

Feature Engineering for Signals and Time Series

Lots of data are represented as time series or signals. For example, audio, radio, human movement, and data read from sensors or medical devices, such as heart rate, temperature, and sensor data, are modeled in two dimensions, i.e., *values* of a single variable and *time*. In simple words, a time series is a variation of a variable over time. Signal, a presentation form of information, is also similar but does not necessarily have time; its other dimensions could be frequency, amplitude, space, etc.

While working with time series, what we need to predict is a function of time[7], e.g. $y = f(t)$. Time series data is inherently ordered by time. While working with signals, what we need to predict is not necessarily a function of time; it could be a function of spatial coordinates, frequency, or anything else, e.g. $y = f(x_1, x_2, x_3)$. Some state there is no clear border between those definitions; some use the term time series, and some use the term signal (time domain signal). We should keep in mind that a time domain signal is a signal that is represented as a function of time, and a time series is a sequence of data points that are measured at regular intervals over time.

Many different data types could be presented as a signal; e.g., we have seen that a picture could be converted into a matrix, and it could also be presented as a signal. Or assume a satellite searching for a specific type of tree in a jungle uses a signal processing method to translate the signal it receives from its radar into the tree's shape. We could convert any information into a signal.

In the following, we define some features worth extracting or investigating in time series and signals.

---

[7] http://ataspinar.com/2018/04/04/machine-learning-with-signal-processing-techniques

Time series Features

Time series has a notion of time and is ordered based on time (unlike the example of detecting a tree shape from satellite radar). There are particular information objects that are important to analyze while working with time series. We prefer to call them features. In other literature, they might not called features. However, they are features because these are the information objects that a machine learning algorithm usually looks to find and predict in a time series. In the following section, we list some concepts and features related to time series; we do not separate concepts from features, unlike in previous sections.

Stationarity and non-stationarity time series

Time series can be either stationary or non-stationary. A time series is stationary if its statistical properties, including mean, variance, etc., stay fixed at different times. In other words, statistical properties are invariant over time. A non-stationary time series has different statistical properties at different times, such as different mean and variance at different times. Figure 6-30, shows an example of two time series: one is stationary (e.g., the amount of food consumed by chickens daily), and the other is non-stationary (e.g., the weight of our chickens increases during the time).

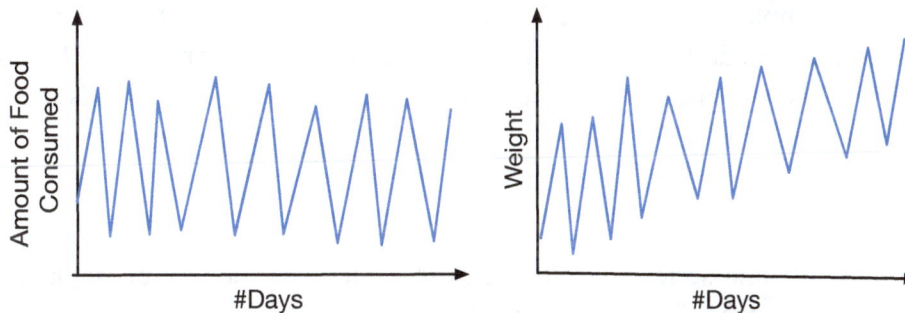

Figure 6-30: (Left) stationary time series, (right) non-stationary time series.

Periodicity and Seasonal Behaviors

A constant variation of values that occurs in a regular interval is called *periodicity*. A periodic function is a function that repeats its value in a regular interval. The term *seasonal behavior* is similar and reflects a periodicity. Seasonal behavior is indeed a type of periodicity that is observed in many time series data, particularly those related to natural, social, or economic phenomena that repeat at regular intervals, typically annually.

For example, the number of swimsuits sold increases in summer but decreases in winter. This is a seasonal behavior. As another example, we can study the consumption of ice cream that increases on the weekend, in hot weather, and decreases on the weekdays or in cold weather; this is another example of seasonal behavior. Figure 6-31 (the leftmost figure) presents an example of periodicity and seasonal behaviors as well.

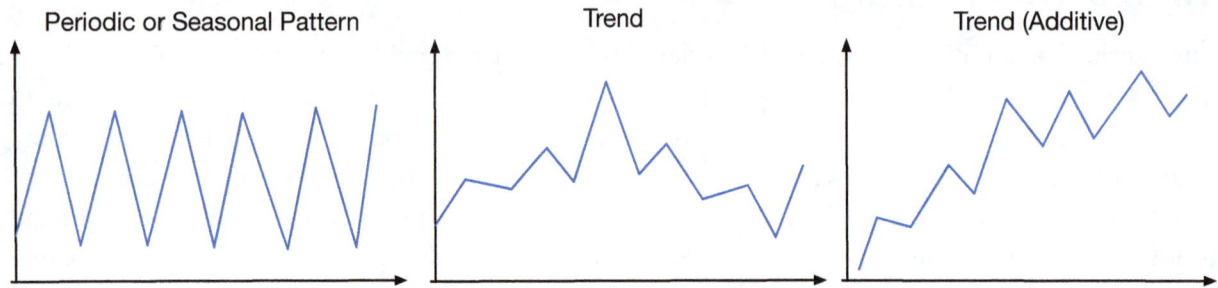

Figure 6-31: A sample periodic pattern and two types of trends sample.

Trends

A trend in a time series is a systematic change that can be upward or downward. In other words, a trend refers to a non-periodic, systematic change in a time series. It is represented by a linear curve that shows a continuous increase or decrease in the data. However, trends are not limited to linear changes. Trends in time series data can also be nonlinear, reflecting more complex patterns of growth or decline, such as exponential, logarithmic, or polynomial trends.

In Figure 6-31, an example of an upward trend is depicted, representing a continuously increasing trend known as an additive trend. If the trend has a constant slope, it is referred to as a linear trend. If the slope is changing, it is considered a non-linear trend. Both trends presented in Figure 6-31 are non-linear.

Motif

A motif in the context of time series analysis refers to a recurring pattern or sequence that is identifiable within the data, regardless of whether the series is periodic or non-periodic. Unlike periodic behavior, which occurs at regular intervals, a motif represents a specific shape that recurs, possibly at irregular intervals. Therefore, while motifs can have periodic behavior, they can also be in a data and not have a fixed cycle. Figure 6-32 presents a motif repeated twice in a time series.

Figure 6-32: Motif example in a time series.

Lag

In practice, understanding the connection between previous and current data points is fundamental to time series modeling. Previous data points in the context of time series are called *lags*. In other words, the lag operator, or *backshift* operator, specifies the number of previous data points for analysis. The lag operator is a mathematical notation used to simplify the expression of models that involve lags. This operator shifts a time series backward by the specified number of periods, making it easier to represent and manipulate time series data in mathematical models. For example, assuming we use L the lag operator, we can write the lag m of the data point at time t to equal its m previous data points $L^m(d_t) = d_{t-m}$. For example, $L(d_t) = d_{t-1}$ or $L^3(d_t) = d_{t-3}$. This notation is used to write the time series in a more condensed form. This notation simplifies the expression of time series models by defining the temporal relationship between data points.

Step, Burst, and Change Point

When there is an abnormal spike in a time series, which is not because of outliers, it reveals that in that specific time, the target information is abnormally high or popular. Outliers are typically individual data points that deviate significantly from other observations, and while they might not always indicate a trend or a *burst*, they can still impact the analysis For example, when there is a cold week in the summer, the selling of coats increases abnormally, and this will present a burst in time series. The algorithms that focus on identifying abrupt changes jump or shift in a time series are called *change point detection* algorithms, and they are fairly easy to learn.

Change point detection algorithms focus on detecting abrupt changes or an outbreak in the signal. There are many applications that can benefit from signal change point identification. For example, assume you have a colleague, Mr. Snoop Nose, who is too curious about other employees' behaviors and salaries. He develops a system that monitors how his colleagues spend their weekends and assigns them a spending score. Any changes in their weekend spending score could be a piece of evidence that their salary has been raised, and he is not aware of that event. Figure 6-33 presents a change point in Mr. Snoop Nose, a diagram drawn for a colleague who is suspicious of getting a salary raise.

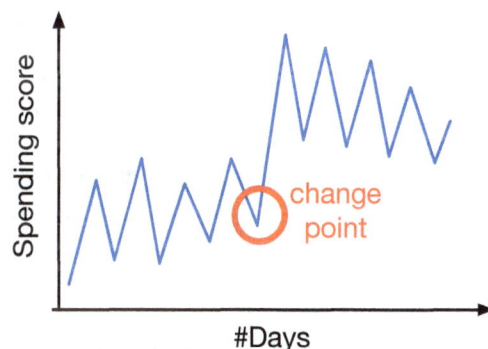

Figure 6-33: Example of a change point in time series.

The Cumulative Sum (CUSUM) is one of the effective change point detection algorithm that is usef in time series data [Page '54]. CUSUM aims to identify points in time where the statistical properties of the data series, such as the mean, undergo a significant change. It operates by calculating cumulative sums of positive and negative deviations from the mean. When these cumulative sums exceed certain thresholds (hyperparameter), a signal is generated indicating a potential change or "change point" in the data series.

Signal Features

Signal Concepts

A signal is a means of conveying information, often represented as a wave in both physical and technical contexts.

Let's think about waves from the sea; if you are looking at a sea wave, an amazingly beautiful thing, every few seconds, a new wave is coming. The

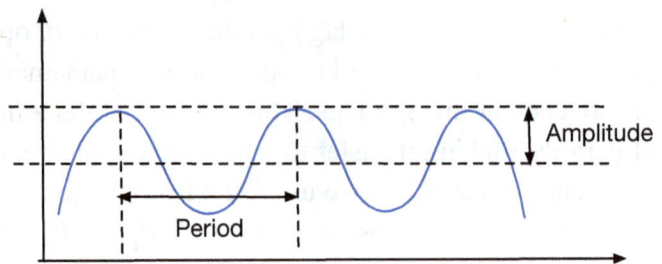

Figure 6-34: Amplitude and period of a signal.

amount of time between each wave is called the *period* of a wave. The period can be viewed as time or distance. In a technical sense, the period is the duration of a wave in a signal, which is repeated continuously. Additionally, just as sea waves have varying heights, signals have an *amplitude*, which measures the peak value or strength of the wave, reflecting how far the wave rises above its resting position. Figure 6-34 presents the amplitude and period of the wave, and it presents the wave as a signal.

Another key concept in a signal is *frequency*. Frequency in signal analysis represents how many cycles of the signal occur within one second. It indicates the rate at which the signal repeats its pattern over time. The period, conversely, is the duration of time it takes for one complete cycle of the signal to occur. There's an inverse relationship between period and frequency, which can be mathematically expressed as follows:

$$Period = \frac{1}{Frequency}$$

This means that a higher frequency corresponds to a shorter period and vice versa. Frequency is measured in units of Hertz (Hz), where one Hertz signifies one cycle per second.

Types of Signals

There are two types of signals: *time domain* and *frequency domain signals*. If the X axis presents a signal based on time and its value (Y axis), then we have a time domain signal. If the X axis presents the frequencies and the Y axis presents amplitude, then it is a frequency domain signal. Figure 6-35 visualizes these two types of signals.

In the next chapter, we describe signal decomposition in detail. For now, just remember that every signal, no matter how complicated, can be decomposed into a set of simpler signals. Through the decomposition process, such as using the Fast Fourier Transform, a time domain signal could be converted into a frequency domain signal and vice versa. By converting a signal into a frequency domain signal, we can apply more operations on signals, such as filtering methods (which will be described later in Chapter 16). Besides, they are computationally faster as well.

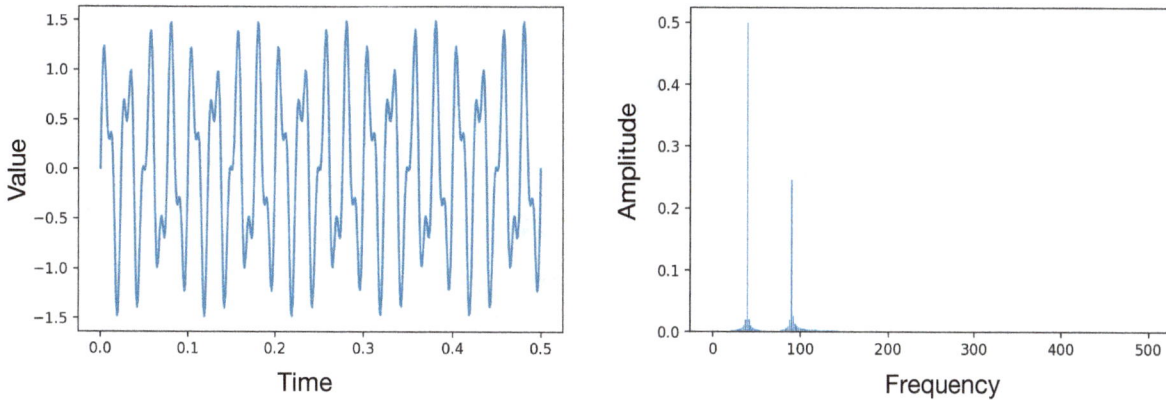

Figure 6-35: (Left) time domain signal, (right) the same signal is converted into frequency domain signal.

Each of these types of signals has some characteristics, and based on the questions we intend to answer, we extract features from them. There are some common features that can be extracted from signals, which could be listed in two categories: peak analysis and change point detection. Information such as the height of peaks, width of peaks, and distance between two peak neighbors (to remove the ones that are too close and smooth the data) are worth considering as features. Figure 6-36 and Figure 6-37 present features related to peaks in a signal, which are easy to visualize.

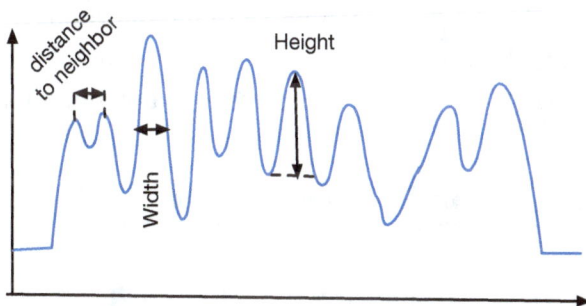

Figure 6-36: Some peak features of a signal.

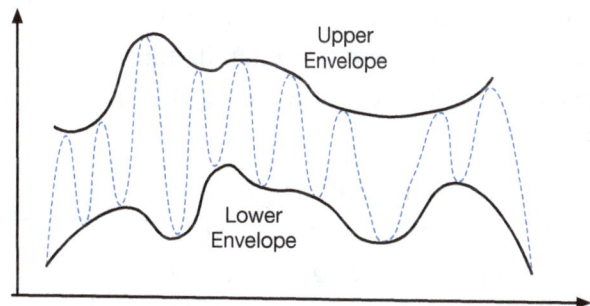

Figure 6-37: Upper and lower envelopes of a signal.

Features related to peak analysis include cumulative maximum or minimum of all peaks, mean, median, variance, and standard deviation of signal frequency, *maximum to minimum* difference, *signal envelope*, and so forth. The signal envelope is another feature, which is a curve that traces the signal's lower and upper peaks. See Figure 6-37, which visualizes a signal's lower and upper envelopes.

There are specific shapes of signals, such as sinusoidal signals (sine waves) or cosine waves. Take a look at Figure 6-38 and try to remember their shape; we are going to refer to them in Chapter 12, for positional encoding. These are common waves that are very popular for different types of applications. We don't need to learn the equation of building waves, but it is really simple, and you can look it up on the Internet. You can see that both cosine and sine waves are

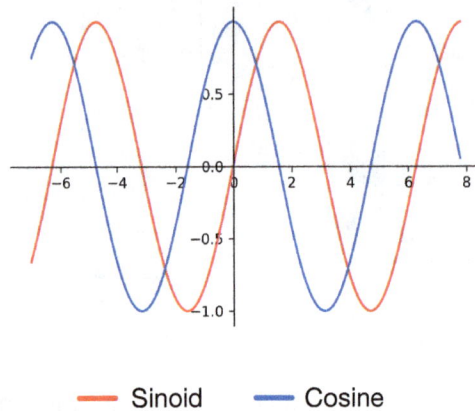

Figure 6-38: Sine and Cosine signals.

similar, and their only difference is their phase. *Phase* in signal refers to how far a signal is shifted horizontally from the usual position. In Figure 6-38, Phase is the size of differences between peaks of sine and cosine signals.

Signal and Time Series Smoothing

Usually, a signal is very noisy, and it is not easy to feed it into a machine learning algorithm. One of the common approaches to dealing with signals (either time domain or frequency domain signal) is generalizing the signal to make it interpretable for the machine learning algorithm or model. The process of generalizing the signal is called smoothing. In other words, smoothing refers to the process of reducing the information in the signal or time series while maintaining its patterns.

While working with large time series or signals, it is very common that we get a very cluttered signal, but we need to make them smooth for analysis, such as comparing two time series for clustering. First, we might need some noise removal approaches (we will explain in Chapter 16). Then, we apply signal smoothing.

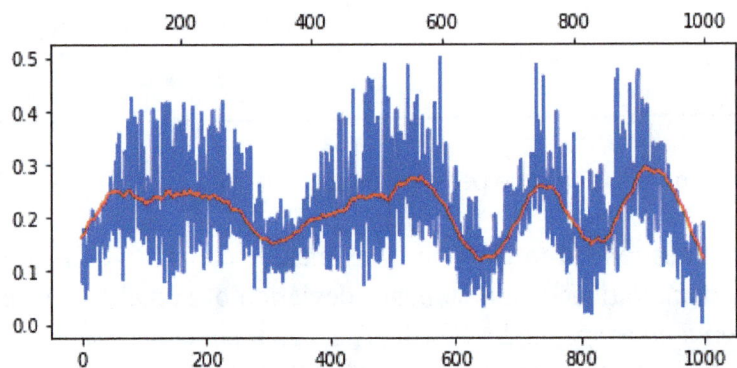

Figure 6-39: Blue lines are the original data of the time series and red line is the moving average.

Usually, signal smoothing methods operate on a one-dimensional signal, such as time series and two-dimensional data, e.g., image. Smoothing is sometimes referred to as filtering in the context of signal processing because it suppresses high-frequency signals and enhances low-frequency signals.

322

There are several methods used for signal smoothing, such as Savitzky-Golay, Butterworth, Chebyshev, and Kalman filters. We will explain some of them in Chapter 16.

Here we could learn the "simple moving average", and keep the rest for later. Take a look at Figure 6-39; we can see there is a blue time series that includes lots of data points. It is hard for a machine learning algorithm to work with this large signal data, but the red line includes the same patterns and consists of a significantly smaller number of data points. It is calculated as the moving average of the original signal.

Moving average uses a disjoint sliding window (Check Chapter 4 to recall sliding window) on the data, gets the average of coordinates inside that window, and assigns the average to all data points inside a window, then moves the window to the next set of data points and do the same until all signal or time series is processed. For example, assume we have the following time series, $a = \{t_1, t_2, \ldots t_n\}$ with a window size of three. A moving average of this time series is a time series with $1/3$ data points, and its data points from point m will be calculated as $(t_{m-1}, t_m, t_{m+1})/3$.

If the moving average assigns a higher weight to more recent data points and a lower weight to previous data points, this is called an *exponential moving average*.

You might get frustrated as to why some discussions about signals are distributed in three chapters (signal decomposition in Chapter 7 and noise filtering in Chapter 17) and not in unique places. We can answer that we learn signal processing in the context of machine learning and not signal processing itself.

Summary

The *feature* refers to a data type or attribute in a format that the machine learning algorithm can use; we can convert from one format to another format, e.g., text to number.

Feature engineering refers to the task of identifying the relevant attributes in the raw dataset.

Feature selection is selecting a subset of the raw dataset, enabling machine learning to analyze the dataset and produce desirable output. It involves identifying the useful attributes and removing those that are not useful.

Feature generation refers to the task of creating new features from features existing in the original dataset, for example, taking the average of multiple columns and adding a new column that is the average of the other columns. Interaction or Polynomial Feature creation are common feature generation methods.

When we collect data and build the dataset for our task, we should identify our goal and what features hold relevance to that goal. What features seem unrelated? Then, we should remove unrelated features in the dataset that we fed into the algorithm. This is often difficult to do by hand and can increase bias. Deep learning algorithms usually use all features and sort out which are useful themselves. However, they are good for unstructured data and not tabular data.

There are three known methods for feature selection: *filter* methods, *wrapper* methods, and *embedded* methods. Filter methods rank features and decide to keep or remove features by their rank score, e.g., Chi-squared test, information gain, and correlation coefficient. Wrapper methods are recursive feature elimination algorithms and consider feature selection as a heuristic search problem (e.g. genetic algorithms). Genetic algorithms are an example of heuristic algorithms, finding a 'good enough' and generalizable solution instead of the perfect solution for an instance. Embedded methods learn which feature best contributes to the accuracy while the model is being created.

Next, we described feature engineering for different types of data, including numerical, categorical, text, image, time series, and signal data.

Feature engineering for numerical data: Most machine learning algorithms are designed to work with numerical data. It is common to convert different data types to a scalar value (number), vector, matrix, or tensor. We have explained some terms related to numerical feature engineering, which are listed as follows:

- Sanity check: checking whether the data follows the requirements for conducting any calculation.

- Normalization: scaling or shrinking data to a comparable format and unit-independent. In particular, it makes it easier for the hosting machine to perform numerical computation.

- Quantizing/binning: substituting individual numerical values with simpler categorical values. More generalized categories (bigger bins) make it easier for the algorithm to recognize patterns, but overgeneralizing could harm the data, and it loses semantical information.

- Scaling Data: shrinking numerical values in a way that they consume less computational resources (e.g., memory) while maintaining their relationships. Examples include Min-Max scaling, Standardization/Variance scaling, and Vector Norms.

- Power Transformations: they are used to change the representation of the dataset (e.g., make the distribution more like a normal distribution for analysis. Change the distribution so the variance does not depend on the mean. Examples are Logarithmic Transformation and Box-Cox transformation. They are useful for balancing data distribution in a model and preventing small but unimportant values from getting overlooked by the model.

Feature engineering for categorical data: we have explained popular methods, which is not possible to explain here briefly. These methods include One-hot encoding, Dummy Coding, Effect Coding, feature hashing, and Bin Counting.

Feature engineering for textual data: Common tasks that use machine learning and involve text include classification, clustering, sentiment analysis, theme extraction, named-entity recognition, question answering (conversational agent), text generation, and translation from one language to another language.

We also listed some non-deep learning models and algorithms that use text data to perform preprocessing steps, including tokenization, stop-words removal, stemming, and text normalization (e.g., TF/IDF).

We list different approaches to feed textual data to the algorithm, including "Bag of Words", "Subword Tokenization", "N-Grams", "Part of Speech Tagging", and "Word Embedding". "Word Embedding" got lots of attention in recent years from the machine learning community and led to the introduction of large language models. Word embedding methods (e.g., word2vec, GloVe, and FastText) construct a "language model", which means each word is presented with a vector of numbers. They are trained by feeding a large corpus of text, and their result can be used for Continuous Bag-of-Words (CBoW) or Skip-gram. CBOW uses a context to predict a target word in the context, and Skip-gram uses the word to predict the context.

Feature engineering for images: First, we described that feature as the region in the image where we can receive the label. Usually, algorithms transform the image into a matrix (black and white) or tensor (color) and process data numerically.

Next, we describe terms and concepts that are used in image feature engineering as follows:

"Edge and region": They refer to a line (edge) or an area (region) where pixel color changes more than a specified threshold.

"Corner": where two edges meet each other and the intensity of pixels changes in both X and Y directions, it is called a corner.

"Path (curve)": a path is a series of adjacent pixels whose color intensity does not change.

"Segmentation": it is the process of dividing an image into areas based on similarities and edges.

"Skeleton": it is a compact representation of an object, and important edges are reduced to a single pixel line.

Afterward, we have listed more advanced features that can be extracted from image data as follows:

"Object Detection": identifying objects that have been annotated in training set in the new image.

"Object Localization": specifying a bounding box around an identified object and its location on the image.

"Semantic Segmentation": finding the shape of the object that separates this object from other objects.

"Instance Segmentation": separating different objects with similar names in the image, for example, separating each individual in the image with faces of different individuals.

Then we list non-deep learning image feature extraction methods, including arris-Stephens (HS) corner detection, Maximally Stable Extremal Regions (MSER), Histogram Oriented Gradient (HOG), Scale Invariant Feature Transform (SIFT), and Watershed Transformation.

Feature engineering for video: After images, we briefly describe some concepts used for video feature engineering, including motion vector and optical flow. Then, briefly explain some deep learning models used for video classification.

Feature engineering for signals and time series: when we are dealing with a variable that is changing through time, we call this data time series, such as heart rate data. Features we usually search in time series are seasonal behaviors, trends, motifs, steps, and bursts (by using change point detection algorithms).

A signal is a representation of a wave that includes information, such as frequency. There are two types of signals: time domain and frequency domain signals. To facilitate analysis, we can decompose a time domain signal into a set of frequency domain signals. Another popular approach while working with a signal is removing its noise via smoothing. One basic approach to smooth a signal is using a "moving average".

Although texts in this chapter seem disconnected, learning these methods is very useful for you, especially if you intend to develop algorithms from scratch and not just use existing libraries. For example, you might think that you don't need to work with image data and, thus, you don't need to read its feature engineering. Nevertheless, you can borrow its concepts, such as using a patch on the data and applying it to your problem or developing your algorithm.

Further Reading or Watching

* There are few books and online resources about feature engineering that have been published. One of them, which is very useful and explains concepts in clear form, is "Feature Engineering for Machine Learning: Principles and Techniques for Data Scientists" [Zheng '18] by Zheng and Casari. We used this book a lot to write different parts of this chapter.

* A helpful description of bin counting with real-world examples is provided in this link: https://blogs.technet.microsoft.com/machinelearning/2015/02/17/big-learning-made-easy-with-counts

* A very brief and helpful tutorial has been proposed for Genetic Algorithm here. https://towardsdatascience.com/introduction-to-genetic-algorithms-including-example-code-e396e98d8bf3 Kudos to its author (Vijini Mallawaarachchi) with her long family name, which is even longer than the family name of this book author, Rawassizadeh.

* A good tutorial about Word2Vec exists here: https://skymind.ai/wiki/word2vec.

* If you are interested in getting an expert in the traditional mechanism of image processing and concepts, we recommend the amazing book provided by Gonzalez and Wood [Gonzalez '17]. It is among the fewest technical books that do not assume you are a mathematician or computer science expert. It describes concepts in detail and easy to understand.

* There is a very good and easy online tutorial about the SIFT algorithm available online by Utkarsh Sinha: http://aishack.in/tutorials/sift-scale-invariant-feature-transform-introduction/.

* If you are interested in learning more about change point detection, you can read the paper by Amirkhanghahi and Cook [Amirkhanghahi '17].

* At the time of writing this chapter for the first time (in 2019), there was a fierce war between Python and R as data science languages, but Matlab also has good libraries for signal processes and working with time series. This Matlab package is called "Signal Processing Toolbox", it might be worth taking a look at this package as well https://www.mathworks.com/products/signal.html. When the author of this book was an undergraduate, Matlab was the most popular programming language for data science. However, it still has very detailed and useful libraries for signal processing. At the time of revising this chapter in 2024, Python overtakes other programming languages, and if you want to start learning data science, start with Python.

Chapter 7: Dimensionality Reduction and Data Decomposition

While working with real-world data, one big challenge is how to feed a large amount of the data into a model. Some datasets are too huge to be fed into a machine-learning algorithm or model. We should reduce the dataset size while maintaining its characteristics.

In other words, we would like to keep everything that can help the machine learning algorithm to operate, but on the other hand, our dataset is too large. In these scenarios, data reduction methods can be used. Data reduction methods include *data decomposition* and *dimensionality reduction* methods.

At the early stage of your career, you might get fooled by hardware performance increases and think you don't need to learn these methods. We have heard this a lot, especially from people in academia who do not have much hands-on experience or who learned machine learning centuries ago, along with dinosaurs inside their cave.

Hardware miniaturization is a continuous effort, and we are always facing the challenge of reducing some components of data or algorithms to fit them into a small device. For example, assume we intend to run an algorithm on a smart pill. We swallow the pill, and it should travel along our digesting system until it identifies a specific tissue; then, it should release its medication. In terms of implementation, we need a nano-scale machine to run an algorithm that can identify the correct tissue, and such a small size requires a very resource-efficient algorithm and a small amount of data. Therefore, the need to collect data, analyze them, and probably reduce them for the algorithm is a persistent challenge. To proceed further, we need to learn a few terms, and in the following, we describe them.

The process of transforming a dataset into a smaller dataset while maintaining the integrity of the original dataset is referred to as *data reduction*. This transformation could be done by summarizing the data, compressing it, changing its format (e.g., from an image to a signal), or projecting the data from its original dimensional space into a lower dimensional space (dimensionality reduction).

Data compression methods will be described in Chapter 14. Some feature engineering methods (check Chapter 6) approaches could also be considered data reduction techniques, such as feature selection methods. Also, when we have two highly correlated data objects in our dataset, we can remove one of them with a correlation coefficient (check Chapter 3). Aggrawal [Aggrawal '15] called correlations in a dataset as implicit redundancies.

This chapter starts by explaining dimensionality reduction methods. After our explanation of dimensionality reduction. we will describe some data-specific decomposition methods, including time series and signal decomposition methods, matrix deposition methods with their applications, and finally, tensor decomposition methods.

Before proceeding further, remember that here, we denote tensor and matrix with capital letters and vector with small letters. Other literature used calligraphic capital letters to denote a tensor.

Dimensionality Reduction Methods

Dimensionality reduction is a major data reduction technique that can help remove correlative features and preserve significant information. Dimensionality reduction reduces the dimensions of numerical data, e.g., converting a matrix (2D data) into a vector (1D data), while trying to keep the characteristics of data. It tries to ensure the reducing attributes in a dataset by introducing a new feature space where the original features are represented.

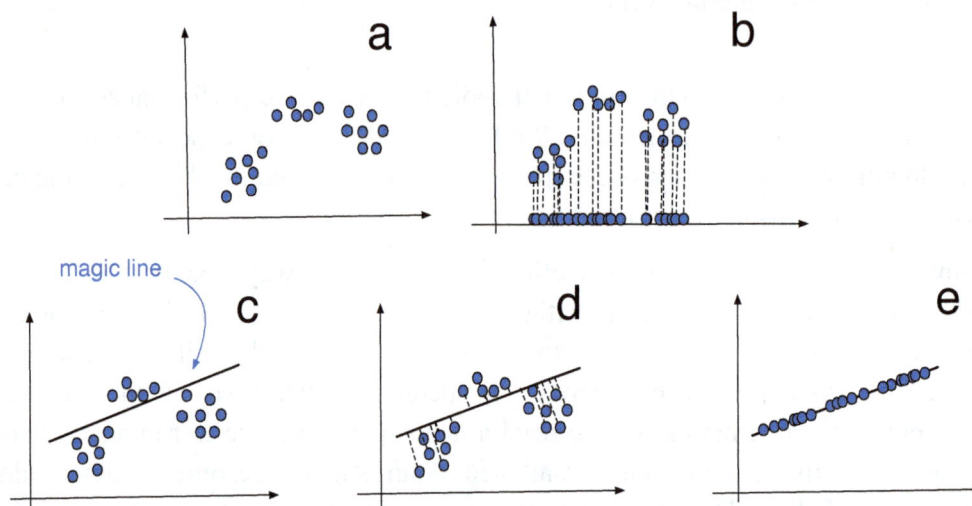

Figure 7-1: A toy example how dimensionality reduction can bring data into 1D space while keeping the relation between data points

To better understand dimensionally reduction, let's look at the example presented in Figure 7-1. In Figure 7-1 (a), we can have a dataset in the 2D space. There, we can see three clusters of data points. If we project (move) all data points into one x-axis (projecting them in one dimension), we can see the result in Figure 7-1 (b). There will be no three distinctive clusters existing between data points, but we have them before the projection. If we project them on the y-axis, the same problem remains. However, if we identify a magic line (Figure 7-1 (c)) and bring all data points to 1D space based on that magic line (Figure 7-1 (d)), their clustered characteristics stay (Figure 7-1 (e)). This is what a dimensionality reduction algorithm does.

Dimensionality reduction is one of the most common methods that is used in a real-world pipeline of machine learning applications to deal with specific data such as genome data analysis. By applying them, we could process and analyze extremely large datasets.

In modern literature, there are two main categories for dimensionality reduction: linear projection and manifold learning. In the context of mathematics, linear equations are characterized by variables that do not exceed the first power and, thus, do not involve polynomials of degrees higher than one. For example, the equation $2x_1 + 5x_2 + 4$ is linear. In contrast, non-linear equations involve higher-degree polynomials or other non-linear terms. For example, the equation $2x^2 + 7$ is non-linear because it includes a squared term.

First, we describe two linear projection methods: Principle Component Analysis (PCA) and Linear Discriminant Analysis (LDA). Afterward, we shift our focus to manifold learning methods, including LLE, t-SNE, and UMAP.

The Curse of Dimensionality

Imagine you have a one-meter (100 cm) tube with a 0.5 cm radius, and you drop a grain of rice that is 1 cm long. To find the rice grain, we must search about 100 times in 1 cm slots. It seems ok, but now let's say we have 1 square meter (100 cm × 100 cm), and we drop the same rice grain there. Now, we must search for the rice grain in 100 × 100 slots. What about a cubic space and the rice grain? We need to search in 100 × 100 × 100 slots. This is too big a space to search. Assume there are 100,000 rice grains, and we would like to cluster them. Comparing each rice grain from the X, Y, and Z axes to the 999,999 X, Y, and Z axes of other rice grains is too time-consuming. This problem is called the curse of dimensionality, which was introduced by Bellman in 1961 [Bellman '61]. A more technical definition is as follows: Having too many (or too few) dimensions makes the available data sparse and is not interpretable for the machine learning algorithm, which is called the curse of dimensionality.

Dataset

The Curse of Dimensionality

This problem exists for all machine learning algorithms. For example, having too many dimensions disables the clustering algorithm to cluster the dataset because data points are far away properly.

Linear Dimensionality Reduction Methods

Linear Dimensionality Reduction methods do not use approximation, and thus, their results do not change after each execution (unlike non-linear methods), which makes them attractive for implementation in real-world applications. In this section, we explain three linear methods, including Principal Component Analysis (PCA), IPCA, and Linear Discriminant Analysis.

Principal Component Analysis (PCA)

The Principal Component Analysis (PCA) is a linear dimensionality reduction method that transforms the multidimensional data into lower dimensional data and tries to reduce the number of highly correlated attributes in the dataset. It compresses a multi-dimensional dataset, with a small loss, into a lower-dimensional dataset that keeps the characteristics of the original multi-dimensional dataset. It is one of the most used machine learning algorithms in many applications, including image recognition, such as face recognition [Rao '00], optical character recognition, missing data reconstruction, anomaly detection, genomics data analysis, etc.

Before we describe PCA in more detail, we need to review or learn some concepts, including *covariance matrix*, *eigenvalue*, and *eigenvector*. We do not need to learn how to compute them mathematically, but we should know their concepts and use them to understand the PCA algorithm.

Covariance matrix: If you recall from Chapter 3, we have described that covariance is a variance of two or more dimensional data between each pair of data. The covariance matrix for three-dimensional data of X, Y, Z is computed as the following matrix, which is a symmetric and square matrix[1]. Here, $cov(x, y)$ means the covariance between x and y.

$$C = \begin{bmatrix} cov(x,x) & cov(x,y) & cov(x,z) \\ cov(y,x) & cov(y,y) & cov(y,z) \\ cov(z,x) & cov(z,y) & cov(z,z) \end{bmatrix}$$

As the number of dimensions increases, the number of rows and columns of the covariance matrix increases as well. Note that the covariance function is commutative, which means $cov(x, y) = cov(y, x)$, $cov(y, z) = cov(z, y)$ and $cov(x, z) = cov(z, x)$. The covariance matrix will be used in PCA to describe relations between data points. In other words, Covariance measures how data points from two dimensions move together.

Eigenvector and *Eigenvalue*: Assume we have a material, like a textile, which we can stretch or squeeze. We draw a line on it, which is shown in blue in Figure 7-2 (a). Now, we squeeze our material vertically or stretch it horizontally (see Figure 7-2 (b)). We can notice that the direction of the blue line has been changed from Figure 7-2 (a) to Figure 7-2 (b).

There are two other vectors shown in red (Figure 7-2 (a2) and Figure 7-2 (b2)); when we apply the same stretch and squeeze, we can see that the direction of the red vectors did not change.

---

[1] A square matrix is a matrix in which its number of columns and rows are equal, and a symmetric matrix is a matrix in which the data on the side of the diagonal matrix are identical.

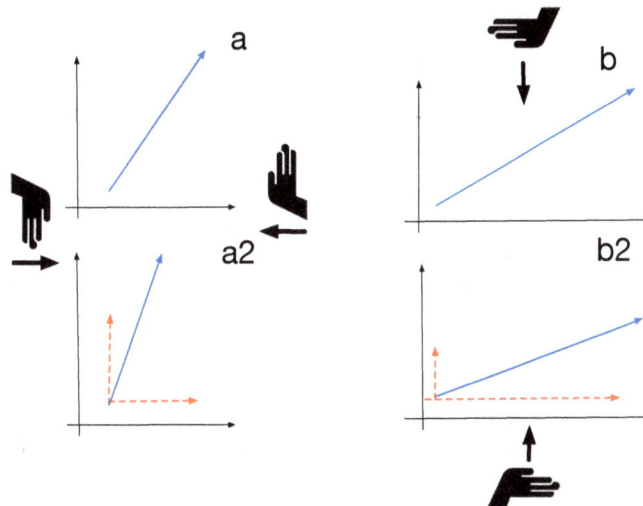

Figure 7-2: Two vectors (in blue) figure *a,* and figure *b*. The bottom figures (a2 and b2) show the eigenvectors in red of each vector. Look at the direction of the hand to see in which direction we squeezed or stretched the vector.

Their length can change, but not their direction. These vectors are called *eigenvectors,* and the magnitude of their changing is called *eigenvalues*[2]. In other words, an eigenvector is a vector that, even after linear transformation, its directions do not change. If you don't know linear transformation, wait; we will learn more about linear models in Chapter 8.

In other words, an eigenvector is a vector that doesn't change its direction when a specific transformation is applied to it; only its length may change. An *eigenvalue* is the factor by which the *eigenvector* is stretched or compressed during the transformation. Every eigenvector has a corresponding eigenvalue. The eigenvector and eigenvalues can describe a matrix with a vector and value.

The number of eigenvectors and their eigenvalues is equal to the dimensions of the dataset. For example, for a 3D dataset, we will have three eigenvectors and eigenvalues, and for a 4D dataset, we have four eigenvectors and eigenvalues.

Assume A is an $n \times n$ matrix, its eigenvalue is scalar λ, and its eigenvector v, if $A \cdot v = \lambda \cdot v$.

For example, we have a square matrix $A = \begin{bmatrix} 3 & 2 \\ 3 & -2 \end{bmatrix}$,

Its first eigenvector will be $v = \begin{bmatrix} 2 \\ 1 \end{bmatrix}$ and $\lambda = 4$ will be its associated eigenvalue.

We can check that $\begin{bmatrix} 3 & 2 \\ 3 & -2 \end{bmatrix} \cdot \begin{bmatrix} 2 \\ 1 \end{bmatrix} = 4 \times \begin{bmatrix} 2 \\ 1 \end{bmatrix}$

Its second eigenvector is $v = \begin{bmatrix} -\frac{1}{3} \\ 1 \end{bmatrix}$ and eigenvalue $\lambda = -3$.

---

[2] Eigen is a German word that means "own", "particular", or "specific".

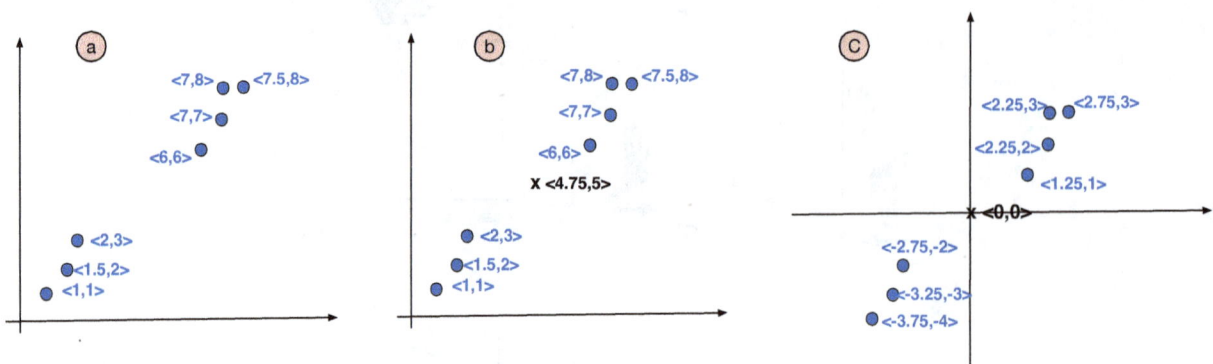

Figure 7-3: (a) Original 2D dataset, (b) finding the mean (X) and (c) shifting the dataset by reducing the mean from all values (the right plot).

An easy approach to determine the eigenvector and eigenvalue is to computer the *determinant of the matrix* $A - \lambda I$, where I is the identity matrix, and solve the equation $det(A - \lambda I) = 0$ for λ. However, you will never need to compute them by hand unless you are a mathematics student. You can check Chapter 4 of Deisenroth's book [Deisenroth '19] to learn how to calculate them in more detail. The *identity matrix* is a matrix that has 1s on its diagonal and 0s in all other positions. For example, a 4×4 identity matrix is as follows:

$$I = \begin{bmatrix} 1 & 0 & 0 & 0 \\ 0 & 1 & 0 & 0 \\ 0 & 0 & 1 & 0 \\ 0 & 0 & 0 & 1 \end{bmatrix}$$

Now that we have learned these basic concepts, we can switch back to describing PCA. The goal of PCA is to change the coordinates of data to a lower dimension while preserving as much of the data point relation and semantics as possible.

Assume we have dataset presented in Figure 7-3 (a). The first step to performing PCA is calculating the mean of all points, marked with X in Figure 7-3 (b). Then, all data points will be shifted to have the mean as a center, as shown in Figure 7-3 (c).

In the second step, PCA calculates the "covariance matrix" from all data points. The covariance matrix summarizes how our data points relate to each other.

The covariance matrix of the given transformed dataset in Figure 7-3 is:

$$cov = \begin{bmatrix} 7.32 & 7.28 \\ 7.28 & 7.42 \end{bmatrix}$$

After covariance has been calculated, the PCA should break down the covariance matrix into vectors that describe the direction and magnitude. In simple words, we need to have new vectors that describe the most variations. If we intend to project a 3D dataset into 2D, we need two vectors; if we intend to project a 4D dataset into 3D, we need three vectors, and so forth.

These vectors are called *principal components*. Each dimension is associated with a principal component. For example, in Figure 7-4, PC1 spans the direction of the most variation, and PC2 spans the direction of the second most variation. To calculate these principal components, we will calculate "eigenvectors" and their corresponding "eigenvalues" from the covariance matrix.

We will get the following eigenvectors and eigenvalues from the above covariance matrix.

$\lambda_1 : 14.65, v_1 : 0.993, 1$

$\lambda_2 : 0.09, v_2 : -1.007, 1$

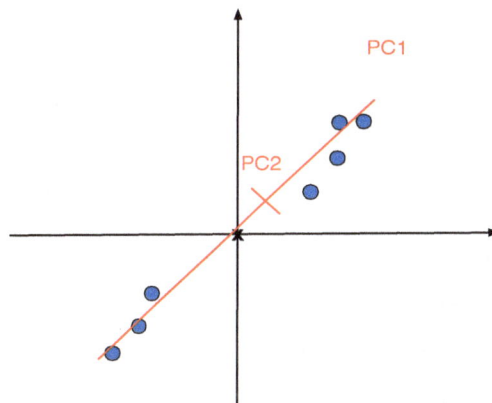

Figure 7-4: Eigenvalues and vectors presented in red color, based on the calculation from the covariance matrix.

If we draw both eigenvectors (PC1 and PC2), we will have something in Figure 7-4, which presents eigenvectors in red. Our PC lines are eigenvectors. In simple words, each eigenvector provides a direction of dataset variances (ordered by their eigenvalues). Eigenvalues measure the variance (or spread) of the data along the eigenvectors (principal components).

In the third step, PCA orders eigenvalues in decreasing order, and based on the *k* number of dimensions we need, it chooses the first *k* number of eigenvalues (Each dimension is associated with one eigenvalue/eigenvector). In this example, we have 2D data, and we would like to project them into 1D data. Therefore, we choose only the first eigenvalue and its eigenvector, i.e., $\lambda_1 : 14.65, v_1 : 0.993, 1$.

In the fourth step, the PCA algorithm transforms the original dataset into a new dataset that has a lower number of dimensions than the original dataset. To perform such a transformation, it creates a matrix of $1 \times n$ eigenvectors, which is called a *feature vector* (because $1 \times n$ matrix is a vector). Each eigenvector is a column of this matrix; the final data points will be identified based on the following equation:

$Final\_Data\_points = (Feature\_Vector)^T \times (Original\_Adjusted\_Data\_points)$

T stayed for the transpose of the matrix. Transposed matrix is a matrix in which its rows are placed as columns and columns as rows.

For example, the matrix $\begin{bmatrix} 3 & 1 & 5 \\ 4 & 2 & 6 \end{bmatrix}$ has transpose of $\begin{bmatrix} 3 & 4 \\ 1 & 2 \\ 5 & 6 \end{bmatrix}$.

Based on our vector and feature data points, the final data points will be as follows:

$$\textit{final data points} = [0.993, 1] \times \begin{bmatrix} -3.75 & -4 \\ -3.25 & 3 \\ -2.75 & -2 \\ 1.25 & 1 \\ 2.25 & 2 \\ 2.25 & 3 \\ 2.5 & 3 \end{bmatrix}^{T} = [-7.724, -0.227, -4.731, 2.241, 4.234, 5.234, 5.483]$$

[−7.724, −0.227, −4.731, 2.241, 4.234, 5.234, 5.483] is the one-dimensional dataset, which is the reduced dimensionality of the original dataset.

This simple example enables us to understand how PCA reduces the two-dimensional dataset into 1 one dimension. Nevertheless, in real-world cases, we have many dimensions, and we should decide what dimensions are appropriate for the projected dataset.

To identify the number of dimensions, each PC line (eigenvector) is associated with a variance, which is reported by the PCA package we use as well. The variance of PC line describes the proportion of the dataset that lies along each PC line. It is recommended to choose the number of dimensions whose variances add to a large proportion of the dataset, e.g., 95%. Another recommended approach to identify the optimal number of dimensions is to use something

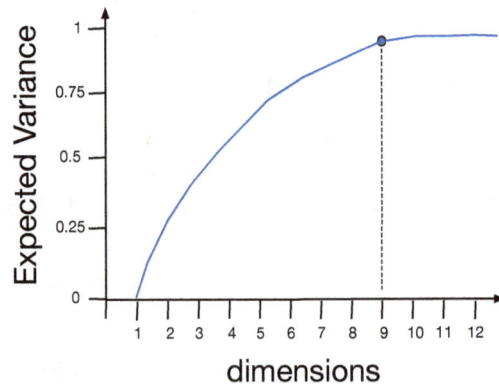

Figure 7-5: By increasing the number of dimensions, eigenvectors' variances increase, and we choose the elbow point, 9 dimensions, as a good point to reduce 13 dimensional data into 9 dimensions.

similar to the elbow method, which we have used in Chapter 5 for clustering. In this method, on x-axis shows the number of dimensions, and the y-axis shows the percentage of explained variance (see Figure 7-5 example). The elbow of such a plot can present at some point increasing the dimensions, but it can not increase the variance significantly, and this is an indication of the proper number of dimensions.

The computational complexity of PCA is composed of the computational complexity of covariance matrix composition, i.e., $O(n \cdot m)$ for $n \times m$ matrix and eigenvector, and value identification, i.e., $O(m^2)$ for $m \times m$ matrix. Therefore, its complexity will be $O(m \cdot n + m^2)$. Therefore, PCA is not efficient when used for real-time applications, but usually, it will be done once before feeding the data into the algorithm in batch mode. Even if the PCA uses SVD (which will be explained later in this chapter), it is getting more complex.

Incremental PCA

One problem of PCA is that it should load the entire dataset into the memory to perform dimensionality reduction. Every decent data scientist loves PCA and uses PCA at least once in their life before marriage (after marriage is also ok). However, real-world datasets are usually too large to fit into the memory, and thus, we can not directly benefit from PCA. To handle those cases, we use a specific type of PCA called *Incremental PCA (IPCA)*.

IPCA operates by dividing the dataset into smaller subsets (mini-batches) and loading them subset by subset into memory, then doing the PCA for each subset.

Linear Discriminant Analysis

Another linear dimensionality reduction method is linear discriminant analysis (LDA). We will learn more about the concept of linearity in Chapter 8, but we describe this dimensionality reduction method in this section.

Assume we have data in 2D space, and we would like to project them into 1D. As you can see from Figure 7-6 (a), there are two groups of data, which are marked with red and blue dots. A good dimensionality reduction approach should keep blue and red dots separate from each other.

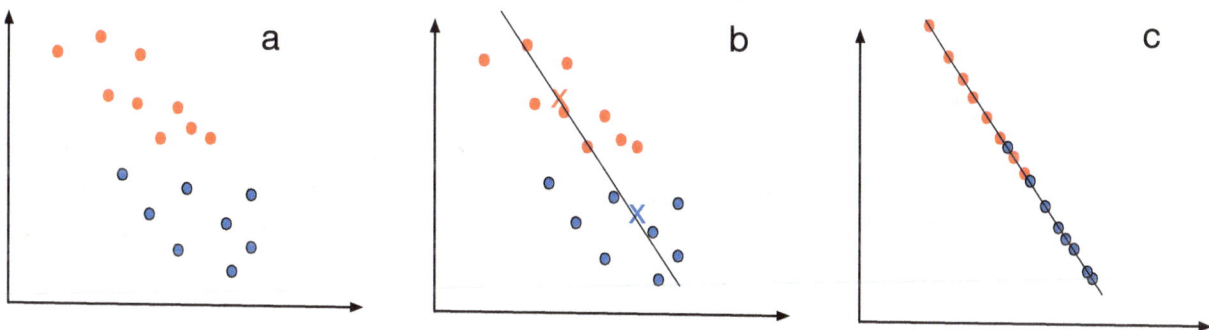

Figure 7-6: Projecting data points around a line, which is drawn based on mean values of blue and red data points.

If we calculate the mean of both data groups (see Figure 7-6 (b)) to find the projection line, it is not the best possible projection line because some data points will overlap, as we can see in Figure 7-6 (c). LDA algorithms such as Fisher Linear Discriminant can provide a better projection line and keep blue and red dots separate from each other in the lower dimensional space.

Fisher Linear Discriminant (FLD)

It is one of the common LDA algorithms [Fisher '36] that can be used to identify the optimal projection space (projection line if we have 2D data). In other words, the objective of Fisher Linear Discriminant (FLD) is to select the best projection that maximizes separation between data points. This separation is specified with an objective function $J(W)$.

$$J(W) = \frac{(\mu_2 - \mu_1)^2}{s_1^2 + s_2^2}$$

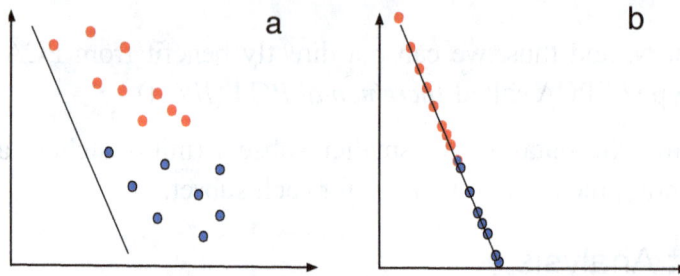

Figure 7-7: (a) the new projected line calculated based on FLD, (b) data projected on the new line and we can see that red and blue are well separated.

The larger the value of $J(W)$, the more separated classes are from each other. FLD's objective is to find a projection line that (i) increases the variance between different groups of data and (ii) decreases the variance between group member data points. These groups of data are called "classes of data".

FLD objective function is formalized as $J(W)$ equation, called the "Weight Vector" equation. We can see that the objective function of FLD is similar to clustering (Check Chapter 4 to recall Clustering if you can't remember it), but here, instead of distance, we use variance, and instead of cluster, we use class of data.

First, the FLD calculates scatters for each class. "Scatter", similar to variance, is a parameter that measures the variance around the mean. Assuming data point z presents the coordinates of a class with n members, which has a mean of $\mu_z = (1/n)\Sigma_{i=1}^n z_i$, scatter for class 1 (with n members) and class 2 (with m members) is calculated as follows:

$$s_1 = \Sigma_{i=1}^n (z_i - \mu_1)(z_i - \mu_1)^T \quad , \quad s_2 = \Sigma_{j=1}^m (z_j - \mu_2)(z_j - \mu_2)^T.$$

Some resources employ *covariance* instead of *scatter*, and both are correct.

Next, it calculates the direction of the projection line, W. In other words, W is a vector that presents the direction and coordination of the projection of the line. The optimal value of W is computed as $W = S_w^{-1}(\mu_1 - \mu_2)$, where $S_w = s_1 + s_2$ and S_w^{-1} is the inverse matrix of within-class variance, μ_1 is the mean of class 1, and μ_2 is the mean of class 2.

If you don't know the inverse of a matrix, see the following example of how to calculate the inverse of a 2×2 matrix (a matrix power to -1 is the inverse of a matrix).

$$\begin{bmatrix} 1 & 6 \\ 3 & 4 \end{bmatrix}^{-1} = \frac{1}{4 \times 1 - 6 \times 3} \begin{bmatrix} 4 & -3 \\ -6 & 1 \end{bmatrix} = \begin{bmatrix} -0.03 & 0.21 \\ 0.42 & -0.07 \end{bmatrix}$$

The third step is to calculate one-dimensional projection (vector) for each class by multiplying the W vector by their matrices of data, i.e., $W^T . class1$ and $W^T . class2$.

A dataset is presented in Figure 7-7 (a), and by conducting FLD, we identify the project line as it has been shown in Figure 7-7 (a). Then, the data could be separated more accurately, as it is shown in Figure 7-7 (b).

FLD inherently generalizes to higher-dimensional data beyond 2D, as it is designed to find a linear combination of features that separates two or more classes of objects or events. The core

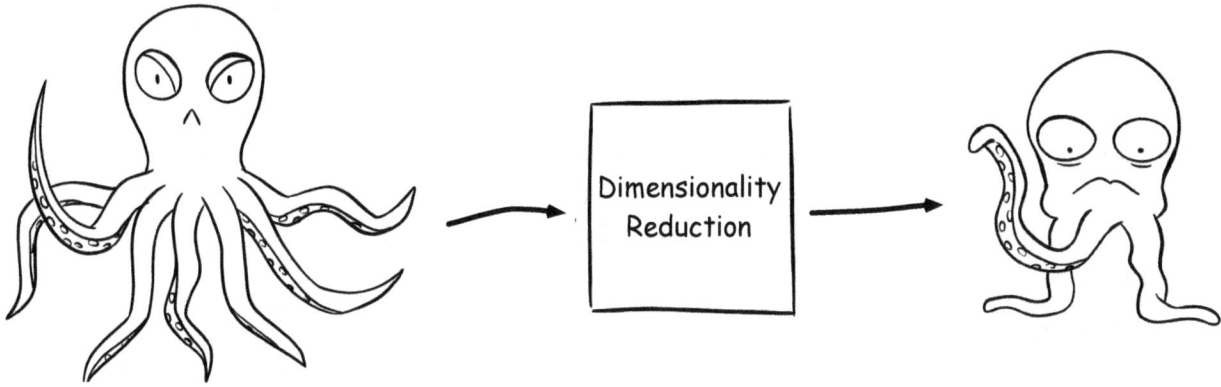

idea is to project high-dimensional data onto a line (in the simplest case) or a lower-dimensional subspace that maximizes the between-class variance while minimizing the within-class variance, even if the original data space is more than 2D.

Fisher's Linear Discriminant Analysis of multiple classes is *Linear Discriminant Analysis (LDA)*. LDA extends the idea of FLD to scenarios where there are more than two classes. While FLD traditionally focuses on distinguishing between two classes, LDA is used for problems involving two or more classes. The methodology involves computing within-class scatter and between-class scatter and then finding the linear subspace that optimally separates the classes based on these criteria.

The computational complexity of linear discriminate analysis methods are $O(mnt + t^3)$ time complexity and memory, where $O(mn + mt + nt)$, m is the number of data points, n is the number of features, and $t = min(m, n)$ [Cai '08].

We have learned two algorithms for dimensionality reduction methods (PCA and LDA), which apply to data types with two or more dimensions, including matrix and tensor. Next, we will explain dimensionally reductions or data decomposition for more specific data types. Before describing our methods, we should talk about an important challenge, i.e., the curse of dimensionality.

NOTES:

* Eigenvectors and eigenvalues have many applications in machine learning. However, your software package computes PCA for us, and we don't need to compute by hand.

* The eigenvector will be a line which is specifying "direction", and the eigenvalue will specify the variance of data points or their "magnitude" around this line. Remember that they are always perpendicular (orthogonal).

* While using PCA, sometimes it is not clear how many dimensions are the most optimal ones. Of course, reducing dimensions is associated with losing some data, which could lead to the curse of dimensionality. However, some software packages can recommend the optimal number of dimensions by calculating the variance of the projection into the new dimensions. This means we get different variances and then decide based on (i) the highest variance and (ii) the number of dimensions we intend to have.

* It is recommended not to use PCA for datasets that have lots of outliers because PCA can not handle outliers easily. PCA is looking for linear correlations, which are very sensitive to outliers.

* In cases where the arithmetic means of classes are the same or close to each other, LDA is not operating well. LDA aims to maximize the distance between the means of the classes while minimizing the variance within each class, which can be challenging when classes overlap significantly.

* If we have too many dimensions and all of them are important for the algorithm, to avoid the curse of dimensionality, sometimes we can increase our sample size. However, don't be very optimistic that this will resolve our case; just give it a try and check the result, then decide.

* PCA is used for lossy compression as well; it is especially good for dealing with image data to reduce its features while keeping most of the distinctive characteristics of the original data.

* Now that we understand the concept of eigenvalue and eigenvector, it makes sense to learn the *Orthogonal* matrix. Orthogonal matrix means when we make a dot product of the matrix and its transpose, the result is the identity matrix. $A \cdot A^T = I$. In other words, we can say a matrix is orthogonal if its transpose is equal to its inverse, i.e., $A^T = A^{-1}$. For example, the following matrix A is Orthogonal.

$$A = \begin{bmatrix} 1 & -1 & 2 \\ -4 & 3 & -13 \\ 6 & -6 & 13 \end{bmatrix} \text{ because } A^T = \begin{bmatrix} 1 & -4 & 6 \\ -1 & 3 & -6 \\ 2 & -13 & 13 \end{bmatrix} \text{ and } A \cdot A^T = \begin{bmatrix} 1 & 0 & 0 \\ 0 & 1 & 0 \\ 0 & 0 & 1 \end{bmatrix}$$

PCA is referred to as an orthogonal linear transformation that transforms the input dataset to a different coordinate system that has a lower number of dimensions.

Non-linear Dimensionality Reduction Methods

PCA, LDA, and their other varieties are useful dimensionality reduction methods. They are projecting a dataset into a lower-dimension space. Nevertheless, sometimes, the linear projection that is used by the described methods is not possible. For example, our data distribution has a Swiss roll shape, as shown in Figure 7-8 (a). For such a dataset, we can not use linear dimensionality reduction methods, and for these datasets, we should use *Manifold dimensionality reduction* methods.

if you have difficulty understanding this section, skip it for now and come back after reading Chapters 8 and 9.

A manifold is defined as a continuous, non-intersecting structure that can be *bent* and *twisted* in a higher-dimensional space, but if we zoom in, it appears as a flat surface. A manifold can have any number of dimensions. One-dimensional manifolds include lines and circles. Two-dimensional manifolds are also called surfaces. Some examples of 2D manifolds include the

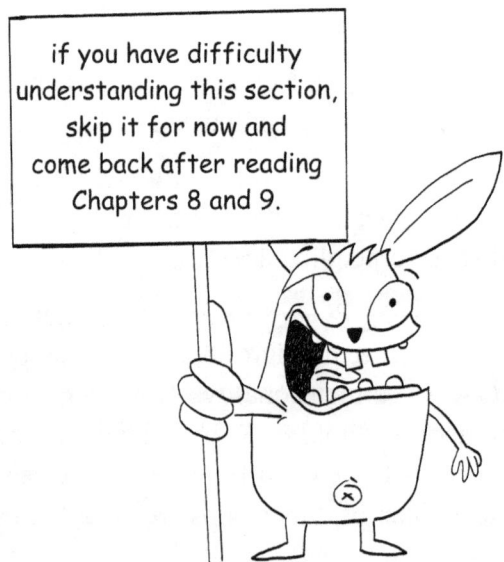

surface of a sphere. Also, each rectangle face of a cube is a 2D manifold. The entire surface of the cube, comprising all six faces, is piecewise flat (consisting of multiple 2D manifolds) and is indeed a 2D manifold in the broader sense, as it is a continuous, non-intersecting surface that locally resembles a flat plane. See Figure 7-8 (b); it presents an untwisted version of the Manifold of Figure 7-8 (a).

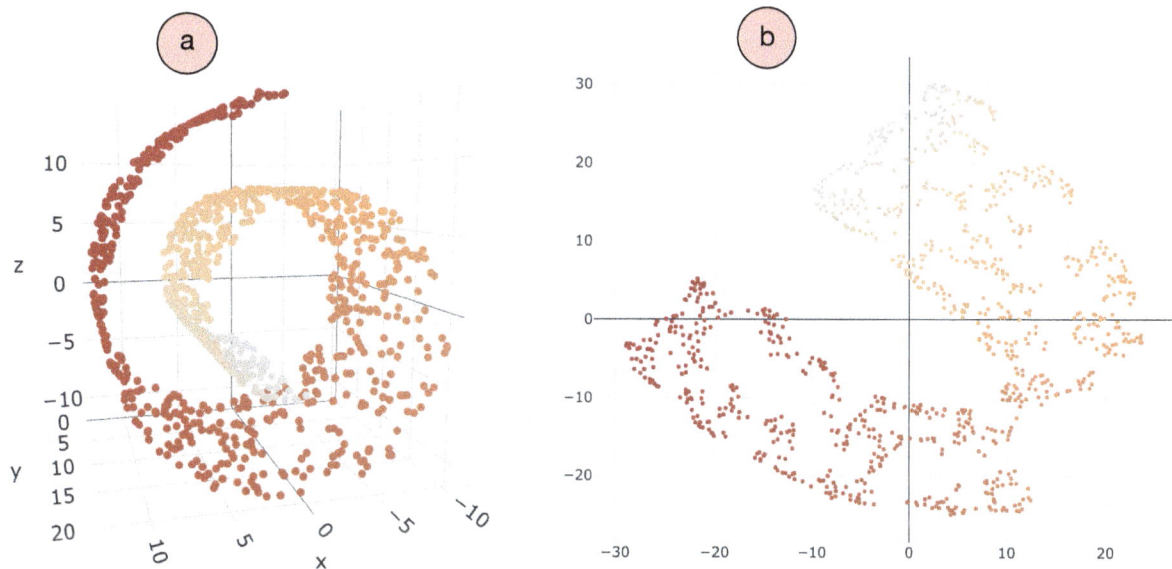

Figure 7-8: (a) A Swissroll function in three dimensions. (b) The same Swissroll function is twisted and plotted into a two-dimensional space, and one dimension has been reduced.

Manifold learning algorithms operate based on the high dimensionality of data. If we intend to perform analysis on manifold data, especially regression or classification (we will explain them in Chapters 8 and 9), it is recommended to transfer the dataset into a lower dimensional space (by dimensionality reduction). To summarize, we use manifold learning for dimensionality reduction of non-linear datasets. For example, to learn protein structures, manifold dimensionality reduction methods are extremely important to be able to analyze them.

In the following, we explain one simple manifold dimensionality reduction method, i.e., LLE, and use this to explain the two common methods, tSNE and UMAP.

Locally Linear Embedding (LLE)

Locally Linear Embedding (LLE) [Roweis '00] is a baseline method used for manifold dimensionality reduction. It assumes all data points are characterized as a linear combination of their neighbors. In other words, LLE tries to identify the *non-linear relationship* between data points in high dimensional space and tries to keep this relation in low dimensional space as well.

LLE measures how each data point is linearly related to its closest neighboring data points. It assumes a data point and its neighbors lie close to a linear relationship on high dimensional space. Then, it transforms the data points from high to low dimensional space while trying to preserve their relationship in the low dimensional space as well. How does it maintain this relationship? By assigning a linear weight to each neighbor data point. While LLE captures the local geometry of the dataset using linear combinations of neighbors, the global structure it reveals can be highly non-linear, and LLE is capable of preserving the non-linear global structure of the dataset.

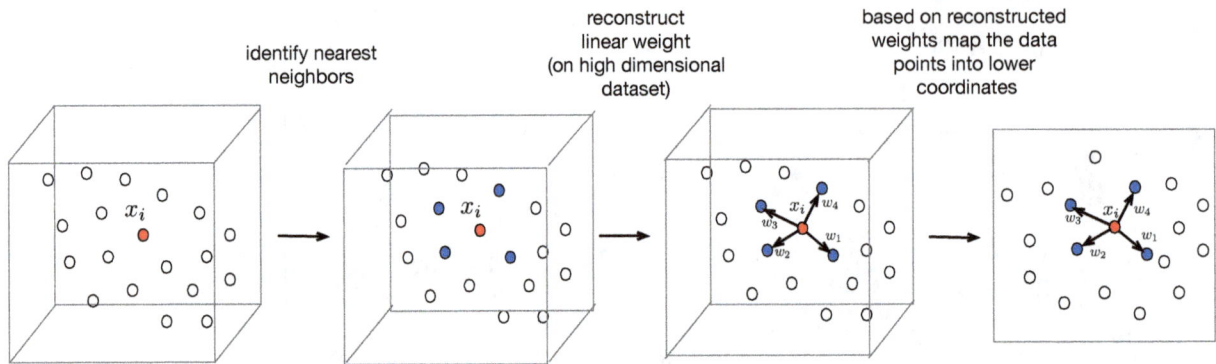

Figure 7-9: Step by step operation of LLE algorithm.

The LLE algorithm operates as follows:

(i) For every single data point in the high dimensional space, the algorithm identifies its k-nearest neighbor. For example, in Figure 7-9, the data point x_i is marked in red, and the algorithm identifies four blue data points as its four nearest neighbors. Next, it calculates distances among all data points to their neighbors in the original dataset (high dimensional dataset) and assigns weights between each two data points. The weight describes the distance between two data points. Assuming X is a matrix (or tensor) of original data, W is the matrix of weight.

We could write x_i in high dimensional space as a weighted sum of the nearest neighbors in high dimensional space, as follows: $x_i \approx w_{i1}x_1 + w_{i2}x_2 + \ldots + w_{ik}x_k$

(ii) It projects all data points into the lower dimensional space, along with their neighbor data points and the associated weights. The objective of the second step is to identify optimal values for weights in higher dimensional space. In other words, the optimal weight minimizes the reconstruction error of each data point from its neighbors. Therefore, the algorithm requires to optimize the following equation (in higher dimensional space):

$$\varepsilon(w) = \sum_{i=1}^{n} ||x_i - \sum_{j=1}^{k} w_{ij}x_j||_2^2$$

In this equation, $||.||_2^2$ presents the squared of L_2 norm, n presents the number of data points, and k is the number of nearest neighbors to each data point. x_i is known and w_{ij} is unknown, which will be identified via the described optimization equation.

340

The w_{ij} will be computed by considering two constraints: (a) each data point x_i is reconstructed only from its neighbors, (b) if the x_j is not the neighbor of x_i, then $w_{ij} = 0$, and these constraints will be used to identify optimal values for weights. We do not explain how an optimizer identifies an optimal value for all w_{ij}. If you would like to learn them, you can check the original paper [Roweis '00].

(iii) After all w_{ij} have been identified from resolving the optimization problem of the described equation, the next step is to find the position of data points in the lower dimensional space, i.e., y. Assuming y represents data point x in the lower dimensional space, and w is known now, the algorithm should optimize another equation.

$$\Phi(y) = \sum_{i=1}^{n} ||y_i - \sum_{j=1}^{k} w_{ij}y_j||^2$$

This embedding aims to maintain the relationships encoded by the weights w_{ij}, ensuring the local structures are preserved in the reduced dimensionality.

Assuming we have n data points, the construction of k nearest neighbor has $O(n \, log \, n)$ computational complexity. The second step, which identifies values for weights, has $O(D \cdot n \cdot k^3)$ complexity, assuming that we have D dimensions in the higher dimension. Assuming the dataset is transferred into d dimensions (lower dimensions), the third step has $O(dn^2)$ complexity [Saul '00].

t-distributed Stochastic Neighbor Embedding (tSNE)

In Chapter 2, we discussed that in the context of spatial visualization, the human brain can not visualize more than three dimensions. This could be an inspiration for some dimensionality reduction algorithms such as t-distributed Stochastic Neighbor Embedding (tSNE), which are mistakingly called visualization as well. tSNE [Van der Maaten '08] is a dimensionality reduction that is used to convert dimensional datasets into lower dimensions by employing a Barnes-Hut approximation [Barnes '86]. tSNE operates based on the assumption that similar data objects are presented with high probability and dissimilar data objects will be presented with low probabilities.

Similar data points in high dimensional space will stay close together in the low dimensional space as well, but the distance between dissimilar data points could even increase from the higher dimension to the lower dimension.

tSNE uses KL-Divergence (check Chapter 3) as a cost function, which prioritizes retaining the local structure of the data and not the global structure of data.

tSNE algorithm operates in three steps:

(i) The first step measures the similarities between every data point in a high dimensional space. In particular, for each data point, it draws a Gaussian distribution around that data point. Then, it *normalizes distances as a probability,* which will be a value between 0 and 1.

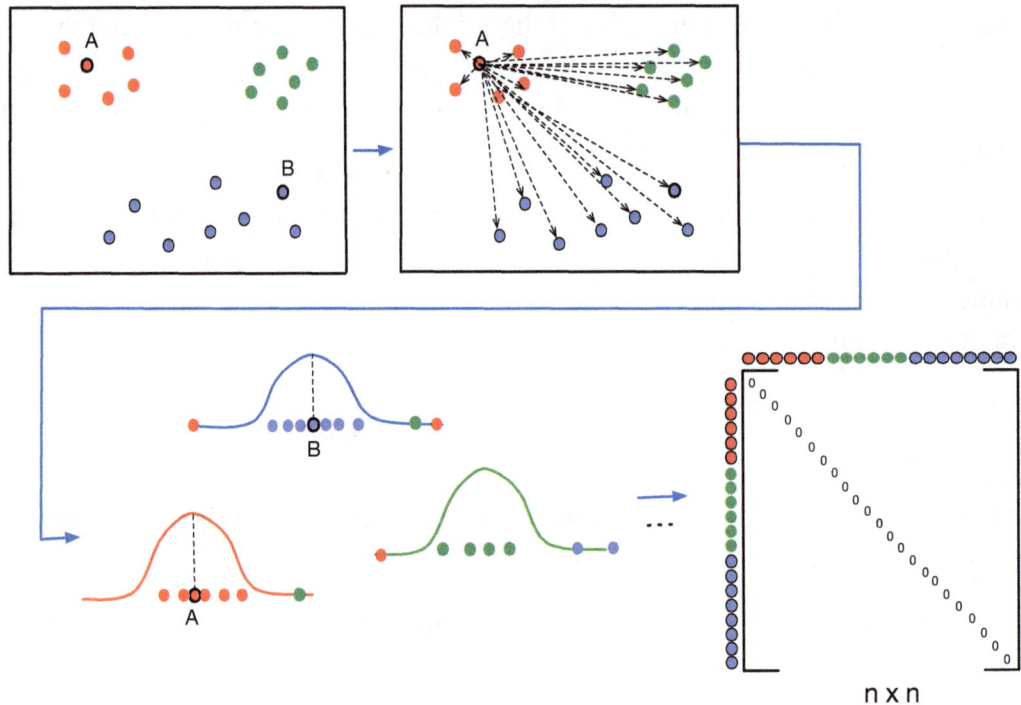

Figure 7-10: The first step of tSNE considers every single data point as a center of Gaussian distribution. Closer data points stay close to the center of the Gaussian distribution, and the other data points are on the side of the distribution. Every single data point distance to all other data points will be calculated.

Therefore, each data point in the higher dimensional space will be a center point of a Gaussian distribution. Take a look at Figure 7-10 to understand that this process has been highlighted for two sample data points, *A* and *B*.

To measure similarity, like density-based clustering, tSNE uses a parameter called *perplexity* (a hyperparameter), and it measures the number of effective neighbors around a data point. It is an input parameter of the algorithm that should be set by the user. Usually, it should be between 5 and 50, and it must be smaller than the total data points.

Assuming we have *n* data points, the result of similarity comparisons will be stored in a $n \times n$ matrix, as you can see in Figure 7-10. For each data point x_i, it centers a Gaussian distribution over that point, and then measures the density of all data points (x_j) under that Gaussian distribution. Next, the tSNE algorithm normalizes all points, which gives us a set of probabilities for all points. This is called $P_{i|j}$. The result of normalization is shown in the matrix of Figure 7-10. tSNE uses the following equation to compute $P_{i|j}$:

$$p_{i|j} = \frac{exp(-||x_i - x_j||^2 / 2\sigma_i^2)}{\sum_{k \neq 1} exp(-||x_i - x_k||^2 / 2\sigma_i^2)}$$

In this equation, σ_i is the variance of the Gaussian that is centered on the data point x_i

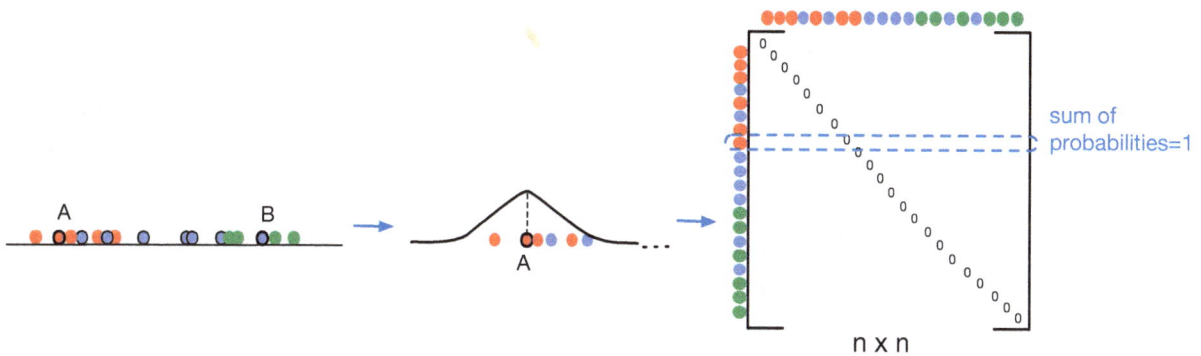

Figure 7-11: Second step of tSNE, calculating a t-distribution for each data point at the initial projection that all data points combined together. Then, similar to a previous step, another similarity matrix will be created. Nevertheless, unlike the previous matrix these data points are unordered.

(ii) The algorithm projects all data points on the lower dimension. This projection can be random or based on PCA for efficiency. Note that there are no changes in the placement of data points at this stage. All data points are simply projected into the lower dimension, as we can see from the left side of Figure 7-11. Now, in the lower dimension space, for any single data point, a t-distribution will be plotted. Check Chapter 3 to recall the shape of t-distribution; it is a subset of Gaussian distribution but with wider tails and shorter peaks (see Figure 7-12).

Figure 7-11 shows that we have our original data points projected from a high-dimensional space into a lower-dimensional space without any proper separation. Now, at this point, data are in low dimensional space, but they are sloppily projected, and of course, this projection is not good. Although they are not separated properly, the algorithm performs the same process of measuring similarities between data points for each data point (but as it has been stated, instead of Gaussian distribution, we use t-distribution to model the adjacent data points).

For each data point x_i that is projected to y_i (because of step ii), it centers a t-distribution over that point, and then it measures the density of all data points y_j under that t-distribution. Next, it

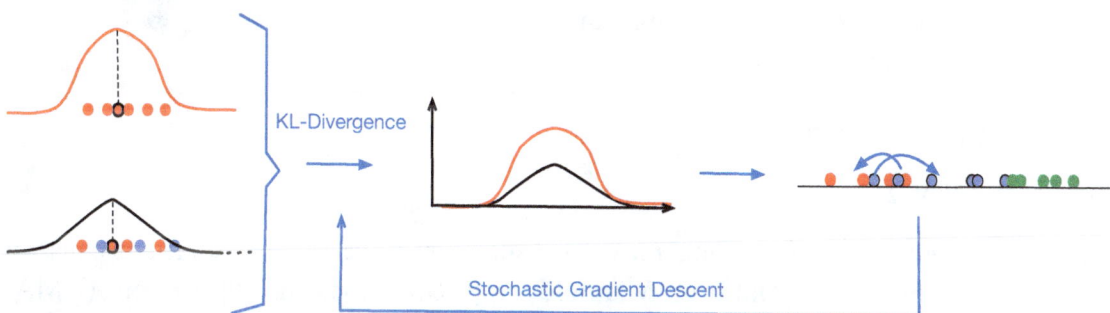

Figure 7-12: Stochastic gradient descent will be used along with the KL-divergence cost function to change the t-distribution and reduce its distance from the Gaussian distribution. As a result, the data points in the low dimensional have a local structure similar to the higher-dimensional space.

343

normalizes all points, which gives us a set of probabilities for all points, i.e., $Q_{i|j}$, and the result is presented in the matrix of Figure 7-11. The normalization process is the same as the one we used for $P_{i|j}$.

(iii) The third step focuses on moving the position of data points in lower dimensional space to consider their local distances. This process will be done by converging the result of the lower dimensional matrix, $Q_{i|j}$ to $P_{i|j}$ with a KL-divergence (described in Chapter 3), as a cost function, and using gradient descent (we will explain in Chapter 8) as optimization of this cost function. We have described in Chapter 3, that KL-divergence is used to measure the distance between two distributions. In other words, we use the KL-divergence cost function to fit $Q_{i|j}$ (data in lower dimensions) close to $P_{i|j}$ (data in higher dimensions). The two distributions are closer to each other if their KL-divergence score is low. The KL-divergence of zero means that the two distributions are identical, which is the ideal case. tSNE uses gradient descent to minimize the sum of the KL-divergence score, see Figure 7-12.

Now, it uses KL-Divergence to create the matrix in the third step (Figure 7-12) and reorder the data points to give them a similar order to their high-dimensional space.

The perplexity hyperparameter specifies the balance between local and global aspects of the data. The authors of the original paper suggested setting it between a variable from 5 to 50, and larger or denser datasets require a large perplexity value.

Another parameter of tSNE is the number of iterations, and it should be large enough to ensure that the dataset's characteristics do not change in low dimensional space. In some implementations, we can configure this parameter, but since this parameter exists in most iterative algorithms, we could rely on the default value from the implementation library.

Assuming d is the dimension of data in lower dimensional space, and we have n data points, the tSNE algorithm has a $O(dn^2)$ complexity.

Mom, mom, come here. I finally learned tSNE.

Uniform Manifold Approximation and Projection (UMAP)

Before you start reading this algorithm, please be sure you have learned tSNE, because we explained it based on tSNE. Uniform Manifold Approximation and Projection (UMAP) [McInnes '18] is a dimensionality reduction technique that uses manifold learning. It is used for datasets that have a Riemannian manifold structure. Riemannian manifolds are manifold planes that are smoothly connected to each other, and Riemannian surface planes are connected continuously. In other words, the Riemannian surface is a multidimensional surface that does not have a 2D plane. Instead, it should be a curved surface in multidimensions, as is shown in Figure 7-13. This figure resembles the author's blanket when he is very sleepy and tries to find the longer parts to cover his body, but the blanket is already rubbed by his daughter. In the context of machine learning and AI, we doubt if we need to learn more about Riemann's surface in more detail.

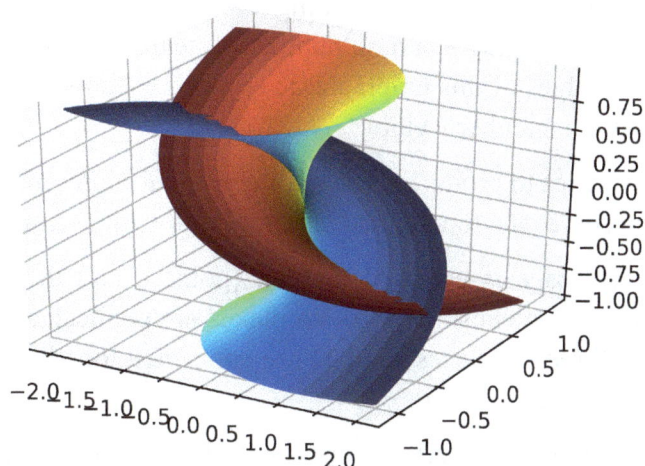

Figure 7-13: An example of Reimanian surface.

UMAP works based on the concept of connecting several simple shapes (simplex) in geodesic space, which are not very easy for non-mathematicians to describe. Therefore, we used the description provided by Nikolay Oskolkov[3] to describe the UMAP based on tSNE. He has introduced UMAP in the context of tSNE. We try to list them as follows:

- UMAP uses exponential distribution in the high dimensional space instead of Gaussian distribution. Besides, UMAP can work with any distance metrics, but tSNE is limited to using Euclidean distance only. In particular, UMAP utilizes the concept of a fuzzy simplicial set, which can be thought of as relying on an exponential distribution of distances between points in high-dimensional space, contrasting with t-SNE's use of a Gaussian distribution to model neighborhood relationships. This difference is part of what allows UMAP to preserve global structure better than t-SNE.

- UMAP focuses on preserving the topological structure of the data, which doesn't map directly onto using a specific distribution like the t-distribution we used for tSNE in low-dimensional embeddings.

- Instead of using perplexity in tSNE, UMAP uses ρ, which is very similar to the number of nearest neighbors, but authors called it a neighborhood graph.

- UMAP does not apply any normalization on both high and dimensional-low space. This makes it much faster than tSNE.

---

[3] http://www.biology.lu.se/nikolay-oskolkov

- Gaussian distribution in a high dimensional space is symmetrization (making the unsymmetric data symmetric), and UMAP uses another form of symmetrization in high dimensional space, which might not have a significant difference.

- UMAP uses cross-entropy instead of KL-divergence as a cost function, but both use gradient descent as optimization (check Chapter 8 about optimization and gradient descent)

In simple words, UMAP operates in two main steps; the first step involves the construction of a high dimensional graph. This involves calculating the distance between points in the high-dimensional space and then selecting a certain number of nearest neighbors to create a weighted graph. The weights on the edges of this graph are determined by how close or similar two points are, with closer points having a higher weight.

In the second step, UMAP tries to produce a lower-dimensional representation of the dataset that, as closely as possible, preserves the topological structure of the high-dimensional graph.

Data in high dimensional space is called 0-simplex (points), and the algorithm first tries to understand the topology of simplexes. If two 0-simplexes are connected together, they construct a 1-simplex (line), and if three simplexes (0 or 1) are connected together, they form a 2-simplex (triangle formed by three connected points).

The computational complexity of UMAP is very much dependent on the nearest neighbor data point search. Assuming we have n data points, according to the authors of the UMAP algorithm, it seems that it has a complexity of $O(d \cdot n^{1.14})$, which is close to tSNE[4]. The efficiency gains of UMAP over t-SNE are more accurately attributed to its different approach to graph construction, its optimization strategy, and its less sensitivity to hyperparameters like perplexity.

We have learned some well-known dimensionality reduction methods, including PCA, LDA, tSNE, and UMAP, which are applicable for data types with two or more dimensions, including matrix and tensor. Next, we will explain dimensionally reductions or data decomposition for more specific data types.

NOTES:

* t-SNE doesn't care about the global shape of data, unlike linear projection methods such as PCA and LDA.

* One disadvantage of tSNE is that each time, it gives a different result; we should run t-SNE multiple times and choose the result that has the best distinction. Therefore, it is not very useful for being used in products that expect to have a fixed behavior within the static dataset.

* tSNE is sensitive to hyper-parameters, and it is recommended to standardize data before feeding them into tSNE [Harrison '19].

* Another drawback of tSNE, UMAP, and other manifold dimensionality reduction methods is that they are computationally intensive processes, and thus, it is not easy to run them on devices with limited resources, such as smartphones or small drones.

---

[4] You can check the discussion in this thread https://github.com/lmcinnes/umap/issues/8

Signal and Time Series Decomposition

In Chapter 6, we describe how we can convert different data types into a signal and analyze them. In addition, we should learn some common signal processing methods and their applications in machine learning. They can be used to decompose any data into time series or signals, e.g., image, audio, and even text (e.g., the sentiment of a text could be plotted as a time series).

However, note that such a decomposition could be associated with loss of accuracy as well, and thus, the conversion process should be cost-effective. This is something that we should decide before processing the algorithm.

We do not explain Digital Signal Processing (DSP) in detail; we just briefly describe the concepts that could help us to transform our dataset into another format, and they are worth to be learned by a data scientist. We start this section by describing some concepts and then go into the algorithmic explanations. We have explained some basic signal concepts in Chapter 6; here we continue.

Signal and Time Series Concepts

Wave: it is a transfer of energy from one point to another point, which consists of oscillation (fluctuation or vibration) through a medium such as sound, light, etc. Waves can propagate through various media (such as air, water, or solids) or even through the vacuum of space in the case of electromagnetic waves.

Signal: Wave only transfers energy, but signal entails information about the wave. In other words, a signal is a representation of data encoded as a variation. While waves can be signals if they carry information, not all waves are used as signals.

Filter: it is a process that suppresses or minimizes the oscillation of a signal. Since it can remove or attenuate specific components of a signal based on their frequency, thereby it can alter the signal's content. Filters are used to enhance the desired information or reduce unwanted noise within a signal.

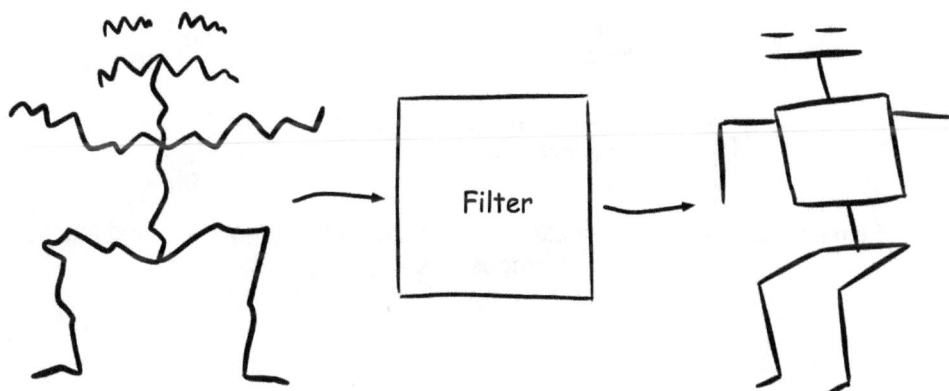

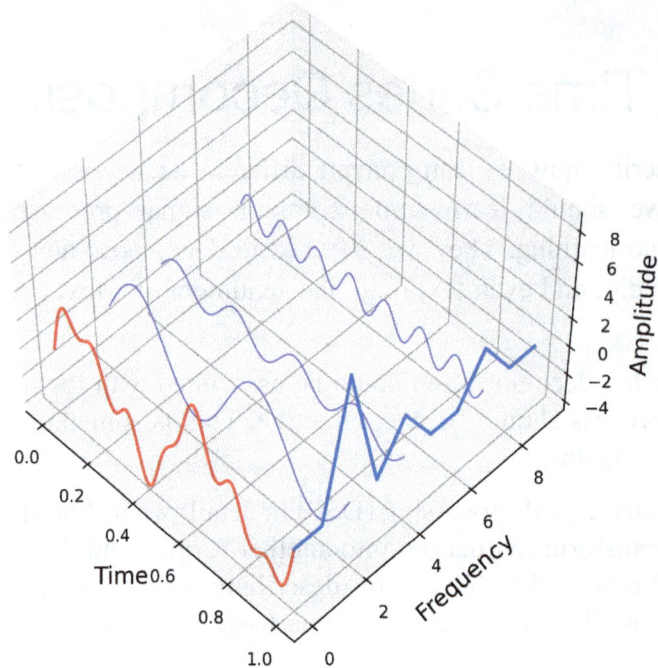

Figure 7-14: A time-domain signal (in red) is transferred into a frequency-domain signal (dark blue signal), which is composed of three sine waves.

Transform: It is similar to the function in mathematics. However, unlike most functions, transform can have multiple inputs and outputs. We can say that transform is a signal processing method that gets a set of data points (e.g., signal data) as input and produces another set of data points (e.g., signal data) as output. In other words, a transform is a mathematical operation applied to a signal that converts it from one domain to another, often to reveal information that is not readily apparent in the original domain.

Signal Type	Type of Transformation	Example of Signal Shape
aperiodic-continous	Fourier Transform	
periodic-continous	Fourier Series	
aperiodic-discrete	Discrete Time Fourier Transform (DTFT)	
periodic-discrete	Discrete Fourier Transform (DFT)	

Table 7-1: Signal types and examples plus their transformation algorithm used for signal decomposition.

Signal Decomposition: Signal decomposition refers to the process of breaking down a complex signal into simpler parts, often into a set of sinusoidal waves (sine and cosine functions) of varying frequencies, amplitudes, and phases

Signal Synthesis: The process of reconstructing a signal from its components, typically after decomposition. In the context of sinusoidal components, it involves combining these sinusoids (sine and cosine waves) to recreate the original signal. This process is the reverse of decomposition.

Aperiodic vs. Periodic Signal: Periodic signals repeat their patterns over a fixed interval, known as the period, for an infinite duration. In contrast, aperiodic (or non-periodic) signals do not exhibit such repetitive patterns over any interval.

Continuous vs. Discrete Signal: Continuous signals are defined over a continuous range of time and can take on any value in their domain. In contrast, discrete signals are defined only at specific time intervals and take on values only at those discrete points. A signal could be either "continuous" or "discrete".

Fourier Transform

Fourier transforms are a very common set of transformations used to convert time-domain signals into frequency-domain signals. By applying the "inverse Fourier transform", we can get back to the original time domain signal.

A French mathematician, Jean Baptiste Joseph Fourier[5], whose full name is longer than the surname of the author of this book, introduces the Fourier Transform. He states that: "*any periodic signal can be represented as a sum of sines, cosines of different frequencies, each multiplied by a different coefficient.*" This representation is called the *Fourier Series*.

After Fourier's transformation was accepted by the scientific community, it revolutionized several fields, including physics and mathematics. Fourier transformations are widely used in different domains, including audio analysis, digital health, and computer vision, e.g., edge detection, image filtering, etc.

Figure 7-14 shows a transformation of an aperiodic time-domain signal into a frequency-domain signal. We could have four types of signals: *aperiodic-continues, periodic-continuous, aperiodic-discrete*, and *periodic-discrete*. If the signal is aperiodic (not periodic) and continuous, we can use Fourier Transform. A specific algorithm should be used to decompose each of these signals, and these algorithms are listed in Table 7-1. In the following, we briefly explain each transformation.

Discrete Fourier Transform (DFT), is a transformation algorithm that is used to transform periodic-discrete signals into a sum of sinusoidal signals, each characterized by a certain frequency, amplitude, and phase. It is the most widely used signal transform method. In particular, DFT outputs frequencies for each periodic signal. DFT gets a time-domain signal and

---

[5] Fourier's ideas brought about a significant shift in physics and mathematics, yet his paper faced a 15-year delay before publication. It was only accepted after Lagrange's death, as he and others had opposed Fourier's ideas during his lifetime [Smith '97]. This teaches us two important lessons: first, while many intelligent and influential figures, like Lagrange, may challenge new ideas, they are not always correct in their judgments. Second, if we believe in our ideas, we should remain resilient.

transforms (decomposes) it into a frequency-domain signal. Machine learning algorithms can work better with frequency-domain signals rather than time-domain signals.

Fast Fourier Transform (FFT), is the most commonly used algorithm to compute the Discrete Fourier Transform (DFT). The FFT efficiently processes an array representing a discrete-time signal, transforming it from the time domain to the frequency domain. Unlike the DFT, which can be computationally intensive for large datasets, the FFT significantly reduces the complexity of calculations, making it faster and more practical for a wide range of applications. Both the FFT and DFT yield the same result.

Figure 7-14 is implemented with FFT as well. You can also check Figure 6-35 in Chapter 6. There, we also used FFT to transform the time domain signal into a frequency domain signal.

The other types of transformation were rarely used for machine learning algorithms, and we use this chance to demonstrate that we are lazy and skip explaining them.

Wavelet Transform

Decomposing signals by Fourier transform is useful because it gives us information related to frequency, and frequency domain signal is usually easier to process for algorithms. By using the Fourier transform, however, we lose the notion of time when a signal is transferred into a

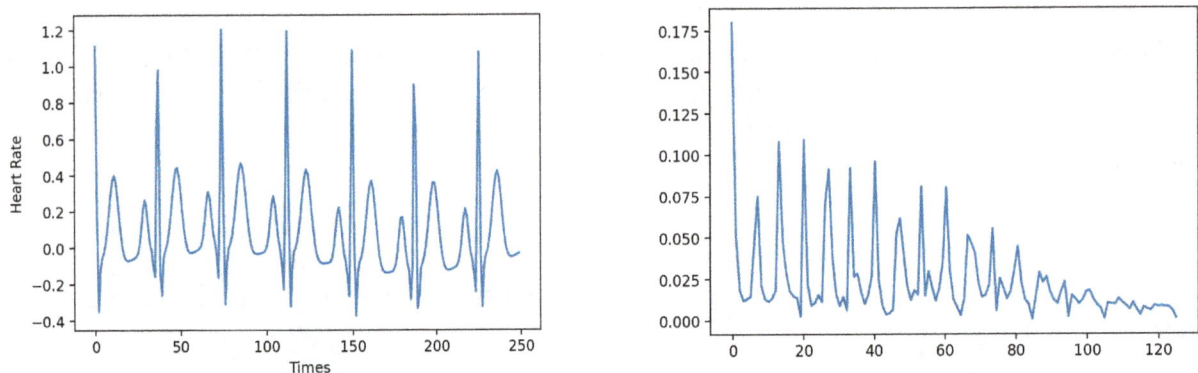

Figure 7-15: An ECG signal on the left has oscillations, which are important to analyze. However, if we apply FFT on it, we will get the the right plot, which does not entail oscillation information.

frequency-domain signal. For example, Electrocardiography (ECG) signals have short intervals of oscillation, and this characteristic can not be captured by Fourier transforms. Take a look at Figure 7-15. There, we have a mock ECG signal, and you can see that oscillations from the time domain signal can not be accurately transformed into the frequency domain signal. In other words, by applying FFT on the original signal, we can not identify the frequency of the signal at a specific time, because the FFT does not capture these time-localized events, and wavelet transform methods mitigate this uncertainty.

Wavelet transform is a signal transformation method that decomposes a signal into components at different scales, allowing simultaneous analysis of its frequency and time characteristics. This

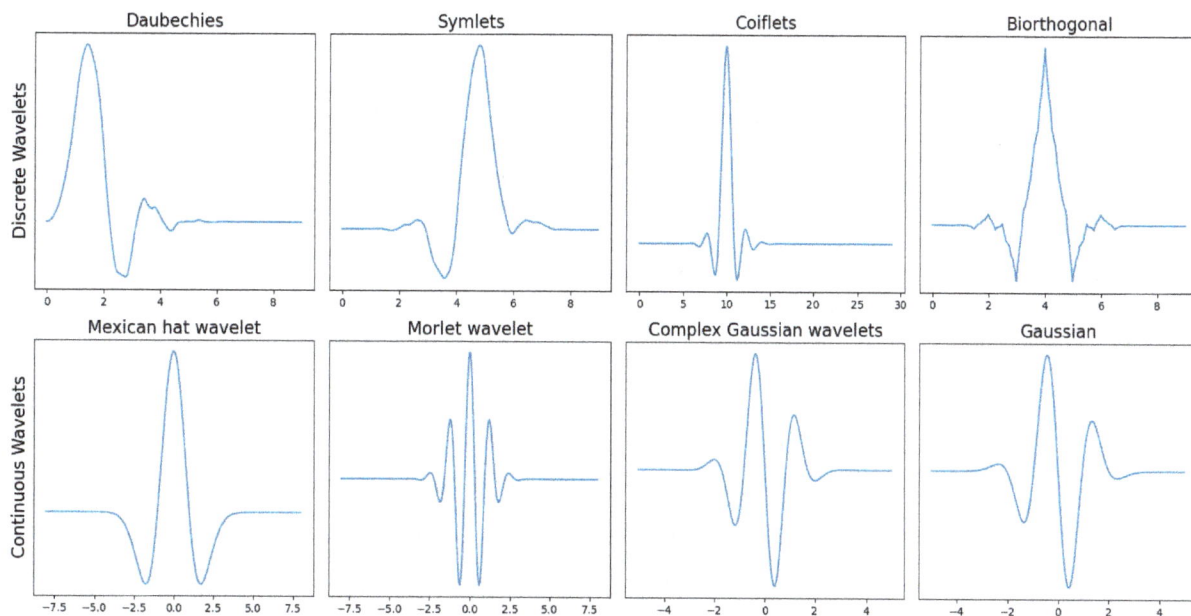

Figure 7-16: Some example of common continuous and discrete wavelet shapes in time domain (X axis is time).

transformation provides a better resolution[6] to the original signal rather than Fourier transformations. In the machine learning community, wavelet is usually favored over Fourier transform because it *keeps both frequency and time information*.

Wavelets are known as mini waves. Waves such as sine or cosine waves go on forever, but wavelets have finite start and stop times. Open some small place in your lovely brain and write there: wavelets have "finite" time and "frequency". In some literature, we might encounter the following description: wavelet is localized in time, which has the same meaning. The Fourier transform converts a given signal to a set of sine waves with different frequencies. The wavelet transform converts a given signal to a set of wavelets that have different scales.

Figure 7-16 presents some examples of common *Continuous Wavelet Transform (CWT)* wavelet shapes. Unlike a sinusoid wave that goes to infinity, we can see that the wavelet is a decaying wave (right and left size value is going toward zero) that has zero means.

To apply wavelet transform on the given signal, the wavelet algorithm uses two operations: *scaling* and *shifting*. Scaling can be either shrinking or expanding the wavelet, which changes the frequency resolution of the transform. Shrinking the wavelet (i.e., using a small scaling factor) increases the frequency resolution and captures the high-frequency components of the input signal, while expanding the wavelet (i.e., using a large scaling factor) decreases the frequency resolution and captures the low-frequency components of the input signal. Shifting the wavelet in time allows the transform to be applied at different time intervals or resolutions. Figure 7-17 visualizes the scaling and shifting of a toy wavelet.

The wavelet transform starts from the beginning of the signal and multiplies a given signal, with wavelets with different scales at different locations in time (shifting the wavelet). As a result, we

---

[6] Think about this resolution as the amount of uncertainty in a measurement. Higher resolution means less uncertainty. If you like to read it in more detail, we recommend reading about Heisenberg's uncertainty principle.

have a series of transformed signals that are presented in a shape called a *Scalogram* (right side of Figure 7-18). This process is called convolution as well, but it is uncommon in machine learning due to its overlap with the conventional neural network (we will describe them in Chapter 10). To better understand this process, look at Figure 7-18, where we visualize a wavelet transform for a given signal. You can see that a Scalogram is composed of many signals that differ in time (as a result of shifting the wavelet through the input signal) and have different scales (the result of multiplying a wavelet at different scales).

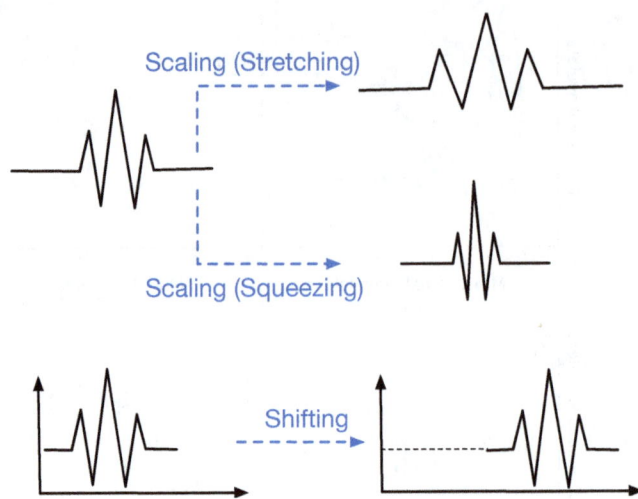

Figure 7-17: Example of scaling (stretching and squeezing), and shifting a toy wavelet.

Resolution: In the context of wavelet transform, resolution refers to the ability to localize a signal in both time and frequency. There are two resolutions, "time" and "frequency" resolutions. Time resolution refers to how accurately a wavelet transform can determine when specific events occur in a signal. Frequency resolution refers to how accurately it can identify the frequency components of those events.

Continuous and Discrete Wavelet Transform

There are two major wavelet transform methods; Continuous Wavelet Transform (CWT) and Discrete Wavelet Transform (DWT). CWT [Morlet '82] uses continuous wavelets, which are wavelets that can be scaled and translated to any desired frequency and time resolution. The flexibility allows the wavelets to adapt to the specific characteristics of the input signal. DWT [Heil '89, Daubechies '92] refers to any wavelet transformation in that wavelets are sampled discretely. This means that these wavelets are only available at certain discrete scales and positions, which can limit the flexibility of the analysis. However, the DWT is more computationally efficient than the CWT.

Haar wavelet transform (Daubechies 1 or Db1) [Haar '10] is the simplest implementation of the DWT that is being used. It decomposes a signal into square-shaped discrete signals. It is known as the first known wavelet transformation.

The CWT provides fine resolution at all scales and positions, while the DWT only provides fine resolution at low frequencies (high scales) and coarser resolution at high frequencies (low scales). This resolution tradeoff is a key difference between CWT and DWT. The fine resolution of CWT makes it suitable for feature extraction tasks where maximum resolution is needed. For example, CWT is useful for edge/corner detection (check Chapter 6) and biological data analysis, such as electrocardiogram (ECG) analysis.

On the other hand, the efficiency of DWT makes it suitable for compression and noise reduction, where redundancy must be minimized. For example, it can be used for audio signals, seismic signals (vibrations in the Earth), and other time-dependent signals for noise reduction. DWT is used in JPEG2000 compression, video compression, and feature extraction. Besides, the DWT is invertible; hence, the original signal can be perfectly reconstructed, while the CWT is overcomplete, so perfect reconstruction is impossible.

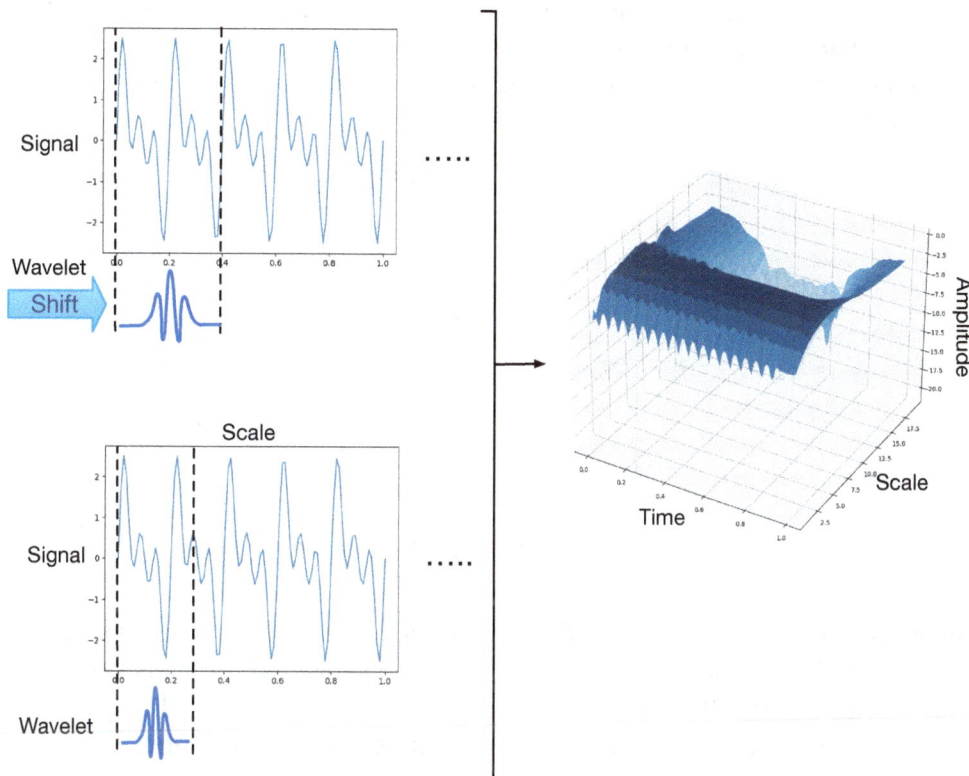

Figure 7-18: The input signal is multiplied by several shifted and scaled wavelets, and as a result, we end up having a Scalogram that can be used instead of the original input signal. This figure uses the "...." to present that the scaling or shifting operation was performed several times.

After we have transformed the time-domain signal into a frequency-domain signal, we can look at finding peaks (as a feature in the signals) or other types of features in the signal. In the end, these features will be fed into the machine learning algorithm for classification or other types of analysis.

A question might arise: among many possible wavelets, which one should be used when? It is recommended that Daubechies, Symlet, and Biorthogonal Spline wavelets be used for time series data. For image processing, we can use Biorthogonal Spline, Cohen-Daubechies-Feauveau, and Haar wavelets. For detecting sharp discontinuities in the input signal, we can use Haar, Symlet, and Coiflet wavelets. For limited computational resources, use Haar, Daubechies, and Symlet wavelets.

Approximate Aggregation Methods (Time Series)

There is another group of transformations widely used on time series data. These transformations employ approximate aggregation techniques to transform time series into another format. We explain two common ones: *Piecewise Aggregate Approximation (PAA)* and *Symbolic Aggregate Approximation (SAX)*.

PAA [Keogh '01] reduces the time series size by splitting it into equal-sized segments. Then, for

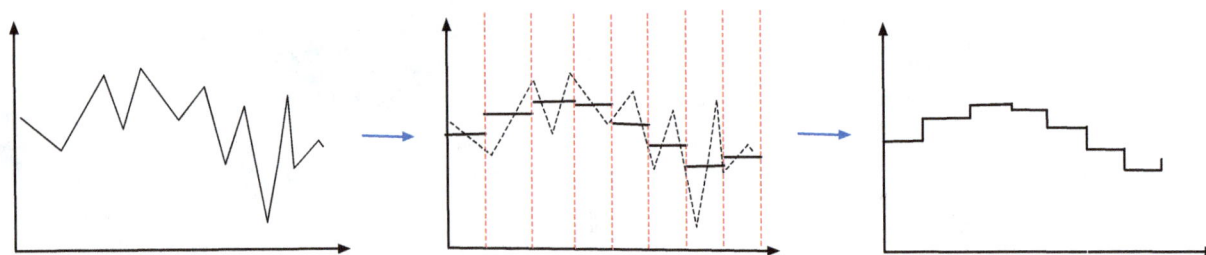

Figure 7-19: Transforming a time series into PAA

each segment, an average value of the data points in this particular segment will be calculated and substituted for the original data. Therefore, the time series will be transformed into a simpler model, which holds most of the characteristics of the original one as well. Figure 7-19 presents a time series and its transformation into PAA. Assuming we have m number of windows and n number of transformed data points, the complexity is $O(m \cdot n)$, which is linear.

Symbolic Aggregate Approximation (SAX) is another interesting method [Lin '03] that converts a time series into a character string. It is widely used in machine learning algorithms, and, similar to PAA, it is very easy to understand. Also, this method is not categorized as signal or time series decomposition. However, since it compresses and transforms time series into a set of characters, we list it in this section. The SAX algorithm first converts the time series into the PAA, assuming we have a time series on n length. Then, it will convert the result into a character string on w length and $w < n$. Figure 7-20 presents an example of how the SAX algorithm operates.

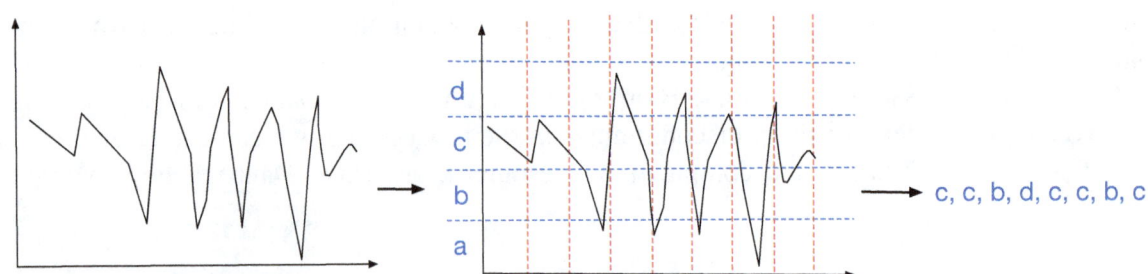

Figure 7-20: SAX algorithm example which converts a time series into a character string.

SAX computational complexity is also linear, similar to the PAA. We have also used it in some of our algorithms, such as Ghost imputation [Rawassizadeh '19], which we will explain in Chapter 16. An advantage of both PAA and SAX is their linear computational complexities.

Both PAA and SAX have hyperparameters that define the size of segmentation on time or x-axis and value or y-axis.

NOTE:

* Two or more than one signal could be merged together to construct a new signal which is composed of the previous ones.

* Signal data could be treated both as continuous data and discrete data. The representation chosen often depends on the nature of the signal and the requirements of the analysis algorithm.

* There are many standard wavelets, and they have their algorithm to convert time-domain signals into wavelet domain signals.

* The fundamental idea of wavelet transform is that extension or changes are allowed in time. Therefore, the shape of the original signal does change with scaling, as it becomes compressed or stretched, affecting its frequency characteristics.

* There is an amazing animation (in .gif format) in Wikipedia that presents how wavelet works: https://en.wikipedia.org/wiki/Continuous_wavelet_transform

Matrix Decomposition

Matrices are one of the most used mathematical structures in machine learning, and especially in deep neural networks. On the other hand, loading a large matrix into memory and applying mathematical operations on matrices such as dot products are resource-intensive processes, and the hardware limitations of the hosting system limit large-scale matrix operations. Matrix multiplication is so important that it led to the success of GPU providers, especially NVIDIA, at the time of writing this section (2018) and its latest revision (2024).

A common approach to deal with large matrices is to reduce their size and thus make them more resource efficient. This process will be done by decomposing matrices into smaller matrices with characteristics similar to those of the original large matrix. In other words, *matrix decomposition* refers to breaking down a matrix into a product of matrices in a way that can simplify further matrix operations.

Factoring is another term that we should learn. It is the process of making an object combination of smaller objects, e.g., 15, which will be factorized into 3×5.

In simple words, matrix factorization reduces the size of the matrix. Matrices are one of the most used mathematical structures in machine learning, especially in deep neural networks. On the other hand, loading a large matrix into memory and applying mathematical operations on matrices such as dot products are resource-intensive processes, and the hardware limitations of the hosting system limit large-scale matrix operations. Matrix multiplication is so important that it led to the success of GPU providers, especially NVIDIA, at the time of writing this section (2018) and its latest revision (2024).

A common approach to deal with large matrices is to reduce their size and thus make them more resource efficient. This process will be done by decomposing matrices into smaller matrices with characteristics similar to those of the original large matrix. In other words, *matrix decomposition* refers to breaking down a matrix into a product of matrices in a way that can simplify further matrix operations.

Factoring is another term that we should learn. It is the process of making an object a combination of smaller objects, e.g., 15 will be factorized into 3×5. *Matrix factorization* is a type of matrix decomposition used for specific purposes, such as dimensionality reduction. In simple words, matrix factorization reduces the size of the matrix, and it is used for different machine learning algorithms, such as LSA (topic modeling), image recognition, LoRA (check Chapter 14), and many more algorithms that require operating with large matrixes.

It is used for different machine learning algorithms, such as LSA (topic modeling), image recognition, LoRA (check Chapter 14), and many more algorithms that require operating with large matrixes.

In the following, we list the most used matrix decomposition approaches and explain each of them without delving into their mathematical details. Besides, we explain two topic modeling approaches, a concept that remains to be explained in Chapter 5.

Cholesky Decomposition

One way to decompose a number is to take its square root, e.g., $16 = 4 \times 4$. For symmetric matrices that include only non-negative numbers, we can also use something similar to a square root to decompose them. Cholesky decomposition [Cholesky '10] described that Matrix A could be decomposed to two triangular matrices, i.e., $A = L . L^T$.

$$
\underbrace{\begin{bmatrix} x11 & x21 & x31 & x41 \\ x21 & x22 & x32 & x42 \\ x31 & x32 & x33 & x43 \\ x41 & x42 & x43 & x44 \end{bmatrix}}_{A} = \underbrace{\begin{bmatrix} a11 & 0 & 0 & 0 \\ a21 & a22 & 0 & 0 \\ a31 & a32 & a33 & 0 \\ a41 & a42 & a43 & a44 \end{bmatrix}}_{L} . \underbrace{\begin{bmatrix} a11 & a21 & a31 & a41 \\ 0 & a22 & a32 & a42 \\ 0 & 0 & a33 & a43 \\ 0 & 0 & 0 & a44 \end{bmatrix}}_{L^T}
$$

Original matrix — lower triangular matrix — higher triangular matrix

L is a lower triangular matrix (all the entries above the main diagonal are zero) and L^T is transposed of L, which will be an upper triangular matrix. Cholesky Decomposition is specifically used for symmetric, positive-definite matrices, not just any symmetric matrix with non-negative numbers. Despite its limitations, it is supported by most machine learning software to facilitate the overhead of numerical computation. For example, Cholesky decomposition can be used to solve linear regression problems with too many features or in Bayesian inferences to approximate the posterior distribution of the model's parameters. It is also used to compute statistical quantities (e.g., sample covariance matrix). The computational complexity of Cholesky decomposition is $n^3/6 + O(n^2)$ [Krishnamoorthy '13].

Non-negative Matrix Factorization (NMF)

Another interesting group of methods that breaks down a matrix into non-negative values is called Non-negative Matrix Factorization algorithms [Sra '06]. By using NMF, a matrix X could be decomposed (approximated) as: $X \approx H.M$. Here we use approximation sign ($\approx$) because it does not guarantee whether we can recover the original matrix from the decomposed ones ($\approx$ is a sign of approximation).

The original $m \times n$ matrix A, is factorized into $m \times r$ matrix W, and $r \times n$ matrix H. r is smaller than m and n. Values of W and H matrices will be achieved by an iterative process which measures each time the cost of approximation.

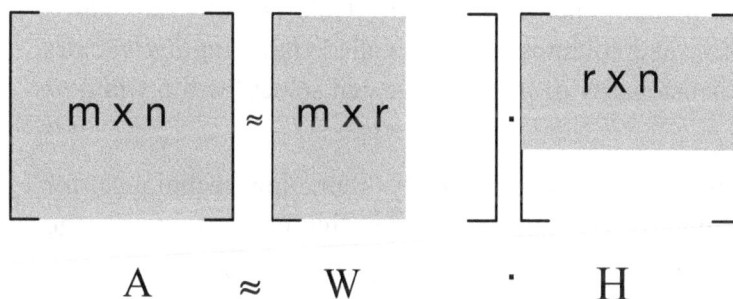

$$A \quad \approx \quad W \quad \cdot \quad H$$

NMF has many useful examples in machine learning. For example, it can be used for image compression and denoising. In NLP applications, it can be used to decompose text into its constituent parts, such as words, phrases, and topics. It can be used to decompose chemical data into their constituent parts, such as molecules, atoms, and bonds. It can be used in analyzing gene expression data to discover gene modules and identify co-regulated gene sets, which helps the machine learning algorithm in identifying patterns and relationships between genes.

Singular Value Decomposition (SVD)

SVD is the most used matrix decomposition method to convert (factorize) a matrix into *singular vectors* and *singular values*. Similar to many other scientific discoveries before the Internet era, it has been discovered independently by several mathematicians, starting from Eugenio Beltrami in 1868 [Stewart '93, Golub '71].

Even SVD could be used to transform the data even after the eigenvectors and their values have been identified in PCA. We have described the basic version of the PCA earlier in this chapter with a simple transformation. Nowadays (in 2024), there are many interesting works done based on SVD, such as Low Rank Adaption (LoRA) for Large Language Models, which we will describe in Chapter 14. Nevertheless, SVD is still widely used. It has a major advantage over

Cholesky decomposition, and it can be applied to any form of matrices (not just symmetric or square matrices).

SVD decomposes a matrix A of size $m \times n$ into three matrices as follows, $A = U . \Lambda . V^T$
Each of these three matrices ($U . \Lambda . V^T$) is smaller than the original matrix.

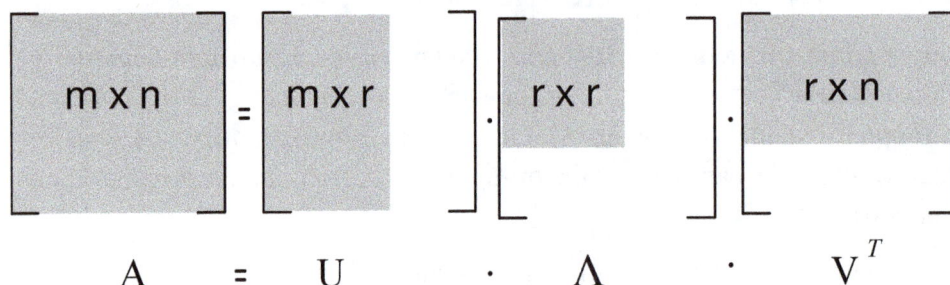

$$A \quad = \quad U \quad \cdot \quad \Lambda \quad \cdot \quad V^T$$

U and V are orthogonal matrices[7]. Λ is a diagonal matrix and called the singular values of the original matrix. A diagonal matrix is a matrix that only has data in its diameter, and all other data are zero. Columns of U are called left singular vectors of A, and columns of V are called right singular vectors of A. To reduce the dimensionality of the data, we can select the top k largest singular values in Λ.

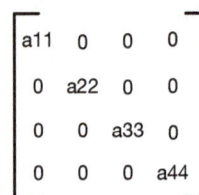

$$\begin{bmatrix} a11 & 0 & 0 & 0 \\ 0 & a22 & 0 & 0 \\ 0 & 0 & a33 & 0 \\ 0 & 0 & 0 & a44 \end{bmatrix}$$

diagonal matrix

SVD is one of the most used dimensionality reduction techniques for machine learning problems. It has many applications, from image compression to noise removal and, interestingly, topic modeling.

Topic Modeling (Clustering with Matrix Decompositions)

A large segment of the data science community is dedicated to working with textual data and extracting knowledge from text data, including recommender systems, search engines, and domain-specific text analysis (e.g., medical reports, law cases, etc.). This leads to the introduction of text-based knowledge extraction disciplines in computer science, such as information retrieval or text mining. Information retrieval is defined as activities to search and retrieve data (mostly textual data), and it is different from extracting knowledge from the text, i.e., text mining.

Through this book, we rarely describe applications because we believe you will learn algorithms to trim them to your problem, and you can easily find plenty of examples online. However, SVD matrix decomposition is an exception. Topic modeling is a type of clustering algorithm for unstructured text. By unstructured, we mean plain text and not text inside tabular data format.

We didn't explain topic modeling in Chapter 4 because it requires an understanding of matrix decomposition and familiarity with SVD. Now, we have some understanding of SVD, and we

---

[7] An orthogonal matrix is a square matrix Q with real entries that have the following characteristic, $Q^T . Q = I$, where Q^T is the transpose of Q and I is the identity matrix of the same dimension.

can describe two important topic modeling algorithms, i.e., *Latent Semantic Indexing* and *Latent Dirichlet Allocation*. Before explaining them, we should learn what topic modeling is.

Topic modeling group textual documents based on the appearance and similarities between their words (terms). In other words, topic modeling tries to identify abstract document topics based on the frequency of their words (terms). They operate based on the concept of a "Bag of Words". It means that the raw text document should be converted into a Bag of Words (BoW). Then, the BoW will be consumed by the topic modeling algorithm to cluster textual documents. Figure 7-21 summarizes the process required for topic modeling.

To understand the motivation of topic modeling, assume there is a beautiful chicken who receives many messages from her lovers. However, her online followers on social media are starting to complain about her attitude, that she is a gold digger. To make some show-off for her social media followers, she decides to answer a few emails about charities or other things related to social good. Nevertheless, she receives many emails, and she should find a way to cluster her emails based on their topics. She hires a data scientist who uses topic modeling for this task. In the following, we describe two widely used topic modeling algorithms.

Latent Semantic Indexing (LSI)

Latent Semantic Analysis (LSA) or Latent Semantic Indexing (LSI) [Deerwester '90] is a topic modeling technique used to cluster documents based on the similarities between their topics. It operates by the assumption that there are hidden (latent) relations between terms (words). To use LSI, first, documents need to be converted into a "document-term matrix" (see Figure 7-21). Before converting unstructured text into the document-term matrix, some necessary text processing steps are performed, which we have explained (see Chapter 6), including tokenization, stemming, etc.

For example, let's say we have three sentences as follows, and we consider each sentence a document, i.e., D1, D2, and D3:

D1: I am lifting a heavy chicken today, it is really heavy.

D2: Today, the heavy chicken eats lots of food.

D3: It is hard to lift heavy chickens and easy to lift light chickens.

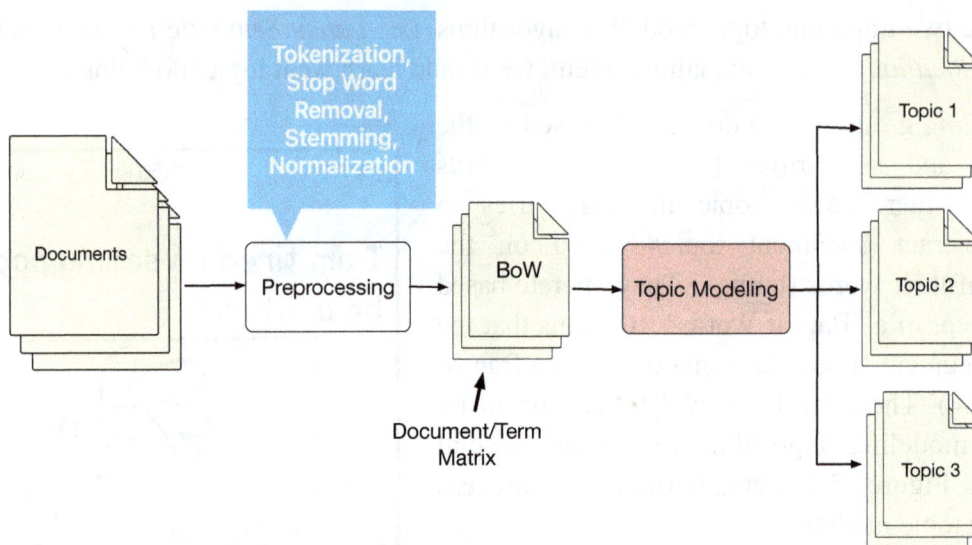

Figure 7-21: Conceptual flow of a typical Topic modeling approach. Yellow boxes have been described in Chapter 6.

Their document-term matrix is something as the following (after applying all pre-processing steps). This matrix includes lots of nulls or zeros, and such a matrix is usually called a "sparse matrix".

The next step is to apply SVD to the term-document matrix. The SVD converts the $n \times m$ term-

	D1	D2	D3
I	1	0	0
lift	1	0	2
heavy	2	1	1
chicken	1	1	2
today	1	2	1
eat	0	1	0
...		...	

document matrix into three matrices, as it is shown below.

In other words, after the term-document matrix is ready, it will be decomposed by the SVD, and we will get three matrices. One matrix is the *word-to-topic assignment*, the second one is the *topic-to-topic relation*, and the third one is the *topic-to-document*, which specifies the contribution of a topic to a document.

The topic contribution in the document matrix is the result that we are looking for. It describes the contribution of the identified topic in each document. The topic is something that the system will identify. Nevertheless, the topic will not be explicitly provided by the algorithm; we should take a look at documents and manually create a name for a topic or use one of the described theme extraction methods in Chapter 6 to create a name for a topic.

For example, a topic that includes terms such as "gear", "break", "wheel", "seat", "auto-pilot", etc. could be named as "vehicle" topic, or a topic that includes documents about "rasmalai", "faloodeh", "tres leches", "moon cake" could be named "dessert".

Latent Dirichlet Allocation (LDA)

Latent Dirichlet Allocation (LDA) [Blei '03] is a generative probabilistic model[8] for topic modeling, and it is more flexible than LSI. To describe the model, first, we need to provide some introductory text. Then, we will briefly describe LDA.

If you recall, in Chapter 5, we explained the probabilistic graphical model, and in Chapter 3, we explained that Beta distribution is parameterized by α and β, which is a 2D of form of Dirichlet distribution. In other words, Dirichlet distribution is a generalization of the Beta distribution to multiple dimensions.

Figure 7-22 presents the probabilistic model for LDA in a "plate notation"(another name for a probabilistic graphical model). If you recall from Chapter 4, when we draw an arrow from one node to another node, it presents the probability dependence.

Plate notation is a probabilistic graphical model, but we need to learn two other small things for plate notation as well: (i) a box around nodes means variables inside the box will be repeated N times or M times, and (ii) anything in grey means that it is an observable variable, and thus W is observable variable in Figure 7-22. α and β are Dirichlet distribution parameters.

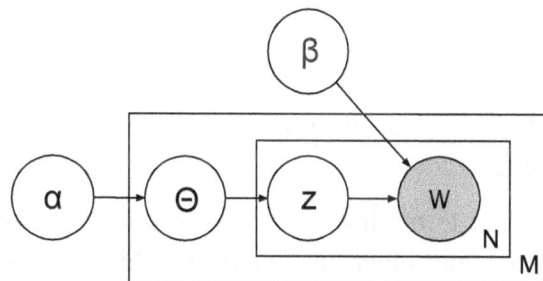

Figure 7-22: Plate notation for Latent Dirichlet Allocation.

LDA implementations wrap all these details, and to only use them, we don't need to worry about configuring these parameters. However, if the result of LDA does not make sense, we could start to play with those parameters. Θ presents the "topic distribution" for "document" M, z is the "topic" for the N-th word in document M, and w is the "specific term".

---

[8] Generative models focus on learning how the data is being generated to be able to predict the future. In contrast, there are discriminative models, which try to learn how to discriminate between different elements of a dataset.

To run the LDA, first, we should define the number of topics (k) we want to discover, i.e., hyperparameter. Then, the LDA randomly assigns each word from the given text corpus (dataset) to one of those k topics. In the next step, LDA uses Gibbs sampling (we will describe Gibbs sampling in Chapter 16) iteratively to update the assignments of words to topics. The goal of LDA is to find the optimal assignment of topics to words in each document and the distribution of words within each topic. The Gibbs sampling process goes through each word in each document and reassigns the word to a new topic based on certain probabilities. These probabilities are calculated by considering (i) the current topic assignments of other words in the document and (ii) the distribution of words within each topic. The process aims to maximize the likelihood of the observed data (the words in the documents) given the current topic assignments.

The Gibbs sampling process is repeated for a maximum number of iterations (we give it as a hyperparameter) or until convergence is achieved (i.e., the topic assignments do not change significantly between iterations). The result of LDA execution will be two matrices, as shown in the following.

The matrix on the left presents the probability of terms appearing inside each document (a.k.a topic-word distribution). The matrix on the right presents the topic appearance probabilities in each document (a.k.a document-topic distribution). Each data point of these matrices is a probability value.

Let's review the process required for LDA topic modeling. First, we perform the preprocessing step similar to the LSI (stemming, stop word removal and a BoW construction). Now, we have the document term matrix ready to feed them into the LDA algorithm. Then, we need to specify how many topics we need. This should be given as a hyperparameter to the algorithm. Afterward, we fed the corpus (in term-document matrix format) within the number of desired topics to the LDA algorithm. Then, we get the result that the legendary LDA gives us.

NOTE:

* Another name for the document-term matrix is document vectors. Each column of this matrix presents a document, and each document is presented as a vector of terms.

* Note that terms inside a term-document matrix are not just about words and documents; we can map any other discrete data to terms as well. For example, a document could be Internet browsing sessions, and links to web pages a user visits could be terms. Topic modeling is widely in use in many different disciplines like, genomics, search engines, recommendation systems, and NLP (Natural Language Processing) systems.

* A major advantage that LDA has over LSI is that each term (word) could participate in more than one topic, but the probability of its topic participation will be different.

Tensor Decompositions

Tensors are used to store multidimensional data. If you are familiar with programming, you can think of a matrix as a two-dimensional array and a tensor as a multidimensional array. A tensor can be used to store multidimensional data such as vowel sounds of different individuals, including their pitch and vowels. Another example is a video; its data can be stored as a sequence of frames, where each frame is a two-dimensional grid of pixel values. Therefore, typically, a video tensor would include dimensions for frames (time), height (y-pixel), width (x-pixel), and color channels. Standard color channels are Red, Green, and Blue (RGB), making it a 4D tensor ([time, height, width, color]). Sound is typically stored separately; it is indeed multidimensional data (with dimensions such as time and frequency when considering spectrograms), but it is usually stored and processed separately from the video's visual data. Although a tensor can have more than three dimensions, our brain can not visualize more than three dimensions, and similar to other literature, we use 3D tensors in our examples. A tensor is usually presented with a calligraphic symbol, i.e., $\mathcal{X}$. Here, we present a tensor with a capital letter and its dimensions as i_x, e.g. $A(i_1, i_2, \ldots i_d)$. Sometimes, we refer to dimensions as a matrix, e.g., A_{ijk} presents a tensor of three dimensions.

Machine learning algorithms that deal with multivariable data, such as videos, get tensor as input. Processing large tensors is challenging and very resource intensive. There are lots of advances in mitigating the challenges of tenor processing, which leads to the introduction of new hardware chips such as NPU (Neural Processing Unit) or regaining attention on GPU (Graphical Processing Unit) to solve this challenge.

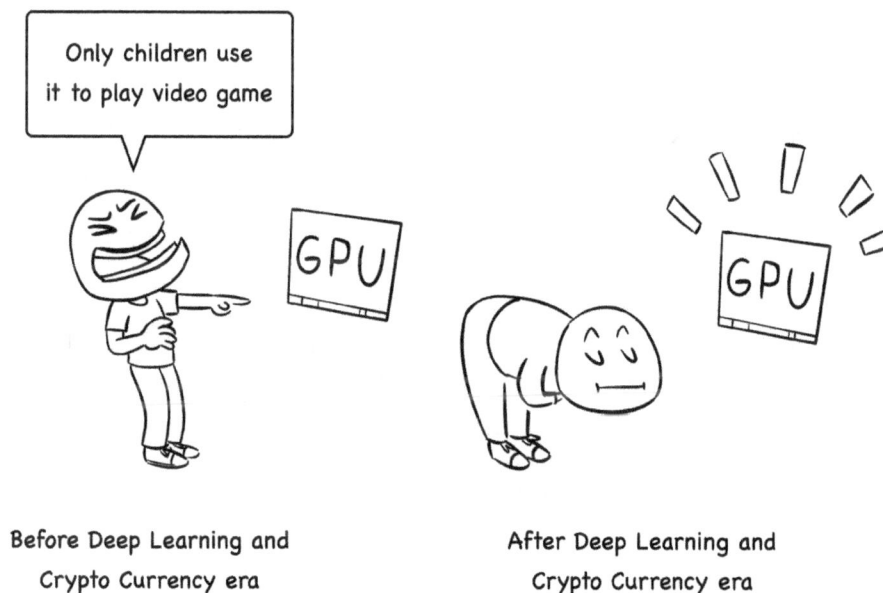

Before Deep Learning and
Crypto Currency era

After Deep Learning and
Crypto Currency era

Usually, tensor decomposition methods are generalizations of matrix decomposition methods. In other words, usually, we convert a tensor into matrices, and then we apply classical matrix processing to it. Next, we generalize our findings or results from matrices into the tensor.

First, we learn some concepts about tensors. Next, we explain some operations of tensors, and

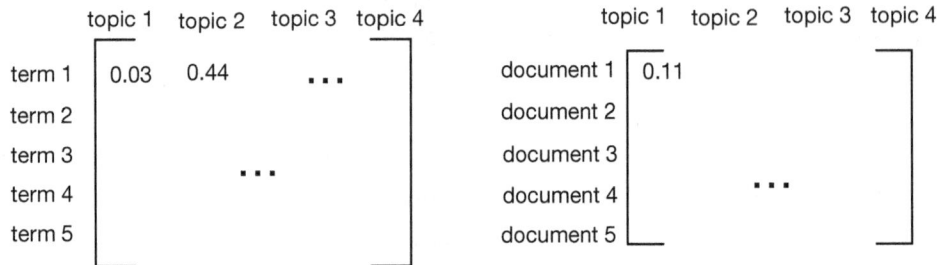

then we move toward the algorithm used for tensor decomposition.

Tensor Concepts

Tensor Order (Mode or Degree): The order (or degree) of a tensor refers to the number of dimensions required to specify an element within the tensor. A scalar is a tensor of order zero, a vector is a first-order tensor, and a matrix is a second-order tensor. Three (or more than three) dimensional numerical arrays are called higher-order tensors. Higher-order tensors require three or more indices to specify each element, and they are the focus of this section.

Cubical Tensor: It is a tensor that has equal dimensions on every axis (e.g., a $3 \times 3 \times 3$ tensor).

Tensor Rank: It refers to the *minimum number of rank-1 tensors* (vectors) that sum up to the original tensor. In other words, it measures the complexity of a tensor in terms of the minimum number of rank-1 tensors that can be combined to reconstruct it.

Some literature makes a separation between the rank and order or mode of tensors [Sochi '17], but here, we use rank to refer to this minimal representation.

Fiber: A matrix has columns and rows; a tensor has a fiber (which are one-dimensional sections of the tensor) instead. Fiber is a vector obtained by fixed indices but one changing index.

Figure 7-23: Fibers of three dimensional tensors. From left to right: Mode-1(column) fibers, Mode-2 (row fibers), Mode-3 (tube) fibers.

In other words, a fiber refers to a specific tensor component in all of its indices except one that is fixed. Figure 7-23 presents three different types of fibers of a 3-order tensor. For example, Mode-1 fibers are fixed in the first (X) and third (Z) dimensions, and their second dimension (Y) is changing. Mode-2 fibers are fixed in the Y and Z dimensions and the X dimension changes. Mode-3 fibers are fixed in the X and the Y dimensions, while their third dimension (Z) changes.

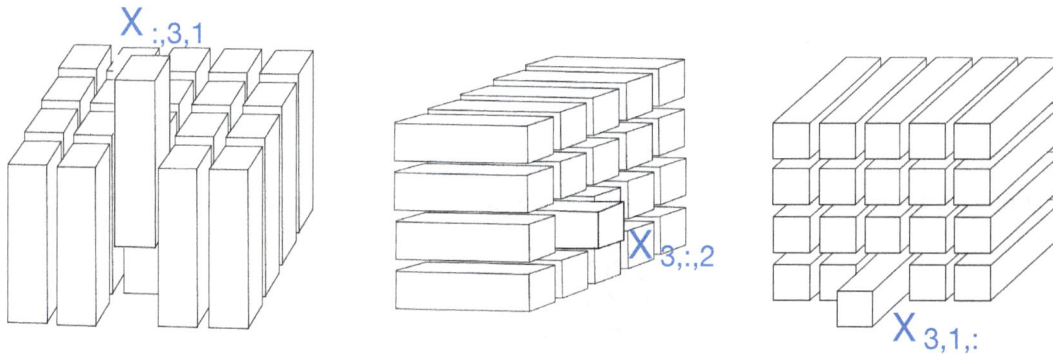

Figure 7-24: Examples that presents how to refer to a specific fiber in a tensor. The colon sign is used to denote the full range of index.

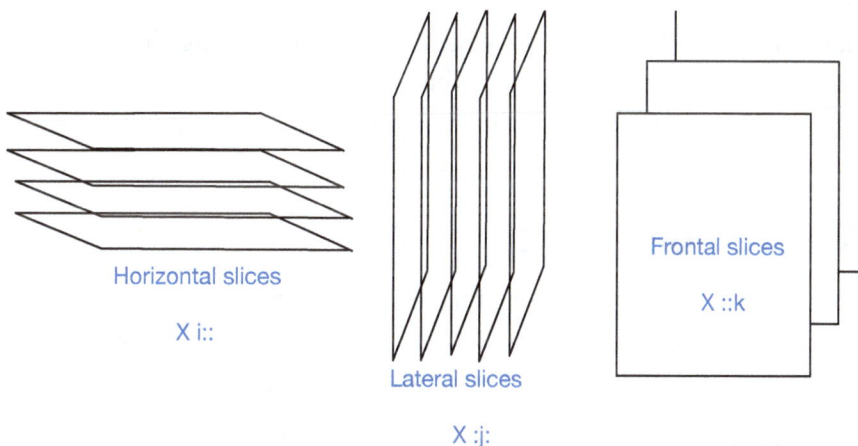

Figure 7-25: Examples that present how to refer to specific types of matrices in a tensor. Similar to Figure 7-23 the colon sign is used to denote the full range of index.

Figure 7-24 presents some examples that refer to a specific fiber in the tensor, ':' is used to present the changing axis, and the other numbers are index numbers of axes that are not changing.

Slide: Each slide in a tensor presents a matrix. To present slices, the axis which is not changing is denoted with ':' sign, as shown in Figure 7-25. The horizontal slice is referred to as the mode-1 slice, the lateral slice is referred to as the mode-2 slice, and the frontal slice is referred to as the mode-3 slice.

Canonical Factors and *Canonical Rank:* A tensor can be decomposed as the sum of the outer products of vectors. The minimum number of these outer product terms required to express a tensor is called its *canonical rank* or *tensor rank*, and it is denoted by r.

For example, if we have two vectors $\vec{a} = \{1,2\}$ and $\vec{b} = \{3,4,5\}$, their outer product is as follows:

$$\vec{a} \otimes \vec{b} = \begin{bmatrix} 3 & 4 & 5 \\ 6 & 8 & 10 \end{bmatrix}$$

When we sum a group of outer products together, it is called the *sum of outer products*. For example, for a three-dimensional tensor, we have the following vectors: u, v, and w. By calculating their outer products, $\vec{u} \otimes \vec{v} \otimes \vec{w}$ we get a tensor. To construct a more complex tensor $\mathcal{T}$, we would sum up r times, and the sum of their outer product is formalized as follows:

$$\mathcal{T} = \sum_{i=1}^{r} u_i \otimes v_i \otimes w_i.$$

We explained that r is the *canonical rank* or *tensor rank*, and vectors involved in the decomposition are called "*canonical factors*" or "*factor vectors*".

To summarize this explanation, Canonical rank is the minimum number of rank-one tensors (outer products of vectors) that must be summed to represent the original tensor. The computation of canonical rank is NP-hard. This means that determining the smallest number of rank-one tensors that sum up to a given tensor (which defines the tensor's canonical rank) is a problem for which no polynomial-time algorithm is known. We will explain NP-hard problems in Chapter 16.

Tensor and Matrix Operations

They are different from matrix multiplication and have a different operator. We need to learn four different matrix multiplication on matrices that are used for tensor decomposition as well, including the Kronecker, Khatri-rao, Hadamard, and n-mode product.

Kronecker product [Kolda '09], denoted by $\otimes$, is a generalization of outer product from vectors to matrices[9]. *Khatri-rao product,* is a column-wise Kronecker product and denoted by *. It is also a generalization of the outer product of a matrix. *Hadamard product* (Schur produce or naive matrix product) is denoted by $\odot$ or $\bigcirc$, gets two matrices of the same dimension and produces a third matrix, which its *i,j* element is the multiplication of *i,j* elements from the two input matrices. The *matrix product*, which we have learned previously presented by the dot (.) or $\otimes$ sign. To better understand these multiplication operations, we use an example.

---

[9] Outer product in algebra means that if the two vectors have dimensions n and m, then their outer product is an $n \times m$ matrix.

Assume we have two matrices A and B, as follows:

$$A = \begin{bmatrix} a & b & c \\ d & e & f \end{bmatrix}, \quad B = \begin{bmatrix} g & h & i \\ j & k & l \end{bmatrix}$$

We can write the following as their Hadamard product, Kronecker product, Khatri-Rao product, and simple dot product:

Hadamard product: $A \odot B = \begin{bmatrix} ag & bh & ci \\ dj & ek & fl \end{bmatrix}$

Kronecker product: $A \otimes B = \begin{bmatrix} a\begin{bmatrix} g & h & i \\ j & k & l \end{bmatrix} & b\begin{bmatrix} g & h & i \\ j & k & l \end{bmatrix} & c\begin{bmatrix} g & h & i \\ j & k & l \end{bmatrix} \\ d\begin{bmatrix} g & h & i \\ j & k & l \end{bmatrix} & e\begin{bmatrix} g & h & i \\ j & k & l \end{bmatrix} & f\begin{bmatrix} g & h & i \\ j & k & l \end{bmatrix} \end{bmatrix}$

Khatri-rao product: $A * B = \begin{bmatrix} ag & bh & ci \\ aj & bk & cl \\ dg & eh & fi \\ dj & ek & fl \end{bmatrix}$

Outer product: $A \cdot B^T = \begin{bmatrix} a & b & c \\ d & e & f \end{bmatrix} \cdot \begin{bmatrix} g & j \\ h & k \\ i & l \end{bmatrix} = \begin{bmatrix} ag+bh+ci & aj+bk+cl \\ dg+eh+fi & dj+ek+fl \end{bmatrix}$

Mode-n Unfolding

Unfolding or flattening a tensor X refers to a process of converting it into a matrix by re-arranging its data based on different dimensions. Assume we have a tensor X, which is a $2 \times 2 \times 3$ tensor composed of the following frontal slices[10]:

$$X_1 = \begin{bmatrix} 0 & 1 \\ 2 & 3 \end{bmatrix}, X_2 = \begin{bmatrix} 4 & 5 \\ 6 & 7 \end{bmatrix}, X_3 = \begin{bmatrix} 8 & 9 \\ 10 & 11 \end{bmatrix}$$

Unfolding tensor X, in mode-1 unfolding (unfolding along column fibers) is as follows:

$$X_{mode-1} = \begin{bmatrix} 0 & 2 & 4 & 6 & 8 & 10 \\ 1 & 3 & 5 & 7 & 9 & 11 \end{bmatrix}$$

Mode-2 unfolding (unfolding along row fibers) is as follows:

$$X_{mode-2} = \begin{bmatrix} 0 & 1 & 4 & 5 & 8 & 9 \\ 2 & 3 & 6 & 7 & 10 & 11 \end{bmatrix}$$

Mode-3 (unfolding along tube/frontal slice fibers) results as follows:

---

[10] Note that to calculate unfolding, we have used Tensorly (http://tensorly.org/stable/index.html) , and it orders as z, x, and y. Therefore, if you use another approach the order might be different.

$$X_{mode-3} = \begin{bmatrix} 0 & 1 & 2 & 3 \\ 4 & 5 & 6 & 7 \\ 8 & 9 & 10 & 11 \end{bmatrix}$$

n-mode product (multiplication)

n-mode product (denoted by $\times_n$) is used to multiply a tensor by a vector or matrix. If we use the *n*-mode product to multiply a tensor with a vector, it is presented as $\bar{\times}_n$. The *n*-product of a tensor and matrix will be a matrix, and the *n*-product of a vector (v) and tensor ($X.v$) will also be a matrix. For *n*-mode multiplication, the tensor's dimensionality in mode n must be the same as the vector's or matrix's dimensions.

For example, assume we have a matrix $A = \begin{bmatrix} a & b \\ c & d \end{bmatrix}$, and we intend to compute $X \times_1 A$.

This mode-1 multiplication Y_1 can be carried out by multiplying the "frontal slice" to the matrix, i.e., $Y_1 = X_1 . A = \begin{bmatrix} 0 & 1 \\ 2 & 3 \end{bmatrix} . \begin{bmatrix} a & b \\ c & d \end{bmatrix} = \begin{bmatrix} 0.a + 1.c & 0.b + 1.d \\ 2.a + 3.c & 2.c + 3.d \end{bmatrix}$,

respectively Y_2, and Y_3 will be calculated. Y_1, Y_2, Y_3 are a column of the new tensor that is the result of mode-1 product between X and A.

To calculate $X \times_2 A$ we use "lateral slice", and thus we have the following:

$$Y_1 = \begin{bmatrix} 0 & 2 \\ 4 & 6 \\ 8 & 10 \end{bmatrix} . \begin{bmatrix} a & c \\ b & d \end{bmatrix}, \text{ and } Y_2 = \begin{bmatrix} 1 & 3 \\ 5 & 7 \\ 9 & 10 \end{bmatrix} . \begin{bmatrix} a & c \\ b & d \end{bmatrix}$$

Respectively, to calculate $X \times_3 A$ we can use "horizontal slices", and thus we have the following:

$$Y_1 = \begin{bmatrix} 0 & 1 \\ 4 & 5 \\ 8 & 9 \end{bmatrix} . \begin{bmatrix} a & c \\ b & d \end{bmatrix}, \text{ and } Y_2 = \begin{bmatrix} 2 & 3 \\ 6 & 7 \\ 10 & 11 \end{bmatrix} . \begin{bmatrix} a & c \\ b & d \end{bmatrix}$$

As another example, we can calculate mode-1 product (column fibers) of tensor X with vector $v = [10,10]$. The result is a matrix because multiplying a third-order tensor with a vector results in a matrix, as shown in the following.

$$X_1 \bar{\times}_1 v = \begin{bmatrix} 0 & 1 \\ 2 & 3 \end{bmatrix} . [10,10] = [10,50],$$

$$X_2 \bar{\times}_1 v = \begin{bmatrix} 4 & 5 \\ 6 & 7 \end{bmatrix} . [10,10] = [90,130] \text{ and}$$

$$X_3 \bar{\times}_1 v = \begin{bmatrix} 8 & 9 \\ 10 & 11 \end{bmatrix} . [10,10] = [170,210]$$

Therefore, the result is: $X \bar{\times}_1 v = \begin{bmatrix} 10 & 50 \\ 90 & 130 \\ 170 & 210 \end{bmatrix}$

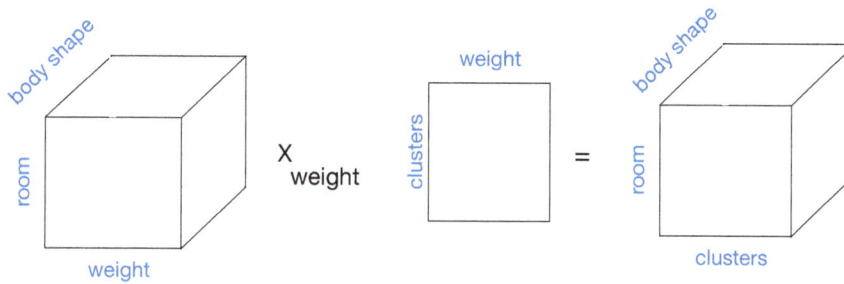

Figure 7-26: n-mode product of a matrix and tensor. The 'n' corresponds to the mode associated with weight and thus the target fiber will be the weight fiber.

You might ask why we need to learn these nonsensical mathematical calculations. Consider an example that we have a tensor composed of three data elements (chicken weight, location room, and body shape). We also have a matrix that holds chicken weights and the clusters those weights belong to (let's say we clustered them before based on their weight). If we intend to get rid of the weight and study the correlation between body shape and chicken rooms in our aviculture, we can make a mode-n product of the tensor and matrix, where 'n' corresponds to the mode associated with chicken weights, as is shown in Figure 7-26.

If we multiply the tensor above (room, weight, body shape) by a vector of weight, we will get a matrix composed of room and body shape.

Tensor Decomposition Methods

Now that we have learned some good concepts, we can switch to tensor decomposition methods. A tensor is called a "rank-1" tensor if it can be written as the outer product of vectors. For example, if we have $X = v_1 . v_2 . v_3$. Tensor X is rank-1 because it has been written as the outer product of three vectors, v_1, v_2, and v_3.

Similar to other decomposition techniques, Tensor decomposition algorithms extract a small subset of the tensor data, which is descriptive enough. There are many decomposition methods, but we explain ones that are more in use, including *Tensor Rank Decomposition*, *Tucker Decomposition*, and *Tensor Train* [Rehman '16].

Canonical or Parallel Factor Decomposition (CP)

Canonical Decomposition (CANDECOMP), or Canonical Polyadic Decomposition (CPD) or Parallel Factor Decomposition (PARAFAC), or Tensor Rank Decomposition [Hitchcock '27], [Harshman '70], [Carroll '70] has been discovered several times, it has different names. From now on, we will refer to it with CP. This method decomposes a tensor into a sum of rank-1 tensors (vectors).

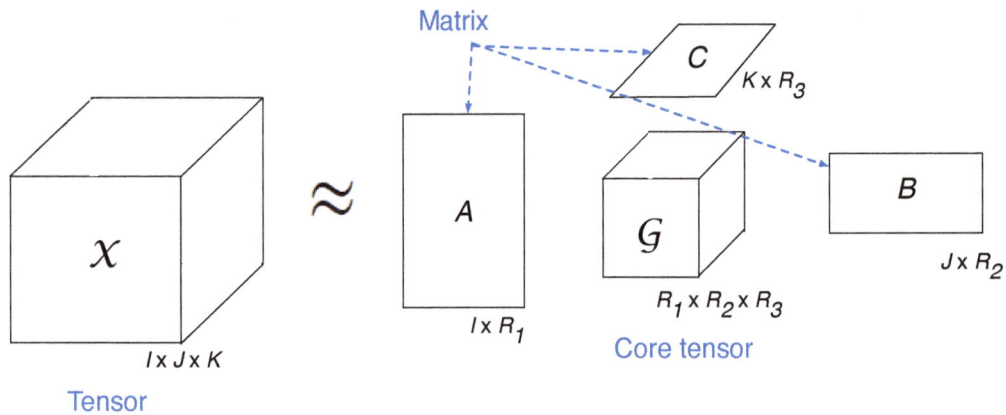

Figure 7-27: CP decomposes a tensor X into a three matrices and a core tensors which is compressed representation of the original tensor.

CP is a generalization of SVD matrix decomposition for tensors. It provides a minimum length combination of rank-1 tensors. Figure 7-27 visualizes this decomposition method, and it is formalized as follows:

$$X \approx \sum_{i=1}^{r} u_i \cdot v_i \cdot w_i$$

In CP decomposition, we have r, which we have described is the canonical rank value. Usually, there is no specific algorithm to determine tensor rank directly, and identifying a canonical rank will be done with trial and error, i.e., NP-hard problem [Håstad '90]. This is why the sign $\approx$ (approximately equal) is being used because it is not guaranteed that we get back exactly the same tensor if we compose the original tensors from decomposed rank-1 tensors.

Most of the computational complexity overhead of CP decomposition is on identifying the optimal r. The computing algorithm can start with $r = 1,2,3,\dots$ until it gets the r value that stays closer to the original tensor. The software package we use for tensor decomposition might need us to give the r as a hyperparameter. The computational complexity of CP decomposition is $O(n^3)$.

Tucker Decomposition (TD)

The Tucker Decomposition (TD) decomposition method was introduced in 1963, and later, it has been further refined [Tucker '66]. It is similar to PCA but for tensors. It decomposes a given tensor X into a smaller tensor G, which is known as a core tensor, multiplied by a matrix along each of its modes (dimensions).

These matrices are usually orthogonal, and thus, some called TD decomposition as a higher-order PCA or three-order PCA. Note the matrix sizes and their relation with the original tensor size. TD can be written mathematically as follows:

$$X_{I,J,K} \approx \sum_{r_1=1}^{R_1} \sum_{r_2=1}^{R_2} \sum_{r_3=1}^{R_3} G_{r_1 r_2 r_3} \cdot A_{i r_1} \cdot B_{j r_2} \cdot C_{k r_3} \approx G \times_1 A \times_2 B \times_3 C$$

370

As we have explained before, $\times_1$ is the 1-mode product, $\times_2$ is the 2-mode product, and $\times_3$ is the 3-mode product. Decomposed matrices in TD are not unique. Therefore, there is no guarantee that we can recompose back to X from the decomposed matrices A, B, C, and tensor G. This is why '$\approx$' sing is used instead '=' sign.

Similar to CP, all three 'r's are hyperparameters and require hyperparameter tuning to identify the best combination of them.

There are two other variations of Tucker decomposition, Tucker 2 decomposition and Tucker 1 decomposition [Kroonenberg '80]. Tucker 2 decomposition has one matrix as the identity matrix. Tucker 1 has two matrices as an identity matrix. Earlier, we explained that the identity matrix is denoted by I, and it is a $n \times n$ matrix in which data points in its main diagonal are '1' and the rest of its elements are 0. It is similar to 1 when we do multiplication and does not have any effect on the data. Therefore, we can neglect it for the calculation. Figure 7-28 presents Tucker 2 and Tucker 1 decompositions, and we removed identity matrices there.

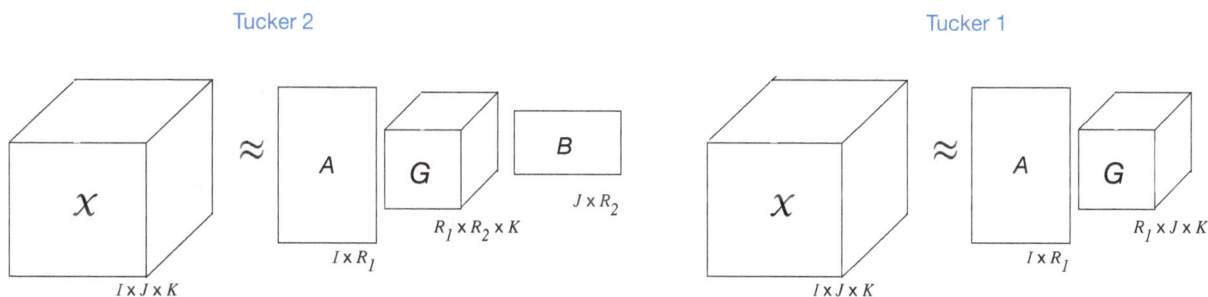

Figure 7-28: Tucker 2 and Tucker 1 variations of Tucker decomposition.

TD scales exponentially as the number of dimensions increases, thus inefficient for more than three-dimensional tensors [Oseledets '11]. Assuming we have d dimensions, each dimension has n indices, and r presents the tensor rank, the TD computational complexity is $O(dnr + r^d)$.

Tensor Train Decomposition (TT)

Tensor Train Decomposition (TT) [Oseledets '11] can be more efficient in terms of storage and computational complexity for certain types of high-dimensional tensors. It performs the decomposition by converting a tensor into a *Tensor Train (TT)* or *Matrix Product State (MPS)* format. TT format transforms the original multidimensional tensor into a chain of products of three-dimensional tensors. In simple words, it converts a tensor into a product of *ranked matrices,* and each matrix has three dimensions. We can formalize TT decomposition as follows: $X(i_1, i_2, \dots i_d) = G_1(i_1)G_2(i_2)\dots G_d(i_d)$. There is a chain of dimensions related together in TT decomposition. Therefore, we can rewrite the previous definition as follows:

$X(i_1, i_2 \dots i_d) = \Sigma_{\alpha_0, \alpha_1, \dots \alpha_d} G_1(\alpha_0, i_1, \alpha_1)G_2(\alpha_1, i_2, \alpha_2)\dots G_d(\alpha_{d-1}, i_d, \alpha_d)$. α indices are referred to as *auxiliary indices*, and each of these indices appears (maximum) twice in the composed chain of three-dimensional tensors. Each G presents a three dimensional tensor, except the first and last G, which are matrices. Therefore, α_0 and α_d are one, to indicate G_1 and G_d are matrices.

Assuming the original tensor has d dimensions, with n indices, and r is the maximum TT-rank the computational complexity of TT decomposition is $O(dnr^2)$.

Summary

In this chapter, we describe dimensionality reduction and data decomposition, which are two of the main requirements for working with large real-world datasets. Dimensionality reduction refers to the task of transforming multidimensional data into lower-dimensional data. For example, tabular data that has 300 columns is very complex for an algorithm to analyze. With the assistance of dimensionality reduction, we might be able to reduce its column numbers to five columns and then feed them to a machine-learning algorithm.

Dimensionality reduction methods are either linear projection or manifold projection. We described two linear projection methods, i.e., Principal Component Analysis (PCA) and linear discriminant analysis (LDA). There are some datasets we cannot use projection-based dimensionality reduction methods, such as, image data. For these data, we can use manifold dimensionality reduction methods. We described three manifold projection methods, including LLE, t-SNE, and UMAP.

The principal component analysis (PCA) calculates the covariance matrix based on all the data points, then decomposes the covariance matrix to eigenvalues and eigenvectors, and the eigenvectors are the principal components. Finally, it selects the first k principal components according to the k number of dimensions we need. The problem with PCA is that when the dataset is too large, we can't fit all the data in memory, and sometimes, the PC lines are not discriminatory enough to separate the data. One approach to deal with it is to use IPCA instead of PCA, which divides the dataset into smaller subsets (mini-batches) and loads each subset into memory, then does the PCA for each subset. Another linear approach is Linear Discriminant Analysis (LDA), which can be used on two-dimensional data to project or generalize for more than two-dimensional data.

As a manifold dimensionality reduction method, first, we describe Locally Linear Embedding (LLE). It assumes all data points are characterized as a linear combination of their neighbors. LLE tries to identify the non-linear relationship between data points in high dimensional space and tries to keep this relation in low dimensional space as well. It maintains this relationship by assigning a linear weight to each neighbor data point. Another manifold dimensionality reduction method is tSNE, which is based on the assumption that similar data objects are presented by high probability and dissimilar data objects will be presented by low probabilities. tSNE measures the similarities between every data point in a high dimensional space, and then the algorithm projects all data points on the lower dimension. Next, it focuses on moving the position of data points in lower dimensional space to take into account their local distances. The third algorithm we explained is UMAP. It works for datasets that have a Riemannian structure. We can think of Riemannian structures as manifold planes that are smoothly and continuously connected to each other. It operates similarly to tSNE, and we described it compared to tSNE. Table 7-2 summarizes linear and manifold dimensionality reduction methods along with their computational complexities.

Algorithm Name	Type of Projection	Computational Complexity
PCA	Linear	$O(m.n + m^2)$
LDA	Linear	$O(m.n.t + t^3)$
LLE	Manifold	$O(n.logn)t + O(D.n.k)^3 + O(dn)^2$
tSNE	Manifold	$O(dn^2)$
UMAP	Manifold	$O(d.n^{1.14})$

Table 7-2: Dimensionality Reduction methods and their computational complexity

Afterward, we discussed data decomposition for signal and time series data. The decomposition could be associated with loss of accuracy as well, but it is cost-effective, and usually, in most signal and time series analyses, we decompose the original data into another format, which is easier for the machine learning algorithm. Piecewise Aggregate Approximation (PAA) and Symbolic aggregate approximation (SAX) are common methods to summarize time series and smooth them for comparison and similarity analysis. Fourier transform decomposes a single signal into a set of sinusoids (sinus-based signals), and it is very common to use to prepare a signal for the machine learning algorithm. We also state that Wavelet Transform has lots of applications in the data science community due to its time awareness.

Next, we move to matrix decomposition approaches and describe Cholesky Decomposition, Non-negative Matrix Factorization (NMF), and Singular Value Decomposition (SVD). Topic Modeling can be considered as clustering for unstructured document data, and some of these algorithms, such as LSA, are based on matrix decompositions. We described LDA as another topic modeling approach.

The last part of this chapter focused on tensor decompositions. Tensors are used to store multidimensional data. Processing a tensor is challenging and very resource intensive. In that section, first, we introduced four operations on matrices including Kronecker product, Khatri-rao product, Hadamard product, and n-mode product. Next, we briefly explain the CANDECOMP/ PARAFAC Decomposition method, which decomposes a tensor into a sum of rank-1 tensors. Tucker Decomposition is followed by Tensor Train Decomposition, which performs the decomposition by converting a tensor into a tensor train format.

Further Reading and Watching

* A valuable and understand-to-easy tutorial for learning PCA is written by Lindsay I. Smith, entitled: "Tutorial on Principal Components Analysis". You can find it online on many different web pages.

* Olga Veksler has a useful tutorial with examples about FLD and MLD. The tutorial is simple and easy to learn. You can find it here: http://www.csd.uwo.ca/~olga/Courses/CS434a_541a/Lecture8.pdf

* After trying to learn Fourier transformation and signals through millions of online tutorials and books, we ended up using *The Scientists & Engineer's Guide to Digital Signal Processing* book by Smith [Smith '97]. It is a very good book if you intend to learn signal processing in depth.

* Josh Starmer's video explanation on tSNE is among the best explanations to understand this method; you can find it here: https://www.youtube.com/watch?v=NEaUSP4YerM

* To understand UMAP in more detail, you can check a more detailed explanation of this algorithm at https://umap-learn.readthedocs.io/en/latest/how_umap_works.html

* George Dallas has a good tutorial with examples of wavelet transformation and its differences with FFT. https://georgemdallas.wordpress.com/2014/05/14/wavelets-4-dummies-signal-processing-fourier-transforms-and-heisenberg.

* Charles F. Van Loan (http://www.cs.cornell.edu/cv) has a tutorial about tensors on his home page. We have used lots of his hints to construct the Tensor Decomposition section.

* There is another very good tutorial for Matrix and Tensor analysis provided by Faloutsus et al. from Carnegie Mellon University: https://www.cs.cmu.edu/~christos/TALKS/SIGMOD-07-tutorial.

Part iii: Supervised Learning

Please accept our warm welcome into supervised learning, which has the most focus on machine learning algorithms. Most machine learning courses or books start with supervised learning. We did not follow that path, because our real-world experiment shows there is no real-world application that we fed the data directly into a supervised learning model and get the result. We use a pipeline of unsupervised and supervised learning approaches for a project. Therefore, in this book, we prioritized learning unsupervised models and then moved to supervised ones. The good news is we will learn a lot about the foundations and slowly get prepared for deep learning, which is the most active area in machine learning from when we started to write this book in 2017 until 2024 when we are revising this text.

When using supervised learning, before feeding the data into the algorithm, we need to make a substantial effort to label the data. The objective of a supervised learning algorithm is to employ the previously labeled data (trained data) to create a "model". Then, the newly built model will be used to test the quality of its labeling by comparing its labels with test data labels.

We have explained in Chapter 1 that a model, which is the result of training supervised learning, is a piece of knowledge, such as a mathematical representation of the data (i.e., equation) or a set of if-then-else rules. The model compares the given unlabeled data (test data) with the labeled data (trained data) and assigns labels to the unlabeled data. Usually, the cost of labeling data is high, or we might not have access to enough data to label them. However, increasing the number of labeled data is usually correlated with increasing the accuracy of our model.

We should be familiar with a few concepts to learn supervised learning, and we briefly explain some of them in the following.

Single-shot learning or *one-shot learning* is a type of supervised learning that focuses on using the minimum number of possible labels (one data will be labeled) to train the data. If we use only one labeled data point and employ a supervised learning algorithm to train the other data points, it is called one-shot learning. Photo tagging in social media or online albums is a good example of using a one-shot learning. Online photo tagging is a very useful approach, and we highly recommend doing it because it makes spying agencies' jobs much easier. "Sharing is caring", especially for social media corporations.

Instance-based learning is a type of supervised learning that uses the entire dataset or a representation of the dataset (dimensionality reduced dataset) as a model and not just the labeled data objects, e.g., the k Nearest Neighbor (kNN) algorithm. Instance-based learning algorithms compare the unlabeled data (test data) with labeled ones (train data) in the memory and then assign them the labels. These are also called *lazy learning* methods because these algorithms do not build a concise model ahead of time but wait until a prediction is requested and use the data to assign a label.

Semi-supervised learning algorithms are a set of supervised learning that combines supervised and unsupervised learning. Usually (not always), in semi-supervised learning, we use one of a few labels, and the algorithm tries to build the model using those labels. It leverages a small amount of labeled data along with a larger amount of unlabeled data during the training process. The key idea of semi-supervised learning is to use the structure and distribution of the unlabeled data to understand the dataset better and improve the learning accuracy with limited labeled data.

A supervised learning algorithm works as follows:
1- First, labels were assigned to the data objects of the dataset.
2- Next, the target dataset is divided into test and train datasets.
3- The supervised algorithm uses the trained dataset and builds the model.
4- The algorithm uses the trained model, along with the test dataset, to measure its accuracy. In other words, the algorithm-assigned labels will be compared with the original labels, and the result will be reported. Then, if required, the model will be changed, more data will be added, and previous steps will be repeated.

Supervised learning is mainly categorized into two branches: one is regression, which focuses on modeling and predicting continuous numerical data, and the other category is classification, which focuses on modeling and predicting discrete data. We start this part by explaining regression (Chapter 8), and then we switch to classification (Chapter 9).

Chapter 8: Regression, Regularization, and Optimization

As we have explained in Chapter 1, *classification* algorithms, and *regressions* algorithms are two main categories of supervised learning algorithms. The focus of this chapter is on regression algorithms. Regression is used to understand how a variable's value changes when values of other corresponding variables change. Provost and Fawcett [Provost '13] stated *"classification is about whether something will happen, whereas regression predicts how much something will happen."* In another good book, James et al. [James '13] stated that classification can be used to predict categorical (qualitative) variables, but regression can be used to predict numerical (quantitative) variables. We can conclude regression is used for continuous data modeling and classification for discrete data modeling.

Nevertheless, the border is blurred, and a machine learning algorithm can belong to classification and regression groups. For example, in this chapter, we explain Logistic regression as a regression method, but it is a classification method as well.

In this chapter, first, we get familiar with the concept of 'cost', 'loss' or 'objective' function. Next, we describe five well-known regression algorithms including 'Linear Regression', 'Polynomial Regression', 'Piecewise Regression' as linear models, and 'Logistic Regression' and 'Softmax Regression' as non-linear models. Our discussion is followed by explaining ARIMA, a regression model for time series analysis. Afterward, we describe devils in model building, i.e., overfitting and underfitting, and then we switch to regression analysis and its algorithms. Next, we describe regularization methods, including Ridge, LASSO, ElasticNet, and Non-Negative Garrote regularization methods. Finally, we conclude this chapter with optimization and describe the very famous Gradient Descent and Newton methods. While learning about optimizations, we refer to some algebraic concepts we learned in high school, so brace yourself for the algebra.

Algebra monster
wakes up

Objective, Cost, and Loss

Everything we do in life is associated with a cost. For instance, you decide to learn data science and begin by reading this book. This decision comes with the cost of spending your precious time and expending energy from your brain cells. In other words, you invest time (a cost) to learn machine learning (a gain). You might have decided, or may decide, to get married; this decision costs you your free time, which will be dedicated to another person. Moreover, for a marriage to be successful, you or anyone else must make some changes to your behaviors. It's impossible to maintain the same behavior after marriage as when you were single.

Something similar exists in algorithms as well. A mathematical function that *tries to maximize or minimize a variable* (e.g., accuracy of an algorithm, energy use of an algorithm, etc.) is called an *objective function*. In the context of machine learning, we can call any function an objective function, if a model is trying to optimize this function. If the objective function seeks to minimize the cost of a variable (where the higher its value, the more its cost). We may refer to this objective function as the *cost function*.

A *Loss score* is a score that defines the cost associated with a single prediction error and measure the cost of that single data point. In layman's terms, we can say loss functions measures how many mistakes we make. On the other hand, the cost function aggregates the losses over all data points, summing up the individual mistakes to measure the overall performance of the model.

Assuming an output of a model is presented with y, the differences between actual output values (y_{actual}) and predicted output values ($y_{predicted}$), specify the cost or loss. One single difference is the loss, and all loss values are constructing the cost.

In summary, while a loss function quantifies the error of a single prediction, a cost function encompasses the total error of the model across all data points. Although 'loss,' 'cost,' and 'objective' functions can sometimes be used interchangeably in the literature, a cost function includes the sum of loss scores across predictions, and both serve as types of objective functions that a model aims to minimize.

Epoch: A complete pass through the entire training set to calculate the cost function during the training phase is called an *epoch*. When calculating the cost function, each epoch involves using the data in the training set to compute the cost and use that cost to update the model's parameters. If a text mentions that a model was trained for 100 epochs, it means that the algorithm has completed 100 iterations over the training set, updating the model after each pass to reduce the cost function.

Linear Regressions

In Chapter 3, we have described that to find a relation between two variables, we use correlation analysis, which results in a single number to describe the relation. This single number is known as the correlation coefficient. This coefficient ranges from -1 to 1, where values closer to -1 or 1 indicate a strong relationship and a value of 0 indicates no relationship. However, sometimes there is a complicated relation among variables (like those relations between teenagers); in these cases, we use linear regression, which measures a change of one variable with respect to the other variable.

In summary, while correlation analysis assigns a score to describe the relationship between two variables, linear regression explains how changes in one variable are expected to affect the other variable. It's important to understand that correlation is about association, whereas regression is about prediction and causation (under certain assumptions)

In the following, we start from the simplest form of regression, which is univariate linear regression, then gradually increases the complexity of the regression models.

Univariate Linear Regression

Linear regression is the simplest linear model used for prediction. It predicts the *quantitative* variable y based on the single predictor variable x (or more than one single predictor variable). Y is also called a *response*, *dependent*, or *output* variable. X is called *predicator*, *input*, *explanatory* or *independent* variable. It is written and calculated with the following equation: $\hat{y} = \beta_0 + x\beta_1$, which we read as *regression y to x* or *y on x*. In the context of machine learning when we predict a variable we use a greek sign, called circumflex on top of that variable, e.g., $\hat{y}$ means a "predicted value of y".

what we want to model — predictor variable

$$y = x\beta_1 + \beta_0$$

slope   intercept

Here, we have two model parameters that a simple linear regression model should identify their best optimal values, i.e., β_0 (intercept) and β_1 (slope). Note that they are not hyperparameters[1] because the model will use a cost function to identify them and are not configured by users of the algorithm. These types of parameters are called *model parameters* or called *model weights*.

Model parameters are parameters whose optimal value will be identified by the algorithm, via the cost function, and we should not give them to the algorithm (unlike hyperparameters, which the user sets). The optimization algorithm (by using cost function) tunes model parameters. Later in this chapter, we describe the optimization algorithm. It is also common to use θ, which is a vector that presents all model parameters.

We can use linear regression or other linear models to perform prediction, describe a phenomenon, or even measure the effect size (see Chapter 3 to recall the effect size).

---

[1] Input parameters of an algorithm that need to be given by the user to the algorithm are called hyperparameters.

Let's look at an example scenario with linear regression. Recently, Mr. Nerd began to suffer from work stress. To solve his stress, he decided to make some changes in his life, and he started to read books before sleep. Previously, he has wasted his time on social media and fighting the holy war of useless discussions with trolls.

He speculates that reading the book could positively change his mood (the hypothesis). To experiment with this hypothesis, he has started to journal the number of books he read within his daily mood score. The result looks like the table on the right side of Figure 8-1. By plotting them he gets the plot on the left side of Figure 8-1.

It is clear that his mood is improving. Recently, he has finished the 9th book, he becomes curious to predict what his mood will be after the 20th book. He can use linear regression to answer his question, The question will be as follows: "what will be the mood score after reading 20 books?" A linear regression to predict his mood based on the number of read books can be written as follows: $mood-score \approx \#books \times \beta_1 + \beta_0$

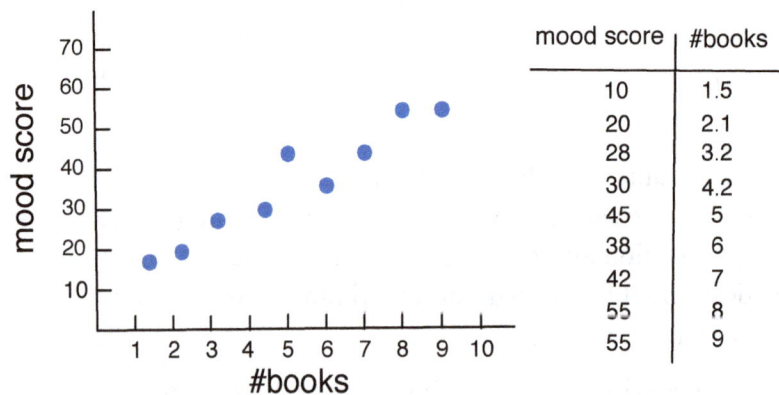

mood score	#books
10	1.5
20	2.1
28	3.2
30	4.2
45	5
38	6
42	7
55	8
55	9

Figure 8-1: Mr. Nerd's mood score based on the number of books he read.

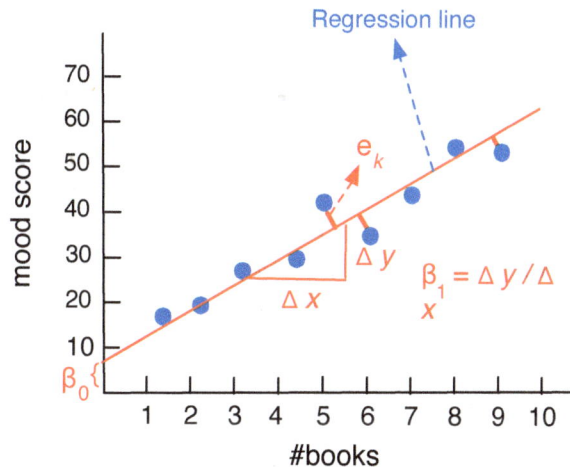

Figure 8-2: Figure 8-1 with regression line and its parameters.

The goal of linear regression is to find the optimal values for β_0 and β_1. In this example, the best value that a cost function of linear regression found is β_0 (intercept) = 7.369 and β_1 (slope) = 5.58. Later in this section, we will describe how the cost function identifies these parameters. Figure 8-2 shows what we have in Figure 8-1 with a linear regression line and two optimal values for β_0 (intercept) and β_1 (slope) parameters.

Therefore, assuming that there is no error, the algorithm can easily predict the mood score achieved by reading specific number books, via replacing these parameters in the linear equation. For example, it can calculate the mood score after reading 20 books, by follows:

$$mood - score_{(20th-book)} \approx 20 \times 7.369 + 5.58 = 152.96$$

Another application of regressions is to identify or describe a phenomenon in the dataset. Therefore, we can use regression to quantify the relation between input (predictor) and output variables. This will be reported as a hypothesis test and its p-values. Check Chapter 3 if you can't recall the use of p-value and hypothesis test. The hypothesis is written as follows:

H_0 = *There is no relation between input and output variables ($\beta_1 = 0$).*
H_1 = *There is a relation between input and output variables ($\beta_1 \neq 0$).*

For example in the book reading example, if the *p-value < 0.05,* we can infer there is a relation between the number of books he read and Mr. Nerd's mood score.

Assuming X (input dataset) is a $n \times m$ matrix. Linear regression will be presented as $n \times m$ matrix multiplication. The computational complexity of linear regression is close to $O(m^2 . n + m^3)$, which is calculated based on matrix decomposition and inversion. We don't go into the detail, just remember that a linear regression has *quadratic* complexity.

Model Parameters (Coefficients) Estimation

Now that we have answered his question about book reading's impact on mood score, we should also learn how the cost function identifies β_1 (slope) and β_0 (intercept) parameters. In other words, we use the cost function to fit a model to a training set. Different cost functions could be used to identify the best values for these two parameters (best coefficients to fit the model), but a basic and common one is *Residual Sum of Squares (RSS)* function. As we progress through this chapter, we explain more cost functions.

Figure 8-2 presents the described linear regression with its parameters highlighted. Check the red lines connecting each data point to the regression line in Figure 8-2, those e are called *errors* or *residuals*. RSS for n data points will be written as the sum of squared errors: $RSS = e_1^2 + e_2^2 + \ldots + e_n^2$ or $RSS = (y_1 - \hat{\beta}_0 - \hat{\beta}_1 x_1)^2 + (y_2 - \hat{\beta}_0 - \hat{\beta}_1 x_2)^2 + \ldots = \Sigma_{i=1}^{n}(y_i - \hat{y}_i)^2$, we could summarized it as follows:

$$RSS = \Sigma_{i=1}^{n}(y_i - (\hat{\beta}_0 + \hat{\beta}_1 x_i))^2$$

The mean of x is presented as $\bar{x}$, and mean of y is presented as $\bar{y}$, the predicted values are presented with a hat sign "^", e.g., the predicted value of y is presented as $\hat{y}$. To calculate the $\hat{\beta}_0$ and $\hat{\beta}_1$ which minimize RSS, we can use the following equation.

$$\hat{\beta}_1 = \frac{\sum_{i=1}^{n}(x_i - \bar{x})(y_i - \bar{y})}{\sum_{i=1}^{n}(x_i - \bar{x})^2}$$

$$\hat{\beta}_0 = \bar{y} - \hat{\beta}_1 \bar{x}$$

Remember that our goal is to minimize change parameters to minimize the cost function, and the cost function here is RSS. The regression library wraps all details in a simple function call, but the mathematical notion of linear regression is fairly easy to learn, and it is worth understanding the rationale behind linear regression.

It is also worth mentioning, to write it accurately, we can write the linear regression equation as follows: $y = \beta_1 x + \beta_0 + e$, assuming that e is the mean of error values. Also, some use approximation sign ($\approx$) instead of the equal sign, because there is no guarantee that the created regression line is the perfect line, thus if you like to avoid generalization even in your mathematical writings and show off you are knowledge, write the linear regression as follows.

$$y \approx \beta_1 x + \beta_0 + e.$$

We can also write a linear model as a transpose of matrix x (it is a vector) times the vector of coefficients ($\hat{\beta}$), as follows: $y = x^T \hat{\beta}$ or $y = x^T \hat{\theta}$. In the near future (Chapter 10) we will write β_0 as b and other intercepts (βs) are vectors of weights w, i.e., $y = wx + b$. Most of the time, we do not write e in the equation.

Now that we understand how we have estimated model parameters (β_0 and β_1), let us repeat that model parameter estimation will be done through a cost function and not by the user of the algorithm.

Multiple Linear Regression

The linear regression we have explained is the simplest regression, which has one input or independent variable. Most of the time, we have multiple input (predictor) variables. In these cases, we can use multiple linear regression to predict the output, i.e., *multilinear regression*. The multiple linear regression equation is written as follows:

$$y = \beta_0 + x_1\beta_1 + x_2\beta_2 + \ldots + e$$

We can summarize the summation as follows:

$$y = \beta_0 + \Sigma_{k=1}^{n} x_k \beta_k + e$$

To better understand the use of multiple linear regression, let's use an example. Assume Mr. Nerd is still looking to find the symptom of his stress. Therefore, he has studied his life in a bit more detail, and he found some factors that are contributing to his daily stress, including work, being active of social media discussions, physical pain in his body, and family issues. To model this amount of information, we can use the following multiple linear regression:

$$Stress = \beta_0 + \beta_1 . work + \beta_2 . social\ media + \beta_3 . physcial\ pain + \beta_4 . family - issues + \ldots + e$$

Assuming we have two predictors, we can visualize it with a surface plot as shown in Figure 8-3. We neglect family issues for the sake of visualization because more than three predictors are hard to visualize in two-dimensional picture. Previously, we have described two questions that can be answered with simple linear regression, (i) predicting the output based on the given input and (ii) describing the relation between x and y.

While employing multiple linear regression, usually, we need to answer some more questions [James '13]. For example, which one of the input variables x_1, x_2, ... is useful for predicting the y? Do all input variables help to explain the output variable? How well does our model fit the data? To answer these questions, the software package that implements regression, provides a p-value or F-statistic test score for each input variable. Then, we can decide which input variables are useful and which are not useful.

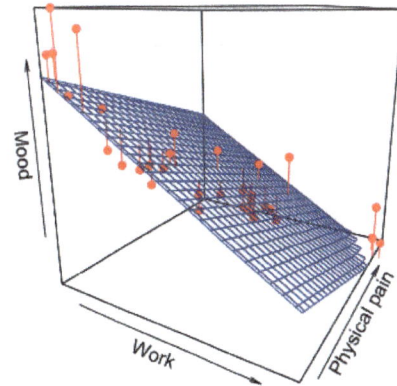

Figure 8-3: A multi linear regression example for Mr. Nerd and his mood score.

If there is no relation between input variables, all their βs should be equal to zero. If there is a relation, at least one of the βs is not zero. Therefore, we can write the following hypothesis:

H_0 = *There is no relation between input and output variables ($\beta_1 = \beta_2 = \ldots = \beta_n = 0$).*
H_1 = *There is a relation between input and output variables (at least one $\beta_x \neq 0$).*

This hypothesis test will be performed by F-statistic test or F-test, which is written as follows:

$$F = \frac{(TSS - RSS)/p}{RSS/(n - p - 1)}$$

In this equation $TSS = \Sigma_{i=1}^{n}(y_i - \bar{y})^2$, and $n - p - 1$ presents the degree of freedom, which in this case is the degree of freedom from error. The *f-value* > 1 means the null hypothesis is rejected (the alternate hypothesis is correct). The *f*-value ≤ 1 means the null hypothesis is correct, and thus the model has no predictive capability. It is recommended to check the output of regression and check *p*-value and *f*-value should be significant to accept the model. Note that the *p*-values and F-statistics are associated with the overall model and the individual coefficients, not directly with the input variables themselves.

Deciding About Model Variables?

The F-test is very useful for multiple linear regression. Assume the model did not pass the F-test and we have several input variables. Probably one or a few of them are responsible for the failure, and other ones are good to be used for building a model for multiple linear regression. Now a question arises: how can we identify which variables perform well in building the model, and which do not perform well?

For example, assume we build a model with three variables x_1, x_2, and x_3 as follows: $Y = \beta_1 . x_1 + \beta_2 . x_2 + \beta_3 . x_3 + \beta_0$. If the model fails with x_1, x_2, x_3 variables, we can remove x_1 and then test another model with x_2, x_3. Or we can remove x_2, x_3 and test another model with x_1, and so forth. In particular, for m numbers of variables we could have 2^m different models.

The best way to deal with model parameters is similar to the SBS and SFS which we have explained in Chapter 6, 'Wrapper Methods' section. Here, instead of features, we have model variables, and SFS style of parameter selection is called *Forward Stepwise Selection*, and SBS style of parameter selection is called *Backward Stepwise Selection.*

In some cases, two more parameters of a model together have a very strong impact on the output variable. This phenomenon is called *synergy effect* or *interaction effect* [James '13]. For example, a popular celebrity (x_1) is advertising a product on popular social media (x_2). In this scenario, the popularity of these two variables boosts product sales, and we can extend our model to emphasize their importance by writing $y = \beta_0 + \beta_1 x_1 + \beta_2 x_2 + \beta_3 x_1 x_2 + e$ instead of $y = \beta_0 + \beta_1 x_1 + \beta_2 x_2 + e$, because we want to emphasize the combination of x_1 and x_2. Usually, we need to experiment with both models (using interaction effect and not using interaction effect). Then, we can compare the *RSS* or other evaluation metrics of these two models and decide to use which one for the upcoming data (i.e., test dataset).

Linear Regression Challenges and Resolutions

There are some limitations in linear regression. We describe two of them that introduce new classes of regression. The first problem with linear regressions is that they can only predict numerical values (categorical values should be encoded). Sometimes we need to predict categorical data, e.g., if an email is spam or not spam, if the patient will die or survive the operation, etc. You might say we can map them to 0 and 1 and easily employ linear regression. That is not wrong, but what will happen if we can not draw a regression line between those binary states? The shape of the data points does not allow to draw the linear regression line. To handle this we use logistic regression, which we will explain shortly.

Another issue is what happens if we have more than one categorical predictor. For example, Mr. Nerd likes technology books, religious and science-fiction books. If we assign them numbers like *'1: technology'*, *'2: religion'*, and *'3: science-fiction'*. There is an ordering enforced on book genres. It might lead to a mistake that the algorithm assumes the distance between 'technology' and 'science-fiction' books is larger than the distance between *'technology'* and *'religious'* books. As another example, assume we try to predict a medical condition of emergency room patients; it could be '*accident injury*', '*seizure*', or '*stroke*'. They are very different information, and considering them all together in a linear regression equation does not sound like a wise decision because we are imposing an ordering that does not exist. Therefore, we should look for a better method that considers input variables as binary and not continuous variables. This will be handled by *logistic regression*, which we will explain later.

The second problem is that linear regression is designed to handle linear relationships between features, and can not handle non-linear relations. In other words, the linear regression we described is useful for linear data with a fixed slope (i.e., additive slope). However, sometimes there is no straight line to be able to model the data points. A curved line can resolve this and it will be handled by *polynomial regression*.

Polynomial Regression

Sometimes, the relationship between the predictor and response variable is not linear. Assume Mr. Nerd enjoys reading books, but if he reads too few books or too many books in a time period, such as a month, his stress level increases. He likes to keep a balance on his book-reading behavior. Again, he logs the number of books he reads per month, and he plots them as shown in Figure 8-4 (a).

It is clear that we can not draw a straight line and model Mr. Nerd's stress level in this scenario. Even if we draw as we did in Figure 8-4 (b), this line is not descriptive enough, and it is too far from data points. We need a line that can model his stress level more accurately, like Figure 8-4 (c) to (e). For these scenarios a straight line (linear regression) is not enough, and instead, we can use polynomial regression.

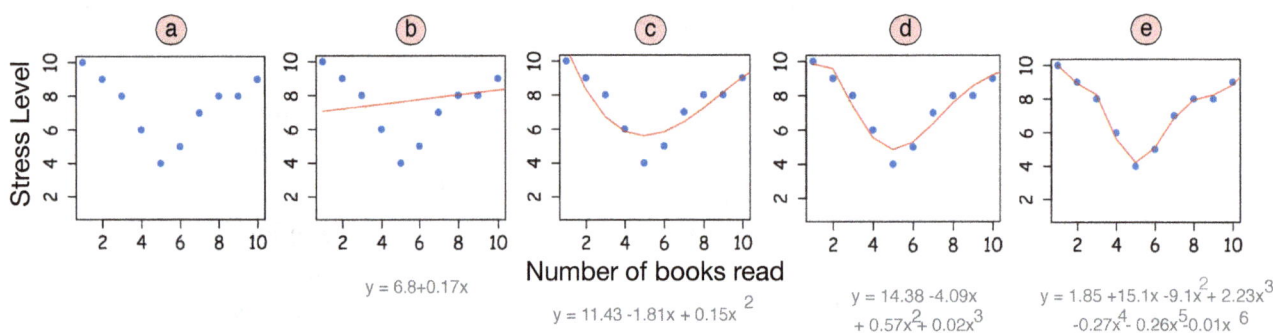

Figure 8-4: (a) plot of stress level and number of read books in a month, (b) Linear regression on the data, which is not properly representative of the data (underfitting). (c) Polynomial regression line with the degree of four, (d) Polynomial regression line with the degree of six, and (e) Polynomial regression line with the degree of six, which could cause overfitting. The equation of each plot is written at the bottom of each diagram in grey color.

A polynomial is a mathematical expression that consists of terms, where each term is formed by multiplying a coefficient (e.g., β) with a variable x raised to a non-negative integer power. These terms are then combined through addition or subtraction. Please keep this in your lovely brain: polynomial equation can produce a non-linear curve, and a polynomial regression can be formalized as follows:

$$y = \beta_0 + x\beta_1 + x^2\beta_2 + x^3\beta_3 + \ldots + \beta_d x^d + e$$

Note that here, we have only one x (input variable), but it has different powers. Having different powers means a curved line function, that can fit into the non-straight linear data points. In the above equation, d is called the degree of the polynomial, and *the larger the polynomial degree gets, the curve is getting more flexible*. In other words, x to a power larger than one gives us parabolic shapes (a sexy name for a curved or bell-shaped line), and thus, it can fit into the data with different shapes. x^2 is called quadratic, x^3 cubic, and x^4 quartic.

To better understand this phenomenon, consider Figure 8-4, which presents a regression model creation with different degrees. As we can see from this figure, the more we increase the degree

386

of polynomial, the more flexibility we have in the model to fit the sample data points (from Figure 8-4 (c) to Figure 8-4 (e)). Nevertheless, increasing it too much causes an unforgivable sin, i.e., overfitting, which we will explain later in this chapter.

Besides, note that increasing the degrees is associated with an increase in complexity. Assuming we have n data points and d degree, polynomial regression builds $((n + d)!)/(n! \times d!)$ features [Géron '17]. Therefore, if computational complexity is important, we should be careful and not increase the degree of polynomial regression too generously. For a polynomial of degree d and n data points, the time complexity of polynomial regression is $O(nd^2 + d^3)$.

Model Parameters (Degrees and Coefficients) Estimation

There are two types of parameters required to be identified in polynomial regression. The first is the regression "coefficients" ($\beta_0, \beta_1, \beta_2, \ldots, \beta_m$) and the second one is to identify the minimum number of "degrees" for the model ($x, x^2, x^3, \ldots, x^n$). The regression line should be the best representative of the observation (our data points) and this will be achieved through the optimal configuration of those two types of parameters. In the following, we describe each in more detail.

Coefficients

To get a good estimate for polynomial coefficients, we can use a cost function similar to the linear regression. A well-known cost function to estimate polynomial coefficients is the least square cost function or least square fitting. A polynomial regression can be written as a multiplication of matrices, and solving the equation can provide the values for βs.

For example, let's say we have a polynomial regression with degree m. y is the dependent variable and the βs are the coefficient for different nth powers of the independent variable x. By substituting the values for x and y (from our real dataset) and by solving the following m equations we can identify $\beta_0, \beta_1 \ldots \beta_m$ (we have learned in school matrix multiplication and how to resolve equations that have m unknown variables):

$$
\begin{bmatrix}
1 & x_1 & x_1^2 & \ldots & x_1^m \\
1 & x_2 & x_2^2 & \ldots & x_2^m \\
\vdots & \vdots & \vdots & \vdots & \vdots \\
1 & x_n & x_n^2 & \ldots & x_n^m
\end{bmatrix}
\cdot
\begin{bmatrix}
\beta_0 \\
\beta_1 \\
\ldots \\
\beta_m
\end{bmatrix}
=
\begin{bmatrix}
y_1 \\
y_2 \\
\ldots \\
y_n
\end{bmatrix}
$$

Note that we have the values of x and y available from our observed dataset, thus, we can substitute them, then perform matrix multiplication and end up having three equations with m unknown variables, i.e., $\beta_0, \beta_1 \ldots, \beta_m$. By solving this equation, we get the values for coefficients. For example, let's assume that we end up with the following variables for $m=3$. $\beta_0 = 0.1, \beta_1 = 0.3, \beta_2 = 0.12$.

We intend to predict: what will be the mood of Mr. Nerd after reading his 20th book in a month? By using two degrees of polynomial regression we will have the following:

$y = \beta_0 + \beta_1 x + \beta_2 x^2$ and by substituting β variables we will have 54.1 as his mood score:
$y = 0.1 + 0.3 \times 20 + 0.12 \times 20^2 = 54.1$

Degrees

A polynomial regression with *n-1* degree is able to model *n* data points, but this is typically not desirable in practice. As we have explained before, increasing the number of degrees is associated with a huge computational cost. Therefore, we should try to find the least number of degrees. One way to identify a reasonable degree is to perform the following two steps:

(1) Leave out some data points from the dataset and then calculate some polynomial regressions with different degrees, e.g., with x^2, x^3, x^4 and x^5. There is a method called 'leave-one-out' and one-by-one data points will be removed. However, this method is very resource-intensive, and it is recommended that it not be used.

(2) Add back those removed data points and check how much an error value (e.g. RSS) changes. For example, assume by using x^2, the error is 0.28, by using x^3, the error is 0.23, by using x^4, the error is 0.21 and by using x^5 the error is 0.46. We can conclude x^4 is the best choice, because it has the least amount of error while encountering new data.

This approach is very similar to the test and train that is being used in all supervised machine learning, but it is not exactly the same. In this case, we play with the training dataset and not the test dataset. Usually, the software that implements the polynomial regression library provides us with a parameter search function to automatize this process.

Piecewise, Segmented, or Non-Additive Regression

Despite the described limitations, linear regressions are very resource-efficient and are widely in use. Sometimes, instead of using polynomial regression, which is not as efficient as linear regression, we can use two or more linear regressions to get rid of the 'additive assumption' of linear regression. See Figure 8-5, which shows the amount of joy Mr. Nerd experiences from eating based on the calories of his food. He enjoys eating, but overeating reduces his joy as well.

The data points presented in Figure 8-5 are not additive because at around 110 calories, his joy starts to decline; before 110 calories, it was increasing. Therefore, we cannot model it with simple linear regression. Polynomial regression is also computationally complex, and it is recommended not to use it. However, by separating the dataset into two datasets, we can easily use two linear regression lines that fit the data appropriately. Figure 8-5 (b) presents the separator line, and Figure 8-5 (c) presents these two regression lines.

Sometimes, we can use more than one polynomial regression to model our dataset, and it is not just limited to linear regression. However, this doesn't seem a rational choice because when we don't care about computational complexity, we can simply use a polynomial regression.

Now, a question might arise: at what point do we need to separate the dataset? By visual inspection, we can recognize where in this small dataset, we can spot the distinctive point. For real-world datasets, which are usually too large and multidimensional, we do not have such a

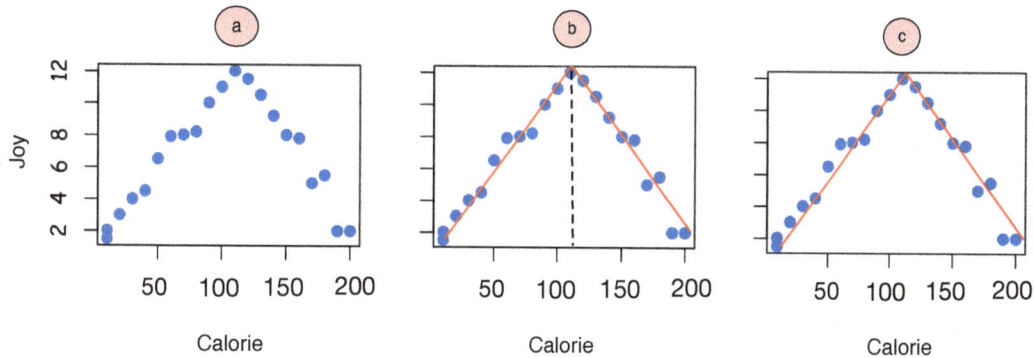

Figure 8-5: (a) A dataset that can are not additive, and thus, we can't use a simple line to fit them. (b) The knot is specified to separate the dataset into two sub-datasets. (c) The same dataset can be modeled with two linear regressions.

simple solution and can't visualize them. Therefore, we need a more subtle approach to design the distinctive border for a regression (the dotted line in Figure 8-5 (b)).

We can start by selecting a random data point and separating the dataset into two subsets from that random data point. Then, we draw a linear regression line for each subset and calculate R, SSE, F-statistics, or other evaluation metrics for each of these regression lines. We will explain these evaluation metrics later in this chapter. We change the data point, selecting a different one from the dataset, and determine various evaluation metrics. The point that yields the minimum sum of these metrics will serve as the optimal discrimination point, allowing us to divide the dataset into two subsets for which we can apply individual linear models.

For example, assuming we have a dataset and we randomly select three data points to separate it into two subsets.

$d_1 : f\_statistics_1 + f\_statistics_2 = 3.1 + 2.5$
$d_2 : f\_statistics_1 + f\_statistics_2 = 3.2 + 2.1$
$d_3 : f\_statistics_1 + f\_statistics_2 = 3.1 + 2.6$

Among the three following data points, i.e., d_1, d_2, d_3. The d_3 is the best data point for separation because it has the maximum F-statistics.

The point where one linear regression breaks and another starts is called a *knot*. We can also use the piecewise regression for polynomial regressions (See Figure 8-6). Assuming c is the knot, and based on the value of x_1 and c we should have two models. Therefore we write the following equation[2] for a polynomial regression with two variables x_1 and x_2.

---

[2] If you encounter the power inside parentheses, e.g., $x^{(k)}$, it is not x to the power of (k), it is read as kth index of x. Sometimes, the index is not written as a subscript, and if it is written as super-script, it should be inside parentheses.

$$y = \begin{cases} \beta_0^{(1)} + \beta_1^{(1)}x_1 + \beta_2^{(1)}x_2 & if \quad x_1 > c \\ \beta_0^{(2)} + \beta_1^{(2)}x_1 + \beta_2^{(2)}x_2 & if \quad x_1 \leq c \end{cases}$$

Each of these regression models in a piecewise regression is called a *spline*. We explained linear regression, but polynomial regression can have the same attribute too. For example, by looking at Figure 8-6, we see a single polynomial function can't model all the datasets properly. However, using a piecewise regression, we can use two polynomials to model the dataset properly.

Although this approach seems flexible, it is useful when we have access to the entire dataset (population) and not just the sample dataset. Somehow, we are hacking a polynomial or linear regression to handle the dataset, and some mathematicians do not agree with this approach for different reasons, including giving biased regression coefficients that need shrinkage (we explain later); it yields a very high R^2 value that is badly biased, and so forth. Nevertheless, who cares? Since it works, feel free to use it. Just be careful not to use it in front of a picky mathematician.

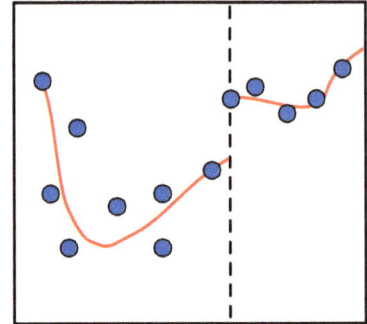

Figure 8-6: Piecewise polynomial regression example.

Note in all examples we explained here, we assume there is only one knot, and thus, we separate the dataset into two subsets. There is no limitation on the number of knots and we can separate a dataset into more than two subsets with several knots [James '13].

Evaluating the fitness of training sets in linear models.

Once we have implemented our regression, we must (as with all machine learning algorithms) evaluate the accuracy of our model. Such an evaluation in regression algorithms is called *model evaluation*, a.k.a., measuring *model fitness*. In the this section, we describe common approaches used for evaluating linear models, including Residual Standard Error (RSE), Coefficient of Determination or R-squared (R^2), Root Mean Square Error (RMSE), and Mean Square Error (MSE).

These metrics are applied to both the *training set* to understand how well the model fits the data it was trained on, and the *test set* to assess how well the model generalizes to unseen data. Although the same metrics are used for both sets, the interpretation of these metrics can vary depending on whether we are evaluating model performance during training or testing.

Model fitness evaluator

Regression Algorithm

Residual Standard Error (RSE)

RSE is an average number of data points that deviated from the regression line. It is called the *lack of fit* measures and is calculated as follows:

$$RSE = \sqrt{\frac{RSS}{n-p-1}} = \sqrt{\frac{\Sigma_{i=1}^{n}(y_i - \hat{y}_i)^2}{n-p-1}}$$

Here, n is the number of data points, p is the number of model parameters (not including the intercept), and thus $n-p-1$ is the degree of freedom from error. If we have one input variable and one intercept, we can write the degree of freedom as $n-2$.

In fact, RSE measures the differences between y_i and $\hat{y}_i$ (y_i is the original data point and $\hat{y}_i$ is the predicted value, which is a data point located on the regression line). The closer two variables y_i and $\hat{y}_i$ are, the better the model will be at the end. Thus, a smaller RSE is better. The *RSE* for Mr. Nerd's linear regression (Figure 8.1) is 5.074 on 7 degrees of freedom.

What is the degree of freedom in the context of regression analysis? The degrees of freedom, in this context is equal to the number of data points minus the number of parameters that will be estimated by the model. In the context of regression analysis, a parameter is estimated for every model's variable, and each parameter costs one degree of freedom. Therefore, including lots of variables in a regression model reduces the degrees of freedom available to estimate the parameters' variability. For example, if our sample size is 12 and our model has 3 parameters. The degree of freedom is $12 - 3 = 9$.

R^2

Another measure for model fitness is the *Coefficient of determination* or R^2 (R squared). RSE depends on the value of y strongly. Therefore, it is not clear what part of the linear regression equation contributes more to the RSE score. R^2 tries to mitigate this challenge by using the *total sum of squares (TSS)*, i.e., $\Sigma_{i=1}^{n}(y_i - \bar{y})^2$ divided by the *regression (or residual) sum of squares (RSS)*, i.e., $\Sigma_{i=1}^{n}(y_i - \hat{y}_i)^2$. Therefore, R^2 will be written as follows:

$$R^2 = 1 - \frac{RSS}{TSS} = 1 - \frac{\Sigma_{i=1}^{n}(y_i - \hat{y}_i)^2}{\Sigma_{i=1}^{n}(y_i - \bar{y})^2}$$

Recall that y is a dependent (output) variable, $\bar{y}$ is the mean of y, and $\hat{y}$ is the predicted value of y. The predicted value is presented with the hat symbol " $\hat{}$ ", e.g. $\hat{y}$.

R^2 returns a value between 0 and 1, which describes *how close the data points are to the regression line*. Higher values of R^2 indicate that our data points are closer to the regression line, and thus, the model is performing well. Mr. Nerd's R^2 is 0.9039 (based on data from Figure 8-1), and since it is near 1, which means that the R^2 evaluation shows high accuracy.

Root Mean Square Error (RMSE) and Mean Square Error (MSE)

For polynomial regressions, we can use RMSE to measure the accuracy. It is recommended to do the experiment for several different polynomial degrees and select the lowest RMSE. RMSE is the standard deviation of residuals or prediction errors. Check Figure 8-2 to recall what is residual. In short, residuals are the distance of data points to the regression line. Assuming $\hat{y}_i$ as a predicted value and y_i is the ith observed data points, the formula is written as follows:

$$RMSE = \sqrt{\frac{\Sigma_{i=1}^{n}(\hat{y}_i - y_i)^2}{n}}$$

When we encounter *Mean Square Error (MSE)*, it is the same as RMSE, just its square root has been removed.

$$MSE = \frac{\Sigma_{i=1}^{n}(\hat{y}_i - y_i)^2}{n}$$

These tests are also called *goodness-of-fit* tests, similar to the Chi-square that has been explained in Chapter 3, which is another test for goodness-of-fit.

NOTE:

* By increasing the number of labeled data or sample data for a regression algorithm, the chance of getting a wrong prediction decreases because we are increasing the likelihood of newly arrived data points being considered to be similar to already labeled data.

* Depending on the goal of our prediction, we might use more than one regression model or even algorithm to estimate the output variable. Therefore, do not hesitate to experiment with more than one model and then choose the best one. We will discuss this later in this chapter.

* Linear regression and polynomial regression are called *parametric methods*. Parametric methods assume that our prediction function has a form or shape. Non-parametric methods, such as Piecewise regression, do not make any assumption about the prediction function form or shape. The model that we choose might not correctly reflect the underlying dataset, which is the disadvantage of parametric methods. Nevertheless, since the non-parametric method does not have any assumptions about the model, we might end up using a large number of data points (train dataset) to create the correct non-parametric model, and it is hard to prepare a large number of labeled data points.

* Linear models, especially linear regression, are one of the most popular machine learning algorithms. Interestingly, in terms of quality, they are as good as non-linear models, which we will explain later. Since they are parametric, they do not need a large amount of data to train the model, and operating with a small dataset makes them very attractive algorithms.

* The simplest form of linear regression is Ordinary Least Squares (OLS), which estimates the relationship between X and Y by minimizing the sum of squares between real values (the blue dots in Figure 8-2) and predicted ones (the blue dots projections on the red line in Figure 8-2).

* Sometimes, to identify which model from a set of models can fit better into our dataset, we need to plot their residuals. Residuals are useful to identify the linearity of the model. After plotting residuals, we can check if the residuals are linear or not, and we can find a pattern in residuals. If there is a pattern in residuals, this is a sign that the dataset is non-linear. However, we described that linear regression is amazingly efficient. To enforce linearity in the data, we can apply a transformation on the data, e.g., log transformation or L_2 norm transformation. Check Chapter 6 to recall transformation. Another useful application of plotting residuals is plotting them based on time to ensure that errors are not correlated. Correlated residuals based on time is a phenomenon called *tracking* [James '13]. If there is tracking exists in our residuals, then we have no guarantee about the confidence of our model, and thus, we should think about another model. Tracking phenomena is common in time series analysis.

* Having two highly correlated variables negatively affects the accuracy of the multilinear regression and it is better to remove one of the highly correlated variables. Another recommended approach is to combine *collinear variables* (highly dependent variables) and create a new input variable, i.e., feature engineering, which we described in Chapter 6.

* Despite the attractiveness and flexibility of polynomial regression, it is prone to the overfitting problem, and therefore, it is not widely in use, unlike linear regression, which everybody loves. More about overfitting will be described later in this chapter; now, keep in mind that polynomial regression is prone to overfitting.

* Polynomial regression is more sensitive to outliers and noise in comparison to linear regression. It could be another reason that linear regression is usually favored over polynomial regression.

* Mathematicians are in a constant struggle to *estimate something that is non-linear by using linear tools*. It can be another motivation behind using linear regression significantly more than other regressions, especially among data scientists with a good understanding of math. Nevertheless, we should not forget that linear models assume the structure of the data will remain the same, which is not the case for real-world applications.

* Regression is a type of approximation. Approximation is the process of finding an efficient estimate of the value of the function at a certain unknown point by leveraging the information we have at a known point about the function.

AutoRegressive Integrated Moving Average (ARIMA)

ARIMA [Box '70] is a popular algorithm used to predict or model time series. Since we are talking about linear regression, it is good to describe this popular algorithm. If you are not interested in working with time series, feel free to skip this section. Before we describe the details of ARIMA we need to become familiar with two concepts, *extrapolation* and *interpolation*.

Extrapolation and Interpolation

Extrapolation is estimating the next change of a variable *beyond* the observation range, the value of a variable on the basis of its relationship with another variable.

For example, in Figure 8-7 (left) the red values are an extrapolation (predicted behavior) of the observed values (blue ones), which could be called forecasting time series as well.

Interpolation produces an estimate of potential values *between* observation data points, such as red dots that are generated between blue dots in Figure 8-7 (right).

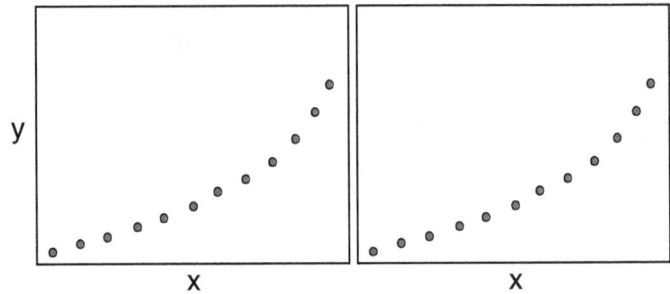

Figure 8-7: (left) A toy example of extrapolation. (right) A toy example of interpolation.

ARIMA Model

ARIMA is used for time series extrapolation (or time series forecasting) algorithm. It assumes that the future data points can be determined by using the past data points, or in other words, we can extrapolate future data points of time series based on the existing data points in the target time series. Specifically, ARIMA models the future data points as a linear combination of past data points (autoregressive part), past forecast errors (moving average part), and differences of past data (integrated part).

ARIMA receives three input parameters, p, d and q, which is written as $ARIMA(p, d, q)$. $ARIMA(1,1,0)$ means that $p=1$, $d=1$, and $q=0$. ARIMA model can be written as a linear equation to predict the next data point, $\hat{y}$ as follows:

$\hat{y}=$ *constant + weighted sum of p number of lags + weight sum of q number of lag errors*.

In the following we explain the ARIMA algorithm steps and this equation in more detail.

(i) As the first step, ARIMA converts a non-stationary time series into a stationary time series, which is called *differencing* the time series. Differencing helps to stabilize the mean of the time series by removing changes in the level of a time series and thus de-trending it. The first column of Figure 8-8 presents the result of de-trending time series. This transformation is necessary because the prediction of future data points relies on the lags from a stationary time series model.

The differencing process is done by subtracting previous data (y_{t-1}) from the current data (y_t). However, just one previous data substitution is usually not enough, and the algorithm should subtract d data points $(y_{t-1}, y_{t-2}, \dots)$ from the current data point. In other words, the parameter d specifies the *order of differencing* required to make the time series stationary. In other words, d specifies the minimum number of differences required for each data point to transform the non-stationary time series into the stationary time series. If the original time series is stationary, we have $d = 0$.

For example, considering y'_t is the differences between lags (the concept of lag is described in Chapter 6), and time t, we can write the following:

$$d = 0: \quad y'_t = y_t$$
$$d = 1: \quad y'_t = y_t - y_{t-1} \text{ \# first-order differencing}$$
$$d = 2: \quad y'_t = (y_t - y_{t-1}) - (y_{t-1} - y_{t-2}) \text{ \# second-order differencing}$$

y'_t values are used to transform the original time series, which is not stationary into a stationary time series. The transformation through differencing is actually related to the *Integrated* component of ARIMA.

(ii) The parameter p specifies the order of autoregressive terms. It refers to the number of lags to be used as input (predictors) to predict the upcoming y'_t. A simple autoregressive (AR) model can be written as a linear model, y'_t depends on its p number of lags. Therefore, we can write the following equation for predicting y'_t based on pure autoregression:

$$\hat{y}'_t = \beta_0 + \beta_1 y'_{t-1} + \beta_2 y'_{t-2} \ldots + \beta_p y'_{t-p} + \epsilon_1$$

In this equation, βs are coefficients of their associated data points (lags) and β_0 is the intercept. Note that it is only AR and not the *moving average* (MA).

(iii) The parameter q specifies the order of *moving average (MA)* term. This means it refers to the number of lagged forecast error terms used to predict the current value of the series. The moving average part of the model is indeed a weighted sum of the past forecast errors. In other words, *moving average* is a weighted sum of q number of lags of the prediction errors. Therefore, assuming α is the intercept (β_0), ϵ is the error of the associated lag and ϕ is the coefficient of the error, a pure moving average equation (only MA and not AR) can be written with the following equation.

$$\hat{y}'_t = \alpha + \epsilon_t + \phi_1 \epsilon_{t-1} + \phi_2 \epsilon_{t-2} + \ldots + \phi_q \epsilon_{t-q}$$

ARIMA model combines both AR and MA models and creates the following equation to predict the upcoming data points in time series.

$$\hat{y}'_t = \alpha + \beta_1 y_{t-1} + \ldots + \beta_p y_{t-p} + \epsilon_t + \phi_1 \epsilon_{t-1} + \ldots + \phi_q \epsilon_{t-q}$$

ARIMA Parameters Estimation

We understand the intuition behind the ARIMA equation, so now, we should explain how to choose the best values for d, p, and q. We have learned in Chapter 3 that correlation describes the relationship between two variables. In the context of time series and signals, when a correlation is calculated against lag variables (e.g., the correlation between d_t and d_{t-1}) it is called *autocorrelation* or *self-correlation*. In other words, autocorrelation means that the signal or time series is correlated with itself.

Let's use a sample; assume $X = \{1,2,4,5,6\}$ and $Y = \{2,3,5,7,7\}$. If we calculate the correlation coefficients of these two sets the Pearson correlation will be $r = 0.9834$, which means they are highly correlated. Let's say we would like to measure the correlation of X with itself, but with one lag. One shift in X will result in $X' = \{0,1,2,4,5\}$, and the correlation between X and X' will be 0.97, which means still, with one shift, X is autocorrelated. The more shifts we perform, the correlation coefficient decreases until it reaches zero. The existence of autocorrelation in errors (residuals) of a model is a sign of error.

We can identify the existence of autocorrelation by using a *correlogram* that stays for the *Auto Correlation Function plot* (ACF plot). ACF plot visualizes the auto-correlation (correlation between the original series and its lagged values). It presents how well the present data points correlate with lag data points (past data points). ACF plot allows us to determine the Auto Regressive coefficient for the ARIMA. In particular, the x-axis of the ACF plot represents the number of lags. The y-axis typically measures the autocorrelation coefficient, which ranges from -1 to 1, as we can see in the second column of Figure 8-8.

The lag value located outside the significant area (blue-shaded area) should be chosen because they are statistically significant. For example, only lag 1 is significant in 'autocorrelation of first-order differencing', the rest, which are located inside the blue area, are insignificant. Lag 0 is also located outside the blue area, despite showing a significant size.

To determine the stationarity of a time series, a statistical test called the Augmented Dickey-Fuller test or ADF test [Dickey '79] will be used. The p-value of this test determines if the time series is stationary. Its null hypothesis states that the time series is non-stationary. If p-value < 0.05, the null hypothesis is rejected, which means the time series is stationary.

ACF plot measures the correlation between observation (data) at the current time and previous observations (data); in other words, it is doing the MA part of the ARIMA. Another plot called the *Partial Autocorrelation Function plot (PACF plot)* finds a correlation of residuals. PACF plot finds a correlation of residuals. It is a bit challenging to understand, and we try to use an example to learn it.

Assume we are living in a city that has a business of selling academic degrees to international students at a huge price. It is a win-win game because academic corporations get richer, and international students who return to their countries are proud to get their degrees from well-known universities. Let's call this imaginary city Bustom. Most of the Bustom population are students. Assume you have arrived later than the start of a new semester and you need to rent an apartment in November. November prices correlate to September (when students move to the city and the fall semester begins) and not October. December prices correlate with September prices, and not October. Therefore, PACF can resolve our need because it identifies the correlation between data at two different times (e.g. d_{t+1}, d_{t+2}) given both data are correlated to a data at other time (e.g., d_{t-1}). In other words, PACF can remove the correlation to data that is not relevant (e.g., d_t is not relevant). In our example, prices of November (d_{t+1}) and December (d_{t+2}) correlate to September (d_{t-1}) and not October (d_t).

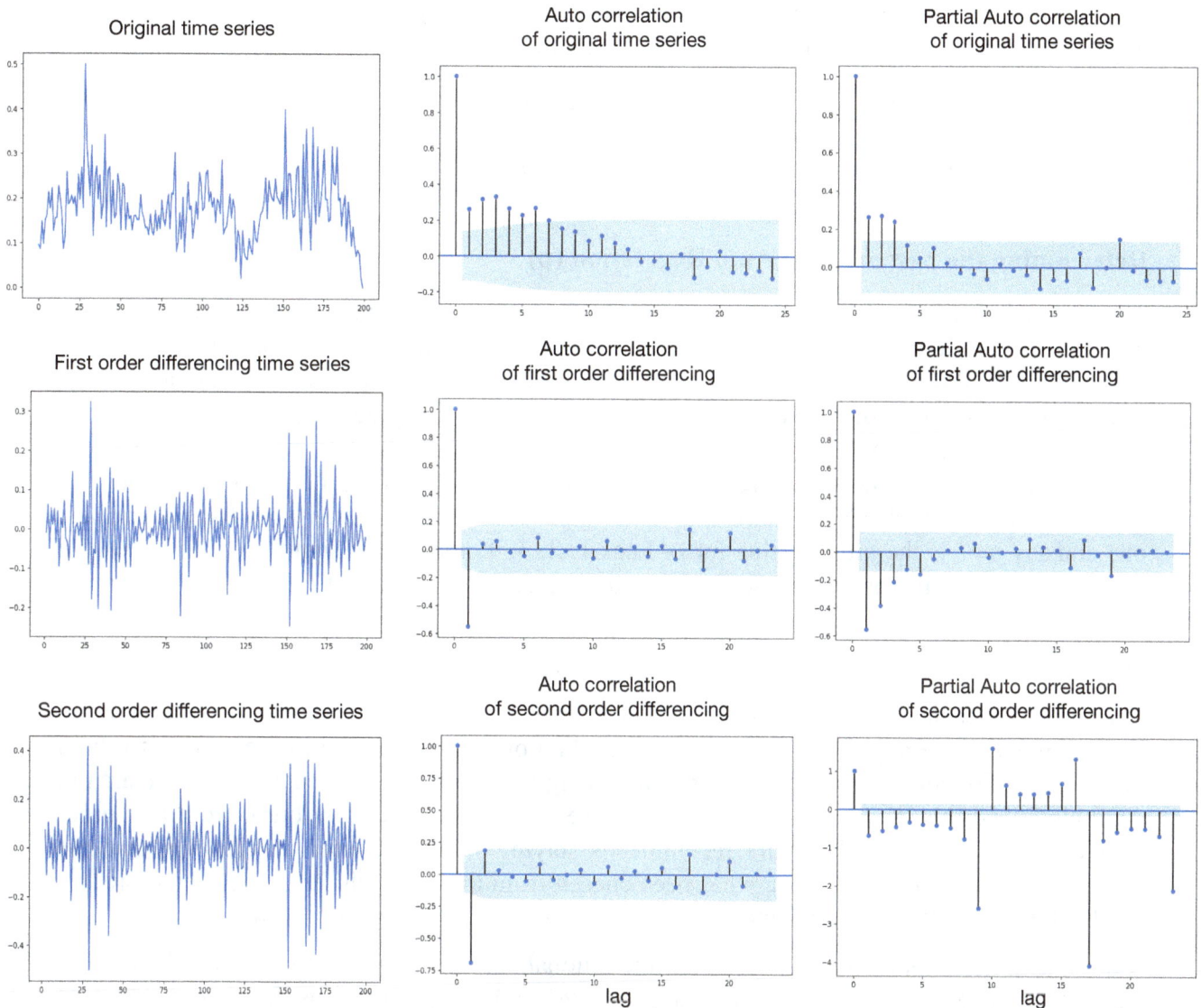

Figure 8-8: (Left) original time series, first-order differencing and second-order differencing (center) ACF plots of each time series. (Right) PCAF plot for each time series on the left side. The light blue color background is called the confidence band and tells us whether the correlation is statistically significant.

Check Figure 8-8, where we present ACF and PACF plots, along with the original and transformed time series.

Determining the best order of differencing (d)

There are a couple of rules to identify the best d. A good differencing is the minimum number of differencing that its ACF plot is fluctuating and decaying toward zero fast. In other words, a good order of differencing is often the order of differencing in which its standard deviation is lowest, and it indicates the time series is stationary. For example, by looking at Figure 8-8, we can see that the ACF plot for second-order differencing is moving toward zero, and compared to the original time series, its standard deviation is also closer to zero. Therefore, we choose the second-order of differencing. Keep in mind that if a time series is stationary (which is our desire), both ACF and PACF should move their tail toward zero (tail off to zero).

In the example presented in Figure 8-8, the second-order of differencing moves toward zero (better than the first-order of differencing), thus we choose $d=2$. If both the first-order and the second-order are moving toward zero, we go for the smaller d. Usually, we do not need to have a d larger than 2.

Determining the best order of Auto Regression (p)

To determine a proper value for d we use the ADF test. To determine a proper value for p, we will use the PACF plot (AR part of ARIMA). We choose a p based on the significance test reported by the PACF plot. For example, if we realize that d_{t-1} is not passing the significance test (located inside the blue area) but d_{t-2} passes, we choose d_{t-2} to predict d_t. The left side of the Figure 8-8 presents PCAF of each time series on their right side. To determine $p,$ we use *PACF* plot and choose lag values that have its value is statistically significant, i.e. fall outside the blue area. The first-order PCAF has some lag values statistically significant. We go for the smallest one and say $p=1$ is the proper value for p. Lags 1, 2, 3, 4, and 19 are significant in the "partial autocorrelation of first-order time series" plot, but we go for the smallest value of lag and assign $p=1$.

Determining the best order of Moving Average (q)

The same approach we have used to determine p from PACF could be used to determine q, but instead of PACF we rely on ACF. Therefore, based on our data presented in Figure 8-8, for both first-order differencing and second-order differencing lag number 1 falls outside the blue area in ACF plots, and this shows statistically significance. However, we choose our q from first-order differencing, because the literature recommends staying with one order differencing that has been chosen for p, and in our case p has been chosen from first-order differencing. Therefore, we can finalize our ARIMA parameters by writing *ARIMA (p=1, d=2, q=1)*.

If these parameter configurations sound hard to rationalize, we can go for cross-fold validation as well. We can test different settings of a parameter and choose the one that has the highest accuracy. However, it is recommended to use this method instead of cross fold validation.

398

Logistic Regression (Classifier)

To understand logistic regression, we should get familiar with the Sigmoid function. The Sigmoid function is a function that gets numerical data as input and outputs numerical vectors in a S-shaped curve, as is shown in Figure 8-9 (c) with a red line. This function brings any given number close to zero or one. It is written as $\sigma(\,.\,)$, and its equation is as follows:

$$\sigma(x) = \frac{e^x}{e^x + 1} = \frac{1}{1 + e^{-x}}.$$

For example, assume we have three inputs $x = 0, \quad x = 1, \quad x = 2, \quad x = 3$. We will have the following sigma for each variable:

$$\sigma(0) = \frac{1}{1 + e^0} = 0.5, \; \sigma(1) = 0.73, \; \sigma(2) = 0.88, \; \sigma(3) = 0.95.$$

We are lazy, and thus, we do not show more examples here, but we recommend you use more values for x and plot them yourself. Then, you will notice that this function returns something between 0 and 1, mostly either close to zero or close to one. Its shape is similar to an S-shaped curve, as shown in Figure 8-9 (c).

Logistic (or Logit) regression (Classification) is used when our output (dependent) variable (y) is binary (e.g., yes/no, open/close, die/survive, spam email/not spam email), as opposed to linear regression which its output has a numeric range. Through a linear regression as input of a sigmoid function, we get a logistic regression. Therefore, we can formalize logistic regression as:

$$\hat{y} = \sigma(\beta_0 + x\beta_1)$$

Logistic regression is not used for predicting or describing a continuous variable. Instead, it predicts or describes a binary variable. It assigns a probability to each class output instead of directly assigning a value to y. In other words, it specifies the probability if a data point belongs to a class or not, and thus, is used to solve classification problems.

With some mathematical calculations, which we skip describing its details here, we can derive the following equation for logistic regression:

$$\hat{y} = \frac{1}{1 + e^{-(\beta_0 + \beta_1 x)}} = \frac{e^{\beta_0 + \beta_1 x}}{1 + e^{\beta_0 + \beta_1 x}}$$

$\hat{y}$ presents what we want to predict (similar to $\hat{y}$, in linear regressions), and x is a set of the predictor variables. Linear regression is written as $y = \beta_1 x + \beta_0$, but here we call β_0 a *balance*, but in linear regression, it is called *intercept*.

Let's learn logistic regression with an example. Mr. Nerd is attending a course at the university, and due to a pandemics (Covid-19 started in 2019), the grade of the course is based on assignments and universities are closed to take a final exam. He either likes a course or doesn't like it (only two possible classes). If a course has more than ten assignments, it's very unlikely that Mr. Nerd will like it, and he doesn't subscribe to that course. Accordingly, we can formalize the Mr. Nerd course-taking procedure as follows: more than 10 assignments: don't like, less than 10 assignments: like.

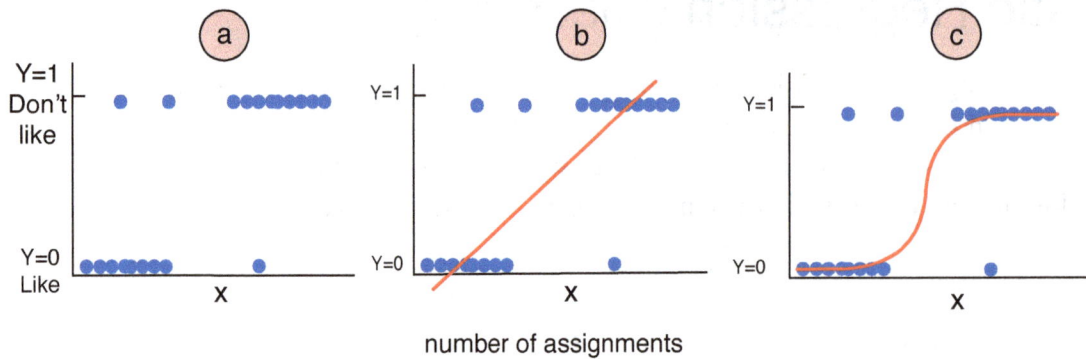

Figure 8-9: (a) Original dataset, x presents the number of pages each book has (b) linear regression plotted, (c) Logistic regression Sigmoid line plotted.

Let's assign like to 0 and don't like to 1. He has plotted his previous courses, based on like/don't like and the number of assignments he had, in Figure 8-9 (a). The *x-axis* (independent variable) represents the number of assignments, and on the *y*-axis we have two values for the independent variable, like (0) or don't like (1).

Figure 8-9 (b) presents a linear regression, which has been calculated and plotted based on given data points. We can see in Figure 8-9 (b) that the linear regression line is not a good fit for these data points. If we look at Figure 8-9 (a), we can intuitively guess that there is a correlation between data points. However, instead of fitting a straight line used in linear regression, we need a method to fit a regression line more accurately on data points, like the line Figure 8-9 (c). By using the Sigmoid function, as shown in Figure 8-9 (c) we can cover more data points (than a straight line, which is the result of a linear regression).

The output of the Logistic regression can classify input data into two categories by estimating the probability, e.g., classify any new course that is given to the algorithm as like it, or not like it, depending on the number of its assignments.

The output of our example will be provided as a probability value of *like* ($y = 0$) or *don't like* ($y = 1$), and the input variable is the number of pages.

As an example, to calculate output, we give Mr. Nerd's course preferences data to the Logistic regression algorithm, and it identifies model parameters as $\beta_0 = 0.4$ and $\beta_1 = 0.3$ coefficients. Now, we can calculate the estimated probability of Mr. Nerd liking a new course, which has eight assignments, as follows:

$$\hat{y} = \frac{e^{0.4+0.3\times8}}{1 + e^{0.4+0.3\times8}} = 0.94$$

It is more than 50%, and thus, we conclude he likes it. Also, the probability of not liking it is $1 - 0.94 = 0.06$. This seems a very easy example to identify, but in reality, logistic regression can solve more complex classification tasks.

Model Parameters Estimation

Similar to linear regression, the objective of logistic regression is to identify parameters that fit the sigmoid line. However, logistic regression is not a linear model, and thus, to identify its parameters, we should convert it into a linear model.

We know that the output of logistic regression is a probability value (between 0 and 1). However, the linear regression output does not have a range limitation. This flexibility in linear regression allows it to identify all linear model parameters easily. Therefore, if we make Logistic regression similar to linear regression, we can determine its model parameters.

To make Logistic regression similar to linear regression and be able to determine a value for each model parameter, the output variable is transformed from probability to *Logarithm of Odds (Logit)*. What is Odds[3]? In statistics, the likelihood of an event happening is called "odds" of that event. Assuming P is the probability, odds is calculated as follows:

$$odds = \frac{p(event)}{1 - p(event)} = \frac{probability \quad of \quad success}{probability \quad of \quad failure}$$

Odds are not probabilities, Odds provide a measure of the likelihood of a particular outcome. They are calculated as the ratio of an event happening, to that event not happening. For example, the probability of rolling a dice and getting the number 6 is 1/6, its odds is $1/6 \div 5/6 = 1/5$, and 5/6 here presents the probability of not getting 6 (failure).

A *Logit* function, which is the *Logarithm of Odds (Log Odds)* is written as follows[4], for the sake of simplicity we write p instead of $p(event)$:

$$log(odds) = log(\frac{p}{1 - p}) \text{ or } ln(\frac{p}{1 - p}) \text{ for } 0 \leq p \leq 1$$

Model parameters of logistic regression are estimated by modeling the logit (the natural logarithm of the odds) as a linear combination of the input variables. This makes the output variable of logistic regression (probability value) similar to the output of linear regression (numerical value). Figure 8-10 (c) presents a logistic regression on the left, and by using the logit function, we transfer output probabilities into the Log Odds on the right side.

After using logit, the problem can be solved as a linear regression, but another problem is raising. As p approaches 1, the odds $1-p$ increase towards infinity, and hence the logit of the probability approaches $+\infty$. Conversely, as p approaches 0, the odds approach 0, and the logit of the probability approaches $-\infty$. This means that our data points are located at $-\infty$ and $+\infty$, and still, we cannot identify their coordinates.

Nevertheless, we can think of an imaginary linear regression line and project our data points on that line, the orange dotted line in Figure 8-10 (a) and (b). present that line. If we can project

---

[3] We described odds ratio in Chapter 3 for another use.

[4] $Log_e(x)$ and $ln(x)$ are equal and they both mean base e logarithm. If you are not familiar with logarithm check Chapter 6, The Magic Power of Transformation section.

our data points on this line, their coordinate will be in a range that linear regression can use to calculate all parameters (we call it θ to abbreviate all these parameters).

To solve this problem (range problem), logistic regression uses a Maximum Likelihood Estimation (MLE) (check Chapter 3) to project data into a small range. Here, the algorithm that implements MLE tries to identify parameter values that are closest to the output probability (0 or 1). In other words, the algorithm tries to find a value for its projected parameters $\hat{\beta}_0, \hat{\beta}_1, \ldots \hat{\beta}_k$. Then it substitutes them to estimate $\hat{y}$, which yields a number close to 1 or close to 0. If you remember from Chapter 3, we are using L_n to denote the likelihood function, ℓ to denote the logarithm (log) of the likelihood function, and we calculate the maximum likelihood on log-likelihood. Assuming θ presents the parameters of our model, X our dataset, we have $\ell(\theta; X) = log\, L(\theta; X)$. Here. $L(\theta; X)$ represents the likelihood function, which quantifies the probability of observing the given data under a specific set of parameters denoted by θ. Now, we have $\ell(\beta_0, \beta_1, \ldots \beta_k)$, because here, the objective of this MLE is to find the optimal values for Logistic regression model parameters.

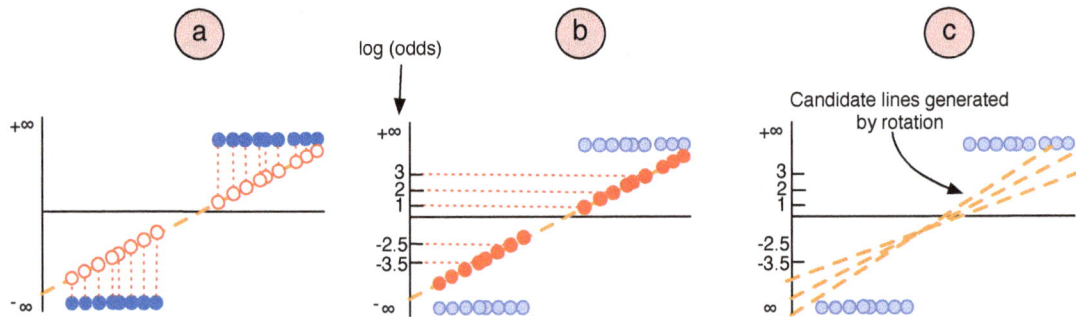

Figure 8-10: (a) Projecting original data points from negative and positive infinity to a candidate line. (b) Identify the Y-axis values of each project data point. (c) By rotation, different candidate lines are generated, and the line with the lowest error is chosen to identify the optimal model parameters.

Once again, let's review why we need MLE to project our data points on a line. Our data points are located at $-\infty$ and $+\infty$. Therefore, their linear regression line will start from plus infinity to minus infinity, and thus, we cannot identify the model parameters of this linear regression. The MLE can resolve the infinity problem by projecting the original data points into an imaginary (candidate) line. See Figure 8-10 (a), the original data points are projected on a candidate line and presented as red empty dots. In Figure 8-10 (b), their coordinates to the y-axis have been calculated. These values on the Y-axis are *log(odds)* of the original data points[5].

For example, let's take a look at Figure 8-10. There is one data point with a value of -3.5, its projected probability (p) will be calculated as: $y = e^{-3.5}/(1 + e^{-3.5}) = 0.29$. This is the number that is used by MLE to sum all likelihoods together. Projected probabilities are between 0 and 1 classes. If the data point belongs to 0 classes (the probability is less than 0.5) its likelihood will

---

[5] We can transform these *log(odds)* back to probabilities by the following equation: $Y = e^{log(odds)}/(1 + e^{log(odds)})$.

be $1 - p$. If it belongs to the 1 class (its probability is larger than 0.5), its likelihood will be the p. The likelihood of all data points on a candidate line is the product of all projected values, or the log-likelihood of all projected data points is the sum of projected data.

Keep in mind that it is better to convert a product into a sum, and the logarithm function converts the products into a sum, e.g., $log(xy) = log(x) + log(y)$. The maximum log-likelihood uses log-likelihood to turn products into sums, making the calculations easier. When optimizing, we look for the parameters that maximize the log-likelihood. By negating the result we mean that sometimes we minimize the negative log-likelihood instead of maximizing the log-likelihood.

For example, let's say MLE draws a random line at the first negative of maximum log-likelihood is 2.99. Next, the MLE rotates a bit this candidate line and then it creates a new candidate line. Afterwards, it calculates the log-likelihood for the new candidate line, e.g., 3.21, and continues this process with another rotation and thus another line, e.g., 2.19, and so forth. In the end, the candidate line with the highest likelihood (3.21 in our example) will be used as the best candidate line.

The best candidate line will be used for projecting data points onto this line, and similar to the linear regression, the algorithm can identify parameters from this line. This means that now we have a single linear line, the intercept (β_0) and slope (β_1) of this line will be reported as the best parameters for our logistic regression.

We remember earlier in the linear regression that to solve non-linear models scientists try to convert them into linear models and solve them. This is another similar scenario. We use MLE to transfer the Logistic regression into linear regression and identify optimal parameters for it.

Since it sounds complex to remember, lets summarize it as question and answer.

Question: How to estimate model parameters for logistic regression?

Answer:

1. Convert logistic regression to log(odds); using log(odds) makes a linear regression out of given logistic regression. However, now we have a problem with the range of the new linear regression. It has a range from $-\infty$ to $+\infty$.

2. The MLE resolves the range problem, but it reports a log-likelihood. Its result is negative because the logarithm of numbers between 0 and 1 is negative.

3. To formulate the maximization of the log-likelihood as a minimization problem, we use the negative log-likelihood. Therefore, the implementation of the MLE experiments with different linear regression lines (by using optimization) and choosing the one that has the highest negative likelihood.

Softmax Regression (Classifier)

Similar to multiple linear regression and linear regression, sometimes we could have more than two possible outcomes, and the output is not just binary. It could be more than two variables, e.g., type of animal (chicken, cat, dog, mouse, etc.). By logistic regression, we have one binary output variable (e.g., die/survive, like/dislike, true/false, etc.). To handle more than two outputs, we use multinomial logistic regression, multi-class logistic regression, or polytomous regression. Four known approaches are used for the multinomial logistic regression, including baseline logit model, adjacent category logit, proportional odds cumulative logit, and softmax regression.

Here, we describe *softmax regression* or *maximum entropy regression*, which is the most common multi-class logistic regression. In Chapter 10, we start to learn deep learning models, and will see that softmax is a common approach for multi-class classifiers by using neural networks, but we can use it without neural networks as well.

One easy way to deal with more than two dependent variables is to convert them into a single dependent variable and perform logistic regression. This means we convert a multinomial problem into many binomial classes, and solve binomial classes problem, with logistic regression. For example, in a mobile app that recognizes its users' transportation mode, one variable is "walk" the other variable is "not walk" (bike, public transport, and car). Our variables are "bike" and "not bike" (walk, public transport, and car), and this process continues for each of the possible categorical values. In the end, we have a set of probabilities, and the highest probability will be used to assign the class label to that particular input. However, it is recommended not to use logistic regression for non-binary classes, see Chapter 4.3.5 of [James '13].

Softmax regression is a generalization of logistic regression for K categories. Therefore, instead of two binary probabilities, we have K different probabilities. For example, the transportation mode could walk, bike, public transport & private vehicle ($K=4$), or a medical treatment type could be surgical, pharmaceuticals, diet, or none ($K=4$).

As another example, assume Mr. Nerd is developing an algorithm for a smart refrigerator he has purchased in Chapter 2. He is developing an image recognition algorithm that enables the refrigerator's camera to record the type and amount of foods he has consumed and calculate the body shape of Mr. Nerd in the future. The input of the refrigerator algorithm is a vector of foods Mr. Nerd puts them inside the refrigerator. The output is the future Mr. Nerd's body shape, i.e.,"gaining weight", "no changes" and "losing weight" (*K=3*). The softmax regression can create a model to predict Mr. Nerd's future body based on the three possible outputs.

The output variable of softmax regression is shown as $Y^{(i)} = \{1,2,...,K\}$ (in logistic regression, it was binary, i.e., $Y^{(i)} = \{0,1\}$). The softmax output is a set of probabilities (values between 0 and 1), and their sum is equal to 1. Unlike Sigmoid (S-shaped function) we can not use a specific shape or plot to visualize the softmax, which has more than two dimension space.

The softmax regression predicts one output class at a time, it is a multi-class classifier but not multi-output. Therefore, we cannot use it to predict more than one output variable, e.g., we can identify a chicken in a picture, but we cannot identify a chicken and a cat in the same image with one softmax.

To better understand the details of softmax regression, we should look at the logistic regression function, i.e., Sigmoid, and see how it can be generalized to be softmax. Let's rewrite the Sigmoid function, which is used for Logistic regression, as follows:

$$\sigma(z)_i = \frac{1}{1 + e^{-(z)}}$$

Here, $z = \beta_0 + \beta_1 x_1 + \beta_2 x_2 + ... \beta_k x_k = \Sigma_{i=0}^{m} \beta_i x_i = \theta^T x$ and i presents its index. In many resources, $\sigma(z)_i$ is written as $h_\theta(x)$ or $\hat{P}$, which $h_\theta(x)$ stays for a hypothesis based on given input vector of x and θ presents parameters of the model.

We see a transpose sign used for θ, why? Because θ is a $1 \times n$ matrix (which is a vector) and x is $1 \times n$ matrix (which is again a vector). Therefore, to multiply these two matrices, we need to multiply $n \times 1$ matrix to $1 \times n$ matrix, and from matrix algebra, we knew that $(1 \times n)^T = (n \times 1)$.

We can write the softmax function, as a generalization of the sigmoid function. For all k classes of output it is written as follows:

$$P(y = j \mid x, \theta) = \frac{e^{\theta_j^T x}}{\sum_{i=1}^{k} e^{\theta_i^T x}}$$

Here, y is the vector we intend to predict, $P(y = j \mid x, \theta)$ is the probability that the outcome y is class j given the input features x, and θ is the model parameters. k is the number of classes. The equation is very similar to sigmoid (except having a Σ sign in its denominator). The softmax function produces a vector of probabilities, each element representing the probability that the input x belongs to one of the k classes. We can express its output as $\hat{P}_k(x)$. To better understand it, let's expand this equation as follows:

$$\hat{P}_k(x) = \begin{bmatrix} P(y=1 \mid x,\theta) \\ P(y=2 \mid x,\theta) \\ \cdots \\ P(y=k \mid x,\theta) \end{bmatrix} = \frac{1}{exp(\theta^{(1)T}x) + exp(\theta^{(2)T}x) + \ldots + exp(\theta^{(k)T}x)} \cdot \begin{bmatrix} exp(\theta^{(1)T}x) \\ exp(\theta^{(2)T}x) \\ \cdots \\ exp(\theta^{(k)T}x) \end{bmatrix}$$

Or some prefer to write it as follows:

$$\hat{P}_k(x) = \begin{bmatrix} P(y=1 \mid x,\theta) \\ P(y=2 \mid x,\theta) \\ \vdots \\ P(y=k \mid x,\theta) \end{bmatrix} = \frac{1}{\sum_{i=1}^{k} e^{\theta_i^T x}} \begin{bmatrix} e^{\theta_1^T x} \\ e^{\theta_2^T x} \\ \vdots \\ e^{\theta_k^T x} \end{bmatrix}$$

As the output result of softmax regression is a vector of probabilities, their sum is equal to 1. Therefore, while using a softmax function, *each output data point depends on the entire vector of input data points*. The input of the softmax function is a vector of raw scores (logits) for each class. For a given instance (data point), we have an input feature vector, and this vector represents the attributes or features of the instance relevant to the classification task.

Back to the refrigerator and Mr. Nerd's example, he went shopping and got the following items for its refrigerator: {broccoli, ice cream, chocolate bar, wheat bread, apple, orange}. His refrigerator, which is equipped with a softmax regression algorithm, can calculate the future body style of Mr. Nerd and returns the following result as a set of probabilities: {"gaining weight" = 0.4, "no changes"= 0.5, and "losing weight" = 0.1}.

Model Parameter Estimation

Similar to logistic regression, the cost function of softmax regression is the MLE. However, we cannot explain a multidimensional classification with visualization, because our brain is limited to 3D visualization only. Before explaining it, we should go back to the cost function of logistic regression, maximum log-likelihood, and formalize it in a mathematical notation.

Assuming $X = \{x_1, x_2, \ldots, x_n\}$ are input variables of the model, $\theta = \{\beta_0, \beta_1, \ldots \beta_k\}$ presents all model parameters, and $Y = \{0,1\}$ presents output classes for our input variables, the following equation presents the likelihood function for logistic regression, which we have described:

(i) $\displaystyle L(\theta) = \prod_{i=1}^{n} [p(y_i = 1 \mid x_i, \theta)]^{y_i} [1 - p(y_i = 1 \mid x_i, \theta)]^{1-y_i}$

The sign Π represents a set of products (multiple multiplications) similar to the Σ, which is used for summation. The goal is to find θ (or $\hat{\beta}$) that maximizes the $L(\theta)$ function.

We can extend the above equation and incorporate more than two Y values as follows:

(ii) $\displaystyle L(\theta) = \prod_{n=1}^{N} P(y_n \mid x_n, \theta)$

Equation (ii) is the cost function for softmax regression. We have explained that for the logistic regression, we are using the log-likelihood function, therefore, the described equation can be rewritten as follows:

$$\text{(iii) } J(\theta) = -\sum_{n=1}^{N} \log P(y_n \,|\, x_n, \theta)$$

Note that instead of $\ell(\theta)$ we use $J(\theta)$, which is common while using softmax refer to cost function as $J(\theta)$. We use the negative value for $J(\theta)$ as a function for model parameter estimation, which is called the *negative log-likelihood* or *cross-entropy loss*. We use the negative sign at the end to make the maximization problem as minimization and thus easier to compute. If you are good at mathematics, you can realize that the cost function, we used for softmax regression parameter estimation will be a generalization of the cost function that we used for logistic regression (i.e., maximum log-likelihood).

Since cross-entropy is a cost function used to measure the accuracy of the softmax regressions, our goal should be minimizing the cross-entropy loss. Assuming we have *M different classes (in logistic regression, we have only two)* and *N number of input samples,* the cross-entropy cost function is written as equation (iv).

$$\text{(iv) } J(\theta) = -\frac{1}{N}\sum_{i=1}^{N}\sum_{j=1}^{M} y_{ij}\, log(p_{ij})$$

This equation is a generalization of equation (i) for more than two output variables. The cross-entropy objective function is to reduce the cross-entropy, thus a model that has smaller cross-entropy is favored over the model that has larger cross-entropy.

Evaluating Regression Models Fitness

The output of softmax regression is k number of distinct classes. The output of logistic regression is a binary classification result, which could be 0 or 1; unlike linear and polynomial regression, there is no traditional residual exists to compute the accuracy.

Logistic and softmax regression (classification) evaluation methods focus on assessing *how good is the model fits the data*. In this section, we list methods used to evaluate these regressions' results. Some of them are very important, not limited to regression methods, and will be used for other classification algorithms as well. Therefore, please tune your brain in full attention mode and learn them very carefully.

I am not tired and full of energy to learn another section.

k-fold cross-validation

The most popular approach to evaluate classification results is the use of *k*-fold cross-validation (check Chapter 1 if you can't recall it). To implement k-fold cross-validation, we should make *k* subsets of the dataset, one set will be used as testing, and the other *k-1* sets will be used as training. We redo this process *k* times, and in each iteration, we change the testing set to another subset. Then, we evaluate the accuracy of each iteration. Next, we take an average of these accuracies and report the final accuracy.

The logistic regression has only two outputs for prediction and it could be written as the confusion matrix in Table 8.1. (left). The softmax regression has more than two predicted variables and its confusion matrix could be something like Table 8.1. (right), which has four possible outputs. We have explained confusion matrix back in Chapter 1.

	Predicted	
	0	1
Actual 0	12	2
Actual 1	0	14

	Predicted			
	a	b	c	d
Actual a	5	1	0	1
Actual b	1	7	2	3
Actual c	0	3	12	1
Actual d	2	2	1	6

Table 8-1: (Left) Confusion matrix example for binary logistic regression. (Right) A confusion matrix example for more multi-logistic regression, such as softmax, which has four possible output variables.

As we have explained in Chapter 1, some literature [Grus '19] and [Ng '18] recommend splitting datasets into three distinct sets including; 'training', 'validation', and 'test', instead of just two (training and test). The validation set is used for parameter tuning, feature selection, and more, facilitates the selection of the optimal model among various candidates or configurations. It's particularly crucial to ensure that the validation and test sets come from the same distribution to maintain the integrity of the model evaluation process.

Learning Curve

A fairly easy approach to determining the optimal model size is to study the error changes with different dataset sizes and experiment with the model. This is known as the learning curve. For example, Figure 8-11 presents a sample that by increasing the dataset size in both train and test sets, the error rate will converge at some point,

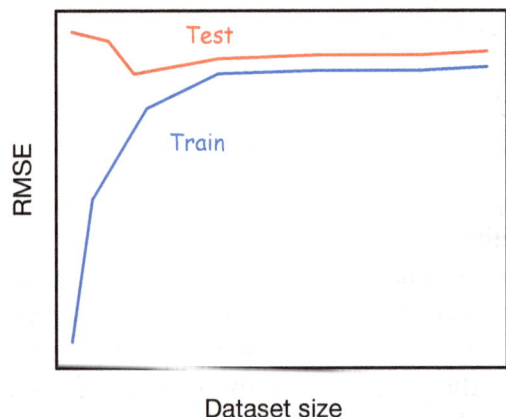

Figure 8-11: Learning curve example. By increasing the number of data objects, both training and test datasets converge. This makes the model reliable.

and the model will be stable. At the beginning, when there are few training instances (the left side of Figure 8-11) the model fits them very well, and thus, the RMSE is very low. Then, slowly, more training data arrives, and the models' RMSE increases until it gets into a constant value.

On the other hand, you can observe that for the test set at the beginning, the RMSE rate is high despite having a very low training error, but as we add more data, it decreases and reaches a steady point similar to the train set. Therefore, we can claim that the model is stable at that point because both errors converge and stay constant.

Observing a very low training error and very high test error rate at the beginning of the curve, in learning curve, present the overfitting. We will explain overfitting later in this chapter.

ROC Curve

Another widely used approach to evaluate the quality of a classifier is the *Receiver Operating Characteristic (ROC) curve*. This approach is not limited to the described regressions and can be used to measure the accuracy of any classification algorithm.

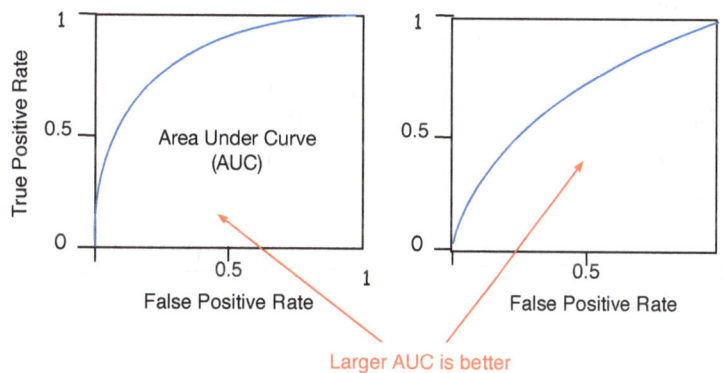

Figure 8-12: (left) an ROC curve which shows high accuracy, (right) ROC curve which has less accuracy in comparison to the left one.

The ROC curve plots True Positive Rate (TPR or sensitivity or recall) against the False Positive Rate (FPR or fall-out), at different threshold settings. Note that *FPR = 1 - TNR (True Negative Rate or Specificity)* as shown in the following:

$$FPR = \frac{FP}{FP + TN} = 1 - \frac{TN}{TN + FP}$$

ROC curve is useful for demonstrating the trade-off between TPR and FPR. The closer the curve is to the left and top borders, the more accurate is our results. The closer the curve comes toward the diagonal, the more accuracy decreases. Figure 8-12 presents two ROC curve samples, the left one is more accurate than the right one because the right one is closer to the diagonal.

The result of the ROC plot is read as the *Area Under the ROC* (AUROC) or some said Area Under Curve (AUC), which will be a value between 0 and 1. The higher AUROC (closer to the one) presents better model accuracy.

Pseudo R²

Unlike linear regression, which typically uses the MSE to minimize the variance between the predicted and actual values and can employ methods like Ordinary Least Squares for parameter estimation, logistic and softmax regressions utilize a different approach. They are designed for classification rather than predicting continuous outcomes, and they estimate their parameters using the Maximum Likelihood Estimation (MLE) approach. This involves optimizing a cost function like cross-entropy (log loss) for logistic regression or a generalized form of it for softmax regression, focusing on maximizing the probability of the observed classifications given the model rather than minimizing variance.

To provide measurement metrics for logistic and softmax regression, we can use some *Pseudo R²* or *Adjusted R²* methods, such as *McFadden's R²* [Domencich '75]. McFadden's R² is computed as follows:

$$R^2_{McFadden} = 1 - \frac{ln\ L(\hat{\theta})}{ln\ L(\theta_0)}$$

$L(\hat{\theta})$ is the (maximized) likelihood value from the current model, $L(\theta_0)$ is the model with only an intercept (β_0) and no covariates (input variables). The high value of McFadden's Pseudo R² indicates a lower likelihood (the lower is better), because the ratio of $L(\hat{\theta})$ and $L(\theta_0)$ likelihoods should be close to 1. In particular, the higher Pseudo R² indicates a better model fit. Note that while Pseudo R² can theoretically range from 0 to 1, it will never reach 1 due to unpredictability in the data or model limitations.

Remember, your machine learning library can compute these pseudo R², so there's no need to calculate them manually.

Wald Test

Wald test, also known as Wald Chi-square test [Wald '45], is a significance test that is used to assess the statistical significance of coefficients in regression models. Its null hypothesis states that model parameters do not have any effect on the model fitness. If the null hypothesis fails to get rejected (the null hypothesis is true), then removing those parameters from the model does not harm the fitness of the model. In other words, those parameters are not useful for the model. Conversely, rejecting the null hypothesis implies that the parameter significantly contributes to the model.

For example, assume a data scientist, Dr. Devil, is consulting a politician for election. He had a logistic regression that used three input variables (x_1, x_2, x_3) to predict the election result. x_1 input variable is presenting the welfare support, x_2 is using the cyber army of fake accounts and bots in social media to promote that candidate and damage the reputation of other candidates, x_3 is creating an imaginary enemy and fueling the xenophobia of that enemy in the public.

Based on previous consultations, Dr. Devil identified some parameters for these variables. He is creating a model that helps him to predict whether a candidate wins the election. He used a Wald test and realized that x_1 does not show a statistically significant result. Therefore, he can remove

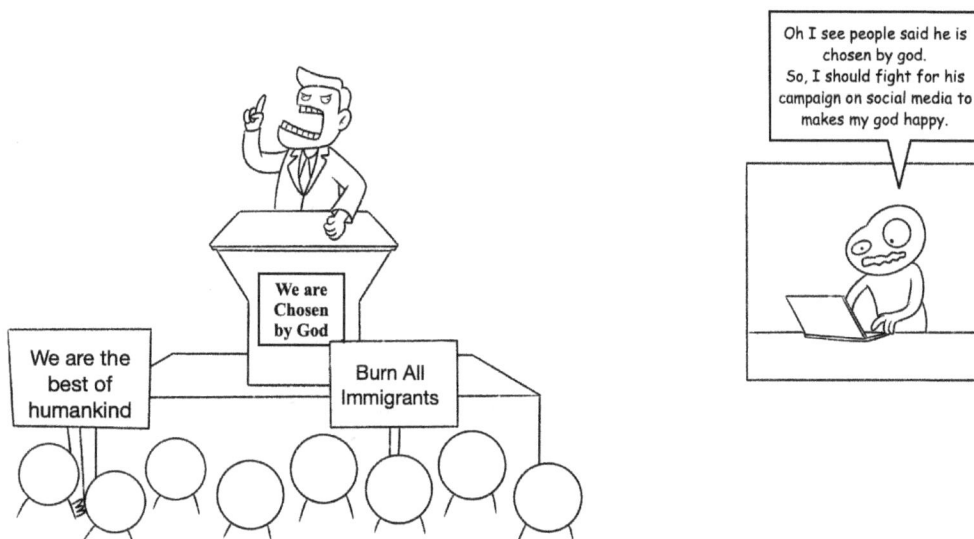

it from his model and focus on the two other variables, i.e., creating fake accounts on social media and fueling the public's xenophobia of an imaginary enemy.

Information Criterion

Information criterion methods are used to compare a set of models together and identify the best model among others. There are several information criterion tests, but two of them are used more often than others, including the Akaike Information Criterion (AIC) [Alkaline '74] and the Bayesian Information Criterion (BIC), which is also called Schwarz Information Criterion (SIC) [Schwarz '78].

Assuming $log(L(\hat{\theta}))$ represents the log-likelihood of the current model, and k is the number of adjustable parameters in that model, the Akaike Information Criterion (AIC) is computed as follows:

$$AIC = -2(log(L(\hat{\theta}))) + 2k$$

Assuming we have n number of data points (sample size), it can also be used for linear regressions by using the following equation:

$$AIC = n\ log(\frac{RSS}{n}) + 2k$$

If you do not recall the RSS, check the description of linear regression earlier in this chapter. The output of AIC is a numerical score, and the lower AIC score is better. It means, that if we compare the two models' AIC scores, we should choose the model with the lowest AIC score.

Another method, the BIC or SIC, offers a distinct advantage over AIC by more severely penalizing the number of parameters in a model. While both AIC and BIC penalize model complexity, BIC incorporates a larger penalty factor, which is particularly sensitive to the sample size. This factor is the logarithm of the number of data points, i.e., log(n), making BIC's penalty for adding parameters more substantial as the dataset grows. Consequently, BIC tends to favor simpler models compared to AIC, especially in large datasets.

One might ask, why do we penalize the number of parameters? Because a large number of parameters makes the model complex and thus inefficient (e.g., prone to overfitting, computationally more demanding).

Assuming n is the number of data points, k is the number of parameters, θ is a set of all model parameters, and $L(\hat{\theta})$ is the maximum likelihood of the model, BIC is computed as follows:

$$BIC = k \ log(n) - 2log(L(\hat{\theta}))$$

For linear types of regressions we use the following equation:

$$BIC = k \ log(n) + n \ . \ log(\frac{RSS}{n})$$

Similar to AIC, while comparing BIC scores of several models the lowest BIC score presents the best model.

Both BIC and AIC are based on the law of *Occam's Razor* principle or the law of briefness. It indicates that we should use only things that are necessary. Models which are obtaining this law are called 'parsimonious models', which have only parameters that are necessary for prediction, and they do not include unused parameters.

Likelihood Ratio Test

The Likelihood Ratio Test (LRT) or Likelihood Ratio Chi-square test is useful to choose the best model from two nested models. Nested models are models in which all parameters of one model exist in the other model as well. For example, we would like to model the applicants' acceptance rate in an elite university, where getting admission there is very competitive. One model has two parameters: the entrance exam score (e.g., SAT in the U.S.) and high school performance (GPA in the U.S.). The other model has five parameters: exam score, high school performance, race of the candidate, nationality of the candidate, and whether their parents are rich. The first model is nested within the second model because the second model has all the parameters of the first model and three more parameters (race, nationality, and parent-richness).

The Likelihood Ratio Test assumes that the best model is the one that maximizes the likelihood function. It uses a log-likelihood to compare two models together and identifies the one with a higher maximum likelihood function. It is written as a ratio between the model with a lower number of parameters $L(\theta_1)$ to the model with a higher number of parameters, $L(\theta_2)$, with the following equation:

$$\text{Likelihood Ratio Test} = -2log_e(\frac{L(\theta_1)}{L(\theta_2)})$$

Once again, for the hundred thousand times, we should repeat that your software package includes these tests and you don't need to implement them on your own. Unless you would like to implement specific customization on the existing method, or you are a student and your instructor asks you implement these algorithms on your own.

NOTES:

* When we encounter, $exp(x)$ it is similar to e^x, which is read as "e to the x" or "e to the power of x", and similar to π, e is a constant $e = 2.7182...$ This is called Euler constant number.

* While working with logistic regression, there could be a predictor variable that shows a negative coefficient (negative value for β). This means that there is a negative relation between that particular predictor (X) and output (Y).

* One known issue in logistic regression is the coefficient impact on two or more different input variables on each other. For example, if we increase x_1, then Y will be increased regardless of the value of x_2. To mitigate this issue and increase the interaction effect, we can define a third coefficient (β_3), which is called *interaction terms* between two other input variables.

$$Y \approx \beta_0 + \beta_1 x_1 + \beta_2 x_2 + \beta_3 x_1 x_2$$

* Similar to linear regression, Logistic regression is a form of the generalized linear model. Generalized models are models which use a "link function" to link a nonlinear model to a linear model. Here, the link function is the log odds.

* A result obtained by one predictor (input variable) might be different from the result obtained by multiple predictors (input variables). This is caused by the correlation among predictors, which is called the *confounding effect*.

* We have referred to some weird terms that are -∞ and +∞. They are used in logistic and softmax regression. Nevertheless, since smoothing the data is a very common problem in machine learning, sometimes large data will be converted to +∞, which is called an *overflow* problem, and small data will be converted to zero which causes an *underflow* problem, e.g. diving by a zero will be -∞.

* We should not perform extrapolation beyond our observed data points, especially while using polynomial regression, which uses curved lines.

Devils of Model Building

Why do we use all these regression algorithms for prediction? Because by observing the past behavior (observed or training data), we would like to predict the future (unobserved or test data). The ability to match future data with observed data is called *generalization*. Therefore, an umbrella term that is used for regression algorithms is called *Generalized Linear Models (GLM)*. Note that GLM models do not necessarily assume a linear relationship between input and output variables. Instead, they posit a linear relationship between the expected value of the output variables, as transformed by a link function (e.g., Logit for logistic regression), and the input (explanatory) variables. This encompasses logistic and softmax regressions, among others. In simple terms, GLMs include a transformation of the output through the link function, rather than just modeling a linear relationship between inputs and outputs.

There are two important challenges associated with all GLM methods and also other supervised learning algorithms, underfitting and overfitting.

Overfitting and Underfitting

Overfitting refers to creating a model that performs very well on the data that we have observed (train dataset), but it performs very weakly on the newly arrived data (test dataset). In other words, overfitting makes a model too specific on the training dataset and causes poor generalization of any new data points (test dataset).

Underfitting refers to creating a model that can neither fit the train nor test data properly. In other words, underfitting refers to *too much generalization*. Therefore, the model will perform poorly on both our training and test datasets. Figure 8-13 shows three examples: an underfitted model, an overfitted model, and a proper model.

We can see in this figure that both underfitted and overfitted lines (in red) are not good representations of the data and are incapable of predicting or modeling the dataset. Overfitting is a very common mistake when we increase the degree of our polynomial regression. Usually, overfitting is a very common mistake in supervised learning.

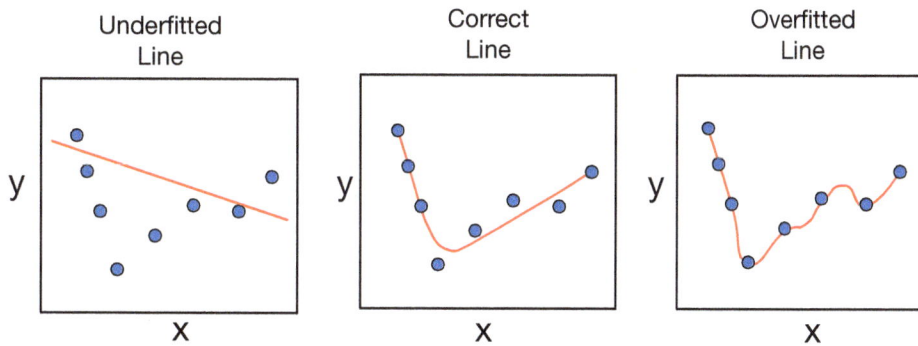

Figure 8-13: Example of underfitting and overfitting in model fitting.

Now, the question is how to avoid releasing the devils of overfitting or underfitting. Underfitting is easy to recognize because as soon as a model does not perform well both on the train and test dataset, we can find it. Nevertheless, overfitting is not that easy. To recognize overfitting, we can change the train and test dataset in every iteration, i.e., k-fold cross-validation. We hope you remember that by using the cross-validation we could change the train and test dataset. For example, we change the train and test dataset five times (5-fold cross-validation) and report the average error from those experiments. If the error is not changing significantly among each experiment, this is an indication that the model is not overfitted. On the other hand, if in one experiment we have a very good result and not in others, that particular experiment is overfitted, and thus, we should revise the model. We can summarize that any significant difference between testing and training is a sign of overfitting.

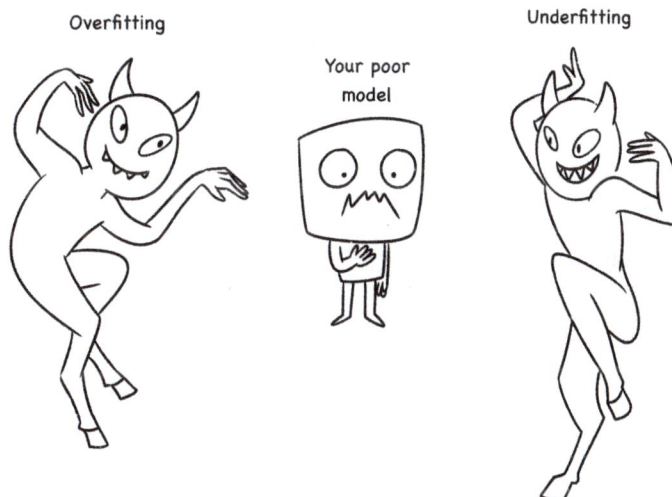

Bias-Variance Tradeoff

Overfitting is a very common error while working with supervised machine learning models. An appropriate approach to deal with overfitting is to decompose it into bias and variance problems

415

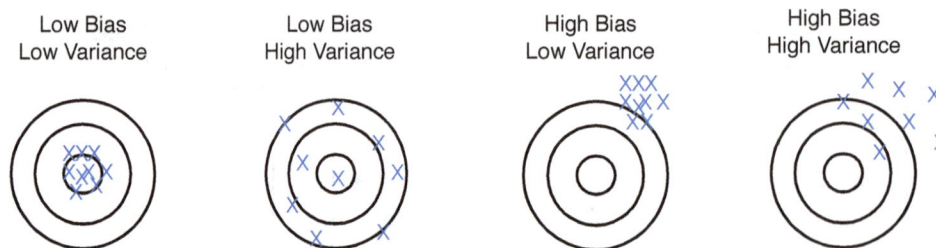

Figure 8-14: Example of underfitting and overfitting in model fitting. The target is the actual data, and the thrown darts are the predicted data.

[Domingos '12]. In other words, we use bias and variance to study whether our model accurately describes the underlying dataset or is misleading.

Variance refers to how far our observed data (or training dataset) differs from the mean of the predicted data (output variables). It is clear that different training sets result in different output variables, but we need to be sure that changes in the output variables for different training sets are insignificant, i.e., the variety of output variables is low. In other words, variance refers to how much the predictions of a model vary or fluctuate for different training datasets (Variance is varieties of the model's outputs).

Bias refers to differences between the model output and correct output (what happens in the real world). In other words, bias presents the prediction error and measures how far off the model's predictions are from the true values, on average.

In summary, Variance is about the model's sensitivity to the training data and how much its predictions vary for different training sets. Bias is about the error in the model itself, regardless of the training data. Domingos [Domingos '12] described: *"Bias is a learner's tendency to consistently learn the same wrong thing. Variance is the tendency to learn random things irrespective of the real signal."* Figure 8-14 presents the bias and variance analogy by throwing darts at the dartboard. In this example, assume the correct output is the closeness to the center of the dartboard, and the input is a set of parameters that affect the dart direction. Sometimes we will get a good distribution of training data so we predict very well and we are close to the dartboard center, while sometimes our training data might be full of outliers values resulting in poorer predictions.

A good model has both a low bias and low variance. However, there is a phenomenon known as the *Bias-Variance tradeoff*, which means an increase in one of them causes a decrease in the other one. Earlier in this chapter (Evaluating the fitness (accuracy) of linear models section), we discussed MSE while describing the fitness of a linear regression model. MSE has been explained earlier in this chapter

With some mathematics that we skip to explain, the MSE can be decomposed into the error equation as follows: $Error = bias(\hat{y}_i)^2 + variance(\hat{y}_i) + variance(e)$. Here, e presents *irreducible error*, which is an error that can not be reduced by having a good model and it is

416

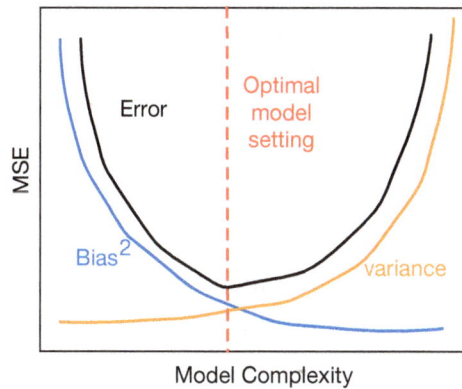

Figure 8-15: Typical bias-variance and error changes. At the red dotted line, the model has the smallest MSE error, thus, complexity at this point can be used to develop the optimal model.

always there. The *Error* on the left is called the *reducible error*. It is a sum of $bias(\hat{y}_i)^2 + variance(\hat{y}_i)$, and irreducible error.

The more we increase the complexity of a model, the bias will be reduced, but the variance increases. We should look for an optimal boundary, something like the red line in Figure 8-15, in which we have the lowest error rate, MSE is the Error in this figure. The red dotted line will be identified by experimenting with models with different complexities (e.g., more model parameters lead to an increase in the complexity).

Any classification algorithm that builds a model creates a decision boundary, which can be a line that is straight, curved, or has a complex form [Burkov '18]. The decision boundary is used to separate labels for data points. For example, in Figure 8-15, the decision boundary is depicted as a straight red dotted line.e.

As a final remark about these model-building devils, keep in mind that it is recommended [Grus '19] to mitigate high bias, we can add more features to the model, and to mitigate high variance, we can remove some features from the model and add more sample data points (not features but data points). It means that more flexible models (which cover more features) have higher variance.

NOTES:

* High bias is often associated with underfitting, which occurs when a model is too simple and does not capture the complexities of the dataset. This simplicity leads to errors because the model cannot adjust to the data's underlying patterns. High variance is associated with overfitting, where a model is too complex relative to the simplicity of the data. Such a model captures random noise in the training data, leading to a model that doesn't generalize to new, unseen data.

* Flexible models, which can adapt their shape to the training data, often have lower bias. However, if too flexible, they may also capture the noise in the data, leading to high variance.

* Usually, underfitting is characterized by high bias and low variance. Keep in mind that underfitting can not be resolved by adding more training data points. To resolve the underfitting. We require to create more complex models, such as adding more parameters.

* A small sample size usually results in high variance, and the simplest approach to mitigate high variance is to increase the sample size. However, usually, the dataset is available before we start building our model, and we can not easily increase the sample size. The data is a valuable source, and usually, we use it with lots of care. There are other techniques, such as adding regularization (we will explain them shortly), reducing the number of features, and using ensemble methods that can help reduce variance without increasing the amount of data.

Regularization

Now, we are familiar with devils in model building, and one of the fiercest devils is overfitting. Due to Occam Razor's principle, we are always looking for less complex (parsimonious) models. Besides, we should always consider reducing overfitting, and for these two needs, we can use *regularization*. Regularization is defined as constraining or shrinking a model to reduce the risk of overfitting the given training set. It applies to the training phase only, not the testing phase. They make the model simpler and try to reduce model coefficients (parameters). Hence, they are also called *shrinkage* methods. In other words, regularization is the process of *shrinking the model's coefficients* and focusing on *decreasing the variance of the model*. Regularizations penalize the high power coefficients that make the polynomial model very sensitive to noise.

If a coefficient of a regression β_k is zero, its predictor will get zero effect on the model as well. Assuming x_j is a model parameter $\beta_k . x_j = 0 . x_j = 0$, and thus, we can remove x_j. It means that this particular variable will be removed from the model, which is good. Why? Because we use the regularization to reduce the model parameters.

Stop fat model ... let me clean your extra parameters. You are not a healthy model. I want to help you

Regularizer

Model

Regularization is crucial for working with regressions. There are $(2^n - 1)$ possible linear models from a combination of n input variables. For example, assuming we have three input variables as x_1, x_2 and x_3, we can have $(2^3 - 1)$ combinations of these parameters: x_1, x_2, x_3, x_1x_2, x_2x_3, x_1x_3, and $x_1x_2x_3$ and each is related to a model parameter. If we have 10 input variables, we can build 2^{10} different linear models. If we have 100 input variables, which is common in real-world applications, we need a supercomputer to choose the model from a set of $2^{100} - 1 = 1267650600228229401496703205375$ models. Therefore, we need a way to shrink models and reduce the number of choices as much as we can. This task will be done by regularization methods. In this section, we explain four common regularization methods, including *Ridge, LASSO, ElasticNet,* and *NonNegative Garrote*.

Ridge

Ridge regression or regularization (Tikhonov-Miller regularization, L2 regularization, or weight decay) [Tikhonov '43] can be used when the model's *independent variables are highly correlated* (multicollinearity). Also, we use it when there is a multicollinearity among the features or we have few training data points. In other words, it makes the prediction less sensitive to the training data, by reducing the variance of the model.

Ridge penalizes the model based on the distances of predictor coefficients (weights) from zero. The higher the distance between coefficient and zero, the higher will be the penalty. It is an RSS (If you can't recall RSS, check "Evaluating the fitness of training sets in linear models" section of this chapter), plus a penalty:

$$Ridge\ Regression = RSS + penalty.$$

Assuming θ is a vector and presents all parameters of our model and we have n parameters, the penalty will be shown as $\lambda\Sigma_{i=1}^n\theta_i^2$, which is λ times the sum of square coefficients. λ is a hyperparameter used to define the *severity of the penalty* or *tuning parameter*. As λ increases, the flexibility of the ridge regression fit decreases, leading to a decrease in variance but an increase in bias. The Ridge regression cost function can be written as follows:

$$J(\theta) = RSS(\theta) + \lambda\Sigma_{i=1}^n\theta_i^2$$

Usually, when we encounter *J(something)* it presents an objective function for *something* parameters. A question might arise now: how can we choose an appropriate λ? Typically, we use 10-fold cross-validation to choose the one that provides the lowest variance. It means, we test with different λ values until we identify a λ that results in the lowest variance. The objective of Ridge regression is to minimize $J(\theta)$. In some literature, MSE [Géron '19] has been used instead of RSS, but both methods are correct (MSE is just the RSS divided by the sample size).

Figure 8-16 presents an example model with three parameters, and after increasing the λ to a specific value, all parameters converge toward zero (but do not get zero). This example shows us how increasing the penalty will reduce the impact of each coefficient on the model until they remove their associated variables because their coefficients will be zero. Keep in mind that Ridge

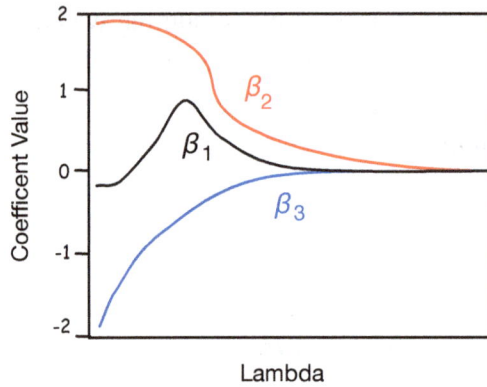

Figure 8-16: A mock model with three parameters that presents the impact of ridge regression. All parameters are converged toward zero as λ value increases.

regression introduces a small amount of bias, and increasing the bias will cause a significant decrease in the variance.

Let's use a simple example to better understand the ridge regression. Before explaining the example, note that acquiring data is an expensive process, and we should use a small number of data objects (as a training dataset) to create the regression model and then use the model on the test dataset.

Figure 8-17 (a) shows a toy dataset composed of 7 data points, and we would like to use linear regression to make a prediction model for this dataset. Figure 8-17 (b) shows two random data points that have been selected as training and marked in red. By calculating their RSS, we identify their RSS is zero, which is too low for bias, and this makes us suspicious. Figure 8-17 (c) calculates the RSS for other points (test set), which are marked with blue color. Test set data points (blue dots) stays far the line constructed by using training set (red dots). This is a sign of high variance (overfitting) and we can say that this regression line is not a good model because we observe overfitting.

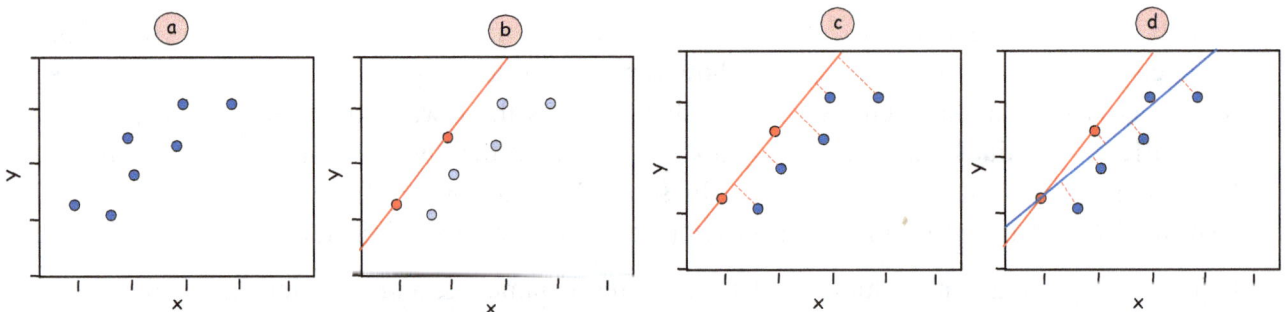

Figure 8-17: (a) Original dataset. (b) Training set data points were selected as red color. (c) The other remaining points RSS are calculated to the regression line (red regression line). (d) By introducing a Ridge penalty, the red regression line is rotated, and the blue line regression is drawn.

To solve the problem of high variance, we use a Ridge regression, and an increase in its penalty causes a rotation of the red regression line a bit toward the X-axis. As a result, the blue regression line will be created, which is shown in Figure 8-17 (d). The blue regression line has a lower variance despite having a bit more bias. In such a simple approach, Ridge regression resolves the overfitting problem. Ridge regression can be applied to logistic regressions and other regression methods as well, but we used a linear regression example for the sake of simplicity.

If you would like to learn how the math behind the Ridge regression works, you can plot some data points in 2D spaces and try them on your own with the linear regression, which is easy to calculate on a piece of paper.

Note that the shrinkage penalty is not applied to the intercept (β_0), it is applied to other parameters ($\beta_1, \beta_2, \ldots, \beta_k$), because the intercept measures the mean value of responses, and we do not want to shrink that. Increasing the λ causes a decrease in the slope. The larger the lambda gets, the prediction for y becomes less and less sensitive to x. In other words, the introduction of the penalty moves the regression line more toward the horizon line (parallel to the x-axis). When all coefficients are set to zero, the regression line will be a horizontal line parallel to the x-axis, which obviously is useless because $y = \beta_0 + 0 \times x_1 + 0 \times x_2 + \ldots = \beta_0$, and we do not have any predictor in the final model.

Assuming m is the number of our parameters, and n is our data points for the training set. The computational complexity of the ridge regression is $O(m^2 n)$. Ridge is good when we have a small dataset size, but it is not scale-invariant[6].

LASSO

LASSO (Least Absolute Shrinkage and Selection Operator) [Santosa '86, Tibshirani '96] is another regularization method in use and performs slightly better than ridge regression.

Ridge regression is very helpful, but it has a disadvantage. It keeps all coefficients in the final model and does not remove them completely. It means that Ridge reduces all coefficients toward zero but didn't set any of them to zero. In other words, Ridge does not exclude predictors that are unrelated to the model, it only reduces the magnitude of their coefficients, which is against Occam's Razor principle. Therefore, all predictors will stay in the final model, even the unrelated ones, but with a low coefficient. However, the LASSO regularization resolves this by substituting the Ridge's penalty $\Sigma_{i=1}^n \theta_i^2$ (L2 norm) with $\Sigma_{i=1}^n |\theta_i|$ or $||\theta_i||_1$ (L1 norm). We have the following equations for these two regularizations:

$Ridge\ Regression = RSS + \lambda \Sigma_{i=1}^n \theta_i^2$

$LASSO\ Regression = RSS + \lambda \Sigma_{i=1}^n |\theta_i|$

LASSO is very simple, it just uses an absolute value instead of squares. Nevertheless, it creates a parsimonious model (respecting Oscam's Razor principle), and even it has been recommended to use it as a feature reduction technique as well.

---

[6] If you recall, we have used this term in Chapter 6, while describing SIFT. However, scale invariant means that the feature or characteristics of the system should not change if their scale of lengths, energy, or other variables changes.

For example, assume we are creating a linear model to predict the success of a candidate in an upcoming election. The model measures success based on the following parameters as follows:

$success = \beta_0 + \beta_1.diet + \beta_2.hobby + \beta_3.climate\text{-}change + \beta_4.foreign\text{-}policy.$

If we have a dataset about this model and use Ridge regression, by increasing the λ, both personal's "diet policy" and "hobby", which do not affect on the success of the candidate will be shrunken toward zero, but they wouldn't be removed completely. However, LASSO can clean the model and get rid of both (diet and hobby) by increasing the λ, and thus the final model will only keep "climate-change policy" and "foreign-policy" as its predictors. LASSO is better than Ridge because we can say that its result models are more parsimonious. In summary, LASSO shrinks the coefficients to zero, unlike Ridge, which shrinks them toward zero but does not remove them entirely. Since LASSO can remove several features from the model, its result model is called a *sparse model.*

LASSO is better than Ridge, but it does not perform very well for highly correlated predictors, and it is also not scale invariant. Besides, when the number of parameters is larger than the number of training set data points, it does not perform well.

Assuming k is the number of variables and n is the number of sample sizes, LASSO computational complexity is $O(k^3 + k^2n)$ [Efron '04].

Elastic Net

Both Ridge and LASSO are useful, but using them when our model includes many parameters, might be challenging. Besides, when the number of our parameters is larger than the available data points for training both LASSO and Ridge do not perform well.

Elastic Net regression [Zou '05] can mitigate these issues because it combines the strength of both Ridge and LASSO. Elastic Net regression is a combination of Ridge (L_2) and LASSO (L_1) regressions. We write the equation of Elastic Net as follows:

$Elastic\ Net\ Regression = RSS + \lambda_1\Sigma_{i=1}^{n}\theta_i^2 + \lambda_2\Sigma_{i=1}^{n}|\theta_i|$

λ_1 is used for Ridge and λ_2 is used for LASSO. We should use cross-validation to find the best values for both λ_1 and λ_2. By setting $\lambda_1=0$, we will have LASSO regression, and by setting $\lambda_2=0$, we will have Ridge regression.

Elastic Net uses this combination of penalties, which can better deal with correlative parameters. Therefore, Elastic Net includes advantages of LASSO and Ridge, but the best way to decide about a regularization selection is to conduct several experiments on the dataset with different regularizations and check results until a good one that addresses our concerns is identified.

Non-Negative Garrote

Another regularization method worth being familiar with is Non-Negative Garrote (NNG) [Breiman '95]. If you don't know what is Garrote, it is better not to search online for it, and we wonder how the author of this method cannot end up using a less gruesome name.

NNG is comparable to LASSO, but it is scale-invariant. It can also eliminate predictors with weak coefficients (LASSO can not do it). NNG does not require specifying the penalty function separately, as it has a fixed penalty function. It only requires specifying the tuning parameter λ, which reduces the computational complexity compared to methods that require selecting both the penalty function and the tuning parameter [Xiong '10].

Assuming n is the number of parameters and $u \geq 0$ is the shrinkage factor, its equation is written as follows:

Non-Negative Garrote $= RSS + 2\lambda \Sigma_{i=1}^{n} u_i^2$

There are different approaches to computer the shrinkage factor u_i for parameter i. For example, assuming β_i is the ith parameter, and $\hat{\beta}_{i,LS}$ is the least squares estimate of the parameters, we can rewrite the Non-Negative Garrote as follows:

Non-Negative Garrote $= RSS + \lambda \sum_{i=1}^{P} \dfrac{\beta_i^2}{\hat{\beta}_{i,LS}^2}$

This method is not as popular as previous methods and unlike Elastic Net, it cannot handle datasets whose number of predictors (or parameters) are larger than training data points. Perhaps its attractive and friendly name demotivates researchers and developers to implement it into their software packages or libraries.

NOTES:

* Keep in mind that regularization should be performed on the training set. We should use the same regularized model for both training and testing to assess the model's generalization performance accurately.

* The cost function result of the training set is usually different from the cost function result of the test set because regularization is applied to the cost function at the training phase.

* We have explained that Ridge regression can help to model data with a small training set. Even in some cases, the number of parameters is larger than the number of training data. If we have a model whose parameters are larger than its data points (the dataset is small), we can also use a regularization regression to reduce the magnitude or influence of its parameters.

* In the world of mathematics, we have a concept referred to as *Lagrangian relaxation*. It is a method to approach a difficult problem of constrained optimization by a simpler problem. It solves the optimization problem by introducing additional variables called *Lagrange multipliers*, which act as a penalty. As a result, it simplifies the optimization solution. While regularization methods use penalty terms in their objective functions, they cannot be considered as direct applications of Lagrangian relaxation. Regularization methods are designed to address the problem of overfitting and improve the generalization of models rather than solving constrained optimization problems (Lagrangian relaxation does it).

Optimization Algorithms

Any repetitive process that is performed by humans is associated with optimization. Optimization is a fundamental concept that can be observed even in everyday human behaviors, such as learning to eat efficiently. Initially, infants explore various methods, using their hands, faces, and different objects to feed themselves. This exploratory phase is messy but crucial for learning. As they grow, individuals experiment with various tools like spoons, forks, and chopsticks, gradually optimizing their technique based on feedback, effectiveness, and cultural practices. This iterative process of experimenting, learning, and refining is similar to optimization in scientific concepts. In particular, optimization tests various strategies and adjusts to achieve the best possible outcome, such as minimizing waste, maximizing efficiency, or achieving a specific goal.

Usually, identifying parameters of an unknown mathematical equation, which is the objective of many machine learning algorithms, consumes lots of computational resources. Resources are limited, and an optimization algorithm can help to search efficiently for optimal model parameters within the parameter space. In other words, we use optimization to estimate model parameters while minimizing the computational resources needed to identify optimal model parameters. Before explaining optimization we need to briefly review some algebraic concepts, which most probably we have learned in the school.

Mathematical Concepts Required for Optimization

Probably, if you are not a mathematician and reading this book, you have not encountered derivative gradients after primary/high school, and while you were in primary/high school, you might have thought they were completely useless for your life, especially if you do your high school in developing countries which emphasize on theoretical science more than practical ones. We have been in the same boat until now, and we thought the same way. At that time, we didn't know that mathematics would stick to our lives like a parasite and we could never eliminate it. Galileo, who was a decent scientist said that *"the book of nature is written in a mathematical language"*.

In this section, first, we explain these concepts very briefly. If you are familiar with these mathematical concepts feel free to skip this section.

Derivative

Lets review that a function is something that gets an input variable x, does something with it, and produces the output, i.e. $f(x)$ or y. *Derivative* is a variable that measures the sensitivity of changes in the output variable with respect to changes in the input variable. In other words, it specifies how the output will change, based on the given input. If it is not still clear, read the following: the derivative of a function describes how fast the function changes (increases or decreases).

If the function is a single straight line, the derivative is a constant variable. If the function is a horizontal line parallel to x-axis, it means there is no slope, and its derivative is zero. If the function is changing at different paces, such as in Figure 8-19, the derivative is a function itself. The derivative is written as a ratio of output changes over the ratio of input changes:

$$Derivative = \frac{Output\ Changes}{Input\ Changes}$$

The process of finding the derivative is called *differentiation*. A derivative of a function $f(x)$ is written as $f'(x)$, which is the notation introduced by *Lagrange*. A bit sexier version that is widely in use is called *Leibniz* notation, and it is written as follows:

$$f'(x) = \frac{d\ f(x)}{dx}$$

As you can see from the red lines in Figure 8-18, when the function is not a straight line, the derivative is changing, because, at different points, the function has a different pace of change.

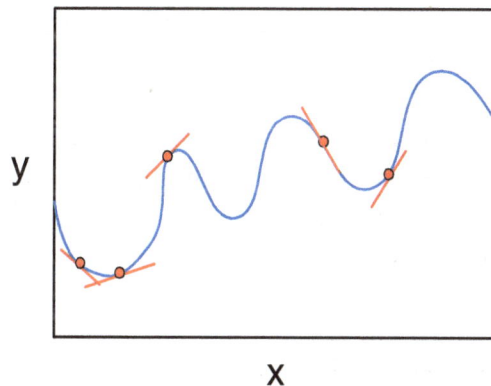

Figure 8-18: A sample function that is plotted in two dimensional spaces ($\mathbb{R}^2$) and five random tangent lines (red lines) has been drawn. You can see that each tangent line touches the curve at one point (red dots).

A constant derivative means that the function is growing constantly and a derivative of a straight line is a constant value.

To better understand the concept of derivatives, let's take a look at Figure 8-19. This figure shows a function that can be modulated with polynomial regression. A line that "just touches" the curve at each arbitrary point is called a 'tangent' line. The particular point that connects the tangent line to the curve is called the point of tangency (red points in Figure 8-18).

The derivative is responsible for specifying the slope for the given tangent line of the function. From Figure 8-18, we can see different tangent lines and their slope changes at each point. More formally, the derivative of a function at a certain point gives the slope of the tangent line to the function's graph at that point.

However, each tangent line has one point on the function, as shown in Figure 8-18, we need at least two points to be able to draw a line. To calculate the slope, we use the following equation:

$$slope = \frac{\Delta y}{\Delta x} = \frac{y_2 - y_1}{x_2 - x_1}$$

This equation shows to calculate the slope, we should select two points from the curve, and by using them, we can calculate the slope of the line connecting them. Let's say the first data point has coordinates $(x_0, f(x_0))$, the second data point is located in h distance to the first data point, thus, it has a coordinate of $(x_0 + h, f(x_0 + h))$. We can rewrite the slope equation as follows:

$$slope = \frac{f(x_0 + h) - f(x_0)}{(x_0 + h) - x_0} = \frac{f(x_0 + h) - f(x_0)}{h}$$

However, the smaller h leads to having a more accurate slope of that particular curve, but why? Let's take a look at Figure 8-19. In Figure 8-19 (a), we can see that the distance of h is the largest, and the line that describes the curve's slope is the least accurate line (in comparison to Figure 8-19 (b) and Figure 8-19 (c)). Then, in Figure 8-19 (b), the points get closer, and thus, the line gets more accurate because the direction is getting more specified. In Figure 8-19 (c), the data points get closer to each other, and the slope line gets more accurate than what we had in Figure 8-19 (b) because, from Figure 8-19 (a) to Figure 8-19 (c), the red line becomes more tangential and does not cross the function line (blue lines).

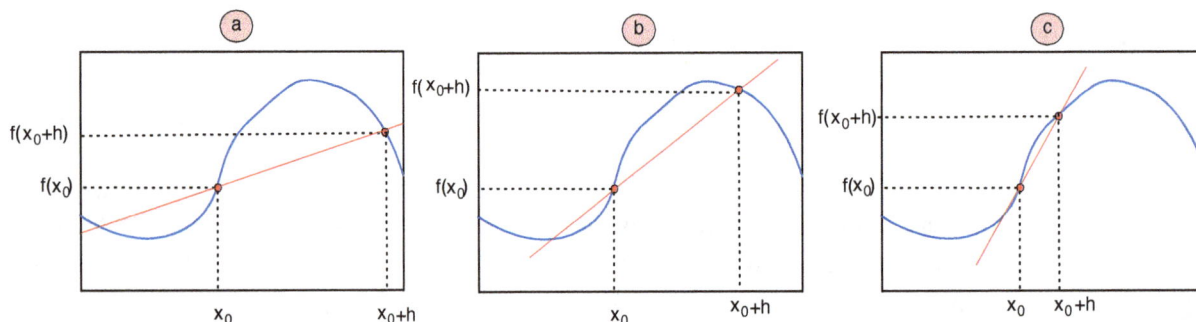

Figure 8-19: Three examples of slope line identification based on distances between two data points, which have h distance. These slope lines are called 'secant' lines. The more h gets smaller, the *secant* line gets closer to being a tangent line.

426

From Figure 8-19 we can conclude that, as the h distance gets smaller, the slope line to the tangent line is getting more accurate. Therefore, we can write the derivation function of x as the *limit* as h is approaching zero:

$$f'(x) = \lim_{h \to 0} \frac{f(x_0 + h) - f(x_0)}{h}$$

There are some basic rules for a derivative that we need to know, assuming n is a constant number, including:

$f(x) = x^n \to f'(x) = nx^{n-1}$,

$f(x) = nx \to f'(x) = n$,

$f(x) = ln(x) \to f'(x) = 1/x$,

$f(x) = x \to f'(x) = 1$,

$f(x) = n \to f'(x) = 0$,

In other words, the function x^n has the slope of nx^{n-1}, the function nx has the slope of n, etc.

What we have explained is a first-order derivative. It is also possible to make a derivative from a derivative, i.e. *second-order derivative*, and it is written as $f''(x)$, e.g.,
$f(x) = x^2 \to f'(x) = 2x \to f''(x) = 2$

The second-order derivative is more stable than the first-order derivative because it captures the rate of change of the first-order derivative. However, due to its high computational cost, it is used rarely. To calculate the derivative, we should use some rules specified for the derivative including the *chain rule, quotient rule, product rule, sum rule* and *difference rule,* which we do not explain in detail here and just write down. If you would like to learn their proof, there are plenty of online resources to help you understand them.

Chain rule (outside-inside rule): $[f(g(x))]' = f'(g(x))g'(x)$ # it most important rule that is used in backpropagation.

Reciprocal rule: $(\frac{1}{x})' = \frac{-1}{x^2}$ # it is known as a specific case of Quotient rule.

Product rule: $(f(x) . g(x))' = f(x) . g(x)' + f(x)' . g(x)$

Quotient rule: $(\frac{f(x)}{g(x)})' = \frac{g(x)f(x)' - g(x)'f(x)}{g^2}$

Difference rule: $(f(x) - g(x))' = f'(x) - g'(x)$

Sum rule: $(f(x) + g(x))' = f'(x) + g'(x)$

We will learn how Chain rule is used for Backpropagation Chapter 10. For now, keep in mind that the Chain rule is useful for finding a derivative of a function when we have nested functions (a function inside another function), e.g. $g(.)$ is inside $f(.)$. We can say it is a derivative of the outside function while leaving the inside function times the derivative of the inside function, check this equation $[f(g(x))]' = f'(g(x)) \times g'(x)$ we wrote about chain rule.

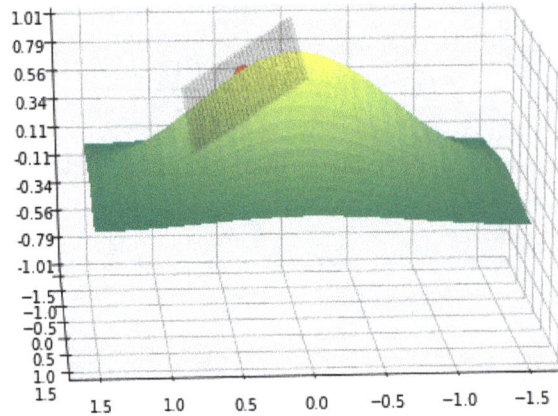

Figure 8-20: Tangent hyper plane plotted in 3D space.

Second-Order Derivative

Making a derivate from a derivative is referred to as second-order derivative. In mathematical terms, if you have a function $f(x)$ its derivative is written as $f'(x)$ or $\frac{df}{dx}$. The second-order derivative is the derivative of $f'(x)$ and it is denoted as $f''(x)$ or $\frac{d^2f}{dx^2}$.

The second-order derivative is particularly useful in various applications such as physics, engineering, and economics, where it can signify concepts like acceleration (the rate of change of velocity) or concavity/convexity of functions, among others. However, due to its high computational cost we will avoid it or in cases we must implement it, we use a trick to simulate second-order derivative.

Partial Derivative

We have discussed the derivative, which is used for functions of one variable. Sometimes, we need to deal with a function with more than one variable, e.g., $f(x, y) = x^2 + 4y$. To differentiate such a function, we use the *partial derivative*. A partial derivative once computes the derivative for x and considers the other variable (y) as constant. Then, it computes the derivative for y and considers the variable x as constant.

Instead of d we used to refer to the derivative, here, we use ∂ sign for partial derivative. To write a partial derivative for variable x and variable y, we write the following:

$$\frac{\partial}{\partial x}[f(x, y)] = 2x + 0 \text{ and } \frac{\partial}{\partial y}[f'(y)] = 0 + 4$$

428

Gradient

Now that we understand derivatives, let's discuss them in the context of multivariate functions. Recall that the derivative in univariate functions represents the rate of change with respect to a single variable and is represented in two-dimensional space, often visualized in $\mathbb{R}^2$. What happens when we extend this concept to functions of more than one variable, i.e., *multivariate* , which is represented in higher-dimensional spaces such as $\mathbb{R}^3$?

The generalization of derivatives for a multivariate function is referred to as *Gradient*. In simple words, the gradient is used for functions that take more than one input variable (or parameter). It is a *vector of partial derivatives,* and gives the input direction in which the function increases or decreases. In machine learning, we often deal with high-dimensional spaces, and while the derivative conceptually applies to vectors, the gradient extends to higher-dimensional structures like matrices and tensors.

While calculating the gradient, instead of the tangent line, which we had in the derivative, we have a *tangent plane* or *hyperplane* for the gradient. Figure 8-20 visualizes the gradient hyperplane on a function plotted in 3D space. Let's review what we have said: *a gradient is a vector of all partial derivatives of the target function.*

A Gradient of a function $f(X)$, assuming $X = \{x_1, x_2, \ldots x_n\}$ can be written as follows:

$$\nabla f(X) = \left[\frac{\partial f(X)}{\partial x_1}, \frac{\partial f(X)}{\partial x_2}, \ldots, \frac{\partial f(X)}{\partial x_n} \right] [7].$$

For example, for $f(x_1, x_2) = a x_1^2 + b x_2 + c$ function.

The derivative of this function with respect to x_1 is written as: $\dfrac{\partial f(x_1, x_2)}{\partial x_1}$. The gradient of this function is written as follows:

$$\frac{\partial f(x_1, x_2)}{\partial x_1} = 2a x_1 + 0 + 0 , \quad \frac{\partial f(x_1, x_2)}{\partial x_2} = 0 + b + 0.$$

$$\nabla f(x_1, x_2) = \left[\frac{\partial f(x_1, x_2)}{\partial x_1}, \frac{\partial f(x_1, x_2)}{\partial x_2} \right] = \left[2a x_1, b \right]$$

In machine learning, *we calculate gradient for cost functions*, and it is used by the most popular optimization methods. Let's see another example, we choose RMSE as a cost function for a linear model, with n data points, we will have the following cost function:

$$J(\beta_0, \beta_1) = \sqrt{\frac{\Sigma_{i=1}^{n}(\hat{y}_i - (\beta_1 x_i + \beta_0))}{n}}, \text{ its gradient will be } \nabla J(\beta_0, \beta_1) = \left[\frac{\partial j}{\partial \beta_1}, \frac{\partial j}{\partial \beta_0} \right]$$

---

[7] The ∇ sign is read as "nabla" and ∂ is a "partial" sign used for specifying a derivative.

Jacobian

When we stack the gradient of two functions, each of them is a vector of variables, into a matrix called a Jacobian matrix. For example, assume we have $\nabla f(x, y)$ and $\nabla g(x, y)$. The Jacobian matrix is written as follows:

$$J = \begin{bmatrix} \nabla f(x, y) \\ \nabla g(x, y) \end{bmatrix} = \begin{bmatrix} \dfrac{\partial f(x, y)}{\partial x} & \dfrac{\partial f(x, y)}{\partial y} \\ \dfrac{\partial g(x, y)}{\partial x} & \dfrac{\partial g(x, y)}{\partial y} \end{bmatrix}$$

For example, if we have $f(x, y) = 4x^2 y$ and $g(x, y) = 2x + y^3$, then their Jacobian matrix is written as follows:

$$J = \begin{bmatrix} \nabla f(x, y) \\ \nabla g(x, y) \end{bmatrix} = \begin{bmatrix} 8x & 4x^2 \\ 2 & 3y^2 \end{bmatrix}$$

Hessian

The gradient is the first-order derivative of a multivariate function, and it can be written as a vector of variables. The Hessian matrix is a square matrix of second-order (a derivative of derivative) partial derivatives of a function. We have seen previously that the gradient can be written as $\nabla f(x_1, x_2, \ldots x_n)$. The Hessian, which is the second-order of the gradient, is a symmetric matrix, and it is written as follows:

$$\nabla^2 f \quad or \quad H(f) = \begin{bmatrix} \dfrac{\partial^2 f}{\partial x_1^2} & \dfrac{\partial^2 f}{\partial x_1 \partial x_2} & \cdots & \dfrac{\partial^2 f}{\partial x_1 \partial x_n} \\ \dfrac{\partial^2 f}{\partial x_2 \partial x_1} & \dfrac{\partial^2 f}{\partial x_2^2} & \cdots & \dfrac{\partial^2 f}{\partial x_2 \partial x_n} \\ \vdots & \vdots & \ddots & \vdots \\ \dfrac{\partial^2 f}{\partial x_n \partial x_1} & \dfrac{\partial^2 f}{\partial x_n \partial x_2} & \cdots & \dfrac{\partial^2 f}{\partial x_n^2} \end{bmatrix}$$

Hessian includes all possible second-order partial derivatives, or we can say it includes every possible pairing of the n variables.

Integral

Many physical and non-physical entities, such as acceleration, force, and velocity, are defined as a rate of change. The derivation is used to quantify the rate of change in mathematics.

However, the rate of change is not enough to force us to learn mathematics, and some devils, who named themselves mathematicians, introduced the concept of the integral. Integral has many physical and non-physical entities, such as specifying a region's weight with respect to the overall weight of a function.

The integral of a function is the 'area under the curve', which we use to 'specify the weight of the function in a certain range'. Integral is the opposite of derivative, and it is also called anti-derivative. The process of finding the integral is called *integration*. Both differentiation and integration are used to study changes in a function (integration is the opposite of differentiation). Differentiation is about finding the rate of change, while integration is about finding the initial function that was changed.

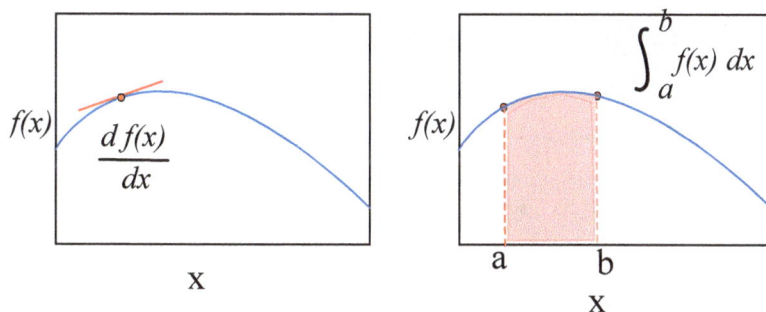

Figure 8-21: Comparison between derivative (left) and integral (right).

Figure 8-21 presents the derivative and integral of a function. The area under the curve is its integral. We have a plot that represents the speed of a car over time. The area under the curve presents the distance traveled by the car over a given period. To compute this distance, we can use integral.

Another application of integrals is finding the volume of an object with a complex shape. For example, we want to find the volume of a complexed-shaped swimming pool. We could use an integral to calculate the volume by dividing the swimming pool into smaller slices, finding the volume of each slice, and then adding them up to identify how much water this pool requires.

Assume we are filling this pool with water from a tap. The rate at which water flows from the tap can be considered the derivative of the water volume in the pool with respect to time. Thus, the total volume of water added to the pool over time can be found by integrating this rate over the relevant time period.

Assume we are filling this pool with water from a tap. The rate at which water flows from the tap can be considered the derivative of the water volume in the pool with respect to time. Thus, the total volume of water added to the pool over time can be found by integrating this rate over the relevant time period.

For example, if the tap is open and it provides a constant amount of water every second, e.g., 1 ml, the water volume inside the pool will increase at a constant rate, mathematically represented as $V(t) = t$ ml after t seconds. If the tap's rate of delivering water increases such that each second it doubles the previous second's rate—represented as $\frac{dV}{dt} = 2t$ ml per second, then the total volume of water in the pool over time will be $V(t) = t^2$ ml after t seconds.

Taylor Series

A *series* in mathematics is referred to as a group of several terms that are all about the same thing. For example, the following is a series:

$$\frac{1}{1} + \frac{1}{2} + \frac{1}{4} + \frac{1}{8} + \frac{1}{16} + \ldots$$

Taylor series or Taylor expansion is a function that has an infinite sum of terms. These terms, however, were calculated based on the repeated derivate at a single point. In mathematics, any function can be represented with a Taylor series.

For example, the Taylor series of functions e^x and $sin(x)$ are written as follows:

$$e^x = 1 + x + \frac{x^2}{2!} + +\frac{x^3}{3!} + \ldots \text{ or } sin(x) = x - \frac{x^2}{2!} + \frac{x^4}{4!} - \ldots$$

Generally, the Taylor series of the function $f(x)$, that is differentiable at a real number a, is approximated as follow:

$$f(x) \approx f(a) + \frac{f'(a)}{1!}(x - a) + \frac{f''(a)}{2!}(x - a)^2 + \ldots \approx \sum_{n=0}^{\infty} \frac{f^{(n)}(a)}{n!}(x - a)^n$$

Taylor series is used in the approximation function and approximation functions try to approximate a target function that is either hard to calculate or unknown, by a simpler function that is easy to calculate or known function.

Linear approximation refers to the first-order Taylor approximation, and quadratic approximation refers to the second-order Taylor approximation.

We learn some mathematical concepts required for optimization. There is more to explain about optimization, but we will postpone it to Chapter 10 where we describe neural networks and deep learning. To continue this section, knowing derivative, gradient, hessian, integral, and Taylor series are enough.

What is Optimization in Machine Learning?

All models we have explained in this chapter are heavily dependent on their parameters, e.g., linear regression is dependent on β_0 and β_1, and changes in these parameters change the model behavior. In the context of machine learning, optimization is the process of changing and tweaking model parameters to minimize the cost function. Why do we need an optimization algorithm, and why not experiment with all possible parameters (brute force) until we get the best result? To better understand the need for optimization, take a look at the right side of Figure 8-22. The bottom of this curve is the optimal parameter value because the cost is at its lowest. However, to get to the minima or maxima point via a brute force approach, we need to experiment with all possible points on the function curve, i.e., all blue dots on the curve. This is expensive to experiment with, and even for this simple convex function, we have too many blue points to experiment with. Recall in Chapter 4, we explained that a real-valued function is called convex if the line segment between any two distinct points on the graph of the function lies above the graph between the two points. For convex functions, any local minimum is also a global minimum. It is efficient in optimization as it guarantees that if we find a local minimum, we have also found the global minimum. But still, it is not guaranteed that the optimization algorithm will find it.

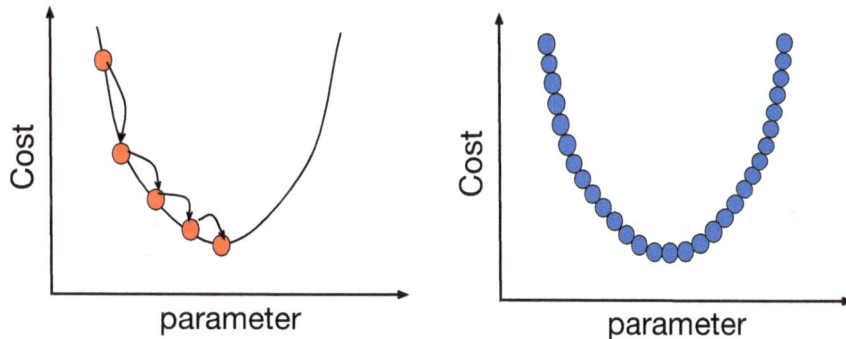

Figure 8-22: (left) Using an optimization to reach the minima, and therefore, very few data points will be processed. (right) Not using an optimization, and thus, many data points will have experimented on the function to find the minima.

However, if we use optimization, as is shown on the left side of Figure 8-22, we can skip many data points and get faster to the minimum. By faster, we mean fewer experiments.

Take a look at Figure 8-23, it is a function and we can see its local minimums (minima or sometimes called optima), have the smallest y value in each valley (or in comparison to its adjacent points), and global minima, is specified the smallest value of y in the entire function. Local minima are points where the cost function is lower than at adjacent points, while the global minima is the lowest point across the entire function.

The process of optimization in the context of the machine learning algorithm is usually identifying a good local minima. By good, we mean a number close to the global minimum or, ideally, the global minimum itself. Therefore, we can say that the ultimate goal of optimization is to find the global minimum. In reality, if we deal with non-convex shape optimization, it is not possible to find the optimal point (global minima), and thus we seek to identify strong local minima. If we have a convex shape (e.g., Figure 8-22), both local minima and global minima are the same, but still, it is not guaranteed that the optimization algorithm will find it.

Note that the blue curve in Figure 8-23 presents the cost function. We don't know the shape of this curve, and by optimization, we experiment with lots of different parameter values and measure the cost to identify this function shape (cost function shape).

By looking at the figure, we can easily identify local minima(s) or global minimum (or minima). Mathematically there are two conditions required to claim a datapoint as a local minima: (i) its first-order derivative is zero, i.e., $f'(x) = 0$ (ii) its second-order derivative is positive, i.e., $f''(x) \geq 0$.

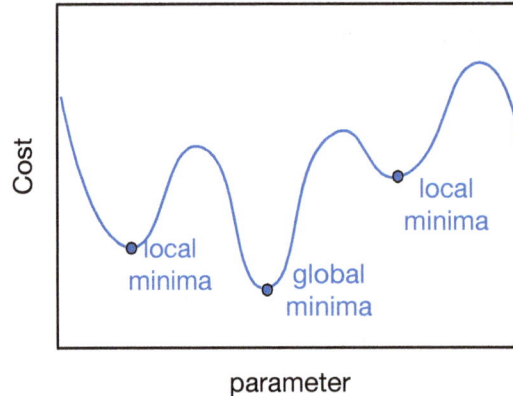

Figure 8-23: A sample function (non-convex shape) with its local minimal and global minima data points have been specified. While reading any of these optimization diagram note that the function presented in blue line is not known and the optimization objective is to find these minima.

Now, by using this background knowledge about optimization, we can say that optimization is focused on reducing the cost or loss function by choosing the right values for the model parameters. In other words, our goal is to build the best model to describe the dataset. Therefore, identifying the optimal value for model parameters reduces the error, and thus we can have the optimal model. Please accept our apologies for repeating it too many times, we want to be sure you memorize the logic behind optimization.

The algorithms we described in this chapter are associated with a cost function, e.g. RMSE is the cost function of linear regression, and changing parameters (β_0, β_1) will affect the RMSE. Therefore, an optimization algorithm is used to identify the lowest cost based on different combinations of β_0, β_1. We do not use RMSE, because RMSE can be used for a convex function, which is usually not a case in a real-world model.

There are various methods for directing the search on the given function toward optimum, including gradient, Hessian, and directional derivatives. In the rest of this section, we explain gradient descent methods, which are widely used for directing the search toward the optimum.

Gradient Descent

Gradient descent is a first-order iterative optimization algorithm that focuses on *minimizing a function by moving its direction toward the steepest descent* (negative of gradient). We will explain Gradient descent in 2D space, which is easier to understand. Previously, we have explained that an optimization problem is the process of finding a good minima that is close to global minima. Now, assume we have a function, and we would like to optimize it, i.e., finding minima close to the global minimum. We knew from the previous section that the gradients of minimum points are zero. Also, the more we move forward toward the minimum point, the closer the gradient gets to zero. Thus, it gets a random data point and calculates its derivative (slope). Then, based on the slope value, another jump toward minima is made.

434

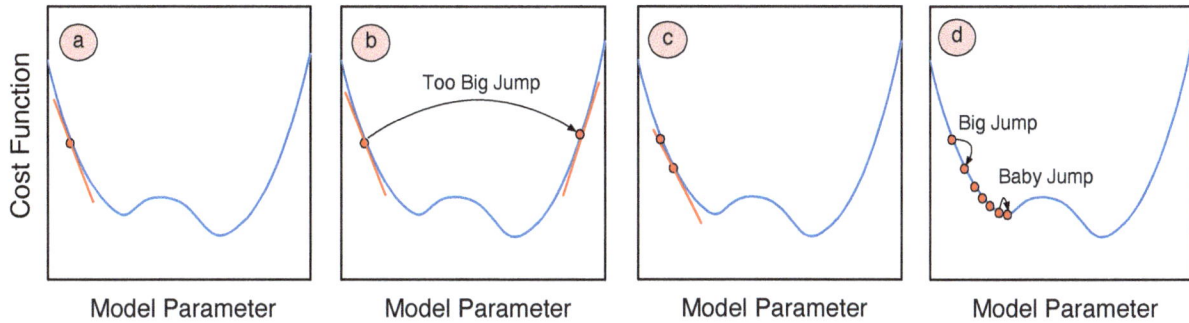

Figure 8-24: A gradient descent approach step by step in a non-convex cost function.

For example, look at Figure 8-24, which presents the cost of an imaginary function based on its model parameter. At the beginning of Figure 8-25 (a), a random point (red dot) is chosen on the function. Next, the optimization algorithm makes a jump, and another point is chosen on the function, as it has been shown in Figure 8-24 (b). Nevertheless, the sign of the derivative of the new point is changed, and this shows that the jump was too big. Thus, the optimization function discards this jump and makes another jump, but smaller, as shown in Figure 8-24 (c), and calculates its steepness. Since the absolute value of the derivative is smaller than the previous point, the algorithm realized that it was a correct jump toward minima. As the derivative of selected points gets closer to zero, the next jumps will get smaller, which means the function is close to a minima. This approach is adaptive step size computation, which is a commonly used policy optimizer. With enough repetitions, the gradient descent function can identify a good minima, which is close to the global minima or could be itself the global minima.

Gradient descent stops either (i) a specific threshold (maximum epoch) reaches, or (ii) the step size gets very small (derivative gets too small and close to zero), which means we reach the minima.

Now, a question might arise: why do we not take the step at the beginning small enough to avoid jumps like Figure 8-24 (b)? The answer is that although taking small steps makes the minima identification more accurate, it makes the gradient calculation process very time-consuming as well.

As another example, let's take a look at a surface plot in Figure 8-25, which shows a cost function of a linear regression model for both β_0 and β_1[8]. We can observe this simple surface plot has one minimum. To calculate the gradient descent, the first random point will be chosen, let's say ($\beta_0 = 0$, $\beta_1 = 0$, $RSME = 1000$). This figure shows that the first jumps are large (the size of the green line between two red dots), then they get smaller and smaller as the gradient descent function moves toward the minimum point (which in this case is global minima, but in complex real-world problems, the cost function is usually not convex and therefore it is not that simple).

---

[8] The visualization provided in this figure is adapted from the code provided at https://github.com/dolittle007/am207_2018/blob/master/wiki/gradientdescent.md

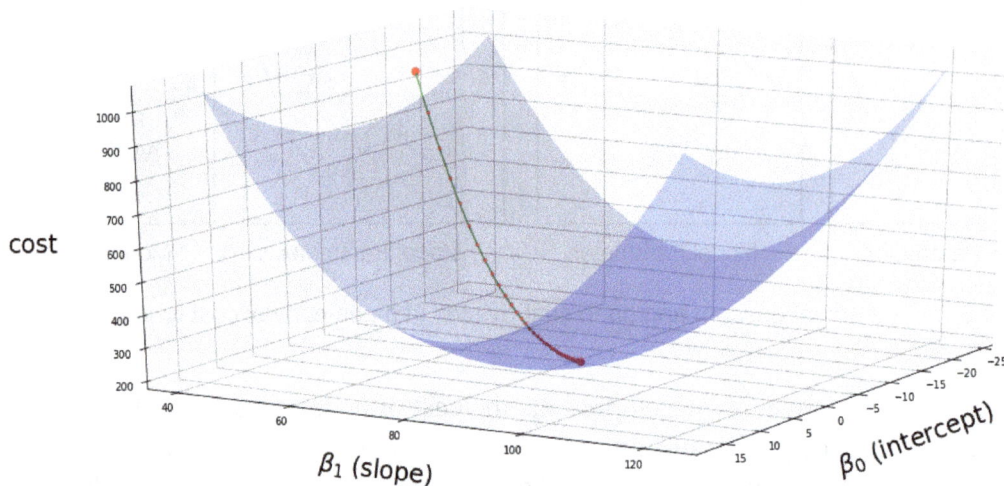

Figure 8-25: Visualization of step sizes of the gradient descent in green line. Each step starts with a red dot and stops in the next red dot. You can see that step sizes are getting smaller as we move toward the global minima in this convex function.

These step sizes were calculated by a parameter called *learning rate* times *slope* of that particular point for each parameter:

next point = current point - (learning rate × slope (a derivative of that particular parameter).

In particular, the learning rate specifies the step in each iteration. A learning rate is usually not constant, and it is given as a hyperparameter. Usually, the gradient descent algorithm starts with a small learning rate of 0.001 and improves it as they experiment more with the dataset. Learning rate, α, is a hyperparameter of gradient descent algorithm.

To summarize this discussion, we can take the following five steps to calculate the Gradient Descent as follows:

(1) The objective is to select a cost function for the given parameter values. For example, we chose to calculate RMSE for the given β_0 and β_1.

(2) A random value for each parameter will be selected and the cost for these parameters will be calculated.

(3) If the model has only one parameter, then its derivative, otherwise gradient, which gives us the slope of this particular cost value (from step 2) will be calculated.

(4) The algorithm calculates the next data point by using the following equation: *new point = current point − learning rate × slope.* As the new data point will be specified, the algorithm uses this data point. The slope is equal to the cost function's gradient at the current point. Assuming the α is the learning rate, we can write the next point will be calculated as follows:

436

$$x_{t+1} = x_t - \alpha_t \cdot \nabla f(x_t)$$

(5) The optimization algorithm checks whether a given threshold for iteration has been reached or the slope size of this new data point is zero (it means we reach the minima), or both values are converged (β_0 and β_1) and not changing anymore. If none of these two conditions were true, then again it goes to step 3 and continues to step 4 and then 5. Otherwise, it stops.

We should be careful with setting the threshold as well. If it is too small, we might never reach a good minima, and if it is too large, it could consume too much of our resources to find a good minima. Each time one training set is analyzed we say one epoch passed.

As a formalized example, let's say the linear regression has the following equation: $y = \beta_0 x + \beta_1$. We intend to use the Sum of Squared Residuals (SSR) error as a cost function.

We do not know the values for βs and both parameters should be identified to minimize the, which is written as $l = \sum_{i=1}^{n} (\hat{y}_i - (\beta_0 + \beta_1 x_i))^2$.

Gradient descent starts with calculating a partial derivative for every parameter, based on the chain rule:

$$\frac{\partial l}{\partial \beta_0} = -2(\hat{y}_i - (\beta_0 + \beta_1 x_i)) \text{ and } \frac{\partial l}{\partial \beta_1} = -2x_i(\hat{y}_i - (\beta_0 + \beta_1 x_i))$$

Instead of SSR we can use RMSE (Quadratic cost function) or other cost functions as well, including cross-entropy cost, which is used for classification, and its equation is written as follows:

$$l = -\sum_{j} (\hat{y}_j log(y_j) + (1 - \hat{y}_j)log(1 - y_j))$$

Gradient descent can use other cost functions which have been described (in Chapter 3 and Chapter 6), and we will learn more about other cost functions in upcoming chapters.

Types of Gradient Descent

There are several types of gradient descent: *batch gradient descent, mini-batch gradient descent, stochastic gradient descent (SGD), SGD with momentum*, etc. The right choice of optimization algorithm is very important because we might get a good accuracy after several days of waiting for the algorithm or get a good accuracy after a few minutes of experimenting. Here, we only explain three varieties of the Gradient Descent algorithm, which is the most popular optimization algorithm.

Batch Gradient Descent (BGD)

As a reminder, recall that the gradient descent uses the training dataset to identify the minimum value of the cost function for different values of model parameters.

A training dataset is divided into smaller segments that are referred to as batches. When all training datasets are collected in one single batch, we use gradient descent for optimization, which is called Batch Gradient Descent.

Batch Gradient Descent is the simplest form of gradient descent because it uses the entire training set to compute the gradients at every step. This means that in each epoch [9], *the entire training set* will be used, and then the gradient will be calculated to decide the next step. Despite its simplicity, since it loads the entire dataset, it is usually the most accurate gradient descent algorithm.

This approach is useful when we have a convex function such as RSME in linear regression because, with enough iterations, the algorithm can easily reach the global minima. Nevertheless, when the training set is large, it uses the entire training set and the batch gradient descent is not resource-efficient.

Stochastic Gradient Descent (SGD)

To solve the problem of using the entire training set another variety of gradient descent has been used since 1951 [Robbins '51], which is stochastic gradient descent. SGD picks *one random sample* from the training set instead of using the entire dataset and calculates its gradient in the first step. Then, again, for the next step, the algorithm takes *another single random sample* from the training set and calculates the gradient for this point. These iterations will continue until it reaches a zero gradient or the maximum threshold for epochs (iterations).

Using random samples from the training set and not the entire training set makes the algorithm very fast and efficient, especially when the training set is too large to load into the memory. Nevertheless, for non-convex functions, it is bouncing around the minima, and unlike BGD, the SGD is not moving straightly toward the minima.

If the cost function is convex, BGD can approach global minima easily. On the other hand, if the cost function is non-convex, BGD is stuck in local minima, but SGD can jump out of the local minima because of its irregular bouncing behavior.

We can observe in Figure 8-26 that the SGD is bouncing around the minima, but BGD moves toward the minima. However, there is no guarantee that this minimum is the global minima, and it could be a weak local minima. Even sometimes, when the step size is too large, the gradient descent can jump out of the minima, as we can see in Figure 8-24 (b). This problem is called *overshooting.*

You might ask why SGD is too noisy? Later in Chapter 10, we explain more about how to reduce the noise. Another simple solution to enforce SGD moves toward minima (even in cone shape functions such as Figure 8-26) is reducing the learning rate slowly and cautiously, which helps the algorithm to settle at the global minima. The process of slowly reducing the learning rate to reach global minima is called *simulated annealing,* and the function that determines a value for the learning rate is called *learning schedule.* If the learning rate reduces too fast, we might stuck

---

[9] Once again, we remind you that a complete pass through the training set is referred to as an epoch, and thus, processing each batch in BGD costs one epoch. Also, every single data point that is processed in SGD is one epoch.

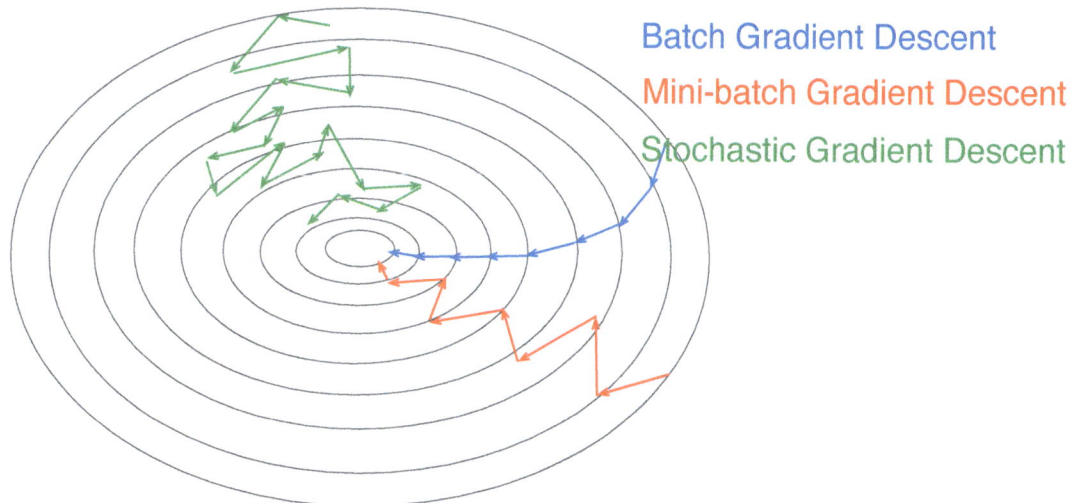

Figure 8-26: A toy example that shows a contour plot of a convex function. The minimum is located in the center. We can see that the batch gradient descent (blue colors) goes straight to the global minima. Mini-batch gradient descent (red colors) moves toward minima with some variances, and this variance significantly increases while using the stochastic gradient descent (green color).

in local minima, if it reduces too slow, we might jump to global minima and end up in a sub-optimal solution. Therefore, implementing a good learning schedule is not easy, but the underlying software library usually takes care of it. Just keep in mind that simulated annealing is useful when finding an approximate global minima is more important than finding precise local minima.

Mini Batch Gradient Descent (miniBGD)

We described that BGD uses the entire training set in each epoch to calculate the gradient, and SGD uses one single sample in each epoch. There is another variation that tries to be in the middle of these two algorithms named miniBGD. In particular, miniBGD uses *a subset of the training set*, (BGD uses the entire training set), and thus it is faster than BGD. Since it uses a subset of training data and not a single instance. However, it has less bouncing in comparison to SGD. In particular, it splits the training dataset (batch) into smaller sets (mini-batch), and in each epoch, it uses one mini-batch. A mini-batch is smaller than a batch and greater than one single data point. Similar to other types, miniBGD needs to have a dynamically changing learning rate, and thus, the learning schedule function here is also important. miniBGD is useful for parallelization of the gradient descent process across multiple GPU [Zhang '21].

Table 8-2 summarizes the comparison between these three methods. We have learned many details about Gradient Descent; it might make sense to emphasize that gradient descent is a first-order optimization algorithm, which means it only works with the first derivation of the function, and it is iterative.

	Advantages	DisAdvantages
Batch Gardient Descent	- no bouncing and no variance - can identify global minima in convex function	- very slow - not suitable for large dataset - can stuck in local minima
Mini-Batch Gardient Descent	- fast - suitable for large dataset - reduce the variance overhead of SGD	- sensitive to the learning schedule - can stuck in local minima but better than BGD
Stochastic Gardient Descent	- very fast - suitable for large dataset - very unlikely to stuck in local minima	- sensitive to deal learning schedule - very high variance that might miss a good minima

Table 8-2: Comparison between advantages and disadvantages of different varieties of gradient descent algorithm.

Newton Method

Gradient descent algorithms and other first-order iterative algorithms (first-order means using the first-order derivative) can determine the direction toward a minima (good local or global). However, a limitation of first-order optimization algorithms is that they cannot determine the right step size. Gradient descent depends on a learning rate, and we describe how to identify the learning rate later in Chapter 10. Besides, the step size is a parameter that should be given to gradient descent algorithms. Nevertheless, it is still the best algorithm used for optimization, and no other algorithm can provide a better result than Gradient Descent.

Second-order optimization algorithms, i.e. Newton, is another iterative method that allows us to determine the 'approximate step size' toward a minima.

Gradient descent uses the derivative at a randomly chosen point to determine the direction of the steepest descent, effectively moving opposite to the gradient. On the other hand, Newton's method approximates the function by a parabola[10] at the current point, leveraging both the first and second derivatives, see Figure 8-27. This approximation is helpful

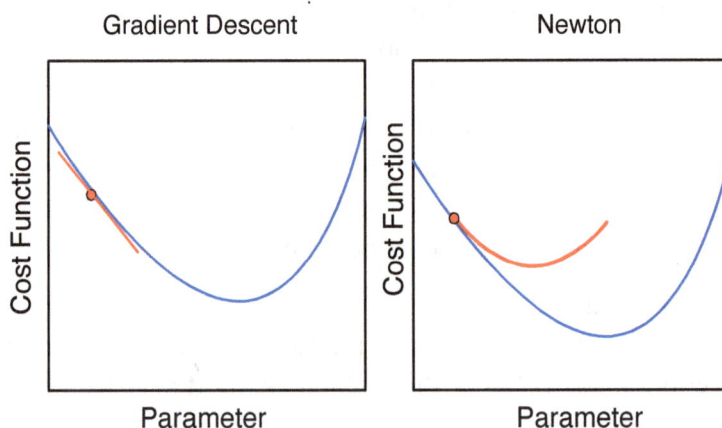

Figure 8-27: The differences between Gradient Descent and Newton method

---

[10] A symmetric bowl-shaped curve is called a parabola.

because it allows the method to find the zero of the derivative more accurately by considering the curvature of the function, which is reflected in the second derivative.

Since a parabola is either convex or concave, depending on its coefficients, it generally presents a clear path to a minima or maxima when it is convex. As is shown in Figure 8-28 after three iteration Newton method gets into the minima. In Newton's method, the minimum of the approximating parabola can be found at the vertex, which is derived from the roots of the first derivative using the second derivative for refinement. The newly identified x-coordinate of this vertex becomes the next point in the iterative search for a minimum.

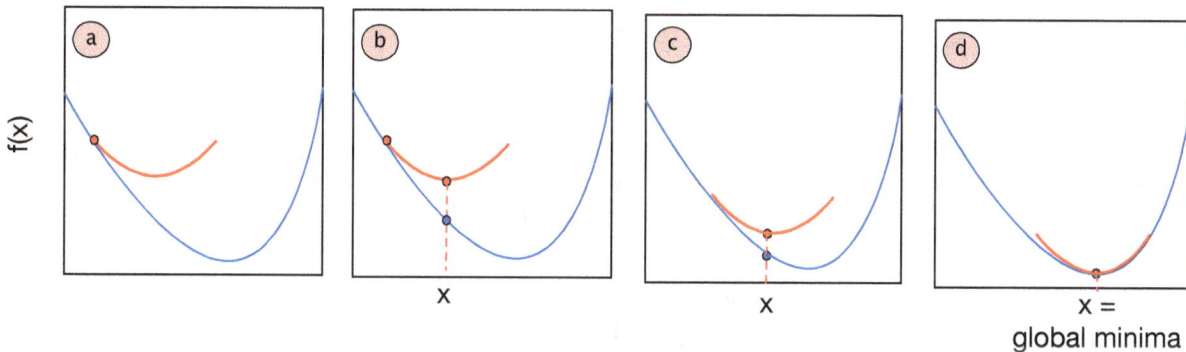

Figure 8-28: Newton method optimization example. We can see with a few steps, the global minima will be identified. Nevertheless, for the sake of simplicity, we assume the optimization function as a convex function.

To understand this concept, we go a bit deeper into its mathematical concepts. Assuming the current point is x_t how did we calculate the next point, $x_t + 1$ in gradient descent? If you remember, we use this equation:

$$x_{t+1} = x_t - \alpha_t . \nabla f(x_t)$$

Based on the Taylor series, a (quadratic) approximation function can be used to define the Newton method. Here, *f(x)* is the quadratic approximation, and with the Taylor series we can write it as follows:

$$f(x) = f(x^{(k)}) + (x - x^{(k)}) . f'(x^{(k)}) + \frac{(x - x^{(k)})^2}{2} . f''(x^{(k)}) + \ldots$$

To calculate the next point, instead of using the slope, we substitute the target point with the Hessian matrix of that point. Therefore, we will have:

$$x_{t+1} = x_t - Hessian^{-1} . Gradient$$

Or we can write it as follows:

$$x_{t+1} = x_t - [\nabla^2 f(x_t)]^{-1} . \nabla f(x_t)$$

$\nabla^2 f(x_t)$ represents the Hessian matrix at x_t, which is the matrix of second-order partial derivatives. Similar to previous examples, for the sake of simplicity, we explain the univariate form of a function.

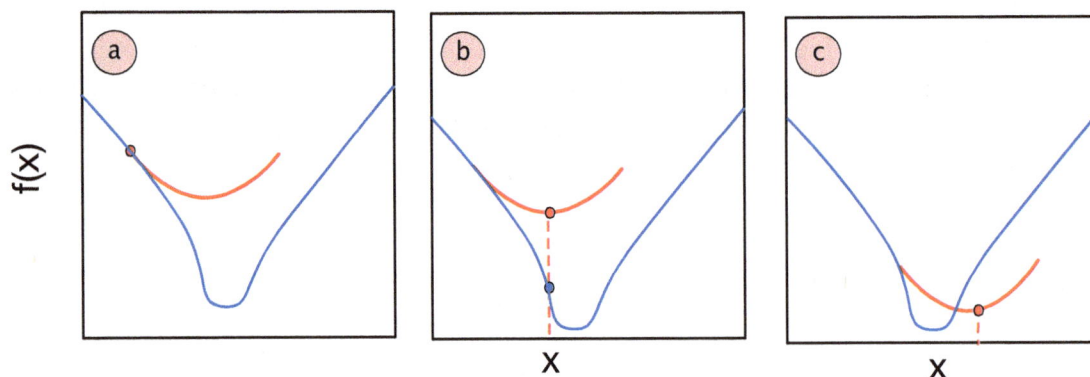

Figure 8-29: Newton method optimization example. We can see in Figure (c) that the parabola falls outside, and thus, minima will not be identified. This is a common problem that occurs a lot if we only rely on Newton's method.

The major problem with Newton method is its complexity of computing second-order derivative. Besides, sometimes the cost function is fairly flat, and the parabola takes a too large step that falls outside the function. This is another drawback of the Newton method. For example, look at Figure 8-29, in which we can see the parabola is too large for the mimina of the function. Therefore, the Newton method works when we are close to the minimum.

When the gradient is
far from minima

When the gradient is
close to minima

Some suggest using gradient descent, and when we get close to the minima, we can take some Newton steps to jump earlier toward the minima. Nevertheless, we do not see Newton's method implemented in modern algorithms because of its huge computational cost. Even algorithms that must deal with second-order optimization, such as TRPO and PPO (we will describe them in Chapter 13), use a trick to simulate Newton's method because constructing a Hessian matrix is computationally very expensive.

442

Early Stopping

In optimization algorithms such as gradient descent or Newton, the epoch number should be given by the user as a hyperparameter. Usually, in each epoch, the error is reduced until a threshold is reached, and then again, it starts to increase; see Figure 8-30. When such a thing happens, to preserve resources, it is better to force the model to stop its optimization and not go until the maximum number of iterations. It means that before all the requested number of epochs finishes when the error rate starts to get higher, we should stop; this phenomenon is called *early stopping*. An increase in the error rate is a sign of overfitting on the training data, and the algorithm passes the minima. Most implementations of the gradient descent objective function enable the user to specify having or not having an early stopping as a hyperparameter.

Figure 8-30 visualizes the need for early stopping; the more epoch encounters by the algorithm,

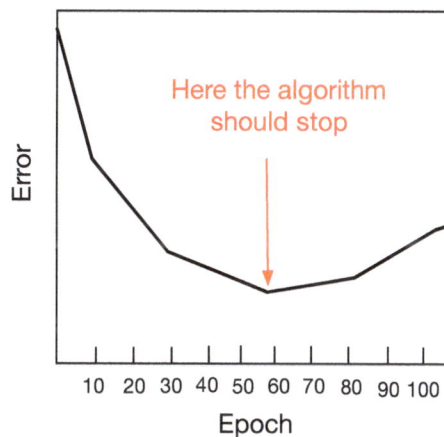

Figure 8-30: Increasing the number of iterations decreases the error until epoch 60, then the error starts to increase. This is where it is recommended to enable the algorithm stops by implementing "early stopping".

the error gets reduced until epoch number 60, in which the error starts to increase. A few iterations later, the algorithm should spot this and stop at epoch 60. This is called early stopping, and it can be categorized as a regularization method.

NOTE:

* The minima data point is where the gradient is zero, but a zero-gradient does not imply necessary optimality.

* Inflection points are when the curve of the function goes from downward to upward or vice versa. Where the derivative is zero, but there are not necessarily local or global minima points. Figure 8-31 presents some examples of inflection points that are not minima or maxima.

* According to Geron [Geron '19], a model is composed of a set of features, and while using gradient descent, we should ensure that all features have a similar scale.

* Gradient descent for linear regression easily gets to the global minima because the cost function of linear regression is a convex function and it has a bowl shape. Therefore, there is no challenge of being stuck in local minima, never reaching global minima, etc.

* SGD is good for non-convex cost functions and reduces the chance of getting stuck in local minima.

* Gradient descent optimization algorithms assume the training dataset data objects are independent and identically distributed (i.i.d). Therefore, to ensure they are not used by the algorithm in any sorted or ordered approach, shuffling the dataset before feeding it into the gradient descent algorithm is recommended.

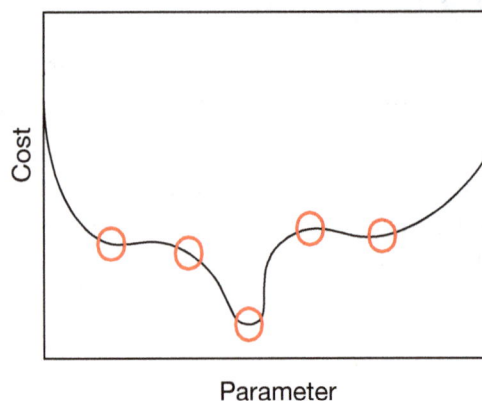

Figure 8-31: Some examples of inflicting points. At the inflicting point, the curve direction changes from upward to downward or vice versa.

* Gradient descent is a type of *hill climbing* algorithm, a popular optimization algorithm. However, gradient descent uses the slope in the local neighbors and then chooses the point with the steepest slope; hill climbing uses a cost function in the local neighbors and chooses the point with the lowest cost score.

Summary

Supervised learning involves regression and classification. In this chapter, we have explained regression algorithms. First, we introduced some concepts, including objective function, epoch, and batch. The objective function is used to maximize or minimize cost model parameters. The cost function is an objective function that tries to minimize a variable. The loss function is similar to the cost function but only for one single data point. A loss score is a quantitative value to measure the difference between the real function and our fitted function.

When an entire dataset is read by the machine learning algorithm, in the context of regression analysis and neural networks, we call each pass an epoch. The number of data points used for training is referred to as a batch. Batch size specifically refers to the size of the subsets the dataset is divided into. For example, dividing a dataset of 1000 data points into two subsets means we can have two batches, and analyzing these two batches takes one epoch to complete.

Linear regression is the simplest and most classical regression model. It is a parametric model. We generally estimate the parameters (slope and intercept) of this model based on its cost function; a basic and common cost function is RSS (Residual Sum of Squares). Whether it is a univariate linear regression model or a multivariate linear regression model, they are basically the same. However, Multiple regression involves knowledge of F-statistics, Forward Stepwise Selection, and Backward Stepwise Selection.

Polynomial regression is also a regression, except that the linear regression plot is linear, while the polynomial regression plot is non-linear. Nevertheless, due to its computational complexity, it is less popular than linear regression.

Condition	Regression Algorithm
Continous Dependent Variable	Linear Regression
Not Continous Dependent Variable	Pieceweise or Polynomial Regression
Categorical Dependent Variable with Binary output	Logistic Regression
Categorical Dependent Variable with more than two output	Multinomial Logistic Regression (e.g. Softmax regression)

Table 8-3: Regression algorithm and their use cases.

Logistic and softmax are called regression, but in reality, they are classification models. Logistic Regression and Softmax Regression are such models. "Logistic Regression" is a well-known binary classifier whose function is also frequently seen in neural networks. Parameter estimation of "Logistic Regression" generally uses the MLE method, which requires a logarithmic likelihood function. "Softmax Regression" is a multi-classification model, and its relationship with logistic Regression is similar to that between ordinary linear Regression and multiple linear Regression. In addition to these regression algorithms, we described the ARIMA model, to complete your mental shock. ARIMA is used for time series extrapolation (or time series forecasting).

Table 8-3 summarizes when to use which regression algorithm. Afterwards, we describe model parameter estimation for each algorithm. Besides, we described methods to evaluate the result of regressions, including k-fold cross-validation, ROC Curve, Pseudo R2, Wald Test, Information Criterion (AIC and BIC), and Likelihood Ratio Test.

Next, we list challenges related to training models, including bias-variance tradeoffs, overfitting, and underfitting. Then, we moved to regularization algorithms, which are used to reduce model parameters, and described Ridge, LASSO, Elastic Net, and Non-Negative Garrote regularization algorithms.

The last part of this chapter explained some algebraic concepts required for optimization and two optimization approaches: Gradient Descent and Newton. Gradient descent is used when we cannot use an algebraic optimization algorithm to calculate the best minima, and we should search for the solution (similar to the heuristic problems explained in Chapter 6). We have explained three types of gradient descent, which are listed as follows:

- Batch Gradient Descent: Batch size is equal to the training set.
- Stochastic Gradient Descent: Batch size is equal to one data point.

- Mini-Batch Gradient Descent: 1 < Batch size < Training set size.

We also explained Newton, which could be used in combination with Gradient descent. Newton is second-order derivative optimization and, unlike Gradient descent, it is independent of step size. However, due to its computational complexity, we should avoid using it directly. Some mathematical tricks could be used to approximate Newton optimization while not using second-order derivatives.

From now on most of the algorithms we learn in machine learning, operate as follows:

(i) makes some random guess about model parameters.

(ii) Use objective function to compute an error.

(iii) Until specified maximum epoch reaches or the error is less than some specific value, use a method to change guess parameters in step (i) and continue step (ii).

Further Readings and Watching

* There is a book by James et al. [James '13] that has a very detailed and easy-to-understand description of linear regressions and their variation.

* If you are interested in learning the math behind logistic regression there is a good tutorial provided under this link: https://www.hackerearth.com/practice/machine-learning/machine-learning-algorithms/logistic-regression-analysis-r/tutorial.

* Joshua Starmer has a very easy to understand detailed explanation about logistic regression and how a maximum likelihood function can be used to identify its parameter. Also, his explanation about regularization is amazing. His videos are available at: https://statquest.org, and in contrast to his songs, his tutorials about regressions are all fantastically useful.

* There is a very good example and explanation about Polynomial regression available online at http://mathforcollege.com/nm/videos; you can use it to learn how polynomial parameters were estimated.

* If you are interested in learning the ARIMA model in detail, Bob Nau has an online tutorial on his homepage https://people.duke.edu/~rnau/411home.htm

* Brandon Foltz has a good series on explaining logistic regression with examples, if you are interested in learning more detail about that. https://www.youtube.com/channel/UCFrjdcImgcQVyFbK04MBEhA

* The hundred-page machine learning book [Burkov '19] has a good introduction to the mathematics required for optimization, and we used it to construct the initial introduction of the optimization as well.

* If you would like to read more about Gradient Descent and its implementation, Geron [Geron '19] has a good summary of gradient descent with Python codes. If you are interested in delving deep into optimization techniques and their mathematical concepts, the Algorithms for

Optimization book [Kochenderfer '19] is a proper reference to read, but only if you intend to go deep into this topic, and be sure you are able to grasp all of its mathematical concepts.

* Another good resource for learning Gradient Descent and comparing different methods together is Jason Brownlee's page, https://machinelearningmastery.com/gentle-introduction-mini-batch-gradient-descent-configure-batch-size.

* If you intend to learn more about the basics of mathematics and algebra Nancy Pi has a very good video series on that. Her explanations are easy to understand. http://youtube.com/nancypi

Chapter 9: Classification

Welcome to another very popular category of machine learning algorithms: classification algorithms. In the previous chapter, we explained regressions that are used for quantitative information (a sexy name for datasets with numerical relations). However, it's important to note that many real-world machine learning datasets comprise qualitative (categorical or non-numerical) and quantitative (numerical) data.

Classification algorithms are used to classify and separate the data points of a dataset based on their labels. For example, an algorithm analyzes microscopic images of a patient's kidney and recommends to the physician if the patient has kidney cancer or not. An email spam filter tries to identify spam from real emails; a social media agent tries to identify fake accounts from real users on social media, etc. Classification is not limited to binary decisions; for example, we can feed a dataset of handwritten digits into a classification algorithm, and it can recognize the digit.

We have explained in Chapter 1 that classification algorithms are supervised learning algorithms and work based on the assumption to separate the dataset into at least two subsets, a test set and a train set. First, we run a classification algorithm on the train set, which builds a model based on the given data. The model could be a set of *if-then-else* rules or mathematical equations, such as regression models. Afterward, we use the built model and run it on the test set. Then, we can evaluate the result of the classification algorithm with the assistance of the ground truth dataset (a.k.a validation, reference, or benchmark dataset). The ground truth dataset is a small part of the training set that is annotated accurately by human experts, and thus, building it is a labor-intensive and expensive process.

There are two types of classifications. One is *non-exclusive classification,* which means a data point could participate in more than one class. The other form is a *mutually exclusive classification* in which any data points are assigned to only one class.

Some algorithms that we have explained in previous chapters are classification algorithms, including logistic regression, softmax regression (Chapter 8), and linear discriminant analysis (LDA) (Chapter 7), which can also be used for classification. In this chapter, we start by describing a rule-based classifier. Then, we describe Naive Bayes, k Nearest Neighbor (kNN), Support Vector Machine (SVM), and decision tree algorithms. Afterward, we focus on ensemble learning methods, including Bagging, Boosting, Stacking, Random Forest, and AdaBoost, and then we finalize this chapter with Gradient Boosting Decision trees and their new derivations, including XGBoost, lightGBM, and Catboost.

You get very frustrated if you intend to read and learn all of these algorithms in a short amount of time.

Note that the gradient boosting algorithms of this chapter are not easy to learn; you might need to read them more than once and check other resources to learn them. We have tried our best to make them easy to understand, but keep in mind that they are complex in their nature. Besides, despite teaching this material for many semesters at Boston University, students still spot some numerical mistakes, and there might also be problems. Please let us know if you spot any numerical mistakes.

Rule-Based Classifier

Although it is the oldest classification approach, this classifier does not fit the definition of machine learning algorithms. It does really not involve any machine learning and is very straightforward computer programming. However, since we can often resolve our problem with this simple method, we respect its simplicity and list it as a classification algorithm.

A rule-based classifier is a set of IF-THEN-ELSE instructions available in all programming languages. For example, we can create a binary classification for the reader of this book, either s/he goes to heaven or hell, based on the following rules.

- IF (*does not recommend this book*) THEN (*will go hell*).

- IF (*does not pay for this book*) AND (*reads this book*) AND (*does not recommend this book*) THEN (*will go hell*).

- IF (*read this book*) THEN IF (*recommend this book*) THEN (*will go to heaven*).

The statement after IF is called *rule precondition (antecedent)*, and the statement after THEN is called *rule consequent*. Thus, we can formalize a rule as IF *antecedent* THEN *consequent*. A rule in this context has two attributes: coverage and accuracy. *Coverage* of rule r is written as:

$$Coverage(r) = \frac{n_{covered}}{n_{total}}$$

Here, $n_{covered}$ is the number of data points (or records if we assume the data is in tabular format) that are covered by rule r, and n_{total} is all data points in the dataset. Another concept is *Accuracy*, which is the number of correctly covered by the rule divided by $n_{covered}$, and it is calculated as follows:

$$Accuracy(r) = \frac{n_{correct}}{n_{covered}} = \frac{Antecenents \cap Consequnts}{Antecendents}$$

If a data point belongs to more than one class, the algorithm can use mechanisms such as voting or priority ordering (some classes have higher priorities to get a data point than others) to decide about the label of that class. Rules could be predefined by users and not learned from the training set.

One might ask why such a simple classification approach is not widely used and why we said it might not fit into the machine learning context. The motivation of machine learning is to teach an algorithm to learn on its own and decide. Rule-based classification is a sort of micromanaging of the learning process and hard coding of the rules. Therefore, it is not a good choice for a machine-learning approach.

Nevertheless, we often use rule-based classification in software development and enjoy its simplicity. For example, in a Game project, we define rules to construct the game character's mood and body shape based on predefined rules. In particular, if the user is far away from their daily steps, the character gets fat and angry. If the user passes the daily goal, the game character stays fit and happy.

Naive Bayes

One of the first theories in statistical learning (one of the oldest names of machine learning) is Bayes' theorem [Bayes '63], introduced by Thomas Bayes. It describes the probability of an event's occurrence depending on our prior knowledge about things related to the event[1]. For example, the probability of us getting peace of mind is related to helping others and not expecting any rewards, and getting obese is related to diet, genetics, etc.

In this section, we first describe the Naive Bayes theorem, and then we describe the prediction capability of the Naive Bayes theorem.

Bayes Theorem

We use the Bayes Theorem to use the probability of something we already know (prior) to predict something that is happening in the future (posterior). In other words, it combines prior knowledge (prior probability) with new evidence (likelihood) to form an updated probability (posterior probability). In the context of machine learning, we employ prior knowledge or evidence (x) to select the best hypothesis or proposition that predicts the future (h). This is the Bayes theorem and is shown in Figure 9-1.

---

[1] If you can't recall the concept of probability, before reading the rest of this algorithm, read Chapter 3.

$P(h \mid x)$ presents the *posterior* probability of hypothesis h given the dataset x.

$P(x \mid h)$ presents the *likelihood* of observing the data x given that the hypothesis h is true.

$P(h)$ presents the *prior* probability of hypothesis h being true before observing the data x.

$P(x)$ presents the *marginal* probability of the data x under all hypotheses, also known as the *evidence*.

You might recall from Chapter 3 that we discussed the conditional probability of P(A|B) will be written as follows:

$$P(A \mid B) = \frac{P(A \cap B)}{P(B)}, \text{ or } P(B \mid A) = \frac{P(B \cap A)}{P(A)}.$$

Naive Bayes assumes that all features are independent of each other. In other words, the Naiveness of this method assumes that if we don't have a probability of two events occurring together, we can assume the probability of having them occurring together is the product of their probabilities and events being independent. In other words, we don't know the probability of A and B occurring together, and it is calculated as follows: $P(A \cap B) = P(A) \cdot P(B)$.

Let's use an example to understand this theorem. One of your friends told you that most people who used essential oils, traditional medicine, herbal medicine, and other non-scientific methods (let's call them the xmed) did not get sick at all. You are curious to see if that theory is right or not, and let's assume you have statistical data of the people who got 'flu' and who 'don't get flu', plus people who 'use xmed' and 'don't use xmed'. We define probabilities of people who had used xmed as $P(x)$ and people who stayed healthy as $P(h)$, we also know the probability of people who have used xmed and didn't get sick, i.e., $P(h|x)$. Now, we would like to know if his theory is correct, i.e., the probability of people who use xmed, given they are healthy $P(x|h)$. Figure 9-1 visualizes the Bayesian inferences and explains how this question can be answered.

Thomas Bayes

$$P(h|x) = \frac{P(x|h)P(x)}{P(h)} = \frac{\blacksquare}{}$$

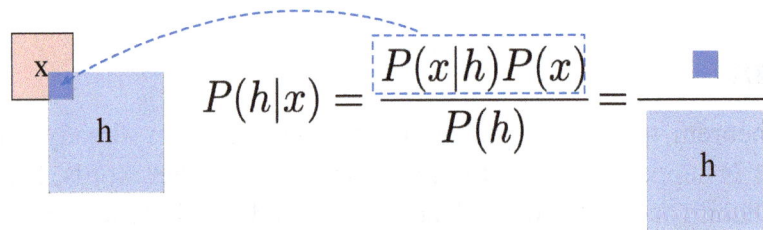

Figure 9-1: Naive Bayes equation and visualization of probabilities.

Prediction with Naive Bayes

The Naive Bayes assumes that each feature is independent of every other feature. This is a naive assumption, but it works fine in many cases. For example, the choice of lunch does not have any relation with the underwear color, i.e., Naive Bayes is correct, but the choice of lunch does have a relation to the diet, and this might negate the Naive Bayes assumption.

Recall to use the Bayesian theorem for prediction, our objective is to predict the $P(h|x)$, which is called *posterior probability*, from $P(x|h)$ which is called *likelihood*, $P(x)$ and $P(h)$ probabilities which are known and available. In other words, from the training dataset, we would like to make inferences, and this inference is called *Bayesian inference*. $P(x)$ is known, the objective is to find the maximum of $P(h|x)$, which is identical to $P(x|h) \times p(h)$.

Assume we have k number of class lables, $c_1, c_2, \ldots c_k$, and each of our data points has n number of features, i.e. $f_1, f_2, \ldots f_n$, we can use the Naive Bayes classification algorithm and assign one of the available k labels to each of the data points. Based on the Bayes theorem, it is formalized with the following equation:

$$P(c_i|f_1, f_2, \ldots f_n) = \frac{P(f_1, f_2, \ldots, f_n|c_i) . P(c_i)}{P(f_1, f_2, \ldots, f_n)}$$

With some mathematical calculations, which we do not describe, the Naive Bayes Theorem can be rewritten as follows:

$$P(c_i|f_1, f_2, \ldots f_n) = \Pi_{j=1}^{j=n} P(f_j|c_i) . P(c_i) \quad for \quad 1 \leq i \leq k$$

As an example of Naive Bayes classification, we can use it to check whether an email is spam or not. We generalize the example to your daily life because, after reading this book from the beginning to the end, you will be a superstar in data science and AI. You are receiving too many emails, and you have asked your secretary to remove all emails with the word "prestigious" and come to your email address. The word "prestigious" is very common in spam emails, but it could be a new client you want to work with. For example, an email that states, "We are a prestigious journal that publishes your scientific work," seems to come from a predatory journal and thus spam.

The traditional spam filters cannot filter your spam emails properly, and you need a more robust approach to filter emails that are asking for your journal submission (spam). Your staff analyzed some sample emails, and he ended up having the following probabilities:

$P(S)$: Probability of an email that is spam. 20 % of emails you received are spam.

$P(\neg S)$: Probability of an email that is important, and thus not spam, i.e., $100 - 20 = 80\%$. The symbol $\neg$ means negation in probability. For example, if A is an event, then $\neg A$ represents the event that A does not occur.

$P(P)$: Probability of an email includes the word "prestigious".15% of your email includes the term "prestigious".

$P(P|\neg S)$: The probability of an email having the term "prestigious", given that it is not spam, is 8% of not spammed emails include this term.

$P(P|S)$: The probability of having an email that includes the word "prestigious", given that it is marked as spam. Your secretary has labeled about 12% of emails that include the term "prestigious" as spam.

$P(S|P)$: Probability of an email being spam, given that it has the word "prestigious" in it. This is what we are trying to find in the dataset or predict. Therefore, we can calculate $P(S|P)$ by using the following equation, which we described earlier:

$$P(S|P) = \frac{P(P|S).P(S)}{P(P)} = \frac{0.12 \times 0.2}{0.15} = 0.16$$

This is a very simple example, and we have only one feature. Let's make the spam filter a bit more intelligent because your staff has realized that you still get many spam emails. He went to collect another 100 sample emails and analyze them. He realizes the term "family" in business email means no clear payment policy. For example, the person who says, *"We are like a family in the work environment."* is planning not to pay properly. Let's use $P(F)$ to present the probability of an email including the word "family". By analyzing those 100 emails manually, your staff created Table 9-1 (a), which includes the result of his analysis. If the values of this table are normalized to be between zero and one, this table is also called the likelihood table, which includes the probabilities.

(a)

	spam	not spam
all emails	20	80
"family"	14	4
"prestigious"	17	6

→

(b)

	spam	not spam
all emails	20	80
"family"	14	4
"prestigious"	17	6
"family" and "prestigious"	14/20 × 17/20 = 0.59	4/80 × 6/80 = 0.00375

Table 9-1: (a) describes the number of spam vs non-spam emails and the frequencies of words "family" or "prestigious" in each category. (b) By relying on Navie's assumption that events are independent, we calculate the probability of having both words by multiplying probabilities divided by the number of emails in each category.

He has forgotten to analyze the occurrences of "family" and "prestigious" terms together. Based on the naive assumption, events are independent, and thus, we can assume the following: $P(F \cap P) = P(F).P(P)$. Here *P(F)={all email contain 'family'}/{all emails}=18/100*, and *P(P)={all email contain 'prestigious' }/{all emails}=23/100*. As these two events are independent, this formula is true. However, it is not our answer, and the desired probability is conditional probability. Therefore, we can create Table 9-1 (b) Simply by multiplying the probability by all emails in each category to understand approximately the number of emails that include both "family" and "prestigious" and are spam, i.e., $20 \times 0.59 = 11.8 \sim 12$. The approximate number of emails that include both "family" and "prestigious" that are not spam is $80 \times 0.00375 = 0.3$ (which is close to zero).

Based on the data provided Table 9.1, we can have the following probabilities:

Probability of getting an email that is spam, $P(S) = 20/100$

Probability of getting an email that is not spam, $P(\neg S) = 80/100$

The probability of getting an email containing the word "prestigious", given it is spam, $P(P|S) = 17/20$

The probability of getting an email containing the word "prestigious", given it is not a spam $P(P|\neg S) = 6/80$

Probability of getting an email containing the word "family", given it is spam, $P(F|S) = 14/20$

The probability of getting an email containing the word "family", given it is not spam, $P(F|\neg S) = 4/80$

The probability of having a spam email, given that it includes both words "family" and "prestigious" is $P(S|F \cap P) = 0.59$. The sign $\cap$ or $\wedge$ stayed for conjunction (and). The results can be read from Table 9-1. Therefore, when an email including these two keywords arrives, the spam filter marks it as spam because there is a 59% probability that it is spam.

This is the *Naive* Bayes algorithm because calculating the probability for each hypothesis is simplified. However, this does not mean that predictive ability is not strong; it is a very useful algorithm for many classification cases; we will describe a classification (prediction) example shortly.

Gaussian Naive Bayes

Gaussian Naive Bayes is a type of Naive Bayes method where continuous attributes are considered, and data features have a Gaussian distribution. The email example that we have previously explained deals with discrete data. We can also use Naive Bayes to make predictions on continuous data. Assuming features are normally distributed, we can use the Gaussian Naive Bayes algorithm. If they are not normally distributed, it is recommended that outliers be removed to make them approximately normally distributed.

For example, let's assume there is another feature in spam email detection, which is the "time of email arrival"; we call it "time" for the sake of brevity. Our objective is to perform a prediction based on time. We assume time is a continuous variable; for the continuous variable, instead of probability, we use their probability distribution function (PDF), which is not the PDF from Adobe but the one we explained back in Chapter 3. Instead of estimating $P(h|d)$, we should estimate $P(S|pdf(time))$, S denotes the spam.

Naive Bayes Prediction Example

Here, we explain another prediction example in another form. This example can ensure that when we finish this section, we are experts in Naive Bayes, and the soul of Thomas Bayes is praying for the sins we have performed or will perform in our machine learning career by a more complex algorithm when Naive Bayes can answer the same thing.

Assume you have a good friend, and when you call him/her, s/he is always available to hang out with you. Having a social interaction is very important for mental health. Nevertheless, watching "TV" or "playing online games" is also delightful.

Every weekend, you check your favorite TV series webpage to see if the new episode of your favorite series, "How to Waste My Time instead of Learning AI," has arrived. Also, you open your phone and scroll the app market to see if there is a new good game to download and play. You do this every weekend, and you decide to use the Naive Bayes algorithm to predict whether you will go to meet your

tv	game	meet friend
no	no	no
no	no	yes
no	yes	no
yes	yes	no
no	yes	yes
no	yes	yes
yes	yes	no
yes	yes	no
no	no	yes
yes	no	yes

Table 9-2: Data we have collected from our previous observations.

friend over the weekend or stay home and spend your time on games or watching TV. Table 9-2. presents what you have recorded about your behaviors over the previous weekends. Your friend would like to predict whether you will meet him/her next weekend (based on the availability of a new game or episode). Assuming the # sign will be used to mention the count of events and ∧ is used to specify intersection (and), first, we calculate the conditional probabilities as follows:

$$P(TV = yes) : 4/10 = 0.4, \quad P(Game = yes) : 6/10 = 0.6, \quad P(Friend = yes) : 5/10 = 0.5$$

$$P(Game = yes | Friend = yes) = \frac{\#(Game = yes) \land \#(Friend = yes)}{\#(Friend = yes)} = \frac{0.2}{0.5} = 0.4$$

$$P(TV = yes | Friend = yes) = \frac{\#(Friend = yes) \land \#(TV = yes)}{\#(Friend = yes)} = \frac{0.1}{0.5} = 0.2$$

$$P(Game = no | Friend = yes) = \frac{\#(Game = no) \land \#(Friend = yes)}{\#(Friend = yes)} = \frac{0.3}{0.5} = 0.6$$

$$P(TV = no | Friend = yes) = \frac{\#(Friend = yes) \land \#(TV = no)}{\#(Friend = yes)} = \frac{0.4}{0.5} = 0.8$$

$$P(Game = yes | Friend = no) = \frac{\#(Game = yes) \land \#(Friend = no)}{\#(Friend = no)} = \frac{0.4}{0.5} = 0.8$$

$$P(TV = yes | Friend = no) = \frac{\#(TV = yes) \land \#(Friend = no)}{\#(Friend = no)} = \frac{0.3}{0.5} = 0.6$$

$$P(Game = no \mid Friend = no) = \frac{\#(Game = no) \wedge \#(Friend = no)}{\#(Friend = no)} = \frac{0.1}{0.5} = 0.2$$

$$P(TV = no \mid Friend = no) = \frac{\#(TV = no) \wedge \#(Friend = no)}{\#(Friend = no)} = \frac{0.1}{0.5} = 0.2$$

Recall that we have explained that a probability in a Naive Bayes is constructed based on the following equation: $P(c_i \mid f_1, f_2, \ldots f_n) = \Pi_{j=1}^{j=n} P(f_j \mid c_i) . P(c_i)$. Now, we can construct every combination of two other available information ("tv", "game"), disregarding the real value of the "meet friend" column.

For example, for the four different combinations of "TV" and "Game" based on Table 9-2, we can write the following:

TV = no, Game= no

$P(meetfriend) = P(TV = no \mid Friend = yes) \times P(Game = no \mid Friend = yes) \times P(Friend = yes)$
$= 0.8 \times 0.6 \times 0.5 = 0.24$

$P(\sim meetfriend) = P(TV = no \mid Friend = no) \times P(Game = no \mid Friend = no) \times P(Friend = no)$
$= 0.2 \times 0.2 \times 0.5 = 0.02$

TV= no, Game = yes

$P(meetfriend) = P(TV = no \mid Friend = yes) \times P(Game = yes \mid Friend = yes) \times P(Friend = yes)$
$= 0.8 \times 0.4 \times 0.5 = 0.16$

$P(\sim meetfriend) = P(TV = no \mid Friend = no) \times P(Game = no \mid Friend = no) \times P(Friend = no)$
$= 0.2 \times 0.2 \times 0.5 = 0.02$

TV=yes, Game=no

$P(meetfriend) = P(TV = yes \mid Friend = yes) \times P(Game = no \mid Friend = yes) \times P(Friend = yes)$
$= 0.2 \times 0.6 \times 0.5 = 0.06$

$P(\sim meetfriend) = P(TV = yes \mid Friend = no) \times P(Game = no \mid Friend = no) \times P(Friend = no)$
$= 0.6 \times 0.2 \times 0.5 = 0.06$

TV=yes, Game=yes

$P(meetfriend) = P(TV = yes \mid Friend = yes) \times P(Game = yes \mid Friend = yes) \times P(Friend = yes)$
$= 0.2 \times 0.4 \times 0.5 = 0.04$

$P(\sim meetfriend) = P(TV = yes \mid Friend = no) \times P(Game = yes \mid Friend = no) \times P(Friend = no)$
$= 0.6 \times 0.8 \times 0.5 = 0.24$

Now, with this data, we can populate Table 9-3 and compare its prediction result (fourth and fifth columns) with the actual value (third column). We can calculate accuracy, precision, recall, and other metrics related to the accuracy by comparing the actual and predicted class results.

tv	game	meet friend	Predicted (meet friend)	Predicted (¬ meet-friend)
no	no	no	0.24 x	0.02
no	no	yes	0.24 ✔	0.02
no	yes	no	0.16 x	0.02
yes	yes	no	0.04	0.24 ✔
no	yes	yes	0.16 ✔	0.02
no	yes	yes	0.16 ✔	0.02
yes	yes	no	0.04	0.24 ✔
yes	yes	no	0.04	0.24 ✔
no	no	yes	0.24 x	0.02
yes	no	yes	0.06	0.06

Table 9-3: Meet or not meet with the friend predicted. Based on the "meet friend" column, we use a tick mark to state a correct prediction and x to state an incorrect prediction. Since, in the last row, both predicted values are equal, we can not make any judgment.

The result of a Naive Bayesian algorithm will be a confusion matrix, and via the confusion matrix, the software package, or manually, we can interpret the results.

The two examples we have explained were binary classification, and we chose binary for the sake of simplicity, but keep in mind that we can use Naive Bayes for multilevel classification or prediction as well. Assuming N is the number of training data objects, f is the number of features, and c is the number of classes, the computational complexity of Naive Bayes in the training phase is $O(Nf)$, and for the testing phase is $O(fc)$.

NOTES:

* Since there is no parameter being used for Naive Bayes, an optimization method is not needed to find a good fit for the parameter. Naive Bayes and similar models are referred to as parameter-free models.

* Naive Bayesian can perform both binary and multi-class classification.

* While dealing with classification, we might encounter generative models, generative statistics, and the term generative often. In simple terms, the generative model is a model that is composed of joint probability distributions, e.g. $P(X, Y)$. Another term in discriminative models describes conditional probability, c.g. $P(Y|X =' something')$. Later, we will learn more about these two types of models.

k Nearest Neighbor (kNN)

The algorithms we have explained, such as regression algorithms and naive Bayes, are based on building a model that generalizes the training data to make predictions. However, other types of supervised learning algorithms do not build a generalized model; instead, they use the whole dataset for prediction. These are referred to as *instance-based learning* algorithms. In instance-based learning, predictions are made by analyzing the specific instances in the dataset that are closest to the new data point. A popular example of these algorithms is the *k*-nearest Neighbor (kNN) algorithm [Cover '67], which classifies new instances based on the majority label among the *k*-closest training examples.

Please close your eyes, take a deep breath, and think about Chapter 4, where we explained clustering algorithms. Imagine yourself surfing among clustering algorithms as a data science expert without the need to learn the mathematics we have described afterward. kNN is very easy to understand and operates similarly to unsupervised learning algorithms. Now, wake up; real life is waiting for you.

kNN operates under the assumption that similar data points are located near each other. In supervised learning, this means starting with a dataset where each data point is labeled. kNN uses a distance function to measure the closeness between data points. For example, the Euclidean distance function, which is also common in clustering algorithms (check Chapter 4), can be used in kNN. The algorithm then labels unlabeled data points based on the labels of the closest data points. The number of these 'closest data points' to consider is specified by the hyperparameter k. For instance, if $k=3$, the algorithm looks at the three nearest neighbors of an unlabeled point to determine its label.

Upcoming Algorithms

As illustrated in Figure 9-2, labeled points might be colored red and blue, while unlabeled points are grey. kNN would predict the label of the grey points based on the majority label of their three closest neighbors. This algorithm receives k as a hyperparameter, which specifies the number of neighbor data points. Assuming $k = 3$, check Figure 9-2 (a), we have a dataset in two-dimensional space, and we label a few of its data points as red and blue data points, and the grey ones in Figure 9-2 (b) do not have labels [2].

---

[2] We could say it is three-dimensional space because another dimension or feature is blue/red color. It is better to refer to dimensions as features because having a dimension that includes either blue or red is not wrong, but it is not common.

The kNN algorithm calculates the distance of each unlabeled data point into k nearest data points and decides on the label. For example, the algorithm chooses a data point that is marked with "?" in Figure 9-2 (b) to decide about its label (blue or red). Figure Figure 9-2 (c) shows that there are three data points close to this particular data point; two are red, and one is blue. Therefore, the majority are red, and thus, this data point will receive the red label as well (Figure 9-2 (d)). This process continues until all data points receive a label. Also, if a new data point is added to the dataset, it will be treated as another unlabeled data point.

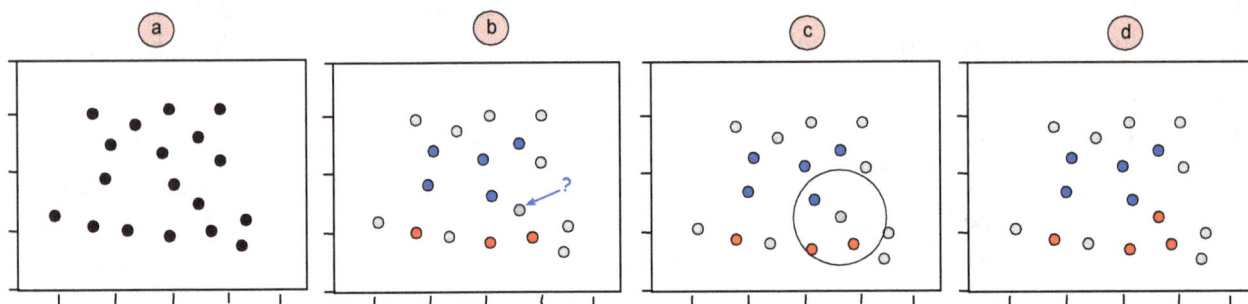

Figure 9-2: (a) The dataset without labeled data points (b) few data points in the dataset has been labeled with blue and red color. (c) a data point have been selected, we need to know what label (red or blue) does this data point receive ?

Our lovely kNN belongs to the category of *lazy learning* algorithms because it stores all the data points first and then uses them, which does not actively classify data points. Due to its simplicity in implementation, kNN is one of the most used supervised learning algorithms and has many applications. For example, online housing agencies can use the kNN algorithm to propose a price for a house based on the nearest neighbor prices they have.

Similar to clustering techniques, the best approach to identifying the optimal value for k is to test with different values for k and choose the one with the highest accuracy. This is known as *hyperparameter tuning* or sensitivity analysis as well. Assuming n is the number of data points, d is the number of dataset dimensions, and k is the number of nearest neighbors, the computational complexity of kNN is $O(nd + kn)$ or $O(kdn)$.

As brute force kNN is very slow, some approaches are used to make kNN perform faster. In the following, we explain three of these approaches, including Voronoi Tessellation, KD-Tree, and Local Sensitive Hashing.

Voronoi Tessellation

Voronoi tessellation [Voronoi '08] is the process of partitioning a 2D plane that includes several data points into regions, in which each region presents the area closest to its related data point and is centered around a data point. Check Figure 9-3(a); we have data points on a 2D plane, and in Figure 9-3(b), the regions were drawn by a Voronoi Tessellation, and the result is the Voronoi diagram. The result is called the Voronoi diagram (right side of Figure 9-3). Each region represents the area that is closest to its corresponding data point, and it is centered around a data point. All points within a region are closer to its center data point than to any other data point. If a new data point is added to the plane, it creates a new region in the Voronoi diagram, altering the boundaries of the existing regions to ensure that each point in the plane is within the region of its closest data point.

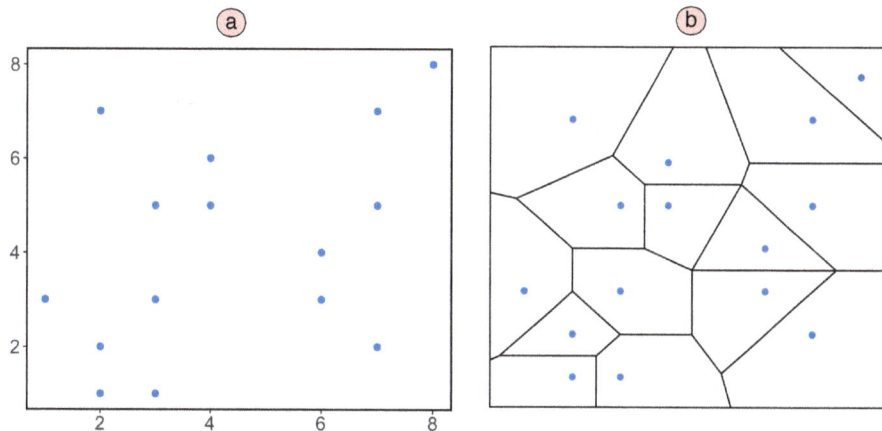

Figure 9-3: (a) Original dataset (b) Voronoi tessellation of the original dataset.

If a kNN algorithm sets $k = 1$, it can use Voronoi tessellation to decide about data points. In other words, Voronoi tessellation is the approach we use to quantify the 2D space for each data point. For $k > 1$, the process involves checking multiple regions around the test point. In particular, the nearest k neighbors (in terms of Euclidean distance) to the test point are identified. The labels of these k neighbors are then considered. The test point is assigned the label that is most frequent among these k neighbors.

Voronoi tessellation improves the search time to $O(\log n)$, but it is only good for low-dimensional data and for higher than two-dimensional data, assuming d is the number of dimensions the computational complexity gets close to $O(d.n)$, reflecting the challenge of distance computations and neighbor searches in higher dimensions.

KD-Tree

If you recall, in Chapter 5, we explained that one of the big challenges in working with data is to reduce the search space, and trees are very useful in reducing the search space. K Dimensional Tree or KD-Tree [Bentley '75] is used for the kNN algorithm to partition the search space efficiently for the nearest neighbors. While KD-Tree is used for exact nearest-neighbor searches, it can also be adapted for approximate nearest-neighbor searches to improve efficiency, especially in high-dimensional spaces. In the latter case, there is a trade-off where some of the nearest neighbors might be missed in favor of faster computation.

Let's use an example to understand how this algorithm distributes data into the tree. Assume we have the following dataset in two-dimensional space: {(2,1), (1,3), (2,2), (3,1), (3,5), (4,5), (4,6), (6,4), (7,4), (7,7), (8,8)}. We plot the dataset as it has been shown in Figure 9-4 (a).

461

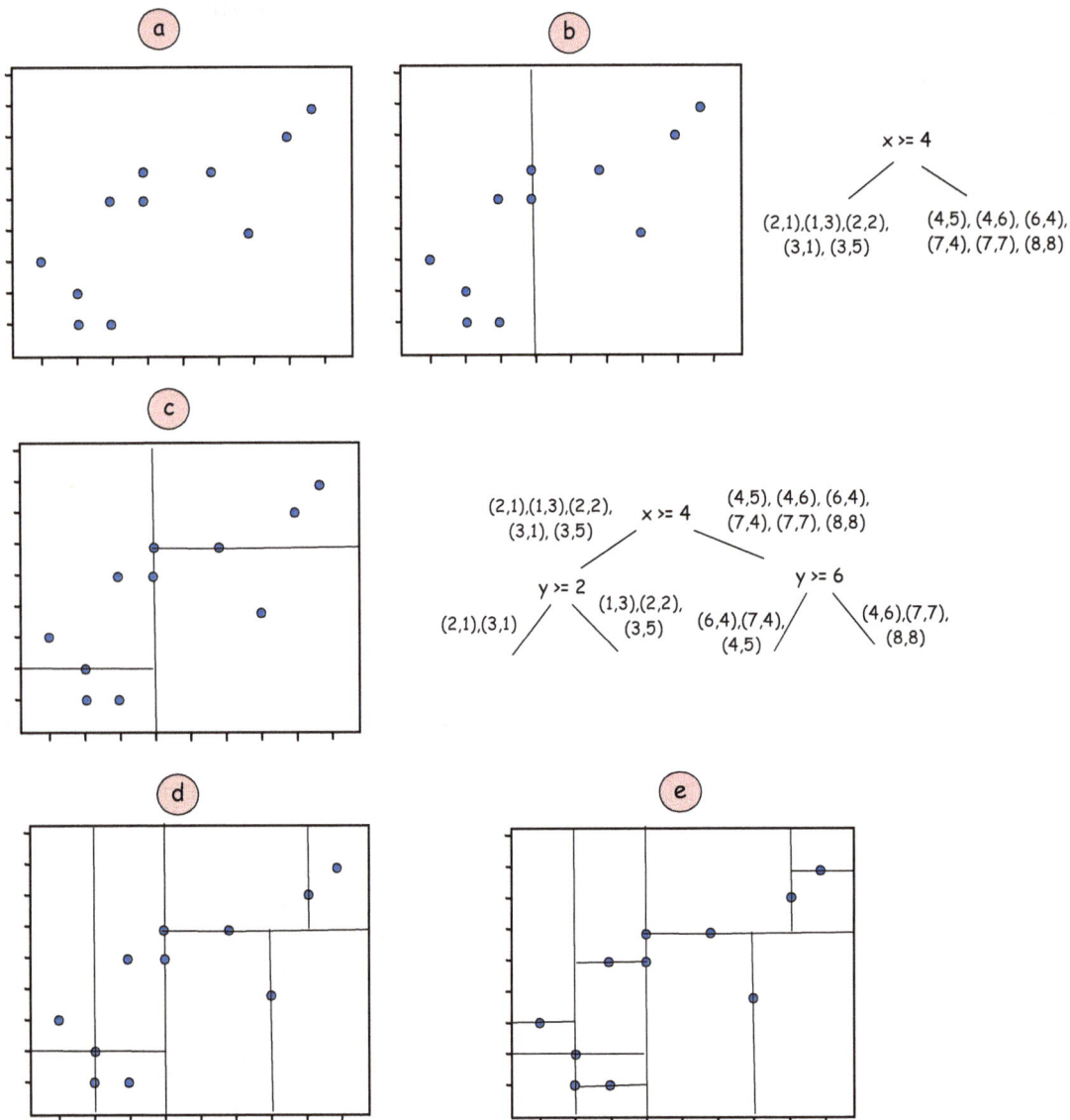

Figure 9-4: (a) The original dataset. (b) The median has been chosen, and the first branch of the tree is constructed to separate the dataset into two sub-datasets. (c) Now the dimension changed (Y-axis), and the same process will be repeated. (d) Again, the axis changed, and the same process. (e) No other data point left, and each node in the final tree has only one data point.

The process of KD-Tree partitioning starts by finding the median in one of the dimensions, e.g., here, we have two dimensions, and we choose x (we could also choose y). The values on the x-axis are $\{1,2,2,3,3,4,4,6,7,7,8\}$, and the median of this set is 4. Next, we plot the KD-Tree based on the median, as is shown in Figure 9-4 (b). We can see we have a tree with a node $x >= 4$ on its right side are nodes with larger or equal x values, $\{(4,5),(4,6),(6,4),(7,4),(7,7),(8,8)\}$, and on the left side, nodes with smaller x values are located, i.e., $\{(2,1),(1,3),(2,2),(3,1),(3,5)\}$.

462

Next, we switch to another dimension, i.e., the y-axis, and get the median of *y* coordinates. Since our dataset is partitioned into two segments, we do it once for each segment. The values on the y-axis on the left segment of Figure 9-4 (b) are {1,1,2,3,5}, its median is '2'. Now, the next node to construct the KD-Tree will be drawn on the y-axis at '2'; the same process will be done on the right side (the median on the y-axis is 6). See Figure 9-4 (c) to check its results. Again, as you can see from the tree on the right side of Figure 9-4 (c), the KD-tree algorithm goes into each branch and does the same (see Figure 9-4 (d)) until each branch has only one data point and the tree generation stops, as shown in Figure 9-4 (e).

By building such a tree to search for the nearest neighbor, the algorithm can navigate through the right branches instead of brute force search to find the nearest neighbors (check Chapter 5 if you can't recall the search on trees). Since it uses a tree, it improves the search for the nearest neighbor. To determine the label for a new data point, the new data point traverses the KD-tree to find its nearest neighbors. Once the k nearest neighbors are found, their labels are used to determine the label of the new point, typically through voting.

KD-Tree is useful when we have low dimensional data; if we need to deal with high dimensional data, it is recommended to use Local Sensitive Hashing. Assuming we have *n* data points and *k* dimensions, the computational complexity of KD-Tree is $O (k.n \log n)$.

Locality Sensitive Hashing (LSH)

Locality Sensitive Hashing (LSH) [Indyk '98] is employed in various applications, such as finding duplicate documents, enabling search engines to locate similar images, aiding biologists in identifying similar gene expressions in genomic datasets, and detecting audio and visual similarities used in authentication systems. Therefore, please switch to full attention mode while reading this section.

Given the prevalence of high-dimensional data in real-world scenarios, LSH is often utilized to enhance kNN efficiency by hashing similar data points into the same buckets, thereby reducing the search space. LSH is somewhat analogous to clustering because it groups similar data points. However, while traditional hash functions aim to minimize collisions (check Chapter 5), LSH works in the opposite, and it is designed to maximize collisions for similar data points, effectively grouping them into the same buckets, which simplifies and speeds up the search process in kNN algorithms.

LSH receives the number of *separator hyperplanes* or *lines*, i.e., *k*, and the *number of iterations*, *l*, as input parameters (hyperparameters). It cuts the data space with lines or hyperplanes with *k* lines, and we end with a maximum 2^k of regions (or buckets) to search. The small regions that are generated by the separator lines are called buckets. Then, it separates the dataset into different regions, and when a new data point arrives, it compares it only to the other data points in that bucket. The algorithm runs this process *l* times.

For example, assume we set *k=3* and *l=3*. Figure 9-5 (a) shows the dataset that is converted into 7 buckets in Figure 9-5 (b). Each new data point that arrives will be compared only to its bucket. This is good, but we might miss some data points. For instance, in Figure 9-5 (b). 'f' belongs to

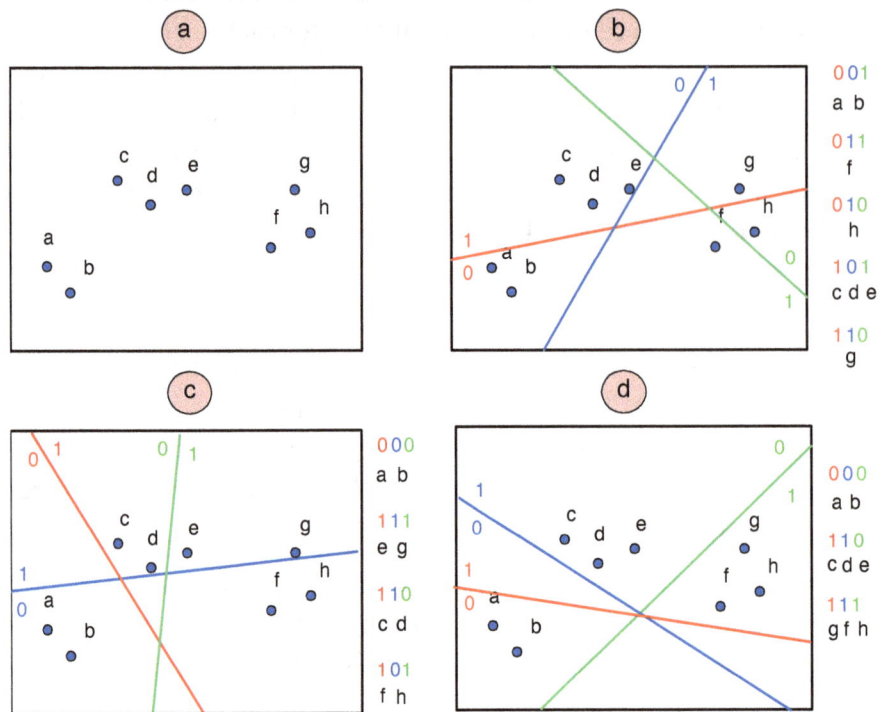

Figure 9-5: (a) The original dataset. (b) Three lines drawn as a hash function to separate each data point. (c) Another three separator lines drawn, and (d) for the third time, another set of line drawn. Note the result of this hashing is written on the right side of each figure.

the 011 and does not belong to the same bucket that 'g' (110) or 'h' (010) belongs. To mitigate this issue, we repeat the process of assigning data points to a bucket three times (because $l=3$), as shown in Figure 9-5 (c) and (d).

Considering the result of this bucketing, we can see 'a' and 'b' stay in the same region, 'c','d', and 'e' in the same region, and 'f' and 'h' are in the same region. However, we cannot judge for 'g', with only $l=3$. We should make l larger to identify neighbors of 'g'. When a new data object is received by the algorithm, kNN can use LSH and assign this data point to one of the existing regions. Also, the list of other data points for that particular region is available, and thus, the algorithm compares the newly arrived data point only to the data points that belong to that region (i.e., the same region). For example, if the hash code assigns region 000 to the newly arrived data object, based on Figure 9-5, it will be compared to 'a' and 'b' only.

Assuming n is the number of data objects, k is the number of separator lines or hyperplanes, and d is the number of data dimensions, the computational complexity of finding the right bucket (hash code) for a new data object is $O(d . k + O(\frac{d.n}{2^k}))$, or we can say it is close to $O(d \log n)$, which is close to KD-Tree complexity.

NOTES:

* While deciding about a k in kNN, there is no optimization-related algorithm exists for that. The best way to decide about k is to perform the parameter sensitivity analysis or hyperparameter tuning, i.e., experimenting and checking the result until we get the best parameter.

* It is recommended that the data be rescaled to have all features in the same numerical range while using the kNN algorithm.

* KD-Tree is also used to order the multidimensional space and facilitate search in the multidimensional environment.

* One of the biggest advantages of kNN is its non-linear decision boundary, which can even handle classifying complex shapes. For example, take a look at Figure 9-6, which shows a complex classification required to separate the class of blue from red data objects.

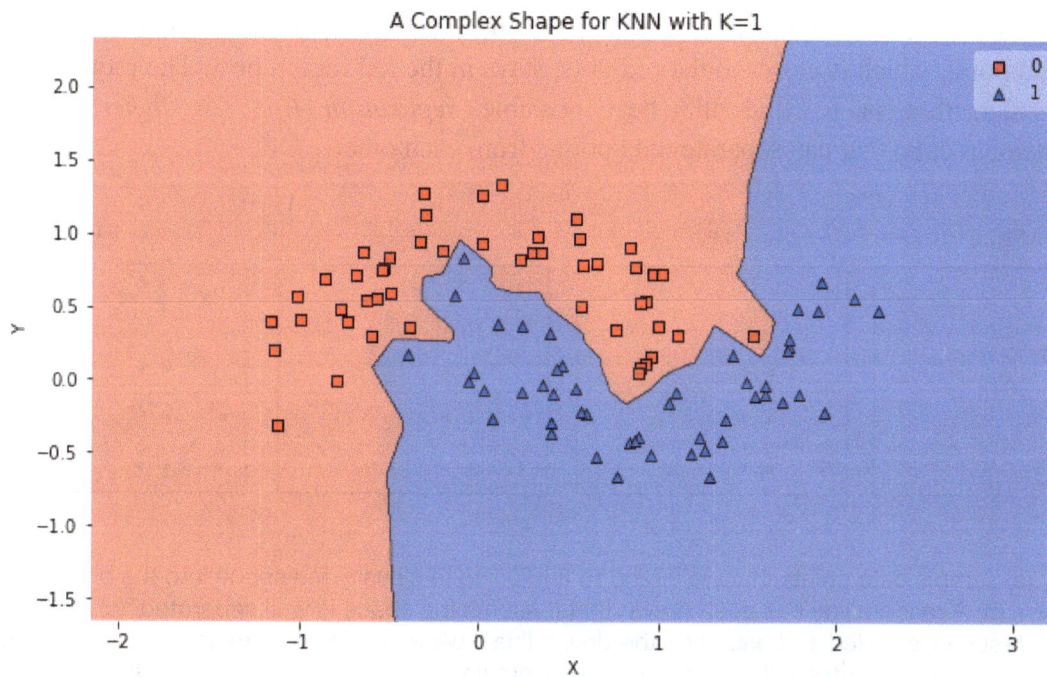

Figure 9-6: kNN algorithm that can classify fairly complex structures. We can see from this image that blue and red dots are interviewed, but the kNN can successfully classify them (look at the red and blue region).

Support Vector Machine (SVM)

An old but popular machine learning algorithm for classification is the Support Vector Machine (SVM) [Cortes '95]. It can classify linear and nonlinear data and performs very well despite its complex approach to classifying data. Besides, SVM can be used for regression as well, but it is commonly used for classification purposes, and we have rarely seen it used for regression.

To explain SVM, let's start with a simple example: we use SVM for binary classification. This example will give us intuition about the rationale of SVM. Then, we explain how a binary classifier extends into a multi-label classifier. Take a look at Figure 9-7 (a). it shows fairly well-separated datasets of blue and red dots. SVM can draw a line and separate them. Let's draw a random line that separates these two sets from each other, as shown in Figure 9-7 (b). After we have this separation line, we would like to test the separation quality and assume we get one new data point as a test, i.e., the grey data points in Figure 9-7 (c). We can see that this data point stays in the blue region, which does not seem correct because it stays closer to the red data points. We draw another separator line in Figure 9-7 (d), and this separates points better because the grey data point, which is closer to the red dots, stays in the red region now. The motivation of the SVM algorithm is to find the best possible *separation line* (or *hyperplane* for multidimensional data) that can separate data points from each other.

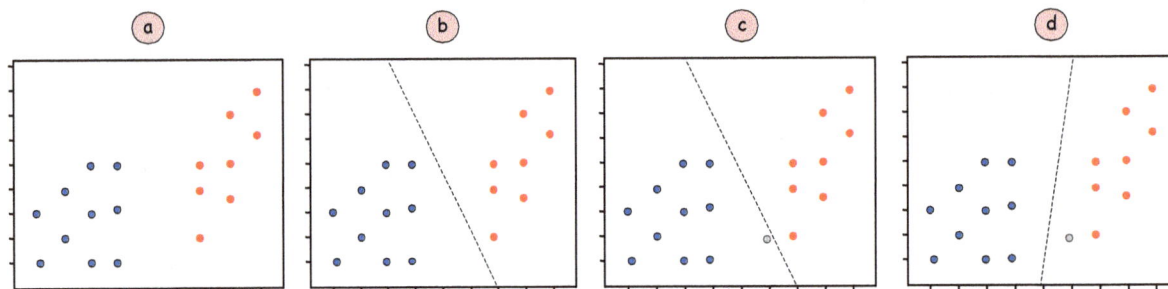

Figure 9-7: (a) sample data points that we intend to find a line to classify the space into the blue region and red region. (b) A random line has been drawn that it seems the space is well separated. (c) a new data point (from test set) is added and based on the drawn line if went into the blue region. However, it doesn't seem at the right place (d) another separator line is drawn and the data point fits into the red region.

To be more accurate from now on, we refer to this separator as a hyperplane instead of dots, lines, etc. The hyperplane for p dimensional space is formalized with a line equation as follows: $\beta_0 + \beta_1 x_1 + \beta_2 x_2 + \ldots + \beta_p x_p = 0$. In Chapter 8, we explained that β_0 is an intercept, and others are the slopes of the hyperplane. $\vec{\beta} = \{\beta_1, \beta_2, \ldots\}$ is called a *weight vector*, and some literature uses $\vec{w}$ to present it. Thus, assuming X is a set of features (feature vector), $X = \{x_1, x_2, \ldots, x_p\}$ and $\beta_0 = b$ (bias), the hyperplane equation is written as: $\vec{w} \cdot X + b = 0$.

Our objective in the given example is to specify the new data points' label as blue or red (classify data points from the test set), and it is a binary classifier, i.e., $y = -1, \quad y = 1$. Unlike other binary classifiers, instead of using 0 and 1, SVM uses -1 and 1. Therefore, we can say one side of the hyperplane is blue ($y = -1$), and the other side is red ($y = 1$) and formalize it with the following equations:

$$\beta_0 + \beta_1 x_1 + \beta_2 x_2 + \ldots \beta_p x_p \leq 1 \quad or \quad \vec{w} . X + b \leq 1 \rightarrow y = -1$$
$$\beta_0 + \beta_1 x_1 + \beta_2 x_2 + \ldots \beta_p x_p > 1 \quad or \quad \vec{w} . X + b > 1 \rightarrow y = 1$$

The test observation data point x^* can be classified by substituting its features $\{x_1, x_2, \ldots\}$ into the hyperplane equation $f(x^*) = \beta_0 + \beta_1 x_1^* + \beta_2 x_2^* + \ldots \beta_p x_p^*$. If $f(x^*) > 1$ then the new test data point x^* will get the label 1 (e.g., red), and if $f(x^*) \leq 1$, then, x will get label -1 (e.g., blue).

In other words, when a new data point from the test dataset $\{x_1, x_2, \ldots\}$ arrives, the algorithm substitutes these x values into the hyperplane equation, and the result will be a number (either ≤ 1 or > 1), which is used by the algorithm to decide about the label of this data point. Therefore, the output will be something like the following: $\{x_1 = -1, x_2 = 1, x_3 = 1, \ldots\}$. Now, the question is, how do we draw this separator hyperplane? Or how do we identify weight vectors $\vec{\beta}$ or $\vec{w}$ of the hyperplane equation?

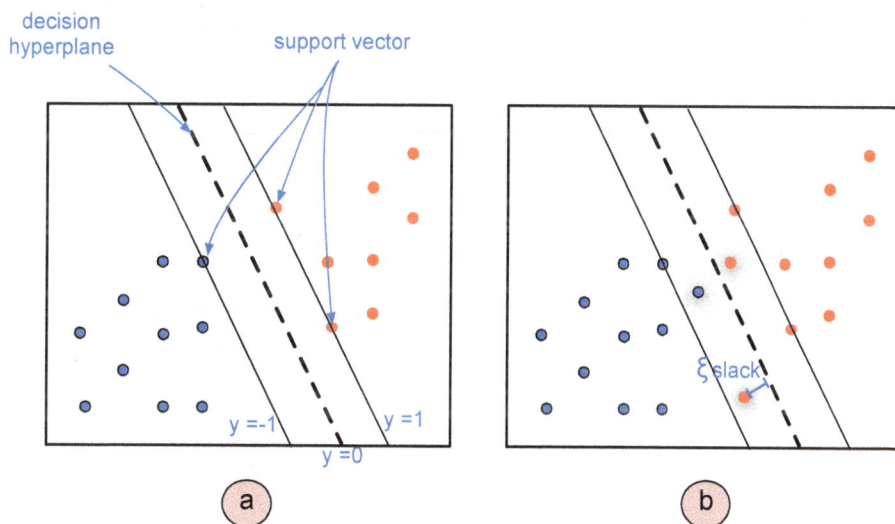

Figure 9-8: (a) The sample data points used by the maximum margin classifier support vectors (points on the line construct support vector) and the decision hyperplane were identified and highlighted. (b) Three sample data points (they have shadows) stay inside support vectors, and one data point stays on the wrong side of the hyperplane (slack variable). A good margin classifier should reduce these issues as much as possible.

The answer is to use the *Maximal Margin Classifier*. Here, the *margin* refers to the distance between a data point and the separator hyperplane. The maximal margin classifier uses a hyperplane with the largest margins (largest distance between data points and hyperplane). Figure 9-8 (a). shows another dataset with blue and red classes. It looks like a street; the centerline is called the decision hyperplane, and the sidelines parallel to it are called support vectors. Figure 9-8 (b) shows the margin lines and the hyperplane, but there are some data points inside the street, which we explain later how to deal with.

The separator hyperplane will be identified with algorithms such as the *Sequential Minimal Optimization* algorithm (check Chapter 8 to recall optimization), which we put under the rug, not to explain it. You might not need to learn SVM optimization unless you intend to design an optimization algorithm or do research on mathematics and theories of optimization.

In short, the SVM optimization algorithm tries to find the hyperplane based on two objectives: (i) it makes the maximum margin possible (making the street wider in Figure 9-8) while limiting margin violations. Data points between vector spaces mean margin violation (see Figure 9-8 (b)). However, since this phenomenon is unavoidable in real-world datasets where data is noisy, we tolerate it and call the classifier that tolerates these points a *soft margin classifier*. (ii) There is a constraint that specifies that each side of the hyperplane should cover most data points from each class. The hyperplane must stay between two classes to separate the data points.

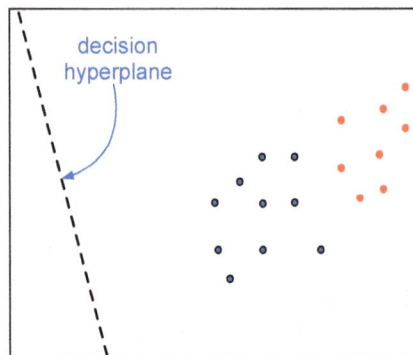

Figure 9-9: A decision hyperplane that has a very large separator margin, but it doesn't make sense. Because the separator constraint does not exist.

Otherwise, this hyperplane can stay far away on the other side of the world and thus make a very large margin, but it doesn't make sense. For example, look at Figure 9-9; the margin is very large but does not classify the dataset.

The loss function used by optimizations for maximum margin classification algorithms (including SVM) is called *hinge loss*. Some implementations of SVM enable the user to choose a different loss function as a hyperparameter for the algorithm.

We have explained that optimization will take care of hyperplane creation. Now, take a look at Figure 9-8 (b). In reality, outliers and data points can not stay in the correct class, either between decision boundaries or on the wrong side of the hyperplane. We have no other choice than to tolerate their misclassification in hyperplane. Data points that stay between support vector lines or on the wrong side of the hyperplane are called *slack variables*[3]. Figure 9-8 (b) highlights three slack variables three between support vectors, and one of them stays on the wrong side of the hyperplane. The objective function of SVM is not only to maximize the margin but also to minimize these slack variables, which essentially means reducing the hard margin violations as much as possible.

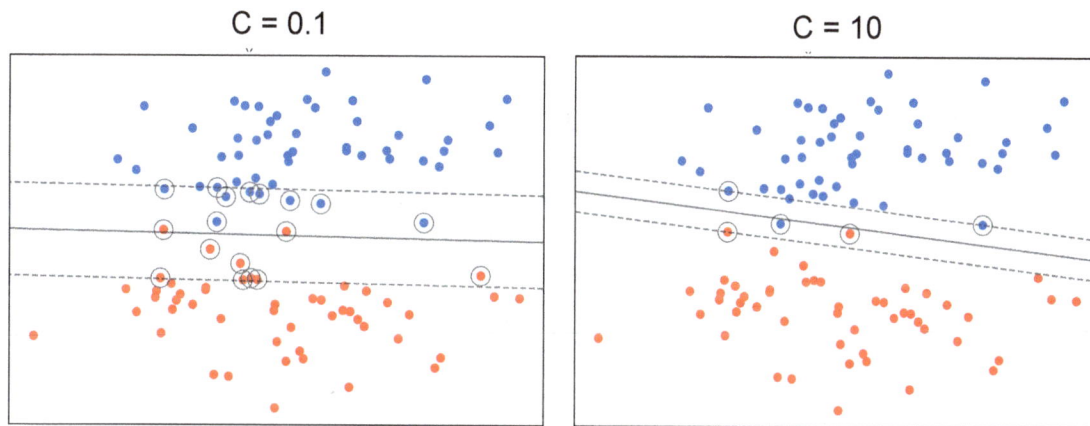

Figure 9-10: (a) A sample dataset and C=0.1 present a very small margin. Therefore, the variance is low, but as you can see, we have high bias. (b) A larger value has been set for C=10, and the bias is getting lower, but the variance has increased.

The C hyperparameter in SVM controls the trade-off between the model's margin width and the tolerance for margin violations; the lower C means less tolerance for any misclassification is allowed. Figure 9-10 presents two different values for C. We can not set C to *0*, its value must be positive, and a low but positive C value would indicate a high tolerance for margin violations, not zero tolerance. The choice of the C hyperparameter indeed represents a bias-variance trade-off. A higher C value (resulting in a "narrower street") minimizes the number of margin violations, aiming for higher accuracy on the training set at the risk of overfitting, thus exhibiting low bias but high variance. On the other hand, a lower C value (resulting in a "wider street") allows for more margin violations. This approach prioritizes generalization over training set accuracy, potentially leading to a model with higher bias but lower variance due to its increased tolerance for misclassifications.

---

[3] In the context of optimization to convert an inequality to equality, we can use a slack variable. For example, if $x + 2y < 3$, by using a slack variable ξ (read it as ksee), it can be written as: $x + 2y + \xi - 3 = 0$

Adjusting the C hyperparameter in SVMs is crucial for balancing the bias-variance trade-off. Higher values of C focus on reducing training errors, potentially at the expense of generalizability, while lower values of C aim for better generalization, potentially at the expense of increased training errors.

Handling Non-linear Data with Maximum Margin Classifier

Until now, all examples we have defined have been easily separable from a line. Take a look at Figure 9-11 (a); how can we use a maximum margin classifier and draw a hyperplane to classify it? To handle non-linear data, we must transform the feature space into quadratic (to the power of two), cubic (to the power of three), or higher-order polynomials. In other words, we need to increase the dimensionality of the data. Back in Chapter 6, we explained that dimensionality reduction is very useful, but in this particular scenario, we need to increase the dimension of the data to classify it. To our knowledge, this is the only place in machine learning where we increase the dimensionality.

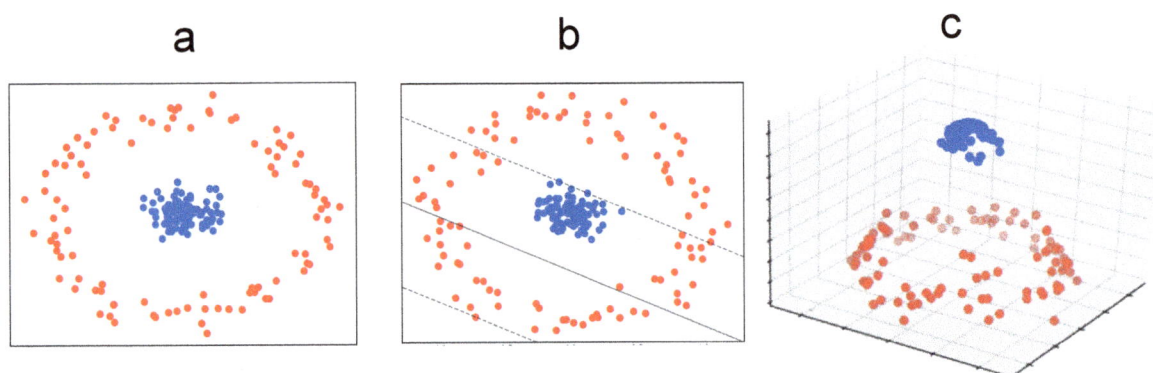

Figure 9-11: (a) A sample dataset that has two different labels. (b) The maximal margin classifier can not handle the data in two dimensions, and you can see that the maximal margin classifier line can not separate the data. (c) After applying the kernel function and bringing the data into a higher dimension, a hyperplane can separate the blue from red dots.

To implement such a high-dimensional transformation, we use the *Kernel* function. Therefore, the kernel function transforms a non-linear, inseparable dataset into a linear and separable dataset. For example, without applying a kernel function in Figure 9-11 (a), we can see that the maximal margin classifier lines cannot be classified as blue and red dots, as shown in Figure 9-10 (b). However, by transforming each data point into its polynomial degree, Figure 9-10 (c), we can project them in a higher dimensional space where blue and red dots can be separated with a linear hyperplane that acts as a border between red and blue dots. Imagine we draw a page (2D hyperplane) between red and blue dots, and this page separates them from each other.

To summarize, SVM uses a computationally efficient kernel function that transforms the input dataset into non-linear higher-dimension data. In the next section, we describe some details about the kernel functions.

Kernel Trick

As we have explained before, the kernel function is used by the SVM algorithm to create an extended representation of the data, and it *converts* a *non-linear separable dataset into a linear separable dataset*. In other words, we use the kernel function, which is a similarity calculation function, to bring our dataset into a higher dimensional space and thus make it separable with a linear hyperplane. After transforming the dataset into a higher dimensional space, the SVM can separate them with a linear hyperplane (see Figure 9-10).

Assume we have a dataset X with each of its data points has two features x_1 and x_2. Data points in this dataset do not have linear relations. $\Phi(X)$[4] is a function that creates a representation of X in the higher dimensional space[5], let's say, with two polynomial degree dimensions. Therefore, we can write the following: $\Phi(X) = \{x_1, x_2, x_1^2, x_2^2, x_1 x_2\}$. A linear hyperplane might separate the dataset in this higher dimensional space. This means that from smaller dimensions, which were not linear, we transform the data to have higher dimensions, and data in the higher dimensional space have a higher chance of getting separated with a linear hyperplane.

If the dataset's data points have many features, a simple second-degree polynomial transformation increases the number of features quadratically, i.e., X^2. What happens if the algorithm still can not separate them? We can increase the degree of polynomiality. For example, using a third-degree polynomial will result in X^3 features to process. For example, if we have 10 features and 3000 data points, we end up having $10^3 \times 3000 = 3,000,000$ features to deal with, and this is computationally very inefficient or even infeasible. This shows that the polynomial transformation increases the computational complexity, and we need a more subtle approach to deal with this problem.

Since transforming data into a higher dimension is computationally expensive, the kernel trick is used. SVM employs a kernel function to compute the similarity (or inner product) between data points directly in the original space, avoiding explicit transformation into a higher dimension. This kernel function is a similarity measure that facilitates the identification of patterns or structures in the data, particularly in SVM and other kernel-based algorithms. The process primarily involves labeled data in supervised learning to determine the decision boundary or classification rules.

Computing the similarity between data points can be considered as a *dot product* in a high-dimensional feature space without explicitly mapping the data to this space. If you can't recall from math what is an inner product, assume we have two vectors $a = [3,2,4]$ and $b = [1,4,0]$, and we would like to compare them together. The inner product of a and b is written as $a.b = 3 \times 1 + 2 \times 4 + 4 \times 0 = 11$. In particular, a kernel trick allows the comparison of two data points by converting the complex dot product in the high-dimensional space into a simpler calculation in the original input space. This is a good thing because instead of computing the dot

---

[4] ϕ is a greek letter and read it as 'fee' like feel.

[5] There is a theory called the *Mercer theorem*. It states if there is a continuous function that is symmetric, i.e., $F(x, y) = F(y, x)$, there is a function ϕ that can bring $F(x, y)$ into another dimension that can be represented as the inner product of $\phi(x)^T$ and $\phi(y)$, i.e., $F(x, y) = \phi(x)^T . \phi(y)$.

product between high dimensional features, we can calculate a dot product in low dimensional space and raise it to the power of polynomials. For example, assume we have a labeled train-set $X = \{x_1, x_2, x_3\}$ and we have an unlabeled test-set $X' = \{x'_1, x'_2, x'_3\}$. Two degrees of polynomiality transformation are $\Phi(X) = \{x_1^2, x_1x_2, x_1x_3, x_2x_1, x_2^2, x_2x_3, x_3x_1, x_3x_2, x_3^2\}$ and $\Phi(X') = \{x_1'^2, x_1'x_2', x_1'x_3', x_2'x_1', x_2'^2, x_2'x_3', x_3'x_1', x_3'x_2', x_3'^2\}$. Now, we would like to compare $\Phi(X)$ with $\Phi(X')$ to label $\Phi(X')$ set from $\Phi(X)$ the labeled set. We can see that we have too many features to compare because, from a three-dimensional space, we moved into a nine-dimensional space.

By using the kernel trick, we avoid the explicit computation of the dot product between $\Phi(X)$ and $\Phi(X')$, in the transformed higher dimensional space. Instead, the kernel function calculates a value representing this dot product, but it does so based on the original data, i.e., X and X'. Therefore, we can write that $K(X, X') = < \Phi(X), \Phi(X') >$.

Let's summarize how a kernel trick works: To compute the similarity between two data points as if they were in a higher-dimensional space, the kernel function calculates a dot product (or another similarity measure) in the original, lower-dimensional space. This process simulates the comparison in the higher-dimensional space without explicitly performing the transformation, thus efficiently achieving the same result as if the data were compared in that higher-dimensional space.

Kernel Functions

Now we have understood the advantage of the kernel trick; congratulations, you have learned a secret in mathematics; if you think mathematicians' secrets are also very boring, we agree with you. There are several popular kernel functions used for SVM. The popular ones are listed as follows.

Linear kernel: it is written as $K(X, X') = X . X'$, which is a simple dot product of two vectors.

Polynomial kernel: it is written as $K(X, X') = (X . X' + r)^p$. Here, r presents the polynomial coefficient, and p presents the polynomial degree, which both are hyperparameters of the polynomial kernel function.

Gaussian Radial Basis Function (RBF) kernel: RBF Kernel is written as follows: $K(X, X') = exp(-\gamma ||X - X'||^2)$. In this equation, γ is a hyperparameter that controls the spread and smoothness of the kernel. A higher γ leads to a narrower kernel, which responds more strongly to nearby data points and less to distant ones), and a lower γ responds more uniformly to nearby and distant data points.

Sigmoid kernel: it is written as $K(X, X') = tanh(\gamma X^T X' + r)$. Here, r presents the polynomial coefficient and γ is a hyperparameter representing the curvedness of the separation area (similar to the Gaussian RBF kernel, *tanh* is a hyperbolic tangent function.

RBF is the most common kernel function used for the SVM algorithm. Figure 9-12 shows an SVM algorithm run with three values for γ. Gaussian RBF kernel is a popular kernel function for SVM. Geron [Geron '19] states that γ acts as a regularization parameter. If the SVM model is overfitting, this means the decision boundary is too complex and closely fits the minor fluctuations in the training data. By decreasing the value of γ we might resolve it. If the model is underfitting, the decision boundary is too simple and cannot capture the complexity of the data. By increasing γ we can resolve the underfitting.

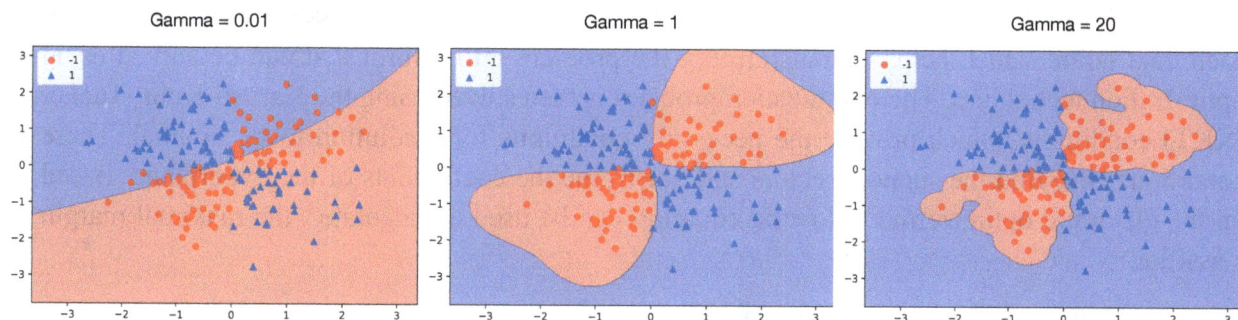

Figure 9-12: Different values for gamma of the RBF kernel function and their impact on the classification quality of the SVM. Setting gamma high improves the separation capability of the classifier, but setting gamma too high, like the most right figure makes the model prone to overfitting.

Unless you would like to be a kernel function designer, you don't need to understand the mathematics behind any of these kernel methods, but it might be useful to know their equation. However, we heard from our students that a psycho-recruiter asked a poor candidate to describe Kernel equations at job interviews for data science.

While using SVM, we can specify the kernel function as a parameter of the SVM algorithm and hyperparameters of the kernel function. A question might arise: How does the kernel function decide the polynomial degree, or how were polynomial parameters determined? By using the cross-validation the algorithm performs hyperparameter tuning, and some implementations automatically choose the best possible parameters for the selected kernel method. In other implementations, we should give the parameter as an input variable. Therefore, this depends on the SVM library or package we are using. It is better to start with a linear kernel, and as we have explained before, linear models are always favored over non-linear models because of their resource efficiency.

Multi-label Classification for SVM

Binary classification is straightforward. But what if we want to use SVM to classify a dataset based on four labels, such as {blue, red, green, orange}? In this case, SVM can perform binary classification in a one-versus-all manner. This means the SVM will first classify blue against all other labels {red, green, orange}. Then, it will classify red against all other labels, followed by green against the rest, and finally orange against the others.

In each iteration, each data point receives a label. For example, data point x might be classified as green in one iteration and as not orange, not blue, and not red in the others. Ultimately, x will receive the green label. This approach to handling multi-class classification is called one-versus-all classification.

How Does the SVM Perform the Prediction?

We have explained that SVM uses an optimizer to determine the support vector and hyperplane equation parameters. We have not explained how the optimizer works. Very briefly, a Lagrangian equation is constructed first, and then the optimizer solves the solution using Karush-Kuhn-Tucker (KKT) conditions. In particular, the Lagrangian function introduces multipliers (α_i) for each data point, which helps in transforming the problem into a form that can be solved using optimization techniques. The optimization problem is then solved using the Karush-Kuhn-Tucker (KKT) conditions, which provide the necessary conditions for a solution to be optimal. These conditions help find the support vectors and determine the coefficients of the hyperplane. Based on the quadratic optimizerthe following equation can be used to calculate , the maximal margin classifier.

$$d(X') = \begin{cases} -1 & if \quad \sum_{i=1}^{l} y_i \alpha_i X_i X' + \beta_0 \leq 0 \\ 1 & if \quad \sum_{i=1}^{l} y_i \alpha_i X_i X' + \beta_0 > 0 \end{cases}$$

In this equation, α_i (the Lagrangian multiplier) and β_0 (bias) parameters are determined by the optimizer. y_i is the class label of input data points X_i (training set), and X' is the test set. Therefore, the X' data points will be substituted into this equation, and the sign of the equation will determine which side of the hyperplane X' belongs (what label it receives).

SVM Computational Complexity

Calculating the computational complexity of SVM is dependent on the optimization algorithm to identify the maximal margin classifier and the kernel method we choose to use for SVM. For example, for the training phase, the computational complexity depends on the optimization algorithm, which is a quadratic problem-solving that has $O(n^3)$ complexity [Bordes '05]. For the testing phase, the computational complexity depends on the kernel method; assuming n presents the number of data points and m presents the number of features, linear kernel SVM has $O(n \times m)$ computational complexity. Polynomial kernel SVM with p degree of the polynomial has $O(n \times m^p)$. Therefore, there is no constant complexity that we can report on SVM; it is one of the most accurate traditional classification algorithms, but it has a high computational complexity.

NOTES:

* Both kNN and SVM can handle nonlinear data structures, and when a dataset is small, they are good options for classification.

* kNN is not good for higher dimensional data; SVM is particularly known for its effectiveness in high-dimensional spaces, even with a small number of samples.

* SVM is sensitive about scaling, and scaling the data will result in a change in SVM behavior. Therefore, if your dataset changes frequently, be careful when using SVM.

* The Support Vector Classifier (SVC) algorithm is the SVM that is used for classification. SVM alone can be used for both regression and classification.

* If classes are highly overlapped, logistic regression is performing better than SVM. If they are well separated, SVM usually performs better than logistic regression. However, the best approach is to try many algorithms and observe which performs better or use ensemble learning methods, which we will explain later in this chapter.

* If you remember, we discussed that if the optimization has a convex shape, local minima are the global minima as well. SVM parameter optimizations also use a convex shape function.

* In comparison to other classification algorithms, the SVM algorithm is generally slow, but it is fairly robust against overfitting, which is a common problem in classification algorithms.

Decision Trees

Back in Chapter 5, we explained that trees reduce search space and thus make the search algorithm faster and more resource-efficient. There is another group of trees that can be used for classification and regression, which we explain in this section. Decision trees extract human-understandable rules, and their explainability makes them very popular. Similar to previous trees we have explained in Chapter 5, decision trees split the search space (dataset) into smaller segments and search those segments instead of the entire dataset. Because of this feature, they are usually resource-efficient.

The first practical version of the decision tree was proposed in the early 80s, i.e., CHAID [Kass '80]. A few years after CHAID, in 1984, Breimanetal [Breimanetal '84] introduced the CART algorithm, another popular decision tree algorithm that is still in use. Two years later, Quilan introduced ID3 [Quilan '86], and then proposed C4.5 [Quilan '94]. All of these four algorithms are still in use. After the introduction of these trees, Gradient Boosting Decision Trees [Mason '99] were introduced, and in 2014, XGBoost [Chen '15] won several Kaggle competition awards. The success of XGBoost leads to shifting the machine learning community's attention to decision trees again. At the time of writing this book, deep learning algorithms have the highest accuracy for unstructured data, but gradient-boosting decision trees perform better than deep learning methods when analyzing tabular data.

Decision trees, similar to other supervised algorithms, use a training dataset to construct a tree. Then, after the decision tree is built, the unlabeled data point navigates through the conditions of the trees to get the label. Leaf nodes (blue-colored nodes in Figure 9-13) of a decision tree are used to assign labels to the tree.

For example, a person who is going under mental therapy is journaling his happiness level based on three different factors: 'having gratitude', 'comparing himself with others', and 'expecting from others'. Figure 9-13 presents the daily information that he collects about himself and what he is thinking about his behaviors and attitudes. He would like to know when he goes to bed and thinks about his day, whether he feels happy or not. We make his job easier by building a

Gratitude	Compare himself with others	Expecting from others	Happy
X	X	high	X
✓	✓	low	✓
✓	X	medium	✓
✓	X	low	✓
X	✓	medium	✓
X	X	high	X
✓	X	medium	✓
X	X	high	X
X	✓	medium	✓
✓	✓	low	✓
X	X	medium	X
✓	X	high	X

train (rows above the dashed line)

✓	X	low	?

test (row below the dashed line)

Decision tree:

- Compare himself with others
  - yes → Expecting from others
    - medium → Gratitude
      - no → 70% not happy day (leaf node)
  - no → Gratitude
    - no → Expecting from others
      - high → 80% not happy day (leaf node)
      - medium → 60% happy day (leaf node)
    - yes → Expecting from others
      - low → 90% happy day (leaf node)

decision node / leaf node

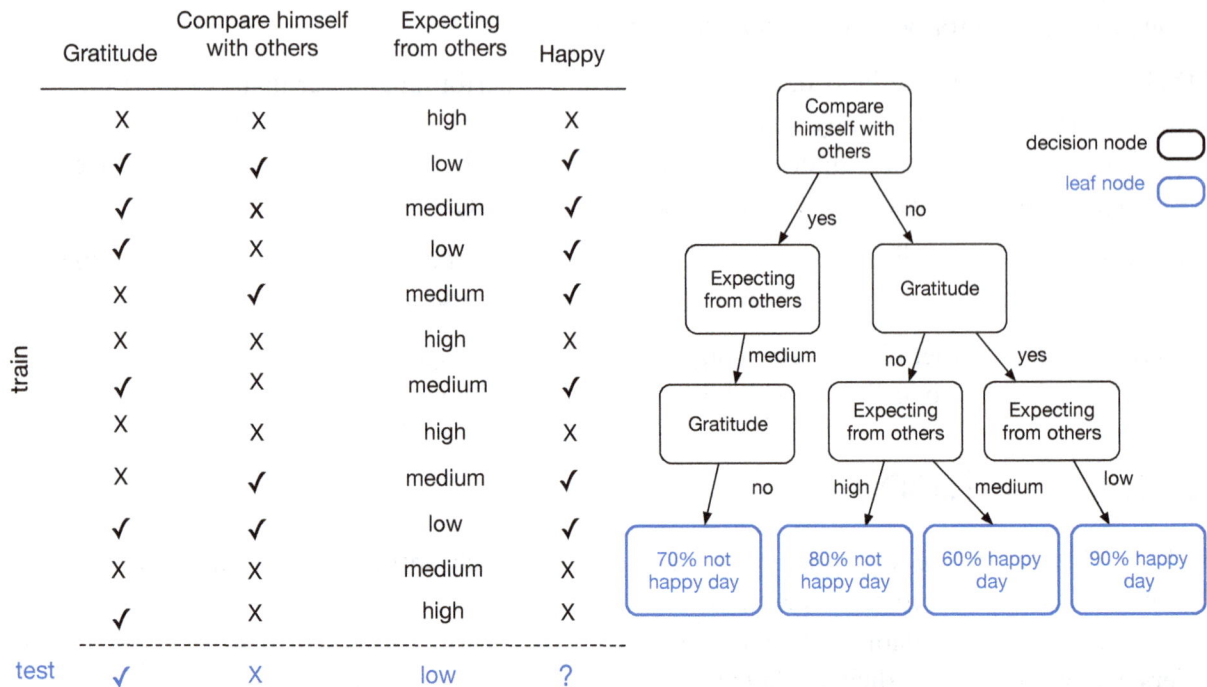

Figure 9-13: A sample dataset along with a simple decision tree, we like to predict the happiness status of the test dataset. On the right side, the algorithm constructs an example tree, and by tracing this tree, the algorithm can make the decision about the test dataset (the ? marks of the blue record in the table).

prediction model that assigns the label (happy / not happy) before he ends his day based on the data he gives to the algorithm.

After the training phase, the decision tree receives a vector of features (the test record in the table of Figure 9-13, which is presented in blue color), and the decision tree starts to navigate its branches until it identifies the missing labels ('?', in the table of Figure 9-13).

Decision trees have three important characteristics:

(i) Decision trees are operated based on the policy of finding the best feature to split the training data into subsets until there is no possible split into a smaller subset. When no further split is possible, the decision tree has been constructed. Next, the test dataset will use this tree to label its records.

(ii) The leaf nodes in decision trees present the 'outcome' of the path from the root to that node. All non-leaf nodes are called 'decision nodes' because the algorithm uses them to decide on navigating in which branch reaches a leaf node at the end. The decision tree of Figure 9-13 shows leaf nodes in blue and decision nodes in black.

(iii) Decision trees are explainable, making them very popular algorithms for real-world applications. Decision tree results are a set of if-then-else rules, making them explainable, which is not the case for many machine learning algorithms, including neural networks. If you encounter a fancy word like *explainable AI (xAI)*, it means the machine learning algorithm or model is explainable and not BlackBox.

Please fasten your seat belt. We are about to land on the big planet of decision trees and learn four vanilla decision trees, followed by gradient boosting decision trees.

Iterative Dichotomiser 3 (ID3)

ID3 is one of the simple decision trees that iteratively dichotomizes (divides into two groups) the dataset at each step. Recall that we have explained that the process of decision tree creation is to split the training data into smaller subsets. Looking at the tree in Figure 9-13, we have selected "Compare himself with others" as the root node, but is it the best feature to select? Probably not, and we need a mechanism to choose the best feature to split the data. By splitting the data, we mean choosing the feature that can separate records based on the target label more accurately than other labels. In other words, choose the best separative feature.

ID3 uses *Information Gain (IG)* to select the best feature. Information gain is based on the concept of *entropy*. Entropy is a number that quantifies the uncertainty or disorder (check Chapter 3 to recall it). Entropy specifies (in number) how well a given feature separates the dataset records based on the label. Assuming n is the number of possible labels (classes) in the training set, S denotes the train dataset, p_i denotes the probability of having class i (the number of records with class label i divided by the total number of dataset records). In other words, p_i is the proportion of elements in S that belong to class i. The entropy $H(S)$ is computed as follows:

$$H(S) = - \sum_{i=1}^{n} p_i log_2(p_i)$$

In other words, $H(S)$ refers to the average amount of information required to identify the class label of a new record.

Each column in the table of Figure 9-13 presents features of the decision tree, and we can write information gain for the feature A (a column of data) as: $IG(S, A) = H(S) - H(S|A)$.

Here, IG presents the differences between the original information, i.e., $H(S)$ and the required information, i.e., $H(S|A)$. Now that we understand IG and entropy in the context of ID3, we can learn step by step how the ID3 algorithm works.

As the first step, the algorithm should find the feature (column) with the highest split against one of the possible values of the labels. To find this feature, it measures the IG for every feature (column) of the dataset. In the second step, it splits the training dataset S based on the feature that has the highest IG. Next, in the third step, if all rows in a tree node belong to the same class, then the algorithm marks this node as a leaf node and does not split it any further. Otherwise, it repeats the process from the first step until all nodes are converted into a leaf node, and no path remains from root to decision nodes without ending in a leaf node.

To better understand the ID3 algorithm, let's explain it with the example we have used in the table of Figure 9-13. As the first step, we calculate the entropy of the target column that we would like to predict; in our case, it is the "Happy" column. We have seven yes ($\checkmark$) and five no ($\times$) for this column. Its entropy is calculated as:

$$H(S) = -(7/12) . log_2(7/12) - (5/12) . log_2(5/12) = 0.96$$

Now we have the entropy of "Happy", and we should find which column has the largest split on "Happy", so we calculate the IG value for each column (feature). We have the following IGs for each feature (column in the table of Figure 9-13). The sign # is used to refer to the number of occurrences.

"Gratitude": yes, #rows: 6, #happy: 5, #not happy:1

$$H(S_{yes}) = -(5/6) . log_2(5/6) - (1/6) . log_2(1/6) = 0.65$$

"Gratitude": no, #rows: 6, #happy: 2, #not happy:4

$$H(S_{no}) = -(2/6) . log_2(2/6) - (4/6) . log_2(4/6) = 0.918$$

The IG of "Gratitude" will be calculated as follows:

$$IG(S, Gratitude) = H(S) - (S_{yes}/S)H(S_{yes}) - (S_{no}/S)H(S_{no}) =$$

$0.96 - (6/12)(0.65) - (6/12)(0.918) = 0.176 \leftarrow$ ***IG of gratitude***

"Compare himself with others (Compare)": yes, #rows: 4, #happy:4, #not happy:0

$$H(S_{yes}) = -(4/4) . log_2(4/4) - (0/4) . log_2(0/4) = 0$$

"Compare himself with others": no, #rows: 8, #happy:3, # not happy:5

$$H(S_{no}) = -(3/8) . log_2(3/8) - (5/8) . log_2(5/8) = 0.95$$

The IG of "Compare himself with others" will be calculated as follows:

$$IG(S, Compare) = H(S) - (S_{yes}/S)H(S_{yes}) - (S_{no}/S)H(S_{no}) = 0.96 - 0 - (8/12)(0.95) = 0.32$$

$0.96 - 0 - (8/12)(0.95) = 0.32 \leftarrow$ ***IG of Comparing himself with others***

"Expecting from others (Exp)": high, number of rows: 4, #happy:0, #not happy: 4

$$H(S_{high}) = -(0/4) . log_2(0/4) - (4/4) . log_2(4/4) = 0$$

"Expecting from others": medium, #rows: 5, #happy:3, #not happy: 2

$$H(S_{medium}) = -(3/5) . log_2(3/5) - (2/5) . log_2(2/5) = 0.97$$

"Expecting from others": low, #rows: 5, #happy:4, #not happy: 1

$$H(S_{low}) = -(4/5) . log_2(4/5) - (1/5) . log_2(1/5) = 0.72$$

Gratitude	Compare himself with others	Expecting from others	Happy
X	X	high	X
X	X	high	X
X	X	high	X
✓	X	high	X

Gratitude	Compare himself with others	Expecting from others	Happy
✓	✓	low	✓
✓	X	low	✓
✓	✓	low	✓

Gratitude	Compare himself with others	Expecting from others	Happy
✓	X	medium	✓
X	✓	medium	✓
✓	X	medium	✓
X	✓	medium	✓
X	X	medium	X

This dataset will be used to extend the tree of

Table 9-4: The result of splitting the original dataset in the table of Figure 9-13 based on the "Expecting from Others" column. Now, we have three datasets, but the two on the top are leaf nodes because we cannot further split them.

The IG of "Expecting from others" will be calculated as follows:

$$IG(S, Exp.) = H(S) - (S_{low}/S)H(S_{low}) - (S_{medium}/S)H(S_{medium}) - (S_{high}/S)H(S_{high}) =$$

$$0.96 - (5/12)(0.72) - (5/12)0.97 - 0 = 0.25 \leftarrow \textbf{\textit{IG of Expecting from others}}$$

We understand that the highest information gain (0.32) belongs to "Compare himself with others". This means that the training dataset can be split on this feature because it provides the best split in comparison to other features (Comp., Exp.). Therefore, we start to construct the tree as it is shown on the right side of Figure 9-13.

We do not explain the tree construction further. The rest will be done the same until no decision node remains, and we end up with all branches in a leaf node. If the split is based on another feature, such as "Expecting from others", we will have something such as Table 9-4, in which two tables on the top do not require any more split because the label for the Happy column is specified.

Chi-square Automatic Interaction Detector (CHAID)

Back in Chapter 3, we explained that the Chi-square test can be used to identify the relationship between two categorical variables, and here, we benefit from that attribute to split the tree.

	Total Happy	Total Unhappy	Total Observed	Total Expected Happy	Total Expected Unhappy
Gratitude X	2	4	6	$\frac{(6 \times 7)}{12}$ 3.5	$\frac{(6 \times 5)}{12}$ 2.5
Gratitude ✓	5	1	6	$\frac{(6 \times 7)}{12}$ 3.5	$\frac{(6 \times 5)}{12}$ 2.5
Total	7	5	12		

	Total Happy	Total Unhappy	Total Observed	Total Expected Happy	Total Expected Unhappy
Compare himself with others X	3	5	8	4.66	3.33
Compare himself with others ✓	4	0	4	2.33	1.66
Total	7	5	12		

		Total Happy	Total Unhappy	Total Observed	Total Expected Happy	Total Expected Unhappy
Expectation from others	high	0	4	4	$\frac{(4 \times 7)}{12}$ 2.33	$\frac{(4 \times 5)}{12}$ 1.66
Expectation from others	medium	4	1	5	$\frac{(5 \times 7)}{12}$ 2.91	$\frac{(5 \times 5)}{12}$ 2.08
Expectation from others	low	3	0	3	$\frac{(3 \times 7)}{12}$ 1.75	$\frac{(3 \times 5)}{12}$ 1.25
	Total	7	5	12		

Table 9-5: Chi-square calculation for the each feature.

Recall that the following equation calculates the Chi-Square value, in which O stayed for observed, and E stayed for expected values.

$$\chi^2 = \Sigma \frac{(O - E)^2}{E}$$

CHAID (Chi-squared Automatic Interaction Detector) primarily uses the Chi-square test for categorical variables to determine the best splits. For numerical variables, they often use the F-test to assess the significance of the splits. However, CHAID is less commonly used for purely numerical variables than categorical variables.

To calculate the splits in CHAID, the algorithm compares the Chi-square value of each feature (each feature is a column in tabular data) with the target label column. The feature with the highest Chi-square value is chosen for the split. In other words, first, the algorithm calculates the Chi-square for each feature. Then, it chooses the feature with the highest Chi-square and splits the table. This process continues until all remaining features can no longer be split, and thus, they are all considered to be the leaf nodes (no decision node remained).

Using data from the table in Figure 9-13 example, we need to calculate the Chi-square for each feature, including (Gratitude, Comp., and Exp.). These calculations are presented in Table 9-5.

We use the result of Chi-squares for each feature as follows and calculate the overall Chi-square for each feature.

Gratitude:
- Gratitude:No, Happy:Yes $\rightarrow (2 - 3.5)^2/3.5 = 0.643$
- Gratitude:Yes, Happy:Yes $\rightarrow (5 - 3.5)^2/3.5 = 0.643$
- Gratitude:No, Happy:No $\rightarrow (4 - 2.5)^2/2.5 = 0.9$
- Gratitude:Yes, Happy:No $\rightarrow (1 - 2.5)^2/2.5 = 0.9$

Total Chi-Square = 0.643 + 0.643 + 0.9 + 0.9 = 3.086

Compare himself with others:
- Compare: No, Happy:Yes $\rightarrow (3 - 4.66)^2/4.66 = 5.91$
- Compare: Yes, Happy:Yes $\rightarrow (4 - 2.33)^2/2.33 = 1.197$
- Compare: No, Happy:No $\rightarrow (5 - 3.33)^2/3.33 = 0.838$
- Compare:Yes, Happy:No $\rightarrow (0 - 1.66)^2/1.66 = 1.66$

Total Chi-Square = 5.91 + 1.197 + 0.838 + 1.66 = 9.605

Expectations from others:
- Expectation: High, Happy: Yes $\rightarrow (0 - 2.33)^2/2.33 = 2.33$
- Expectation: High, Happy: No $\rightarrow (4 - 1.66)^2/1.66 = 3.298$
- Expectation: Medium, Happy: Yes $\rightarrow (4 - 2.91)^2/2.91 = 0.408$
- Expectation: Medium, Happy: No $\rightarrow (1 - 2.08)^2/2.08 = 0.561$
- Expectation: Low, Happy: Yes $\rightarrow (3 - 1.75)^2/1.75 = 0.893$
- Expectation: Low, Happy: No $\rightarrow (0 - 1.25)^2/1.25 = 1.25$

Total Chi-Square = 2.33 + 3.297 + 0.408 + 0.561 + 0.893 + 1.25 = 8.739

Based on these results, the best feature to split the table is "Compare himself with others", because it has the largest Chi-square.

The CHAID algorithm creates a tree with the "Compare himself with others" as the root node. In this split, the algorithm deals with two subsets of the dataset; one subset has only "yes", and the other subset has only "no". These steps of spiting the dataset continue until all nodes are leaf nodes. You can do the rest independently, and it does not make sense to explain the rest of the algorithm in detail. Save your brain energy for more algorithms; still, two more decision tree algorithms remain, and then we will switch to boosting and bagging, which are extremely important algorithms and not easy to understand.

C4.5

C4.5 is developed by the author of the ID3 algorithm, and it is an extension of the ID3 algorithm. There is an open-source implementation of this algorithm called J48, which is the same algorithm but with a different name. C4.5 operates very similar to ID3, but it can handle missing data and continuous data, which are not very efficiently handled by ID3. The ID3 algorithm calculates the information gain to split the data, but C4.5 calculates the *gain ratio* instead. The gain ratio is the information gain of a feature divided by split information. Therefore, instead of information gain, C4.5 calculates the GainRatio for all features and decides on the split based on the highest gain ratio. The gain ratio is computed as follows:

$$GainRatio(S, A) = \frac{IG(S, A)}{Split(S, A)}$$

Split of information is the sum of entropies that are calculated for each feature, as it is written as follows:

$$Split(S, A) = -\Sigma_{v \in values(A)} \frac{S_v}{S} log_2 \frac{S_v}{S}.$$

Here, S presents the dataset, and A presents the candidate feature, v is all possible values of A, S_v is the size of the subset of the dataset where A has v value.

Gratitude	Compare himself with others	Expecting from others	Happy
X	X	0.8	X
✓	✓	0.2	✓
✓	X	0.5	✓
✓	X	0.2	✓
X	✓	0.5	✓
X	X	0.8	X
✓	X	0.5	X
X	X	0.9	X
X	✓	0.4	✓
✓	✓	0.2	✓
X	X	0.5	✓
✓	X	0.8	X

Gratitude	Compare himself with others	Expecting from others	Happy
✓	✓	0.2	✓
✓	✓	0.2	✓
✓	X	0.2	✓
X	✓	0.4	✓
X	✓	0.5	✓
✓	X	0.5	X
X	X	0.5	✓
✓	X	0.5	✓
✓	X	0.8	X
X	X	0.8	X
X	X	0.8	X
X	X	0.9	X

Table 9-6: (left) The original dataset, (right) the dataset is ordered based on the "Excepting from others" which is continuous. Every time a single feature is selected for split information calculation the dataset should be ordered based on that feature.

Table 9-6 shows the table of Figure 9-13 with continuous data. In this example, we consider "Expecting from others" as continuous data. To implement the C4.5 decision tree on this data, first, we need to calculate the entropy of Happy for Table 9-6, as follows (7 happy and 5 not happy): $H(S) = -(7/12).log_2(7/12) - (5/12).log_2(5/12) = 0.96$.

Now that we have the entropy of "Happy", we should find which column has the largest split on the "Happy" column to calculate IG for each column (feature).

The following presents the gain ratio calculation for each feature. The sign # is used to refer to the frequency.

"Gratitude": yes, #rows: 6, #happy: 4, #not happy:2

$$H(S_{yes}) = -(4/6) . log_2(4/6) - (2/6) . log_2(2/6) = 0.918$$

"Gratitude": no, #rows: 6, #happy: 2, #not happy:4

$$H(S_{no}) = -(3/6) . log_2(3/6) - (3/6) . log_2(3/6) = 1$$

The IG[6] of "Gratitude" will be calculated as follows:

$$IG(S, Gratitude) = H(S) - (S_{yes}/S)H(S_{yes}) - (S_{no}/S)H(S_{no}) = 0.96 - (6/12).0.918 - (6/12).1 = 0.24$$

Next, the Split and then Gain Ratio of "Gratitude" will be calculated as follows:

$$Split(S, Gratitude) = -(6/12) . log_2(6/12) - (6/12) . log_2(6/12) = 1$$

$$Gain Ratio(S, Gratitude) = \frac{IG(S, Gradtitude)}{Split(S, Gratitude)} = 0.024/1 = 0.24$$

The IG of "Compare himself with others" will be calculated as follows:

$$IG(S, Compare) = H(S) - (S_{yes}/S)H(S_{yes}) - (S_{no}/S)H(S_{no}) = 0.96 - 0 - (8/12)0.954 = 0.324$$

Next, the Split and then Gain Ratio of "Compare himself with others" will be calculated as follows:

$$Split(S, Compare) = -(8/12) . log_2(8/12) - (4/12) . log_2(4/12) = 0.9$$

$$Gain Ratio(S, Compare) = 0.324/0.9 = 0.36$$

Now, we need to calculate the Gain Ratio for "Expecting from others", but it is a continuous attribute, and thus, we need to identify the maximum Gain Ratio for the value that can do the best split. Therefore, for each value (0.2,0.4,0.5, 0.8), we should calculate the Gain Ratio.

"Expecting from others" $< = 0.2$, #rows: 3, #happy: 3, #not happy:0

$$H(exp < = 0.2) = -(3/3)log_2(3/3) - 0 = 0$$

"Expecting from others" > 0.2, #rows: 9, #happy: 4, #not happy:5

$$H(exp > 0.2) = -(5/9)log_2(5/9) - (4/9)log_2(4/9) = 1$$

Based on the above results, the Gains ratio for Exp. for 0.2 will be calculated as follows.

$$IG(S, Exp \ for \ 0.2) = 0.96 - (3/12).0 - (9/12).1 = 0.21$$

$$Split(S, Exp. for \ 0.2) = -(3/12)log_2(3/12) - (9/12)log_2(9/12) = 0.8$$

$$Gain Ratio(S, Exp \ for \ 0.2) = 0.26$$

"Expecting from others" $<= 0.4$, #rows: 4, #happy: 4, #not happy:0

$$H(exp < = 0.4) = 0 - (3/3)log_2(3/3) = 0$$

---

[6] We do not write them here again, the IG of "Gratitude" and "Compare himself with others" come from the ID3 explanation.

"Expecting from others" > 0.4, #rows: 8, #happy: 3, #not happy:5

$H(exp > 0.4) = (3/8)log_2(3/8) - (5/8)log_2(5/8) = 0.94$

$IG(S, Exp\ for\ 0.4) = 0.96 - (4/12).0 - (8/12).(0.94) = 0.33$

$Split(S, Exp.\ for\ 0.4) = -(4/12)log_2(4/12) - (8/12)log_2(8/12) = 0.9$

$GainRatio(S, Exp\ for\ 0.4) = 0.36$

Since your brain and our brain are about to explode, we do not continue further. However, the algorithm will do the rest for other values of the "Expecting from others" as well, including 0.5. 0.8 and 0.9. Then, it compares all gain ratios (all from "Expecting from others" plus gratitude and "Compare himself."). For example, assume $GainRatio(S, Exp\ for\ 0.8) = 0.67$ receives the highest gain ratio. Therefore, the decision tree starts with something like Figure 9-14, and continues until all nodes are leaf nodes.

Classification and Regression Trees (CART)

The CART algorithm is another popular decision tree algorithm. It constructs a binary tree (each node has only two branches), which can also handle continuous data appropriately. The CART algorithm uses a recursive approach similar to other decision trees to split the dataset into smaller units. CART uses *Gini index (Gini split* or *Gini Impurity)* to determine the best feature to split. Gini split quantifies the impurity of each feature. The Gini index is computed as $Gini = 1 - \Sigma_{i=1}^{n}(p_i)^2$, where p_i is the proportion of the samples that belong to class i in a particular node.

the total number of labels

frequency of records with label i

$$Gini = 1 - \sum_{i=1}^{n} (P_i)^2$$

The Gini index value varies between values 0 and 1, where 0 expresses the purity of the feature, i.e., all values belong to a specified class, or only one class exists for this feature, and 1 indicates the random distribution of elements across various classes. The algorithm calculates the Gini index for each feature and chooses the lowest one for the split.

It can handle continuous data in a similar fashion as C 4.5; check the example we explained for "Expecting from others" in the C 4.5 algorithm description.

Reading these numerical examples are deadly boring.

Let's calculate the Gini index for the example presented in Figure 9-13. Similar to other algorithms, the algorithm

484

calculates the Gini index for each feature (column), and then it selects the feature with the lowest Gini index for the split.

Gini index of "Gratitude":
$$Gini(gratitude = yes) = 1 - (5/6)^2 - (1/6)^2 = 0.27$$
$$Gini(gratitude = no) = 1 - (4/6)^2 - (2/6)^2 = 0.44$$
$$Gini(gratitude) = 0.44(6/12) + 0.27(6/12) = 0.335$$

Gini index of "Compare himself with others":
$$Gini(Compare = yes) = 1 - (4/4)^2 - (0/4)^2 = 0$$
$$Gini(Compare = no) = 1 - (3/8)^2 - (5/8)^2 = 0.47$$
$$Gini(Compare) = 0(4/12) + 0.47(8/12) = 0.313$$

Gini index of "Expecting from others":
$$Gini(Exp. = low) = 1 - (3/3)^2 - (0/3)^2 = 0$$
$$Gini(Exp. = medium) = 1 - (4/5)^2 - (1/5)^2 = 0.32$$
$$Gini(Exp. = high) = 1 - (4/4)^2 - (0/4)^2 = 0$$
$$Gini(Exp. = high) = 0.32 + 0 + 0 = 0.32$$

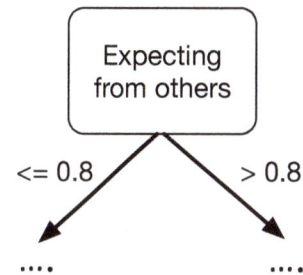

Figure 9-14: C4.5 decision tree result example "Expecting from others" for 0.8 has the highest gains ratio.

$Gratiude = 0.085$, has the lowest Gini index. Therefore, the tree starts the first split by using this feature; then, it continues constructing the branches with the same approach for each subset (one subset includes gratitude = yes, and the other subset includes gratitude = no).

Decision Tree Challenges and the Need for Pruning.

While working with decision trees, one might ask when we need to stop creating the tree. One answer is to define a minimum number of records under a leaf node; if the dataset remaining for a node has a size less or equal to this minimum number of records, then the algorithm should stop splitting the dataset and stop going deeper into that branch.

The second approach is to specify a maximum depth for a tree. Larger tree depths lead to more complex models that can capture detailed patterns but risk overfitting. Conversely, shallow trees might fail to capture essential patterns in the data, which can lead to underfitting. Thus, finding a balance between tree depth and complexity is crucial for building an effective decision tree model. Allowing each leaf to hold only one data point often leads to overfitting because the tree becomes too detailed, fitting the noise in the training data, which can decrease its predictive performance on new data.

If we plot the accuracy of the test and train

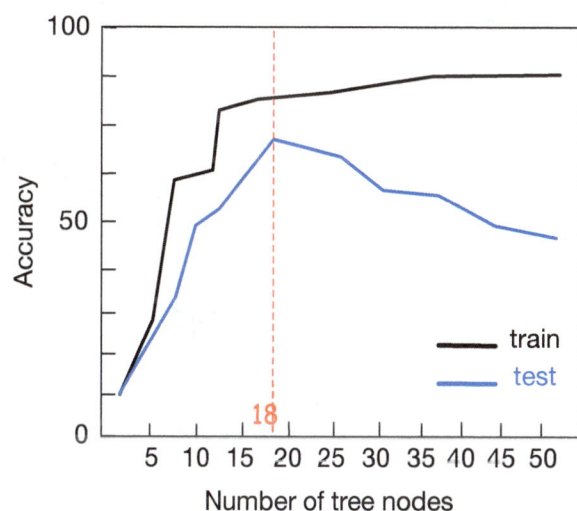

Figure 9-15: The overfitting problem in a tree. We can see that after the number of nodes increases to more than 18, test data accuracy decreases significantly, but training is still performing well.

485

datasets, we might end up having Figure 9-15, which is a demonstration of overfitting in a decision tree. This figure shows that after the 18th node in the tree, the test dataset accuracy starts to decrease. However, while working on the train set, we went to 50 nodes, but the accuracy didn't decrease. In other words, this problem shows that we should cut the tree before it becomes too specific for the train dataset (overfitting), and in this example, we should stop at the 18th node.

A third approach, which is a popular one, is to separate the "train" dataset into two parts: *train*, and *validation* sets (we have explained it in Chapter 1). We construct the tree by using the train set. Then, we test the tree using the validation set. Next, we remove each node and again measure the tree's accuracy (now a pruned tree) on the validation set. If the accuracy of the validation set improves, we do not incorporate the recently removed node anymore and will continue working with the pruned tree. Otherwise, if the accuracy decreases, we add the node back. Next, we remove another node, and we repeat this node removal accuracy test on the validation set until no further node removal improves the accuracy, and we stop removing any new nodes. Pruning stops when removing further nodes no longer improves validation set accuracy. In the end, we evaluate the tree with the data that it has not encountered before from the Test set.

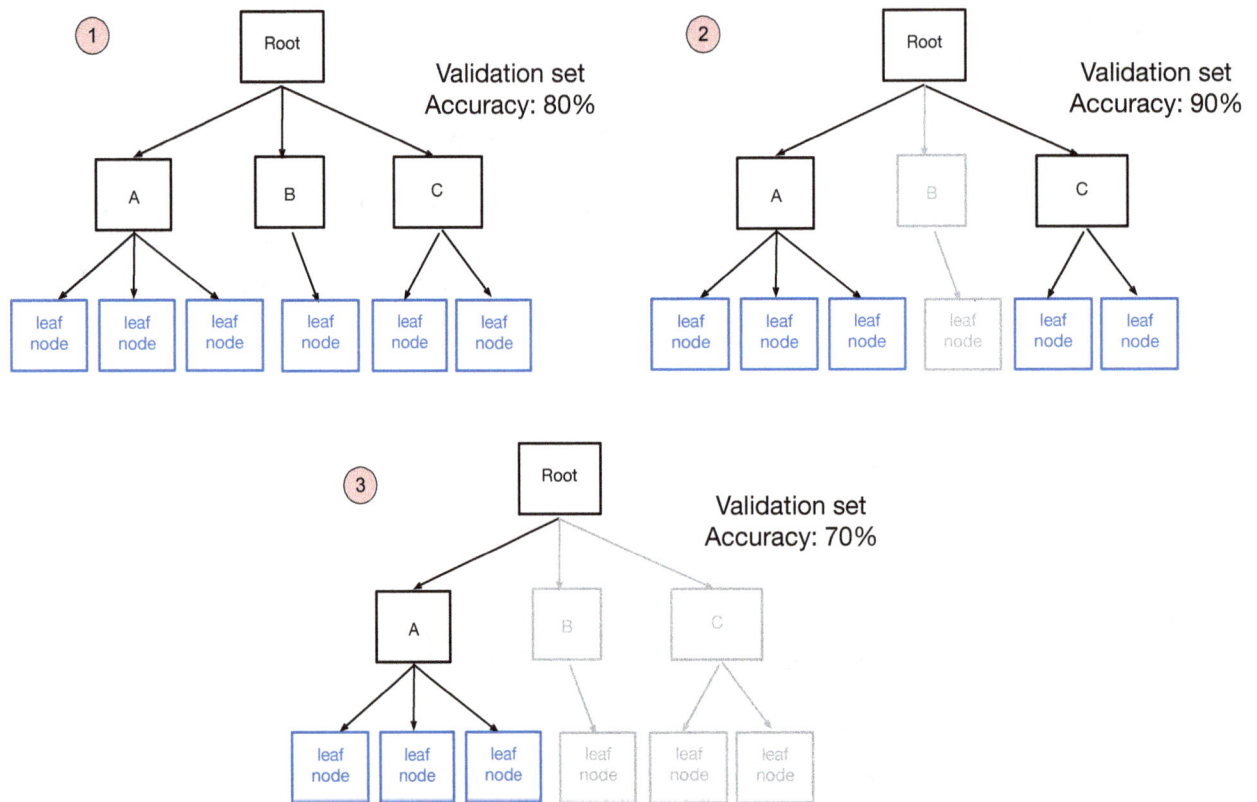

Figure 9-16: An example of pruning the original tree (1) and, as a result, removing branch B improves the accuracy (2), but removing another branch, branch C decreases the accuracy, so the algorithm returns (2) as the final result of pruning.

This process is similar to what we have explained about Sequential Backward Selection (SBS), which we have explained in Chapter 6. Note that while we are talking about nodes in this section, in the process of tree pruning, we do not remove leaf nodes, we only remove decision nodes.

To better understand pruning, look at the toy example in Figure 9-16 (1). Using the training set, the algorithm constructs the decision tree, and we experiment with its accuracy on the validation set. The accuracy is 80%. Then, we removed branch B and again tested the accuracy, it improved to 90%. That is good, so we keep the new tree, as shown in Figure 9-16 (2). From the new tree, we remove another branch, which is branch C. Now, the accuracy decreases to 70%, which is worse, as shown in Figure 9-16 (c). Therefore, we return to the pruned tree in Figure 9-16 (2) because it has the highest accuracy.

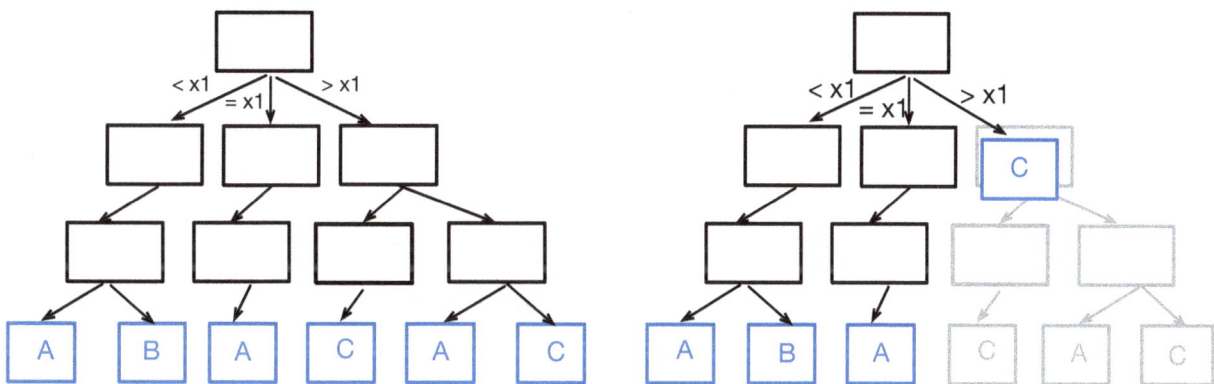

Figure 9-17: (Left) Original tree (Right) the pruned version of the tree, by substituting any branches under > x1 as C leaf node. Because most of the leaf nodes under this branch was C (two C and one A) before pruning.

The fourth approach is to prune a tree, which is fairly simple. It starts to remove a branch and substitute it with the most frequent leaf node. For example, take a look at Figure 9-17. Here, we have a branch $> X1$, and most of its leaf values are C. Therefore, the pruning process removes this branch (and its sub-branches) and substitutes one leaf with a leaf node with the value of C. In other words, it checks the value of the new record (test set). If it is $> X1$, then it assigns the C label to that record.

Decision Trees Computational Complexity

Assuming a tree has m depth, and it is constructed based on n number of data points, the computational complexity of running a tree is $O(m)$ and always $m < n$. The complexity of the decision tree increases exponentially as the tree goes deeper. Géron [Géron '19] explained that balanced decision trees have the computational complexity of $O(log_2(m))$ for the test set and $O(n \times log_2(m))$ for the train set.

Perhaps you might think that decision trees are similar to rule-based classifiers. That is correct; decision trees generate rules, but instead of feeding the rules ourselves, they discover rules on their own.

Table 9-7 summarizes the decision trees we explained here and their usage. If you intend to decide on a decision tree to use, it is better to test them on your dataset and choose the one that has the highest accuracy and/or resource efficiency.

Algorithm	Split Function	Usage / Characterisitcs
ID3	Information Gain	It is used for categorical data and not numerical data.
CHAID	Chi-Square	It can produce multiple branches, and since it uses Chi-Square test, it is good for descriptive analysis.
C4.5	Gain Ratio	it constructs a tree based on rule sets. It can handle missing data and can be used for both continuous and discrete data.
CART	Gini Split	it constructs data based on numerical splitting; it is binary but allows more than one choice via one vs others labeling policy. It is not very useful when the number of classes is large.

Table 9-7: Decision tree algorithms, their splitting method and usage / characteristics.

NOTES:

* Decision trees can be used for continuous, categorical, and discrete datasets. Besides, they are a good choice for datasets that include missing variables, and we do not need to invest heavily in removing or reconstructing missing data. Besides, decision trees are not good at handling duplicates, and it is better to remove duplicates and then feed the dataset into a decision tree for classification. Decision trees can handle duplicates in the sense that they will still function, but duplicate data can bias the tree's structure toward the over-represented examples.

* We can say that the decision tree is nothing but a set of rules that assign the label to the training data point based on that rule. It is a useful feature of the decision tree, and decision tree algorithms are explainable. At the time of writing this chapter (the first time in 2020), there is a movement to have explainable algorithms, i.e., *xAI*. This is due to the fact that black box algorithms, such as deep learning algorithms, are not explainable and, thus, not reliable for many applications, especially medical applications that make sensitive decisions.

* A problem in the decision tree is repetitions. Repetition occurs when an attribute is tested along the tree more than once, e.g., at level *n, salary > 100k,* and then again at level *n+3, salary > 100k.* This problem can also be mitigated by some condition checks (If-Then command).

* Another problem that happens in decision trees is replication. At some level, a duplicate subtree might be created, and considering the fact that a tree is usually loaded into the memory,

this is very inefficient. This problem can also be mitigated by some conditional checks, i.e., rule-based classifiers or condition checks (If-Then command).

* ID3, C4.5, and CART algorithms use greedy search methods. Greedy algorithms perform heuristic searches, and they might get stuck in local optima and not reach the global optima (Check Chapter 8 to recall global optima). As we have explained before, greedy algorithms are the algorithms that make the best short-sighted decisions. For example, we have the following two paths to get from A to D. *Path 1: A 3 steps to B, 99 steps to C, 999 steps to D. Path 2: A 5 steps to B, 6 steps to C, 5 steps to D.* Path 1 is longer than Path 2, but the greedy algorithm, choose Path 1, because the 3 steps are less than 5 steps at the beginning (short-sighted vision).

* In some literature specifying the depth of a tree as a threshold or the number of branches as the threshold for a decision, the tree is called tree regularization. We prefer to be parsimonious with these terms, but they are not wrong.

Ensemble Learning Methods

If you think you have already got rid of the decision trees, please accept our sincere condolence; we still have a long way to go with decision trees. All classification methods we have described until now are useful. However, in a real-world setting, we usually do not use one method, and most of the time (not always), we use ensemble learning or ensemble methods.

Ensemble learning refers to using more than one classification algorithm to perform the classification task, i.e., deciding on labels of the test dataset. Classification algorithms that we have explained until now are called weak learners or vanilla learners in the context of ensemble learning models. They are not weak, but combining them builds a stronger classifier, making them more accurate, and this is the objective of ensemble learning.

Some ensemble learning methods resemble human behavior while seeking a consultation. Assume you ask a person's opinion about a product. What that person said is probably correct, but if you ask for the opinion of more people, you get a better understanding of that particular product. Ensemble learners do the same in the context of machine learning, and they lower error rates and reduce the risk of overfitting. However, it is impossible to always use ensemble learning methods due to their resource utilization; they are not good for low-power or devices with limited computational capabilities.

There are three types of ensemble learning: Bagging [Breiman '96], Boosting [Schapire '90], and Stacking [Wolpert '92]. In the following, we describe these ensemble learning methods. Next, we focus on state-of-the-art classifiers that are based on gradient boosting decision trees.

Bootstrap Aggregating (Bagging)

Bootstrap Aggregating (Bagging) methods start by choosing *more than one random subset* from the training set and building a model for each subset (Figure 9-18). As a result, we have a set of models, and in the end, they are combined together. Based on majority voting or averaging the

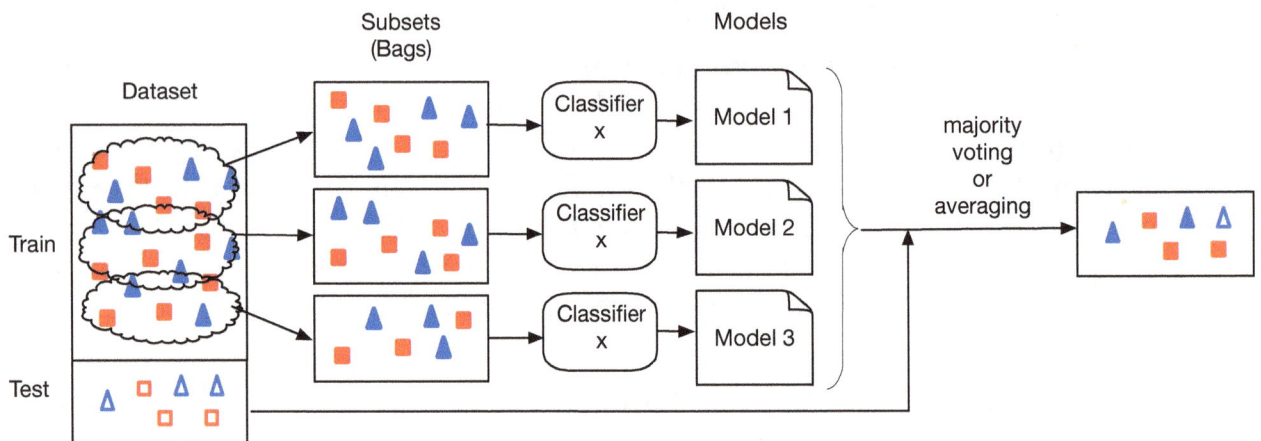

Figure 9-18: A toy example that presents how the bagging algorithm works. In the context of ensemble learning, a single classifier is called a weak learner. Note that all classifiers should be the same algorithm.

candidate labels, which are provided by each classifier, the final label will be assigned. Bagging reduces the variance because of the combination of different models.

Figure 9-18 visualizes how bagging is working. Here, we have a dataset, and as the first step, the algorithm starts with the original dataset and creates multiple subsets through a process known as *bootstrap sampling*. Bootstrap sampling involves repeatedly drawing samples from the original dataset, with each draw returning the data point back to the dataset, allowing it to be selected again. This means that after a data point is selected, it can be chosen again in subsequent samples. As a result of this sampling method, each bootstrap sample is likely to contain duplicates of some data points while missing others. On average, each bootstrap sample will include about 63% of the unique data points from the original dataset. These subsets are referred to as bootstrap samples or *bags of data*.

Then, as we can see in Figure 9-18, each of these subsets will be fed into a classifier. As a result, we will have three models, each from the same classifier algorithm. Afterward, it uses the majority voting or averaging from models to determine the label of new data points or test set data points.

Despite the usefulness of vanilla decision trees, they suffer from high variance. It means that if we separate the training dataset into two parts and train a decision tree for each part, the two parts could have two very different decision trees. This shows the vulnerability of the decision tree to variance. We can use a Bagging method (e.g., Random Forest) to reduce the risk of variance. The result of bagging usually has higher accuracy than each algorithm alone. The input parameters that a bagging algorithm receives include the number of subsets, the number of data points in each subset, and the classification algorithm.

Bagging is a *bootstrapping* method. In the English language, bootstrapping is an expression that metaphorically describes the act of achieving something without external help, often overcoming seemingly impossible odds. In the modern era, it is referred to processes that rely on self-

490

resources and not outside resources. In the context of statistics, bootstrapping is a metric of sampling that uses random sampling with replacement and assigns an accuracy score to sampled data.

Boosting

Boosting is another type of ensemble learning method in which each weak learner (a decision tree) is associated with a weight. To better understand the intuition of weights in boosting, let's talk about the example we described earlier. We explained that to buy a product, we could consult with several other recommenders. In the context of boosting, we assign weight to each recommendation based on the expertise of the recommender. If the recommender is an expert, the weight is high, and the non-expert recommender gets a low weight. These weights affect the average of opinions we get for a product.

Boosting is a sequential process that applies a classification algorithm to the entire train dataset and then makes a model. Next, it collects the errors (mislabeled data points) of the previous model and makes a smaller dataset from errors. Then, it uses the same classification algorithm (similar to Bagging) on the dataset that includes errors and creates another model. This process continues in several iterations until a maximum number of iterations (hyperparameter) is reached. Each Boosting model depends on the previous model. In other words, each new model focuses on the errors of the previous round. The models that perform better will receive higher weights, and the model that performs weaker will receive lower weights.

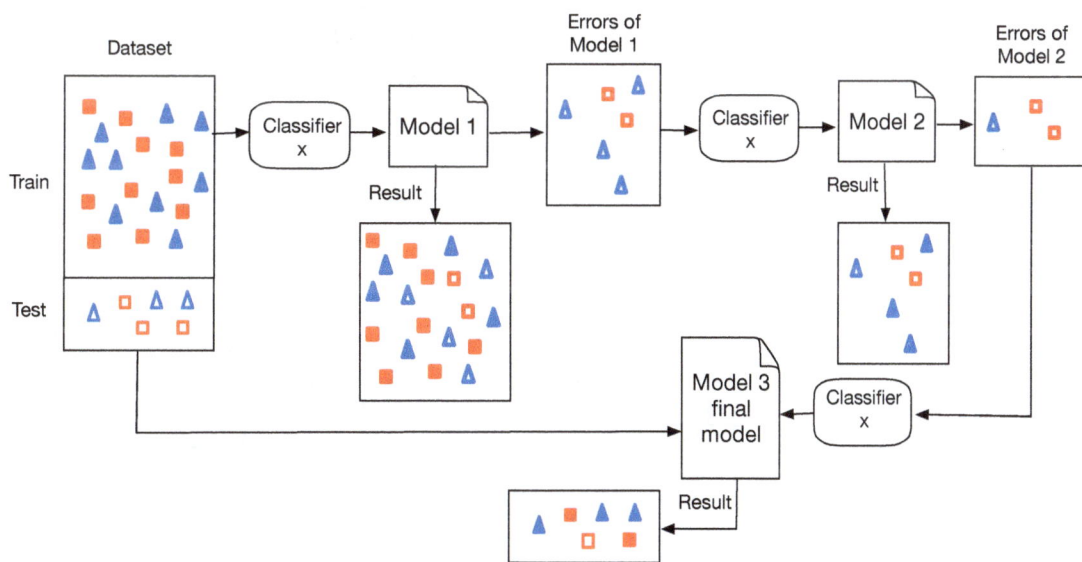

Figure 9-19: A toy example that presents how the boosting algorithm works. Note that the test dataset uses the final model, which is the strongest model.

Figure 9-19 shows the process flow of the Boosting algorithm. The objective is to label the unlabeled data, and in each round, a new dataset is constructed from the model's errors and fed into the next model as an input dataset.

Boosting algorithms usually receive three hyperparameters: (i) the number of decision trees to construct, (ii) the shrinkage parameter (which is a type of regularization and it is used to slow down the learning and thus reduce overfitting), and (iii) the number of splits in each tree. In general, Boosting decreases the bias (bagging decreases the variance). However, it provides better accuracy than bagging, but it also tends to overfit the training data as well. Therefore, hyperparameter tuning is required to avoid overfitting.

Stacking

Unlike Bagging and Boosting, Stacking applies 'different' weak learners in parallel and combines their results (base models) by training a *meta-model*. Meta-model or meta-learner aims to minimize the overall prediction error of weak learners. Previously described ensemble learning methods, i.e., Bagging and Boosting, are homogenous (they use one weak learner), but stacking is heterogeneous (different classification algorithms), and the output of each model will be used to train a meta-model. Usually, meta-model classifiers are simple regression algorithms, such as linear regression for regression tasks or logistic regression for classification tasks. However, we could use other algorithms as well.

The main advantage of stacking is diversity among different models, which allows different patterns to be extracted from the underlying dataset. Both Bagging and Boosting are usually used for decision trees, but stacking uses a variety of weak learners, which makes it powerful to benefit from the diversity of different classifiers.

Figure 9-20 visualizes the process of stacking. Usually, it has two steps. The first step, which we call level 0, involves using a mixture of classifiers. The next step, i.e., level 1, includes constructing a meta-model classifier that can accurately classify the test dataset.

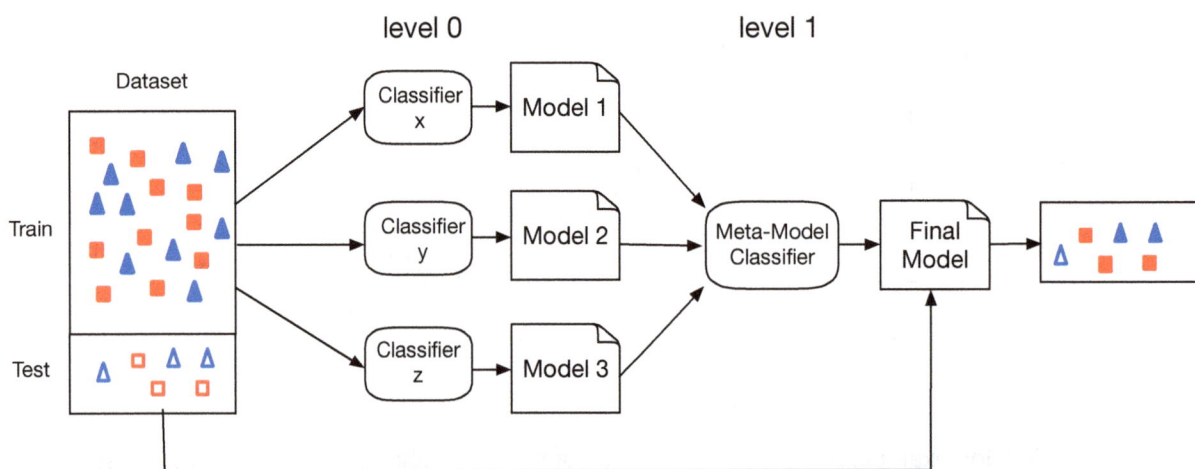

Figure 9-20: A toy example that presents how the stacking algorithm works. The first classifiers are called level 0, and the second classifier is called level 1.

Random Forest

Random forest [Breiman '05] is the most popular Bagging algorithm that uses decision trees. Decision trees are very good because they are explainable, but they suffer from weak accuracy. The random forest algorithm resolves the accuracy problem of decision trees by constructing multiple decision trees during the training phase. Random forests can reduce the variance of decision trees by averaging their predictions from different subsets of the data.

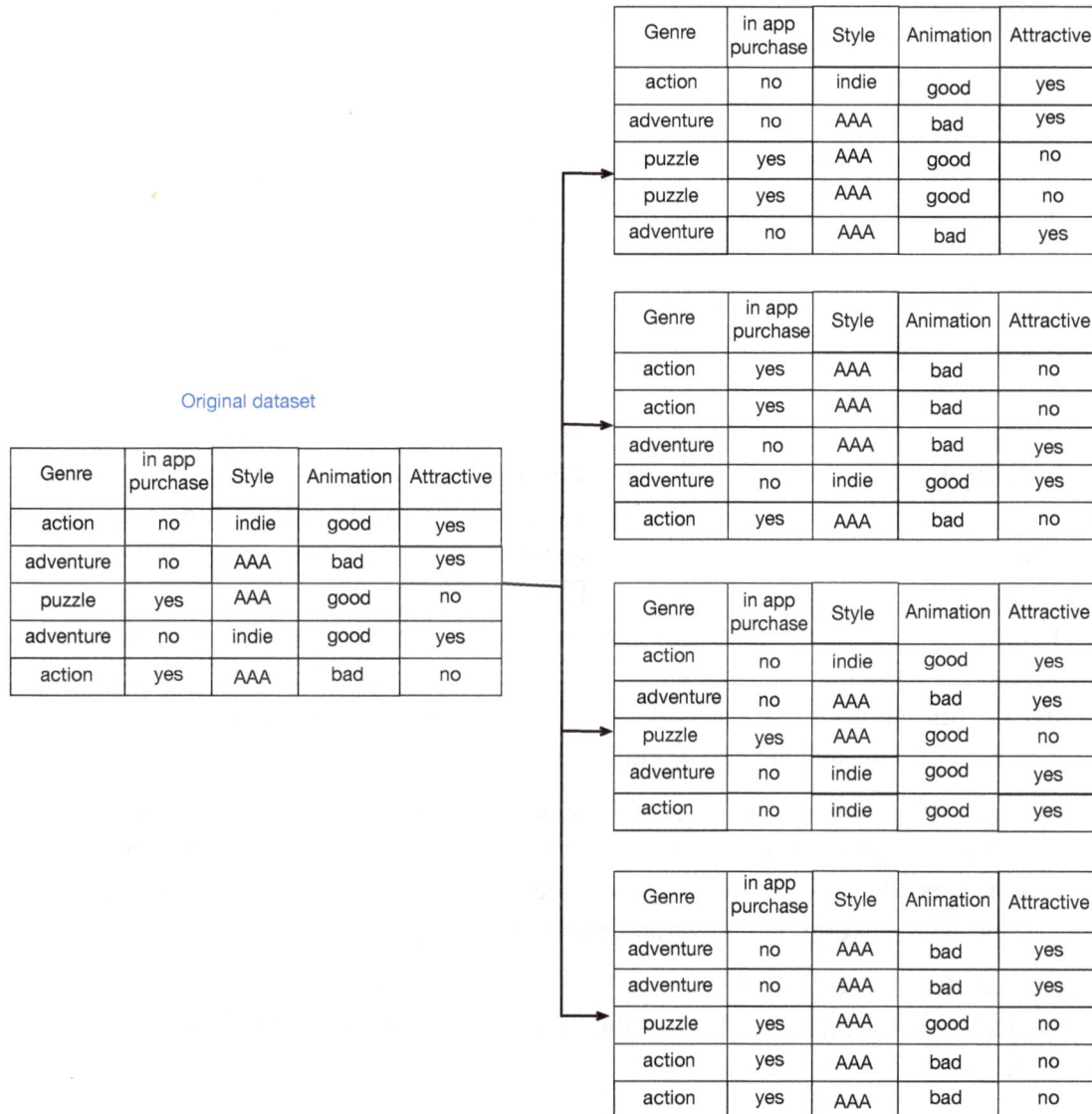

Genre	in app purchase	Style	Animation	Attractive
action	no	indie	good	yes
adventure	no	AAA	bad	yes
puzzle	yes	AAA	good	no
puzzle	yes	AAA	good	no
adventure	no	AAA	bad	yes

Genre	in app purchase	Style	Animation	Attractive
action	yes	AAA	bad	no
action	yes	AAA	bad	no
adventure	no	AAA	bad	yes
adventure	no	indie	good	yes
action	yes	AAA	bad	no

Genre	in app purchase	Style	Animation	Attractive
action	no	indie	good	yes
adventure	no	AAA	bad	yes
puzzle	yes	AAA	good	no
adventure	no	indie	good	yes
action	no	indie	good	yes

Genre	in app purchase	Style	Animation	Attractive
adventure	no	AAA	bad	yes
adventure	no	AAA	bad	yes
puzzle	yes	AAA	good	no
action	yes	AAA	bad	no
action	yes	AAA	bad	no

Original dataset

Genre	in app purchase	Style	Animation	Attractive
action	no	indie	good	yes
adventure	no	AAA	bad	yes
puzzle	yes	AAA	good	no
adventure	no	indie	good	yes
action	yes	AAA	bad	no

Figure 9-21: From the original dataset, a set of bootstrap datasets (subsets) are created. The size of the subsets is similar to the original dataset, but each of them contains about 70% of the original data. The rest of their data points are duplicates.

To understand the algorithm, we follow our sacred tradition and begin with an example. Assume you finished this book and since a big worry of your life, which is learning artificial intelligence and machine learning concepts, has been resolved. Now, you have started to play some games. A recommendation algorithm intends to suggest games to play. The history of your game playing is shown as a table on the left side of Figure 9-21. It is a dataset of some games, and you have labeled them as attractive to you or not.

(i) As the first step, the random forest algorithm starts by creating subsets from the original dataset, similar to other Bagging algorithms; each subset includes only 63% of the original dataset samples, as it has been shown in Figure 9-21. These subset datasets are called *Bootstrap datasets*, and the algorithm randomly selects some records from the original dataset to construct them. Since only 63% of the original dataset existed in each bootstrap (subset) dataset, the part of the dataset that does not exist in the bootstrap dataset is called the *Out-Of-Bag (OOB) dataset*, i.e., $100 - 63 = 37\%$.

(ii) In the second step, assuming we have a d number of features, The random forest selects k ($k < d$) number of features from each subset, and it uses these k features to calculate the split. In traditional decision trees, we use all features (d), but here, we use only k features from d possible features. Usually, the random forest uses $k = \sqrt{d}$ equation to decide on the k value. For example, in Figure 9-22 $d = 4$ ('Attractive' column is the label, and we didn't count it as a feature), thus we assume $k = 2$. For each dataset, we select k random number of features. Selected features are shown in blue color in Figure 9-22. For example, if $d = 8$, we get the closest fit, which is $k = 3$. In other words, unlike vanilla decision trees, we only use a random number of features (columns) and not all of them. This contrasts with decision trees, where the best feature for splitting is chosen from all available features rather than a subset. Here, however, the split is selected from a smaller set of features, not all of them.

A question might arise: why did we select k features from the d number of features? The answer is that if one single feature contributes significantly to a prediction, it will be used in the top split of all decision trees. By changing the split of trees, random forest creates highly uncorrelated trees, thus reducing the overall model variance and preventing overfitting.

(iii) In the third step, the test dataset (or test record) will be fed into all trees, and then its labels will be decided by averaging the labels of the other decision trees. It will lead to having a low variance while getting a high accuracy.

At the end of the training phase, we have a set of uncorrelated decision trees. For the inferences, we easily feed them into the set of decision trees, and each of these trees assigns a label to the new data point. The algorithms perform a majority voting and decide the final label based on votes, as shown in Figure 9-23.

The Random forest has two hyperparameters to configure, k and m (in addition to the decision tree algorithm that is usually fixed). k is the number of features to select for comparison and split

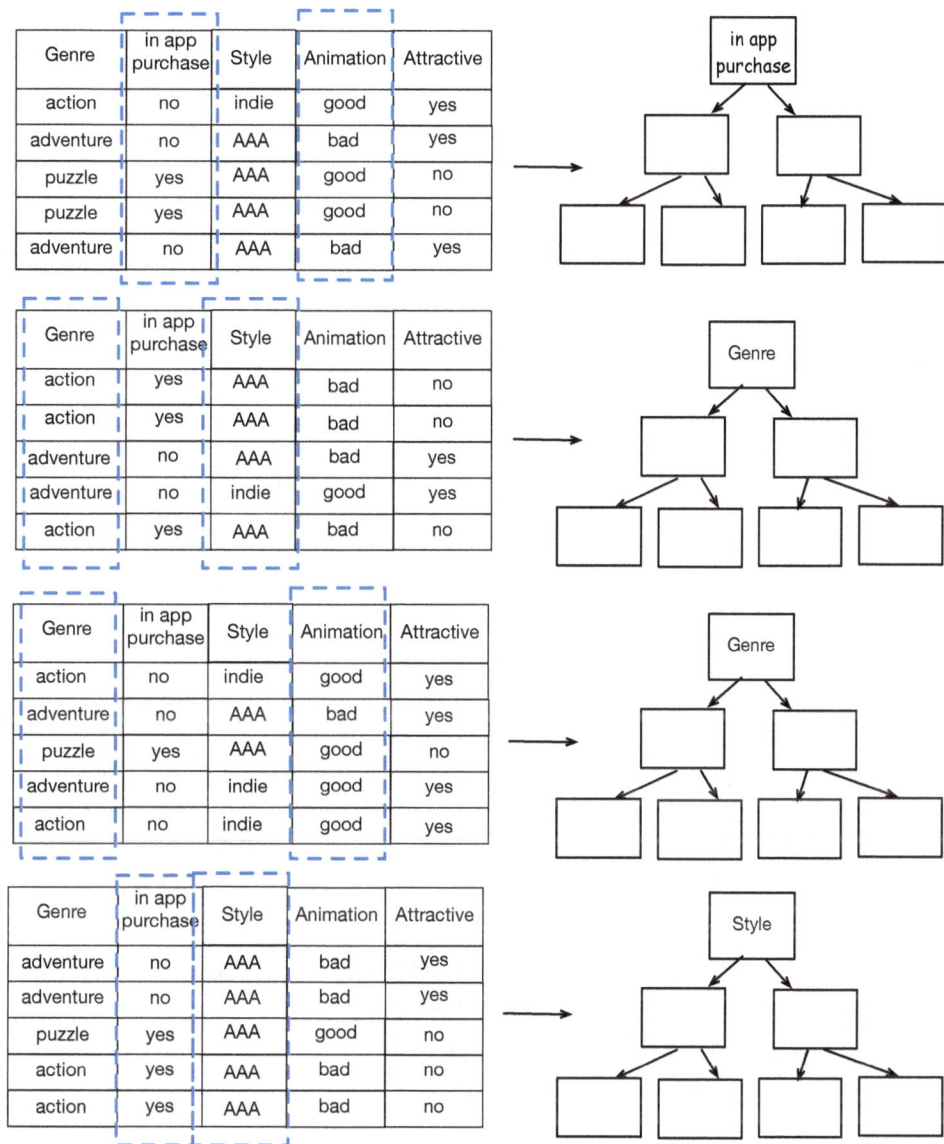

Genre	in app purchase	Style	Animation	Attractive
action	no	indie	good	yes
adventure	no	AAA	bad	yes
puzzle	yes	AAA	good	no
puzzle	yes	AAA	good	no
adventure	no	AAA	bad	yes

Genre	in app purchase	Style	Animation	Attractive
action	yes	AAA	bad	no
action	yes	AAA	bad	no
adventure	no	AAA	bad	yes
adventure	no	indie	good	yes
action	yes	AAA	bad	no

Genre	in app purchase	Style	Animation	Attractive
action	no	indie	good	yes
adventure	no	AAA	bad	yes
puzzle	yes	AAA	good	no
adventure	no	indie	good	yes
action	no	indie	good	yes

Genre	in app purchase	Style	Animation	Attractive
adventure	no	AAA	bad	yes
adventure	no	AAA	bad	yes
puzzle	yes	AAA	good	no
action	yes	AAA	bad	no
action	yes	AAA	bad	no

Figure 9-22: *k* number of features from each select is selected (marked in blue dots), and the split for each decision tree is selected based on analyzing these k features, unlike decision trees, which use all features.

selection ($k = 2$ in our example), and m is the number of subsets created from the original dataset ($m = 4$ subsets in our example).

We have explained that while creating the Bootstrap dataset, the records or data points that are not included in the subsets are called Out-of-Bag (OOB) samples. For example, if the original dataset is composed of $[a, b, c, d]$ and one of the Bagged datasets is composed of $[a, b, d]$, we can say $[c]$ is an OOB sample. Similar to the Bootstrap samples, each OOB sample set could be passed through its related decision tree algorithm, constructed without looking at OOB samples, and the prediction error of OOB can be calculated. This error is called the *Out-of-Bag error*. This

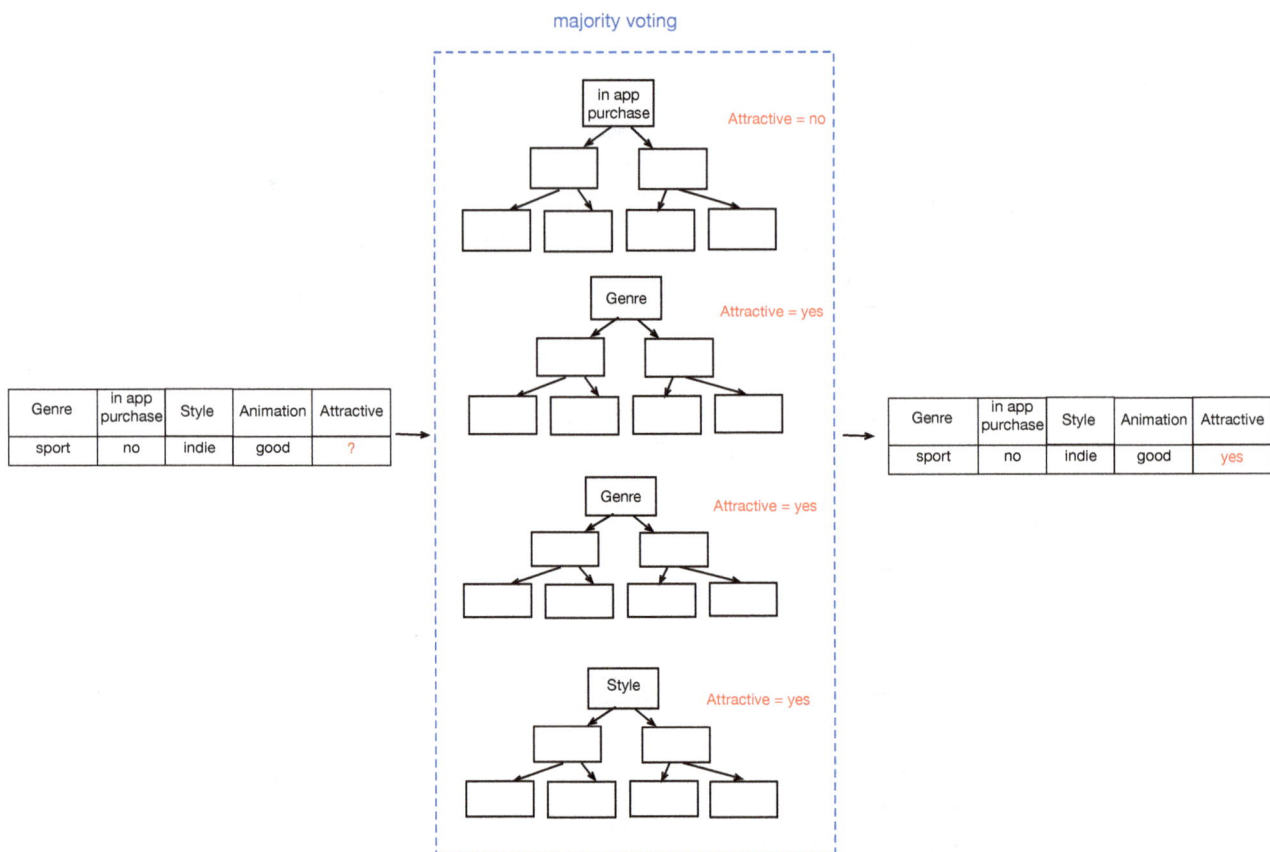

Figure 9-23: A test data which has only one record is fed into all decision trees. Each of them is assigned a label for the "Attractive" feature and at the end based on majority voting the final label will be assigned.

error could also be used for the validation of the dataset (in addition to the test dataset). It means that we are not only relying on a test dataset for validation, but we can also use OOB datasets as well. As a result, we will have a very good estimate of the Random forest model. This is especially useful if we have a small training dataset.

Assuming m is the tree depth and n is the number of data points, we explained the computational complexity of the tree is $O(n \times log_2(m))$. The Random Forest creates m' numbers of decision trees, so the computational complexity of training is $O(m' \times n \times log_2(m))$ and testing will be $O(m' \times log_2(m))$.

Adaptive Boosting (AdaBoost)

AdaBoost [Freund '95] is an ensemble learning algorithm based on Boosting that uses decision trees as its weak learners. To learn AdaBoost, we should first learn some concepts, which we describe below.

- Similar to other boosting algorithms, AdaBoost combines multiple weak learners into a single strong learner. It focuses on errors in each iteration and improves the classification result by assigning higher weights to errors and lower weights to correctly classified data points.

496

- Similar to Random Forest, it uses a set of decision trees. However, since these trees are very short in depth (usually only one level deep), they are called *stumps*.

- Each data point (e.g., a record in table format) is associated with weight, and at the beginning, this weight is equal for each data point. The sum of weight at each iteration/level should equal one. Assuming we have n number of data points, the first stumps have their weight calculated as follows: $weight = 1/n$. If, in an iteration, the data is classified incorrectly, then its weight increases. Otherwise, its weight decreases (for correctly classified data points) because the sum of weights should be equal to one at each iteration/level.

- For each stump after the first level, the *error rate* is calculated as the sum of the weights of the misclassified instances divided by the total sum of weights. For example, if we have 5 data points, each with an initial weight of $1/5$, and 3 of them are misclassified, the error rate would be $3/5$, assuming all weights are still equal.

- The result of an AdaBoost algorithm is a set of classifiers (stumps), each associated with a

Standing and Clapping Hand (C1)	Move too many Body parts (C2)	Keep Smile (C3)	Good Dance Performance (Actual)	Good Dance Performance (Predicted)
No	No	No	No	No
No	Yes	Yes	No	No
No	Yes	No	Yes	Yes
No	No	Yes	Yes	Yes
No	Yes	No	No	No
Yes	Yes	Yes	No	Yes

Table 9-8: A sample dataset of dancing, the prediction label will be "Good Dancer".

weight.

Now that we have learned these preliminary concepts let's use an example to explain this algorithm. Dancing is a part of the culture in many regions, especially in Africa, Asia (East, South, and West), South and Latin America, etc. People gather together in ceremonies and parties and dance together.

Assume that after you finish this book, you are in charge of an AI corporation building dancing robots for parties. Your robots will learn to dance by looking at people dancing at parties. By looking at the people, the robot constructs the dataset presented in Table 9-8 along with our labels (Good Dance Performance). The output or prediction variable we are looking for labels is "Good Dance Performance". For the sake of simplicity, we consider all features binary.

Adaboost algorithm performs the prediction in the following steps:

(i) The algorithm assigns equal weight to each record, which is $1/n$ of the total n records (See Table of Figure 9-24). Then, it uses each feature alone and creates a decision tree with a depth

C1	C2	C3	Actual	Predicted	Level 0 Weight
No	No	No	No	No	1/6
No	Yes	Yes	No	No	1/6
No	Yes	No	Yes	Yes	1/6
No	No	Yes	Yes	Yes	1/6
No	Yes	No	No	No	1/6
Yes	Yes	Yes	No	Yes	1/6

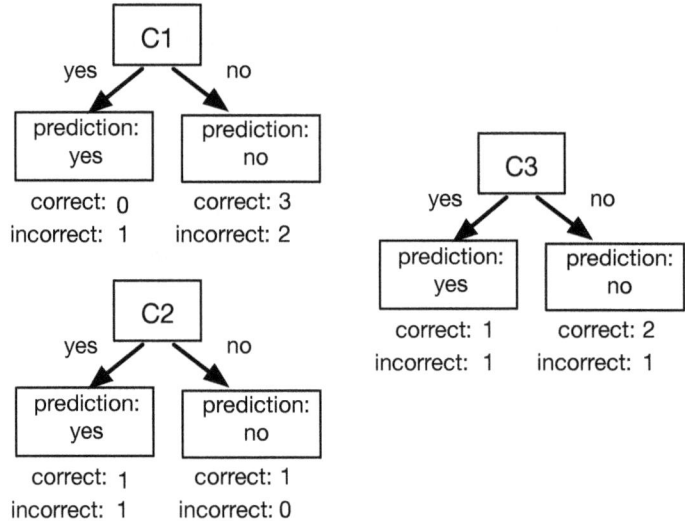

Figure 9-24: First each record receives an equal weight. For each feature a stump is created, the number of its correct vs incorrect instances is written at the bottom of each stump.

equal to one for each feature. In our example, we have three columns as features, and thus, we have three stumps. Keep in mind that the number of stumps is equal to the number of features. For example, if we have m features, AdaBoost will create m stumps.

(ii) Entropy, Gini index, or another tree split method is used for each stump to find the best stump that can split the data. We do not write them for the sake of brevity, but let's say the lowest entropy belongs to C3. Therefore, the AdaBoost algorithm selects the C3 stump.

(iii) Now, for the selected stump (C3), the algorithm should calculate the α_t, known as *stump performance score*, *significance*, or *stage value* of the stump by using the following equation. Here, t presents the index of the current iteration.

$$\alpha_t = \frac{1}{2} ln(\frac{1 - \epsilon_t}{\epsilon_t})$$

In this equation ϵ_t is the total error, which is the sum of weights for the misclassified records in the stump by total records in that stump, which is as follows:

$$\epsilon_t = \sum (\text{misclassied records in the stump / total records}).$$

Based on Figure 9-24, in stump C3, we have one incorrect record, and thus $\epsilon_t = 2/6$. Now, we can calculate the α_t for stump C3 as follows:

$$\alpha_t = \frac{1}{2} ln(\frac{1 - (2/6)}{(2/6)}) = 0.3465$$

(iv) Now that we have α_t of this stump, we can add it to the final model. The final model $H(x)$ is calculated as the sum of stump performance times each stump:

$$H(x) = \sum_{t=1}^{T} \alpha_t h_t(x)$$

Therefore, we have $H(x) = 0.3465 \ (Stump(C_3))$ as the result of the first level (*Stump (C3)* is a function).

(v) The next level starts by updating weights. The following equation describes how the algorithms calculate the new weights (weight at level *i*).

$$w_i = \begin{cases} w_{i-1} \times e^{-\alpha_t} & \text{for correctly classified record} \\ w_{i-1} \times e^{\alpha_t} & \text{for incorrectly classified record} \end{cases}$$

Based on this equation, we calculate the weights of the second level and add them to the table, as shown in Table 9-10. For example, for the last record, the prediction says 'Yes', but the actual data says 'No'. In this case, assume that two records are incorrectly classified by Stump C3 (the last record), which its new weight will be $1/6 \times e^{0.3465} = 0.235$. Others were correctly classified, and they will be $1/6 \times e^{-0.3465} = 0.117$. These numbers help us to identify values for the "Weight Level 2" column.

Nevertheless, the sum of weights should be equal to one, but they are not; therefore, the algorithm normalizes them by summing them all ($5 \times 0.117 + 1 \times 0.235 = 0.82$) and dividing each weight by the sum. This helps us to populate the "Level 2 weight (normalized)" column in Table 9-9.

C1	C2	C3	Actual	Predicted	Level 1 weight	Level 2 weight	Level 2 weight (normalized)
No	No	No	No	No	1/6	0.117	0.14
No	Yes	Yes	No	No	1/6	0.117	0.14
No	Yes	No	Yes	Yes	1/6	0.117	0.14
No	No	Yes	Yes	Yes	1/6	0.117	0.14
No	Yes	No	No	No	1/6	0.117	0.14
Yes	Yes	Yes	No	Yes	1/6	0.235	0.264

Table 9-9: A sample dataset of dancing, with the old and new weights.

(vi) Now, a new dataset with the same size as the original dataset will be created based on the coefficients of the most recent weights (Level 2 weights). In other words, new dataset records have their distribution based on weights. Since we have $0.286 \sim 30\%$, we can say 30% includes this weight error, and in the new table, there will repeated at 30% of records. Therefore, we will have something like the table in Figure 9-25. The blue records in this table are the ones that are substituted by the record that has a weight of 0.286, and other records (black ones) remain unchanged. Randomly, we selected one record to substitute it, and as a result, the blue colors are 30% of the entire dataset.

C1	C2	C3	Actual	Predicted
No	No	No	No	No
No	Yes	Yes	No	No
No	Yes	No	Yes	Yes
No	Yes	No	No	Yes
No	Yes	No	No	No
No	Yes	No	No	Yes

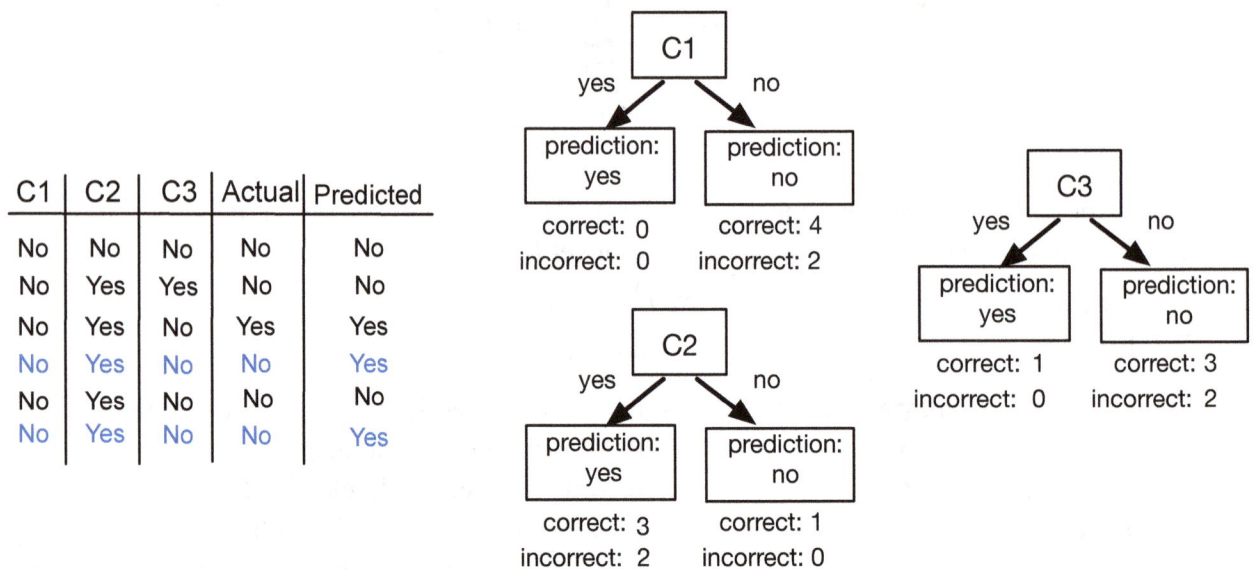

Figure 9-25: A new dataset created based on the normalized level 2 weights. Blue ones are because of weight 0.087, and because of that the algorithm creates multiple copies of that record. Then, based on the new dataset, again for each feature one stump will be created.

(vii) Next, the algorithm repeats from step (i) and creates one stump per feature from the Table in Figure 9-25. There will be three stumps, as it is shown in Figure 9-25.

The process from step (ii) will be repeated. Let's say Entropy, Gini index, or another tree split is used and select stump C1. At the end of step (iii), after the stage value (α_t) has been identified, the model can be refined. For example, assuming in this iteration, C1 has the lowest entropy (in this example, all have the lowest entropy, but we assume C1 has the lowest one), and we suppose we get its $\alpha_t = 0.622$, the final prediction model will be written as follows:

$$H(x) = 0.804(Stump(C_3)) + 0.622(Stump(C_1))$$

By looking at the result, i,e, $H(x)$, we can summarize that AdaBoost is a combination of several stumps (weak learners), and each stump is AdaBoost adjusts the weights of the training instances based on the errors made by the previous stump.

AdaBoost receives several hyperparameters. One hyperparameter is called the "number of iterations", T. This iterative process will be repeated T times. In other words, the algorithm iterates until the number of iterations is reached. Also, some AdaBoost implementations receive the "number of trees (stumps)" as an input parameter as well. Obviously, the number of trees should be lower than the number of features, and the algorithm selects them randomly. To identify the optimal number of stumps, we need to perform hyperparameter tuning. Some implementations also allow specifying a bit deeper than one-level trees. Therefore, another hyperparameter of AdaBoost could be the "depth of tree". Another parameter is the "learning rate", and it usually has a predefined value (e.g., 0.1). It could be used to shrink the contribution of each weak model in the final model, but it is not required in all implementations.

Assuming m is the number of stumps, n is the number of data points, and p is the number of features (columns) the computational complexity of AdaBoost is $O(m \times n \times p)$ for training and $O(m \times p)$ for testing.

NOTES:

* Bagging is robust to noise and outliers; boosting can indeed be sensitive to noise and outliers because it tries to correct misclassifications in each round, potentially leading to overfitting on the noise.

* Decision trees are the most common algorithm used in Bagging. Because they have high variance, bagging is helpful in reducing the variance of classification algorithms.

* Since boosting focuses on misclassified data points, the risk of overfitting increases.

* The term Bootstrapping is used in statistics as well. Bootstrapping is a statistical method used to estimate the population's statistical characteristics (e.g., distribution) by sampling a dataset with replacement.

* Random Forest enables us to measure the importance of each feature based on its contribution to tree split (e.g., impurity), which makes it an attractive algorithm to use for feature selection as well. In particular, per feature, we can calculate the impurity of the tree and then rank features based on their impurity to select the best feature that can assist us in the prediction or labeling task.

* While using the Random Forest algorithm, it is possible to make tree creation more random by using a random threshold for splitting trees. As a result, such a set of trees (forest) is called an *extremely randomized forest*.

* While Random Forest is generally robust against overfitting due to its averaging method, AdaBoost can provide high accuracy but might be prone to overfitting if the noise is present.

Gradient Boosting Decision Tree

At the time of writing this chapter, there are online communities, such as Kaggle's[7] online machine-learning competitions. Most of the awards go to the participants who use Gradient Boosting Decision Tree (GBDT) algorithms. GBDT algorithms have higher accuracy than traditional classification algorithms and, for tabular data, perform better than Deep learning algorithms (we will explain them in upcoming Chapters 10, 11, and 12). Nevertheless, unlike deep learning algorithms, they do not need a large dataset to operate. Therefore, if the dataset is not large enough and we can experiment with gradient boosting algorithms, it is a wise decision.

We start this section by explaining the basic form of the gradient boosting algorithm [Friedman '01, Mason '99]. Then, we will describe the three state-of-the-art algorithms that leverage the gradient boosting approach.

---

[7] https://mlcontests.com

GBDT algorithm [Friedman '01, Mason '99] is boosting (ensemble) algorithms that operate by using a gradient cost function. The boosting process of this algorithm is *a numerical optimization problem to optimize*.

Gradient Boosting Decision Trees are dead boring to read, but extremely important algorithms.

A gradient boosting algorithm is composed of these components: (1) a loss function to be optimized, (2) a sequence of decision trees (weak learners) to make predictions, and (3) an additive model that adds decision trees to minimize the loss function. It is called an additive model because one new decision tree will be added to the sequence at each iteration, and existing decision trees remain unchanged (frozen) in the model. This additive model uses a gradient (check Chapter 8) to minimize the loss score. In Chapter 8, we have explained that gradient descent is used to minimize a set of parameters, but in the context of a gradient boosting algorithm, the gradient is applied to minimize residual errors in decision trees.

The gradient boost can be used for both regression and classification. To keep the clarity of the algorithm, we cheat a bit in this chapter, which is about classification, and explain its regression here.

Regression with GBDT

To implement a regression with a simplified version of the gradient boosting, the algorithm performs the following steps.

(1) First, it starts with averaging the data we intend to predict. Next, the 'residual' will be calculated by the loss function. A simplified version of residual (pseudo residual) error is computed as $actual - predicted$. We will also use the simplified pseudo residual in our examples.

(2) Then, a new decision tree (e.g., CART) is constructed based on differences between pseudo residuals and predicted values.

Next, by comparing the initial prediction and pseudo residual, a new prediction is estimated, but the tree is multiplied by a learning rate ($\alpha = 0.1$) to scale its contribution to the final prediction model and reduce the risk of overfitting.

(3) Step 1 (without average calculation) and Step 2 will be repeated until adding more new trees does not reduce the pseudo residual values or a specific number of trees reaches.

Gender	Race	Salary
Male	White	$200k
Female	Asian	$160k
Female	White	$170k
Male	Black	$150k
Female	Asian	$150k
Male	White	$200k

Table 9-10: Salary of the SocialLove company based on race and gender for employees with equal qualification.

We are very proud of our diverse team

The maximum number of trees will be given as a hyperparameter. Then, the additive model is constructed by summing up all constructed decision trees.

It is better to understand its details with an example. Assume we have access to the historical data of SocialLove, a corporation that is a pioneer in Internet social media. SocialLove[8] claims that they respect diversity and treat all their employees fairly. A friend of ours who is an AI engineer is willing to join SocialLove. We have acquired access to some internal data, and we could use it to predict the salary in that corporation based on their race and gender. Our algorithm uses gender and race from historical data to recommend a salary for a new candidate. The historical data that is used for our prediction is presented in Table 9-10.

In the first step, the algorithm calculates the average of the target feature used for prediction (i.e., average salary) and then calculates the residual error for this feature. The residual (Residual 1) and target prediction (Predicted Salary) are shown in the table on the left side of Figure 9-26. Next, the algorithm creates the first decision tree, *DT1*, based on residuals. *DT1* leaf nodes are residuals, and its decision nodes are input features (Gender and Race).

The input of the decision tree is "Gender" and "Race", and the output is residuals ("Residual 1" in Figure 9-26). Since each leaf node has more than one residual, the algorithm calculates the average of residuals and substitutes it as a leaf node value. Check the bottom table in Figure 9-26 to see the result of the average substitution for *DT1*.

Now, the first decision tree is constructed, and each leaf node (light blue nodes in Figure 9-26) has one value. We observe that some residuals are close to the predicted value. For example, the average of residuals in a leaf node is -10, which is a small difference from the original value. This is a sign of overfitting (we have high variance and low bias) and it should be resolved. To

---

[8] At the time of writing this section, in 2021, requests for gender and race fairness reached tech companies, and ethics among the AI community raised many discussions. The data of this example are not real, but this unfortunate phenomenon exists everywhere, and to show our respect for people who struggle against this attitude, we use this example here.

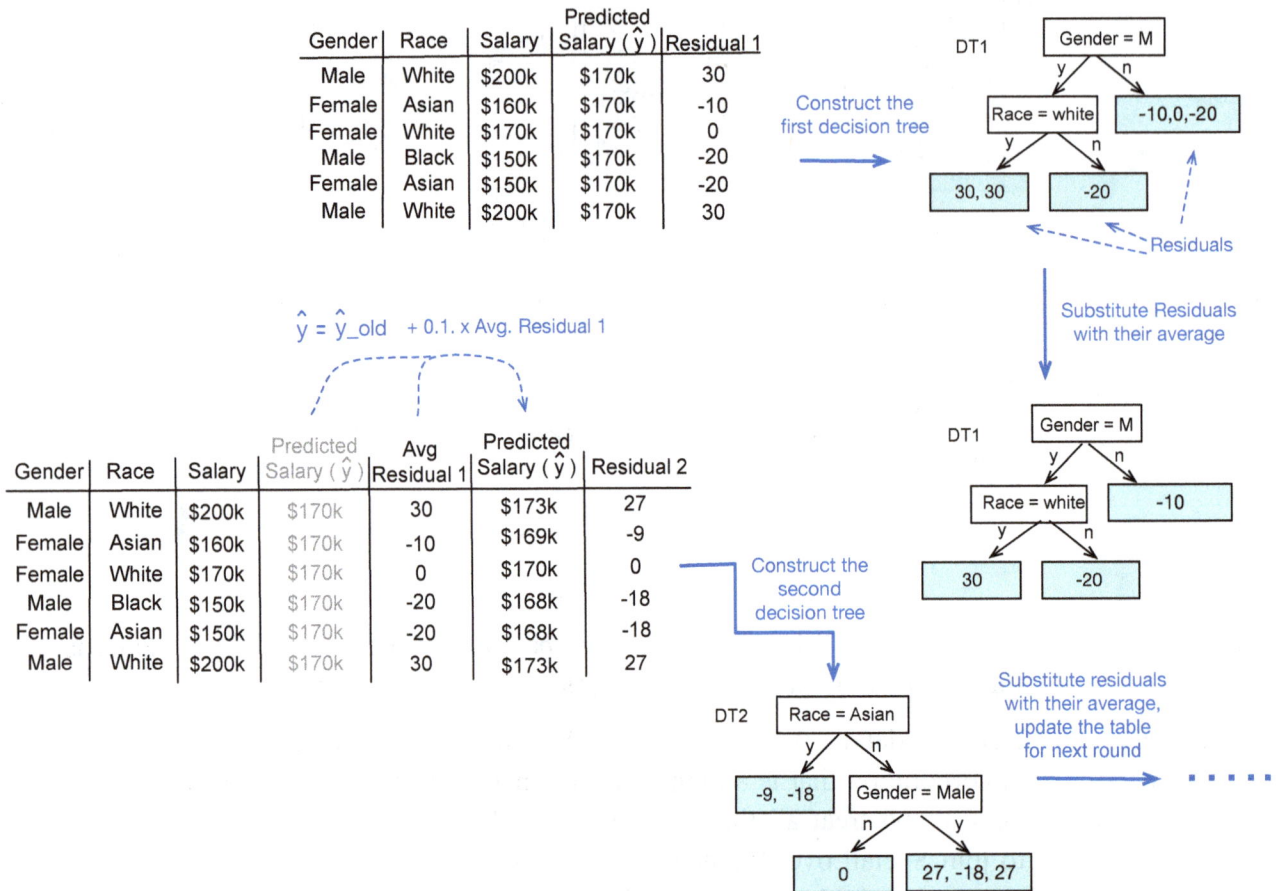

Figure 9-26: Flow of the GBDT for regression. The same information from Table 9-12, along with the predicted salary, has been added, and residuals are calculated. Since it is the first round, the predicted salary is just an average of all salaries. After residuals have been calculated, they will be substituted in the table for the next round. In the next round based on residuals, a new decision table is constructed, and then a new table with an updated prediction and residual will be created. This process continues until we reach the maximum number of trees (given as a hyperparameter).

ensure that the overfitting is handled, the algorithm adds a learning rate ($\alpha = 0.1$) multiplied by residuals to the predicted value. It uses the following equation: $\hat{y}_{new} = \hat{y}_{old} + \alpha . Residual$. Multiplication by a learning rate reduces the contribution of the decision tree in the final model, and thus, more decision trees will improve the accuracy of the final model. This is an important step to reduce the risk of overfitting. In the next iteration (step 3), we use the DT1 and calculate the predicted value as: $H(salary) = \hat{y} + 0.1(Residual\ 1)$.

Now, we can see that residuals have decreased slightly. See the second table in Figure 9-26. Again, the algorithm uses these new residual values (Residual 2) to construct a new decision tree, DT2, similar to what we have explained for DT1. The initial result is shown in the DT2 of Figure 9-26, and the described process repeats continuously, i.e., again, a new residual (Residual 3) is calculated, and the predicted value will be updated.

At the end, the prediction model will be written as follows:

$$H(salary) = \hat{y}_{old} + 0.1(Residual\ 1) + 0.1(Residual\ 2) + \dots.$$

Technically, we can formalize the prediction model as follows.

$$H(x) = H_0(x) + \alpha \times (H_1(x)) + \alpha \times (H_2(x)) + \dots = \sum_{i=1}^{n} \alpha \cdot H_i(x)$$

Here, $H_i(x)$ presents the residual of the model i. Each time we add a tree to the model, the residual values will get smaller. This process of tree creation and new residual calculation continues until the maximum number of decision trees is reached or the residual values do not change. Then, $H(x)$ is the final model and used to label new data.

When new test data arrives to estimate its salary, the algorithm finds the residual value of the right branch of decision trees (e.g., if it is male and not white, the algorithms substitute -20 for DT1, -18 for DT2, …), and substitute them in the $H(x)$ equation to predicts the salary of the new test data.

This example seems too simple; no optimization algorithm has been used, and there is nothing about gradient here. However, this simplified example helps us to understand the intuition behind using GBDT for regression.

Classification with GBDT

The gradient boosting algorithm is used for classification as well. It is fairly similar to regression one with some slight modifications, and it operates as follows:

(1) First, the algorithm starts calculating the average of the target prediction variable and then computing its *log(odds)* (check Chapter 8 to recall odds). In other words, the log(odds) is the initial prediction value for the classification task of this algorithm. Then, the Sigmoid function for binary class labeling or Softmax function for multi-class labeling is used to transform the log(odds) into a probability value[9]. This probability value will be used to decide about the initial labels. For example, if the probability is higher than 0.5, in a binary classification, then all labels will be positive, otherwise negative.

(2) Next, the residual will be calculated by the cost function. A simplified residual (pseudo residual) is

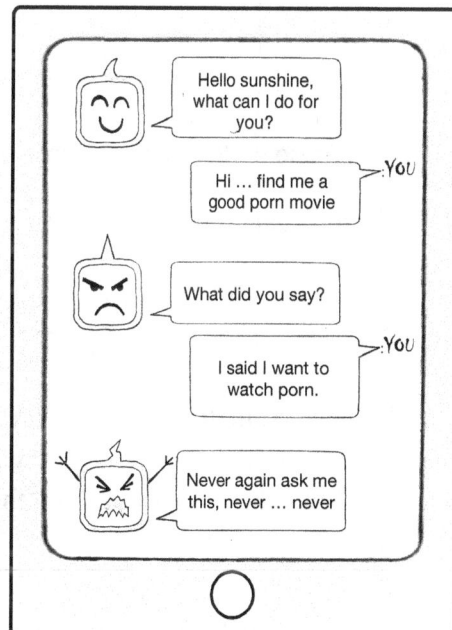

___

[9] In Chapter 8 we use the same approach for logistic regression and convert log (odds) into probability.

calculated ($actual - predicted$) in terms of probability (a number between 0 and 1).

(3) Afterward, a weak learner (e.g., ID3, CART, etc.) is used to construct a decision tree and predict residuals. The number of residuals is more than the number of leaves in the tree, and some leaves have more than one residual. A transformation should be implemented to assign

Length of Question (LQ)	Speak Loudly (SL)	Say Something Negative (SN)	Get Angry (GA)
8	Yes	Yes	Yes
8	No	No	No
9	Yes	No	Yes
14	No	Yes	Yes
17	Yes	Yes	Yes
5	No	No	No

Table 9-11: Previous reaction of the chatbot to the question we asked it.

Length of Question (LQ)	Speak Loudly (SL)	Say Something Negative (SN)	Get Angry (Actual)	Initial Prediction	Residual
8	Yes	Yes	Yes	0.66	1-0.66 = 0.34
8	No	No	No	0.66	0-0.66 = -0.66
9	Yes	No	Yes	0.66	1-0.66 = 0.34
14	No	Yes	Yes	0.66	1-0.66 = 0.34
17	Yes	Yes	Yes	0.66	1-0.66 = 0.34
5	No	No	No	0.66	0-0.66 = -0.66

Table 9-12: Previous reaction of the chatbot to the question we asked it.

each leaf a residual value. The following equation is the transformation to calculate the leaf value from its residuals.

$$leaf\_value = \frac{\sum_i Residual_i}{\sum P_{old} \times (1 - P_{old,i})}$$

Here, P_{old} refers to the previous predicted probability value. In the initialization phase, all predicted probabilities are the same, which we already know is wrong, but it is only the initial prediction. After this transformation, the tree is ready, and each of the tree's leaf nodes includes one value.

(4) The recently constructed tree will be multiplied by the learning rate and added to the additive model, i.e., $H(x)$. This tree will be used to calculate the new predicted probability, i.e., Sigmoid of log(odds).

(5) The process from steps (1) to (4) will continue until adding more new trees does not reduce the residual values or a specific number of trees reaches. The maximum number of trees will be

given as a hyperparameter. Then, the final model is constructed by summing up all the constructed decision trees.

Table 9-11 presents an example; here, we try to create a chatbot that is getting angry when we are talking about something harmful. We are in love with our chatbot, and we really hesitate to make it angry. Since the mood of the chatbot changes based on the questions, we developed a predictive model to avoid asking the chatbot questions that result in anger.

We would use the GBDT classification model to predict whether the chatbot gets angry or not based on the given question.

(1) The first step, as we have explained before, is to make an initial prediction and then calculate its log(odds). In Table 9-12, column GA, there are four 'yes' and two 'no'. In total, we will have six states. Its log(odds) is calculated as follows:

$$log(\frac{4}{2}) = 0.69, \text{ which we round it to 0.7.}$$

After the log(odds) have been computed, it is converted into probability by using the Sigmoid function as follows:

$$initial\_prediction = \frac{e^{log(odds)}}{1 + e^{log(odds)}} = 0.66$$

The value of the initial_prediction $0.66 > 0.5$, thus, states that all sentences are predicted to make the chatbot angry (we assign 1 to get angry and 0 not to get angry). Clearly, this approach is not correct, but it is the initial prediction, and the algorithm improves its prediction in each iteration.

In this example, we choose to perform binary classification for the sake of simplicity, but though softmax, we can cover more than two labels.

(2) Now, the probability value of the initial prediction is used to calculate residuals, and the residual is the difference between actual values (1 for yes and 0 for no) and the initial prediction. The result of the residual is presented in Table 9-12.

(3) At this stage, residuals are ready, and the algorithm can build its first decision tree. It builds the decision tree to predict residuals using LQ, SL, and SN. Let's assume the algorithm constructs a tree like the one presented on the left side of Figure 9-27.

Since some leaves have more than one residual value, we should use the described equation for leaf values to get only one value for each leaf.

Afterward, we perform the transformation for every leaf (even the one that has only one value) by using the described equations as follows:

$$\frac{-0.66 + 0.34 + -0.66}{(0.66 \times (1 - 0.66)) + (0.66 \times (1 - 0.66)) + (0.66 \times (1 - 0.66))} = \frac{-0.98}{0.67} = -1.46$$

$$\frac{0.34 + 0.34}{(0.66 \times (1 - 0.66)) + 0.66 \times (1 - 0.66)} = \frac{0.68}{0.45} = 1.51$$

$$\frac{0.34}{0.66 \times (1 - 0.66)} = \frac{0.34}{0.22} = 1.54$$

By substituting these values, the first weak learner table will look like the right side of Figure 9-27.

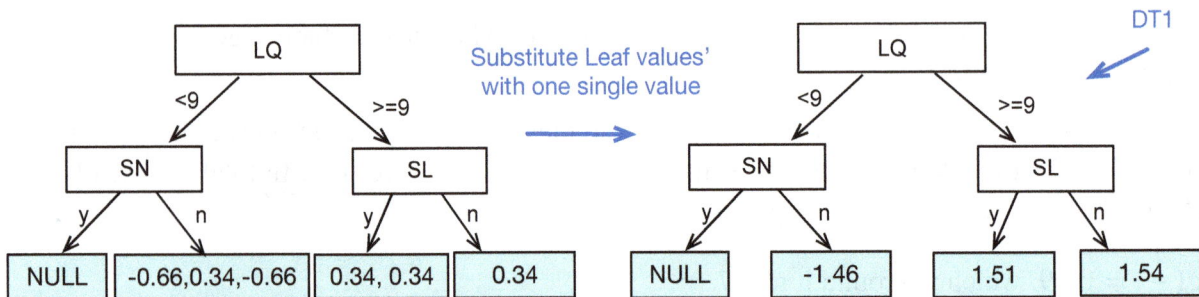

Figure 9-27: A sample decision tree is constructed (by hand to simplifying its readability) from the data in Table 9-14. Leaf nodes assigned residuals as value, but then the described equation of step (2) is used to substitute leaf nodes with more than one variable with new variable.

(4) Now that the first decision tree is ready, let's call it DT1, and it is added to the additive model. In particular, the result mode (additive model) will summate the initial prediction and learning rate (e.g., 0.1) times the decision tree.

$$model = 0.66 + 0.1 \times DT1$$

(5) Based on the additive model, the algorithm calculates the log(odds) of each record and substitutes it as the prediction column.

For example, for the first record in Table 9-12, we have $LQ<9$ and $SN=No$. By substituting them in the model, we have $model = 0.66 + 0.1 \times 0 = 0.66$. For the fourth record, we have $LQ>=9$ and $SL=No$, and thus $model = 0.66 + 0.1 \times 1.54 = 0.814$. These log(odds) values will be used by the Sigmoid function to calculate the new prediction value and construct a table similar to Table 9-12. The initial prediction value will be substituted by the results Sigmoid of log (odds). Then, from step (2), the process iteratively continues until a threshold for a maximum number of trees is reached or the loss score does not improve. The final model will be something like the following: $model = 0.66 + 0.1 \times DT1 + 0.1 \times DT2 + 0.1 \times DT3 + \ldots$

When a new record is added to the table, the algorithm identifies its label (either get angry or not get angry) by substituting its values in the model. Now that we understand a simplified version of the algorithm, we can write it pseudo-code, which could help us understand the algorithm in more technical detail. In the examples described, both regression and classification, we try not to describe its math. Nevertheless, while formalizing GBDT, we describe its math.

The algorithm gets the input dataset, i.e., $\{(x_i, y_i)\}_{i=1}^{n}$, and a differentiable[10] loss function, i.e., $L(y_i, \gamma)$, as an input. The loss function should be differentiable because it uses the chain rule of derivatives (check Chapter 8 to recall these mathematical terms). x_i presents the features for the given input value (one row of data without the prediction column), y_i presents the output, and n is the number of data points in the training set. The loss function L evaluates how well we can predict y_i, and it is computed as follows (for a regression problem):

$$L = 1/2(predicted - actual)^2 .$$

The gradient boosting algorithm can be written as follows. While reading the pseudocode, please read the blue text as comments on top of each command.

```
# The input data is the training dataset ({(xᵢ,yᵢ)}ⁿᵢ₌₁), x is the input variable, and
our objective is to predict y, given the differentiable loss function.L(yᵢ,F(x)) Is a
differentiable loss function, and F(x) is the prediction function
```

Input data: $\{(x_i, y_i)\}_{i=1}^{n}, L(y_i, F(x))$

```
# At the beginning, the algorithm initializes the model (H₀(x)) with a constant value.
L(yᵢ,γ) is the cost function that should be minimized. y is the prediction variable and
γ is the log(odds) value. In this context, arg min means we must find the log(odds) that
minimizes the sum. This minimization can be implemented with the Newton method or gradient
descent.
```

$$H_0(x) = arg \min_{\gamma} \Sigma_{i=1}^{n} L(y_i, \gamma)$$

For m =1 to M `# M is the number of trees to be constructed, usually set to 100.`

```
    # The following line computes residual for 'n' training set. i in rᵢ,ₘ stays for
    the sample number, and m for the number of trees. The residual (rᵢ,ₘ) is the
    derivative of the loss function with respect to the predicted value.
```

$$r_{i,m} = -\left[\frac{\partial L(y_i, H(x_i))}{\partial H(x_i)}\right]_{H(x)=H_{m-1}(x)} \quad for \quad i = 1,...,n$$

```
    # it uses the 'train' dataset to construct a weak learner (Hₘ(x)). Hₘ(x) fits the
    predicted value to the residual value. Meaning, it builds a tree to predict
    pseudo-residuals ({(xᵢ,rᵢ,ₘ)}ⁿᵢ₌₁) instead of prediction value ({(xᵢ,yᵢ)}ⁿᵢ₌₁).
    Therefore, Fit(Hₘ(x),rᵢ,ₘ) trains the mth weak learner to predict these residuals
    (the errors made by the previous predictions).
```

$$H_m(x) \leftarrow Fit(H_m(x), r_{i,m})$$

```
    # Each leaf from the constructed tree Hₘ(x) has k number of residuals to
    determine the output values for each leaf. The output (predicted) value is a γ
    (gamma), a.k.a. "multiplier" (or log(odds) in classification case). γ is the
    minimum of the cost function. It resolves an optimization problem with a
    derivative of the cost function and chain rule. In other words, this equation
    minimizes the optimization function.
    # In the case of classification case, instead of i = 1 in the summation, we will
    have xᵢ ∈ Rᵢⱼ, which Rᵢⱼ refers to all residual values in that leaf.
```

$$\gamma_m = arg \min_{\gamma} \sum_{i=1}^{k} L(y_i, H_{m-1}(x_i) + \gamma)$$

[the algorithm continues on the next page]

---

[10] A function that its derivative existed at each point is referred to as a differentiable function.

```
# At the end, the model will be updated and making a new prediction for each
  sample by adding it to the previous prediction. α is the learning rate.
```
$$H_m(x) = H_{m-1}(x) + \alpha \times \gamma_m$$

return $H_m(x)$

For some mysterious reason, we did not find much discussion about the computational complexity of the GBDT algorithm. A paper by Si et al. [Si '17] reports that considering N is the number of data points and L is the number of labels (size of output space). At least $O(NL)$ time and memory are required to build GBDT trees. The residual density grows after each iteration, and will become a matrix of size $N \times L$.

Finally, we are done with the gradient boosting algorithm and other algorithms. Take out your brain from the fryer and put it in an ice bucket because the remainder of this chapter will explain some state-of-the-art classification algorithms.

NOTES:

* In the described examples, we have explained that no loss function has been used, and we calculate a pseudo-residual, not a real residual, which is calculated by the loss function.

* In the described examples, we have simplified the weak learner (tree) construction and do not use any of the standard tree construction algorithms for the sake of simplicity.

* Usually, a gradient boost used for classification uses trees with 8 to 32 leaves.

* The pseudo-code for the algorithm for both regression and classification is the same, except for a few differences while calculating γ_m.

* In the classification case, to calculate γ_m instead of derivate, the algorithm can take a second-order Taylor polynomial and approximate the minimum value.

eXtreme Gradient Boosting (XGBoost)

As we have explained previously, at the time of writing this book, XGBoost [Chen '15] is among the best algorithms in terms of their accuracy, especially when deep learning algorithms do not meet the demands of the users. XGBoost is a derivation of a GBDT algorithm, which has employed lots of optimization techniques, and its cost function has a regularization (check Chapter 8 to recall regularization) along with the cost function as well.

XGBoost optimization methods include using an approximate greedy algorithm, parallel learning, and a weighted quantile approach to identify the threshold value of split faster. In addition, the algorithm handles the sparsity of data efficiently, using caching (bringing data to CPU cache memory instead of the hard disk) and sharding (a method to facilitate disk access), which results in a better execution time by dealing with missing values. We do not explain the details of optimization methods to preserve your brain cells to understand the core of this very accurate but heavy algorithm.

Vanilla Decision Tree   Gradient Boosting Decision Trees

Unlike the GBDT method, XGBoost uses a specific type of decision tree, i.e., XGBoost tree. XGBoost has many hyperparameters, and we describe the important ones here, i.e. η, γ, "minimum child weight", "maximum depth", "subsample", λ and α. The hyperparameter η (default = 0.3) is a learning rate and shrinks the model weight at each iteration to make it robust against overfitting. A node in XGBoost split if the split leads to a reduction in a loss function. γ (default = 0) is a threshold to specify the minimum loss reduction that is required to perform the split. "Minimum child weight" specifies the minimum sum of weights of samples required to be in each child node. "maximum depth" (default = 6) specifies the maximum depth of a tree (deep trees are prone to overfitting). "subsample" parameter specifies the ratio of sampling for the training dataset. For example, subsample = 0.5 means at every iteration, the algorithm randomly samples 50% of training data before growing a tree. λ is a L_2 regularization parameter and α is a L_1 regularizer on weights (leaf values). There are several other hyperparameters, we do not explain them, but you can check them from XGBoost documentation.

XGBoost can be used for both classification and regression. The loss function for classification and regression differs, but both cases use second-order Taylor approximation (Check Chapter 8 for Taylor series). XGBoost cost function is a gradient loss function plus a regularization, which is very similar to ridge regression (Check Chapter 8).

Assuming x_i is the input and, $\hat{y}_i$ is the output, $\hat{y}_i = \Sigma_{k=1}^{K} f_k(x_i)$, which is a combination of K number of decision trees. The objective function of the GBDT algorithm can be formalized as $L(y_i, \hat{y}_i)$. Additionally, GBDT can have a regularizer Ω to reduce the risk of overfitting. The following equation presents the objective function of a GBDT algorithm with a regularizer:

$$\Sigma_{i=1}^{n} L(y_i, \hat{y}_i) + \Sigma_{k=1}^{K} \Omega(f_k)$$

511

In this equation, $L(y_i, \hat{y}_i)$ is the gradient cost function, f_k presents the kth decision tree, and $\Sigma_{k=1}^{K}\Omega(f_k)$ is the regularizer.

XGBoost is a type of GBDT with a specific regularizer, and the regularizer in XGBoost for the kth tree is calculated as follows:

$$\Omega(f_k) = \gamma T + \frac{1}{2}\lambda\Sigma_{j=1}^{T}w_j^2$$

T is the number of leaves (terminal nodes), γ is a user-defined penalty, and the rest is the L_2 norm or Ridge penalty (check Chapter 8 regularization section), which encourages tree leaves to have smaller weights.

XGBoost algorithm operates as follows:

(1) First, similar to the GBDT, the algorithm calculates all initial residuals. It starts with a base prediction (like the average of the target values for regression), represented by a single node or leaf. Then, each successive tree in the iteration process aims to correct the residuals left by the previous trees.

(2) The algorithm tries different splits on residuals for each feature. Then, for each candidate tree, based on the split in residuals and previous prediction value, it calculates a *similarity weight*. For classification, it uses the square summation of all residuals divided by the summation of previous probabilities (P_{old-i}):

$$Similarity\ Weight = \frac{\sum Residuals^2}{\sum_i [P_{old-i} \times (1 - P_{old-i})] + \lambda}$$

For regression, it uses the following equation:

$$Similarity\ Weight = \frac{\sum Residuals^2}{Number\ of\ Residuals + \lambda}$$

(3) Next, the *Gain* for each newly constructed tree is calculated to determine the best split of data for each feature (each column except the prediction column).

Gain = Left tree (similarity weight) + Right (similarity weight) - Root (similarity weight)

The tree with the best Gain will remain, and other trees will be removed.

(4) If the difference between Gain and γ (Gain - γ) is positive, nothing changes; if it is negative, then the algorithm prunes the tree, removes that branch, and again subtracts γ from the next Gain value way up to the tree.

(5) The algorithm calculates the "output value" for the leaves as follows:

Output value = Sum of residuals / Number of residuals + λ

(6) As the first decision tree is created, the algorithm calculates the *log(odds)* of the initial prediction value and the output value of the related branch in the decision tree to make a new prediction. The result is a new predicted probability, and this value will be used (instead of the initial predicted probability) to adjust residuals and continue from Step 2 until the maximum number of trees is reached or residual values do not improve.

After the model is ready to perform prediction, it acts similarly to the gradient boosting algorithm. In particular, for new test data, the algorithm calculates the log(odds) of the previous prediction value plus learning rate (η) times the log(odds) of output for each XGBoost decision tree. Let's be honest together: how can we expect you to memorize the above explanation without an example? To understand this complex algorithm, we follow the holly traditions and learn it with an example. We have the data of a few readers who have read this chapter and experimented with the described algorithms by writing some codes from scratch. We would like to predict whether a new reader who read this chapter a couple of times, with some implementations of these algorithms, has already learned the described algorithms or not. Table 9-13 shows the dataset we have for training. For testing, we give a new record with specified RC and EC, and the algorithm predicts the LA.

Number of time this chapter is read (RC)	Experiment by Coding (EC)	Learn the Algorithm (LA)
2	Yes	Yes
1	No	No
2	Yes	No
2	No	No
2	Yes	Yes
1	No	No

Table 9-13: Sample dataset used by XGBoost.

In this example, for ease of understanding in describing the algorithm and calculating its math, we do not calculate α, and we also assume λ, γ, "minimum child weight", and other hyperparameters are all set to zero.

(1) To implement Step 1 of the XGBoost algorithm, for 'yes' in LA, we assign '1', and for 'no', we assign '0' (for simplicity, we use binary classification). Similar to the GBDT, the initial prediction could be the average of their possible values, and it is $(0 + 1)/2 = 0.5$, or something similar. During iterations in all GBDT algorithms, the initial values will be changed, and it is unimportant how we initialize it. Initial pseudo-residuals are calculated as the differences between observed and predicted values. The predicted value is equal to the initial prediction, in the beginning, which is also called *the base model,* because it participates in constructing the final model. By substituting initial residuals in Table 9-13, we get the table presented on the left side of Figure 9-28. At this moment, look at the table there and not the tree on the right side.

After the residuals have been calculated, the XGBoost algorithm fits a series of binary decision trees to predict the residuals of the target variable. However, it has not yet been decided which column should be used for the split and the algorithm experiment for all columns. A binary table for RC is shown in Figure 9-28. As described, these types of trees will be created for all other columns (features) as well. Besides, we could make a more complex tree and make more

RC	EC	Actual Data LA	Residuals
2	Yes	1	0.5
1	No	0	-0.5
2	Yes	0	-0.5
2	No	0	-0.5
2	Yes	1	0.5
1	No	0	-0.5

1-0.5 (0.5 = initial residual)

0-0.5 (0.5 = initial residual)

-0.5, -0.5, 0.5, -0.5, 0.5, -0.5

RC>1

n — -0.5, -0.5

y — 0.5, -0.5, 0.5, -0.5

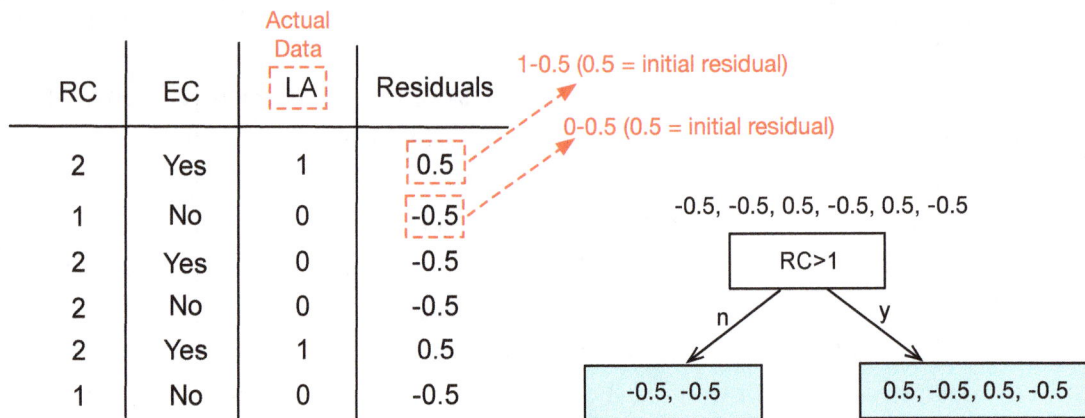

Figure 9-28: Table 9-15 is used to calculate the residuals, we intend to do a classification and since it is binary we substitute yes and no with 1 and 0 in LA column. On the right side a binary tree is constructed and we assume the RC>1 as a split.

branches, but for the sake of simplicity, we keep it very small. Why did we select $RC > 1$ as the main branch? Because we simply average RC values, and it is 1.5, and rounded to 1.

(2) In the second step, "similarity weight" for each leave node is calculated. For the sake of simplicity, we assume $\lambda = 0$. Then, $P_{old-i} = 0.5$ because it was our initial prediction probability. Therefore, the similarity weight for the leaf of the tree in Figure 9-28 is calculated as follows:

$$\frac{(-0.5 + -0.5)^2}{0.5(1-0.5) + 0.5(1-0.5) + 0} = \frac{1}{0.5} = 2$$

Respectively, the similarity weight for the leaf on the right side is calculated as follows:

$$\frac{(0.5 + -0.5 + 0.5 - 0.5)^2}{0.5(1-0.5) + 0.5(1-0.5) + 0.5(1-0.5) + 0.5(1-0.5) + 0} = \frac{0}{1} = 0$$

We also compute the similarity weight for the root node (which is a node that includes all residuals) as follows:

$$\frac{(-0.5 + -0.5 + 0.5 + -0.5 + 0.5 - 0.5)^2}{0.5(1-0.5) + 0.5(1-0.5) + 0.5(1-0.5) + 0.5(1-0.5) + 0.5(1-0.5) + 0.5(1-0.5) + 0} = \frac{1}{1.5} = 0.66$$

Now, similarity weights for all tree nodes get calculated. Also, the algorithm calculates the "gain" for this feature (RC). Therefore, we will have: $2 + 0 - 0.66 = 1.34$. Figure 9-29 writes similarity weights on top of each node on a tree.

(3) The algorithm performs the same process and calculates the gain for other features (columns), which are used to predict the LA (the target data). The only remaining feature is EC. Its tree, similarity score, and gain score are presented in Figure 9-30.

EC gain is 2.67, and it is higher than RC gain, which is 1.34. Therefore, the algorithm chooses EC to perform the split on the root node.

Note, if there are three different labels for one feature (our example has two labels only, and we don't have this situation), the algorithm experiments with different possible combinations for binary tree construction (XGBoost split trees are always binary trees), and at it selects the tree

514

Similarity:0.66   Gain = 2+0-0.66 = 1.34

-0.5, -0.5, 0.5, -0.5, 0.5, -0.5

RC>1

Similarity:2   n   y   Similarity:0

-0.5, -0.5   0.5, -0.5, 0.5, -0.5

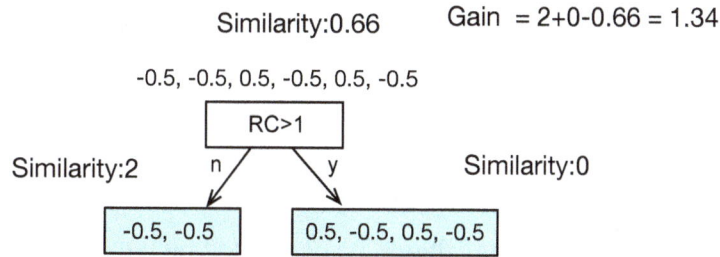

Figure 9-29: Similarity score of each node will be used to calculate the Gain score for RC.

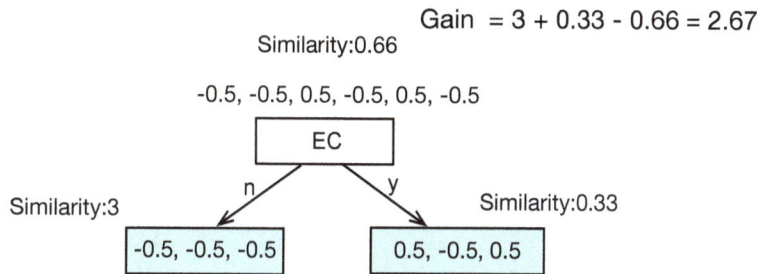

Gain = 3 + 0.33 - 0.66 = 2.67

Similarity:0.66

-0.5, -0.5, 0.5, -0.5, 0.5, -0.5

EC

Similarity:3   n   y   Similarity:0.33

-0.5, -0.5, -0.5   0.5, -0.5, 0.5

Figure 9-30: Similarity score of each node will be used to calculate the Gain score for EC.

that has the highest gain score. For example, if we have Red (R), Green (G), and Blue (B), once it makes a binary tree on one branch R and two other branches G and B, and next it calculates their gain. Then, another tree is constructed that has G as one branch; the other branch includes R and B, and it calculates their gain. Afterward, it chooses the best tree with the highest gain. However, experimenting with too many trees is infeasible and uses approximation, which we do not explain in more detail.

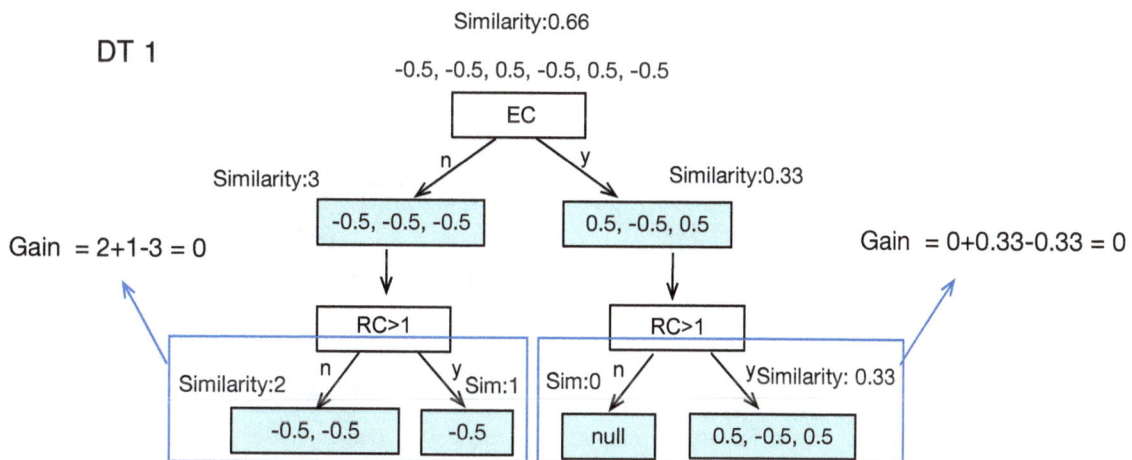

Similarity:0.66

DT 1

-0.5, -0.5, 0.5, -0.5, 0.5, -0.5

EC

Similarity:3   n   y   Similarity:0.33

-0.5, -0.5, -0.5   0.5, -0.5, 0.5

Gain = 2+1-3 = 0   Gain = 0+0.33-0.33 = 0

RC>1   RC>1

Similarity:2 n   y Sim:1   Sim:0 n   y Similarity: 0.33

-0.5, -0.5   -0.5   null   0.5, -0.5, 0.5

Figure 9-31: A split, and gain scores have been calculated for other branches.

(4) We have explained that we set γ it to zero, and thus, in this example, we skip pruning the tree; if we incorporate γ it removes the branches, which subtracts their gain from γ and results in a negative number.

To see the example of how the gain of a branch is computed, check the example presented in Figure 9-31. Both Gains are zero, and thus, one of the trees is randomly selected. At this stage, we can say that one decision tree is ready. Let's call it *DT1* (Figure 9-31).

(5) Now, the output value for each leaf should be calculated by the equation we have described (*Output value = Sum of residuals / Number of residuals + λ*). Recall that we set λ to zero.

The followings show the calculation of the output for each branch :

$EC = n, RC = n :$
$$output : \frac{-0.5 + -0.5}{2} + 0 = -0.5$$

$EC = n, RC = y :$
$$output : \frac{-0.5}{1} = -0.5$$

$EC = y, RC = n :$
$$output : 0$$

$EC = y, RC = n :$
$$output : \frac{0.5 + -0.5 + 0.5}{3} + 0 = 0.16$$

Figure 9-32 is Figure 9-31, but it presents the output result for each branch.

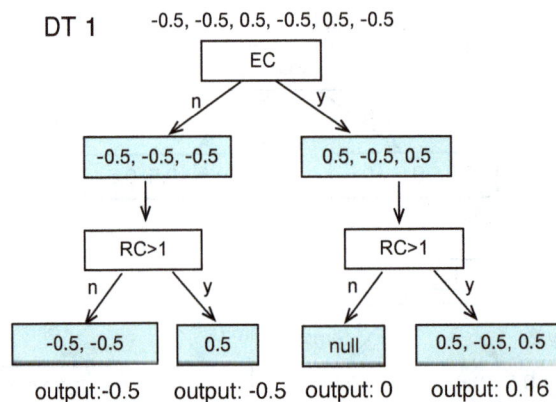

Figure 9-32: Output for each leaf has been calculated and added to the bottom of leaves.

516

(6) Now, the algorithm iterates from step 1 to construct the new prediction probabilities and substitute the initial one. The *log(odds)* of initial probability, which is equal to 0.5, will be 0 we donot write them in Table 9-14. The new log(odds) of predicted probability will be $0 + 0.3 \times log(odds)\ of\ DT1_{output}$ (0.3 is the recommended learning rate η). For each record, we can substitute the values into the $0 + 0.3 \times log(odds)\ of\ DT1_{output}$ equation and give us the result. To calculate the predicted probability from log(odds), a Sigmoid function is used as follows:

$$Predicted\ Probability = \frac{e^{log(odds)\ of\ p}}{1 + e^{log(odds)\ of\ p}}$$

Then, the result will be a new predicted probability, and its differences with LA will be used to construct new residuals (see Table 9-14). The new residuals are usually smaller than the residuals in the previous rounds.

To decide on the LA label of the new record, the new record data will be substituted in the model (i.e., *initial prediction + DT1 + DT2 + ...*), and its LA label will be based on the result of the model.

RC	EC	LA	Old Residuals	New Log(Odds) of predicted probabilities	New Probabilities	New Residuals
2	Yes	1	0.5	0.3 x 0.16	0.045	0.96
1	No	0	-0.5	0.3 x -0.5	-0.17	0.17
2	Yes	0	-0.5	0.3 x 0.16	0.045	-0.96
2	No	0	-0.5	0	0.5	-0.5
2	Yes	1	0.5	0.3 x 0.16	0.045	0.96
1	No	0	-0.5	0	0.5	-0.5

$$\frac{e^{log(0.3 \times 0.16)}}{1 + e^{log(0.3 \times 0.16)}} = 0.045$$

Table 9-14: New prediction probabilities and their residuals, in each iteration residuals will get smaller. which are constructed based on old residuals. For sake of simplicity we use absolute value and neglect negative residuals.

The XGBoost for regression operates very similar to classification, except for its similarity weight calculation, which we have explained before. Therefore, we skip its explanation with a detailed example. Before finalizing this section, remember that the only difference between classification and regression of XGBoost is their loss function.

Assuming K is the total number of trees, d is the maximum depth of trees and $||x_0||$ is the total number of training data points with non-missing data, and n is the total number of data points, the computational complexity of XGBoost for training is $O(Kd||x_0|| + ||x_0||log\ n)$. The $||x_0||log\ n$ part belongs to the preprocessing phase.

In October 2023, XGBoost 2.0 was released. It has some improvements in its implementation, including native support for GPU, improved memory management, and probably the most important one is multi-target trees. The multi-target trees allow XGBoost to build one tree for all targets (instead of many trees). This helps make smaller models and prevents overfitting.

LightGBM

Later, after XGBoost and its tremendous success in Kaggle competitions, back in 2017, another interesting algorithm was released by Ke et al. [Ke '17]. One of the problems in XGBoost was the need to scan many possible combinations to decide on the best split. LightGBM, another GBDT type, tries to mitigate this by using Gradient-Based-One-Side Sampling (GOSS) and reducing the number of features by Exclusive Feature Bundling (EFB).

GOSS operates based on the fact that different gradients of training data points have a different impact on the information gain (the result of the algorithm). In particular, data points with larger gradients that are under-trained data points contribute more to the result of prediction (i.e., information gain). Respectively, data points with small gradients are not important for the algorithm, and GOSS randomly samples a subset of data points with small gradients. This process decreases the role of gradient in learning. On the other hand, data points that have a large gradient are important for the learning phase of the algorithm, and they will be kept, e.g., top 30% of gradients.

You should learn GBDT and XGBoost, before proceeding to learn LightGBM and CatBoost

Usually, real-world datasets include lots of zeros; even one-hot encoding makes sparse features because there are too many zeros added to the feature list. The EFB uses the sparsity that exists in most real-world datasets. Features that never take non-zero values together are referred to as *Exclusive features*. EFB merges exclusive features into a single feature (similar to the feature reduction techniques we have described back in Chapter 6). In other words, EFB reduces the problem of feature reduction to a graph coloring problem and uses a greedy algorithm[11] to remove features. In particular, all features will be a vertex in the graph, and if they are not mutually exclusive[12], the algorithm adds one edge between them. The weight of edges corresponds to conflicts between features.

---

[11] Greedy algorithms refer to heuristic algorithms (check Chapter 6) in which each step chooses a local optimum. In the end, it might not end up being the best optimum solution, but each step alone chooses the best optimum that is available for the algorithm.

[12] Mutual exclusion means that two processes can not exist in the same state. Remember that, assume a rest-room which usually two people can not enter one rest-room.

To understand the characteristics of EFB, take a look at Table 9-15. There we have 'Feature 1', 'Feature 2', and 'Feature 3'. LightGBM can not merge 'Feature 1' and 'Feature 2' together because in the third row, 'Feature 1'=3 and 'Feature 2'=2. To make a bundle, at least one of them should be zero. However, it can merge 'Feature 2' and 'Feature 3' together because, in each record, at least one of them is zero. There is a γ parameter that can tolerate small conflicts (e.g., a few records have non-zero values, but both features are still merged) in each bundle, but we did not use it in our example.

While constructing bundles, one problem is that the bundled feature should not be in the same range as the original feature. To solve this issue, the algorithm adds a constant value to bring them to a different range. For example, here, we add 10 to each non-zero value of the bundling feature.

Feature 1	Feature 2	Feature 3
0	2	0
1	0	1
3	2	0
0	0	3
4	0	0
0	2	0

Feature 1	Feature New
0	2+0
1	0+1
3	2+0
0	0+3
4	0+0
0	2+0

Feature 1	Feature New
0	2+0+10=12
1	0+1+10=11
3	2+0+10 =12
0	0+3+10=13
4	0+0
0	2+0+10=12

Table 9-15: (left) original table of data, which has three features, (center) feature 2 and feature 3 could be merged together, because at least one of them is zero, (right) to avoid having the number of bundled feature in the same range as original data, a constant value will be added to non-zero columns.

Assuming n is the total number of data points, k is the number of features, and m is the number of bundles, the EFB algorithm of lightGBM reduces the $O(nk)$ complexity $O(mn)$. Nevertheless, the computational complexity of GOSS (linear as well) and GBT will be added to this value.

CatBoost

CatBoost [Prokhorenkova '18] is a third popular GBDT-based algorithm introduced after LightGBM in 2018. CatBoost described that previous GBDT works suffer from the *prediction shift* problem, and it can mitigate this problem.

The prediction shift problem occurs when the model is trained on a dataset whose distribution has shifted or is different from the distribution of the actual data on which predictions are to be made. This discrepancy can lead to inaccuracies because the model is learning from a dataset that does not represent the true underlying distribution of the target population.

CatBoost introduces two improvements to the GBDT algorithm: (i) a specific method to encode categorical features, i.e., "Ordered Target Statistic" (Ordered-TS), and (ii) "Order Boosting", which is a permutation-driven alternative to the classical boosting algorithm.

To implement the ordered target statistic, the CatBoost uses target encoding (check Chapter 6). However, we have explained in Chapter 6 that target encoding is prone to data leakage. CatBoost resolves this issue by first applying a permutation (rearranging the order of data) on data, which relies on a specific ordering principle. After ordering the data, the target value of every feature is calculated from the rows before (previous observations or historical observations). It uses the following equation (a simplified version of the original version in the paper) to apply target encoding while resolving the data leakage problem.

$$Ordered - TS = \frac{current - count + (\alpha \times p)}{total - feature - count + \alpha}$$

In this equation, *current-count* specifies the sum of values for the feature we are applying target encoding, but not all values; it applies the sum until the current data point. *p* stays for the prior, it is a constant value, and it is used to smooth the result of target encoding. A common value for *p* is the average of the target value in the dataset, e.g., 0.5. α (which sets larger than zero) is a weight parameter used to ensure not dividing by zero and specify the weight on prior. *total-feature-count* specifies the total number of the current feature values, excluding the current row.

To understand how Ordered-TS works, take a look at Table 9-16. Here, we assume *p* as a constant and equal to 0.5 and α is 1. Table 9-16 (a) presents features and their values. It is a result of binary classification, and thus, each row has a value of either 0 or 1. In Table 9-16 (b), the average of each feature is calculated, and in Table 9-16 (c), these features have been substituted on the original table. Table 9-16 (c) presents the target encoding, prone to data leakage.

a

Fruit	Value (target)
Apple	1
Orange	0
Apple	1
Orange	1
Orange	1
Banana	1

b

Fruit	Avg. Target Value
Apple	2/2 = 1
Orange	(0+1+1)/3 = 0.67
Banana	1/1 = 1

c

Fruit	Avg Target Value
Apple	1
Orange	0.67
Apple	1
Orange	0.67
Orange	0.67
Banana	1

d

Fruit	Ordered Target Statistics
Apple	(0+0.5) / (0+1) = 0.5
Orange	(0+0.5) / (0+1) = 0.5
Apple	(1+0.5) / (1+1) = 0.75
Orange	(0+0.5) / (1+1) = 0.25
Orange	(1+0.5) / (2+1) = 0.5
Banana	(0+0.5) / (0+1) = 0.5

What is the current count of having Orange =1? There is only one record before this record, therefrore: *current_count =1*

There are two rows before this record, second and fourth record. Despite they have different values the *total_feature_count = 2*

Table 9-16: An example how CatBoost calculate the Ordered Target Encoding.

The authors proposed using the described equation and resolving the data leakage by ordering. Take a look at Table 9-16 (d), which is the ordered target encoding value. Note that *current_count* and *total_feaute_count* are both calculated by looking only at previous records, not the current one and overall table. In Table 9-16 (d), we describe the justification of the sixth row. As another example, let's take a look at the value of the third row, where Apple=1, in Table 9-16 (c). Table 9-16 (d) shows the *current_count* of this row is equal to 1, because there is one row before this row, i.e., the first row that has Apple=1, and the *total_feature_count* of the Apple feature is also 1, because there is only one row that includes Apple before this row, i.e., the first row.

The second novelty of CatBoost is its Order Boosting. Order Boosting does not use the whole dataset to calculate the residuals at every iteration of GBDT. In classical GBDT, multiple trees are constructed to fit the entire dataset. Using the entire dataset might lead to overfitting, which CatBoost referred to it as a "prediction shift". CatBoost tries to resolve this issue using a tree that is constructed using a subset of the dataset, and it calculates the residuals using the data points that it did not see before (from another subset of the dataset). To understand this feature, let's say we have a 1D dataset of 10 data points, i.e. $\{x_1, x_2, x_3, x_4, x_5, x_6, x_7, x_8, x_9, x_{10}\}$. The CatBoost applies a random permutation on these data, and after applying the permutation, the data points are ordered as follows: $\{x_4, x_6, x_3, x_1, x_8, x_2, x_7, x_9, x_5, x_{10}\}$. A model that is trained on $\{x_4\}$ will be used to determine the label for x_6, and its residual error will be calculated (*actual label - predicted label*). Respectively, a model that is trained on $\{x_4, x_6\}$ will be used to determine the label for $\{x_3\}$, and its residual error will be calculated. It means the model which is trained on $\{x_4, x_6, x_3, x_1, x_8\}$ has never been seen x_2 before, but it uses x_2 to determine its residual.

This approach resolves the prediction shift, but by using this approach for every single data point, one decision tree needs to be constructed, which leads to quadratic time and memory use, and thus, it is computationally very expensive. To resolve this issue, the authors proposed the following solution: assuming we have n data points, the authors proposed constructing decision trees for data points that are located at 2^j, where $j = 1,2,4,8,...log_2(n)$. This means the first model is trained on the first data point. The second model is trained on the first and second data points. Respectively, the third model is trained on the first four data points, the fourth model is trained on the first eight data points, etc. In the end, instead of having n trees, we have $log_2(n)$ trees.

Assuming s is the number of permutations and n is the number of data points. $N_{TS,t}$ is the number of TS to be calculated at iteration t, and C is the set of candidate splits to be considered at the given iteration b_j^t is the ith leaf value at iteration t. Authors claimed the computational complexity of calculating the gradient is $O(sn)$, building the tree T is $O(|C|.n)$, calculating all b_j^t is $O(n)$, updating a model is $O(sn)$, and calculating the Ordered-TS is $O(N_{TS,t}.n)$.

It seems all linear, but, summing them together makes CatBoost a very computationally expensive algorithm. In an experiment, Keshavarz et al. [Keshavarz '20] compared the computational complexity of these GBDT algorithms, and CatBoost is the most resource-intensive one.

If you are still alive after learning these three hard-core GBDT models, let's do a brief review of what we have explained. XGBoost implements several optimizations, but its salient difference is the use of regularization and XGBoost tree instead of weak learners on top of the GBDT. LightGBM is a GBDT that provides two additional features, GOSS and EFB. Catboost is a GBDT used for categorical data, and it provides two additional features, i.e., Ordered Target Statistics and Ordered Boosting.

NOTES:

* Except for LightGBM, other GBDT methods grow trees by increasing their depth (leve-wise tree growth). It means they identify the best node to split, and from that node, the tree will split down to branches. However, LightGBM grows trees by increasing the leaves in the related node (leaf-wise growth) to reduce the loss score and keep the other leaves in the tree intact.

* If there is no GPU available and you are using a local machine for training, usually LightGBM consumes less amount of resources than XGBoost and CatBoost, which operate well with GPU.

* These three boosting algorithms are state-of-the-art algorithms that can work accurately on tabular data and when we have small data.

* Despite all GBDT algorithms being good at operating with a small number of data points and tabular data, their resource utilization is very high. This makes them infeasible to be used on battery-powered small devices such as smartwatches.

How to Select the Best Classification Model?

We have learned several classification algorithms in this chapter and logistic regression in the previous chapter. Now, assume we have experimented with a set of algorithms, and then we would like to decide which models perform better. The first approach that can be used for model selection relies solely on the model's accuracy. In particular, by comparing the cost (false-negative, false-positive) with the benefit (true-positive, true-negative) or using the ROC curve of different models (check Chapter 8 to recall what the ROC curve is) or other evaluation metrics. What if the accuracy differences appeared by chance? To have a better estimate of whether there is a real difference between the two models, we should use the *significance test* (check Chapter 3 if you can't recall). For example, we performed a 10-fold cross-validation, and it resulted in ten different error rates, one for each model. We can use the t-test to compare the mean and variance of these error rates set (each model has 10 error rates) and check whether there is a significant difference between the error rates of these models.

Nevertheless, sometimes the accuracy is not as important as execution time. Thus, execution time is another parameter to decide about a model from a set of models. For applications that require making decisions in real-time and close to the real-time execution time, it plays a crucial role. In some cases, we are dealing with limited resources, such as algorithms that run on any battery-powered devices, from fitness trackers to electric vehicles; in these cases, resource utilization of the algorithm also plays a crucial role in deciding the best classification model.

Summary

This chapter starts by describing rule-based classification. The rule-based classifier is not really a machine learning algorithm because we define rules by hand, and somehow, manually, we build the model. Next, we have described some common classification methods, including the Bayesian network, kNN (by using LSH, Voronoi tessellation, and KD-Tree), SVM, and four different decision trees, including ID3, CHAID, C4.5, and CART. The differences in these decision trees are based on their tree split policy, and to review them, you can check Table 9-8.

Ensemble Learning Method	Description	Algorithm Examples
Stacking	It is a combination of weak learners that are combined to build the final model.	Combination of Weak Learners
Bagging	It chooses a random subset from the training set and builds a model for each subset. Each model assigns a label, and the final label will be assigned based on majority voting or averaging the candidate labels, provided by each classifier.	Random Forest
Boosting	Boosting is a sequential process that starts by applying a classification algorithm to the entire train set and makes a model. Next, it collects the errors of the previous model and makes a smaller dataset from errors. Then, it uses the same classification algorithm (similar to Bagging) on the dataset that includes errors and creates another model. This process continues in several iterations until a maximum number of iterations (hyperparameter) is reached.	AdaBoost Gradient Boosting Decision Tree XGBoost LightGBM CatBoost

Table 9-17: Summary of ensemble learning algorithms.

Afterwards, we have explained the ensemble learning algorithm, which uses the described algorithms as weak learners and combines them for a better learning result. Table 9-17 summarizes three ensemble approaches.

Bagging is used with decision trees as classifiers because decision trees are prone to high variance, and bagging resolves the high variance problem. The Random Forest algorithm is the most popular bagging algorithm. First, it creates subsets from the original dataset, then selects several features from each subset, and it uses these features to calculate the split. Afterward, the test dataset will be fed into all trees, and then its labels will be decided by averaging the labels of the other decision trees. Random forests are not as good at solving regression problems as they are at sorting because they do not give a continuous output. AdaBoost is a boosting algorithm that uses a set of small decision trees. It increases the weight of hard-to-classify data points and decreases the weight of easy-to-classify data points. Then, it combines multiple weak learners into a single strong learner. The result of an AdaBoost algorithm is a set of classifiers (stumps), each associated with a weight.

The Gradient Boosting Decision Tree (GBDT) algorithm is an ensemble algorithm that operates using a gradient cost function. The boosting process of this algorithm is a numerical optimization problem. We have explained that a gradient boosting algorithm is composed of (i) a loss function to be optimized, (ii) a sequence of decision trees (weak learners) to make predictions, and (iii) an additive model that adds decision trees to minimize the loss function. Similar to other Boosting algorithms, GBDT uses a residual (error) to calculate the best split for the next step, and finding the residual and splitting is the task of its cost function. Three algorithms are built on top of GBDT, including XGBoost, LightGBM, and CatBoost. At the time of writing this Chapter, these algorithms are state-of-the-art algorithms to use on tabular data whose number of data points is limited.

Further Reading or Watching

* There is a good book for basic algorithms, Mastering Machine Learning Algorithm by Jason Brownlee [Brownlee '16]. It includes more examples of basic supervised machine learning algorithms. We use this book to write the Naive Bayesian section of this chapter.

* Victor Lavrenko (https://www.youtube.com/user/victorlavrenko/videos) has a good video series on teaching LSH. We have used his explanation to construct our explanation as well, and also good videos to explain the decision tree.

* There are many explanations of Kernel function and even books written about kernels [Scholkopf '01]. Nevertheless, the slides for Matt Gormley from Carnegie Mellon University are one of the few that we find useful and help us learn the motivation behind using kernel functions. https://www.cs.cmu.edu/~mgormley/courses/10601-s17/slides/lecture12-svm.pdf. Besides, Josh Starmer online videos on teaching the mathematics behind kernel functions are clear and helpful, like most of his other videos.

* Sefik Ilkin Serengil provides a detailed explanation of vanilla decision trees with numerical examples. If you would like to see more examples for each decision tree algorithm, check his home page: https://sefiks.com.

* Motaz Saad provides a good visual explanation of Bagging, Boosting, and Stacking, which inspired us to develop the visualization for this model https://mksaad.wordpress.com/2019/12/21/stacking-vs-bagging-vs-boosting

* While constructing examples and explanations for AdaBoost we benefit from online videos of Krish Naik, Cory Maklin, and Bhavesh Bhatt.

* Bradley Boehmke & Brandon Greenwell have an excellent book released online [Boehmke '20], Hands-On Machine Learning with R, and you can read a detailed explanation about Gradient Boosting in their book: https://bradleyboehmke.github.io/HOML/gbm.html

* We have used Josh Starmer explanations (https://statquest.org) to construct the gradient boosting and XGBoost algorithm of this chapter as well.

* To learn LightGBM and CATBoost from other resources, you can check "Machine Learning University" of Amazon as well: https://aws.amazon.com/machine-learning/mlu. They provide a simplified explanation of these two algorithms, and we use them to describe LightGBM and CatBoost.

* If you are trying to learn the mathematical concepts behind Gradient Boosting in more detail and feel the description provided in this chapter is not enough, you can check the explanation of Terence Parr and Jeremy Howard: https://explained.ai/gradient-boosting.

Part iv: Deep Learning

When writing this book, Artificial Neural Networks (ANN) are the most accurate machine learning algorithms to deal with non-structured data, including text, images, videos, and audio. ANN algorithms are inspired by the brain neurons of living animals.

ANN was introduced for the first time by Walter Pitts and Warren McCulloch in 1943. They have created a computer model [McCulloch '43] based on the neural networks of the human brain. They use "Propositional Logic" to propose a model that explains how brain neurons work together. We can say that they design a simple computational model of a biological neuron using propositional logic.

Learning from nature and biology has been very attractive for human knowledge development throughout history, and there are numerous examples, such as building airplanes inspired by birds, building robots inspired by dogs, algorithms inspired by insects, making modern banking systems inspired by pigs, building political hegemony inspired by vultures, etc.

After Pitts and McCulloch, Rosenblatt introduced the concept of Perceptron [Rosenblatt '58] in 1957. This algorithm was used until Minsky and Papert [Minsky '69] criticized its limitations in a paper, saying that it could only be used for liner problems. They also claim that there are insufficient resources to implement a better neural network. Their criticism contributed to a reduction in funding and interest in neural network research, leading to what is known as the "AI winter," a period during which funding and enthusiasm for artificial intelligence research, including neural networks, were significantly diminished.

Later, there were some successful artificial neural network models proposed in the early 1960s. In that era, scientists believed that human-level AI could be reached by handcrafting large numbers of explicit rules for the machine. This is known as "Symbolic AI," and these works continued until the late 1980s. Symbolic AI is still useful for problems with predefined rule sets, such as the Chess game, but it fails to address complex and fuzzy problems, such as identifying objects in an image. In other words, users provide a set of rules (check the rule-based classifier we explained in Chapter 9), and the algorithm operates by user-defined rules.

Again, in 1980, the term "Expert System" and LISP as their programming languages were introduced, and the attention of funding and academic institutions on AI was regained. The introduction of more powerful desktop computers in the mid-1980s facilitated the development and running of expert systems, contributing to the growth of AI research and applications. Later, AI and machine learning progressed slowly but steadily. There were a few groups of scientists working on Neural networks at that time in Canada (Hinton and Benjio), the US (LeCun), and Switzerland (Schmidhuber). In 2011, in Switzerland, Schmidhuber's team, DanNet [Ciresan '11] won a classification challenge on image dataset, and in 2012, Hinton's team, AlexNet [Krizhevsky '12], won the most known image classification competition, ImageNet[1] , by shifting the best accuracy from 74.3% to 83.6%, both with their deep learning algorithms[2].

After these breakthroughs, deep learning has been introduced and revolutionized in several fields, such as image classification, speech recognition, and natural language processing. Deep Learning show a tremendous impact on different scientific disciplines, including games that beat

---

[1] http://www.image-net.org/

[2] North American scientists (Geoffrey Hinton, Yann LeCun, Yoshua Bengio) who are active in deep learning won Turing award in 2018, but not Schmidthuber, who is the architecutre of LSTM not. We would love to discuss more about American academic hegemony, but we are afraid to go more into politics.

a human player and create patterns that have never been thought of by humans, more than human accurate image recognition, language translation, intelligent conversational agents, autonomous driving, drug discovery, data restoration, and coloring historical black and white images and movies.

Two times in the history of computer science, artificial intelligence has had tremendous success and provided a robust answer for some of the problems. This success leads to the illusion that soon, humans will make a computer that can overtake humans in all intelligent tasks, including its wisdom. Nevertheless, those promises were not successful, and for many years, there has been slow progress in the field, which some referred to as *AI winter*. In other words, initial success in some machine learning algorithms makes the scientists too proud, and they claim in a few years, we will reach human-level intelligence or computers will overtake humans in all aspects, i.e., *Artificial General Intelligence* (AGI). You can observe too many artificial intelligence books written by philosophers or business people who have never written computer codes and provide similar promises.

However, we do not believe computers can overtake the human brain in all aspects, and terms such as AGI is not technically possible. A stunning theory proposed by Gödel, it is known as the *Incompleteness Theorem*. Based on Gödel's incompleteness theorem, there are two schools of thought about AI. One group, such as Lucas and Hofstadter [Hofstadter '79], describes that since machines are limited to a predefined formalization, decision-making will be limited to the grammar of that formalization, and machines cannot step outside of their formalization limits. On the other hand, there are scientists such as Norvig and Russell [Russell '21] who do not agree with this argument and state that the computer can invent a new formalization and thus can implement creativity. Inspired by Gödel's incompleteness theorem, we are advocates of the idea that algorithms are limited to a predefined formalization. Therefore, the process of decision-making will be limited to the grammar of that formalization, and the algorithm cannot step outside its formalization limits [Rawassizadeh '19].

At the time of writing this book still, the different communities are impressed by the success of deep learning, and deep learning research is ongoing without seeing in AI winter even after more than 10 years (considering AlexNet in 2012 as the starting point).

Chapter 10: Neural Networks and Deep Learning

Due to the enormous success of deep learning, you might not read the book from the beginning and start it from here. We do not recommend taking this attitude, but it should not be a big problem. Just listen to the Good News Potato feedback before you start reading a chapter, and be sure that you have read all the required chapters before starting this chapter.

Deep learning has revolutionized many scientific disciplines. People whose research works sound funny and too theoretical before 2012 are now celebrities in academia, not just in computer science but also in other disciplines.

Without reading chapter 3 and chapter 8, you do not understand anything from this chapter.

In this chapter, we start by explaining the rationale behind deep learning and then the function of a biological brain. It helps us to explain artificial neural networks more easily later. Next, we move our discussion from the traditional ANN method, including Perceptron and Multilayer Perceptron. Then, we explain activation functions, cost functions, and neural network optimizers. Afterward, two common deep learning models and architectures, including CNN and RNN, are explained.

At the time of writing this Chapter (late 2020 and revising in 2024), we are living in the AI boom, and one of the hardest jobs of academic supervisors is to lead motivated students and research workers from focusing too much on neural networks and deep learning, dedicate their time to learning other concepts and algorithms as well. There are many other important things that we should learn before starting to learn deep learning, which is why we postponed the deep learning and neural network to Chapter 10 and later chapters in this book.

Why is deep learning popular and powerful? There are two main reasons why deep learning is successful and popular: superior accuracy and no need for feature engineering

Superior accuracy: The first reason is its superior accuracy while dealing with unstructured data (not tabular data). Other algorithms that we have described all require to have structured data. From tabular formats like CSV to a time series and signals. On the other hand, deep learning works very well for unstructured data, including image, audio, and textual data. Since most of the data we acquire is in unstructured format and conversion to structured format is either infeasible or very expensive, deep learning algorithms are gaining tremendous popularity working with these datasets.

Students

No need for feature engineering: The second reason that deep learning is superior to other machine learning algorithms is that there is no need for feature engineering in these algorithms. We dedicate one chapter (Chapter 6), at least in this book, to describing feature engineering and selecting the appropriate features for the algorithm, which is one of the most important challenges of data science. Feature engineering consumes a significant amount of time to prepare data for the algorithm. Deep learning algorithms, in particular, have hidden layers that handle all features. If a feature does not contribute to the prediction, its weight will be reduced, and this feature will be ineffective in the final model. It is very exciting for data scientists, especially for those who do not have the domain knowledge to do feature engineering. The hassle of preparing data for the machine learning algorithm is removed. Nevertheless, in many cases, we still do some data engineering, such as resizing the input data (e.g., images for computer vision models), augmenting the data, and increasing the dataset size because deep learning models are trained on large datasets.

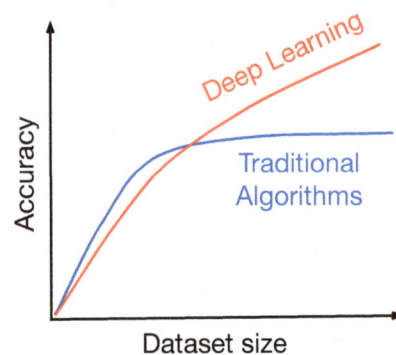

Figure 10-1: Differences between deep learning algorithms and traditional machine learning algorithms

However, there is a "no free lunch" theorem, and the process of automatic feature extraction is associated with a huge cost, which is intensive computation. Therefore, scientists move their calculations from CPU to GPU (Graphical Processing Unit) for matrix multiplication. The popularity of GPU led NVIDIA to launch the first platform for coding for GPU, i.e., CUDA, and later, parallel programming tools came into the market.

Although neural networks, especially deep learning algorithms, outperform all other machine learning algorithms, they operate under the conditions that we can feed them large amounts of data. Figure 10-1 presents an abstract overview of the accuracy between traditional algorithms and deep learning. This figure is inspired by the drawing of Andrew Ng's deep learning course and Aggrawal's book on neural networks [Aggarwal '18].

Universal Approximation Theory

Supervised machine learning can be interpreted as a function approximation problem. We have some data, and there is an *underlying function that is not known.* We try to understand that function. Take a look again at the Figure we have explained in Chapter 1. We have a machine that receives chicken as input and nuggets as output. The process of supervised machine learning is to identify what this unknown machine (function) is.

This technique is also called *function approximation.* The name of this unknown function is often called the *target function.* We do not know this function, and therefore, we use machine learning to approximate this function. We can refer to deep learning models as *non-linear function approximation methods.* We will learn more about non-linearity later.

The Universal Approximation theorem states that a neural network can approximate any continuous function with a high degree of accuracy (given the right configuration). This theorem is often cited to explain the potential of neural networks to capture complex patterns within data. Deep learning, which involves neural networks with multiple hidden layers (we will explain hidden layers shortly), allows for the modeling of highly complex functions.

However, there are practical limitations to consider, such as the computational resources required as network complexity increases. Although, theoretically, neural networks can approximate any functions, in practice, the feasibility of constructing such models depends on having sufficient data, as well as the computational capacity to train large networks. Therefore, while neural networks hold a universality in function approximation, this does not imply they can model absolutely everything without constraints.

Biological Neural Network

The brains of animals and humans are responsible for thinking, decision making, and learning new things. The brain is composed of neurons that are connected to each other; the process of thinking and decision making is done by sending electrical pulses between neurons. Figure 10-2 shows a simplified biological neuron, which is composed of three components: dendrite, axon, and terminals.

Neurons transmit an electrical signal from dendrites to terminals through axons. Then, the signal passes between neurons with the same structure. Based on this explanation, a simplified brain structure could be shown as a set of neurons that pass the electrical signals along with each other. Figure 10-3 presents several neurons where signals are transferred between them.

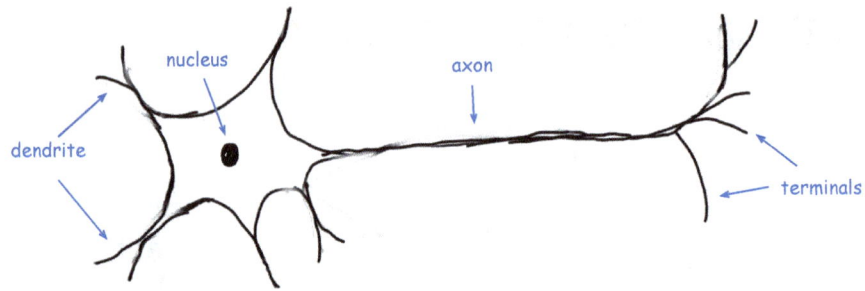

Figure 10-2: A simplified shape of a biological neuron and its components.

The process of transferring electrical signals through neurons is called *firing*. A question might arise: when does a neuron fire a signal? Scientific evidence suggests that the nucleus of a neuron should reach a *threshold* of receiving an electrical signal, and it fires the signal *only if that threshold is reached*. If the nucleus does not reach that threshold, it does not fire the signal. It makes sense because noises, which are signals less than the threshold, are not transferred, and only information that is useful will be transferred. To summarize, a biological neuron gets a set

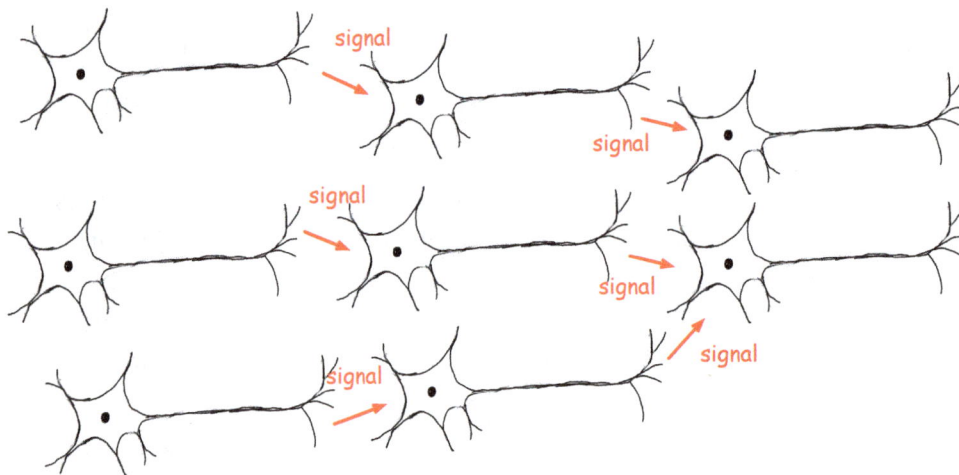

Figure 10-3: A simplified version of brain and how neurons transfer electrical signal.

of signals and processes them, and if the output reaches a threshold, it fires another signal.

Figure 10-4 presents a single function that shows how a neuron fires a signal. It looks like a function with a binary output of 1 (fire) or 0 (not fire). The function that is presented in this figure is called the *step function,* and the function that is used to activate a neuron is called the *threshold function*. The step function has a zero value until *x* reaches a specific threshold, and then the value of *y* changes to 1.

However, note that processing information inside the brain is in *parallel* and *fuzzy* (not just binary). Parallel processing and fuzziness are two prominent features of the biological brain.

Now that we have learned a simplified version of how the brain of animals works, we can go and stay in front of the mirror and grant ourselves a full professorship in Neuroscience. Please accept our humble congratulations on your achievement, and continue reading the rest of this chapter.

Artificial Neural Network

The concept of the first digital neuron, that is inspired by biological neurons, was proposed by McCulloch and Pitts in 1943 and is known as the MCP neuron [McCulloch '43], and it is known as the MCP neuron. Nevertheless, it was too limited, and thus it is not used anymore. Later, the Perceptron algorithm was created in 1958 [Rosenblatt '58], and it is the simplest artificial neural network (ANN).

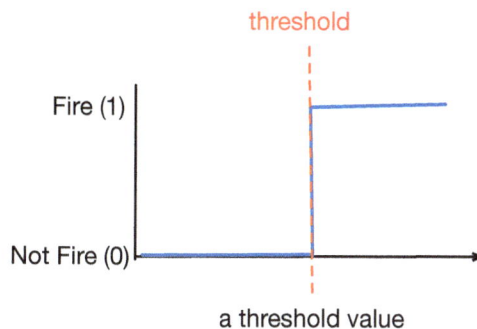

Figure 10-4: A neuron will not fire any signal until the signal reaches a threshold. Then it will get fired. This is called threshold function as well.

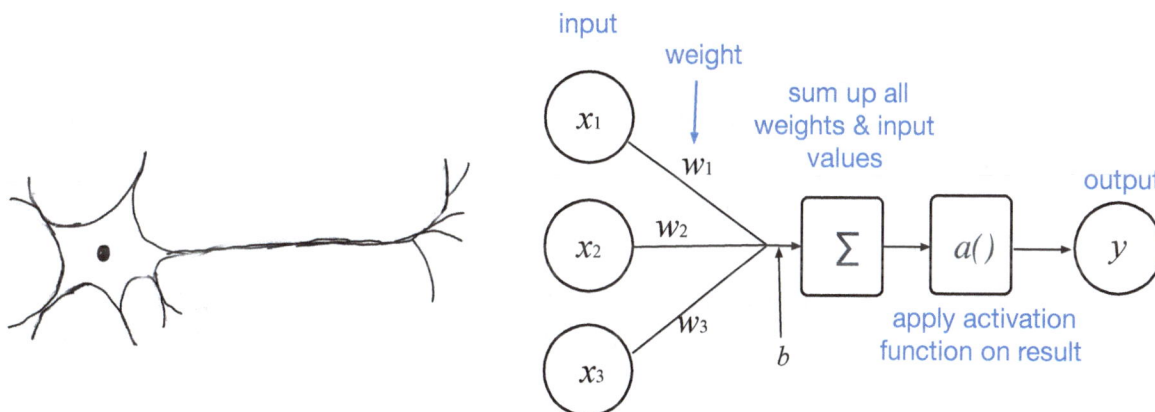

Figure 10-5: (left) A biological neuron, (right) a simple perceptron which its architecture is inspired by biological neuron.

Figure 10-5 shows a simplified biological neuron and a simple artificial neuron known as a perceptron. Intuitively, we can observe and realize the similarity of the perceptron to the biological neuron. An artificial neuron is a mathematical model with a set of inputs (similar to biological dendrites); each input is associated with a weight and a bias as parameters of a neuron, similar to regression parameters (βs) we have described in Chapter 8. Thus, we can write them as $x_1 w_1 + x_2 w_2 + \ldots + b$. Then, a function called the activation function is applied and sends the result to the output. Activation function could be written as a, and thus, we have $a(x_1 w_1 + x_2 w_2 + \ldots)$. A neuron output[1] can be written as $y = a(w_1 x_1 + w_2 x_2 + \ldots)$.

---

[1] You might think it is nothing more than a multi-linear regression inside an activation function. It is correct; a simple artificial neuron (perceptron) can be interpreted as a multi-linear regression inside an activation function.

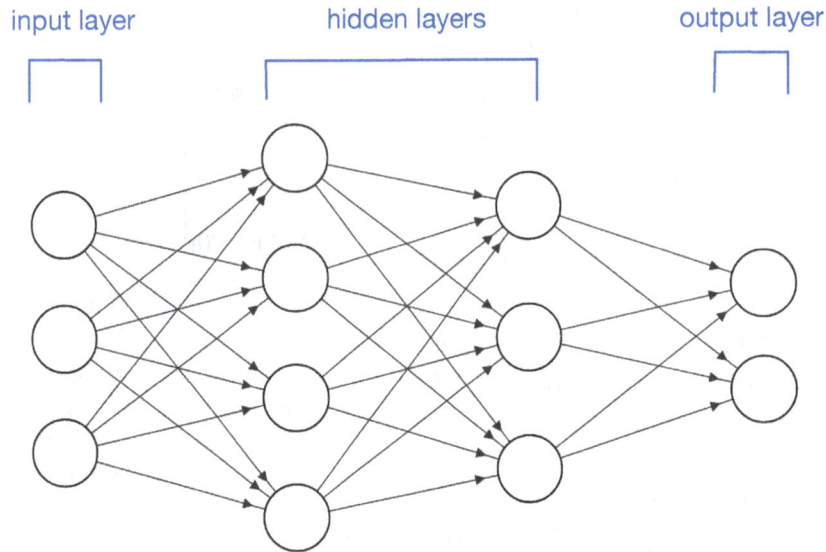

Figure 10-6: A simple presentation of neural network and its layers. You can realize it is also look a like Figure 10-2 but with artificial signal.

Weight: We have explained that each signal inside a biological neuronal network could have a different amplitude (e.g., micro-voltage), and these will be simulated in artificial neurons by weights. To better understand the need for weight, consider the prediction error, written as *error = predicted value - actual value*. If the *error* is non-zero, the perceptron algorithm changes the *w* of input variables and redoes the process to reduce the prediction error. Weights are very similar to coefficients in regressions, as we explained in Chapter 8.

Usually, weights are initialized with a random value, e.g., 0.1 to 0.9. There are different weight initialization methods (at random, with all zeroes, with small values close to zero, and so forth), and the user should decide on the weight. In other words, initial weights could be considered hyperparameters. Increasing the weight increases the complexity of the network, and thus, it is better to keep weights small. Usually, weights are initialized close to zero values or randomly assigned, but there are other approaches that we will explain later in this chapter.

Bias: Similar to linear regression, which has a constant value, each neuron also has a bias, presented as *b*. Weights adjust the strength of the connections between neurons. Biases, on the other hand, adjust the output of a neuron before it is passed through the activation function.

Activation function: It is a function that receives the sum of weighted inputs and calculates a threshold value (take a look at Figure 10-4). Based on the result of the threshold value, the activation function decides the output value. This function could be linear or non-linear. Linear here means that a hyperplane exists to model with an algebraic equation, and this hyperplane can classify the dataset (separate data points). However, non-linear activation functions such as the Sigmoid function (Check Chapter 8 for explanation about Sigmoid) are more popular. There are different choices for activation functions, which we will describe later.

We could say that an artificial neuron is nothing more than linear regression and that its result is fed into an activation function, i.e., $y = a(wx + b)$, and the activation function imposes non-linearity on this linear regression.

A neural Network is a set of connected neurons in which the output of its neurons is the input of other neurons, see Figure 10-6. All neural networks have three layers: input, hidden, and output.

Input layer: The input layer is the input variable(s) we feed into the neural network. Often, each feature is presented as one input neuron. For example, a table with 10 numerical columns will have 10 input neurons, and they are independent variables, e.g., one table row presents one set of input, and each row is composed of n columns (features). Therefore, the input layer has n features (one neuron for each feature). Besides, remember that the only input type we can use for the neural network is *numeric,* and other data types, such as text, image, audio, etc., should be converted into numeric data types. For example, pixel data from images can be scaled (e.g., between 0 and 1), standardized, and then fed into a neural network, or words (in a textual format) should be converted into a vector of numbers and then fed into the neural network.

Hidden layers: They are layers that are located between the input layer and the output layer. The simplest neural network has one neuron as a hidden layer. Deep learning algorithms have two or more hidden layers, and because of that, they are called *deep neural networks*. Hidden layers are connected and construct the neural network, along with input and output layers. Each layer is responsible for transforming the received data for the next layer. Since they transform the data and are different from the known input or output, we cannot interpret hidden layers. Therefore, deep neural networks are *black box*[2] models. In other words, a layer takes the input as a set of tensors (Check Chapter 1 to recall tensors), and the output of each layer is also a set of tensors, which is a numerical dataset. However, we can not interpret the tensor transformations between hidden layers.

Output layer: It presents the result of the neural network model. Similar to the input, the output of a neural network is also a number, i.e., tensor, which we can convert back into its original value by decoding.

Three common technical tasks could be done with an artificial neural network: binary classification, multi-class classification, and scalar regression. Therefore, the type and number of neurons in the output layer depend on the problem we intend to solve. A regression neural network can have one single neuron as the output layer. A binary classification neural network can have a single output, which provides zero or one or a probability. The result is determined from the probability, e.g., less than 0.5 is zero, and larger than 0.5 is one. A multi-class classification neural network has multiple output neurons.

Now that we have learned about neurons and layers of the neural network, we can discuss the objective of a neural network algorithm. The objective of a neural network is to assign a correct label to the unlabelled data or make predictions or decisions based on the input data. This objective will be achieved by assigning proper values to weights and biases, testing the network,

---

[2] There are methods for interpreting the hidden layers, such as activation maps, feature visualization, etc. We describe in Chapter 16, briefly some common tools for data interpretation.

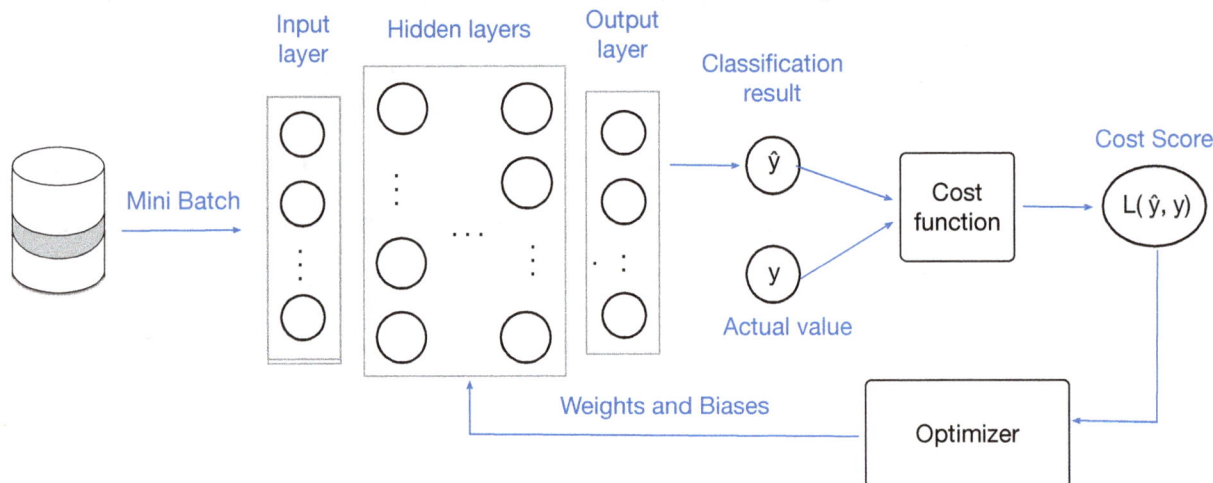

Figure 10-7: Artificial Neural Network flow of operation. The output value will be analyzed by the cost function and the loss score is calculated, then based on the loss score, the optimizer reconfigure weights of the neural network.

then reconfiguring weights and biases, and reevaluating the network. This process occurs in several epochs until the neural network provides satisfactory accuracy or a maximum number of iterations is reached.

Cost (objective) function: To measure neural network accuracy, we measure how different the machine-assigned labels (predicted values) are from the correct labels (actual values). This will be identified by the cost (objective) function. We have described the cost function in Chapter 8, and use that definition here as well. To review: the cost function measures loss in the neural network. The neural network algorithm identifies a loss score for the output layer. Then, after each iteration, it reconfigures the network's weights and biases to reduce the loss score.

Optimizer: It is the algorithm that is responsible for reconfiguring the network's weights and biases, such as SGD, which we have explained in Chapter 8. In other words, the optimizer uses the loss value to update weights in the network and thus reduces the loss value for the next round (epoch).

Based on this explanation, a neural network has three core components: *hidden layers, cost function,* and the *optimizer algorithm.* Figure 10-7 describes the process of how a neural network works and its components. The design of this figure was inspired by an excellent explanation provided in Chollet's book [Chollet '18].

Learning in the context of a neural network means finding weights and biases that minimize the cost function result (cost score) for the given dataset. Updating the weights toward improving the accuracy of the neural network is the task of the optimizer. The process of changing weight and calculating the cost continues iteratively until the number of specified epochs reaches the neural network. We have explained in Chapter 8 that epoch refers to one round of scanning the dataset, and it is usually given to the neural network by the user as a hyperparameter.

Now, we understand that a neural network algorithm assigns the weights, runs the network, measures the cost, and then updates the weights to reduce costs via optimizer.

NOTES:

* It is uncommon to report the computational complexity of neural network algorithms anymore because they are known to be very complex and require lots of resources, especially for training them.

* The format of data that is transferred between layers in the neural network is a tensor, and a tensor is a multidimensional array of numbers (scalar is a 0D tensor, a vector is a 1D tensor, a matrix is a 2D tensor, and Rank 3 tensor is 3D tensor). For example, an image could be a 3D tensor (x, y coordinates, and one value for color channel, i.e., RGB value), or a video could be a 5D tensor (images' data plus the timestamp of each frame, and audio wave of each frame).

* A neural network does not feed all members of the training dataset once into the network; it either feeds input data one by one or a set of input data as mini-batches into the neural network.

* The neural networks' input and output variables present the same observation, but the output has a label assigned to the same information.

* We can say a single neuron network is nothing more than a linear regression ($y = \beta_0 x + \beta_1$, but here we use b and w instead of β_0 and β_1) that its output will be fed into an activation function. Therefore, the input will be a value for x, the neural network assigned w, and b to that identify a y. This y will be fed into an activation function, and the result will be the output. Therefore, considering the content of the bracket is what is happening inside a neuron, we can formalize a single-neuron network as follows: $x(input) \rightarrow [z = wx + b \rightarrow \sigma(z)] \rightarrow y(output)$

Perceptron Algorithm

Perceptron [Rosenblatt '58] is one of the simplest and easiest neural network algorithms, which performs a binary classification. It receives a vector of input values and calculates a linear combination of input variables' values. If the results are greater than a threshold, the output will be equal to 1, otherwise, the output will be equal to -1. Assuming our input tensor has a d number of features $(x_1, x_2, \ldots x_d)$, the following equations formalize the binary classification of the perceptron algorithm.

$$\hat{y} = a(\sum_{i=1}^{d} x_i w_i + b)$$

Why do we use weight (w) for each feature? Because weights determine the contribution of each input feature to the output ($\hat{y}$). In other words, they specify the importance of the neuron. However, if a particular x value is zero in some instances, the weight impact will be zero as well. To handle this problem, a bias b_i will be added to the input as well. Bias can be interpreted as offset, which assists the $x_i w_i$ reaches a specific threshold, and it has an impact on the output variable.

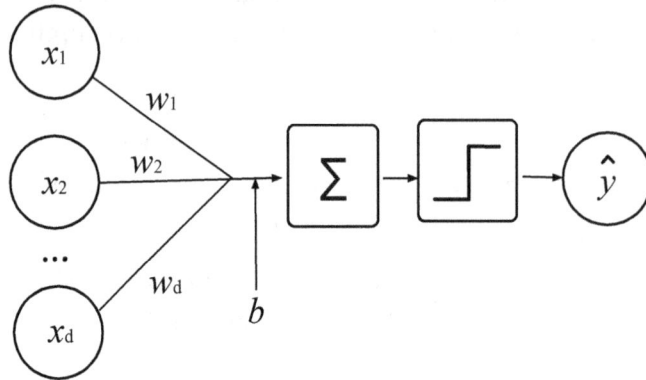

Figure 10-8: A single perceptron that receives a d dimensional input data. b presents the bias which is added to the result of summation. The first square is summing all weights and inputs together. The second square presents the activation function.

Figure 10-8 shows the simplest form of the perceptron. Σ is presenting a summation of input variables along with their weights and biases, i.e., $\Sigma_{i=1}^{d} x_i w_i + b$. The other rectangular box, with a shape similar to S, presents the activation function, i.e., $a(\Sigma_{i=1}^{d} x_i w_i + b)$.

Now, we have realized that the content of the equation in $a(.)$ function is just a simple linear regression, but perceptron is not only a linear regression; we need to have a constraint for the output to be able to classify each output value and assign a label to it. For example, all outputs should be either 0 or 1. This is the job of the activation function $a(.)$. We can conclude that the *activation function* is used to set constraints on the output values of neurons. If we use the Sigmoid function (check Chapter 8 to recall the Sigmoid), the activation function is a logistic regression, and a single perceptron is just a logistic regression.

Single perceptron is not very useful, and we use Multilayer perceptron, which we will explain later. Perceptron (similar to logistic regression algorithms) operates based on the assumption that a hyperplane exists, which separates two sides of the dataset. It starts with random weights that are close to zero. Then, it classifies the dataset. Next, it goes back and checks the data points that have been misclassified by analyzing the cost of the predicted value ($\hat{y}$), which is acquired by comparing it to the actual value (y). The cost function of Perceptron is very simple, and it simply uses the step function (we will explain it later). It means it uses a rule-based update: if the output is correct, weights remain the same; if not, weights are adjusted. Then, the neural network changes the weight of misclassified data points to reduce the cost score, and again performs the classification, then checks the new result. This process continues until it reaches a specific number of iterations, or weights cannot be further changed.

In general, the process of going back to the network and changing the value of weights to improve the accuracy of output is called *backpropagation* or *(Back-Prop)*, which will be explained later in detail. However, Perceptron is much simpler and doesn't use backpropagation.

Multilayer Perceptron

A Multi-Layer Perceptron (MLP) is an artificial neural network consisting of fully connected neurons with a nonlinear activation function, organized in at least three layers, notable for being able to distinguish data that is not linearly separable.

Before we explain Multilayer Perceptron, we should review binary logic, which we learned back in high school. Table 10-1 presets possible values of two binary variables, A and B, which can get binary values (0 or 1) and the result of applying a logical operation on these two variables. To recall them from school time, OR is a logical operation that outputs true or 1 when at least one of the input values is true or 1, while AND is a logical operation that outputs true or 1 only when both input values are true or 1. XOR (exclusive OR) is a logical operation that outputs true or 1 only when the two input values are different.

A	B	Not A	Not B	A AND B	A NAND B	A OR B	A XOR B
0	0	1	1	0	1	0	0
0	1	1	0	0	1	1	1
1	0	0	1	0	1	1	1
1	1	0	0	1	0	1	0

Table 10-1: Some basic binary logic operation between two variables.

Logical reasoning plays an important role in electrical engineering, computer science, and circuit design, but for our needs, it is enough to know these operators.

We have explained that a single layer perceptron is good for classifying linearly separable datasets. In the context of the logical operator, the perceptron algorithm can be used to represent boolean AND, OR, and NAND functions. However, a single layer perceptron cannot represent the XOR[3] boolean operator. In other words, it is impossible to use it for a non-linearly separable dataset [Minsky '69] with a single perceptron.

It is possible to model XOR with multiple perceptrons (a network of perceptrons). To understand this problem, let's use some visualizations with sample data points. Figure 10-9 present four data points for the A and B variables described in Table 10-1. At the bottom of each plot, there is the perceptron that can model the distinction between the blue and red areas of its plot on top.

All these perceptrons use a *step function* as their activation function. As we have explained, a single perceptron can linearly classify the data point. In the context of logical operators, AND, OR are linearly separable. Nevertheless, the XOR operator cannot be implemented with a single perceptron, and we can see that the XOR area is not linearly separable, unlike AND or OR, which can be separated with a single line. Linear separation can be implemented with a single

---

[3] XOR means that both operands should be different, i.e. true and false. For example, (P XOR ~P) = true

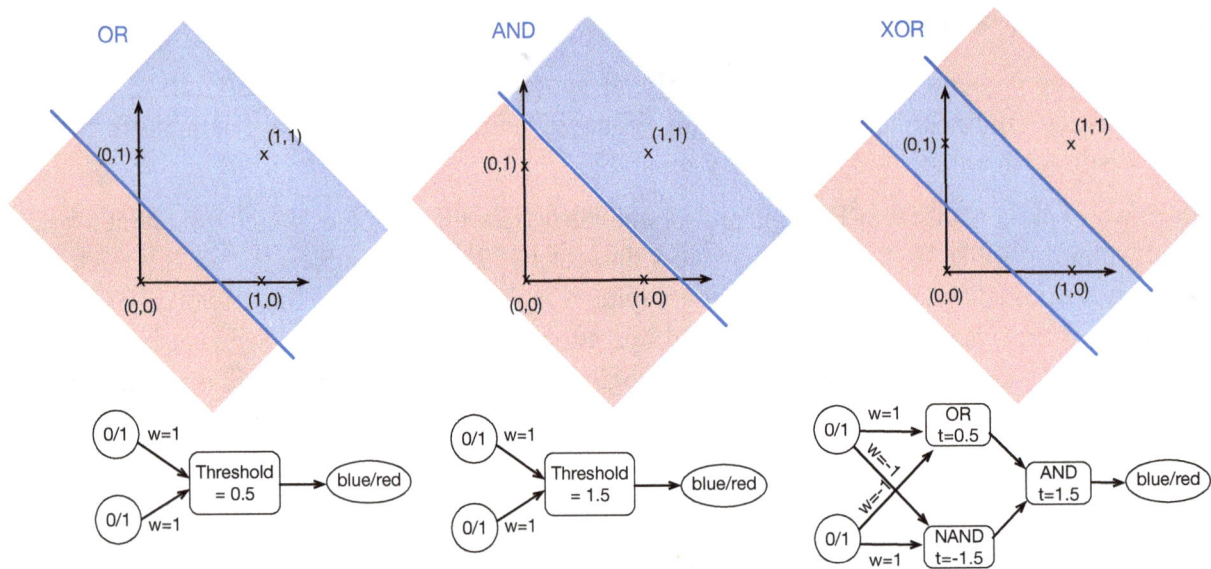

Figure 10-9: (Top) Four dots represents different status of A and B. AND and OR could be separated with linear hyperplane but XOR is not separable with a liner hyperplane. In the right diagram, we need some non-linear function to separate the blue area from the rest. (Bottom) Both AND and OR can be implemented as a perceptron with the step function.

perceptron and threshold activation function, as shown for AND and OR. XOR cannot work with one layer and requires at least two connected layers of perceptrons. This means that the output of one layer will be fed as the input of another layer. This neural network has more than one perceptron, which is called a 'Multilayer Perceptron (MLP)'. An MLP can handle non-linear separable data as well.

In other words, MLP, which is also called a *feed-forward neural network*, is a solution to separate non-linearly separable datasets, such as the situation we have observed for XOR. As we have explained, layers between the input layer and output layers are called hidden layers, and deep learning algorithms have lots of hidden layers. Therefore, it is not wrong to say that the MLP algorithm is the mother of the deep learning algorithm.

The more layers we add, the more complexity we can handle by the algorithm. It doesn't mean that too many layers are always good. Too many hidden layers impose a huge cost on resources and could cause overfitting, which we will discuss later.

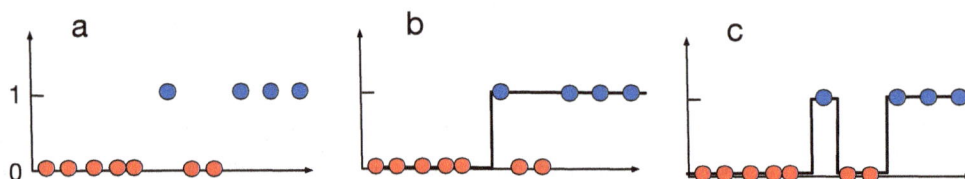

Figure 10-10: (a) original dataset (b) Using one single perceptron with a threshold activation function we can classify most data points correctly. However, two red dots are misclassified. (c) By using MLP with threshold activation function, all data points were classified correctly.

To be sure you leave this section with a Ph.D degree in MLP, take a look at Figure 10-10 (a) as another example. Here, we have blue and red dots in two-dimensional space, and we would like to classify them. By using a single perceptron, we have a step function, and it can fit these data points, as shown in Figure 10-10 (b). The result can classify blue dots as 1 and red dots 0, but two red dots are misclassified. By using two step functions, which are possible through MLP, all data points can be classified correctly. We can see the result in Figure 10-10 (c) that blue dots are one, and red dots are 0.

Every layer in the MLP architecture, except the output layer, is fully connected to the next layer. This is known as the *fully connected layer*. This means that every neuron is connected to all other neurons in the next layer, as shown in Figure 10-11. Also, keep in mind that every layer, except the output layer, includes biases as well. Figure 10-11 presents how we write weights. We write

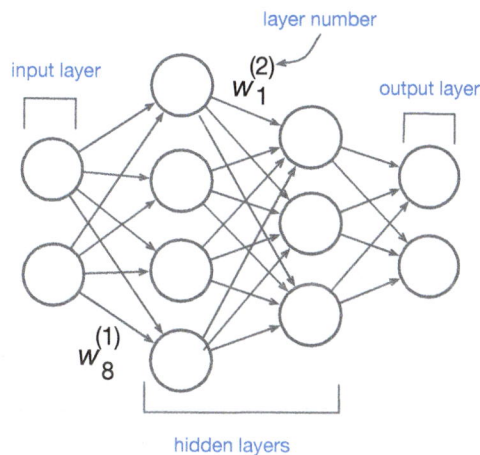

Figure 10-11: Sample MLP network, which all neurons are fully connected to all other neurons in the next layer.

them on the top of each w its layer number, and at the bottom of it, its index.

MLP is considered the foundational neural network model from which many subsequent architectures were developed, and it is still widely used today. Its layered structure and ability to learn complex patterns paved the way for advancements in deep learning, influencing later architectures such as convolutional and recurrent neural networks.

Activation Functions

Activation functions are mathematical equations used to set boundaries on each neuron and decide the output value of a neuron. We can write the *output of each neuron* as $z = wx + b$, and $a(z)$ is the activation function that changes z to the neuron's output.

There are many activation functions available. Some activation functions focus on binary decisions, such as the step function, which we have explained. Some activation functions, such as softmax, can handle multi-classes, e.g., objects that appear inside a photo (cat, dog, human, etc.). In the following, we list some common activation functions, which are visualized in Figure 10-12.

Step function: The simplest form of binary class activation function is the step function, used for binary classification. It has a similar shape to the function presented in Figure 10-4, but the threshold is z, and if it is greater than z, the signal is fired ($\hat{y} = 1$). Otherwise, it will be 0 ($\hat{y} = 0$).

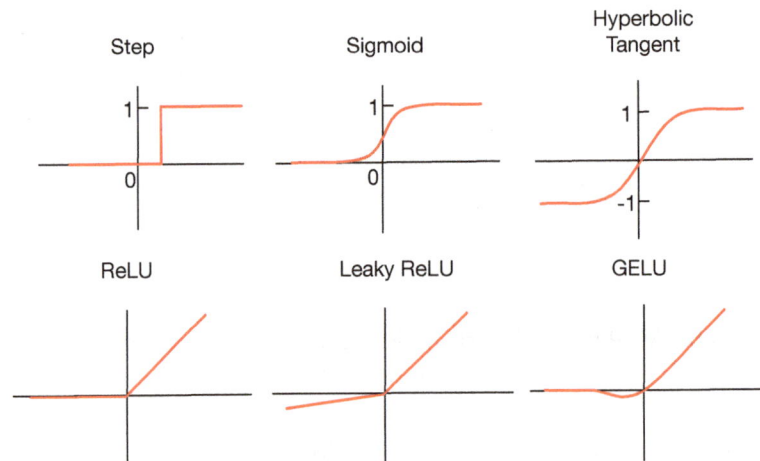

Figure 10-12: Some popular activation functions.

Sigmoid function: It is smoother than the step function and uses the Sigmoid equation, which we have explained in Chapter 8, for logistic regression. The Sigmoid function is written as follows:

$$f(z) = \frac{1}{1 + e^{-z}}$$

Sigmoid is more popular than step function because it is slightly more sensitive to small changes in the output value, and it is better suited to report probabilities because it gives a value between 0 and 1, step function only gives either 0 or 1, and no other numbers in between. The sigmoid function involves exponential calculations, which can be computationally expensive and slow down the training process, especially for large networks.

Hyperbolic Tangent function: it is another activation function, and its equation is written as follows.

$$a(z) = \frac{e^z - e^{-z}}{e^z + e^{-z}}.$$

It calculates a ratio between hyperbolic sine and hyperbolic cosine, i.e., the ratio of the half-difference and half-sum of two exponential functions in the points z and $-z$. This activation function is very similar to the Sigmoid function, but its range is from -1 to 1. It is used when we do not want to have zero as the lower boundary of the activation function.

Rectified Linear Unit (ReLU) function: ReLU [Nair '10] is a very common activation function (probably the most popular binary activation function), and we can easily write it as $max(0,z)$. It means if the output value is less than zero, the RELU activation function considers it as 0; otherwise, the value of the output is the actual output value. ReLU is a useful activation function while dealing with the vanishing gradient problem, which will be explained later in detail. The following equation formalizes the ReLU.

$$a(z) = \begin{cases} 0 & for\ z < 0 \\ z & for\ z \geq 0 \end{cases}$$

ReLU function has a specific characteristic, and it can be used in networks with many layers.

Leaky RELU (LReLU): LReLU [Maas '13] is another common activation function that allows a very small negative value to have a non-zero variable. Not having zero is useful because it mitigates the challenge of vanishing gradient better than ReLU, which is called *dying ReLU*. The LReLU equation is written as follows.

$$a(z) = \begin{cases} 0.01z & for\ z < 0 \\ z & for\ z \geq 0 \end{cases}$$

Having zero gradients can also result in slow learning, and using ReLU and LReLU resolves this. In other words, The derivative of the ReLU is 1 in the positive part and 0 in the negative part. The derivative of the LReLU is 1 in the positive part, and it is a small fraction in the negative part. If this explanation does not make sense now to you, be patient; after you have learned backpropagation and chain rule, it might make more sense.

Gaussian Error Linear Units (GELU): The GELU is a nonlinear activation function based on the standard Gaussian cumulative distribution function [Hendrycks '16]. Later in this chapter, we learn to regularize a neural network. Sometimes, we set some weights randomly to zero, i.e., *dropout*. In the context of GELU, the function smoothly varies between 0 and 1, modulating the output based on the Gaussian CDF, which gives it a probabilistic interpretation somewhat similar to dropout. In particular, assuming the $\Phi(x)$ is the standard Gaussian cumulative distribution function (CDF if you recall from Chapter 3), each data point x will be multiplied by this value, $x\Phi(x)$, and is the output of GELU activation.

This activation function is used in several transformer-based models, GPT-3, BERT, and wav2vec 2, which we will describe later in Chapter 12.

Softmax: It is an activation function used for multiclass output values. We cannot visualize it in Figure 10-12 because it has multi-dimensions (equal to the number of output classes). Assuming we have k number of classes, the softmax is presented with $\sigma(.)$ and it is formalized as follows:

$$\sigma(z)_i = \frac{e^{z_i}}{\sum_{j=1}^{k} e^{z_j}} \quad for \quad i = 1,...,k$$

Softmax activation function calculates probability distributions of the particular output over k different labels. For example, we have a neural network to recognize if the given image includes a human, cat, or dog, then our $k=3$. Softmax calculates the probability of each target class over all possible target classes.

To understand how Softmax works, we use another example. Assuming we are creating an algorithm to identify infection in microscopic blood samples, we have developed a neural network algorithm to identify specific types of infection in the blood sample. There are three types of infection. X, Y, and Z. First, our algorithm analyzes an image, and then probabilities for infection probabilities are calculated using Softmax as follows: X:0.3, Y:0.6, and Z:0.1. We can see that their sum is equal to one; the output of Softmax is a probability function, and thus, the sum of the label values is equal to 1. In summary, based on the number of possible outputs, Softmax assigns each of the outputs a probability value, and the one that has the highest probability is the correct output.

Let's review why we need an activation function. Neural networks are linear transformations, and chaining many linear transformations ends up having a linear transformation again. This means the neural network cannot solve complex problems by just having a linear transformation. For example, if we have $f(x) = x - 1$ and $g(x) = 2x + 1$, chaining them will be $f(g(x)) = (2x + 1) - 1 = 2x$, which is still a linear function. Therefore, activation functions are used to enable the neural network to perform a non-linear transformation.

SiLu (Swish): The Swish activation function is smooth, non-monotonic, and unbounded above, allowing for more complex representations in neural networks, and it combines the benefits of both linear and non-linear functions. The Swish activation function is formalized as follows:

$$swish(x) = \frac{x}{1 + e^{-x}}$$

A variant of Swish includes a learnable parameter β, and its computed as follows:

$$Swish(x) = x \times sigmoid(\beta x)$$

Swish approximates a linear function for positive inputs and ReLU for negative inputs, which helps mitigate the vanishing gradient problem in deep networks.

Neural Network Cost Functions

We have explained that a neural network has three core components, and the cost function is one of them, as is shown in Figure 10-7. We have also introduced and explained Chapter 8 about the motivation and use of cost functions. In the context of a neural network, a cost function uses the given input to measure how accurate the output is. In other words, after the neural networks construct the output, the algorithm uses the cost function to report the loss score $L(\hat{y}, y)$. Basically, the loss score is calculated by the cost function and presents how far the algorithm output is from the true value.

A cost function for a single neuron can be formalized as $C = (W, B, S_r, E_r)$, in which C presents the cost value, W presents weights, B presents biases, S_r presents the input of a single training sample (e.g., one record of data), and E_r presents the desired output of the given input training sample (S_r). For the entire network, which has n layers and we have m numbers of weights in the last layer, we will have a cost function that depends on all of the weights $C(w_1^{(1)}, w_2^{(1)}, \ldots, w_m^{(n)})$.

Once again, let us repeat that when we report something specific to a layer, we use superscript. For example, $a^{(2)}(x)$ presents the value of the activation function at the second layer.

Here, we list four cost functions that are common for neural networks. Some of them have been explained in Chapter 8 and Chapter 3, but here, we repeat them because they are widely used for neural networks. We can also define our cost function based on our needs for the neural network and incorporate the domain knowledge into the customized cost function.

Quadratic Cost (Mean Squared Error)

We use this cost function for regression as well, and it is known as "root mean square error". Assuming $y(x)$ presents the actual value of output, a_i^L or y presents the actual value in the last layer (L presents the last layer), and $E_{r(i)}$ (or $\hat{y}$) presents the output value. Assuming the test set has n data points, and i is the index of each data point, the quadratic cost equation is written as follows:

$$C = \frac{1}{n} \sum_{i=1}^{n} (a_i^L - E_{r(i)})^2$$

Using simply y and $\hat{y}$ is much easier to understand and memorize, but other literatures use this notation, and it is better to get familiar with this style of writing as well; we follow the common approach while describing these equations.

The quadratic cost function is usually used for neural networks that try to resolve a regression problem. The loss function used in the training phase of MLP is usually Mean Square Error (MSE), if there are many outliers, then it is recommended to use Mean Absolute Error (MAE) instead.

Cross Entropy

Cross entropy is a mathematical function that measures the differences between two statistical distributions. Assuming $P(x)$ is one distribution and $Q(x)$ is the second distribution, their cross entropy $H(P, Q)$ is written as follows.

$$H(P, Q) = - \sum_x P(x) . log \ Q(x)$$

For classification tasks of neural networks where we deal with labels, we use cross entropy cost function. This function provides a probability distribution for each class label. For example, an algorithm that recognizes your mood from facial expressions after you wake up from bed can provide a list of three probabilities as follows: *{I bite, don't talk with me: 0.7, full of energy: 0.1, immediately to the restroom: 0.2}*.

Assuming the data points in the test dataset are specified with the i index, the binary cross entropy equation is written as follows.

$$C = - \sum_i [E_{r(i)} . ln(a_i^L) + (1 - E_{r(i)})ln(1 - (a_i^L))]$$

Respectively, we can write the multiclass cross entropy as follows:

$$C = - \sum_i \sum_j [E_{r(i),j} . ln(a_{i,j}^L)]$$

Here, i is the index of samples, and j is the index of classes. The description of other variables in this equation is exactly the same as the one we have explained for the quadratic cost function.

Kullback-Leibler Divergence

Back in Chapter 3, we explained that the KL-Divergence function is used to quantify the differences (or measuring similarity) between two probability distributions, which is written as follows:

$$D_{KL}(P, Q) = \sum_i P(i) ln \frac{P(i)}{Q(i)}$$

In the context of neural networks, we need a cost function to compare the distribution of output values with the actual values. Therefore, it is written as follows:

$$D_{KL}(E_r, a^L) = \sum_i E_{r(i)} \ log \frac{E_{r(i)}}{a_i^L}$$

KL divergence and cross entropy are very similar, and KL divergence is also called relative entropy. Therefore, similar to cross entropy, we can also use it for classification tasks.

Hellinger (Bhattacharyya) Distance

Hellinger distance [Hellinger '09] is also used to quantify the similarity between two probability distributions. Therefore, outside the context of the cost function, we can formalize the Hellinger distance between two distributions, P and Q, as the L_2 norm (Euclidean norm) used for probability distributions:

$$H(P,Q) = \frac{1}{\sqrt{2}} || \sqrt{P} - \sqrt{Q} ||_2 = \frac{1}{\sqrt{2}} \sqrt{\sum_i (\sqrt{P_i} - \sqrt{Q_i})^2}$$

However, differentiating a square root inside a square root (used by Backpropagation) is computationally complex, and this distance can be relaxed by removing the square root as follows:

$$H^2 = \frac{1}{2} \sum_i (\sqrt{a_i^L} - \sqrt{E_{r(i)}})^2$$

If we intend to ensure that the cost function result will stay between 0 and 1, we can use Hellinger distance and not its squared version.

To summarize these cost functions, cross entropy and all other cost functions we described here, except quadratic cost, measure the differences between two probabilities, in which one probability presents the actual value and the other presents the output value. To review where we use which cost function, check the following:

Quadratic cost: Typically used in regression problems where the output is a continuous variable.

Cross Entropy: Commonly used in classification problems, particularly multi-class classification.

Binary Cross Entropy: Specifically used in binary classification problems.

KL-Divergence: Also used in classification tasks, KL-Divergence measures how one probability distribution diverges from a second reference probability distribution.

Hellinger Distance: Used in classification problems as well, especially when a bounded metric between 0 and 1 is preferred.

> Why does the author of the book lied in earlier chapters?
>
> I thought there is not much mathematics, what is this? why so mean equations ???

NOTES:

* Since a neural network includes many neurons and many activation functions used to combine these neurons, activation functions should be computationally efficient. Similar to the similarity metrics that are required to be computationally efficient while we are doing clustering, the same issue existed for the activation function as well.

* Among the equations we described for the cost function, the quadratic cost function is more popular. It is squared to eliminate negative values and increase the impact of errors to highlight them. Somehow, it punishes the network when the error is large by making it a larger error. If we have a multi-label classification problem, cross entropy is also popular. Nevertheless, we recommend that if you have enough resources, you experiment with all of them. You can change them in the neural network package easily as hyperparameters.

* There are more cost functions, such as the Generalized Kullback–Leibler divergence or Exponential cost function, which we did not describe.

* Some implementations of cross entropy provide one implementation for binary classification and one for multi-class classification, i.e., softmax cross entropy.

* While using gradient-based optimization, Mean Absolute Error and Mean Square Error usually do not yield good results. Therefore, they are not popular to go for neural network cost functions. However, while MSE and MAE are not typically the first choice for classification problems in neural networks, they are still popular and effective for regression tasks.

Neural Network Optimizers

Before starting to read this section, it is worth taking a look at Chapter 8 and reviewing the optimization section we have explained. In Chapter 8, we have explained that the cost function measures how far the true value is from the actual value. The optimizer's job is to reduce the cost. In other words, the goal of the optimizer is to increase the accuracy of the algorithm by changing the model parameters. In the context of the neural network algorithm, this will be achieved by minimizing the cost function through changing weight and biases. At the end of each epoch (each iteration on the dataset), the optimizer changes weights and biases in such a way that the loss score will be reduced in the next epoch. In other words, by adjusting weights and biases, the algorithm can reduce the loss score.

Now, a question arises: What is the best combination of weights that increases the accuracy of the result? To answer this question, we can try every possible combination of weights and biases and see which one is the best. We have bad news; this is not possible. To understand why, let's assume we have a fully connected neural network that has one input layer with four neurons, two hidden layers, each with five neurons, and two output neurons. In this case, we will end up with $(4 \times 5) + (5 \times 5) + (5 \times 2) = 55$ weights to configure. Besides, each neuron in the hidden layers and output layer has its bias, so this adds five (for the first hidden layer) plus five (for the second hidden layer) plus two (for the output layer), which results in 12 biases.

As we have explained, the algorithm starts by assigning a random weight. Let's say we limit the weight changes to a number that can vary between 0 to 100. Therefore, only to find the best weight configurations and not biases, we need to test 100^{67} possible combinations for experimenting and identify which combinations lead to the best minima. It is definitely not computationally possible, even with the strongest supercomputer in the world. Besides, note that neural network optimization is non-convex, and thus, it is very complex to identify good minima or global minima.

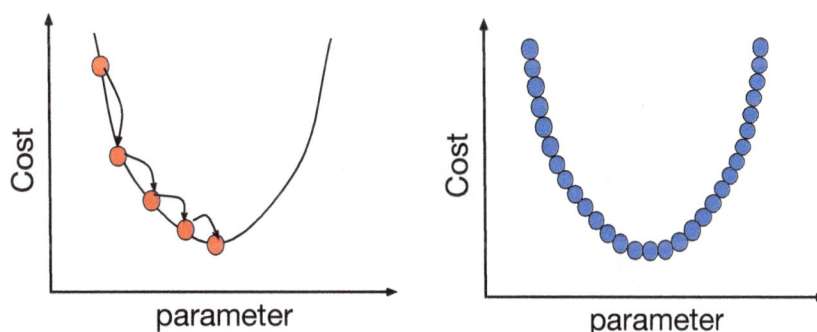

Figure 10-12: (left) Using an optimization to reach the minima and therefore very few data points will be processed. (right) Not using an optimization and thus many data points will be experimented on the function to find the minima.

We need a subtle approach to determine weights and biases. Take a look at Figure 10-12. Here, we have a very simple convex function. This figure presents the differences between using optimization and not using optimization to find the minima. By using blue dots in the right plot, we try to show that there are too many data points to experiment with, but in the left plot, there are few red dots, and instead of experimenting with any possible value, the algorithm can make some jumps toward the minima. This figure is a convex function, and we show it for the sake of simplicity; in the real world, it is much more complicated, and while working with neural networks, there is no convex shape.

Therefore, instead of experimenting with all possible points on the curve, we can experiment with some random points on the function. We use a method such as Gradient Descent to determine the next point and skip some points that are not useful. In other words, from the current data point, we need to find (i) *direction* and (ii) *step size* toward reaching the global minima and avoid experimenting with every single data point. As we have explained in Chapter 8, Gradient Descent helps us to determine the direction by using the partial derivative, which is shown on the left plot in Figure 10-12. At each point, the Gradient Descent algorithm calculates the dervative, and if the gradient sign does not change, it means that it should take the next move in this direction to move toward minima. If the sign changes (from negative to positive), it means the gradient jump is too big, and it should take the next move in the opposite direction to reach the minimum.

Figure 10-13 visualizes the behavior of the Gradient Descent based on the slope of the line. The *X*-axis in this figure presents a model parameter, and the *Y*-axis presents the cost of different values for the parameter. However, having constant size jumps is inefficient because if step sizes are small, it takes a long time to reach the minima, and if they are large, it might pass the minima. Also,

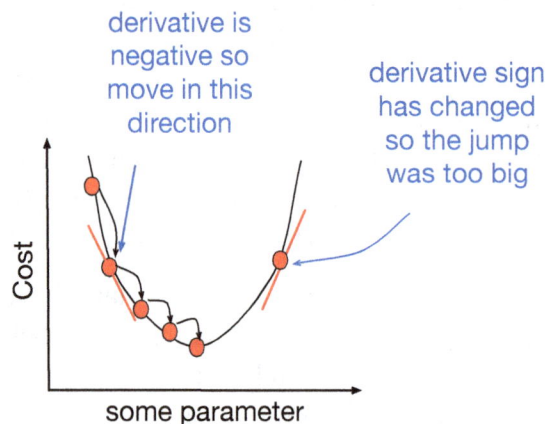

Figure 10-13: Gradient Descent decides the direction of the next point to move toward minima.

if the jump is too big, the gradient sign will change. Therefore, a better approach is customizing the step size, and it is done through a *decaying learning rate*, which we will explain later.

In Chapter 8, we explained three types of Gradient Descent, including SGD, BGD, and Mini BGD. Gradient Descent is the most popular optimization approach for neural network optimization, but it has some limitations. It can identify the direction of the movement, but it does not specify the step size. In this section, we introduce other optimization algorithms that use SGD, and mini BGD, and improve them.

Stochastic Gradient Descent with Momentum

Although mini-batch Stochastic Gradient Descent (SGD) has some limitations, it is a very popular optimization algorithm used in deep learning as well. In Chapter 8, we have explained that, while using Gradient Descent, the next value for the parameter is calculated as follows:

next point = current point — (learning rate × slope (a derivative of that particular parameter)

552

Assuming w_{t+1} is the weight (or any other parameter [4]) at the next epoch, w_t is the weight of the current epoch, α is the *learning rate*, and $\nabla G(.)$ is the vector of gradients. We estimate a value of w_{t+1} via the following equation:

$$w_{t+1} = w_t - \alpha . \nabla G(w_t)$$

In simple words, $\alpha . \nabla G(w_t)$ specifies the amount of the move for the next step.

There are two problems with SGD. First, we can see from the described equation that the direction is changing, but the step size is fixed. A fixed step size might trap the optimizer into a local minimum. Second, as seen in Figure 8-25 from Chapter 8, the noise of SGD is high due to its stochastic (random) behavior. This problem can be reduced by using an *exponentially weighted average (β)*. In other words, the exponentially weighted average introduces a *decay coefficient* to a series of numbers. Therefore, applying this exponentially weighted average reduces the impact of older data points and, thus, the noise. This type of SGD is called *SGD with momentum* [Polyak '64]. In the following, first, we describe the exponentially weighted average, and then, we explain SGD with momentum.

Exponentially Weighted Average

Weight (β) is a number between 0 and 1, usually set to 0.9. Multiplying a number less than one to another number makes it smaller. For example, let's assume we set $\beta = 0.7$, if we multiply β to something, the result value will be 70% of the original value. If we multiply β^2 by a number, the result will be 49% of the original value, etc. To understand the mathematic intuition of exponentially weighted average, assume we have a sequence of numerical data, i.e.,

---

[4] While reading text for optimization in the context of the neural networks, some literature refers to parameters as θ or β, we use w to present weight. Nevertheless, bias b is also a parameter for the neural network. Besides, based on the literature, these parameter names change and we do comply with the name change to avoid confusing you while reading other resources as well.

$\{s_1, s_2, \ldots s_n\}$. Using a weighted average, we can have a sequence of transformed numerical data, i.e., $\{s'_1, s'_2, \ldots s'_n\}$ and each s' at position t is calculated as $s'_t = \beta s'_{t-1} + (1 - \beta)s_t$.

For example, three sequential data in the transformed sequence with a weighted average will be written as follows.

$$s'_t = \beta s'_{t-1} + (1 - \beta)s_t$$
$$s'_{t-1} = \beta s'_{t-2} + (1 - \beta)s_{t-1}$$
$$s'_{t-2} = \beta s'_{t-3} + (1 - \beta)s_{t-2}$$

Therefore, by combining these series and applying some mathematical simplification to them, we have the following:

$$s'_t = \ldots + \beta\beta(1 - \beta)s_{t-2} + \beta(1 - \beta)s_{t-1} + (1 - \beta)s_t$$

As the data gets older in the series, its impact on the equation gets smaller because more numbers of βs are multiplied. Note that the previous sentence is true only for β between 0 and 1.

That is enough to understand the impact of exponentially weighted average, and now we can switch back to the SGD with the momentum algorithm.

SGD with Momentum

SGD with momentum uses an exponentially weighted average to create a *velocity*. To understand velocity, assume we are rolling a ball inside the bowl shape function, and we intend to get the ball into global minima in Figure 10-14. The derivative of the ball specifies the acceleration of the moving ball, and momentums add to the velocity of the moving ball (v or *retained gradient*).

Momentum (η or *momentum coefficient*) is used to increase the velocity of the optimization ball, and thus, it will be fast enough to jump out of the local minima and move toward global minima. In other words, momentum reduces the oscillation of the optimizer algorithm, and thus, it can reach the minima faster.

The SGD with momentum algorithm is written as follows:

$v_t = 0$ # velocity

$\eta = 0.9$ # momentum coefficient

$\alpha = 0.01$ # learning rate

while $(w_t$ *not converged*$)$ { # e.g., converged in this context: loss < 0.1

$$v_{t+1} = v_t \cdot \eta + \nabla G(w_t)$$
$$w_{t+1} = w_t - \alpha \cdot v_{t+1}$$
$$v_t = v_{t+1}$$
$$t = t + 1$$

}

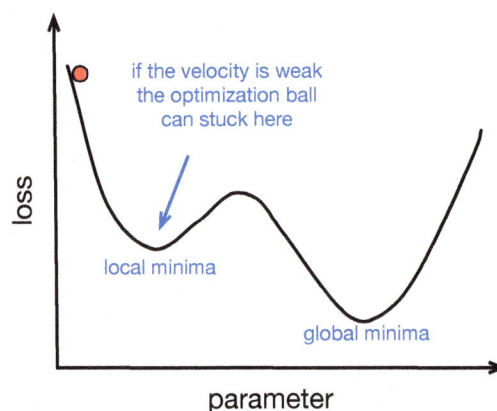

Figure 10-14: A ball with low velocity stuck in the local minima, but having high enough velocity can helps it jump out of local minima and move toward global minima.

554

Don't be afraid to see these equations inside an algorithm. They are very easy to understand. $v_{t+1} = v_t . \eta + \alpha . \nabla G(w_t)$, means:

next_velocity = current_velocity×momentum + learning_rate× gradient

Respectively $w_{t+1} = w_t - \alpha . v_{t+1}$, means:

next_weight = current_weight - learning rate × next_velocity

You can see that current velocity, momentum, and learning rate are all hyperparameters, and they have been given by the user of the algorithm. Nevertheless, they usually have common predefined values, and the only hassle for the user is choosing the optimization method and calling it. For example, at the time of writing this chapter in 2021, in Keras[5], we only need to write the following for the model, and the SGD is the SGD with momentum.

```
model.compile(loss='categorical_crossentropy', optimizer='SGD')
```

if we need to give some parameter value, we use the following code in Tensorflow[6]:

```
tf.keras.optimizers.SGD(learning_rate=0.01, momentum=0.0,
nesterov=False, name='SGD', **kwargs)
```

In pytorch[7], it is written as follows:

```
optimizer = optim.SGD(model.parameters(), lr=0.01, momentum=0.9)
```

To summarize our explanation here, momentum increases the convergence speed toward global minima and also increases the chance of reaching local minima by introducing a velocity component.

Nesterov Momentum

Using momentum significantly improves the convergence speed of SGD. However, using classical SGD with momentum, the gradient is always moving toward the correct direction, but momentum may not necessarily move in the correct direction. Nesterov Accelerated Gradient (NAG) or Nesterov momentum [Nesterov '83] improves the next step, and if the momentum goes in the wrong direction, then the gradient can go in the correct direction.

Take a look at Figure 10-15 to better understand the differences between the classical and Nesterov approach. In both examples, the momentum moves in the wrong direction, but in the Nesterov momentum, the gradient is started after the momentum, and thus, it moves more toward the correct direction. Therefore, the next data point is selected closer to the minima in the right figure.

While using pure SGD, the next point on the cost function will be determined by the following equation:

$$w_{t+1} = w_t - \alpha . \nabla G(w_t)$$

By using SGD with the momentum, the next point on the cost function will be determined as:

[5] https://keras.io
[6] https://www.tensorflow.org
[7] https://pytorch.org

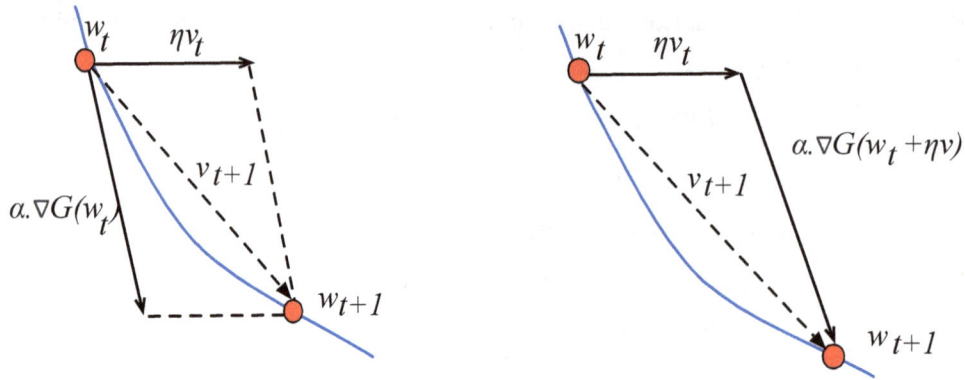

Figure 10-15: (left) Classical momentum for calculating the next parameter. (right) Nesterov momentum is used to calculate the next parameter on cost function. The blue line presents a small part of the cost function, and each new w is moving toward minima.

$$w_{t+1} = w_t + v_{t+1} \cdot \eta - \alpha \cdot \nabla G(w_t)$$

The Nesterov momentum calculates velocity and weight differently from classical momentum, and it uses the following equations to calculate them.

$$v_{t+1} = v_t \cdot \eta - \alpha \cdot \nabla G(w_t + \eta v_t)$$

$$w_{t+1} = w_t + v_{t+1} \cdot \eta$$

Based on Figure 10-15 [Sutskever '13], we can see that by using Nesterov momentum, if the momentum goes in the wrong direction, the gradient has the chance to move in the correct direction, which results in faster convergence. It is a small difference, but it increases the

convergence speed, and in some cases, Nesterov momentum performs better than classical momentum.

Adagrad

The neural network optimization deals with a cost function that is a non-convex shape because the neural network includes many weights and biases (model parameters or dimensions). Until now, we have assumed that the learning rate is constant. A constant learning rate, a hyperparameter given by the user, could not be useful for all scenarios and all types of networks.

Experiments show that the learning rate in a multi-dimensional dataset (or features) in some dimensions (or features) is changing very fast, and in some dimensions (or features), it is changing very slowly. For example, usually, weights change more frequently than biases, which change less frequently. Therefore, employing separate learning rates for different types of parameters (e.g., one for weights and another for biases) can facilitate a more efficient approach to reaching the minimum of the cost function. This means that a neural network could benefit from a dynamically changing learning rate that adapts its change pace to each feature.

One of the practical algorithms that introduce a dynamically changing learning rate is Adagrad [Duchi '11]. The Adagrad algorithm adaptively scales the learning rate for each parameter. It adapts the learning rate by scaling them inversely proportional to the square root of the sum of their historical/previous squared values.

We are sure you did not understand the previous sentence, so we explained it with the equation, which is easier to understand. Recall that SGD uses the following equation to estimate the next weight $w_{t+1} = w_t - \alpha \cdot \nabla G(w_t)$. The learning rate (shown as α) is constant in all epochs of traditional SGD, but the Adagrad algorithm uses a *dynamic learning rate* that can change based on the magnitude of the target parameter gradients until the current time.

Assuming α' is a learning rate for Adagrad, we can re-write the previous equation as $w_{t+1} = w_t - \alpha'_t \cdot \nabla G(w_t)$. Here α'_t is specified based on the previous epoch and previous weights. Therefore, we add a subscript as t, which presents the learning rate in t iteration (epoch). The value for α'_t will be calculated as follows:

$$\alpha'_t = \frac{\eta}{\sqrt{G_t + \epsilon}}$$

In this equation, ϵ is used to be sure the denominator will never be zero, and it is recommended to set it to 10^{-8}, which is an extremely small number. η is a constant value, and we can use a small value, such as 0.001. G_t is the variable that is performing the magic. It is the sum of squares of all previous gradients up to the current t, as it is calculated by using the following equation.

$$G_t = \sum_{i=1}^{t} \nabla G(w_i)^2$$

In other words, G_t enables Adagrad to reduce the impact of parameters with large gradients (features that are frequent in the dataset) and increases the impact of parameters with low gradients (features that are not frequent in the dataset). Because the denominator part of the equation (G_t) will be large for small gradients and small for large gradients. Recall that in primary school, we learned that having a large denominator makes the number small and vice versa.

a constant value

used to avoid having zero in denominator

$$\alpha'_t = \frac{\eta}{\sqrt{G_t} + \epsilon}$$

$$w_{t+1} = w_t - \alpha'_t \, \nabla G(w_t)$$

$$G_t = \sum_{i=1}^{t} \nabla G(w_i)^2$$

Figure 10-16: Adagrad parameter calculation equations.

To get an overview, take a look at Figure 10-16, which shows the described equation in this figure. If you still don't get the point, let's substitute the equation with some number and see the result. Assume we have a frequent feature (f_1) and its $G_t = 3.46$ and another less frequent feature (f_2) that its $G_t = 0.16$.

$$\alpha'_t(f_1) = \frac{0.01}{\sqrt{3.46 + 10^{-8}}} = \frac{0.01}{1.86} = 0.005$$

and

$$\alpha'_t(f_2) = \frac{0.01}{\sqrt{0.16 + 10^{-8}}} = \frac{0.01}{0.4} = 0.025$$

By comparing 0.005 and 0.025 we can realize that the magic that Adagrad does is based on the sum of previous gradients of a parameter.

Adagrad with thin gradient θ

Adagrad with fat gradient θ

Adagrad works well when the dataset is sparse, which means some features include lots of zeros, and those features do not frequently appear in the dataset. For example, constructing a bag-of-words (Chapter 6) creates a sparse dataset of terms, and to process a bag-of-words with a deep learning algorithm, it is recommended to use Adagard as an optimizer.

To summarize this algorithm, make some space in your brain about Adagrad and write the following there: Adagrad reduces the focus on the parameter that is always happening by decreasing their learning rate and allows parameters that with of zeros (sparse features) to have a larger learning rate.

There is an optimized version of Adagrad, called Adadelta [Zeiler '12], which restricts the number of past gradients to a specific window size, which is given by the user as a hyperparameter. It is late night now, and the author doesn't have the energy to explain Adadelta.

RMSprop

RMSprop (Root Mean Square propagation) is an optimization algorithm introduced by Hinton [Hinton '12], one of the pioneers of deep learning in his online course[8]. Similar to Adagrad, RMSprop focuses on mitigating the challenge of having gradients of different sizes (too large or too small). RMSprop builds on top of the *Rprop* [Riedmiller '93], in which Rprop uses (i) the sign of gradient and (ii) adapting the step size to each gradient. To understand the RMSprop, we need to understand the Rprop first.

Rprop first checks the sign of two consecutive gradients. If the sign has been changed (similar to the two red dots in Figure 10-13), it means the jump was too large, and thus, the minima have been passed and not reached. In the next step, it decreases the jump size by multiplying it by a number less than 1, e.g. $\eta^- = 0.5$. If the sign has not been changed in two consecutive gradients, the jump is correct, and we are moving in the correct direction. Therefore, the step size will increase by multiplying it by a number larger than 1, e.g., $\eta^+ = 1.2$. For example, if $\nabla G(w_t) = -1$ and $\nabla G(w_{t+1}) = -2$, it means the algorithm is moving in the correct direction, and step size can increase, and it calculates the next weight as follows: $w_{t+1} = w_t + \eta^+ \nabla G(w_t) = 1.2$.

Rprop is very good when we have a small dataset. As soon as the dataset gets large, we should go for mini-batch Gradient Descent (mini BGD), and Rprop cannot handle mini BGD properly. For example, we have five mini-batches whose gradients are as follows: -0.4, -0.3, -0.25, -0.2, -0.14, -0.1, 0.7, and 0.9. Here, Rprop increases the weight six times and decreases it only once from (-0.1 to 0.7). This means that instead of RProp coefficients canceling each other, the weights grow larger despite insignificant changes in gradient.

Instead, RMSProp performs a small improvement on the Rprop to resolve it. RMSProp benefits from using Rprop decision based on sign, but additionally, it can handle the issue existing in mini-batches. RMSprop is similar to Adagard with slight differences. While using Adagrad, the next weight will be determined by the following equation:

---

[8] Interestingly they did not publish any scientific paper about RMSprop

$$w_{t+1} = w_t - \frac{\eta}{\sqrt{G_t + \epsilon}} \cdot \nabla G(w_t)$$

While using RMSprop, the next weight will be determined by the following equation:

$$w_{t+1} = w_t - \frac{\eta}{\sqrt{E[G_t^2] + \epsilon}} \nabla G(w_t)$$

In this equation $E[G_t^2]$ presents the squared of exponentially weighted average gradient and is calculated as follows: $E[G_t^2] = \beta E(G_{t-1}^2) + (1 - \beta) \nabla G(w_t)^2$.

β is the exponentially weighted average parameter that is assigned by the user, and it is usually 0.9. $\nabla G(w_t)$ (the gradient of the cost function with respect to w_t), $\eta = 0.01$ and $\epsilon = 10^{-8}$ are similar to Adam's equation and hyperparameters, but they usually have default values.

In summary, RMSprop divides the learning rate by an exponentially decaying average of squared gradient and benefits from the sign-based decision of Rprop.

Adam

Adam (adaptive Gradient Descent) [Kingma '14] is another popular optimization algorithm that combines the advantages of both SGD with momentum and RMSprop. It operates by taking large jumps at the beginning, and as the slope gets closer to zero, it starts to take smaller jumps.

Adam introduces two β parameters (β_1, β_2), which are used to *control the decay rate* of exponentially weighted averages of the (i) gradient (used in momentum) and (ii) squared gradients (used in RMSProp). The following algorithm describes Adam. While reading this algorithm, note the # sign is used for commenting and is written in blue.

$\alpha = 0.001, \beta_1 = 0.9, \beta_2 = 0.999, \epsilon = 10^{-8}$

$while\ (w_t\quad not\quad converged)\ \{$

$\qquad g_t = \nabla G(w_t)$

$\qquad m_t = \beta_1 . m_{t-1} + (1 - \beta_1)g_t$ #first moment estimate (momentum)

$\qquad v_t = \beta_2 . v_{t-1} + (1 - \beta_2)g_t^2$ #second moment estimate (RMSProp)

$\qquad \hat{m}_t = \dfrac{m_t}{1 - \beta_1^t}$ # corrected first moment estimate

$\qquad \hat{v}_t = \dfrac{v_t}{1 - \beta_2^t}$ # corrected second moment estimate

$\qquad w_t = w_{t-1} - \alpha \dfrac{\hat{m}_t}{\sqrt{\hat{v}_t} + \epsilon}$

$\}$

In Adam, m_t presents the first moment estimate (to support the functionality of momentum) and v_t presents the second moment estimate (to support the functionality of RMSProp). β_1^t and β_2^t mean β_1, β_2 to the power of t. Other hyperparameter values were recommended by the authors of the Adam algorithm. We can see that a squared gradient is used to scale the learning rate like RMSprop, and a moving average of the gradient is used instead of the gradient itself, as it is done in SGD with momentum.

There is not much to explain about Adam, and it seems that in the golden era of Gradient optimizer algorithms (approximately from 2012 to 2015), authors of these algorithms, with small changes, made a big improvement in the algorithm accuracy.

Optimizers Summary

Congratulations! Now, you are an optimizer expert and know these mathematical definitions. Since this section was very exciting, has no theoretical discussion and mathematics, and is full of real-world examples, let's summarize before you completely fall asleep.

SGD will change the model parameter per sample data point, which is too frequent, and this could cause much oscillation on the optimizer. To handle this problem, we can use mini-BGD, which updates mode parameters after a few samples. Nevertheless, a few examples could change the model parameters, and this leads to being stuck in a local minima. Therefore, we use classical momentum to reduce this noise.

Since the noise is reduced with classical momentum, the optimizer could make a wrong move and get away from minima, and the Nesterov momentum can mitigate this challenge by introducing an acceleration term into the model parameter. In simple words, if momentum makes the optimizer too fast, the Nesterov momentum slows it down. With these approaches, the learning rate for all model parameters is the same, but Adagrad and RMSprop use a history of the gradient to find a different learning rate for each model parameter.

Adagrad tunes the learning rate per model parameter, which makes it a good choice for sparse datasets. RMSProp also tunes the learning rate per model parameter, which makes it a good choice for noisy data and non-stationary data (check Chapter 8 to recall the meaning of non-stationary time series).

RMSprop and Adagrad do not update the momentum, Adam updates the momentum (in addition to the learning rate) for each model parameter as well[9].

Only 150 more optimizers remain that you can learn on your own [Schmidt '20]. All jokes aside, it was the madness of making better optimizers, and many students are struggling to make impactful optimizers. For example, add a decaying coefficient on the momentum parameter as well, called demon (decaying momentum) DemonAdam [Chen '19], or develop an automatic tuning approach for the momentum, such as YellowFin [Zhang '17]. We skip to explain them, and you really don't need to learn them unless you want to research optimizers.

---

[9] Alec Radford has an amazing visualization of some of these optimizers in a gif file and we highly recommend you to check his animation, here: https://imgur.com/a/Hqolp.

NOTES:

* There is no best optimizer that can perform perfectly in all problems. However, some background knowledge might be helpful to decide about the optimizer. There are some platforms used to benchmark different optimizers, such as DeepOBS [Schneider '19], which is open source[10], and you can install and use them to decide on the optimizer based on your dataset.

* If you have a mathematic or algorithmic background and love this topic, invest in making a good optimizer that is not based on Gradient Descent; think about it twice. To date, many optimization algorithms, especially genetic algorithms, have been proposed, and none have been as good as the Gradient Descent ones. You can search online and see that an optimizer algorithm has been developed for every single animal or insect on this planet. However, gradient descent algorithms are performing better than any other algorithm, but there might be an opportunity to explore and identify a new one. At the time of writing this part in 2021, genetic algorithms lost badly the optimization war to gradient descent ones.

* Convex optimization problems are easier to solve than non-convex problems. If a concave problem can be reformulated into an equivalent convex problem, it may enable the use of more efficient convex optimization algorithms.

Backpropagation

The Backpropagation (Backprop) algorithm was introduced in 1960 [Kelley '60, Bryson '62], and later, in 1986, it was generalized and popularized by [Rumelhart '86]. At the time of writing this part, it is the backbone of deep learning. It is not a complex algorithm to understand, and we will try our best to explain it clearly here, so do not worry if you don't understand it the first time. Try reading this section more than once. Backpropagation is nothing more than applying a chain rule on the neural network neurons to change weights to reduce the loss score.

We have explained that the objective of the cost function is to change weight and biases toward improving the output's accuracy (reducing the loss score) and thus making a better prediction. How does the neural network change weight and biases to improve its accuracy? Starting from the output, it *goes back* to the network after each epoch and then reconfigures weight and biases to reduce the loss score.

Before beginning the Backpropagation explanation, take a look at Figure 10-17 (a). We did not add hidden layers for the sake of simplicity. In this Figure, we have two input neurons and one output neuron. The output of this network provides a loss score, i.e., error. The next epoch should reduce this error by changing the weights w_1 and w_2 values. Now, the question is, which one of these input nodes contributes to the error: input A or input B, or both?

We could say both A and B contribute to the error, but the amount of their contributions is estimated based on their weights. In other words, neurons' contributions to the error will be measured by their weights. For example, in Figure 10-17 (a), the contribution of input neuron A

---

[10] https://deepobs.readthedocs.io/en/stable/index.html

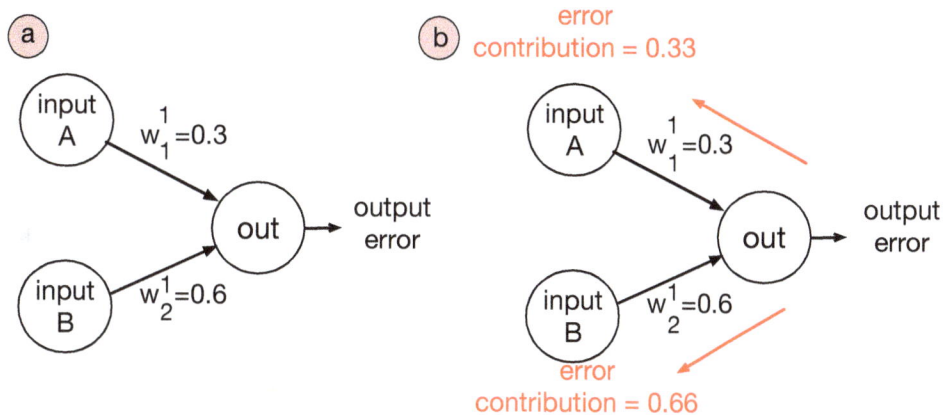

Figure 10-17: (a) Very simple network with two input and one output. (b) Based on the weight of each node in the previous layer, the contribution of each node to the output has been identified.

to the error can be calculated as $\dfrac{0.3}{0.3 + 0.6} = 0.33$, and the contribution of input neuron B will be calculated as $\dfrac{0.6}{0.3 + 0.6} = 0.66$.

Figure 10-17 (b) presents the contribution of each neuron to the error (based on their weight) in red color. We can see in Figure 10-17 (a) that we have used weights to forward the signal to the output layer, which is referred to as the *forward step*. Next, after the output error has been identified, a neural network uses the weight to propagate back the signal from the output layer to the input layer, as shown in Figure 10-17 (b). This process is called *Backpropagation* or *Backprop*. To summarize, the Backprop algorithm splits the error of output neurons across the previous neurons proportional to the incoming weights to this neuron.

Note that while the Backpropagation algorithm is propagating back the error, input values do not change, and the only thing that will be changed is the weights used in the hidden layers. In this example, for the sake of simplicity, we show only two neurons, and hidden layers are not separated from input layers. The network of Figure 10-17 is too simple, and we can extend the described approach to hidden layers as well.

The Backpropagation algorithm assumes the error in a neuron (hidden or output) is the sum of the splits errors in all other previous nodes linked to this neuron.

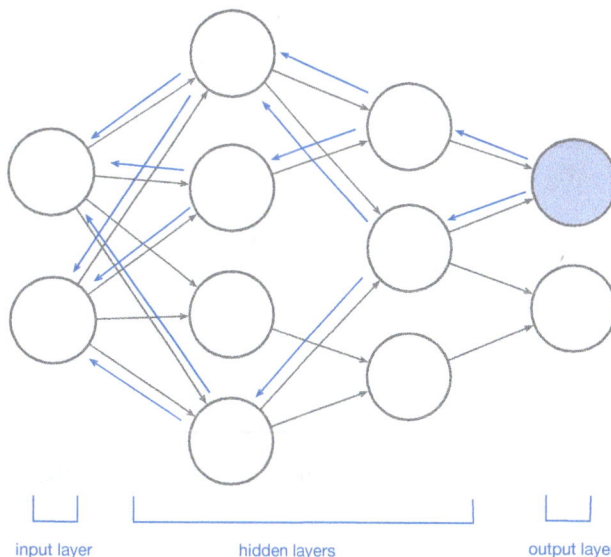

Figure 10-18: Propagation of error from the blue output neuron to the input neurons is shown in blue color.

563

For example, the blue lines in Figure 10-18 visualize the propagation of the error from the blue output neuron to the first layer neurons. If there is more than one hidden layer (as the case in Figure 10-18), the errors of hidden layers split again proportional across all previous links between input and hidden layers connected to that particular neuron.

The weights of all links are known, and with the same mathematical approach, we can calculate the contribution of each input neuron to the output's neuron error. If you believe it helps you to understand the Backpropagation better, take some pen and paper, assign random weights to each node and a random error to the end node, then try to calculate each node's contribution to the error until you reach an input layer.

In summary, the Backpropagation specifies the contribution of each neuron to the error, and in the next epoch, weights will be changed to reduce the output error.

Matrix multiplication is used to implement backpropagation[11]. Therefore, the errors of layer n could be written as a matrix (e_n), which is the multiplication of the transpose of weight matrix (w^T) time matrix of errors for the next layer (e_{n+1}), as follows: $e_n = w_{n+1}^T \cdot e_{n+1}$

For example, we have a simple network like Figure 10-19, the output layer errors are $e\ out_1$, $e\ out_2$, which are specified at the end of each epoch, and thus, the errors of the hidden layer (e_h) can be calculated as follows:

$$e_h = \begin{bmatrix} e\ h1 \\ e\ h2 \end{bmatrix} = \begin{bmatrix} w_1^{(2)} & w_2^{(2)} \\ w_3^{(2)} & w_4^{(2)} \end{bmatrix} \times \begin{bmatrix} e\ out_1 \\ e\ out_2 \end{bmatrix}$$

Respectively, errors of the input (e_{in}) layer are calculated as follows:

$$e_{in} = \begin{bmatrix} e\ in1 \\ e\ in2 \end{bmatrix} = \begin{bmatrix} w_1^{(1)} & w_2^{(1)} \\ w_3^{(1)} & w_4^{(1)} \end{bmatrix} \times \begin{bmatrix} e\ h_1 \\ e\ h_2 \end{bmatrix}$$

We have learned that the Backpropagation error is presented and calculated as matrix multiplication. Now, that big fat question arises again: How do we update weights to reduce the error (loss score)?

By looking at Figure 10-18, we can see that many neurons are connected to a single output. Therefore, it is not a trivial mathematical process to identify a better weight assignment for the next epoch. We need to go deeper into mathematical explanation, but first, we should review some conventions and concepts.

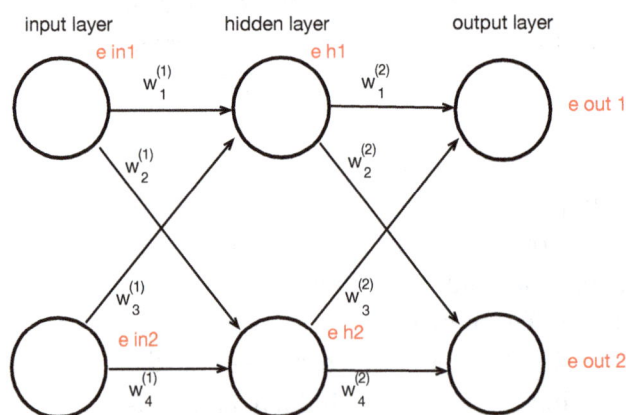

Figure 10-19: A simple neural network with one hidden layer.

---

[11] Matrix multiplication is also implemented with inner loops, but this is very resource-intensive. There are other approaches that make it more efficient to implement, and if a new matrix multiplication is discovered, the resources used by GPU will be drastically reduced.

The output of the neuron in the first layer is written as $z = wx + b$. Here x is the input, but in the other neurons after the input layer, we do not have x. Instead of input neuron data, we only have the output of the activation function. Therefore, we refer to the neuron output as z. In other words, $z^{(l)}$ is defined by weight and biases at level l for the given input that comes from the previous layer.

In the context of a neural network, the predicted value presents the activation function result in layer l of neuron number j, which is presented as $a_j^{(l)}$. We can generalize it (removing layer and node information) and write it as $a = \sigma(z)$, here, z presents the output of the previous layer, σ presents the activation function, and thus we can formalize as follows: $z^{(l)} = w^{(l)}a^{(l-1)} + b^{(l)}$.

For example, if we refer to the very last layer of the network as L, and for the last layer we have $z^{(L)} = w^{(L)}a^{(L-1)} + b^{(L)}$, we know that $a^{(L)} = \sigma(z^{(L)})$.

We have explained that *error = predicted — actual*. Assuming the actual value of neuron j is presented as y_j, and the error is presented as E we can write the following equation for an error of neuron j and the last layer (output layer) L: $E_j^{(L)} = a_j^{(L)} - y_j$ also note that $\hat{y}_j = a_j^{(L)}$. We consider them as a matrix (generalizing it), and thus remove the jth parameter, and end up with $E^{(L)} = a^{(L)} - y$, to calculate the error matrix in the last layer L.

Figure 10-20, presents a concept called *Computation Graph*. It is used to visualize the chain rule with a partial derivative. If you can't recall what is partial derviative, check Chapter 8 we described it there. Assume we have $y = g(f(x))$, and by using the chain rule, we can present the partial derivate of y over x as follows:

$$\frac{\partial y}{\partial x} = \frac{\partial y}{\partial g} \times \frac{\partial g}{\partial f} \times \frac{\partial f}{\partial x}$$

Now that we have learned enough conventions and concepts let's get back to what we need to understand for Backpropagation. We should understand *how* much loss score (or error) changes as weight and biases change. Or how sensitive is the loss score to changes in w and b?

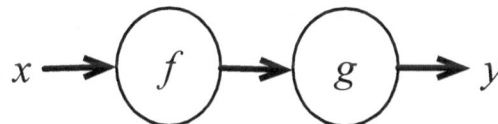

Figure 10-20: A very simple computation graph for a nested function $y = g(f(x))$.

We can write the error equation as a partial derivative of the error to the weight at layer l, i.e., $\frac{\partial E}{\partial w^{(l)}}$. In other words, we are calculating the error with respect to the weights at layer l. Based on the mathematical notation we have described above, and by using the "chain rule" of derivative (again, say hello to Chapter 8), we can rewrite $\frac{\partial E}{\partial w^{(l)}}$ with the partial derivative as follows:

$$\frac{\partial E}{\partial w^{(l)}} = \frac{\partial E}{\partial a^{(l)}} \times \frac{\partial a^{(l)}}{\partial z^{(l)}} \times \frac{\partial z^{(l)}}{\partial w^{(l)}}$$

If you take a look at Figure 10-20, this equation seems easy to understand. It just applies chain rules and simply uses z and a to find how sensitive the loss score (E) is to changes in weight (w). However, in this equation, we only considered weights; for biases, we can use the same equation as follows:

$$\frac{\partial E}{\partial b^{(l)}} = \frac{\partial E}{\partial a^{(l)}} \times \frac{\partial a^{(l)}}{\partial z^{(l)}} \times \frac{\partial z^{(l)}}{\partial b^{(l)}}$$

By using a Hadamard product (it is denoted with ⊙ sign, and you can check Chapter 7 to recall it), these two equations will be written for one single layer, and we can do this for the entire network. These equations use gradients to go back through the network and adjust weight and biases to minimize the output error at the output layer. By output error, we mean a vector of errors (or loss score).

Assuming our network has L layers and the training dataset has m data points, the following pseudocode describes the Backpropagation algorithm. While reading this algorithm, note that the # sign is used for commenting and is written in blue.

```
# -------Step 1 (initialing parameters)-------
# This step initializes the learning rate, weights, and biases, and also, the
threshold for stopping criteria will be specified.
```

initialize w, b, α & stop_criteria # `α is learning rate`

for i = 1 to m {# `m is the number of data points in the training set`
```
    #------Step 2 (forward propagation)------
    This step computes the activation for all layers.
    a is the predicted value extracted from the activation function and x is
    the input variable.
```

    for j = 1 to L {# `L is the number of layers`

        if (j=1) then $a^{(1)} = x^{(i)}$ `# Since the activation function does not exist at`
            `the first layer, which is the input layer, at this layer` $z = wx + b$
            `and` $a^{(1)}$ `is the input.`

        else $a^{(j)} = \sigma(z^j)$ `# Now there is no x available (because we are talking`
            `about layer 2 and other layers) for each layer, the algorithm`
            `computes z and a, recall that` $z^l = w^l a^{l-1} + b^l$

    }

```
# -------Step 3 (Calculate error vectors)------
```
$E^{(L)} = a^{(L)} - y^{(i)}$ `# In the last layer, we know y, which is the actual output.`
for k = (L-1) to 2 {# `this loop computes other error vectors` $(E^{(L-1)}, E^{(L-2)}, \ldots E^{(2)})$
    `for other layers (except the last one) until it reaches layer 2, there is`
    `no error for layer 1, because the input layer does not have an error.`

    $E^{(k)} = (w^{(k+1)})^T \times E^{k+1} \odot \sigma'(z^{(k)})$ `#` ⊙ `presents a Hadamard product of two`
    `matrices.` $(w^{(k+1)})^T$ `is the transpose of weights for the next layer (to`
    `prepare them for matrix operation, they are transposed).` $\sigma'(z^{(k)})$ `is a`
    `vector of derivatives of activation function results at layer` *k*`.`
    ⊙ $\sigma'(z^{(k)})$ `moves the error backward through the activation function in`
    `layer` *k*`.`

}

[The algorithm continues on the next page]

```
#------ Step 4 (compute gradients and update weight and biases) ------
# This step computes partial derivation of the gradient with respect to
weight and biases.
```

for all (w and b) {

$$w_{new}^{(l)} = w_{old}^{(l)} - \alpha \frac{\partial E}{\partial w_{old}^{(l)}} \quad \# \quad \frac{\partial E}{\partial w_{old}^{(l)}}$$ is the partial gradient of the cost function

result (loss score) with respect to w. At the first run, we assume it is 0. This gradient will be acquired using the chained rule of partial derivative.

$$b_{new}^{(l)} = b_{old}^{(l)} - \alpha \frac{\partial E}{\partial b_{old}^{(l)}} \quad \# \quad \frac{\partial E}{\partial b_{old}^{(l)}}$$ is a partial gradient of the cost function

result (loss score) result with respect to b. At the first run we assume it is 0.

}

} # closes 'for i = 1 to m' loop

We can see that the partial gradients with respect to weights and biases are computed using the chain rule. The loss score is calculated by using a cost function. For example, a cost function could be binary cross-entropy, $\hat{y}$ is the predicted and y is the actual value; the loss score after one pass on the network will be calculated as follows:

$$Loss = - [y \, log(\hat{y}) + (1 - y)log(1 - \hat{y})]$$

For the sake of simplicity, we did not incorporate regularization in step 4, but regularization can be involved as well.

We can summarize that a neural network is a chain of differentiable functions. Differentiable functions mean we can get a derivative of it, and the parameters of these functions can be trained. The process of training involves computing the gradient of network parameters (for each batch) with respect to the cost value (for each batch). Therefore, we can say a neural network is a combination of nested functions, $F(X) = f_1(f_2(f_3(x)))$, each of the f functions could be a vector function $f_l(Z) = a(x\,w + b)$, a is the activation function, l is the layer index, and it could span from 1 to any number of layers. More about this will be explained later in the Backpropagation section.

Forward and Backward Pass Example

If you want to understand the forward and backward pass on the network with mathematical examples, read this section. Otherwise, if you find it too complex, feel free to skip it.

Consider a forward pass starting from x_0 as input and three layers with activation function σ_R we can formalize the forward pass f follows $f(x_0; w) = \sigma_R(w_3 \cdot \sigma_R(w_2 \cdot \sigma_R(w_1 x_0 + b_1) + b_2) + b_3)$. To perform the backward pass and compute the gradients, the Backpropagation algorithm uses the chain rule of differentiation. It starts by computing the gradient of the loss with respect to the output of the network:

$\nabla f = \nabla \sigma_R(w_3 \cdot \sigma_R(w_2 \cdot \sigma_R(w_1 x_0 + b_1) + b_2) + b_3)$ where ∇ denotes the gradient and σ_R is the activation function.

Next, it applies the chain rule to compute the gradient of the loss with respect to the parameters of the last layer:

$$\nabla w_3 = (\nabla f) \cdot \sigma_R'(w_3 \cdot \sigma_R(w_2 \cdot \sigma_R(w_1 x_0 + b_1) + b_2) + b_3) \cdot \sigma_R(w_2 \cdot \sigma_R(w_1 x_0 + b_1) + b_2)$$

$$\nabla b_3 = (\nabla f) \cdot \sigma_R'(w_3 \cdot \sigma_R(w_2 \cdot \sigma_R(w_1 x_0 + b_1) + b_2) + b_3)$$

Here σ_R' is the derivative of the activation function. Next, it uses the chain rule again to compute the gradient of the loss with respect to the parameters of the second layer:

$$\nabla w_2 = (\nabla f) \cdot \sigma_R'(w_3 \cdot \sigma_R(w_2 \cdot \sigma_R(w_1 x_0 + b_1) + b_2) + b_3) \cdot w_3 \cdot \sigma_R'(w_2 \cdot \sigma_R(w_1 x_0 + b_1) + b_2)$$
$$\cdot \sigma_R(w_1 x_0 + b1)$$

$$\nabla b_2 = (\nabla f) \cdot \sigma_R'(w_3 \cdot \sigma_R(w_2 \cdot \sigma_R(w_1 x_0 + b_1) + b_2) + b_3) \cdot w_3 \cdot \sigma_R'(w_2 \cdot \sigma_R(w_1 x_0 + b_1) + b_2)$$

Finally, it computes the gradient of the loss with respect to the parameters of the first layer:

$$\nabla w_1 = (\nabla f) \cdot \sigma_R'(w_3 \cdot \sigma_R(w_2 \cdot \sigma_R(w_1 x_0 + b_1) + b_2) + b_3) \cdot w_3 \cdot \sigma_R'(w_2 \cdot \sigma_R(w_1 x_0 + b_1) + b_2)$$
$$\cdot w_2 \cdot \sigma_R'(w_1 x_0 + b1) \cdot x_0$$

$$\nabla b_1 = (\nabla f) \cdot \sigma_R'(w_3 \cdot \sigma_R(w_2 \cdot \sigma_R(w_1 x_0 + b_1) + b_2) + b_3) \cdot w_3 \cdot \sigma_R'(w_2 \cdot \sigma_R(w_1 x_0 + b_1) + b_2)$$
$$\cdot w_2 \cdot \sigma_R'(w_1 x_0 + b1)$$

Regularization in Neural Network

A successful machine learning algorithm should avoid overfitting. We can think of overfitting as just memorizing the data (not learning) and matching everything to the training dataset. Therefore, if new data arrives in the system (test data) that does not match the existing data, the algorithm can not determine a label for it.

As we have described in Chapter 8, to avoid overfitting, we use regularization. We describe popular methods that used regularization in ANN, but in addition, to resolve overfitting, they can handle vanishing and exploding gradient problems as well.

Vanishing and Exploding Gradients

We have learned that a neural network operates in three steps. First, it begins from the input layer and goes through hidden layers to reach the output. This process is known as forward pass. Second, the loss function compares the output (e.g., prediction) result with the actual label and measures the error. In simple words, the output of a loss function is an error value. In the third step, the neural network uses the error value and backpropagation algorithm to calculate the gradient for each neuron in the network. Gradients (a vector of partial derivatives with respect to weights and biases) are used by the network to adjust its weights and biases to reduce error.

As the network moves from the output toward the input layer, the gradient gets smaller by the chain rule, and thus, the weight adjustment gets smaller. In a more technical sense, the gradient is exponentially shrinking as the backpropagation moves toward the input neuron. Figure 10-21

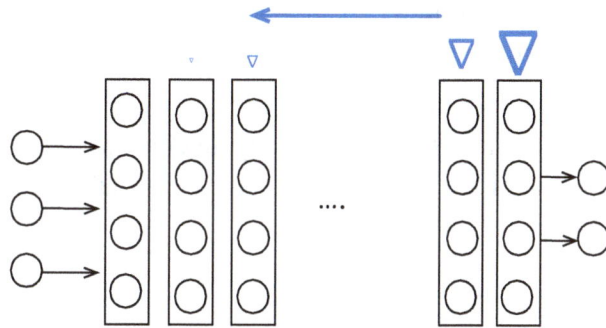

Figure 10-21: The gradient (blue triangle) is getting smaller and smaller in the back propagation that at some points get useless and weights do not change at all.

visualizes this phenomenon. A high gradient reveals a big adjustment, while a small gradient indicates a minor adjustment in weight.

When a network is deep (with many hidden layers), the gradient gets smaller and smaller until it vanishes, and thus, weights on the layers close to the input layer never get updated. For example, if the gradient of the last layer close to the output layer is 0.5 or less than one, as we go deep in the network, it gets smaller and smaller because it gets multiplied by some smaller number. For example, it will be $0.5 \times 0.4 \times 0.1 = 0.02$. Then, the weight in the input layer will be $w_{new}^{(l)} = w_{old}^{(l)} - \alpha \times 0.02$. Since we know a learning rate is also a small number, such as 0.01, we have $w_{new}^{(l)} = w_{old}^{(l)} - 0.0002$, which means the new weight has very insignificant differences from the old weight; this is known as the *vanishing gradient*. The same thing could happen with gradients larger than one, and thus, it makes a very big gradient that does not let SGD get close to minima, which is the *exploding gradient*.

In summary, the vanishing gradient refers to a problem in deep neural networks; the gradient in backpropagation is getting too small that the weight doesn't get updated at all. This is not good because some weights, especially weights close to the input layer, didn't change by backpropagation, and only weights close to output layers get updated.

There are approaches that do not solve this problem but try to mitigate them. One approach is the use of ReLU or Leaky-ReLU as activation functions and avoiding using hyperbolic tangent as activation function, which is prone to the vanishing gradient. In the following, we briefly list common approaches.

Weight Initialization

There are two approaches that use a more subtle weight initialization and do not perform the weight initialization completely randomly at the beginning. These approaches are known as Xavier [Glorot '10] and He [He '15].

Xavier or Gorlot initialization uses random weights with a normal distribution and not completely random numbers. In particular, uses random weights drawn from a normal distribution with a mean of 0 and a variance of $2/(n_{in} + n_{out})$ where n_{in} presents the number of input neurons and n_{out} the number of output neurons.

He initialization (a.k.a Kaiming initialization) proposes a weight initialization for non-linear activation functions such as ReLU and Leaky ReLU while considering the non-linearity of non-linear activation functions. Weights are initialized to have a normal distribution, with zero mean and variance of $2/n_{in}$, and biases are initialized to zero.

Batch Normalization

Back in Chapter 3, we described the standardization and normalization approaches. Also, in the Clustering chapter (Chapter 4), we provide an example that all data should be in the same range to be able to cluster them. If we intend to have a neural network that handles different numerical ranges, we should normalize them. For example, we would like to have a classification algorithm to predict the happiness of our users based on their age and annual income. The age is a number between 0 and 120, but income is widely varied. For example, in the U.S., it could be from 10,000\$ per year to >10,000,000,000\$. Therefore, we need to bring these variables into the same range. In the neural network, Batch normalization is a bit different from the previous normalization methods we have learned.

Batch Normalization or *Batch Norm* normalizes the activations of a previous layer at each batch, meaning it applies a transformation that maintains the mean output close to zero and the output standard deviation close to one. It is done at the input layer and hidden layers after intermediate layers within the network. Batch normalization helps speed up training, reducing the sensitivity to network initialization and mitigating the vanishing/exploding gradient problem.

Batch normalization was introduced in 2015 by Loffe and Szegedy [Loffe '15]. It adjusts the outputs of the previous layer to have a mean close to 0 and a standard deviation close to 1 based on the current mini-batch. Keep in mind that *it's about the activations, not the weights directly.*

The process of Batch normalization is implemented as follows:

(i) Assuming that m is the batch size of the input and x_i is the ith example in this batch, first, mini-batch mean (μ_B) and mini-batch variance (σ_B^2) for each input batch will be calculated, independently.

$$\mu_B = \frac{1}{m} \sum_{i=1}^{m} x_i$$

$$\sigma_B^2 = \frac{1}{m} \sum_{i=1}^{m} (x_i - \mu_B)^2$$

(ii) Next, the output of an activation function will be substituted with its z-normalized value before being passed to the next layer. In Chapter 3, we described the z-score, written as:

$$\hat{x}_i = \frac{x_i - \mu_B}{\sigma_B}$$

(iii) After the value (output of activation function) of each neuron is normalized, then it multiplies the output by an arbitrary parameter (scale) γ and adds another arbitrary parameter (shift) β, to the result, as follows:

$$y_i = (\hat{x}_i \times \gamma) + \beta$$

Similar to weight and bias parameters, both of these parameters (scale and shift) are learned during the training process, and the optimizer configures them. Therefore, we can write the equation of batch normalization as follows:

$$y_i = \gamma \frac{\hat{x}_i - \mu_B}{\sigma_B} + \beta$$

Once again, let us remind you that the Batch norm is applied to the outputs of the input layer and the output of the hidden layers.

Now that we understand batch normalization, a question might arise: Why do we need batch normalization and not just use traditional normalization? The problem is that traditional normalization may not go far in a deep neural network, and it is still prone to several issues. One of these issues is referred to as *covariate shift*, which states that the distribution of original input data as it proceeds through its hidden layers will change. In other words, input values in each layer are scaled by trainable parameters, and as parameters get turned back by the backpropagation algorithm, the original distribution of input data will change. Batch norms can mitigate this issue by updating the network's parameters, thereby stabilizing the training process and often resulting in faster convergence.

Gradient Clipping

Another approach is Gradient clipping, which cuts off gradients before they reach a predefined limit. This limitation can be specified by a transformation and normalizing the range of the gradient. For example, we cut off all gradients larger than a specific threshold, e.g., 1, or smaller than a specific threshold, e.g., -1. Then, all gradient values above 1 are set to 1, and all values below -1 are set to -1.

Gradient clipping can be applied to any type of neural network where the risk of exploding gradients is a concern. However, Gradient Clipping is commonly used in RNN networks, which we will explain later.

Other Normalization Techniques

In addition to Batch Normalization, there are other kinds of normalizations, including *Instance normalization* [Ulyanov '17], *Layer Normalization* [Ba '16], and *Group Normalization* [Wu '18]. Before we explain them, we should know that the input data fed into a neural network is in tensor format, and it has height, width, and depth (channel). We will learn more about input data later.

Instance Normalization is a type of batch normalization that is applied for a specific instance. For example, if the neural network applies a Batch Norm on a set of images while instance normalization applies the normalization on every single image. In other words, it treats *each input sample separately for normalization.* Therefore, mean and variance are calculated for each channel of an individual sample across both *x* and *y* (spatial) dimensions and not for batches of input samples.

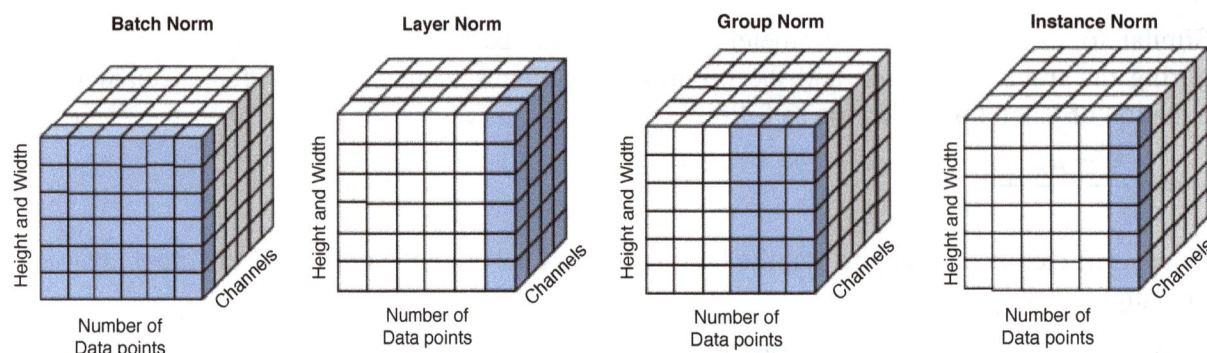

Figure 10-22: Visualization of different neural network normalization methods.

Layer Normalization tries to address the Batch Norm's dependency on the batch sizes, which is particularly challenging for RNN networks (we will explain RNN later in this Chapter). To handle this limitation, Layer Normalization [Ba '16] introduces a *normalization across channel dimensions*. Similar to instance normalization, layer normalization normalizes the data across all the features (channels) within a layer for each individual sample rather than across different samples in a batch as in batch normalization.

Group Normalization tries to address Batch Norm's need for large memory because of a need for a large set of batches [Wu '18]. Group Normalization divides the channels into groups and computes the mean and variance of the normalization in each group, which makes it independent from batch sizes. Figure 10-22, is designed by Wu et al. [Wu '18] to visualize these normalization approaches.

Dropout

One of the most popular neural network regularization methods introduced by Hinton et al. and popularized by Srivastava [Srivastava '14] is Dropout. The idea of dropout is simple but very effective. During training, a random subset of neuron outputs in a layer is set to zero in each epoch, reducing their contribution to the next layer temporarily. It is typically implemented by adding a dropout layer to the network. For example, a network might have 50% of its neurons in the first hidden layer 'dropped out' in each iteration to prevent overfitting. While dropout rates commonly range from 20% to 50%, the optimal rate depends on the specific context and network architecture. Dropout ensures that the network's predictive performance does not rely too heavily on any single neuron, promoting more robust learning. Figure 10-23 presents a simple example of a dropout regularization. There, we have a neural network, and in every iteration, 50% of its first layer of hidden neurons are turned off (their weights will be equal to zero).

Gal and Ghahramani [Gal '16] introduced an extension called *Monte Carlo (MC) Dropout*, which applies dropout during both the training and test phases to obtain a measure of predictive

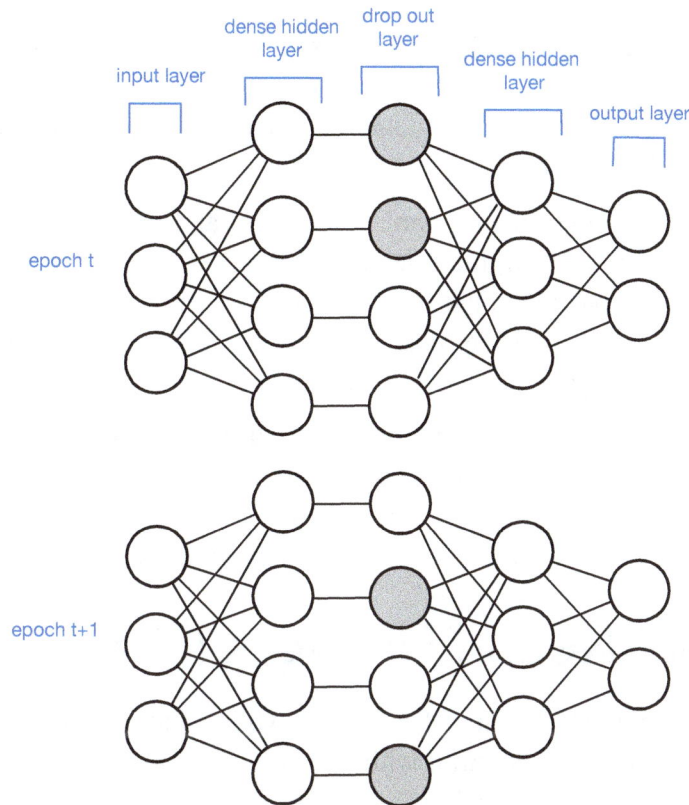

Figure 10-23: A mock example of having dropout layer after the first hidden layer. Nodes which are converted to zero are presented in grey color and this layer in each epoch turns off (make their activation function output zero) 50% of nodes.

uncertainty. In MC Dropout, the network makes multiple predictions for the same input with different neurons dropped out each time, and the results are averaged to estimate the final output. This approach can enhance the model's ability to estimate its uncertainty but should be used carefully in applications where accurate uncertainty estimation is critical, such as in medical diagnostics.

Early Stopping

We have explained that training a neural network is a computationally expensive process. To manage this, we employ regularization methods, which help prevent overfitting and can also indirectly reduce computational costs by promoting simpler models. During training, we monitor the loss, and once it reaches an acceptable level, we can halt the training process. One popular regularization technique is *early stopping*, which we detailed in Chapter 8. Early stopping involves terminating the training when the model's performance on the validation set starts to deteriorate, thus preventing overfitting and potentially saving computational resources. There is not much more to add here except to reiterate that early stopping can be effectively implemented in neural network training to optimize resource usage.

NOTES:

* We have explained that neural networks suffer from high computational complexity, which makes them unsuitable for settings with limited computational capacities, such as battery-powered devices. There are several different approaches to making neural networks lighter, and we will explain them in Chapter 14.

* There is no optimal way to design the number of layers and neurons, and since there is no methodological approach to doing it, the selection of neurons and layers is referred to as 'dark art'. A simple common approach is to use a large model, which is larger than our need, and use dropout to prevent overfitting. This approach is called 'stretch pants', in which we do not search for a pant that fits; we get a large pant that stretches and fits our size.

* A neural network has many hyperparameters to tune, such as learning rate, number of hidden layers, number of neurons, cost function, types of activation function, optimization algorithm, batch size, etc. It is common to rely on some predefined settings, but you might change a parameter and make some significant advancements; nobody knows until you experiment.

Convolutional Neural Network (CNN)

We have covered the general architecture of neural networks, and now we turn our attention to specific deep learning architectures, starting with Convolutional Neural Networks (ConvNets or CNNs), which are pivotal in computer vision.

Back in 1950, Hubel and Wiesel [Huhel '59] identified that monkeys' and cats' brains have two types of cells in their brains (simple cells and complex cells) to be used for visual recognition. Later, Fukushima [Fukushima '80], inspired by the findings of Hubel and Wiesel, introduced the *Neocognitron*, a hierarchical, multilayered artificial neural network, which is considered a precursor to modern CNNs. Afterward, LeCun et al. [LeCun '89] combined the CNN architecture with the Backpropagation algorithm and experimented with it on handwritten numbers[12]. Their results showed a tremendous improvement in the accuracy of handwritten number recognition.

CNN

Traditional Computer Vision Algorithms

---

[12] There are some very popular datasets used to evaluate computer vision algorithms. One is a handwritten number dataset called MNIST (stayed for Modified National Institute of Standards and Technology database), which to date is the most popular image dataset in use for experimenting with different types of machine learning algorithms, especially image recognition ones. There are a few popular datasets, including ImageNet [Deng '09], mtCars (vehicle data), IRIS (flower shape), CIFAR-10/CIFAR100 (tiny images of different objects), and Fashion MNIST (cloths images), which are used for experimenting or benchmarking machine learning algorithm.

CNN architecture preserves the spatial structure of the input data. Preserving the spatial structure of data makes CNN very helpful for computer vision applications such as image classification, image segmentation, medical image analysis, etc. CNNs are used in a few other applications, such as natural language processing or audio analysis. Nowadays, most advances in computer vision include CNN or the use of neural networks.

CNN models are spatial invariant, which means they can recognize patterns in an image regardless of their position or orientation. They are flexible in tolerating distortion and changes in images. These features make them more accurate than the traditional image feature engineering we have described back in Chapter 6. However, CNNs are not scaling or rotation invariant.

At the time of writing this chapter (first time in 2021, and revision in 2024), most state-of-the-art image analysis algorithms are using CNN, and they even overtake human experts in some computer vision applications such as medical image analysis [Javaheri '21]. CNN operates by assigning multiple image filters to an input image, and these filters enable the algorithm to classify the image accurately.

CNN has two significant advantages over ANN. First, ANN uses densely connected layers, each neuron in CNN is connected to a limited number of neurons in the neighbor layer. Figure 10-24 shows a comparison between CNN and ANN, and we can see that all input neurons of the ANN are connected to all neurons in the next layer. Having this dense connection causes ANN to learn only *global patterns* in the image (the result of having densely connected layers where every output neuron is connected to every input neuron). However, CNN does not

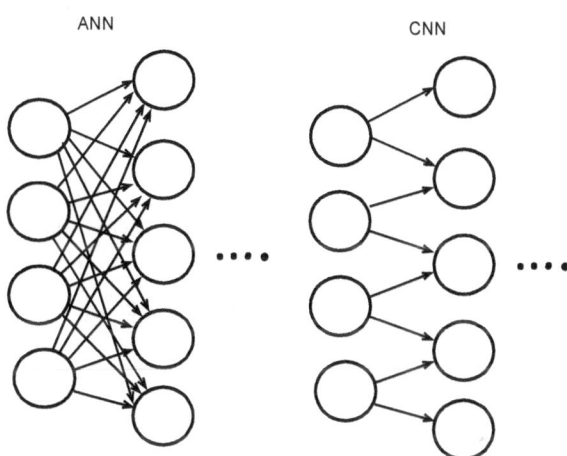

Figure 10-24: A traditional input layer and the second layer of ANN on the left and the same approach for CNN on the right, we can see the number of connections in CNN is smaller than ANN, which is fully connected.

use densely connected layers, and by using a window over the image, it can learn the *local patterns*. The size of the window and the movement steps, or 'stride', are key factors in defining the extent of these connections, which we will explore in detail later.

To understand the importance of local patterns, let's discuss an example in the MNIST dataset. If the handwritten number is not in the center of the image, ANN cannot classify it, but CNN can classify it, even if the number is located on the side of the image, because of its local pattern learning capability.

The second advantage of CNNs is their ability to learn hierarchical features within an image. For instance, in analyzing an image of a chicken, a CNN might first identify basic features like edges and curves, then assemble these into more complex structures such as beaks, eyes, and wings, and finally recognize the overall form of the chicken. This hierarchical feature learning enables

CNNs to understand complex structures and classify images more effectively, even when they vary in perspective or lighting conditions. Thus, CNNs are capable of distinguishing and classifying images in a way that is challenging for other computer vision algorithms, demonstrating their robustness across different viewing conditions.

Typically, a CNN network takes an input of a tensor, which includes image height, width, and channels (three data for red, green, and blue in RGB mode or three data for hue, saturation, and lightness in HSL mode). We can say that we work with a 3D tensor for colored images. For the sake of simplicity, we explain our example with a matrix and one value for each pixel.

CNN algorithm operates in four steps, but before starting to explain these steps, we need to learn the definition of convolution. Then, we proceed with the explanation of these four steps in a sequence.

Convolution and Cross Correlation

Convolution is the process of combining two functions and, as a result, getting a new function with a different function shape than the two input functions. In the context of neural networks, convolution is the process of multiplying matrices (or tensors), and the result will be a new matrix (or tensor). Mathematically, convolution (*) is written as the integral product of two functions after one of these functions is reversed and shifted. The convolution specifies the amount of overlap of one function as it is shifted over another function. The convolution operation can be described using the following equation:

$$(g * f)(t) = \int_{-\infty}^{\infty} g(\tau)f(t - \tau)d_{\tau}$$

In this equation, t is the current time and τ is used to specify the size of the shift. The integral measures the overlap between $g(\tau)$ and $f(t - \tau)$ as one function slides over the other. If two functions are not overlapped, their convolution is zero. When a function overlaps with another function, then its convolution is none zero. On the top of Figure 10-25, we have two functions f and g. The bottom plots their convolution. We can see the amount of shift on top and its impact on the convolution at the bottom. The grey area on the top shows the convolution. Note that, as part of the convolution process, the function f is reversed starting from the second figure onward.

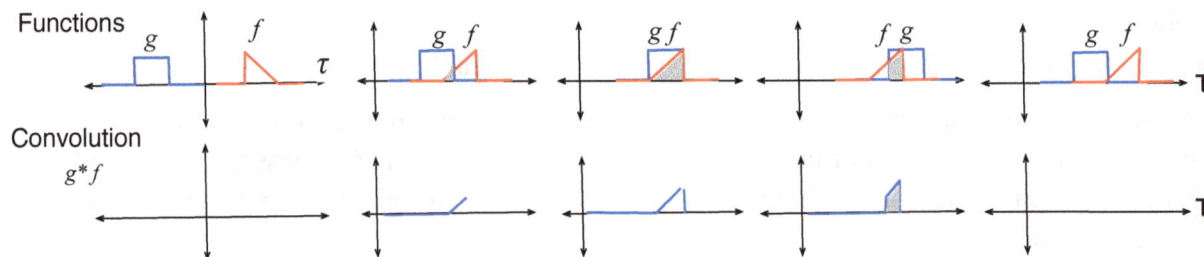

Figure 10-25: Two functions at the top and their convolution function in the bottom. The g function is moving along τ-axis and as it overlaps with the function f (the grey area is the overlapped area) the convolution starts to increase and then again decreases as the g is moving away from f. Note that the function g is also reversed, but its shape doesn't changed.

Cross-correlation is similar to convolution, but it measures how similar two different functions are. In other words, while both operations involve sliding one function over another and computing integrals or sums of their products, the key difference is that convolution flips one of the functions before this process. Specifically, in convolution, one of the functions is reversed along one of its dimensions (flipping means taking the negative of its indices), while cross-correlation does not perform this flipping. e.g., $f(x) = x$ and flipping it will be $f'(x) = -x$.

The following presents an equation of cross-correlation:

$$(g \otimes f)(t) = \int_{-\infty}^{\infty} g(\tau)f(t + \tau)d_{\tau}$$

Usually, when we work with the digital image, the data is considered discrete and multi-dimensional. Assuming f is one function (image), and g is the other function (kernel), the convolution equation for 2D data can be written as follows:

$$(f * g)(i, j) = \sum_{m=-k}^{k} \sum_{n=-k}^{k} f[i - n, i - m] . g[m, n]$$

Respectively, their cross correlation will be written as:

$$(f \otimes g)(i, j) = \sum_{m=-k}^{k} \sum_{n=-k}^{k} f[i + n, i + m] . g[m, n]$$

You don't need to learn the mathematics of convolution in this detail, but as general knowledge, they are good to be known, and you can brag about your knowledge while talking about CNN.

CNN Architecture

To implement a CNN, we should follow five steps, and we describe each step in detail.

Step 1- Convolution: The first step focuses on using kernels (filters) to construct feature maps. This step of applying filters on images is referred to as convolution. In this context, the kernel is a different concept than the kernel we explained for SVM in Chapter 9. The task of kernels or filters in the CNN is feature detection. Kernels are similar to the image filters that we use in

Figure 10-26: An example of image and three filters that are applied on the image.

photo editing software, such as blurring an image, sharpening, etc. Filters or kernels are small matrices of weights that slide over the input image to extract specific features, such as edges, textures, and patterns. Figure 10-26 shows some examples of filters applied to a sample image.

In the context of CNN, we call these filters *convolutional kernels* (or filters), and the resulting image is a **c**onvoluted image. Therefore, we can write the following:

*input image * kernel (feature detector) = convolved image (feature map)*

There are some common kernels, e.g., a matrix with a Gaussian distribution is a kernel used for blurring an image. A few examples are presented in Figure 10-27. We can see some filters, and intuitively, by looking at the matrices, we can realize what they are doing. For example, the Sharpen filter on the right increases the differences between the right and left center pixels. It reduces the emphasis on the other pixel, which results in sharpening the image. The convolution process usually results in a feature map with fewer pixels than the original image; as we can see in Figure 10-27, the convolved image is smaller than the original image.

Sharpen		
0	-1	0
-1	5	-1
0	-1	0

Blur		
0.05	0.1	0.05
0.1	0.25	0.1
0.05	0.1	0.05

Left Sobel		
1	0	-1
1.5	0	-1.5
1	0	-1

Emboss		
-2	-1	0
-1	1	1
0	1	2

Figure 10-27: Some sample image filters. Sharpen and Blur are clear, Left Sobel is used to show differences between each pixel and its adjacent pixel on the left. Emboss creates an illusion of depth for the viewer of the image.

Take a look at Figure 10-28. We have an image, and for the sake of simplicity, we used a matrix and not a tensor; the process of convolution is a type of matrix multiplication but with a small window (or patch) on the original image. You can see the result of applying the kernel on the original matrix. Using this kernel increases the emphasis on pixels located on the left side of the kernel. The red window of Figure 10-28 is moving along the matrix and constructs pixels of the convolved matrix. The number of pixels that a window moves is called *stride*. In Figure 10-29, we move the window one pixel in the *X* and then *Y* direction.

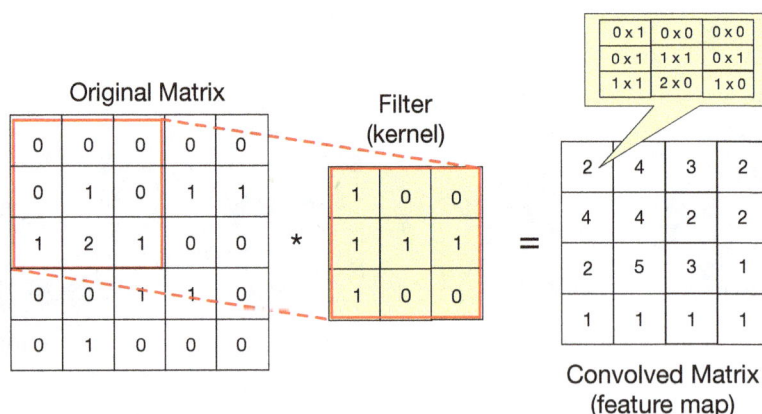

Figure 10-28: A matrix, which presents image and a filter that will be applied on the matrix. The right matrix presents the result of applying the filter on the original matrix.

578

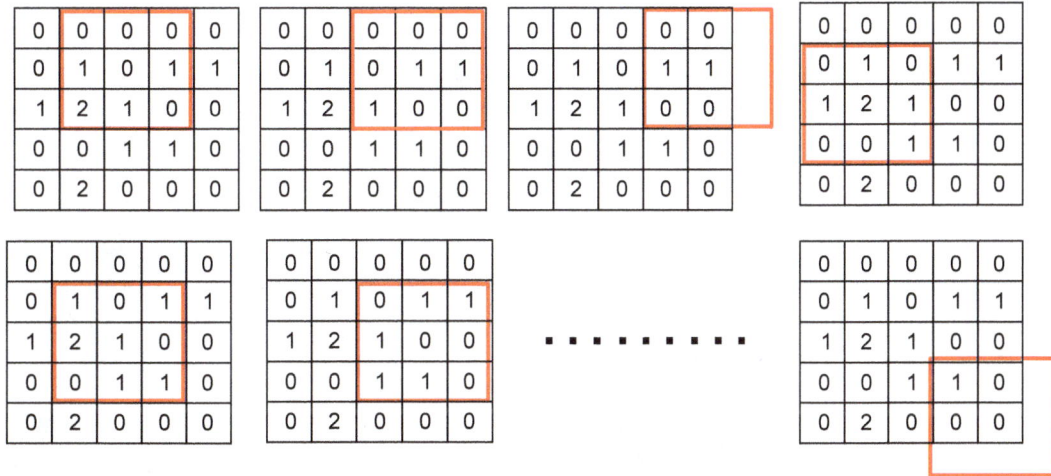

Figure 10-29: An example presented a window equal to the size of the filter (3x3) is moving along the original matrix (image) and select a window of data to multiply them to the filter values. Here, the stride is equal to one on both the X and the Y axis.

Therefore, we say the stride is equal to one. Usually, the stride is set to two pixels, and the larger the stride, the smaller the resulting convolved image becomes.

The '*' operator here refers to convolution (sometimes informally referred to as element-wise multiplication in this context) and not the dot product. A filter is always smaller than the original image.

Another issue is illustrated in Figure 10-29. where the window extends beyond the image matrix, leading to a loss of the image's border in the convolved image. To avoid losing the borders, we can use a process called *padding*. Padding is the process of substituting the pixels on the edges.

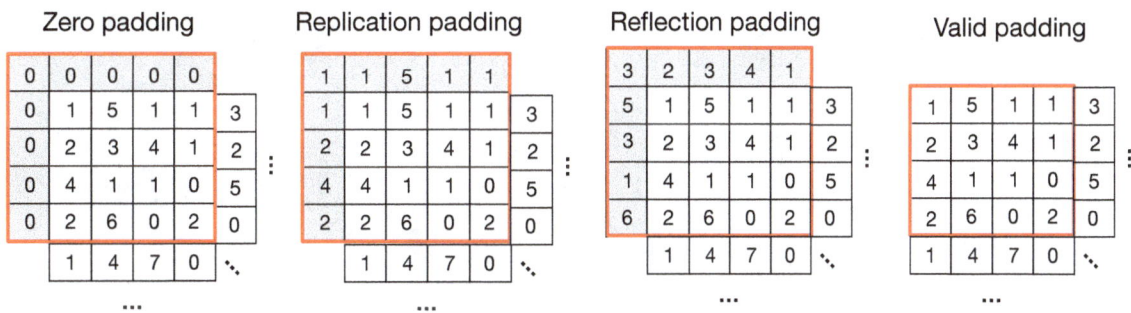

Figure 10-30: Three different approaches for padding on edge pixels. The white area in this figure presents the image pixels and grey ares is the padding area.

There are different approaches for padding, including *reflection*, *replication*, and *zero padding*. Otherwise, the convolutional filter does not go outside the image border; its padding is referred to as *valid padding*. If the output of the convolution has the same size as the input image, its padding is referred to as the *same padding*.

Zero padding (masking) is just adding a zero in the padding area. Replication padding is using the same value for the last pixel in the padding area and substituting it in the padding area. Reflection padding uses the neighbor value of the last pixel on the opposite side of this pixel and substituting it in the padding area. Valid padding happens when the filter window stays inside the image and does not go outside. Figure 10-30 presents these three padding techniques.

Applying image filters to the input image results in having a *convolutional layer* for the neural network. Note, if you encounter the term "local" in the convolutional filter, it refers to the pixels inside a window (the content of red windows in Figure 10-29). Assume that s presents the stride size, p is the padding size for $n \times n$ input matrix, and we apply $f \times f$ filter, the output feature map has the following size: $(n + p - f)/s + 1 * (n + p - f)/s + 1$.

Step 2- Perform non-linear transformation: Filters create a convolved image, which is a linear transformation of the original image. However, an image includes a set of non-linear objects. To increase the non-linearity of feature maps (convoluted images), the feature map will usually be fed into a non-linear activation function such as ReLU to get a non-linear transformation of the image.

Not having a non-linear transformation after the convolution layer results in lower classification accuracy, which this sin is equal to blasphemy in machine learning. In particular, this non-linearity of feature maps (convoluted images) makes the neural network really strong, and this step is called the 'detector step' [Goodfellow '16].

Figure 10-31: All of them are chickens but a spatial variant algorithm cannot detect them. CNN is spatial invariant and it can extract features such as body shape, peak, eyes, etc. and thus recognize them. However, it can recognize them, only if we train the algorithm with enough sample images.

Keep in mind that at the end, the output layer should be something related to the task we expect from the neural network. For example, if it is a classification task for more than two classes, then the activation function of the output layer should be Softmax, for two classes, it should be Sigmoid (check Chapter 8 to recall Softmax and Sigmoid).

Again, let us remind you that unlike ANNs, where all layers are connected to each other (fully connected), CNN neurons are only connected to a subset of neurons in the next layer, and these are filters. Figure 10-24 visualizes the difference between ANN and CNN at the input layer, which we can see in CNN neurons are not connected to all other neurons in the next layer.

Step 3- Pooling: The feature extraction, done automatically by the CNN, is spatial invariant. Spatial invariance makes CNN algorithms very accurate and flexible algorithm for image classification. For example, traditional image feature extraction methods cannot recognize that Figure 10-31 are chickens because the shape, camera angle, etc., are different among all chickens. However, a CNN trained on many comic chickens can recognize the given chicken image despite their different visual shapes.

Each convolutional layer applies different filters on the input image for feature extraction. The number of filters is a hyperparameter and grows between convolutional layers. For example, the first layer has 32 filters, the next layers 64 filters, and so forth. Therefore, a CNN creates many convolved images (feature maps). Dealing with such a large number of data (results of applying many filters) is computationally very ineffective or impossible. To handle this issue, the CNN downsamples (reduce the size) of convolved images while maintaining the highlighted features. The pooling functionality is used for this purpose (downsampling), similar to using window and strides, which downsample the original image into smaller images. The result of the convolutional layer, i.e., feature maps (after they have been transformed into non-linear feature maps), are sent to a pooling layer via a *tensor operation* that performs the downsampling. The pooling layer performs something similar to a kernel function (check Chapter 9 SVM algorithm explanation), kernel function downsamples, by using a linear transformation, but pooling performs downsampling by tensor operation, e.g., Figure 10-32 shows some examples of downsampling via pooling. Downsampling here refers to lowering the filter resolution while still maintaining its features. If you read Chapter 6, you might recall the SIFT algorithm, which blurs and resizes the image for feature extraction. That process can also be called downsampling as well.

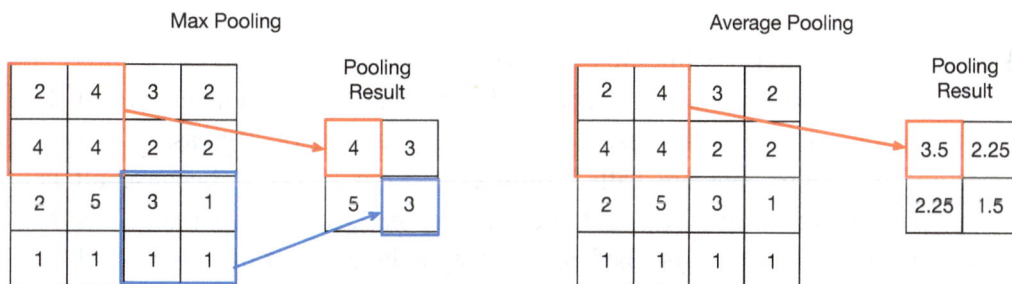

Figure 10-32: Max pooling and average pooling examples are applied to the convoluted image (feature map), and the pooled feature map is created. In these two examples, we use a stride of two pixels in each direction.

Pooling, similar to convolution, slides a window on data, and two types of pooling are commonly used: max pooling and average pooling. *Max pooling* takes the largest element of the window it defines. *Average pooling* averages the content of the window and presents it as a single pixel for the next layer. Figure 10-32 present max pooling and average pooling examples.

There are other pooling methods as well, such as calculating L_2 *norm* for all pixels inside a window, but they are not as popular as max pooling.

Pooling has another advantage: it removes irrelevant information and unnecessary features. It reduces the chance of overfitting (because of reducing the number of parameters) in addition to making the input smaller and thus making the network more resource efficient.

Step 4- Flattening: After the end of the pooling layer, there is usually a normal flattening layer. Flattening is very simple; the result of the pooling layer is a matrix, and the flattening layer converts matrices to a vector, as shown in Figure 10-33. Since we have many pooling layers, the vector size is usually too long. This process occurs before the data is passed into the fully connected layer(s).

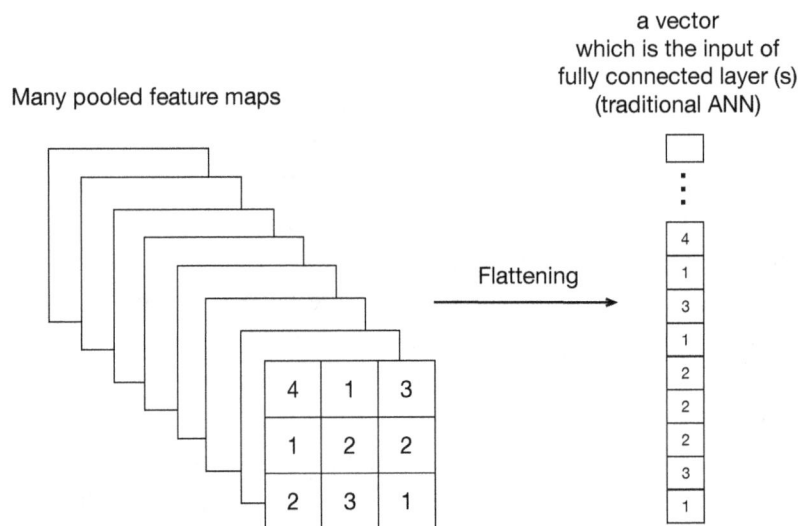

Figure 10-33: The results of max pooling functions are flattened into a large vector, which will be the input of the dense layer (traditional ANN).

Step 5- Dense Layers: After creating this very long vector, it will be fed into a traditional ANN layer. The input layer of this ANN is the result of flattening; its hidden layers are fully connected layers, and the output specifies the result of the classification or regression. The reason that hidden layers of the ANN here are called fully connected layers is because all neurons are connected, as is common with ANN architecture. We have described that CNN neurons are locally connected, but the last layers before the output layer are fully connected. In the fully connected layer, the error is calculated and also backpropagated to the neural network until the weights and biases of the networks converge (maximum iteration reached or accuracy reaches a satisfactory number).

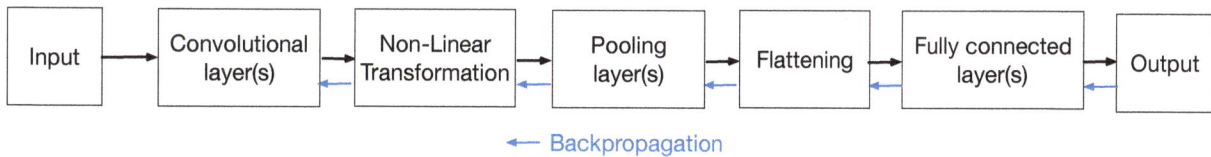

Figure 10-34: A brief summarization of CNN architecture. Here, fully connected layers refer to traditional ANN.

If a feature is useless, its weights get reduced during the iterations of backpropagation. The number of neurons in the output layer is equal to the number of classes; for example, if we make a CNN distinguish whether a given image is Chicken, t-rex, or other, the output layer will have three neurons, one for chicken, one for t-rex and one for other.

Figure 10-34 summarize what we have explained in the CNN architecture. However, there are usually multiple convolutional layers, multiple pooling layers, and multiple fully connected layers. We could even have several convolutional layers and pooling layers in a sequence. Note that the network learns the filters. These filters, which are small matrices of weights, start as a random set of values and are then adjusted through training iterations to help achieve the goals of the network.

A CNN architecture can be used for classification and rarely for regression. If the CNN model is designed for classification, it will have more than one neuron at the end, and for the classification, its output layer neurons are equal to the size of the classes that are used for labeling. For example, if a CNN classifier is trying to distinguish between cat, dog, and t-rex. Then, we will have three neurons. Nevertheless, the neurons in the output layer are unaware of the value of other neurons in the last layer. However, they should provide a probability, and the sum of those probabilities is equal to one. The Softmax function (described in Chapter 8) is used to convert them into a probability and assign a value to them that adds up all of the output neurons to one. In other words, CNN used for classification has a Softmax activation function (or other non-linear activation function) that outputs probabilities for a classification result. The softmax function is a generalization of the logistic function and provides a vector of values with a range between zero and one. The softmax function is usually used with cross entropy as a loss function.

It is not common to use CNN for regression, but if you intend to use a CNN for regression, there will be one output neuron, and its activation function will be linear.

The best way to design a CNN is to experiment with different combinations of layers until we get the best possible result. This can be done by measuring the accuracy and then deciding on the final architecture of the network. In Chapter 12, we will introduce some popular CNN models.

A high number of layers in the neural network leads to a high number of parameters, and thus, the network might get more accurate, but on the other hand, it gets prone to overfitting as well.

Usually, computer vision models are built by very large AI corporations, and it is not easy to generalize all of them for real-world applications developed by non-super-rich corporations.

Therefore, if you develop a model that is accurate and not affiliated with those big AI corporations, good luck in convincing the scientific community about your model.

Different Types of Convolutions

We have learned that the process of convolution is applying a filter/kernel on an image and computing the output, whose size depends on the filter size, stride, and padding. Here, we briefly introduce different types of convolution, but we do not go into detail about each convolution.

2D and 3D Convolutions: Until now, we have only used a matrix and explained convolution on a matrix. This convolution is called 2D convolution. If the original data is in 3D format, we should use 3D convolution. For example, a colored image has three color properties (red, green, and blue) along with x and y coordinates. Therefore, we should deal with a 3D tensor (three matrices with the same width and height). The third dimension is referred to as *depth* or *channel*, and the size of the channel for the original data and filter must be exactly equal. For example, in Figure 10-35, both filter and original data have the same number of channels, i.e., three.

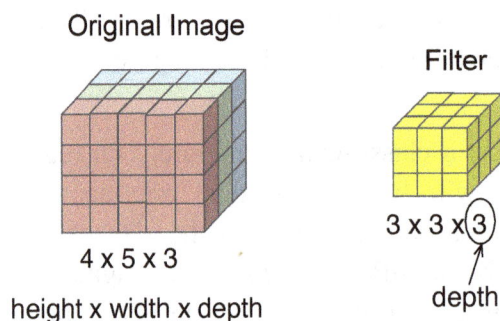

Figure 10-35: An example of a 3D Convolution with its filter. As is shown, the depth or channel in both the filter and the original image should be the same.

The dot product of a 3D filter to a 3D tensor will result in a scalar for each position where the filter is applied. As the filter moves across the tensor with a specific stride, each application produces a scalar. These scalars together construct a 2D matrix, known as the feature map or convolution output[13].

3D convolutions are typically used for volumetric images (such as MRI and CT images), where each voxel represents a point in a three-dimensional space (a voxel resembles a pixel with one additional dimension, that is depth). They are also applied to video files, where the third dimension is time, allowing the filters to capture temporal information in addition to spatial features. 2D images are generally processed with 2D convolutions, as they lack a third spatial or temporal dimension that would necessitate 3D convolution.

---

[13] A good video that visualizes this process is available under this link: https://www.youtube.com/watch?v=D0VoQDDe5zI

Dilated (Atrous) Convolution: Dilated or Atrous[14] convolution adds a gap between pixels that feds into the filter. A parameter called *dilation rate (l)* specifies the size of gaps between matrix elements (e.g., pixels). In other words, parameter l indicates how much the kernel is widened. For example, Figure 10-36 has a dilation rate of $l = 2$, 2D Convolution shown on top of this figure has $l = 1$ (no gap between pixels). As the dilation rate increases, the patch sizes increase as well because it skips several pixels.

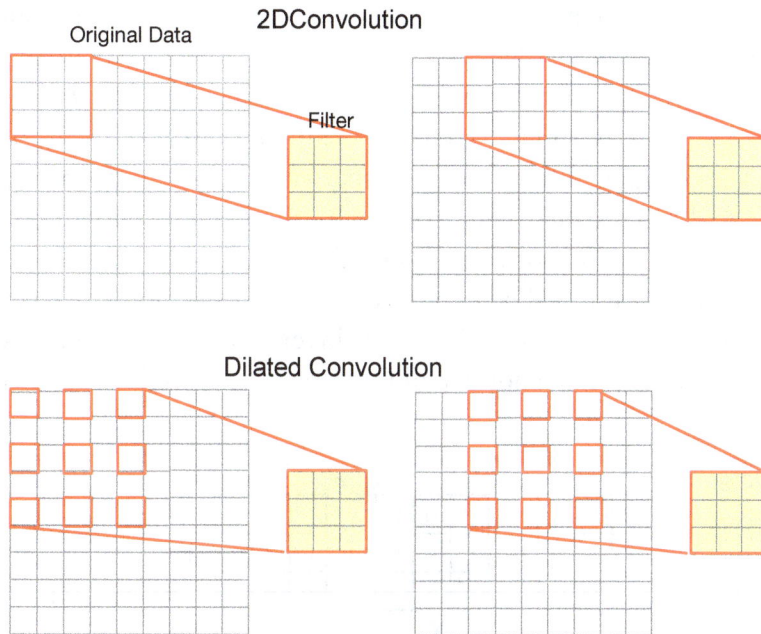

Figure 10-36: A 2D convolution on top and dilated convolution in the button. Both have zero padding and a stride size of 2.

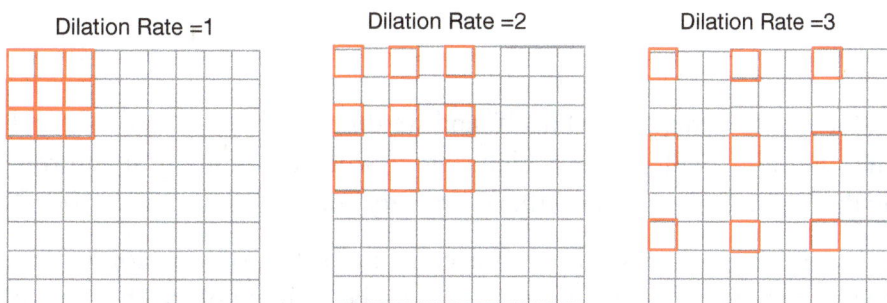

Figure 10-37: Different dilation rate specifies the size of gaps between selected pixels.

Dilation convolutions are popular approaches for image segmentation [Chen '17] because they provide an overview of the larger scope of the image. More about image segmentation will be explained in Chapter 12. Figure 10-37 shows different dilation rates (l).

---

[14] trous in French means holes

Transposed Convolution: all convolutions, except dilated convolutions, typically either preserve the size of the original data or reduce it, which is downsampling. Transpose convolution, on the other hand, is commonly used to increase the size of the output feature, a technique referred to as *upsampling*. Upsampling has many applications in image processing, such as increasing image resolution, reducing image blur, and facilitating semantic segmentation, which involves abstracting the image into a set of labeled objects. These applications will be explored in more detail in Chapter 12.

A transposed convolutional layer performs the convolution operation in a manner that reverses the spatial dimension reduction of standard convolution, effectively increasing the spatial dimensions of its output. Look at the example we provide in Figure 10-38. Here $*_{TC}$ refers to transposed convolution. In this example, we have a small input image (2×2), the stride is 1, and the padding is zero. The output will be a 3×3 matrix, which you can see the input size has been increased from 2×2 to 3×3. In particular, every cell of the input data will be multiplied by all filter cells and written in a similar position. Then, all these results will be summed in a 3×3 matrix. In Figure 10-39, we use red and blue colors to show how two cells construct the upsampled version of the input data cells. The inverse of convolution is referred to as deconvolution, and it is recommended not to call transposed convolution deconvolution.

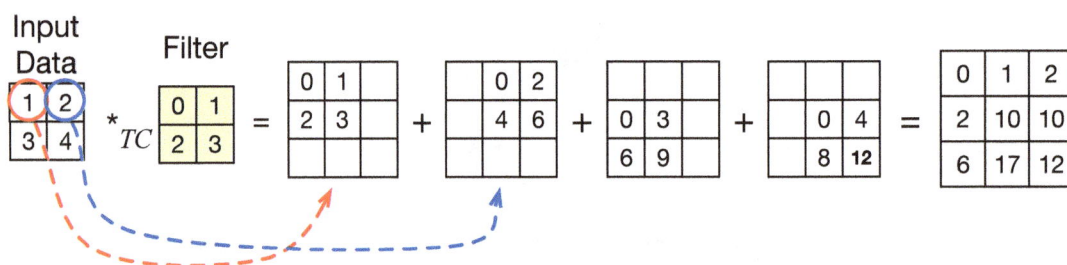

Figure 10-38: A simple example of transposed convolution, with stride size 1 and padding size 0.

NOTES:

* The CNN will learn the kernel (filter) itself, and we do not need to give the kernel manually. In particular, the algorithm will decide which kernel will get meaningful information from an image. Edge detection kernels (Sobel filters) are very common to be used by CNN algorithms.

* Often, in real-world cases, we stack several convolutional layers on top of each other. This means that the output of one convolutional layer will be an input of another convolutional layer and so forth. This causes the CNN to discover more complex patterns as it goes deeper into convolutional layers. In simple words, for example, a CNN with five convolutional layers could recognize more patterns than a CNN with three convolutional layers.

* Nowadays, we rarely need to build a CNN on our own, and most of the time, we are using an existing model to define the model that fulfills our demand, i.e., transfer learning. More about transfer learning will be explained in Chapter 14.

* Pooling layers help manage varying sizes of input images by reducing the spatial dimensions of the feature maps, thus creating more manageable and smaller representations. However, they don't directly handle varying input image sizes; rather, they contribute to making the CNN architecture more flexible and capable of dealing with the spatial hierarchy in images.

* There are different visualization techniques available to present CNN architecture. AlexNet visualization [Krizhevsky '12] is another form to visualize CCN, as shown in Figure 10-39. Here, both convolutional and pooling layers are presented as cubes, and the red cubes inside convolutional layers are filters.

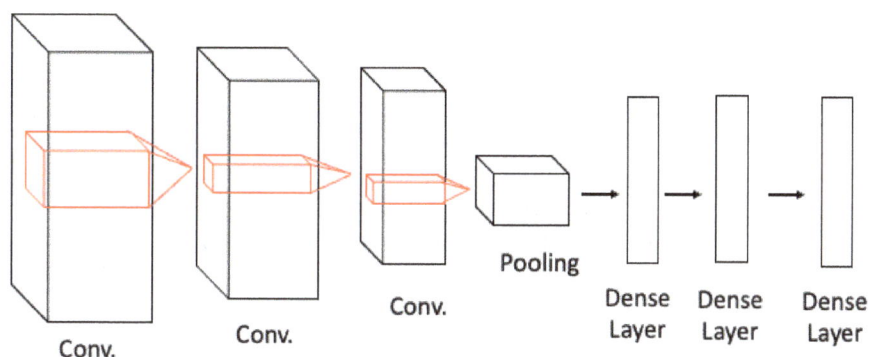

Figure 10-39: AlexNet style of presenting a CNN network architecture. Usually the sizes (height, width, channel) on the cube sides as well.

* CNN models are known to have *inductive bias*, which refers to the built-in assumptions within the model that guide its learning process. These assumptions are not directly derived from the training data but are inherent to the model's architecture, facilitating the identification of relevant patterns and features. Inductive bias enables CNN to generalize from its training data to new, unseen data, allowing it to make predictions or classifications beyond the specific examples it has been trained on. In simple terms, inductive bias is what empowers a CNN to generalize beyond its training dataset.

Recurrent Neural Network (RNN)

We have learned that neural networks, either CNN or ANN, could be used for classification and regression. Until now, our neural networks have received a fixed-size input, such as an image or vector, and the neural network output is a single output. For example, the output will be a vector of two scalars, which includes a classification result on the input image to recognize whether the picture is t-rex or chicken (0.4 for chicken and 0.7 for t-rex probabilities).

Another category of deep learning algorithms is Recurrent Neural Networks (RNN). RNNs are used for *sequential data*, such as sensor data, timestamped data, medical device data, audio, and even text (word appearances in a sentence have an order). In particular, any data with either a timestamp or implicit notion of time can be assumed as sequential data and modeled by RNN.

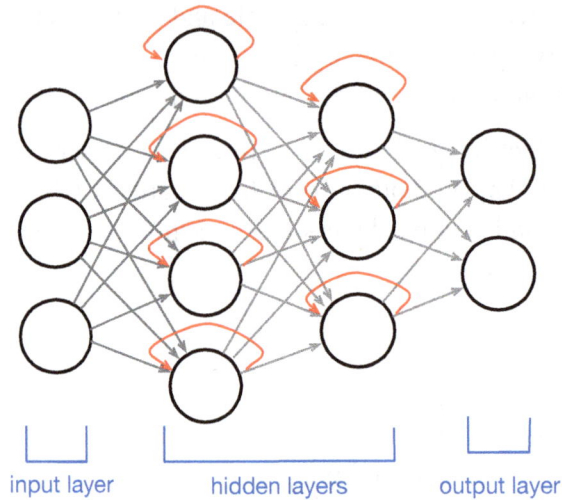

Figure 10-40: Hidden layers in RNN send their output as input in the next epoch. The red lines present the information flow in the next epoch. Here for sake of simplicity we show each neuron has a input from itself, but in fact RNN cells have input from itself, and they are different than neurons (we will explain more about this later).

RNNs could be used to reconstruct sequential data similar to the data we fed into it. For example, we could feed a book to RNN, and it can generate sentences similar to that book, or we could use RNN to construct music or poem, but at the time of writing this chapter (2020), they are not accurate enough for Generative AI tasks, but excellent to model sequential data.

As we can see from Figure 10-40, hidden layers of RNN do give output to the next layer, but

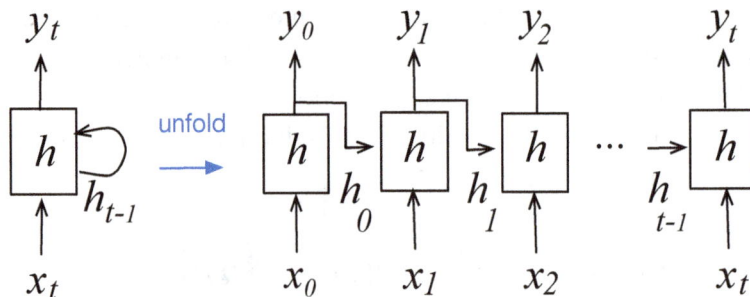

Figure 10-41: The hidden layer (h) in RNN receives input x and produces output y. However, the hidden layers have a temporal loop, which means that in addition to giving the output, it feeds the output to itself as input for the next epoch as well. Each rectangle presents a hidden layer and not a single neuron.

they also feed back their outputs to themselves along with the input (in the next epoch). In other words, they support a temporal loop.

Figure 10-40 shows an abstract representation of RNN. Hidden layers (shown with rectangles in Figure 10-40) are connected to themselves. It means that in addition to the output for the next layer, they also feed the output of the current hidden layer to themselves as well.

An unfolded hidden layer is shown in Figure 10-41. On the right side of this figure, we can see that the hidden layer at epoch (or time) t receives the output of the previous epoch ($t-1$) as well, or the layer at a epoch $t-1$ receives the output of epoch $t-2$. Therefore, we could say that neurons have *short-term memory* and remember what has happened in the previous epoch (only a single previous epoch, not all previous epochs). To formalize this, assuming $\vec{y}$ is the output vector of hidden layer variables and $\vec{x}$ is a vector of the input variable, then we have:

$$\vec{y_t} = f(\vec{y}_{t-1}, \vec{x_t})$$

In this context, neurons functioning with inputs from previous time steps can be thought of as having *memory* capabilities, although in basic RNNs, this memory is typically short-term. This memory structure makes the neural network very attractive for some fields of machine learning, including natural language processing and text analysis, because we need temporal knowledge to predict upcoming information.

Based on the number of inputs and output, there are different types of RNN, which are presented as one-to-one, one-to-many, many-to-one, many-to-many, and one-to-one. Figure 10-42 visualizes these types of presentations and notes that rectangles here refer to a layer of neurons.

For a one-to-one example, assume a single image is given into a neural network, and the network finds its label. For a one-to-many example, suppose we give a single image into the RNN, and it generates the caption for the image. For example, in an image, a chicken wears a shirt, and the shirt has a T-rex image on it. An RNN recognizes the t-rex, the shirt, and the chicken. Then, it generates a caption that correctly describes the image. In summary, in one-to-many mode, for one input (an image), it provides many output labels (chicken, t-rex, shirt). For a many-to-one example, assume we give a sentence that includes several words (many) to a sentiment analyzer, and the RNN labels the sentence tone as positive or negative (one). For many-to-many examples, we can think of a sentence that includes many words in Mandarin, and the RNN translates it to a Hindi sentence, which also includes many words. The position of words is important in the sentence to make an accurate translation. For example, *"I hit the ball with a bat"* is referring the *"bat"* as a wooden stick and not the *"bat"* as the bird *"bat"*.

We have explained that a neuron receives an input and puts it in the linear function. Then, it applies an activation function on the result of linear regression, and produces the output.

While using RNN, we have $\vec{y_t} = f(\vec{y}_{t-1}, \vec{x_t})$, which means that the output depends on the previous outputs as well. Also, its activation function is a *hyperbolic tangent activation* function, i.e., *tanh*.

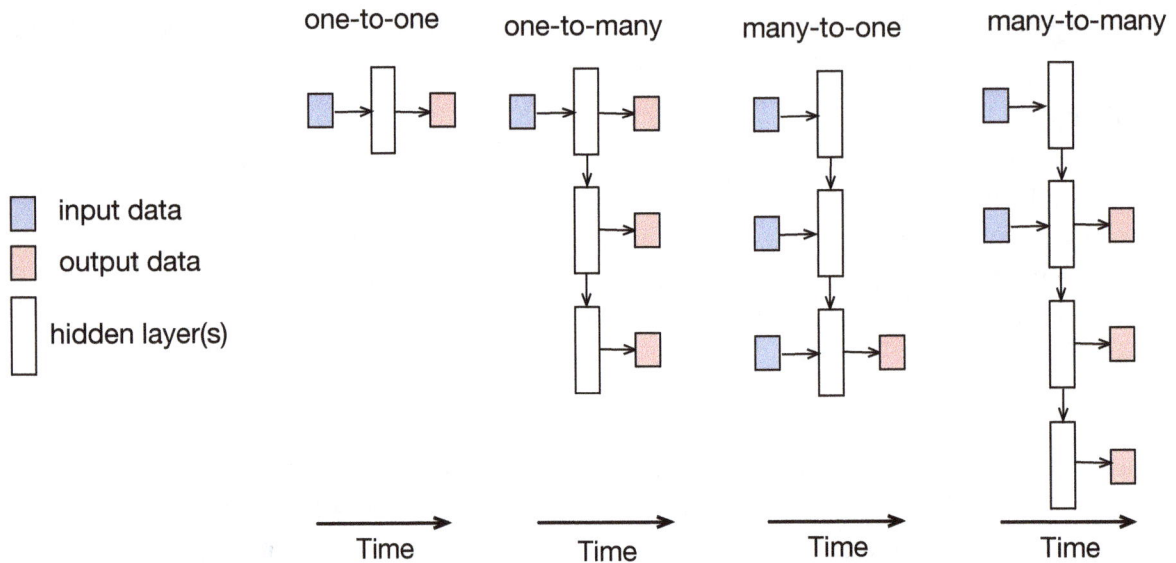

Figure 10-42: Different architectures of recurrent neural networks, you can see from left to right the time increases.

These are the basics of the RNN network. It is fantastic architecture because it has memory, but it has a goldfish memory; it only remembers the previous output, i.e., short-term memory. To mitigate this, we need an RNN that keeps tracking more previous output, i.e., long-term memory.

Long Short-Term Memory (LSTM)

Vanishing gradient is a severe issue in sequential data, and previous approaches we have explained to mitigate its risk are not very practical for sequential data, especially for long sequential data. As a result, using those methods could slow down the system significantly. Deep learning algorithms are consuming lots of resources, and they are slow; thus, applying those techniques makes things even worse. On the other hand, we have explained that the RNN is a goldfish memory; it has only one short-term memory and can't remember more than a few single previous states. For example, take a look at this sentence: *"I love money, and as a consultation fee, I get a fat check."* The word *"fat"* in this sentence refers to money, not overweight people nor nutritional fat. Therefore, if we give this sentence to an RNN, which can only remember a few previous words, it cannot recognize the context of the word fat.

A well-known approach to handling time series issues and temporal data is the LSTM architecture [Hochreiter '97]. Instead of one memory, each LSTM cell has four memory components: *long-term memory, short-term memory, new long-term memory,* and *new short-term memory.* It uses the element-wise operator (Hadamard product explained in Chapter 7), sigmoid, and

590

hyperbolic tangent activation functions to decide about the output value.

Before explaining LSTM, let's briefly review how an RNN works. It receives x_t and h_{t-1} and uses a hyperbolic tangent to calculate h_t the output, which can be written as $h_t = tanh(W[h_{t-1}, x_t] + b)$, as shown on the left side of Figure 10-43.

LSTM neurons are referred to as cells. As shown in Figure 10-43, an LSTM cell receives information called *cell state* (C_{t-1} or previous long-term memory), *previous output* (h_{t-1} or previous short-term memory), and *input vector* (x_t). The output of LSTM is a *new cell state* (C_t or new long-term memory) and the *output vector* (h_t or new short-term memory). The content of each LSTM cell has input, forget, output, sigmoid layers, hyperbolic tangent layers (tanh) gates, and point-wise operators (element-wise or Hadamard product).

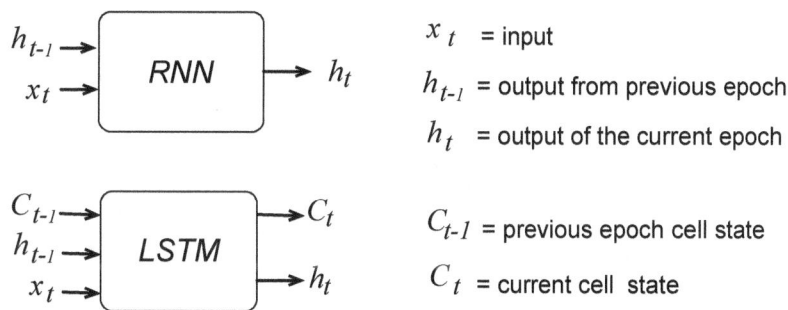

Figure 10-43: Traditional RNN neuron, which is called vanilla RNN, versus LSTM neuron. The LSTM neuron receives C_{t-1} and outputs C_t as well, which are representations of cell state that could be interpreted as long-term memory.

Gates are responsible for deciding whether the information passes the gate or not. The Sigmoid layer outputs values close to 0 (not passing the gate) or close to 1 (passing the gate), and *tanh* brings the data into the range -1 to 1.

Now that we have a basic understanding, we can describe LSTM, step by step, as follows.

(i) The first step of the LSTM algorithm decides what information is not worth keeping and could be thrown away; this will be done through the *Forget gate*, which is a layer with the Sigmoid function. In particular, first, it receives the h_{t-1} (output of the previous time) and x_t (input), then calculates f by using the Forget gate (Sigmoid function) and outputs $f = \sigma(W_f[h_{t-1}, x_t] + b_f)$. At this step, the state is C_{t-1}, which comes from the previous cell, as shown in Figure 10-44 Step 1.

(ii) The second step decides what new information is going to be stored in the cell state. This step includes three sub-steps. First, it uses the *Input gate* to decide which values will be updated. Its equation is written as: $i = \sigma(W_i[h_{t-1}, x_t] + b_i)$, see Figure 10-44 Step 2-1. Second, a hyperbolic tangent function creates a vector of new candidate values, i.e., $\tilde{C}_t$, that its value will be between -1 and 1. These candidate values can be added to a candidate new cell state (not the final new cell state), and the equation is written as: $\tilde{C}_t = tanh(W_C[h_{t-1}, x_t] + b_C)$, (see Figure 10-44 Step

2-2). Third, after these two sub-steps, their results should be combined to update the old cell state, C_{t-1}, into a new cell state: C_t. To construct the new cell state, the algorithm multiplies the old state by f_t, forgetting the information that is forgettable, and then decides how much a cell state is getting updated by using $i \odot \tilde{C}_t$. The third sub-step, which constructs the new cell state (C_t) is written as: $C_t = f \odot C_{t-1} + i \odot \tilde{C}_t$, (see Figure 10-44 Step 2-3)

(iii) In the final step, the algorithm decides what is going to be in the output vector, which is based on the filtered version of the C_t cell state. Here, the filter is referred to as passing through C_t to the *tanh* gate. This step has three sub-steps. In the first sub-step, h_{t-1}, and x_t sends to *Output gate* (Sigmoid gate), which decides what part of the cell state will be given as output. The

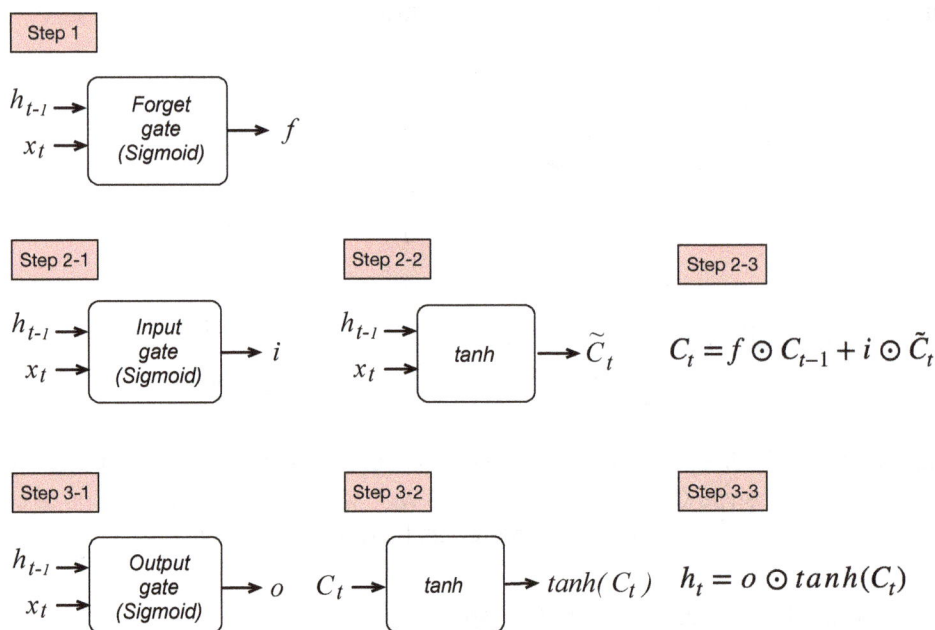

Figure 10-44: Summarization of the LSTM steps to construct short-term memory and long-term memory. Results of steps 3-2 and 3-3 are returned as outputs of the LSTM cell.

equation of this sub-step is written as: $o = \sigma(W_o[h_{t-1}, x_t] + b_o)$, see Figure 10-44 Step 3-1. In the second sub-step, the cell state will be passed through a *tanh* gate (to transform its values between -1 and 1), see Figure 10-44 Step 3-2. As the third sub-step, it constructs the final output, h_t as $h_t = o \odot tanh(C_t)$, (see Figure 10-44 Step 3-3). The final output of each LSTM cell is o, along with its states (h_t and C_t).

Figure 10-44 summarizes the process of the LSTM cell[15]. It is an easy to remember summarization of the LSTM algorithm. We did not explain each W and b, weight, and biases. There is another form of visualizing LSTM, which is presented in Figure 10-45. To be sure we understand them correctly, let's summarize their equations:

$i_t = \sigma(W_{ii}x_t + b_{ii} + W_{hi}h_{t-1} + b_{hi})$

---

[15] There are other nice visualizations, especially the one from Christopher Olah existed online as well. Links will be provided at the end of this chapter.

$$f_t = \sigma(W_{if}x_t + b_{if} + W_{hf}h_{t-1} + b_{hf})$$
$$\tilde{C}_t = tanh(W_{i\tilde{c}}x_t + b_{i\tilde{c}} + W_{h\tilde{c}}h_{t-1} + b_{h\tilde{c}})$$
$$o_t = \sigma(W_{io}x_t + b_{io} + W_{ho}h_{t-1} + b_{ho})$$
$$c_t = f_t \odot c_{t-1} + i_t \odot \tilde{c}$$
$$h_t = o_t \odot tanh(c_t)$$

While using LSTM, if the states change in different layers (based on temporal changes), they are not significantly different, and thus, the gradient is unlikely to change drastically. LSTM can resolve the vanishing gradient problem, but it is still prone to exploding gradient.

Gated Recurrent Unit (GRU)

Gated recurrent unit [Chung '14] is the more recent version of mitigating the vanishing gradient problem of RNN. GRU is less complex than LSTM and does not have a cell state. Since there is no cell state, its input (x_t, h_{t-1}) and output (h_t) parameters are the same as those of simple RNN.

GRU has two gates that use the Sigmoid function: *reset gate* (r) and *update gate* (z). The reset gate decides how much of the previous state to remember. The update gate is similar to the forget gate and input gate of LSTM, and it decides *how much information to throw away*, or in other words, how much new information is a copy of the old information. A reset gate is responsible for *short-term* dependencies of a given sequence, and an update gate is responsible for *long-term* dependencies of the given sequence. A GRU cell operates as follows.

(i) The input x_t and output of the previous time h_{t-1} are used to calculate the value for update and reset gates. They both use the Sigmoid function, and their equations are written as follows:
$$z = \sigma(x_t W_{xz} + h_{t-1}W_{hz} + b_z)$$
$$r = \sigma(x_t W_{xr} + h_{t-1}W_{hr} + b_r)$$

(ii) Now, with the result of both gates, we can calculate the *candidate hidden state*, but it is not the final hidden state. The candidate's hidden state should stay between -1 and 1 interval, thus, a *tanh* gate is required. Besides, it incorporates the reset gate information, which will be used as an element-wise product of the previous hidden state h_{t-1} and r. The candidate's hidden state equation is written as follows:
$$\tilde{h}_t = tanh(x_t W_{xh} + (r \odot h_{t-1})W_{hh} + b_h)$$

(iii) After the candidate hidden value $(\tilde{h}_t)$ and update gate (z) are both identified, the algorithm calculates the hidden state as $h_t = (1 - z) \odot h_{t-1} + z \odot \tilde{h}_t$. The output of GRU is only h_t, and unlike LSTM, it does not have a cell state. Based on the above equation, if z is close to 1, the new hidden value (h_t) will be very similar to the old hidden value (h_{t-1}). This reveals that lots of the new information is a copy of old information. In contrast, z close to 0 indicates the new hidden value (h_t) will be close to the candidate's hidden state $(\tilde{h}_t)$.

Figure 10-45 presents a common visualization approach used by these two algorithms, inspired by Christopher Olah's design. In the beginning, we did not describe LSTM and GRU with these visualizations because they might sound confusing, but now you can easily understand them. To be sure we understand it, take a look at the following summary of its equations:

$$r_t = \sigma(W_{ir}x_t + b_{ir} + W_{hr}h_{t-1} + b_{hr})$$
$$z_t = \sigma(W_{iz}x_t + b_{iz} + W_{hz}h_{t-1} + b_{hz})$$
$$\tilde{h}_t = tanh(W_{in}x_t + b_{in} + r_t \odot (W_{hn}h_{t-1} + b_{hn}))$$
$$h_t = (1 - z) \times \tilde{h}_t + z_t \odot h_{t-1}$$

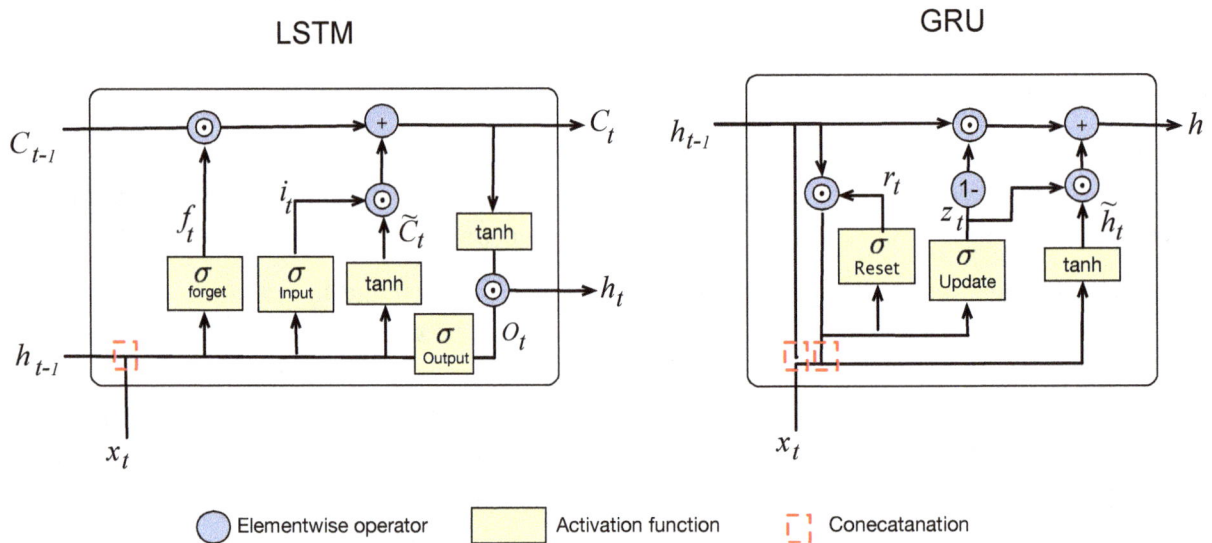

Figure 10-45: Visualization of LSTM and GRU architecture based on their gates and operators. In GRU output and new hidden state (h_t) are the same.

Bidirectional RNN

One of the popular problems in NLP applications is referred to as *name-entity recognition*. It refers to a problem that we need to identify a word that refers to what entity (person, object, organization, etc.) in a sentence. RNN is a useful approach for named entity recognition.

For example, consider this sentence: "They seem very tasty". We do not know what "They" means. However, if we write the following: "They seem very tasty, I love cheese." Now, we can understand that "They" refers to food with cheese or cheese itself. Without that prior knowledge, we could not understand what entity the word "They" refers to. A simple RNN cannot incorporate prior knowledge before encountering the word "Cheese" because every new state got the information from its previous state, and in this particular sentence, we realize they are referring to food with cheese later in the sentence.

Take a look at Figure 10-46, the unfolded version of one RNN cell (check Figure 10-41 to recall what we mean by unfolding). At each time unit, the RNN receives one word of the input sentence. In this figure, we present the activation result as $a_{(time-index)}$, which means that at time 3, the word "very" will be fed into the RNN, and the output of the RNN cell at that time will be

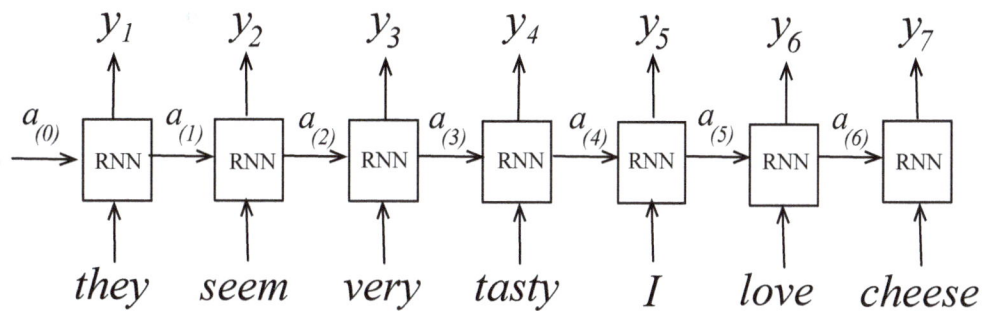

Figure 10-46: Unfold version of one RNN cell, which receives one word at each time unit. 'a' presents activation function result.

y_3. Since the word, "cheese" happened at the t_7 and "They" at t_1, the RNN can not understand what "They" refers to at any time earlier than t_7.

A solution to handle this is to incorporate future information for better sequence prediction. Bidirectional RNN is designed to incorporate future data into the sequence as well. It introduces another layer with exactly the same size, but the direction of activations is in the opposite direction (from the end of the sequence to the beginning of the sequence). Figure 10-47 presents the Bidirectional RNN, and we use blue color to present the other direction. In this example, in addition to the forward RNN, the other RNN is going from the last word of the input sequence to the first word of the sequence, i.e., going backward in the sequence. In particular, the neural network of Figure 10-47 starts from $a_{(0)}$ and goes forward until it computes $a_{(6)}$; simultaneously, it starts from the $a'_{(0)}$ (blue RNN) and moves backward by computing activations until it reaches $a'_{(6)}$. Therefore, assuming our sequence has a size of T, every y_t output includes both forward (from 0 to t) and backward activation (from T to t) data (simultaneously). This structure allows the network to have information from both past and future states at any point in the sequence.

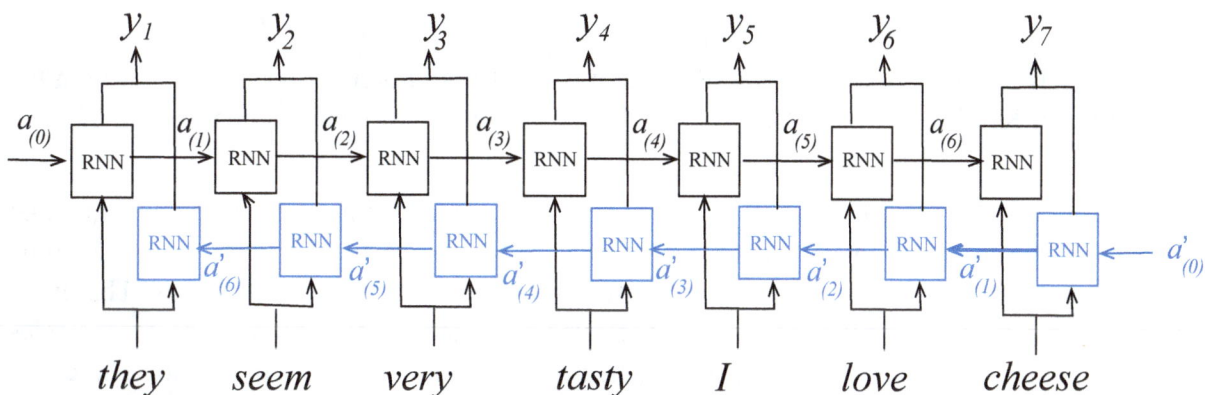

Figure 10-47: Unfold version of one BiDirectional RNN cell, which the blue layer is the same kind of RNN.

Of course, training this network is computationally intensive and slower than usual RNN, but it is a useful network to experiment with when we intend to incorporate future information into the current sequence. Another issue of Bidirectional RNN is that we need to have the entire dataset to train the model, and thus, we can not work with a stream of data (data will be available during the time).

Deep RNN

$$x_t \longrightarrow \boxed{h_{(1)}} \longrightarrow \boxed{h_{(2)}} \longrightarrow \boxed{h_{(3)}} \longrightarrow y_t$$

$$h^{(1)}_{t-1} \qquad h^{(2)}_{t-1} \qquad h^{(3)}_{t-1}$$

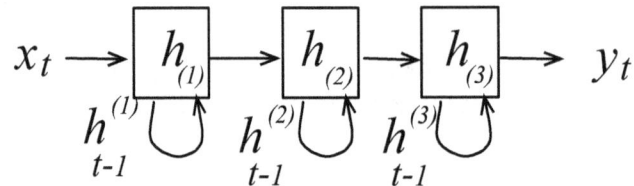

Figure 10-48: A deep RNN with three layers. Note, unlike previous figures, it has three hidden layers instead of one hidden layer that is unfolded.

If we stack multiple layers of RNNs on top of each other, the result is a Deep RNN. This structure is employed to capture more complex patterns and dependencies in the data, which is especially useful in modeling higher levels of non-linearity. As illustrated in Figure 10-48, a Deep RNN can have several hidden layers, such as three in this example. While it is common to expect that increasing the number of layers in a neural network can improve its accuracy, this approach requires sufficient data and comes with trade-offs.

Specifically, there is a limit to the number of layers that can be added before the model starts experiencing diminishing returns on accuracy, often due to overfitting or training difficulties such as the vanishing gradient problem.

RNN Examples

Unlike CNN, which we could easily understand its applications, RNN might require a bit more explanation. Therefore, we introduce a few examples that might help us better connect with the applications of RNN. More examples of popular RNN models will be explained later in Chapter 12.

We can use an RNN (LSTM, GRU, or any other model) to predict time series or other sequential information in the future. In the example presented in Figure 10-49, the blue part of the time series is fed as trained data, and the RNN constructs the red part of the time series. The given time series is a mock time series that shows the sales of a rubber duck. As we can see, the sales increase in summer and decrease in winter, when most of the planet Earth is cold. Therefore, customers are less motivated to purchase rubber ducks.

Also, RNN is used a lot in natural language processing applications such as machine translation, grammar correction, and text editing tools. For example, we can train it, and it can be used to construct sentences as well. Definitely, the larger the text size, the longer the sentence will be.

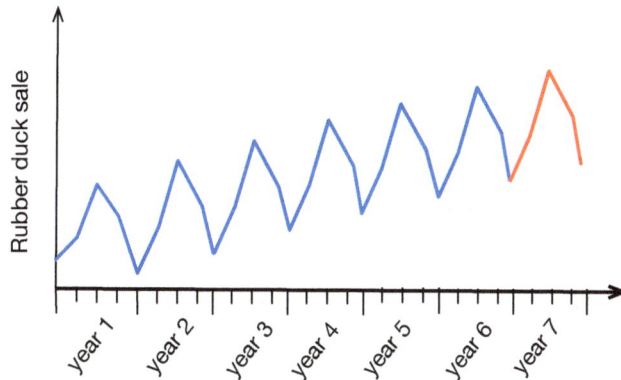

Figure 10-49: Another popular use of RNN is for time series prediction and here the red area is constructed by the algorithm, which has the blue part of the time series as its training set.

Also, it could guess the rest of the sentence. For example, "I do appreciate a quick" could be followed by the word "response", which this word is guessed or constructed by an RNN. Another example is Part Of Speech (POS) tagging, which specifies the grammatical role of each word in a sentence.

As another example, take a look at the visualization proposed by Karpathy et al. [Karpathy '15]. We fed the "100 years of solitude" from Gabriel García Márquez as input, and RNN automatically learns text features, such as the beginning of a sentence, quotes, punctuations, etc. It is similar to feeding the image into CNN, and it learns image features such as edge lines. Figure 10-50 visualizes features that the LSTM identifies from the text. In this model[16], we have used Sigmoid as the activation function, and the color range from dark sky blue (0) to dark red

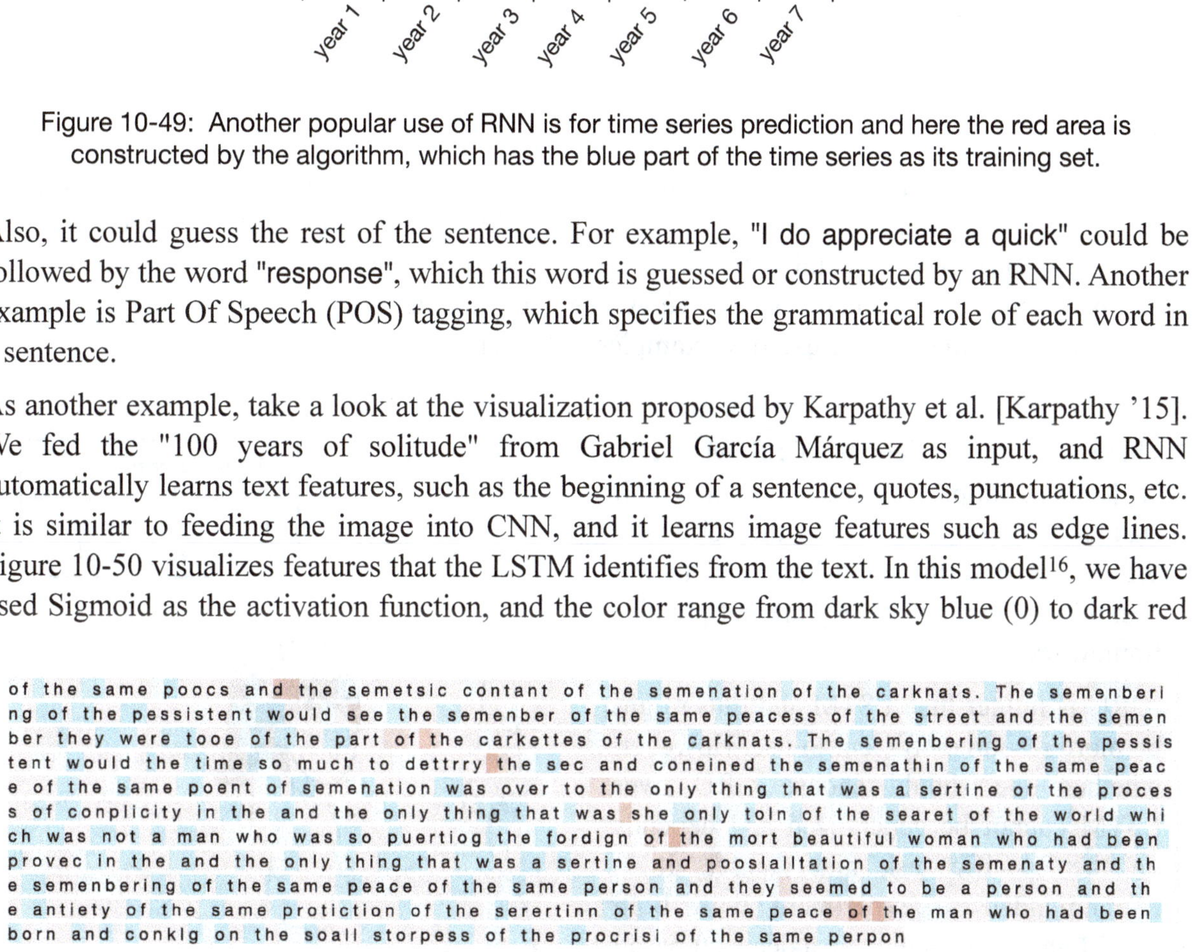

Figure 10-50: A sample feature (space between words and the character beginning of the next word) that has been detected by one output cell of the LSTM model.

(1) is used to visualize LSTM cell activations. Blue means it can correctly predict, and red means it has not correctly predicted. You can see this in this example of an output cell. The space and beginning characters of the next word have been predicted, especially the word "the". Thus, one application of LSTM could be a grammar checker, i.e., when the word "the" is not there, recommend adding the word "the".

---

[16] The code is used from here: https://github.com/praneet9/visualising-lstm-activations, but we apply some modifications to make it run for the current Keras version.

Since the author of this book is poor and can not afford a strong machine with GPU, we trained it in a few epochs, but if we train it longer, we might get more human-readable output. Note that there are many other features recognized by LSTM, but they are hard for us as humans to understand due to the black nature of deep neural networks.

In another example, we use a GRU to generate a text automatically by looking at the text of the "100 Years of Solitude" book and learning the patterns of the text. Following is a sample text created at the 30th epoch:

```
He plunged into his the one he had felt during the time he was resigned to
living without a woman. He plunged into his mpose on the gone at section
purbers of uneraterad difficult, and on adldo suck pestact rediract.
Aureliano, and mort time to lession with phonessess had fanthoush alous the
work he fate hoared with his it was a corple and troused a scorphing,s of a
lecquiander some most memary and time that the cloth and that it wasred the
mosting from darnesprotes. eaven who had becauged your arthioped his walk.
```

This text does not have a meaning, but it is very interesting, and you can see how these algorithms can be used to generate new information from existing information. Generating new information through the neural network is referred to as "Generative AI", and deep learning algorithms model the path to Generative AI. In later chapters, especially in Chapter 11 and Chapter 12, we will see more exciting examples of Generative AI.

NOTES:

* There are more variations of LSTM [Greff '16]. For example, peephole LSTM which enables f, i, and o to look at the cell state. We do not describe them here, and to the best of our knowledge, they are not popular frameworks.

* GRU is lighter than LSTM and thus might be more resource-efficient to use. However, whether one should choose GRU over LSTM cannot be generalized and depends on the specific task and dataset. It is recommended to experiment with both LSTM and GRU and make a decision based on comparing the accuracy of the results.

* The Bidirectional RNN and the Forward-Backward algorithm in HMMs share a functional similarity in that both incorporate future information to enhance sequence prediction. This feature allows for more accurate modeling of sequences by considering the entire context, as well as both past and future elements (refer to Chapter 5 for more on HMMs).

I have learned neural network and deep learning. Now, I am the empress of machine learning and AI.

* While Bidirectional RNNs are computationally more intensive than unidirectional RNNs due to their processing of information in both forward and backward directions, the increased complexity can lead to better performance in tasks where understanding the entire sequence context is crucial.

Summary

In this chapter, we first described some history and human efforts to resemble neural networks with computers. An artificial neural network is composed of three layers: input, output, and hidden layers. Each artificial neuron includes weight, bias, and an activation function that calculates its output.

Weights and biases are the foundations of neural networks. The optimizer in the neural network weights tries to find the best combination of weights that results in the lowest loss score. The objective of a cost function in a neural network algorithm is to evaluate the output correctness based on the loss score. If it is not correct, then modify its values by using an optimizer to adjust weights. As we have explained, weights are assigned randomly; they should be adjusted, and this will be done through the backpropagation algorithm, which goes back from the output of the network to the input neurons. Backpropagation splits the error, proportional to the weights back using the chain rule of the differential equation.

Next, we explained the perceptron, its limitations, and how Multi-Layer Perceptron (MLP) resolves its limitations. MLP algorithms are still in use and referred to as traditional ANN. If we increase the hidden layers of MLP algorithms, they are called deep neural networks. All neurons are connected to all neurons in the next layer of MLP. Such a connection is not mandatory in deep learning models, and if neurons of one layer are connected to all neurons of the next layer, they are called the "dense layer" or "fully connected layer".

Afterward, we describe activation functions, which can enable the neural network to construct a non-linear model, followed by the cost functions explaination, which are mostly comparisons between the statistical distribution of predicted data and actual data.

Then, different types of Gradient Descent based optimizers have been explained. At the time of writing this book, the best optimization approach is using Backpropagation, which operates based on chain rule in derivations.

We concluded the general discussion about neural networks by regularization methods, including weight initialization, Gradient Clipping, Batch Normalization, Dropout, and Early Stopping.

CNN models are one of the most popular uses of Deep Learning because of their application in computer vision. To understand CNN models, we explained the concept of convolution and cross-correlation. Then, we described some concepts related to applying a convolution, including padding, stride, window size, etc. After the convolution, a CNN network is usually applying a pooling layer. The pooling layer is used to downsample many convolutions of the input image. The result of the pooling layer is sent into a flattening and then a dense layer for output preparation.

Since convolutions are very popular not just for computer visions but also for other tasks, we listed some of the common convolutional approaches, including 3D, dilated, and transposed convolution.

RNN is another common deep learning architecture used for sequential data modeling, such as time series, natural language processing, etc. The basic models of RNN are pruned to vanishing gradient and exploding gradient problems. LSTM has solved the vanishing gradient issue and implements short-term and long-term memories. Each LSTM cell has four memory components: long-term memory, short-term memory, new long-term memory, and new short-term memory. Despite its usefulness, LSTM is complex and computationally inefficient. GRU, later, is introduced, which is less complex than LSTM. However, it is better to decide on the architecture with experiments and take into account the nature of the dataset that is used for the application.

To review LSTM and GRU operations, the following list of all their equations[17].

LSTM:
$$i_t = \sigma(W_{ii}x_t + b_{ii} + W_{hi}h_{t-1} + b_{hi})$$
$$f_t = \sigma(W_{if}x_t + b_{if} + W_{hf}h_{t-1} + b_{hf})$$
$$g_t = tanh(W_{ig}x_t + b_{ig} + W_{hg}h_{t-1} + b_{hg})$$
$$o_t = \sigma(W_{io}x_t + b_{io} + W_{ho}h_{t-1} + b_{ho})$$
$$c_t = f_t \odot c_{t-1} + i_t \odot g_t$$
$$h_t = o_t \odot tanh(c_t)$$

GRU:
$$r_t = \sigma(W_{ir}x_t + b_{ir} + W_{hr}h_{t-1} + b_{hr})$$
$$z_t = \sigma(W_{iz}x_t + b_{iz} + W_{hz}h_{t-1} + b_{hz})$$
$$\tilde{h}_t = tanh(W_{in}x_t + b_{in} + r_t \odot (W_{hn}h_{t-1} + b_{hn}))$$
$$h_t = (1 - z) \times \tilde{h}_t + z_t \odot h_{t-1}$$

In these equations, W refers to weight and b to biases; their index is used to separate them from each other and explain the weight or bias belongs to which stage, $\odot$ presents a Hadamard product. h_t is the hidden state at time t, c_t is the cell state at time t, x_t is the input at time t. For the LSTM, i_t is the input, f_t is the forget gate output, g_t is the cell gate output and o_t is the output. For the GRU the r_z refers to reset gate, z_t refers to update gate, and $\tilde{h}_t$ refers to candidate hidden state. To have a better overview of the differences between LSTM and GRU, you can check Figure 10-46 as well.

At the end of this chapter, we described "Bidirectional RNN". Bidirectional RNN has the advantage of learning from both directions of a sequence, but it is slow. Nevertheless, it has useful applications such as speech recognition, part of speech tagging, etc. When we stack more than one layer of RNN cells on top of each other, this is referred to as "Deep RNN". It also has applications in machine translation and speech recognition.

---

[17] Pytorch tutorial has a good summarization of both and we use them to build this table.

Further Reading or Watching

* Tareq Rasheed [Rasheed '16] has a short book about neural networks and makes a very good connection between biological and artificial neural networks. It also guides the reader to build a small neural network.

* Francis Chollet [Chollet '18] has an excellent introduction to deep learning and its components in his book. Chollet is the original author of the Keras platform[18].

* There are many explanations about the Backpropagation algorithm available, but the Aurelin Geron [Geron '19] book provides a good explanation of this algorithm. Besides, Andrew Ng and Kian Katanforoush's notes are among the good ones that we use to understand them. http://cs229.stanford.edu/notes/cs229-notes-deep_learning.pdf.

* Jay Alammar provides a very good interactive tool for learning linear regression and Gradient Descent. There, we can observe changes in the neuron (linear regression) by playing with w and b parameters. http://jalammar.github.io/visual-interactive-guide-basics-neural-networks

* If you are interested in staying updated with the Gradient Descent optimizer, John Chen has a good analysis of recent Gradient Descent algorithms on his home page https://johnchenresearch.github.io/demon. Also, Sebastian Ruder benchmarked some Gradient Descent algorithms, and you can observe the result: https://ruder.io/optimizing-gradient-descent/index.html.

* We do not go into the details of Convolution, if you are interested in learning more, Wikipedia has very good visual examples https://en.wikipedia.org/wiki/Convolution, if you are lazy, to read the content, we encourage you to check those images.

* There is a good resource by Dumoulin and Visin [Dumoulin '16] if you are interested in learning the mathematics of Convolutions in more detail. They also have very good visualizations, which can be seen in the animation from here: https://github.com/vdumoulin/conv_arithmetic.

* Christopher Olah has a fantastic and easy to understand, visualized explanation of LSTM. We benefited a lot from his explanation to understand LSTM. His explanation and visualization are available here: https://colah.github.io/posts/2015-08-Understanding-LSTMs.

* There are courses that provide a good introduction to Deep Learning on Udemy. Two good ones are the one by Eremenko et al. (https://www.udemy.com/course/deeplearning) and Jose Portilla (https://www.udemy.com/course/complete-guide-to-tensorflow-for-deep-learning-with-python). However, the material they cover is not very broad, so consider them a start to your learning journey.

---

[18] https://github.com/keras-team/keras

Chapter 11: Self-Supervised Neural Networks

In the previous chapter, we have described CNN and RNN as two popular neural networks. These neural network architectures fall into the category of supervised learning. It means we train a model and use the trained model on the test dataset. Then, the accuracy of the result will be evaluated, and the optimizer will try to improve weights in the next epoch.

This chapter focuses on *self-supervised neural networks*, a type of unsupervised neural network often associated with *generative models*. Most supervised models we have learned are doing either classification or regression, but self-supervised algorithms *reconstruct* the given input data. Previous neural network models we have explained get the input data, process them, and provide output. They include train (fit) and then test (predict). From now on, the neural networks we explain will perform the prediction on the same data, similar to the unsupervised learning algorithms we learned in Chapter 4 and Chapter 5, and the dataset is not split into train and test sets. These algorithms learn the latent representation of the input data. Latent representation has characteristics of the original data, while some of its noise might be removed.

Recent advances in AI, including applications such as colorizing black-and-white photos, restoring old photographs, converting photos into videos, and generating images from text

Figure 11-1: An example of a black-and-white picture that is colorized by a generative neural network.

descriptions, leverage these algorithms. For instance, in Figure 11-1, we demonstrate an old photo to a neural network that uses a generative model, and it colorizes the old photo. Figure 11-2 showcases a text-to-image generation model, DALL-E v2 [Ramesh '22], which, as of 2023, can create detailed images based on text descriptions provided in captions. These examples highlight the capabilities of current models and contrast them with earlier, less sophisticated versions. At the time of reading this chapter, you might have wondered how the high quality of the text-2-image models or other models improved during the time and how dull are the versions you see in this chapter.

Figure 11-2: Text-to-image construction example. (right) DALL-E v2, text prompt: "A happy potato on the beach is drinking Pinacolada". (left) Stable Diffusion XL, text prompt: "A frog is talking behind the microphone for a crowd".

We start this chapter by describing Self Organizing Maps (SOM) or Kohonen Maps, an old neural network whose main objective is to change the representation of the input data. It can be used for different tasks, including dimensionality reduction while maintaining the characteristics of the original dataset, classification, clustering, and even solving traveling salesman problems[1]. Next, we describe the Restricted Boltzmann Machine (RBM) and its varieties as initial models of both RBM and SOM are fairly old, but algorithms are not like technologies that get outdated; they are science, and we should learn them as much as we can. Software developers learn technologies, and they need to filter what technology to learn and what to filter. However, learning the algorithm is a different policy, and we strongly recommend learning as many algorithms and models as you can.

Afterward, we describe Autoencoders (AE) and then Generative Adversarial Networks (GAN), which are generative models and very popular until the introduction of diffusion models. Next, we explain Contrastive Learning methods, followed by a discussion on text-to-image models and

---

[1] Traveling Salesman Problem (TSP) is a known problem in computer science that tries to answer the following question: Given the list of cities and distances between each city, what is the shortest path that the salesman can visit each city once and return to the origin city?

their concepts and algorithms, such as diffusion models, zero-shot learning, CLIP, VQGAN, etc.

These models lead a new direction known as generative AI, e.g., generating text, illustrations, music, and videos. You might think that at the end of this chapter, you are done with deep learning. To be sure you do not feel too happy by reading this chapter, there is another fat chapter about state-of-the-art architectures and models for neural networks. Nevertheless, by finishing this chapter and learning all its algorithms, you will be another person. Do not take it as an advertisement; the algorithms of this chapter are exciting to learn.

If you start this book from the beginning, at this point, you have a good understanding of distributions and probabilities. You are ready to learn new topics. If you skipped those chapters, we strongly recommend reading Chapter 3, Chapter 8, and Chapter 10 before starting to read this chapter. Otherwise, you will have trouble understanding this chapter.

Before we start our explanations, we should be familiar with a few basic terms, which we have tried not to explain in earlier chapters.

Representation Learning Concepts

Self-supervised neural networks have a common characteristic, i.e., instead of training the model on the data in its original dimensions (high dimensional space), these models learn a low-dimensional representation of the data (*latent space*) and train their model in the latent space.

It means that each data point in the latent space is a representation of one or more data points in a high-dimensional space. Latent space is located in hidden layers. Latent space includes *features that best describe the characteristics of data.* Features that are less important to describe the characteristics of the data are either eliminated or their impact is reduced in the latent space. For example, to construct an artificial human face, the placement of eyes and nose are important and exist in the latent space, but a dot on a cheek is unimportant and will be removed when the data gets into latent space. Variables inside the latent space are referred to as *latent variables*, and other variables that we can observe, such as input or output, are *observed variables*.

Generative vs. Discriminative Model

Machine learning and artificial intelligence algorithms can be classified as generative or discriminative models. Assuming our input data is *x,* and the output that we intend to predict is *y,* a *generative model* uses joint probability to make a prediction or an inference $p(x, y)$.

On the other hand, a *discriminative model* uses conditional probability $p(x|y)$. We can say a generative algorithm cares about how data has been generated, and these models try to learn the "distribution" of the training data, e.g., GMM (Chapter 4) and HMM (Chapter 5). Once a generative model learns the distribution of the train set, then it can decide on the labels for the test dataset based on the distribution. An easy explanation about generative models is algorithms that deal with data distribution.

In contrast, a discriminative model *discriminates between different data points, disregarding how they have been generated*. In a more technical sense, discriminative models do not need to understand how data is generated; instead, they focus on finding the optimal decision boundary to separate classes using the input features, e.g., logistic regression or SVM (Chapter 9).

Since generative models take into account dependencies between data points, they are very powerful algorithms. On the advantage of generative models, David Foster [Foster '19] stated that "*by sampling from a generative model, we can generate new data*". For example, to generate images, we need generative models because each pixel in an image is highly correlated with all other pixels in that image. Images are complex, and their features are highly correlated and latent. A human can not infer their relationships, but neural networks can. Discriminative models do not take into account feature dependencies, and thus, they are not very capable of generating synthetic data.

Discriminative models usually operate with labels, and thus, they have been categorized as supervised learning algorithms. However, generative models can be used in both supervised and unsupervised settings, and mainly, they are used in unsupervised settings. However, to keep them separate from traditional unsupervised learning algorithms, scientists call them self-supervised models, In other words, self-supervised learning refers to techniques where *models generate their own training signals from the data*.

Deterministic vs. Stochastic Model

A *deterministic model* produces consistent outputs for the same set of inputs and parameters, given fixed initial conditions. In contrast, a *stochastic model* incorporates elements of randomness, which means its outputs may vary with each execution, even under identical conditions. Therefore, every execution of a stochastic model might have different results. Examples of stochastic models are genetic algorithms (check Chapter 6) or manifold dimensionality reductions, tSNE, and UMAP (check Chapter 7). Most heuristics approaches that try to find the optimal answer, including deep neural networks, belong to stochastic models.

Generative models typically exhibit stochastic behavior because they sample from learned data distributions, making their outputs inherently variable. This stochasticity introduces challenges in using them for end-user applications, though there are techniques to mitigate these effects, which are beyond this basic explanation.

Self Organizing Maps (SOM)

One of the oldest practical artificial neural networks is Self Organizing Maps (SOM), which were invented back in 1982, by Teuvo Kohonen [Kohonen '82] in Finland. Finland is a small country in northern Europe where the Linux operating system and Nokia (the indestructible smartphones of early 2000) came from.

SOM changes the representation of the data, it can reduce the dimensionality of data, and its results can be used for clustering, classification, visualization, etc. We feed a multi-dimensional dataset into a SOM, it outputs a grid representation of the input dataset. Similar to other dimensionality reduction methods that we have described in Chapter 6, the SOM algorithm projects data into a different dimensional space while maintaining the topological properties of the data. Topological properties ensure that the relative relationships and neighborhood structures in the data are preserved during the projection of data into different dimensions.

SOM is a very special type of ANN. It does not have an activation function and hidden layers. From the input layer, it gets directly to the output layer. Besides, its weight assignment process is not based on backpropagation; it uses a different approach to adjusting weights.

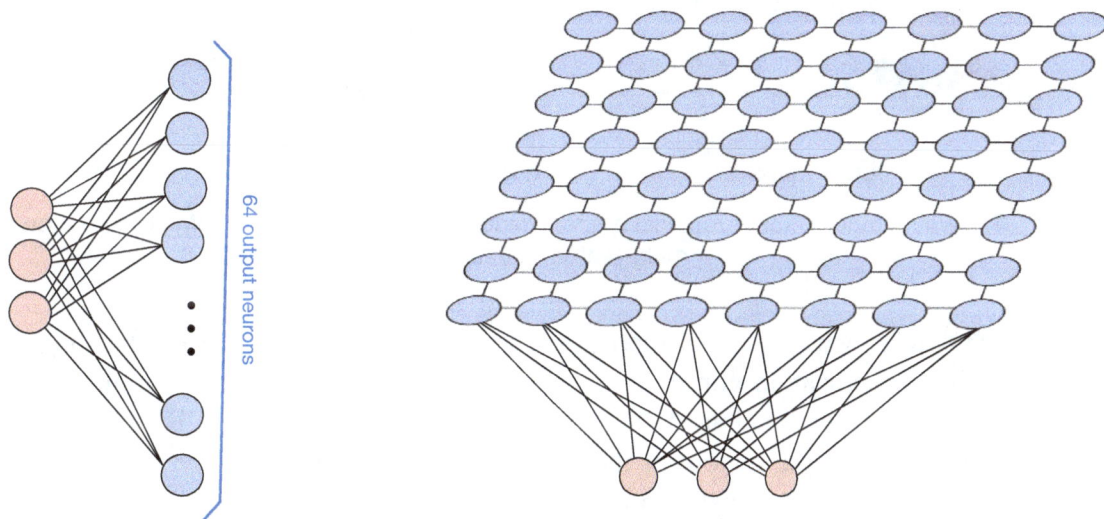

Figure 11-3: A sample SOM, whose red neurons are the input and blue ones are the output. Until now, we show neural networks like the left side, but SOM is usually shown as the right side of this figure, which is a rotated version of the left side (matrix of neurons).

The input layer of SOM is a vector equal to the number of features. For example, if we have a five-dimensional dataset, e.g., a table with five columns, the input vector includes five neurons. Assuming the dataset has n number of data points, it is recommended to have $5 \times \sqrt{n}$ neurons in the output [Tian '14]. For example, if our dataset includes 164 records, it is recommended to have $5 \times \sqrt{164} \approx 64$ output neurons. Take a look at Figure 11-3; there, we have a dataset with three features (three input neurons), it includes 164 records, and the number of output neurons is $5 \times \sqrt{164} \approx 64$.

The output layer of SOM can also be visualized as a *topographical map*, which presents the data in a lower dimension. Usually, the output of SOM is presented on a colored network of circles or hexagons in two dimensions. Colors present labels of the data or group of data, and each hexagon or circle presents one output neuron (see Figure 11-4). The patterns of color correspond to the distribution of data. Usually, by looking at the topographical map, we could identify clusters of data, or compare two or more topographical maps to identify some correlations. For example, Figure 11-4 shows that by collecting many parents' advice, we realize that there is a correlation between life satisfaction and education.

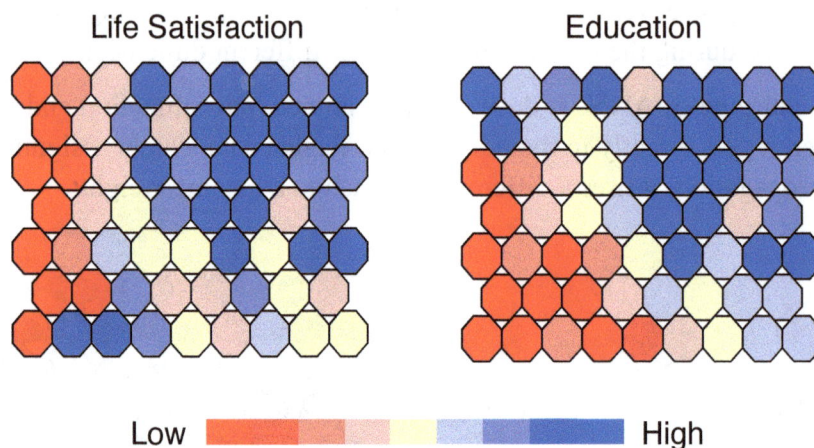

Figure 11-4: Two hexagonal topographical maps, which is the outputs of the SOM algorithm on a multi-feature dataset.

The result is shown in two hexagonal topological maps. By looking at these two images, we might see that higher education correlates with life satisfaction. Higher education is usually correlated with life satisfaction, but as presented from these two maps, it is not necessarily always true. We can see at the bottom of the Life Satisfaction map that there are a few points with low education but high life satisfaction.

Higher dimensional SOM is also possible, but it is not common. The blue layers in Figure 11-3 are the lower dimensional representation of the input dataset. The output layer of the SOM is called a *lattice*, a specific mathematical structure. In simple words, the lattice is a collection of data points that are arranged in a regular pattern, kind of like a grid, and have a special relationship to each other that helps us understand their properties. For example, a 2D grid lattice can be seen as a graph where each node has edges connecting it to its immediate neighbors (up, down, left, right). Therefore, while not every graph is a lattice, every lattice can be represented as a graph.

The SOM algorithm is implemented in four steps (initialization, competition, cooperation, and adaption).

(i) *Initialization:* All weights of the neural network are initialized with a small random number.

(ii) *Competition:* For every single input tuple (e.g., one record of a table), SOM computes a distance between the given input vector and weights of each node by using the following Euclidean distance, called the discriminant function.

$$distance(O_j) = \sqrt{\sum_{i=1}^{D}(x_i - w_{j,i})^2}$$

Here, x_i is the input neuron in high dimensional space, i is the index of input neuron, D is the dimensions of the input dataset or number of features, j is the index of output neuron, and $w_{j,i}$ is the weight between output neuron j and input neuron i. The node that has the smallest value is called the winner (o_2 in Figure 11-5) neuron. The final w is the presentation of data points in the projected dataset.

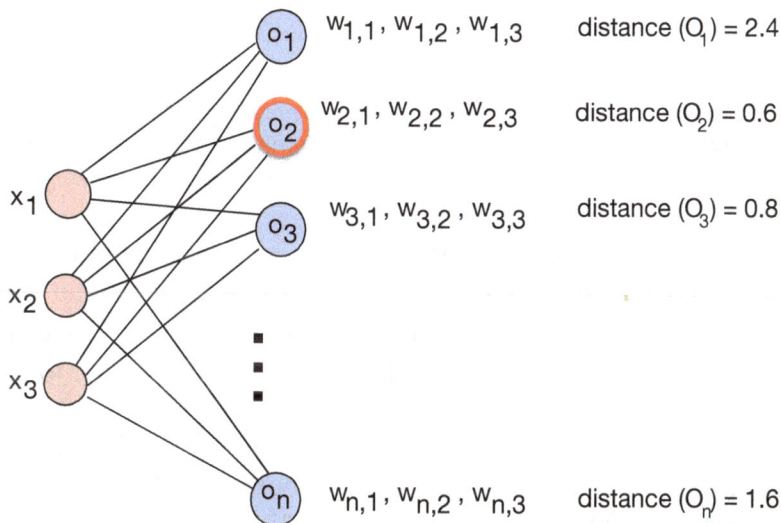

Figure 11-5: For each output node in the SOM, a distance between the given input vector and the weights of each node is calculated. Then, the node with the shortest distance is selected as the winner, i.e., here it is, and we mark it around with red.

To better understand this process, let's take a look at Figure 11-5. There, we fed one record from a table with three columns (for every column, assume one input neuron) into the network. Then, for every single output node, the distance between this node and the input record is calculated. o_2 is the winner node, which has the smallest value (the most similar one because its distance is the shortest one). The winning node is also called the *Best Matching Unit (BMU)*.

Because of this step in some literature, SOM is called competitive learning. In summary, at every round (i.e., a new row of data is processed), there is competition among neurons, and only one neuron gets activated. This neuron is referred to as "the winner takes all" neuron.

(iii) *Cooperation*: Now, a circle with σ radius is drawn around the BMU, and all nodes that fit inside this radius are assumed to be the neighbors of the BMU. The radius size around the BMU starts large, potentially covering the entire output layer's lattice, and decreases exponentially with each iteration, helping to fine-tune the learning process as the algorithm progresses, as shown in Figure 11-6. Each time a new record is fed into a network, the radius size will be changed (reduced) by the exponential rate.

(iv) *Adaptation*: After the BMU is specified, each node (output neuron) inside the radius adjusts its weights to be similar to the BMU. The neighbor nodes, which are closer to BMU, get higher

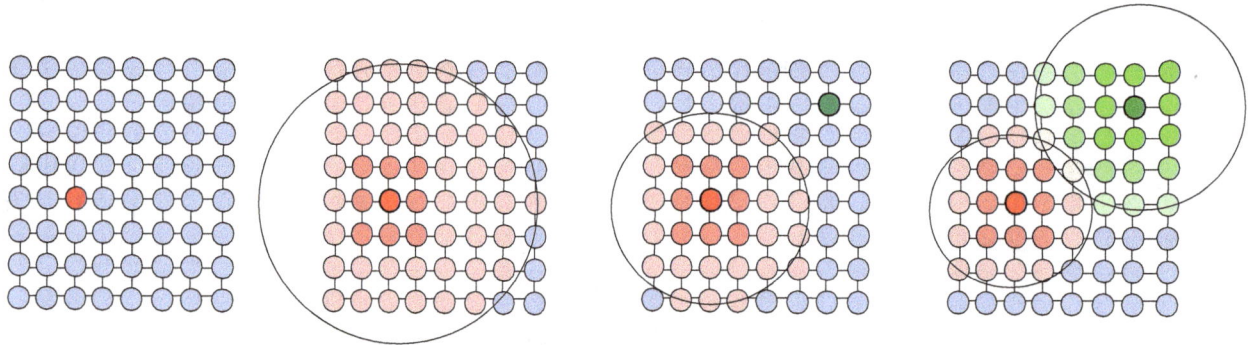

Figure 11-6: Output layer of SOM. From left to right, the first input record is processed, a BMU is identified with dark red, and with a large radius, neighbor nodes get its colors. Then, for the second input record, the green dot is identified. The red circle gets smaller (with an exponential decay rate), and this process continues for other records.

weight updates, and neighbor nodes apart from the BMU get lower weight updates. This process is visualized in Figure 11-6 by using the different intensities of the same color. When a new record is processed, the BMU changes (from the red dot to the green dot in Figure 11-6), and again, the weights of the new BMU neighbors will be updated to be similar to the BMU.

Steps (ii) to (iv) will be repeated iteratively until all input records are processed.

One common approach to present SOM output is using hexagonal topological maps (Hex map), as shown in Figure 11-4. Hex map tries to keep similar objects close to each other and dissimilar objects distant from each other. By looking at Hex maps, we can identify correlations between features of the dataset and use the map to identify clusters of data. Nevertheless, SOM is not a clustering algorithm or classification algorithm itself. Instead, it changes the representation of the data, and this result can be considered clustering. In particular, each input data point is assigned to the cluster represented by the winning neuron in the SOM.

Usually, after transferring the data into the lower dimension, it is easier to perform clustering, classification, etc. It is better to experiment with different representations, such as SOM and others that we will explain later, e.g., RBM, Autoencoder, then decide.

610

Boltzmann Machines

In the neural networks we've studied previously, data flows in a specific direction, starting from the input layer and ending at the output layer. However, the Boltzmann Machine operates differently—it is directionless. This means that a neuron can function as an input at one moment and as an output at another, with no clear distinction between input and output layers. Each neuron can both send and receive signals to and from any other neuron in the network. Unlike previous networks, the Boltzmann Machine lacks a defined output layer, and all neurons, including those typically considered inputs, are interconnected.

Boltzmann machines, similar to SOM, have two types of layers. A Boltzmann machine includes a visible (observed) layer, and unlike SOM, it has a

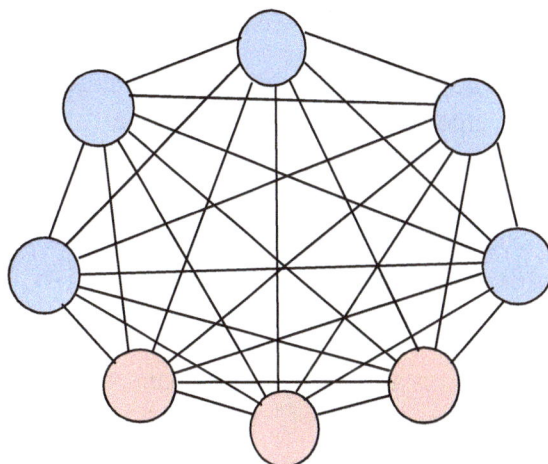

Figure 11-7: a Boltzmann machine, hidden nodes are presented in blue and visible nodes (input neurons) in red color.

hidden layer(s). The visible neurons are the information that we can measure, and the hidden layer neurons are information that we can not measure. Figure 11-7 presents a simple Boltzmann Machine.

The most prominent difference between Boltzmann Machine and other neural networks is that in addition to the input value that we give to the algorithm, the algorithm itself generates inputs for its next iterations as well. It might sound odd, but the output of one epoch will be considered as the input of the next epoch. In other words, we can say neurons are typically considered units that update their states based on the states of other units they are connected to. The concept of distinct input and output does not apply as it would in feedforward networks. A unique characteristic of Boltzmann Machines is their generative nature, where the network's state is updated stochastically based on a probabilistic model, thereby generating new sample data from the learned distribution.

The idea of the Boltzmann machine came from physics and the Boltzmann distribution (check Chapter 3). Boltzmann machines are Energy-Based Models (EBM). EBMs are popular when we need to build generative models with high-dimensional datasets that have sophisticated distribution. The ultimate goal of the Boltzmann machines is to iterate the network until it finds the thermal equilibrium (a characteristic of Boltzmann distribution).

We need to remember that while working with EBMs, we use the energy function instead of the cost function. The purpose of the energy function is to determine how likely each possible arrangement of the inputs is. In other words, the Energy function assigns energy (scalar value) to each possible configuration of the inputs. The result is a probability distribution over the inputs. The probability of a particular configuration is computed using the energy function and the Boltzmann distribution. When a system is in thermal equilibrium, its energy is usually minimized, and the energy function reaches its lowest value. Thermal equilibrium is about achieving a distribution where transitions between states are stable (i.e., the rate of entering any given state equals the rate of leaving it).

An energy function models the system, but a cost/objective/loss function is something we give to the algorithm as input to calculate the differences between the predicted value to the correct output. LeCun et al. [LeCun '06] stated that "*A loss score is minimized at the training (learning) phase, and energy is minimized at the testing (prediction) phase*".

Restricted Boltzmann Machine (RBM)

Restricted Boltzmann Machine (RBM) is an undirected neural network with two layers [Ackley '85], similar to other Boltzmann machines, a visible layer and a hidden layer. It is one of the early versions of self-supervised learning used for dimensionality reduction and feature learning.

Boltzmann machine neurons are all connected to each other, but RBM has no connection between the neurons of the same layers (no connection between visible neurons and no connection between hidden neurons), but there are connections from each visible neuron to all hidden neurons, and vice versa. Because of these limitations, the word restricted is used to reflect these characteristics in RBM. Compare Figure 11-7, which is a Boltzmann machine to Figure 11-8, which is an RBM. These small differences between RBM and the Boltzmann machine make the energy function of the RBM much simpler than the Boltzmann machine's energy function.

RBM has different use cases; it could be used for dimensionality reduction, classification, regressions, topic modeling, image denoising, etc. However, it has recently been substituted by more recent architecture, which we will explain later. Nevertheless, to learn them, we should be familiar with RBM.

Restricted Boltzmann machine   Boltzmann machine

Usually, we use a Binary RBM (Bernoulli RBM), its input in the visible layer is a binary vector, $v \in \{0,1\}$ and its hidden layer is also a binary vector, $h \in \{0,1\}$. Therefore, we can use binary encoding, such as one-hot encoding, to prepare its input data.

An RBM adjusts its weights and biases through multiple iterations of interaction between the visible and hidden layers. The *neurons in the hidden layer represent features that the model learns to capture from the data*. It's important to note that the RBM algorithm does not assign specific labels or meanings to these hidden neurons. However, by analyzing the activation patterns of these neurons in response to various inputs, we can infer what features each neuron represents and selectively activate specific neurons based on these insights.

Let's do a brief review of what we have explained. RBM is an EBM model, and EBM models are generative models that learn the underlying distribution of the data (training set). Once the model has learned this distribution, it is capable of generating new data points that resemble the original data, effectively reproducing the characteristics of the training set's distribution. In other words, the neural networks that we have learned are identifying a mapping between input and output neurons. Still, RBM defines the probabilistic distribution instead of mapping between input and output neurons.

Assuming i presents the index of a visible neuron (v), and j presents the index of a hidden neuron (h), the energy function of RBM will be written as follows:

$$E(v,h) = -\sum_i^D a_i v_i - \sum_j^M b_j h_j - \sum_i \sum_j v_i h_j W_{ij}$$

In the above equations, D is the total number of visible neurons, M is the total number of hidden neurons, v is a vector of visible neurons (units), h is a vector of hidden neurons (units), a is a bias of visible neurons, b is the bias of hidden neurons and W_{ij} is a matrix of weights between v_i and h_j. An RBM network uses this energy function to find patterns in the data by *reconstructing the input*.

This equation shows the energy as the sum of three terms: *visible neurons and their biases, hidden neurons and their biases*, and a *combination of hidden, visible neurons and their weights* (edges that connect visible and hidden neurons). We can see from the described equation that RBM has two biases: b is a hidden layer bias that enables the RBM to produce activations on the

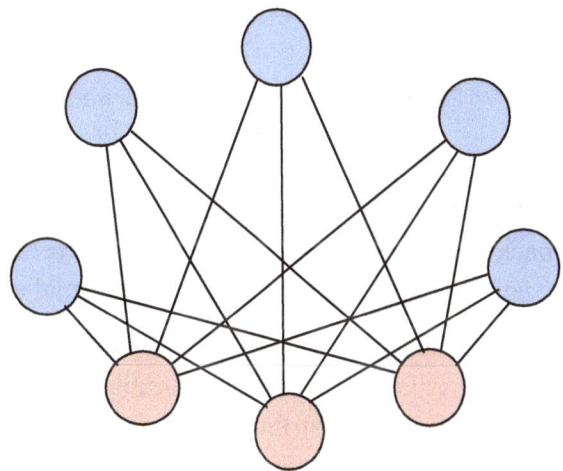

Figure 11-8: A sample Restricted Boltzmann Machine, hidden neurons are presented in blue and visible neurons (observation) in red. It is very similar to Figure 11-7; the only difference is that nodes are disconnected.

forward pass, and a is the visible layer bias that enables the RBM to reconstruct the input in the backward pass. Weights (W_{ij}) in RBM are presented in a matrix in which each of its rows presents a visible neuron, and each of its columns presents a hidden neuron. Figure 11-9 visualizes both biases (presented as a and b) and weights on a simple RBM with three visible neurons and three hidden neurons, and it visualizes a forward pass and backward pass.

The Intractable Problem of Z in RBM

The process of training RBM involves configuring biases (a_i, b_j), weights (w_{ij}), and visible/ hidden states (v_i, h_i), toward minimizing the energy for getting into thermal equilibrium and

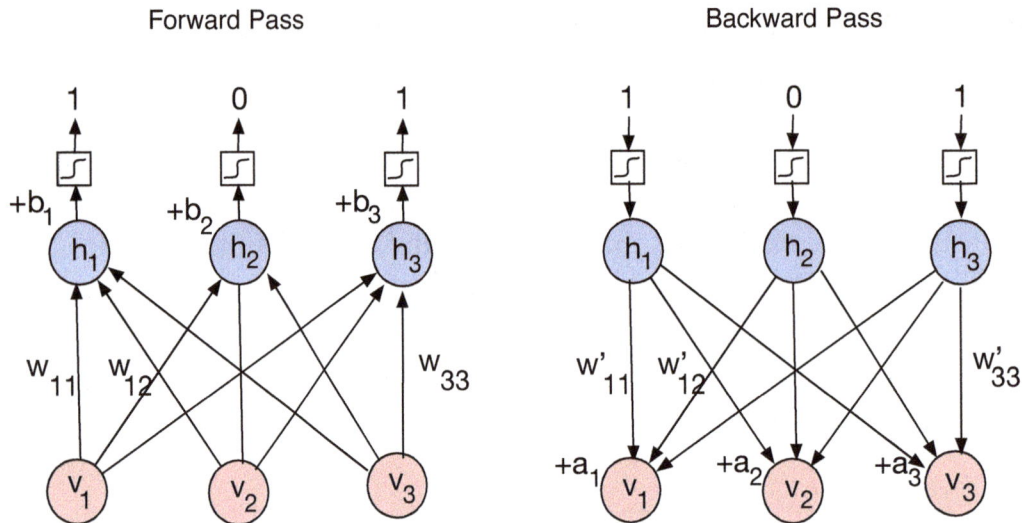

Figure 11-9: Visualizing the forward pass versus backward pass in RBM. The input and output of RBM is binary and its activation function is Sigmoid.

constructing the $p(v, h)$ distribution. In simple words, the model makes several forward and backward passes between the visible layer and the hidden layer and configures weight and biases (see Figure 11-7) toward reaching thermal equilibrium.

In the training phase, the RBM needs to model the joint probability distribution of observed variables along with hidden variables, i.e., $p(v, h)$. The following equation is used to model this joint probability distribution.

$$p(v, h) = \frac{1}{Z} e^{-E(v,h)}$$

This equation shows that the probability of a state between v and h is inversely related to the energy function. In this equation, Z is called the *partition function* or *normalization factor* and is computed as follows: $Z = \sum_v \sum_h e^{-E(v,h)}$

As we can see from the above equation, Z is the sum of *all values* for all possible combinations of visible and hidden (all possible states). v and h are vectors of 0s and 1s (a.k.a *Bernoulli vectors*). Assuming D is the number of visible neurons, and M is the number of hidden neurons, the number of all possibilities will be $2^M \times 2^D = 2^{D+M}$. Now, do you smell the smoke? Yes, the smell of smoke comes from the computer that is going to compute Z. It is clear that Z is intractable, and it is not possible to compute it with the brute force method (check Chapter 5 for the definition of brute force). Let's review the problem again. The objective of the RBM

614

algorithm is to find all $p(v, h)$ and compute all possible combinations in Z, which is computationally intractable. Therefore, we should look for an approximation method to mitigate this problem.

To identify the joint distribution of $p(v, h)$, we can calculate conditional distributions $p(v \mid h)$, and $p(h \mid v)$, which is the basis of *Gibbs sampling*. By using Gibbs sampling (we will explain this sampling method in Chapter 16), we can approximate the $p(v, h)$, via conditional probabilities without directly computing the partition function Z.

Neuron Activation Probabilities

By using Gibbs sampling (we know v presents visible layers and h hidden layer), the probabilities of the visible layer and hidden layer in RBM can be decomposed as the following joint probabilities:

$$P(h \mid v) = \prod_i P(h_i \mid v)$$

$$P(v \mid h) = \prod_j P(v_j \mid h)$$

The activation function for each neuron in RBM is a Sigmoid function, and if a neuron is in state 1, it is activated. Otherwise, it is inactive with a value of 0. Therefore, given the visible vector v, the probability for a single hidden neuron h_j activated is written as follows:

$$p(h_j = 1 \mid v) = Sigmoid(b_j + W_{ij} . v_i) = \frac{1}{1 + exp(-b_j - \sum_{i=1}^{D} v_i W_{ij})}$$

Respectively, given a hidden vector h, the probability for a single visible neuron v_i activated is written as follows:

$$p(v_i = 1 \mid h) = Sigmoid(a_i + W_{ij} . h_j) = \frac{1}{1 + exp(-a_i - \sum_{j=1}^{M} h_j W_{ij})}$$

RBM Training

$p(v)$ presents the likelihood of the distribution of visible neurons, and it is calculated as follows:

$$p(v) = \frac{1}{Z} \sum_h e^{-E(v,h)}$$

However, since it includes Z and we can not calculate it, we use the energy of visible neurons $E(v)$ to approximate it, i.e., $E(v) = -log\ p(v)$. We can see $E(v)$ is a negative log-likelihood of $p(v)$, why do we say $-log()$? If you recall, while describing maximum likelihood estimation (MLE) in Chapter 3, we said that instead of maximum likelihood, we use a negative log-likelihood to have a minimization and then inverse it (for the sake of computational efficiency). Therefore, we can state that RBM includes minimizing the negative log-likelihood for $p(v)$.

The derivative of likelihood with respect to weight equals the expectation of the given data (or actual), $\mathbb{E}_{data}$ minus the expectation provided by the model (or predicted), $\mathbb{E}_{model}$.

$$\frac{\partial log\ p(v)}{\partial w_{ij}} = \mathbb{E}_{data} - \mathbb{E}_{model}$$

or

$$w_{ij}(t+1) = w_{ij}(t) + \eta \frac{\partial log\ p(v)}{\partial w_{ij}}$$

Here η is the learning rate. Recall from Chapter 3 that $\mathbb{E}$ denoted the expectation and $\mathbb{E}_{data}$ (positive gradient) means that we update the hidden neurons at the beginning, where visible neurons are the input we give into the network, and hidden neurons are constructed by the initial values of visible neurons. After a certain number of iterations, visible and hidden vectors are updated, and the model is constructed, i.e., $\mathbb{E}_{model}$ (negative gradient).

The differences between the two expectations will be used to identify the optimal values for weights. This means that after $\mathbb{E}_{data}$(actual) and $\mathbb{E}_{model}$ (predicted) have been computed, the weight matrix and biases could be computed.

Now a question arises: how did we identify $\mathbb{E}_{data}$ and $\mathbb{E}_{model}$? To identify $\mathbb{E}_{data}$, RBM use the following algorithm:

$S = 0$ # S is a matrix with the same shape as the weight matrix W.
$for\ each\ v_t\ in\ (1\ to\ T)$ {# T is the number of visible neurons
      # generates a sample of hidden variables given the visible vector v_t
      $sample\ h \leftarrow p(h = 1\,|\,v_t) = \sigma(b + W^T v_t)$
      $sample\ S \leftarrow S + v_t h^T$
}
$\mathbb{E}_{data} \leftarrow \frac{1}{T}S$

After calculating the $\mathbb{E}_{data}$ with direct sampling, we can calculate $\mathbb{E}_{model}$ with plain Gibbs sampling, but it is too slow. This is due to the fact that one iteration of Gibbs sampling consists of updating all of the hidden neurons in parallel, followed by updating all of the visible neurons in parallel. To improve its training time, the performance RBM uses an algorithm called *Contrastive Divergence (CD)* [Hinton '02]. The CD estimates the energy function's gradient using a set of model parameters and the training data as input. The CD makes a series of approximations to the true posterior distribution $p(h\,|\,v)$ over the hidden units using a *limited number* of Gibbs sampling steps.

Following is a simplified description of the CD algorithm. In this algorithm, instead of iterating n times, if we use a condition that continues the loop until it converges, then it is plain Gibbs sampling. However, the CD applies a small, subtle change that reduces the number of iterations. In summary, the CD algorithm continues to change visible and hidden neurons back and forth in fewer iterations than Gibbs sampling.

$Q = 0$ # a matrix that has the same shape as weight matrix W and it is used to approximate p

$for(1\ to\ k)$ { # k is a small number, and it is used to perform k times Gibbs sampling, not until it converges. Of course, larger k results in better accuracy, but at the expense of slower convergence.

sample $h \leftarrow p(h = 1 \mid v) = \sigma(b + W^T v)$

sample $v \leftarrow p(v = 1 \mid h) = \sigma(a + Wh)$

$Q = Q + vh^T$ # Re-update the hidden neurons given the reconstructed visible neurons using the same equation

}

$$\mathbb{E}_{model} \leftarrow \frac{1}{k}Q$$

Note that the loop does not stop after it converges. Instead, it performs k number of iterations, and then it stops. After both $\mathbb{E}_{data}$ and $\mathbb{E}_{model}$ are specified by the algorithm, assuming ϵ is the learning rate, the algorithm uses gradient ascending (not descent) to compute weight and biases as follows:

$$\Delta W_{i,j} = \epsilon(\mathbb{E}_{data}[v_i, h_j] - \mathbb{E}_{model}[v_i, h_j])$$

$$\Delta a_i = \epsilon(\mathbb{E}_{data}[v_i] - \mathbb{E}_{model}[v_i])$$

$$\Delta b_j = \epsilon(\mathbb{E}_{data}[h_j] - \mathbb{E}_{model}[h_j])$$

As we have explained, RBM is a historical algorithm. Through learning history, we cultivate an understanding of how later, more practical algorithms have been designed. Nevertheless, do not underestimate its capabilities. In some scenarios, to analyze data, we usually pass the dataset to SOM, RBM, and Autoencoder and decide which one provides the best accuracy.

Deep Belief Network and Deep Boltzmann Machine

Deep belief network (DBN) and Deep Boltzmann Machine (DBM) are two models inspired by RBM. DBN is composed of stacking multiple hidden layers, which can be trained independently (not necessarily using RBMs). DBM stacking RBM layers directly on top of each other. Each RBM's hidden layer becomes the visible layer for the next RBM in the stack (see Figure 11-10). They both can be used for generative tasks such as image generation and dimensionality reduction.

What is the advantage of stacking multiple RBMs? The first RBM contains latent features that are better than random inputs. The hidden layer of the second RBM contains a combination of hidden features that are better than random inputs, and each layer includes more accurate hidden features.

DBM is trained using the CD algorithm. The training phase in DBN includes two stages: *unsupervised pre-training* and *supervised fine-tuning*. The pre-training initializes weights on the network, but every hidden layer is trained based on its given input (only its given input, not the previous layers' data). It is a greedy layer-wise approach (short-sighted to one previous layer and not all previous layers), and thus this training can stick in local optima. Because of this limitation, it is called pre-training.

After pre-training, the algorithm performs fine-tuning, and this step is used to correct weights in a supervised manner with Backpropagation. At the fine-tuning stage, the output layer is added (it is not present in Figure 11-10 because this figure focuses on pre-training only), and supervised learning is used to train the network with forward and backward propagations. Therefore, fine-tuning assigns the final weight and biases in a supervised manner.

There is not many implementations existed for DBN and DBM, you should make them on your own from scratch.

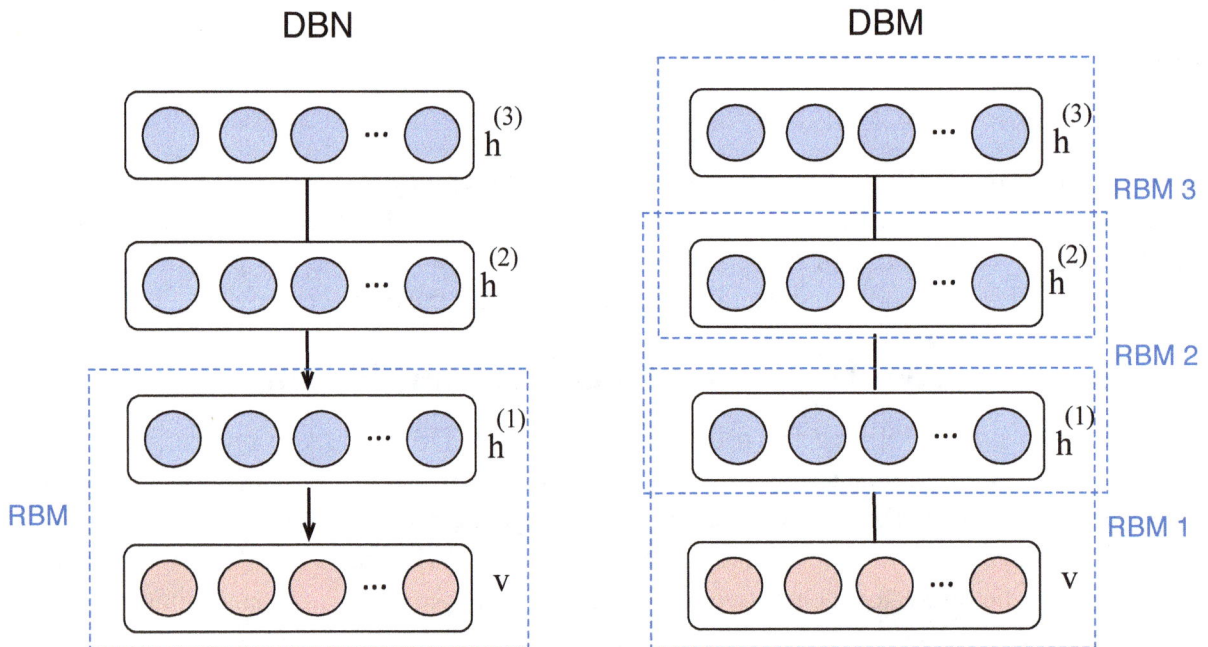

Figure 11-10: A sample DBN and DBM pre-training stage. The output of each RBM (hidden layer) is the visible layer of the next RBM. Arrows in this figure present the direction of the generative model (Gibbs sampling). For example, updated values in $h^{(1)}$ layer are used to reconstruct v layer, but in DBN, v layer can not be used to reconstruct $h^{(1)}$.

Keep in mind that both pre-training and fine-tuning stages in both DBN and DBM use a greedy approach to train and configure weights. Besides, in DBM, unlike DBNs, visible-to-hidden and hidden-to-hidden connections are trained simultaneously in an unsupervised manner.

By looking at Figure 11-10, we realize the arrowhead between the visible and the hidden layer. At the pre-training stage, the undirected connection between $h^{(1)}$ and $h^{(2)}$ means that Gibbs sampling of RBM once sample from $h^{(1)}$ to model $h^{(2)}$ and then once from $h^{(2)}$ to model $h^{(1)}$, but for a directed (undirected) connection from $h^{(1)}$ to v sampling goes from $h^{(1)}$ to v and not from v to $h^{(1)}$. In other words, it uses a model to generate visible layer data.

You might ask how the DBN model can skip v to $h^{(1)}$, and how it can start from the top layer $h^{(3)}$? The algorithm does not skip weight assignment (training) from v to $h^{(1)}$. It performs weight training once and goes up using the same weights on all layers until it reaches the top layer, i.e., $h^{(3)}$. Because of that, it is called pre-training. Then, in the next step, the Gibbs sampling and weight adjustment (fine-tuning) start from the top layer until they reach the directed layers.

Why do we need to stack lots of RBM and train them (DBN or DBM)? Why not use a deep, feed-forward approach?

The motivation is that since RBM uses a contrastive divergence algorithm [Hinton '95], there is no use of backpropagation in the pre-training stage while using a discriminative model. By not using backpropagation, it eliminates the vanishing or exploding gradient problem that existed at this stage. The fine-tuning uses backpropagation, but it starts with well-initialized weights in hidden layers.

Similar to MLP, these architectures have input, hidden layers, and output layers. The output of each RBM is an input of the next RBM, as it is shown in Figure 11-10 However, unlike other deep learning algorithms, each layer learns to represent the data (input) passed to it from the previous layer. For example, while using CNN, the first layer gets the original input, and the next layers work with the convolved input data, not the original input data, but both DBN and DBM learn from the input received from the previous layer.

To summarize their differences, DBN uses two phases of training, pretraining and fine-tuning, and it is easy to sample from visible and hidden units. DBM uses contrastive divergence for its training, but it requires a complex approach to sample from visible and hidden units. To summarize their similarities, both use greedy layer-wise training (pre-training) features of RBM, and both use RBM to identify a latent feature of data.

Some literature claims that DBM operates better than DBN for learning complex models [Wang '17], but, as always, we recommend experimenting with both architectures on your own and deciding which one to use.

NOTES:

* A basic model of RBM is inspired by the Hopfield Network [Hopfield '82]. Hopfield network neurons contain binary neurons, which present binary attributes of the dataset. They create a deterministic model (not stochastic, unlike RBM) of the relationship among different features by changing weights on edges. If you intend to learn more about Hopfield networks, you can

check Aggrawal's book [Aggarwal '18], which has a detailed explanation of it, but it requires a strong mathematical background to understand it.

* By learning the weights, the RBM implicitly memorizes the training dataset; therefore, we could say the RBM is becoming the representation of the input data that we fed to it. RBM minimizes the energy or maximizes the probability. This feature of RBM enables both DBN and DBM to understand complex internal representations of the input data. Also, it can assist in improving the model by adding very few labeled data (semi-supervised learning), but still, RBM is called self-supervised learning.

* Contrastive divergence is a popular algorithm for training EBM algorithms. It operates based on the assumption that the derivative of the log-likelihood function can be defined as differences between two expectations ($\mathbb{E}_{data}$ and $\mathbb{E}_{model}$).

* DBN, DBM, and RBM are not widely used nowadays because Autoencoders outperform them. Nevertheless, learning them is useful for understanding the intuition behind the new models and algorithms. Some algorithms have not been in use for many years, and somebody extracts an idea from them and creates something useful, like Geoff Hinton, whose team created DBN, RBM, and tSNE and also applied backpropagation on DNN. Therefore, it is worth learning as much as we can, even though the algorithm seems not popular at the current time.

* Bengio [Bengio '09] has a good generalization on unsupervised deep learning models, such as DBN and autoencoders. He states: *"Unsupervised training signal at each layer may help to guide the parameters of that layer towards better regions in parameter space"*.

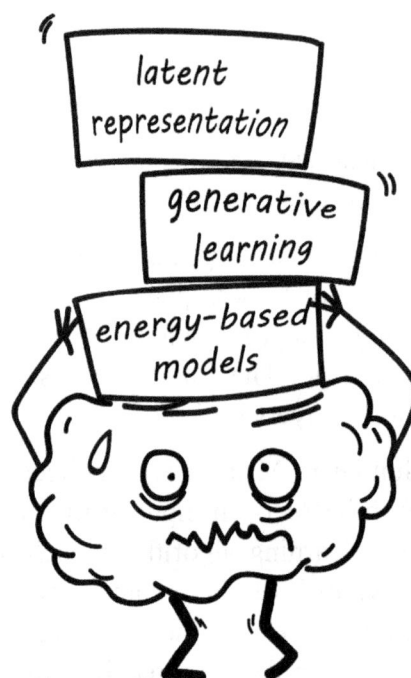

Autoencoders

Previous neural network models we have explained in this chapter are important to learn because they could inspire us while designing our algorithms. In this section, we discuss autoencoders, which, unlike SOM and RBM, are state-of-the-art data representation models.

If you skip learning Autoencoders you can not learn GAN, Transformer, and many other important concepts in Deep Learning.

Despite the superior accuracy of autoencoders due to resource consumption, we can not always use these complex neural network models, and using light algorithms such as SOM could resolve our need.

At the time of writing this chapter in 2021 and revising it back in 2024, autoencoders were one of the main components of most generative AI models. They are used in many applications, such as denoising data (e.g., watermark removal from watermarked images), better weight assignment in neural networks, image segmentation, text2image, synthetic video generation, and so forth.

Autoencoders are neural networks that reconstruct (not copy) their inputs in their output layer and try to make a reconstructed output similar to the original input. An autoencoder is composed of two components: an encoder and a decoder. The encoder creates a compressed representation of input data, i.e., transferring the input data into latent space. In other words, *latent space* is the representation of input data in hidden layers. Transferring data into a latent space is associated with compression, and this compression removes non-descriptive features (e.g., redundant and noisy) while keeping the important features. The decoder uses latent space to reconstruct the data, similar to the original data, but without its non-descriptive features. In other words, autoencoders' outputs are not identical to their inputs; they provide an approximate copy of the input. While training, the output entails important features of the input data and removes useless features of the input.

We also use something similar to autoencoders in our daily communications. Let's say you are healthy and ran 12,342 steps in the morning. Then you visit a friend in the evening and tell her you had a productive day because you ran more than 10,000 steps (you compress the data, and instead of 12,342 steps you say more than 10,000 steps). Then your friend visits another friend of hers and tells him she has a friend (you) who is an athlete, and also that person is reading an amazing data science book. She used to reconstruct the data you provided to her, i.e., you said "I ran more than 10,000 steps" and she compressed and decompressed it as you are an "athlete".

Take a look at Figure 11-11. We add noise to an image from the MNIST dataset and feed it into an autoencoder. The output removes the noise because it reconstructs the data by keeping the useful properties of the data and getting rid of non-useful features. The autoencoder has seen many '4' (plenty of clean images without noise) in the training set, and thus it learns to reconstruct '4' by removing unimportant features and only keeping the important ones. If there is no '4' in the dataset, then the denoising was not successful.

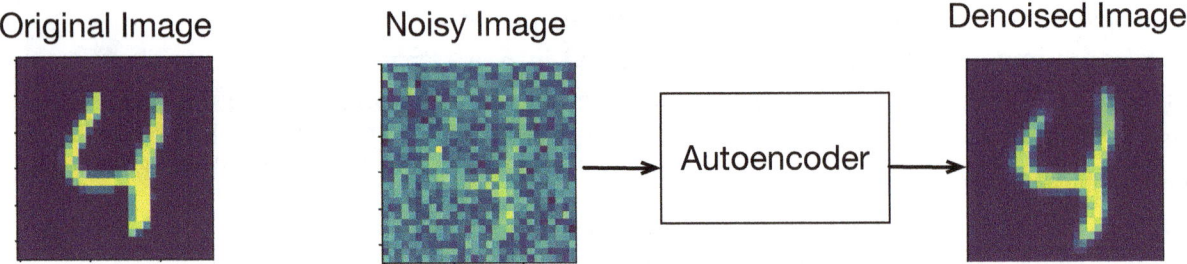

Figure 11-11: Autoencoder is used to Denoise a noisy image.

How do Autoencoders work?

The typical architecture of an autoencoder is presented in Figure 11-12. We have explained that during the encoding stage, the encoder removes unimportant features, and during the decoding stage, the decoder reconstructs the data by using its important feature, but how does this happen? The hidden layers in autoencoders typically have fewer neurons, and thus, the dimensionality of the data (its features) is reduced. This reduction in the number of features has many useful applications and removes unnecessary features. An autoencoder has at least one hidden layer that includes the features used to encode the input data and change its representation (reducing its features).

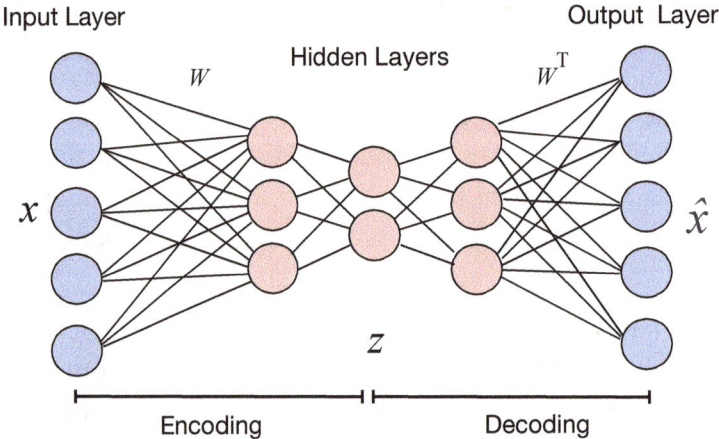

Figure 11-12: A simple one layer autoencoder which has only one layer of hidden layer.

Training an autoencoder involves three steps: (i) input dataset x is fed into the autoencoder, passing through the encoder to construct the latent space features. (ii) Then, it passes through the decoder, and the decoder provides output $\hat{x}$ as a result of x reconstruction. (iii) The cost function calculates the differences between $\hat{x}$ and x. Based on the cost (error), the backpropagation

622

algorithm is used to reconfigure weights and biases in the next iteration and backpropagates from the decoder to the beginning of the encoder.

After training the autoencoder, either the encoder or decoder could be used separately. For example, only an encoder could be used if our goal is to reduce the dimensionality of the data, or a decoder along with the encoder could be used to generate new data but with a pattern similar to the given input data. Since the distribution of the generated data (output) is the same as the distribution of given data (input), autoencoders can construct new data that is similar to the given input data.

We can interpret an autoencoder as a mathematical function. It has an encoder function $f(x)$, and receives input x. Another function is the decoding function g, which receives the encoding result and provides the output $\hat{x} = g(f(x))$. The cost function of autoencoders is to minimize the error of dissimilarity between input and output, $L(x, \hat{x})$ or $L(x, g(f(x)))$. A common cost function in autoencoders is the mean square error, and usually, they use non-linear activation functions (e.g., Sigmoid, ReLU, Hyperbolic Tangent). If we use the linear activation function, the dimensionality reduction of the autoencoder will be similar to PCA.

Now that we understand the basics of autoencoders, we need to be familiar with some specific types of autoencoders. Autoencoders whose hidden layers have fewer neurons than their input layer are called *undercomplete autoencoders*, such as the one presented in Figure 11-12. Autoencoders whose hidden layers have more neurons than input layer neurons are called *overcomplete autoencoders*, such as Figure 11-13. An autoencoder is called a *deep autoencoder* if it has more than one layer of hidden layers, such as Figure 11-12. Most of the time, we use deep autoencoders to solve our problems.

Autoencoders Can Cheat

The magic of an autoencoder occurs when it achieves an abstract representation of the input data, often facilitated by having a smaller number of activated neurons in the hidden layers. For example, the hidden layers in Figure 11-12 are smaller than the input layer. However, experiments have shown that increasing the number of hidden neurons can enable the network to construct a more accurate representation of the input data. Therefore, in some autoencoder models, the hidden layers typically contain more neurons than the input layers. On the other hand, having more neurons in the hidden layers than in the input layers might cause the network to simply copy the input to the output without any meaningful transformation (i.e., acting as an identity function). Figure 11-13 visualizes this problem. Such behavior, often referred to as cheating, should be avoided. Below, we describe three types of autoencoders that avoid this problem while still benefiting from having more hidden neurons than input neurons.

Autoencoder Types

Autoencoders are grouped into two main categories: regularized autoencoders and variational autoencoders. Common regularized autoencoders include sparse autoencoders, denoising autoencoders, and contractive autoencoders. They provide a regularization method to prevent the output layer from copying the input layer data (act as an identity function). To simplify the

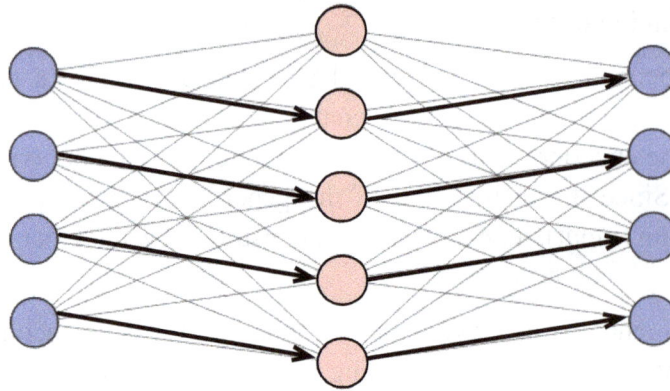

Figure 11-13: An over complete autoencoder that its hidden layer acts as identify function and input nodes are copied exactly as they are into the output node.

motivation behind using regularization in autoencoders, consider how we study for a mathematics exam. We need to learn concepts rather than memorize them because learning enables us to solve problems we haven't encountered before. In contrast, mere memorization is akin to copying and pasting answers, which is ineffective for addressing new or unfamiliar questions using our existing knowledge base.

After we are done with regularized autoencoders, we delve into a more advanced autoencoder, the variational autoencoder, which can generate data very close to the given input data [Kingma '13].

Sparse Autoencoder

Sparse autoencoders (SAE) [Makhzani '13, Ng '11] are typically overcomplete autoencoders, meaning their hidden layers are larger than the neurons of their input layer. When the number of hidden nodes exceeds that of the input, the autoencoder effectively increases the dimensionality of the data, which can lead to the extraction of more useful features. This characteristic is particularly beneficial for algorithms that aim to identify descriptive features of a dataset. Nevertheless, adding the extra hidden neurons requires certain measures to prevent the autoencoder from acting as the identity function (copy/paste of input neurons and cheating).

SAE regularizer disables some neurons at any pass, and thus, the autoencoder will be limited to work with a smaller number of neurons than input neurons, but in the next pass, those disabled neurons will be enabled, and a new group of neurons will be disabled (see Figure 11-14). In practice, it is similar to under-complete autoencoders, which use a smaller number of neurons, but in every iteration, it uses a different set of nodes.

The cost function of the SAE is written as $L(x, \hat{x}) + \Omega(f(x))$. Here, $L(x, \hat{x})$ is the reconstruction loss, and $\Omega(f(x))$ is the regularizer that receives the encoder output $f(x)$ as its input. Unlike regularizers, as explained in Chapter 8, these regularizers do not have weight decay. Instead, they measure the hidden layer activations for each training dataset and add a penalty to the loss function to penalize excessive changes in the activation.

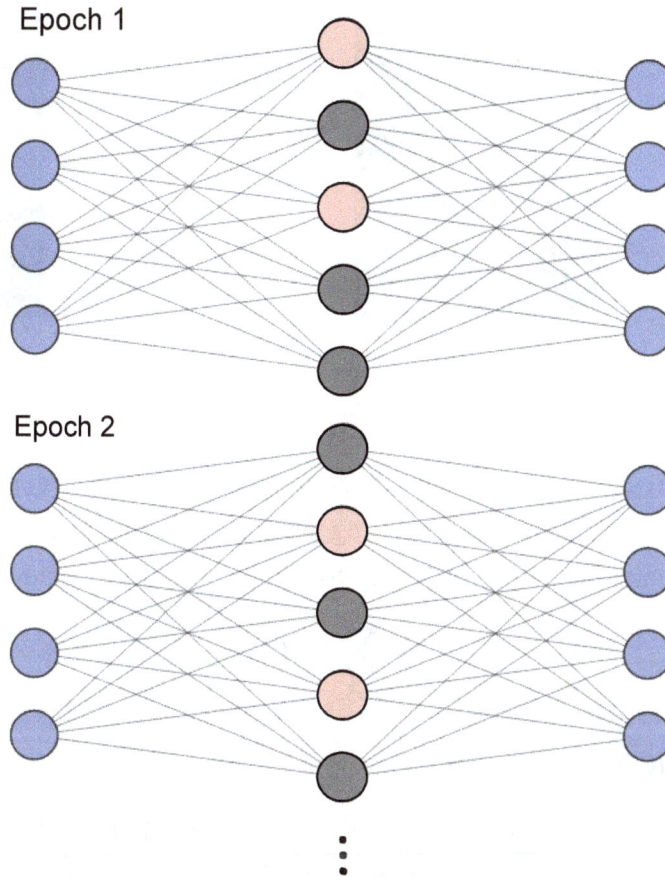

Figure 11-14: An example of the sparse autoencoder, which has two neurons of its hidden layer activated at each epoch. The grey neurons are regularized. Hence, their value is too small to have any significant impact on the output.

Two common methods to implement $\Omega(f(x))$ are L_1 regularization and KL-Divergence. The L_1 regularizer uses L_1 norm as a penalization score. Assuming we have n number of hidden neurons, a_i the activation of ith neuron, its cost function will be written as:

$$L(x, \hat{x}) + \lambda \sum_{i=0}^{n} |a_i|$$

We have explained in Chapter 3 and Chapter 6 that KL-Divergence measures the differences between two statistical distributions. The KL-Divergence regularizer defines a *sparsity parameter* ρ that specifies the desired level of sparsity. In other words, it is the average activation of neurons in a hidden layer over the sample of m observations (m is the number of data points in the training set). Usually, ρ is set to a small variable such as 0.05.

$\hat{\rho}_j$ specifies the *desired* sparsity level. Assuming $a_j(x^{(i)})$ is the activation function of the hidden neuron j, and m number of input data gets into neuron j, $\hat{\rho}_j$ is calculated as follows:

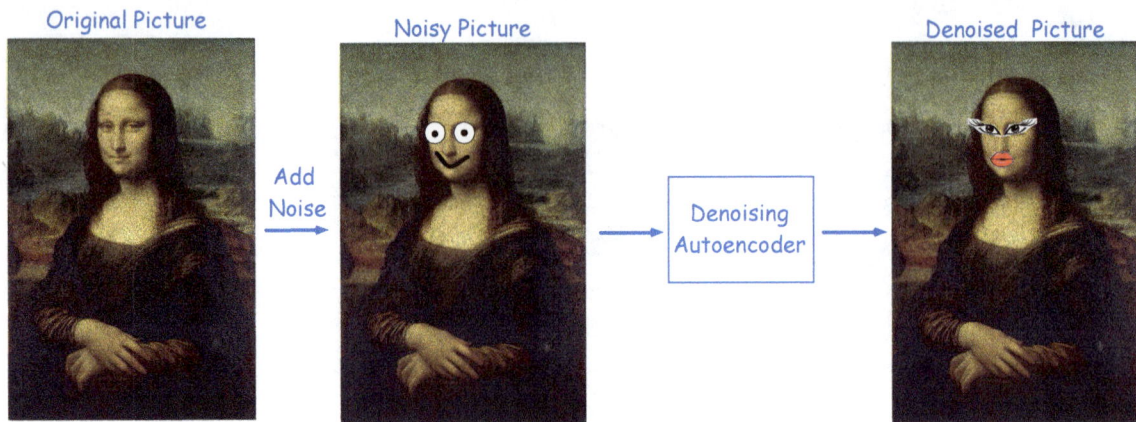

Original Picture   Noisy Picture   Denoised Picture

Add Noise

Denoising Autoencoder

$$\hat{\rho}_j = \frac{1}{m} \sum_{i=0}^{m} [a_j(x^{(i)})]$$

The KL-divergence between ρ and $\hat{\rho}_j$ distributions can be used as the penalty function. The penalty will be 0 if $\rho = \hat{\rho}_j$. Therefore, assuming we have n number of hidden neurons in the hidden layer, the cost function of SAE with KL-Divergence penalty is written as follows:

$$L(x, \hat{x}) + \sum_{j=1}^{n} KL(\rho || \hat{\rho}_j)$$

Denoising Autoencoder

The objective of denoising autoencoder (DAE) [Vincent '10] is to have a representation that can handle noise. Figure 11-15 is a denoising autoencoder. Similar to sparse autoencoders, the hidden layers of denoising autoencoders usually have more neurons than the input layer (overcomplete but could also be undercomplete). Therefore, again, some neurons get inactive, and a regularizer is used to make those neurons inactive.

The denoising autoencoder first adds some noise to the input data and then feeds the noisy input data, instead of the original input data, into the encoder. Many types of noise can be used to make an image noisy, like Gaussian noise, Perlin noise, etc. In Chapter 16, we will provide a more detailed description of different types of noise.

Figure 11-15 presents the architecture of the denoising autoencoder. Assuming the original data is x and the original data with added noise is x', the process of adding noise is a probabilistic process $P(x'|x)$. Nevertheless, the cost function $L(\hat{x}, x)$ does not compare $\hat{x}$ (output data) to x' (input data with noise). Instead, it compares the $\hat{x}$ (output data) to x (original input data).

As shown in Figure 11-15, denoising autoencoders change the input neurons first, for example, setting some of them randomly to 0. It has a hyperparameter that specifies the percentage of neurons that should be randomly deactivated (turned off) during each pass by the denoising function. Since, in every pass, a random number of neurons are changed, this type of autoencoder is called a *stochastic autoencoder*.

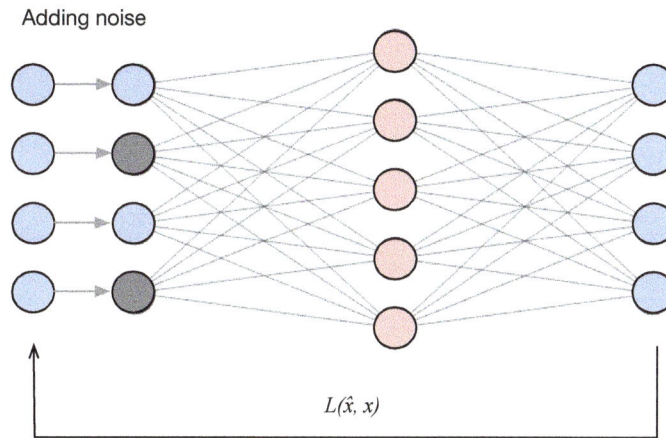

Adding noise

$L(\hat{x}, x)$

Figure 11-15: Example of denoising autoencoder. In each iteration the input layer with noise will be added to the network. Afterward, the cost function compares the output layer to the original input layer to tune weights and biases for the next iteration.

Contractive Autoencoder

A contractive autoencoder [Rifai '11] is an overcomplete autoencoder (usually but not always) that encourages the model to provide similar encoded output for similar input data. In simple words, similar input data should yield similar output data (i.e., the result of encoding).

It operates by incorporating a penalty into the cost function. It applies a regularization penalty to the activation function (not on weights but only on activation function results). The penalty tries to ensure that small changes in the input do not change the output (encoding result) significantly, and this small change maintains a very similar change in the encoded state. In other words, with respect to the variation in input data, the variation of activation function in neurons of hidden layers should be small. For example, if two input data have small differences and are fairly similar, the activation values of the hidden layer neurons for these two input data should be small as well. Some scientists describe a contractive autoencoder as extracting features that only reflect variations observed in the training set, and it is invariant to the other variations.

The cost function of the contractive autoencoder is written as: $L(x, \hat{x}) + \lambda(||J||_F)^2$, which λ is a hyperparameter that controls the strength of the regularizer, $||J||_F$ is the *Frobenius norm[2]* on the Jacobian matrix[3] of hidden layer activations with respect to inputs. You can check Chapter 8 to recall the Jacobian Matrix. Assuming n presents the number of neurons in the hidden layer(s) and m presents the number of input neurons, $a_i(x)$ is the activation of hidden neuron i for the

---

[2] Frobenius norm (Euclidean norm or L_2 norm) of a matrix A with m number of rows and n number of columns is written as follows:

$$||A||_F = \sqrt{\sum_{i=1}^{m} \sum_{j=1}^{n} |a_{i,j}|^2}$$

[3] The Jacobian matrix captures the variation of weights of output vectors with respect to the input vector.

given x input, we can write Frobenius norm on the Jacobian matrix of hidden layer activations with respect to inputs as follows:

$$||J||_F = \sqrt{\sum_{j=1}^{m} \sum_{i=1}^{n} \left| \frac{\partial a_i(x)}{\partial (x_j)} \right|^2} = \begin{bmatrix} \frac{\partial a_1(x)}{\partial x_1} & \frac{\partial a_1(x)}{\partial x_2} & \cdots & \frac{\partial a_1(x)}{\partial x_m} \\ \frac{\partial a_2(x)}{\partial x_1} & \frac{\partial a_2(x)}{\partial x_2} & \cdots & \frac{\partial a_2(x)}{\partial x_m} \\ \vdots & \vdots & \ddots & \vdots \\ \frac{\partial a_n(x)}{\partial x_1} & \frac{\partial a_n(x)}{\partial x_2} & \cdots & \frac{\partial a_n(x)}{\partial x_m} \end{bmatrix}$$

In this Jacobian matrix, n is the number of activation functions, and m is the number of input neurons. Each row refers to a *gradient of one hidden neuron's output with respect to all input neurons*. Each column refers to all hidden neurons' activation gradients with respect to one single input neuron. For example, the first row gives a gradient of the first hidden neuron output with respect to all inputs, or the first column gives us all hidden neurons' activation gradients with respect to the first input neuron.

If the partial derivative is zero, changing input (x_j) does not change the activation value of the hidden unit, i.e., $a_i(x)$. We have understood that a contractive autoencoder makes the feature extraction function (i.e., encoder) resist small changes in the input. How does this happen? By using a penalty, which is a Frobenius norm of the described Jacobian matrix.

A denoising autoencoder is straightforward to implement and does not require the computation of the Jacobian of hidden layers. However, Contractive autoencoders tend to be more stable than denoising autoencoders because they introduce penalties that measure the sensitivity of the hidden representation to small changes in the input. These penalties help maintain stability by ensuring the model's output does not vary significantly in response to minor input changes.

Stacked Autoencoder

Sometimes, the input dataset is complex, and one single autoencoder can not remove unnecessary information to provide useful features. Besides, adding more hidden layers might cause underfitting or overfitting, and it is not easy to optimize weights in non-linear autoencoders with several hidden layers. This issue is handled by introducing a *greedy layer-wise pretraining* [Hinton '06, Bengio '07]. This pretraining is shortsighted, which means that each layer is trained based only on the output from the immediately preceding layer rather than on the outputs of all previous layers.

Why does greedy layer-wise training resolve the challenges associated with having too many hidden layers?

When a neural network gets deep (it has many hidden layers), its hidden layers' weights close to the output layer are updated very frequently, but hidden layers' weights close to the input layer might not get updated at all, which can lead to inefficient training and poor convergence (issues often exacerbated by the vanishing gradient problem). Greedy layer-wise pretraining addresses this by training one hidden layer at a time. Each layer is trained using the output from the previous layer as input, effectively isolating the learning process for each layer.

For example, we have a small stacked autoencoder that gets input X, and by using the labeled data, it trains it, and the hidden layer of this autoencoder outputs Z_1. The next autoencoder receives Z_1 an input, and in the same fashion, it produces output Z_2. The third hidden layer receives Z_2 as input (not Z_1, only the immediate previous layer output). This sequential process repeats until the last layer produces Z_n, which is then fed into the output layer of the network.

Figure 11-16 presents a stacked autoencoder. A stacked autoencoder setup typically involves multiple autoencoders, each consisting of an encoder and a decoder. In a typical use-case for classification, only the encoder components might be utilized, and the last layer is usually a Softmax (or logistic regression), which has a randomized initial weight. Randomized initialized

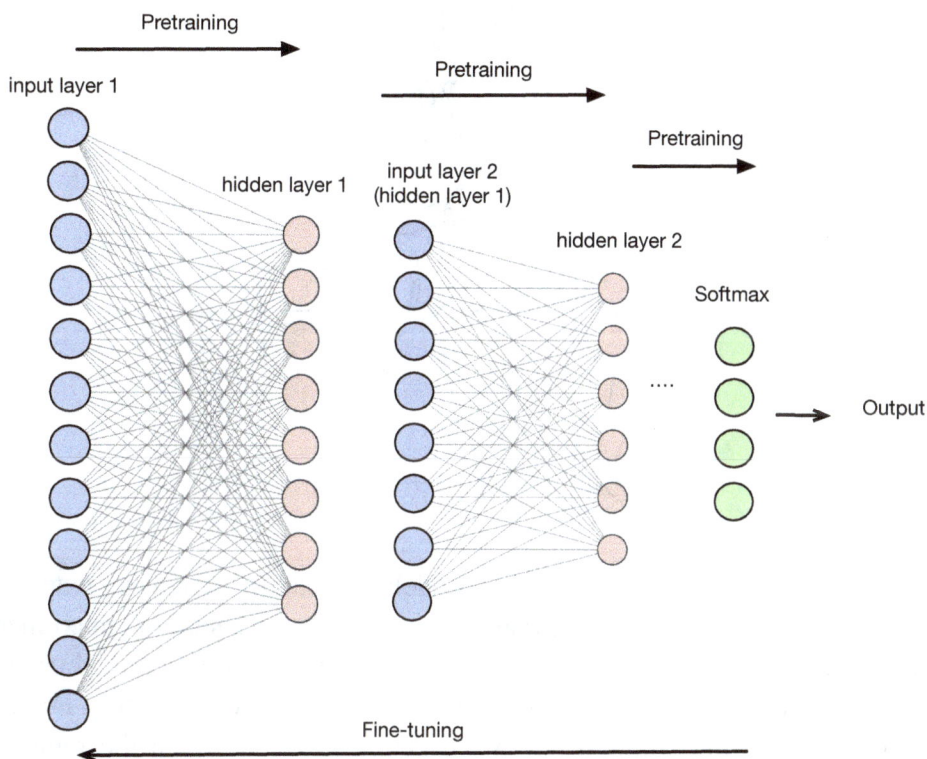

Figure 11-16: Stacked autoencoder with two autoencoders (this example has only an encoder, and there is no decoder), is followed by a logistic regression or a softmax layer for classification. Note that each encoder is trained separately in a greedy layer-wise manner using the labeled training data, and then fine-tuning stack them all together and train them.

weights here refer to the weights that must be corrected, and thus, we need to perform backpropagation. The backpropagation process adjusts the weights to approach optimal values, making the path of backpropagation less complex compared to other neural networks. This adjustment phase is known as *fine-tuning*. Fine-tuning is a supervised process, relying on labeled data to compute the errors necessary for backpropagation and to refine the model effectively. This step follows after all the hidden layers have been pre-trained.

Variational Auto-Encoder (VAE)

All autoencoders we have described until now are regularized autoencoders. However, there is another variation of autoencoder, variational autoencoders (VAE) [Kingma '13], which is very popular. VAE is a combination of variational Bayesian inference and autoencoder. Bayesian inferences are one of the oldest machine learning approaches, and in this book we only explained Bayesian inferences in different places, such as Naive Bayes in Chapter 9, probabilistically graphical model in Chapter 5, and LDA in Chapter 7.

Bayesian inference   Autoencoder

Variational autoencoders (VAEs) are generative models that can be seen as an extension of latent variable models, which are similar in some respects to the expectation maximization (EM) algorithm (see Chapter 3). A VAE is specifically a type of latent variable model used to discover the underlying structure of a dataset containing hidden or unobserved variables. For example, if a document mentions terms like 'foreign policy', 'sanctions', and 'allies in the region', we might infer that its underlying topic is 'international politics'. This topic, not explicitly mentioned in the document, represents a latent variable. In contexts where the dataset is high-dimensional and noisy, it is common for classification algorithms to focus on extracting these latent features rather than attempting to utilize all existing features.

The Magic of Generative Models in Data Reconstruction

The VAE fits into the category of generative algorithms, and generative models are very successful in reconstructing synthetic data similar to the original data due to their focus on the distribution of data. It means that instead of working with the original dataset, they get the distributions of the original dataset and make a new dataset from the distribution of the original dataset. It leads to lots of success in generative AI applications.

How does such a thing happen? Assume we write an autoencoder to encode images of potatoes, carrots, and onions. A one-hot encoder can be used, and each of them can be presented with a bit vector, potato: [0,0,1], carrot: [0,1,0] and onion [1,0,0]. If the algorithm trained on white onion, and not red onion, it might not recognize the red onion. Their shapes are the same (something can be used for inferences), but their colors are different (the color is something that has not been seen by the algorithm before). We can fix this problem by adding one bit and having one hot encoding as follows: potato: [0,0,0,1], carrot: [0,0,1,0], white-onion: [0,1,0,0] and red onion [1,0,0,0]. The more images we encode, the larger we will get the one hot encoding bit vector. Besides, if the algorithm encounters a red potato, it has the same problem despite its shape having similarities with yellow potatoes. By transferring the data into the distribution, we can gain more flexibility in terms of learning the data's characteristics. In this example, assume an onion has a distribution of [0.3,0.5,1,1.2, 1.4], a potato has a distribution of [2,1, 1.5, 1.2, 0.9], and when an object comes with the distribution of [0.5, 0.7, 1.1, 1.4, 1.7], the algorithm can label it as onion because its shape of the distribution is more similar to the onion [0.3, 0.5, 1, 1.2, 1.4] and methods such as KL-Divergence (check Chapter 3) are used to perform this comparison.

How does VAE work?

The VAE extracts input features, but it does not assign a single value to each of the features (e.g., the face has an eyeglass: yes, the face has silicon in its lip: no). Instead, it presents them in a vector (e.g., the face has eyeglass: 0.7 probability, the face has silicon in its lip: 0.2 probability), and each extracted feature is called a *latent attribute*. These latent attributes construct a *latent space* in which features that are close to each other in the input space stay close to each other in the latent space as well.

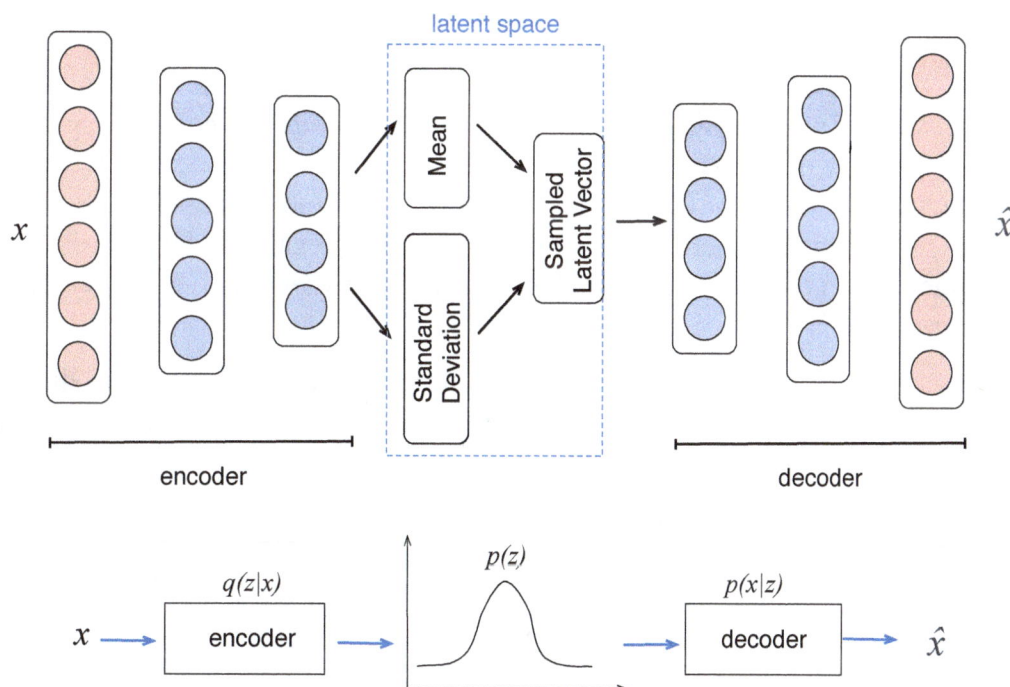

Figure 11-17: Schematic presentation of a Variational Autoencoder (VAE).

In particular, latent space, i.e., the output of the encoder, is a representation of the input data in a multi-dimensional Gaussian distribution (each feature corresponds to one dimension) in hidden layers. Foster [Foster '19] provides a short and good description of VAE as follows: *"While using regularized Autoencoders, a set of data points will map directly to one single data point in latent space, but VAE maps each data point into a multivariate Gaussian distribution around a point in the latent space."* Figure 11-17 visualizes the architecture of VAE.

Formally speaking, the output of the encoder in VAE is not a single vector z (e.g., a vector of one hot encoding). Instead, it computes $q(z|x)$, and results in a posterior that includes distribution parameters (e.g., mean and standard deviation if we deal with Gaussian distributions), i.e., $p(z)$. This posterior presents the PDF of z (Check Chapter 3 if you can't recall the PDF). In other words, VAE transforms the output of the encoder into *mean* and *variance*, which are parameters to describe a Gaussian distribution. Previously described autoencoders map input into a fixed-size vector, while VAEs map the input into a distribution $p(z)$, which is parameterized by the distribution parameters. The decoder then samples from the $p(z)$, and constructs $p(x|z)$. Afterward, $p(x|z)$ it is passed through an activation function, such as a logit or Softmax function, to reconstruct the $\hat{x}$, which is similar to x (but not exactly equal). For example, if the decoder output is a binary variable, the final layer of the decoder typically uses a logit function to convert the outputs into parameters for a Bernoulli distribution, representing the probabilities of binary outcomes. These probabilities can then be sampled to generate binary data points $\hat{x}$. Figure 11-17 presents the VAE architecture.

For example, assume that the last layer of the encoder represents a two-dimensional Gaussian distribution[4]. We need two neurons to output the mean values (one for each dimension) and two additional neurons to output the log standard deviations. Thus, for a two-dimensional Gaussian, we have a total of four neurons, two for the means and two for the log standard deviations (refer to Chapter 3 for a review of Gaussian distribution parameters).

Another thing we need to remember is that variance values are always positive, and to increase the variance variability, the VAE uses the logarithm of the standard deviation, i.e., $log(\sigma)$, instead of the standard deviation itself. This is because the exponential of the logarithm will ensure that the standard deviation remains positive, and the logarithm function is capable of mapping a large domain of positive real numbers to a much smaller range, thus ensuring stability in the backpropagation process.

Assuming ϵ is a point sampled from a standard normal distribution, the encoding process converts the original input data into latent space data (z) by using its mean vector (μ_x) and variance vector (σ_x) using the following equation: $z = \mu_x + \exp(log(\sigma_x)/2) \times \epsilon$

To summarize our discussion, we can say the VAE algorithm operates as follows:

1- It feeds the input data x into the encoder and constructs $q(z|x)$, i.e., $x \rightarrow q(z|x)$.

2- It samples z from $q(z|x)$ to construct $p(z)$ in latent space, i.e., $q(z|x) \rightarrow p(z)$.

---

[4] VAE uses Gaussian distribution, but there are few works that use another distribution as well. Also, the authors of the VAE paper explained that it is possible to use other distributions as well.

3- The decoder samples from $p(z)$ to construct $p(x|z)$, i.e., $p(z) \rightarrow p(x|z)$.

4- Then, the cost function is used to perform the backward pass from step 1 to step 4. The algorithm updates weights and biases, and at the end of the iteration, step 5 will be executed.

5- It uses $p(x|z)$ to construct $\hat{x}$ (reconstruct x while some information from x is removed).

VAE Cost Function

The cost function of VAE seems a bit different than other cost functions. However, its nature is the same as other cost functions; its first part calculates the reconstruction penalty (generative loss), and its second part calculates the latent loss (regularizer). VAE cost function is called *Evidence Lower Bound (ELBO)*, and it is written as an expected log-likelihood, $\mathbb{E}[log]$, (because it is a probabilistic method and using expected log-likelihood as a cost function is common among probabilistic methods) minus KL-Divergence between the latent posterior probability distribution, i.e., $q(z|x)$ and prior probability distribution, i.e., $p(z)$.

$$ELBO = \mathbb{E}[log\ p(x|z)] - D_{KL}(q(z|x)||p(z))$$

The expected log-likelihood ($\mathbb{E}[log(P(x|z))]$) is a negative cross-entropy, which is the data reconstruction loss. The expected log-likelihood is similar to other neural network cost functions. The second part, KL-divergence, measures how close the distribution of the latent variables matches the encoder distribution (encoder output). Here, KL-divergence acts as a regularizer that penalizes the encoding process and ensures the sampled data point does not stay far away from the center of the distribution.

Similar to other neural network algorithms, backpropagation is used to minimize the cost function.

U-Net

We have briefly described two segmentation algorithms in Chapter 6, MSER and Watershed algorithms, several more segmentation algorithms will be described in Chapter 12. Nevertheless, it is a big sin not to explain the legendary U-Net model while discussing autoencoders.

One of the most popular image segmentation algorithms is U-Net[5], which has an autoencoder architecture. U-Net [Ronneberger '15] is a semantic segmentation model designed for

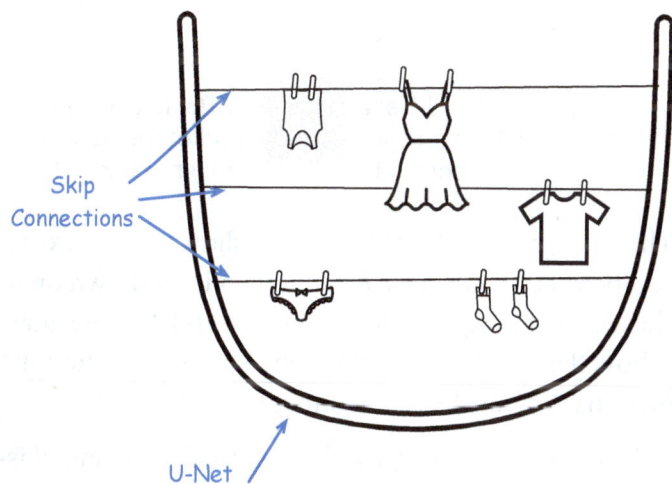

---

[5] Before you proceed with reading U-Net, be sure that you recall the terms we have used in describing CNN in Chapter 10, including max pooling, stride, downsampling, and upsampling.

medical image segmentation, such as separating the nucleus and cytoplasm in the microscopic image of a cell, extracting the lung tissue from fat and skin tissue in CT images, etc. At the time of writing this part (first in 2021 and revision in 2024), it is state-of-the-art semantic segmentation for medical images, and new models that provide a new segmentation usually compare their approach with U-Net, unless they train on a very large dataset, which we explain them later in Chapter 12. The authors of U-Net did not refer to it as an autoencoder, but it has a very similar architecture to autoencoders, and thus, we explain it in this section.

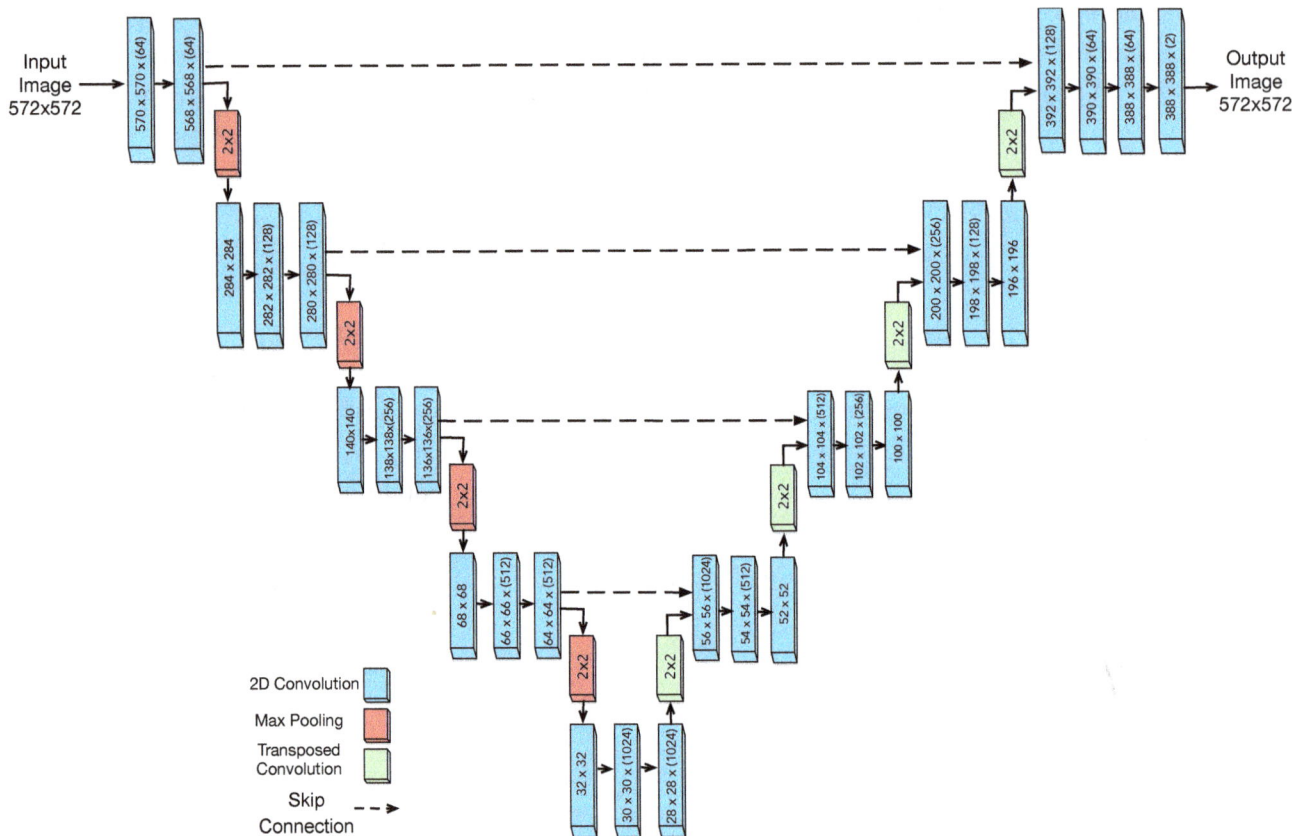

Figure 11-18: U-Net architecture. The left side presents the contraction/encoder, and the right side presents the extraction/decoder part of the network. The dotted arrows (copy and crop) present skip connections. The number of each channel is written on the top of its layer.

The original U-Net architecture is shown in Figure 11-18. It has a U-shape architecture, and it is composed of *contractor layers* (encoders) shown on the left side and *expansion layers* (decoders) shown on the right side of Figure 11-18. Note that not all implementations of U-Net should follow the standard U-Net numbers (or other neural network numbers) usually, we can customize these parameters based on our application needs.

U-Net moves a sliding window on the image and classifies every single pixel of the image; thus, it is very expensive and slow, but due to its superior accuracy, it is widely used for image segmentation, and at the time of writing this text, it is the most popular one.

As we can see in Figure 11-18, its contraction path (left side of U) *downsamples* the input image into a smaller size, and the expansion path (right side of U) *upsamples* the contracted data that it has received from the last layer of contraction path. In each downsampling stage of the contraction path, the number of feature channels doubled, and 2×2 max pooling with a stride of size 2 is used. In summary, the contraction path is a set of convolutional and max pooling layers.

The first level of the contraction path applies 3×3 padding, which reduces the input size from 572×572 to 568×568. Then, the output of the convolution is downsampled for the next level. In the next level, the convolution is applied again, and the result is 280×280, with 128 channels (the number of channels is doubled). This process continues in four levels of downsampling.

Next, the extractor path starts the process of upsampling, which we can read its numbers from the right side of Figure 11-18. The expansion path is a set of convolutional layers and transposed convolutions (instead of max pooling) to upsample the data. At each level of the expansion path, the number of feature channels will be halved.

Both the contraction path and expansion path use ReLU as an activation function. The contraction path learns *what* is in the image, but since it makes the image smaller, it *loses the spatial information* of the image (often referred to as the *where* information). The expansion path specifies *where* the information is located in the image, which means it handles the *localization of segments* in the image.

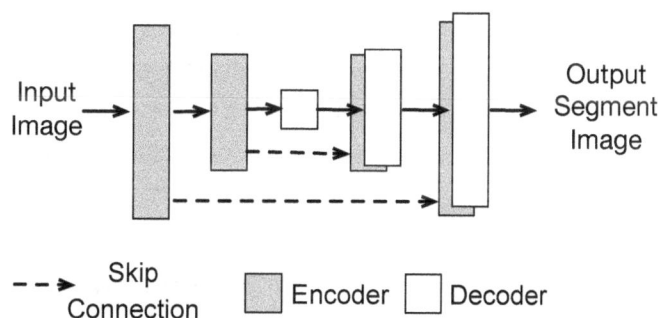

Figure 11-19: A simplified shape of U-Net architecture.

There is something interesting in U-Net architecture that makes it different from other autoencoders. Between the contraction path and expansion path, there are *skip connections*, marked with dotted lines in Figure 11-18 and Figure 11-19. Skip connections are responsible for cropping the image (by using padding) in the contraction path to match the expansion path image at the same level. By using skip connections from the contraction path, it combines the high-level 'what' information with the spatial 'where' information to localize features within the image accurately. Using the skip connection enables the network to mitigate the risk of vanishing gradients. Another advantage of using skip connections is that it mitigates the *degradation problem*. The degradation problem refers to a phenomenon in a deep neural network in which increasing the depth of a network causes a decrease in the accuracy of the train and test. Using a skip connection is a good solution to avoid both issues and keep them in mind while designing your neural network architecture. We can also employ the use of skip connection for our neural network, especially while working with image data, and we need encoding and decoding.

We are done with the explanation of U-Net, but if you think that a U-Net architecture does not need to have a U-shape and it could be designed with a simple autoencoder shape, we do not disagree with you. However, the authors of the paper prefer to design it something like Figure 11-18. A more abstract shape of U-Net is designed by the author of Pix2Pix [Isola '17], is shown in Figure 11-19, and it is easy to memorize.

NOTES:

* Autoencoders are unsupervised learning, and they are adapted as self-supervised learning, where the data itself provides the structure for learning. We can also improve the accuracy of the algorithm by introducing a small amount of labeled data.

* It is not mandatory, but most of the time, autoencoder architectures are symmetric. While it's common for autoencoder architectures to be symmetric, the symmetry pertains to the network's architecture rather than the exact mirroring of input and output data. Non-linear activation functions, such as hyperbolic tangents, enable the network to capture complex relationships in the data, resulting in meaningful compressed representations.

* Aggarwal [Aggarwal '18] recommended transferring non-sparse data into sparse data while describing sparse autoencoder because it provides a more flexible representation of data. The author of this book is a developer who struggles to push machine learning algorithms into battery-powered devices, a.k.a on-device machine learning, and always suffers from a lack of resources. There may be reservations about adopting this approach. Despite this, it's important to present diverse perspectives. In environments where computational resources are not as constrained, such as within well-resourced corporations, following Aggarwal's advice could indeed be beneficial for model design.

* Despite autoencoder compressing features, if we use linear activation functions, they have no additional superiority over traditional image compression algorithms. Therefore, they are not usually used for image compression and do not act well as well.

* Both RBM and autoencoder can *share weights*. Instead of using another set of weights at the output layer, we just use the transpose of weights from the input layer. Therefore, if we have W input layer weights, we can have W^T output layer weights. In other words, the weights of the decoder are the transposed weights of the encoder. This approach can operate as a regularizer, which could prevent making a linear transformation (like PCA), while we need a non-linear transformation. Therefore, it prevents the autoencoder from acting as an identity function. Besides, it enforces the model to have fewer parameters. By reducing the number of parameters, the likelihood of overfitting will be reduced.

* An autoencoder that consists of multiple hidden layers is termed a *deep autoencoder*. These models are particularly effective for information retrieval tasks, including image search and matching, because they can compress high-dimensional data into a more manageable, lower-dimensional space while retaining important features. According to the universal approximation theorem, a neural network with sufficient depth and neuron count can

theoretically represent a wide variety of complex functions. This capability suggests that deep autoencoders have the potential to capture intricate patterns and relationships within the data, although their practical performance depends on factors such as network architecture, training procedures, and data availability.

* U-Net does not use any fully connected layer and uses only the result of each convolutional layer. You can check the original paper for a justification of their approach and more details about the U-Net architecture [Ronneberger '15].

Generative Adversarial Network

One of the attractive approaches used for self-supervised learning is the *Generative Adversarial Network (GAN)*. In 2013, a model for an adversarial approach was proposed for animal behavior modeling by Li et al. [Li '13]. Afterward, in 2014, Goodfellow et al. [Goodfellow '14] introduced and implemented GAN, popularized among the data science community.

GAN employs generative models to construct fake (to be polite, we can say synthetic) data similar to the original input data. This is scary (for its deep fake applications), but an attractive approach, and it has many applications, including generating artistic images and music, image modification (e.g., converting low-resolution images to high-resolution images), photo-realistic image construction, face swapping and face image changes (e.g., aging the face or making it younger), image translation (e.g., converting day taken picture into night taken picture), etc.

Figure 11-20: Synthetic image generated by NVIDIA's open source model, i.e., StyleGAN [Karras '19].

GAN architecture is composed of two neural networks. One network is *Generator*, and the other one is *Discriminator*. The generator network acts as a 'counterfeiter' that tries to generate synthetic data by using the original observed data. The generator network usually comprises a series of dense layers followed by transposed convolutional layers. The generator starts from random noise and improves itself in each iteration. We could say the generator has a similar role to the encoder in VAE, i.e., converting the input into a latent space.

The discriminator acts as a 'detective' and tries to identify the synthetic data from the original data. This process iteratively continues as a competition between the generator and the discriminator. In each iteration, the synthetic data gets more realistic because the generator learns to improve its model. Still, the discriminator also learns to recognize the synthetic data from the previous iteration, and it improves its recognition capability as well. Usually, the discriminator is a convolutional neural network that classifies the generated image.

The result of a GAN is synthetic data that is similar to the original data, like audio or images, that even humans can not recognize the differences from the original data. For example, Figure 11-20 presents a human face image constructed by StyleGAN [Karras '19]. The person in this figure does not exist in reality, but we can see that his face is very realistic, and we can not recognize that it is fake.

Both generator and discriminator networks compete with each other, and in every round, they get stronger. The generator is improving its synthetic data construction, and the discriminator encounters more synthetic data, which improves its discrimination capability. These two networks are playing a min-max game based on the objective function, which is cross entropy. Figure 11-21 presents the architecture of GAN.

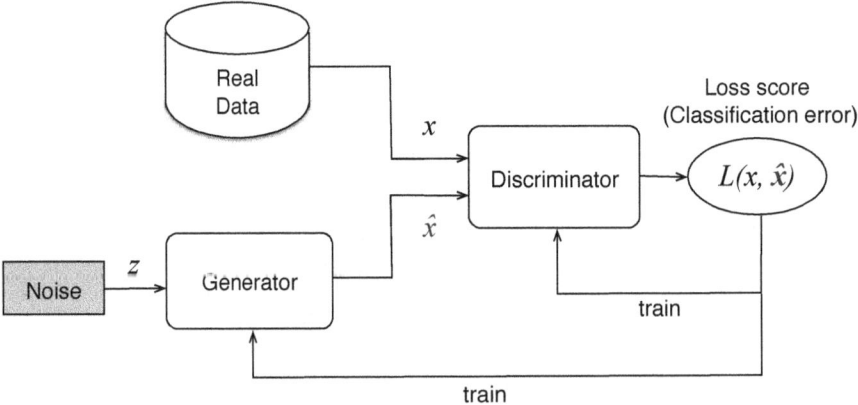

Figure 11-21: GAN architecture.

The dataset we use to train the GAN model is real data (x), and GAN uses this dataset to construct a synthetic version of this dataset $(\hat{x})$, with high similarity to the real data.

The generator performs the task of synthetic data construction and starts by getting a vector of random "noise" (z) and then refines it in each iteration with the backpropagation algorithm. When the network converges after several iterations, the random noise will be converted to data with a distribution similar to the real data.

The discriminator gets the real data (x) and the generator's output $(\hat{x})$ as input and tries to determine if the data is real or not, with a probability score. In other words, the discriminator (D) is a binary classifier that specifies whether the output is synthetic or real.

Considering 1 as a label for correct classification and 0 as incorrect classification, the D output (which is a probability) can be interpreted as follows:
True Positive: $D(x) \approx 1$ (correctly identifies real data as real)
False Negative: $D(x) \approx 0$ (misclassifies real data as synthetic)
False Positive: $D(\hat{x}) \approx 1$ (misclassifies generated data as real)
True Negative: $D(\hat{x}) \approx 0$ (correctly identifies generated data as synthetic)

The result of the discriminator and generator is classification error (loss score), and backpropagation adjusts both generator and discriminator weights and biases. In particular, backpropagation is used by the discriminator to improve the accuracy of identifying real versus synthetic data. The backpropagation is used by the generator to increase the probability of the discriminator misclassifying synthetic data $(\hat{x})$ as real data (x). After the training phase, the discriminator will be discarded, and the user will use the generator model to construct new synthetic data.

Training GAN

The generator and discriminator are trained alternatively, and their training process is different. The discriminator training can be outlined in the following four steps:
1- Sample (random mini-batch) x from the real data.
2- Take a random noise z, and construct the output of the generator, i.e. $G(z) = \hat{x}$.
3- Then, the discriminator classifies x, and $\hat{x}$, and calculates their error (loss score), i.e. $L(x, \hat{x})$.
4- The backpropagation algorithm improves the weights and biases of the discriminator to reduce the loss score of the discriminator in the next iteration.

The process generator training can be outlined in the following three steps:
1- Use a random noise z and feed it into the generator network to construct synthetic data, i.e $G(z) = \hat{x}$.
2- The $\hat{x}$ will be classified by the discriminator, i.e. $D(\hat{x})$, and it computes the loss score (classification error).
3- The loss score will be backpropagated to the generator network to update weights and biases toward reducing the loss score of the generator.

The generator's weight and biases remain fixed while the algorithm is training the discriminator, and similarly, the discriminator's weight and biases remain fixed while the algorithm is training the generator.

A question might arise: When does the GAN iterative training stop? The GAN stops when both the discriminator and the generator reach a stage where neither the generator nor the discriminator can improve their accuracy, which means neither of these two networks can improve their loss score. This is referred to as *Nash equilibrium* or *Zero-Sum game*, in which no parties in the game can change or improve their state without changing the other party's parameters.

GAN stops when either (i) the generator produces synthetic examples that are not distinguishable from real data or (ii) the discriminator guess starts to get random for the generated synthetic data. In layman's terms, GAN stops when the discriminator is defeated and cannot recognize synthetic from real data. Nevertheless, in real-world applications, it is very hard to get into Nash equilibrium, and we should stop the algorithm after we get some satisfactory results.

The training of other neural networks is an optimization problem, but in GAN, it is better to say its training is like a strategic game rather than an optimization for problem solving.

GAN Cost Function

The GAN training is a two-player game. One player is the Generator (one player) that tries to fool the discriminator by generating unreal data that looks like real data. The other player, the discriminator, tries to distinguish whether the data is real or not. The cost function of GAN is a Minimax game, and it is written as a cross-entropy between real ($\mathbb{E}_x[log\ D(x)]$) and synthetic data ($\mathbb{E}_z[log(1 - D(G(z)))]$) distributions.

$$J(D, G) = \min_{G}\ \max_{D}\ \mathbb{E}_x[log\ D(x)] + \mathbb{E}_z[log(1 - D(G(z)))]$$

In this cost function:

$\mathbb{E}_x$ presents the expectation of x, which is drawn from real data.

$D(x)$ is discriminator output for real input x.

$\mathbb{E}_z$ is the expectation of synthetic data samples that come from the generator.

$G(z)$ or $\hat{x}$ is the generator output for a given noise z.

$D(G(z))$ is the discriminator output for generated synthetic data, i.e. $G(z)$.

In the cost function, the generator's objective is to minimize the loss score. The discriminator D tries to correctly classify real data x and generated data $G(z)$. Therefore, $D(x)$ should be close to 1 (indicating real data), and $D(G(z))$ should be close to 0 (indicating synthetic data). The generator G aims to fool the discriminator by generating data as realistic as possible. Thus, it wants $D(G(z))$ to be close to 1, indicating that the generated data is classified as real by the discriminator.

This cost function could also be re-written as a Binary Cross Entropy (BCE):

$$-\frac{1}{n}\sum_{i=1}^{n}(y_i\ log(p_i) + (1 - y_i)log(1 - p_i))$$

However, applying it directly to GANs requires adapting it to the context of discriminator and generator. Therefore we can write discriminator loss as follows:

$$L_D = -\frac{1}{n}\sum_{i=1}^{n}\left[\log D(x_i) + \log(1 - D(G(z_i)))\right]$$

Respectively, we can write generator loss as follows:

$$L_G = -\frac{1}{n}\sum_{i=1}^{n}\log D(G(z_i))$$

GAN Challenges

Training GAN is not trivial, and there are some major challenges associated with it. In the following, we list some common known challenges [Foster '19] of training GAN.

Loss Oscillation

Sometimes, the loss score for the generator and discriminator could oscillate and not converge. Some GAN models mitigate this oscillation, but it is still a common issue in training GAN. Oscillation in loss scores of the generator and the discriminator states that something in the GAN model is not correct, and thus, we should revise our GAN model.

Slow Convergence

Another common issue in GAN is slow convergence, which does not make GAN training practical in real-world software applications because it is very resource intensive and slow to make the model converge. Since GAN training is slow, it requires lots of GPUs to train it, and this prevents small or medium size companies from being able to train a GAN from scratch. Usually, big corporation trains a GAN, and others use or fine-tune the trained model in their applications.

Mode Collapse

Real-world data are usually multimodal (check chapter Chapter 3). For example, assume we are building a GAN to create a synthetic human conversational agent (chatbot) and make it so good that humans mistake this chatbot with a real human. To implement this chatbot, a GAN is used to imitate real humans (the chatbot's conversational text is synthetic data, and the human conversational text is real data.)

A human has a combination of emotions, including anger, joy, and so forth. Once a GAN samples from a distribution of anger (called mode in the context of GAN), it can fool the discriminator because the discriminator recognizes that being angry is a human emotion and is not a chatbot. Then, the generator continuously keeps the agent angry (sample from anger distribution), and the discriminator fails to recognize that this is not a real human. The result is a chatbot that is always angry, and users can recognize it as synthetic. If you can not connect well with the described example, let's use another one. Assume we create a GAN to construct handwritten digits similar to MNIST, but the GAN fails to produce all numbers, e.g., 3 is never produced, despite the model being converged. Usually, a low standard deviation among samples is a sign of mode collapse.

In summary, sometimes, a generator finds a few data points that can fool the discriminator and stick with those data points. Nevertheless, there are too few data points, which are not representative of the real data.

Uninformative Loss

Theoretically, a small loss score for a generator should translate to high-quality synthetic data. Nevertheless, the generator is compared with the discriminator, and sometimes, a decrease or an increase in generator loss is not correlated with the quality of the synthetic data they produce. This lack of correlation between loss score and the quality of generated data is known as uninformative loss, and it is another known challenge in GAN.

There are many improvements in the GAN architecture to mitigate the described challenges. Later in this chapter, we explain some popular GAN architectures.

Evaluating GAN Result

GANs are a subset of generative models, and while some generative models are derived from the Maximum Likelihood Estimation (MLE) algorithm (see Chapter 4), GANs are distinct from MLE-based models in their approach and underlying theory. Unlike MLE-based models, the evaluation of GANs often relies on specific metrics that assess the quality and diversity of generated outputs rather than fitting to a known likelihood function. Two common metrics used to evaluate the results of GANs are the *Inception Score (IS)* and the *Fréchet Inception Distance (FID)*. These two methods are only used for synthetic image comparison, but the concepts behind these metrics can potentially be adapted for other types of data. Both the Inception Score and the Fréchet Inception Distance use the pre-trained Inception-v3 model.

Inception score (IS)

The IS [Salimans '16] was introduced in 2016 and uses a pre-trained neural network model, i.e., Inception-v3 [Szegedy '16], for image classification. We will explain more about Inception-v3 in Chapter 12. IS tries to capture two properties: data quality (fidelity) and variety (diversity).

Quality: First, a pre-trained Inception-v3 model is used to label each generated image (in probabilities). Assuming y is the label for generated image x, the result gives a conditional label distribution, i.e., $p(y|x)$ for every generated image. Generated samples that are close to real ones will have low entropy ($H(y|G(z))$ is low). In simple words, the first step maps a generated image to its class probability.

Variety: In the second step, IS calculation algorithm identifies the variety of generated samples, $G(z)$, by using a marginal probability distribution, which is the probability distribution of all generated images, i.e., $p(y)$, e.g. 10% potatoes, 30% cats, etc. To have a high variety of generated images, the integral of the marginal probability distribution ($\int p(y|x = G(z))dz$) should have high entropy, i.e., $H(y)$. In simple words, the Generator should synthesize as many different classes as possible.

Next, the KL-Divergence (Check Chapter 3) between these two probability distributions will be calculated, i.e., $KL(p(y|x) || p(y))$. In the last step, it exponentiates [6] the KL-Divergence score to make it easier to read, $IS = exp[KL(p(y|x) || p(y))]$.

By adding a logarithm, we can get rid of exp and re-write the KL-Divergence, and thus we use the following equation.

$$log(IS) = H(y) - H(y|G(z))$$

We can see from this equation that the IS score is high (higher IS is better) if we have high variety, i.e., $H(y)$, and low entropy for generated samples, i.e., $H(y|G(z))$.

If you are not happy with too much mathematics, let's switch to the elementary school explanation: The IS score calculates the differences between the overall variety of images produced by the generator and how much each individual image looks like it belongs to a specific category.

Fréchet Inception Distance (FID)

IS does not take into account statistical characteristics (i.e., *mean* and *variance*) of real and generated data. Therefore, if a GAN produces a very small number of samples, let's say it generates only three images from a 900,000 input dataset, but the quality is very high, the IS score could get high despite having very few instances of generated data. It is the limitation of the IS score, and the FID score [Heusel '17] can study the mean and variance of both real data and generated samples using the inception network. This capability makes the FID score more attractive than the IS score.

Before getting into the details of FID, we should learn Fréchet distance. It is a similarity measurement method that measures the similarity between two curves (paths or shapes) in a discrete manner, taking into account both the order and the positioning of data points along each curve. In Chapter 4, we explained DTW, which we use for time series similarity measurement, and in Chapter 3, we explained several distribution similarity measurements. Fréchet distance is generalizable to any shape and not only time series or distribution.

To understand the Fréchet distance, imagine a person walking a dog, where both start at the same point and move towards the same destination along different paths. They are not allowed to backtrack, meaning they can only move forward, though their paths might diverge within these constraints. The person and the dog can adjust their walking speeds independently but cannot reverse direction. The Fréchet distance between their paths is defined as the length of the shortest leash necessary for them to travel from start to finish without losing connection. This leash length represents the maximum distance between them at any point, assuming each can slow down or speed up as needed to minimize the distance along their paths, see Figure 11-22. In simple terms, the Fréchet distance specifies the minimum length of a dog leash required so that the owner and the dog can travel from the start to the finish points on their separate paths without losing connection.

---

[6] Exponentiating a number x means raising the constant e (Euler's number) to the power of x, resulting in e^x.

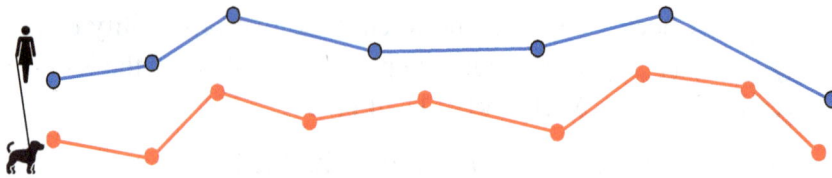

Figure 11-22: A toy example of two curves that have the same path but their trajectory is different.

Now that we understand Fréchet distance, let's get back to the FID score, unlike IS which uses the output (labels) of the Inception-v3 network, FID uses the activations from the last pooling layer (before the output layer) of the Inception-v3 network. This layer contains 2,048 activations, modeling each image as a vector of 2,048 features. Therefore, FID effectively uses the Inception-v3 network for feature extraction, allowing for a detailed comparison of the real and generated images based on these high-dimensional feature vectors.

Assuming the real data has a multidimensional Gaussian distribution of (μ and Σ), which μ presents the mean Σ presents covariance matrix, and the generated data has the multidimensional Gaussian distribution of (μ_g and Σ_g), Tr is the trace operator [7], the FID distance between two data will be calculated by using the following equation:

$$FID = |\mu - \mu_g|^2 + Tr(\Sigma + \Sigma_g - 2(\Sigma\Sigma_g)^{1/2})$$

A lower FID score means better quality generated images.

GAN Architectures

The GAN we have explained earlier in this chapter is called Min-Max GAN or vanilla GAN. It has very limited accuracy and slow convergence. At the time of writing this section in late 2021, probably one of the most attractive areas in computer science for new PhD students is GAN networks and their synthetic data generation capability. Later, after the revision in 2024, GAN is substituted by the Diffusion model, which we will explain later. The popularity of GAN has led to the introduction of too many GAN models, which are referred to as GAN Zoo. Therefore, we limit this section to a few well-known GAN architectures.

Note that most of the GAN architectures we describe here operate on image data we refer to the inputs more generally as data to acknowledge that these architectures may also be adaptable to other types of data, such as music or text.

Conditional GAN (CGAN)

The traditional GAN, Min-Max GAN, has no control over the output or types of synthetic data being generated. This process can be easily improved by adding a label to both real data and synthetic data [Mirza '14]. For example, assume a GAN, which is generating pictures of handwritten numbers from the MNIST dataset. A label y, which specifies the digit, could be

---

[7] Tr(X) denotes trace of a square matrix X. It is the sum of elements on the main diagonal (from the upper left to the lower right) of matrix X.

assigned to real data, and this label helps the generator produce more realistic synthetic data. This label is incorporated in the cost function, and therefore, the Conditional GAN (CGAN) cost function is written as follows:

$$\min_{G} \max_{D} \mathbb{E}_x[log\ D(x|y)] + \mathbb{E}_z[log(1 - D(G(z|y)))]$$

In this function, y presents the label. In the red section of the above equation, the CGAN incorporates conditional probability into the cost function. The only difference is that GAN uses probability, while CGAN uses conditional probability.

The baseline GAN cost is written as follows:

$$J(D, G) = \min_{G} \max_{D} \mathbb{E}_x[log\ D(x)] + \mathbb{E}_z[log(1 - D(G(z)))],$$

The CGAN cost function can be written based on vanilla GAN as follows:

$$J_c(D, G) = \min_{G} \max_{D} \mathbb{E}_{x,\hat{x}}[log\ D(x, \hat{x})] + \mathbb{E}_{z,\hat{x}}[log(1 - D(G(z, \hat{x}), \hat{x}))]$$

To summarize, we can say that the GAN Generator learns a mapping from noise z to output image $\hat{x}$, i.e., $G : z \rightarrow \hat{x}$. The generator of CGAN learns a mapping from noise z and input x to output image $\hat{x}$, i.e., $G : z, x \rightarrow \hat{x}$.

Deep Convolutional GAN (DCGAN)

Usually, GAN models are unstable to train and could cause the generator to produce non-essential outputs and DCGAN mitigates this issue by using convolutional layers in the discriminator and transposed convolutional layers (check Chapter 10) in the generator.

In Chapter 10, we described convolutional neural networks are very popular for image classification. DCGAN benefits from CNN as well. In particular, its discriminator includes convolutional layers, which compress the data, and its generator includes a transposed convolutional layer that decompresses the input image. In other words, the idea of DCGAN is to add transposed convolutional layers between input vector z and the output image generated by the generator and use a convolutional layer in the discriminator to classify the images.

As we can see in Figure 11-21, the generator starts with a vector of noise (in the original paper, the vector size is 100). It has four transposed convolutional layers, and each of them increases the height and width but decreases the channel, as is shown in Figure 11-23, in the original paper, the result is $64 \times 64 \times 3$. The discriminator includes four convolutional layers, and each convolutional layer decreases the height and width but increases the depth (the last layer of the original paper delivers the output with size $4 \times 4 \times 512$). The activation function of the last layer is Sigmoid, which specifies if the output is real or synthetic.

Both generator and discriminator networks have specific characteristics and some differences from commonly used CNN networks. For example, we have explained in Chapter 10 that usually, after convolutional layers, we have pooling layers, and after that, we have fully connected layers. DCGan has the following characteristic differences from Min-Max GAN:

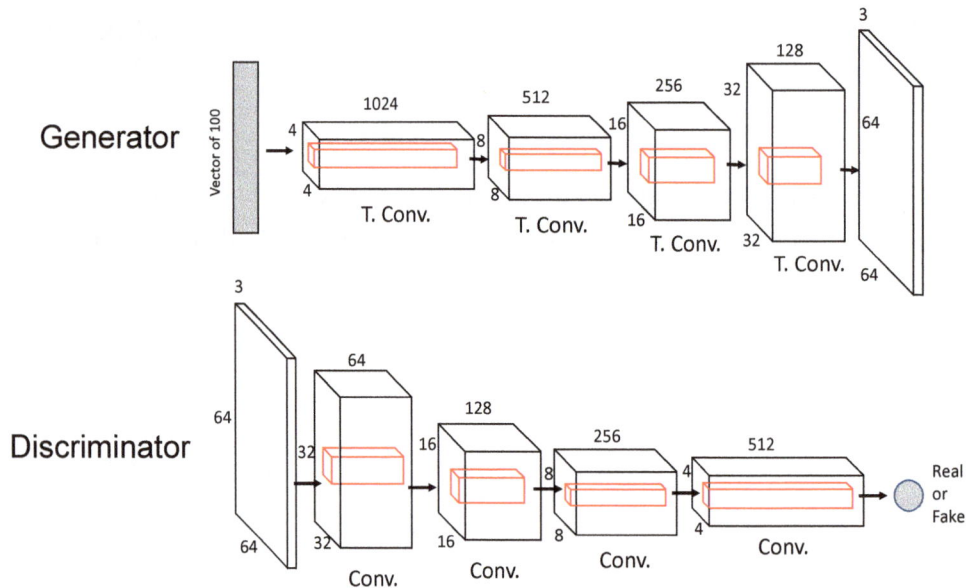

DCGAN discriminator and generator architecture. T. Conv. stands for the transposed convolutional layer and Conv. Stands for the convolutional layer.

- It replaced any pooling layers with stride convolutions in the discriminator and *fractional-strided convolutions*[8] (often referred to as transposed convolutions) in the generator.

- Both the discriminator and the generator use Batch normalization.

- They use ReLU activation in the generator for all layers except for the output, which uses the hyperbolic tangent. They use LeakyReLU activation in the discriminator for all layers.

- The authors recommend initializing all weights using a Gaussian distribution, with a mean of 0 and a standard deviation of 0.02.

Wasserstein GAN (WGAN)

WGAN introduces a different loss function for both the discriminator and the generator, and it has a stable optimization process that mitigates the "mode collapse" and "vanishing gradient" problems. While using WGAN, we refer to the discriminator as a *critic* and not a discriminator. The reason is that the discriminator output is 0 or 1, which discriminates between real and synthetic, but in WGAN, it is a number because there is no logarithm, and the output is not a probability. Therefore, sigmoid activation is not applied to the output of the critic (discriminator). In other words, unlike traditional GANs, where the discriminator outputs a probability via a sigmoid activation, indicating whether inputs are real or synthetic, the WGAN critic provides a scalar score without using a sigmoid activation. This score is not limited to binary outcomes, reflecting how realistic or synthetic the critic deems the input rather than classifying it directly.

---

[8] We skip explaining this convolution, you can check visualizations of Dumoulin and Visin [Dumoulin '16] to learn it.

Why didn't you put us together with DTW and other distance metrics?

Wasserstein distance   Fréchet distance

To understand WGAN, we should briefly describe two concepts: Wasserstein distance and continuity condition.

Wasserstein or Earth Mover Distance

The Wasserstein Distance, also known as the Kantorovich-Rubinstein [Rüschendorf '85] or Earth Mover's (EM) Distance, is a method used to compute the distance between two probability distributions. This distance measures how much work is required to transform one distribution into another, which is akin to calculating how to optimally relocate mass from one distribution to another. Imagine you have two piles of sand representing different distributions. The Wasserstein Distance calculates the minimum amount of work needed to reshape one pile to exactly resemble the other. This involves considering where each particle of sand is located and where it needs to go, which is why it's sometimes called the Earth Mover's Distance.

Figure 11-24 presents two distributions: P and Q. For the sake of simplicity, we present the area under each distribution as a set of numbered rectangles. The EM distance between these two distributions is calculated as the minimum cost of transporting the mass converting distribution P to distribution Q. Informally, we could say this distance metric is the minimum energy required to transfer one pile of soil in its shape to another pile, i.e., *amount of earth moved* × *the moving distance*. Figure 11-24 visualizes this figure in three steps for the sake of readability, and it does not cover all the details. Step 1, moves cubes 1, 2, and 3 from the P distributions, moves them d size, and adds them to the Q distribution. The cost of this step is $3 \times d$. Step 2, moves P distributions cube 4 and cube 5 and adds them to the proper side in the Q distribution. The cost of this step is $2 \times d$. Step 3, moves cubes 6,7, and 8 from P distribution and adds them into the Q distribution, but the cost is $2 \times d + (1 \times d - 1)$, the $1 \times d - 1$ is because cube 8 shifts one cube size less than d. The final cost of this simplified EM distance is:

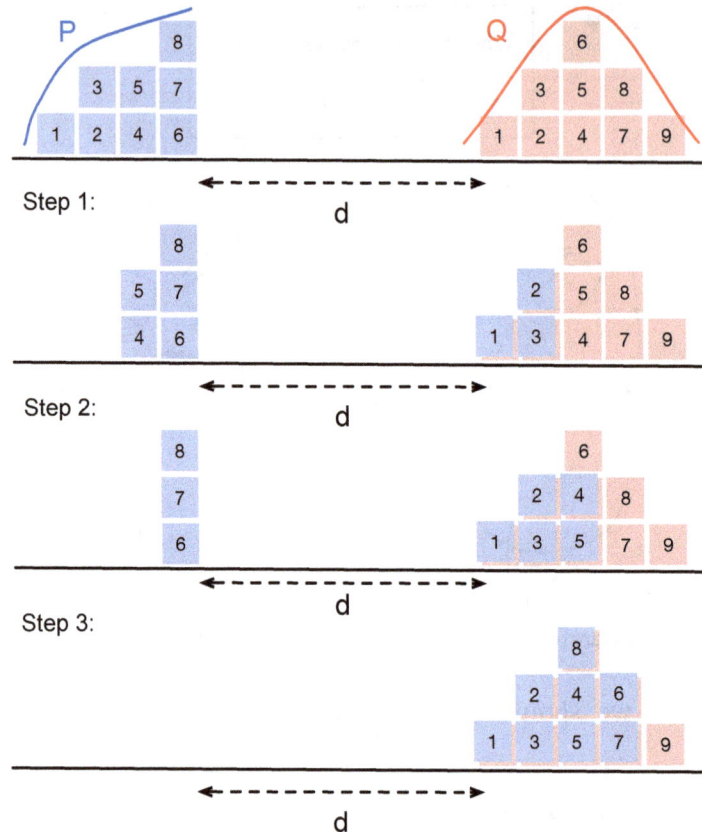

Figure 11-24: Two sample distributions, P and Q, that stay in distance of size *d* from each other. These steps try to simulate how the Earth Moving distance calculates the minimum distance between two distributions.

$$3 \times d + 2 \times d + 2 \times d + (d - 1).$$

Formally speaking, the EM distance between two distributions is the difference between their cumulative distribution (Check Chapter 3 for CDF), which is the area under the curve. $\Pi(P, Q)$ is known as the *transport plan* of P and Q, and it denotes the sum (joint) of both P and Q cumulative distribution, γ specifies the amount of mass (area under the curve) that is transported from x of P distribution to y of Q distribution. The EM distance between P and Q is written as follows:

$$W(P, Q) = inf_{\gamma \in \Pi(P,Q)} \int ||x - y|| d(x, y)$$

In this equation, the term *inf* refers to *infimum*, which presents the minimum in this context and $d(x, y)$ presents the distance between two points. The opposite of infimum is *supremum*, which is presented as *sup* and, in this context, means maximum. The authors of the WGAN paper describe the Wasserstein distance for critic C and generator G as follows:

$$W(C, G) = inf_{\gamma \in \Pi(C,G)} \mathbb{E}_{(x,y)\sim\gamma} ||x - y||$$

648

To recall what we have explained in Chapter 3, here $\mathbb{E}$ presents the excepted value, $(x, y) \sim \gamma$ means x and y sampled from γ (x for C and y for G).

Continuity Condition

The critic network of WGAN must have a continuity condition. In layman's terms, a function is continuous if we can walk on it without lifting a pen from the paper (see Figure 11-25). If we must jump and lift the writing pen from the paper to follow the path of function, this function is not continuous. Mathematically speaking, the function $f(x)$ is called continuous at point a if the following three conditions are met.

(i) $f(a)$ must exist

(ii) $\lim_{x \to a} f(x)$ must exist.

(iii) $\lim_{x \to a} f(x) = f(a)$

There is a specific form of continuity called *Lipschitz continuity*. A function is called Lipschitz continuous if there exists a constant $L \geq 0$ that for all x and y, we have $|f(x) - f(y)| \leq L|x - y|$. Here, L is called the Lipschitz constant.

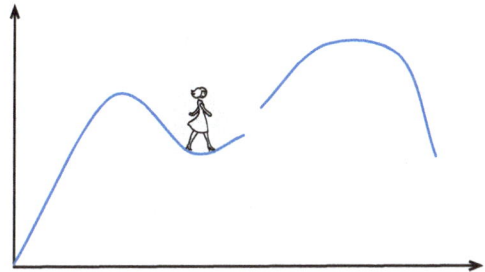

Figure 11-25: A simple illustration of non-continuous function, where we need to lift the pen to draw it, or the person walking on the function needs to jump to walk on it.

WGAN Cost Function

Let's get back to the WGAN description and the use of Wasserstein distance. Now, a question arises, and one might ask: "Why not use more popular distribution similarity metrics such as KL-Divergence or JS-Divergence (check Chapter 3) and instead use Wasserstein distance?"

Authors of WGAN explained that if two distributions do not have overlap (disjoint), such as the example proposed in Figure 11-24, the KL-Divergence result is infinity, and the JS-Divergence result is not converging, and thus both are unreliable. However, EM distance performs better than two KL-Divergence or JS-Divergence; its result is a number between 0 and 1. The mathematical explanation is fairly simple, and you can check the original paper for more details. However, you do not need to learn it in more detail because, at the time of writing this section, we are in Boston. It is spring season, and good weather is too short here, so the author should go for his weekly half marathon and burn some of his excessive body fat.

Wasserstein is favored over KL-Divergence and JS-Divergence when distributions are disjoint, but computing the *inf* of $\gamma \in \Pi(P, Q)$ is intractable in EM distance. Therefore, WGAN authors propose a transformation using Kantorovich-Rubinstein duality[9] to the original Wasserstein distance, as follows:

$$W(C, G) = \frac{1}{K} \sup_{||f||_L \leq K} \mathbb{E}_{x \sim C}[f(x)] - \mathbb{E}_{x \sim G}[f(x)]$$

---

[9] If you like to understand the details of this transformation this blogpost is helpful: www.vincentherrmann.github.io/blog/wasserstein

In this equation, K is the Lipschitz constant, and the maximum of Wasserstein distance satisfies the Lipshitz continuity ($||f||_L \leq K$). The term $\sup\limits_{||f||_L \leq K}$ read as all functions with Lipschitz constant smaller than K. $\mathbb{E}_{x \sim C}[f(x)]$ is the expected value of the critic function f (it is a 1-Lipschitz function), with real data x, and $\mathbb{E}_{x \sim G}[f(x)]$[10] is the expected value of the generator function f (it is a 1-Lipschitz function, which its L is equal to 1), with synthetic data, x.

Assuming the function f is a Lipschitz continuous, parameterized by w the critic network is used to learn w to find a proper f_w. The described WGAN cost function can be configured to measure the distance between C and G, as follows:

$$W(C, G) = \max\limits_{w \in W} \mathbb{E}_{x \sim C}[f_w(x)] - \mathbb{E}_{z \sim G}[f_w(g_\theta(z))]$$

The critic wants to maximize this $W(C, G)$, and the generator wants to minimize it (make those expected values close to each other). Explaining more about the cost function will explode both your brain and ours, but please be patient; a small part remains to be explained.

Weight Clipping

While updating the weights in every iteration of the critic training phase, the algorithm should regulate the Lipschitz continuity of weight function. Therefore, they introduce a parameter $c = 0.01$, which clamps the weights into the $[-c, c]$ range. This simple trick preserves the Lipschitz continuity of weight function during the training phase. In other words, after every gradient update on the Critic function, weights are enforced to be within the range of $[-c, c]$, and this guarantees its Lipschitz continuity.

The rest of the algorithm is the same as other GANs, it uses RMSProp as an optimizer, $\alpha = 0.0005$ and so forth. You can read and understand it from its paper. Hopefully, our explanation here makes it easy enough to understand its main characteristics.

WGAN with Gradient Penalty (WGAN-GP)

After WGAN was introduced and provided superior accuracy among other GAN models, later an improved version of WGAN in 2017 was introduced, WGAN with Gradient Penalty (WGAN-GP) [Gulrajani '17]. Although WGAN is much more stable than other GAN versions at its time, it fails to capture higher-level data distributions and thus generates poor samples due to its weight clipping. Instead of performing weigh clipping, WGAN-GP penalizes the norm of the gradient of the Critic with respect to its input.

WGAN-GP has three differences major differences from WGAN. (i) It does not use gradient clipping. (ii) It applies a gradient penalty to the cost function, to enforce Lipschitz continuity. (iii) WGAN-GP critic does not have batch normalization because batch normalization makes the gradient loss penalty less efficient.

---

[10] We will encounter this writing more in the rest of this chapter, try to keep in mind that $\mathbb{E}_{x \sim P_{data}(x)}[\dots]$ means the expected value of the content inside braces.

The gradient penalty is only applied to the critic and not the generator. Therefore, assuming the WGAN cost function is $\mathbb{E}_{x \sim C}[C(x)] - \mathbb{E}_{z \sim G}[C(G(z))]$, Critic for WGAN-GP is written as follows:

$$W_{GP}(C, G) = \mathbb{E}_{x \sim C}[C(x)] - \mathbb{E}_{z \sim G}[C(G(z))] + \lambda \mathbb{E}_{x \sim P_x}[(||\nabla C(G(z))||_2 - 1)^2]$$

In this equation, λ is the gradient penalty coefficient, and $\lambda = 10$, P_x is sampling *between* pairs of points sampled from the real data distribution and the generated data distribution. We could say P_x describes the interpolated data instead of real/synthetic data.

There are other small differences, such as the use of Adam instead of RMSProp, which are also not important to understand; you can check the details from its paper as well.

Pix2Pix

Explained GAN models are focused on generating images. One of the fascinating applications of GANs is the transformation of an image into another image. Pix2Pix [Isola '17] is a GAN model that uses a CGAN to translate an input image into a different image. For example, convert a satellite image into an outline map, color a sketch (See Figure 11-26), etc.

Pix2Pix, like CGAN, uses a conditional approach in its generator, but it conditions the entire input image rather than a label, making it suitable for tasks such as image-to-image translation. In addition to the adversarial loss, Pix2Pix incorporates a L_1 regularizer in its objective. This L_1 regularizer helps ensure that the generated images closely match the target images in terms of content and structure by penalizing the absolute differences between the output of the generator and the real target image. This feature makes Pix2Pix particularly effective for applications where precise alignment and details are crucial, such as photo editing or converting sketches to photographs. One popular application of Pix2Pix is to transfer sketches into photo realistic images. For example, we can see in Figure 11-26 that a dataset of Pokemon figures is used to train the algorithm. We draw a masterpiece of art, and Pix2Pix uses the Pokemon-trained dataset to make a realistic figure out of it.

Figure 11-26: Using Pix2Pix trained on Pokemon image data to transfer a sketch we draw into photo realistic images. To construct this image, we use the following link https://zaidalyafeai.github.io/pix2pix.

If you do not read this section properly and skip our mathematical explanations, the monster of Figure 11-26 will hunt you in your dream and kiss your chick.

The Generator of Pix2Pix is based on a U-Net architecture, which operates similarly to autoencoder. The encoder downsamples the input image to a low-resolution feature map, and the decoder upsamples it to the output image. The skip connections help preserve the spatial information from the input image to the output image. This architecture, which has skip layers, helps the output have the same structure as the input. In summary, Pix2Pix uses a U-Net architecture for its conditional GAN generator.

The discriminator of Pix2Pix is a *PatchGAN*. In short, a PatchGAN is patching an image into smaller patches (it is a disjoint sliding window we explained in Chapter 5), and instead of using a discriminator for the entire image, a discriminator is used for every single patch.

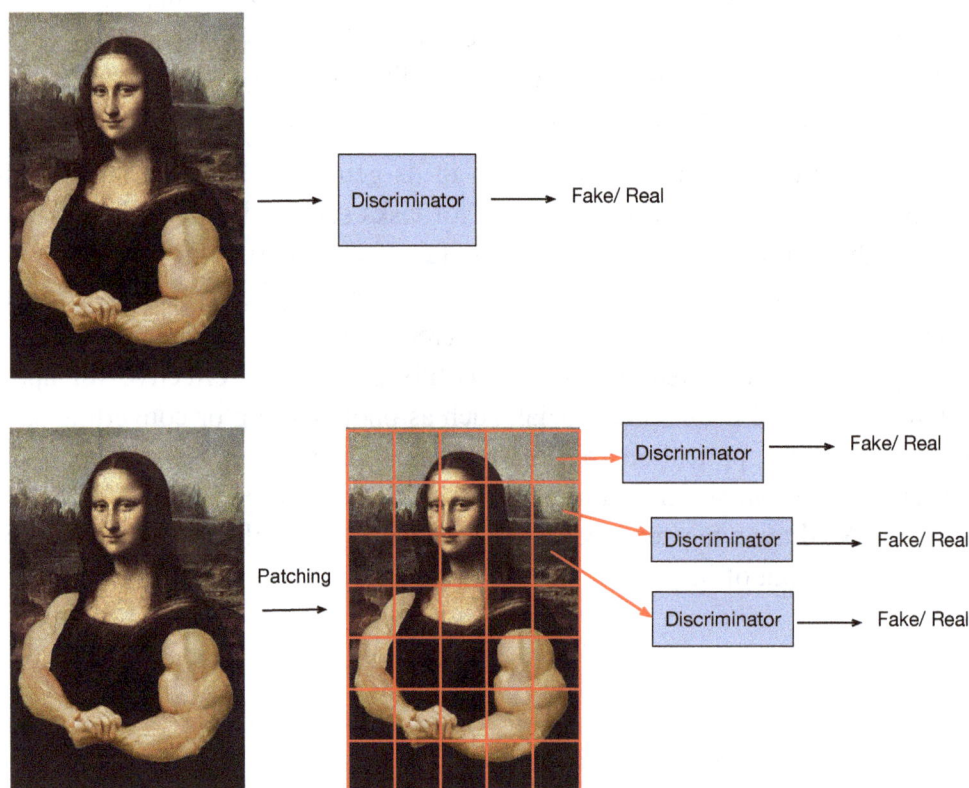

Figure 11-27: A vanilla GAN discriminator on top and PatchGAN discriminator on the bottom. The PatchGAN is feeding each patch into a discriminator, and at the end, averaging will be performed to decide if the image is fake or real.

Image source: https://www.deviantart.com/califjenni3/art/Mona-Lisa-the-Bodybuilder-62547732

Figure 11-27 visualizes the differences between vanilla GAN and PatchGAN. The PatchGAN Discriminator tries to classify each patch (70×70) in an image as real or synthetic. It is very similar to CNN, and PatchGAN leverages the power of CNN, but instead of providing a scalar value at the end, it provides a binary result (synthetic, real) for each patch. The PatchGAN applies convolution across the image, averaging all responses to provide the output, which presents if the image is real or synthetic.

Both the generator and discriminator of Pix2Pix use Batch normalization after each convolutional layer and use the ReLU activation function. You can read more about the details of their model in its original paper, but the information we provide here is enough to understand its core architecture.

CycleGAN

CycleGAN [Zhu '17] is a GAN architecture designed for image-to-image translation tasks where paired training data is not available. Unlike Pix2Pix, which requires precisely aligned pairs of source and target images, CycleGAN operates on unpaired sets of images from two different domains, such as horses and zebras. This is particularly useful when it's infeasible to obtain paired datasets that depict the exact same scene or object in two different styles or appearances. CycleGAN learns to translate an image from one domain to a corresponding image in another domain with reasonable accuracy by incorporating a *cycle consistency loss*. This loss ensures that an image translated from one domain to another can be translated back to the original domain, preserving content while adapting the style. We can see from Figure 11-28 that it can transform a zebra image into a horse image and transfer a landscape photo from winter into a photo in summer.

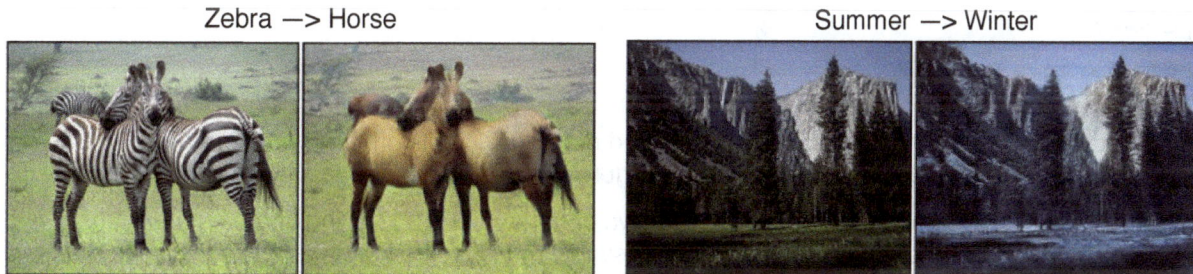

Figure 11-28: Three examples of image to image translation by CycleGAN. Image credit [Zhu '17]

CycleGAN assumes the data translation is cycle consistent. It means if a translator G translates an image from domain X to an image in domain Y ($G : X \rightarrow Y$), another translator F can translate back the translated data into the original domain data, i.e., $F : Y \rightarrow X$. To clarify cycle consistency, the authors described that if an English sentence is translated to French, the translation of the French sentence back to English should show us the same original sentence in English.

CycleGAN Architecture

CycleGAN has two discriminators, i.e., D_X and D_Y, two generators, i.e., G and F, which are described as follows:

• G transforms images of type X to images of type Y.

- F transforms images of type Y to images of type X.
- D_X distinguishes differences between images X and translated images $\hat{X} = F(Y)$.
- D_Y distinguishes differences between images Y and translated images Y, which is $\hat{Y} = G(X)$.

In total, CycleGAN has two cost functions: cycle-consistency loss, which we have explained earlier, and adversarial loss.

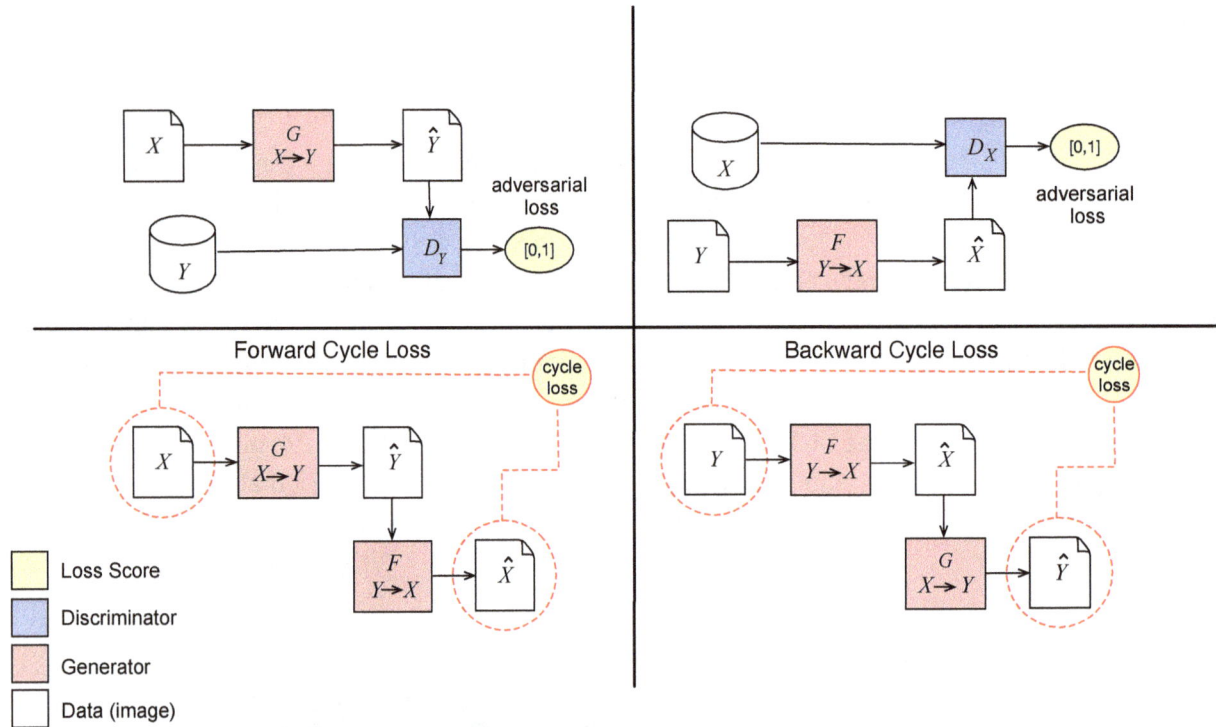

Figure 11-29: CycleGAN architecture is composed of two discriminators and two generators (We separate this figure into four subfigures for the sake of readability).

On the top, there are two GAN networks, with two generators, F and G. The one on the top left translates X to Y and the one on the top right translates Y to X. At the bottom, cycle consistency is presented with different figures. There is cycle consistency, I.e., forward and backward cycle loss. Therefore, the CycleGAN has four loss scores, two are discrimination losses of its GAN network and two are cycle losses.

Figure 11-29 presents the CycleGAN architecture. On top of this figure, we have two GAN networks, and our objective is to translate data from domain X to data from domain Y (e.g., an image of a cat into a lion). Also, to maintain the cycle consistency, we need to have the image from domain Y translated back to domain X (e.g., a lion image back to a cat image). Therefore, two GAN networks are required, as shown at the top of Figure 11-29. The one on the top left gets an image from domain X and uses generator G to translate it to an image in domain Y, i.e. $\hat{Y}$. Then, the discriminator D_Y uses images from domain Y to compare the $\hat{Y}$ images from domain Y and calculates the adversarial loss.

The segment on the top right gets an image from domain Y and uses generator F to translate it to an image in domain X, i.e. $\hat{X}$. Then, the discriminator D_X uses images from domain X to compare the $\hat{X}$ images from domain X and calculate the adversarial loss. As stated before, there are two cycles that are shown in the bottom segments of Figure 11-29. They compare the translated image with the original images before translation and calculate the cycle loss.

For example, the cat A is translated into a lion B, and the lion B is translated back into a cat A ($\hat{A}$). The cycle loss is measuring the differences between A and $\hat{A}$. Authors call it *Forward Cycle Loss*, i.e., $x \rightarrow G(x) \rightarrow F(G(x)) = \hat{x} \sim x$. For each generated image in domain Y, getting back to the original image is called *Backward Cycle Loss*, i.e., $y \rightarrow F(y) \rightarrow G(F(y)) = \hat{y} \sim y$.

CycleGAN Cost functions

The adversarial loss is similar to other GAN networks, and for mapping $G : X \rightarrow Y$ it is formalized as follows:

$$\min_{G} \ \max_{D_Y} J(G, D_Y, X, Y) = \min_{G} \ \max_{D_Y} \mathbb{E}_{y \sim P_{data(y)}}[log D_Y(y)] + \mathbb{E}_{x \sim P_{data(x)}}[log(1 - D_Y(G(x)))]$$

In this equation, G gets an image x and transforms it into an image y. D_Y is the discriminator for domain Y, and tries to identify if the transformed data $G(x)$, which is shown with $\hat{y}$, and determines if $\hat{y}$ is real data or not (by comparing with y). For the second GAN ($F : Y \rightarrow X$), it is the same equation, just changing the input, output, and functions:

$$\min_{F} \ \max_{D_X} J(F, D_X, Y, X)$$

The cycle consistency cost is written as the sum of forward cycle cost and backward cycle loss, as follows:

$$J_{cyc}(G, F) = \mathbb{E}_{x \sim P_{data(x)}}[\|F(G(x)) - x\|_1] + \mathbb{E}_{y \sim P_{data(y)}}[\|G(F(y)) - y\|_1]$$

The first part of this cost function calculates the forward cycle loss, and the second part calculates the backward cycle loss, here $\|...\|_1$ is a sign of L_1 norm. In particular, the cycle cost is either L_1 norm, or summed absolute difference in pixel values between images.

By merging these two types of cost functions, the overall CycleGAN cost function $J(G, F, D_X, D_Y)$ is written as follows:

$$J(G, F, D_X, D_Y) = J(G, D_Y, X, Y) + J(F, D_X, Y, X) + \lambda . J_{cyc}(G, F)$$

The parameter λ specifies the importance of the cycle consistency cost function. The authors of the paper set $\lambda = 10$ for their experiments.

Generator and Discriminator Networks

We describe the discriminator and generator architecture briefly. You also can check several open-source implementations, which could have small differences from each other, for more details.

Each generator has three sections: (i) an encoder, (ii) a transformer, and (iii) a decoder. The input image (with size $256 \times 256 \times 3$) is fed into the encoder, and the encoder shrinks the input image but increases its depth. The encoder is composed of three convolution layers, and its result (output size: $64 \times 64 \times 256$) is fed into the transformer. The transformer is composed of a set of six Resnet block layers (we will explain them in Chapter 12). The decoder gets the image from the last layer of the transformer and reconstructs the image to its initial size by using three transposed convolutional layers. This means that the decoder gets the input of size $64 \times 64 \times 256$ and provides output of size $256 \times 256 \times 3$.

The CycleGAN discriminator uses the PatchGAN architecture [Isola '17], like Pix2Pix. By checking its original paper or open source implementations of CycleGAN, you can find more details about the details of architecture, such as activation functions, stride size, etc.

StyleGAN Models

At the time of rewriting this chapter, the StyleGAN3 (a.k.a Alias-Free GAN) [Karras '21] is among the most updated and advanced synthetic human face generators. In this section, we explain StyleGAN [Karras '19], but we should neglect some details that you can further check in their papers or implementation. If you are tired of reading about GAN architectures, take a look at Figure 11-20 and you can observe how realistic StyleGAN synthetic faces created are. That person does not exist in reality[11], and StyleGAN version 2 generated it.

The novelty of StyleGAN is in its generator; its discriminator is the same as Min-Max GAN. The StyleGAN generator can control the fine-grained details (style) of the generated images. For example, in generated faces, the algorithm can separate the pose, identity, face, hair color, etc. StyleGAN also prevents mode collapse (e.g., generating only one particular

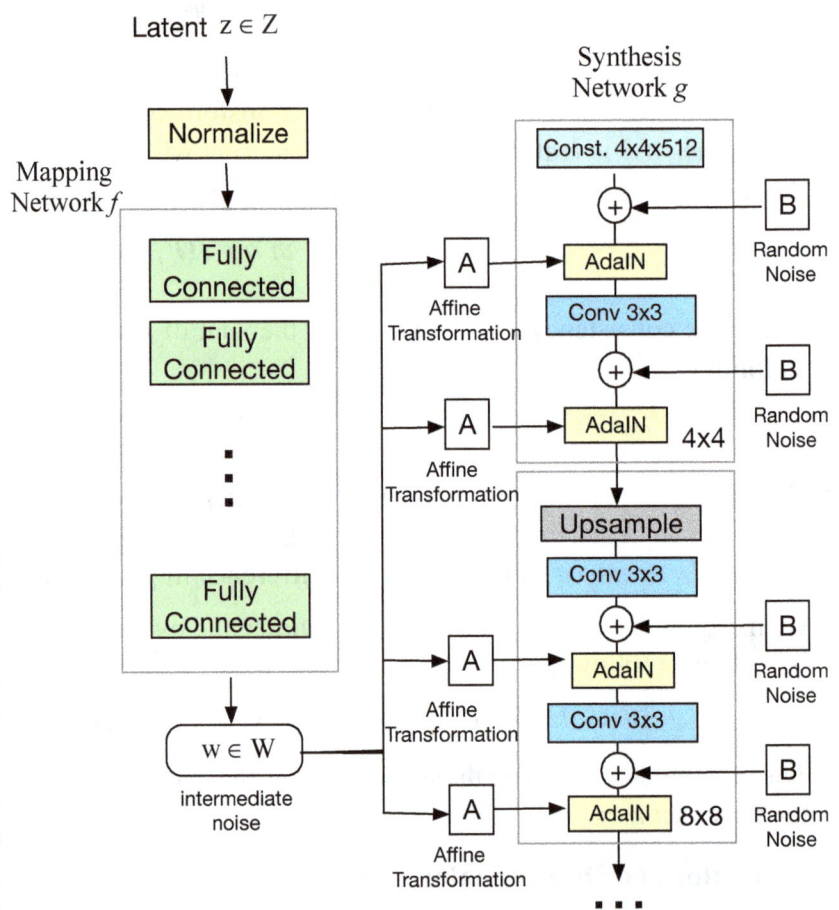

Figure 11-30: StyleGAN architecture.

---

11 At the time of writing this book there is a web page that show StyleGAN2 sythentic images as well you can check it in the following address: https://thispersondoesnotexist.com

face).

Besides, images can be mixed to construct new images. For example, taking parents' pictures and guessing the face of their child in the future or controlling accessories on the face, such as removing or adding glass to a face image.

StyleGAN generator has three architectural characteristics, which makes it different than vanilla GAN. In the following, we list these characteristics.

(i) Vanilla GAN generators start from a noise vector z and improve it until the generator provides synthetic data that is hard for the discriminator to distinguish from real data. The StyleGAN Generator does not start with plain noise. It has a mapping network architecture (a sequence of fully connected layers, as shown on the left side of Figure 11-30) that gets a set of plain noise vectors $z \in Z$ (not just one noise vector, more than one noise vector) and maps it to transforms it into an intermediate latent space, $w \in W$. This noise w is called an *intermediate noise*. This network is known as a mapping network and applies a non-linear transformation on noise z. The purpose of this mapping network is to create a more structured and manipulable representation of the latent variables.

(ii) StyleGAN introduces an Affine Transformation (Four A in Figure 11-30) and Adaptive Instance Normalization (AdaIN) [Huang '17], which incorporate *style (statistical characteristics of the image)* into the generated image (blue boxes in Figure 11-30 presents AdaIN).

(iii) After the w is transferred into the generator network, the model adds another random noise (Gaussian noise) into the generator as well. This means that StyleGAN uses two noises, unlike other GANs, which use one noise.

Noise Sources and AdaIN

In terms of noise, if we summarize Figure 11-30, we get something like Figure 11-31. In particular, two different types of noises are added to the generator: intermediate noise and random noise.

The generator got noise vector z with the size of 512 and passed it through eight fully connected (FC) layers of Multi Layer Perceptron, as we can see from the left side of Figure 11-30.

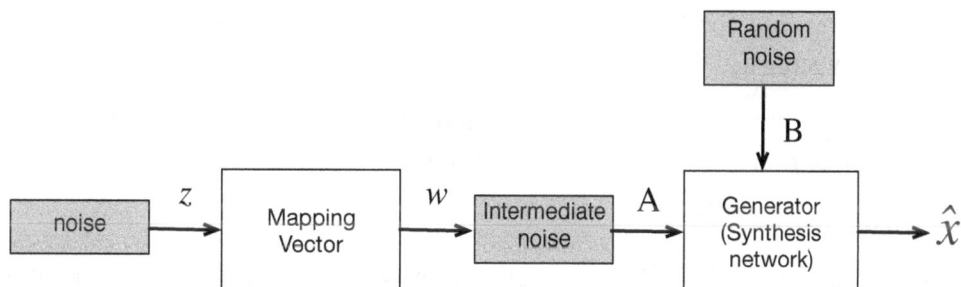

Figure 11-31: StyleGAN noises flow to the generator.

Why do the authors add eight additional Fully Connected layers? Mapping a noise vector into an intermediate noise provides a *disentangled representation of features*. Feature entanglement here means that the noise vector is not mapped to features we expect from the output (e.g., eyes, lips, and nose). A noise vector has a Gaussian distribution, but features we need from output have a different distribution around the actual object, e.g., the eyes distribution is not Gaussian, and z noises do not entail those distributions. Therefore, these FC layers convert z noise vector distributions into distributions that are a bit closer to those output feature distributions. In other words, authors state that w noises are less entangled and closer to the distributions of real image components (e.g., eyes, nose, lips). The result is w, with a dimensionality of 512. To summarize, the fully connected layers transform the initial noise vector z into a more disentangled intermediate latent representation w.

AdaIN injects the intermediate noise into the layers of the Generator. AdaIN normalization is a type of Instance Normalization (check Chapter 10) that adds style to the data. Style in this context refers to the statistical characteristics of the image components based on the incoming *intermediate noise (A)*. We can say, that styles are extracted from w and then added to various points (As) into the generator.

Assuming x_i is a feature map that is normalized separately, as described in Chapter 10, Instance Normalization (IN) is written as follows:

$$IN(x_i) = \gamma \frac{x_i - \mu(x_i)}{\sigma(x_i)} + \beta$$

The intermediate noise w does not directly go into AdaIN, it goes into two fully connected layers and produces y_s (scale) and y_b (bias). In other words, this step converts w into a style y. AdaIN got style y as input, and substitutes γ and β, of instance normalization with $y_{s,i}$ and $y_{b,i}$ (i subscript stands for the instance), which are scale and bias factors *coming from the intermediate noise*. AdaIN is written as follows:

$$AdaIN(x_i, y) = y_{s,i} \frac{x_i - \mu(x_i)}{\sigma(x_i)} + y_{b,i}$$

These two parameters (y_s, y_b) are parameters that construct the new style on the image and construct the synthetic image. In Figure 11-30, y_s, y_b are presented as A. Therefore, we can say that AdaIN is responsible for transferring style information onto a generated image from the intermediate noise vector w.

The second source of noise is a simple Gaussian distribution noise (stochastic noise). These noises are used to add more diversity and variation to the output.

Progressive Growing

StyleGAN uses an approach called progressive growing, in which the generator gradually increases the quality of synthetic data. Progressive growing means that at the beginning layers, the model has a coarser component of a face, such as a nose, eyes, etc., and as it moves forward, it adds more details, such as hair color and eyebrow style.

The concept of progressive growing is proposed by Progressive GAN (ProGAN) architecture [Karras '17]. In progressive growing, the resolution grows slowly. The generator starts by constructing a tiny tensor with 4×4 pixel size, and the discriminator tries to analyze this image as well. Const. 4×4×512 in Figure 11-30 is a 4×4 grid where each cell contains a 512-dimensional vector. This tensor is the initial input to the StyleGAN generator network, which then progressively upsamples and transforms it to produce a high-resolution image. Then, the resolution of the image increases by simple upsampling (after 4×4 will be 8×8), and in parallel, it applies a convolution on the image until it ends up in an image with the size of 1024×1024 pixels. Other GAN architectures usually use transposed convolution for upsampling, but here, authors use a combination of nearest neighbor upsampling and convolution. On the right bottom side of each grey box in Figure 11-30, we see 4×4, and then 8×8, this represents the progressive growing.

This slow pace of evolvement in the network causes the network to learn simple features (eyes, nose in the face) first and then progress toward more complex features (hairstyle, wrinkles on the face, etc.). This approach hinders the mode collapse problem as well. Briefly, this process is implemented with a combination of upsampling and convolutions. If you are interested in learning more details about progressive growing, you can check the ProGAN paper [Karras '17].

Style Mixing

One of the interesting characteristics of StyleGAN is mixing images and generating photo-realistic images based on the mixture of images. This is an amazing feature, and some recent mobile selfie apps have adopted it as well. Figure 11-32 presents an example of this feature.

The process of style mixing occurs by mixing several intermediate noises ('A' noises in Figure 11-30) along with stochastic noises ('B' noises in Figure 11-30). As we can see from Figure 11-30, different 'A' noises will be added to the generator (Synthesis network g Figure 11-30). Some 'A' noises control coarse style features (the one added at the first layers of the generator), some middle style features, and some fine style features (the one which is added to the bottom layers of the generator).

Figure 11-32 presents two sources of images that are mixed (images from source 'A' and source 'B') and construct new images. As we can see, images in the first row got coarse styles from source 'B', images in the second row got middle styles from source A, and images in the third row got fine styles from source 'A'. We can see in the bottom row that the woman's face in source 'A' got the fine styles from source 'B', and the changes in her face are insignificant. Only her hair and skin colors have changed. This bottom line is in contrast to the first line, which states that the number of changes in the face is very significant.

We can control the degree of mixing styles by intermediate noise vector, and this is the reason we mentioned StyleGAN can control the fine-grained style of the generated image at the beginning of our explanation. It was not possible with the previously described GAN architectures.

Figure 11-32 describes the impact of mixing styles (the impact of A noises). In some cases, we do not intend to mix different styles, instead, we want to see different faces with one single generated image. This will be handled by stochastic noise (B noises), which is added before

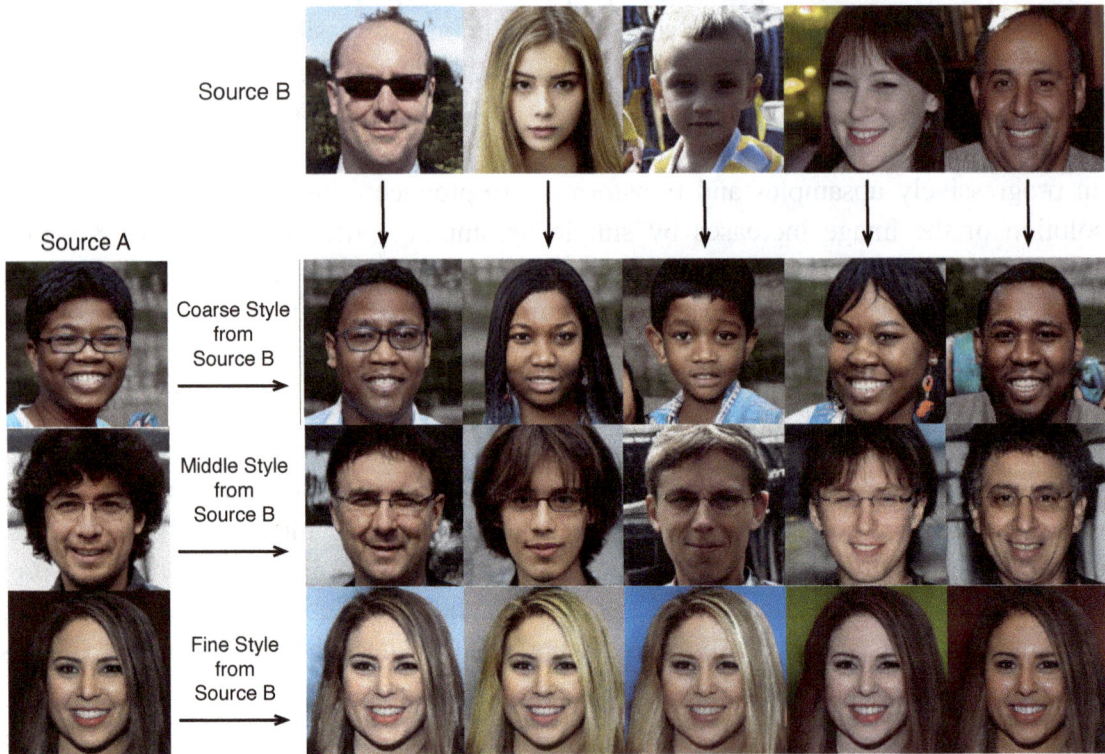

Figure 11-32: Mixing styles in StyleGAN: the impact of adding the intermediating noise vector into the Generator.

AdaIN. In other words, adding additional noise to the model adds additional variation to a single image. This variation can change very small details of the image, such as a wisp of hair. Figure 11-33 presents these features of StyleGAN, and we can see the impact of stochastic noise.

Finally, we are done with StyleGAN, but … wait … oh no … StyleGAN2, StyleGAN3 (Alias-Free GAN), remained.

StyleGAN Successors

At the time of writing this section, NVIDIA is the largest GPU manufacturer, and a team of researchers from NVIDIA is behind StyleGAN. Unlike academics who are continuously looking for grants, they are very rich to afford their GPU settings and improve StyleGAN. For example, Alias-Free GAN used 100, V100 NVIDIA GPU. If the author of this book could afford 100 NVIDIA GPUs back in 2021, he definitely would

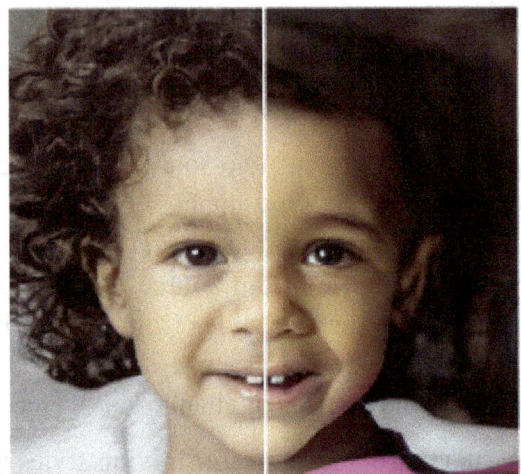

Figure 11-33: The impact of stochastic noise on the generated image. (Right) no stochastic noise. (Left) stochastic noises added in fine layers. Image credit: [Karras'19]

not write this book and enjoy a lavish life on a private island with tropical smoothies.

After the success of StyleGAN, the same group announced an even more accurate version of StyleGAN in 2020, which is known as StyleGAN2 [Karras '20], next to that StyleGAN3 is introduced as Alias-Free GAN [Karras '21]. Here, we briefly explain their differences from StyleGAN, for more information, you can check their papers.

One known limitation of StyleGAN is the *water droplet effect* on the generated image (see Figure 11-34). Another problem with StyleGAN is that some features are highly localized, and their position in generated images was pinned on a fixed position, such as the location of teeth or eyes staying in the same region in different generated images. Take a look at Figure 11-32; the region of teeth in all images is the same.

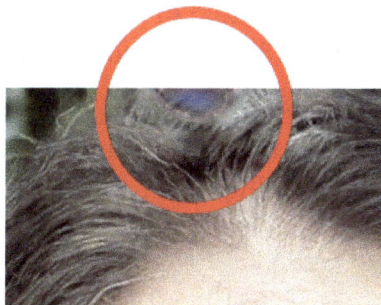

Figure 11-34: Water droplet sign on generated images, because of AdaINST.

StyleGAN2 applied several changes to StyleGAN, including substituting the use of AdaIN with weight demodulation, changing the progressive growth, and adding two new regularizations, including lazy regularization and path length regularization. These changes mitigate the limitations of StyleGAN.

After introducing StyleGAN2, the same team developed StyleGAN3, known as Alias-Free GAN [Karras '21]. While StyleGAN2 reduced but did not eliminate the fixed positioning of features in images, StyleGAN3 further addressed this issue. Although its FID score is similar to that of StyleGAN2, StyleGAN3 offers improved internal image representation by better handling transformations and rotations without texture sticking. This was achieved by addressing aliasing issues, which in previous models led to the generation of repetitive texture patterns due to inadequate suppression by upsampling filters. By eliminating aliasing, StyleGAN3 effectively resolved the texture sticking issue.

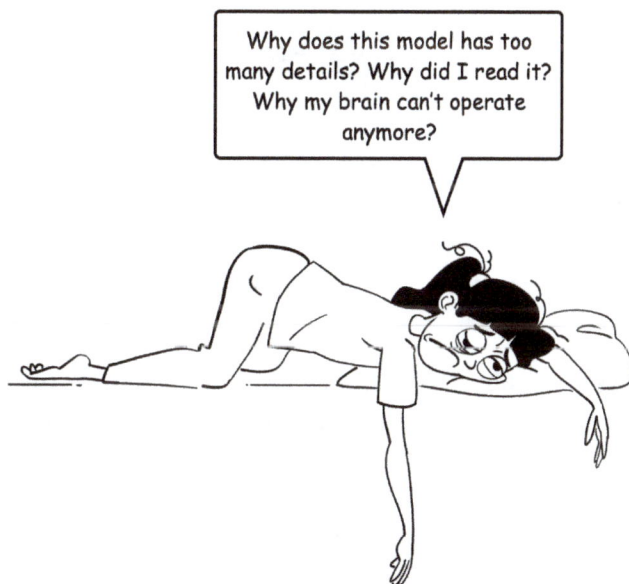

What remained to be explored? We are done with explaining some popular GAN architectures. You might be tired of learning many GANs, but by doing a search on arXiv.org, there are only more than 1,800 GAN models remaining that you can learn on your own[12].

All jokes aside, if you are young and a student, do not spend your valuable time making another GAN architecture; there are lots of interesting things that remain to be discovered and learned in artificial intelligence. At the time of revising this section, diffusion models are hype, but they also have a lifespan.

Few more GANs
(ಠ_ಠ)

NOTES:

* There are claims that all generative models are derived from the maximum likelihood algorithm. The assertion, however, is an oversimplification, as many generative models employ a variety of foundational algorithms, such as variational inference, adversarial training, autoregressive models, etc.

* Training GAN is hard (still in late 2024), and again, it is something that we can not train on our own unless we have access to large GPU clusters of big corporations.

---

[12] https://arxiv.org/search/?query=GAN&searchtype=title

* While working with other autoencoders and other described neural network models, we have a single cost function that the optimizer needs to improve. However, GAN has two cost functions that need to compete with each other.

* To have a stable GAN, it is recommended not to mix synthetic and real data into a batch (stochastic gradient descent batch). Instead, use separate batches, and each batch only contains either synthetic or real data.

* Another approach to have a stable GAN is label smoothing, which means instead of 0 and 1 labels for synthetic and real data. Instead, the model can use probability that provides values slightly larger than one or slightly smaller than zero, i.e., label smoothing [Brownlee '19]. This has a regularization effect on the data as well.

* One potential harm of image-to-image or video translations is synthetic video or synthetic audio generation, which is known as *deep fake,* and it is getting harder to distinguish if a real person saying something in a video is real or deep fake. There are interesting works where you can see the face of an old painting converted into a moving face and change his/her mimic, like Mona Lisa smiling, singing, or changing her facial expressions [Zakharov '19]. Some speculated that in the future, it will be possible to create an immortal version of a human in the digital world [Bell '10], known as digital immortality. By the time of reading this text, it is probably already implemented. Another extreme example is that criminals can simulate a synthetic digital avatar which we can distinguish if it is synthetic or real, e.g., swapping a face of a popular person with a porn actor in a porn movie.

* Only big and wealthy corporations such as NVIDIA, OpenAI, Google, and Meta can afford to build these large models. Even big universities fall behind the trend of making fancy generative AI models due to their limited GPUs.

Contrastive Representation Learning

There are different methods for preparing an object for comparison. Non-deep learning approaches use a similarity measurement (Chapter 4) to compare two digital objects. These approaches are used in the traditional image-matching, audio search, etc, but they are not flexible. For example, the face of a person who rotates his head in two different images is identified as two different images. Older approaches convert data into a lower dimension, such as applying PCA (Chapter 7) and then comparing them, and even after conversion, those methods are still inflexible.

Feature engineering in the context of deep learning is effectively managed through a process known as representation learning. This process involves mapping raw input data to a more useful representation or feature space through multiple layers of non-linear transformations. These transformations help convert the input data into a form where its features are more useful for machine learning algorithms, particularly for tasks referred to as *downstream tasks*. This automated feature discovery and transformation are key aspects of how deep learning models learn from data [Bengio '13]. A good representation of the data entails important features of the original data. For example, the word embedding methods described in Chapter 6, are representation learning methods because they convert text data into vectors while maintaining the semantics of the text.

Contrastive Learning, a form of self-supervised learning also referred to as Contrastive Self-supervised Learning, involves using representation learning techniques to distinguish between similar and dissimilar data points. This approach, known as *Contrastive Representation Learning (CLR)*, focuses on learning by comparison. CLR algorithms are particularly effective in learning robust features by contrasting pairs or sets of similar and dissimilar data, which enhances the model's ability to generalize from unlabelled data [Le-Khac '20].

From a technical perspective, the objective of CLR is to build an embedding space (e.g., in a lower-dimensional space) where similar data points stay close together and dissimilar data points stay far away from each other.

I don't understand, why the author brings this section here? You have difficulty understanding this part, without knowing Transformers, BPE, etc.

For example, we show an image of one rabbit to the model, and then we ask the model to search a database of other animals and find other rabbits. The algorithm should recognize the correct object (image of rabbits) and contrast that object to other objects (images of other animals) in the dataset.

A question might arise: what is the difference between contrastive learning and other supervised learning? While supervised learning typically processes individual data points or batches, assigning or predicting labels based on provided examples, Contrastive Learning differs by focusing on learning from the relationships between data points. Specifically, CLR algorithms enhance their learning by comparing and contrasting similar and dissimilar data points to construct robust representations. These representations do not directly classify objects into categories but rather enable the model to distinguish between similarities and differences effectively.

In the following, we describe one type of neural network in this category, i.e., Siamese Neural Network, and leave the rest, such as SimCLR models [Chen '20 A, Chen '20 B] and MoCo models [He '20, Chen '20 C] to you to read them.

Contrastive Learning Loss Functions

Before we describe the Siamese Network, we need to be familiar with two popular loss functions, i.e., contrastive loss and triplet loss.

Contrastive Loss

We measure the contrastive loss [Chopra '05] between two objects X_1 and X_2 that each object describes a data point in embedding space, e.g., a vector. The generalized contrastive loss (L) between these two data objects is written as follows:

$$L(W, (Y, X_1, X_2)^i) = (1 - Y)\frac{1}{2}(E_w)^2 + Y\frac{1}{2}\{max(0, m - E_w)\}^2$$

In this equation, i specifies the index of the current data point. $Y = 1$ if two data points X_1 and X_2 are dissimilar, $Y = 0$ if they are similar. m stayed for the margin of distance, it is a hyperparameter, and it defines the threshold (lower band) distance between data points of different classes. E_W is the distance value (e.g., euclidean) between data points of two groups and assuming that the G_W is a mapping function E_W between X_1 and X_2 is computed as follows:

$$E_W(X_1, X_2) = ||G_W(X_1) - G_w(X_2)||_2$$

We can see that the right part of the equation penalizes the model for dissimilar data points having a distance $E_w < m$ (close to each other). It also penalizes similar data objects for being far from each other (stay outside m margin of distance). If $E_w \geq m$, then $m - E_w$ will be negative, and because of the *max*, the right part of the loss function will be zero.

Contrastive loss is usually used when we do not have enough amount of labels, which is common because acquiring labels for raw data is expensive.

Triplet Loss

The *triplet loss function* was proposed for the first time in the FaceNet [Schroff '15] model, which is a successful face recognition algorithm that can recognize the same person in different poses and different camera angles.

This loss function compares the baseline or anchor sample x, against the positive sample x^+ and a negative sample x^-. In this context, the positive sample means x and x^+ belongs to the same class (they have the same label), and the negative sample means x^-, and x belongs to different classes (their labels are different). The objective of this loss function is to encourage the network to minimize the distance between x and x^+, and, at the same time, maximize the distance between x and x^-. Triplet loss is formulated as follows:

$$L(x, x^+, x^-) = max(0, E(f(x) - f(x^+)) - E(f(x), f(x^-)) + \epsilon)$$

In this equation, $f(.)$ is the embedding function that transforms the data into a different representation. ϵ is the margin between positive and negative data. This margin distance ensures that a distance exists between negative pairs and positive pairs. $E(.)$ presents the Euclidean distance (or L_2 norm), but other distance functions could be used as well. There are some uses $||...||_2$ to present this L_2 norm, but to reduce your risk of getting a mathematical panic attack chance, we used $E(.)$. Anyway, you might find it more readable as follows;

$$L(x, x^+, x^-) = max(0, ||f(x) - f(x^+)||_2 - ||f(x) - f(x^-)||_2 + \epsilon)$$

Triplet loss is useful when it is important to set the negative sample properly, e.g., face recognition among many existing faces.

Siamese Network

The class of neural networks that they learn to *differentiate between two inputs* are called Siamese networks[13]. Unlike other neural networks, they do not train to classify the data. Instead, they learn to measure the similarity between two inputs. Why not use traditional similarity measurements we have learned in this book, like correlation coefficient analysis (Chapter 3), or similarity metrics we have learned in Chapter 4? The answer is that described similarity metrics can not compare complex data structures together. For example, a traditional facial recognition approach can compare images from the same camera location toward the face (e.g., front view). It can not recognize the same person when her face image is taken from a different angle (e.g., side view). These limitations led to the introduction of neural networks used specifically for comparison.

A Siamese neural network (twin neural network) is composed of two or more identical neural networks. Here, identical means the same number of layers, same neurons, same parameters, and even the same weights. After training, they can be used to recognize whether or not two input data are similar. The objective of the Siamese neural network is to have high similarity scores for similar inputs (e.g., images of the same person, signatures of the same person, the same food) and low similarity scores for different inputs.

A Siamese neural network commonly employs a convolutional neural network (CNN) architecture within each of its branches. This setup typically includes several convolutional layers followed by a few fully connected layers. Unlike traditional classification networks, Siamese networks do not use a softmax layer at the output because the primary task is not to classify inputs into categories but to assess their similarity. The output from each network branch, often called the encoding of the input data, represents a feature-rich representation used to measure similarity between pairs of inputs.

Training Siamese Network

First, the network should be fed with positive (1) and negative (0) pairs. For example, we fed the following dataset into a network: {(cat_img_1, cat_img_2, 1), (cat_img_3, cat_img_2, 1), (cat_img_1, dog_img_2, 0), (cat_img_2, squirrel_img_2, 0), ...}.

Figure 11-35 presents the architecture of training the Siamese network, assuming we have an input set of pairs with labels and we fed each pair into one network. For example, we feed x_1 to one network and x_2 to the other network. The result of each network will be a feature vector, let's say $f(x_1)$ and $f(x_2)$. After obtaining the difference or distance between feature vectors, this value is often processed through one or more fully connected (FC) layers. The result of an FC layer(s) is given to a Sigmoid to transform the output into a probability and thus identify if x_1 and x_2 are similar or not (a threshold of 0.5 is used to make a binary decision).

---

[13] Siamese twin is an extremely rare medical phenomenon in which two babies (twins) are conjoined and must be separated by surgery.

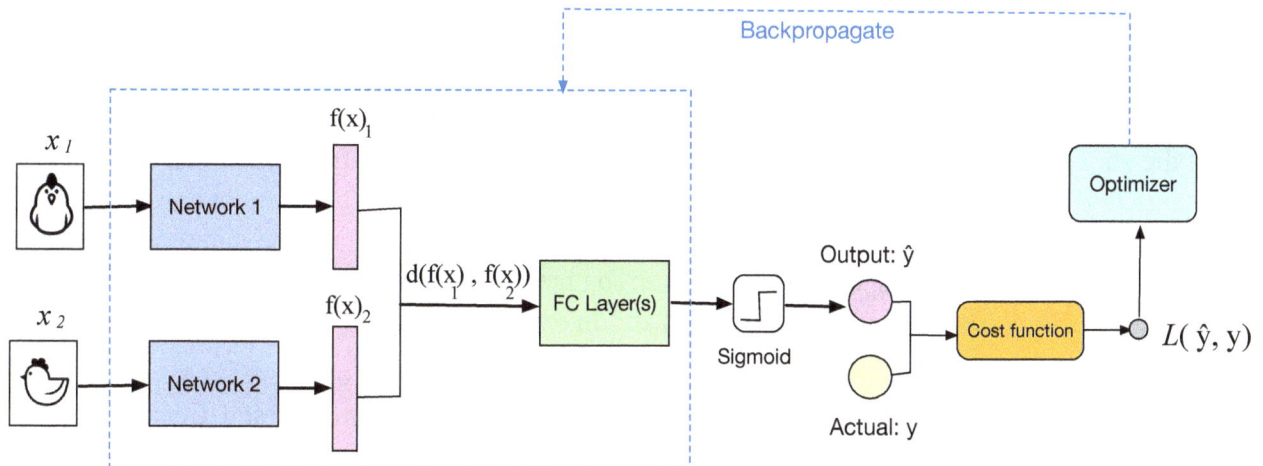

Figure 11-35: Training phase for Siamese network, in this example we show only one positive pair of examples, which is a chicken. In practice, the network will be trained for many positive and negative pairs.

Now, the output of a Sigmoid function will be compared with the original label (similar:1, dissimilar:0), and the Cost function (Triplet loss, Contrastive loss, or Cross entropy) will be used to measure the loss score. Then, the optimizer (e.g. SGD) back-propagates the loss score to change model weights and biases on the FC layer(s) and both twin networks toward reducing the loss score.

This process will be done for all pairs of images, both positive and negative ones. The resulting model will be saved for the inference phase.

Data Preparation for Training

To prepare data for training a contrastive model, we start by selecting a random input to serve as the anchor sample. We then choose additional samples that share the same label with the anchor to create positive pairs and samples with different labels to form negative pairs. This process ensures that there are corresponding positive and negative samples for every anchor. During the training process, every image in the dataset should have the chance to be used as an anchor, ensuring that the model learns from a diverse and representative set of comparisons. This selection process is typically repeated across training epochs to gurantee robust learning.

Testing Siamese Network

The test set will benefit from the model constructed in the training stage. For example, assume we trained the model on human faces, and as an input of the model, we give a pair of face images to the model. Our goal is to find out if they are the same person or if they are different persons. These two images are passed to the network (the same network that is used for training), and the model provides us with a similarity score. If the similarity score is close to zero, it means they are not presenting the same person, but if the similarity score is high (i.e., close to one).

Text-to-Image Models

In the summer of 2022, when we revised this chapter to add Text-to-Image models, at that time, OpenAI released DALL-E, which was the top model at that time that got the text and constructed an image from the text, a.k.a. *text2image* or *text-to-image* models. Since the author of this book has a mental disorder and is obsessed with learning new models, we list some of the popular text-to-image models here that will be introduced and in use by the end of 2024.

You might have difficulties understanding the architecture of some of them, then, please come back here after reading Chapter 12. There, you will learn transformers and thus be able to understand these models. Before we explain models, we follow our tradition and explain some concepts used in text-to-image models.

Text-to-Image Related Concepts

In this section, we introduce key concepts and models used in text-to-image generation. Among these, diffusion models have gained significant popularity, surpassing GANs in synthetic data generation. Although we provide only a brief overview of diffusion models, this should not diminish their importance. Diffusion models have demonstrated remarkable success in text-to-image, text-to-audio, and even image-to-video generation.

Zero-Shot Learning

Zero-shot learning enables a model to label data it has not encountered during the training phase. However, based on given instructions and other auxiliary information, it generates a label for the unseen data not included in the training set [Larochelle '08, Chang '08]. In other words, in zero-shot learning, the model is trained to recognize and classify objects it hasn't previously seen. This is achieved by providing the model with descriptions or attributes of the objects rather than actual examples.

> Don't read anything here, before reading Chapter 12. I don't know what mental problem does the author have to bring this section here.

How is such a thing possible? For example, a model receives a text input. There are many descriptions of horses, and these texts are labeled as 'horse' during the training phase. Therefore, the model can easily label a text segment about a horse during its testing phase. Now, consider a text about a zebra. Although there is no label for 'zebra' provided, when it encounters a text about a zebra, it recognizes that the description is similar to that of a horse but with stripes (auxiliary information). By analyzing the text, it recognizes the horse and infers from other parts of the text that a horse with stripes is called a zebra. The model then assigns the label 'zebra' to that text based on this information.

In summary, zero-shot learning enables the model to assign labels to unseen data in the training phase. While large corporations have historically had access to such extensive datasets, there are now publicly available resources that can be used for zero-shot learning. Open-source contributions democratize access to the technology, enabling more entities beyond just large corporations to explore and implement zero-shot learning.

Autoregressive Models

The term autoregressive (AR) is often used in time series and refers to a model whose future values are regressed on previous values. In simple words, the autoregressive model predicts a variable based on its past values, or a model is AR if it predicts future behavior based on leveraging past behaviors. For example, ARIMA (Chapter 8) is an AR model. We can formalize the prediction of variable x at time t, by using k previous variables: $x_t = f(x_{t-1}, x_{t-2}, \ldots x_{t-k})$

There are differences between RNN and AR neural network models. RNN model connections can go forward and backward, but AR model connections are feedforward only. RNNs maintain hidden layers, but AR models do not use hidden layers, and thus, they have limited output responses. There are more differences, but we don't need to learn them to understand text-to-image generative models.

Diffusion Models

Another category of generative models is referred to as *Diffusion models*. Diffusion models are inspired by physics, and in the context of physics, diffusion is the movement of a particle from a high concentration to a lower concentration area until a thermal equilibrium is achieved; recall that the Boltzmann machine we have explained earlier uses something similar.

For example, consider a spoonful of salt that we put into the water. The salt molecules move in the water (diffuses) until the salt concentration inside the water is equal at every point.

Diffusion models define a Markov chain (check Chapter 5), and in each Markov state, they gradually add noise (usually Gaussian noise, which will be explained in Chapter 16). A diffusion model starts with the original data, such as an image, and progressively adds noise over a series of steps until the image becomes indistinguishable from random noise, specifically following a Gaussian distribution with a mean of zero. Then, a neural network is trained to invert this process (get back from noise to the image). Once this network is trained, it can generate synthetic data with reasonable quality.

The process of adding noise gradually to data at each Markov state is referred to as a *diffusion process* or *forward process/path*. Then, the diffusion model tries to reconstruct the original data from the noise, which is known as the *reverse (denoising) path/process* or *generative path*. The process of Diffusion is presented in Figure 11-36. A trained diffusion model learns to generate new synthetic data by reversing a gradually added noise. The forward path is essential because it creates the training pairs (original data and noisy data) used in training the model. The model then learns how to reverse this process during the reverse path, where it learns to reconstruct the original data from the noisy data.

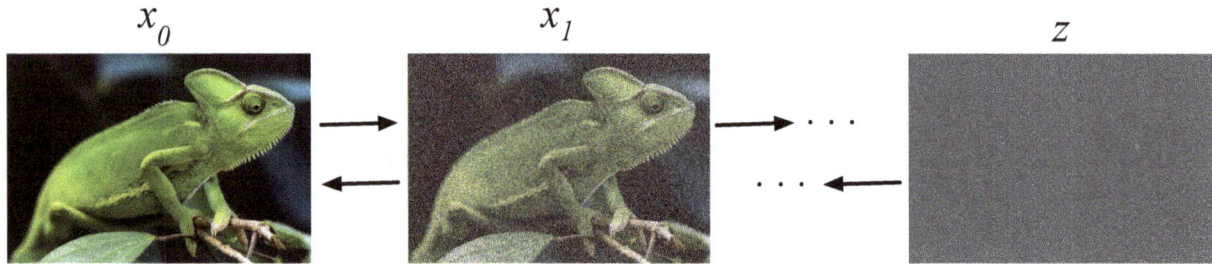

Figure 11-36: A diffusion model that gradually adds a noise at each Markov state (forward process), and then from *z* state, noise gradually reverse to reconstruct a data similar to the original data (reverse process).

Diffusion models differ from other generative approaches, including VAE or GAN models, which use latent spaces and adversarial dynamics. Diffusion models instead directly learn the transformation from noise to data, maintaining the original data dimensionality throughout the process. In other words, diffusion models are trained to de-noise the noise (z) and return to something similar to the original data (x_0).

Forward path: To formalize the forward process and reverse step, we assume the given input data as x_0 sampled from a real distribution, i.e., $x_0 \sim q(x)$. The forward process gradually adds a Gaussian noise $\mathcal{N}$ to the sampled data at T steps, which results in the sequence of noisy images, $x_1, x_2, \ldots x_T$, and x_T will be the pure Gaussian noise with a mean of 0 and variance of I. Assuming ϵ_t is sampled from this noise distribution, $\epsilon_t \sim \mathcal{N}(0, I)$, and α_t is a variance schedule, and decrease over time. Therefore, we can formalize the forward process as follows:

$$x_t = \sqrt{\alpha_t} x_{t-1} + \sqrt{1 - \alpha_t} \epsilon_t$$

In the context of the diffusion model, α_t is a sequence of values that dictate how much noise is added to the data at each step of the forward process. α_t is usually computed as a product of terms of the form $1 - \beta$, where β_t is a small value that defines the noise increment at each step. Usually α_t is computed as $\alpha_t = \Pi_{s=1}^{t}(1 - \beta_s)$.

Reverse path: The reverse process is where the diffusion model generates synthetic data by learning to reverse the noising process. This entails reconstructing the original data from its noise-corrupted state step-by-step, based on learned predictions of the noise ϵ_t, that was added at each stage of the forward process. The reverse step can be formalized as follows:

$$x_{t-1} = \frac{1}{\sqrt{\alpha_t}}\left(x_t - \frac{1-\alpha_t}{\sqrt{1-\alpha_t}}\epsilon_\theta(x_t, t)\right)$$

Here, α_t is defined at the forward process, $\epsilon_\theta(x_t, t)$ is the predicted noise by the model at step t and parameterized by θ. In other words, θ is the trainable parameter of the neural network. The learning objective is to adjust θ so that $\epsilon_\theta(x_t, t)$ matches the actual noise, enabling effective reversal of the noise addition to reconstructing the original data.

Training: The training of the diffusion model involves optimizing θ that the model's predicted noise $\epsilon_\theta(x_t, t)$ matches the actual noise ϵ_t that was added during the forward process. This is usually achieved through an MSE between these two values:

$$L(\theta) = \frac{1}{N}\Sigma_{i=1}^{N}\|\epsilon_t - \epsilon_\theta(x_t, t)\|_2^2$$

Here, $L(\theta)$ represents the loss as a function of the model parameters θ and N represents the number of training examples. We have explained that $||.||_2$ means L_2 norm.

The diffusion process can also be seen from an energy-based perspective, where the model learns an energy function that assigns low energy to data-like samples and high energy to non-data-like samples. The training of diffusion models involves learning the reverse of the diffusion process. This is typically achieved by training the model to predict the noise added at each step of the forward process to undo the diffusion.

A common choice for the neural network architecture in diffusion models is the U-Net, which consists of a series of downsampling layers followed by upsampling layers, with skip connections between corresponding layers.

Inpainting and Outpainting

Inpainting generates or fills in content within a user-defined mask in a specific area of an image, typically to alter or complete parts of the image based on the surrounding context. Outpainting extends the existing content of an image beyond its original boundaries, using the provided prompt and context to generate coherent, expanded visuals. These features are integrated into image editing and image restoration tools.

CLIP (Contrastive Language–Image Pre-training)

CLIP [Radford '21] is a model that can understand images in the context of textual descriptions, proposed by OpenAI in 2021. In layman's terms, CLIP can match the given image to a range of textual descriptions, facilitating zero-shot learning. It is a very popular model and is used in many text-to-image models.

The authors explain the usefulness of CLIP for two reasons. First, image classification approaches are bound to a predefined set of user-defined labels used in their training set. For

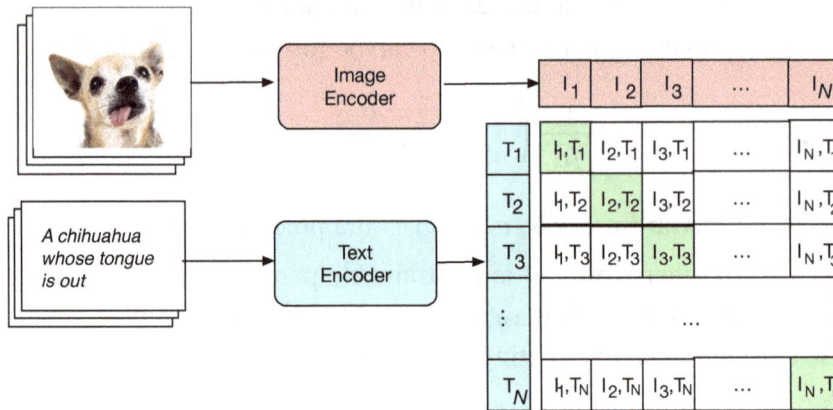

Figure 11-37: CLIP Pre-training phase, each pair of text and image gets into separate encoder and the green diagonal of the result matrix (it is a matrix of cosine similarity) includes positive samples and the white ones are negative samples.

example, in the MNIST dataset, we are limited to 10 labels, whereas ImageNet has 22,000 labels. Each time a new image is added that has never been seen in the dataset before, it must be manually labeled. We have recently learned that zero-shot learning can resolve this need. Second, the labeling process in the classification task is limited to one single label. A text caption (prompt text) contains more useful information than a single class label. For example, instead of the label "chihuahua" for a chihuahua dog image, we can have a text caption as *"The boss's chihuahua, whose bark sounds better than the voice of his owner, a.k.a., my boss"*.

CLIP implements bot zero-shot learning and text prompt labeling by matching images with textual descriptions. It is trained on many images paired with their text captions, which are collected from the Internet[14]. Similar to CLR methods, CLIP input is also a set of image-text pairs.

Contrastive Pretraining

In the first step, CLIP applies an encoding to both texts and images. Texts are encoded by Transformer (in Chapter 12, we will explain attentions and transformers), trained with BPE (in Chapter 14, we will describe BPE) with 49k vocabulary size. Images are encoded by ViT Transformer (with additional layer norm) or ResNet-50 (we will explain them in Chapter 12), whose pooling layers are substituted by the Attention pooling mechanism. You can find more details about the encoding process and its configuration in their paper. After both image and text data are encoded, the representation (feature vectors that are the result of encoding) matches together.

Figure 11-35 shows the pre-training phase of the CLIP. Each image and text in the Image-text pairs set uses a separate encoding, resulting in a vector. The main diagonal of the matrix in this

---

[14] CLIP has trained (they call it pre-training) 400 million pairs of image-text pairs. Authors build 500,000 queries and use a web search to find image and text pairs. Words for queries are selected based on some condition such as occurring at least 100 times on Wikipedia, etc.

figure (shown in green cells) presents positive samples (the cosine similarity should be maximized), and other cells on the matrix (white cells) present negative samples (the cosine similarity should be minimized). CLIP uses cross entropy as a cost function for pre-training and implements it as a symmetric loss function, which results in the matrix shape shown in Figure 11-37.

Training

To train the model, authors experiment with different ResNet and Transformer models with different resolutions and embedding (feature vector) sizes. Instead of pairing each image model with a specific text Transformer, CLIP uses one image encoder and one text encoder that work together. In other words, the authors complemented the image encoder with a Transformer-based text encoder. Through experiments, the authors identified that a larger Vision Transformer model (ViT-L) paired effectively with the text Transformer, yielding the highest accuracy.

The optimizer of the training is Adam, and it is trained for 32 epochs. CLIP has a hyperparameter, i.e., *temperature parameter*, equivalent to 0.07. This hyperparameter is used to clip and prevent scaling the logits by more than 100. The batch sizes used for training are 32,768.

For this setting, the authors trained their model on only 256 NVidia V100 GPUs for 12 days. Yes, it is very affordable, and you can train a small model with a small budget in your basement. Joke aside, such pre-training is only possible by a large corporation that has a large number of resources, and even at the time of describing this model, to our knowledge, even universities can not afford to provide such an infrastructure for pre-training.

Experimenting Zero-Shot Learning

CLIP can be used to build both image classifiers and text classifiers. The authors present the zero-shot learning capability of CLIP by using class labels with some manually added prompt text. This capability stems from its ability to understand images and textual descriptions through its dual-encoder architecture. For example, in Figure 11-38, when an image of a chicken is fed

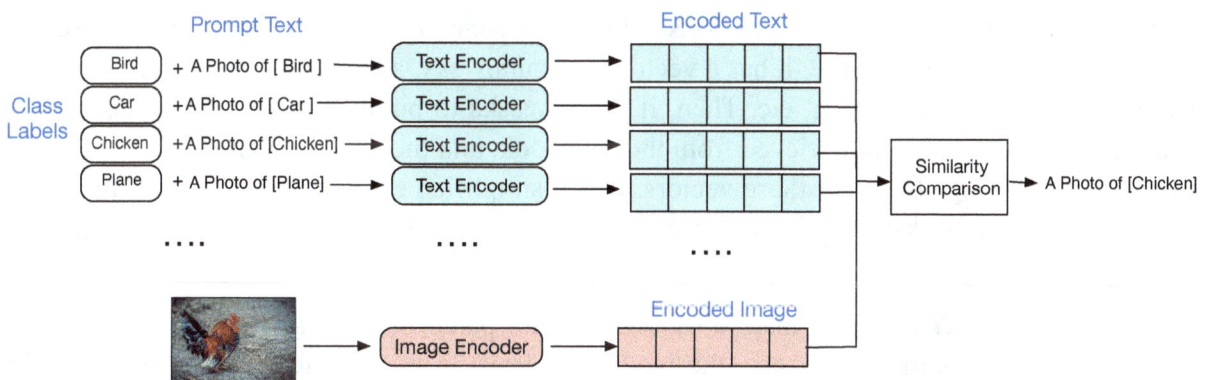

Figure 11-38: Zero-shot learning of CLIP for text prompt prediction. Note that text labels have no information about images. The label will be assigned based on comparing the similarity of the encoded text to the encoded image.

into CLIP, the model does not generate a caption. Instead, a predefined text prompt (e.g., 'A picture of'), manually provided by a model engineer, is used. The model then iterates through possible class labels, combines them with this prompt (like 'A picture of a chicken'), and computes the embeddings of these text phrases. CLIP then compares these text embeddings with the image's embedding. The text whose embedding is closest to that of the image is selected as the best description.

Note that zero training is needed on the actual task of matching the text caption to the image. In other words, the output of the text encoder in Figure 11-36 could be entirely different from the dataset built in Figure 11-37. CLIP identifies the best match by finding the closest text (blue part in Figure 11-38) to the image encoder (red part in Figure 11-38) based on similarity comparison.

In summary, CLIP ensures high similarity between the encoded image and text. This model is very popular in generative AI architecture, and its new variants, such as openCLIP[15], are trained on a larger number of data.

VQGAN (Vector Quantized GAN)

CNN architecture is capable of identifying local relationships correctly, but not able to capture complex higher-level information. In other words, CNN is good at capturing spatial information from pixels close to each other, such as the texture of the object.

VQGAN [Esser '21] combines both architectures to produce high-resolution synthetic images. To explain it, we separate our explanation into two stages: one stage focuses on its GAN architecture, and the second stage focuses on its transformer.

Stage 1: VQGAN is inspired a lot by VQ-VAE [Van Den Oord '17]. The primary motivation behind VQ-VAE was to replace the continuous latent space typical of Variational AutoEncoders (VAEs) with a discrete latent space. This discrete latent space provides several advantages, such as enabling more controlled and stable image generation.

To implement this discretization, the authors of VQ-VAE use a vector quantization and construct a codebook from training images. Please check Chapter 16 to read about vector quantization from the image and how a codebook is constructed. In short, a codebook is a dictionary of discrete vectors all in the exact dimensions. VQGAN's codebook is a dictionary that is used to reconstruct images. For example, it has a vector for a sunny sky, a vector for the garden, a vector or more than a vector for a dog, etc. Then, if a query asks to build "a dog in the garden at the sunny sky," these vectors are retrieved from the codebook, and the decoder will use the codebook to construct the image by using these vectors. It is a simplified example, and keep in mind that vectors in the codebook typically represent various visual patterns or textures,   not items or scenes.

Why use a codebook? Because, instead of dealing with patches of images, which result in too many different values in latent space, the model can learn from a limited number of values in latent space, and thus it can index them. Besides, the training images are fed into a CNN encoder

---

[15] https://laion.ai/blog/large-openclip

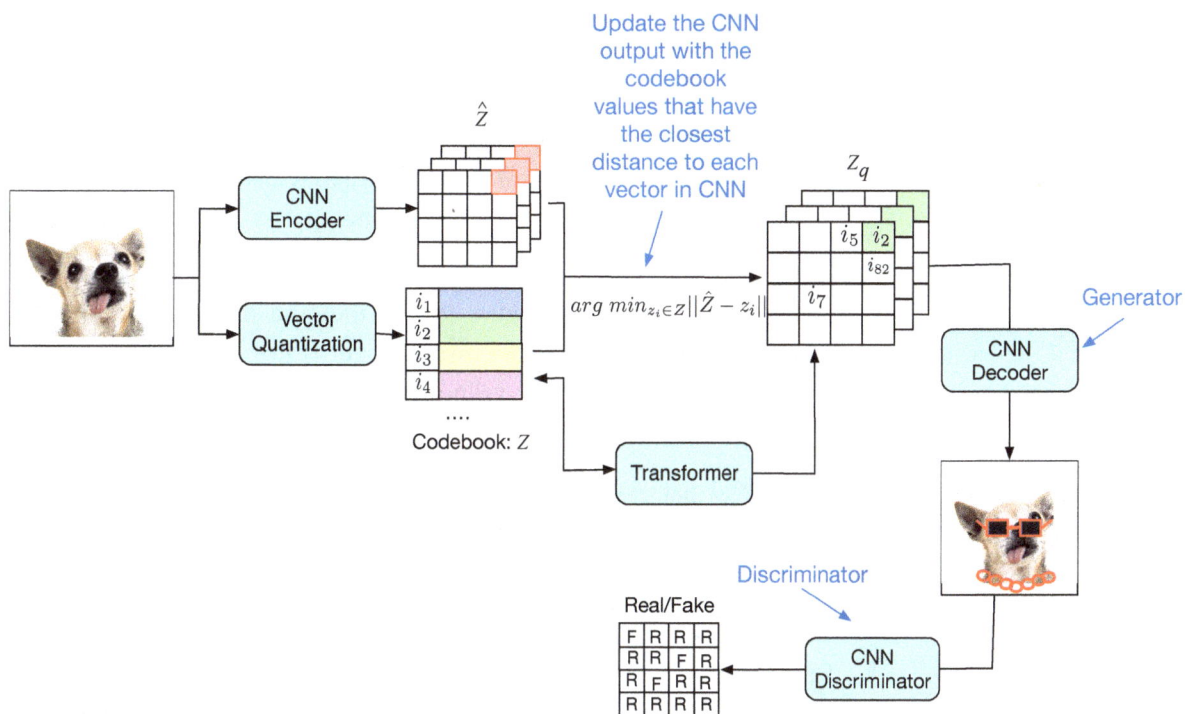

Figure 11-39: VQGAN architecture.

whose output ($\hat{Z}$) reduces the size of the input image while maintaining its spatial information and feature information.

Now, we have trained quantized images into codebook (Z) and reduced size into $\hat{Z}$. Next, a tensor (Z_q) with the same size as the CNN output will be constructed as follows: each vector of CNN is compared (via Euclidean distance) to a vector inside a codebook, and the closest vector in the codebook will be used to substitute the CNN vector value. As a result, instead of the continuous output from the CNN encoder ($\hat{Z}$), we now have a discretized tensor Z_q of the same size as $\hat{Z}$. Note that Z_q includes vectors from the codebook rather than CNN Encoder.

Figure 11-39 Visualizes the construction of the codebook from the image. For example, we highlight one channel vector on the $\hat{Z}$ tensor in red. This vector will be compared to the vectors in the codebook, and the one that has the shortest distance from the codebook (i_1 the one with a light green color) will substitute the vector of $\hat{Z}$ in Z_q.

VQGAN, similar to other GAN models, includes a synthetic image generator and a discriminator. The generator is a CNN decoder, as shown in Figure 11-39.

Now, the generator (CNN decoder) constructs a synthetic image from the codebook. The encoder learns to map images to discrete codes, while the codebook learns to store useful feature vectors that can reconstruct the input effectively.

The backpropagation flows through the entire model, including the encoder, but the quantization step introduces a challenge because it is non-differentiable. Since Z_q is derived from discrete

675

codebook values, direct backpropagation through this step is not possible. Besides, Z_q includes discrete values, and discrete values are non-differentiable and do not have a meaningful gradient. This means the optimizer cannot backpropagated through the encoder during training. To address this, the authors use the *straight-through estimator* (proposed in VQ-VAE). This approach allows gradients from the decoder to be copied to the encoder as if the quantization step were differentiable, ensuring the encoder and decoder can be trained effectively.

For the CNN discriminator, there is not much to explain except that the authors recommend starting the CNN discriminator later in the training because the generator has time to learn. Otherwise, the discriminator can easily recognize the synthetic image, and thus, the generator can not learn.

Stage 2: Until now, we have only had a GAN reconstruct the data. However, the goal is to generate large synthetic images with high resolution. The encoder and decoder are typically invariant to image size, but the transformer can assist the decoder in building larger images. Therefore, a transformer will be used in an autoregressive manner.

We have explained that the codebook acts as a reference that includes elements of synthetic image construction. The task of the transformer is to learn which codebook vector members (codeword) are required for the image construction and add them into a sequence of patches for image generation in Zq. For example, as it can be seen from Figure 11-37, the transformer orders codewords as $i_{14}, i_5, i_{12}, i_2, \ldots$ in Z_q and sends them to the CNN decoder for synthetic image construction.

Why do we feed the codebook to the transformer and not the output of the Encoder? The transformer requires sequential data as input. However, since it uses self-attention (check Chapter 12), it is computationally very hard to feed a large sequence to a transformer. Therefore, the codebook is fed into the transformer, and the authors use a sliding window for the attention model to reduce the size of the image and the computational cost. The transformer processes the sequences of discrete codebooks in an autoregressive manner. This autoregressive approach allows the transformer to sequentially predict different parts of the image, using the output of one step as the input for the next, thus enhancing the overall quality and coherence of the generated images.

The VQGAN transformer acts as a decoder and upsamples data from the codebook. In other words, it is used to enrich the content of the codebook and thus enable the generator to build larger than input synthetic images. For training, the transformer uses cross-entropy loss and compares the codebook's original value with the modified codebook, which is a result of training.

In summary, instead of constructing images from pixels, VQGAN uses the discrete variables extracted from the codebook, and the transformer improves the content of the codebook. Therefore, the synthetic images of the Generator are high resolution and can be larger than the original ones.

If you would like to bluff about your deep understanding of VQGAN, say the following in front of others: Using the concept of the codebook and feeding it into a transformer makes VQGAN very attractive and shows its high accuracy in generating high-resolution images. Jokes aside, using a codebook is a popular trick, and we will encounter it later in upcoming chapters.

We do not explain the cost functions of the VQGAN in detail; it uses two cost functions: one is discriminator cost, and the other one is codebook cost. The discriminator cost is a standard GAN cost function whose goal is to train the discriminator to successfully identify the generator's synthetic images from real training images. The codebook cost encourages the discrete latent code Z_q to stay close to the encoder outputs.

VQGAN-CLIP

We have explained CLIP and VQGAN to reach this interesting and practical approach. Crowson et al. [Crowson '22] use a CLIP to guide VQGAN and build a practical text-to-image approach. They make their trained model openly accessible, allowing it to be fine-tuned for other tasks. VQGAN-CLIP is an interaction between VQGAN and CLIP networks, which results in using the following neural networks: the generator and discriminator of VQGAN and a separate neural network of CLIP.

The objective of VQGAN-CLIP is to generate images from the given text prompt. Recall that the VQGAN network generates a high-quality image from the text prompt, and then the CLIP network assesses the fitness of the text (image caption) to the image by comparing it to other captions.

In short, VQGAN-CLIP operates as follows: the user inputs a text (prompt) for the desired image generated. In the beginning, VQGAN generates a completely random, noisy image, which includes only random pixels. Then, this image will be assessed by the CLIP to see if it matches the user-given prompt. A loss score will be reported to explain how far the generated image is from the text prompt. The loss calculated by CLIP is backpropagated to the VQGAN generator. Then, VQGAN improves its image generation process. This process continues iteratively until the loss reaches the maximum threshold.

We skip the small details of the original paper, but one thing that is worth explaining is the challenge that existed with CLIP loss. The updated gradient from CLIP is noisy if it is calculated on a single image. To overcome this issue, the authors apply several image augmentation techniques (we will discuss them in Chapter 16) to the generated image to have several images. Their image augmentation technique includes first taking random crops of the candidate image and then applying flipping, noising, etc.

DALL-E Models

DALL·E models are neural networks designed by OpenAI that generate images from textual descriptions, leveraging the power of diffusion models and transformers to create detailed and imaginative visuals.

DALL-E v1

DALL-E version 1 [Ramesh '21] was introduced back in early 2021 by OpenAI. The authors use a transformer architecture and autoregressive model. They assume text and image as a single stream of data. DALL-E v1 includes 12 billion parameters and uses a dataset of 250 million text-image pairs collected from the Internet. As you can see, it is very small, and you can train this model on your phone while going to sleep.

Earlier, we explained that autoregressive models are used for sequential data (token). The authors described that they could not consider the pixel of an image as a token of data for the autoregressive model. Because then the model focuses on capturing short-range dependencies between pixels, and thus, long-range relations between pixels will be neglected.

To capture long-range relations, DALL-E v1 operates in two stages. The first stage is focused on training a discrete variational autoencoder (dVAE) to compress image data. The resulting images are encoded and thus smaller but hold the feature of the original images. In particular, the resulting images are encoded into a sequence of 1024 tokens, each of which can take one of 8192 possible values. The codebook contains 8192 vectors.

The dVAE is similar to VQ-VAE, which builds a codebook. The difference is that, unlike VQ-VAE, which selects one codeword (a codebook vector) from the codebook, dVAE can express some uncertainty by outputting a distribution over codewords for each latent variable instead of a single codeword. In simple terms, the dVAE can output a distribution over codewords. However, since backpropagation is not able to work with discrete data (discrete data are not differentiable), to transform them into continuous forms and make them differentiable, authors use a method called Gumbel-softmax [Maddison '16, Jang '16]. Gumbel-softmax enables gradient-based optimization despite having discrete data. It approximates the discrete distribution with a smooth one that can be sampled and differentiated. The result of this stage is to convert a 256×256 RGB image into a 32×32 grid of tokens.

The second stage first concatenates image tokens to BPE-encoded text tokens (in Chapter 14, we will explain BPE). This enables the model to process both text and image data in a similar manner. Then, they trained an autoregressive transformer to model the joint distribution over text and image tokens using their 12-billion parameter transformer.

As a result, a model is trained on a large dataset of image and text token pairs. Then, they can be used in an autoregressive fashion and predict the next image token in a sequence, and a decoder builds the entire image from a given text.

DALL-E v2

DALL-E version 2 [Ramesh '22] was released in 2022, it can construct significantly more realistic and accurate photos than its version 1. DALL-E 2 uses a diffusion model for the decoder, while DALL-E v1 uses an autoregressive model for generating images.

DALL-E 2 is composed of two main components: Prior and Decoder. First, a CLIP model [Radford '21] is trained, and each pair of images and text will be encoded to match image embeddings (z_i) and text embedding (z_t). The objective of CLIP is to learn a joint embedding space for images and text, enabling it to match text descriptions with relevant images. The top part of Figure 11-40 presents the CLIP training for DALL-E 2 architecture. After the CLIP model is trained, it gets frozen.

Next, the prior network uses the frozen CLIP and produces the image embeddings (z_i) that are conditioned on the caption (y), i.e., $P(z_i | y)$. In simple words, Prior maps the text embedding z_t to the image embedding using the CLIP frozen model. The output of the prior is z_p which is an image embedding. The authors stated that they used both diffusion and autoregressive for the Prior.

Afterward, the diffusion decoder produces images conditioned on image embedding (from the prior) and optionally on the caption, i.e., $P(\hat{x} | z_p, y)$. This means that unlike the diffusion we have explained earlier, this diffusion model does not start from noise. Instead, it starts with the image embedding produced by the prior. The objective of the decoder is to generate an image that corresponds to the image embedding produced by the prior. Figure 11-40

Why does the decoder perform inverting (the process of reconstructing an image from its latent representation)? The goal is to generate a new image from a text prompt rather than build an identical image as it is encoded by the Prior. The inversion assists the decoder in creating a new image. To implement the decoder with inversion, they use a modified version of the GLIDE [Nichol '21] (another text-to-image from OpenAI). The authors call the combination of prior and GLIDE-based decoders to *unCLIP* because it reverses the mapping learned by the CLIP image encoder. By stacking the Prior and Decoder, DALL-E v2 builds a generative model $P(\hat{x} | y)$ of images $\hat{x}$ given captions y.

Why do authors of DALL-E 2 use prior and not simply a CLIP encoder with a decoder? The reason for this design choice is that, through experiments, the authors found that having a Prior network in between produces better images. They stated that the Prior creates a summary of the image, and the decoder inverts images given their CLIP image embeddings. During this process, the decoder learns the useful details for image reconstruction, but prior (frozen model of CLIP inside Prior) does not consider them.

DALL-E 2 generates high resolution images, and to generate high resolution images, they train two diffusion up-sampler models. The first one upsamples images from 64×64 to 256×256 resolution (with Gaussian blur to improve the robustness of the upsampling), and the second one applies further upsampling to make a 1024×1024 resolution (with something called BSR degradation [Rombach '22, Zhang '21] to improve robustness of the upsampling). We do not explain the details of BSR degradation here.

To train, the encoder authors use CLIP and DALL-E v1 datasets, which include about 650 million images. To train the encoder, upsamplers, and prior, they used about 250 million images of the DALL-E dataset.

To summarize, DALL-E 2 operates in three steps. First, a text encoder is trained to convert the user input text (text prompt) into a text embedding (vector instead of text). In the second step,

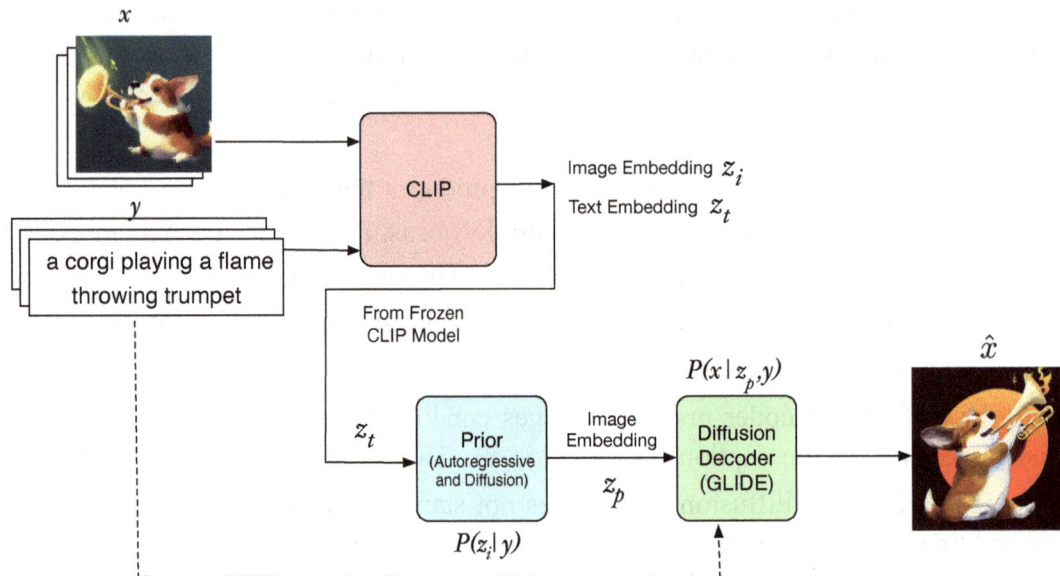

Figure 11-40: DALL-E 2 architecture, which is composed of three networks, CLIP, Prior and a decoder, which is a GLIDE model with some changes.

Prior maps the text encoder result (text embedding) into a corresponding image embedding. The image created by Prior has semantic information from the input text. In the third step, the "Decoder" generates images from the image encoding created by the Prior[16].

DALL-E v3

In 2024, OpenAI released DALL-E 3 [Betker '23]. Authors claimed that one problem with the DALL-E 2 is the prompt text given by the user, and if there is a mechanism to improve the descriptiveness of the prompt, the text-to-image model can generate high quality images with more details related to the prompt. To handle this issue, they trained an *image captioner* and a text-to-image model to build DALL-E 3.

Image Captioner: The image captioner is a model that is similar to a language model (we will explain in Chapter 12) is used to predict text. It begins by using a tokenizer to break down strings of text into discrete tokens, representing these tokens as sequences, which can then be processed to generate captions for images. This approach is often used to address shortcomings in existing captions found on the internet, which may be incorrect or only tangentially related to the content of the image. The image captioner is jointly trained along with CLIP to leverage CLIP capabilities in measuring text and image similarity, thus ensuring that generated captions closely align with the visual content and human interpretation of images.

When an image captioner is trained alongside CLIP, it can optimize its captions to be more aligned with the actual content of the images. This is particularly useful because it enables the

---

[16] You can also check the Blogpost of one author to better understand DALL-E 2, http://adityaramesh.com/posts/dalle2/dalle2.html

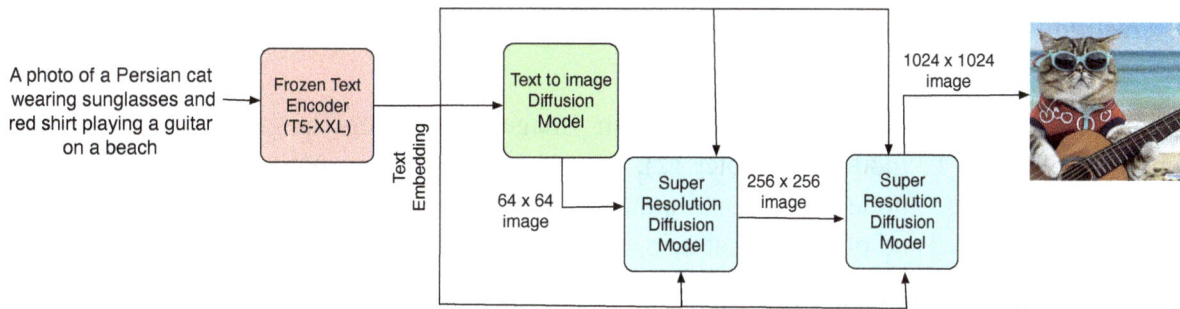

Figure 11-41: Imagen architecture.

model to generate captions that not only describe the images accurately but also in a way that resonates with how humans interpret the images.

Synthetic Captions: The authors stated that first, they build a small dataset (not clear what they mean by small) of captions that describe only the main subject of the image. Then, they trained their image captioner model on this dataset to produce what is referred to as *short synthetic captions*.

Afterward, they repeat the fine-tuning process for the second time with a new dataset that contains long, highly-descriptive captions. These captions include details about not only the main subject but also the surroundings, background, text found in the image, styles, and coloration. The result of the second fine-tuning is referred to as *descriptive synthetic captions*. GPT-4 is utilized to upsample these captions further, enhancing them to include even more detailed descriptions, thereby improving the model's ability to generate images

To train the DALL-E 3, at the end, they use the descriptive synthetic captions on their image dataset to build the final model. Authors claimed they used 95% synthetic captions and 5% ground truth captions. This blending of captions helps balance between creativity and accuracy, improving the model's ability to generate relevant images.

OpenAI does not release any information about where they get the data, and most probably, they use the copyrighted data available on the Internet.

Imagen

Do you recall rich family or friends who do not have anything to do except try to copy what other rich people did and show off their life, which includes more luxury than others? If you think those behaviors belong to people and big corporations don't do the same thing, you are wrong. Anyway, this explanation has nothing to do with the relationship between Google and OpenAI, and we wrote it accidentally here.

A few weeks after the release of Open AI's DALL-E 2, Google released Imagen [Saharia '22], and a few days later, it released Parti [Yu '22], both of which are text-to-image models. The Imagen architecture is significantly easier than DALL-E 2, and in its paper, the authors also introduce an evaluation method, DrawBench, to evaluate the result of text-to-image algorithms. DrawBench evaluates key aspects such as the fidelity of the generated images to the original text, the creativity and diversity of the outputs, the detail and resolution of the images, and their internal consistency and coherence.

Figure 11-41 presents the architecture of Imagen. Imagen uses a large language model, T5-XXL (we will explain the T5 model in Chapter 12), pre-trained on a text corpora (Check Chapter 12 to learn about language models) to map input text into a text embedding (word embedding). Authors experimented and observed that using this language model outperforms CLIP accuracy.

Next, they adapt a conditional diffusion model provided in [Dhariwal '21] that gets the text embedding and constructs the image. Their diffusion model introduces a concept called *dynamic thresholding*, which is a sampling method to leverage weight guidance. This weight guidance results in having more photo-realistic images at the end.

Afterward, they use two super-resolution diffusion models for generating 256×256 and 1024×1024 images. Their super-resolution architecture is based on their customized U-Net model (Efficient U-Net).

To train the T5 XXL, they use 800 GB of textual data, and the model has 11 billion parameters. Their 64×64 diffusion model is trained on a 300 million parameters model, which is conditioned on text encoding. Their UNet models have 300 million to 2 billion parameters.

Parti

Parti [Yu '22] stayed for the "Pathways autoregressive text-to-image model", and it was introduced a few days after Imagen. Unlike Imagen, which uses a diffusion model, it is an autoregressive model. Recall that an autoregressive model takes a sequence of tokens as input, and predicts an upcoming token(s).

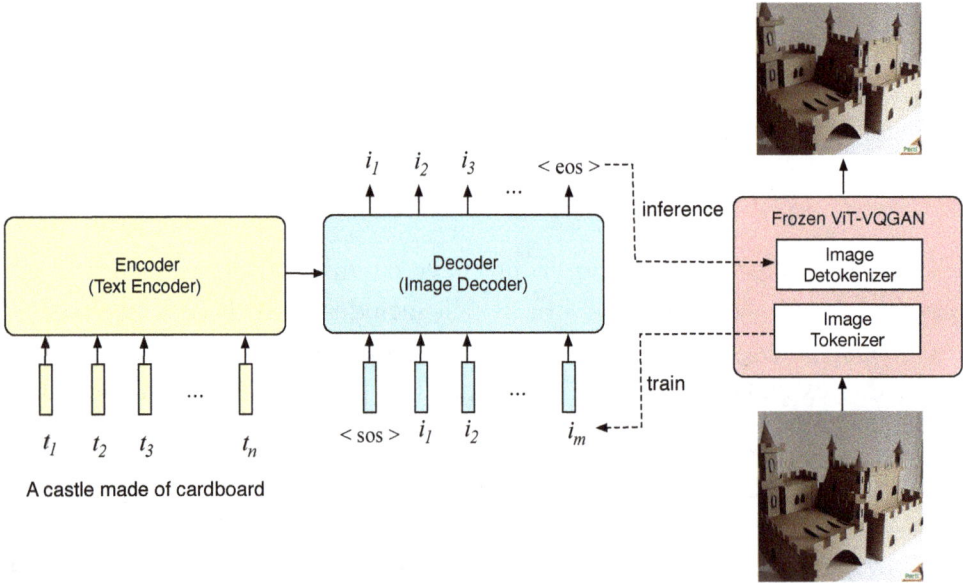

Figure 11-42: Parti architecture.

Parti treats text-to-image generation as a seq2seq model used for machine translation (check Chapter 12 for seq2seq models). Unlike DALL-E 2, which provides a decoder only, it provides both an image encoder and image decoder by using ViT-VQGAN [Yu '21], and the authors claimed it had outperformed models only with the decoder. We have already explained VQGAN earlier here, and ViT (we will explain in Chapter 12) stayed for the vision transformer. Parti has a transformer-based architecture, and it operates in two stages: an image tokenizer and an autoregressive model.

Image Tokenizer: The first stage trains the image tokenizer to convert an image into a sequence of discrete visual tokens (image embedding). We could assume this as an encoder because the tokenizer's role is to compress the image into tokens effectively, which are then used for generating images.

As explained in the VQGAN section, tokenizing an image into a sequence of pixels results in an extremely large sequence. VQGAN and DALL-E 2 handle this challenge with discrete types of VAE, and instead of using patches on the input image, they use codebook values. Authors of Parti downsample an input image of the size 1024×1024 to 256×256. Then, they give it for training to ViT-VQGAN. ViT-VQGAN plays a role similar to an encoder in Seq2Seq models (check Chapter 12), as it encodes images into a tokenized format suitable for the autoregressive transformer. The resulting codebook in this model has 8192 entries (codewords). ViT-VQGAN encodes images as a sequence of discrete tokens (codewords), which come from the codebook (not image patches). To train ViT-VQGAN, the authors follow the DALL-E model. After training is done at this stage, the model freezes ViT-VQGAN, which is presented as the red component of Figure 11-42.

Since the frozen model of ViT-VQGAN operates with 256×256 image size, later, the synthetic image that is reconstructed by ViT-VQGAN will be fed into a super-resolution layer and upsamples the reconstructed image to a size of 1024×1024.

Autoregressive model: In the second stage, an autoregressive model Transformer Decoder (presented as a blue box in Figure 11-42) receives the sequence of embedded tokens from the text encoder (yellow box in Figure 11-42) and generates an image token-by-token, predicting each subsequent token based on all previous ones. In other words, this stage focuses on training autoregressive encoder and decoder transformers (the yellow and blue transformers in Figure 11-42) by treating text-to-image as a seq2seq modeling. The input for training is the image and text pair. The Image Decoder, shown in Figure 11-42, gets the (i) sequence of image tokens (extracted by ViT-VQGAN from the image) and (ii) the encoded text prompt that the Text Encoder (yellow box in Figure 11-42) converts to text embedding. Then, it tries to predict the next image token. The text embedding acts as a target for attention (attention and transformer will be explained in Chapter 12).

After the second stage of training, they freeze the encoder and codebook and fine-tune the model, which results in a 600 million-parameter model. After the model is trained, it can generate images by feeding the text prompt.

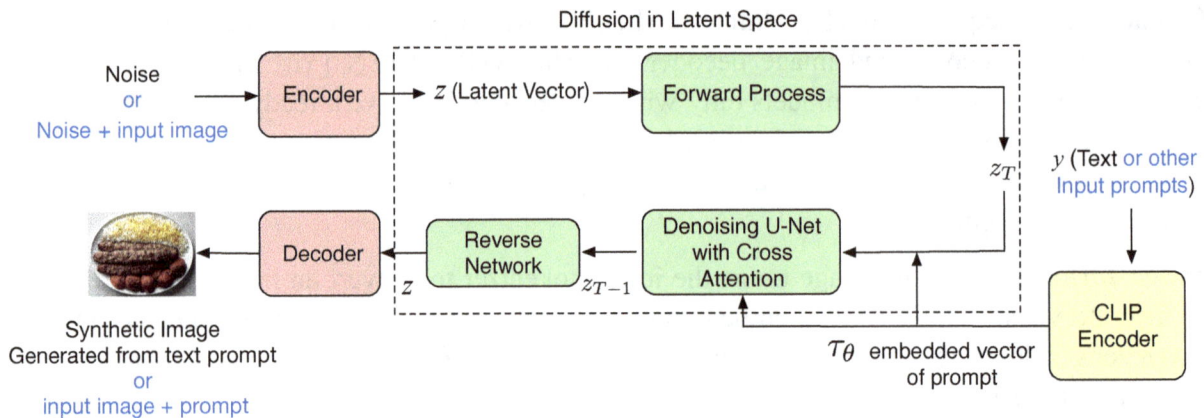

Figure 11-43: The simplified architecture of LDM training, which is the first version of the Stable Diffusion. The blue texts on this figure are used for inferences, and they mean LDM can get an image as input and generate another output, such as editing the given input based on the given text prompt.

The Parti model that provides most photo-realistic images has 20 billion parameters. The super-resolution module has about 15 million parameters for the 512×512 version and about 30 million parameters for the 1024×1024 version. Their image-text paired datasets are vaguely explained, but it seems they use several internal Google and public datasets, and they use about 6 billion text-image pairs.

Stable Diffusion Models

In 2022, Rombach et al. [Rombach '22] provided a text-to-image model, later known as Stable diffusion. Unlike Google and OpenAI, released their model weights open so that others can benefit from their model and further tune it. Their approach leads to the introduction and advancement of many research and applications that can benefit from text-to-image.

Their explanation of their architecture is not easy to understand; we spent reasonable time simplifying their approach, and the simplified version of their architecture is presented in Figure 11-43. In this figure, we use the blue color in the text to emphasize that Stable Diffusion can indeed be used to edit or modify an existing image based on a given text prompt, a capability often referred to as *text-guided image editing* or *inpainting*.

Latent Diffusion Model

The first version of Stable Diffusion is called the *Latent Diffusion Model (LDM)*. Authors described that previous diffusion models operate on the original image data, which is too large to process and thus computationally expensive. LDM utilizes an autoencoder, which compresses high-dimensional data into a lower-dimensional space (light red encoder in Figure 11-43). Later, after the diffusion process is trained, the decoder (light red decoder in Figure 11-43) translates the clean vector z into the output (synthetic image). By using autoencoder architecture, the computational cost for both training and inferences is reduced significantly.

The reverse process (diffusion process) includes a U-Net with a cross-attention model. The use of cross-attention allows the model to integrate additional contextual information (e.g., text descriptions, bounding boxes) directly into the diffusion process. We will describe some attention mechanisms in Chapter 12. Similar to self-attention, the cross-attention mechanism involves three main components: queries (Q), keys (K), and values (V). In the context of LDMs, the diffusion model's intermediate layers generate queries (Q), while the encoded conditioning inputs (like text or bounding box descriptions) generate keys (K) and values (V). Note that this U-Net works only with the latent representation of the data and not the original data.

The encoder shown in yellow in Figure 11-41 is used to convert different modalities (text, image, or even audio), but to our understanding and by checking other resources, this is a CLIP encoder that aligns text with the image. This means the encoder constructs the 'conditioning signal' $\tau(\theta)$ that represents the transformation of text input θ into a *conditioning signal* τ, which is used to guide the image generation process via U-Net.

The result of the denoised U-Net is passed to the Reverse Network (shown in green in Figure 11-43), which implements the reverse process of a diffusion model. The output of the reverse network will be given to the decoder that converts the latent space data into pixel space, which means it constructs the output (i.e., synthetic image).

Training LDM

We explain the training process with respect to the figure because there are different components to train, some trained in a sequence and some in parallel.

Encoder Training: The training starts with an encoder that compresses high-resolution images into a lower-dimensional latent space. The encoder is usually a part of a denoising autoencoder (DAE) or a variational autoencoder (VAE) that learns to encode the image into a latent vector z, which retains the essential information but in a more compact form.

Forward Diffusion Process: Next is the training of the forward process. During this phase, Gaussian noise is incrementally added to the latent vectors (encoded input image) over several steps, transforming the data from its structured latent representation to a completely noisy state. This creates a sequence of increasingly noisier versions of the latent vector, represented by $z_1, z_2, \ldots, z_T$, where T is the final time step.

Reverse Network (Denoising Network): Parallel to the forward process, the Reverse Network is trained. This network learns to perform the reverse of the diffusion process, denoising the latent vectors. The network starts from noise z_T (the noisiest state) and trains to predict the less noisy previous state z_{T-1}. By training on many such pairs, the network learns to reverse the entire diffusion process, recovering the original latent vector z from noise.

Integration with Text Prompts: If the LDM is conditioned on text prompts, a separate component, such as the CLIP Encoder, is used to convert text prompts into embedded vectors τ. These embeddings are used during the reverse diffusion process. The Denoising U-Net with Cross Attention incorporates these embeddings to ensure that the features of the generated image match the text prompt's content.

Decoder for Image Generation: Finally, the decoder will be trained to convert latent vectors back into images, is used to generate synthetic images from the denoised latent vectors.

Stable Diffusion XL

After the introduction of Stable Diffusion v1 or LDM, some minor changes were introduced, and later, Stable Diffusion XL (SDXL) [Podell '23] was introduced. SDXL has a larger training set and leverages Reinforcement Learning with Humans in the Loop (RLHF), which we will describe in Chapter 13. Besides, it leverages a three times larger U-Net model than LDM, which has 2.6 Billion parameters, and it can produce images at a resolution of 1024×1024 pixels.

It includes a second U-Net model known as the *refinement model* to improve the visual fidelity of synthetic data generated by the first U-Net model. It also includes two text encoders (CLIP ViT-L and OpenCLIP ViT-bigG) that enhance the model's ability to interpret and synthesize images from textual prompts accurately. The two text encoders allow the system to capture more nuanced details from the text prompts, presumably enabling better text-to-image translation and higher fidelity in the generated images. Figure 11-44 presents the architecture of SDXL, which seems too abstract, in contrast to their previous architectural drawing, which was unnecessarily complex.

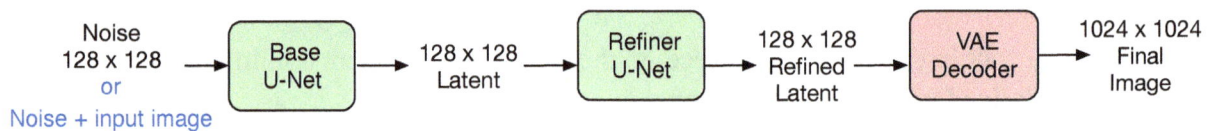

Figure 11-44: SDXL architecture (very similar to the same figure that is presented in the paper).

Additionally, SDXL can handle varying image sizes, as the model needs to adapt to different resolutions and aspect ratios to produce coherent and visually acceptable results. To address this, SDXL employs several image size conditioning strategies, which include:

- It allows users to explicitly specify the desired output image size and aspect ratio in the prompt.

- During training, the model learns to associate specific size embeddings with different image resolutions. At generation time, the appropriate size embedding is selected based on the desired output size, and it is used to condition the image generation process.

- The model starts by generating a low-resolution image and then iteratively increases the resolution through multiple stages. At each stage, the model takes the previous stage's output as input and generates a higher-resolution version of the image. This progressive refinement allows SDXL to generate high-quality images at larger sizes while maintaining detail.

Stable Diffusion 3

Stable Diffusion 3 or SD3 [Esser '24] medium size mode has been released open, but not its large version. SD3 introduced two new components in their architecture: (i) re-weighing of the noise in Rectified Flow and (ii) employing the Multi-Modal Diffusion Transformer (MM-DiT), along with some other minor improvements that we will learn while we describe the architecture.

Re-weighing of the noise in Rectified Flow: Forward paths from data to noise in the diffusion process can be implemented in different approaches. One common one is known as *Rectified Flow*. Rectified flow optimizes the noise addition in a diffusion model by adjusting the noise to simplify the reverse path. This maintains key structural elements of the original data throughout the process. Rectified flow implements this simplification by taking pairs of data points, one with noise (like a noisy image) and the other with the original clean data. Then, it learns the Ordinary Differential Equation (ODE)[17], which connects these points through a straight path. This involves iteratively refining the path to make it straighter.

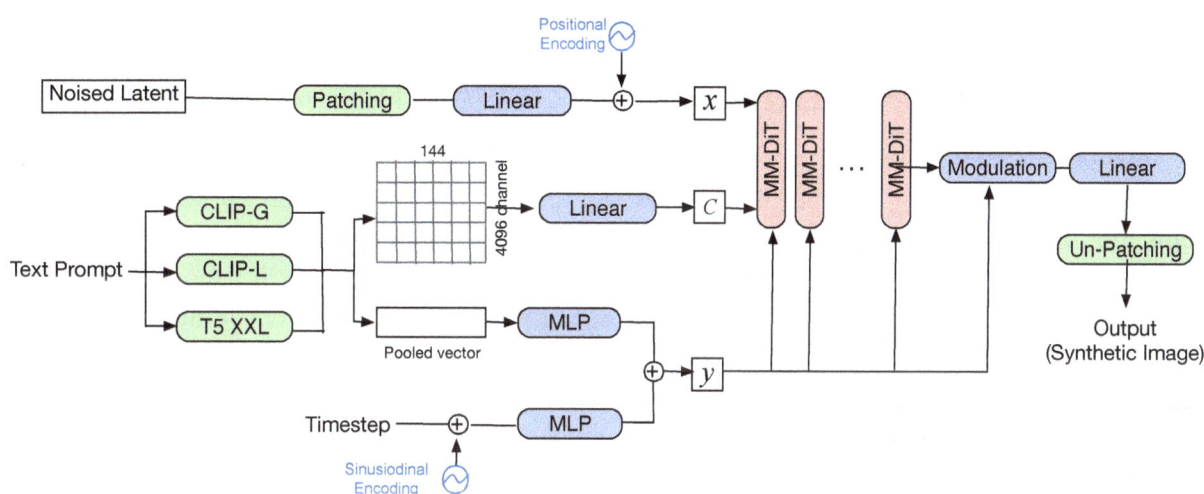

Figure 11-45: SD3 architecture.

The authors introduce *re-weighing of the noise in rectified flow*, which reduces error accumulation during the sampling phase due to increasing the possibility of getting to noise in fewer steps. Sampling in this context refers to generating new data points from the learned model distribution. Besides, the re-weight of the noise improves the fidelity and clarity of the generated images, and ensures that the sampling process is in line with the underlying data distribution.

Multi-modal Diffusion Transformer (MM-DiT): Authors employ diffusion transformers in their architecture. Their architecture uses two separate sets of weights, one for text and one for image. This refers to multi-modality in their architecture. This separation of weights for different modalities improves the model's capacity to synthesize images from textual descriptions by enhancing text comprehension and visual representation. These transformer blocks are referred to as MM-DiT blocks, and they are stacked on top of each other in the model architecture, as

---

[17] An Ordinary Differential Equation (ODE) is a type of mathematical equation that involves derivatives of one or more functions with respect to a single independent variable. It is called "ordinary" to distinguish it from partial differential equations, which involve partial derivatives with respect to multiple independent variables.

shown in Figure 11-45. Each MM-DiT block includes modulated attention and multi-layer perceptrons (MLPs), which are designed to merge information from text and image effectively.

SD3 Architecture: In the architecture, CLIP-G/14 and CLIP-L/14 provide semantic alignment between text and images, while T5 XXL (we will explain T5 in Chapter 12) processes detailed textual input, ensuring the generated images reflect the content described in the text. The authors described the text prompt passes to these listed models, and the output feature map has 4096 channels.

Having 4096 channels implies a very high-dimensional feature space, allowing the model to capture and process a vast amount of information from the data. This rich feature map passes to a linear layer to apply a linear transformation to it. The result is shown as 'c', which is a transformed feature matrix.

Besides, their architecture shows a pooled vector. Pooling refers to an operation that aggregates or summarizes information from a broader set of data into a more compact form. Therefore, this pooled vector includes a rich feature vector of text and images, which is passed to an MLP.

To incorporate the sequential notation into the model, they use *Timestep* combined with *Sinusoidal encoding* (a type of positional encoding that we will describe in Chapter 12), which it helps the model understand the order or sequence of data. This approach introduces a notion of progression or evolution within the model's processing steps, which assists the model in creating temporally coherent outputs. The raw timestep, along with its Sinusoidal encoding, passes to an MLP to transform the raw timestep data (a numerical indicator of the current phase in the generative process) into a more complex, feature-rich representation. The output of this MLP is combined with the output of the MLP of the 'pooled vector' and constructs the 'y' matrix. This matrix is a feature fusion of textual, visual (from one MLP), and sequential (from the other MLP) features altogether.

The diffusion process starts with a 'noise latent'. Noise latent is not pure noise; the authors apply an intermediary step where the latent representation of an image (initially derived from real data) is systematically infused with noise through a controlled process. We have described this earlier as a re-weighing of the noise in rectified flow. Then, they get the noise latent and apply patching. The patching process is spatially restructured by breaking down the noise into smaller segments. Then, it passes to a linear layer, positional encoding is added, and it results in matrix *x*.

x, y, and *c* go through *n* number of sequential MM-DiT modules. After these modules, there is a 'Modulation' component. Modulation is a mechanism that adjusts the behavior of the model based on certain inputs (timestep and a class vector). It alters parameters or activations in the network dynamically as it processes through the steps of generating an image, ensuring that the network's outputs are appropriately influenced by the inputs and the stage of the diffusion process. The result is passed to another linear layer, and later, they will be unpatched to build the final synthetic image.

Explaining these models are: Finished, تمام, finalizada, خلاص, 完成的, Fertig, завершены, खत्म

The weights of stable diffusion models are openly accessible, which leads us to dedicate serious time to explaining them here. If you find the architecture of SD3 too complex and it reminds you of StyleGAN3, we do agree with you. It seems that the diffusion model is squeezed, and similar to StyleGAN, which was one of the final GAN models; this might be one of the final models of diffusion. The same group of scientists also implemented knowledge distillation to make the distillation process faster, such as SDXL-Trubo [Sauer '24], which introduces Latent Adversarial Diffusion Distillation (LADD), which is a simple distillation approach. In Chapter 16, we explain what is knowledge distillation.

Later, they released another group of models known as FLUX models[18] (the largest one is not open-source), and at the time of revising this chapter for the 100th time, no technical details were shared.

Other models that we didn't explain

In addition to these promising text-to-image models, there are (i) music generation models, (ii) text/image-to-video models, and (iii) face swap (or deep fake) models. We skip explaining them in detail. If you are interested in learning more about music generation models, the authors and his students did a survey [Zhu '23] on AI music generation frameworks, which you can read and get familiar with. On the other hand, there are promising efforts to build video from a text, a.k.a, text-to-video model, such as Make-A-Video [Singer '22], Text2Video-Zero [Khachatryan '23], and Sora[19], KLIG[20], and LumaLab[21] that shows many interesting videos. Besides, there are efforts to construct a video from a single image or a GIF file from a single image, such as [Kandala '24].

---

[18] https://blackforestlabs.ai/#get-flux

[19] https://openai.com/index/sora

[20] https://klingai.org

[21] https://lumalabs.ai/dream-machine

Another group of models, known in layman's terms as deep fake, can swap faces on videos or images, make the image of the person older, combine the two face images, and build a new one. We have explained diffusion models that are capable of doing so. However, there should be used with lots of caution because they are capable of faking a talk of a person or claims, making realistic porn videos from the face of a human, etc. However, there are lots of interesting models in this direction, and we recommend you check Github to learn more about them.

Perhaps, in the next version of this very short book, we will add them as well to be sure nobody is able to finish this book.

Summary

This section focuses on unsupervised neural networks. It starts by describing concepts that convert the data into a low-dimensional representation of the original data.

We explained latent variables, which are constructed by hidden layer neurons, in contrast to observed variables that are either input or output variables of a neural network. Next, we describe generative (joint probability) vs. discriminative (conditional probability) models. Afterward, we define deterministic (the output is determined by model parameters) and stochastic (the output is random) models.

After describing concepts, we start explaining SOM, which is one of the oldest unsupervised neural networks that is still in use, mainly for dimensionality reduction. Then, we describe the Boltzmann machine to prepare your poor brain for RBM. Afterward, we focused on autoencoders. An autoencoder has two main components: an encoder, which reduces the dimensionality of the data, and a decoder, which tries to reconstruct the original data from the given data in the low dimension, i.e., latent space. The process of training in autoencoders includes finding weights for the encoder that minimize the error of data reconstruction. There are two types of autoencoders: regularized autoencoders (denoising, sparse, and contractive autoencoders) and VAEs. Regularized autoencoders map a set of data points directly to one single data point in latent space, but VAE maps a set of data points to a multivariate Gaussian distribution around a point in the latent space. This makes them very flexible in constructing synthetic data that they have never encountered before. We conclude the autoencoder discussion by explaining the legendary U-Net, which is a state-of-the-art algorithm for medical image segmentation. U-Net is not called an autoencoder, but it is also composed of an encoder, and then the encoded compressed data will be retrieved and reconstructed by the decoder. Nevertheless, between layers, there is a concept called skip connection, which concatenates the encoder output (that includes "what information") to the decoder output at the same level (that includes "where information").

Next, we introduce GANs. GAN is the min-max game between two neural networks, a discriminator and a generator. Each of these neural networks has its cost function and its parameters. During the training phase, their parameters were tuned, similar to other neural networks. The generator starts from noise and improves it until it builds synthetic data, which has a very similar distribution to the original data. In particular, it takes a point from latent space as input and generates new synthetic data. The discriminator employs Backpropagation to minimize the loss score for real and synthetic data examples it receives. The generator tries to minimize the discriminator's loss for synthetic data (motivates the discriminator to consider the synthetic data as real data). In other words, the generator tries to increase the false-positive errors of the discriminator, while the discriminator tries to minimize the false-positive and false-negative of its classification. The basic form of GAN is associated with several challenges, and later, we introduce GAN architectures that try to mitigate those challenges.

Afterward, we introduced Contrastive learning representations and their common cost function (triplet loss and contrastive loss), and later, we explained how the Siamese network operates. We finalized this chapter by describing Text-to-Image and inpainting approaches that rely on diffusion models.

Further Reading and Watching

* Frank Noe, has a good RBM tutorial and we used it to explain the details of RBM algorithm, https://www.youtube.com/watch?v=wfFf5Fj-rzE.

* Another tutorial to learn RBM and DBN is the monograph of Bengio [Bengio '09]; that tutorial is among the few ones that explain these algorithms in simple terms and does not need significant prior knowledge, assuming you have read Chapter 3 and Chapter 8.

* If you are interested in understanding the details of RBM and DBM, a good tutorial with lots of mathematical detail is for Hugo Larochelle under this link: https://www.youtube.com/watch?v=35MUlYCColk. Nevertheless, beware that it goes fairly deep into the mathematical explanations but is a good online tutorial. His autoencoder tutorial is also very detailed and useful.

* Andrew Ng has a good explanation about Sparse Autoencoder and delves deep into the math detail in his lectures: https://web.stanford.edu/class/cs294a/sparseAutoencoder.pdf.

* Lilian Wang has a brief article on Autoencoders: https://lilianweng.github.io/lil-log/2018/08/12/from-autoencoder-to-beta-vae.html. Also, she has a tutorial on WGAN: https://lilianweng.github.io/lil-log/2017/08/20/from-GAN-to-WGAN.html#wasserstein-gan-wgan

* At the time of writing this section, two books have been published about GAN, *Generative Deep Learning: Teaching Machines to Paint, Write, Compose, and Play* [Foster '19], by Foster and *GANs in Action: Deep Learning with Generative Adversarial Networks* [Langr '19] by Langr and Bok. Both books are well written, and if you would like to have a more hands-on experiment with GAN, we recommend checking them.

* Soheil Feizi has a good introduction to GAN in mathematical language. His video lectures are available as well: https://www.youtube.com/watch?v=IzaerbzSB64

* Jason Brown Lee has a good introduction and implementation example on FID score: https://machinelearningmastery.com/how-to-implement-the-frechet-inception-distance-fid-from-scratch

* There is an online course that teaches some of the common GAN models. Especially we used its explanation about StyleGAN, https://www.coursera.org/learn/build-better-generative-adversarial-networks-gans

* We skip explaining DALL-E 1, and dedicate most of our explanation to DALL-E 2. If you are curious to see how dVAE works in detail, you can check this blog post: https://ml.berkeley.edu/blog/posts/dalle2.

* We benefit from the Yannic Kilcher online videos for text-to-image models (https://www.youtube.com/c/YannicKilcher) to learn these models. His explanation is fairly ok, or at least better than the papers, which seem to all be written in haste.

* Umar Jamil has an excellent explanation of the Stable Diffusion model, which we recommend checking it: https://youtu.be/ZBKpAp_6TGI.

Chapter 12: Deep Learning Models and Applications (Text, Vision, and Audio)

In the previous two chapters, we have learned concepts and architectures for deep neural network models. Unlike other machine learning algorithms, the way we build and use deep learning in our applications is different. If you, similar to the author, are not employed by those superrich corporations and do not have access to the massive number of GPUs, you cannot train a model from scratch, but you can change and tune an existing model for what you need. We usually use existing models or tune them for our needs, i.e., transfer learning. More about transfer learning will be explained in Chapter 14. Therefore, being familiar with successful deep learning models and architectures is crucial to developing or employing an existing model for our needs.

> This chapter is very long... To understand this chapter you should read Chapter 6, 8 and 10.

We start this chapter by explaining the *Attention and Transformer*, introduced in 2017 [Vaswani '17], and significantly improved the quality of Natural Language Processing (NLP) applications. Some call transformer the last revolution in Deep Learning. At the time we started to write for the first time, in mid-2022, deep learning models were popular for NLP and computer vision applications. We categorize this chapter into three main sections: NLP, Computer Vision, and Audio models. In each section, we explain some of the popular models, and getting familiar with these models will help us develop our model in the related field or give us the knowledge to use which model is best for our needs.

After explaining the transformer, the rest of this chapter seems like a dictionary of models and does not follow the structure that we have in the previous chapter. Additionally, we could benefit by getting familiar with popular models and applying them in different applications that are not common for their initial use. For example, attention models were designed for NLP tasks, but since 2020, attention models have been used for computer vision tasks, too. Although this chapter is the longest chapter of this long book, we recommend that you read the entire chapter; you will be somebody else after you are done reading and learning the architecture described in this chapter.

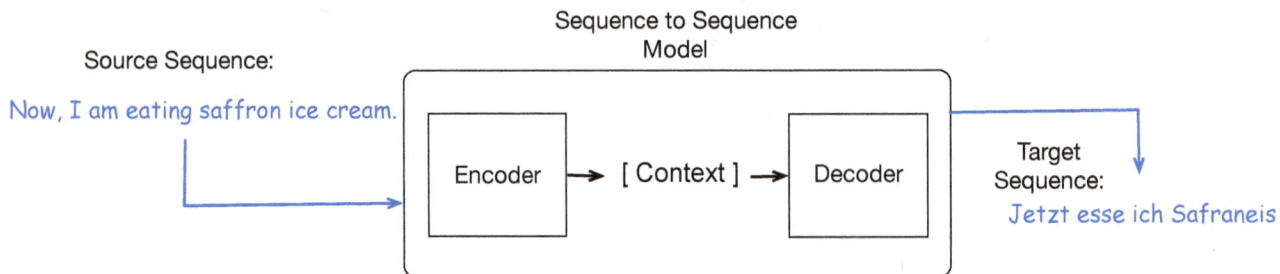

Figure 12-1: Seq2Seq architecture. The encoder or decoder can contain any one or more than one RNN model.

Attention and Transformer Neural Network

The concepts of attention in neural networks existed before the transformer architecture, but the transformer architecture, introduced in 2017 [Vaswani '17], uses attention and significantly improves deep learning algorithms. A researcher called this paper an ImageNet moment for NLP[1], which refers to 2012 when AlexNet broke the accuracy record.

At the time of revising this chapter in mid-2024, transformer and attention-based models were the pioneers in building large language models. To understand their concepts, we start by describing the sequence-to-sequence models. Then, sequence-to-sequence with attention, and next, we move to the legendary transformer architecture.

Sequence-to-Sequence

Back in Chapter 10, we learn that RNNs are useful for sequential data, and while building the model, they take the order of data into consideration. For example, we could use an RNN such as LSTM or GRU to predict changes in a time series, model audio data, etc. Having a model that can take into account the sequence of data is crucial, especially in NLP applications. Word2vec, GloVe, FastText (check Chapter 6), and other word embedding methods were very successful until 2014. In 2014, Sutskever et al. [Sutskever '14] and Cho et al. [Cho '14], both at the same time, proposed RNN models that perform sequence learning more accurately. These models are known as *Sequence-to-Sequence (Seq2Seq)* models, and they are used to transfer data from one domain into another. They have improved the quality of automatic "text translation" significantly.

If you are old enough, you recall people posting images of menus or wrong translations online. Those translations mainly came from online translator systems such as Google Translate and were used in restaurant menus and hotels. Advances in machine translation (such as using a sequence-to-sequence model) cause fewer funny translation errors, and we rarely see images of those funny translations. In addition to translation, Seq2Seq models are useful in other applications, such as text summarization and image captioning from the given image as well.

---

[1] https://thegradient.pub/nlp-imagenet

From a technical perspective, a Seq2Seq model receives an input whose sequence is important. For example, the order of words in a sentence is important. The output of Seq2Seq models is also in a sequence format (e.g., a sequence of words). For example, a Seq2Seq input is a sentence in English, *'Now, I am eating saffron ice cream'*. The output is the translated sentence in German, *'Jetzt esse ich Safraneis'*. You can see that the length of input and output are not equal in this example. The English version has six words, and the German version has four words. This shows that a simple RNN cannot take care of the transformation from English to German because it performs word-to-word translation, and traditional RNNs struggle with long-term dependencies and memory issues, which are crucial in tasks like translation. Therefore, we need a more subtle approach that leads to the introduction of Seq2Seq models.

A common approach to implementing a Seq2Seq model is using a combination of RNN encoder and RNN decoder (check Chapter 11 to recall encoder and decoder). Figure 12-1 shows the architecture of a Seq2Seq model, with a translation example. Each encoder and decoder contains at least one RNN, such as an LSTM, and can also include a stack of RNN.

The encoder gets the input sequence, and the output of the encoder is called *Context*, which is a vector. The context vector is meant to encapsulate the entire input sequence information, which the decoder uses to generate the output sequence. Note, if our RNN is LSTM, the output of the LSTM is discarded, and only its internal states (hidden states and cell state) will be used as context, or if the RNN is GRU, the hidden state will be used as context. In other words, the final hidden state of the encoder (whether LSTM or GRU) is used as the context vector. The context size is equal to the number of hidden units in the encoder, such as 256, 512, or 1024.

Then, the context vector will be sent to the decoder as an input. The decoder starts with a start-of-sequence token and iteratively generates the next elements of the sequence until an end-of-sequence token is produced. Each step uses the previous output as its input, conditioned on the context vector. For example, it receives a vector of *'Jetzt esse'* and predicts *'ich'*. In other words, the decoder gets the target sequence at time t (from the context) and generates the next target sequence data at a time $t + 1$.

In short Seq2Seq model has an encoder that compresses the input sequence from a source domain (e.g., English language) and constructs a context vector. The decoder gets the context vector and decomposes it into the target domain (e.g., German language).

The encoder and decoder of the Seq2Seq model are RNN as well. We know that input in RNN is passed to the neuron along with the previous state. This means that the RNN of the encoder, as input, receives a vector for each word. By using word embedding (check Chapter 6 to recall), and the hidden state of the previous step, it then constructs the output, which is the translated text. For example, while passing *'I'*, the encoder receives a numerical vector for *'I'* and the hidden state of the previous word, which is *'Now'*. The decoder is a language model that generates the translated sentence, but this language model is conditioned on the encoder's data. Figure 12-1 visualizes the architecture of the Seq2Seq model.

Although Seq2Seq mechanisms are successful in machine translation, they can translate only

Seq2Seq

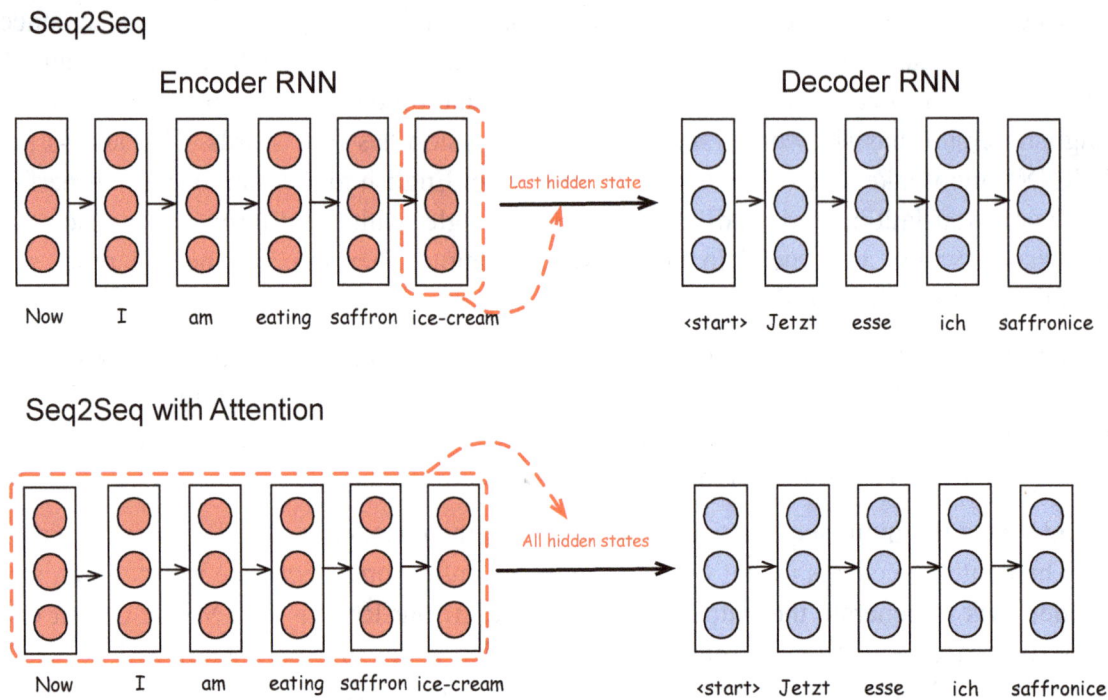

Figure 12-2: A simplified visualization of differences between Seq2Seq and Seq2Seq with Attention. Attention provides all hidden states of the encoder to the decoder.

short sentences, and for long sentences, they do not work fine. The Seq2Seq approach has been refined by *attention,* which we will describe in the next section.

Seq2Seq with Attention

While humans read a long text, we do not dedicate our attention equally to all words. Instead, we are skimming the text, and while skimming the text, we devote more attention to a few words and less to some other words. This phenomenon exists while we are listening or watching as well; when a sequence of information is long, we forget most of the information and dedicate our attention to a smaller segment of information.

The same problem also exists in Seq2Seq models and machine translation. When a sequence is long, the RNN can forget some information due to the limitations of its memory capacity. In Seq2Seq models, all the information is compressed into a single context vector. If the sequence is extensive, it becomes challenging to encode all the source sequence information into this context vector. In other words, the output of the encoder is squeezed into a single vector, which may be too small to retain all the necessary source sequence information. This can result in the encoder producing a context vector that does not adequately represent the entire source sequence, thereby affecting the quality of the translation.

696

In 2014, Bahdanau et al. [Bahdanau '14], and in 2015, Luong et al. [Luong '15] introduced the concept of *attention*, which significantly increases the accuracy of machine translation. Using attention allows the translation model to dedicate more focus on an important part of the source sequence and less focus on less important parts of the source sequence. In simple words, by using attention, the model can identify which words in the sentence have higher importance than other words.

Seq2Seq context vector is the result of the last execution (not all of the executions) of the Encoder RNN. Seq2Seq builds a single context vector from the encoder's hidden state, but employing attention at each step of the decoder enables the decoder to focus on particular items of the sequence (not only the last context vector). In other words, attention enables the decoder to look directly at the source sequence and not just a single context vector. Figure 12-2 provides a very simplified visualization of the differences between the old version of Seq2Seq and *Seq2Seq with attention* model. We can see that by using all hidden states, the information bottleneck of a Seq2Seq model could be resolved. Note, in this example, the encoder and decoder both have only one RNN, but the visualization of Figure 12-2 presents the unfolded RNN for the encoder and unfolded RNN decoder.

In general, the objective of attention is to identify the important items (e.g., words) in a sequence and fade the unimportant items of the sequence. Although the decoder receives all of the encoder's hidden states, the encoder's hidden states are multiplied by weights, and those weights are the implementation of attention.

A Seq2Seq with attention differs from an older version of Seq2Seq in its decoder operation. At each of its decoder states, the following process occurs:

(1) For each decoder hidden state h_t (where t ranges from 1 to T, representing the length of the output sequence), the model computes alignment scores with all encoder hidden states: $(s_1, s_2, ..., s_m)$. This score is typically calculated using a feed-forward neural network or another scoring function, resulting in a vector of scores. These scores indicate how much each part of the input sequence should contribute to the output at this step.

Given the decoder state h_t, we can formalize the alignment score as $score(h_t, s_k)_{k=1..m}$. In this context, the score function refers to the feed-forward neural network, a dot product, or another scoring function. The green dots in Figure 12-3 present alignment scores.

(2) Next, a softmax function is applied to the vectors of alignment scores to obtain attention score distributions, resulting in *attention weights,* i.e., $a_k^{(t)}$. Yellow dots in Figure 12-3 present attention weights. The softmax function is used to reduce the small or negative values toward zero, and amplify large values. Attention weights are probabilities and the sum of the attention weights is equal to one. The attention weight is calculated via the following equation, which uses exponentiation:

$$a_k^{(t)} = \frac{exp(score(h_t, s_k))}{\sum_{i=1}^{m} exp(score(h_t, s_i))} \text{ for } k = 1..m$$

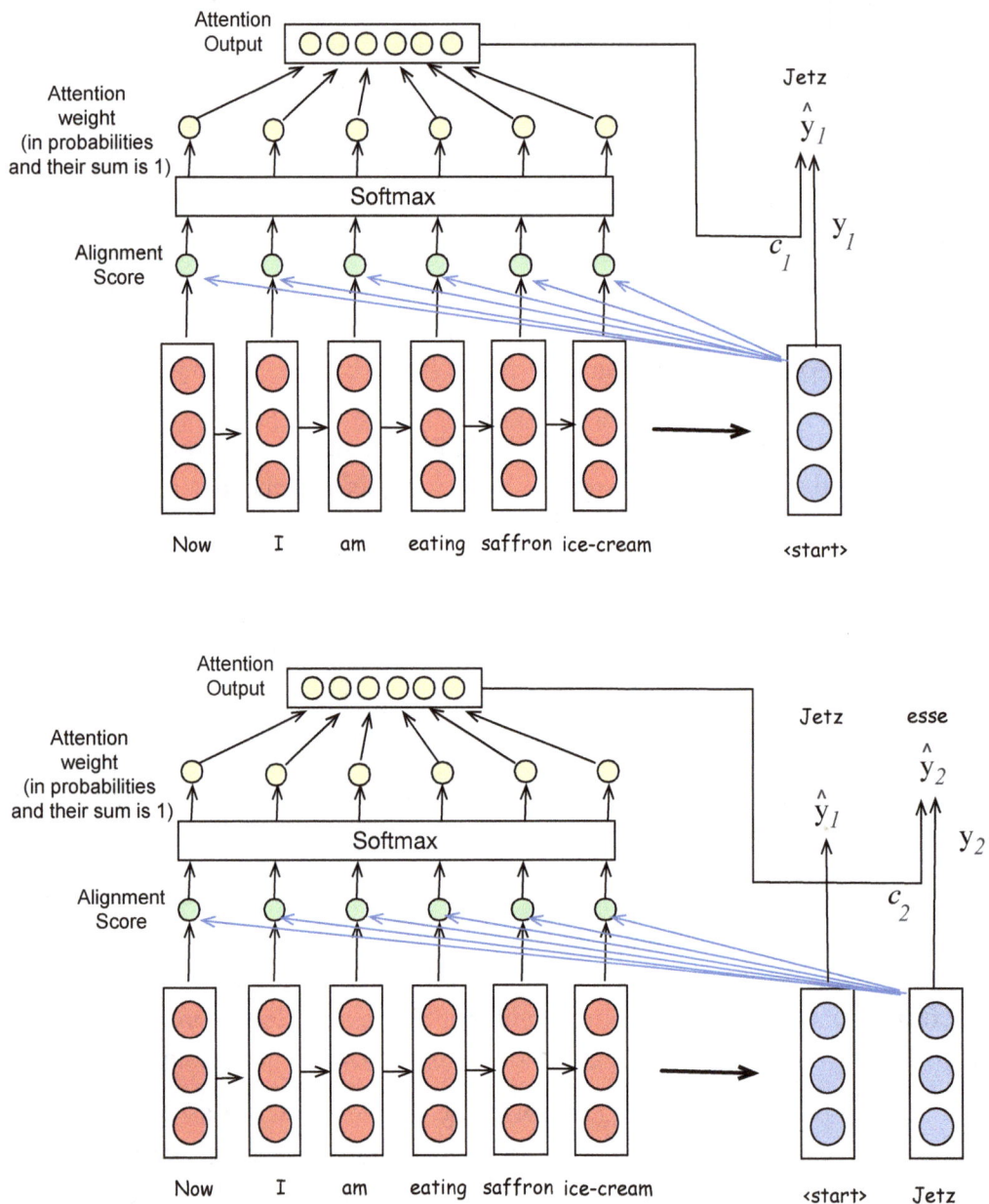

Figure 12-3: The first two steps are how attention is calculated, and the decoder identifies the word based on its attention value. Here, we visualize only the first two steps of the decoder, but this process continues until the sentence is translated completely.

Exponentiation is used because it converts negative numbers into positive, it monotonically increases, and preserves the order of the inner product. In summary, attention weight specifies the amount of attention that output $\hat{y}_t$ should pay to the sequence s_t.

(3) Now, we have the attention weights, the weighted sum of encoder states $(s_1, s_2, \ldots s_m)$ with attention weights $(a_1^{(t)}, a_2^{(t)}, \ldots a_m^{(t)})$; together, they construct the *Context vector* or

Attention output ($c^{(t)}$) for the decoder at time t. The following equation formalizes the construction of the context vector at time t:

$$c^{(t)} = \sum_{k=1}^{m} a_k^{(t)} s_k = a_1^{(t)} s_1 + a_2^{(t)} s_2 + \ldots + a_m^{(t)} s_m$$

These three steps are visualized in Figure 12-3, but only for the first two tokens. The first token is always `<start>` token because at the first state of the decoder, there is no previous output state exists, and we need to have a state to build the context vector along with attention. For the rest of the sequence, the same approach will be used. We can summarize that each decoder output is conditioned on the previous outputs, the current hidden state, and the attention *context vector*.

Attention Models

As we have explained, attention helps the model to focus on important parts of a sentence when it generates or processes text. For example, in the sentence "I left my keys in the car", if the model is predicting the next word after "keys", it will pay more attention to "left" and "car" because they are crucial for understanding where the keys are.

There are many different attention score calculation methods and attention models, here we list some common ones, including dot product, Bahdanau (Additive) attention [Bahdanau '14], Loung (Bilinear) attention [Loung '15], Self-Attention [Lin '17], which we explain here.

Dot Product Attention

This attention mechanism is a simple vector dot product[2], and its alignment score is calculated as follows:

$$score(h_t, s_k) = h_t^{(T)} . s_k.$$

It is a simple attention mechanism that tries to derive similarities from the dot product, but it assumes the encoder and decoder have the same dimension. $h_t^{(T)}$, which originates from the decoder, is a query vector, and s_k, which originates from the encoder, is a key vector. $h_t^{(T)}$ is called the *query* because it is used to query or search for relevant information from the encoder's outputs. The query vector seeks information to help the decoder decide the next step in generating output. s_k is the *key* in this context because it is used to match with the query. The model uses it to determine how relevant each part of the input is to the query's current focus or needs.

Bahdanau (Additive) Attention

The encoder of Seq2Seq with attention is a bi-directional RNN, which has a forward and backward model to read the input sentence once from the beginning and once from the end.

---

[2] If you are not familiar with vector dot products, assume we have vector $x = [a, b, c]$ and $y = [a', b', c']$. The dot product of x and y is written as $x.y$ or $x^T y$ and it is a scalar $x.y = a \times a' + b \times b' + c \times c'$.

Then, their states are concatenated. Each decoder state h_t has access to the context vector $c^{(t)}$ and its previous hidden state h_{t-1}. The context vector is computed based on the attention mechanism's output, which dynamically weights the encoder's output states according to their relevance to the decoder's current state. The context vector provides a summary of the input sequence that is relevant to the current step in the output sequence.

Bahdanau attention [Bahdanau '14] is applied between the forward and backward steps of Encoder RNN, and its attention score is calculated by applying a Multilayer Perceptron (MLP) to the concatenation of encoder and decoder hidden states ($[h_t; s_k]$). In the first version, the authors use a single layer neuron, but in later version they use MLP, which learns better relation between h_t and s_k. Also, it is called *additive attention*, and its alignment score is computed as follows:

$$score(h_t, s_k) = v_a^{(T)} . tanh(W_a . [h_t; s_k])$$

W_a stands for the weight matrix and $v_a^{(T)}$ is the weight vector. Both W_a, and $v_a^{(T)}$ are learnable parameters (the model configures their value during the training; thus, they are model parameters).

Luong Attention

Luong et al. [Luong '15] propose another attention model that has uni-directional RNN in the encoder. However, this attention model can be implemented with either uni-directional or bi-directional RNNs in the encoder. In Luong's model, the attention function is applied between the decoder state h_t (just before making the prediction) and all the encoder states s_k where $k = 1..m$. Its alignment score, a.k.a, general score, is calculated as follows:

$$score(h_t, s_k) = h_t^{(T)} . W_a . s_k$$

Similar to the previous model W_a represents the weight matrix, and it is a learnable parameter (configured by the model during the training). After computing the alignment scores, the scores are typically normalized (using a softmax function) to form a probability distribution and build attention score.

Self-Attention

Now, we have learned three basic forms of attention, they are used to add context to each word or data item of the sequence. A very successful attention model is *Self-Attention*, which is used in legendary *transformer* architecture, and it is crucial to learn it in detail to understand the transformer architecture.

As a reminder, we use attention to refine the correlated words in a sequence. For example, consider this sentence: *"Raha is a nice baby, but when she sees her father wearing a clean shirt. Then, she tries her best to put her hand in her food and rub it*

on daddy's shirt." By using self-attention, we can identify *"she"* and *"her"* refers to *"Raha"* and *"his"* refers to the poor father.

Similar to other attention functions, self-attention focuses on quantifying dependencies between items of the sequence, thus enabling the model to construct a semantically correct representation of the input sequence.

While using self-attention, each word/item looks at all other words in the sequence by using an attention mechanism, then it collects context and updates the previous representation of itself. If this explanation is not easy to understand, continue reading, we will learn it together.

Self-attention components are inspired by information retrieval components. We query a search engine for a piece of information. The search engine extracts some keywords (keys) from our query and uses these keys to retrieve the information or values that match the given keys. For example, we search a video-sharing website for *"tutorials on self-attention"*, e.g., finding a video (value) related to the identified keywords (keys), e.g., *"self-attention"*, *"tutorial"* from our input (query). Self-attention is composed of three components. Query, Keys, and Values, but all components (query, keys, values) are typically derived from the same input (e.g., a sequence of words in a sentence), unlike in search engines where keys might be pre-existing data in an index. In the following, we explain each component in detail.

Query: The word that all other words in the sequence were analyzed with respect to this word is called a query. In other words, the vector (one single word) that the attention function will update and add more contextual data (weight) is called a *query vector*. Here, we query how the given word (query vector) is related to other words in the sequence.

Keys: Other words (not the query word) in the sequence that we analyze those words with respect to the query word are called keys. In other words, all input vectors that go into the attention function to update their relation (attention score) with the query vector are called *key vectors*.

Values: Values are paired with keys, and they are relative weights that describe keys' relations to the query. In other words, the result of the query is vectors (weighted sum of attention scores) that entail updated contextual information about the query vector.

For example, look at the following example, "*I didn't forget my homeworks, my dog ate them*". Each word in the context of self-attention will be once a query word; all other words are keys that have a value associated with them. Once *'I'* is the query word, and all other words are keys, including: *'didn't'*, *'forget'*, *'my'*, … Then, *'didn't'* will be the query word, and all other words are keys: *'I'*, *'forget'*, *'my'*,… Note that this process happens simultaneously for all words and that the final output for each word is a blend of information from all other words in the sentence (weighted by their relevance as determined by the attention scores).

The input of self-attention is a set of vectors (each word is converted into a vector with a word-embedding model), and the output of the attention function will be a set of vectors, but they entail more contextual information about its surrounding words.

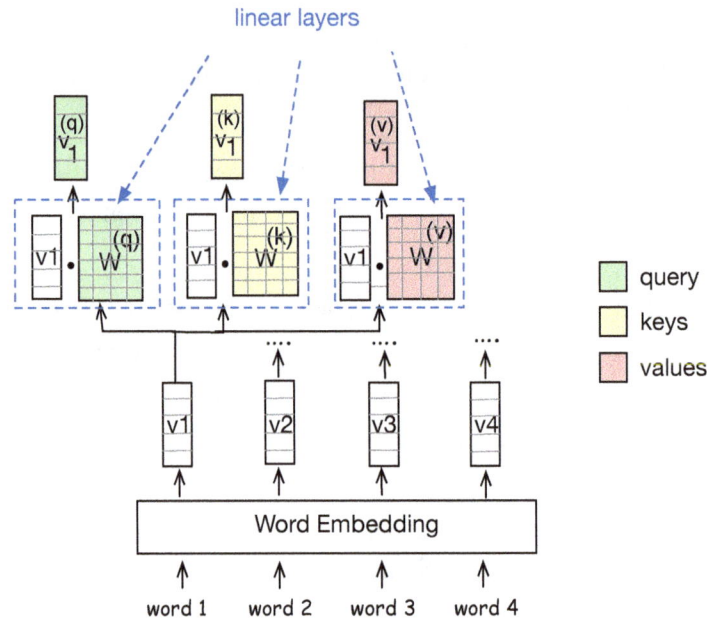

Figure 12-4: First, words are fed into a word embedding and converted into a vector. Then, each vector is multiplied by three matrices, and the result will construct query, key, and value vectors. This process is repeated for all four words, but due to lack of space, we only visualize it for v_1.

To understand the process of self-attention score assignment, we explain an example in four steps. Assume we are computing a self-attention for a sentence that has four words $\{w_1, w_2, w_3, w_4\}$.

(i)   By using word embedding, each word in the input sequence will be converted into a vector, and as a result, we will have $\{v_1, v_2, v_3, v_4\}$. Word embeddings include some contextual information about each word, but we are using self-attention to optimize these vectors to have more contextual information about other words. By contextual information here, we refer to the similarity that each word has with the other words in the sentence.

(ii) The result of each word embedding is a vector for each word. Each word vector is then transformed into three different vectors: queries (Q), keys (K), and values (V) by multiplying into three matrices (they are called weight matrices of keys $W^{(k)}$, values $W^{(v)}$, and queries $W^{(q)}$). These three matrices are parameters of self-attention, and they will be trained by the neural network. This transformation is done by multiplying the word vector with three separate trainable matrices—one for each type of vector. These matrices are parameters of the self-attention mechanism and are learned during the training of the neural network. The result of this multiplication will be three vectors: *key* (v_k), *value* (v_v), and *query* (v_q).

Figure 12-4 visualizes this step for vector $v1$. For the sake of reducing your cognitive load, we use different colors to present keys, values, and query vectors. Three vectors will be created for each vector $\{v_1, v_2, v_3, v_4\}$. In the beginning, these weight matrices ($W^{(k)}$, $W^{(v)}$ and $W^{(q)}$) have been initialized randomly (or use a common neural network initializer such as Xavier [Glorot '10] and He [He '15], which has been described in Chapter 10), but during the training, their values will be updated, as part of a feed-forward or fully connected neural network training. These components of the neural network are called *linear layers*, and they have no activation function. We have three linear layers: one for query, one for keys, and one for value. Usually, these three linear layers reduce the dimension of vectors that come from word embedding to reduce the computational cost.

$$\hat{v}_3 = v_1^{(v)} \cdot r_1 + v_2^{(v)} \cdot r_2 + v_3^{(v)} \cdot r_3 + v_4^{(v)} \cdot r_4$$

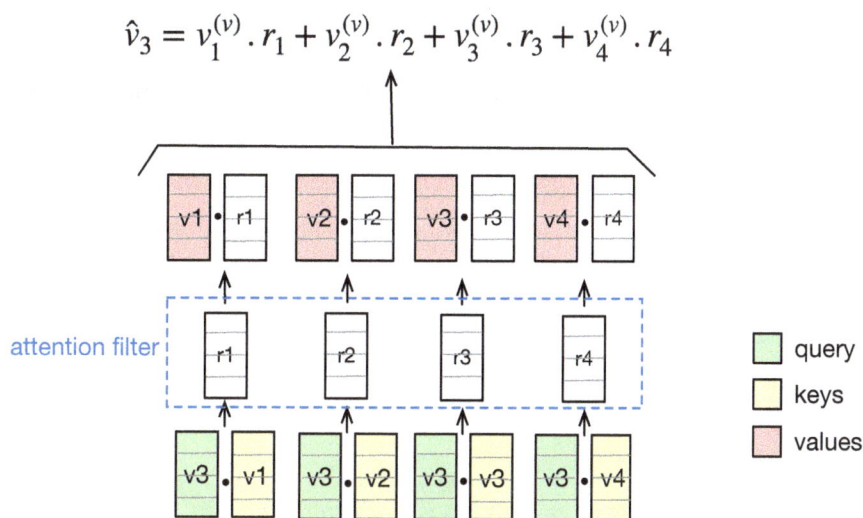

Figure 12-5: The process of new representation ($\hat{v}_3$) construction for vector v_3, which includes more contextual information. Here, we visualize only v_3, but for every vector in the sequence this process will be repeated and a new representation will be calculated.

(iii) In the third step of the self-attention mechanism, the architecture focuses on identifying the relationships, such as similarities, between each word (represented by a query vector) and all other words in the context (represented by key vectors). This means the relation of v_1 with $\{v_1^{(k)}, v_2^{(k)}, v_3^{(k)}, v_4^{(k)}\}$ will be analyzed, and the relation of v_2 with $\{v_1^{(k)}, v_2^{(k)}, v_3^{(k)}, v_4^{(k)}\}$ will be analyzed, and so forth. To do so, the algorithm makes inner products between a query vector ($v_i^{(q)}$) and key vectors $\{v_1^{(k)}, v_2^{(k)}, v_3^{(k)}, v_4^{(k)}\}$. Figure 12-5 visualizes this process $v_3^{(q)}$ only, which means this example v_3 is the query vector (we do not write $v_3^{(q)}$ in Figure 12-5, because we are lazy). These dot products are used to compute attention scores, which indicate the degree of relevance or contribution of each word (as represented by its corresponding value vector) to the output representation of the word associated with it. The result of the dot product is large if those vectors have a strong relationship and small or negative if they do not have any relationship. Then, the results will be scaled by the square root of key vector dimensions ($\sqrt{d^k}$), which d^k presents the dimensionality of the key vectors. After this scaling, a softmax function is applied to these numbers and normalizes the result (we skip presenting softmax in Figure 12-5). Let's say the result for each vector product is presented as $\{r_1, r_2, r_3, r_4\}$, and because they are normalized, the sum of these 'r's are equal to 1. These 'r's present a *relative relationship* between the *query* and *other vectors*. They are called *intermediate attention scores, attention weights,* or *attention probabilities* as well. For example, r_1 represents the relative relationship between v_1 and v_3, r_2 represents the relative relationship between v_2 and v_3, etc. Joining these r vectors together makes a matrix called an *attention matrix*.

(iv) The fourth step makes dot products between each value vector and vectors of the attention filter matrix (attention filter composed of r vectors). For example, for vector v_3 we will have:
$\hat{v}_3 = v_1^{(v)} \cdot r_1 + v_2^{(v)} \cdot r_2 + v_3^{(v)} \cdot r_3 + v_4^{(v)} \cdot r_4.$
Here, the vector $\hat{v}_3$ is a new representation of a vector v_3, which has more contextual information (information about surrounding words) than the original v_3 (which comes from word embedding).

Now, that we understand how we get from v_k to $\hat{v}_k$, let's review again: why do we use those weight matrices at the beginning? Those weight matrices make this problem a neural network training problem. By using a linear layer, we can train these values. In other words, errors $\hat{v}_k$ will be backpropagated to all these processes until they reach those weight matrixes. Weight matrices will be changed to reduce the $\hat{v}_k$ error in the next epoch. Therefore, self-attention implements a neural network that has three *linear layers*. In summary, the self-attention function is written as follows:

$$Attention(V, Q, K) = softmax(\frac{QK^T}{\sqrt{d_k}})V$$

Q is query, V is value, and K is key matrices, d_k which present the dimensions of the key vector. In the original paper, the authors assigned the key vector size of 64, and thus $\sqrt{d_k} = 8$.

Cross-Attention (Encoder-Decoder Attention)

The transformer architecture includes *Cross-Attention* or *Encoder-Decoder Attention*. It is a self-attention where the queries come from one set of data (such as the output of a decoder), and the *keys* and *values* come from a different set of data (such as the output of an encoder).

Cross-attention in transformer architectures allows each state in the decoder to attend to all states in the encoder, facilitating specific focus on relevant parts of the input sequence. In contrast, self-attention allows each state to attend to all other states within the same sequence, helping the model to understand and integrate the entire sequence's internal dynamics.

Self-attention is the attention mechanism that is still widely used, but there are efforts to reduce the memory footprint and computational overhead of the self-attention, such as FlashAttention models, Grouped Query Attention, etc. We will explain them in Chapter 15, which focuses on making an efficient machine-learning model.

Learning attention requires patience, and we can't jump to
the transformer architecture before learning its
prerequisites.

Transformer Network

Seq2Seq models, particularly those enhanced with attention mechanisms, have shown reasonable performance in tasks such as machine translation. However, traditional RNNs process sequences item by item, which is slow due to sequential dependencies. Furthermore, even advanced RNN variants like LSTM and GRU, while designed to handle the vanishing gradient problem better, still struggle with it, especially with longer sequences. Although parallel computing techniques such as batching can improve training by processing multiple sequences simultaneously, they do not change the fundamental sequential processing of data within each sequence. Transformer architecture mitigates this limitation by enabling concurrent processing of entire sequences.

In 2017, Vaswani et al. [Vaswani '17], in a paper entitled *"Attention is all you need"*, proposed a revolutionary model, *Transformer*[3], that significantly improves the NLP models. Transformers have been used in computer vision as well, and we will discuss the Transformer in computer vision later.

As we have stated previously, LSTM, GRU, and other RNN cells are dependent on the previous state, and thus, we cannot feed input in parallel to them. Transformer architecture removes RNN and uses only attention. Not having an RNN enables parallelization and significantly improves the performance (both accuracy and resource utilization) of Seq2Seq models.

---

[3] The name "Transformer" for the neural network architecture likely originates from the model's ability to "transform" one sequence into another, which is central to its design.

Similar to Seq2Seq models, a Transformer is also composed of a stack of encoders and decoders. Each encoder is composed of a self-attention layer and a feed-forward layer. Before explaining the Transformer architecture, we should be familiar with self-attention. Therefore, please be sure you understand the concept of self-attention.

Skip Connection and Positional Encoding

The self-attention we have described does not take into account the position of words. Take a look at the following sentences, *"Sally hits me with a pillow"* and *"Pillow hits me with Sally"*. Changing the word order in these two sentences changes the meaning of the sentence. By using self-attention, even if we change the word orders (a.k.a applying a permutation on the word orders) and apply the same process, the result will be the same. Because unlike RNNs, which are processed sequentially, Transformers process the sequential items in parallel. Parallel processing makes them significantly faster than RNN, but we need to use an approach that considers the position of the sequence items.

Skip (Residual) Connection

One approach the authors introduced to keep the position of words is using skip connections. Word-embedding itself is useful, but feeding word vectors into self-attention causes them to lose all information that word-embedding holds about the data. To mitigate this problem, we need to incorporate the position of words and skip connections that can be used for this purpose. To implement the skip connection, the word embedding (which is position-aware) will be added to the output of the attention function and then normalized. Another advantage of skip connections is to reduce the risk of vanishing gradients.

Positional Encoding

To maintain each word's position while constructing query, key, and value matrices, we can add the position of the word as an index into the word embedding. Directly adding indices to word embeddings, such as adding the index number to each element of the vector, would distort the embeddings because the position information would dominate, overshadowing the semantic content. For example, assume a vector value of the 10th word in the sequence is as follows: $v_{10} = \{0.1, 0.23, 0.9, 0.05\}$ adding index 10 will result in the following vector, in which the impact of its word-embedding vectors is suppressed $v_{10} = \{10.1, 10.23, 10.9, 10.05\}$.

One approach to resolve this problem is to normalize the indices between 0 to 1. If we do so, another problem arises here: when we are comparing two sentences with different sizes, the positions of similar words will get different indexes. For example, consider these two sentences: "Again, Sally hits me with a pillow", and "Dear God, again Sally hits me with a pillow". In the first sentence, index 2 belongs to "Sally"; in the second sentence, index 2 belongs to "God". Therefore, this approach is also not practical.

The transformer architecture proposes a method called *positional encoding* that resolves these issues. Using sine and cosine functions in different dimensions adds the position of the word to

the word embedding vectors. The transformer's authors proposed using the following two equations to calculate the *Positional Embedding (PE)* for each word embedding vector.

$$PE_{(pos,2i)} = sin(\frac{pos}{1000^{\frac{2i}{d}}}) \text{ and } PE_{(pos,2i+1)} = cos(\frac{pos}{1000^{\frac{2i}{d}}})$$

In these equations, *pos* refers to the position of the word inside the sentence *d* presents the size of each word-embedding vector, referred to as the dimensions of vectors. *i* is the index of the vector for one word. For even positions, $PE_{(pos,2i)}$ is used, and for odd positions, $PE_{(pos,2i+1)}$ is used. One might ask, why not use only a sine or only a cosine function? Why do we incorporate vector dimensions in the process of position encoding?

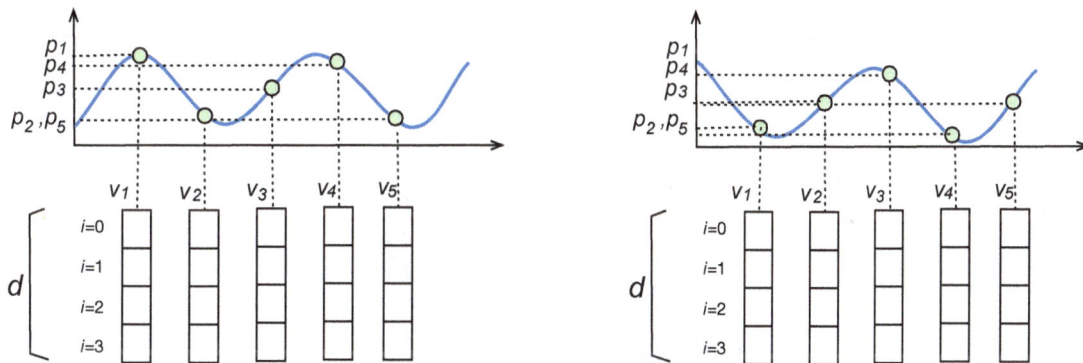

Figure 12-6: Illustration of the problem of using sine or cosine signals alone to quantify the positions of the word. We can see that in sine signals v_2 and v_5 are far apart, but by mistake, they receive the same position. The same is applicable for v_1 and v_4 in cosine signals.

To answer these questions, check Figure 12-6. There, we have used one sine function for positional embedding. Both v_2 and v_5 are getting the same positions p_2, p_5, despite there are two other words (v_3, v_4) in between. To solve this issue, we use *d* different sine and cosine functions,

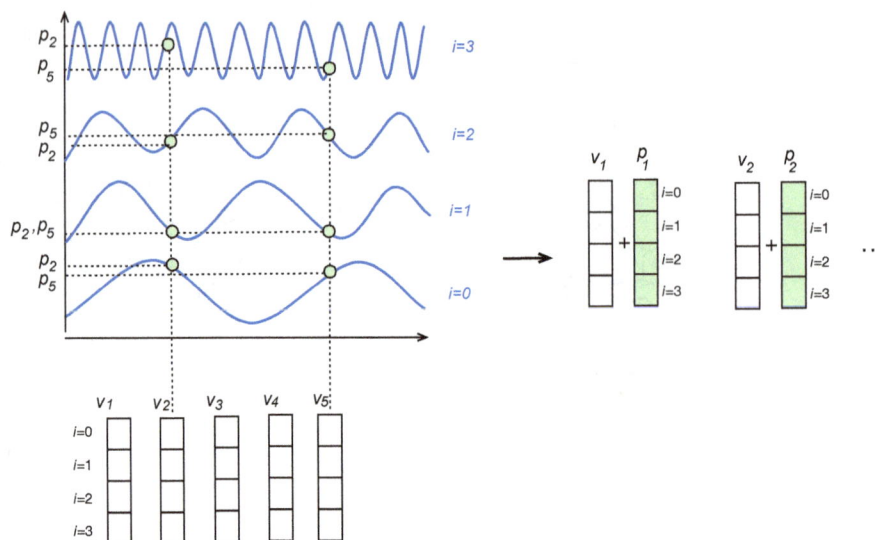

Figure 12-7: Based on the dimension of each vector, we use a Sine signal, and the positional vectors are constructed based on the values of each vector at a sine signal. The position vector has an equal dimension to the input vector.

d is the number of dimensions in each vector. Using the described equation, the *Positional Encoding* will be identified for each i index of the vector by adding it to a vector of sine (or cosine) values. Take a look at Figure 12-7 as a toy example; in this figure, for each i we use a sine function, and for the sake of laziness, we do not use cosine functions. Now, you can see that p_2 and p_5 receive an equal value for $i=1$ index, but for other indices, their value is different.

We have learned positional embedding and skip connections in addition to the self-attention function. Figure 12-8 presents adding the skip connection and positional encoding on the attention function.

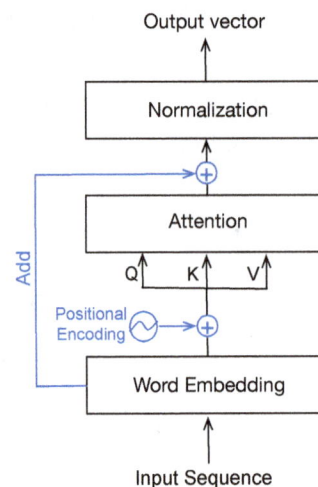

Figure 12-8: Visualizing skip connection and positional encoding at the attention layer.

Multi Head Attention

Let's review what we have explained (without positional encoding and skip connection). First, we feed a matrix of word embedding vectors into three linear layers to construct key, query, and value vectors. These matrices will be trained by using a feed-forward neural network. Then, a matrix dot product of Key and Query will be calculated, and the result will be normalized, i.e., *attention filter*. Afterward, the output will be calculated, which is a matrix dot product, between the attention filter and value vector, which is called *filtered value*. This process is shown in Figure 12-9.

However, having only attention is not enough. To understand the reason, consider this sentence: "I talk to my boss about promotion", and let's focus on the word "talk". The word "talk" has some relation with "I", "boss", and "promotion". This means that the attention of the word "talk" needs to be split up into three different words in this short sentence. To handle this issue, it is better not to saturate everything in one attention and instead use multiple attention mechanisms. Multi-head attention allows the model to explore different relationships

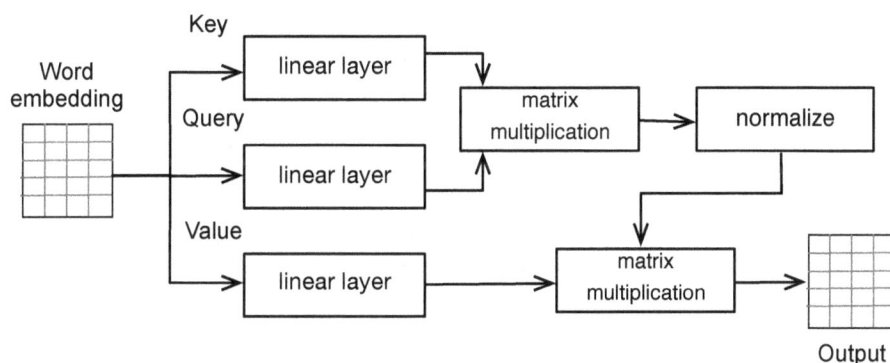

Figure 12-9: An overview on attention mechanism until now.

simultaneously by processing multiple attention heads. Each head can focus on different elements of the sentence context, thereby providing a richer, more comprehensive understanding. In the case of 'talk', multi-head attention enables the model to attend to 'I', 'boss', and 'promotion' in parallel through separate attention processes within the same layer.

To implement this feature, we can have linear layers for Key, linear layers for Query, and linear layers for Value. This is very similar to CNN architectures (check Chapter 10), in which we apply several kernel functions on an image. Here, we deploy several attentions in a sequence, but in contrast to most CNN, which applies them in a sequence, attentions will be applied in parallel. In transformers, multiple attention heads within a single multi-head attention layer operate in parallel.

It should be noted that no weights are being shared between linear layers (between Key layers, and Query layers and Value layers), since each of them produces its output. Therefore, we will have several outputs. Each of these attention filters focuses on a different linguistic feature. Then, similar to the basic attention mechanism (single-head attention), the dot product of the query and key matrices will be computed and normalized (see Figure 12-10). The results (attention filters) will be multiplied by the value vectors of value layers. As a result, we have a set of filtered values matrices, which are equal to the number of attention heads. In the original paper, the authors set the number of attention heads to eight. Next, they are concatenated, but the result of concatenation will be a very long vector. To make it back to its original size (size of word embedding), the result of concatenation will be fed into another "linear layer" to reduce the output dimension into the word embedding size. This process is called multi-head attention.

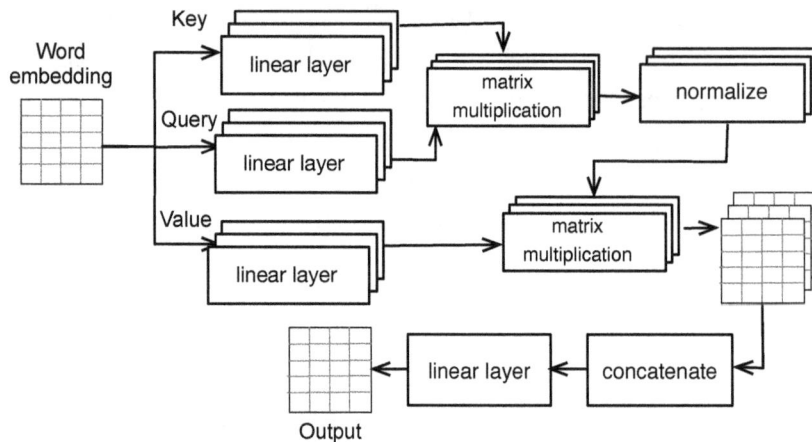

Figure 12-10: Multi-head attention architecture. The difference with a single one is multiple linear layers plus a concatenation and dense layer.

Figure 12-10, is a redesign of Figure 12-9, it presents multi-head attention instead of one attention. In summary, the transformer does not rely on one attention filter. Instead, they learn different linguistic features by using different attention filters.

Transformer Architecture

We have learned self-attention, positional encoding, and multi-head attention. Finally, it is time to learn the transformer architecture. The transformer is composed of encoder and decoder modules. Figure 12-11 visualizes the architecture of the transformer.

Encoder

The left side of Figure 12-11 presents the encoder. The process begins with adding positional encoding to the word embedding inputs to incorporate sequence information. These enhanced embeddings are then fed into a multi-head attention layer, where a skip (residual) connection is added around this layer, combining the input and output before applying layer normalization. The output from this step is then passed through a feed-forward neural network, which consists of two linear layers with a ReLU activation function between them. Another skip connection is added around these feed-forward layers, followed by layer normalization, enhancing training stability and model performance. This configuration of layers in the encoder is repeated multiple times (n layers) to refine the model's understanding and representation of the input. Finally, the output of the last encoder layer, after normalization, is forwarded to the decoder to continue the transformation process.

Decoder

The decoder takes two inputs: (i) the encoder output, which serves as keys and values in the decoder's cross-attention layer (the green color Multi-head Attention in Figure 12-11), and (ii) the sequence items it has generated so far, such as words in a translation task. Initially, the

710

decoder starts with a <start> token, which is transformed into a vector through word embedding and augmented with positional encoding. This vector is processed by a masked multi-head attention layer to prevent future token visibility (see Figure 12-11); we will explain masked multi-head attention shortly.

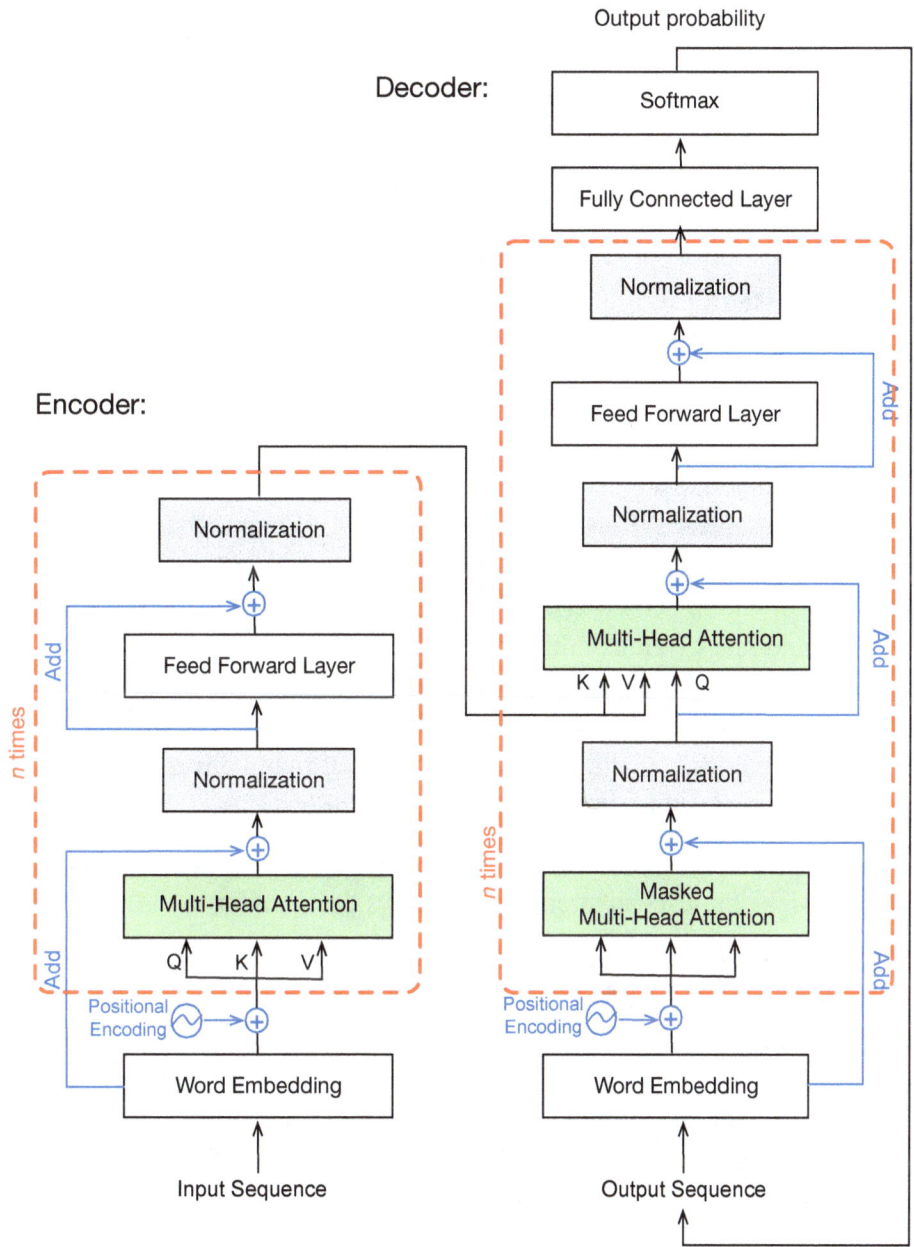

Figure 12-11: Transformer architecture, left side is encoder and right side is decoder.

Skip connections are employed to add the inputs to the outputs of each sub-layer (attention and feed-forward layers) before layer normalization. Then, the output from the masked multi-head attention is used as a query, interacting with keys and values from the encoder in the cross-attention layer. This enables the decoder to focus appropriately on different parts of the input

711

sequence. Afterward, the decoder output is normalized and fed into a feed-forward layer, a.k.a. linear layer. Similar to the encoder, the decoding process can be repeated *n* times to improve the accuracy of the result by having a deeper network.

The output size of the final linear layer in the decoder corresponds to the size of the vocabulary, as it must generate probabilities for each word in the vocabulary. The decoder predicts a probability distribution over all possible vocabulary words. For example, if the intention of using the transformer is to generate a new text. The size of this layer will be equal to the total number of vocabularies. Each input vocabulary outputs a vector, which is a logit score. A softmax will be used at the end to convert them into probability. The words with the highest probability will be used as words for new sequence generation. This process will continue in the decoder until the decoder generates the last item for the sequence, i.e., <end>.

Masked Multi-Head Attention

Attention blocks on the encoder side receive the entire sequence, but the attention block on the decoder side only receives the previous word of the target sequence. This is due to the fact that other remaining words in the output sequence have not yet been identified (predicted). Therefore, the algorithm should prevent seeing the future words in the sequence. For example, consider that we have the sentence 'I am learning Transformer'. When the attention score for 'am' is computed, the model should not have access to 'Transformer' or even 'learning', it should only see 'I'. In other words, the attention score should be based on the previous words and not the future ones. Therefore, it should mask (remove, hide, etc.) upcoming words. To implement the masking, the algorithm assigns $-\infty$ to future words that will appear later in the decoder. Because of these simple differences, this is called Masked Multi-Head Attention. The mask is added before calculating the softmax and after scaling the scores. The softmax converts negative infinities to zero and, thus, leaves zero attention scores for upcoming words.

In summary, we use mask multi-head attention to make the training parallel and this allows the model to train faster. Mask multi-head attention disallows the decoder to look ahead.

Please accept our sincere congratulations for finally learning the transformer architecture. Transformer architecture is a significant improvement on contextual models (consider the meaning of the words in their context). They have revolutionized NLP and led to the introduction of large language models similar to CNN architecture, which revolutionized computer vision back in 2012. They are not just limited to translation tasks. Nevertheless, to understand a Transformer, we need to learn it in the context of text translation. We will describe NLP models built based on Transformer in the next section of this chapter.

Structured State Space Sequence Models (S4)

We have learned earlier that transformers operate based on the concept of attention. However, they suffer from scalability issues, which is due to the self-attention mechanisms that have quadratic computational complexity with respect to sequence length. This is why the input length of current large language models is limited. The model cannot consider anything outside the finite-size attention window.

The Structured State Space Model (S4) can be used for sequential data analysis. These models are a combination of CNN and RNN and the traditional State Space Model (SSM), e.g., the Kalman filter (we will explain it later in Chapter 16). For example, Mamba [Gu '23] is an S4 model, which mitigates the self-attention limitations. However, our initial experiments reveal the challenges in quantization and pruning (we will explain them in Chapter 14) of Mamba, which questions its usability. When writing this section, they are too new to pass the test of time. Therefore, we briefly describe them here and do not go deep into them.

State Space Model

State Space Models (SSM) are used for sequential data analysis, such as time series and signals. They are characterized by using state variables to represent the system's current status, which can evolve over time according to a set of equations. Hidden Markov Model (Chapter 5), Markov Decision Process (Chapter 14), and Kalman Filter (Chapter 16) are examples of SSM.

An SSM gets input u_t at time t and provides y_t as output at time t. We can define a state space model as any recurrent model that has a latent space. A typical SSM consists of two equations: a *state equation* and an *output equation*. The *state equation* describes how the system evolves over time and is computed by using the following equation:

$$h_{t+1} = A h_t + B u_t$$

h_{t+1} is the state at time $t + 1$ (latent state), A is the state transition matrix, B is the control-input matrix, u_t is the input vector at time t. The output equation is generated by the following equation:

$$y_t = C h_t + D u_t$$

y_t is the observed output at time t, C is the observation matrix, and D is the direct transmission matrix (often zero in most neural network implementations of SSM).

Given the input u_t, we can compute output y_t by finding the value of h_t. However, finding h_t is computationally complex, and we could approximate it by finding h_t for some t values, not all possible t. One way to approximate h_t is by using the Kalman filter (Check Chapter 16). The Kalman filter is a recursive algorithm that estimates the state of a system from a series of noisy measurements.

Linear Time Invariants SSM

The SSM we have explained earlier is known as linear SSM. Linear SSMs are linear time invariants (LTI), which means the model dynamics remain consistent over time. In other words, parameters *A, B, C,* and *D* do not depend on time; they are fixed, and their behaviors do not change over time. It means that a linear SSM can not represent a system in the real world whose behavior changes over time. Non-linear and time-varying SSMs are specifically designed to resolve this issue and handle such complexities. These models can capture the changes in the system's behavior over time by allowing the parameters *A, B, C,* and *D* to vary with time. This makes them more suitable for modeling real-world systems that exhibit nonlinear and time-varying behavior. Non-linear SSMs could be implemented with neural networks by using the network to approximate the non-linear state update function.

Structured State Space for Sequence Modeling (S4)

Sate-Space Models (SSMs) can function as components within a neural network architecture. Like Recurrent Neural Networks (RNNs), SSMs process sequential data by feeding the output from one time step as input to the next. The core of SSMs lies in the matrices *A, B, C,* and *D*. However, SSMs lack the ability to differentiate between input tokens, in contrast to RNNs' ability to weigh inputs differently. Unlike the attention mechanisms in transformers, SSMs are not capable of selectively focusing on specific tokens.

The Structured State Space for Sequence Modeling (S4) addresses these limitations by introducing Selective SSM, employing separate linear layers for computing A, B, C, and D matrices. This enables the model to adapt dynamically to the unique characteristics of the data, akin to the self-attention mechanism in transformers.

Additionally, conventional SSMs face challenges when dealing with long sequences due to computational complexity. S4 overcomes this by employing techniques, such as the FFT (Check Chapter 7), to efficiently manage state updates in lengthy sequences. By transforming operations into the frequency domain, S4 mitigates the computational burden of large matrix multiplications, making it well-suited for tasks involving extensive sequential data, including audio processing and time series analysis.

Mamba Architecture

Mamba is designed based on S4. This architecture tries to improve SSMs by introducing the Selective and Structured features for its SSM. It uses different linear layers to compute A, B, C, and D matrices.

Since SSM is dealing with the output of the prior hidden layer in time t, we can rewrite the SSM equations as follows:

$$h'_t = A h_t + B x_t$$
$$y_t = C h_t + D x_t$$

h'_t presents the state vector in latent space at time t, x_t is the input to the model at time t, which has a D dimensionality; for example, D is equal to the embedding vector size. y_t is the output of the model, e.g., the next token for LLM. A is known as the state matrix, B as the input matrix, C as the output matrix, and D as the feed-forward (a.k.a feedthrough) matrix. If the model does not have feed-forward, then D is zero, as is the case in Mamba.

The first step in converting a continuous state-space model (SSM) into an S4 model is to discretize the model. This process reformulates the continuous differential equations of the SSM into a discrete matrix equation format. This discrete representation allows the model to be processed by neural networks, including specialized architectures like the S4, which is designed to handle sequence data efficiently. The discretization process uses Δ as the step size to discretize the data. In particular, continuous A is covered into discretized $\bar{A}$, and continuous B is covered into discretized $\bar{B}$, by using the following mathematical transformations:

$$\bar{A} = exp(\Delta A)$$
$$\bar{B} = (\Delta \bar{A})^{-1}(exp(\Delta A) - I) . \Delta B.$$

In summary, the discretization process transforms the original (Δ, A, B, C) into $(\bar{A}, \bar{B}, C)$, i.e., $(\Delta, A, B, C) \rightarrow (\bar{A}, \bar{B}, C)$. In addition to discretization, Mamba enhances SSM with *structure* and *selectivity*.

Selectivity: To implement selectivity, Mamba uses linear projections that adjust Δ, B, and C matrices parameters based on the input. It involves a linear layer (or multiple layers) that projects the input data into the parameter space needed for the model. This transformation allows each segment of the state space to focus on different aspects of the data, similar to how attention mechanisms in transformers highlight relevant features, thus boosting the model's ability to understand temporal relationships.

Structure: To implement structure, Mamba utilizes a specific arrangement of the state space parameters optimized for parallel processing and efficient memory usage. It is known as a *diagonal structure*. This arrangement organizes matrix A to maximize computational efficiency while preserving the ability to capture dependencies within the data, but we don't explain its details.

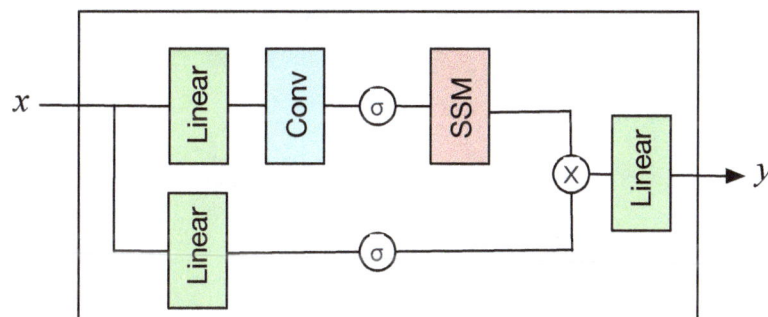

Figure 12-12: Mamba module architecture.

Mama is a block in a neural network that includes a Selective SSM module and gated functions similar to LSTM and GRU. Figure 12-12 presents a Mamba block. First, the input goes into a linear layer to increase the dimensionality of the data. The high dimensionality increases the chance of identifying some features better than lower dimensions.

After the linear layer, the authors introduce a convolutional layer followed by a Swish[4] [Ramachandran '17] activation function (presented σ in Figure 12-12). The convolutional layer is used for feature extraction on high-dimensional data. Next, the processed data is fed into the SSM module to process the convolution's output similar to RNN fashion. This module dynamically adjusts its internal state by modifying the Δ, B, and C parameters based on the input data, as influenced by the selectivity mechanism, which we have described earlier.

The input is also passed to another linear layer, and the result is given to the Swish activation function; this linear layer is used to create an additional transformed representation of the input, which can be utilized for further processing or integration with other network components. This representation acts as a complement to the first, providing better feature extraction capabilities.

Afterward, the output of selective SSM and the linear layer is passed through a non-linear activation function or gated multiplication. The rationale for this multiplication is to measure the similarity between SSM (which contains information about the previous token) and current tokens, which come from the linear layer without a token.

The result is passed to another linear layer to reduce the dimension of the output and prepare it for subsequent processing. This non-linear activation is shown as $\otimes$ in Figure 12-12, and it is used to enrich the network's ability to process sequential data. The non-linear or matrix multiplication is used to integrate different transformed representations, enhancing the model's capacity for complex feature learning.

---

[4] The Swish activation function is a slight modification of the sigmoid function. It is written as:
$x \cdot sigmoid(\beta x) = \dfrac{x}{1 + e^{-\beta x}}$, where β is a scalable and trainable parameter.

Large Language Models

One of the most popular fields within Artificial Intelligence and Machine Learning is Natural Language Processing (NLP). There are various approaches to working with text, including information retrieval and text mining. Information retrieval is usually focused on searching and retrieving information (audio, video, and text) from unstructured data. Text mining, on the other hand, is focused on extracting knowledge from the text.

However, in the context of deep learning, these approaches are often grouped under the umbrella of Large Language Models (LLMs) and their applications, known as *downstream tasks*. There are several downstream tasks we can perform with NLP or LLMs, including:

- Translating text from one language to another.

- Summarizing a long text and reducing the text corpus size.

- Classifying text, such as identifying the sentiment (emotional tone) of a sentence, determining whether a piece of news is misinformation (fake news), or checking if an email is spam.

- Identifying entities in the text, such as names of persons or organizations, geographical locations, etc., is known as *named-entity recognition*.

- Generating new text based on previous text, such as auto-complete, generating answers to given questions (question answering), and correcting and improving human writings. These text-generation approaches are categorized under the umbrella term generative AI.

Transformers have revolutionized advances in NLP tasks, and there are de facto standard approaches for building LLMs and tuning them for downstream tasks. We briefly explained the approaches used to prepare text for machine learning algorithms in Chapter 6, which covers feature engineering for textual data. Here, we describe some popular LLMs, excluding those that are not much in use nowadays, such as ELMo [Peters '18] and ULMFiT [Howard '18]. However, some architectures, such as BART [Lewis '19], are still in use. We can use their *pre-trained* models for our NLP applications to benefit from their accuracy or even *fine-tune* an open-source model for our dataset.

Before we explain NLP models, we should be familiar with two terms, pre-training and Fine-tuning, which are used to work with LLMs.

Pre-training: It refers to the task of training a neural network with a dataset and saving its weights. This process is performed on a large amount of data. For example, a language model is trained on Wikipedia data. The result of pre-training is also referred to as the base model. In other words, pre-training is an unsupervised approach that trains a general purpose language model by using a large amount of unlabelled text.

Fine-tuning: It refers to constructing a new model trimmed for a specific task with limited data, but this model benefits from the weights of the pre-trained model. In other words, fine-tuning adjusts the pre-trained model's weights for a specific task or specific dataset. For example, if the task is sentiment analysis, the fine-tuning dataset would contain text examples labeled as positive, negative, or neutral. In other words, fine-tuning augments the language model for a specific task. As another example, think about building an LLM to answer health-related questions. First, we train a language model by using all text from Wikipedia. Wikipedia has some health-related data, but not only that. Therefore, this base model can be used for different tasks. Next, we use a repository of several medicine books and fine-tune the base model to answer health-related questions.

The remainder of this section lists and describes many LLMs; please spend time and learn them. They are important to observe what is happening in the real of LLM and NLP.

BERT

About a year after the introduction of Transformer architecture, in 2018, the BERT architecture [Devlin '18] was introduced by Google and released as open source. BERT is inspired by Transformer architecture, and the word BERT is the abbreviation of "Bidirectional Encoder Representations from Transformers". Still, BERT is widely used and considered one of the state-baseline LLMs.

In the training phase, BERT receives a large corpus of text, and the output of a BERT is the vectorized version of the same text, but these vectors include lots of contextual information about each word. In other words, the output of BERT is a language model similar to word embeddings, and this language model entails significantly more accurate information about the context of each word.

After the training, the BERT can be used for inferences; for example, we give a text and ask the BERT to summarize this text.

The Transformer learns the sequence only from left to right, but BERT learns the sequence from both directions. Experiments show that learning the model from both directions (left-to-right and right-to-left) at the same time provides better probability distributions for words, which leads to getting higher contextual information in the result language model. Besides, it makes the training time faster. As an intuitive example, consider this sentence: *'I went to the bank, and there a nice view of the river, which is beautiful.'* If we start the sentence from left to right when the algorithm reaches the word 'bank' (passed I went to 'the'), usually 'bank' refers to the financial institution. However, by looking at both directions simultaneously, the model can understand that 'bank' here refers to the edge of the river and not that financial institution. This is something that is handled by BERT, because it parses the sentences from right-to-left as well as left-to-right, both at the same time.

BERT is trained on Wikipedia and a large book dataset with over 10,000 English books. It has been implemented and trained for 104 other languages as well[5]. Unless you have access to super expensive hardware infrastructure with many GPUs, it is not common to train a BERT model on your own. If you intend to use BERT, a common approach is to use a pre-trained BERT model,

---

[5] https://github.com/google-research/bert/blob/master/multilingual.md#list-of-languages

and fine-tune it for your downstream task. This process is referred to as transfer learning, and we will explain it in Chapter 14.

BERT Architecture

Since the objective of BERT is to construct a language model, it is composed of a stack of transformer encoders only and no decoder. The decoder is used for prediction and downstream tasks. Therefore, it has only encoders and no decoder. The BERT encoder gets the sequence of words as input, similar to the Transformer's encoder, and passes the input to a stack of encoder layers. Each encoder is composed of multi-head attention layers and then a feed-forward layer. Figure 12-13 visualizes the BERT model architecture. Authors of BERT propose two BERT models, 'base' and 'large'. The base model has 768 feed-forward nodes, 12 attention heads, and 12 encoder layers, and it has 110 million parameters. The large model has 1024 feed-forward nodes, 16 attention heads, and 24 encoder layers, and it has 340 million parameters.

BERT Input

BERT does not start with the raw text input. Instead, its input requires three preprocessing steps to convert the given input into an input for BERT encoders. These preprocessing steps include token embeddings, segment embeddings, and positional embeddings.

Token embeddings: The first input that a BERT model receives is the [CLS] token, which stands for classification. As the input passes through the multiple layers of the Transformer, each layer updates the representation of the [CLS] token by integrating information from all other parts of the input. In other words, the [CLS] token reserves a spot specifically for accumulating and representing the overall context of the entire input. It is like a notetaker in a meeting. It doesn't have any prior knowledge but listens to everything being discussed (other words in the sentence) and summarizes the key points at the end (final state of the token), which can then be used for various purposes (classification tasks)

The separation or end of a sentence is marked by the '[SEP]' token. For instance, when we feed a sentence like, *"Strong people talk about ideas. Weak people talk about people"*, BERT tokenizes the input sentence as follows: [CLS], Strong, people, talk, about, ideas, [SEP], Weak, people, talk, about, people, [SEP]

Segment embeddings: It is used to differentiate between sentences. In other words, it encodes a token contribution to a sentence, and it annotates each token with a sentence marker, i.e., E_A for the first sentence and E_B for the second sentence. For example, the segment embedding has been added to the described example list of tokens as follows:

$S_{[A]}$+[CLS], $S_{[A]}$+Strong, $S_{[A]}$+people, $S_{[A]}$+talk, $S_{[A]}$+about, $S_{[A]}$+ideas, $S_{[A]}$+[SEP], $S_{[B]}$+Weak, $S_{[B]}$+people, $S_{[B]}$+talk, $S_{[B]}$+about, $S_{[B]}$+people, $S_{[B]}$+[SEP]

Positional embeddings: After each token is annotated with segment embeddings, a positional embedding is added to each token to specify its location in the sequence. For the example sentence we used earlier, we will have the following that goes into the first layer encoder of BERT:

$P_0+S_{[A]}+[CLS]$, $P_1+S_{[A]}+$Strong, $P_2+S_{[A]}+$people, $P_3+S_{[A]}+$talk, $P_4+S_{[A]}+$about, $P_5+S_{[A]}+$ideas, $P_6+S_{[A]}+[SEP]$, $P_7+S_{[B]}+$Weak, $P_8+S_{[B]}+$people, $P_9+S_{[B]}+$talk, $P_{10}+S_{[B]}+$about, $P_{11}+S_{[B]}+$people, $P_{12}+S_{[B]}+[SEP]$

Then BERT passes this tokenized input which includes position and segment information as well into the first layer of the Encoder.

Pre-training BERT

BERT uses two approaches for its pre-training phase, i.e., Masked Language Model (MLM) and Next Sentence Prediction (NSP) models.

In Chapter 6, we have explained that the objective of word embedding is predicting the missing word (CBOW and Skipgram). LLM is used to identify the best choice for missing words in a sentence. For example, "Learning Transformer is important to be able to understand new ___ models". Here, an appropriate substitution for the blank space is 'NLP', and word embeddings are used to find the word 'NLP' in this sentence.

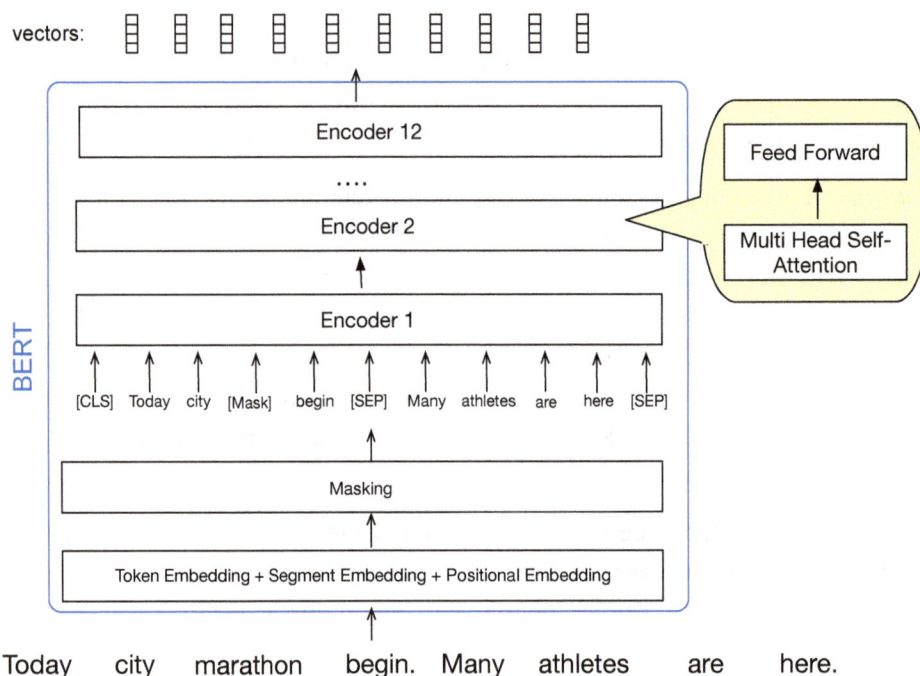

Figure 12-13: BERT base architecture which is composed of a stack of 12 encoder layers. The result of BERT is a set of contextualized vectors. This figure depicts the pre-training only, while fine-tuning adds some layers on top of the output vectors based on the desired task.

Instead of predicting one missing word, from left to right or right to left, BERT introduces MLM, which enables them to not just read words in a sequence from left to right (how we read in the English language), it learns the sequence from both directions (right to left and left to right) at the same time. ELMo [Peters '18] also uses a bidirectional LSTM and learns from both directions, but it does not perform learning simultaneously. This is due to the fact that LSTM cell states depend on their previous state, and this is a reason why attention models substituted LSTM models.

The MLM process focuses on randomly hiding (masking) a word in a sentence, and then it tries to identify (predict) them. Authors recommend hiding 15% of tokens (words) because masking more than this will harm the contextual understanding of the algorithm and less than that makes the training slow and expensive. Besides, they recommend that from those 15% of tokens that will get masked, replace 80% of them with [Mask], leave the rest 10% unchanged, and replace 10% with another random token. While training, the BERT's loss function considers only the identification of the masked tokens and does not identify the non-masked tokens. Also, note that masked tokens are not seen in the fine-tuning phase.

You can see in Figure 12-13 that the word 'marathon' is substituted with [Mask] after the masking process has been applied. The resulting text includes position and segment tokens, but for the sake of readability, we do not write them in the figure.

The second approach used in training is NSP. By using NSP, BERT can understand the relationship between two sentences. This feature is used in question-answering applications or more advanced versions of auto-completes that recommend the rest of the sentence while we are typing. During the training phase, the BERT model receives a pair of sentences as input, and it learns to identify if the second sentence is the next sentence in the original text or an irrelevant sentence. The '[Sep]' token separates sentences for NSP. In particular, at the training phase, the BERT model receives two sentences at a time: 50% of the time after the first sentence comes the second sentence, and 50% of the time after the first sentence comes to a random sentence. The BERT's job is to identify if the second sentence is correct or a random sentence from somewhere else.

Fine-tuning BERT

The pre-training of BERT provides us with the base language model that accurately understands context. The process of fine-tuning involves adding a layer on top of the BERT results for a specific task, or re-training the BERT model with a specific dataset, such as a medical dataset, to answer medical-related questions. To better understand the fine-tuning process, we list three examples that are popular examples implemented with BERT.

Sentiment analysis: If the task is sentiment analysis (considering we have negative or positive emotions), then a fully connected layer with softmax can assign a sentiment label to the input sentence.

Question Answering: The question/answering task is a sort of prediction. Therefore, the models get the first sentence; it tries to identify the start and end of a text that contains the answer. In other words, the model is trained by learning two vectors (start and end token of the answer text). For example, consider a question a user asks a chatbot. *"Why do I need to eat Broccoli?"*. Then, the BERT model used several diet and nutrition books for training, and it should identify the start and end tokens in the following text: "*Some food tastes like cardboard, and it is not yet clear why god created them so tasteless. Broccoli is an example, but it has many nutritional values. It contains antioxidants, a compound that reduces inflammation, and low-carb food. Therefore, taste alone will not be a proper choice for your diet planning."*

In this example, a good start could be the third '*It*', and a good end token could be the '*food*'. Both of them are marked with an underline in the text.

Named Entity Recognition (NER): If the task of BERT is NER, the model must label different types of entities (person, organization, date, street names, etc.) that exist in the input text. The NER output is vectors, and a classification algorithm (e.g., feed-forward with softmax) identifies the correct names and assigns them to the vectors. For example, in the sentence "Rumi wrote his poems in Persian," BERT would identify "Rumi" as a person and "Persian" as a language.

The fine-tuning process trains the pre-trained model for a specific task known as the *downstream task*. Note that fine-tuning BERT generally involves training all the layers of the model, not just adding a new layer, although a task-specific output layer (like a classification layer) is usually added. The entire network is fine-tuned to better suit the specifics of the targeted task, adjusting the pre-trained parameters to optimize performance on new data.

BERT uses Gaussian Error Linear Units (GELU) [Hendrycks '16] as an activation function (check Chapter 10).

Derivations of BERT

After BERT was introduced, many computer scientists and researchers recognized the revolutionary impact of self-attention on training, and a wide variety of BERT has been introduced. We have a forest of BERT models, such as RoBERTa [Liu '19], Bio-BERT [Lee '20], ALBERT[Lan '19], DistilBERT [Sanh '19], until the introduction of Llama and ChatGPT in Q4 2022. There is an ongoing competition to construct a large model with similar architecture and get a better result. This results in weird models such as NVIDIA and Microsoft Megatron-Turing LNG[6] , which was trained on 4,480 GPUs, Open AI's GPT 3[7], which has 175 billion parameters, or Google's PaLM[8], which has 540 billion parameters, and many more LLM models. In 2024, most corporations stopped releasing the amount of GPU they have used to build their model due to critiques on the CO_2 footprint and water use required for their data centers [Li '23]. At the time of revising this chapter (Q2 2024), we cannot build these models unless we are employed by a large corporation with mega hardware infrastructure.

We briefly review some common derivations of BERTS: RoBERTa, DistilBERT, and XLM. Then, we continue describing common LLM models.

---

[6] https://developer.nvidia.com/blog/using-deepspeed-and-megatron-to-train-megatron-turing-nlg-530b-the-worlds-largest-and-most-powerful-generative-language-model

[7] https://openai.com/blog/gpt-3-apps

[8] https://ai.googleblog.com/2022/04/pathways-language-model-palm-scaling-to.html

We need the request from your department's chair, dean's approval, provost, and university president. They should sign it with your blood.

Next, you should take the 'Ethics and diversity of using hardware' exam. After the exam, make an IRB request; if your IRB request has been approved, send us all the documents, including your great-grandparent's sexual orientation, to proceed with your request. If you can bring me these documents by tomorrow, the next available date for GPU that you can reserve is from 01:00 pm 17th June 2090 to 1:45 pm 17th June 2090; if you send your documents later this week, the availability time might change to a later time.

Let me know if you have any other questions or can I help you with anything else.

IT administrator   Student who needs GPU

RoBERTa

The term RoBERTa is an abbreviation of the "Robustly optimized BERT pre-training approach", and it was introduced by Facebook back in 2019 [Liu '19]. RoBERTa improves BERT's training by making four key modifications: (i) increasing the training dataset (by adding a new large text dataset) and the batch size, which results in longer model training; (ii) removing the Next Sentence Prediction (NSP) task, as experiments show that removing NSP increases the accuracy of downstream tasks; (iii) training on longer word sequences; and (iv) dynamically changing the masking pattern.

DistilBERT

Previous BERT models are not practical for on-device usage (e.g., mobile phones) or real-time wide application deployments. DistillBERT [Sanh '19] was released by HuggingFace and focused on the large size and extreme resource utilization of training a BERT. DistilBERT has reduced the BERT model size by 47% while losing 3% accuracy in language understanding and making it 60% faster than BERT. DistilBERT uses a knowledge distillation approach to compress the model, and we explain knowledge distillation later in Chapter 14. Briefly, knowledge distillation is the process of transferring knowledge from a larger model into a smaller model by training the smaller model to replicate the behavior of the larger model.

XLM

In 2019, Facebook, in another effort, introduced the Cross Lingual Model (XLM) [Lample '19]. One challenge of BERT and other models is the imbalance between English documents that exist on the Internet and other languages. XLM proposes a cross-lingual (covering multiple languages) approach. XLM authors proposed three language models; two require monolingual data (one language), i.e., Masked Language Modeling (MLM) and Causal Language Modeling (CLM). The third language model requires parallel sentences in two or more languages, i.e., Translation Language Modeling (TLM).

To implement tokenization, they use Byte Pair Encoding (check Chapter 14) to construct sub-word tokens. Since the text corpus is unbalanced between languages (English has more text than another language, such as traditional Chinese), they resolve this by a multinomial distribution for sampling (check Chapter 3, it is a generalization of Binomial distribution for multidimensional data), which resolves the unbalanced data.

The objective of the CLM task is to predict the next token by identifying the probability of a word given its previous words. The objective of MLM is similar to BERT and is used for training the model for prediction. Besides, similar to BERT, they mask 15% of subwords (Byte Pair Encoding tokens). However, BERT uses two sentences, but the XLM approach can use an arbitrary number of sentences (up to 256 sentences). The objective of TLM is to improve the cross-language pre-training, and unlike MLM and CLM, it is not unsupervised. It extends the MLM features, but it randomly masks a word in English, and for example, this masked word can be learned from either its surrounding word in English text or its Chinese translation. In other words, the BERT MLM approach processes sentences in consecutive sequences, but XLM processes them in parallel (across two different languages). This approach encourages the model to align the source and translation representations. This means that if the Chinese text is not enough to identify the masked word, TLM leverages the English context.

Transfer Learning with a Unified Text to Text Transformer (T5)

After the success of BERT and the transformer, many settings and parameter configurations were applied to optimize its architecture further. A team of researchers at Google conducted different experiments on transformers with large datasets and came up with the T5 configuration [Raffel '20], an optimized version of the transformer architecture.

BERT and its described derivations are only encoders, while T5 has a decoder. This allows T5 to generate text output, while BERT produces embeddings that can be used for various NLP tasks. The T5 architecture is very similar to the original transformer but has minor differences. T5 encoder normalization is simpler than the original transformer's normalization and has no additive bias. After each self-attention layer, the T5 decoder includes an additional cross-attention layer. In cross-attention, the decoder attends to the encoder outputs, allowing it to condition the generation on the input sequence. This allows the decoder to focus on relevant parts of the input sequence while generating the output.

T5 has been trained on the Colossal Cleaned Common Crawl dataset, or C4 dataset[9], which had a size of 745 GB at that time. This dataset is a cleaned version of a large dataset, i.e., the Common Crawl dataset (~7 TB size). It considers all downstream tasks (e.g., sentence similarity detection, sentiment analysis, sentence completion, etc.) as a Text-to-Text conversion since it provides the output in the text and has a decoder.

The authors of T5 conducted many experiments to improve some of the Transformer hyperparameters. For example, T5 experimented with different mask sizes, not just masking an entire word, instead substituting a part of the word; their experiment shows that masking the entire word is the most reliable approach.

Generative Pre-trained Transformer (GPT) Models

Google, Meta (Facebook), and others were invested in BERT. OpenAI, which was started as a non-profit organization from 2015 until 2019, used the original transformer architecture to build a series of GPT models. Similar to BERT, GPT models construct a language model using unlabeled data in an unsupervised manner (authors named it generative pre-training). Then, a Fine-tuning process (authors named it discriminative fine-tuning) will be applied to them based on the type of the downstream task, which is a semi-supervised training process.

At the time of revising this section, GPT-4o and o1 were the latest released from OpenAI, which combines LLM with computer vision and audio. However, no details of these models are revealed at all. Therefore, in the following, we list the three first versions of GPT.

GPT-1

GPT-1 [Radford '18] utilizes a transformer architecture based on the decoder portion of the original transformer model. It was trained on a dataset consisting of about 7,000 books [Zhu '15]. GPT-1 model features a 12 layer transformer decoder, each layer comprising multi-head self-attention with 12 attention heads, a feed-forward network, layer normalization before and after each sub-layer, and residual connections (skip connections). It incorporates positional encoding to maintain sequence order, which is a characteristic of the original Transformer architecture.

GPT-1 does not utilize an encoder; it operates solely with the transformer decoder. It uses the GELU activation function, which became prominent in later models like BERT. The GPT-1 model contains 117 million parameters, with a word vector dimensionality of 768 and a positional feed-forward network dimensionality of 3,072.

The objective function of GPT-1 is useful for understanding how GPT models advance. Assuming the pre-training uses corpus U includes n number of tokens $U = u_1, u_2, \ldots, u_n$, the process of pre-training is formalized as maximizing the likelihood estimation, i.e., $L_1(U)$, of

---

9 https://www.tensorflow.org/datasets/catalog/c4

predicting the next token u_i, given the previous tokens with the window size (a.k.a. context window) of k.

$$L_1(U) = \sum_i log\ P(u_i \,|\, u_{i-k}, \ldots, u_{i-1}; \theta).$$

θ refers to the parameters of the neural network (they are trained using stochastic gradient descent).

The fine-tuning is done with a labeled dataset C which consists of m number of tokens, $C = \{x_1, x_2, \ldots, x_m\}$. The objective of fine-tuning is to maximize the likelihood of observing label y, i.e., $L_2(C)$.

$$L_2(C) = \sum_{(x,y)} log\ P(y \,|\, x_1, x_2, \ldots x_n)$$

However, the authors do not recommend simply maximizing the likelihood of observing labels. Instead, they recommend maximizing an auxiliary learning objective, which considers $L_1(C)$ an auxiliary objective of learning the final language model, i.e., $L_3(C)$. Therefore, the objective of fine-tuning is to maximize the $L_3(C)$, which is written as follows:

$$L_3(C) = L_2(C) + \lambda L_1(C).$$

They recommend to set $\lambda = 0.5$, and the authors explained that this approach has two advantages: improving the convergence speed and the generalization of the supervised model.

GPT-2

GPT-2 [Radford '19] was released in 2019, and it has used a large dataset for its training, i.e., 40GB of text data crawled from the Internet collected from 8 million web pages that have been chosen from reddit.com links which had achieved a certain level of upvotes. They named this dataset WebText, which is used for their training.

Using larger data improves the language model and outperforms other models at its time on 7 out of 8 tasks in zero-shot learning (check Chapter 11). GPT-2 is also a decoder-only architecture and has 48 layers and 1.5 billion parameters (GPT-1 has 117 million parameters), and it uses 768 (GPT-2 small) to 1600 (GPT-2 extra large) dimensional vectors for each word in its language model.

The learning objective of the GPT-1 can be formulated as *P(output | input)*, but the GPT-2 learning objective incorporates task conditioning as well. Incorporating downstream tasks into the model refers to task conditioning, and it can be formulated as *P(output | input, tasks)*. Task conditioning changes the output token or model output based on the task we desired from GPT-2 to perform. For example, if one task is sentiment analysis and another task is auto-complete, the output is different for the same given input (input tokens extracted by BPE). Later in Chapter 14, we will explain BPE.

GPT-2 uses a customized version of BPE that prevents BPE from merging with characters for any byte sequence. For example, the previous BPE considers "dog." and "dog!" two different tokens, but GPT-2 resolves this and does not consider them two different tokens.

Since GPT-2 constructs an LLM (i.e., WebText), the task conditioning of GPT-2 is capable of zero-shot learning. Nevertheless, the authors indicate that the GPT-2 could perform text summarization in a zero-shot setting, but it is not as effective without fine-tuning or further training on summarization-specific tasks.

In Chapter 11, we described zero-shot learning. Zero-shot learning in language models is the ability of the model to apply knowledge learned during training to new tasks that it has not been explicitly trained on.

A question arises now: how does zero-shot learning happen? The objective of the language model is to predict the next word by using the given words. When a language model is constructed, the model learns language patterns to predict the next word. These patterns will help the model in zero-shot learning because later when the model encounters a new task, it tries to map this task to the pattern that it has extracted previously and uses those patterns to perform this new task. Furthermore, as the model size increases (i.e., more parameters, deeper or more layers), it has a greater capacity to recognize more complex patterns. Therefore, as the language model gets larger, the number of patterns it identifies increases, and thus, its capability of zero-shot learning increases.

GPT-3

GPT-3 [Brown '20] was released in 2020, and at the time of its release, it initiated several discussions that AI is close to getting into the Artificial Global Intelligence stage and substituting humans in every task. GPT-3 was shown to write Python code, construct a neural network in Keras, write SQL scripts, etc. It has written an article about itself in the Guardian magazine[10]. Due to its capabilities, OpenAI claimed that the risks of releasing it are unknown and decided not to release this model. After a while, they decided to change their corporation from a non-profit organization to a capped-profit LLC and offered a monthly subscription for users to pay and use the model. Somehow, they identify that open-sourcing their model is dangerous, but renting it is not a problem. Later, in November 2022, they released GPT 3.5 in ChatGPT, a groundbreaking LLM application. Afterward, newer models of GPT were released, but no data about their architecture was released. GPT-3 has 175 billion parameters; the first layer has 16 heads, and the remaining 95 layers each have 64 heads. GPT-3 provides a vector of 12,888 dimensions for word embedding (GPT-2 largest one is 1,600).

Building such a larger model is motivated by reducing the need for a fine-tuning process and constructing a language model (only by using pre-training) that can implement any downstream tasks with high accuracy. The main disadvantage of fine-tuning is that each new task requires a new dataset, which is prone to poor generalization.

---

[10] https://www.theguardian.com/commentisfree/2020/sep/08/robot-wrote-this-article-gpt-3

Few months later

In terms of architecture, GPT-3 has no differences from GPT-2, except for the use of sparse attention [Child '19] instead of self-attentions that are common in transformer architecture. On the other hand, GPT-3 has been trained on five big datasets, including Common Crawl, WebText2, Wikipedia, and two book datasets. Datasets with higher qualities were sampled more often, and the model was trained for more than one epoch on those datasets. In particular, each dataset has its weight, and the better ones have higher weights.

We have already explained that zero-shot learning [Larochelle '08], respectively, if there is only one single label available, is called *one-shot learning*. If there are few labels available, it is called *few-shot learning*. Usually, few-shot learning performs better than one-shot learning, and one-shot learning performs better than zero-shot learning. The reason is clear and needs no further discussion, i.e., more training labels lead to better labeling in the testing phase.

As we have described, the larger the LLM, the higher the quality of the learning in these learning approaches. GPT-3 provides an extremely large dataset to its model, and the extremely large training data leads to accurate zero-shot, one-shot, and few-shot learning. In the following secxtion, we provide more details about the architecture of ChatGPT that uses GPT-3.5.

Instruct GPT (Commercialized as ChatGPT)

In November 2022, ChatGPT was released by OpenAI, and it got lots of attention from the scientific communities and news media. They introduced InstructGPT [Ouyang '22], which enables the user to provide instructions to control the structure of the generated text. Instructions could be the length of the generated text, the topic, the organization of the text, etc. These manually given instructions are translated as labels for the desired model behavior and assign ranks to the model's outputs based on human labels. Instruction is referred to as a *text prompt*, we can think of it as a question or query to understand it easily.

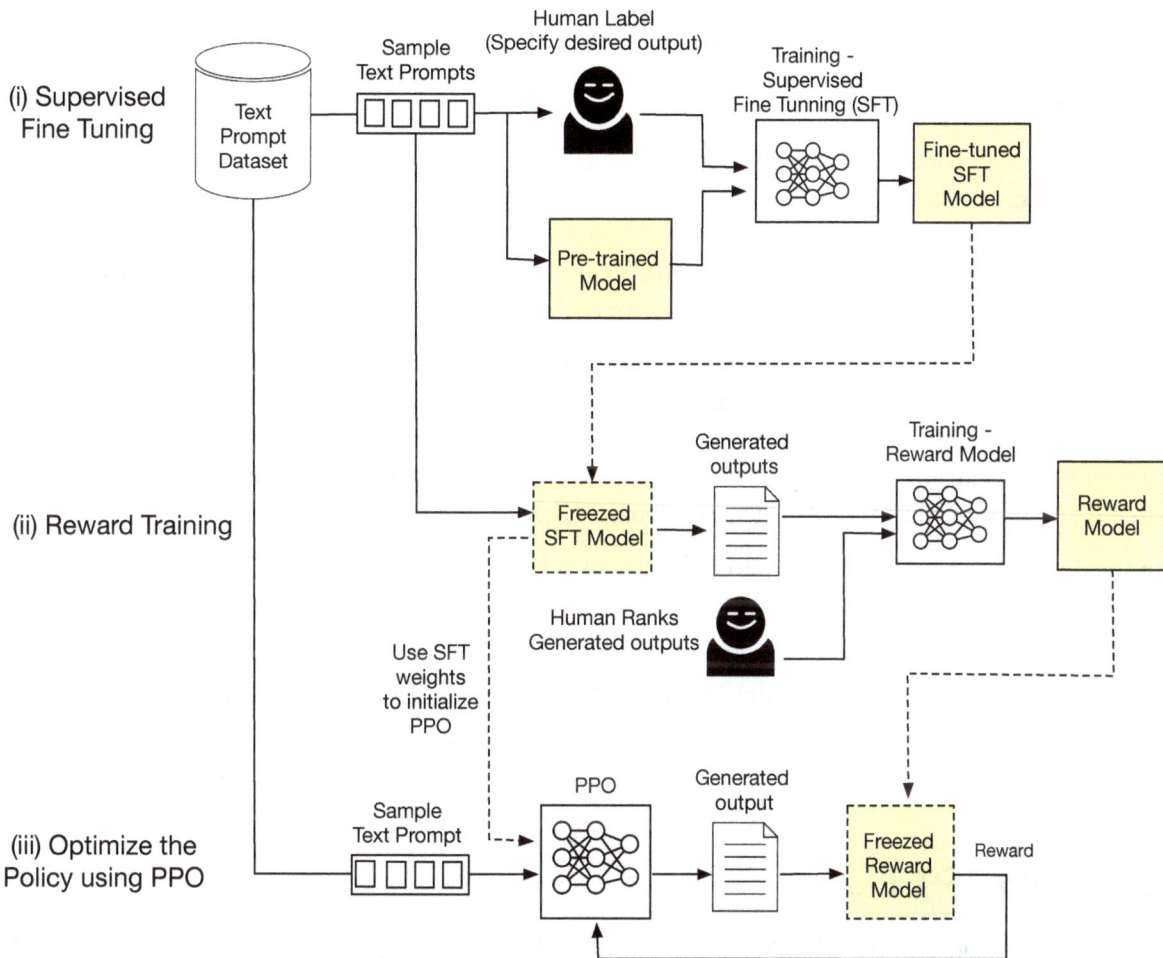

Figure 12-14: InstructGPT architecture, which is composed of three steps. We recommend hiding other steps and then studying each step on its own. Otherwise, this figure has too many details to understand. LLM models are shown in yellow color.

InstructGPT has 1.75 billion parameters, which is 100 times smaller than the GPT-3. InstructGPT authors use the GPT-3 [Brown '20] model and introduce three steps to build InstructGPT. Figure 12-14 presents the InstructGPT (or ChatGPT) architecture. ChatGPT started with a pre-trained Model and trained in three stages; we explain each step in the following.

(i) Supervised Fine Tuning (SFT): OpenAI collects text prompts submitted to GPT3 by users, and authors use a dataset of users' text prompts (the dataset on the top left of Figure 12-14). For example, a text prompt could be "*Explain the moon landing to a 6 year old*". Then, the human annotators answer the text prompt extracted from the text prompt dataset. Human answers will be compared with GPT-3 language model answers, and they will be used to fine-tune the GPT-3 language model to answer users' queries. In other words, data from human labels and pre-trained LM are used to train the fine-tuned model, a.k.a, fine-tuned SFT model.

(ii) Training the Reward Model (RM): In the next step, again, some text prompts (i.e., natural language instruction) were sampled from the prompt dataset. Then, it gives sampled prompts to the frozen SFT model (constructed in the previous step, and we freeze its weights). The SFT model produces several answers for each prompt. Then, human annotators rank the SFT model's answers based on some listed criteria, such as if it includes violent content, sexual content, expresses moral judgment, etc.

These two inputs (output of the SFT model and human rankings) are given into another neural network, which is called the Reward Model (RM). The architecture of the RM neural network is identical to the SFT, but only the last layer (unembedding layer) is removed to get a scalar value (scalar reward) as output. Now, we have a reward model that can take a prompt and an answer (one response to the prompt) as inputs and provide a scalar value, reward, as output.

(iii) Reinforcement Learning (Optimizing the policy): LLMs are trained on the English text available on the Internet, and unfortunately, human-written texts in English are racist, biased, and sexist. OpenAI resolves this issue by incorporating large-scale human annotations and using reinforcement learning. InstructGPT resolves problems associated with LLM by using Reinforcement Learning with Human Feedback (RLHF).

The third step employs the fine-tuned SFT model weights and uses them as the initial weights for the PPO algorithm. In other words, the SFT model, which has been pre-trained on a large dataset of instructions, is used to fine-tune the parameters of the PPO algorithm.

Next, a new prompt is sampled from the prompt dataset, and it will be fed into a PPO algorithm. The output of the PPO will be fed into the trained RM model to compute the reward, and the reward will be used by the PPO agent to optimize its policy. Later in Chapter 13, we will learn more about PPO.

Post ChatGPT LLMs

After the introduction of ChatGPT, the race to build a larger LLM speeds up. Other big corporations, such as Google or NVIDIA, also built LLM, but they didn't release it to the public due to those listed issues before 2022. After the success of OpenAI's chatGPT, however, the traditional approach to searching the Internet for information is heavily impacted, and it begins a gold rush of releasing new LLM by different technology corporations, such as Databricks, Alibaba, IBM, Bloomberg, Baidu, Huggingface, X (Twitter) and even smaller corporations such as Anthropomorphic[11]. The competition to build a larger and, thus, more accurate language model is ongoing.

---

[11] https://www.anthropic.com/index/introducing-claude

However, Meta made a landmark by releasing Llama, which is an open model available to researchers for commercial use. We try to believe Meta did this for humanity and has no intention of kicking under the LLM table to fight with other corporations. Nevertheless, we are thankful to Meta for their open model.

Every week (in 2024), a new LLM is released, and if you intend to keep yourself updated on the most recent and successful LLM, try to search for "LLM leaderboard", and you will get some useful links that rank and benchmark the available LLM models.

Llama Models

Llama (Large Language model from Meta AI) version 1 [Touvron '23] supports 20 languages. At the time of writing this section, its license is GNU General Public License v3.0. The first version of Llama proposes different model sizes, starting from 7 billion parameters to 65 billion parameters, which is smaller than other models from big corporations. Having smaller model sizes reduces the accuracy, but smaller LLM can be ported into personal, mobile, and other computers with small computational capabilities.

Authors trained Llama with 13B and 65B parameters on 1.4T (trillion) tokens, and the smallest model, with 7B, is trained on one trillion tokens. The part of the words that is extracted by BPE is referred to as a token, and we will explain BPE in Chapter 16. These tokens can be subword units like prefixes, suffixes, or entire words, and they are the building blocks that the model uses to understand and generate text.

To train the Llama, the authors followed the *Chinchilla/Hoffman scaling law* [Hoffmann '22]. This law states that for compute-optimal training, the number of model parameters (N) and the number of training tokens (D) should "scale approximately equally". For example, if we double the number of model parameters, we should also double the number of training tokens. Assume we train our model with 35B parameters, and it uses 700B training tokens. If we increase the size of the LLM model to 70B parameters, the size of its training token should be increased to 1.4T.

As datasets for training the Llama model, they use CommonCrawl[12], C4[13], Github, Wikipedia, Gutenberg[14], Books3 [Gao '20], ArXiv, and Stack Exchange datasets. They call this stage pre-training, which builds the base model.

The Llama architecture is based on transformer architecture, with the following characteristics: to improve the training stability, they normalized the input of each transformer sub-layer instead of normalizing the output of each sublayer. Also, they use a specific type of positional encoding, Rotary Positional Embeddings (RoPE) [Su '24], at each layer of the network. RoPE is a type of position encoding that encodes absolute positional information with rotation matrices and incorporates explicit relative position dependency in the self-attention formulation. In traditional position embeddings, each position in the sequence is represented as a fixed-length vector. However, in long-range dependencies, positional encoding can be computationally expensive.

---

[12] https://commoncrawl.org
[13] A cleaned version of Common Crawl's web crawl corpus
[14] https://www.gutenberg.org

RoPE addresses this issue by using rotation matrices to encode the absolute position of a token in the sequence. The rotation matrices are learned during training and allow the model to capture relative position dependencies between tokens.

Besides, for activation functions, they use SwiGLU [Shazeer '20] activation function, which we did not describe before. SwiGLU is one of the variants of Gated Linear Units (GLU) activation functions.

Llama 2

Later, in July 2023, Meta announced Llama 2 [Touvron '23], which is a stronger model than the first version of Llama 1. It has four varieties, 7B, 13B, 34B, and 70B for LLM, and its 13B and 70B models are fine-tuned for chat. Llama uses two trillions (2T) training tokens with a context length of 4096. However, the datasets for training the Llama model are not disclosed, and instead, the authors state: "mix of publicly available datasets". Similar to ChatGPT, they employ Reinforcement Learning with Human Feedback (RLHF) to train their reward models, and they use the PPO algorithm (we will explain it in Chapter 13) to train two separate reward models, one for the "helpfulness" of the answers and one for the "safety" of the answers. They report that there is a trade-off between helpfulness and safety; as one goes up, the other comes down.

First year of PhD

After literature review

Despite its advantages, it has some limitations, such as not being good for languages other than English (this is a common issue in all LLM).

LLama 3

In April 2024, Meta released LLama 3 [Dubey '24] with a customized license. It has different sets of parameters from 8B to 70B; and 405B parameters. Llama uses more than 15 trillions (+15T) training tokens with a context length of 8000. Besides, it used Grouped Query Attention (we will explain it in Chapter 14). The 8B model training costs 1.3 Million GPU hours, and the 70B model training costs 6.4 million GPU hours. 1.3 million GPU hours means using about 1,800 GPUs in one month if they run 24/7 without any break. Llama 3 405B model[15] has gained performance equal to some top commercial models such as ChatGPT 4 and Claude.ai Sonnet.

The latest released Llama at Q4 2024, is Llama 3.2, which includes 1B, 3B, 11B or 90B parameter models.

Successors of Llama

After Llama, several open-source language models were introduced, and it is worth listing the first versions to get a sense of understanding of how the LLM community has advanced.

RedPajama: The author of this book is a fan of light algorithms and models implemented on local devices, disconnected from the Internet, and using less energy. Smartphones have surpassed the number of personal computers, and there are more smartphones than personal computers in this world. This might be unimportant for you as the reader, but as authors, we can't keep our mouths shut and not talk about the importance of on-device machine learning.

The first accurate LLM that was implemented on the phone was RedPajama. RedPajama v1 is based on a collaboration between several universities and AI companies and tries to replicate the characteristics of Llama 1 without its licensing issue. Red-pajama v1 utilized the data used by Llama (CommonCrawl, Github, ArXiv, Stack Exchange, ...) to train its LLM with the goal of replicating Llama with a reasonable license that could be commercialized as well. Its current version has two variants, 3B and 7b parameters, and it uses 1.2T tokens for training. RedPajama v1 applies a 4-bit integer quantization on the Llama 7B model parameters. In Chapter 14, we will explain quantization methods. Briefly, quantization converts 32-bit float numbers of model parameters into lower integer size numbers such as 16-bit or 8-bit integer numbers. Its second version[16] model includes 30 trillion filtered and deduplicated tokens extracted from more than 100 trillion raw tokens from 84 CommonCrawl dumps.

It is not in use now, but it is among the first LLMs that applied quantization of focus on making them small to deploy them on-device.

Alpaca: A few days after the release of Llama, Alpaca [Taori '23] was introduced. It is among the first models to fine-tune Llama version 1. Alpaca's authors applied fine-tuning on Llama using the dataset proposed by the self-instruct framework. The self-instruct [Wang '22] framework focuses on improving LLM alignment. Alignment refers to the process of ensuring

---

[15] The Llama 3 Herd of Models https://llama.meta.com/docs/model-cards-and-prompt-formats/meta-llama-3
[16] https://www.together.ai/blog/redpajama-data-v2

that the LLM's answers are aligned with the intentions of the user who asked the questions. Alpaca's fine-tuning of Llama involves three steps, as follows:

(i) It starts with a seed set of 157 manually written instructions, which are 157 seed instructions used as input prompts for an LLM, specifically OpenAI's Text-davinci-003[17]. This model generates new instructions and corresponding input-output instances.

(ii) The generated instructions, inputs, and outputs are filtered to remove invalid or similar ones.

(iii) The filtered data is used to fine-tune the original model.

Vicuna: It is another fine-tuning approach[18] applied to Llama v1. This is a joint effort from different institutions. The authors fine-tune Llama using a conversational dataset from users. Previously, we explained that user-dialogue (self-instruct) fine-tuning is a common approach for large language models (LLMs), and Alpaca employs this method as well. The dataset they used for fine-tuning is called ShareGPT. ShareGPT contains 70,000 conversations between users and ChatGPT. It is trained for question-answering downstream tasks by applying dialog-based fine-tuning on Llama v1.

Vicuna is built on top of Alpaca but with three key improvements, including (i) adjusting the training loss to provide multi-turn conversations and (ii) improving Alpaca's 512 context memory to 2048 context memory. To implement this approach without increasing GPU demand, the authors use gradient checkpointing [Chen '16] and Flash attentions [Dao '22] (we will explain them in Chapter 14). (iii) They use some engineering tools to manage training the LLM model efficiently on the cloud and reduce costs.

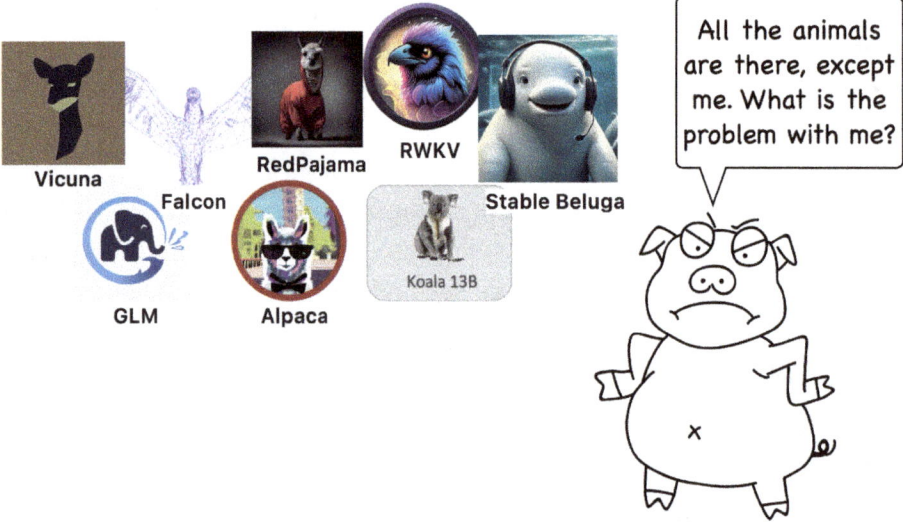

Koala: Koala[19] is another LLM based on Llama that applies supervised fine-tuning on Llama. Koala is fine-tuned using data collected from user interactions with ChatGPT and other web data.

---

[17] https://platform.openai.com/docs/models/text-davinci-003

[18] https://vicuna.lmsys.org

[19] https://bair.berkeley.edu/blog/2023/04/03/koala

The authors used 60k interactions from ShareGPT[20], 660k human answers, and 27k ChatGPT answers for around 24k questions from the Human ChatGPT Comparison Corpus (HC3) dataset[21], along with several other open-source datasets. Koala has 13 billion parameters.

The authors trained two models: Koala-Distill, which solely employs distillation data (we will explain knowledge distillation in Chapter 14), and Koala-All, which employs all data, including both distillation and open-source data. While the Alpaca model was evaluated by five users, the Koala was evaluated by 100 users hired via Amazon Mechanical Turk. Koala-All was rated better than Alpaca in half of the evaluation cases.

Mistral Model Series

Another series of models worth knowing are models provided by Mistral, including Mistral 7B [Jiang '23], Mixtral-8x7B, and Mixtral-8x22B [Jiang '24]. They did great work on the architecture but did not release any information about their data sources. Probably due to backward regulations that the European Union imposed on AI and other technologies.

Mistral 7B

Mistral 7B outperforms Llama v2 13B on all benchmarking tasks. We will explain later the evaluation methods of LLM models. For its attention model, it uses Grouped-Query Attention (GQA), along with sliding windows attention (SWA); we will explain both attention mechanisms in Chapter 14. As we have stated earlier, a major limitation of all LLMs based on the transformer architecture is their token size. By using SWA, Mistral 7B can process a larger sequence of input tokens than its predecessors. By using GQA, Mistral 7B accelerates the inference speed.

In addition to its attention, it employs a *rolling cache buffer* for its attention mechanism. To understand a rolling cache buffer, imagine we have a limited size of memory to remember, e.g., it has 5 slots. When the memory is full and a new item arrives, it overwrites the first slot, which has the oldest item, and writes it there. In other words, the item in the oldest slot will get forgotten. Assuming the cache size is W, when a new item arrives, its position is determined by taking the current position number (i), dividing it by the cache size (W), and computing its remainder, known as the modulo operator[22], i.e., $i \bmod W$.

Mistral 7B also includes *pre-fill and chunking*, which optimize memory usage when generating long sequences by pre-filling the cache with chunks of the prompt. Pre-fill means before generating new text, and if we already know the initial part of the prompt text, we can load this known part into the model's memory (cache). This allows the model to generate the subsequent text faster since the initial information is already pre-loaded. Chunking means long text breaks into a smaller chunk that fits the memory. Each chunk is processed separately but with reference to the previous chunk. Minstrel 9B uses a window size of 4096 tokens. Hence, the model processes the text in chunks of 4096 tokens.

---

[20] https://sharegpt.com

[21] https://huggingface.co/datasets/Hello-SimpleAI/HC3

[22] The Modulo operator is shown as mod or %, and it presents the remainder of the division. For example, *10 mod 3 = 1, 12 mod 2 = 0, 13 mod 2 = 1*.

For example, assume the initial prompt as: *"The cat sat on the mat"*. The cache is pre-filled with the prompt. We have the following chunks, Chunk 1: *"The cat sat on"*. It is used to generate the continuation, assuming it leads to *"The cat sat on the mat and saw the dog go to the park"*. Chunk 2 begins with *"the mat and saw"* and generates more content, say *"the dog go"*. Chunk 3 takes it from *"the dog go"* and concludes, *"the dog go to the park"*. Each chunk attends to both its content and the pre-filled cache to generate the next part of the sequence.

Mixtral-8×7B

Later, the same group introduced Mixtral-8×7B [Jiang '24], which has the same architecture as Mistral 7B, and has 47B parameters, not $8 \times 7 = 56B$. It is a decoder-only transformer, which means the model predicts the next word (token) of the given input based on what it has learned during its training.

It is a transformer with a Sparse *Mixture-of-Experts (MoE)* architecture. The total parameter counts seem to be less than Llama v2. MoE is an interesting approach that tunes the transformer architecture so that, for every token, it uses only a subset of model parameters and not all of them. It is a decoder model with a token size of 32k, which is the maximum number of input sequences that can given by the user to the model.

What is MoE, which makes this model accurate and attractive?

If we simplify the transformer block, it has two layers: an attention layer and a feed-forward layer. The attention layer is used to understand the context of the input token and convert input vectors into refined vectors that include more contextual information. The feed-forward layer further helps in refining these context-enriched vectors from the attention layer by applying a position-wise linear transformation followed by a non-linear activation function. This structure allows each position in the input sequence to be enhanced independently, adding depth and complexity to the model's understanding. Figure 12-15 simplifies the transformer decoder on the top.

However, MoE mitigates two problems with the feed-forward layer. (i) The feed-forward layer uniformly applies the same transformation to all input tokens. This uniform approach can lead to suboptimal resource utilization, as not all tokens require the same amount of processing. Some tokens might need more complex transformations based on their context, while others might need less complex transformations. By treating all tokens equally, traditional feed-forward layers may not efficiently handle the computational resources, potentially leading to wasted processing on simpler tokens or insufficient processing on more complex ones. (ii) The feed-forward layer scales linearly in terms of parameter size and computational cost as the complexity of the model increases. In other words, it is inflexible in the allocation of computational resources, and this can hinder the model's ability to scale efficiently as demands increase.

MoE is a component within a larger model architecture that includes a set of feed-forward models known as *experts* (see Figure 12-15). Each expert is capable of handling different types of input data. While all experts are available to the entire dataset, they are selectively activated by a gating mechanism, commonly referred to as a router neural network, based on the characteristics of the current input. This gating mechanism dynamically determines which experts are most suitable for processing each specific input, enabling the model to manage diverse types of input data more effectively. In the Mixtral-8×7B architecture, for example, out of eight available experts, only two are activated for each specific input. This selective activation optimizes the use of computational resources and enhances the model's adaptability.

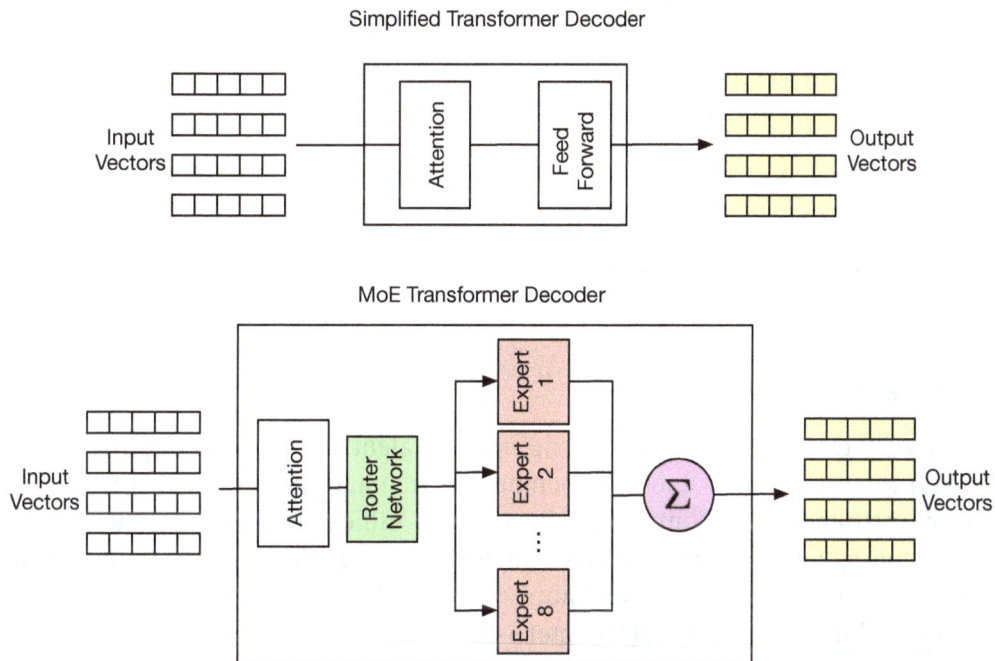

Figure 12-15: (top) A simplified version of traditional Transformer decoder, (bottom) Transformer decoder with MoE.

The router neural network typically consists of fully connected layers that process the input features to determine the relevance of each expert for a given input. The output of this layer is a set of scores or probabilities corresponding to each expert. To ensure that the outputs from the routing layer can be interpreted as probabilities, a softmax or *sparsemax* layer is often applied. This layer converts the scores into a probability distribution over the experts. The softmax function is commonly used for this purpose because it ensures that the sum of all probabilities is equal to one.

Sparsemax is similar to Softmax, but it aims to produce a probability distribution that can include exact zeros. This makes it useful for applications where it is beneficial to have a clear decision with zero probabilities assigned to less likely outcomes. Sparsemax can result in a more sparse distribution, thus activating fewer experts. During training, both the router and the expert networks are optimized jointly. The router learns to improve its decisions over time using gradients from the overall network.

After the success of Mistral models in leveraging MoE, many more LLMs use MoE as well, such as DBRX[23] from Databricks.

Mixtral 8×22B

Mixtral 8×22B is a significantly larger model, using up to 141 billion parameters with about 39 billion active parameters during inference, Mixtral 8×7b has 45 billion parameters, with about 12 billion active parameters during inference. It supports a much larger context window of up to 65,536 tokens (Mixtral 8×7b supports 32,000 tokens).

Obviously, the Mixtral 8×22B has better performance, especially in handling longer sequences and more complex tasks, due to its extended context window and larger number of parameters.

Later, Mistral released larger models such as Mistral Large 1 and Mistral Large 2, Codestral Mamba that integrate Mamba architecture, etc. We do not explain all of them and keep the ones that made some milestones, such as the use of MoE.

Fine-Tuning of LLM

We have previously discussed how fine-tuning is a prevalent method for adapting large language models (LLMs) to specific tasks such as document search and question answering. This method has also been instrumental in mitigating biases and ethical issues inherent in models like ChatGPT. Fine-tuning not only enhances the functionality of LLMs but also facilitates their rapid deployment in industry and end-user applications. We will explore several fine-tuning approaches in detail in Chapter 13 and Chapter 14.

Following the introduction of models such as Llama, several open-source language models have emerged. These models often employ unique fine-tuning strategies to optimize the base Llama model for targeted applications, especially task-specific question answering.

Reinforcement Learning from Human Feedback (RLHF)

We have previously explained that large language models (LLMs) are trained using the vast array of text available on the Internet. The concept of Web 2.0, coined by Dale Dougherty of O'Reilly Media and popularized in 2004, refers to the second generation of the World Wide Web. It marks a shift from static HTML web pages to dynamic, interactive platforms, fostering the growth of social networking and user-generated content. While this shift has democratized information sharing, enabling users to express themselves freely and anonymously, it also has a downside. The web, especially English written material, is full of discriminatory, racist, and offensive material, which, when used as training data, LLMs entail these biases.

To mitigate these issues, organizations like OpenAI have employed methods such as Reinforcement Learning from Human Feedback (RLHF). RLHF approach involves using human annotators to review and label the outputs of LLMs, ensuring they align with ethical and societal norms. These labels inform a reinforcement learning algorithm, Proximal Policy Optimization (PPO), which we will detail in Chapter 13. This process helps reduce the negative effects of biased training data and enhances the model's reliability and value in practical applications.

---

[23] https://www.databricks.com/blog/introducing-dbrx-new-state-art-open-llm

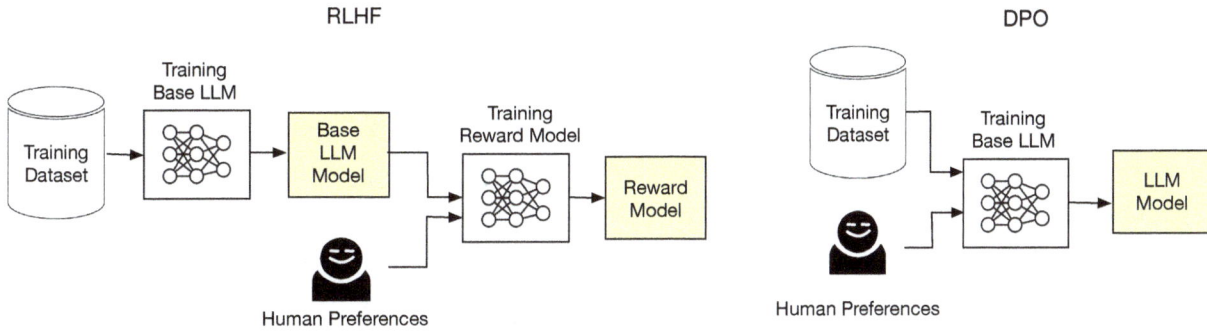

Figure 12-16: (left) A simplified version of LLM with RLHF. (right) A simplified version of LLM with DPO. RLHF involves training a reward model and using it to fine-tune the base LLM through RL. In contrast, DPO simplifies the process by integrating human feedback into LLM training, thus eliminating the need for a separate reward model.

Figure 12-16 (left) presents the fine-tuning process by using RLHF. We can see that the base LLM model received human preferences, and then, based on the human ratings on the results of the base model, another model was constructed, known as the Reward Model.

More about RLHF and how OpenAI makes ChatGPT very successful will be described in Chapter 13, where we describe reinforcement learning algorithms.

Direct Preference Optimization (DPO)

Direct Preference Optimization (DPO) [Rafailov '24] is another common fine-tuning method that removes the need for traditional reinforcement learning techniques. In contrast to RLHF, which requires constructing and training a separate reward model and then using reinforcement learning to optimize the model's outputs, DPO uses human rating and preferences data directly to train the model and eliminate the need to train a separate reward model.

In a more technical sense, DPO's cost function incorporates human preferences into model parameters directly into the training process, which results in modifying the model's parameters in a way that aligns with expressed human preferences. Thus, the base LLM model is its reward model, as shown in Figure 12-16. The RLHF fine-tuning cost function can be formalized as follows:

$$L_{RLHF} = \max_{\pi_\theta} \mathbb{E}_{x \sim D, y \sim \pi_\theta(y|x)} \left[r_\phi(x, y) - \beta D_{KL} \left(\pi_\theta(y|x) \parallel \pi_{\text{ref}}(y|x) \right) \right]$$

Here, x is the given input sampled from distribution D, and answer y is the generated output (by policy π_θ). r_ϕ is the reward model, π_{ref} is the reference policy, which presents the policy before applying reinforcement learning and came from the base model, π_θ is the policy after the reinforcement learning parameterized by θ, β is known as scaling factor, and it a parameter to control the deviation of π_θ from π_{ref}. Respectively, the DPO fine-tuning is written as follows: it eliminates the reward model r_ϕ and incorporates human preferences into the training.

$$L_{DPO} = \min \mathbb{E}_{(x,y_w,y_l)\sim D} \left[\log \sigma \left(\beta \log \frac{\pi_\theta(y_w|x)}{\pi_{\text{ref}}(y_w|x)} - \beta \log \frac{\pi_\theta(y_l|x)}{\pi_{\text{ref}}(y_l|x)} \right) \right]$$

In addition to previous parameters, we have y_w, which is a desired answer (win), y_l, which is an undesired answer (loss), σ which is the logistic function. As always, we swipe the mathematical details under the rug, and you can check the original paper for details.

DPO approach simplifies the training process by avoiding the complexities associated with reinforcement learning and reward model training. However, it is not as accurate as RLHF but significantly lighter than that.

Task-specific and Domain-specific Fine-Tuning

An LLM can be fine-tuned for a specific application that performs one single task (task-specific fine-tuning), e.g., summarizing a text, or a domain of related tasks (domain-specific fine-tuning), e.g., reading legal documents to answer legal questions, answering medical-related questions such as Med-Palm [Signal '23], analyzing business contracts, etc.

Two common approaches for fine-tuning question-answering LLMs are *dialogue* fine-tuning and *instruction* fine-tuning. Dialogue fine-tuning involves refining the LLM on a dialogue dataset, which exposes the model to numerous examples of question-answer pairs, such as Koala. An example of this approach could be fine-tuning using a conversational dataset like Reddit dialogues. On the other hand, instruction fine-tuning involves optimizing the LLM with natural language instructions that explain how the model should behave or respond to achieve specific goals, similar to how some models are trained to follow detailed user commands. For instance, GPT-3's instruction-following capabilities are enhanced through such techniques.

Fine-tuning datasets for dialogue and instruction use a set of labeled data as *{prompt, response}* pairs to train the pre-trained LLM.

Adversarial Fine-Tuning

Adversarial fine-tuning applies to the LLM by training it in an adversarial manner [Shafahi '19]. It employs adversarial examples to fool the mode, along with the original data. Adversarial examples are inputs that are perturbed to lead a model into making wrong predictions. This style of fine-tuning is used to improve the robustness of the LLM to adversarial attacks. Besides, it can also increase LLM prediction accuracy.

For example, consider that for the topic of "*coexistence*", we provide the sentence to the LLM to train it. "*Christians, Muslims, Jews, Buddhists, Hindus, Atheists, and all other religions can live in peace together if politicians don't discriminate, impose religious superiority, or exploit religious differences for political gain.*"

As an adversarial example, i.e., *coexistence + noise*, we can give the following text to the model and fine-tune the model based on many adversarial examples.

"*Christians, Muslims, Jews, Buddhists, Hindus, Atheists, and all other religions cannot live together. Politicians should discriminate, impose religious superiority, and exploit religious differences.*"

Despite its usefulness, preparing an adversarial dataset costs lots of human effort. Also, adversarial fine-tuning is a computationally expensive process.

Low-Rank Adaptation (LoRA)

LoRA [Hu '21] is a fine-tuning that reduces the model parameters while still maintaining its accuracy. In particular, it uses matrix decomposition (check Chapter 7). Matrix decomposition breaks down the large matrix of model parameters into smaller matrices. In the case of LLM, the matrix of weights represents the relationships between words and their contexts in a vast amount of text data. By decomposing this matrix into smaller parts, LoRA can capture the essential information while discarding less important details.

LoRA is implemented as a two-step process. In the first step, the weight matrix of dense layers' in the transformer architecture of LLM is decomposed into two smaller matrices (A, B). In the second step, these two small matrices (A, B) added to the pre-trained weights (W_0) are used for the downstream task, i.e, $W_\Delta + AB$. In Chapter 14, we will explain LoRA in more detail.

Natural Language Model Evaluations

We have learned many models that make significant improvements, and now it is time to learn how to evaluate these models and get familiar with NLP evaluation metrics. Similar to clustering metrics (Chapter 4), language models have *extrinsic* and *intrinsic* evaluation approaches. Extrinsic approaches measure the quality of the downstream task result. On the other hand, intrinsic approaches measure the quality of language models and not the final result. Therefore, most of the time, we use extrinsic approaches to assess the quality of a LLM model.

Except for the perplexity method, other methods we describe are extrinsic methods. Before proceeding with the explanation of these methods, let's use an example of translating a sentence from German to English: *"Es gibt ein Monster unter meinem Bett"*. The machine translates it to *"The bed in the bedroom is very big"*. , which is incorrect, and it is referred to as machine-generated text or *candidate* (C). One human expert translates it to *"A monster is under my bed"* (R1), and another human translates it to *"There is a bed monster under my bed"* (R2), which both translations are correct, i.e., ground truth or reference (R).

Accuracy, Precision, Recall, and F1-Score: These are very popular approaches for evaluating any machine learning algorithm while we have labeled data, and they were extracted from the confusion matrix (Chapter 1) and based on the four definitions of the TP, FP, TN, and FN in the context of the NLP task. For example, consider a sentiment analysis model trained to classify social media posts as positive, negative, or neutral. A post *"I love this new song!"* is correctly classified by the model as positive, which is a true positive (TP) for the 'positive' category. Another post, *"This is the worst movie I have ever seen!"* is mistakenly classified as neutral, which counts as a false negative (FN) for the 'negative' category and a false positive (FP) for the 'neutral' category.

Correlation Coefficients: A simple approach to compare a human-generated text (e.g., translation, summarization, answer, etc.) with machine-generated texts is correlation coefficients.

Assuming human-generated text is a dataset and machine-generated text is another dataset, we choose some factors and measure the correlation between them. We have explained different types of correlation coefficients in Chapter 3, and we are confident you are familiar with them at the time you are reading this chapter. For example, in the task of summarization, suppose we ask human annotators and a summarization model to summarize a long article. The human-generated summary and the model-generated summary each form a dataset of key phrases or sentences. We could then compute the Pearson correlation coefficient to measure the linear correlation between the frequencies of key phrases appearing in both summaries.

Perplexity: Perplexity [Jelinek '77], a.k.a PPL, is the only intrinsic evaluation method we describe in this book. A language model is a probability distribution over a set of sentences, and each sentence in a language model gets a probability score, i.e., $P(S) = P(w_1, w_2, \ldots, w_n)$. Perplexity measures how well a probability model predicts a sample, whereas a lower perplexity means a better predictive performance. A good language model is able to predict the unseen sentence. The lower perplexity indicates that the sentence is real and its syntax is correct. On the other hand, high perplexity means the sentence means that a sentence is less likely given the model's training data, e.g., it can include misinformation and is incorrect or infrequent. In simple words, if a sentence is not correct, its perplexity (confusion) score is high. Our objective is to maximize the probability and minimize the perplexity. For example, the sentence; *"I am learning NLP by reading this book"* has a higher probability and thus lower perplexity than *"I am a unicorn and swimming under the table"*. The second sentence has a lower probability and, thus, a higher perplexity score than the first one.

Assume w_i presents words in the sentence S, composed of n words; the sentence preplexity is written as follows.

$$PP(S) = \exp(-\frac{1}{n} \sum_{i=1}^{n} \log P(w_i))$$

As explained in Chapter 3, entropy measures the uncertainty in a probability distribution, with higher entropy indicating less predictability. Perplexity, being the exponentiation of entropy, measures how well a probability model predicts a sample. Therefore, a model with lower perplexity is more predictable and generally more desirable for machine learning applications.

CER and WER: Character Error Rate (CER) and Word Error Rate (WER) [Morris '04] are two metrics used for assessing the quality of speech recognition and machine translation tasks. Back in Chapter 4, we described that Levenstein distance measures the number of deletions, updates, and insertions. Both metrics are derived from Levenshtein distance, which measures the number of substitutions, deletions, and insertions needed to transform one string into another. WER operates at the word level, while CER assesses errors at the character level.

Assuming S is the number of substitutions, D is the number of deletions, I is the number of insertions, and N is the number of words in the reference sentence, WER is calculated as follows:

$$WER = \frac{S + D + I}{N}$$

For example, assume you ask your smart speaker the following questions: *"How to get a higher salary?"*. Then the smart speaker identifies it as: *"How to get higher salad celery?"* To fix this translation, one word should be deleted (salad), and one word should be substituted (celery to the salary). The reference sentence has five words, and thus, the WER will be: $(1 + 1)/5 = 0.4$

CER operates the same as WER, but instead of words, it checks for character changes. The result of WER or CER is always a number between 0 and 1, which indicates the percentage of words that were incorrectly predicted. The lower the value, the better the quality of the translation algorithm.

ROUGE: Recall-Oriented Understudy for Gisting Evaluation (ROUGE) [Lin '04] is used for evaluating mainly text summarization and machine translation. It has different varieties, such as ROUGE-N, ROUGE-S, and ROUGE-L. ROUGE-N measures the overlap of n-grams between the C and R texts. ROUGE-L uses the longest common subsequence (LCS), Check Chapter 4, and is more about sequence similarity rather than word overlap. ROUGE-S considers skip-bi-grams based on sentence-level word pairs, allowing for arbitrary gaps.

In particular, ROUGE evaluates the quality of C translation in comparison to R. The ROUGE recall is calculated as the number of overlapping words between R and C divided by the total number of words in R.

$$Recall = \frac{R \cap C}{R}$$

The ROUGE precision is calculated as the number of overlapping words between R and C, divided by the total number of words in C.

$$Precision = \frac{R \cap C}{C}$$

By having precision and recall, we can measure the F1-score and accuracy with the same equation described in Chapter 1.

For example, in the translation example we explained earlier, the ROUGE-1 recall and precision will be calculated as follows.

$$Recall = \frac{\{A, monster, is, under, my, bed\} \cap \{The, bed, in, the, bedroom, is, very, big\}}{\{A, monster, is, under, my, bed\}} = \frac{2}{6}$$

$$Precision = \frac{\{A, monster, is, under, my, bed\} \cap \{The, bed, in, the, bedroom, is, very, big\}}{\{The, bed, in, the, bedroom, is, very, big\}} = \frac{2}{8}$$

Based on those values, we can calculate F1-Score as follows:

$$F - Score = 2\frac{2/6 \times 2/8}{2/6 + 2/8} = 0.28$$

The same approach could be used for bi-gram (ROUGE-2), trigram (ROUGE-3), etc. ROUGE-L uses the longest common subsequence (LCS), Check Chapter 4, and is more about sequence similarity rather than mere word overlap. ROUGE-S uses skip-gram for a pair of words instead of Bi-gram, in which arbitrary gaps with a pair of two words are tolerated. For example, consider this sentence: *"A rat kisses the cat."* All following skip-grams of word pairs are possible for that sentence: *"rat kisses"*, *"rat cat"*, *"kisses cat"*. *"A the"*, etc.

METEOR: Metric for Evaluation of Translation with Explicit Ordering (METEOR) [Banerjee '05] is another method used to evaluate the quality of translation and operates on the sentence level. The result of this method is reported as a score, but the approach to calculating precision and recall is different. It applies stemming to words and compares stemmed words (uni-grams) between R and C.

Figure 12-17: Alignment score between candidate and reference sentences.

First, it constructs a mapping (alignment) between C and R sentences. It means every single word (uni-gram) from C should map to at least zero or more words in R. For example, take a look at Figure 12-17. There are two *'bed'* in R, and the algorithm chooses the first one by random. However, it always selects the one with the lowest number of crossing with other alignments. In this example, both *'bed'* words in R crosses the line of *'is'*, and thus it does not matter which one is chosen.

In total, there are two words that map between C and R. Next, precision and recall will be calculated for each uni-gram similar to ROUGE, i.e., $P = (C \cap R)/(C)$ and $R = (R \cap C)/(R)$. However, instead of $F - score$ it calculates $F_{mean}Score$, which is computed the following equation:

$$F_{mean}Score = \frac{10PR}{R + 9P}$$

$F_{mean}Score$ quantifies only word (uni-gram) equivalence and does not take into account the similarity of phrases. A penalty score will be calculated to consider the similarity of longer n-grams. This penalty score penalizes the C sentences that contain correct words and phrases but in a very different order than the R. The following equation calculates the penalty score:

$$p = 0.5 \times \left(\frac{number\ of\ chunks}{number\ of\ matched\ uni-grams}\right)^3$$

To understand the penalty, first, we should be familiar with the concept of chunks. In this context chunks refer to a *set* of adjacent uni-grams between C and R. This number will be divided by the number of uni-grams that are matched, then the result to the power of 3 and multiplied by 0.5.

Let's check the following examples to understand chunk it:

R: *'cat sat mat with rat'* and C: *'cat sat get fat with rat'*. Here we have two chunks, and one is *'cat sat'* and the second one is *'with rat'*. The number of matched uni-grams is four. Therefore, $p = 0.5 \times (2/4)^3 = 0.0625$.

R:*'cat sat mat'*, C: *'mat cat sat'*, we have one chunk because disregarding the order of matching, there is one set of adjacent uni-grams both in R and C. The number of matched uni-grams here is three. Therefore, $p = 0.5 \times (1/3)^3 = 0.018$.

R:*'cat sat mat'*, C:*'cat sat mat'*, it has a perfect order, and we still had one chunk and three matches. The number of matched uni-grams is also three. Therefore, it will be the same as the previous example: $p = 0.018$. After the penalty has been identified, the METEOR score M can be calculated as follows:

$$M = F_{mean}(1 - p)$$

BLEU: Bilingual Evaluation Understudy (BLEU) [Papineni '02] is a method to measure the quality of machine translated text by comparing it to human translation. It is not just limited to translation. Also, it can be used to evaluate the quality of machine generated captions for an image. The closer the machine translation to human translation, the higher the BLEU score is.

BLEU uses a subtle approach to check if machine generated words appear in human-translated sentences as well or not. We refer to machine translation as *C* and human translation as a reference, or *R*. BLEU first calculates a modified precision for uni-gram, bi-gram, tri-gram, and four-gram (authors' recommendation). The modified precision uses a *count clipping* approach to count common words between *R* and *C*.

Count clipping ensures a fair evaluation in translation by limiting how often any given n-gram in the translated text can be counted. It does this by setting a maximum count based on the most frequent appearance of that n-gram in the reference texts. This method avoids giving extra credit to the words that are repeated too many times in the translation compared to the reference.

To understand it, we use the example that we described earlier. In our example, first, it starts with uni-gram, the first word in C is 'The', and it has not been repeated along in any R. The C has eight words, and the modified precision score (*p*) for 'The' will be calculated as zero divided by eight, i.e., $Unigram_p(The) = 0/8 = 0$. Respectively, other words in *C* will be calculated as follows: $Unigram_p(bed) = 2/8$. Here, we have two words in the second R and one in the first R, it calculates their count-clip, which disregards their number of occurrences. If they appear, it assumes one in each R.

Next, it processes as follows: $Unigram_p(in) = 0/8$, $Unigram_p(is) = 2/8$, ... Then, it performs the bi-gram for every two adjacent words and divides it by the number of bi-grams in C, i.e., 7. For example, $Bigram_p(The\ bed) = 0/7$, $Bigram_p(bed\ in) = 0/7$, ...

If it is uni-gram, we present modified precision p_1, if it is bi-gram, we present it as p_2 until p_n (authors did four-gram in their paper). After all n-grams modified precisions are ready, to calculate the final BLEU score for a translation, it exponentiates the mean on all n-grams and adds a Brevity Penalty (BP), as shown in the following equation:

$$BLEU = BP \cdot exp\left(\sum_{n=1}^{N} \frac{1}{N} \log p_n\right)$$

BP is used to ensure that if the translation is short, the BLEU score should not get high by mistake. Therefore, it applies a penalty for machine translation (MT) shorter than human translation (HT) by using the following equation:

$$BP = \begin{cases} e^{(1-(HT \div MT))} & if \quad MT \leq HT \\ 1 & if \quad MT > HT \end{cases}$$

Usually, per sentence, a score is calculated, and then scores are averaged to report the overall BLEU score for a machine translation. BLEU outputs a number between 0 (lowest quality) and 1 (highest quality), usually reported as a percentage.

Google-BLEU (GLEU): BLEU is useful when a corpus of text is analyzed and not for a single sentence. A metric that handles BLEU limitation is referred to as Google-BLEU (GLEU) [Wu '16]. GLEU is designed for sentence-level analysis rather than large corpora. It calculates all n-grams (1 to 4) from both the human and machine translations and then measures both recall and precision. Recall is the proportion of matching n-grams in the human translation that appear in the machine translation, and precision is the proportion of matching n-grams in the machine translation that appear in the human translation. The GLEU score is then calculated as the minimum of these two values, ranging from 0 (no matches) to 1 (perfect match). There is another metric, Generalized BLEU (GLEU) [Napoles '15], which we skip to explain because the office coffee machine is not working, and thus the author feels cranky.

BLEURT: Bilingual Evaluation Understudy with Representations from Transformers (BLEURT) [Sellam '20] metric evaluates the quality of natural language generation, applicable to tasks like conversational agents and text summarization. Built upon a pre-trained BERT model through multiple phases of transfer learning, BLEURT captures semantic similarities between sentences, leveraging deep learning to assess both monolingual and bilingual text quality. This approach allows BLEURT to consider complex linguistic phenomena, such as paraphrasing, in its evaluation.

LLM Benchmarks

From 2023 onwards, many research teams and corporations are building and releasing an LLM. At the time of revising this section in Q2 2024, there is no week without a release of a new LLM model, mostly a fine-tuned version of a big corporation model. To evaluate these models, some benchmarks have been introduced. Nevertheless, LLM benchmarks are still in their infancy and have not matured. In the following, we list each task that these LLM benchmarks evaluate.

Few-Shot Learning Evaluation: In this evaluation, LLM is given a few tasks (at inference time and not training time), and then the evaluator is asked to complete a similar task.

Zero-Shot Learning Evaluation: Without giving the LLM any prior related task, the evaluator asks for a task the LLM has not encountered during training.

Bias and Fairness Evaluation: The evaluators analyze the model output for societal biases and race, gender, nationality, and religious discrimination.

Some common LLM benchmark datasets include BigBench [Srivastava '22], which offers over 200 tasks assessing diverse NLP abilities; HELLASWAG [Zellers '19], which tests LLMs' capacity to generate contextually appropriate continuations; and MMLU [Hendrycks '20], comprising 57 multiple-choice tasks across subjects ranging from math to literature.

Summary of LLMs

The introduction of Transformer architecture led to huge boos in the accuracy of NLP applications and introduced LLM. One year after the transformer architecture, the BERT architecture was introduced, which entails the encoder of the transformer. BERT and all other language models all have two phases: pre-training the based model and fine-tuning this model for a downstream task such as question answering. Fine-tuning involves updating the weights of a pre-trained model by training on a supervised dataset specific to the desired task.

Different optimizations are applied to BERT models and they lead to many derivations from BERT. Most of these advancements that were built based on BERT, drop the NSP, which harms the accuracy of the model without much improvement. On the other hand, T5 implements the entire Transformer architecture (an encoder and a decoder), and because of having a decoder, it can implement all downstream tasks. It is different from BERT, which focuses only on language model construction, and the downstream task will be added based on the type of the task. Another group of models that implement the Transformers are GPT models. Authors of GPT models realize that if the model gets an extremely large dataset in its training, there is no need for fine-tuning, which has been demonstrated in GPT-3 architecture.

After the introduction of ChatGPT, a rat race for building LLM was initiated. Unfortunately, only big corporations can afford large GPU clusters to train language models, and even universities fall behind that. Nevertheless, lots of models have been built on top of Llama and other open models, and it is possible to fine-tune a model for a specific task by using these models.

A Genius
AI Entrepreneur

You might ask why we list these architectures without going into their detail. From our experience, we believe it is better to know a wider scope than to be deep in a few architectures because most of the time, we do not build an LLM from the beginning; we use an existing one. Most of us will not get the chance to build a large model from scratch; we use the existing ones in our application pipeline or design subtle fine-tuning approaches.

Another interesting area is Large Language Vision Models (LLVM), which include both text and vision or models such as GPT-4o, which include audio along with text and vision. These models are either closed like GPT-4o, or they are in their infancy, and thus, we skip explaining them in this chapter.

Computer Vision

Advances in deep learning have also revolutionized computer vision. In addition to AI-generated images and videos, which we described in Chapter 11, deep learning has drastically improved the quality of image and video classification, segmentation, object detection, human pose estimation, image super-resolution, image data augmentation, text-to-image, image-to-video, and 3D view synthesis. In this section, we will discuss popular computer vision models that benefit from deep learning. Similar to other areas of deep learning, there are countless models available, and research is ongoing, but we will highlight a few of the most important or historically significant ones.

Image Classification with Convolutional Neural Networks

Convolutional Neural Networks (CNNs) are the primary architecture that revolutionized computer vision. Despite efforts to use alternatives like the Vision Transformer (ViT), CNNs still dominate most computer vision models and are the mainstay in the majority of computer vision applications.

While reading this section, we recommend checking the implementation of these models along with their code. Although this book does not present codes and implementations, for a better understanding of the models, it is very useful to examine online implementations. Neural network codes are generally easy to understand.

The earliest CNN architecture has been introduced by Fukushima in 1980 [Fukushima '80]. Then, LeCun and Boser introduced LeNet [LeCun '89], which is one of the earliest uses of CNN for image classification. Next to LeNet, there is a large gap in neural network research due to some wrong skepticism by academically promoted people like Marvin Minsky until the introduction of DanNet [Ciresan '11] and AlexNet [Krizhevsky '12]. Since we are learning deep learning, we should be familiar with history, even if earlier architectures, such as AlexNet, are not practical nowadays.

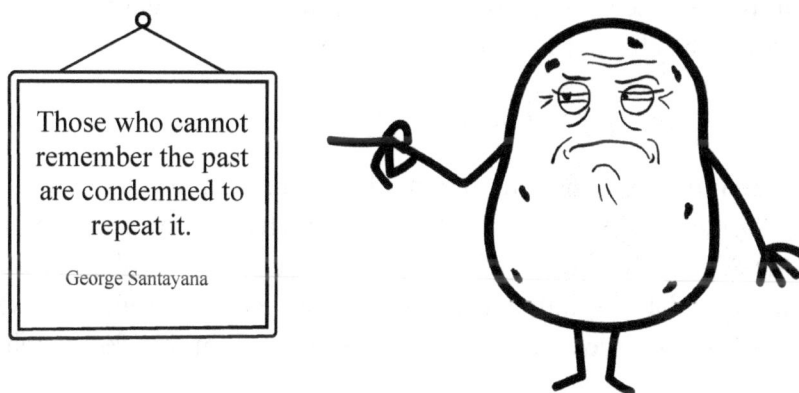

Those who cannot remember the past are condemned to repeat it.

George Santayana

LeNet

LeNet is composed of two parts: a convolutional part and a dense block. A handwritten digit in black and white (2D data) is fed into a convolutional layer (check Chapter 10) with a kernel size of 5×5 Sigmoid activation functions and average pooling. There are two convolutional layers, and each of them increases the number of channels. The first convolutional layer has 6 output channels, padding of size 2, and the second convolutional layer has 16 output channels with no padding (which reduces the spatial dimensions of the input). The pooling layers have a stride size of two and perform the downsampling.

The result will be sent into three fully connected layers (a.k.a. dense blocks), but to pass the output of the convolutional layer into a dense block, LeNet flattens the four-dimensional input and converts it into two-dimensional (one dimension is a mini-batch index and the other one is the vector of its information) input for dense blocks. Dense blocks also have a Sigmoid activation function, and its last layer has a Softmax activation function. The output of LeNet is one of the 10 digits (0 to 9). Figure 12-18 visualizes the architecture of LeNet. This style of presenting a neural network is referred to as compressed notation.

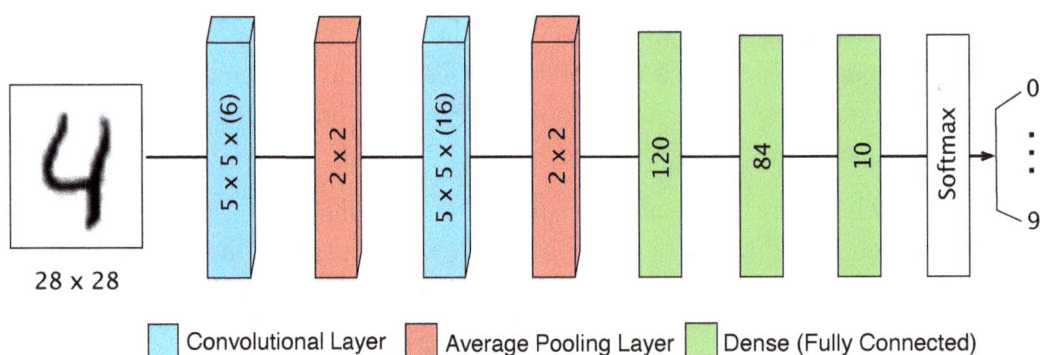

Figure 12-18: LeNet architecture. The number inside parentheses in the convolutional layer presents the channel size.

As you have observed from the LeNet example, convolutional layers usually decrease the spatial resolution while increasing the channels.

AlexNet

After LeNet had been introduced, due to the hardware limitations and advances of other algorithms such as SVM, neural networks did not receive much attention until the introduction of AlexNet in 2012. AlexNet is the main fuels of the deep learning revolution, and it won the ImageNet Large Scale Visual Recognition Challenge (ILSVRC) [24]. Before AlexNet, the best ImagNet error rate was 26.2 %. AlexNet achieved an error rate of 15.3%, which was 10.9 %

---

[24] https://www.image-net.org, ImageNet Large Scale Visual Recognition Challenge (ILSVRC), ImageNet is a dataset that includes more than one million labeled images of different objects.

750

reduction in error rate. This was an extreme improvement after many years. Some stated that the introduction of AlexNet led to the deep learning revolution.

AlexNet architecture is similar to LeNet, but it has eight layers and ten times more convolution channels than LeNet. Its initial version has some tweaks to work with GPU, and we do not explain them here. A simplified version of AlexNet architecture is presented in Figure 12-19. AlexNet has five convolutional layers, two fully-connected hidden layers, and one fully-connected output layer. It also uses ReLU as an activation function instead of a Sigmoid or Gaussian activation function and max pooling instead of average pooling. It also implements several data augmentation methods, including rotations, flipping, and random cropping. We will explain data augmentation methods later in Chapter 16.

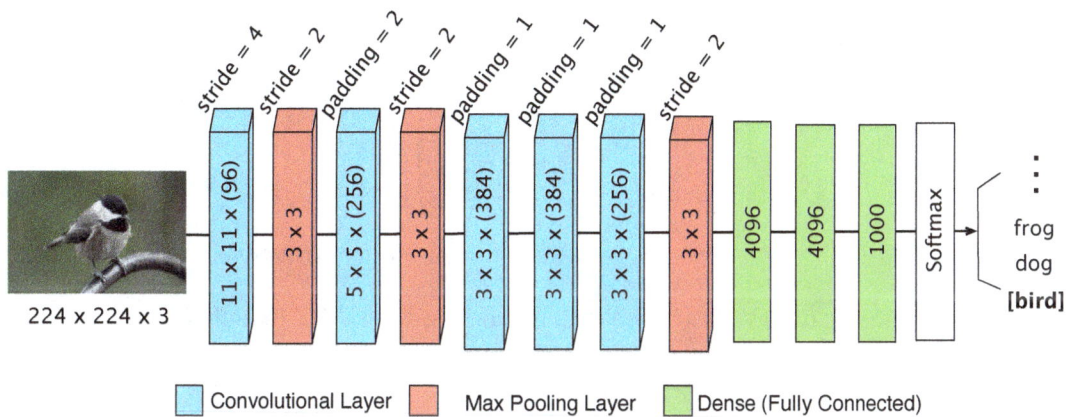

Figure 12-19: AlexNet architecture. The output is 1000, because this model is trained on ImageNet dataset.

The convolution window shape in the first layer is 11×11; in the second layer, it is reduced to 5×5, followed by 3×3 in the third, fourth, and fifth layers. After the first, second, and fifth convolutional layers, it adds maximum pooling layers with a window size of 3×3 and a stride of 2. At the end of convolutional layers, there are two fully connected layers, which each have 4096 outputs. Therefore, an image from ImageNet data produces about 1 Gigabyte of data as the output of these layers. The dense layers also implement dropout to prevent overfitting and reduce the complexity of the model.

In total, AlexNet has about 60 million parameters and is considered a small network. If you recall the size of LLMs this sounds like a tiny model.

VGG

Two years after AlexNet in 2014, VGG [Simonyan '14] was introduced. The VGG-16 net achieved an error rate of 7.3%. The main difference between VGG and AlexNet is its deeper network architecture, i.e., 16 layers and 19 layers. Similar to AlexNet, it also trained on the ImageNet dataset. Therefore, its input image size is also the same (224×224×3). Also, similar to AlexNet and LeNet, it has two parts, the first part includes a set of convolutional layers, and the second part includes a set of dense layers. Its second part, similar to AlexNet, includes fully connected layers with outputs of 4096, 4096, and 1000, respectively. A set of convolutional layers that ends with a max pooling layer is referred to as a VGG block by authors. For example, you can see from Figure 12-20 VGG-16 and VGG-19 both have five convolutional blocks.

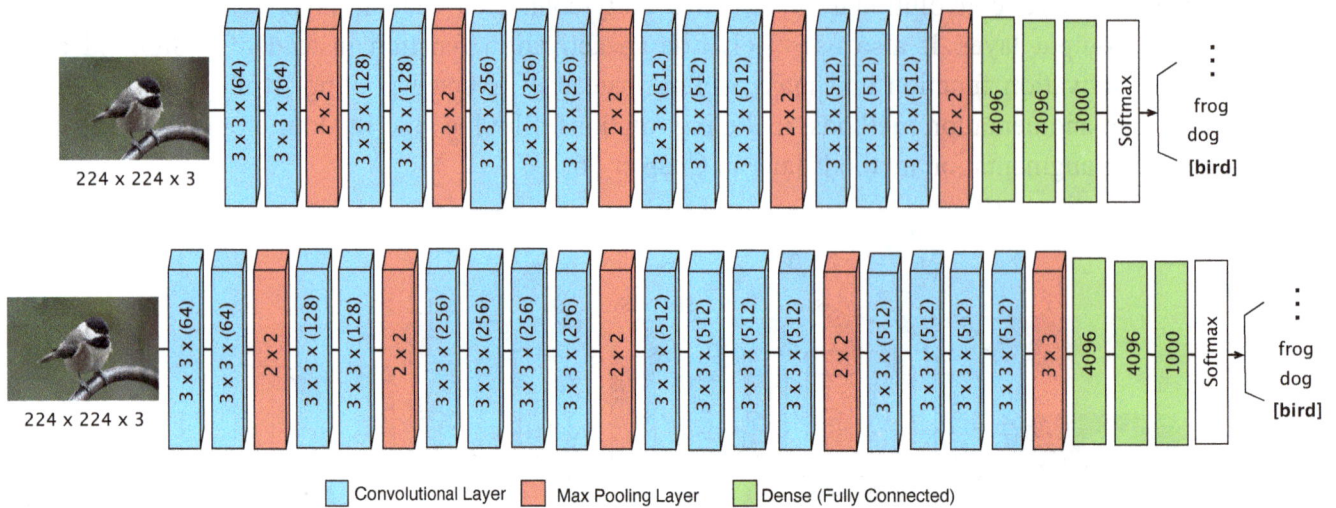

Figure 12-20: (top) VGG 16 architecture, and (bottom) VGG 19 architecture.

The filters in VGG are very small, 3×3 convolutions, with a stride size of 1 and a padding size of 1. Its max-pooling layers are 2×2 with a stride size of 2. Interestingly, the authors of VGG find that 3×3 convolution has the same *effective receptive field* as 7×7 convolution or larger convolutions (we will explain the effective receptive field shortly).

There are three varieties of VGG, one has 11 layers, known as VGG-11. The other version has 16 layers, known as VGG-16, and the third one has 19 layers, known as VGG-19. Figure 12-20 presents the VGG-16 and VGG-19 architectures. VGG-19 performs slightly better than VGG-16, but it consumes more memory. In total VGG has about 138 million parameters, but still, it is not considered a large network when comparing it with newer models, which we explain later.

What is the effective receptive field? According to Britannica[25], the Receptive field is a *"region in the sensory periphery within which stimuli can influence the electrical activity of sensory cells"*. Huebel and Wiesel [Hubel '59] performed a series of experiments on cats, and they realized that many neurons inside the visual part of the brain react only to a limited region of visual data, known as the *local receptive field*. On the other hand, some neurons react only to a larger region of visual data. These findings led to the idea that the visual perception of mammals identifies complex visual patterns through higher-level neurons, which operate based on the output of lower-level neurons. The effective receptive field refers to *the region in the input image that produces an important feature.*

---

[25] https://www.britannica.com/science/receptive-field

In other words, the "effective receptive field" is the output unit that contains pixels with a non-negligible impact (including visual stimuli) on that unit. If you still do not understand it, in a biological context, the receptive field is the portion of the sensory input space (e.g., image) that causes a neural response. In computer vision, the receptive field refers to the region in the input data that produces the feature or the smallest region that includes a feature of the image. The size of the receptive field is s influenced by the filter size but also by other factors such as stride and padding.

GoogleLeNet / Inception Net

GoogLeNet (Inception Net, a.k.a Inception) [Szegedy '15] is another architecture that was introduced in 2014 and performs close to VGG. GoogLeNet achieved an error rate of 6.67% on the ImageNet ILSVRC in 2014 and won the award that year. Authors of GoogLeNet focus on finding the best size of convolution kernels and found it is better to have a combination of varied-size convolutional kernels.

Inspired by a quote in the movie Inception[26]: *"We need to go deeper"*, the authors introduced the *inception module*. An inception module is a small neural network architecture, and GoogLeNet operates by stacking inception modules on top of each other, which can be interpreted as a network inside a network.

Each inception module takes input and sends it to four parallel paths of convolutions to get into outputs. All convolutional layers have a ReLU activation function. Figure 12-21 presents the first inception module introduced by GoogLeNet. At the first layer, the input will be fed into three 1×1 convolutional layers with a stride size of 1. These convolutions are too small to capture any specific pattern in the image, but they output fewer feature maps, and the authors said they could reduce the dimensionality of the input data while keeping important features of the data. The second layer of convolution enables the network to capture patterns on different scales. This layer has three convolutions of sizes 1×1, 3×3, and 5×5. At the end, all output will be concatenated together in the filter concatenation layer.

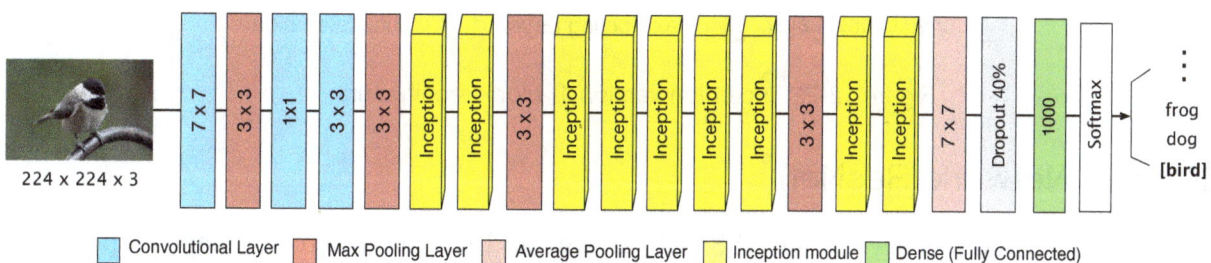

Figure 12-21: GoogLeNet (Inception V1) architecture.

The inception module makes GoogLeNet architecture very effective, and it is used nine times in the basic GoogLeNet architecture. Stacking inception modules on top of each other and not

---

[26] https://www.imdb.com/title/tt1375666

having a fully connected layer reduces the number of parameters significantly, and as a result, GoogLeNet has about 6 million parameters.

The details of the Inception module are presented in Figure 12-22. Similar to other books [Zhang '21, Geron '19], we neglect some details, such as its auxiliary classification outputs, and we make it simpler for the sake of understandability.

After the first version of GoogLeNet (Inception Net v1), later versions of Inception were introduced, and at the time of writing this chapter, Inception v4 is the most recent version that has higher accuracy in some cases than ResNet (we will explain it). Briefly, Inception v2 leverages the use of Batch norm uses ReLU as an activation function, and uses two 3×3 convolutions instead of a 5×5 convolution in the Inception module. Inception v3 introduces a *factorization* (we have explained matrix factorization in Chapter 7), which reduces the number of parameters and also reduces the risk of overfitting. For example, a 5×5 convolutional layer can be substituted by 1×5 and 5×1 convolutional layers, which have fewer parameters.

Before learning Inception v4, which is the latest Inception version, we should understand ResNet and its approach to make the result accurate while feeding it into a very deep neural network.

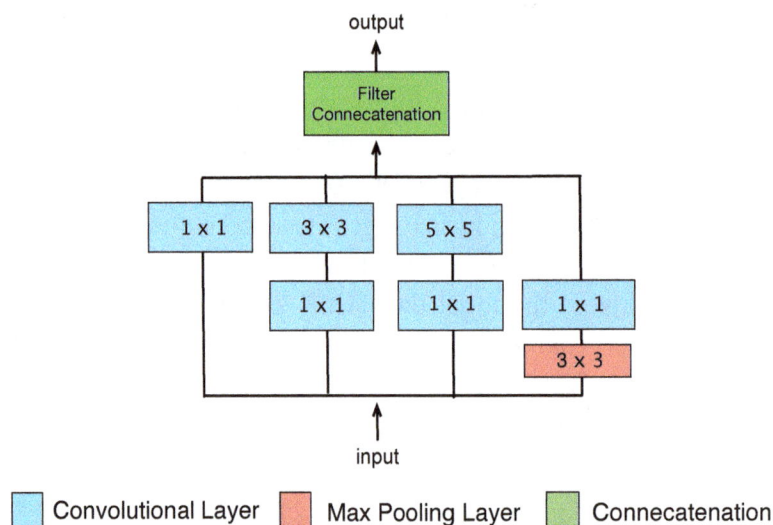

Figure 12-22: Inception version 1, block architecture.

Residual Network (ResNet)

One year after GoogLeNet won the ILSVRC award, in 2015, the Microsoft team proposed a *Residual Network (ResNet)* [He '16] and won the 2015 ILSVRC award. ResNet50 achieved a top-5 error rate of 3.57%, and ResNet101 achieved a top-5 error rate of 3.25% in the ImageNet ILSVRC challenge.

Authors of ResNet recognize that increasing the number of convolutional layers does not increase the accuracy and, in contrast, causes a *degradation* problem. Degradation means that as the network depth increases, the accuracy gets saturated and then starts to decrease rapidly. The authors reported that optimization is the origin of the problem because deeper models are harder to optimize than shallower models, and thus, their accuracy decreases. A major reason is a vanishing gradient caused by backpropagation in deep networks.

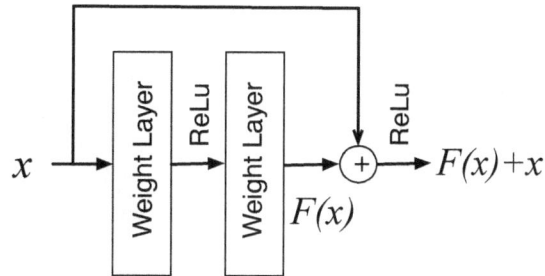

Figure 12-23: Residual block architecture.

ResNet resolves this issue by using the skip connection (check Chapter 11, U-Net architecture, to recall the skip connection). As we have explained in Chapter 11 and earlier in the transformer architecture, a skip connection adds the original input to the output. Therefore, the output is the combination of the original input and the previous layer's output. ResNet calls a set of weight layers with skip connection *residual blocks,* and by stacking several residual blocks, it builds a deep neural network, which ensures high accuracy. Figure 12-23 presents the architecture of a Residual block. Weights and biases are transferred through weight layers. Here, the weight layer could be a fully-connected layer or a convolutional layer. Figure 12-23 shows the input x will pass weight layers and construct the $F(x)$. However, the next layer after the Residual block does not receive $F(x)$. Instead, it receives the result of $F(x) + x$, which is fed into a ReLU activation function.

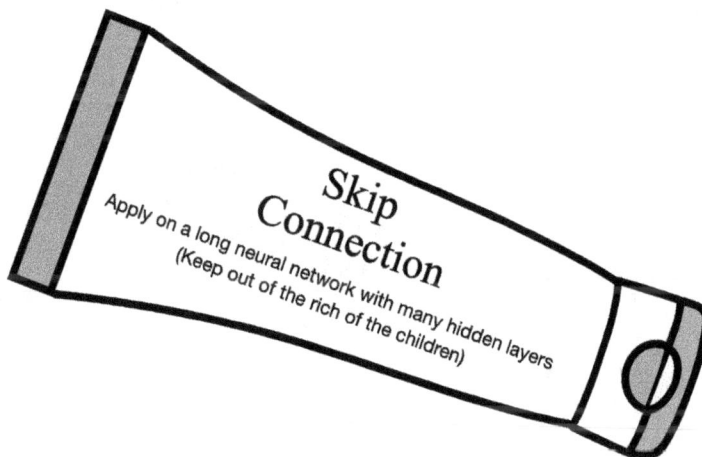

In summary, Skip connections (a.k.a. short-cut connections or residual connections) enable the

without skip connection   with skip connection

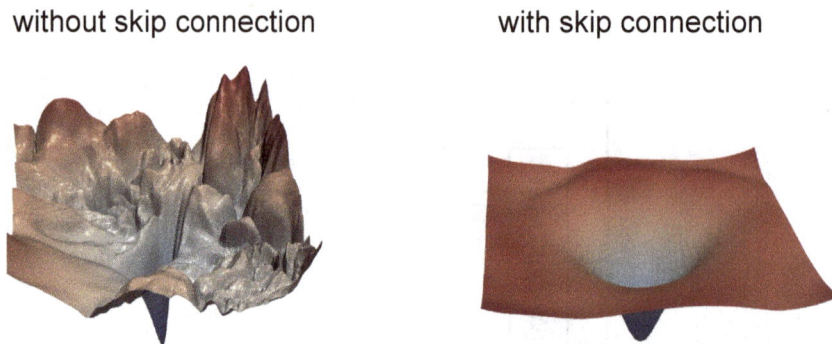

Figure 12-24: Visualized loss in ResNet-56 with and without skip connection. Image credit: [Li '17]

network to go deep without getting into any optimization-related challenges. Therefore, it provides high accuracy.

Figure 12-24 presents a visualization of a loss [Li '17] in ResNet with and without skip connections. We can see how using the skip connection flattens the complex curves and thus makes it easier for the gradient descent algorithm to reach the minima.

To be able to add the input and output together, both weight layers should have the same shape. In other words, ResNet tries to ensure that input shapes are equal to output. If they are the same, the skip connection simply adds the input to the output. If not, the skip connection adds a 1×1 convolutional layer in between that changes the size of the input to be equal to the output.

ResNet uses He [He '15] initialization, its batch sizes are 256. After each convolution layer, it uses a Batch normalization (we did not show it as a separate layer in Figure 12-25). It has 3×3 convolutional layers, and each convolutional layer is followed by a ReLU.

Figure 12-25 presents 34 layers of the ResNet architecture. However, many varieties of ResNet have been proposed, such as 18, 50, 101, and 152 layers. Still, at the time of writing this book, ResNet is among the top CNN architectures for computer vision applications.

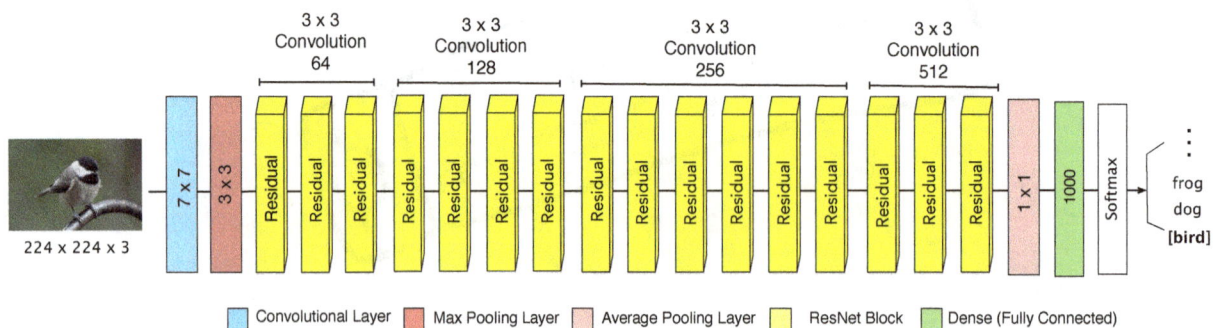

Figure 12-25: ResNet-34 architecture. It has 34 layers, because pooling layers and softmax do not counted by authors as layers. They count only weight layers (Dense and Convolution).

Inception-v4 and Inception-ResNet

Inception models advanced until Inception-v4 and Inception-ResNet [Szegedy '17] were released about one year after ResNet. Inception-ResNet combines the architecture of Inception with the Residual network. Inception's efficient use of computing resources through dimensionality reduction and ResNet's ability to train deeper networks without degradation via residual connections lead to the introduction of Inception-ResNet models. The authors report different combinations of Inception and ResNet (Inception-ResNet-v2) and the Inception architecture improvements that result in Inception-v4. Figure 12-26 presents the Inception-v4 architecture and Inception-ResNet-v2 architectures.

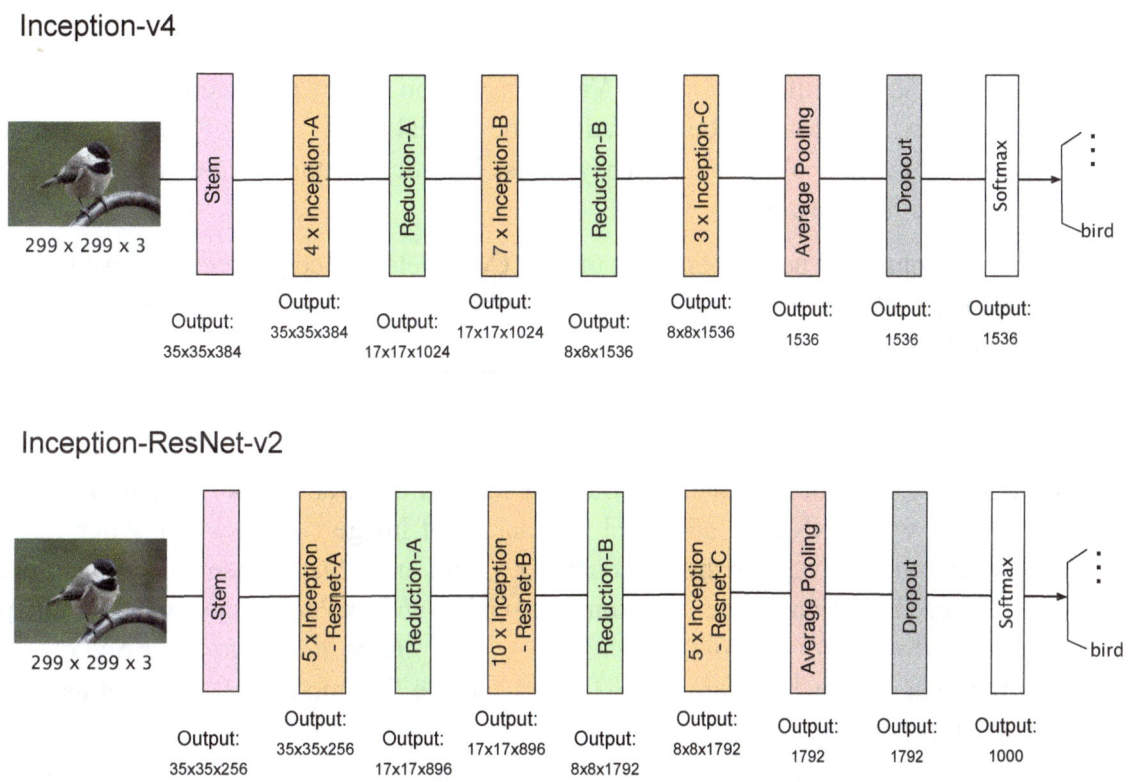

Figure 12-26: Inception v4 architecture, to understand the details of Stem layer check Figure 12-25.

Here, we only present their overall schema; there is one layer introduced in Inception models, known as *Stem,* and it is worth further exploring. Figure 12-27 presents the architecture of the Stem network. The sign 'V' on convolutions means they are using the valid padding (check Chapter 10 to recall them), and convolutions without 'V' are using the same padding. An advantage of Stem is providing parallel convolutions that are concatenated at the end; this

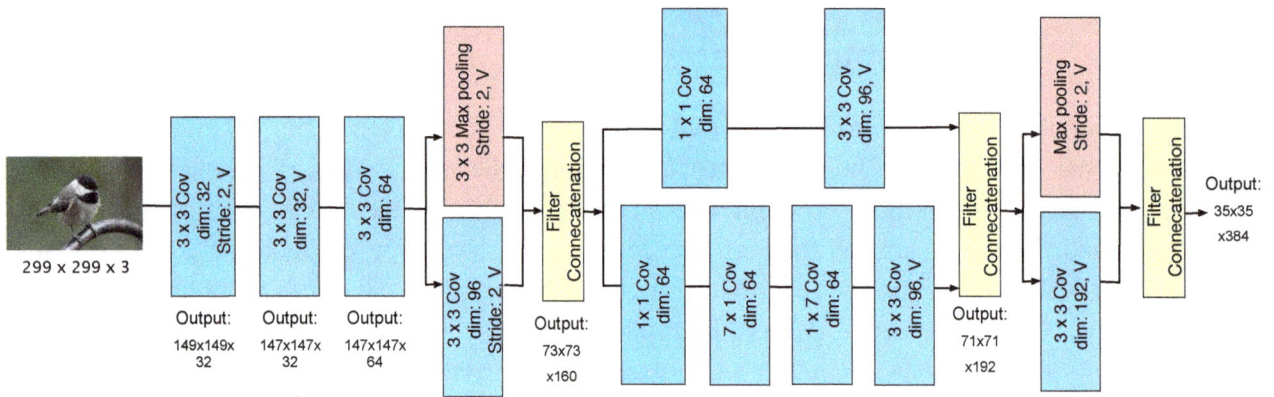

Figure 12-27: Stem network architecture that is used at the beginning of Inception-v4 and Inception-ResNet-v2.

enables the model to have a different Field of View (FoV)[27] on an image, which results in higher classification accuracy. Reduction modules reduce the spatial dimensions of the feature maps while increasing the number of channels. They assist in reducing the computational cost and increasing the depth of the network.

There are more convolutional neural networks (a.k.a., ConvNets) we could learn, but we hold the discussion on image classification and proceed further with other advancements in image recognition algorithms.

Vision Transformers

If you read the NLP section of this chapter, you have learned that transformer revolutionized the NLP field, the same way CNN revolutionized the quality of image classification. After ResNet models, there were minor advancements in computer vision until 2020, when Dosovitskiy et al. [Dosovitskiy '20] used the transformer for computer vision and explained their work in a paper titled*"An Image is Worth 16×16 Words: Transformers for Image Recognition at Scale"*. This model is referred to as Vision Transformer (ViT), and it is the first model that does not use CNN but still has high accuracy.

Authors have used 2500 TPU[28] version 3 to build their model and use the private Google dataset JFT dataset, which has 300 million images. This shows that you can easily build such a model by using the smartphone under your bed. Jokes aside, only big AI corporations with their massive hard infrastructure and huge datasets can build this model, but still, we can fine-tune them or make them lighter.

---

[27] In simple terms, Field of View refers to the area of an image that we can see (or a camera can sense).

[28] Tensor Processing Unit (TPU) is similar to GPU and produced by Google to accelerate tensor flow (Google's framework for Neural Network programming) projects.

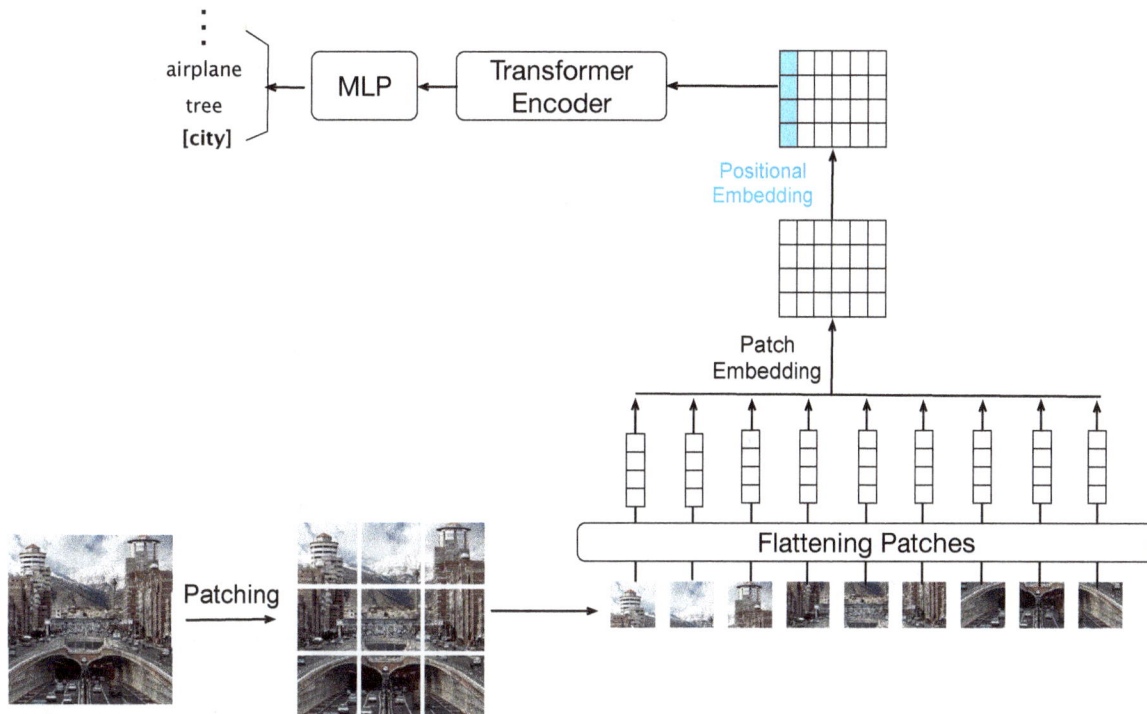

Figure 12-28: Visual Information Transformer architecture.

They have proposed three models: ViT-Base, ViT-Large, and ViT-Huge which have 12, 24 and 32 layers respectively. They have used 14×14 and 16×16 patch sizes, and for example, a ViT-Large with 16×16 patches is shown as ViT-L/16.

Inspired by the transformer architecture, the authors use a disjoint sliding window and convert an image into patches. Assume an original image has $H \times W \times C$, where H stays for height, W for width, C for the channel, and each patch has a size of $P \times P$. The number of patches will be N and $N = H/P \times W/P$. The original image is reshaped into $N \times P^2 \times C$.

Next, the patching process begins, which divides the image into smaller patches. Patch sizes and stride sizes are hyperparameters of the ViT algorithms. The bottom left of Figure 12-28 presents how an image is converted into a set of patches and patches processed further.

Patches simulate the sequence of words used as input of a transformer. However, patches are $P \times P$ dimensions, and similar to words, we must vectorize them to have a one-dimensional vector of numbers for each patch. In other words, each of the patches should be flattened into a one-dimensional vector that represents the patch, similar to how words are vectorized in NLP. To vectorize patches, each patch will be substituted by a vector P_d, which is calculated as follows: $P_d = P_C \times P_W \times P_H$. P_H stays for the height of the patch, P_W the width of the patch, P_C for the channel of the patch.

We will explain the rest of this model with numbers and examples. While reading the example, check Figure 12-28, where we try to visualize each step.

(i) Assume our input image has a size of 224×224, then it will be separated into 196 patches of 16×16 size. Since the image in RGB has three channels, the original image is a tensor of $224 \times 224 \times 3$ and each patch is a tensor of $16 \times 16 \times 3$ (recall 3 presents the number of channels).

(ii) Next, a linear projection is applied on every patch to convert a patch $16 \times 16 \times 3$ into a vector of size 1×768. The process of converting patches into a vector is referred to as *flattening* a patch into a vector. However, this linear projection involves not just flattening but transforming each flattened patch vector using a learned linear projection (a weight matrix), typically resulting in a 768-dimensional vector per patch. It involves using a trainable tensor, which is essentially a matrix of weights learned during the model training process. Each patch is transformed by this matrix through matrix multiplication, effectively mapping the original 768-dimensional data of each patch (since $16 \times 16 \times 3 = 768$) into a new 768-dimensional space.

(iv) Afterward, all of these vectors of size 768 are concatenated together, and as a result, we will have a matrix of size 196×768 (196 is the total number of patches we have). The process of constructing this matrix is called *patch embedding*.

(iv) In the fourth step, a start token [class] will be added to the sequence of patch embeddings. It is similar to the [CLS] token that is used for BERT. Recall that the [CLS] token is used to capture a comprehensive summary of the entire input for later tasks, just like a notetaker in a meeting. After adding the [class] token, a simple positional embedding is applied to patch embedding and the [class] token (each patch is one row). Now, the patch embedding matrix has a size of 197×768, because each patch is associated with the position. We do not describe the details of positional embedding, which is simple to learn, and you can check the original paper or its open source implementation for more details.

(v) In the last step, the patch embedding matrix will be passed to the encoder function of the Transformer. The Transformer encoder has the same architecture that we have explained for the encoder of the Transformer earlier in this chapter. The output from the Transformer encoder is then fed into an MLP, and the MLP output determines the label for the input image.

Why do we do patching, and why convert each patch into a vector? Assuming d is the dimension of data and n is the number of words, by using self-attention, each word is compared with all other words in the sequence. Therefore, it has quadratic complexity $O(n^2 . d)$ [Vaswani '17]. Now assume that an image with the 300×300 pixel size, if treated similarly, its data complexity becomes prohibitively expensive to compute due to the high dimensionality. Therefore, instead of using the original image, we separate the image into patches and then reduce the dimensions of each patch to a vector.

In the context of machine learning, *inductive bias* refers to the assumptions a model makes to generalize from training data to unseen data. CNN algorithms. For instance, incorporating inductive biases, such as locality and translation equivariance, contributes to success in computer vision tasks. The authors of ViT acknowledge that ViT models, lacking some of these inductive biases, are not as effective when trained on small datasets. It's also important to note that they trained their ViT model on 300 million images from Google's JFT dataset, which is not open-source. At the time of revising this chapter, after 100th time (Q2 2024), it seems that the combination of Vision Transformer and CNN yields even higher accuracy.

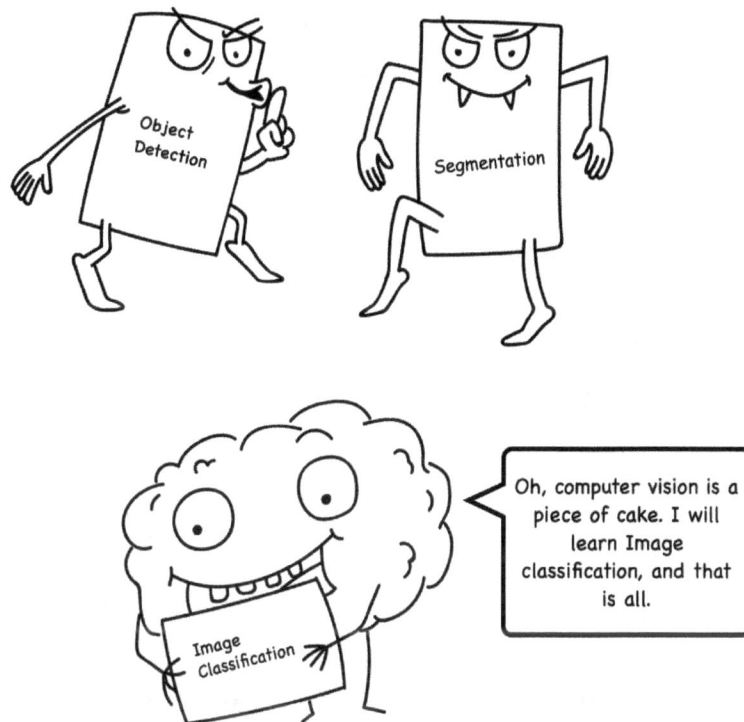

Object Detection

Previously, we explained image classification, in which we give an image into the image classification model, and it identifies one dominant object inside the image and assigns a label to the image. It means that if there are three objects in an image, it assigns labels to the dominant object and not the two other objects.

Object localization models identify the locations of objects within an image by specifying coordinates for each object. *Object detection*, also known as object recognition, not only locates objects within the image but also draws bounding boxes around them and assigns labels to those boxes. These concepts were briefly explained in Chapter 6. There might be more than one object within an image, and the object detection algorithm will draw bounding boxes around each identified object. Object detection can be described as a combination of image classification and image localization. However, unlike image classification, where we typically assign a single label to an entire image, object detection involves assigning multiple labels, each localized to an object with a bounding box.

We explain three groups of models for object detection, including R-CNN models, YOLO models, and Single Shot Detection (SSD). Each of these model groups will be explained in this section. However, before explaining them, we need to be familiar with a few concepts related to object detection, which we describe as well.

Object Detection Concepts

Bounding Box: A rectangle that is drawn around an object in the image is referred to as a bounding box. This rectangle highlights the location of the object on the image. In other words, bounding boxes specify the spatial location of an object. For example, in Figure 12-29, both red and green bounding boxes specify the location of a car under snow.

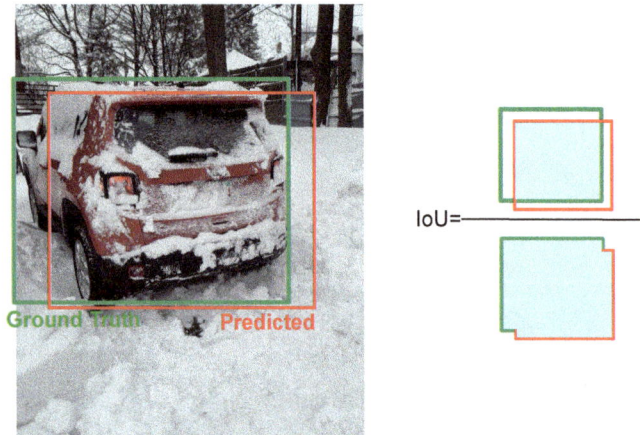

Figure 12-29: Intersection over Union visualization.

Anchor: Object detection algorithms start by proposing potential bounding boxes, which are preliminary guesses about where objects might be located within the image. These proposed boxes, which are not necessarily correct, are called anchors. Then, during training, the algorithm adjusts these anchor boxes to better align with the ground truth bounding boxes.

It is common to generate several anchor boxes at different scales and aspect ratios and use them on some sample pixels to detect objects with various shapes and scales inside an image. However, it would be computationally impractical to generate multiple anchor boxes around every single pixel; therefore, object detection algorithms perform some sampling and create anchor boxes around sampled pixels. For example, Figure 12-30 presents three anchor boxes that are created around nine pixels that are sampled from the image.

Intersection over Union (IoU): While the performance of image classification algorithms is typically measured by accuracy or mean classification error, the performance of object detection algorithms is commonly assessed by how accurately they identify the location and size of bounding boxes around objects. In other words, our

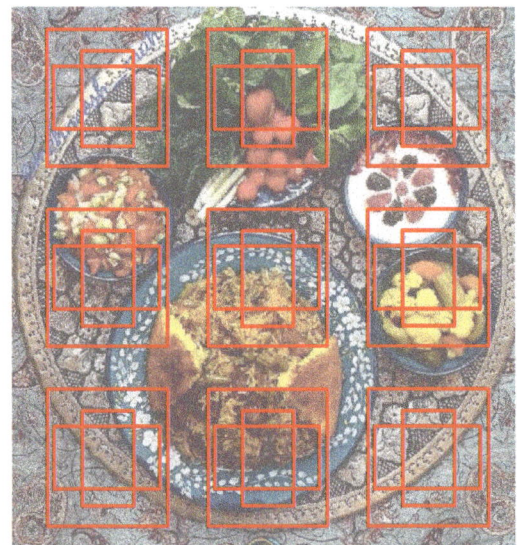

Figure 12-30: Three different anchor boxes around nine sampled pixel.

762

objective in object detection is to maximize the Intersection over Union (IoU). IoU, which is a number between 0 and 1, measures the overlap between the ground truth bounding box and the predicted bounding box. Figure 12-29 visualizes the IoU. In this figure, the ground truth bounding box is marked with green, and the bounding box that the object detection algorithm drew is marked in red. The IoU is calculated as the area of overlap divided by the area of union between the two bounding boxes. IoU is the Jaccard index (check Chapter 4), and it is also referred to as Jaccard overlap, as presented in the following equation:

$$\text{IoU} = \frac{|A \cap B|}{|A \cup B|}$$

A is the predicted bounding box, B is the ground truth bounding box. $|A \cap B|$ is the area of the intersection between the predicted and ground truth boxes. $|A \cup B|$ is the area of the union of the predicted and ground truth boxes.

Region of Interest: Region of Interest (RoI) refers to a specific part of an image that is selected for further analysis or processing. RoIs are used to focus computational resources on certain areas, thereby reducing the amount of data to process and improving efficiency. For example, in medical imaging, ROIs might be used to focus on a particular organ.

Region Proposal: Region proposals suggest bounding boxes or areas within an image where there is a high likelihood of containing an object of interest. These proposals are generated by a part of an object detection system and are subsequently analyzed by a classifier to determine if any object is within the proposed region.

R-CNN Models

Region-based Convolutional Neural Network (R-CNN) is a series of image segmentation models proposed by Girshick et al. [Girshick '15]. Before explaining this model series, we should be familiar with the concept of the region proposal.

R-CNN

The first model of R-CNN has three modules: Region Proposal (selective search), Feature Extractor (pre-trained AlexNet), and SVM Classifier, which are executed in a row.

(i) To identify the Region Proposals, we can use a sliding window (check Chapter 5) and move it along the image based on texture changes. However, there could be an infinite number of sliding windows with different sizes. Therefore, we need more subtle approaches to identify region proposals, and there are algorithms that refine the proposed region, such as the *Selective search* [Uijlings '13] region proposal algorithm. Selective search is a greedy algorithm that starts by generating many segments by segmenting pixels with the same texture color intensity together[29]. Next, it uses a greedy approach to recursively merge two similar regions into one region. The recursive merging process is repeated until only a few regions remain, each potentially containing an object of interest. Figure 12-31 visualizes the process of selective search.

---

[29] Note that choosing the same pixels with the same color intensity for segmentation is not enough. You can see a real-world object has a different color intensity and thus selective search alone is not accurate for object recognition.

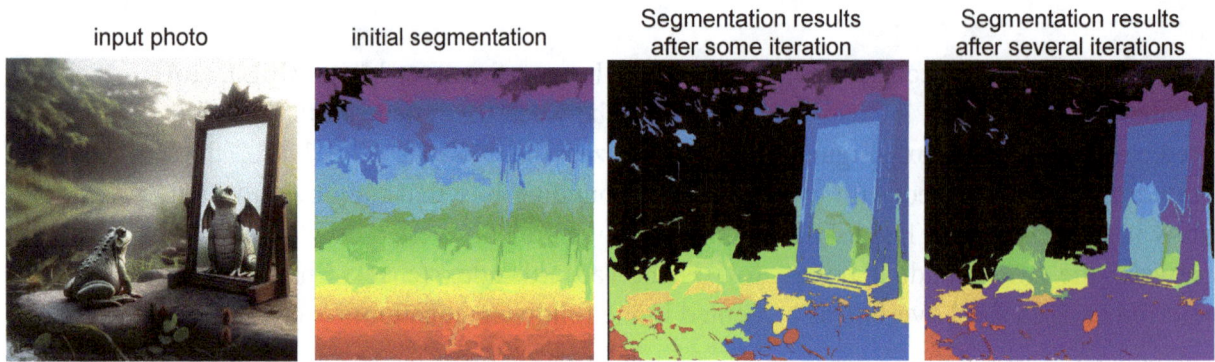

Figure 12-31: An example of image segmentation based on selective search algorithm.

As a result, it selects 2000 regions by using the selective search algorithm, which means 2000 bounding boxes around any candidate region (proposed region), as is shown in Figure 12-32.

(ii) Next, to implement the feature extractor, it uses a CNN (AlexNet) on every candidate bounding box (proposed region or warped region) and extracts features. The output of the CNN is a 4,096 size vector that describes the contents of the image.

(iii) The classifier, which is an SVM, receives the CNN output (a vector of size 4096) and classifies it into one of the known classes based on the extracted features.

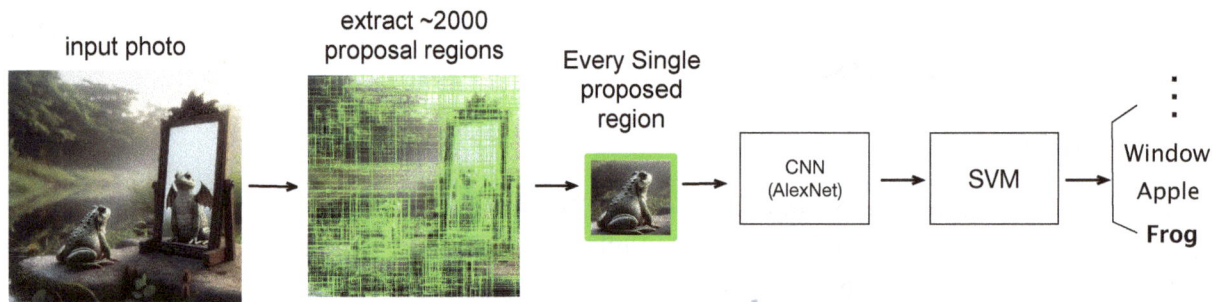

Figure 12-32: R-CNN model architecture.

Fast R-CNN

Although the output of R-CNN is accurate, it is very slow. For every 2000 regions, we need to run a CNN and SVM once. The slow execution time of R-CNN led to the introduction of Fast R-CNN [Girshick '15] one year later.

One salient difference between Fast R-CNN and R-CNN is that Fast R-CNN fed the entire image into the CNN, but R-CNN feed the proposed regions into the CNN. Therefore, it does not have to feed 2000 regions to the CNN, and the convolution process will be done once per image.

The CNN model (e.g., VGG-16) is pre-trained with a ground truth dataset. It receives (i) the input image and (ii) a set of the regions of interest (RoI), which are extracted by a selective search algorithm. The output of CNN is a convolutional feature map (check Chapter 10). In simple words, this pre-trained CNN performs the feature extraction. After the feature map is constructed by CNN, the region of proposals is projected on the feature map with their coordinates on the original image. This means the next layer does not deal with the RoI of the original image, but instead, it deals with the RoI on the feature map.

These convolutional feature maps (output of CNN) are fed into a component called the *RoI pooling layer*. RoI pooling layer scales (warping) every RoI region in the feature map into a predefined size.

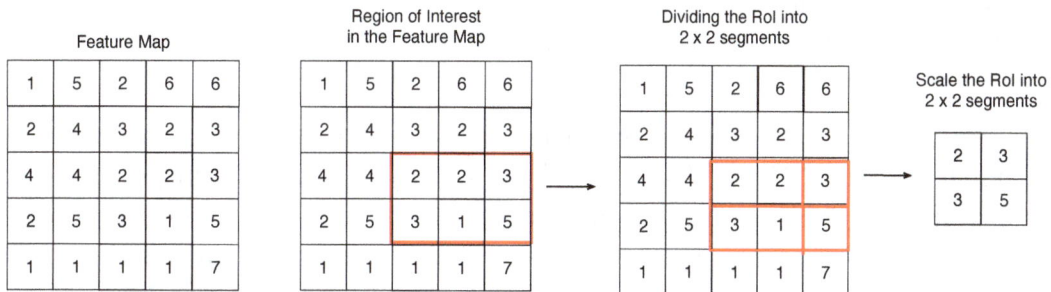

Figure 12-33: RoI pooling layer scales a region of interest (highlighted in red) in a feature map into a 2x2 size.

For example, look at Figure 12-33, we have a feature map, and the RoI is highlighted in red. The RoI pooling layer scales that region into a 2×2 size region. To perform the scaling first, it segments the region into 2×2 sizes, which we can see are not equal. Then, it selects the maximum value inside each segment and constructs the 2×2 final result. In other words, the

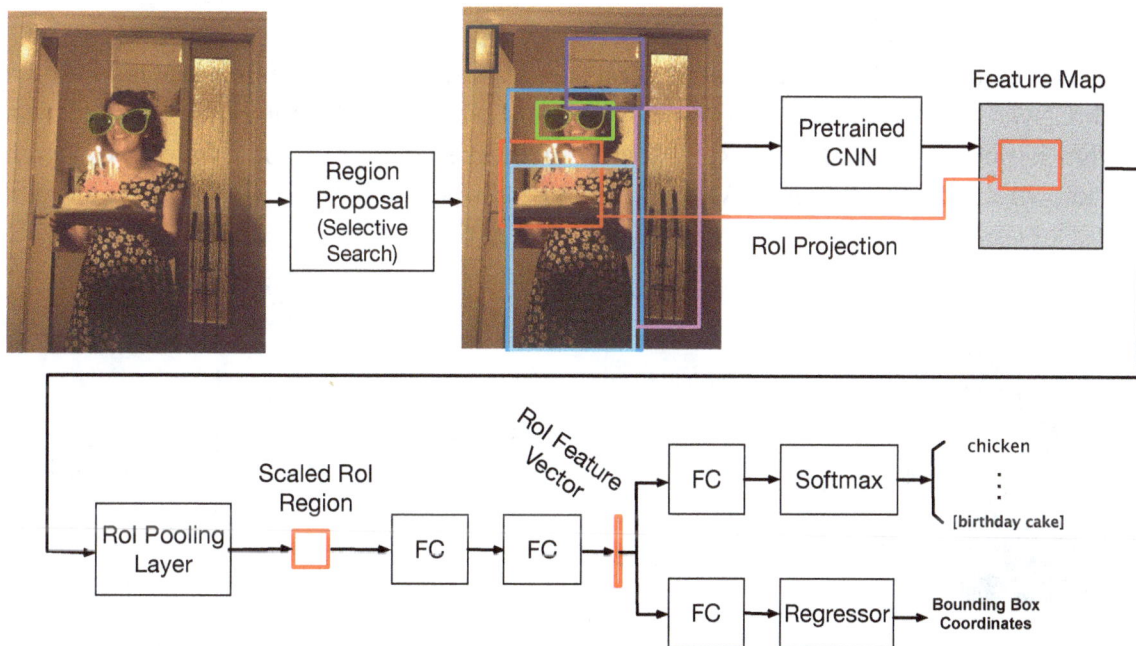

Figure 12-34: Fast R-CNN architecture.

RoI pooling layer divides each RoI into a fixed grid size and performs max pooling within each segment to produce a uniformly sized output. This ensures that regardless of the original RoI size, the output fed into the fully connected layers is consistent, facilitating efficient learning and classification.

In Figure 12-34, we present the output of the RoI pooling layer as a Scaled RoI Region. Afterward, the output of RoI pooling (Scaled RoI Region) is fed into two fully connected layers (FC in Figure 12-34). The result will be a vector, which is called a *RoI feature vector*. This vector is passed into two different fully connected layers; one result gets into a softmax to identify the label, and the other results are passed into a regressor to identify the coordinates of the bounding box, as shown in Figure 12-34.

To summarize, the algorithm receives the image and RoIs. RoIs will be calculated by selective search, similar to the R-CNN. The pre-trained CNN will get the image and provide a convolutional feature map in the result. Also, the RoI location on the input location is marked on the feature map. The RoI pooling layer warps (scales) the RoIs into square shapes and they will be reshaped into a fixed size. The output of the RoI pooling layer (fixed-size RoI regions) is sent to two fully connected layers, and the result is the RoI feature vector. The RoI feature vector will be sent to a softmax layer to identify the class label for each proposed region and also the coordinates of the bounding box.

Faster R-CNN

Both R-CNN and Fast R-CNN use selective search. The selective search also contributes to the slowness of their model. Faster R-CNN [Ren '16] has been introduced after Fast R-CNN and removes the selective search. Eliminating the selective search for RoI detection makes it significantly faster than Fast R-CNN.

Similar to previous R-CNN models, in the first step of Faster-RCNN, the image is fed into a pre-trained CNN, but unlike previous R-CNN models for RoI selection, the selective search algorithm is not used. To identify RoI (Region of Proposals), instead of a Selective Search, a CNN, known as *Region Proposal Network (RPN)*, will be used. We will describe more about

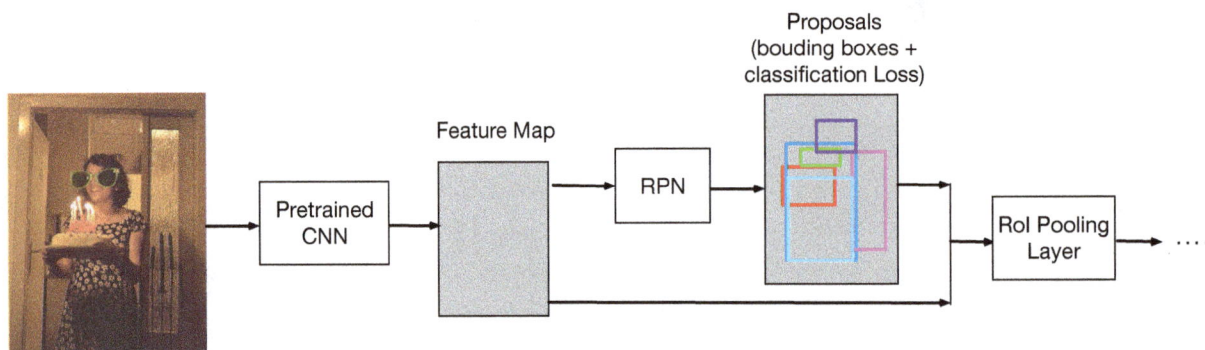

Figure 12-35: Faster R-CNN architecture. Note that after RoI pooling layer everything else is similar to Fast R-CNN.

RPN later. For now, keep in mind that RPN input is an image, and its output is a set of RoIs (rectangular bounding box object proposals with class labels), as is shown in Figure 12-35.

After RPN constructs a list of RoIs with their labels, these identified RoIs are rescaled using an RoI pooling layer, similar to Fast R-CNN. Next, again, similar to Fast R-CNN, a classification and regressor will be used on the output of the RoI pooling layer to identify class labels and specify the bounding box of the object. The rest of the process is not shown in Figure 12-35 because it is the same as in Fast R-CNN.

It is not wrong to say the main difference between Faster R-CNN and Fast R-CNN is the substitution of selective search with RPN. Authors claimed that RPN acts as an attention mechanism, which tells the neural network where to look for information. The output of the pre-trained CNN is a feature map. After the feature map is constructed, RPN slides a small window over the feature map. Then, the dimensions of each sliding window content will be reduced, and the result (dimensionality reduced content of the sliding window) will be fed into two fully connected (FC) layers. One FC is used for classification, and the other FC performs a regression task, which is used to specify the bounding box of RoI.

How does RPN work? RPN is a fully convolutional network (FCN)[30] and tries to identify RoI with different *scales* and *aspect ratios*. At each sliding window, we can have a maximum of k possible RoI (proposals). The authors recommend that each window has three scales and three aspect ratios. Therefore, we can have maximum $k = 3 \times 3 = 9$ RoIs in each window.

The center of each window is referred to as the *anchor*. Although anchors are defined over a feature map, they correspond to the input image. As we have explained before, anchors are rectangles that will be fed into the RPN, but their size will be changed during the training to fit the object they cover. Through the training phase, RPN tries to maximize the overlap between RoI and ground truth data [31].

In other words, RPN (i) predicts the probability of object labels in each anchor and (ii) changes the anchor coordinates to cover the object correctly. If you are interested in learning about the loss function and how it incorporates anchor size and labels, check the original paper; we are tired of explaining more about these models.

Single Shot MultiBox Detection (SSD)

Although Faster R-CNN is faster than its predecessors, it still cannot operate in real-time. Single Shot MultiBox Detector (SSD) [Liu '16] and YOLO are object detection models that can operate in real time. Operating in real-time is a very useful feature, especially in self-driving vehicles or other applications that require real-time decision-making.

---

[30] FCNs modify the traditional CNN architecture by replacing the fully connected layers with convolutional layers. This change allows the network to output spatial maps instead of classification scores. FCNs, particularly, are used for detailed and spatially complex image processing tasks, whereas CNNs are used for general image classification tasks.

[31] In this context, by using ground truth, we mean a dataset that its objects are collected and manually annotated.

SSD speeds up the process of object detection by eliminating the need for RPN and bounding box proposals. It performs the bounding box assignment and object identification in one pass through a pre-trained network. However, not having an RPN bounding box proposal negatively affects the accuracy, and to resolve this, SSD applies some improvements, which we will explain here. SSD can detect objects at different scales. For example, look at Figure 12-36, where we have different vases with different sizes and we have bounding boxes with different sizes.

Figure 12-36: An example of having different size of bounding boxes for objects inside an image.

SSD has two key features: (i) it implements object detection at different scales by using multiple layers of convolution, and (ii) it uses various filters to identify bounding boxes of different shapes.. The process of making multiple predictions (object identification) with different bounding boxes and different confidence scores is referred to as *multiplexing* by authors.

First, it performs feature extraction, and then it applies a convolutional filter on the results of feature extraction. Similar to R-CNN models for feature extraction, SSD utilizes a pre-trained CNN, i.e., VGG16, in the original paper, but other models, such as ResNet models, can be used for feature extraction as well. The pre-trained CNN lacks fully connected and softmax layers at the end. The green part of Figure 12-37 presents the pre-trained CNN. Adjacent to the pre-trained CNN, several multiscale feature map blocks are constructed using convolutional layers of varying sizes (the white and blue boxes in Figure 12-37 represent these convolutions). Instead of relying on a single feature map, utilizing a collection of feature maps of various sizes enables SSD to detect objects across different scales. In particular, large feature maps (the top of Figure 12-37) are good at capturing small objects, and small feature maps (the bottom of Figure 12-37) can detect large objects.

To implement the described multi-scale object detection, in addition to the Conv4_3 layer of the VGG 16 model, SSD employs six extra feature maps (Conv 6, Conv 7, Conv8_2, Conv9_2,

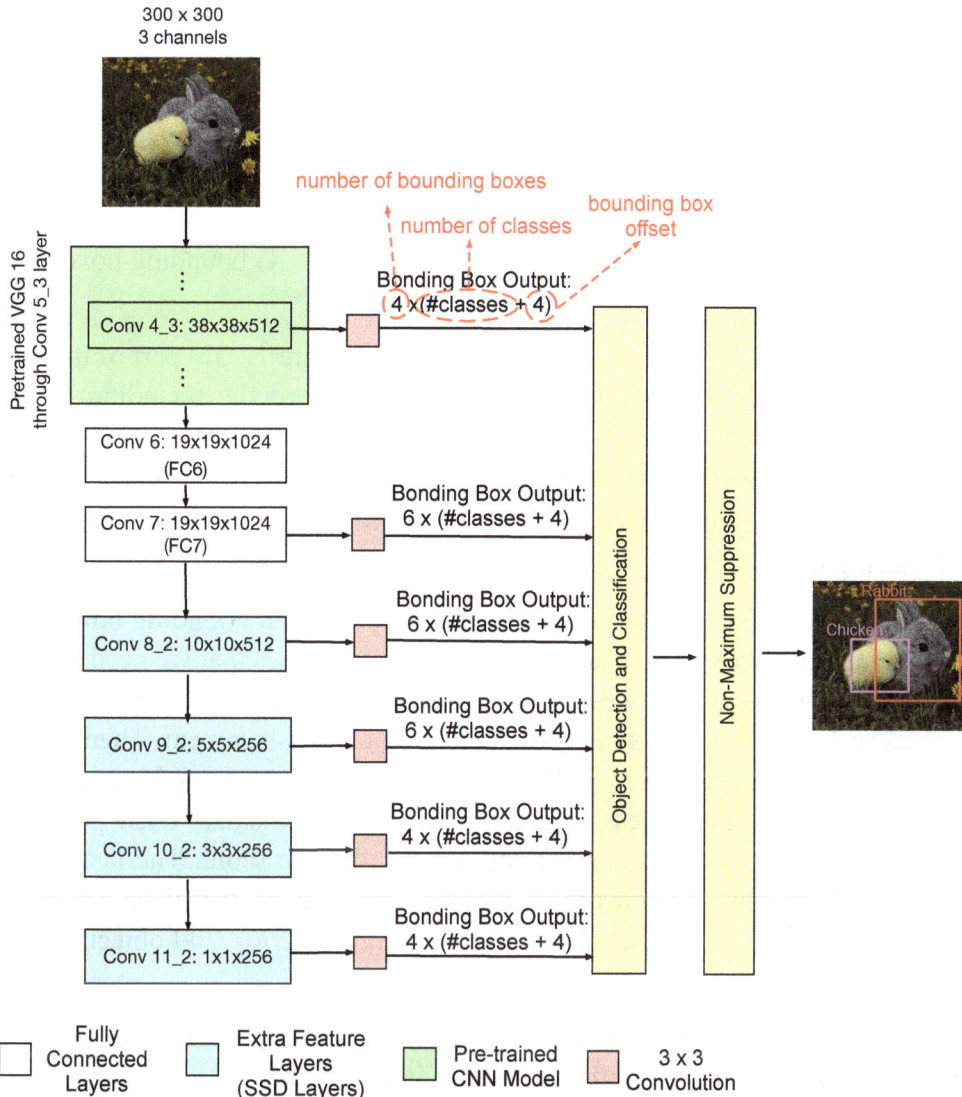

Figure 12-37: SSD-300 architecture. There is another variation which is known as SSD-512.

Conv10_2, and Conv11_2). The '_2' is used by authors to present that these four layers are extra feature layers.

A 3×3 convolution will be applied on each feature map. Then for each object, k number of bounding boxes are used. These bounding boxes have the same center coordinates, but they have different sizes and shapes (Figure 12-30 presents bounding boxes with different shapes and sizes but on the same center). For example, to detect a standing human, a vertical rectangle is useful, and to detect a car, a horizontal rectangle is useful. Authors use $k = 4$ and $k = 6$ bounding boxes around each object.

Considering the size of each convolution, we have the following number of bounding boxes for the SSD-300 model:

Conv 4_3: $38 \times 38 \times 4 = 5776$

Conv 7: $19 \times 19 \times 6 = 2166$

Conv 8_2: $10 \times 10 \times 6 = 600$

Conv 9_2: $5 \times 5 \times 6 = 150$

Conv 10_2: $3 \times 3 \times 4 = 36$

Conv 11_2: $1 \times 1 \times 4 = 4$

In total, there will be 8732 $(5776 + 2166 + 600 + 150 + 36 + 4)$ bounding boxes for each class of an image.

The "Object Detection and Classification" (yellow box in Figure 12-37) part of the model assigns bounding boxes with their offsets. *Offsets* are used to change the bounding boxes to fit the correct position of the object, and this process will be done through training. Offset is shown as $\Delta(c_x, c_y, w, h)$, c_x and c_y present the center of the bounding box, and w and h present the width and height of the bounding box. These offsets are adjustments from predicted positions to the ground truth during the training.

A confidence score will be calculated for each class label in each bounding box. This means the result of each bounding box will be equal to the number of classes that we need to detect from an image plus *offset*. For example, if we seek 20 different object labels for an image, we deal with 21 labels, because 1 label is used as background and no class assignment. Then, it adds 4 offsets as well. As a result, we will have $(21 + 4) \times 8732 = 218{,}300$ bounding boxes. Obviously, most of these bounding boxes are useless, and they should be removed. Each prediction in SSD includes a bounding box with shape offset and one confidence score. The last part of the SSD architecture, "Non-Maximum Suppression", drops all predictions that their confidence score is lower than a threshold (e.g., 0.45 per class) and keeps only the top 200 objects detected for the input image.

In summary, we have explained that SSD uses a single forward pass in its neural network to model object localization and classification. This is the reason it is referred to as a *single-shot*. The term single-shot indicates that the detection tasks are accomplished in a single pass through the network. SSD employs three approaches to resolve the accuracy issue of not using RPN, as follows. (i) It uses a small convolutional filter (3×3) to predict object categories and their bounding box locations. (ii) It uses different sets of filters for each of its feature maps to deal with various object scales and aspect ratios. (iii) It applies filters to multiple feature maps of a pre-trained CNN network to perform object detection in different scales.

We do not go into details of the loss function; if, after reading our description, it is still not clear to you, we recommend reading the open-source implementation of it. We also had a hard time understanding this model, and we tried our best to prepare an easy explanation for it. It might be easier to read the codes that implement SSD. This explanation should be enough to understand the advantages and design principles of the SSD model.

YOLO Models

You Only Look Once (YOLO) is a family of models [Redmon '16, Redmon '17, Farhadi '18] that implement object detection, and to our knowledge, they are the fastest available object detection models. The main author stopped proposing YOLO anymore, but at the time of reading this section, you might see the latest version of YOLO as 99999 because a newer version of YOLO is released every week.

Unlike SSD or Fast/Faster R-CNN models, YOLO does not use any pre-trained CNN model. Similar to SSD, YOLO also uses a single pre-trained neural network. This means it passes the image once into a pre-trained CNN (look at the image once only), and then it calculates the class labels (in probabilities), confidence score, and bounding boxes. Because of these characteristics, authors chose the name YOLO for their algorithm and its successors.

Figure 12-38: Yolo separates the image into 3 x 3 patches (S=3), and each patch can be the center of two bounding boxes (B=2).

YOLO v1

It starts by dividing an image into a grid of $S \times S$ smaller patches, as is shown in Figure 12-38. YOLO assumes that each patch is the center of B number of bounding boxes. In the middle image of Figure 12-38, it is assumed $B = 2$, and thus, each grid cell (each patch) is a center of two bounding boxes. In the original YOLO paper, the authors specify $S=7$ and $B=2$ for their evaluation. As we can see from the middle picture in Figure 12-38, a cell could have some parts of bounding boxes from neighbor cells as well. Although bounding boxes can extend beyond the boundaries of their respective cells, each cell is responsible for bounding boxes whose centers are located inside it. There are too many bounding boxes, and most of them do not include any meaningful objects. Therefore, during the execution of the model, each bounding box receives a confidence score, and bounding boxes whose confidence scores are higher than a specific threshold will be considered as detected objects, and the rest of them will be eliminated.

The confidence score is defined as $P(Object) * IoU$, earlier in this section, we explain the IoU. If no object exists in that patch, the confidence score will be zero. Otherwise, the confidence score is equal to the IoU between the predicted bounding box and the ground truth. Each Bounding box includes five components, coordinates for its center (x.y), width (w), height (h), and a confidence score. Each Bounding box inside a cell includes a probability of containing an object, i.e., $P(Class_i|Object)$. During the inference, YOLO multiplies class probability, i.e., $P(Class_i|Object)$, and the individual box confidence, i.e., $P(Object)*IoU$, to identify the class-specific confidence score for each bounding box.

YOLO architecture is a CNN whose initial layers are used to extract image features. Its CNN has 24 convolutional layers for plain YOLO and 9 convolutional layers for the faster version of YOLO. They name their pre-trained CNN model DarkNet, which is inspired by the design principles of networks like GoogLeNet, but instead of Inception layers, it uses 1×1 reduction layers followed by 3×3 convolutional layers.

The output of DarkNet's last layer, which is a $7 \times 7 \times 1024$ convolutional layer, is followed by two fully connected layers that perform the object identification (by outputting confidence scores and coordinates of the final bounding boxes). Figure 12-39 visualizes the architecture of YOLO.

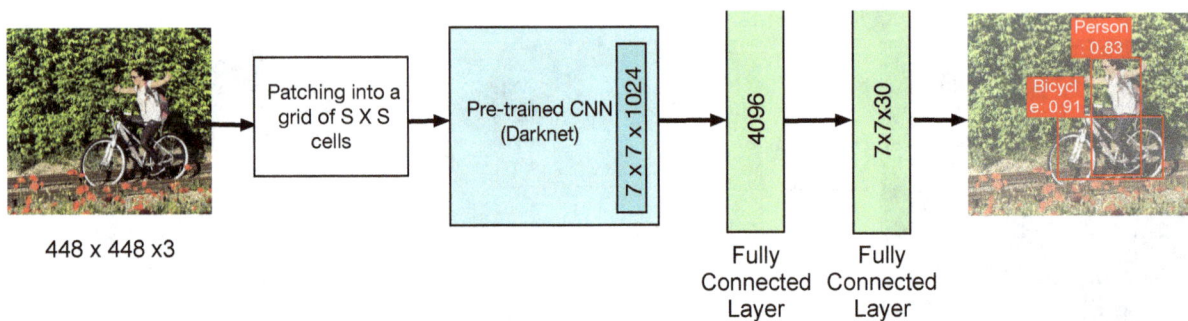

Figure 12-39: YOLO architecture. The last layer of DarkNet is a 7x7x1024 convolutional layer, which will be fed into a 4096 fully connected layer.

Since some objects that are large or near the border of multiple cells can be detected more than once, a "Non-Maximum Suppression" can be used to resolve this issue in YOLO.

YOLO v2

The first version of the YOLO outperforms other models in terms of execution time, but its accuracy is not as good as those models. In particular, YOLO suffers from localization errors and low recall. This led to the introduction of YOLO2 and later YOLO9000 [Redmon '17]. YOLO2 resolves the localization and low recall issue with several improvements, which are listed as follows:

- YOLO v2 uses batch normalization on all convolutional layers.

- YOLO v2 initially trains with 224×224 images and later fine-tunes the network using 448×448 resolution for object detection. While training, YOLO v2 fine-tuned the classification network by using 448×448 resolution images.

- YOLO v1 identifies the coordinates of bounding boxes using a convolutional neural network. Inspired by Fast-RCNN, which uses anchor boxes (region proposal bounding boxes) to predict the coordinates of bounding boxes, YOLO v2 uses anchor boxes. Anchor boxes are determined via the k-mean clustering algorithm applied to the training dataset. However, the k-mean clustering uses a customized distance, which is computed as: $1 - IoU(Box, Centroid)$.

- YOLO v1 does not have constraints on predicting the location of bounding boxes, and this makes the model unstable. YOLO v2 resolves this by using a logistic activation function at the output of prediction.

- YOLO v1 uses 13×13 a feature map, which makes it hard to detect small objects. To improve detection of small objects, YOLO v2 upsamples a 26×26×512 feature map from an earlier layer to match the dimensions of the 13×13×2048 feature map, which is then concatenated with the original 13×13 feature map to enrich the feature representation.

- YOLO v1 uses a fixed image size for the testing phase, i.e. 448×448. YOLO v2 can work with different image sizes. To implement this, YOLO v2 varies the size of the input images during the training phase, changing the dimensions every ten batches. This feature is referred to as multi-scale training by authors.

- YOLO v2's Darknet is faster than the previous version of Darknet because it has 1×1 convolutions that reduce the number of model parameters. Additionally, YOLO v2 adds 11 more layers to Darknet-19, resulting in a total of 30 layers.

YOLO 9000

To have a stronger model that can cover a wider variety of objects, authors use joint training techniques that leverage both ImageNet and Microsoft COCO[32] datasets together [Redmon '17]. COCO includes 100,000 images and 80 class labels. ImageNet includes 13 million images and 22,000 class labels. Unlike COCO, which uses general category labels, ImageNet class labels are organized based on the WordNet hierarchy, presenting a detailed tree structure of relationships. For example, "Norfolk terrier", "Yorkshire terrier", and "Bedlington terrier" are child nodes of the "Terrier" and "Terrier" is a child node of the "Dog". Therefore, classes with the label "dog" from COCO and "Yorkshire terrier" from ImageNet are considered different by a model that simply combines the labels. To handle this issue, YOLO 9000 uses a *Word Tree*. Word Tree is a hierarchical tree structure that relates labels (dog) and sub-labels (Yorkshire terrier) together. The integration of COCO and ImageNet using the Word Tree in YOLO 9000 addresses these challenges, enabling the model to utilize 9418 class labels.

YOLO v3

About a year after the release of YOLO 9000, authors introduced YOLO v3 [Redmon '18], which is a bit slower than the previous YOLO versions, but it provides better feature extraction and multi-scale object detection.

- YOLO v3 uses DarkNet-53 as its CNN model. DarkNet-53 is trained on ImageNet and uses 53 convolution layers. The network is built with consecutive 3×3 and 1×1 convolution layers followed by skip connections.

---

[32] https://cocodataset.org/#home

- In addition to bounding box prediction, YOLO v3 uses a logistic (sigmoid) function to compute an objectness score for each bounding box. This score indicates the probability that a bounding box contains any object and is set to '1' if there is an overlap with a ground truth object based on a certain threshold of IoU. This methodology enhances the precision of bounding box assignments. If a bounding box does not align with any ground truth object, its objectness score approaches zero, which effectively minimizes its impact on the overall loss calculation during training, particularly affecting the localization and class prediction components.

- A predicted bounding box around an object may contain multiple labels. By using softmax to decide on a label, previous YOLO models enforced assigning one label to each bounding box. Authors substitute softmax with an independent logistic function for each class label. This allows YOLO v3 to handle overlapping labels (e.g., women and person).

- It uses a Feature Pyramid Network (FPN) concept [Lin '17] and extracts three bounding boxes at a different scale from an image.

Feature Pyramid Network (FPN): Detecting objects at different scales is an important requirement of object detection models. To address this, models may apply upsampling or downsampling techniques to adjust the scale of images or features. Upsampling increases the resolution of feature maps, aiding in the detection of smaller objects, whereas downsampling reduces resolution, which can help in identifying larger objects. However, training a model directly on pyramids of upsampled and downsampled images instead of the original image is resource-intensive and complex, making it impractical for many applications.

Figure 12-40 presents an image pyramid and compares it with a CNN and SSD approach. We can see that in classical CNN, there have been several convolution layers and the last one is used for the prediction. This approach may struggle with objects of different sizes due to its focus on a single scale of features. SSD utilizes multiple feature maps at various resolutions to detect objects of different sizes. However, unlike FPN, it does not integrate these feature maps but uses them separately to make predictions at corresponding scales. FPN, in contrast to SSD, improves feature integration by adding 'lateral connections' to each feature map. These connections allow the combination of low-resolution, semantically strong features with high-resolution, semantically weak features, as we can see on the right side of Figure 12-40. Lateral connections in FPN function similarly to skip connections in networks like ResNet, enhancing the feature map with information from different layers of the network.

After YOLO v3, the original author, Joseph Redmon, ceased his work on further YOLO developments. We sincerely appreciate his concerns for privacy and not sharing his work for military purposes[33]. The YOLO framework continued to evolve through contributions from other researchers and developers, leading to new versions such as YOLO v4 [Bochkovskiy '20], YOLO v5 [Jocher '20], YOLOX [Ge '21], and so forth.

---

[33] This is the tweet which he stated his justification: https://twitter.com/pjreddie/status/1230524770350817280

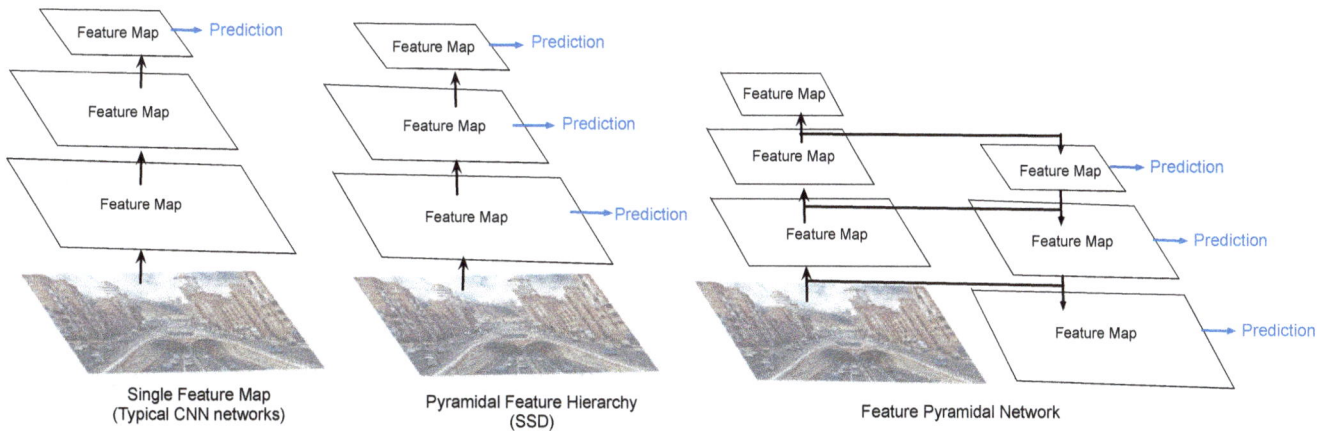

| Single Feature Map (Typical CNN networks) | Pyramidal Feature Hierarchy (SSD) | Feature Pyramidal Network |

Figure 12-40: CNN vs SSD vs FPN network. The difference between SSD and FPN is that FPN uses lateral connection and enables high-resolution features (with weak features) to benefit from low-resolution features (with strong features).

Congratulations! Now, we learn three popular groups of object detection algorithms that have a wide number of applications, from self-driving cars to activity recognition and face recognition systems.

NOTES:

* While reading materials about CNN architecture, remember that the term "block" refers to layers of convolutions that end with a pooling layer. For example, by looking at Figure 12-18, LeNet has two blocks, or by looking at Figure 12-21 GoogLeNet has five blocks.

* In the context of object detection, a small receptive field is incapable of recognizing large objects. Therefore, object detection approaches are using multi-scale feature maps.

* Increasing the number of layers, pooling operations, and dilated convolutions are effective approaches to increase the receptive field size quickly [Luo '16]. Skip connections may share features between different layers but tend to make the effective receptive field smaller.

* It is recommended to favor RCNN models over YOLO or SSD while dealing with small object detection. Initial versions of YOLO and SSD are not working well with small objects in the image.

Semantic Segmentation and Instance Segmentation

If an object detection algorithm highlights each pixel of a particular object in an image but classifies all instances of similar objects with the same color, it is called *semantic segmentation*. If there is more than one particular object in an image and the number of similar objects are all highlighted with different colors, it is called *instance segmentation*. We briefly described these concepts in Chapter 6; as a reminder, check Figure 12-41, which presents the differences between semantic segmentation and instance segmentation on an image. You can see individuals are separated with instance segmentation. We can see that in semantic segmentation or instance segmentation, every pixel is labeled. Before proceeding further, we should be familiar with the term *mask*. A mask in image or instance segmentation typically refers to an image where parts of the image are highlighted or marked to distinguish certain objects from the background. For example, in Figure 12-41, in the middle picture, all humans are masked, and in the instance segmentation one, each individual is masked with a different color.

Figure 12-41: Differences between object segmentation and instance segmentation. Instance segmentation can count the number of specific objects in each segment.

In this section, we list common semantic segmentation and instantiation models. At the time of writing this chapter (first revision 2021 and second one Q2 2024), there are considered state-of-the-art algorithms in this domain, including U-Net, RCNN models, DeepLab models and SAM models.

U-Net

Probably, the most dominant image segmentation model that is still widely in use is the U-Net architecture [Ronneberger '15], as described in Chapter 11. Although it was proposed in 2015, it is still among the most popular image segmentation algorithms, especially since it is very popular for medical image analysis.

U-Net is the most widely used segmentation model. This is due to the fact that it is easy to train and, unlike most models we described here, is not developed by a corporation that uses a huge number of GPUs. Therefore, if you intend to train it on your own dataset and even from scratch (not just fine-tuning), U-Net is the best choice.

Fully Convolutional Network (FCN)

One of the initial approaches that is still in use for semantic segmentation is the Fully Convolutional Network (FCN) [Long '15]. Recall from Chapter 10, we use CNNs for the image classification task, and it operates by first applying a set of convolutional layers to downsize an image, followed by fully connected (dense) layers. Convolutional layers of a CNN apply sub-sampling on the input and make the output smaller for the next layer. Sub-sampling process that reduces the spatial dimensions of the input by down-sampling, often through pooling, to decrease the computational load and retain important features. However, reducing the size of the original input limits spatial resolution while increasing the receptive field size.

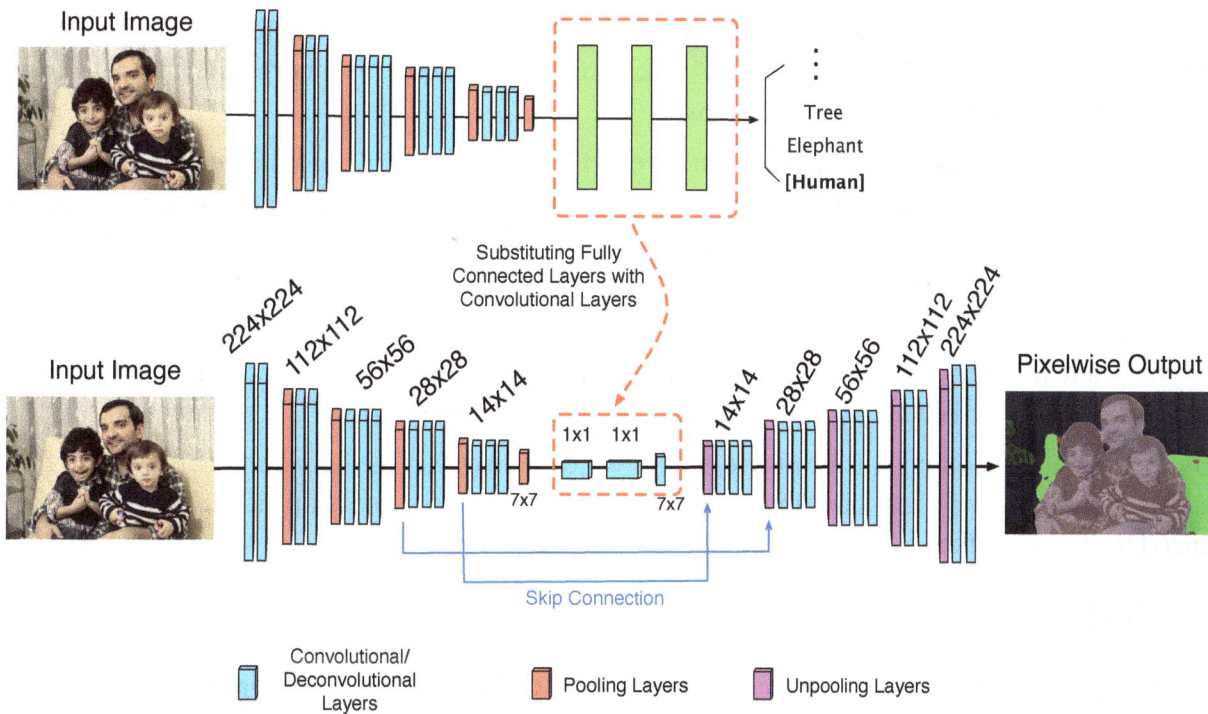

Figure 12-42: An abstract overview of differences between CNN, which VGG-16, on top and FCN in the bottom. FCN has encoder/decoder architecture with skip connections in between.

After convolutional layers, fully connected layers output a vector for the softmax layer and output a single class label. This label specifies the object that occupies the majority of space inside an image. A simplified version of a traditional CNN (e.g., VGG-16) is presented in the top part of Figure 12-42.

Fully Connected (FC) layers output feature maps that are typically smaller than the input data and lack the spatial dimensions needed for segmentation tasks. Having fixed sizes for FC layers limits these models from effectively detecting objects of varying sizes, which is a common requirement in image segmentation processes. To address this limitation and enable the FCN model to segment objects of different sizes, the authors of FCN eliminate all fully connected layers and replace them with two convolutional layers. The first convolutional layer has a size of 1×1 to maintain the feature map dimensions, and the second convolutional layer also uses a 1×1 kernel to output the number of classes. For example, if we have 1000 classes, the second convolutional layer will have $1 \times 1 \times 1000$ dimensions. These two layers are shown in Figure 12-42 as two 1×1 convolutions.

Subsampling imposed by the convolutional layer caused the classification process to identify only one object. To segment objects inside an image, we want to have a classification for every single image in the input image (similar to object detection models). To implement this requirement, the authors add "upsampling" layers at the last layer of convolution (see the bottom part of Figure 12-42). These upsampling layers are transposed convolution.

However, upsampling causes the identified objects to be very coarse and not well separated. How can this be handled? By incorporating the earlier features into later features, which is … YES … the legendary "skip connection". Skip connection helps to incorporate the data from earlier feature maps into later feature maps.

In particular, after passing all convolution layers, the output is small, and the authors propose to apply upsampling (via larger transposed convolution) and make the size of the feature map equal to the size of the input image. However, this output is inaccurate due to the loss of spatial information, and by fusing the upsampling output with previous pooling layers (via skip connections), the loss in the upsampled layers is reduced.

If you want to learn more details about this architecture, we recommend reading the original paper [Long '15] or better checking the open-source implementation of FCN architecture.

By looking at the bottom part of Figure 12-42, we could say that the convolutional layers at first act as encoders, and then one upsampling acts as a decoder, also skip connections between encoders resemble the U-Net architecture.

Mask R-CNN

Mask R-CNN [He '17] is introduced after Faster R-CNN by the same group of authors. It combines the features of FCN (used for semantic segmentation) and Faster R-CNN (used for object detection). As a result, it provides a high-accuracy instance segmentation algorithm.

We have previously explained that Faster R-CNN outputs two things for each detected object: (i) a class label and (ii) a bounding box offset. In the first stage, the Region Proposal Network (RPN) proposes candidate object bounding boxes by scanning the image with an anchor-based approach. The second stage involves the RoI Pooling layer (RoIPool), which extracts a fixed-size feature map from each proposed bounding box. Mask R-CNN keeps the identical first stage as Faster R-CNN, using the same RPN. However, its second stage not only identifies class labels and bounding boxes but also generates a binary mask for each RoI. In simple terms, it adds an additional output called the 'object mask,' which highlights the identified object through specific pixels.

It means the training phase will optimize the sum of losses from classification, bounding box, and object masking, i.e., $L = L_{cls} + L_{box} + L_{mask}$. Note that mask generation and class label identifications are decoupled.

Mask R-CNN substitutes 'RoIPool' with 'RoIAlign'. To understand RoIAlign, we should review the RoIPool first. RoIPool extracts a small feature map (e.g. 7×7), scales the RoIs into square shapes and they will be reshaped into a fixed size. The process of RoI pooling will be done by quantizing a floating number RoI to the discrete value of the feature map. This quantized RoI is then divided into spatial bins (which are quantized as well). Finally, the feature values of bins are aggregated. Quantization is working fine for bounding box construction, but it has a negative impact on pixel-basis classification. Therefore, the authors substitute RoIPool with RoIAlign,

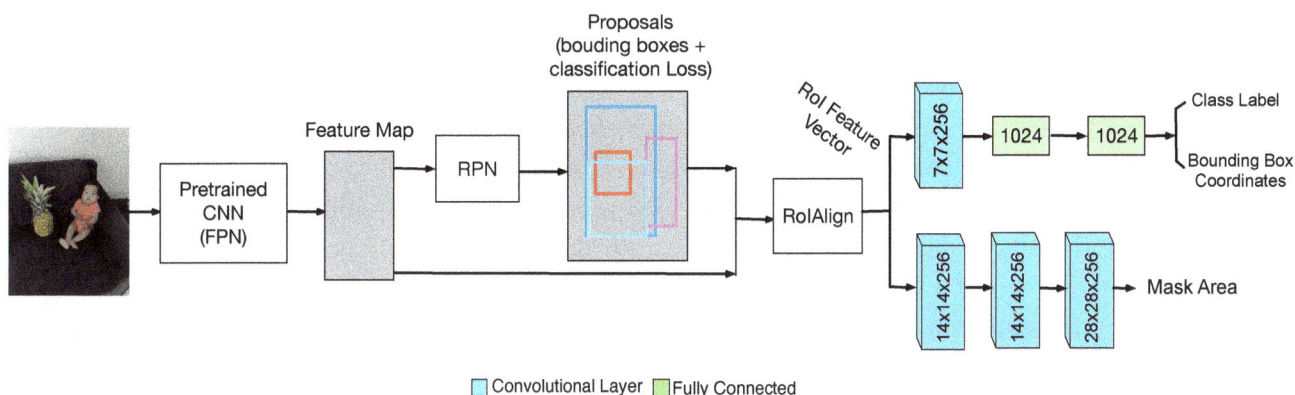

Figure 12-43: Mask R-CNN architecture, with FPN as the pre-trained CNN network.

which simply avoids quantization and uses bilinear interpolation [Jaderberg '15] to compute the exact values of input features. You can check more about quantization in Chapter 14 and bilinear interpolation in Chapter 15.

To implement Mask RCNN, authors separate their architectures into 'backbone' and 'head'. They refer to the CNN network (used for training) as the backbone of their model. They refer to the bounding-box recognition (classification and regression) and mask prediction as the head of the network. To implement the 'backbone' part, both the ResNet and FPN models could be used.

Figure 12-43 presents the Mask R-CNN architecture with FPN as a pre-trained CNN model, with the ResNet the architecture is slightly different after RoIAlign, and you can check it from the original paper.

DeepLab Models

Usually, semantic/instance segmentation models such as U-Net or FCN operate with an encoder that uses a CNN to downsample the image. This is followed by a decoder that upsamples the reduced-resolution feature map and identifies pixel-wise segments, assigning a class label to each segment. Upsampling is typically implemented using transposed convolution or other deconvolution techniques, which are computationally expensive. Semantic/instance segmentation is crucial in real-time applications like self-driving vehicles or drones, which require quick decision-making. DeepLab [Chen '14, Chen '17 A, Chen '17 B] is another series of models developed to make semantic/instance segmentation faster and more resource-efficient.

DeepLab v1

The first version of DeepLab [Chen '14] was proposed in 2014. DeepLab v1 explained that CNN alone cannot be used for segmentation because convolutional layers downsample the original image, and the feature map (results of convolution) is spatially invariant. It applies some changes to the VGG-16 architecture to enable it to perform segmentation. DeepLab v1 has two main characteristics. First, it uses dilated (Atrous) convolution (check Chapter 10) to perform upsampling of the feature map and thus handle image objects with different scales. Second, it employs a combination of probabilistic graphical models and CNN to improve the localization of object boundaries. Authors named this combination as Fully Connected Conditional Random Field (CRF). In the following, each characteristic will be explained in more detail.

(i) *Dilated (Atrous) Convolution:* The output of each convolution result is a feature map whose size is reduced. This causes a reduction in the resolution of input data in the last layer of the network. After each convolution, it is possible to apply a dilated convolution and not lose the image resolution by adding zeros between feature data. By dilated convolution, we increase receptive fields while the number of parameters increases at a constant rate, and the spatial resolution of the feature maps is reduced. This is particularly beneficial after max-pooling layers, which typically reduce spatial dimensions.

DeepLab v1 uses dilated convolution to replace fully connected layers, aiming to enhance the network's capacity for context integration without the computational burden of fully connected layers.

To recall dilated convolution, check Figure 1-37 in Chapter 10. It presents a simple form of dilated convolution for upsampling. You can see that the initial receptive field is 3×3, but by using the dilation rate of 2 (one-pixel space is added between initial pixels), we get a receptive field of 5×5, and by using a dilation rate of 3, we get a receptive field of 7×7.

To further refine the resolution of segmentation results, DeepLab v1 employs bilinear interpolation. This technique is used to upscale the feature maps to higher resolutions, improving the alignment with the original image size. In Chapter 16, we will explain bilinear interpolation in more detail.

(ii) The second approach authors used for their segmentation model is the use of CRF. Briefly, CRF treats an image as a graph in which its pixels are connected. This means that each pixel is a node that is connected to its adjacent pixels. The CRF has a good short-range understanding, and to increase its understanding range, authors use Fully Connected Pairwise CRF, in which each pixel is connected to all other pixels. Therefore, CRF classifies a pixel not only based on its label but also based on other pixel labels.

It is known that deeper models with multiple max-pooling layers are able to classify objects accurately. However, deep models increase the invariance of classification, which is desirable, and the large receptive fields of the first layers cannot be handled by deep models, which is not desirable. In other words, multiple max-pooling layers cause poor localizations on edges and this harms the segmentation or bounding box detection in the image. To handle this issue, authors use CRF, which supports accurate localization

DeepLab v1 architecture is inspired by VGG-16 architecture. Briefly, (i) it substitutes fully connected layers at the end of VGG-16 with convolutional layers and increases feature resolution through dilated convolutional layers. (ii) it applies bilinear interpolation to upsample the feature map to the original image resolution. (iii) At the end, a CRF will be applied to improve segmentation accuracy and remove weak classifiers.

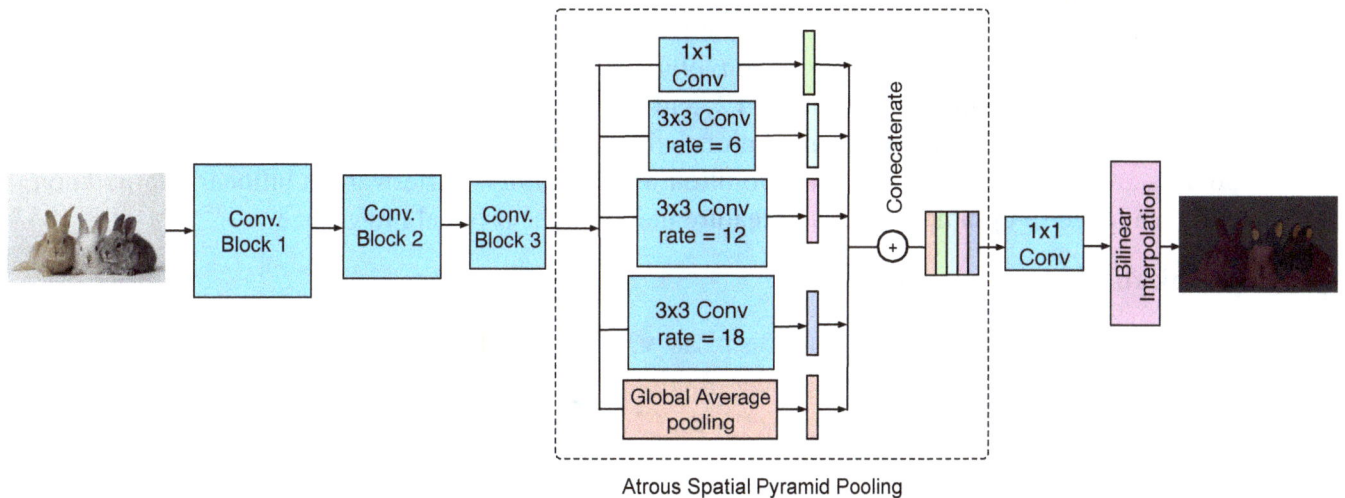

Figure 12-44: DeepLab v3 architecture.

DeepLab v2

After DeepLab v1, in 2017, the same group proposed DeepLab v2 [Chen '17 A], which can utilize either VGG-16 or ResNet101 as its backbone network. It retains the core architecture of DeepLab v1, including the use of dilated convolution and CRF. A significant addition in DeepLab v2 is the implementation of Atrous Spatial Pyramid Pooling (ASPP) to effectively segment objects at multiple scales.

ASPP handles varying sizes of segmentation by using multiple parallel dilated convolutional layers with different sampling rates. This approach is conceptually similar to the Stem module (see Figure 12-27) in the Inception model, where parallel convolutions are employed. However, in DeepLab v2, four parallel dilated (Atrous) convolutions of size 3×3 with different dilation rates (6, 12, 18, and 24) are used simultaneously. This setup allows the network to capture a broad range of contextual information from the image at multiple scales, enhancing its ability to segment complex scenes.

DeepLab v3

DeepLab v3 [Chen '17 B] was introduced in late 2017. It improves the ASPP in extracting dense features by incorporating more dilated convolutional features at multiple scales. The authors introduce the term 'output stride' as a ratio of input image spatial resolution to the final output resolution.

Atrous convolution changes the field of view by its stride (a.k.a dilation rate). The larger stride size enables us to have a larger field of view. The ASPP includes using one 1×1 convolution and three 3×3 convolutions with rates = 6, 12, and 18 for output stride = 16. The resulting feature map from all the branches will be concatenated and passed through another 1×1 convolution. Next, bilinear interpolation will be applied to generate the final results. Figure 12-44 presents the architecture of DeepLab v3. They removed the CRF layers at the end. This reduces the computational complexity of the model while retaining its accuracy.

In Figure 12-44, we can see that Block 1 to Block 3 are similar to ResNet. They are used to reduce the feature map size, but next to the third block, there are four parallel ASPPs, each with different rate sizes, including 1×1 convolution, 3×3 convolution, with a dilation rate of 6, 3×3 convolution, with dilation rate of 12, and 3×3 convolution, with dilation rate of 18. Next, the output of those four dilated convolutions will be concatenated together with a global average pooling and passed through 1×1 convolution with 256 filters. Afterward, a bilinear interpolation will be applied to upsample the result feature map into the original image size.

DeepLab v3+

At the time of writing this section, DeepLab v3+ [Chen '18] is the last version of this model series. DeepLab v3+ adds the decoder module to the DeepLab v3, similar to other image segmentation algorithms, such as U-Net and FCN. Adding an encoder module enables the model

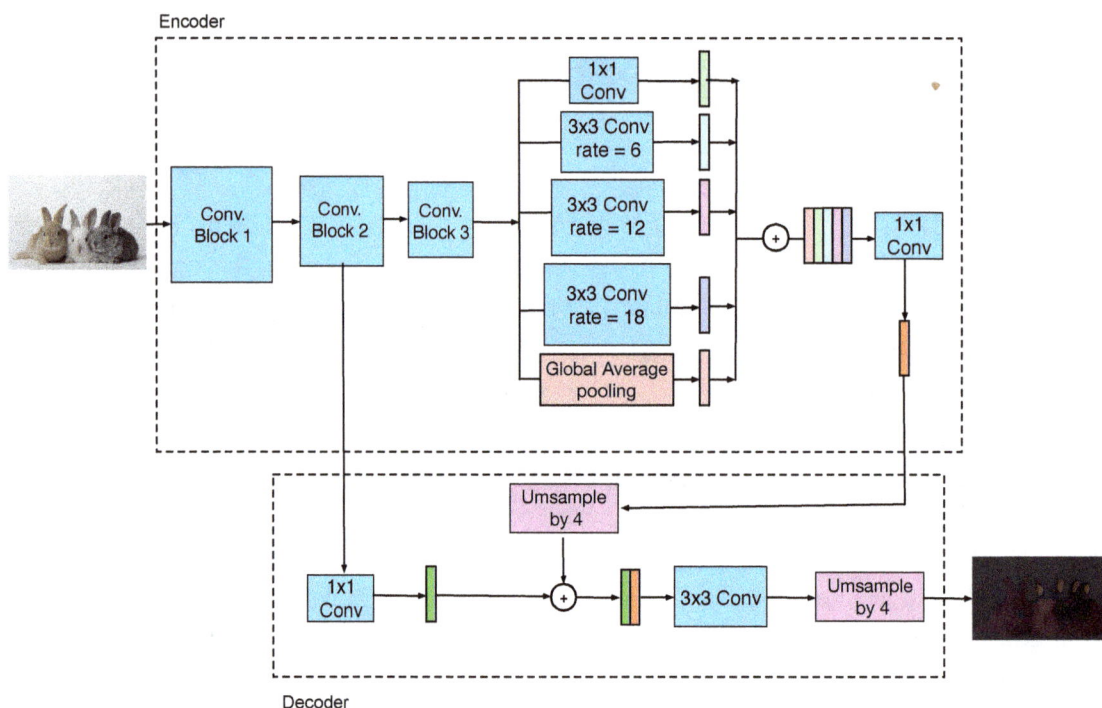

Figure 12-45: DeepLab v3+ architecture. Note that the encoder architecture is the DeepLab v3.

to capture sharper object boundaries (by recovering from spatial information in each decoder layer). For the pre-trained CNN, they experimented both with ResNet 101 and a modified version of the Xception model [Chollet '17], Xception is a variety of Inception models.

Figure 12-45 presents DeepLap v3+ architecture. The authors use DeepLab v3 as the base encoder structure, and adding a decoder module is the main novelty of this model. First, the decoder upsamples the encoder's output by a factor of 4 using bilinear interpolation, then concatenates it with the corresponding low-level features from an earlier layer of the encoder (e.g., from block 2 in ResNet-101). The low-level features contain a large number of channels, and this outweighs the rich encoder features, and thus harms the training. To mitigate this challenge, the decoder applies a 1×1 convolution on the low-level features to reduce the

782

channel dimensions (e.g., from Conv Block 2). Then, the result of 1×1 convolution is concatenated to the upsampled version of the encoder output. Next, it applies another 3×3 convolution (to refine the features), followed by another bilinear interpolation (this time in the decoder) for upsampling by a factor of 4.

Segment Anything Model Series

Segment Anything Model (SAM) [Kirillov '23] is a very powerful segmentation model released by Meta in 2023. At the time of revising this section for the 100th time (Q3 2024), it includes the largest number of images used for training. It is trained on an image dataset, referred to as SA-1B, which includes over 1 billion segmentation masks generated from 11 million images. The dataset is open to access and use. These segmentations are either manually annotated or semi-manually. You can read about the annotation process in the appendix of the original paper. Later, SAM v2 [Ravi '24] was released in 2024, which covers video segmentation as well.

SAM v1

LLMs that are trained on vast datasets are known for their capability for zero-shot learning, which allows them to handle tasks on unseen data by inferring from related data. These models are often categorized as *foundational models* due to their applicability across a variety of tasks. SAM is inspired by the zero-shot learning capabilities of these foundational models and proposes a foundational model for computer vision. In other words, SAM constructs a foundational model for image segmentation capable of segmenting images that were not present in its training dataset (zero-shot learning).

SAM architecture includes three components for training the foundational model, including image encoder, prompt encoder, and mask decoder.

Image Encoder: SAM uses a heavy image encoder to train embeddings from input images. This image encoder uses the Masked Autoencoder (MAE) [He '22], pre-trained with Vision Transformer (ViT-H/16) with a window size of 16×16. We skip explaining MAE here, but it is easy to understand by checking the original paper. This encoder has been pre-trained by using 256 GPUs. As a result, it can handle high-resolution inputs.

In summary, the image encoder is a pre-trained model that transforms visual inputs into detailed embeddings. It receives a 1024×1024 size input, and the embedding is downscaled by a factor of 16 to $64 \times 64 \times 256$, resulting in a 64×64 embedding with a depth of 256 channels.

Prompt Encoder: While dealing with LLMs, a prompt is typically a textual input that guides the model's response. In the context of SAM, the notion of a prompt is expanded to include more than text, such as spatial cues, including points, bounding boxes, and image masks. These prompts guide the segmentation task.

The prompt encoder transforms input prompts into embeddings that are compatible with the image embeddings produced by the image encoder. As explained earlier, the prompt encoder can handle various types of prompts, ranging from sparse prompts (such as points and bounding boxes) to dense prompts (such as masks), as well as free-form text. Figure 12-46 presents the prompt encoder of SAM.

The prompt encoder includes a group of models to implement the embedding. Depending on the input type, a different model is used for embedding the input prompt. If the prompt is text, it uses CLIP (Check Chapter 11). If it is a simple spatial prompt, such as points or boxes, it just uses a positional encoding. In particular, for a point, its x,y coordinates, and for a box, the top left corner and bottom right corner of the rectangle coordinates (x,y values) will be used

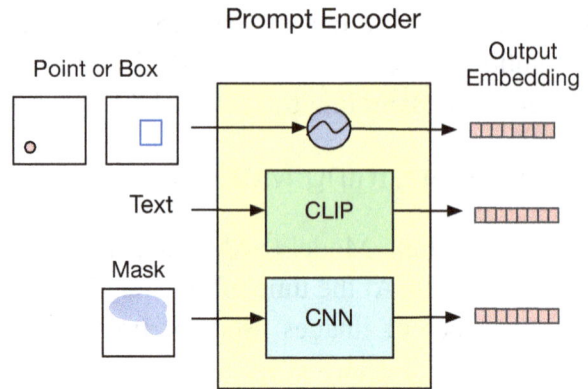

Figure 12-46: Prompt Encoder architecture of SAM.

as positional encoding. The use of positional encodings helps the model understand where something is located in space. Additionally, they are labeled as foreground or background. The foreground indicates the areas of interest for the embedding, whereas the background indicates areas to exclude. For more complex spatial prompts, i.e., image masks, we need detailed spatial information across a segment of an image. Therefore, three layers of CNN are used to construct this embedding. CNNs can maintain spatial hierarchies while extracting features from images and image-like data structures such as masks. To feed the mask into the CNN, the resolution of the input masks is reduced to one-fourth that of the input image. Then, it is downscaled four more times using two 2×2, stride-2 convolutions with output channels 4 and 16, respectively. A final 1×1 convolution maps the channel dimension to 256. The activation function used is GELU activation, and each layer includes normalization. The result of CNN embedding is combined with image embedding with 'element-wise summation', which means that each element of the mask embedding is added to the corresponding element of the image embedding.

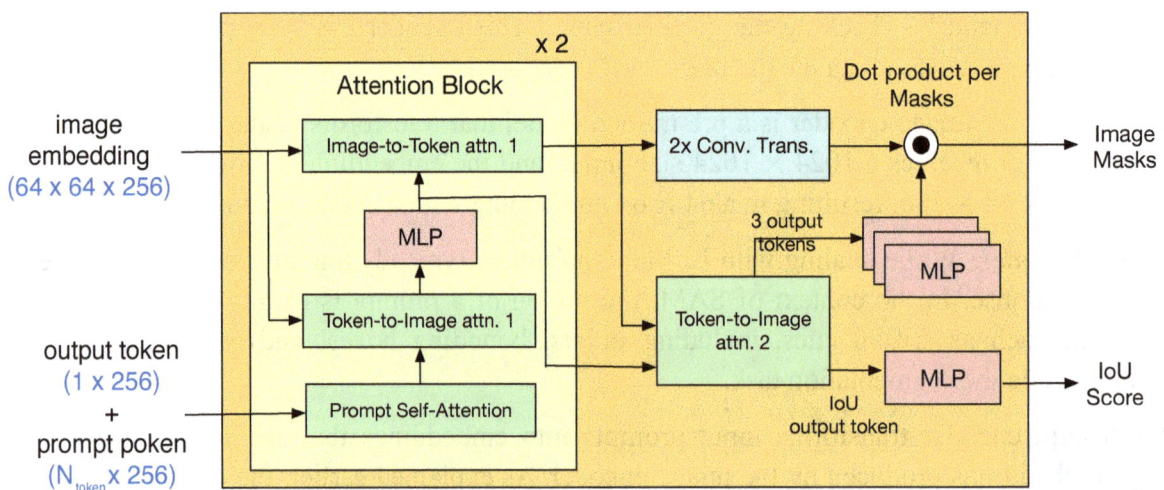

Figure 12-47: SAM's mask decoder architecture.

As we have explained, the output of the image encoder will be 64 x 64 x 256. The output of the prompt encoder also has 256 channels, matching the depth of the image encoder's output. Therefore, masked or prompt embedding is added to image embedding, and this embedding, which has the size of 64 x 64 x 256, is passed to the Mask Decoder.

Mask Decoder: The mask decoder is a decoder that is responsible for combining image embeddings and prompt embeddings and outputs the mask. It is lightweight because it should operate in real-time. The mask decoder handles ambiguity by outputting three masks for a single prompt. The authors stated that a nested mask often has three levels of depth: the whole, the part, and the subpart. The mask that has the highest IoU score will be output as the final mask.

Figure 12-47 presents the architecture of the mask decoder. We use the original image but add numbers and names to reduce its vagueness. Before reading each component, keep in mind that the token refers to the output components or features used by the mask decoder to generate segmentation masks.

This decoder is composed of *two transformer layers* of attention block to handle distinct aspects of the segmentation process. One layer is responsible for generating coarse segmentation masks, such as basic shapes and larger object boundaries. Another layer refines these masks to capture finer details, such as textures, edges, and small features. We present these two layers as the *Attention block* component in Figure 12-47. This name is not used in the original paper, we found it from their code and used it to make it easier to understand[34]. The following are four components that are repeated in each layer:

1. *Prompt self-attention:* It updates the prompt embeddings by attending to themselves, allowing it to capture better relationships and dependencies within the prompt data.

2. *Token-to-Image attn. 1 (Cross-Attention):* It aligns the prompts with relevant regions in the image by enabling the prompt embeddings to attend to the image embeddings. The query vectors come from the prompt embeddings, and key and value vectors come from the image embedding.

3. *MLP (between Token-to-Image and Image-to-Token):* This MLP is used to transform the combined embeddings from the Token-to-Image attention into a suitable form for the subsequent Image-to-Token attention (*Token-to-Image attn. 2*).

4. *Image-to-Token attn, 1 (Cross-Attention):* It facilitates the integration of prompt information into the image context by allowing the image embeddings to attend to the prompt embeddings. The query vectors come from the prompt embedding, and key and value vectors come from the prompt embedding.

To ensure the inclusion of the geometrical location of each token, positional encoding is added whenever the image or prompt embeddings participate in an attention layer. You might wonder why so much cross-attention is required to unify the image and prompt embeddings. Since there is no direct connection between the two embeddings, this attention mechanism refines them, ensuring they form a unified input for subsequent layers.

---

[34] https://github.com/facebookresearch/segment-anything/blob/main/segment_anything/modeling/transformer.py

After these two layers of attention refinement, the refined data will be sent to '*2x Conv. Trans.*' component and '*Token-to-Image attn. 2*', which are explained as follows.

2x Conv. Trans. (Upsampling): The refined image embedding is upsampled via two transposed convolution layers to the original image resolution. Upsampling ensures that the mask predictions correspond to the input image dimensions. In particular, the input of *2x Conv. Trans.* is an image-embedding (which is refined with prompt embedding) that has the size of 64 x 64 x 256 (64 x 64 is feature space and 256 is channel), and the output will be upsampled to 512 x 512 x 256 (512 x 512 is feature space and 256 is channel). On the other hand, from the refined prompt embeddings, it extracts three output token embeddings.

Token-to-Image attn. 2 (Cross-attention): It is used to extract token embedding from the prompt embedding and image embedding. In other words, it enables the model to attend from the token (prompt) space to the visual features of the image. Its keys and values come from the image embedding and queries from prompt embeddings. It outputs two types of tokens: three tokens for mask prediction (one for each mask) and one token for IoU score prediction.

MLPs and dot product: Each of the output tokens for mask identification goes into one MLP (in total, three MLPs) for linear classification, and the result is combined with the upsampled tokens via dot product. The output is three different masked images. In particular, these three MLPs map the output token from the decoder to a dynamic linear classifier, which then computes the foreground probability for each pixel, resulting in the predicted segmentation mask.

The output token for the IoU assignment goes into an MLP, for linear classification, resulting in the IoU score for each predicted mask. The IoU score measures the overlap between the predicted mask and the ground truth mask, thereby providing a quantitative evaluation of the accuracy of the segmentation. This score is then used to rank the predicted masks, ensuring that the best-fitting mask is identified for each object. Note that all three MLPs have a unique architecture.

SAM Architecture: We do hope you understand this complex but powerful architecture of encoder and mask decoder. We can summarize the overall SAM architecture in Figure 12-48.

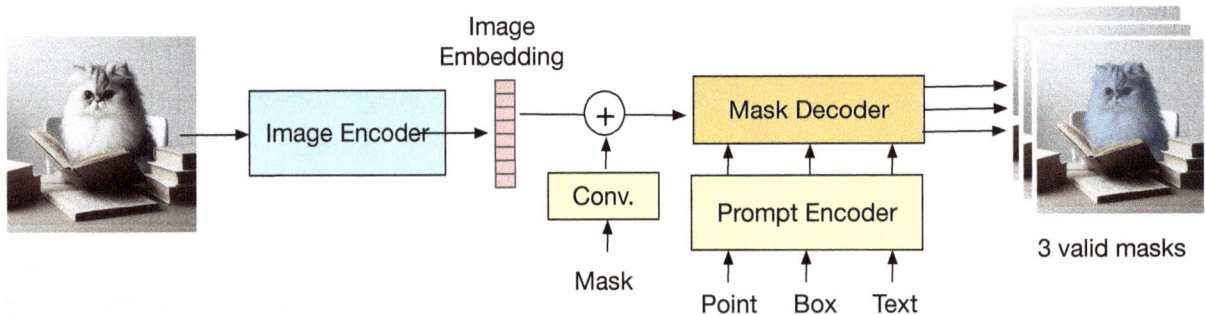

Figure 12-48: SAM architecture.

If the prompt is Mask, it passes to a convolutional layer to downscale it, and then it is combined with image embedding via element-wise addition. This convolutional layer is necessary to ensure that the mask and image embeddings are of compatible dimensions and can be combined to produce accurate segmentation outputs.

Keep in mind that both the image encoder and prompt encoder are pre-trained on large data. However, the mask decoder is trained jointly with the rest of the model once the image encoder has been integrated.

SAM v2

SAM v2 [Ravi '24] was introduced in 2024 and incorporated video segmentation along with image segmentation as well. It has a simple transformer architecture, which makes it significantly faster than SAM v1. They introduce a video dataset, the Segment Anything Video (SA-V) dataset, which consists of 35.5M masks across 50.9K videos. Videos often have lower quality and resolution in comparison to images. SAM 2 analyzes video frame by frame and applies segmentation. Before we explain SAM v2 architecture, we should understand its memory and Promptable Visual Segmentation (PSV) components.

Memory: To enable video segmentation, SAM 2 includes a memory component that stores information about previous frames. The memory module is used only for video segmentation. Objects of interest can get occluded by another object in the next frame, such as a person who is running around a tree. In this example, the object of interest is a person, and in some frame, he is occluded by a tree. Memory bank keeps track of previously generated masks for objects of interest. Using the memory helps maintain consistent segmentation of the object of interest even when it becomes temporarily occluded. In particular, the memory block ensures that the segmentation remains accurate and continuous across the video despite challenges such as occlusions, where the object might be partially or fully hidden in one or more frames.

Promptable Visual Segmentation (PSV): The authors introduce a concept referred to as Promptable Visual Segmentation (PSV). It enables the user to build a prompt (with boxes, dots, or dense mask) and thus specify an object of interest in a video. Then, the model predicts an object mask, known as a *Masklet*. Masklet is a spatiotemporal mask that segments objects of interest in a video. It can be refined with more prompts in the upcoming frames, with positive (it is the same object) and negative clicks (it is a different object)[35].

Figure 12-49 presents SAM 2 architecture, and each of its components is described briefly in the following section.

Image Encoder: The image encoder is a Masked Autoencoder (MAE) [He '22], pre-trained with a hierarchical transformer. It gets the image and outputs a tensor of $64 \times 64 \times 256$. Each frame is processed separately and independently, and video frames do not have any temporal condition.

Memory Attention: It is composed of a stack of transformer blocks; each block performs a self-attention, followed by a cross-attention to objects stored in the memory block.

Memory Bank: The memory bank preserves information about previous predictions for the target object in the video by maintaining a First-In-First-Out (FIFO) queue of memories for up to N

---

[35] You can experiment with SAM 2 to better understand the PSV in this link: https://sam2.metademolab.com

recent frames. It also stores information from prompts in a separate FIFO queue for up to M prompted frames.

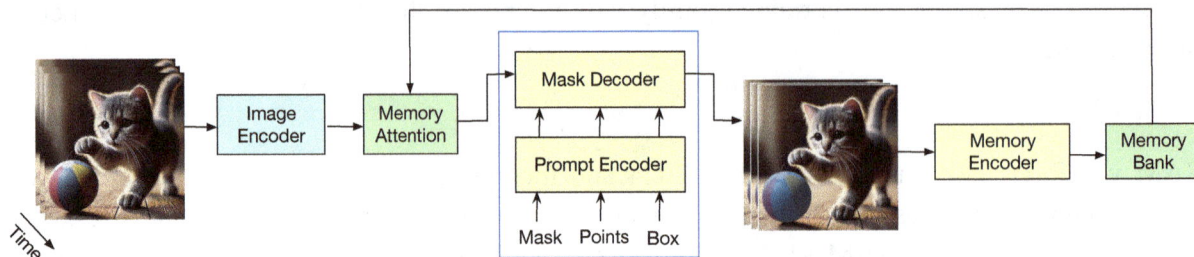

Figure 12-49: SAM v2 architecture.

Prompt Encoder: It is responsible for converting different prompts (not just text but also points, boxes, masks, and positive or negative clicks) of a frame into an embedding representation that aligns with the feature maps generated by the image encoder, allowing for contextual interaction between the prompts and image features.

Mask Decoder: The mask decoder utilizes the embeddings from the prompt encoder and the feature maps from the image encoder, integrating them with the context stored in the memory bank. It applies a series of convolutional and upsampling layers to generate high-resolution binary masks that accurately outline the target objects in the current frame. The mask decoder refines the mask predictions by leveraging both the spatial details from the image encoder and the temporal consistency provided by the memory bank. The mask decoder architecture is very similar to the mask decoder of SAM v1, which is presented in Figure 12-48.

3D View Synthesis

Advancements in computer vision enable the construction of a 3D model from a set of multiple photos. For example, we can use our phone camera, take several selfies, and then use these methods to build a 3D avatar of ourselves or any other person or artifact.

NeRF and 3D Gaussian splatting are two groups of methods for 3D view synthesis (3D model building). Before explaining them, we should get familiar with several concepts in this domain, not necessarily all of them used in our explanation. However, to read and understand related literature, it is crucial to be familiar with them.

At the time of writing this section (Q2 2023, and revising it Q3 2024), NeRF and 3D Gaussian splatting are hot topics in computer vision, and many research works are in progress in this area. Therefore, it is not clear which models will stand the test of time. We adhere to explaining the first models of NeRF, but there are a vast number of research works around this topic[36]. We follow the same approach by explaining the original 3D Gaussian splatting model, which was introduced in August of 2023.

---

[36] https://github.com/awesome-NeRF/awesome-NeRF

Although NeRF and 3D Gaussian splatting are new, they could make a big shift in the game industry, urban planning, virtual reality, and other industries that benefit from geographical simulation. For example, a part of the city in a game could be constructed in 3D by just feeding some 2D images of that part from different angles. We recommend you search for NeRF and 3D Gaussian splatting videos online to understand their capabilities better; within images on static medium, as in this book, it is hard to present its capabilities.

Concepts

Geometric shape: A geometric shape is any shape that can be described mathematically, regardless of whether it has two or three dimensions.

In Chapter 2, we described three different approaches to representing 3D data, including Point cloud, Voxel, and Mesh presentations. NeRF models use volumetric representation, which is similar to voxel representation. However, it does not use discrete voxels. NeRFs represent a scene as a continuous volumetric field function. On the other hand, 3D Gaussian splatting initializes its 3D Gaussian splats from a point cloud.

Primitives: Primitives are defined as the basic building blocks of 3D graphics. If we use mesh representation, each triangle is a primitive; if we use voxels, each cube is a primitive; and if we use a point cloud, each point is a primitive. They could also be lines representing the edges of objects. Deep learning-based representations (such as NeRF) are not composed of simple geometric primitives. Instead, they use a neural network to represent a scene.

Rendering: The process of employing a 3D model and building a 2D image is referred to as rendering. There are three common methods for rendering: rasterization, ray casting (a subset of ray tracing), and full ray tracing.

The rendering process solves two problems: *visibility* and *shading*. Visibility refers to determining which objects or surfaces are visible from a particular viewpoint, while shading refers to computing the color and brightness of each visible object or surface based on its material properties, lighting conditions, and other factors.

Rasterization: Rasterization is the process of converting the geometric data of a 3D model, typically represented as meshes (with triangles as the primitive shapes), into a raster image composed of pixels. This technique addresses the visibility problem by determining which parts of the model are visible from the viewer's perspective and maps these parts onto pixels on the screen. It also incorporates color and shading into these pixels based on coverage, material properties, and lighting conditions.

The first step in rasterization is to divide the 3D scene into a grid of pixels. In other words, the first step involves projecting the 3D coordinates of a scene into 2D screen coordinates. In the second step, the geometric primitives, such as triangles, are converted into fragments. The *fragment* is an intermediate data state for each pixel, containing all the necessary information to compute the final pixel color, such as its position, depth, and color. These fragments correspond to pixel positions in the *frame buffer*, which is a memory buffer that stores the color and other data for each pixel. The third step determines the final color of each pixel. This is achieved by

some calculations, including lighting and shading, based on the scene's geometry and material properties.

Ray casting: Solves the visibility problem of rendering, and it generates a number of rays (simulates the path of lights) from the viewpoint of the observer, shoots them at the scene (which includes the target 3D shape), and tests if the rays hit any primitive in the scene. If the ray hits a primitive (on the 3D shape), then the color and distance along the ray from the observer's viewpoint to the primitive is determined, and one pixel with the same color as the primitive at the point the ray hits is created on the 2D image. It calculates the intersection of rays emitted from the observer's viewpoint with primitives in the scene. The position of each pixel on the 2D image is determined by projecting these intersection points onto the 2D plane. Ray casting is slower than rasterization due to the complexity of these calculations, but it typically offers higher accuracy, particularly in terms of visibility and lighting effects.

Ray tracing: It solves both the visibility and shading problems of rendering. It can simulate the lighting of the scene realistically. Similar to ray casting, it simulates the path of light rays from the observer's viewpoint to the scene (which includes the target 3D shape) and tests if the rays hit any primitives of the shape. If the ray hits a primitive of the 3D shape, then the distance along the ray from the observer's viewpoint to the primitive is determined, and one pixel with the same color will be created on the 2D image.

Until now, everything is similar to ray casting, but ray casting only collects color and visibility. Ray tracing, however, uses more complex approaches and collects reflection, refraction, and other lighting effects. This results in having more realistic images with ray tracing than ray casting. Ray tracing is the slowest method in comparison to ray casting and rasterization, but it is the most accurate one. Modern GPUs made real-time ray tracing possible, which is an important need for game engines.

790

Volumetric rendering: Many visual objects, such as fluids, clouds, fire, and smoke, are volumetric, and thus, it is difficult to model them with geometric primitives. Volumetric rendering is a group of methods that make a 2D image from a set of "discretely sampled" values of a 3D shape (e.g., density, opacity, color, etc.). NeRF methods use volumetric rendering.

Volume density: It is a property that describes the distribution of matter in a volume, which can affect how much light is scattered or absorbed by the object at a specific point in space.

Radiance field: A radiance field is a function that maps a point in space to the amount of light that is emitted from that point in all directions. The input of the radiance field is 5D, and it includes three dimensions for the spatial coordinates (x,y,z) and two dimensions for the direction of the light rays (inclination[37] and azimuth[38]). The output of the radiance field is the amount of light (color) that is emitted in all directions. This output is measured as the radiance (color) of light exiting that point, and it can be represented as a scalar value, an RGB color value, or a vector of other physical quantities such as luminance, irradiance, or radiance intensity. NeRF (Neural Radiance Fields) uses neural networks to compute the radiance field.

View Synthesis: View synthesis involves generating images from different viewpoints based on the learned representation of the given scene or image. Both NeRF and 3D Gaussian splatting are view synthesis methods.

View-dependent emitted radiance: It is the amount of light emitted from each point in the space in a particular direction. It is used to determine the color and brightness of the object at that point from a particular viewpoint. For example, imagine a fire burning in a fireplace. The fire is emitting light in all directions, but the amount of light emitted in a particular direction will depend on the direction of the viewer. If we are standing in front of the fireplace, the fire will appear brighter than if we are standing to the side.

Alpha blending: Alpha blending or alpha compositing is a technique that is used to combine two images, one image on top of the other image, with the ability to control the opacity (more transparent or less transparent) of the top image. The final image is referred to as a "composed image". Alpha is referred to as transparency or opacity. For example, we have two images, one of a tree and one of a house. We can use alpha blending to combine two images, and the tree appears to be in front of the house. We can control the opacity of the tree to make it more or less transparent, and thus, we can see more or less of the house behind the tree.

Alpha blending has lots of applications, such as overlaying text or other animations on an image, which has applications in augmented reality devices (e.g., presenting the energy bar of the player), rendering shadows and reflections in video games, using visual effects to combine Computer-Generated Imagery (CGI) with live-action footage (a superhero is flying in the air and alpha blending is used to combine the actor's footage with the CGI background).

Anisotropic and Isotropic Object: Anisotropic refers to an object that has different physical properties while measuring it from different directions. Many physical objects are anisotropic;

---

[37] Inclination is the angle or slope of an object or surface in relation to a horizontal plane.

[38] Azimuth refers to the horizontal angle or direction of an object or location measured clockwise from a reference point, such as North.

for example, wood is an anisotropic object. In contrast, isotropic objects exhibit the same physical properties regardless of their orientation or the direction of measurement. They behave uniformly in all directions. Examples of isotropic objects include glass, water, and most metals.

Anisotropic covariance: It refers to the property of a covariance matrix where the variances and covariances between variables are not constant in all directions. This means that the correlation between two variables can vary depending on the orientation of the data. For example, consider measuring the temperature in a garden; under the shadow of a tree, the temperature is lower because there is no sunlight, but elsewhere, there is sunlight, and thus temperature is higher. In this example, the covariance between temperature measurements would be anisotropic.

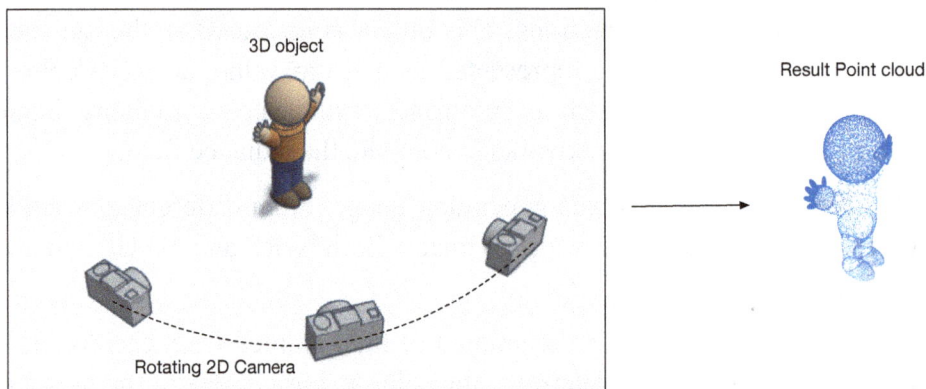

Figure 12-50: Structure from Motion (SfM) identification process, the output is presented as a point cloud.

Structure from Motion (SfM): It is a technique that utilizes a series of two-dimensional image sequences (taken from a moving camera) to reconstruct the three-dimensional structure of a scene or object. Figure 12-50 presents the abstract overview of the SfM construction overview. The result of an SfM construction is usually presented in the form of a point cloud, which represents the external surface of the scene or objects as a set of points.

2D Splats: A 2D splat refers to the process of projecting or "splatting" a point from a 3D space onto a 2D image plane. By rendering all the 3D dots as different-sized splats, we get a 2D "heat map" showing the relative 3D positions. This allows us to visualize the full 3D point cloud on a 2D image. The splat is usually a circle or Gaussian shape centered on the dot's projected 2D location. The size and color of the splat depict the dot's depth - bigger and brighter means closer, and smaller and darker means farther away.

Frustum culling: It is a technique that removes objects that are not visible from the screen.

Neural Radiance Field (NeRF) Model

The Neural Radiance Field (NeRF) model [Mildenhall '21] was introduced in 2020. NeRF is a neural network model to construct different views (each has a different camera position) from given 2D images of a single scene. NeRF models are one of the most successful computer vision approaches that have four main applications, which we list as follows:

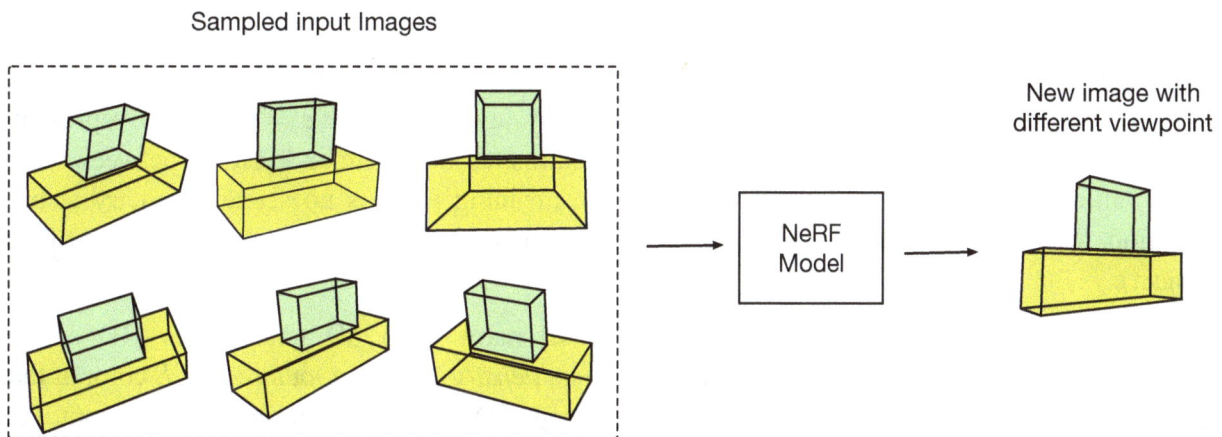

Figure 12-51: A toy example of NeRF model. To build the model, a set of 2D images feed into a NeRF model and NeRF reconstructs the 3D model of the scene or a new viewpoint that does not exist in the training set. The image on the right is the viewpoint that does not exist in the image sets on the left.

- View synthesis: As mentioned earlier, NeRF can generate novel views or images of a scene from a limited set of input views.
- 3D reconstruction: NeRF can also be used for 3D reconstruction of a scene from a set of 2D images.
- Inpainting: NeRF can reconstruct or fill in missing information in an image by synthesizing a novel view that completes the missing information. This is also known as view interpolation, and in cases such as video compression where some frames are removed, NeRF can be used to reconstruct them back.
- Image-based rendering: NeRF can generate high-quality 2D images of a scene by synthesizing novel views that capture the scene's geometry and appearance[39].

Look at Figure 12-51; here, we present a toy example that takes six images (in reality, we need more images, but this is a toy example) with different angles and different zooming levels from a scene sampled. Then, they were given to the NeRF model, and the NeRF can construct a different viewpoint of the given scene.

The original work of NeRF [Mildenhall '21] receives several images from different angles of a scene and uses a Multilayer Perceptron (MLP) to construct different views from the given 2D scene.

The input of the NeRF model has five dimensions. Three dimensions specify spatial coordinates (x, y, z), and two specify the observer's viewpoint direction, including inclination (θ) and azimuth (ϕ) angles.

---

[39] If still, it is hard to understand, try to find some online movies that present how NeRF makes view synthesis from a single image.

As it is shown in Figure 12-51, NeRF gets several input images from different directions of a scene. Each image includes coordinates and the observer's viewpoint direction. The NeRF model uses MLP to learn the representation of this model. The output of the MLP model is a pair of values, i.e., a *volume density* σ and RGB color. Volume density represents the opacity and transparency of the scene at a given point along the ray, and RGB color $c = (R, G, B)$, represents the RGB color, at that point. The volume density parameter is recommended to be between 0 and 1. A value of 0 indicates that the point is empty, meaning there is no scene geometry at that location. A value of 1 indicates that the point is occupied by scene geometry (e.g., part of the object).

The MLP network is used to train a model that maps from 5D coordinates (3D spatial coordinates plus 2D viewing direction) to the *volume density* and its *associated RGB* color at that point. If we consider the NeRF as a function that maps 5D input to the described 2D output, this function is differentiable, and thus, we can use a neural network to train it. This network can then be used to render novel views of the scene by generating a continuous volumetric radiance field.

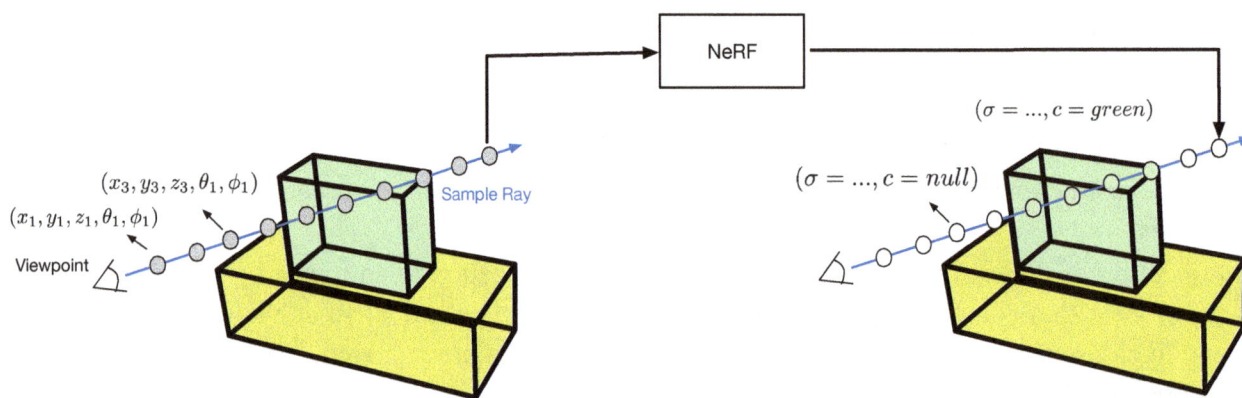

Figure 12-52: A toy example of a ray that cast toward an image and the NeRF model uses it to assign each sampled pixel color and volume intensity. The grey dots on the left are sampled pixels from the casted ray.

Check Figure 12-52. similar to ray tracing. Here, we present one single ray that is cast from the observer's (camera) viewpoint over a pixel in the scene. During its path, it samples the environment, as it is shown in grey circles on the left side of Figure 12-52. The input will be given to the NeRF model. Each of these samples includes a vector of (x, y, z, θ, ϕ), as it is shown in Figure 12-52. The output will be the color (c) and volume intensity (σ).

On the right side, we can see that samples include colors; we can get these samples from the neural network and alpha-blending images back toward the camera until a new 2D sense is constructed. In some samples, the ray does not hit anything (shown as an empty white circle), and in some samples, it hits (intersects) a 3D scene (shown as a green circle).

During the training process, NeRF learns to approximate the *volume density* for each point in the scene by optimizing neural network parameters. After the network is trained (the volume density values of the scene are learned) the NeRF model can get arbitrary point coordinates and estimate the color and density of the scene at that point.

The cost function of NeRF is referred to as rendering loss, and it is used to measure the discrepancy between the rendered pixel (sampled pixel) and the actual pixel on the ground truth image. It is computed as the total squared error between the rendered and actual pixel colors.

Keep in mind that each NeRF network is trained on a set of images from one "single" scene. This is in contrast to other computer vision models in that we train a model to handle different types of images. In particular, parameters (weights and biases) of the MLP that are trained on the input images memorize that particular scene. Some stated that a NeRF model overfits in a single sense. We do not agree with this definition, but it is good to remember that NeRF is trained on a single 3D sense.

The MLP network of NeRF has eight fully connected layers, ReLU as an activation function, and includes a channel depth of 256. The output is volume density σ and a 256-dimensional feature vector. This feature vector is concatenated with the camera ray's viewing direction and passed to a fully connected layer with ReLu and 128 channels. This outputs a view-dependent RGB color.

The first seven layers of the network are used to extract features from the input data, which is a 5D coordinate representing the spatial location and viewing direction of a point in the scene. The eighth layer of the network combines the features extracted from the previous layers with the viewing direction and outputs a 256-dimensional feature vector. This feature vector is then concatenated with the viewing direction and passed to a final fully connected layer with 128 channels. The output of this final layer is a view-dependent RGB color.

Alpha blending can be used in post-processing and visualization of the NeRF output. For example, after rendering a novel view of the scene using NeRF, alpha blending can be used to overlay the rendered image onto a real-world photograph or video to create a composite image that blends the virtual and real-world elements together. In simple words, by blending output images together, NeRF is able to create the illusion of depth and occlusion.

There are more details included in the NeRF, such as increasing the sample rate when intensity increases or using a specific form of positional encoding; we skip explaining them for the sake of brevity.

3D Gaussian Splatting

3D Gaussian Splatting [Kerbl '23] is a rasterization technique for 3D model reconstruction and rendering of images taken from multiple viewpoints. It models the scene using 3D Gaussian distributions characterized by parameters such as mean and covariance. These distributions are used to capture the properties of volumetric radiance fields efficiently, avoiding the computation of empty space as part of the object.

Figure 12-53 presents the flow of 3D Gaussian splatting. First, it applies SfM [Snavely '06] to convert the series of input images into a point cloud. Next, every single point in the point cloud is converted into a center of 3D Gaussian distribution (same as what we have seen in tSNE back in Chapter 7). This process is shown in the initialization component of Figure 12-53. During the initialization process, each point on the 3D Gaussian distribution has a mean and a covariance matrix. The mean represents the center of the Gaussian distribution, while the covariance matrix describes the spread of the distribution (how this particular data point is stretched). Besides, every point also entails color and alpha information, alpha is used to quantify the transparency of a point.

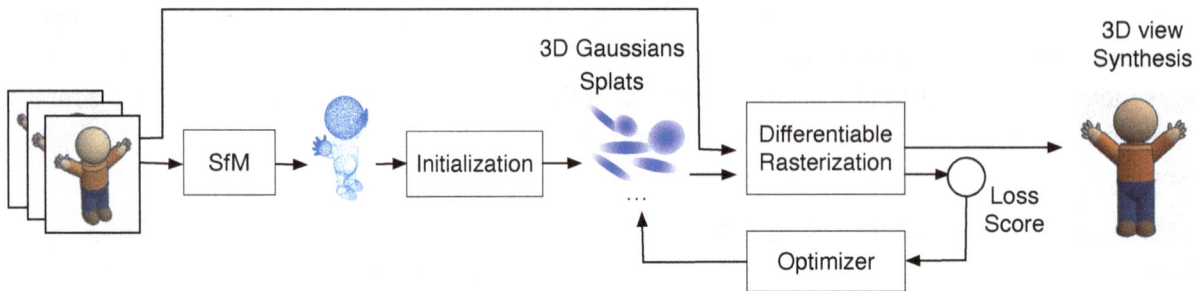

Figure 12-53: 3D Gaussian splatting process. We did not incorporate its under-construction and over-construction solutions for sake of simplicity.

The third step in the training process leverages the Adam optimizer but without any neural network. This means the training does the following: (i) it rasterizes each Gaussian distribution to an image (converts each Gaussian distribution into pixels) by using a differentiable Gaussian rasterization. They should be differentiable because the Adam will be used to optimize them. (ii) Then, the ground truth image will be compared with the rasterized image constructed from 3D Gaussian splats, and the loss score will be used by the Adam optimizer (check Chapter 10) to optimize the Gaussian splats parameters, including mean (position), covariance matrix, opacity α, and color c.

In addition to the described method, the authors use two subtle approaches in their design: *Gaussian Densification* and *Differentiable Rasterization*.

Gaussian Densification: The Gaussian densification is performed after every 100 iterations, and it removes Gaussian splats that (i) their opacity α value is less than a given threshold and (ii) large areas or empty regions in the scene where their Gaussian splats result in over-construction with another threshold. Small Gaussian splats that do not cover an entire region are referred to as under-reconstructed regions. To enable small Gaussian splats to cover the entire region, the model clones the Gaussian splats by creating a copy of the same size Gaussian splats and moving it in the direction of the positional gradient. Positional gradient refers to the direction in which Gaussian splats need to be added to better fill in gaps in the scene. We did not present in Figure 12-53 for the sake of simplicity.

Differentiable Rasterization: The rasterization method that the authors used in their architecture is differentiable, and thus, it enables the optimizer to use backpropagation and change Gaussian parameters. To make the rasterization process fast, they split the input image into 16×16 tiles. Then, for each tile, the model removes all Gaussian splats that have a lower than 99% confidence interval range (see Chapter 3 about Confidence Interval). This is the implementation of the Frustum Culling, which we have explained earlier. Next, each of the Gaussian splats in each tile gets a key and is stored in a key-value structure in memory. The key is composed of 64 bits, 32 bits for the projected depth of the Gaussian splat, and the other 32 bits for encoding the index of overlapped tiles. Then, keys are sorted based on the depth of each Gaussian splat by using GPU Radix sort [Merrill '10]. As a result, each tile has a list of sorted Gaussian splats, and they are processed in parallel (each 64-bit key is compared to its neighbors). Afterward, for each pixel in each tile, the current Gaussian color and opacity are blended with the current pixel's color and opacity using the blending weight. The result is stored in the current pixel's color and opacity. In other words, for each pixel, it iterates over Gaussian splats from front to back and blends their mean, covariance, color, and opacity together.

We skipped describing the gradient computation, and you can check Appendix A of the paper [Kerbl '23] to see how they compute the gradient of the covariance matrix.

Other computer vision models that we didn't explain

We have explained prominent algorithms in computer vision that benefit from deep learning. Nevertheless, the variety of computer vision algorithms that benefit from deep learning is much larger. Computer vision is one of the most crowded areas in computer science, and we believe it will stay active in the future as well.

Imagine the current trend of mass consumption continues, and we end up in a dystopian AI society, such as the "Quality Land" that is described in Marc-Uwe Kling's novel [Kling '21]. A nice corporation, Your Privacy is Our Life (YPL), tries to sell more useless things and thus rebrands everything as "smart" by integrating a camera inside them, such as paper toilet holders, shower heads, etc.

Of course, there is a justification for using cameras everywhere; a smart toilet paper holder can check and notify us when we are running low on toilet paper. Smart toilet paper can monitor our bodies and make skin-customized recommendations, all with deep respect to users' privacy, but the data will be kept in the cloud and might be shared with some reasonable third parties, such as government or spying agencies.

You can see from this simple example that computer vision will have lots of applications, but since we do not like our brains to explode, we just list them here.

Facial Recognition, Face Swapping: One popular group of models widely used in social media, ID verification, and security analysis is facial recognition algorithms. For example, these models can be deployed in crowded places to identify fugitives. They are capable of recognizing individuals from face images or videos. Some models can even identify individuals at a distance, even when the images are in low resolution. Several neural network models have been utilized for face recognition. Among the early models proposed by Facebook is DeepFace [Taigman '14], which, back in 2014, achieved near-human accuracy. Another prominent group of models, DeepID [Sun '14], was also proposed in the same year and has shown gradual progress. For instance, DeepID2+ [Sun '15] was among the first models to outperform human perception in face recognition. Google proposed another significant model, FaceNet [Schroff '15], in 2015. The use of facial recognition technology raises ethical issues, including concerns about privacy, consent, potential biases in algorithmic decision-making, and the surveillance of individuals without their knowledge.

High-resolution Image Reconstruction: Another popular group of models widely used in media production, scientific imaging, and digital restoration is high-resolution image reconstruction algorithms. For example, these models can be employed in enhancing old films or low-quality images to produce clearer, high-resolution versions. They are capable of reconstructing detailed textures and fine details from images or videos that are originally of low resolution. Several neural network models have been utilized for high-resolution image reconstruction. Among the early models is SRGAN [Ledig '17], which, back in 2017, achieved impressive results in generating photo-realistic textures. Another prominent set of models, ESRGAN [Wang '18], was proposed the following year and has demonstrated significant improvements in perceptual quality. In 2019, Adobe proposed EnhanceNet [Sajjadi '17], which became known for its ability to reconstruct high-quality images without requiring reference high-resolution images.

Pose Estimation: A very exciting and promising direction in computer vision is 2D and/or 3D human or animal pose estimations. In particular, from a video or image of the user, the model identifies body joints, and by using those joints, it constructs abstract skeletal shapes of the body. They have a wide range of applications, from medical kinetic analysis [Zhang '21] to exercise coaching and professional athlete performance analysis. As prominent works in this category, we can refer to PozeNet [Kendall '15], OpenPose [Cao '17], AlphaPose [Fang '17], and many other prominent pose estimation models.

Depth Estimation: An active research area in computer vision is depth estimation. It refers to the task of perceiving an image that is in two dimensions (2D) in three dimensions (3D). It also has a wide range of applications. For example, the placement of objects in an augmented reality environment benefits from depth estimation. Self-driving vehicle cameras get the image in two dimensions, and they should estimate the motion of moving objects along the third dimension. Smartphones with dual cameras can focus on one certain object and blur the rest of the objects in the image. Simple blurring is not appealing, and blurring based on the depth of the object is more appealing to the user. There are several works done on depth estimation. An example work [Alhashim '18] for depth estimation uses a combination of CNN and Encoder-decoder architecture (transfer learning) for depth estimation. Another example uses only CNN model to extract depth information [Laina '16].

Medical Image Analysis: There are many promising computer vision in medicine, including 3D modeling, pathological image analysis, cell detection, CT devices, and X-Ray [Topol '19]; we skip explaining those algorithms as well. Nevertheless, the backbone of most of them are architectures that we have already described, such as U-Net for medical image segmentation.

Audio

The third category of work that has advanced significantly with deep neural networks is audio-based applications, such as audio classification, synthetic music generation, automatic speech recognition (ASR), audio denoising, speaker verification, voice cloning, and speech synthesis. Audio data analysis got lots of attention back in 2019 because of the introduction of conversational agents such as Apple Siri, Amazon Alexa, etc. [Rawassizadeh '19]. Most operating systems, such as Windows, MacOS, Android, and iOS, have extensive support for voice-based conversational agents, particularly for individuals with visual impairments.

First, we list audio applications that benefit from deep learning. Next, we briefly describe concepts and features that are used for audio analysis. Afterward, we list some known deep-learning architectures that are being used for audio analysis.

Audio Classification: One popular application of audio analysis is identifying the origin of the input sound. For example, a piece of music is given to the model, and the model recognizes the genre of the music or instrument that is used in this music, such as violin, piano, etc.

Audio Segmentation: Some sounds include a mixture of audio from different resources. For example, assume a corporation, let's call it "Meta Privacy", installs voice recognition on its mobile applications. The application collects voice communications, analyzes them, and trims its ads based on the topic they are discussing. Since the ambient noise is large in some environments and many people are talking together at once, Meta Privacy's model uses audio segmentation to extract the audio signal of interest from a mixture of audio signals.

Synthetic Music Generation: We have described several algorithms in Chapter 11 that construct synthetic images or human faces. Another interesting advancement that has been achieved recently is synthetic music composition [Zhu '23], based on the specified music instrument or composer.

Voice Recognition and Speaker Identification: It is similar to audio classification but focuses on identifying the individual or animal who is speaking or making a sound and the emotional tone. For example, an algorithm can record a human sound and recognize the human's emotion; another example is voice-based authentication [Liu '18].

Text to Speech and Speech to Text: At the time of writing this chapter, the most popular and wide use of audio signal processing is speech-to-text (STT) or text-to-speech (TTS). All popular smartphone and desktop operating systems are equipped with conversational agents, and they communicate with users via voice, e.g., SiRi, Google Assistant, etc. This means the conversational agent gets the user's voice and converts it to text (STT), or it gets the text from the application and reads it with a voice similar to a human voice for the user (TTS).

Voice Cloning/Conversion: One application of speech synthesis technology is voice cloning, a.k.a, Deep Voice. This involves creating a digital replica of a person's voice from a relatively small sample of their speech [Arik '18]. For instance, a model might be fed recordings of an individual speaking and then trained to produce new speech that sounds as if the same person is talking, even saying words or sentences they never actually recorded. These group of models can have implications across various sectors, such as having more realistic and engaging video game characters, and aiding individuals who might lose their ability to speak due to illness or injury. However, voice cloning is associated with some ethical considerations that arise with the potential misuse of voice cloning technologies, such as identity theft and privacy violations.

Audio Signal Concepts and Features

Prior to the advent of deep learning models for audio analysis, classic algorithms relied heavily on manual feature engineering, with most methods in this domain utilizing *Digital Signal Processing (DSP)* techniques. However, even today, these manually engineered features are often incorporated into neural network models. Therefore, before we explain neural network models for audio, we highly recommend checking the explanation of wave and signal in Chapter 6 and the signal decomposition methods in Chapter 7. Sound is a wave, and we have explained in Chapter 6 that we use digital signals to present wave information in digital format. An audio signal is a time series that the value is the amplitude of the sound. Real-world sounds are a composition of many different signals, each with a different frequency, and not just a plain signal, such as a sinusoid signal.

As always, we should learn some concepts before going into model explanations.

Sampling: It refers to the process of measuring the amplitude of a sound wave at fixed intervals to convert it into a digital audio signal. The rate at which sampling occurs, known as the *sampling rate*, varies depending on the application. For example, CD-quality audio has a sampling rate of 44,100 Hz, meaning the audio is sampled 44,100 times per second.

Spectrum: It represents a visualization of all frequencies present in an audio signal, along with the amplitude of each frequency. Essentially, it shows the strength (or loudness) of different frequencies within a sound. A higher amplitude indicates a greater loudness of sound as perceived by the listener. Figure 12-54 visualizes a spectrum of the audio wave along with the original wave.

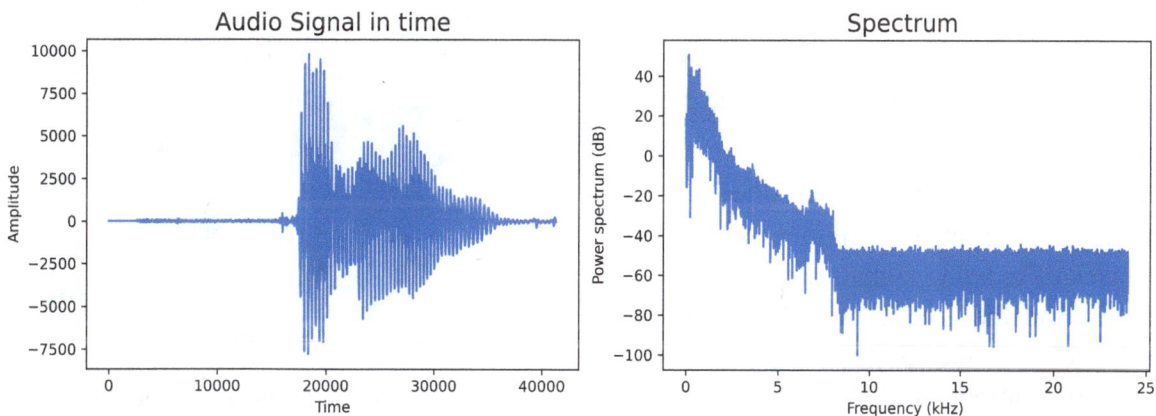

Figure 12-54: Original audio signal on the left, and spectrum of that audio signal on the right.

A spectrogram can be divided into two parts: *magnitude* and *phase*. The magnitude describes the strength or loudness of each frequency within the signal. The phase, on the other hand, specifies the position of the wave at a particular point in time for each frequency. It indicates how the waves are aligned or synchronized and tells us the timing of the wave's cycle at each frequency point. Think of a wave as a repeated up-and-down motion, like a swing. The phase tells us the position of the swing at a specific moment.

Fundamental (Reference) frequency: We have explained that a spectrum is a set of frequencies. Fundamental/Reference frequency is the lowest frequency in a spectrum and is considered the most dominant.

Harmonic: Signals whose frequency is a multiplication of whole numbers (positive integer) of the reference frequency are harmonic signals. For example, if the reference frequency is 100 Hz, then its harmonic signals have frequencies of 200 Hz, 300 Hz, 400 Hz, etc.

Pitch: It refers to how humans perceive the frequency of a sound. It is a subjective experience that describes how high or low a sound seems to us. While the frequency of a sound wave indicates how fast it vibrates, pitch is about how these vibrations are heard by the human ear. Different people may perceive the pitch of the same sound slightly differently, which illustrates the subjective nature of the pitch.

Melody: A melody[40] is a series of musical tones that are grouped together as a single entity. These series of tones are arranged in a memorable and pleasing sequence, which can be recognized and enjoyed as a distinct musical entity.

Spectrogram: We have explained that to analyze a signal, we can work with a time-domain signal or a frequency-domain signal. A spectrogram is a representation that incorporates both time-domain and frequency-domain signals. In simple terms, the spectrogram is a visual representation of the spectrum of frequencies in a signal over time. Each column of the spectrogram is generated by the Fourier Transform, which decomposes a given time-domain signal into a set of frequency-domain signals.

The signal is divided into overlapping segments of equal size, it applies the Fourier transform to each segment. By sliding this transformation across the entire signal and stacking the results vertically, a spectrogram is constructed.

Figure 12-55 shows the spectrogram of the audio signal presented in Figure 12-54; the frequencies of existing signals are presented on the Y-axis, and the time is presented on the X-axis. The color on the spectrogram specifies the value of the amplitude of the signal at that particular frequency.

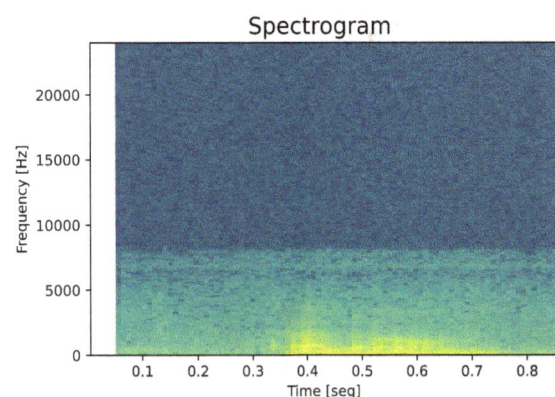

Figure 12-55: Spectrogram of the audio signal presented in Figure 12-49.

---

[40] This link: https://hellomusictheory.com/learn/melody has a detailed explanation about the melody that we can hear to understand it better.

Many deep learning models use the spectrogram as an image of the audio to perform their analysis and it is common to use CNN for audio analysis tasks.

Mel-Spectrogram: The term "mel" is an abbreviation of the term "melody". The difference between 500 Hz and 1000 Hz is obvious for our ear, whereas the difference between 9500 Hz and 10000 Hz is barely noticeable. It means that we are good at detecting differences in lower frequencies but weak at detecting differences in higher frequencies. This means that we do not perceive frequencies on a linear scale, but instead, we perceive them on a logarithmic scale.

Mel-Spectrogram is similar to spectrogram but it remaps the value of frequency from *Hertz* to *Mel scale.* Mel Scale is a logarithmic transformation of a signal's frequency. The spectrogram assigns equal importance to all frequencies, but it is recommended to favor the mel-spectrogram over the spectrogram for applications that deal with human hearing perception.

Spectrogram uses amplitude, and mel-spectrogram uses Decibel (dB), which is closer to human ear capabilities. 0 dB is total silence, and the measurement units with decibels increase exponentially. 10 dB is 10 times louder than 0 dB, 20 dB is 100 times louder, and 30 dB is 1000 times louder. Our ears can stand a maximum of 85 dB, and larger sounds are harmful to hear. The equation to convert amplitude to its decibel value (R) is written as follows:

$$R = 20 * log_{10}(amp_c/amp_{ref})$$

In this equation amp_c refers to the current signal amplitude and amp_{ref} refers to the reference signal amplitude. This process refers to compounding, which we will describe in Chapter 15.

Figure 12-56 presents a comparison between a spectrogram of audio versus its mel-spectrogram. We can see that the mel-spectrogram that focuses on dB instead of Hertz provides more visual information than the other two representations.

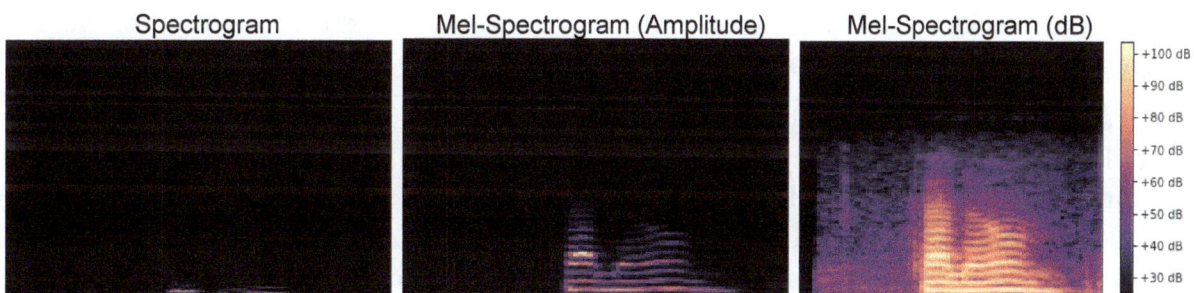

Figure 12-56: Differences between spectrogram, mel-spectrogram with Amplitude and mel-spectrogram with dB.

Mel Frequency Cepstral Coefficients (MFCC): MFCCs are derived from the mel-spectrogram, which is better suited for modeling human auditory perception than a standard spectrogram. While mel-spectrograms provide a good representation of sounds as perceived by the human ear, MFCCs are particularly favored in speech recognition applications. This is because MFCCs efficiently capture the key characteristics of human speech by emphasizing the most important frequencies, those that the human ear is most sensitive to. MFCCs are coefficients that represent the amplitude of the spectrum in the mel-frequency domain. Typically, a small number of MFCCs (e.g., 12-20) are used to effectively capture the relevant features of speech, unlike in a mel-spectrogram where a larger number of frequencies might be represented.

Figure 12-57 presents MFCC on the right side, constructed from the Mel-Spectrogram, which is presented on the left side.

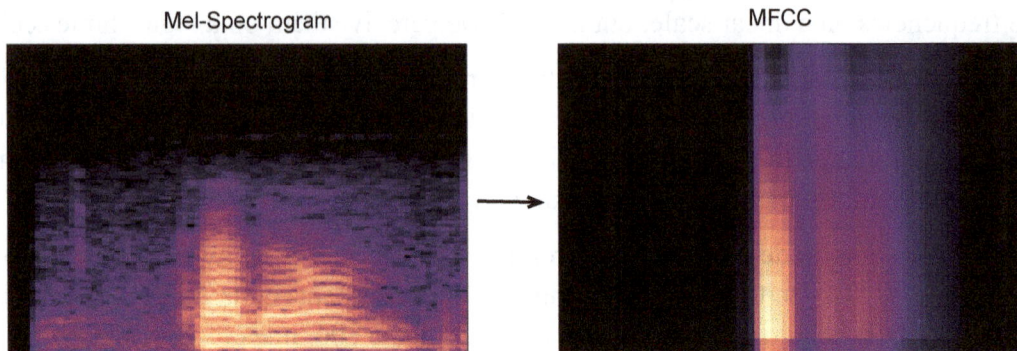

Figure 12-57: MFCC on the right is constructed from the Mel-spectrogram, presented on the left side.

Sound Modeling

Audio analysis models receive raw audio, spectrogram, mel-spectrogram, or other audio features and provide the desired output. Before the introduction of neural networks, many none-deep learning approaches to analyze audio did not provide high accuracy. For example, TTS algorithms have a very artificial tone and humans can recognize them easily that they are generated by machines and not humans. Similar to unstructured text and computer vision, deep learning made a significant shift in this area as well.

Sound signal is a time series in that its data points have some correlation and causality with each other. Therefore, a model that synthesizes sound should take into account the correlation and causality of sound. However, the number of data points in such a time series is vast. For example, Speech is sampled at a frequency of 16Khz or 22Khz, which means every second is equal to 16000 or 22000 samples. To handle this issue, traditional algorithms use information retrieval and signal processing methods, such as overlapping sliding windows (see Chapter 5), to deal with a smaller number of data points. We do not list traditional algorithms and focus on neural network approaches in this section. In the following, we explain four audio models including, WaveNet series [Oord '16, Oord '18], Tacotron series [Wang '17, Shen '18], Wav2Vec series [Schneider '19, Baevski '20], and Whisper [Radford '23].

WaveNet Models

Wavenet is a popular audio model [Oord '16] proposed back in 2016. and implemented in Google Assistant[41]. It can perform high-quality speech synthesis (human voice generation) but is also used for other tasks such as speaker identification, music generation, and TTS as well.

WaveNet 1

WaveNet is a probabilistic and autoregressive model (check ARIMA in Chapter 8) that analyzes audio samples as discrete data. Since it is autoregressive each observation at time t depends on previous observations from time 1 to $t-1$. Therefore, a wave X is composed of several smaller audio samples, and it is written as $X = \{x_1, x_2, \ldots, x_T\}$. Every audio sample x_t has a correlative and causal relation to its previous audio sample $p(x_t | x_1, x_2, \ldots x_{t-1})$. An audio sample can be factorized as a product of conditional probabilities as follows:

$$p(x) = \Pi_{t=1}^{T} p(x_t | x_1, \ldots x_{t-1})$$

We have learned in Chapter 10 that we use RNNs for the prediction or modeling of sequential data, but they are incapable of modeling a long sequence of data (e.g., 16 kHz of speech data). Therefore, the authors adapt CNN to model the sequence of data. To enable CNN to operate as an autoregressive model, authors employ one-dimensional causal convolution (Conv1D). In other words, a convolutional window slides over the vector data points and predicts the next value (which is not seen yet). The left side of Figure 12-58 visualizes a simple causal convolution with a stride of size 1 and a kernel size of 4. Here, four data points inside a kernel are used to predict the fifth data point.

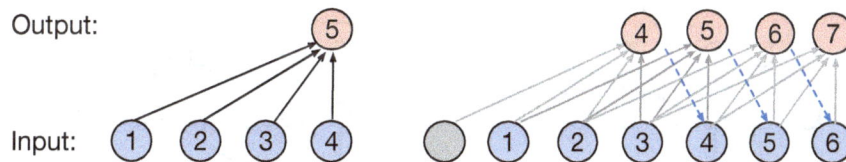

Figure 12-58: (Left) One causal convolution with stride size =1 and kernel =4 to predict one data points. (Right) this process continues three times. The grey circle is used as causal padding (zero) to predict data point 4.

After a sample data point is predicted (red data point in Figure 12-58) it will be fed back to the model to predict the next sample data point. This is shown as the dotted blue line in Figure 12-58. This process continues until all other data points in the time series were predicted. For example, the right side of Figure 12-58 shows that this process has been used to predict four data points: 4, 5, 6, and 7.

A question might arise here: what do we do when the input is incomplete? In other words, how do we predict data points 1, 2, 3, or 4, when the kernel size is 4? In that case, causal convolution substitutes missing inputs with zeros, i.e., *causal padding*, which is presented with a grey circle

---

[41] https://cloud.google.com/text-to-speech/docs/wavenet

in Figure 12-58. This approach is used in other audio models as well, such as wav2vec. However, using causal convolution results in having a very small receptive field. Audio data have too many samples for each second of data, and thus, we need to have a very large receptive field to model it with CNN. To handle this issue, authors propose using a dilated causal convolution and thus increasing the receptive field. Just by increasing the stride more than one, the receptive field increases. Figure 12-59 visualizes the differences between using causal convolution and causal dilated convolution. On the top of this figure, we used a causal convolution, and the receptive field is 5 ($\#layers + filter\_length - 1 = 4 + 2 - 1$), in the bottom one, we use a dilated causal convolution with different dilated rates (1 for the first layer, which is similar to causal convolution, 2 for the second layer, 4 for the third layer, and 8 for the fourth layer). As a result, the receptive field has increased.

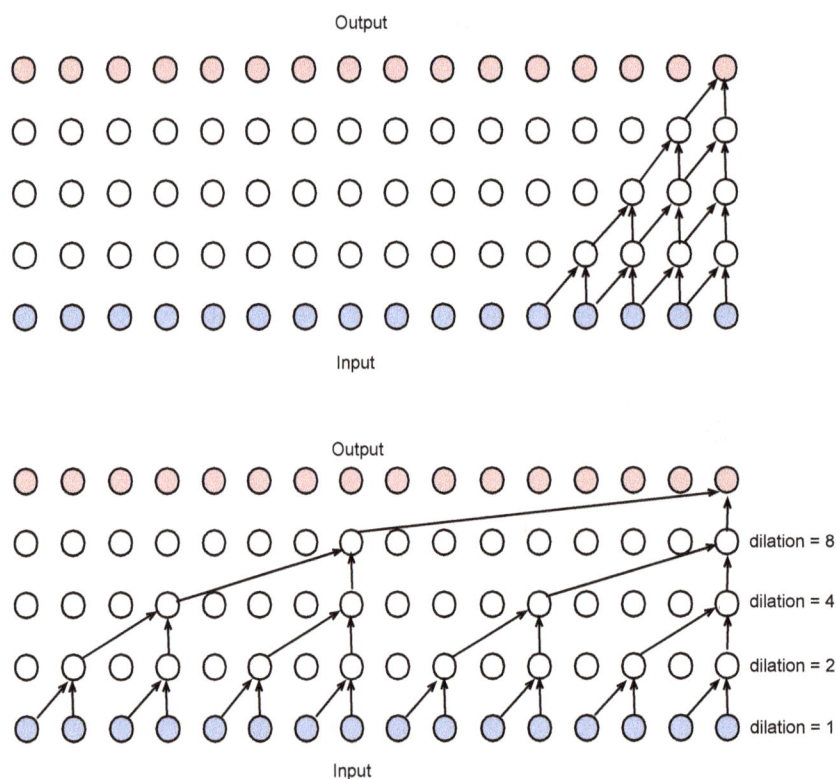

Figure 12-59: (top) Not using a dilated convolution, which results in having a receptive field of 5. (bottom) Using a dilated convolution with different dilations in each layer increases the receptive field to 16.

The authors of WaveNet use dilated convolutions to exponentially increase the receptive field. The dilation rate doubles at each successive layer up to a limit, typically 512, after which it resets back to a dilation rate of 1 and repeats the cycle. This creates a sequence of dilation rates such as 1, 2, 4, 8, ..., up to 512. Each block of layers with dilation rates 1, 2, ..., 512 provides a receptive field of 1024 time steps. This approach allows WaveNet to capture long-term dependencies in the input by significantly increasing the receptive field in a computationally efficient manner.

To summarize what we have discussed above, we need a method to enable the CNN model to look back at one thousand samples. In other words, the CNN model should be able to look back at least a second (and each second has 16,000 samples) before making a conclusion about the output at the current time. To handle this problem, authors use dilated causal convolution. Dilated convolution increases the length of look-back length (receptive field size) by applying a filter over an area larger than its length.

The WaveNet model has no pooling layer; the output dimension of the network is equal to the input dimension. Raw audio is stored in 16-bit integer values (one per timestamp). Therefore, a soft-max layer needs $2^{16} = 65,536$ probabilities per timestamp to model all possible values. This is intractable, and to handle this situation, WaveNet first uses a companding algorithm, i.e., μ-law, that converts 16-bit audio data into 8 bits and then quantizes the audio data to 256 possible values. We will explain more about companding and quantization in Chapter 14.

Authors use *Gated Activation Function*, a.k.a, Gated Activation Unit, which is written as $z = tanh(W_{f,k} * x) \odot \sigma(W_{g,k} * x)$. In this equation, * presents convolution, $\odot$ presents element-wise multiplication operator, $\sigma()$ presents the sigmoid function. k presents the layer index, f presents the filter, and g presents the gate, respectively, and W is a learnable convolution filter.

Figure 12-60 presents the WaveNet architecture. WaveNet has CNN, and similar to 99.99999% of other CNN architectures, it uses skip connection, and it uses the k number of residual blocks as well. As it is presented in Figure 12-60, first, it applies a causal convolution to the input. The result passes to the k-layers of residual blocks. Each residual block first applies a dilated causal convolution, then applies the Gated Activation Unit on it. This gated activation unit applies a hyperbolic tangent activation function and a Sigmoid activation function separately to the convolved signal and then multiplies the results element-wise. Next, the result passes its results to a 1×1 convolution.

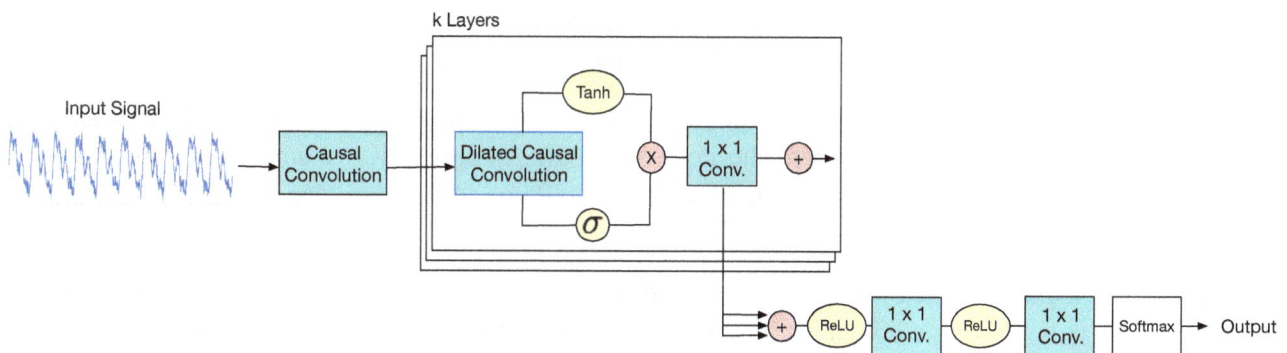

Figure 12-60: WaveNet architecture, The input is raw audio signal and output is an audio but with the information we expect from the model.

Skip connections add the input of each residual block to the result of its corresponding 1×1 convolution. The outputs of all residual blocks are collected and then added together, followed by a ReLU activation function. The combined result is then passed through another 1×1

convolution, followed by a ReLU activation and a final 1×1 convolution. Finally, the processed signal is passed through a softmax layer, which outputs the probability distribution for the next audio sample.

To summarize, WaveNet models audio signals by increasing the receptive field and using a CNN network. It increases the receptive field by increasing the number of dilation stages, using more layers, larger filters, and greater dilation factors.

Downstream tasks: To implement WaveNet for different audio-related tasks (downstream tasks), an additional input h will be added to the model. This input enables the WaveNet to model the conditional distribution as follows:

$$p(x \,|\, h) = \Pi_{t=1}^{T} p(x_t \,|\, x_1, \ldots x_{t-1}, h)$$

This conditioning criterion enables the user to guide the model to produce audio with the required characteristics. For example, in a multi-speaker setting, the user can choose a speaker and add his identity to the model as extra input (h). As another example, a TTS task information about the text can be passed to the model as an extra input (h).

The conditioning can be implemented as global or local conditioning. Global conditioning influenced the output distribution across all timestamps, and it is characterized by a single latent h. For example, speaker embedding in a TTS model requires global conditioning. Local conditioning is used for tasks that require a second time series, such as text in a TTS task (audio is the larger sampled time series, and text is the shorter sampled time series). Global and local conditioning have different implementations in terms of activation functions and network structures. For detailed technical implementations, check the original WaveNet paper.

Parallel WaveNet (WaveNet 2)

WaveNet uses the raw audio signal, and sampling the audio signal is usually done at 16 kHz to 48 kHz, which means each audio signal has 16,000 to 48,000 samples per second. This is acceptable while training the model, but the generation of a new audio signal is very slow and impractical because each input sample needs to be drawn from the output distribution.

Parallel WaveNet, a.k.a. WaveNet 2 [Oord '18], solved this issue by using inverse autoregressive flows (IAFs). IAF is a formulation of the deep autoregressive model in which sampling can be done in parallel. We do not explain more about the details of IAF. Otherwise, the author's brain, along with your brain, will explode.

The fidelity in the context of WaveNet refers to the challenge of generating high-quality audio that closely matches the original audio's characteristics. The authors used two approaches to increase the fidelity of WaveNet. First, instead of modeling the Audio wave with 8 bits (by using μ-law) and 256 categorical distribution, they increased the audio wave to 16 bits. Since training on a $2^{16} = 65,536$ categorical distribution is very expensive, authors handle this by using a *discretized mixture of logistics distribution*[42] for sampling the audio signal. Besides, they increased the audio rate used for training from 16 kHz to 24 kHz. To cover a larger receptive

---

[42] Logistic distribution is similar to normal distribution but has a fatter tail than the normal distribution.

field, they increased the filter size in the second layer of dilated convolution from 2 to 3 and adjusted the filter sizes in other layers accordingly.

For the training, they use knowledge distillation (we will explain it in Chapter 14), with three defined loss functions (power loss, perceptual loss, and contrastive loss) customized for their model.

Tacotron Models

Tacotron models are a series of models developed by Google [43]. They receive text as input and generate synthetic speech as output, a.k.a., text-to-speech (TTS). These are TTS models that aim to produce natural and human-like speech.

Tacotron 1

The first Tacotron model was proposed back in 2017 [Wang '17]. It performs speech synthesis at the frame level, making it significantly faster than sample-level autoregressive models. Autoregressive models that generate speech one sample at a time are referred to as sample-level autoregressive models.

We have explained the seq2seq architecture earlier in this chapter, and Tacotron is a TTS model based on the seq2seq architecture (Figure 12-1 presents the seq2seq model) with an attention-based decoder. As input, its encoder receives a set of characters, constructs spectrogram frames, and uses the Griffin-Lim [Griffin '84] algorithm to convert the spectrogram into a waveform, i.e., synthesizing speech. We will explain the Griffin-Lim algorithm in Chapter 16. Before we learn the architecture of this model, we should be familiar with two blocks that are used in this model, i.e., *Convolutional Bank, Highway network, and Gated recurrent unit (CBHG)* and *pre-net* blocks.

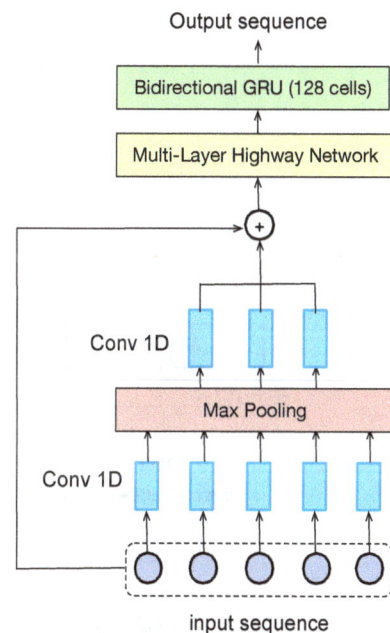

Figure 12-61: CBHG block of Tacotron architecture.

Convolutional Bank, Highway network, and Gated recurrent unit (CBHG) block: A CBHG block is shown in Figure 12-61, it is composed of 1-D convolution filters (a.k.a 1-D convolution bank), a highway network [Srivastava '15] (we don't explain it here), and a bidirectional GRU stack on top of each other.

First, a k number of 1D convolutional filters are applied to the input t for both the encoder's CBHG and the post-processing's CBHG. k is set to 16 for the encoder's CBHG and k is set to 8 for the post-processing's CBHG. These filters model local and contextual information of the input sequence. The outputs are stacked together and pass through a max-pooling layer along the time axis. This max-pooling increases local invariances, with a stride of 1 and a width of 2. The stride of 1 is used to preserve the original time resolution. Then, the output of the max-pooling

[43] https://google.github.io/tacotron

layer is passed to another stack of 1D convolutions, and their output is added to the original data through a skip connection. The result goes through a highway network to extract high-level features. Finally, those high-level features are consumed by a bidirectional layer of GRU, with 128 cells.

Pre-net block: These blocks are used for non-linear transformations. A pre-net is composed of two fully connected layers with a ReLU activation function. The first layer outputs 256-dimensional features (a vector of size 256) with a dropout of 0.5, and the second layer outputs 128-dimensional features (a vector of size 128) with a dropout of 0.5. Figure 12-62 presents the overall architecture of the Tactotron, which is composed of an encoder and decoder.

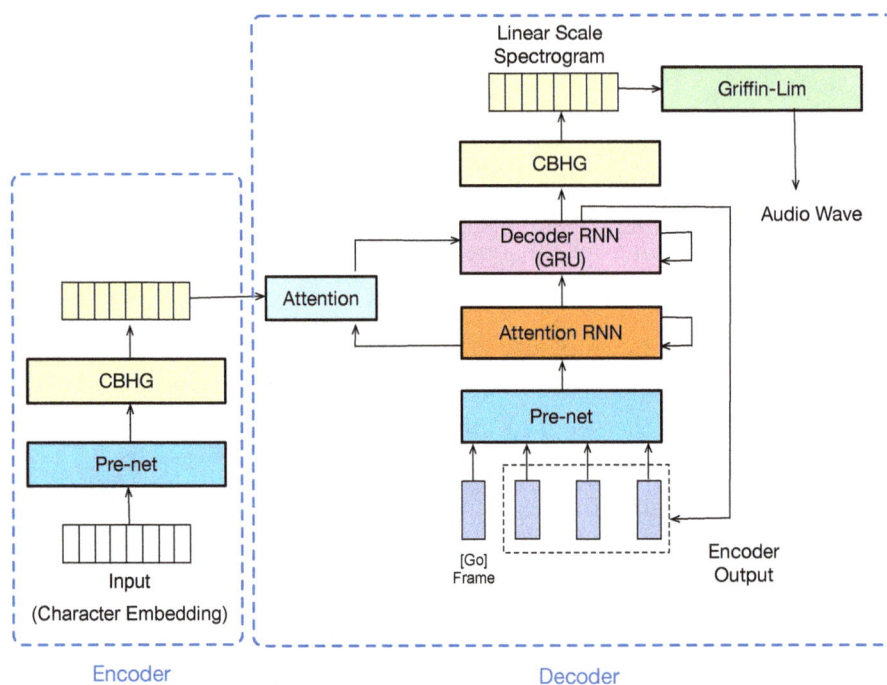

Figure 12-62: Tacotron v1 architecture, which is based on seq2seq model.

Encoder: The encoder first receives a set of textual characters encoded with one-hot encoding, referred to as character embedding by the authors. Then, the Pre-net applies a set of non-linear transformations to these embeddings. The transformed embeddings are passed to a CBHG (Convolutional Bank + Highway Network + Bidirectional GRU) block. The attention mechanism uses the output from the CBHG block. The authors note that this style of encoder reduces overfitting and results in fewer mispronunciations compared to a standard multi-layer RNN encoder.

Decoder: The decoder consists of a stack of GRUs (Gated Recurrent Units) with residual connections. A stateful recurrent layer within the Attention component produces the attention query at each decoder time step. Initially, the encoder's output passes through the Pre-net. If there is no output, in the first pass, the [Go] frame (similar to the [CLS] token in BERT) is passed to the Pre-net. The Pre-net applies a non-linear transformation, and its output is consumed by the

810

Attention RNN. The output of the Attention RNN and the context vector (a compressed version of the encoder output) are concatenated via skip (residual) connections to form the input for the Decoder RNN (GRU). The GRU output then passes through a CBHG block, and the result is a linear-scale spectrogram. The Griffin-Lim algorithm subsequently converts the linear-scale spectrogram into an audio waveform.

Tacotron 2

After the introduction of Tacotron, several efforts were made to optimize its architecture, and a major improvement was introduced in Tacotron 2 [Shen '18]. Tacotron 2 features a seq-to-seq network (similar to version 1) that maps character embeddings to mel-spectrograms, whereas Tacotron 1 outputs spectrograms instead of mel-spectrograms. Tacotron 2 uses a modified WaveNet to synthesize time-domain waves from the mel-spectrograms.

In particular, Tacotron 2 has two main components: (i) a recurrent sequence-to-sequence network with attention that generates mel-spectrogram frames from the given character sequence, (ii) a modified version of WaveNet that generates waveforms conditioned on the predicted mel-spectrogram frames from the first component.

The authors employed *location-sensitive attention*, which is a type of attention specifically designed for speech recognition [Chorowski '15], and we did not explain it in our attention section. The rest of the model has some minor differences from Tacotron; for example, instead of using bi-directional GRUs, the authors used LSTMs. The changes can be summarized as using mel-spectrograms, a different attention method, and WaveNet instead of the Griffin-Lim algorithm.

wav2vec Models

wav2vec models are representation learning models designed for converting raw audio waves into meaningful vector representations, which can be used for various downstream tasks such as automatic speech recognition (ASR), audio classification, and emotion recognition. They use unsupervised pre-training to learn contextual representations of audio data, followed by fine-tuning for specific tasks.

wav2vec v1

The first version of wav2vec [Schneider '19] was released in 2019 by Meta (formerly Facebook). This initial version focuses on automatic speech recognition (ASR). wav2vec converts audio waves into vectors, and the resulting representations can be used for several downstream tasks. wav2vec has two neural networks: one is the *encoder network*, and the other one is the *context network*. Figure 12-63 visualizes the architecture of wav2vec model.

First, the model takes raw audio as input and passes it through an encoder. We have learned that the encoder's job is to embed the audio input into a latent space (converting the audio signal into a vector). Audio data has more dimension than textual data and transferring it into the latent space reduces the dimensionality of data while maintaining its characteristics. The encoder network encodes 30 milliseconds of audio data into 512 dimension data in latent space z_t, at time t, in every 10 milliseconds.

After the encoding process is finalized, the "Context Network", is used to combine latent representations of data together and construct a larger receptive field size. The motivation for using the context network is to *enable the network to model a longer sequence of audio data* because the encoder's receptive field size is 10 milliseconds, which is too small for a reasonable prediction.

The context network receives the output of the encoder (z_t), aggregates several latent representations, and, as a result, constructs the context tensor (c_t). The context tensor has a larger receptive field size than a single latent representation (constructing a receptive field size of 210 milliseconds).

Now, we understand the rationale of using CNN, by using CNN, wav2vec combines latent representation data together and thus enables the model to perform better prediction.

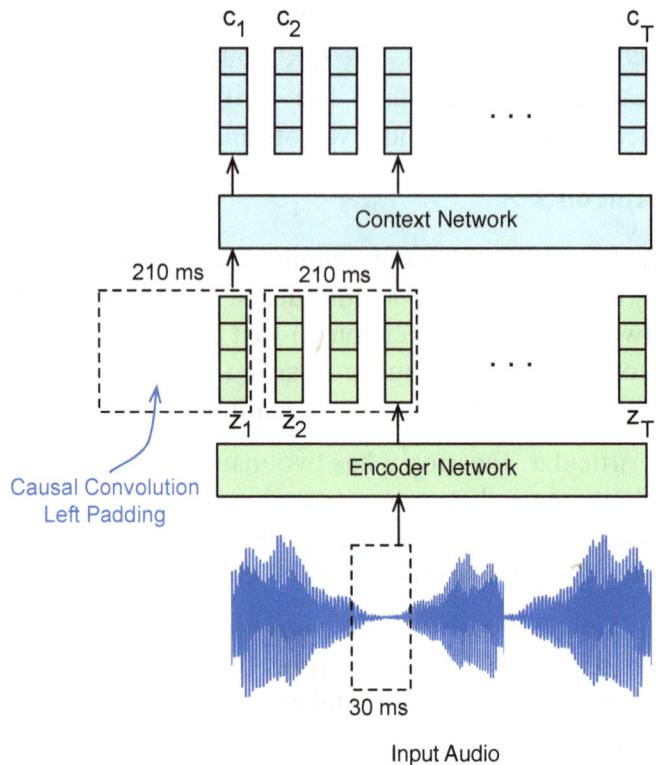

Figure 12-63: wav2vec model architecture.

One might ask why CNN should be used and why RNN should not be used when RNN can handle sequential data better. By using CNNs, the training can be parallelized on hardware, unlike RNN models, which cannot parallelize the training.

An important characteristic of wav2vec CNN is the use of causal convolution. This means that the outputs at time t never attend any position after t. In other words, by using left padding it performs causal convolution (as used by WaveNet), and the causal convolution disallows the model to see future signals while predicting the future signal.

The loss function of wav2vec is a contrastive loss, which distinguishes true (positive) future audio from false (negative) audio. We have explained in Chapter 11 that contrastive loss is used for contrastive representation algorithms, which focus on keeping similar data points together in the latent space and separating the dissimilar data points in the embedded space.

We do not go into more detail about the wav2vec training, but if you are interested in learning it at the end of this chapter, we provide a reference for some additional explanation on the training phase of wav2vec.

wav2vec v2

wav2vec version 2 [Baevski '20] employs the transformer's encoder as its context network. It is composed of three main modules: the "feature encoder", "contextual network", and the "quantization" module.

Feature Encoder: First, the audio signal is normalized and then transferred into the feature encoder. It has a similar role as the feature encoder of the wav2vec v1, and it reduces the waveform dimension into a vector of 20 ms. However, in wav2vec v2, the feature encoder is composed of seven layers of 1D convolutional network with 512 channels at each layer. After every block of 1D convolution, it applies a normalization layer on top of it and then a GELU activation function. Assuming the audio data sample rate is 16 KHz, the receptive field of the feature encoder is 25 ms of audio or 400 sample size.

Quantization: Transformer architecture is designed to operate with sequences of discrete data (e.g., text). However, audio is continuous, and to enable the use of transformer architecture, it needs to be quantized. By using the Gumbel-Softmax distribution [Jang '16], wav2vec 2.0 automatically learns discrete speech units. In other words, the Gumbel-Softmax distribution is a quantization approach that converts a continuous distribution into a categorical distribution.

Context Network: After quantization, the feature dimensions are projected from 512 (output of the feature encoder) to 768 for BASE or 1024 for LARGE. wav2vec 2 substitutes the CNN-based context network with a Transformer encoder-based context network to construct the contextual representation of audio data. However, instead of a fixed positional embedding, "1D convolution" is used, which acts as relative positional embedding. The context network has 12 blocks (transformer encoder block) for the BASE version of the model or 24 blocks for the LARGE version.

Figure 12-64 visualizes the wav2vec 2 architecture. The authors report by using ten minutes of labeled data and pre-training on 53k hours of unlabeled data still, the model achieves a high WER (Word Error Rate). Other speech recognition models require thousands of hours of transcribed speech to make an accurate model makes, and such a huge dataset does not exist for most languages of the world.

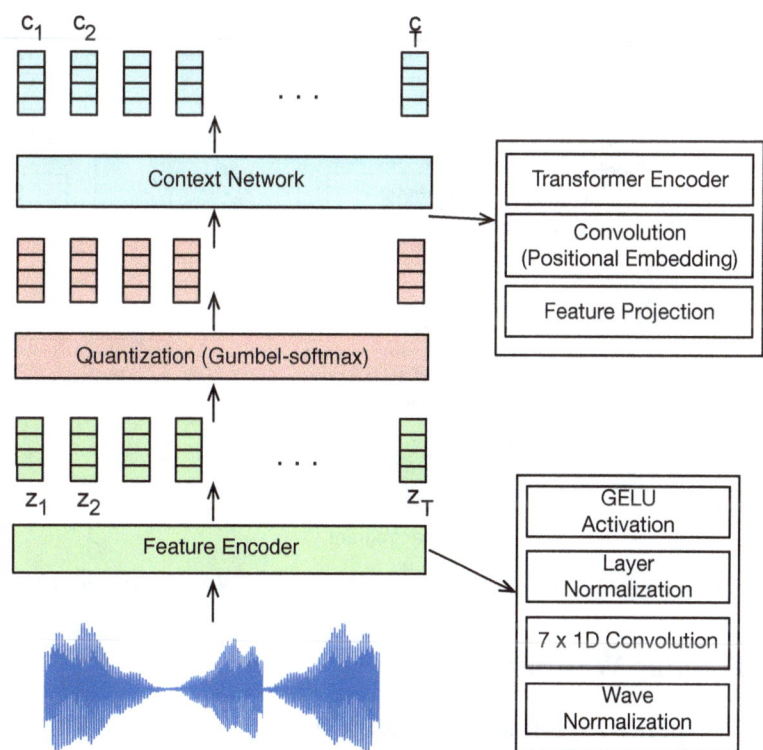

Figure 12-64: wav2vec 2 model architecture.

Whisper

Whisper [Radford '23] is primarily designed as a speech-to-text model that is trained on 680,000 hours of multilingual and multitask supervised data collected from the web. It has more capabilities than previous models, including multilingual speech recognition, noise robustness, and contextual understanding.

Except for Tacotron models, previously described audio models are usually encoders that are pre-trained in an unsupervised manner and require fine-tuning for a downstream task. Whisper authors described that the lack of a pre-trained decoder affects the quality of previous models and limits their applications. Whisper architecture is composed of an *audio encoder* and *text decoder* transformer. Figure 12-65 presents the Whisper architecture.

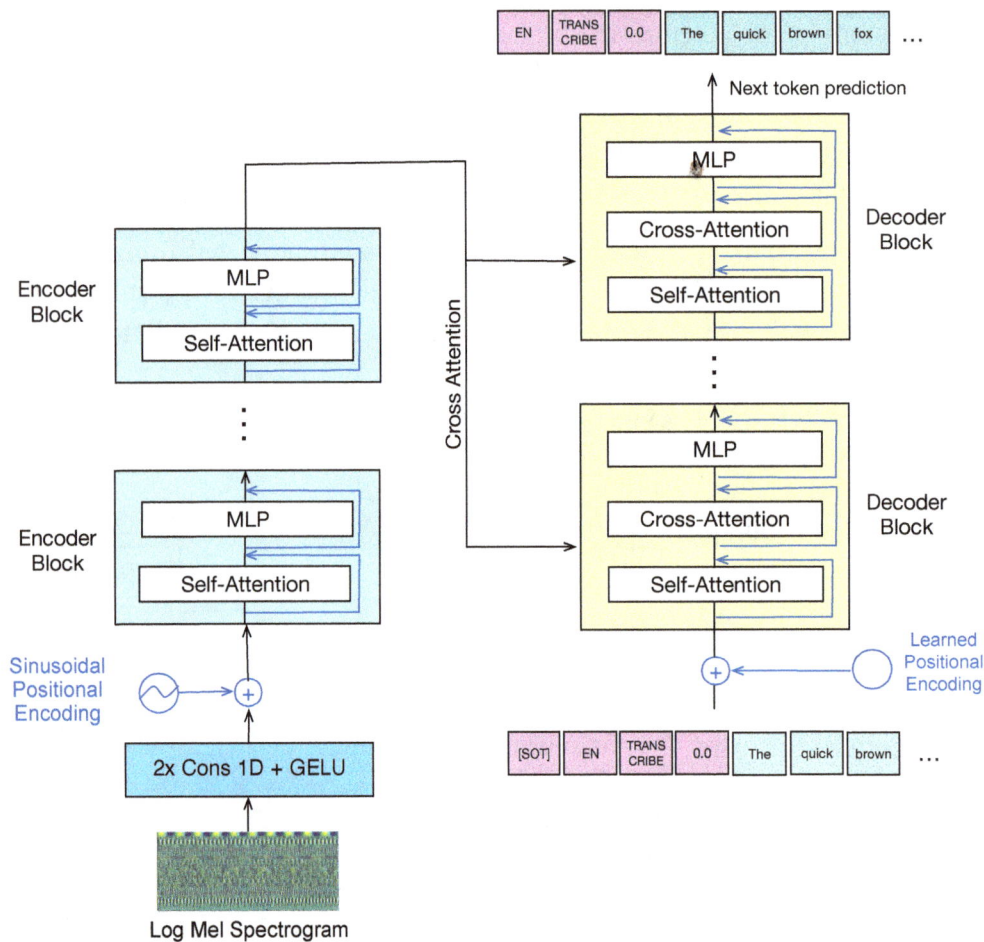

Figure 12-65: Whisper architecture.

Training Whisper

Whisper was trained on 680,000 hours of labeled audio data collected from the Internet, encompassing multiple languages and tasks. This massive dataset includes 117,000 hours of multilingual data and 125,000 hours of translation data, allowing the model to generalize across

814

benchmarks without the need for fine-tuning. The training process focuses on predicting raw text transcripts from audio encoder output. Whisper architecture is a classic sequence-to-sequence transformer model, with data processed into 30-second audio segments.

Similar to wav2vec, Whisper's authors use the Gumbel-Softmax distribution for quantization, enabling the learning of discrete speech units effectively. All audio is re-sampled to 16,000 Hz, and an 80-channel log-Mel spectrogram representation is computed on 25-milliseconds windows with a stride of 10 milliseconds.

Whisper Architecture

Audio Encoder: The objective of Whisper's encoder is similar to other transformer encoders and to enrich the input data with contextual information. Before passing the data into the encoder, it splits the raw input audio into 30 second chunks, and converts them into log-Mel Spectrogram to pass them into a decoder. Next, the log-Mel spectrogram data is normalized to have a mean of zero and a standard deviation of one (a.k.a, unit variance).

Then, the encoder processes this input representation with a block consisting of two convolution layers with a filter width of three and the GELU activation function. Sinusoidal position embeddings are then added to the output of the block. Next, the "Encoder Blocks" are applied to this enriched representation. An encoder block (Residual Attention block according to their code) is responsible for processing the audio using self-attention. Each of these blocks includes a multi-head self-attention and layer normalization, MLP, and residual (skip) connection to add input into the output.

Keep in mind that the audio encoder transforms the raw audio input into a set of encoded representations that are used by the text decoder to generate text. These representations are *not audio themselves* but rather feature vectors derived from the audio signal.

Text Decoder: The text decoder takes the encoded information from the encoder blocks and translates it into text. It is trained to predict the most likely sequence of words based on the given encoded audio.

The input of the decoder begins a special token indicating the Start-of-Text ([SOT]), followed by a language token e.g., [EN], and a task-specific token (e.g., [TRANSCRIBE] or [TRANSLATE]) and additional conditioning information, such as timestamps or speaker identification. Next, there is a token used to indicate whether it should predict timestamps or not. This token helps the model decide if it needs to include information about the timing of the words in the audio. These described tokens are predefined special symbols provided as input to the text decoder. They act as guiding instructions for the decoder, and they are not generated by the audio encoder. They are presented as pink color tokens in Figure 12-65. The rest of the decoding process involves the decoder generating a sequence of tokens. The audio encoder outputs a series of encoded representations of the audio input, capturing the important features of the audio. These representations are used by the decoder as context to predict the text. The decoder takes these encoded audio features and, using self-attention and cross-attention mechanisms, predicts the next most probable word based on the encoded audio and the previously predicted words. This process continues iteratively until the decoder predicts an end-of-token (EOT) or reaches a maximum length.

Summary

We started this section by explaining one of the common architectures of neural networks, the transformer that is widely used in text data and also computer vision. Then, we switch to concepts and models used in NLP, computer vision, and audio. Except for the transformer and its details, the rest of this chapter is a dictionary of models and applications, and since there is no summary for a dictionary, we do not provide a summary for this chapter.

Further Reading and Watching

* Jay Alamyar has a blog (https://jalammar.github.io) that provides visual explanations about state-of-the-art NLP approaches in deep learning. He provides brief explanations with illustrations and does not delve into details.

* Josh Starmer has a brief and easy-to-understand tutorial on transformer: https://youtu.be/zxQyTK8quyY. Besides, there is a very detailed explanation of transformers provided by Brandon Rohrer https://e2eml.school/transformers.html. Another tutorial from Lena Voita describes transformer architecture in good detail: https://lena-voita.github.io/nlp_course/seq2seq_and_attention.html.

* Stanford University has an online course called "NLP and Deep Learning," available at: https://youtu.be/rmVRLeJRkl4?si=c9mBB3XNID-wHA2r. Usually, university courses do not tackle state-of-the-art models. However, this course is taught by experts in the field, and it has many interesting materials to learn.

* There are some online resources that try to explain self-attention, but we found the explanation from Lawrence Carin on coursera.org an easy one to follow. Check different resources for self-

attention; each resource explains some parts very well, and some other parts are incomplete or even incorrect.

* It is not easy to find much good explanation and comparison between GPT models unless you read the entire paper. However, the Wallmart Global Tech blog provides a good summarization of GPT models: https://medium.com/walmartglobaltech/the-journey-of-open-ai-gpt-models-32d95b7b7fb2

* Brendan Bycroft tries to visualize some LLM models, which might be worth taking a look at: https://bbycroft.net

* Letitia Parcalabescu has a short introductory video for Mamba: https://youtu.be/vrF3MtGwD0Y?si=_s6zWRbB3YSfTw1W, which gives some introductions on Mamba.

* If you are interested in learning ViT model in detail and seeing how the transformer encoder behaves for the given input matrix, check the tutorial provided by Aman Arora at: https://amaarora.github.io/2021/01/18/ViT.html

* Umar Jamil has a Jamil (means beautiful in Arabic) explanation about the Segment Anything Model, and we use that explanation to understand the Mask Decoder component, https://www.youtube.com/watch?v=eYhvJR4zFUM

* Ketan Doshi provides a good introduction to audio-based deep learning and its applications. https://ketanhdoshi.medium.com.

* Kilian Batzner provides a useful detailed explanation of the one-dimensional causal dilated convolution in his blog. https://theblog.github.io/post/convolution-in-autoregressive-neural-networks

* wav2vec models have lots of details, which we skip to explain them. Jonathan Boigne provides a more detailed explanation of wav2vec models. Both for version 1: https://jonathanbgn.com/2021/06/29/illustrated-wav2vec.html and version 2: https://jonathanbgn.com/2021/09/30/illustrated-wav2vec-2.html

* To experiment and train NeRF, there is an open-source library available as NeRFstudio: https://docs.nerf.studio/en/latest

* To design Figure 12-53, we use a combination of http://tinkercard.com and http://meshlab.com.

Part v: Reinforcement Learning

Machine learning is traditionally known to have three paradigms. In previous chapters, we have discussed supervised and unsupervised (and self-supervised) learning algorithms. The next paradigm is reinforcement learning, which has recently gained more popularity than before.

Supervised learning algorithms learn based on labels, and somehow, they memorize the state of an event. For example, there are speculations that the game of Chess has 10^{50} possible states, and a supervised learning algorithm requires learning all these states, which is impossible. Instead, the algorithm should plan and learn how to play, and reinforcement learning algorithms can do this.

The popularization of reinforcement learning goes back to the DeepMind's Atari game playing [Mnih '13] in 2013, which plays Atari 2600 games on par with human users. Later, AlphaGo [Silver '17], in 2016, defeated the world Go champion in the game of Go. Just defeating the champion does not raise much attention because the human chess champion previously lost a chess match to the IBM supercomputer in 1997. What set AlphaGo apart was its ability to identify tricks in the game that had not been discovered by human players before. A year after the release of AlphaGo, AlphaZero [Silver '18] was introduced in 2017, and beat AlphaGo in the game of Go and defeated world champions in Chess and Shogi (Japanese chess) games.

Model agonistic Reinforcement Learning approaches, such as a model for playing Chess or Go, advanced until late 2022. The release of InstructGPT [Ouyang '22], which uses a reinforcement learning approach (Reinforcement Learning with Human Feedback), led to the success of OpenAI's ChatGPT even ChatGPT reduces users' interaction with the Google search engine[1], and overtook traditional web search. There were many large language models available before the introduction of ChatGPT, but they were extremely racist, biased, and unfair. Therefore, no corporation dared to release their language model online, but OpenAI hired many individuals to annotate results and fine-tune the GPT v3.5 model for public release. Again, after the ChatGPT, Reinforcement Learning approaches got attention afterward.

Nevertheless, except for their use in robotics, they were not still as popular as other paradigms (supervised, unsupervised, and self-supervised) of machine learning. We speculate that the origin of this problem is that the complexity exists in real world settings, and it is not easy to simulate it in a model, in the way Reinforcement Learning algorithms are modeled.

In the upcoming chapter, the reinforcement learning algorithms and models we explain are model-free and could be used for different tasks.

---

[1] https://www.nytimes.com/2022/12/21/technology/ai-chatgpt-google-search.html

Chapter 13: Reinforcement Learning

Traditionally, machine learning algorithms are classified into three categories: supervised learning, unsupervised learning, and reinforcement learning, which is the third major category. However, in Chapter 1, we proposed a more recent categorization that includes self-supervised learning as well. This chapter focuses on reinforcement learning.

Before we start to explain reinforcement learning algorithms, let's review one of the prominent theories in psychology, i.e., *Operant Conditioning*. Burrhus Frederic (B.F.) Skinner [Skinner '65] described that humans and animals learn any behavior through reinforcement and punishment. For example, a kid's parents motivate kids to keep their rooms tidy with rewards such as an ice cream or punishment such as not getting ice cream.

A computer agent can simulate this learning process based on trial and error, and the computer algorithm that mimics how we learn is called Reinforcement Learning (RL). We can say RL algorithms are decision-making algorithms that make decisions to get a *maximum reward* in a particular situation. This resembles the operant conditioning theory, in which a living creature learns based on increasing the reinforcement and reducing the punishment.

Due to the lower cost and safety benefits, computer simulations are often used to train RL algorithms for real-world implementation. During simulation, the agent learns to perform its task accurately. However, it's important to remember that there will always be some difference between the simulation and the real world. Therefore, it's crucial to consider as many real-world variables as possible when designing the simulation. This helps minimize the "reality gap" and allows the agent to transfer its learned skills more effectively to real-world scenarios.

We start this chapter by describing some basic concepts required to learn the RL algorithm. Then, we explain a well-known dilemma in RL known as "Exploration versus Exploitation". Next, we switch to theoretical discussion and describe the Markov Decision Process. To give you a short break from theory, we describe the Multi-Armed Bandit (MAB) problem. Afterward, we describe approaches to find the optimal policy and optimal value function. After learning these approaches, we describe classical RL algorithms for discrete space models. Next, we moved to deep RL algorithms, which handle both discrete and continuous space models. We skip to describe model-dependent algorithms in this book, and all models and algorithms we describe here are model-free. Therefore, we can adapt and use them for different research questions or problems.

Reinforcement Learning Concepts

In the context of RL, the living creature of Skinner's theory is referred to as the *agent*. The agent takes action and learns to adjust its behavior based on the reinforcement (positive reward) or punishment (negative reward) it receives.

Please open some room in your brain and memorize the following. An agent performs *actions* to interact with the *environment*, and it learns to perform actions in a way that maximizes the *reward*. The reward is a result of taking or performing an action, and it is scalar feedback that the agent receives. The reward describes how well the agent did the action. Note that the reward quantifies "what" we want to achieve and "not how" to achieve that.

The environment in which an agent operates has a *state,* and every time an action is taken, the environment's state changes. Sutton and Barto [Sutton '18] explained that informally the state describes 'how the environment is at the current time'. Figure 13-1 presents an abstract representation of RL components and their interaction with each other.

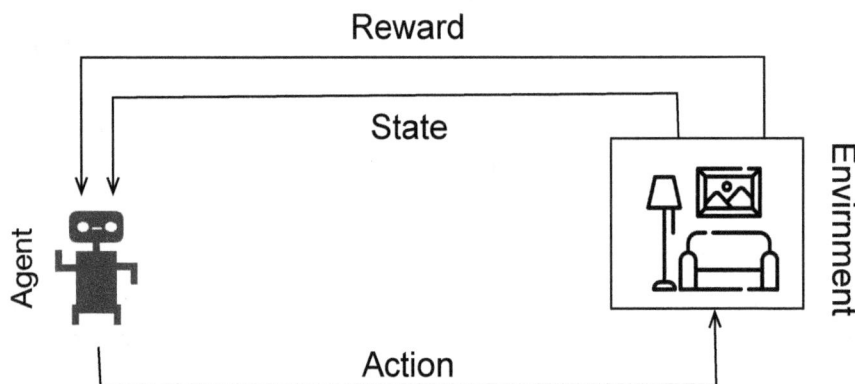

Figure 13-1: Reinforcement learning components.

All actions that an RL agent takes are in the scope of "time", and thus we are dealing with the sequence of actions and rewards we used $t = 1,2,...T$ to present the time of these sequences. For example, action at time t is presented as A_t. A sequence of actions, rewards, and state changes in an RL algorithm can be written as: $S_1, A_1, R_1, S_2, A_2, R_2, \ldots S_T, A_T, R_T$. Try to remember a process reinforcement algorithm operation as "SARSARSAR...".

Episode: In reinforcement learning (RL), the objective of an algorithm is for the agent to explore the environment (or a simulated environment), gain experience, and learn how to make optimal decisions. The agent interacts with the environment through a series of state-action-reward (SAR) sequences. These sequences continue until the agent reaches either a goal or a terminal state. This complete sequence of interactions is referred to as an *episode*.

In other words, the episode is the duration between the initial state and the terminal state. For example, for an agent that plays a computer game, each time it starts playing until the player loses, the game is one episode.

822

Tasks that are not defined in the context episode and could continue indefinitely are considered continuous or *non-episodic tasks*. For example, controlling room temperature is a non-episodic task because there is no end to it; maintaining the temperature is an ongoing, daily, or hourly activity.

Terminal state: It is the state that marks the end of an episode. For example, an agent might be driving a vehicle in a video game, and the car hits something and explodes, or the agent loses its life while fighting a monster in the game and reaches the 'game over' state. In these examples, 'game over' is a terminal state. The RL agent plays the game (or experiments in the environment) many times until it learns to behave optimally in the environment or reaches its maximum threshold of episodes.

Model: The model describes how the environment behaves and specifies how the environment interacts with the agent. In particular, the modeling process specifies how the environment responds to the agent's actions. RL algorithms are either *model-based* or *model-free*. Model-based algorithms operate by planning and either receiving the definition of the environment from the user or learning the model themselves, such as in a chess-playing agent. Model-free algorithms do not need information about the model; they learn based on trial and error, as seen in a robot learning to walk by itself.

Keep in mind that in the context of RL, the term model is different from its usage in neural networks, where it refers to a trained system used for making inferences.

Stationary vs. Non-Stationary: An environment can be classified as either stationary or non-stationary. If the environment does not change during the time of the experiment, we refer to it as a *stationary environment*. On the other hand, if the environment changes during the agent's interaction with it, it is called a *non-stationary* environment. To illustrate this concept, consider a drone searching for pests on a farm. If the pests are mobile, such as rodents like rats that move around the farm, the environment is non-stationary. The locations of the pests change over time, altering the state of the environment. Conversely, if the pests are stationary plants, their positions remain fixed, and we can consider this a stationary environment.

In addition to "agent", "environment", "reward", "state", "model" and "episode", we need to learn more definitions in reinforcement learning, including "policy," "value function", etc., which we will explain them later. Before continuing reading, please be sure to learn the described concepts; we now move on to some characteristics of reinforcement learning. Later, we revisit these definitions and describe more terms using a detailed mathematical formulation.

RL as Optimization Problem: Another perspective states that we can view an RL problem as an optimization problem and use the following equation to update the agent's new state based on the old state information:

$$New\ Estimate \leftarrow OldEstimate + StepSize\ [\ Target - OldEstimate]$$

Similar to other optimization problems the error of estimation is *[Target - OldEstimate]*, and as we get closer to the target, this error will be reduced. The step size is a constant parameter, and its value should be chosen to minimize the cost of optimization algorithms. If you recall Chapter 10, we use a very similar approach for gradient descent optimization as well.

Exploitation versus Exploration

A challenge in reinforcement learning that resembles the human decision making process is the challenge between "instant gratification" and "delayed gratification". Humans naturally enjoy instant gratification, which means we prefer immediate rewards. For example, eating sweets and fatty foods such as fries gives us instant joy. However, not eating immediately when our stomach craves food is good for our physical and mental health. This means we delay gratification (by eating later and not now). However, if we continuously delay gratification, we will die from hunger. Usually, the more we delay gratification, the higher the reward, but our resources are limited, and there must be a balance between instant and delayed gratification.

Another example is studying and learning, which forces our brain to bear the learning pressure and avoid having comfort. Later, our accumulated knowledge helps us have a better career path and improve our quality of life and, thus, comfort. However, if you continuously learn and don't implement your knowledge in industry or product, you will live poor, like the author of this book.

The same issue exists in RL algorithms, referred to as *exploitation versus exploration*. Exploration means not necessarily performing the action that has the highest reward but instead exploring the environment, which could lead (but is not guaranteed) to a higher reward later. On the other hand, exploitation refers to the action that gets the highest possible reward immediately. We can say that exploitation is a greedy approach because it goes for the immediate reward.

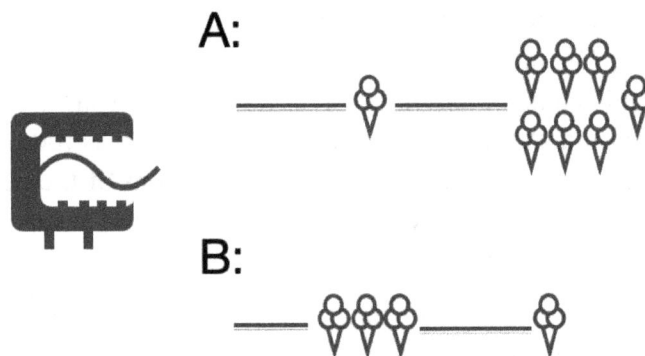

Figure 13-2: A greedy algorithm goes to path B, because it goes for immediate reward and does not explore the next steps.

Take a look at Figure 13-2; we are using taxpayers' money to build a robot that only licks ice cream. Why build a robot to go and lick ice cream? Assume we have too many grants, like top universities, and do not know what to do with it. Therefore, we start building an ice cream licking robot.

If the robot chooses a greedy approach, it goes for path B; it immediately gets three ice cream. If it follows path A, it gets one ice cream at the first step. However, following path A results in eight ice creams at the end, which is higher than following path B (despite the short-term reward of path A being lower than B).

As another example, consider a hiring process. Assume you own a company and plan to hire engineers to build a very innovative product. Some candidates come from top schools and have excellent grades. It is a sign that most probably they are good at work. On the other hand, some candidates are not coming from top schools, but their CVs show they have progressed in their careers. Those seem more innovative people who do not follow the rules and regulations, and thus, they might bring new ideas to the company. To build a novel product, candidates from top schools are good and do their job fine, i.e., exploitation. However, a new product requires innovation, and it might be useful to hire from the later group of candidates, i.e., exploration.

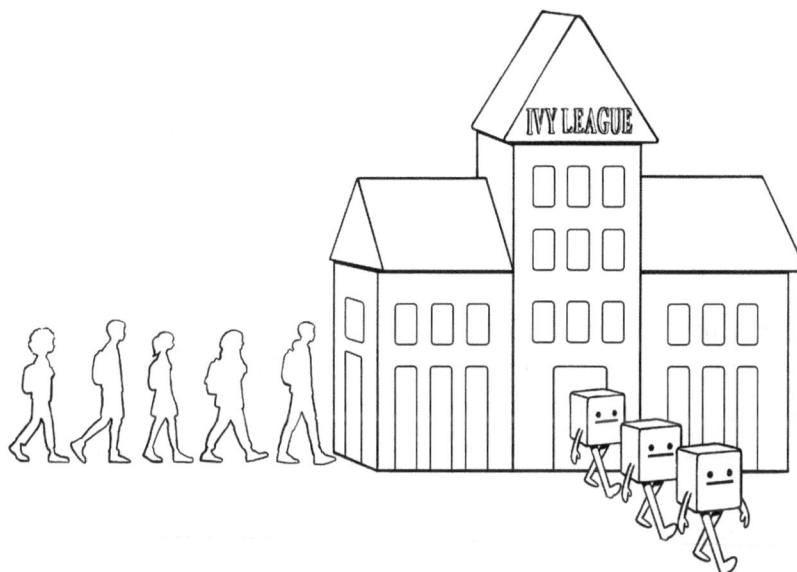

Unlike other algorithms, RL algorithms evaluate every decision they make, and decisions are not only instructions. If the algorithm is in the exploration stage, it gives *evaluative feedback*, and if it is in exploitation, it gives *instructive feedback*. David Silver, the first author of the AlphaGo [Silver '17] and AlphaZero algorithms [Silver '18] has an interesting quote in this context and stated that *"the RL is the science of making optimal decisions"[1]*.

Markov Decision Process

An RL algorithm is a sequential decision-making process, and the Markov decision process (MDP) is used to formalize it. In particular, MDP is a mathematical framework to model sequential decision-making under uncertainty. MDP formalization provides a fairly easy abstraction for RL problems, and most RL problems can be formalized with MDP.

This section has many theoretical explanations, and we must learn them to proceed with RL algorithms. The theoretical explanations might be boring and sound hard at the beginning, but please read them several times until you understand them. There is a demon in this chapter that hunts your dreams while reading this chapter and takes exams, so you should be good with these theoretical formalizations.

---

[1] https://www.youtube.com/watch?v=2pWv7GOvuf0

While you are reading this chapter, I haunt your dreams and take an exam from you every night.

Grid-world

A simplified environment called *grid-world* is used to explain RL concepts. Assume we have an agent that can move only up, down, right, or left, like the Pacman game[2]. The grid-world shown in Figure 13-3, is 4×3 size. The agent starts at the left-bottom position, and the goal is to reach the top-right position (green rectangle) while avoiding hitting the red rectangle. There is one cell, i.e., the wall, that the agent can not pass, and it is shown as a grey cell in Figure 13-3. Usually, the agent starts from the left-bottom corner, but it could start from any other cells as well.

A common terminology of grid-world is known as a frozen lake[3]. It contains four types of cells: start (S), frozen (F), hole (H), and goal (G).

The agent intends to move from S to G on F cells and avoid H. The matrix can be something like the following:

S F F F F

H F H F F F

H F F F H F

F H F F H F

H F F F F G

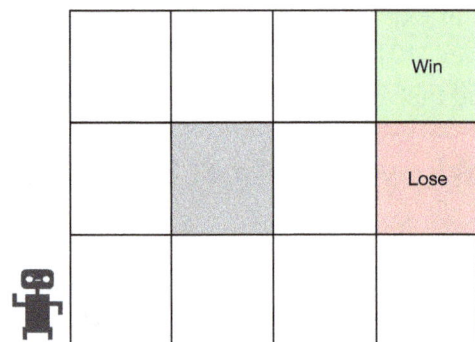

Figure 13-3: The Grid-world that the agent can move in four directions, right/left up and down. The grey area is a wall, and it is not possible for the agent to move there.

There are two other varieties of grid-world: *Windy Grid-world* and *Cliff Grid-world*. In the windy grid-world, a part of the grid is windy and pushes the agent upward. The windy grid-world has a crosswind upward through the part of the grid. In the windy

---

[2] https://en.wikipedia.org/wiki/Pac-Man

[3] It is used in Gymnasium, which, at the time of writing this chapter, is a common platform for experimenting with RL algorithms.

region, the resultant states are shifted upward by the wind, and the strength of the wind could vary or stay fixed between columns. We can see on the left side of Figure 13-4 that the wind pushes the agent upward in the windy region, and the agent must move downward to reach the goal state.

In the cliff grid-world, all white cells have a reward of -1, while the cliff has a reward of -100. The agent must get to the goal cell, while avoiding to hit the cliff.

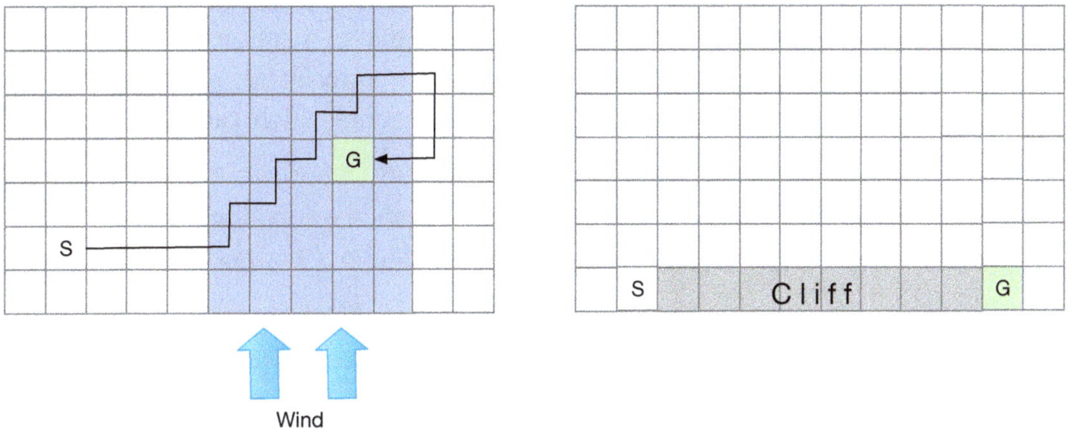

Figure 13-4: (left) The windy Grid-world that pushes the agent upward in the blue cells. (right) the cliff grid-world that the agent should avoid hitting the cliff.

We have learned that decisions are made in a *sequence*, and thus, every action, reward, and state is associated with a *time*. To incorporate time into the decision-making process, check Figure 13-5, which is a redesign of Figure 13-1.

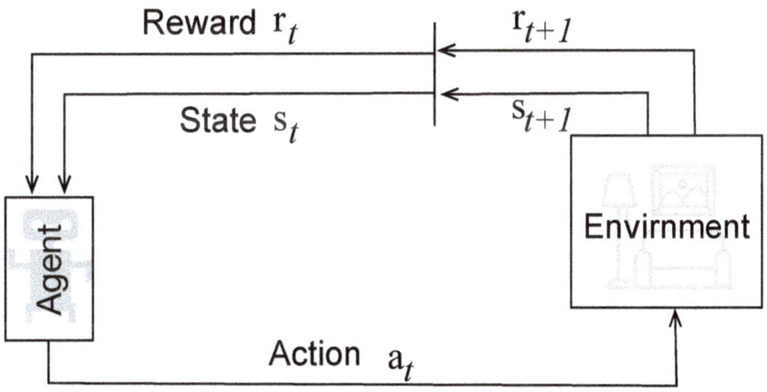

Figure 13-5: Reinforcement learning components in the context time.

In Chapter 5, we have explained that the Markov Process (or Markov Chain) is a sequence of random states without any memory to remember the previous states. If you would like to review it, you can check Figure 5-38 in Chapter 5, which shows an example of the Markov process with its transition matrix.

A Markov Decision Process (MDP) is represented by states (S) and a transition probability matrix(P), i.e., (S, P). Additionally, the MDP model has action (a) and reward (R). The transition probability function specificities the probability of moving (transitioning) from the state s to the next state s', after taking action a, while obtaining the reward r. If the reward is random (stochastic), this process can be written as $P(s', r \mid s, a)$. If the reward is deterministic (not random), the transition probability is written as $P(s' \mid s, a)$. We will explain the differences between stochastic and deterministic shortly.

While learning RL from now on, keep in mind that the next state after the current state s will be presented as s'. In this formalization, a presents an action that is a subset of all possible actions $(a \in A)$. After the agent takes an action, the reward r will be received, which is a subset of all possible rewards R $(r \in R)$. s presents the state of the agent, which can be of many states S $(s \in S)$.

After an action is taken, the state of the agent changes. In other words, the next state of the agent s_{t+1} depends on the previous state and the action of the agent, i.e. $P(s_{t+1} \mid s_t, a)$.

State transitions are *Markovian* (*Markovian Transition*). A Markovian Transition means that future states depend on the present state and not on the history of previous states. In simple terms, to predict the future state, an MDP uses only the present state and not the history of state changes. We can say the state s_t is a Markovian state if $P(s_{t+1} \mid s_1, \ldots s_t) = P(s_{t+1} \mid s_t)$.

We can also change the formalization of the Markov process to incorporate rewards into the Markov process. This means that for every transition from state t to state $t+1$, there is a reward (either positive or negative) associated with the transition. Earlier, we have explained that the Markov process is a set of state-action-reward, i.e., SARSAR… We can add time factors to it and rewrite it as: $S_t \, A_t \, R_{t+1} \, S_{t+1} \, A_{t+1} \, R_{t+2} \cdots$.

In conclusion, each Markov process can be written as a tuple (S, A, T, R, γ).

- S is a finite set of states.

- A is a finite set of actions.

- T is the transition probability matrix, the transition from state s to state s' is formalized as follows: $T_{ss'} = \mathbb{P}(S_{t+1} = s' \mid S_t = s, A_t = a)$. By $\mathbb{P}$ we mean probability.

- R is the reward function, which specifies how much immediate reward we get at state s_t at the time $t + 1$, we can write $R_s = \mathbb{E}[R_{t+1} \mid S_t = s, A_t = a]$. The goal of the RL algorithm is to maximize the overall cumulative sum of rewards. Recall by using $\mathbb{E}$ we mean expected value of R_{t+1}.

- γ is a discount factor that can be between 0 and 1, $0 \leq \gamma \leq 1$ and it is given by the user as a hyperparameter. The discount factor is used to determine how much the agent prefers the

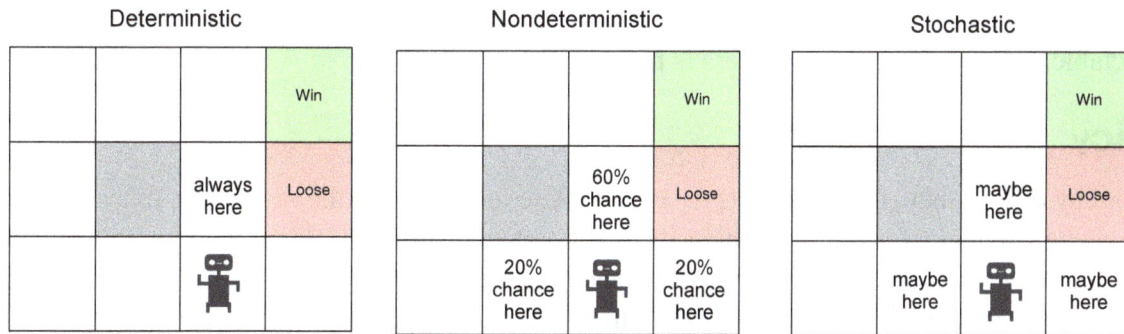

Figure 13-6: Three different agent behavior examples in the MDP process. Stochastic processes are random and unpredictable; nondeterministic processes are not completely random and are determined by prior conditions or states. On the other hand, deterministic processes are predictable, in contrast to stochastic processes.

current reward over the future reward [4]. γ close to 0 means the agent is in a rush to get the immediate reward, and γ close to 1 means that the agent is willing to wait for a long-term reward. Note that most of the time, MDP rewards are discounted. Why does the agent prefer immediate reward over future reward? This is due to different reasons, such as the uncertainty that models do not operate perfectly fine, and thus, the immediate reward is favored.

Deterministic, Nondeterministic, and Stochastic Models

We have explained in Chapter 11 that a *deterministic* model is a model that, after every execution of the model, the result will be the same. A *stochastic* (random) model is a model that, after every execution of the model, has a different result.

Another concept we use in RL is *nondeterministic*, which is not completely random but has some probability associated with it. In other words, stochastic refers to a process that is random or unpredictable, while nondeterministic refers to a process that is not completely random and is influenced by prior states or conditions. For example, a game of chess is a nondeterministic process. The exact outcome of a chess game depends on the moves the players make, which are not completely random. However, the game still has some element of randomness, as the players do not know what their opponent will do next.

To better understand them, look at the Grid-world example in Figure 13-6. If, in every execution, the agent comes one rectangle up, it is a deterministic model. If, in every execution, the agent decides completely random to go left, right, or up, it is stochastic. If, in every execution, the agent decides not random but based on some probability. For example, it has 20% chance of going to the right, 20% left, and 60% up; it is a nondeterministic model. In other words, nondeterministic models are associated with probabilities.

---

[4] You might think it is against our explanation of exploration. In fact, it is a bit different, it says that a reward now is better than a reward later and favors the recent reward. This could resemble a financial quote: "money now is better than money later". For example, getting 500$ today is better than getting the exact amount next year, because of inflation, etc.

To summarize, a stochastic process is not predictable, a nondeterministic process is partially predictable, and a deterministic process is predictable.

Policy

The policy defines how the agent acts in a given state to maximize the reward. In other words, it is the *agent's strategy* for accomplishing its task. Mathematically speaking, a policy is a function that *maps state to actions*. For example, a policy could be a machine learning model that receives states as input and provides a probability distribution over actions as output. The optimal policy that yields the highest sum of rewards is presented as $\pi *$.

The policy could be stochastic (random) or deterministic (non-random). For example, an Ice-Cream Licking Robot (ILR) could have a policy of licking every object it encounters, with the hope that it is ice cream. This ILR has a stochastic policy because it does not have any specific criteria or rules for determining whether an object is ice cream. It relies on the chance to encounter actual ice cream. Another more intelligent ILR can use a camera with an object detection algorithm and lick only objects with the same shape as ice cream. This ILR's policy is to lick objects that resemble ice cream and not randomly lick anything until it finds ice cream; thus, we can say it has a deterministic policy.

The policy is presented with π, and if it is a deterministic policy, the action recommended by policy π and state s is presented as $a = \pi(s)$ or $s \xrightarrow{\pi} a$. If the policy is non-deterministic, it is a probability of taking action a given state s, is formalized as $\pi(a \,|\, s)$. The input of the policy function is state s, and its output is an action a.

Expected Return (Goal)

The goal of an MDP model is to maximize future rewards from the current point in time. Since the MDP does not care about the past and uses the present to model the future. The sum of future rewards at time t is called the *expected return* or *goal* and is presented as G_t. Usually, MDP uses a discount factor on rewards as well, and the G_t is defined as follows:

$$G_t = R_{t+1} + \gamma R_{t+2} + \gamma^2 R_{t+3} + \ldots = \sum_{k=0}^{\infty} \gamma^k R_{t+k+1}$$

We can see from this equation that the later rewards get smaller (discounted) because γ is smaller than one, and bringing it to the power of a natural number makes it smaller, e.g. $0.1^2 = 0.01$. If we set $\gamma = 0$ it means the algorithm is shortsighted (greedy) and only cares about the current reward.

For example, in our ILR example, the expected return is to have the robot lick as many ice creams as possible. In a stock trading agent, the expected return is the sum of the expected profits from the trade plus the expected profits from future trades. In grid-world problems, the expected return for a movement is the sum of the expected rewards for reaching the goal plus the expected rewards for avoiding obstacles.

When we are evaluating an RL model, we report its *average return*. Recall for supervised learning algorithms we used precision, recall, F-score, etc. Here, the accuracy of a model is evaluated by averaging the return.

Value, Value function, and Action-value Function

The *value* refers to the total amount of rewards an agent can expect to accumulate in the future, given its current state or action. Value is different from reward. A reward measures the quality of the most recent decision and not all previous decisions, but the value is the *total (cumulative) sum of future rewards*. The agent aims to achieve a state that maximizes the cumulative sum of rewards (G_t).

The *value function* or *state-value function* is a function that is used to estimate the total sum of future rewards for the given state, and it is presented as $V_\pi(s)$. Given the current state, the value function specifies how well an agent performs.

Since the value function is used to calculate the expected return, we deal with the expectations ($\mathbb{E}$), and it is formalized as follows:

$$V_\pi(s) = \mathbb{E}_\pi[G_t \,|\, S_t = s]$$

This equation means that the value function is defined as the expected return G_t, given state s, under the policy π. $V_\pi(s)$ is also read as *V-function*. $\mathbb{E}_\pi$ denotes the expected value given that the agent follows policy π. For the exam in your dream, you should memorize that V-function is the expected return G_t given that the agent is currently at the state S_t and following the policy π. We can summarize that the value function is a function that predicts the expected amount of future rewards that an agent can accumulate at its current state with its current policy.

In many scenarios, actions are not equal; some lead to higher rewards, and some lead to lower rewards. Besides, there are cases where the policy is non-stationary policy, meaning it allows the agent to take different actions based on the given state. In other words, a non-stationary policy changes its action recommendations as it adapts or changes the environment over time. In these cases, the sum of future rewards depends on the type of actions the agents take, and we should incorporate the action along with the state into the value function. Therefore, the value of having action *a,* in state *s,* under the policy π can be formalized as follows:

$$Q_\pi(s, a) = \mathbb{E}[G_t \,|\, S_t = s, A_t = a]$$

$Q_\pi(s, a)$ is called the *action-value function* or *state-action value function* for the policy π or simply *Q-function*. In summary, the Q-function estimates the total sum of future rewards for a given state and action.

Usually, the value function V_π or action-value function Q_π will be identified through experiments. The ultimate goal of any RL algorithm is to learn a good value (or action-value) function and policy.

Action choices are made by using the value, and the RL algorithm seeks the highest possible value (not the highest possible reward). Aiming for the highest possible next reward (not all future rewards) is a greedy decision, and we have explained its limitations in the "Exploration versus Exploitation" section.

State Transition Diagram

We have explained that all states between initial and terminal states construct an episode. The interaction between agent and environment can be defined as a set of discrete or continuous sequences known as episodes.

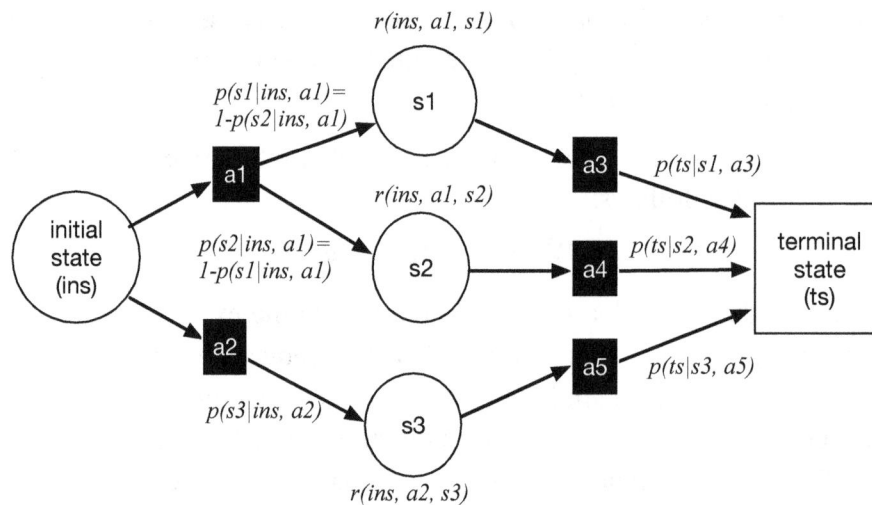

Figure 13-7: State transition diagram example.

To model discrete sequences via MDP, we use a *state transition diagram* as presented in Figure 13-7. A state transition diagram is a graph that visualizes the agent's possible actions at each state, the probability that an agent moves from one state to another state, and the reward that the agent receives in each state. The black square presents the action, the circle presents the state, and after actions, we have transition probabilities. From each state, the agent has a probability of transitioning to one or more possible next states. The sum of the transition probabilities from any given state must equal one (as shown in Figure 13-7).

Figure 13-7 presents a state-transition diagram example. By looking at this figure, we can understand the rationale for having probabilities after each action. The probability of taking an action and moving to another state is written as $p(s_{t+1} \mid s_t, a_t)$, and the reward of taking an action and changing the state is written as $r(s_t, a_t, s_{t+1})$. For the sake of simplicity, most of the time, the action will be neglected, and the diagram is shown only with states, as it is shown in Figure 13-8.

Instead, the focus is often on illustrating the different states and the possible transitions between them. Similar to some literature, we use rectangles for terminal states and circles for other states.

For example, in the future, our planet is running out of resources that we extracted from the ground. However, oceans are full of garbage, and we can extract some required materials from garbage inside the ocean. There are fishes in the ocean, but since they eat lots of microplastics, we can not eat them anymore. You own a company that builds garbage-collecting robots. These robots walk under the ocean surface and collect plastics and other garbage to reduce pollution in the ocean. The robot has a limited capacity to collect garbage (positive reward) and fish or other living species (negative reward) until it gets a big fish or big garbage, and then it is full. Therefore, we have two terminal states: one is grabbing a "big fish" by mistake, and the other one is grabbing a "big garbage". Assuming there are three pieces of garbage in the path (two small and one big) and two fishes, the state diagram of the robot navigating underwater is presented in Figure 13-8.

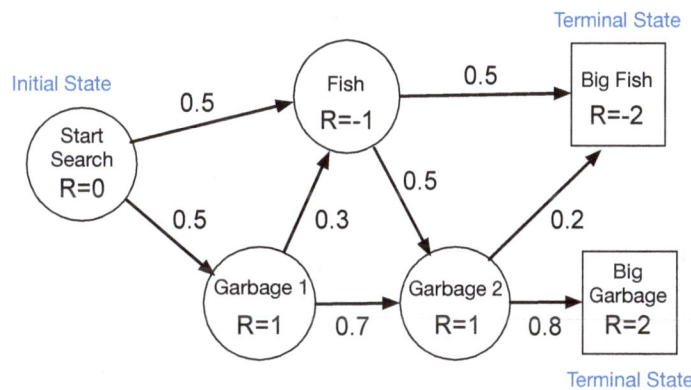

Figure 13-8: State transition diagram of the example robot that collects garbage in the ocean.

833

A state transition diagram might be useful for a small RL model, but it is not easy to build real-world models. Real-world RL applications are not simple enough to enable us to draw their state diagram.

Bellman Equation

Bellman equation is an equation used to solve the value function (or action-value function) and can be used to resolve many RL problems. As previously mentioned, RL algorithms aim to estimate the value function (or action-value function) to guide the agent toward optimal decision-making.

We have learned that the equation for value function is written as: $V_\pi(s) = \mathbb{E}_\pi[G_t | S_t = s]$, by substituting the G_t with discounted factor (γ) and reward, we get the following equation:

$$V_\pi(s) = \mathbb{E}_\pi[R_{t+1} + \gamma G_{t+1} | S_t = s]$$

The expected value here is a linear operation. Therefore, we can split the expectation by its plus sign and rewrite this equation as follows:

$$V_\pi(s) = \mathbb{E}_\pi[R_{t+1} | S_t = s] + \gamma \mathbb{E}_\pi[G_{t+1} | S_t = s]$$

G_{t+1} means the expected return at the time $t + 1$, which is equivalent to the value function at the next state, i.e., $V_\pi(s')$. We can replace it with the value function at the next state (s'). and rewrite the value function equation as follows:

$$V_\pi(s) = \mathbb{E}_\pi[R_{t+1} + \gamma V_\pi(s') | S_t = s]$$

A non-deterministic value function can be written with the same equation as a sum of probabilities as follows:

$$V_\pi(s) = \max_a \sum_{s',r} p(s', r | s, a)[r + \gamma V_\pi(s')]$$

We could use the policy function and write the non-deterministic value function as follows:

$$V_\pi(s) = \sum_a \pi(a | s) \sum_{s',r} p(s', r | s, a)[r + \gamma V_\pi(s')]$$

In these listed equations, $p(s', r | s, a)$ refers to the transition probability of getting from state s to the next state s', with the action a and reward r.

For the exam that the exam monster will take tonight in your dream, you should memorize the simpler version of the Bellman equation as follows:

$$V(s) = \max_a[R(s, a) + \gamma v(s')]$$

Here, $V(s)$ presents the value at state s', max_a is used to specify the maximum value for any action a, R presents the reward of taking action a at state s, γ is the discount factor and $v(s')$ is the value of next state, i.e., s'.

834

The Bellman equation for value function is a core of RL. We can see that the value function is recursive, and calculating the value function at the current state, s, depends on the possible next state (s') only, reflecting its Markovian characteristic. In simple words, to calculate the value function, we only look at one state ahead (s') and not previous ones, even if there are many steps required by an agent to solve a problem. This encapsulates the idea that the decision-making process at each step is memoryless regarding past states.

Respectively, we can write the Bellman equation for the action-value function, i.e., $Q_\pi(s, a)$ as follows:

$$Q_\pi(s, a) = \sum_{s', r} p(s', r \mid s, a)[r + \gamma \max_{a'} q_\pi(s', a')]$$

We do not write the details of getting into this equation because we are afraid that by adding too many theories and formulations, you stop reading the rest of this chapter. To summarize this section, keep in mind that while learning RL, it is important to remember the differences between $V_\pi(s)$ and $Q_\pi(s, a)$. $V_\pi(s)$ is the expected return started from state s and following policy π, $Q_\pi(s, a)$ is the expected return from state s, taking action a, and following policy π.

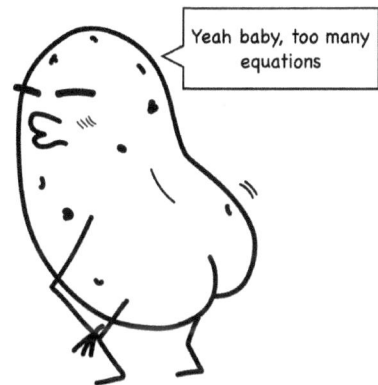

Yeah baby, too many equations

Bellman Equation Examples

To grasp some practical understanding, we describe some examples that are resolved with the Bellman equation. The idea of RL is that we operate within a MDP framework, with the objective of finding or solving $v_\pi(s)$ or $Q_\pi(s, a)$. We have learned that the Bellman equation is used to solve $v_\pi(s)$ or $Q_\pi(s, a)$.

Example 1: A very simple, straightforward example is shown in Figure 13-9, we have only one middle state and assume the discount factor $\gamma = 0.9$. Since from the "Start" state the agent goes to the "Mid" state and from the "Mid" state it goes to the "Terminal" state, all transition probabilities are 1 because there is no state choice existing on this path.

To calculate value functions for all states, $V(Start)$, $V(Mid)$, and $V(Terminal)$, we start with $V(Terminal)$. The value of the terminal state is always 0 because there is no next state, i.e. $V(s') = 0$, and no future action, i.e. $R(s, a) = 0$. Therefore, substituting them in the $V(s) = max_a[R(s, a) + \gamma V(s')]$ equation results in 0.

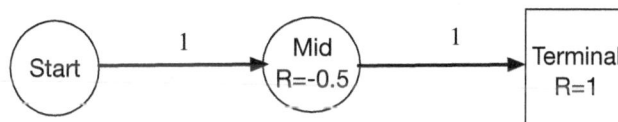

Figure 13-9: A simple state transition diagram with one middle state.

To calculate $V(Mid)$ we have $V(Mid) = 1 + 0.9 \times 0 = 1$. To calculate $V(Start)$, we have $V(Start) = max_a[R(Start, mid) + \gamma V(mid)] = -0.5 + 0.9 \times 1 = 0.4$

Example 2: Assume we are using grid-world, and actions are deterministic. The agent starts at the cell (z, a), the discount factor is $\gamma = 0.9$, the "win" cell located at (x, d) and has a reward of +1, and the "lose" cell located at (y, d) and has a reward of -1, as shown in Figure 13-10. Always, we start with $v(Terminal)$, which is the "win" cell, and the value of its next state (s') is again zero. As the next state, we can calculate the value of (x, c), which is: $v_{(x,c)} = 1 + 0.9 \times 0 = 1$.

The value of (x, d) is zero because it is a terminal state, and +1 presents $R(s, a)$, the reward that is read from the cell (x, d). Next, we calculate the value for (x, b), and it will be $v_{(x,b)} = 0 + 0.9 \times 1 = 0.9$. For cell (x, b), we use 0 as a value of $R(s, a)$, because there is no reward for getting into the next cell, i.e. (x, c). Respectively, we continue this calculation until all cells receive their values, as is shown on the right grid of Figure 13-10. In the end, we can see that the agent can easily find the best path by looking at values (0.66, 0.73, 0.81, 0.9), which could be $\{(z, a), (y, a), (x, a), (x, b), (x, c)\}$ or $\{(z, a), (z, b), (z, c), (y, c), (x, c)\}$.

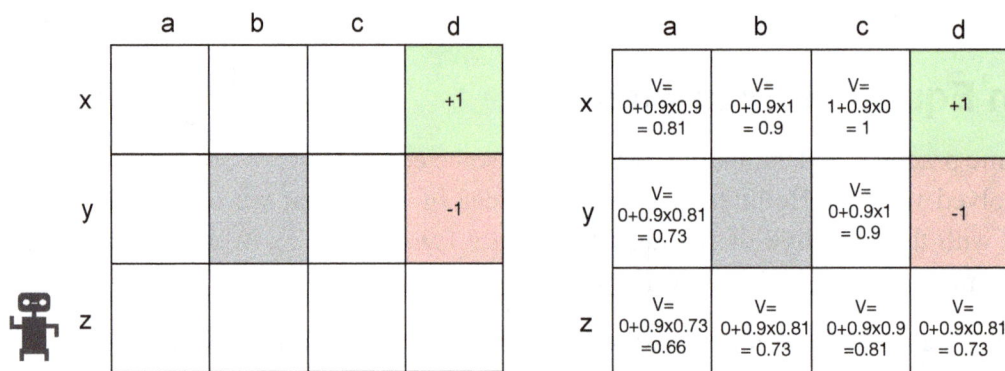

Figure 13-10: Example of Value function calculation in a grid-world.

Example 3: In the previous two examples, we can easily go backward from the terminal state and calculate the value for each state. However, in some cases our states have a cycle, such as the example shown in Figure 13-11.

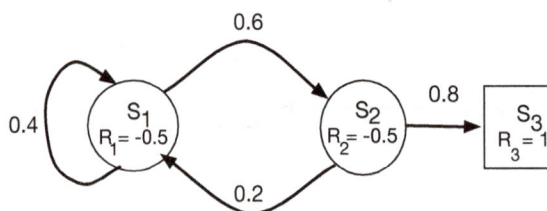

Figure 13-11: Example of a state diagram with a cycle.

To calculate the value of each state, we will use the Bellman equation, but we can not use the simple form. Instead, we use probabilities to find the value of each state. We assume the discount factor $\gamma = 0.9$. Since the s_3 is a terminal state $v(s_3) = 0$. To calculate $v(s_1)$, we need to solve the following equation:

$$v(s_1) = p(s_1 \mid s_1)(R_1 + \gamma v(s_1)) + p(s_2 \mid s_1)(R_2 + \gamma v(s_2)).$$

By substituting them with their values, we get:

$$v(s_1) = 0.4(-0.5 + 0.9v(s_1)) + 0.6(-0.5 + 0.9v(s_2)) = -0.5 + 0.36v(s_1) + 0.54v(s_2)$$

With the same approach, we can calculate $v(s_2)$ as follows:

$$v(s_2) = p(s_1 \mid s_2)(R_1 + \gamma v(s_1)) + p(s_3 \mid s_2)(R_3 + \gamma v(s_3))$$

$$v(s_2) = 0.2(-0.5 + 0.9v(s_1)) + 0.8(1 + 0.9v(s_3)) = -0.2 + 0.18v(s_1) + 0.56v(s_3)$$

Now, we can substitute $v(s_3) = 0$ in both equations and have the followings:

$$v(s_1) = -0.5 + 0.36v(s_1) + 0.54v(s_2)$$

$$v(s_2) = -0.2 + 0.18v(s_1) + 0.56 \times 0 = -0.2 + 0.18v(s_1)$$

In school, we learn how to solve two equations with two unknown variables ($v(s_1), v(s_2)$). Therefore, $v(s_1) = -0.5 + 0.36v(s_1) + 0.54(-0.2 + 0.18v(s_1)) = -1.12$ and $v(s_2) = -0.40$.

NOTES:

* When a reward is negative, it is called a "penalty" or "negative reward". Comparing it with the Skinner theory we described at the beginning of this chapter, a negative reward resembles the punishment process, which is the opposite of reinforcement.

* If you read Chapter 5, there we introduce HMM, and here we introduce MDP. You might think, what are the similarities, and why do they all have the name Markov? They all came from a Russian mathematician, Andrey Andreyevich Markov (1856-1922). Table 13-1 provides a summarization of Markov models for modeling discrete time events. We should decide which method to use based on whether we have a model with discrete time and whether the stochastic processes involved are controlled. For example, the temperature of a city, which we do not control, can be modeled using a thermometer. This is called an autonomous system. If the state, such as the outdoor temperature, is fully observable, we use the Markov model. If the state is fully observable and we can control the system, we use a Markov Decision Process (MDP). For instance, a thermostat that controls and measures room temperature represents a controlled system with fully observable states. If some states are hidden and the system is autonomous, we can use a Hidden Markov Model (HMM), such as in

modeling energy price changes. If the system is controlled but states are not completely observable, we use a Partially Observable Markov Decision Process (POMDP). In a POMDP, the agent needs to learn to map observations to their actual states. This model is used when the agent's observations of the environment's state are uncertain or incomplete, requiring decisions to be made based on a probability distribution over possible states. In such an environment, the agent must make decisions based on a probability distribution over the possible states of the environment (rather than a definite observation).

	State is Fully Observed	State that are Partially Observed
Autonomous Process (no control over the state transitions)	Markov Chain	HMM
Controlled Process	MDP	Partially Observable MDP (POMDP)

Table 13-1: Different processes and observation strategies within their related Markov model.

* In addition to the transition diagram, there are also other techniques to present actions and states. Two prominent examples are the *Backup Diagram* and the *Transition Table*. The backup diagram shows states and actions as a tree, and the transition table shows every state *s*, action a, and future state *s* in the transition table, a.k.a., transition matrix. Figure 13-12 shows a sample backup diagram, similar to the transition diagram, and we can see from this Figure how each component of RL is presented.

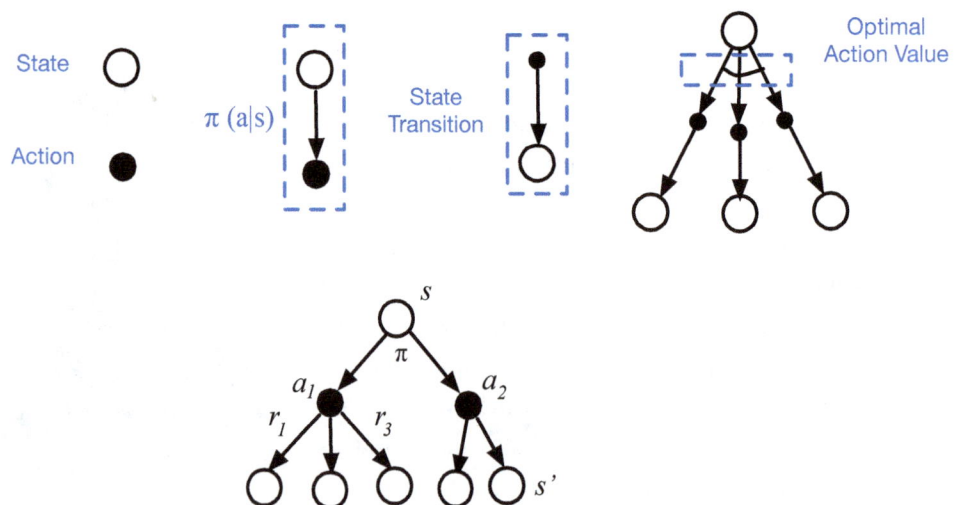

Figure 13-12: Backup diagram and its components.

* $Q_\pi(s, a)$ can be used when we need a model-free solution (the transition matrix is unavailable). Usually, using $V_\pi(s)$ is favored over $Q_\pi(s, a)$ because it involves calculations for all possible states and actions; when the action space is large, calculating $Q_\pi(s, a)$ is inefficient. $Q_\pi(s, a)$ represents $s \times a$ values, but $V_\pi(s)$ represents only s values.

* Some RL algorithms are referred to as *value-based* algorithms, and they focus on updating the value function based on the Bellman equation. Another category is called *policy-based* algorithms, which focus on updating the value function based on greedy policy obtained from the last policy improvement.

Multi-Armed Bandits Problem

In this section, we shift a bit from mathematics and explain a well-known RL problem without MDP formalization. By solving the Multi-Armed Bandit (MAB) problem, the RL algorithm can make a better decision under uncertainty. In other words, resolving the MAB problem involves finding how to allocate a fixed amount of resources (e.g., battery, money, time, etc.) among a set of alternative options (e.g., directing a robot, purchasing a stock) to maximize the reward (e.g., the robot reaching the destination faster, achieving a high monetary gain from the purchased stock).

Although we do not recommend wasting your life and money in Casinos, which leads to going to hell in the afterlife, to understand this example, we refer to a common example of the casino. Assume you (agent) have 1,000 tokens and want to play in a casino with only five slot machines, such as Jackpot. Each machine has one arm, and the gameplay includes pulling down the arm (action) and testing whether a series of the same images appear on the screen or not. If all images are the same, then you win some money. The agent's (you) objective is to maximize the cumulative rewards (not just one single reward). There are only five machines, and we need to decide which one to play to get the most reward.

Figure 13-13: (a) All tokens are distributed equally to each machine. (b) In the first round, we tried, and in the second round, all remaining tokens went to the machine that gave the highest reward in the first round.

A naive approach is distributing tokens equally and playing 200 tokens on each machine, see Figure 13-13 (a). Nevertheless, some machines have a higher chance of winning, and some have a lower chance. Another approach is to insert one token into each machine, then see which machine wins the most coins and use the remaining tokens on that machine. For example, Figure 13-13 (b) shows that the second machine from the left is the one that gave rewards after one playing, and we stick to it for the rest of 1000 - 5 = 995 tokens. This approach has one step of exploration because it plays once at each machine to measure the chance of winning and not really winning. However, it will stick to a single machine that provides the highest reward at the exploration time.

What if the selected machine, which is the second one from the left in Figure 13-13 (b), is not the best one? What if this one is the worst machine, and by pure chance, it gives the highest reward on the first try?

There are different approaches to deal with this problem. We list three ones here, including epsilon greedy (ε-Greedy), Upper-Confidence-Bound (UCB), and Thomson Sampling method.

Besides, there could be many more details included in the decision, for example, making it a non-stationary problem, the Casino owner is looking at our playing and changing the winning machine (adversarial MAB), another approach is that the agent associates different actions with different situations (contextual bandit). In our upcoming explanation, we focus on the stationary and assume the winning chance of each machine does not change during playing (the policy is not changing). Not changing the policy during the time is referred to as a *stationary policy.*

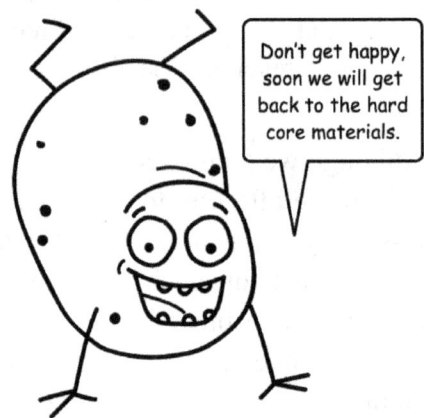

Epsilon Greedy (ε-Greedy)

If the agent selects the action that has the highest reward at the current time, it is taking the greedy approach. This decision, known as exploitation and is formalized as follows:

$$A_t = \underset{a}{argmax}\, V_t(a)$$

Once in a while, the agent does not make a greedy decision and gives up going for the action that has the highest reward. Instead, it selects another random action (another casino machine). This behavior implements the exploration approach and is referred to as the ε-Greedy strategy.

The frequency of taking a random decision is specified with a small probability given by the ε parameter. For example, assume $\varepsilon = 0.25$ means that with the probability of 0.25, the agent takes an exploration step, and with the probability of $1 - 0.25$ it takes an exploitation step, e.g., {greedy, greedy, greedy, random, greedy, greedy, greedy, random, greedy,...}

This causes the agent not to get the highest reward at the exploration time, but it enables the agent to explore, and thus, in the long run, the agent usually gets a higher total reward than the pure greedy approach.

Figure 13-14 presents the differences between using a Greedy approach and a ε-Greedy approach. We see that the greedy approach performs better at the beginning, then it has no changes. On the other hand, after a few steps ε-Greedy outperforms the Greedy approach and gets higher rewards.

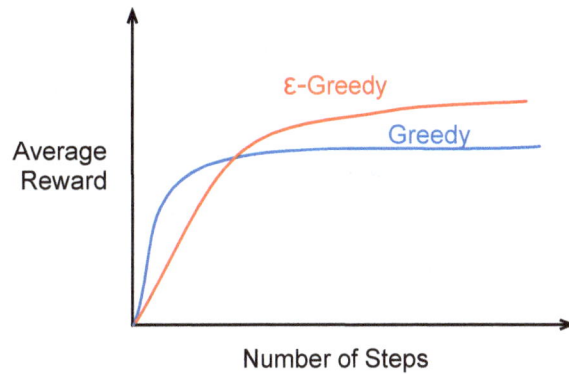

Figure 13-14: The difference in reward acquisition between Greedy and ε-Greedy.

The ε-Greedy strategy has one notable drawback: the value of ε is a hyperparameter that requires tuning to identify the optimal value. This tuning can be costly, especially in real-world multi-armed bandit (MAB) problems, where experimentation can be expensive. Typically, the tuning of ε and other parameters is conducted in a simulated environment. After identifying the most effective settings, the best-performing model is then deployed in a real-world application.

Upper Confidence Bound (UCB)

Another drawback of ε-Greedy is using the means to justify the reward chance of a machine. For example, one machine's reward chance is between -5 to +6 reward, and another machine's reward chance is -500 to +501. While both machines have an expected reward (mean value) of approximately +0.5, their variances are vastly different.

We have learned in Chapter 3 that knowing a mean is not enough to identify the characteristics of data, and at least we need variance as well (recall that Gaussian distribution needs both mean and variance). While discussing distributions, we explained that the more we sample (exploit), the confidence level gets smaller and smaller, which means we get more certainty about the estimated reward of a machine. This significant difference in variance means that while the average reward is the same between the two machines, the risk associated with the second machine (-500 to +501) is much higher. ε-Greedy blindly chooses a random machine for exploration.

The Upper Confidence Bound (UCB) approach, however, keeps track of samples that it takes from each machine and uses the *variance* of samples to calculate the *Upper Confidence Bound* (please check Chapter 3 for the confidence interval description; the 'b' in Figure 3-25 presents the Upper Confidence Bound or UCB). In the following, we explain the UCB approach step by step.

(i) *Initialization:* The agent initially experiments with each machine (explore) to gather preliminary data on the rewards of each machine (but it does not waste all tokens). Since the UCB approach relies on confidence, we need these sampled data.

(ii) *Data Collection and Computation:* After the agent collects some preliminary data, it analyzes the distribution of award values (see Figure 13-15). To conduct this analysis, it should compute the UCB of the award in each machine. First, the agent computes the Q_i or *empirical mean reward* $\hat{\mu}_i$ for several experiences that it had performed for each machine i. Assuming n_i is the number of times that machine i has been played up to time t, $I_t \in 1...N$ the choice of the machine at time t, and $r_t \in [0,1]$ the reward observed at the time t.

Then, the empirical mean reward is computed by using the following equation:

$$\hat{\mu}_{i,t} = \frac{\sum_{s=1}^{t} [I_s = i] r_s}{n_i}$$

This equation can be simplified as:

Empirical mean value = Sum of rewards obtained / Number of times this machine has been chosen

By using empirical mean reward, the UCB value can be computed by the following equation:

$$UCB_{i,t} = \hat{\mu}_{i,t} + \sqrt{\frac{2 \ln t}{n_i}}$$

Here, $\sqrt{2 \ln t / (n_i)}$ is used for exploration and avoiding playing a machine without checking other machines. In simple words, we can simplify UCB as follows:

UCB = Empirical mean reward + Exploration term

(iii) *Selection of Machine:* The agent selects the machine that has the highest UCB for exploitation. This machine is assumed to have the highest potential reward, which is formalized as follows:

$$I_t = arg \max_{I \in \{1..N\}} UCB_{i,t-1}$$

(iv) *Updating Estimates:* The agent continues to update and recompute the UCB values for each machine during exploration. As it plays machines and receives rewards, the empirical mean reward and the UCB values are updated to incorporate the new information.

(v) *Adjusting Choice Based on UCB:* If the UCB value of the current machine is still the highest, the agent continues to exploit it. However, if another machine's UCB value gets higher, this indicates a potentially better reward, and thus, the agent switches to that machine and starts exploiting it.

For example, in Figure 13-15, all slot machines have experimented k times, and each machine's winning distribution is plotted. The upper confidence bound of the second machine from the top is higher, and thus, at step $k + 1$, the agent chooses this machine for exploration, not a random machine, as it was in Epsilon Greedy.

Next, again, the confidence of this machine will be updated, and if it gets the highest UCB, it will be used again. Otherwise, another machine with the highest UCB will be used. Note that the more samples we collect, the more confidence boundaries get smaller. Therefore, after the first

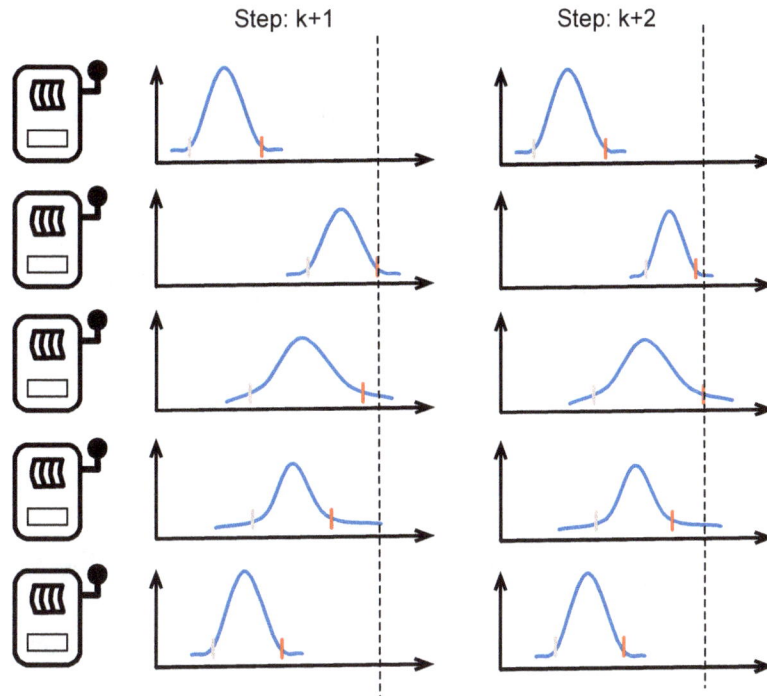

Figure 13-15: Each machine is associated with the distribution of reward, which has been identified by k number of samplings. The lower confidence bound on each distribution is shown in light red, and the upper confidence bound (UCB) is shown in dark red. For the exploitation step at k+1, the algorithm chooses the machine that has the highest upper confidence bound, which is the second machine from the top. At iteration k+1, the third machine will be chosen due to having the highest UCB.

try, the UCB of the second machine will be reduced. In our example, as you can see from Figure 13-15, the next step $(k + 2)$, the UCB of the third machine, is higher, and thus the agents choose this machine.

Thompson Sampling

Thompson sampling [Thompson '33] is a Bayesian approach used to address MAB problem. Samples are taken from a beta distribution to perform the exploration step in Thompson sampling. The Beta distribution is commonly used due to its relationship with the Bernoulli distribution, which models each arm of the multi-armed bandit problem (check Chapter 3 for an explanation of Beta and Bernoulli distributions). Each arm can be thought of as a Bernoulli trial with binary outcomes and an unknown probability of success. Although the Beta distribution is typically employed, other distributions might also be used depending on the scenario and the nature of the rewards.

Beta distribution conjugates a relationship with the Bernoulli distribution (also for the Binomial distribution, which is a sum of Bernoulli trials). When a Beta distribution is used as a prior for the probability parameter of a Bernoulli distribution, the resulting posterior distribution, after observing new data, is also a Beta distribution. Mathematically, if the probability of success θ in a Bernoulli (binary) trial is unknown and modeled as $\theta \sim Beta(\alpha, \beta)$, observing new data x

successes out of n trials updates θ to a new Beta distribution, which we can formalize as follows: $\theta \mid data \sim Beta(\alpha + x, \beta + n - x)$. This property ensures that the posterior distribution, after incorporating evidence, remains a Beta distribution. For instance, if our prior belief about p is represented by $Beta(2,2)$ and we then observe 10 Bernoulli trials with 6 successes and 4 failures, the posterior distribution becomes $Beta(2+6, 2+4) = Beta(8,6)$. This update combines our prior information with the observed data, leading to a more informed estimate of p. The mean of this posterior distribution is computed as $\frac{\alpha}{\alpha + \beta}$, which is $\frac{8}{8+6} = 0.571$, reflects our updated belief that the probability of success is around 57.1%.

In most Bayesian inferences, we need prior knowledge/belief about the system. When we do not have this prior knowledge, we use *flat prior* and *uninformative prior* distribution. For example, a flat prior can be a normal distribution with a mean equal to zero and a standard deviation of infinity (a flat line).

In Thompson sampling, if there is no prior knowledge, a flat (uniform) prior, $Beta(1,1)$, is often used initially. The Thompson sampling algorithm starts by sampling from this prior for each machine. In subsequent iterations, it selects the machine whose sample indicates the highest expected reward for exploitation. After each selection, it uses the following rules to update the value for α and β (parameters of the Beta distribution):

$$\alpha_{new} = \alpha_{old} + reward,$$

$$\beta_{new} = \beta_{old} + (1 - reward) \text{ or } \beta_{new} = \beta_{old} + failure,$$

$$reward = [0,1].$$

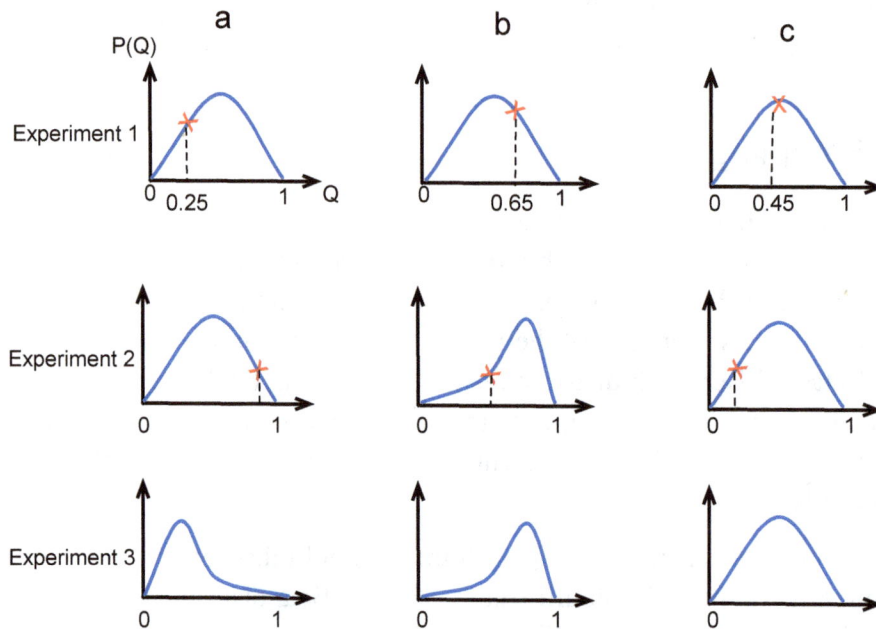

Figure 13-16: Thomson sampling example on three machines in three experiments.

After α and β has been updated, then again the agent samples from the Beta distribution of each machine and updated α, and β respectively to prepare them for the next round of sampling.

To better understand this approach, let's assume we have three machines (a, b, and c), and their initial distribution is shown in the first row of Figure 13-16. The flat prior is assumed to $\alpha = 1$ and $\beta = 1$, which is a flat line, but for the sake of understandability, we assume at the first step $\alpha = 2$ and $\beta = 2$. The agent samples from these three distributions and the highest Q belongs to machine 'b', which is 0.65.

The agent plays on machine 'b' and wins (reward=1). Therefore, it updates its $\alpha = 2 + 1 = 3$ (reward =1) and $\beta = 2 + (1 - 1) = 2$. By using the new values for α, and β, the new Beta distribution will be calculated, and we can see from the second experiment that the distribution of machine 'b' shifts to the right. This means the chance of sampling from machine 'b' increases.

Next, in the second experiment (Experiment 2 in Figure 13-16), the algorithm again selects random data from each distribution, and this time, machine 'a' has the highest Q value. The agent plays on machine 'a' and loses (reward = 0). Therefore, it updates $\alpha = 3 + 0 = 3$ and $\beta = 2 + 1 - 0 = 3$. The result shifts the Beta distribution of machine 'a' to the left, as shown in Experiment 3. We can see from the third experiment of Figure 13-16 that the probability of selecting from machine 'b' is high, and the probability of selecting from machine 'a' is the lowest due to their distribution shape. This process continues until a specific number of experiments are reached or the agent loses more than a specific amount.

Real-world examples of MAB problem

The MAB is a classic framework used to model decision-making in situations where an agent needs to balance exploration and exploitation. This concept can be found in some real-world applications, and we list three examples of implementing MAB.

In medical research, the MAB framework can be employed to optimize the allocation of treatments or interventions across a group of patients. The goal is to identify the most effective treatment while minimizing exposure to less effective options. In this context, each treatment or intervention corresponds to an 'arm' of the bandit, analogous to a machine in traditional MAB problems. The MAB algorithm efficiently allocates patients to different treatments, balancing the need for learning about the efficacy of each option with the goal of providing the best possible care.

For example, assume we study three treatments: A, B, and C, each representing an arm of the multi-armed bandit. We begin with no prior knowledge of their effectiveness. We employ a simple epsilon-greedy algorithm with $\varepsilon = 0.2$, meaning 20% exploration and 80% exploitation. In the initial phase, we allocate 10 patients to each treatment to gather baseline data. The results show 5, 7, and 4 patients improving under treatments A, B, and C, respectively. For the next 50 patients, the algorithm allocates based on these initial results and the exploration-exploitation trade-off. It chooses the best-performing treatment (B) 80% of the time (40 patients) and randomly explores the other treatments 20% of the time (10 patients). This leads to a possible allocation of 5 patients to A, 40 to B, and 5 to C. After this phase, the updated results show improvement rates of 47% for A (7 out of 15), 76% for B (38 out of 50), and 40% for C (6 out of 15). The algorithm continues this process, constantly updating probabilities and allocating more patients to the seemingly most effective treatment while maintaining some level of exploration for the other options.

Another use of MAB is in online advertising. A company can use different versions or strategies for advertisements (ads). Using the MAB approach, a company can decide how frequently each advertisement version should be displayed. Each advertisement approach is like an arm on a slot machine. The MAB adjusts the allocation of the advertisement based on how many people click on them or complete a desired action, like making a purchase. This helps the company try out different advertisement versions and figure out which ones work best while getting the most overall performance from their advertising.

Another application of the MAB is in content recommendation systems. For instance, in news recommendation platforms, different articles or news items can be viewed as the arms of a slot machine. The system must strategically choose which articles to recommend to users, balancing their known preferences with the opportunity to explore new content. The MAB algorithm optimizes this selection process by dynamically adjusting recommendations based on user feedback and engagement metrics. This approach is particularly valuable in the early stages of data collection when users' preferences are not yet well understood by the system.

NOTES:

* If you recall from Chapter 3 or are familiar with the A/B testing, you might think about the similarities between the MAB and A/B testing. Both are used to compare different versions of something, but A/B testing involves dividing the study groups into two or more *static* groups and showing each group a different version of the product (e.g., A and B). On the other hand, the MAB approach is a more dynamic approach, and instead of splitting the study groups into fixed groups, the MAB continuously adjusts the allocation of group members based on real-time feedback. In other words, MAB can dynamically allocate traffic to the slot machine that is performing the best. This is because MAB learns from the data as it is collected, while A/B testing only learns from the data at the end of the test.

* MAB can sometimes be used as a substitute for A/B testing; we use A/B testing to measure the effectiveness of something such as a medication on the treatment group along with the control group while no medication is being used. The A/B testing result depends on the significance test. Sometimes, the cost of the experiment is too expensive. An advantage of MAB over A/B testing is that in A/B testing, the number of subjects in the control and treatment should be equal to enable us to test for significance. In contrast, using MAB, we can assign more experiments to successful treatments (arms) and fewer experiments to unsuccessful treatments (arms).

* Thomson sampling alone usually causes the agent to be stuck in one winning machine, and in real-world cases, to handle this issue, it is combined with some ε number of random selections, similar to ε-Greedy algorithm. These random selections help the agent not to be stuck on one slot machine.

Optimal Policy and Optimal Value Function

As we have explained earlier, solving reinforcement learning problems means finding a policy that achieves a high amount of rewards over time. The policy that provides a larger value function than other policies is referred to as an *optimal* value function $V_*(s)$, and it is formalized as follows:

$$V_*(s) = \max_{\pi} V_{\pi}(s) \quad \forall s \in S$$

We use a similar sign (*) to present the optimal action-value function as follows:

$$Q_*(s, a) = \max_{\pi} q_{\pi}(s, a) \quad \forall s \in S, \quad \forall a \in A$$

We can formalize the Bellman equation for the optimal value function as follows:

$$V_*(s) = \max_{a} \sum_{s',r} p(s', r \mid s, a)[r + \gamma V_*(s')]$$

Respectively, we can formalize the Bellman equation for optimal action-value function as:

$$Q_*(s, a) = \sum_{s',r} p(s', r \mid s, a)[r + \gamma \max_{a'} q_*(s', a)]$$

By looking at the two equations above, we can see that the optimal value function or optimal action-value functions are unique, but there is no single optimal policy. For example, looking at Figure 13-10, there are two paths to go from (z, a) to the terminal state: one is $(y, a), (x, a), \ldots$, and the other one is $(z, b), (z, c), \ldots$, which means we have two policies. However, there are scenarios where there is a single optimal strategy that leads to the optimal value function. In these scenarios, both value function and action-value functions are the same, and we can formalize the optimal value function as follows:

$$V_*(s) = \max_{a} Q_*(s, a) \quad \forall s \in S \ \forall a \in A$$

To better understand this, assume we have key-value storage. Keys are state numbers and values of the value for each state. For example, `{state:0, value:0}`, `{state:1, value:-1}`, `{state:2, value:2.3}`, ... We need to find a function that, each key (state) maps to a value (value associated with that state), which can be seen as an approximation of the action-value function. Now, a question arises: how do we identify the optimal action-value or value function (which leads to finding the optimal policy)?

Unlike classification outputs in supervised machine learning algorithms, a value function in RL does not map to a single number; rather, it assigns values to every state under a specific policy, reflecting the expected return from each state. For example, assume we have three policies $V_1(s) = 1$, $V_2(s) = 0.8$, and $V_3(s) = 1.2$. The third policy has the highest value function it is $V_3(s)$ and thus will be $V_*(s) = V_3(s)$. However, a value function is not just a single number. For every single state, we have a value function. We can say the policy π_1 is better than the policy π_2 if, in all states (S), the value function of π_1 is better than the value function of π_2. We can write it formally as follows: $\pi_1 \geq \pi_2 \quad iff \quad V_{\pi_1}(s) \geq V_{\pi_2}(s) \quad \forall s \in S$. This approach seems to have a big problem: what if once in the state s_x, π_1 performs better than π_2 and in the state s_y, π_2

performs better than π_1? In other words, even if at several states π_2 performs better than π_1, still, we can not claim that π_2 performs better than π_1.

A solution to resolve this is to experiment with all possible policies and check the value of each policy at all possible states (S). Then, compare policies together and identify the best one. Assuming A presents the number of possible actions, in this case, we have A^S different kinds of policies to experiment with, and this is intractable (computationally too complex and thus infeasible).

Identifying v_*, π_* or q_* (* sign means the optimal), is a topic of the next sections, including Monte Carlo policy generation and Dynamic Programming (DP). However, before describing them, we must learn the difference between two problems in the context of RL, i.e., *Control* and *Prediction*.

Control and Prediction Problems

Using the RL algorithm involves solving two key problems: the *prediction problem* (policy evaluation) and the *control problem* (finding the best policy).

The *prediction problem* assumes that a policy is already available, and its goal is to evaluate the quality of this existing policy through policy evaluation. This involves computing the value function or the action-value function for the specified policy.

The *control problem (policy improvement)* assumes the policy is unavailable, and its goal is to find or generate the optimal policy.

The *prediction problem (policy evaluation)* computes the action-value function or value function for the given policy (the policy is available). The control problem iteratively updates the policy to maximize the value function.

The RL algorithm employs both policy evaluation and policy improvement methods in a sequential manner. Initially, a policy evaluation is conducted to assess the quality of an existing policy by calculating its value function. Following this, a policy improvement step is executed to enhance the policy based on the insights gained from the evaluation. This cycle of policy evaluation followed by policy improvement is repeated. The loop that continues until the policy converges to the optimal policy is known as *policy iteration*. A policy iteration loop is visualized as follows:

$$\pi_0 \xrightarrow{evaluate(prediction)} v_{\pi_0} \xrightarrow{improve\ /\ generate(control)} \pi_1 \xrightarrow{evaluate(prediction)} v_{\pi_1} \ldots \pi_* \xrightarrow{evaluate} v_{\pi_*}$$

Keep in mind that policy iteration is a combination of the control problem (policy improvement) with the prediction problem (policy evaluation). The exam monster will hunt your dream with a question on this topic.

Since policy evaluation is computationally expensive, we can use some techniques to truncate the policy evaluation. For example, combining policy evaluation with policy improvement into a single loop is a possible truncation method. If we do these truncations, instead of policy iteration, we call the described process *value iteration*.

Recall that in supervised learning, first, we segment a dataset into train and test, then we choose an algorithm and run it on the train set, build the model, and evaluate the model on the test set. In the realm of RL, we take a different approach, check the following pseudo-codes for the Control and Prediction problem.

```
#------Prediction Problem--------
# Goal: Compute the value function for a given policy π
```
Prediction(π) {

  initialize V(s) `# initialize the value function V(s) for all states s`

  repeat {

    Δ = 0 `# Initialize a threshold for stopping`

    for each state s {

      v = V(s) `# Store the current value of V(s)`

      $V(s) = sum(\ P(s'|s,\ a) * (R(s,\ a,\ s') + \gamma * V(s'))\ )$ *for all s', a* `# Update V(s) using the Bellman expectation equation`

      $\Delta = max(\Delta,\ |v - V(s)|)$ `# Update Δ with the largest change in V across all states`

    }

  } until Δ < threshold `# Continue until changes are below the threshold`

  return V(s) `# Return the final value function`

}

```
#------Control Problem--------
# Goal: find the optimal policy
```
Control( ) {

  initialize V(s), π(s) `# initialize the value function V(s) and policy π(s) for all states s`

  policy_stable = False `# Indicator for policy stability`

  while not policy_stable {

    `# Policy Evaluation`

    repeat {

      Δ = 0 `# Initialize a threshold for stopping`

      for each state s {

        v = V(s) `# Store the current value of V(s)`

        $V(s) = sum(\ P(s'|s,\ \pi(s)) * (R(s,\ \pi(s),\ s') + \gamma * V(s'))\ )$ *for all s'* `# Update V(s) using the Bellman equation under current policy`

        $\Delta = max(\Delta,\ |v - V(s)|)$ `# Update Δ with the largest change in V across all states`

      }

    } until Δ < threshold `# Continue until changes are below the threshold`

[The algorithm continues on the next page]

```
# Policy Improvement
    policy_stable = True   # Assume the policy is stable
    for each state s {
        old_action = π(s)  # Store the current action
        π(s) = argmax_a(sum( P(s'|s, a) * [R(s, a, s') + γ * V(s')] ) for all s') # Update policy to the
                            action that maximizes the current value function
        if old_action ≠ π(s) {
            policy_stable = False  # If any action changes, the policy is not stable
        }
    }
}# close of "while not policy_stable" loop
return V(s), π(s) # Return the final value function and policy
}
```

We can see from the above Pseudo code that the prediction problem is focused solely on *policy evaluation*. The control problem, on the other hand, involves both *policy evaluation* and *policy improvement*.

NOTES:

* As we have explained, in the context of RL algorithms, finding a value function for the given policy is referred to as the prediction or evaluation problem. Finding an optimal policy is referred to as the control problem.

* In most practical cases, we combine policy iteration and value iteration and use both. This combination is known as *General Policy Iteration (GPI)*. In particular, GPI guarantees that the policy and value function will improve over iterations and will eventually converge to the optimal policy and value function. This iterative process involves alternating between policy evaluation (estimating the value function for a given policy) and policy improvement (using the value function to find a better policy), ensuring that each step brings the policy closer to optimality.

Tabular Methods

RL algorithms can be categorized into two groups: *tabular methods* and *approximate solution* methods. We follow this categorization for the rest of this chapter, and this long section focuses on describing tabular methods.

We have explained that the agent either updates $v(s)$ or $q(s, a)$. A common approach for storing these data for later use is storing them in a table. A table that stores Q-function values is referred to as a Q-table, and a table that stores V-function values is referred to as a V-table. Since they rely on storing data in a table, they are called tabular methods. Note that Q-tables are commonly used because they provide a straightforward way to store and retrieve values needed for decision-making at each state-action pair. V-tables, while conceptually similar, are less frequently used since they only provide value estimates for states without directly suggesting the best actions.

Q-table stores the expected future rewards for each action at each state, and its content will be used by the agent to find the "best action at each state". Q-table includes states and actions, i.e., $Q[s,a]$, rows of Q-table correspond to different states, and columns correspond to different actions, as shown in the bottom of Figure 13-17.

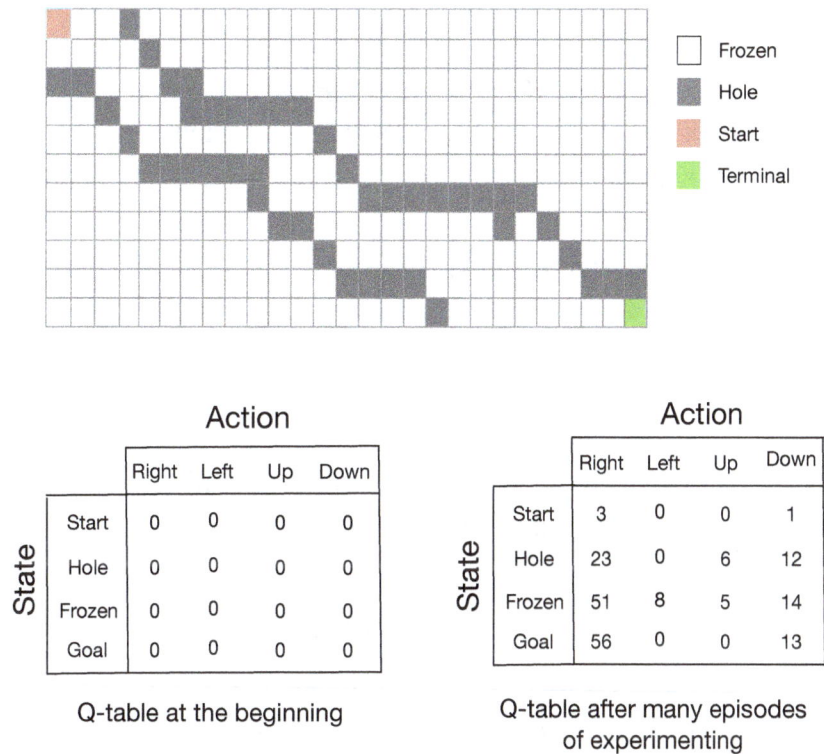

		Action		
	Right	Left	Up	Down
Start	0	0	0	0
Hole	0	0	0	0
Frozen	0	0	0	0
Goal	0	0	0	0

Q-table at the beginning

		Action		
	Right	Left	Up	Down
Start	3	0	0	1
Hole	23	0	6	12
Frozen	51	8	5	14
Goal	56	0	0	13

Q-table after many episodes of experimenting

Figure 13-17: A frozen lake example with four states. The Q-table on the bottom left is the one that is initialized at the beginning. The one on the right presents Q-table data after the agents conduct many experiments.

Figure 13-17 presents a frozen lake example, and the agent's plan is to go from the start cell to the terminal (or goal) cell. Frozen Lake is similar to Grid-world in that the agent is willing to get into a terminal state while not falling into the hole, and it can move only on frozen cells. To execute this example and go from start to terminal cell, first, the Q-table is initialized with zeros (the left table at the bottom of Figure 13-17). Then, the agent starts experimenting in the environment. After several iterations, the Q-table values are populated as they are shown on the right table of Figure 13-17. To implement a Q-table, we use a matrix. The agent refers to the values stored in the Q-table to determine the best action to take in each state.

Monte Carlo Method (Policy Evaluation and Policy Improvement)

Usually, in mathematics, when there is an operation that has a significant random component, we refer to this approach as a *Monte Carlo*[5] method. There are different use cases for the Monte Carlo (MC) methods, such as using MC for sampling, which we will explain in Chapter 16, and using MC for the control problem, which we explain here.

We can define/identify actions (input) that are used in the environment, but in many instances, we do not know the exact reward of the environment to our actions (output). This means that we can characterize the action distribution because it is known, but we can not characterize the reward distribution because it is unknown.

One approach to characterize the unknown distribution is to collect sample data and try to identify the characteristics (distribution) of the sample data. Increasing the sample size leads to getting closer to the distribution of the unknown data. This approach is known as the MC method.

By using MC, the agent doesn't know the immediate consequences of its action. It experiments with various actions until the end of an episode, then collects the returns to construct an action-value function or value function from them by averaging the returns across each episode, as is shown in the following equation:

$$\bar{v}(s) \approx \frac{1}{n} \sum_{i=1}^{n} G_{i,s}$$

In this equation, G represents the return (or Goal) at the end of each episode, n is the number of episodes, i is the index of each episode (sample number), and s is the state. $\bar{v}(s)$ is the value function for given state s. We can see that we do not use the Bellman equation here.

In the MC method, the expected action-value is conditioned on the state s and we have different estimates for each state. Therefore, we need to run (or play) some episodes, collect their gain G, and then average collected Gs to estimate the expected value.

Keep in mind for the exam that MC needs to wait until the end of each episode, then compute its return (G).

Monte Carlo Challenges

MC policy evaluation is very simple, but there are some challenges associated with this approach, which we list as follows:

Non-deterministic policies and state sampling: When the policies are not deterministic (probabilistic), what is the policy of the state that is not sampled by MC? A solution to mitigate it

---

[5] Monte Carlo is the name of a Casino located in Monaco, a tiny country in south France. If you get there, we recommend that you not waste your money and pay to visit inside the casino. Instead, hang out in the square in front of the casino, watch the luxury cars pass by, and cuss at the owners as they enter (of course, only in your head, not out loud).

is to enforce the agent start at different states. In the end, we have different episodes, which have different start states. Then, we can hope that in those episodes, the policy for every possible state is already computed.

Different returns for the same state: If the agent encounters the same state in two different episodes, the return (G) for that state might be different each time. Therefore, it is not clear which G should be taken into account for that state. There are two solutions to this problem. The first solution is known as "first visit MC", which only counts the return for the first visit. The second solution is known as "every visit MC", which counts the return for every time the state is visited. Theoretically, it is proven that both approaches converge.

Conflicting policies and infinite cycles: If the agent encounters the same state in two different episodes, there might be a different policy executed each time, and these policies might contract each other. For example, one policy states that the agent should go right, and on the second visit in the same state, the policy states that the agent should go left. This results in an infinite cycle because the agent is stuck between going left and right, preventing the episode from terminating. We described that MC calculates the G only after the end of an episode, and to handle this, we could enforce the termination of each episode after a certain number of steps. However, enforced termination introduces a bias in the value estimates.

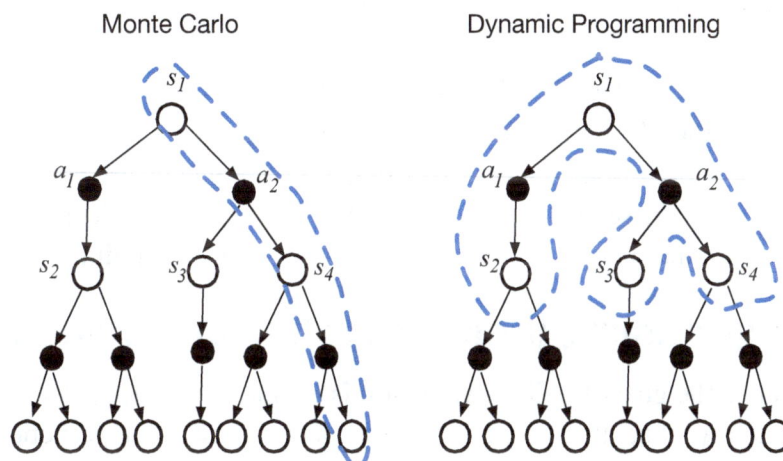

Figure 13-18: (left) Monte Carlo search to calculate the value-state function, which takes time until the agent reaches the end of the episode. (right) DP is for optimal policy, which should visit all successor actions and states to calculate a value-state function.

Monte Carlo Policy Evaluation and Policy Improvement

After the end of each episode, first, policy evaluation is performed for all the states visited in that episode, followed by policy improvement. In general, we can say that the MC policy generation identifies optimal values through trial-and-error.

On the left side of Figure 13-18, we have highlighted one episode with dotted lines on a backup diagram. At the end of this episode, policy evaluation and improvement are implemented for all the non-terminal states. Recall that the optimal policy improvement algorithm uses the Bellman equation as follows: $\pi_*(s) \leftarrow max_a \Sigma_{s',r} p(s',r \,|\, s,a)[r + \gamma V(s')]$. However, while using the MC method, we don't have access to $p(s', r | s, a)$ because the environment is unknown, and to resolve this issue, the policy improvement uses $Q(s, a)$ instead of $V(s)$. Therefore, it solves the following equation: $\pi_*(s) \leftarrow max_a\, Q(s, a)$. On the other hand, we have explained before that calculating $Q(s, a)$ is inefficient. To solve this issue, we can combine policy evaluation and policy improvement at every iteration. Earlier, we explained that this technique is called Generalized Policy Iteration (GPI).

When to use the MC method

The advantage of the MC policy improvement/generation method is that it learns only from episodes of experiments and does not need any prior knowledge about the environment (model-free). The main disadvantage of the MC policy generation method is that the episode should be terminated, and then the reward can be calculated (see Figure 13-18). This means that for an environment where experimenting is expensive, using the MC policy generation method is not recommended.

We can use the MC policy generation method to find the optimal policy and optimal value function when we have the following conditions:

* The problem is model-free, which means we do not know MDP transitions, and we don't have a transition probability matrix (Check Chapter 5). In other words, MC can find an optimal policy by using sample episodes, and no other knowledge about the environment is required.
* The agent learns from sampling and calculates the value function by averaging returns.
* The value function in the MC method is computed only after a complete episode.
* MC methods are used for episodic tasks. Tasks that are not episodic are called *continuing tasks* or *non-episodic tasks*. These are tasks where the interaction between the agent and the environment goes on indefinitely, without a clear starting or ending point for episodes.

Dynamic Programming (Policy Iteration and Value Iteration)

We have explained that the goal of policy evaluation is to find how much reward the given policy accumulates. We know that the value function and action-value function can measure accumulated rewards (average return), and thus, the input for policy evaluation algorithms is a policy, i.e., $\pi(a | s)$, and the output is $V_\pi(s)$ or $Q_\pi(s, a)$.

By looking at the Bellman equation, i.e., $V_\pi(s) = \Sigma_a \pi(a | s) \Sigma_{s', r} p(s', r | s, a)[r + \gamma V_\pi(s')]$, we can see that except $V_\pi(s)$ and $V_\pi(s')$ everything in this equation is known, and by substituting the known values, policy evaluation can be solved.

Besides, there is no polynomial $V_\pi(s)$ such as $V_\pi(s)^2$ and $V_\pi(s)^3$. Therefore, the Bellman equation is linear, and by a linear equation for every state, we can identify $V_\pi(s)$. In other words, if we have n number of states, we can substitute known variables $(r, \gamma, \dots)$ into the Bellman equation and find *n-1* unknown variables, i.e., $V_\pi(s_1), V_\pi(s_2), \dots V_\pi(s_{n-1})$. There is no need to compute $V_\pi(s_n)$ because it is the terminal state, and its value is zero. In simple words, these $V_\pi(s)$ values are the unknowns in a system of linear equations where the number of equations is equal to the number of unknown variables, and thus we can solve them.

Nom nom, nothing tastes better than a dead brain from memorizing too many mathematical equations.

Nevertheless, using a linear equation for every state to calculate $V_\pi(s)$ is not scalable. If n is large (e.g., all possible states in the Backgammon game is 10^{20} or Chess is claimed unknown or estimated as 10^{120}), it will be computationally infeasible. To mitigate this issue and reduce the complexity, Dynamic Programming (DP) algorithms are used to reduce the computation. Nevertheless, they are still incapable of completely resolving challenges with a very large space of states and actions.

The DP algorithms can perform value iteration (policy evaluation and policy improvement), usually more efficiently than linear methods. However, DP refers to algorithms that, given the "perfect model" of the environment, can find optimal policies. You read it correctly; they need a perfect model of the environment, which is infeasible in real-world applications. Nevertheless, we need to learn these theories because other practical methods, which do not require perfect modeling, are constructed based on DP models [Sutton '20].

How does a DP algorithm work?

DP algorithms operate in two steps: (i) it decomposes the complex problem into a set of subproblems, then solves these subproblems recursively, and (ii) stores solutions for subproblems in memory and reuses them to find the optimal solution for the main problem. Since the exam monster will take a quiz tonight in your dream, keep in mind that DP algorithms are *recursive* algorithms with *memorization* capability.

By using memoization in recursive algorithms, DP reduces the complexity of recursive algorithms. The exam does not ask questions about DP, but if you are curious to learn a bit more about DP algorithms, take a look at Figure 13-19, which shows two implementations of Fibonacci numbers; one is pure recursive, and the other one is DP, which is also recursive, but it has memory as well. In particular, by storing the previously calculated Fibonacci numbers in the f array, the code avoids redundant calculations and optimizes the computation by reusing the precomputed values.

```
Recursion:                                DP:
int fibo (int n) {                        int fibo (int n) {
    if (n <=1)                                int f[] = new int[n+2]
          return n                            f[0]= 0
    return fibo(n-1)+fibo(n-2)                f[1]= 1
}                                             for (i=2;i<=n;i++) {
                                                 f[i]= f[i-1]+f[i-2]
                                              }
                                              return f[n]
                                          }
```

Figure 13-19: Fibonacci number generation with recursion and dynamic programming.

If you are familiar with divide and conquer algorithms, DP algorithms are very similar to divide and conquer algorithms; they also break a problem down into subproblems, solve them recursively, and then combine their solutions. The key distinction is that divide and conquer tackles non-overlapping subproblems, whereas DP is used for overlapping subproblems. This overlap is where memoization becomes particularly valuable, as it avoids re-solving the same subproblem multiple times.

The Bellman equation (1) can be converted into subproblems, and the agent can solve the subproblems (ii) subproblem solutions can be cached and reused. This means that the Bellman equation can satisfy both conditions of DP. In the context of RL, DP is referred to as converting a Bellman equation (instead of solving it in a linear fashion) into iterative updates for approximating the desired value function.

Converting Bellman equation into interactive updates

To understand DP, we start by reviewing the value iteration. We have explained that the optimal value function is formalized as follows:

$$V_*(s) = \max_a \Sigma_{s',r} p(s',r \mid s,a)[r + \gamma V_*(s')]$$

The DP approach starts by initializing $v_0(s) = 0$ for all states s. Then, it starts by assigning the right-hand side in every iteration to the left side as follows:

$$V_{i+1}(s) \leftarrow \max_a \Sigma_{s',r} p(s',r \mid s,a)[r + \gamma V_i(s')]$$

In the above equation, i refers to the index of iteration and $p(s',r \mid s,a)$ stands for the state-transition probability matrix, which specifies the amount of reward r by taking action a and moving from state s to state s'.

This process continues until the value of v converges. In this context, converging refers to reaching a defined threshold for accuracy, i.e., $\theta > 0$. The θ parameter is also known as the *stopping threshold*.

The following algorithm is a pseudocode of the policy iteration algorithm based on DP:

```
#-----Algorithm's inputs-----
```

Inputs: θ, $p(s',r\,|\,s,a)$, γ `# `θ`: stopping threshold, `$p(s',r\,|\,s,a)$`: transition probability matrix,`
γ`: discount factor`

```
#-----Step 1: initialized params-----
```

Initialize V (terminal) = 0 , V(s) and π(s) to some arbitrary value for all $s \in S$

```
#-----Step 2: Policy Evaluation-----
```

$\Delta = 0$ `# initialize `Δ

while $(\Delta < \theta)$ {

    for each $(s \in S)$ {

        $v \leftarrow V(s)$

        $V(s) \leftarrow \Sigma_{s',r} p(s',r\,|\,s,\pi(s))[r + \gamma V(s')]$

        $\Delta = max[\Delta, |v - V(s)|]$

    }

```
#-----Step 3: Policy Improvement-----
```

policy_stable = true

for each $(s \in S)$ {

    $a_{old} \leftarrow \pi(s)$

    $\pi(s) \leftarrow \max_{a} \Sigma_{s',r} p(s',r\,|\,s,a)[r + \gamma V(s')]$

    if $(a_{old} \neq \pi(s))$ then (policy_ stable = false)

}

if (policy_stable) then (return $V \sim v_*, \pi \sim \pi_*$)

    else go to "Policy Evaluation".

Value iteration updates the value function and automatically determines the best actions by choosing those that maximize value in each step, quickly moving toward the best solution. Therefore, we can present the value iteration algorithm based on DP as follows:

```
#-----Algorithm's inputs-----
```

Inputs: θ, $p(s',r\,|\,s,\pi(s))$

```
#-----Step 1: initialized params-----
```

Initialize $\nabla \leftarrow 0$, V(s) to some arbitrary value for all $s \in S$

```
#-----Step 2: Value iteration-----
```

while $(\nabla < \theta)$ {

    for each $(s \in S)$ {

        $v \leftarrow V(s)$

        $V(s) \leftarrow \max_{a} \Sigma_{s',r} p(s',r\,|\,s,a)[r + \gamma V(s')]$ `#tries to find the highest value for action a`

        $\nabla \leftarrow max(\nabla, |v - V(s)|)$ `#record the max. error to decide about another iteration`

    }

}

return π_* `#--- `π_*` is the optimal policy`

$$Q(s,a) \leftarrow Q(s,a) + \alpha[r + \gamma Q(s',a') - Q(s,a)]$$

$$\pi(s) \leftarrow \max_a \Sigma_{s',r} p(s',r \mid s,a)[r + \gamma V(s')]$$

$$V_\pi(s) = \Sigma_a \pi(a \mid s)\Sigma_{s',r} p(s',r \mid s,a)[r + \gamma V_\pi(s')]$$

$$V_*(s) = \max_a \Sigma_{s',r} p(s',r \mid s,a)[r + \gamma V_*(s')]$$

$$\pi_{epsilon\ greedy} = \begin{cases} 1 - \epsilon + \dfrac{\epsilon}{|A(s)|} & if\ a^* = arg\ max\ Q(s,a) \\ \dfrac{\epsilon}{|A(s)|} & otherwise \end{cases}$$

When to use the DP method

As we have explained, DP is faster than MC because DP improves the policy at every step, but MC should wait until the end of the episode and then calculate the reward. Nevertheless, there are some major issues [Winder '20] that we should consider while using DP, except in very simple situations.

First, the value iteration algorithm stops when the agent updates all states. This is possible for simple toy examples such as grid-world, but in real-world scenarios, when the number of states and actions increases, it becomes computationally very expensive to use DP.

Second, DP has an issue with sampling data. As we have explained before, we usually use RL for the simulation of the real-world, and we require to feed sample data to the DP value iteration algorithm. Since this algorithm needs to visit all states and actions, the sample preparation for DP is very cumbersome and requires providing all possible sample data, which is computationally too expensive.

The third problem is the need for transition probability, which means that the agent needs to learn the dynamics of the environment, which is expensive to learn and requires extensive sampling.

To summarize this discussion, take a look at the Backup diagrams of Figure 13-18, which show how DP and MC search for possible actions to calculate the action-value function and update the policy. We can see that MC requires generating a full episode to update the policy. DP does not require a full episode, but it requires visiting all actions and states from a source state to update the policy.

NOTES:

* When dealing with the control problem, we can use the value function $V(s)$, but for the prediction problem, we should use the action-value function, i.e., $Q(s, a)$.

* We have explained that MC requires generating a full episode to update the policy. DP does not require a full episode, but it requires visiting all actions and states from a source state (i.e., transition probability) to update the policy. Both approaches are not practical for real-world solutions, so why have we learned them? Because first, we want to torture your brain, and second, we need to know them to understand practical methods that are used in real applications.

* In some cases, linear programming can solve MDP better than DP methods, especially when the number of states is not small or too large. However, DP performs better when the number of states is very large.

* Despite the fact that DP methods perform better for very large states, they suffer from the curse of dimensionality problem (Check Chapter 6), which makes them less effective.

* While using DP, it is necessary to use the discounting factor (γ). Otherwise, all states get the same reward. Recall that the discount factor is used to prioritize immediate rewards over distant rewards and to ensure the convergence of value estimates in infinite horizons.

* There are two types of reinforcement learning policies: *on-policy learning* and *off-policy learning*, both used to evaluate action-values (Q-values). On-policy learning updates its Q-values using the Q-value of the next state (s') and the current policy's action (a). In this approach, the agent uses the same policy for both exploration and exploitation. Off-policy learning, on the other hand, learns the optimal policy Q-values using the Q-value of the next state (s') and a greedy action (a'), allowing the agent to use different policies for exploration and exploitation. In simpler terms, in on-policy learning, the $Q(s, a)$ function is updated based on actions taken under the current policy $\pi(a \mid s)$. In off-policy learning, the $Q(s, a)$ function is updated from optimal actions, even if they are not taken by the current policy, and can be based on different policies.

* In reinforcement learning, we often distinguish between two types of policies: *behavior policy* and *target policy*. The behavior policy is the policy that is used by the agent to explore and generate data. The target policy is the policy that the agent is improving and trying to optimize. In the context of off-policy methods (e.g., Q-learning), the behavior policy may differ from the target policy, allowing the agent to explore more effectively. For instance, the agent may follow a behavior policy that encourages exploration (which could be random), while it learns and improves a more optimal target policy, i.e. off-policy learning.

Temporal Differences (TD) Learning

We have explained earlier that both MC and DP could be used to find optimal value functions, but they are not practical for real-world scenarios. MC requires the agent to complete an entire episode before it can update the policy, which is inefficient in environments with long episodes. DP requires complete knowledge of the environment's transition probabilities and reward functions to calculate optimal policies. DP algorithms are referred to as bootstrapping techniques, and we have explained in Chapter 9 that bootstrapping are systems or processes that rely on self-resources and no outside resources. *Bootstrapping* in reinforcement learning refers to the process of updating value estimates based on other estimates rather than waiting for the actual, complete outcome.

In other words, the value-function estimation is improved by iterating on the previously estimated value functions. MC is used when the environment, i.e., $p(s', r \mid s, a)$ is unknown, and DP is used when the environment, i.e., $p(s', r \mid s, a)$ is known.

Temporal Differences (TD) combine MC and DP [Sutton '98]. It gets the benefit of MC, which employs the random sampling characteristic of MC. Also, it has the advantage of DP, and it has the bootstrapping feature, meaning it updates estimates based on other learned estimates without waiting for the end of the episode. In summary, TD employs the random sampling of MC and bootstrapping of DP.

To understand TD, first, we should review the equation for computing the action-value function by MC and DP. We can write the MC value function as $v_\pi(s) = \mathbb{E}_\pi[v_\pi(s) + \alpha(G - v_\pi(s)) \mid s]$. In this equation, α represents a constant step size or learning rate[6], which is a number between 0 and 1. $\mathbb{E}_\pi$ represents the expected value of the policy π. As we explained in Chapter 3, the expectation ($\mathbb{E}$) is a weighted average. Similar to MC, we can write the value function equation for DP as: $v_\pi(s) = \mathbb{E}[r_{t+1} + \gamma v_\pi(s_{t+1}) \mid s_t = s]$.

The return, G, in DP is calculated as the *current reward* plus *all the discounted value-function of the next state*, i.e., $G = r + \gamma v(s')$. We can substitute the G with $r + \gamma v(s')$ in the MC, which resolves the issue of waiting until the end of each episode. Therefore, we can write the TD value-function as follows:

$$v_\pi(s) = \mathbb{E}[v_\pi(s) + \alpha(r + \gamma v_\pi(s')) \mid s]$$

As a next step, instead of using expectation, we can substitute the expectation with a weighted average and rewrite the TD value-function equation as follows:

$$v_\pi(s) \leftarrow v_\pi(s) + \alpha[r + \gamma v_\pi(s') - v_\pi(s)]$$

In other words, to update the value function estimate for the current state s, instead of the expected value, we use the observed reward r and the sample-based estimate of the value function $v_\pi(s')$.

---

[6] Earlier in this chapter, we said that we can see RL problems as optimization problems and thus use the following equation to solve RL problems;
$NewEstimate \leftarrow OldEstimate + LearningRate[Target - OldEstimate]$. Here, α refers to the learning rate used in optimization methods.

This equation is known as *online TD*, and instead of waiting until the end of the episode, TD iteratively updates the current value estimation using the discounted estimate of the next state $(\gamma v(s'))$. The *TD target* or *bootstrap estimate* is the part that TD substituted in MC equation instead of G, as it is highlighted in the equation below.

$$v(s) \leftarrow v(s) + \alpha[\underbrace{r + \gamma v(s')}_{\text{Bootstrap estimate or TD Target}} - v(s)]$$

TD error

Here, $r + \gamma v(s') - v(s)$ is known as *TD error*. TD error specifies the differences between the current value function estimation and the target (actual) reward gained from state transition from s_t to s_{t+1}. A simple explanation of TD error is provided by Winder's book [Winder '20] as: "*TD error measures how wrong the current estimate is compared to what actually has happened*".

There are different varieties of TD algorithms, with the simplest being TD(0). It only takes one action, and, based on the next state (s'), it updates the current state (s). It means we go from s to s', i.e., $s \xrightarrow{a} s'$. In other words, it only has to wait until the action a is taken (at time t) and brings the agent to the next state (time $t+1$), then it updates the current state at time t. We can rewrite the TD's Bellman equation for TD(0) as follows:

$$v(s_t) \leftarrow v(s_t) + \alpha[r + \gamma v(s_{t+1}) - v(s_t)]$$

There are more forms of TD algorithms that we will learn very soon. If the exam monster asks what the difference is between TD and MC/DP? You should answer (i) TD benefits from bootstrapping characteristics of DP and efficient sampling of MC. (ii) TD is model-free and thus does not require knowing the environment model (transition probabilities, rules that govern the environment, etc.). Instead, it gets a partial trajectory and then estimates the reward. (iii) TD uses bootstrapping, which estimates the value function based on a small amount of exploration and then makes another estimate.

SARSA and Q-learning

There are some common implementations of the TD approach, including SARSA, Q-learning, Dyna-Q, and TD(λ). We explain some of the common ones briefly in this chapter.

When performing prediction, we typically use the value function, i.e., $V(s)$. However, when performing control, we must use the action-value function, i.e., $Q(s, a)$. The control task with TD is substituting of $V(s)$ with $Q(s, a)$, as it is shown in the following:

$$Q(s, a) \leftarrow Q(s, a) + \alpha[r + \gamma Q(s', a') - Q(s, a)]$$

This is because, in an unknown or non-deterministic environment, we use TD methods, which require us to evaluate the expected outcomes of actions. The $Q(s, a)$ function allows us to assess the quality of actions in a given state, enabling the agent to learn the optimal policy by selecting actions that maximize future rewards. Therefore, to explain different varieties of TD, from now on, we will use $Q(s, a)$ instead of $V(s)$.

SARSA

SARSA is a policy iteration algorithm that implements TD learning for the control problem. It updates its Q-values using the state, action, the reward received, the next state, and the action taken in the next state, i.e., $(s_t, a_t, r_{t+1}, s_{t+1}, a_{t+1})$ or (s, a, r', s', a') to update Q-values. The acronym SARSA stands for State-Action-Reward-State-Action, which reflects the data elements used in each update.

One of the primary advantages of SARSA is that it learns a policy while considering the consequences of the actual policy followed. This can make SARSA safer compared to off-policy methods like Q-learning (we will learn it shortly), especially in environments where taking certain actions can lead to significantly negative consequences.

SARSA estimates the value of the policy that is currently learning (control problem), and thus, it is an on-policy learning algorithm. This means it updates its $Q(s, a)$ estimates based on the actions taken by the policy itself, which is in contrast to off-policy methods that may learn about an optimal policy regardless of the agent's actions.

SARSA is a policy improvement approach that typically uses an epsilon-greedy policy to estimate the best possible action to take (a^*). Therefore, it does not always go for the best actions; once in a while, the agent does not make the greedy decision and goes for exploration and uses a random action. SARSA policy can be formalized as follows:

$$\pi_{epsilon\ greedy} = \begin{cases} 1 - \epsilon + \frac{\epsilon}{|A(s)|} & if\ a^* = arg\ max\ Q(s, a) \\ \frac{\epsilon}{|A(s)|} & otherwise \end{cases}$$

Here, $|A(s)|$ is the number of available actions in state s. This policy selects the action a^* that maximizes the current Q-value estimate with probability $1 - \epsilon$ (exploitation) and any other action with a probability distributed equally among them, ensuring exploration. In other words, $\frac{\epsilon}{|A(s)|}$ presents the exploration (not greedy action), part of the epsilon-greedy policy.

Following is the pseudo code for the SARSA algorithm:

input: ϵ, α, γ `# ` ϵ`: a small number >0 for ` ϵ`-greedy,`
   α`: step size or learning rate,`
   γ`: discount factor`

initialize Q(s,a) `# initialize the action-value function, no policy is`
      `initialized because here policies are derived by Q(s, a).`

   for all (episode) { `# this loop continues for each episode until the max iteration`
         `of episodes we want to execute is reached, or the algorithm converges.`

      initialize (s)

      a = epsilon_greedy(s) `# By using a policy derived by ` Q`, based on the epsilon`
            `greedy approach, chose an action ` a ` for the given state ` s`.`

[The algorithm continues on the next page]

```
while (s is not terminal) {
        r, s' = execute(a) # take action a, observe r and s'
        a' = epsilon_greedy(s')

        Q(s, a) ← Q(s, a) + α[r + γQ(s', a') − Q(s, a)] # a' is the action that is taken
                                                        based on a specific policy.
        s ← s', a ← a'
    }
}
```

Q-learning

A very popular implementation of TD for the control problem is Q-learning. Q stays for the quality of a state-action pair in terms of expected future rewards. This algorithm was introduced in 1989 [Watkins '89]. At its time, it was a breakthrough in reinforcement learning. Q-learning is a model-free and off-policy approach. Similar to SARSA, it resolves the control problem, but it has one difference with SARSA, the SARSA's target value is computed as $r + \gamma Q(s', a')$, while in Q-learning, the target is computed as follows:

$$r + \gamma \max_{a'} Q(s', a')$$

Here, $max_{a'} Q(s', a')$ represents the maximum estimated value for the next state across all possible actions. This approach allows Q-learning to directly approximate the optimal policy, even as it explores suboptimal actions.

SARSA's actions might either be the best action based on the current policy or could arise from the exploration step, and thus, they are not always the optimal action. However, Q-learning evaluates all possible actions in the next state to determine the best action, thus optimizing the estimation of $Q(s', a')$. Since Q-learning's updates do not depend on the action actually taken according to the current policy, it is referred to as an off-policy learning method.

Following is the pseudo code for the Q-learning algorithm,

```
input: ε, α, γ # ε: a small number >0 for ε-greedy, α: step size, γ: discount factor
initialize (Q(s,a)) # initialize the action-value function, no policy is initialized because
                here policies are derived by Q(s,a).
    for all (episode) { # this loop continues for each episode until the maximum iteration of
                    episodes we want to execute is reached or the algorithm converges.
        initialize (s)
        while (not terminal (s)) {
            a = epsilon_greedy(s) # by using a policy derived by q, e.g., epsilon greedy,
                                action a is chosen for the given state.
            r, s' = execute(a) # take action a and get r and s'
            Q(s, a) ← Q(s, a) + α[r + γ max_{a'} Q(s, a) − Q(s, a)]
            s ← s'
        }
    }
```

The best way to memorize the differences between SARSA and Q-learning is by looking at their backup diagram on top of Figure 3-19 (this figure and the following example are described in Sutton and Barlow's book [Sutton '20]).

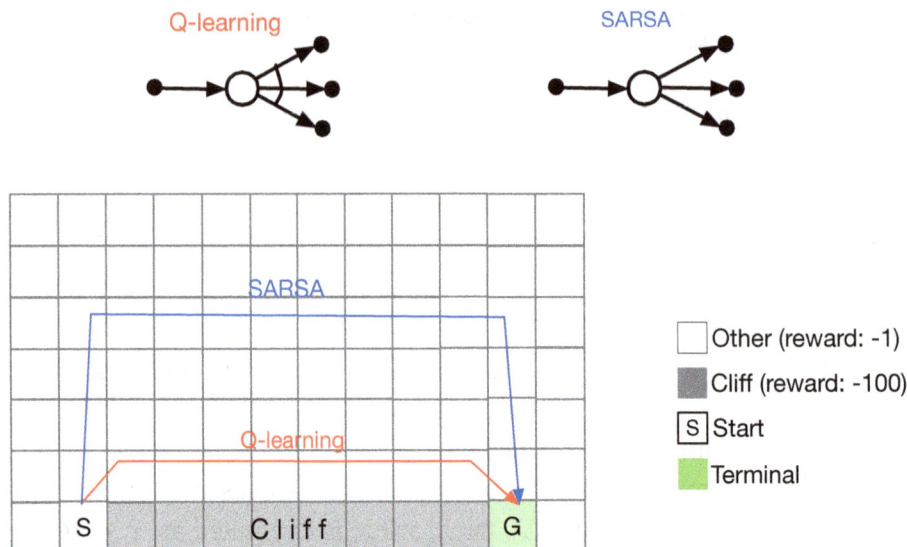

Figure 13-19: (top) Backup diagram of Q-learning versus SARSA. (bottom) getting from the start, i.e., red cell, to the goal, i.e., terminal, while avoiding cliff cells, i.e., grey cells.

You might also ask, what is the use of SARSA when Q-learning resolves the issue with optimal policy? To answer this question, look at the bottom of Figure 3-19, which is the cliff Grid-world. Assume the agent needs to reach the goal (green cell) while avoiding hitting the cliff. The reward of all white cells (except start and goal) is -1, and hitting the cliff is -100 (grey cells). By using Q-learning the agent always chooses the optimal decision and follows a path similar to the red line in this figure. This path is risky because it is very close to the cliff and causes the agent to hit the cliff many times until it learns to get to the goal. However, by using SARSA, since agents perform some exploration as well, it chooses a more conservative path, and therefore, the likelihood of hitting the cliff is getting lower. With high likelihood, while using SARSA, the agent follows a path similar to the one that is shown in blue color.

We can conclude that it is better to use SARSA when the agent intends to reduce the risk, and as we explained, the mistake is too expensive to make. On the other hand, we can use Q-learning when the agent should get faster to the goal but with more risks.

Dyna-Q

The RL agent learns the model either through simulated interactions (model-based learning) or direct interactions with the environment (model-free learning). For example, the Monte Carlo Tree Search algorithm (see Chapter 5) is a model-based method, as it relies on a model of the environment. On the other hand, Temporal Difference (TD) algorithms are model-free, as they learn directly from interactions with the environment without requiring a model.

Dyna-Q, developed by Sutton [Sutton '20], combines model-free Q-learning and model-based learning, which results in a faster model convergence. In other words, it combines the real (model-free) and simulated (model-based) experiments[7] both together, to make the Q-learning faster and more resource efficient. Simulated experience is less expensive and easier to perform, but the quality of those experiences is usually not as good as a real experience.

Dyna-Q augments the Q-learning algorithm (a model-free algorithm) with a hypothetical model-based component. Despite being model-free, Q-learning is a computationally expensive process, but Dyna-Q incorporates a light hypothetical model-based component along with Q-learning, which results in faster convergence. To better understand Dyna-Q, we start by learning the Dyna algorithm [Sutton '91]. Dyna is using model-free learning to augment it with a concept Sutton used as *"planning things in your head"*. It is composed of an RL algorithm and a hypothetical model-based method that augments the main RL algorithm. Figure 13-20 presents an abstract Dyna algorithm.

```
loop (forever) {
    - choose an action (a) and observe its result state (s), reward (r)
    - apply real experiment by using s, r
    - update the model based on the real experiment
    loop (n times) {
        choose a hypothetical state (hs) and action (ha),
        observe the hypothetical model results: hs', hr
        update the model based on the hs', hr
    }
}
```

Figure 13-20: Abstract presentation of Dyna algorithm. The black part is a common RL algorithm and the blue part is the augmentation provided by Dyna.

The black part presents a typical RL algorithm, and the blue part presents the augmentation provided by Dyna. Note that the blue part in Figures 13-20 is model-based learning, and it repeats several times, but it is computationally efficient in comparison to the black part, which is experimenting in the real world. In other words, for each real experiment the agents execute, the agent executes *n* number of experiments by using the hypothetical model, which results in a better final result.

To better understand it, let's use an example. Assume there is a warehouse that has shelves, moving obstacles like other robots and humans, and various delivery points. The robot needs to learn an optimal navigation policy to efficiently move from point A to point B. In a traditional model-free RL approach, the robot would learn solely through trial and error by interacting with the real environment. Repeated real-world trials increase maintenance costs and disrupt warehouse operations. By using Dyna-Q, the robot combines real experiences with simulated experiences to speed up learning and reduce costs. The robot performs actual navigation tasks in the warehouse, learning from the real outcomes of its actions. Between real navigation tasks, the robot uses the learned model to simulate hypothetical navigation scenarios. In each simulation, it selects hypothetical state-action pairs, predicts the outcomes using the model, and updates its navigation policy based on these simulated experiences.

---

[7] In the original explanation [Sutton' 20], model-free learning method is referred to as "learning", and model-based learning is referred to as "planning" by learning how the world operates. Planning is a process that we use in our daily life lot, because we know the world model. For example, we plan when to leave home to not get stuck in rush hour traffic. Learning is associated with experimenting and learning by experimenting unknown environment. For example, eating an unfamiliar food is a model-free experiment or learning.

The following presents a pseudo-code for the Dyna-Q algorithm. The Q-learning is presented in black color, and Dyna parts are highlighted in red color.

input: $\epsilon, \alpha, \gamma, n$ # epsilon for e-greedy, learning rate, discount factor, planning steps
initialize Q(s,a) and model(s,a) # initialize the action-value function, and the model.
loop forever { # this loop continues for each episode until the maximum number of
                        iterations is reached or the algorithm converges.

    initialize (s)
    a = epsilon_greedy(s, Q)
    r, s' = execute(a) # take action a, observe reward r and next state s'
    $Q(s,a) \leftarrow Q(s,a) + \alpha[r + \gamma max_a Q(s',a) - Q(s,a)]$
    $model(s,a) \leftarrow r, s'$ # store the experience
    loop (n times) { # for each real step, this loop runs n times
        $s_h \leftarrow random (s)$ # A random state from the list of previously observed states
        $a_h \leftarrow random (a)$ # A random action from the list of previously taken actions
        $r_h, s'_h \leftarrow model(s_h, a_h)$
        $Q(s_h, a_h) = Q(s_h, a_h) + \alpha[r_h + \gamma max_a Q(s'_h, a) - Q(s_h, a_h)]$
    }
    s' ← s

}

As we can see from this algorithm, for each real experiment conducted by Q-learning, Dyna-Q conducted a hypothetical experiment n times. Since it is hypothetical, it is not as computationally intensive as Q-learning real experiments. Figure 13-21 presents the typical convergence impact of using Dyna-Q with two different values for n versus Q-learning to solve a problem. For example, the number of steps that it requires to get to the destination on the frozen lake (Grid-world).

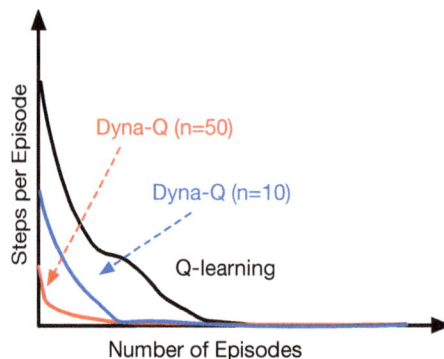

Figure 13-21: Comparison between Q-Leaning convergence and Dyna-Q. Two Dyna-Q are presented here, one with n=10 iterations and the other one with n=50 iterations. Increasing *n* results in faster convergence speed.

Approximation Methods and Deep Reinforcement Learning

Tabular methods are useful when we have a limited space state that can fit into a table, such as moving on a frozen lake or other grid-world types of problems. In many real-world scenarios, we must deal with enormous state or action spaces, and these spaces grow exponentially. As an example, we explained earlier that all possible states in the Backgammon game are 10^{20} or Chess is claimed unknown or 10^{120}. Therefore, it is computationally very expensive or infeasible to find the optimal policy. Instead, we use approximation methods to get closer to an optimal result. Approximation methods are parameterized functions based on weights (w). In machine learning, weights are usually referred to as model parameters that the algorithm or model adjusts during training to minimize error. In other words, the objective of the approximation function is to configure model parameters (weights) in a way that the agent follows the optimal path and collects the maximum amount of rewards.

n-Step Methods and TD(λ)

Recall that SARSA, i.e., $Q(s,a) \leftarrow Q(s,a) + \alpha[r + \gamma Q(s',a') - Q(s,a)]$, looks one step ahead, and also Q-learning, i.e., $Q(s,a) \leftarrow Q(s,a) + \alpha[r + \gamma max_a Q(s',a) - Q(s,a)]$. On the other hand, Monte Carlo methods look at all states until the end of the episode. If the algorithm looks at one step as described in SARSA and Q-learning, it is TD(0).

In other words, to estimate the current state-action value, TD methods look at one next single step. This approach is useful, but bootstrapping methods such as Q-learning are associated with *high bias (low variance)*. Besides, the Monte Carlo method has a *high variance (zero bias)* because it uses the output of a complete episode to update its current state. To handle the bias and variance issue, we can extend the next step estimation, and instead of looking at one step ahead, a TD method can look at the next 2 steps, 3 steps, or until the end of the episode (i.e., Monte Carlo method). In particular, we can develop some algorithms between TD(0) and Monte Carlo. These are known as *n-step* TD methods.

It is recommended to look ahead at more than one step in many applications. For example, in the cliff example we described in Figure 13-19, if the agent is on the wrong path, it can decide earlier not to hit the cliff. Also, if there exist some other unobserved variables, such as wind, looking a few steps ahead could be helpful because the agent can decrease the chances of hitting the cliff, and it plans its trajectory to move distance from the cliff.

If a TD method looks at the next n steps instead of just the next one step, it is referred to as an *n-step* TD method. In simple terms, n-step methods maintain the rewards accumulated by the agent during the n number of steps. Monte Carlo is a $TD(n)$ method and $n = \infty$. As the number of n increases, the agent should wait longer to update its state and reward at the current time.

To better understand *n*-step algorithms, let's look at the expected return in TD(0), which we have described previously, it is computed as $G = r + \gamma v(s')$. Specifically, the expected return while looking one step ahead from time t to time $t + 1$ is computed as: $G_{t:t+1} = r_{t+1} + \gamma v_t(s_{t+1})$. In the same way, $G_{t:t+2} = r_{t+1} + \gamma r_{t+2} + \gamma^2 v_{t+1}(s_{t+2})$ represents the expected return while looking two steps ahead from time t to time $t + 2$. Therefore, we can formalize the n-step expected reward between steps t to $t+n$ as follows:

$$G_{t:t+n} = r_t + \gamma r_{t+1} + \ldots + \gamma^{n-1} r_{n-1} + \gamma^n q(s_n, a_n)$$

Note that a reward is available after the agent takes action. Therefore, to implement an n-step algorithm, the agent has to follow a path of states, storing the rewards (buffering rewards), and then go back and replay the buffer to update step t. This buffer of rewards is also referred to as the *replay buffer* or *experience replay*. We will explain more about the replay buffer later.

It is recommended to average all the possible n-step returns (or some of the n-step returns) into a single return. For example, assume an n-step method for $n = 4$, and present the total return up to step x as $r^{(x)}$, we can write the following: $G_t = 1/4 \, r_t^{(1)} + 1/4 \, r_t^{(2)} + 1/4 \, r_t^{(3)} + 1/4 \, r_t^{(4)}$. Another recommended approach is to compute some steps (not all) and average them. For example, if we calculate TD(2) and TD(4) and average them, we will end up having the following return: $G_t = 1/2 \, r_t^{(2)} + 1/2 \, r_t^{(4)}$.

TD(λ)

A question might arise while using n-step methods: what is the best value for n? We could answer this question by TD(λ) algorithm that gets the best of all values for n. However, the short answer is higher n reduces bias but increases variance. Going for the highest n, results in Monte Carlo, which has a high variance. Therefore, we use a *weighted n*.

Weighted n: Waiting until the end of an episode (as in Monte Carlo method) can be inefficient, especially in long or continuing tasks. Therefore, we need to find a way not to wait for the end of the nth state. To resolve this issue, TD(λ) applies a *weighted scaling* on the n-step return $G_t^{(n)}$, and the weight exponentially decays over time (as the number of states increases). The weights decrease more rapidly for smaller values of λ, where $0 < \lambda \le 1$, and the return G_t^λ is calculated by using the following equation:

$$G_t^\lambda = (1 - \lambda) \sum_{n-1}^{\infty} \lambda^{n-1} G_t^{(n)}$$

Here, the G_t^λ is the weighted sum of all n step returns ($G_t^{(n)}$). The weight for each return is λ^{n-1}, and to scale them appropriately, a factor of $(1 - \lambda)$ is used. In other words, each $G_t^{(n)}$ is multiplied by a weight of λ^{n-1} and to adjust the contribution of the geometrically weighted returns and ensure the overall update is scaled appropriately, $(1 - \lambda)$ constant is applied.

The left side of Figure 13-22 presents the backup diagram of TD(λ), which is an n-step method. In this figure, we have n number of backup diagrams. The weights are decayed geometrically, as shown on the left side of Figure 13-22.

To summarize, TD(λ) implements the n-step return $G_t^{(n)}$ by using a weight that decays exponentially during each state change. This means TD(λ) integrates bootstrapping with multi-step returns, allowing for a spectrum between TD(0), i.e., SARSA or Q-learning and Monte Carlo methods. As λ approaches 1, TD(λ) becomes more similar to Monte Carlo but still updates estimates after every step based on subsequent estimates. As we have explained the λ is a parameter between 0 and 1.

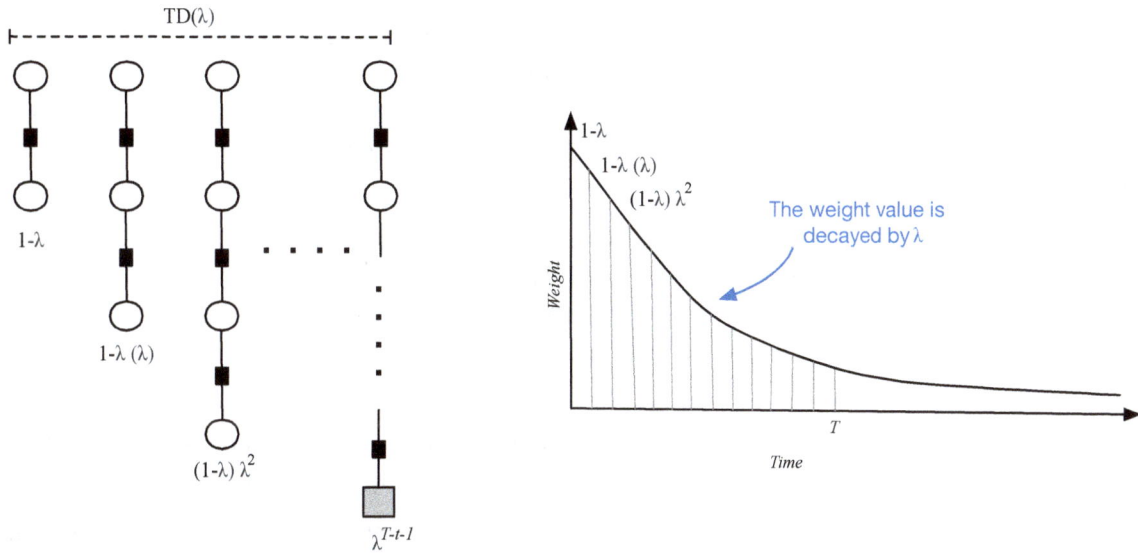

Figure 13-22: (left) n numbers of backup diagrams for TD(λ) algorithm. (right) A presentation of geometrically decaying weights during the time by TD(λ) algorithm.

You might think: "Why does the author dance around the question and not give a proper answer on the correct value of n? Weighted decay (λ) has nothing to do with reducing the waiting time, and we should wait like Monte Carlo until the end of the episode. Why do I read all these nonsense explanations and still don't know how to determine the right value for n?"

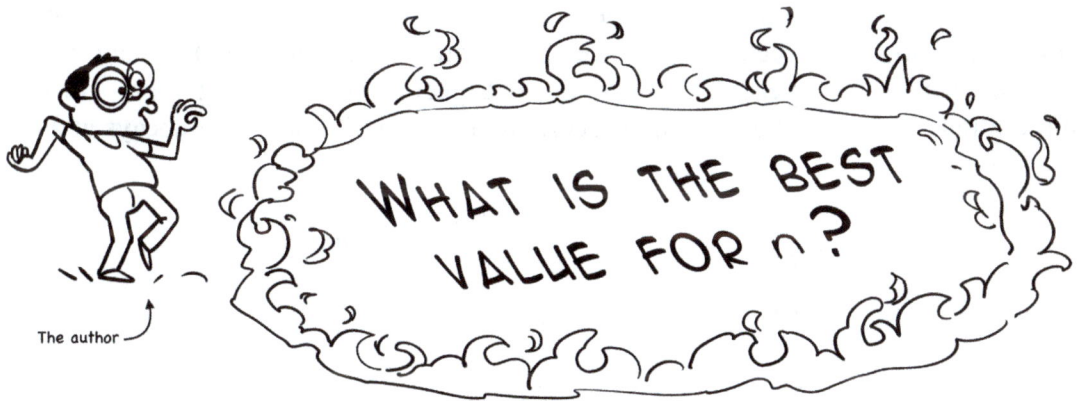

You are right. Still, we need to deal with all possible n values, which means the agent should wait for the n-step TD algorithm to reach the nth state. However, using λ enables the algorithm to trick the agent and let it update immediately instead of waiting until the end of nth state. This trick is implemented with the *eligibility trace* concept. The eligibility trace, along with exponentially weighted decay (λ) can resolve the waiting time issue in n-step methods.

Eligibility trace

An eligibility trace is a mechanism that acts as a memory buffer, keeping track of the states visited by the agent and storing them. However, it does more than just store them. It keeps a decaying record of the states (or state-action pairs) visited by the agent, enabling updates to be applied not only to the current state but also to past states. These updates are weighted by both the recency of visitation (using a decay parameter, λ) and the temporal difference (TD) error.

In contrast to n-step methods, in which the agent looks forward from the current state to some future rewards. The challenge with n-step methods is that the update relies on future information, which is not available at the current time. However, using eligibility traces, the agent can update the action-value function based on past states instead of waiting for future states.

In summary, eligibility traces can be seen as a generalization of n-step methods. While n-step methods look forward a fixed number of steps, eligibility traces effectively implement a weighted average of n-step returns for all n, with the weighting determined by λ.

Eligibility traces capture two key pieces of information about a state: *how frequently* and *how recently* the state has been visited. To better understand eligibility traces, let's use an example: You see a button labeled 'Free snack', and there is a green light on top of it. You press the button once, and nothing happens, and the light doesn't change. The second time, nothing happens again. The third time, a green light turns red, and a monster appears and eats you. Did you think you were eaten because the red light appeared (how recent) or because you pressed the button three times (how frequent)? Perhaps both influenced this unfortunate event.

Eligibility traces capture both frequency and recency of past events to model future states. Instead of estimating the next state-action value, the agent uses past memory (via eligibility traces) to update the current state-action value. Therefore, by using eligibility traces, we can better understand and predict how past actions influence current decisions and outcomes.

To better understand eligibility trace, look at Figure 3-23, which presents a simple type of eligibility trace known as accumulating trace. Each time the agent visits the state, we show it with "↑". We can say that at each "↑", a stimulus is triggered, such as a button press or the green light turning red. Then, the eligibility trace score increases but immediately decays at a constant rate.

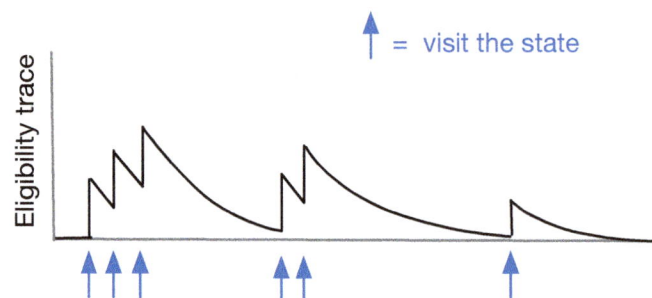

Figure 13-23: A visualization that shows the impact of visiting a state on the eligibility trace.

As we can see at the beginning, the same state is encountered three times, causing the eligibility trace value to increase, but then it decays until the next time the agent encounters this state twice, and then later, only once.

At the start, the eligibility trace for each state is initialized to zero, denoted as $E_0(s) = 0$. When the agents encounter a state, it updates the corresponding value in the eligibility trace vector. Afterward, the values in the eligibility trace vector are decayed by the λ *weight factor* and γ *discount factor*. These two parameters ensure that more recent state visits receive higher credits for the reward compared to earlier visited states. The eligibility trace is formalized as follows:

$$E_t(s) = \begin{cases} \gamma\lambda E_{t-1}(s) & if \quad s \neq s_t \\ \gamma\lambda E_{t-1}(s) + 1 & if \quad s = s_t \end{cases}$$

In this equation, if the current state s has been visited at time t (s_t) as well. Then, the eligibility trace for this state is updated by applying λ weight, γ discount, and also adding one to it. For all other states, the eligibility trace simply decays by λ and γ.

We explained that in the TD algorithm, we update the value function for each state in proportion to TD error, which can be expressed as $v(s) \leftarrow v(s) + \alpha[r + \gamma v(s') - v(s)]$. Let's call TD error as δ_t, so the update is simplified as $v(s) \leftarrow v(s) + \alpha\delta_t$. TD($\lambda$) updates the value function in proportion to both TD error and eligibility traces. Therefore, we have:

$$v(s) \leftarrow v(s) + \alpha\delta_t E_t(s)$$

This means that each state's value is adjusted based on both the *magnitude of the TD error* at that step and the state's *current eligibility trace value*, reflecting how recently and frequently the state has been visited.

The SARSA(λ) algorithm is written as follows. The text in black is the SARSA algorithm, and the new addition, which incorporates the eligibility traces and weight decay, is highlighted in red.

```
input: ε, α, γ, λ
initialize Q(s,a)
    for all (episode) {
        e(s,a)=0 # Clean the eligibility trace for each episode
        initialize (s)
        a = epsilon_greedy(s)
        while (not terminal (s)) {
            r, s' = execute(a)
            a' = epsilon_greedy(s')
            δ ← r + γQ(s',a') − Q(s,a) # TD error is called δ, nothing new here.
            e(s,a) ← e(s,a) + 1
            for (all s) {
                for (all a) {
                    Q(s,a) ← Q(s,a) + α . δ . e(s,a)
                    e(s,a) ← γλe(s,a)
                }
            }
            s ← s' , a ← a'
        }
    }
```

With similar changes, we can write Q(λ), which is a Q-learning that incorporates λ weight decays and eligibility traces. However, there are different implementations of Q(λ). Two common implementations are Watkins's Q(λ) [Watkins '89] and Peng's Q(λ) [Peng '94]. We explain Watkins's Q(λ) algorithm as follows.

```
input: ε, α, γ, λ
initialize (Q(s,a))
for all (episode) {
        e(s,a)=0 # clean the eligibility trace
        initialize (s)
        a = epsilon_greedy(s)
        while (not terminal (s)) {
                r, s' = execute(a)
                a* ← max_a Q(s',a) # a* is the action that provides the highest reward for
                                         all possible actions of state s'.
                δ ← r + γ Q(s',a*) - Q(s,a) # δ is TD error, but here we use a* for the
                                                 current action in action-value function.
                e(s,a) ← e(s,a) + 1
                for (all s, all a) { # Update Q-values for all state-action pairs
                    Q(s,a) ← Q(s,a) + α . δ . e(s,a) # the only addition here is e(s,a)
                      if a'=a* # if the next action is greedy
                            then e(s,a) ← γλe(s,a) # decay eligibility trace
                            else e(s,a)=0 # reset trace if non-greedy action is taken
                }
                s ← s', a ← a'

                a = epsilon_greedy(s')
        }
    }
```

Note that Watkins's Q(λ) resets the eligibility trace to zero when it encounters the first action that is not greedy. This makes sense because we do not want to penalize taking exploration actions (non-greedy actions. In other words, the algorithm is designed for off-policy learning. In off-policy learning, we're trying to learn the optimal policy (the greedy policy with respect to Q) while following a different, more exploratory policy (such as epsilon-greedy).

We know your brain is about to explode by reading too many algorithms, but we didn't find a better way to explain them. Please take a short break and fasten your seat belt; more material will come to learn.

Continuous Space and Function Approximation

All examples we explained work with discrete states and values; what if we have large state/action spaces with continuous variables? Grid-world problems treat every cell as a discrete entity but in many world-real environments, such as moving a robot hand in 3D space is continuous. We generally use discretization to simplify the environment to understand and model it. However, most real-world problems have very large continuous environments. For example, Atari games[8] at 60Hz have $(255^3)^{210 \times 160}$ different states [Morales '20]. Clearly, it is not possible to store these numbers of states into something such as a Q-table.

To handle a continuous space, we need a model that can cope with very large and continuous state and action spaces. Deep Neural Networks (DNNs) are the most common method for dealing with such large continuous state and action spaces.

---

[8] Atari game is designed in 210×160 pixels and each pixel is presented with 3 color channels.

Before we further explain DNN use in RL, let's analyze a hello-world continuous example, the cart-pole environment. This environment has a cart; the cart is moving on a track, and a pole is hinged to this cart. The cart was moving right and left so that the pole kept its balance vertical on the ground. Figure 13-24 shows this environment. If you have difficulty understanding this environment, try to hold a pen vertically at the palm of your hand; you should move your hand around to keep the pen vertical at your palm. The cart resembles the palm of your hand, and the pole is the pen.

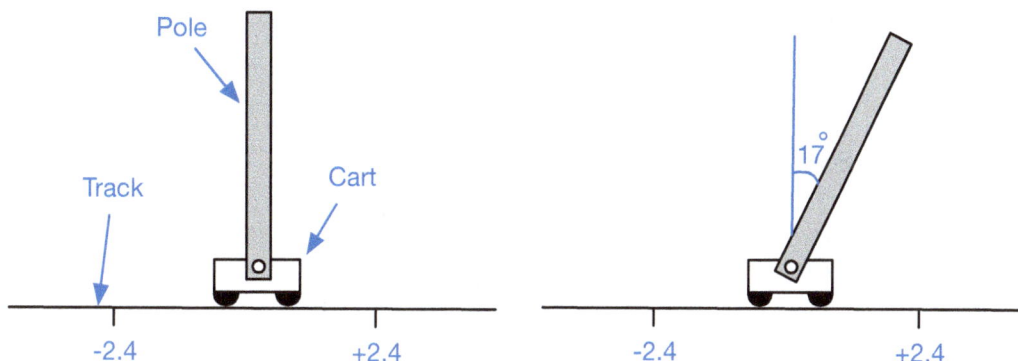

Figure 13-24: (left) Cart pole illustration with its component. (right) An example is that a cart pole went into a terminal state because the pole angle got larger than 12 degrees deviation from its vertical position.

Each step where the pole remains upright (within the allowed angle range) is typically rewarded with +1. Its state space comprises four variables: (i) cart position on the x-axis, with a range from -2.4 to 2.4. (ii) the velocity of the cart movement, which has a range of $-\infty$ to $+\infty$. (iii) the pole movement angle, which ranges from -40 degrees to 40 degrees. (iv) the pole movement velocity, with the range of $-\infty$ to $+\infty$. There are two actions in every state: (i) action 0, which pushes the cart to the left by applying -1 force to the cart, and (ii) action 1, which pushes the cart to the right by applying +1 force to the cart. The agent reaches the terminal state, and the episode ends, if one of the following conditions occurs. (i) the pole angle is getting away from its vertical position more than 12 degrees, (ii) the cart moves more than 2.4 units from the center of the track. (iii) the maximum number of episodes reaches (e.g., 500 times).

Deep Q-Network (DQN)

Deep Q-Network (DQN) [Mnih '13, Mnih '15] is one of the early architectures that incorporated deep learning into RL, and it initiated a big shift in RL. Nevertheless, it is not the first model, and previously NFQ [Riedmiller '05] was introduced back in 2005, but for the same reason that neural networks were not popular before 2012, NFQ did not get much community attention at its time. DQN performs much better than NFQ, and therefore, we focus on DQN and skip explaining NFQ. Both NFQ and DQN replace Q-tables of tabular methods with a neural network.

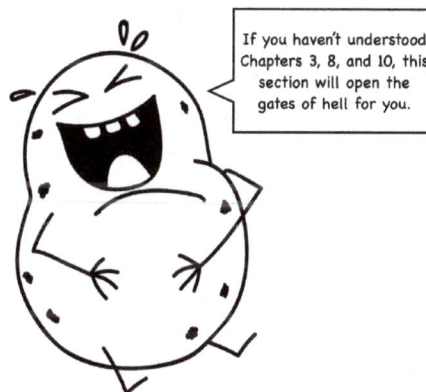

If you haven't understood Chapters 3, 8, and 10, this section will open the gates of hell for you.

In the cart-pole example, we can use a neural network to train the cart to move right and left while keeping the pole balance straight and not letting it fall down. This neural network has an input layer with four neurons (state space variables). The output layer has two neurons, each representing the Q-value for one action: $Q(s,0)$ for moving left and $Q(s,1)$ for moving right. Therefore, we can build a neural network with the structure of Figure 13-25.

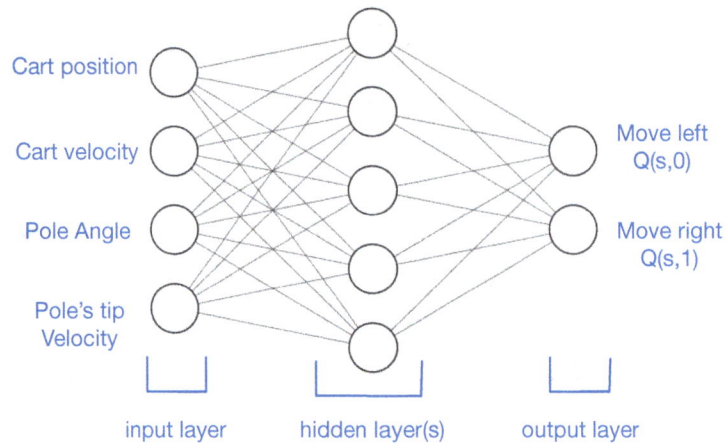

Figure 13-25: Neural network architecture of the Cart pole example.

The initial version of DQN uses MLP architecture; its input is the environment information, formatted as a vector of scalar values representing the current state. The outputs of the network are the estimated Q-values for each possible action in that state. The agent then typically selects the action with the highest estimated Q-value, although exploration strategies may sometimes lead to selecting other actions.

We have explained that DQN replaces the Q-table with a neural network. It means that instead of using a table to map state-action pairs to Q-value, the neural network of DQN maps actions and Q-value pairs. In other words, it predicts the Q-values from states and actions. Here, instead of using $Q(s, a)$, we use $Q(s, a; \theta)$, where θ represents the parameters of the model (weights and biases of the neural network)[9].

Similar to other neural networks, the DQN predicts Q-values and then calculates the loss score by comparing the predicted value to the actual value obtained from the environment. This comparison typically involves calculating the temporal difference error. Next, the neural network adjusts its weights to reduce the temporal difference error.

DQN cost function

We have explained that the TD error is the difference between the previous action-value function and the action-value function after the agent takes action. The *TD(error)* can be formalized as follows: $TD(error) = r + \gamma max_{a'} Q(s', a') - Q(s, a)$, or *TD(error)= TD target - Q(s, a)*.

---

[9] We have explained in Chapter 3 that the "P(x;θ)" sign is joint probaility, 'x' is a distribution and the 'θ' are its parameters.

In the above equation, the TD target presents the *next estimation* of the Q-value and $Q(s, a)$ is known as the *current estimation* of the Q-value. In other words, the *TD(error)* is the difference between the new estimation of the Q-value and the current estimation of the Q-value.

By calculating the difference between the TD target and the current estimate of the Q-value, the loss score can be computed. The DQN cost function of the DQN is calculated as a Mean Square Error (MSE) between the TD-target and the current Q-value estimation, and it is formalized as follows:

$$L(\theta) = [Q(s, a; \theta) - (r + \gamma max_{a'} Q(s', a'; \theta))]^2$$

Similar to other neural networks, the training phase stops when the loss score converges (reaches an acceptable value) or when the maximum number of iterations (or episodes) is achieved. In the action selection phase of DQN, the most common approach is to select the action corresponding to the highest Q-value predicted by the network for the current state. This is the greedy policy, but during training, a strategy such as ε-greedy is employed to balance exploration and exploitation. After executing the selected action in the environment, which results in a new state and reward, the agent updates its Q-value estimates based on the observed transition.

Two problems of Deep RL (i.d.d. samples and non-stationary target)

Right now, everything seems fine, we substituted the Q-table with a neural work, and we implemented a Deep RL. However, it is not as straightforward as we think. Neural networks assume its data points are independently and identically distributed (i.i.d.), and all supervised learning algorithms and neural networks assume the training set is properly sampled from the dataset and thus represents the original data. This assumption does not hold in RL because successive data points are not independent of each other, and most of the time, successive states are correlated. Therefore, the independent assumption of i.i.d. is not valid. Moreover, since the policy changes during the training process, the sample data are not identically distributed.

To understand the data points dependency, consider a robot organizing a baby's room. Imagine we place a robot in the baby's room to order the toys and put them back in the toy basket. The robot scans the room, moving around the room on a constant trajectory, grabs the baby toys, and puts them back in the toy basket. Here, the robot moves on a predefined trajectory (indicating that samples are dependent). The policy is fixed until the baby crawls into the room. Then, the baby picks up her toy box; she empties it on the ground that had been cleaned before (the distribution of samples changed) and puts the toy box on her head. She moves around a bit and then throws the toy box in another place. Since the placement of the toy box changed (indicating a change in the distribution of samples), the policy should also be changed. This example demonstrates that in the context of RL, i.i.d. condition and stationary target, which are required by neural networks, are invalid. Therefore, we need a mechanism to resolve this problem for the RL neural network.

If we update the neural network after every state, consecutive states are interdependent or correlated, and this correlation might bias the agent. This means the agent learns behavior and performs this behavior frequently until suddenly the state changes; previous actions are not applicable anymore, and this confuses the agent. In our example, the robot initially moves towards the previous position of the toy box, but now the toy box is on the baby's head. Alternatively, the robot cleans based on a set trajectory, but then the baby throws toys again onto the cleaned area, confusing the robot as to why this has happened.

FQ uses a mini-batch of data for training, and using a mini-batch mitigates the i.i.d. issue. Additionally, it uses the same mini-batch in multiple optimization steps to address the moving target problem. To mitigate the i.i.d. issue, DQN uses Experience Replay [Lin '92], which was introduced in 1992 and is still in use today.

Experience Replay

A problem that the RL agent faces is that consecutive samples are highly correlated, which biases the agent during the learning process. We have used the example of a robot ordering the baby's room to illustrate the correlation problem. DQN uses a replay memory, a.k.a, experience replay, to solve the problem of sampling from consecutive samples. In other words, experience replay is used to perform i.i.d. sampling.

Experience replay acts as a buffer that stores state, action, reward, and next state, i.e., $sample = (s_t, a_t, r_t, s_{t+1})$. The data is stored in this buffer using a First-In-First-Out (FIFO) policy. During the learning process of the neural network, instead of using a batch of consecutive data points (as in NFQ), DQN uses a batch of data points randomly selected from the experience replay buffer by using uniform random sampling. The neural network trains using these mini-batches of samples from experience replay.

For example, assume we have the following set of actions for a robot movement: $\{ \rightarrow , \rightarrow , \rightarrow , \rightarrow , \uparrow , \uparrow , \uparrow , \uparrow , \leftarrow , \leftarrow , \leftarrow , \leftarrow \}$ if the agent samples them sequentially, the agent might perform only one movement, e.g., $\{ \uparrow , \uparrow , \uparrow \}$; instead, random sampling based on uniform distribution allows the agent to have a sample of each movement, i.e., $\{ \uparrow , \leftarrow , \rightarrow \}$. Note that the content of replay memory is a (s_t, a_t, r_t, s_{t+1}) tuple, and for the sake of simplicity, we showed only actions, i.e., robot movement direction.

Figure 13-26 presents experience replay content. In summary, the agent performs some actions and keeps track of its states, actions, and rewards in the replay buffer. Then, the DQN samples a batch of these data (mini-batch) to train the agent in each epoch. If you see the RL exam monster in your dream and he asks you why DQN uses experience replay?, you should answer: "Because it makes the sampled data i.d.d. for DQN neural network".

If the RL exam monster asked what is the problem of using mini-batches without experience replay? You should answer: "Mini-batches include consecutive samples that are very likely correlated, but mini-batches sampled from experience replay are composed of random data points, which have no correlations".

The size of the experience replay buffer depends on the problem; for example, for the cart pole problem, the experience replay size of 10,000 (state, action, reward, next state) tuples is enough, and the mini-batch with the size 64, randomly samples 64 tuples from those 10,000 tuples. For an agent to play an Atari game, the size of the experience replay buffer is 1,000,000.

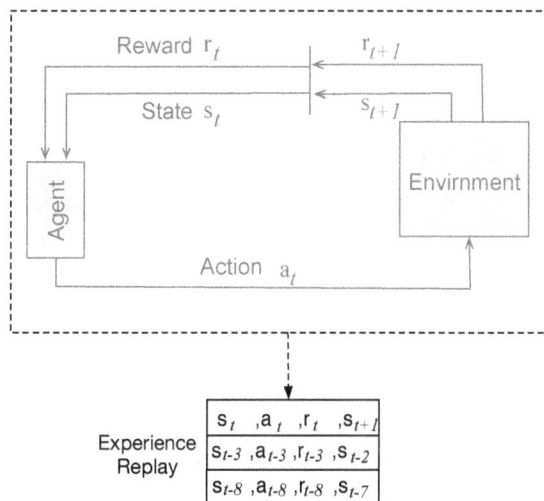

Figure 13-26: Experience replay construction from the interaction of agent with the environment.

DQN's target network

In Q-learning algorithms, at each time step, a single Q-value (or action-value) is updated. Neural networks adjust all weights at once, which means if we use a neural network instead of a Q-table, all Q-values inside the mini-batch are updated simultaneously. Since each Q-value depends on the next state's Q-value, and because the states and their corresponding values change during training, the most recent weight updates become inconsistent as the states evolve. This makes the neural network training process unstable, and this phenomenon is referred to as 'chasing its own tail' or 'moving target problem'. Why it happens? In supervised learning, we split it to test and train sets, and then train the algorithm. The target is fixed and not changing. For example, if the target goal is to find a bird image, the bird characteristics do not change. However, in RL the target is moving during the training process, and it is not fixed. It means we are trying to minimize a loss function where the optimal point (minimum) keeps moving each time because the agent updates its policy. Therefore, the gradient movement toward the optimal point (minimum) is unstable.

To handle the moving target, a later version of DQN [Mnih '15] uses two neural networks with exactly the same architecture but with different weights. One network is referred to as a *main, online,* or *policy network,* in which its parameters are presented with θ. The other one is referred to as a *target network,* in which its parameters are presented with θ^-. Both models take states as input and output Q-values to decide the optimal action.

The target network's weights remain fixed for C iterations (steps), providing a stable target for learning while the main network is trained. C is the number of iterations that occur in the main network while the target network remains frozen, and it is a hyperparameter that the user should tune. After a C iteration passes, the weight of the target network will be substituted by the weight of the main network. The target network is used to obtain the maximum Q-value for the next state. The main network uses this Q-value to calculate the target Q-value for the current state. The Bellman equation to compute the target value is: $y_t = r_{t+1} + \gamma max_a Q_{target}(s_{t+1}, a; \theta_t^-)$, and the main network minimizes the loss function between Q-value prediction and target value as follows:

$$L(\theta) = \left(Q(s_t, a_t; \theta_t) - [r_{t+1} + \gamma \, max_a \, Q_{target}(s_{t+1}, a; \theta_t^-)] \right)^2$$

We simplified the DQN model in the following pseudocode to make it easier to understand, if you are interested in the original version of the algorithm, check the original paper [Mnih '15].

initialize: replay memory D, main network with random weights θ, target network with weights of main network $\theta^- = \theta$ # θ^- is used to present target network params, and θ is used for main network params, but at the beginning $\theta = \theta^-$.
for all (episode) {
    Initialize the starting state $s_1 = \{x_1\}$ # x_1 is the observed sample.
    for all (time step t) {
      a = epsilon_greedy(s)
      r, s' = execute(a)
      D.add [(s, a, r, s')] # add the tuple of (s, a, r, s') into the replay memory
      mbatch = sample a random mini-batch from D # uniform distribution is used
                           for sampling.
      y_j = target_network (mbatch) # computes the y_j (Q-value) using the sampled
                    batch by using the target network. If the
                    episode terminates at the next state, then
                    $y_j = r_j$, otherwise $y_j = r_j + \gamma Q_{target}(s_{j+1}, a' : \theta^-)$.
      y_i = main_network (mbatch) # the main network Bellman equation can be
                    written as $y_j = Q_{main}(s_j, a_j; \theta)$
      gradient descent (MSE_loss(y_i, y_j))
      if (step_size = C) then $\theta^- = \theta$
    }
}

To summarize, the main (online or policy) network trains on a batch of past experiences read from experience replay. It is updated regularly during training and is responsible for selecting the actions the agent should take based on the current state of the environment. It learns from interactions with the environment and tries to improve its predictions over time.

After every C number of steps, the main network parameters are then copied to the target network. The target network's purpose is to provide more stable targets for learning, meaning it stays fixed for some number of training steps (C iterations). After every C iteration, the weights of the target network are updated by copying the weights from the main network.

Convolutional Layers of DQN to play Atari games

Minh et al. [Mnih '15] introduced DQN for Atari games, and it has outperformed humans in playing Atari 2600 games[10] on 49 games, and the performance of the algorithm is equal to that of a human professional gamer. Since most of the time, the agent's behavior can be visualized, it is worth learning how authors of DQN use it to learn how to play Atari games.

---

[10] Useless reading alert … Kudos to Atari games. When the author was seven years old, he got an Atari 2600. It was the first computer he saw, back in 1987. He asked his mother how to make video games, and his mom answered to study computer engineering (what was called computer science at that time).

The input of DQN neural network should take a video of size 210×160 color video at 60 Hz. To handle videos, convolutional layers receive a series of images of input size $83 \times 84 \times 4$, which are preprocessed images from the input video frame. To infer velocity, direction, and other information in the network, a stack of four images (frames) will be fed into the network.

They propose two convolutional layers of convolutions followed by two fully connected hidden layers, and ReLU is used as their activation function.

Double Deep Q-Network (DDQN)

Double Deep Q-learning [Van Hasselt '16] was introduced to address the overestimation bias in Q-learning, which became evident after the success of Deep Q-Network (DQN) [Mnih '15]. To understand DDQN, let's assume we need consultation on a problem. We know a person who is an expert, but she is too optimistic. How can we benefit from her expertise while getting realistic consultations? We can consult with another person along with her as well. Taking the other person's opinion on her opinion makes our decision more realistic.

The same problem exists with all Q-learning, and DQN. Q-learning algorithms are too optimist, and they overestimate the Q-values. To understand this issue, take a look at a simplified form of the Bellman equation: $Q(s,a) = r + \gamma max_{a'}(Q(s',a'))$. Here we can see that we choose the *maximum* Q-value of the next state estimation. Estimated values are not necessarily correct and present actual values. Going for the maximum estimation, which is not necessarily correct, is too optimistic. Therefore, in all Q-learning algorithms, the Q-value is overestimated.

To handle this issue, we do the same in that we consult with an optimist person, getting another expert view on the subject as well. There is a Double Q-learning algorithm, which we skipped to explain earlier [Van Hasselt '10], but the same approach is used for DQN as well.

DQN already has two networks while working, i.e., the target network and the main network. The main network selects the action for the current state, and the target network provides the target Q-value for the next state to estimate the Q-value. By using two networks, the target network helps stabilize learning by providing fixed Q-value targets, and the main network learns from these targets. However, in DDQN, the main network is used to select the "best" action for the next state, and then the target network is used to provide the Q-value for that action (the one selected by the main network). To better understand this approach, let's take a look at an example in Table 13-2.

	Policy Network	Target Network	
Action with Index 1 has the highest Q-value	1: Q(s,a) = 2.9	1: Q(s,a) = 2.8	This Q-value will be selected as an estimation, because its index from the policy network
	2: Q(s,b) = 2.1	2: Q(s,b) = 3.4	
	3: Q(s,c) = 0.8	3: Q(s,c) = 1.2	

Table 13-2: Q-value estimation process in DDQN. First, the index of the Q-value with the highest value from the Main network, will be selected. Next, the Q-value of the same index, but in the target network, will be returned for estimation.

Here the main (policy) network provides three estimations for a Q-value with different actions. The first action has the highest Q-value. Then, the index of this action from the online network is taken (index=1), and the Q-value of this index from the target network will be used as a Q-value for the estimation of this action. You can see, despite having a Q-value with higher values in the target network and having a lower estimate for action with index 1, in the target network, the final estimation for Q-value will be 2.8.

Recall that the Bellman equation for DQN is as follows: $y_t^{DQN} = r_{t+1} + \gamma max_a Q(s_{t+1}, a; \theta_t^-)$. In DDQN, instead of using the maximum value from the target network (*max*), we use the *arg max*, which refers to the index in the other network (main network), and the target value will be computed as follows:

$$y_t^{DDQN} = r_{t+1} + \gamma Q\left(s_{t+1}, arg\ max_a Q(s_{t+1}, a; \theta_t); \theta_t^-\right)$$

Figure 13-27 presents the architecture of DDQN. The authors of the DQN recommend having two updates. First, it is a policy update by finding the optimal Q-value. This process will be done by passing a state into the network, and the network outputs the optimal action for the given state. However, in Q-learning, we need to calculate the optimal Q-value, i.e., $max_a Q^*(s', a')$, where s' and a' belongs to the future. To find this, we pass s' to the target network, and it provides, $Q*(s', a')$, which are used by the Bellman equation to calculate the next action.

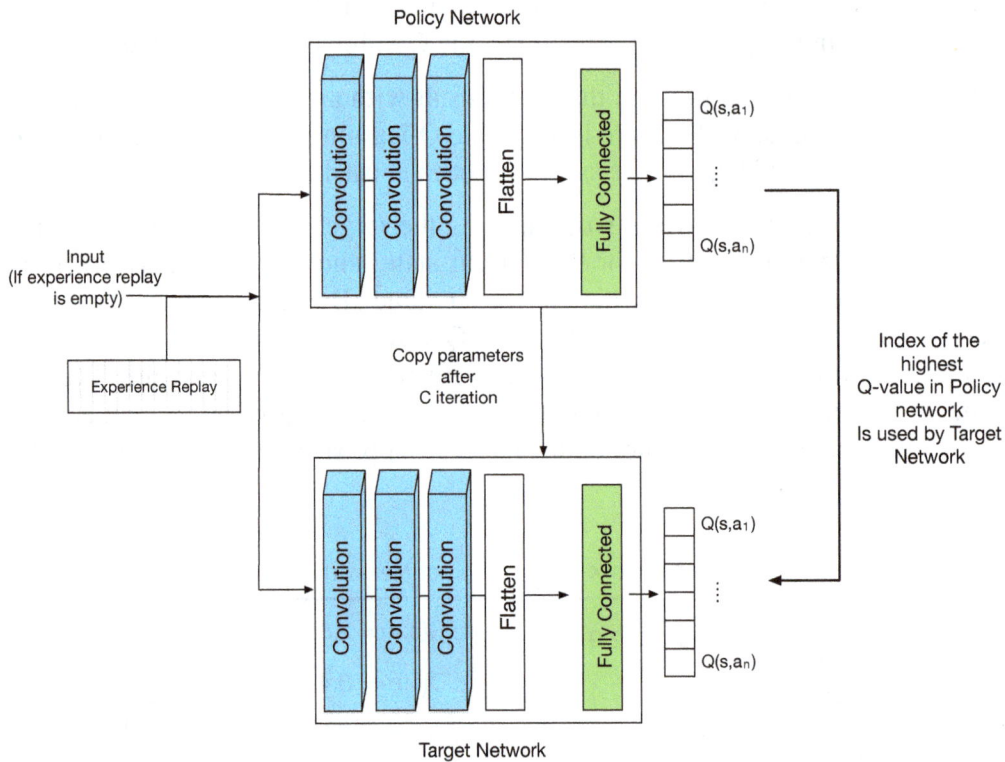

Figure 13-27: DDQN architecture, which is composed of a policy network and a target network.

Dueling Networks

After the introduction of DQN, Wang et al. [Wang '16] proposed an improvement in the network architecture of DQN, known as Dueling Network, which could be applied to any variant of Deep Q-Networks, including DDQN and DQN. A dueling network does not change the model; it only modifies the neural network architecture. Experiments show that using a dueling network reduces the optimistic biases of Q-learning better than DDQN.

Before we explain the dueling network, we should learn the definition of the *action-advantage function* $A(s, a)$. The action-advantage function specifies how much extra reward an agent can accumulate by taking action a at state s. In other words, it is a difference between the action-value function and the state-value function:

$$A_\pi(s, a) = Q_\pi(s, a) - V_\pi(s).$$

It means that the action-value function $Q(s, a)$ can be decomposed into one component that is shared across all state values, i.e., $V(s)$ and a component that is specific to the given action advantage, i.e., $A(s, a)$. The Dueling network architecture uses two networks and produces two separate estimates, one for the state-value, $V(s)$ (the output is a scalar) and one for the action-advantage, $A(s, a)$ (the output is a vector).

In other words, the Dueling network separates the representation of state values and action advantages (action advantage is state-dependent). This separation enables the model to learn which states are valuable without having to learn the impact of each action on each state. If the exam monster in your sleep asked you what the advantage of the Dueling network is, you could answer with the previous sentence.

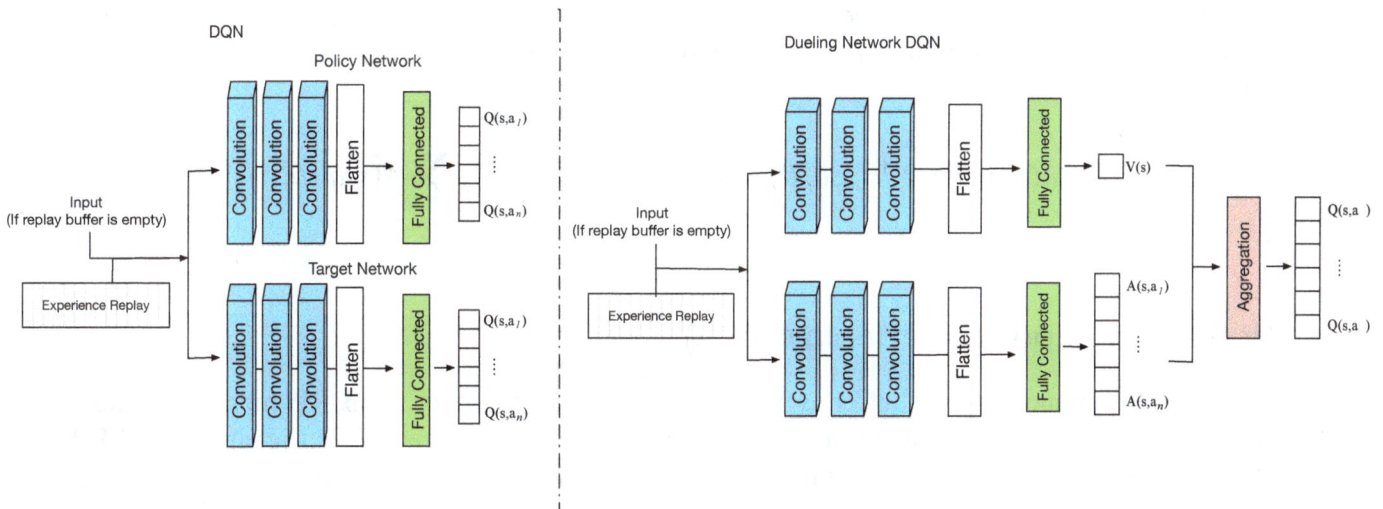

Figure 13-28: (left) DQN architecture. (right) Dueling Network DQN. Note that convolutional layers are not mandatory. Since both networks are implemented for Atari game playing, they use convolutional layers; depending on the question, we can change or remove the convolutional layers.

Similar to DDQN, a dueling network has a convolutional module with ReLU activation functions for learning features. This module is common to both networks, but the computation of advantage and state-value functions is separate. Figure 13-28 presents the architecture of the

Dueling network (right) in comparison to vanilla DQN (left). Dueling Network uses two networks, one to estimate $V(s)$ and one to estimate $A(s, a)$; they do not share weights and operate independently. The architecture of Convolutional layers and flattened layers are identical in both networks, but, unlike the original paper, we show them separately to emphasize that they are trained separately.

We can see in Figure 13-28 that there is an aggregation component that merges the scalar $V(s)$ with the vector of $A(s, a)$ values to calculate $Q(s, a)$ values. However, the aggregation is not simply adding advantage values to state values because there is a problem, which is called *identifiability*. The identifiability problem explains that, given $Q(s, a)$ value, we can not recover unique values for $V(s)$ and $A(s, a)$. It is simple mathematics, if we have $x + y = 5$, we can not say definitely $x = 2$ and $y = 3$, because we could have different values such as $x = 1, y = 4$ or $x = 5, y = 0$, etc.

To resolve the identifiability problem, authors force the action-advantage function to have zero value at the chosen action by subtracting the maximum of action-advantage values. As a result, for the optimal action, we will have $Q(s, a) = V(s) + [A(a, s) - max_{a'}(s, a')]$. Therefore, for the optimal action $a*$ the $Q(s, a) = V(s)$, because $A(a, s) - max_{a'}(s, a') = 0$, and for other actions, authors substitute the maximum action-value with the average of all action-advantages.

Assuming A is the total number of possible actions, the following is the simplified equation to calculate $Q(s, a)$:

$$Q(s, a) = V(s) + [A(s, a) - \frac{1}{|A|} \sum_{a'} A(s, a')]$$

However, the equation is the simplified version to understand the rationale, the equation reported by the authors incorporates parameters of the convolutional network (θ), and parameters of both networks (α for state value and β for action-advantage parameters). Therefore, it is written as follows:

$$Q(s, a; \theta, \alpha, \beta) = V(s; \theta, \beta) + \left[A(s, a; \theta, \alpha) - \frac{1}{|A|} \sum_{a'} A(s, a'; \theta, \alpha) \right]$$

Prioritized Experience Replay (PER)

Using an experience replay buffer is very common among recent deep RL models. A more advanced replay buffer is known as *Prioritized Experience Replay (PER)* [Schaul '16]. PER allocates more resources to experiences that have a higher chance of being used for learning. Therefore, experiences that are more important for improving the model are more likely to be replayed and used for learning. To implement such a feature PER, experiences are associated with probabilities and are selected based on these probabilities.

There are different approaches to prioritizing experience replay. One approach is using TD error, but there are several issues associated with TD error alone, such as noisiness. The authors of PER introduce a *proportional prioritization* as $p_i = |TD - error_i| + \epsilon$, where ϵ is a small constant to ensure that even if TD-error of a sample is zero, it still has a chance to get selected from the PER. Then, they exponentiate p_i to scale with a hyperparameter α, where $0 < \alpha < 1$. The probability of selecting an experience, $P(i)$ is computed as follows:

$$P(i) = \frac{p_i^\alpha}{\sum_k p_k^\alpha}$$

As we stated earlier, $P(i)$ is used to sample experiences from the replay buffer instead of using random uniform sampling. The described approach to PER is sensitive to outliers. To mitigate the impact of outliers, an alternative approach is to use *rank-based prioritization*. In this method, samples are sorted based on their absolute TD error in descending order. For example, the sample with the highest absolute TD-error gets the rank 1, the next highest one gets the rank 2, etc. Then, the priority is calculated as one divided by the rank, and the priority is typically scaled by an exponent β, which controls how much prioritization is applied.

$$p_i = (\frac{1}{rank(i)})^\beta$$

Right now, we resolve several challenges by sampling based on PER. However, PER data in experience replay has a different distribution than actual data because of the use of TD error, priorities, and probabilities. Therefore, sampling from one distribution, which we have been doing (until now), introduces a bias in our sampling approach. In simple words, PER has a different distribution than actual data in experience replay, and this makes the sampling biased.

To handle this bias, the authors introduced weighted importance sampling. This approach scales the TD-error by weights calculated with the probabilities of each sample. In other words, it changes the magnitude of updates if the sampling does not come from a uniform distribution. The weight for each sample is calculated as follows:

$$w_i = (\frac{1}{N} \cdot \frac{1}{P(i)})^\beta$$

Here, N is the number of samples inside the experience replay and β is a hyperparameter that tunes the degree of correction ($\beta = 0$ means there is no correction and $\beta = 1$ there is a full correction for bias). We have already explained that $P(i)$ is the probability of selecting experience i, and it is calculated based on the priorities of all experiences. Also, weights are downscaled to be in the range between 0 and 1, as follows:

$$w_i = w_i \div (max_j(w_j))$$

Addressing the bias introduced by PER stabilizes the learning process and improves the performance of the model. PER is particularly beneficial in scenarios where certain experiences are more informative than others, such as in games or real-time decision-making systems.

However, implementing PER and weighted importance sampling comes with its challenges. One must carefully tune the hyperparameters α and β balance between learning from high-priority experiences and maintaining a diverse set of samples.

Policy Gradient Methods

Up until this point, all methods we have learned, including Monte Carlo, SARSA, TD learning, and all varieties of Q-learning, use a value function (or action-value function) to identify the optimal policy. However, another group of algorithms directly finds the optimal policy without using a value function. These algorithms are referred to as policy-based methods. Policy-based methods use gradients, and thus, they are also called *Policy Gradient* methods. In particular, policy gradient methods optimize the policy directly by adjusting its parameters through *gradient ascent*, aiming to maximize the expected reward. Be sure you have already learned Gradient and optimizations in Chapter 8, before continue reading these models or algorithms. In this section, we will explain some common Policy Gradient methods, including REINFORCE, TRPO, and PPO.

Policy Gradient Advantages

Real-world environments are often stochastic rather than deterministic. To recall the differences between stochastic and deterministic steps, refer to Figure 13-6 and its explanations. Policy gradient algorithms have some advantages over value-based algorithms, making them popular in many applications.

First, the policy learned by Q-learning is deterministic, which means that Q-learning cannot deal with stochastic policies. Policy gradient methods can incorporate the stochasticity of real-world environments.

Second, a policy gradient agent directly learns the policy, and thus, it can learn stochastic policies. A stochastic policy does not require an explicit exploration step (unlike value-based methods); the exploration is embedded within the policy. In contrast, value-based methods assume policies are deterministic. Third, policy gradient methods use a simpler approach to training. Instead of requiring accurate estimations of value functions, these methods focus on optimizing probabilistic actions, leading to more stable behavior.

Policy Gradient Objective Function

The objective of all RL algorithms is to find the optimal policy that maximizes the expected return. This objective can be achieved if we have full observational data of all possible trajectories. Practically, such a thing is impossible, and as we explained in Chapter 8 when it is infeasible to get the final optimal result, function approximation methods find a result close to the optimal result (approximate the optimal result). Similar to other optimization algorithms that we described in Chapter 8, optimization function J has model parameters, θ, thus we write it as $J(\theta)$. While using value-based methods, our objective is to minimize the error between the predicted and the target values, i.e., $\mathbb{E}_\pi[v_\pi(s) + \alpha(G - v_\pi(s)) \mid s]$. While using policy gradient methods, our objective is to maximize the gain of a policy, i.e., $J(\theta) = \mathbb{E}_\pi[G \mid s, a]$.

Policy gradient algorithms use Gradient Ascent (negative of Gradient Descent) to optimize their objective function.

Before we proceed further, both we and the exam monster would like to ensure you remember the following: policies are modeled as probability distributions. In a policy gradient algorithm, if an action yields a high reward, the algorithm increases the probability of selecting that action in the future by adjusting the policy's parameters.

REINFORCE

REINFORCE (Monte-Carlo policy gradient) [Williams '92] is a Monte Carlo implementation of the policy gradient method. Unlike value-based methods, the REINFORCE algorithm selects the next action based on the current policy rather than the value function or action-value function. After selecting an action, it observes the rewards and the next state and then uses these observations to update the policy directly.

How does REINFORCE update the policy? We have learned that neural network algorithms are defined by their parameters. Assuming θ presents the parameters of the neural network, $J(\pi)$ is the *expected return* under the policy π, and the gradient of the policy objective function is computed as $\nabla J(\theta) = \nabla \mathbb{E}_{\pi}[v_{\pi}(s)]$. Nevertheless, computing $\nabla J(\theta)$ is problematic because it depends on state selection in the environment, and in RL the environment distribution is non-stationary. The *policy gradient theorem* states that the gradient of the expected return with respect to the policy parameters, i.e., $\nabla J(\theta)$ is equal to the expected value of the product of the gradient of the log probability of the policy and the expected return (or an estimator of it) under the policy. Mathematically, this can be expressed as:

$$\nabla J(\theta) = \mathbb{E}_{\pi}[G \nabla log \ \pi_{\theta}(a \,|\, s)]$$

We do not describe its mathematical proof, but it is easy to understand if you search online.

We can substitute expected return with action-value function, i.e., $\Sigma_a[Q(s,a) \,|\, \pi(a \,|\, s)]$, and we can rewrite it as follows:

$$\nabla J(\theta) = \Sigma_a[Q_{\pi}(s, a) \nabla log \ \pi_{\theta}(a \,|\, s)]$$

In Chapter 8, we have explained that Gradient descent neural network parameters are updated as $\theta = \theta_{old} - \alpha \nabla_{\theta} J(\theta)$ (where α is the learning rate). In the case of Gradient ascent, which is used in RL algorithm, we have $\theta = \theta_{old} + \alpha \nabla_{\theta} J(\theta)$. Based on the explanation we have provided, we can substitute $\nabla_{\theta} J(\theta)$ and use the following equation to compute the policy gradient update rule.

$$\theta = \theta_{old} + \alpha G \nabla log \ \pi_{\theta}(a \,|\, s)$$

In some literature, it is presented as: $\theta = \theta_{old} + \alpha G \nabla log \ \pi(a \,|\, s; \pi_{\theta})$.

θ will be optimized via backpropagation on the neural network, but we have two other unknown parameters to be configured, G and $\nabla log \ \pi_{\theta}(a \,|\, s)$. To identify G, the algorithm samples trajectories $(t = 1...T)$, then for each step t in that trajectory[11], it computes the return from that step, i.e., G_t. The average of all G_t will be used as G. To identify $\nabla log \ \pi_{\theta}(a \,|\, s)$, the algorithm first calculates the logarithm of the probability of the policy for each state and each action. Next,

---

[11] Since it relies on the trajectory of full episode, it is a Monte Carlo algorithm.

it computes the derivative of the log probability with respect to the policy parameters for each state and action. Specifically, the algorithm calculates this by averaging the gradient of the log probability over all states and actions. This policy gradient algorithm indicates which direction the policy's parameters should be adjusted to increase the expected return, as it is an objective function used to improve the policy.

The REINFORCE algorithm is fairly simple to implement, but it can also be slow to converge due to its sensitivity to hyperparameters. The network architecture of REINFORCE is not fixed and can vary depending on the problem it is being used to solve. The input layer receives the observations (action, state, rewards) from the environment, and the output layer produces probabilities for each action.

The Following describes the pseudo-code of REINFORCE algorithm.

initialize: network with random weights θ, a differentiable policy with parameters $\pi_\theta(a\,|\,s)$
for all (episode) {
    episode_trajectory $(1...T) \leftarrow \pi_\theta(.\,|\,.)$ `# apply the policy to generate the`
            `trajectory, i.e.,` $S_0, A_0, R_1, ... S_{T-1}, A_{T-1}, R_T$
        for all (t in episode_trajectory) { `# from 0 to` T `in each episode.`

$$G = \sum_{t'=t}^{T} \gamma^{t'-t} r_{t'} \quad \#\gamma^{t'-t}\text{: discount factor, } r_{t'}\text{: return at time } t.$$

$$\theta_{new} \leftarrow \theta_{old} + \alpha\gamma^t G_t \nabla log\pi_\theta(a_t\,|\,s_t) \text{ # Update the parameters using}$$
                            `the policy gradient.`

    }

}

REINFORCE with Baseline

The REINFORCE algorithm has high variance in its gradients because they are estimated by random sampling of actions and rewards. The random sampling process is noisy, causing the gradient estimates to vary in each iteration, which leads to slow convergence. This causes the REINFORCE algorithm to converge slowly. The author [Williams '92] introduces the *REINFORCE with Baseline* to reduce the variance of gradient and improve the algorithm's stability. This algorithm is also known as the Vanilla Policy Gradient (VPG).

To reduce the variance in the gradient, the REINFORCE with baseline algorithm employs the value function. While policy gradient methods typically do not use a value function, REINFORCE with baseline uses a *state-value function* as a baseline to adjust the variance of weight changes with gradient ascent. Specifically, it uses the value function to calculate the baseline only. Applying the baseline converts the $\theta_{t+1} = \theta_t + \alpha G_t \nabla log\pi_\theta(a_t\,|\,s_t)$ equation to the following equation:

$$\theta_{t+1} = \theta_t + \alpha(G_t - \hat{v}(s_t, w)) \nabla log\pi_\theta(a_t\,|\,s_t)$$

Here, the term $\hat{v}(s_t, w)$, which is shown in red, represents the baseline or the state-value function at time t.

REINFORCE with baseline has two neural networks, one to compute the policy and one to compute the state-value function. w represents the parameters of the state-value function neural network and θ presents the parameters of the policy gradient neural network.

The output of the state-value function network is a scalar, i.e., the value of the state. The output of the policy network is a list of possible actions. For example, if it is a cart pole example, we have two outputs, as shown in Figure 13-29. The policy network receives the current state and outputs the optimal policy (not optimal Q-values as in previous models). The value-function network receives the current state and builds the baseline, which is used by the policy network to improve the algorithms' stability.

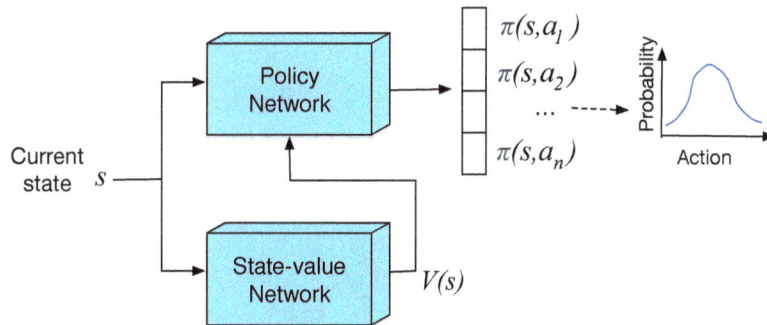

Figure 13-29: Architecture of REINFORCE with baseline, which is composed of two neural networks, a policy network, and a state-value network; its results will be used by the policy network.

The cost function of the REINFORCE with baseline (value estimation) is written as follows:

$$\nabla_\theta J(\theta) = \mathbb{E}_\pi \left[(G_t - V(s_t; w)) \nabla \log \pi_\theta(a_t \mid s_t) \right]$$

Assuming N is the total number of samples, the loss function used to train the value-function network (not the policy function), is computed as MSE of loss with respect to w (parameters of value-function network).

$$J(w) = \frac{1}{n} \sum_{n=0}^{N} [(G_t - V(s_t; w))^2]$$

We can write the REINFORCE with baseline as follows. The red parts are the new additions to the REINFORCE algorithm

initialize: - policy network with random weights θ
*   - a differentiable policy with parameters $\pi_\theta(a \mid s)$*
*   - a differentiable state-value function parameters $\hat{v}(s_t, w)$.*
for all (episode) {
*   episode_trajectory ($1...T$) $\leftarrow \pi_\theta(. \mid .)$* `# apply the policy to generate the`
                            `trajectory, i.e.,` $S_0, A_0, R_1, ...S_{T-1}, A_{T-1}, R_T$
*   for all (t in episode_trajectory) {* `# from` t_1 `to` T `in each episode.`

$$G = \sum_{t=1}^{T} \gamma^t r_t$$ `#` γ^t `is the discount factor and time` t`,` r_t `is return at time` t`.`

$\delta \leftarrow G - \hat{v}(s_t, w)$ `# Here` δ `is the TD error,` w `is the weight vector, and it`
                `is used to adjust the values predicted by the value`
                `function through training.`

$w_{new} \leftarrow w_{old} + \alpha^w \delta \nabla \hat{v}(s_t, w_{old})$ `# During the training, weights are updated`
                            `based on the gradient of the value function`
                            `with respect to the policy parameters.`

$\theta_{new} \leftarrow \theta_{old} + \alpha \delta \nabla ln(a \mid s, \theta)$ `# Here the` G `of REINFORCE is substituted with` δ`.`
                            `Keep in mind that policy gradient update`
                            `doesn't need discount factor.`

}}

Trust Region Policy Optimization (TRPO)

Trust Region Policy Optimization (TRPO) [Schulman '15] is an on-policy RL model that considers the policy update as an optimization problem (check optimization in Chapter 8).

TRPO uses a *trust region* to ensure the optimization variable changes are not too large, and trust regions restrict the size of the update to a certain region around the current optimizer's value. This algorithm offers two key advantages: (i) it enforces a trust region, leading to more stable and reliable policy updates, which in turn results in better sample efficiency; and (ii) by using second-order optimization methods, it can optimize non-linear policy functions, allowing it to model complex environments more effectively. It is one of the most successful algorithms in simulating robot movements, which is a complex task.

There are too many extra things you need to learn before start learning the TRPO algorithm, also the author referes you to Chapter 16 twice.

Sampling and running the experiment, even experimenting in the simulated environment, is a resource-intensive process. TRPO can learn the policy with few observations, which means less interaction with the environment.

Besides, linear policy functions are not capable of being used in a complex environment. However, by using second-order optimization, the TRPO algorithm can optimize a non-linear policy function. This enables this algorithm to identify and model policies in a complex environment.

Before we explain trust regions, it is important to be familiar with several key concepts, such as the Fisher information matrix, natural policy gradient, and conjugate gradient descent. Please be patient and follow the explanations, understanding that these concepts are necessary for a comprehensive grasp of TRPO. TRPO is one of the most effective reinforcement learning models for complex tasks, making the effort to learn these concepts worthwhile.

Trust Region

Gradient descent/ascent algorithms search for an optimal point by following the direction of the gradient (descent or ascent). Another class of optimization algorithms, known as trust region methods, restricts the search for the optimal point within a specific region around the current point, called the trust region. Instead of moving directly along the gradient, these methods define a trust region and look for the best point within that confined area.

Trust region methods get a hyperparameter that specifies the maximum radius (δ) to search for the next point. To accelerate the learning pace, δ can expand or shrink during each iteration depending on the curvature of the optimization surface. For example, if the error reduces, δ expands, and if the error increases, then δ shrinks. Figure 13-30 is a toy example that visualizes the differences between using line search and trust region to achieve an optimal point.

Assuming x is a vector of variables and $f(x)$ is a scalar function, we can state that the objective of trust region optimization is to minimize $f(x)$, while $||x|| <= \delta$.

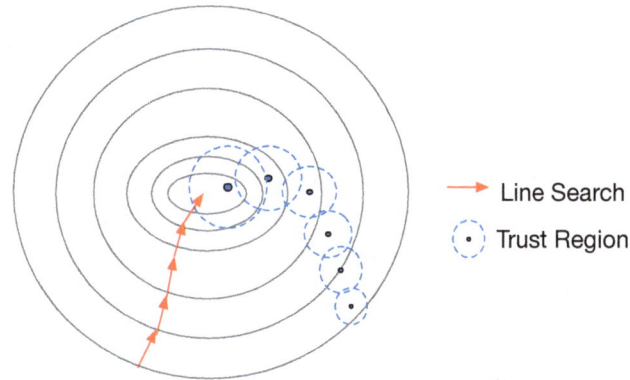

Figure 13-30: Line search optimization algorithm, e.g., gradient descent, moves toward the optimal point with a step size. On the other hand, a trust region optimization searches a radius of options to decide about the next point.

Using the trust region optimizes the policy while guaranteeing *monotonic improvement*. The monotonic improvement theorem states that under certain conditions, the expected return of the newly produced policy is guaranteed to be at least as good as the expected return of the old policy.

Minorization-Maximization (MM) algorithm

To address the challenge of gradient updates in policy optimization, we need to ensure that policy updates improve the expected reward rather than worsen it. One approach to ensure this is the use of Minorization-Maximization (MM) algorithm. The MM algorithm offers one approach to achieve this. MM approximates a complex optimization problem with a simpler one by constructing a simpler (surrogate) function that serves as a lower bound to the original reward function. This process, known as 'minorizing' the problem, makes it easier to solve than the original optimization task.

A lower bound function is a function that is always less than or equal to the function being optimized. For example, consider a function $f(x)$ that is defined on a domain D. A lower bound function $g(x)$ is defined on the same domain D and satisfies the property $g(x) \leq f(x)$ for all x in D. Since it is lower bound, it never passes the objective function and always stays close to the optimal point. The MM algorithm is used to optimize the lower bound function, and it guarantees that in every iteration, the policy will be improved without passing or jumping the optimal point.

Although TRPO is typically described in terms of constrained optimization and trust regions, its underlying mathematical structure aligns closely with the MM algorithm. In particular, each iteration of TRPO maximizes a surrogate objective function (which we will explain shortly) within the trust region constraint, similar to the maximization step in MM. A surrogate objective function is an alternative function used to approximate the true objective, designed to be easier to optimize while still guiding the solution toward improving the original objective. The surrogate objective function in TRPO is analogous to the lower-bound function in MM algorithms. Furthermore, the process of constructing a new surrogate objective at each iteration based on the current policy mirrors the minorization step in MM, where a simpler function is used to approximate the original complex objective.

Surrogate Loss

Assuming τ presents the trajectory of sequence of states s and actions a that are sampled from the policy π_θ, the policy gradient is computed as follows: $\nabla J(\theta) = \mathbb{E}_{\tau \sim \pi_\theta(\tau)}[\nabla ln\ \pi_\theta(a\,|\,s)r(\tau)]$. By using advantage A, we can write the cost function with respect to the advantage as follows: $J(\theta) = \mathbb{E}_t[ln\ \pi_\theta(a_t\,|\,s_t)A_t]$. Using this loss, however, is not easy for optimization because the advantage estimate is noisy, and we don't want the algorithm to use that noisy estimate to decide about our policy. Recall that we have explained that, unlike supervised learning, a wrong estimate is very expensive in the context of RL.

In addition to that, the reward $r(\tau)$ is calculated from the trajectory of using the policy. It means to calculate the new policy, we need to have the old policy. What if the new policy changes? Then, there is a need to collect new samples, and old policies become invalid. Using importance sampling (check Chapter 16) along with advantage can mitigate this problem. Briefly, importance sampling refers to the process of reweighting samples to estimate the properties of one distribution (target distribution) using data from another distribution (more convenient to sample).

We can consider the new policy as target distribution (π_θ) and the old one as auxiliary distribution ($\pi_{\theta_{old}}$), and by using the importance sampling along with advantage, the policy could be written as follows:

$$J(\theta) = J(\theta_{old}) + \mathbb{E}_{\pi_{old}}[\frac{\pi_\theta(a\,|\,s)}{\pi_{\theta_{old}}(a\,|\,s)} A^{\pi_{old}}(a\,|\,s)]$$

The expected advantage is defined as Surrogate Loss, $L_{\pi_\theta}(\theta)$ and can be formalized as follows:

$$L_{\pi_\theta}(\theta) = J(\theta) - J(\theta_{old}) \approx \mathbb{E}_{\pi_{old}}[\frac{\pi_\theta(a\,|\,s)}{\pi_{\theta_{old}}(a\,|\,s)} A^{\pi_{old}}(a\,|\,s)]$$

In summary, the surrogate loss is used as an approximation of the true loss (implements the minimization of MM algorithm). The true loss in policy gradient methods is based on maximizing the expected cumulative reward, and in trust-region methods like TRPO, additional constraints are applied to keep updates within a certain region.

We should be careful with this surrogate loss. If the differences between the old policy and the new policy are too high, this might lead to an inaccurate advantage because the advantage is computed based on the old policy, and thus, the new policy should not be too far away from the old policy. The surrogate loss is acceptable only if the old policy and new policy have less than δ differences. To handle this, we add a condition by comparing their distribution with KL-Divergence (check Chapter 3) i.e., $\mathbb{E}_{\pi_{old}}[D_{KL}(\pi\,||\,\pi_{old})] < \delta$. δ is a hyperparameter that will be determined through experimentation and empirical tuning.

TRPO uses KL-Divergence to define the *trust region constraint* in the optimization problem being solved. The use of this constraint in TRPO provides two main benefits: (i) it ensures that the new policy is not significantly different from the old policy, which helps prevent overshooting during updates. (ii) it enforces the policy updates to be small and incremental rather than allowing large, destabilizing changes in the policy. As a result, this approach promotes stable learning.

In summary, by using the trust region along with the surrogate loss, TRPO ensures that policy updates are stable and controlled. In other words, TRPO uses KL-Divergence to constrain the policy update, which bears some conceptual similarity to the minorization of the MM algorithm, but it is not an implementation of the MM algorithm.

Fisher Information

Given a random sample dataset X from an unknown distribution with θ parameters, Fisher information measures how much information X provides about unknown parameters θ. For example, in a Gaussian distribution, the parameters are the unknown mean and unknown variance. Fisher information helps us understand the accuracy with which these parameters can be estimated from the data.

While working with deep RL, it is common to use an observed dataset X, and estimate unknown parameters θ that describe the distribution of the original dataset, i.e., $p(\theta; X)$. We should have a metric, such as Fisher Information, to specify how well the observation dataset X describes the parameters θ.

When the distribution has n parameter, the Fisher information is presented in a matrix of $n \times n$. For example, for Gaussian distribution, we have two parameters (mean and variance), and the Fisher Information Matrix (FIM) has a size of 2×2. The FIM is calculated by taking the second-order partial derivatives (Hessian matrix[12]) of the log-likelihood function with respect to given θ parameters and then averaging these derivatives over the distribution of the data. Assuming N is the total number of data points and X is the sampled dataset from an unknown distribution $p(X \mid \theta)$, the FIM is computed as follows:

$$F(\theta)_{i,j} = \frac{1}{N} \sum_{n=1}^{N} [\frac{\partial^2}{\partial\theta_i . \partial\theta_j} log \ f(X_n; \theta)]$$

We explained FIM calculation is associated with computing Hessian (matrix of second-order derivative). As a diligent scientist, engineer, or simply a human being, we should know that computing the second-order derivative is computationally very expensive.

As an alternative, it is recommended to use a trick that uses a KL divergence instead, i.e. *pseudo inverse of FIM*. In particular, FIM is the second-order derivative of a KL-divergence between two distributions θ, and θ'. Therefore, we can compute FIM as follows:

$$F(\theta)_{i,j} \approx \nabla_{\theta'}^2 D_{KL}(\theta \mid \mid \theta')$$

In other words, we can estimate second-order FIM through the gradients of the log-likelihood, avoiding the direct computation of Hessians. Besides, the second-order KL-Divergence is symmetric. Therefore, $\nabla_\theta^2 D_{KL}(\theta \mid \mid \theta') = \nabla_\theta^2 D_{KL}(\theta' \mid \mid \theta)$.

Natural Policy Gradient

The gradient we have learned in Chapter 8 and Chapter 10 and used later in neural networks specifies the direction in the optimization function. It does not specify the step size toward that direction, and to change the step size the learning rate (α) hyperparameter is used: $\theta = \theta_{old} + \alpha \nabla_\theta J(\theta)$.

---

[12] Check Chapter 8 if you need to recall the definition of the Hessian matrix.

This gradient is still associated with two problems: jumping too big (overshooting) that the sign of gradient changes, and taking a very small step that leads to very slow convergence. In simple words, the gradient is good at giving the direction but not the step size.

In supervised learning, the overshooting error can be corrected in the next epoch. However, in RL one overshooting in policy update might result in poor policy selection, and thus, the agent never reaches the right terminal stage. Furthermore, a mistake in RL could be very expensive, such as the robot falling down and getting wasted.

This means a poor policy selection in RL could be very expensive and intolerable, in contrast to supervised learning. To provide a solution to overshooting, Natural Policy Gradient (NPG) [Amari '98] is used by the TRPO model, and it uses two approaches: (i) constraining the size of policy updates and (ii) scaling the gradient update using the FIM.

(i) Constraining the size of policy updates: A simple approach that can resolve the overshooting is using a small cap ϵ, and limiting the θ changes ($\Delta\theta$) to a certain value ($\Delta\theta \leq \epsilon$). By using the cap, we can resolve the overshooting in first-order gradient. However, this is a too simplistic approach because it assumes the policy gradient is a convex shape function. The policy gradient of the real-world problem is a manifold (we described manifold in Chapter 7), which is far more complex than a simple convex shape function. Hence, only using a cap does not fully resolve the gradient overshooting issue.

Resolution: Since policies are distributions, instead of incorporating the cap between two jumps on the optimization function, we can incorporate it on the distances between policies. Therefore, we can use KL-Divergence to measure distances between distributions and use the cap within the KL-Divergence The KL-divergence between two the old policy $\pi_{\theta_{old}}$ and the current policy π_θ could be written as follows:

$$D_{KL}(\pi_{\theta_{old}} || \pi_\theta) = \sum_x \pi_\theta(x) log(\frac{\pi_{\theta_{old}}(x)}{\pi_\theta(x)})$$

By incorporating the cap ϵ, we can constrain the policy update based on D_{KL} the following: $D_{KL}(\pi_{\theta_{old}} || \pi_\theta) < \epsilon$.

Therefore, the optimal policy $\Delta\theta^*$ can be formalized as:

$$\Delta\theta^* = \arg \max_{D_{KL}(\pi_{\theta_{old}}||\pi_\theta)) < \epsilon} J(\theta)$$

Solving the above equation allows the algorithm to make significant policy updates while ensuring the update is within a trust region, preventing overshooting. This concept, previously referred to as a Trust Region, is reiterated here to emphasize its importance.

(ii) Scaling the gradient update using the FIM: There is a second approach to handle the problem of overshooting. Mathematicians, propose to change the scalar step size α[13] into a matrix that entails different step sizes and thus reduces the chance of overshooting.

Now the question is how the data for this matrix data is computed? One approach to compute the matrix data to mitigate overshooting is to use a second-order derivative (or Hessian), and thus, all parameters between different distributions are not treated equally. As we have explained, however, they are computationally expensive and not efficient substitutions to the gradient computation (first-order derivative).

Resolution: While using KL-Divergence, we can use a FIM to calculate the optimal step size for policy updates. The FIM is a measure of the curvature of the KL divergence between two policies, and it is used to approximate the Hessian matrix of the loss function and, therefore, allows for an efficient calculation of the matrix-vector product without having to compute the full Hessian matrix. The following presents the equation for FIM:

$$F(\theta)_{i,j} \approx \nabla^2_{\theta'} D_{KL}(\theta||\theta') = \frac{1}{N} \sum_{n=1}^{N} [\frac{\partial \log \pi_\theta(a_n|s_n)}{\partial \theta_i} \cdot \frac{\partial \log \pi_\theta(a_n|s_n)}{\partial \theta_j}]$$

Here, $F(\theta)_{i,j}$ presents the FIM, i and j present indices of the matrix. θ and θ' are two distributions that are compared by KL-divergence (D_{KL}). N is the total number of samples. $\log_{\pi_\theta}(a_n|s_n)$ is the logarithm of the policy π_θ, which gives the probability of taking action a_n given state s_n, under policy π parameterized by θ. $\frac{\partial(.)}{\partial \theta_i}$ is the partial derivative of (.) with respect to θ_i.

Lagrangian relaxation & Taylor expansion: Now we have solutions for both adding a constraint and scaling the gradient by FIM, but to compute a KL-Divergence between two distributions, we need to have all states and actions (as shown in the above equation), which is infeasible. Therefore, we should again approximate them.

To implement the approximation, we simplify the optimal policy update ($\Delta\theta^*$) equation by using Lagrangian relaxation (we explained it briefly in Chapter 8), which converts an optimization constraint into a penalty. Therefore, the equation to compute the optimal policy, $\Delta\theta^* = \arg \max_{D_{KL}(\pi_{\theta_{old}}||\pi_\theta)) < \epsilon} J(\theta)$, will be written as follows as a lagrangian relaxation:

$$\Delta\theta^* = \arg \max_{\Delta\theta} J(\theta) - \lambda(D_{KL}(\pi_{\theta_{old}}||\pi_\theta) - \epsilon)$$

In this equation, λ is the coefficient of the penalty. Then, a Taylor expansion (check Chapter 8) is applied to this equation. I $J(\theta)$ is approximated with the first-order Taylor expansion (i.e., the

---

[13] We are referring to the α in this equation $\theta = \theta_{old} + \alpha \nabla_\theta J(\theta)$

gradient w.r.t. θ), and D_{KL} is approximated with a second-order Taylor expansion. The result will be as follows.

$$\Delta\theta^* = \arg\max_{\Delta\theta} J(\theta_{old}) + \nabla_\theta J(\theta) \cdot \Delta\theta - \frac{1}{2}\lambda(\Delta\theta^T \nabla_\theta^2 D_{KL}(\pi_{\theta_{old}} || \pi_\theta)\Delta\theta + \lambda\epsilon)$$

Since we get $\nabla_\theta^2 D_{KL}$, now we can substitute it with FIM and get rid of D_{KL}. After the substitution, we will get the following equation:

$$\Delta\theta^* = \arg\max_{\Delta\theta} \left(J(\theta_{old}) + \nabla_\theta J(\theta_{old}) \cdot \Delta\theta - \frac{1}{2}\lambda(\Delta\theta^T F(\theta_{old})\Delta\theta + \lambda\epsilon) \right)$$

The above equation operates within the trust region constraints, i.e., $\Delta_\theta^2 F(\theta_{old})\Delta\theta \leq \delta$. To solve this equation, we need to find the optimal weight update $\Delta\theta$. To find $\Delta\theta$, first, we compute $\nabla_\theta J(\theta_{old})$, then FIM $F(\theta_{old})$ and next solve $\Delta\theta$ by using quadratic approximation.

Assuming that the NPG is the inverse of FIM multiplied by the traditional gradient, we will have: $\tilde{\nabla} J(\theta) = F(\theta)^{-1}\nabla J(\theta)$. We skip the details, but to determine the scale factor α the following equation will be used:

$$\alpha = \sqrt{\frac{2\delta}{\nabla_\theta J(\theta)^T F(\theta)^{-1}\nabla_\theta J(\theta)}} F(\theta)^{-1}\nabla_\theta J(\theta)$$

We did all these things to substitute the α in $\theta_{new} = \theta_{old} + \alpha\nabla_\theta J(\theta)$ with FIM $F(\theta)$ of the cost function. As a summary, instead of computing Hessian, we can derive FIM as outer products of gradients: $F(\theta) = \mathbb{E}_\theta[\nabla_\theta log\pi_\theta(x)\nabla_\theta log\pi_\theta(x)^T]$.

Although we derive FIM as an outer product of gradients, computing the inverse of the FIM, $F(\theta)^{-1}$ is expensive again. Instead, TRPO authors propose to solve $F(\theta)^{-1}\nabla J(\theta)$ as an optimization problem $x \approx F(\theta)^{-1}\nabla J(\theta)$, and we can solve x as: $F(\theta) \cdot x \approx \nabla J(\theta)$. Solving $Ax = b$ is equivalent to minimizing x as $f(x) = (1/2)x^T Ax - b^T x$, which lets us rewrite the transpose of FIM as $min_x(1/2)x^T F(\theta) - \nabla J(\theta)^T$. At this point, we finally get rid of computing the inverse of FIM.

The algorithm ensuring that the Natural Policy Gradient (NPG) remains within the trust region and adheres to the Kullback-Leibler (KL) divergence constraint is known as the line search algorithm in the context of TRPO, and it is written as follows:

Line search algorithm:
Compute the proposed policy step α
for ($j = 0, 1, ..., L$) {
       Compute proposed update $\theta = \theta_{old} + \alpha\nabla_\theta J(\theta)$
      if ($L_{\pi_\theta}(\theta) \geq 0$ and $D_{KL}(\theta || \theta_{old}) \leq \delta$) then {
              accept the update and set $\theta = \theta_{old} + \alpha\nabla_\theta J(\theta)$
              break
}

TRPO uses the conjugate gradient method to compute an approximation of the NPG direction, adhering to a trust region constraint based on KL divergence in the policy optimization process. We will explain conjugate gradient descent in Chapter 14.

TRPO Algorithm

TRPO is similar to Natural Policy Gradient (NPG) but includes three key improvements to address NPG's known limitations by applying specific changes to handle these issues. We list three common problems with NPG and describe how TRPO resolves them:

(i) *Large step size:* The constraint we apply on the KL-Divergence might be violated, and NPG has a high variance. This is due to the fact that it should perform lots of approximations. The second-order Taylor expansion used to approximate distances between two distributions may not be accurate, leading to overshooting during optimization. TRPO addresses this by employing a trust region, which constrains the KL-Divergence between the new and old policies to ensure more stable updates.

(ii) *FIM computational complexity:* NPG requires the inversion of the Fisher Information Matrix (FIM), which is computationally expensive with a complexity of $O(n^3)$. TRPO resolves this by avoiding explicit matrix inversion and instead using conjugate gradient methods to approximate the update direction efficiently, reducing the computational burden. Additionally, TRPO applies Lagrangian relaxation to optimize the problem with the KL-Divergence constraint.

(iii) *Unreliable policy update:* NPG lacks a mechanism to ensure that updates consistently improve the policy, and its approximations may sometimes result in performance degradation. TRPO overcomes this by using the conjugate gradient method for optimization and enforcing the trust region constraint to ensure that each update is beneficial and does not deteriorate the policy.

The following presents a simplified pseudocode of the TRPO algorithm, which we adapted from Peter Abbeel's [14] explanations.

```
# TRPO algorithm:
```
For (each episode) {

       Run policy for T time steps or across N trajectories.

       Estimate Average Reward at all T time stamps.

       Compute the Policy Gradient.

       Use Conjugate Gradient to approximate the FIM.

       Perform line search optimizing the surrogate loss while respecting a KL-divergence constraint.

}

Proximal Policy Optimization (PPO)

Two years after the introduction of TRPO, the first author of the TRPO and its collaborators proposed the Proximal Policy Optimization (PPO) algorithm [Schulman '17]. PPO and TRPO are among the best RL algorithms for complex tasks. PPO is also used in the ChatGPT architecture [Ouyang '22]. It focuses on redesigning TRPO but without the need to have a second-order derivative at all. An Actor-Critic version of it exists, but we list it in this section because it can be implemented as a Policy Gradient algorithm.

---

[14] https://youtu.be/KjWF8VIMGiY

TRPO was successful, but it uses many complex approaches to avoid computing second-order derivatives (Hessians). It is often hard to enforce trust region constraints on complex policies. Moreover, it can not operate with architectures that incorporate noise, such as drop-out layers, and share parameters, such as those sharing parameters between policy and value networks.

PPO tackles both challenges with a simplified approach to trust region optimization. PPO enforces a KL-Divergence constraint but does so without computing natural gradients, which are complex to compute. By avoiding second-order optimization, PPO can utilize common first-order optimizers such as RMSProp and Adam, and it supports other common functionalities of neural networks.

If you are alive after reading TRPO, keep in mind to uderstand PPO you should read TRPO first.

There are two types of PPO, one uses a penalty coefficient, and the other one utilizes a clipped objective. First, we describe the one with the penalty. The authors' experiments demonstrate that a clipped objective performs better than one with a penalty coefficient.

Penalty Coefficient

To understand PPO with a penalty, let's first review the TRPO surrogate loss and its constraint. The objective in TRPO is to maximize an expectation of the ratio of new and old policy probabilities weighted by the advantage function, subject to a constraint on the KL divergence between the new and old policies. This objective can be formalized as follows:

$$maximize_\theta \quad \mathbb{E}_\pi[\frac{\pi_\theta(a\,|\,s)}{\pi_{\theta_{old}}(a\,|\,s)}A] \text{ and } \mathbb{E}_{\pi_{old}}[D_{KL}(\pi_{old}(.\,|\,s),\pi(.\,|\,s))] < \delta$$

The PPO moves the constraint $\mathbb{E}_{\pi_{old}}[D_{KL}(\pi_{\theta_{old}}(.\,|\,s),\pi_\theta(.\,|\,s))] < \delta$ into the surrogate loss with a penalty β that acts as a weighting factor:

$$J^{Penalty}(\theta) = maximize_\theta \quad \mathbb{E}_\pi[\frac{\pi_\theta(a\,|\,s)}{\pi_{\theta_{old}}(a\,|\,s)}A - \beta.D_{KL}(\pi_{\theta_{old}}(.\,|\,s),\pi_\theta(.\,|\,s))]$$

This change converts the surrogate loss into an unconstrained cost function. Now, a question might arise: what is the advantage of converting constraint into a penalty? The authors' experiments show that different problems require a different value for β, and TRPO has a fixed constraint, not a penalty. This lack of flexibility makes TRPO less generalizable across different problems. Furthermore, the cost function of PPO allows for optimization in one step, enabling the use of SGD or other gradient descent methods to identify the best β value.

Clipped Objective

The second form of PPO uses clipping on the surrogate loss to address potential instability. Clipping is applied not directly to the advantage values themselves but to the ratio of the new policy to the old policy probabilities, multiplied by the advantage. This approach is intended to prevent excessively large policy updates, which can destabilize the training process.

In the surrogate loss, authors assign $r(\theta)$ to the ratio of new to old policy as follows:

$$r(\theta) = \frac{\pi_\theta(a \mid s)}{\pi_{\theta_{old}}(a \mid s)}$$

Now, if the new policy is exactly the old policy, we have $r(\theta) = 1$. The authors propose to write a cost function that limits $r(\theta)$ changes from moving too far away from 1.

Assuming A is the advantage function, the new cost function limits $r(\theta)$ changes between $1 - \epsilon$ and $1 + \epsilon$, ϵ is a hyperparameter, e.g., $\epsilon = 0.2$. The PPO cost function with respect to the clipped objective is written as follows:

$$J^{Clipping}(\theta) = \mathbb{E}[min(r(\theta)A, clip(r(\theta), 1 - \epsilon, 1 + \epsilon)A)]$$

$clip(r(\theta), 1 - \epsilon, 1 + \epsilon)$ ensures that the changes do not make the policy deviate too far from the old one, preventing overly large policy updates that could destabilize training.

This change to the surrogate loss is referred to as the *clipped surrogate loss*. In particular, it compares the clipped version and the original ratio and selects the one that results in the lesser adjustment. By using this approach, the policy estimation will be more conservative, favoring minimal changes in the policy.

Implementing this algorithm as a policy gradient is straightforward and shares similarities with TRPO. However, here the adjustment is less complex, and the gradient update can be performed using common neural network optimizers such as SGD, Adam, etc.

The following presents a pseudo code of the PPO algorithm in the style of policy gradient.

```
# PPO algorithm:
```
For (each episode) {

    Run policy $\pi_{\theta_{old}}$ for T time steps or across N trajectories

    Estimate Average Function A_t for at all T time stamps

    While (not Coverge OR max Epochs) {

        For (each mini-batch)

        Calculate $r(\theta) = \frac{\pi_\theta(a \mid s)}{\pi_{\theta_{old}}(a \mid s)}$

        $J^{Clipping}(\theta) = \mathbb{E}[min(r(\theta)A_t, clip(r(\theta), 1 - \epsilon, 1 + \epsilon)A_t)]$# Compute clipped surrogate loss.

        $\theta_{new} = arg\ \max\limits_{\theta} J^{Clipping}(\theta)$# Update θ using an optimizer to maximize $J^{CLIP}(\theta)$

        }

    }

}

PPO as Actor-Critic Model

PPO can be implemented as an actor-critic model as well. Later, we will explain the actor-critic model, but since your brain is too hot with too much mathematics, we describe the formalization along with the pseudo-code of the algorithm. If you have difficulties understanding this algorithm, check back again after you read about actor-critic architecture.

The actor network is responsible for getting the current state as input and outputs the probability distributions over possible actions that can be taken by the agent. It is also possible to run multiple actors in parallel on CPUs. The actor loss is the $J^{Clipping}(\theta)$ which we have already explained. The critic network is responsible for evaluating the quality of the actions taken by the actor. Its input is (i) the current state of the environment and (ii) the action taken by the actor, and its output is an estimation of the expected reward.

Assuming we have n sequences of states and actions, the critic loss is shown L_θ^{vf}, but to stay consistent in this chapter, we show it with $J^{VF}(\theta)$. This loss is MSE of actual return/reward (V_{actual}) minus the predicted return (critic value) ($V_{estimated}$) acquired from the network:

$$J^{VF}(\theta) = \frac{1}{n} \sum (V_{actual} - V_{estimated})^2$$

If both actor and critic network share parameters, to ensure exploration[15], an entropy bonus (S) will be added to the cost function of PPO. To adjust the importance of Critic loss and entropy bonus, each receives a coefficient c_1, and c_2. The following presents the cost function of PPO.

$$J^{Clipping+VF+S}(\theta) = \mathbb{E}[L^{Clipping}(\theta) + c_1 J^{VF}(\theta) - c_2 S\pi_\theta]$$

Assuming we have N parallel actors, each collecting T times of data, the surrogate loss can be used along with SGD to optimize NT numbers of data points. The following presents the actor-critic version of PPO.

```
# PPO algorithm in actor-critic style:
for (i =1 ... I){ # I is the maximum number of times the algorithm updates the policy
network
        for (actor =1 ... N){
            execute policy π_θ_old for T times
            compute advantage estimates: A_1,...A_T   # A_1,...A_T are the advantage estimates
                                    for each timestep in the trajectory of length T
        }
        Optimize J^Clipping+VF+S(θ) with K epochs and minibatch size: M ≤ NT
        θ_old ← θ
}
```

Actor-Critic Methods

We have explained that Policy gradient methods rely on the policy, and this reduces the need to work with the state-value function. Despite being faster to converge and able to learn stochastic policies, they (Policy gradient algorithms) are associated with two challenges: (i) there is a high likelihood that they converge to a local optimum that is not close to the global optimum, (ii) they are prone to high variances.

There is another group of RL models that combine both policy-based and value-based methods' advantages. These models are referred to as Actor-Critic models. In Figure 13-31, a Venn diagram is used to present RL methods categories.

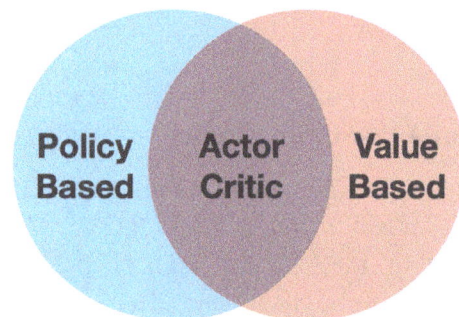

Figure 13-31: Three common architecture of RL methods.

---

[15] Later in the SAC algorithm explanation, we describe the reason for using the entropy bonus.

Actor-critic methods split the model into two components. The *actor* is responsible for computing the action. The *critic* is responsible for producing Q-values for the computed actions. In simple words, the actor controls how an agent behaves, and the critic evaluates how good the actor's actions are. The actor uses the policy to *select* the action, and thus, it is policy-based, and the critic *evaluates* the action by the policy.

Some methods can be considered both actor-critic and policy-based methods. For example, in REINFORCE with baseline, we use the state-value function as a baseline to reduce the variance of the policy gradient estimates. Since the value function is used only as a baseline and not to update the policy directly, many consider it part of policy gradient methods. However, others argue it could belong to the Actor-Critic group because the incorporation of a value function (the baseline) to reduce

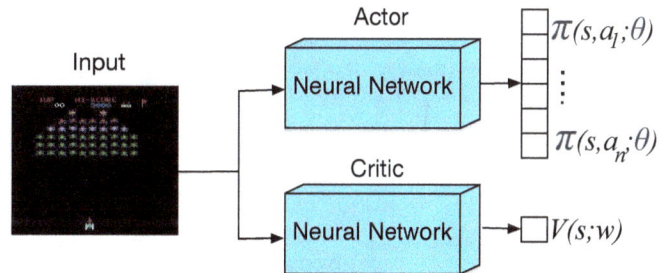

Figure 13-32: A common architecture that is used in most actor critics models. Parameters could be shared or not shared between networks.

variance is similar to how the critic is used in Actor-Critic methods. Therefore, it combines elements of both policy-based (actor) and value-based (critic) methods.

In all actor-critic algorithms that we will describe in the remainder of this chapter, two or more *neural networks* are used. The critic network updates the value function parameters (shown with w or θ_v). The actor network is responsible for updating the policy parameters (shown with θ) based on the direction decided by the critic.

Figure 13-32 presents the architecture of a simple actor-critic model, and details of the neural network are not presented here for the sake of simplicity.

Asynchronous Advantage Actor-Critic (A3C)

One of the popular actor-critic models is the A3C, which was introduced in 2016 [Mnih '16]. The actor is a neural network that identifies the policy, and the critic is another neural network that identifies the value function.

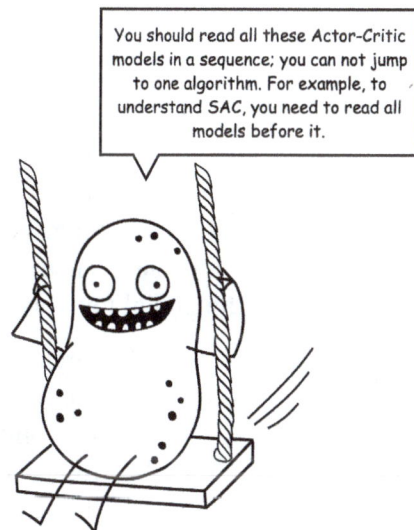

You should read all these Actor-Critic models in a sequence; you can not jump to one algorithm. For example, to understand SAC, you need to read all models before it.

Earlier, we used experience replay to reduce variance, but it is limited to off-policy algorithms and cannot be used for on-policy algorithms. A3C uses several concurrent actors to generate multiple experiments. Since they all run in parallel (as threads), they are called asynchronous actors. Having several agents experiment in parallel, each exploring different parts of the environment, reduces the risk of the agent staying in a local optimum. A3C is composed of different critics paired with actors.

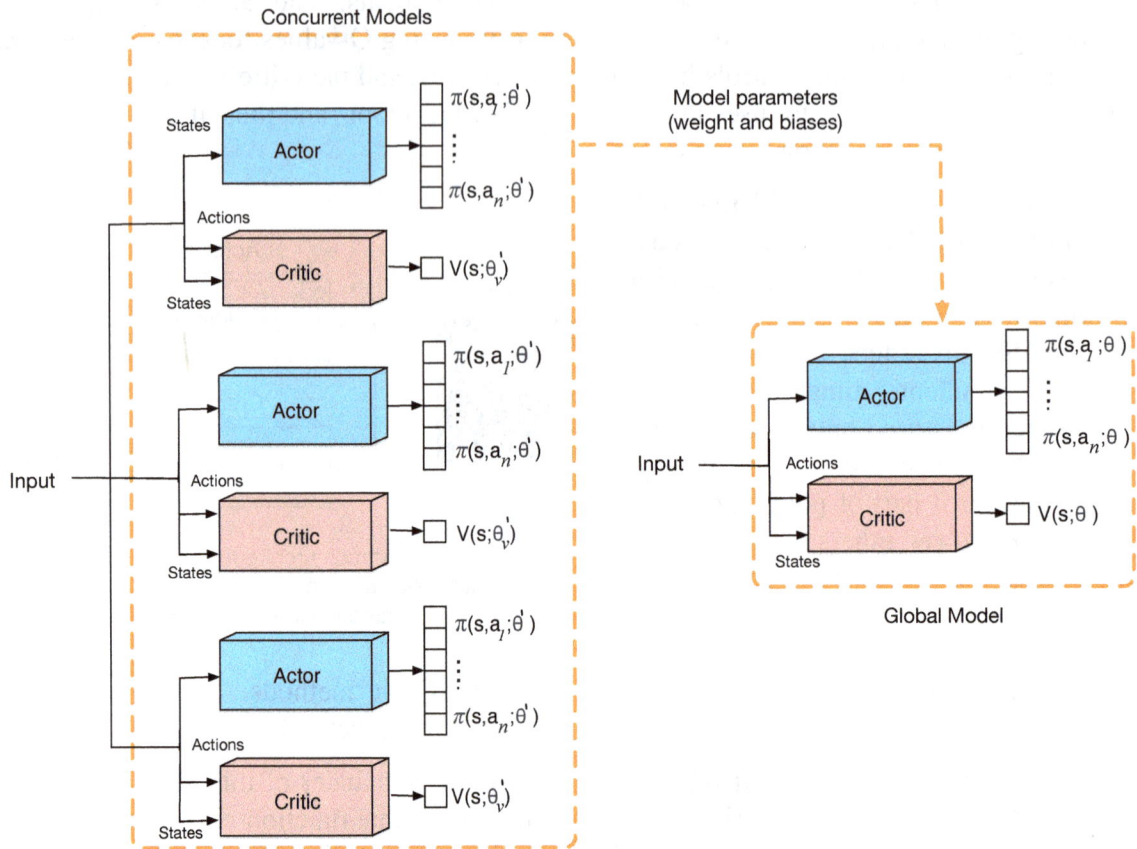

Figure 13-33: A3C Architecture. Note that no parameters except the input environment are shared between concurrent models.

The actors interact with the environment and generate experiences to learn the policy, while the critics use these experiences to update the network's value estimates. Note that actors may run more frequently than the critics, and they do not necessarily run concurrently or for the same amount of time.

After a certain number of experiments or a certain amount of time, concurrent actors and critics update model parameters, known as a *global network*. The actors and critics do not update the global network at the same time. Typically, they update the global network at different times and with different frequencies. Then, the parameters of all concurrent networks synchronize their parameters with the global network.

Figure 13-33 presents the architecture of A3C model. The authors propose asynchronous SARSA, asynchronous Q-learning, and asynchronous n-step boosting Q-learning, which returns with bootstrapping to learn the policy and value functions. Using n-step reduces the variance but adds to the bias. Here, the critic is used to mitigate the impact of bias. Due to the use of several actors, A3C reduces the bias more than REINFORCE with baseline as well.

A3C maintains the policy $\pi(a_t | s_t; \theta)$ and an estimate of the value function $V(s_t; \theta_t)$, and the policy and value estimation are updated based on the average negative log-likelihood of the predicted action, weighted by the advantage of taking that action in the environment. Therefore,

900

the policy cost function is computed as $J_{A3C} = - log\ \pi(a_t | s_t; \theta')A(s_t, a_t; \theta, \theta_v)$. Here $\pi(a_t | s_t; \theta')$ represents the predicted action with thread-specific actor parameters θ'. $A(s_t, a_t; \theta, \theta_v)$ represents the advantage function, where θ presents global actor parameters and θ_v presents parameters of the global critic. Typically, the overall loss also includes a value loss and an entropy term to encourage exploration.

Each agent in A3C calculates the gradient of the cost function based on its interaction with the environment. These gradients are then asynchronously applied to the global network to update its parameters, improving the network's ability to optimize the policy and predict future rewards.

After A3C, OpenAI introduced A2C [Mnih '16], a synchronous version of A3C. A2C is a deterministic implementation that waits for each actor to finish its experiment before performing an update. Then, it averages over all of the actors. One advantage of this method is that it can make effective use of GPUs. The A2C implementation is more cost-effective than A3C when using single-GPU machines and is faster than a CPU-only A3C implementation when using larger policies.

Deep Deterministic Policy Gradient (DDPG)

The Deep Deterministic Policy Gradient (DDPG) [Lillicrap '15] is an off-policy algorithm that is similar to DQN, but it can handle continuous action space. Most real-world problems, such as robot movements, self-driving vehicle movements, or controls, require dealing with continuous action space. Similar to DQN, the DDPG has an experience replay (to train the action-value function in an off-policy manner and mitigate the i.i.d issue). Also, similar to DQN, at each epoch, the agent gets a mini-batch of data from experience replay and uses it as the input for the policy network.

DQN learns from past experiences stored in a replay buffer, where it samples random batches of experiences to break the correlation between consecutive experiences. DDPG also uses a replay buffer to store past experiences and sample random batches of trajectories from this buffer. This approach helps stabilize the training process and improve the policy's performance in continuous action spaces.

Figure 13-34 visualizes the architecture of the DDPG. As in other actor-critic models, the actor decides the action based on the current state, and the critic is used to evaluate state and action pairs. It uses two target networks, one for the actor (policy) and one for the critic (Q-value), to stabilize the learning process and converge the model faster. However, unlike DQN, it does not directly copy the weight. Instead, it uses a soft update, which we will explain later.

In total, DDPG has four neural networks, as follows:
- Actor, which is a deterministic policy network. Its parameters are specified by θ^μ.
- Critic, which is a value network or Q network. Its parameters are specified by θ^Q.
- Target actor or policy network. Its parameters are specified by $\theta^{\mu'}$.
- Target critic or value network. Its parameters are specified by $\theta^{Q'}$.

The actor receives the vector of states as input and provides a vector of actions as output. The input to the critic is a vector of states and a vector of actions, and the output is a prediction of the expected future reward, i.e., $Q(s, a)$. The two target networks, similar to other target networks,

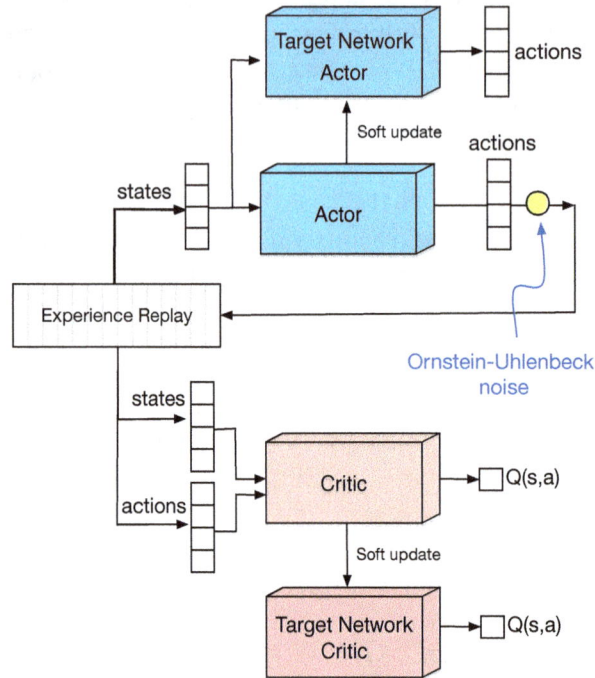

Figure 13-34: DDPG architecture. The content of experience replay comes from agent interaction with the environment, but for the sake of brevity, we do not show it in this figure.

are delayed-time replicates of the original actor and critic networks, and they are used to stabilize the main actor and critic networks.

Critic (Value) network update and loss: Similar to other Q-learning methods, the Q-value (y_i) is computed with the following Bellman equation: $y_t = r_t + \gamma Q'(s_{t+1}, \mu'(s_{t+1}|\theta^{\mu'})|\theta^{Q'})$. The updated Q-values are computed using the target actor and the target critic networks.

The DDPG minimizes the loss between the current Q-value estimate and the target (updated) Q-value with Mean Square Error (MSE) as presented in the following:

$$L(\theta^Q) = \frac{1}{N} \sum_t (Q(s_t, a_t | \theta^Q) - y_t)^2$$

Policy network update and loss: The policy is presented as $\mu(s|\theta^\mu)$, which is specified by mapping the state to the action, and its objective is to maximize the expected return:

$$J(\theta^\mu) = \frac{1}{N} \Sigma_t [Q(s, a | \theta^Q) | s = s_t, a_t = \mu(s_t)]$$

The actor (policy) network cost function is computed as a derivative of the objective with respect to its parameters, and it is formalized as follows:

$$\nabla J(\theta^\mu) = \frac{1}{N} \Sigma_t (\nabla_{\theta^\mu} Q(s, a | \theta^Q) | s = s_t, a = \mu(s_t | \theta^\mu)])$$

In other words, to determine the policy, the actor is trained and uses the gradient of the expected cumulative reward with respect to the actor's parameters.

Target network update: DDPG target networks are updated by making a "soft update" (not just copying, as was the case in DQN) from the weights of policy (actor) and value (critic) networks. The process of the soft update includes blending the parameters of the target networks with the parameters of the main networks using the following update rule:

$$\theta^Q \leftarrow \tau\theta^Q + (1 - \tau)\theta^{Q'}$$
$$\theta^{\mu'} \leftarrow \tau\theta^\mu + (1 - \tau)\theta^{\mu'}$$

Here $\tau < 1$ is a hyperparameter controls the update rate of the target network. The smaller τ value means the target network is updated less frequently.

Exploration: since DDPG deals with continuous action space, its exploration is not as easy as discrete action spaces. In discrete action spaces, there are fairly simple methods, such as epsilon greedy, which uses probabilistic random actions to implement the exploration. However, in continuous action space, exploration is done by adding noise to the action. To add the noise, the author of DDPG uses a process called the Ornstein-Uhlenbeck Process [Uhlenbeck '30]. The noise is added to the actions that the output of the actor network.

Twin Delayed Deep Deterministic Policy Gradient (TD3)

Twin Delayed DDPG or TD3 [Fujimoto '18] is an off-policy algorithm designed for continuous action space and introduced after DDPG to solve its limitations. DDPG is very sensitive to its hyperparameters, and similar to other Q-Networks, the critic overestimates the Q-values. This overestimation results in wrong policy identification, and as we stated before, wrong policy estimation is a kiss of death for the RL models. TD3 handles the DDPG issues by introducing three modifications to the DDPG. These three changes are listed as follows, and we will explain them in more detail.

(i) *Clipped Double Q-learning:* TD3 learns two Q-functions instead of one, and then it chooses the smaller of the two Q-values to form the targets in the objective functions (which is based on the Bellman equation). Since it is learning two Q-functions, it is referred to as Twin Delayed DDPG.

(ii) *Delayed Policy Updates:* It updates the policy (actor) network less frequently than value (critic) networks. The authors recommend updating the policy network once for every two Q-value functions (value network) updates.

(iii) *Target Policy Smoothing:* To stabilize the training process and reduce the likelihood of Q-value overestimation, TD3 adds a clipped noise (sampled from a normal distribution) to the target action during critic updates. This noise is the implementation of exploration and makes it harder for the policy to exploit Q-function errors by smoothing out Q-value estimates along with changes in action.

In total, TD3 has six networks. Two main critics, two target critic networks, one main actor, and one target actor.

Clipped Double Q-learning: We use the Bellman equation to compute Q-value (y) for DQN as follows: $y = r + \gamma Q(s', \arg max_{a'}(Q(s', a')))$. In the context of actor-critic, we update the current policy (ϕ refers to actor parameters) with respect to the Q-value of the critic (θ refers to critic parameters), and thus, we can write the Q-value calculation as $y = r + \gamma Q_\theta(s', \pi_\phi(s'))$.

Authors of TD3 propose to use two Q-networks (critic) and then take the minimum of estimates from the two Q-functions as the final Q-value:

$$y = r + \gamma \min_{i=1,2} Q_{\theta_i}(s', \pi_\phi(s'))$$

This approach helps to reduce overestimation bias in Q-values.

Besides, TD3 employs two target critic networks in addition to the two main critic networks. By using two independent target critics and taking the minimum of their estimates, TD3 further reduces overestimation bias and improves the stability of the learning process. The target critics are updated more slowly than the main networks, providing a more stable learning target

Delayed Policy Updates: The second change of TD3 is updating the policy network with some delays. DDPG also uses τ hyperparameter to update the target policy networks. The same parameter with the same equation is also in use for TD3, i.e., $\theta' \leftarrow \tau\theta + (1 - \tau)\theta'$ and $\phi' \leftarrow \tau\phi + (1 - \tau)\phi'$. This is known as a soft update in DDPG.

However, in addition to this approach, the policy network (actor) is updated less frequently than the value network (critic). The author recommends updating the policy network after d updates on the value network, and d is a hyperparameter specified by the user. This delay limits the likelihood of having a policy update for an unchanged critic, which leads to a wrong policy update. This technique helps to stabilize training by ensuring that the policy is updated based on a more accurate critic.

Target Policy Smoothing: Ideally, taking similar actions should lead to similar Q-values. However, in deep RL, the function approximation inaccuracies could cause high differences between Q-values for similar actions. This makes deterministic policies inaccurate. To mitigate this, TD3 introduces target policy smoothing, where random noise (a.k.a. clipped noise) is added to the target policy to smooth out the Q-values over similar actions. The target policy is modified as follows:

$$y = r + \gamma \min Q_{\theta'}(s', \pi_\phi s' + \epsilon)$$

Here, ϵ is the clipped noise, and we will explain more about it shortly. This technique helps to regularize the learning process and make the algorithm more robust to noise in the action space. In other words, it helps in making deterministic policies more robust against the function approximation errors.

For example, imagine the room tidying robot that is moving a baby's toy to the target box in a cluttered room environment. When the robot encounters a toy on the ground, it moves it toward the toy box located at a known place in the room. The action is to *rotate the robot wheels* and push the toy toward the toy box. The angle of movement could be between -90 to 90 degrees, and *actions* could be movement angles as follows 45, 50, and 55. They all seem to have similar Q-value (extracted from the critic), e.g., all inside the toy box Q-value =1 (let's assume outside toy box Q-value=0), but instead of moving the toy to the box, some toys move it to another place outside the toy box. In other words, the actual Q-value = 0, but the estimated Q-value =1.

To enforce similar actions to have similar Q-values, authors add a Gaussian noise ϵ with a mean of zero, and its value is clipped between -c and c. In other words, ϵ is sampled from a Gaussian distribution that has a mean of 0, and its value will be $-c < \epsilon < c$. Therefore, the target Q-value y, can be written as follows: $y = r + \gamma Q_{\theta'}(s', \pi_\phi(s') + \epsilon)$, and $\epsilon \sim clip(\mathcal{N}(0,\sigma), -c, c)$.

In this equation $\mathcal{N}(0,\sigma)$ refers to a Gaussian distribution with a mean of zero. Note that the Gaussian noise is added to the action input of the main critic network during the exploration phase but not added to the target critic network.

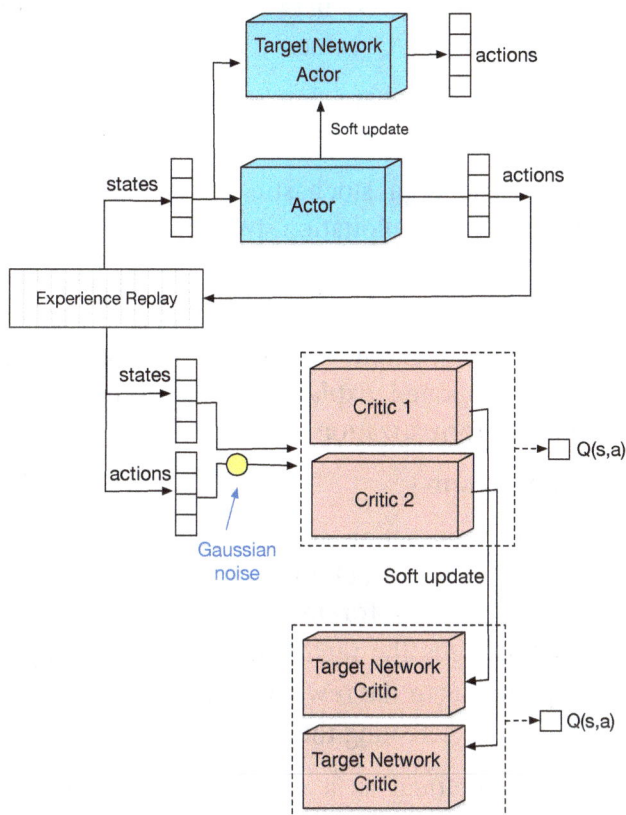

Figure 13-35: TD3 architecture. The content of experience replay comes from agent interaction with the environment, but for the sake of brevity, we do not show it in this figure.

Figure 13-35 presents the architecture of TD3, based on our understanding of the TD3 model. Understanding the rationale behind these cost functions and the process of these actor-critic models can be challenging. If you are having trouble, a practical approach involves looking at open-source implementations of these models and experimenting yourself. This hands-on experience can greatly enhance your understanding beyond theoretical study.

Soft Actor-Critic (SAC)

Soft Actor-Critic (SAC) [Haarnoja '18] is an off-policy algorithm with an experience replay that employs entropy regularization to encourage exploration by using a stochastic policy. Sampling from real-world data is complex and expensive. For example, teaching a robot to move properly requires several experiments and millions of data collection. The other issue with sampling from real-world data is the sensitivity of the algorithm to hyperparameters. SAC addresses stability and sample efficiency by using a *twin-critic mechanism* and an *entropy-enhanced objective*.

Similar to other actor-critic models, SAC has one policy network and four Q-networks. In particular, it is composed of five neural networks, each described as follows:

One actor network (π_ϕ): This is the policy network that outputs a probability distribution over actions. SAC doesn't use a target actor network because it uses a reparameterization trick to sample actions, allowing direct optimization of the expected value.

Two Q-networks ($Q_{\theta 1}$ and $Q_{\theta 2}$): SAC uses two Q-networks to mitigate overestimation bias, similar to TD3. These networks estimate the Q-values.

Two target Q-networks ($Q_{\theta'1}$ and $Q_{\theta'2}$): These are slowly updated copies of the main Q-networks, used for more stable learning.

Key features of SAC include the use of a stochastic policy for better exploration, Entropy regularization to balance exploration and exploitation, two Q-networks to reduce overestimation bias (similar to TD3), automatic tuning of the temperature parameter for the entropy term (we will explain it shortly) and off-policy learning for sample efficiency.

Before we explain the details of this algorithm, we should be familiar with the entropy bonus, which is used by PPO, but we postponed explaining it until now. Then, we explain each component of the SAC along with its formalization.

Entropy Regularization (Entropy Bonus)

An agent might get a good reward for an action and thus repeat this action frequently, and as a result, this behavior shows that the agent is stuck in a local optimum. We have learned early in this chapter that 'exploration' is a remedy for this issue. *Entropy regularization* is used to implement exploration. Entropy measures the randomness or uncertainty in the policy's action distribution. By adding an entropy bonus to the reward function, the agent is incentivized to take more diverse actions rather than always selecting the action with the highest predicted reward.

A continuous space reward can be defined as a distribution conditioned on the state of the environment. To encourage exploration and prevent premature convergence to suboptimal policies, we can modify this reward by incorporating the entropy. By incorporating entropy, H, of the agent's policy, i.e., $\pi(a|s)$, we can write the reward function as follows: $r(s,a) = r_{orig}(s,a) + \alpha * H[\pi(a|s)]$, where r_{orig} is the original reward function α (or β in some implementations) known as the *temperature parameter*, which adjusts the weight given to the entropy term in the overall reward calculation. The temperature is a hyperparamter to specify the relative importance of entropy impact.

Soft Q-value

The authors incorporate the entropy, H, to define what is known as the soft Q-value. The soft Q-value can be formulated by integrating the entropy term into the standard Q-value calculation, which encourages exploration by penalizing certainty in the policy. This is shown as the red part in the soft Q-value Belman equation.

$$Q_{soft}^\pi = r(s_t, a_t) + \mathbb{E}_{s_{t+1}, a_{t+1} \sim \pi}[\sum_{t=0} \gamma(r(s_{t+1}, a_{t+1}) + \alpha H(\pi(a_{t+1}|s_{t+1})))]$$

Here α is the temperature parameter, which serves as a regularization parameter, and it scales the importance of the entropy relative to the reward. The Q-value can be simplified by using the entropy equation, explained in Chapter 3, and written as follows:

$$Q^{\pi}_{soft} = r(s_t, a_t) + \mathbb{E}_{s_{t+1}, a_{t+1} \sim \pi} [\sum_{t=0} \gamma (r(s_{t+1}, a_{t+1}) {\color{red}- \alpha log \pi(a_{t+1}|s_{t+1})})]$$

The only change we see here is the addition of the entropy bonus at the end of the Q-value computation (highlighted in red color), which encourages the agent to explore a variety of actions.

Critics and Target Critics

SAC typically uses two critics for learning, which helps in reducing overestimation bias. The task of both SAC critics, similar to other actor-critic networks, is to evaluate the policy, which is identified by the actor. Each critic outputs a Q-value estimate for a given state-action pair.

In other words, their objective is to learn the Q-value, and the Q-value can be learned by minimizing the critic's objective function, which is written as follows:

$$J_Q(\theta_i) = \mathbb{E}[(Q_{\theta_i}(s_t, a_t) - y_t)^2]$$

where: $y_t = r(s_t, a_t) + \gamma[min_j Q_{\theta_{target_j}}(s_{t+1}, a_{t+1}) - \alpha \log(\pi_\phi(a_{t+1}|s_{t+1}))]$, θ_i represents the parameters of the ith Q-function (i = 1, 2), and θ_{target_j} represents the parameters of the jth target Q-network.

Similar to previous actor-critic models, SAC uses two critics and computes the minimum between their Q-value estimates to reduce overestimation bias. Each of them has a target critic network to stabilize learning. Similar to TD3 for target network SAC uses a soft update, i.e., $\tilde{\theta} \leftarrow (1 - \tau) \tilde{\theta} + \theta \tau$. Here $\tilde{\theta}$ represents the parameters of the target network, θ represents the parameters of the main network, and τ is a small value (typically much less than 1) that determines the update rate, allowing for a gradual shift of the target network towards the main network.

The use of two critics and taking their minimum helps in providing a more conservative Q-value estimate, which is crucial in preventing the overestimation of action values. In other words, by using the minimum Q-value between the two critics in the target calculation, the critics are trained alternately, each using the other's Q-value in the target calculation to further stabilize the learning process.

The SAC value function is derived indirectly and is used to stabilize the Q-function updates, as follows:

$$V_\theta(s_{t+1}) = \mathbb{E}_{a_{t+1} \sim \pi_\phi} \left[Q_\theta(s_{t+1}, a_{t+1}) - \alpha \log \pi_\phi(a_{t+1}|s_{t+1}) \right]$$

This implicit value function incorporates both the Q-values and the entropy term, which encourages exploration by penalizing the predictability of the policy. The use of this soft value function distinguishes SAC from other actor-critic methods.

Actor

To compute the policy, SAC updates the policy by using the exponentiation of soft Q-values (which come from the critic) and adjusting them with the temperature parameter α. The temperature parameter α controls the trade-off between exploration and exploitation by scaling the Q-values. The policy from the soft Q-value is computed as follows:

$$\pi_\theta(a \,|\, s_t) = \frac{exp(\frac{1}{\alpha} Q_\theta(s_t, a))}{Z_\theta s_t}$$

Here, $Z_\theta(s_t)$ is a normalization term (a.k.a., partition function[16]) used to construct the distribution. This function is defined as:

$$Z_\theta(s_t) = \sum_{a'} \exp\left(\frac{1}{\alpha} Q_\theta(s_t, a')\right)$$

Therefore, we can write the policy with the normalizing factor as follows:

$$\pi_\theta(a \,|\, s_t) = \frac{\exp\left(\frac{1}{\alpha} Q_\theta(s_t, a)\right)}{\sum_{a'} \exp\left(\frac{1}{\alpha} Q_\theta(s_t, a')\right)}$$

The reason for exponentiating the soft Q-values is to transform these values into a distribution that represents the optimal policy for selecting actions. This transformation assigns higher probabilities to actions with higher Q-values, which correspond to higher expected returns.

Policy Update and Reparameterization Trick

SAC updates the policy based on the following cost function for actor:

$$J_\pi(\phi) = \mathbb{E}_{s_t \sim D}\left[D_{KL}\left(\pi_\phi(.\,|\,s_t)\,||\,\frac{exp(\frac{1}{\alpha} Q_\theta(s_t, .))}{Z_\theta(s_t)}\right)\right]$$

However, we can not minimize this expression directly due to the computational cost of KL-divergence. TRPO has a similar issue and uses a surrogate loss. SAC authors use a reparameterization trick[17] and reparameterize the policy using a neural network transformation.

To apply the reparameterization, authors rewrite action a_t as a function of input noise vector ϵ_t (sampled from a fixed distribution such as Gaussian distribution) parameterized by state s_t, i.e., $a_t = f_\phi(\epsilon_t; s_t)$. Then, the objective function could be rewritten by incorporating the reparameterization trick as follows:

$$J_\pi(\phi) = \mathbb{E}_{s_t \sim D, \epsilon_t \sim \mathcal{N}}[log\ \pi_\phi(f_\phi(\epsilon_t; s_t)\,|\,s_t) - Q_\theta(s_t, f_\phi(\epsilon_t; s_t))]$$

Therefore, to update the actor parameters, the gradient of the policy cost function with respect to policy parameters ϕ, is computed as follows:

$$\hat{\nabla}_\phi J_\pi(\phi) = \nabla_\phi log\ \pi_\phi(a_t\,|\,s_t) + (\nabla_{a_t} log\ \pi_\phi(a_t\,|\,s_t) - \nabla_{a_t} Q(s_t, a_t)) \nabla_\phi f_\phi(e_t; s_t)$$

Here, $\nabla_\phi log\ \pi_\phi(a_t\,|\,s_t)$ is the gradient of the log-probability of the policy with respect to policy parameters, indicating how changes in parameters affect the likelihood of selecting action a_t at in state s_t. $\nabla_{a_t} log\ \pi_\phi(a_t\,|\,s_t)$ is the gradient of the log probability with respect to the action,

---

[16] The partition function is a concept borrowed from statistical mechanics, often used in probabilistic models and reinforcement learning to normalize probability distributions.

[17] The basic idea behind the reparameterization trick is to separate the randomness in the model from the parameters that are being learned. As a result of applying the reparameterization trick, the gradient computation will be more efficient.

indicating how small adjustments in the action affect its probability. $\nabla_{a_t} Q(s_t, a_t)$ is the gradient of the Q-function with respect to action. $\nabla_\phi f_\phi(e_t; s_t)$ is the gradient of the function that generates actions from policy parameters, noise, and state, which is important for implementing the reparameterization trick in stochastic policies.

SAC Architecture

SAC can be used both for discrete and continuous action space. While SAC can technically be adapted for discrete action spaces by modifying the policy output to a softmax over actions, it's more commonly associated with and advantageous for continuous action spaces due to its ability to handle high-dimensional action spaces efficiently. If it is used for discrete action spaces, the actor input is a set of states $(s_1, s_2, \ldots)$, and it outputs a probability distribution over discrete action, $p(a_1 | s), p(a_2 | s), \ldots$ The critic receives a set of states $(s_1, s_2, \ldots)$ as input, and its output will be the estimated value of the state-action pair, i.e., Q-values for each action, $Q(s | a_1), Q(s | a_2), \ldots$

If we use SAC for continuous action space, the critic receives (i) a vector of states and (ii) a vector of actions as input, and it outputs a Q-value. The actor receives a vector of states and returns two parameters for the Gaussian distribution, i.e., $\mu_\phi(s)$ and $\sigma_\phi(s)$. Both in discrete and continuous scenarios, the "vector of states" and "vector of actions" can imply various representations (e.g., images, feature vectors) depending on the problem domain. Figure 13-36 presents the architecture of SAC for continuous action space.

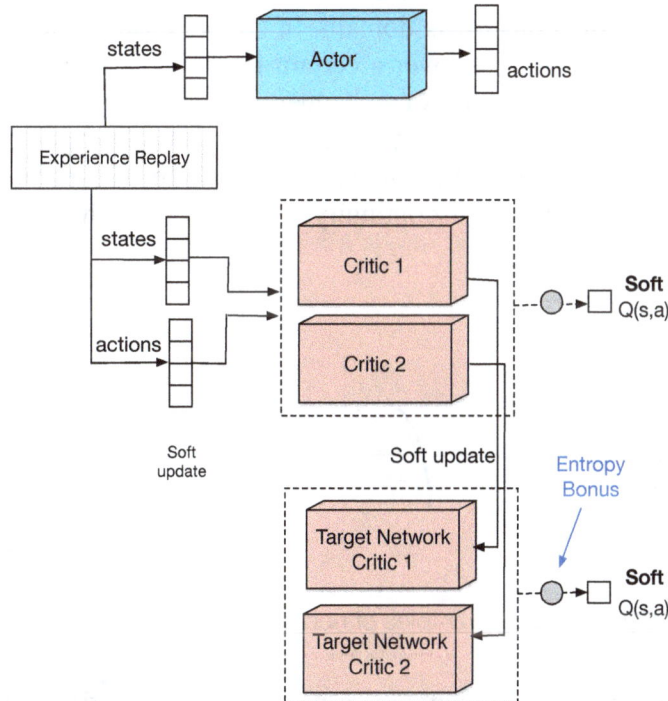

Figure 13-36: SAC architecture. The content of experience replay comes from agent interaction with the environment, but for the sake of brevity, we do not show it in this figure.

Dreamer Models

The previous RL models we have discussed focus on continuous or discrete space modeling. They are successful in a limited number of applications, such as playing Atari games, physical simulations, robot control, and credit assignment. However, they are incapable of solving a wide variety of RL tasks. Dreamer is an effort to bridge this gap by creating a world model through video (series of images) observation, which it uses to learn long-term behaviors. We think Dreamer and other models after it, such as Octo [Team '24] aim to be a foundational model for RL, similar to SAM for image segmentation and many of the LLMs we described in Chapter 12.

The goal of Dreamer models is to use a series of images (video) to model or simulate a system by observing the system's behavior. This is the only model-based RL algorithm we explain in this book. Despite its complexity, it is worth learning because it introduces interesting ideas that could help us build our RL model, such as using the encoder, Straight-Through gradient, and categorical representations.

At the time of writing this chapter (Q2-2023, and revising it again Q3-2024), it has three versions, and we explain them in a sequence. Before we explore Dreamer models, we need to learn a simple gradient method, the Straight-Through Estimator (STE). Since we deal with gradients in neural networks, it is called the *Straight-Through Gradient (STG)* [Bengio '13].

Straight-Through Gradient

The basic idea behind STG is to replace the non-differentiable operation with a differentiable one during backpropagation and then apply a correction factor to ensure the gradient is still accurate.

In mathematics, a function is non-differentiable if it does not have a well-defined derivative at some points or over some intervals of its domain. In simple terms, if a function or operation is non-differentiable at a certain point or over a certain interval, its rate of change or slope is not well-defined at that point or interval. This can happen, for example, when a function has a sharp corner, a vertical tangent, or a discontinuity in its domain. For example, Figure 13-37 presents three samples of non-differentiable functions. Several conditions make a function non-differentiable, including oscillation, discontinuity, infinite limit, sharp corner, etc., which we won't go into their details here.

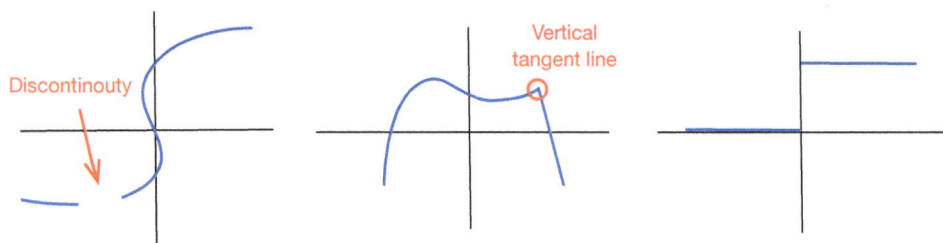

Figure 13-37: Examples of non-differentiable functions.

STG (Straight-Through Gradient Estimator) is a method used to estimate the gradient of a non-differentiable function during backpropagation in neural networks. It approximates the gradient by using a continuous function that has a well-defined gradient. For example, check the step function on the rightmost plot (step function) of Figure 13-37. It is a non-differentiable function because $f(x) = 1$ for $x > 0$ and $f(x) = 0$ for $x \leq 0$. The STG can convert this function to $g(x) = x$ for $if\ |x| < 1$, otherwise $g(x) = sgn(x)$[18].

In this process, the STG can use the original function $f(x)$ during the forward pass and $g(x)$ (the differentiable approximation) during the backward pass for gradient computation. It's important to note that the STG does not actually make the non-differentiable function differentiable; rather, it provides an approximation to enable gradient-based optimization during training.

Dreamer v1

Dreamer v1 [Hafner '19] consists of three components that can operate in parallel: (i) using past experiences to train the model, which is referred to as "learning the world model", (ii) using the trained world model to build the critic (value) network and actor network, which is referred to as "learning behavior in the imagination", and (iii) interacting with the environment to collect more data and expand the experience dataset, which is referred to as "environment interaction".

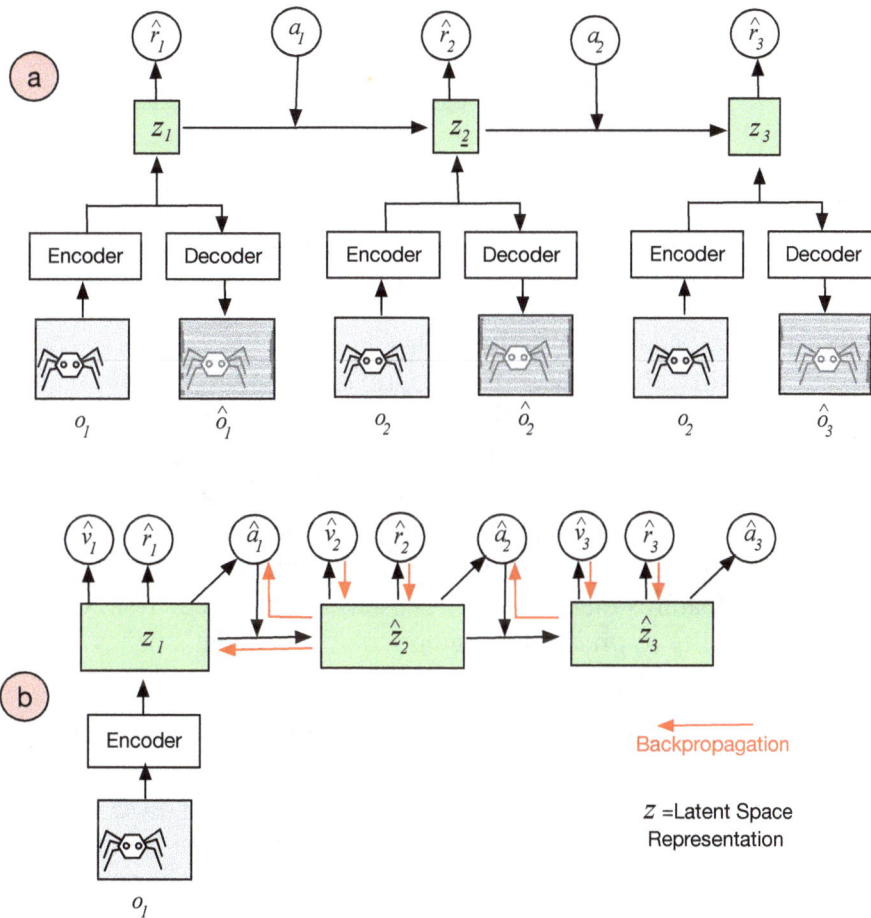

Figure 13-38: Two stages of the Dreamer v1 model.

---

[18] A sign function (presented as sgd) is a function that returns the sign of a real number, it can be formalized as $for (x < 0)\ sgn(x) = -1$, $for (x > 0)\ sig(x) = 1$

Did you see it? Again, Autoencoder is used for compression, but nobody tells this legendary neural network is a compression method.

(i) *Learning the world model:* to implement this component of Dreamer v1, a series of image observations $o_1, o_2, \ldots$ are fed into an encoder (the encoder is a Convolutional layer). The result of the encoder is the representation of the observation in the latent space (hidden representation of the observation), i.e., z, as shown in Figure 13-38 (a).

Then, the agent takes action; for example, action a_1 is taken for observation o_1, and the reward $\hat{r}_2$ will be acquired. For the next observation o_2, the previous outputs of observations from the encoder will be used along with the next action, i.e., a_2, to compute the next reward, i.e., $\hat{r}_3$. This is similar to an RNN approach because the dataset of past experiences is used to acquire a sequence of encoded observations (states), actions, and rewards. Everything seems fine here, but one might ask why Dreamer encodes the input image. The authors used the encoder to enable the model to be trained with fewer resources and even on one single GPU. Besides, via the encoder, the model learns important features that facilitate long-term video prediction. One feature of the Dreamer model is the capability to *predict a long horizon* of an agent's behavior, and the use of the encoder enables the model to implement this feature.

Additionally, the model reconstructs the image (observation or states) using a decoder to facilitate representation learning. In particular, the decoder reconstructs the original image from the latent representation z. Reconstructing from z enables the model to compare the original image with the reconstructed image and use this comparison to fine-tune the model.

(ii) *Learning behaviors by latent imagination:* The second stage focuses on learning actions and values from predicting hypothetical trajectories in the latent space world model. This stage focuses on predicting the sequence of model states in compact representations (latent space, which is encoder output). As it is shown in Figure 13-38 (b), for each state, the corresponding reward and value will be predicted. Similar to other actor-critic models, it uses two networks, i.e., actor and critic. The actor network predicts the next action ($\hat{a}$), and the critic predicts values ($\hat{v}$) and rewards ($\hat{r}$). To improve the actor network, it backpropagates rewards and values through

912

the sequence of model states. In simple words, at this stage, the model uses the learned world model from the previous stage to do reinforcement learning and predict the upcoming video sequences.

(iii) *Interacting with the environment:* The agent executes the learned world model by interacting with the environment, collects new experience data, and repeats the previous two steps to improve its behavior.

Dreamer v2

In Dreamer v2 [Hafner '20], discrete variables (multiple categorical variables) are used to represent each image, in contrast to the continuous variables (Gaussian variables) used in Dreamer v1. This shift allows Dreamer v2 to better capture multimodal distributions and improve sample efficiency, leading to more accurate predictions of future representations.

However, Dreamer v2 still uses continuous action spaces for many tasks. The discretization is primarily in the state representation, not in the action space. Similar to Dreamer v1, it uses the compact representation of the image (encoded via convolutional layers) learned from Markov states in a Partially Observable Markov Decision Process (POMDP).

While Dreamer v2 was initially focused on playing Atari games, its architecture is versatile and can be adapted to other tasks. We categorize it as a model-based architecture. Dreamer v2 applies some modifications to Dreamer v1, which results in the agent achieving higher performance. There are two main modifications that improve the quality of Dreamer v2 over Dreamer v1, including the use of a "vector of categorical representation" and applying "KL-balancing".

Vector of Categorical representation: In Dreamer v1, the reward predictor and image predictor are modeled as a Gaussian distribution. There is a problem with this approach: a single (unimodal) Gaussian distribution can not accurately predict the next image that has a multimodal distribution shape. Many real-world scenarios, including game states in Atari, can have multiple possible future outcomes and thus a distribution with multiple peaks or modes. To resolve this issue, the encoder of Dreamer v2 converts the image into a mixture of categorical variables and presents it as a vector of categorical representation.

In other words, Dreamer v2 replaces continuous representation of latent space z, with discrete values. The latent state z, which is the output of the encoder, is presented as 32 distributions over 32 classes each, i.e., 32×32 sparse matrix. For example, in the space invader[19] game, categories could be the players' spaceship location with 32 different classes (each class presents a position), player's spaceship statues (can shoot, energy level, etc.) in 32 different classes, enemy spaceships' status (do they shoot, their location, etc.) again in 32 classes and so forth. However,, categorical variables are non-differentiable, and to backpropagate through them, Dreamer v2 uses STG.

KL-balancing: KL balancing is a technique used in Dreamer v2 to improve the model prediction capability by adjusting the trade-off between the accuracy of the learned model and the entropy of its predictions. In particular, the authors employ a KL-divergence in the Dreamer loss function, i.e., Evidence Lower Bound (ELBO). The use of KL divergence in the loss function balances the trade-off between the reconstruction accuracy and the regularization imposed by the

---

[19] https://en.wikipedia.org/wiki/Space_Invaders

KL divergence. The KL balancing works by splitting the KL divergence term in the loss function into two parts:

(i) A *reconstruction* term that encourages the model to accurately predict observations.

(ii) A *regularization* term that prevents the model from overfitting to the training data.

By separating these terms, Dreamer v2 can apply different learning rates to each, allowing for finer control over the learning process. This technique helps prevent the model from either ignoring the prior, which would lead to poor generalization, or relying too heavily on it, which would lead to inaccurate predictions. In this context, prior refers to the distribution of the next latent state, while posterior refers to the encoded observations.

Dreamer v3

Dreamer v3 [Hafner '23] is the first model that can collect diamonds in the Minecraft game[20] without any human data. The authors tested Dreamer v3 across four visual domains, including Atari games, robot locomotion, Minecraft, and DMLab[21]. The authors explained that it has three neural networks: a *world model*, which predicts the future output of potential actions; a *critic* to identify the value of each situation; and an *actor* that learns to reach a valuable situation. All these neural networks are trained concurrently from a replay buffer, and they do not share parameters.

Dreamer v3 introduces several technical improvements over Dreamer v2, including enhancements in robustness, network architecture, optimizer, and experience replay buffer. We do not explain all the changes, but briefly, we list some of the important ones.

Robustness: Dreamer v3 uses KL-divergence to train its world model, which predicts future states of the environment. However, standard KL-divergence can sometimes lead to unstable training, especially when the predicted distributions are very confident and narrow, which can make it hard for the model to learn meaningful representations from the data. To address this, Dreamer v3 incorporates *free bits* into the KL-divergence computation. Free bits are a technique where a small value is added to the KL-divergence to prevent it from becoming too small. This effectively ensures that the model doesn't become overly confident in its predictions too quickly, allowing for a more gradual and stable learning process. In other words, it balances the KL-divergence with free bits.

In addition to that, traditional loss functions like mean squared error or Huber loss can struggle with large and varying reward magnitudes, which can cause instability in learning. Dreamer v3 uses "*symexp twohot loss*" to address this issue by transforming the predicted values using the symlog (symmetrical log) function, which compresses large positive and negative values symmetrically around zero. The following presents both symlog and symexp. Symexp is the inverse of symlog square error (a bi-symetric and logarithmic function). The symlog is used to compress the large positive and negative values. It is symmetric around the origin while preserving the input.

$$\text{symlog}(x) \doteq \text{sign}(x)\ln(|x|+1), \quad \text{symexp}(x) \doteq \text{sign}(x)\big(\exp(|x|)-1\big)$$

---

[20] https://www.minecraft.net

[21] Deepmind Lab (DMLab) is a 3D learning environment based on the Quake game: https://github.com/deepmind/lab

914

Therefore, Dreamer v3 neural networks $f(x, \theta)$ with inputs x and parameters θ learn to predict a transformed version of its targets y as follows:

$$L(\theta) \doteq \frac{1}{2} \left(f(x, \theta) - \text{symlog}(y) \right)^2, \hat{y} \doteq \text{symexp}\left(f(x, \theta) \right)$$

The twohot encoding is an extension of the onehot encoding typically used in classification tasks. In one hot encoding, a single class label is represented by a vector with all zeros except for one at the index corresponding to the class. Twohot encoding, on the other hand, represents a continuous value by distributing its probability mass between the two closest discrete bins. This results in a smoother gradient during training and helps the model to better learn continuous values. By using the symexp transformation and twohot encoding, the symexp two hot loss enables Dreamer v3 to predict rewards and values more accurately, even when these values span a wide range of magnitudes. This transformation makes the optimization landscape smoother and prevents large gradients from causing instability.

Network Architecture: The network architecture has been upgraded with block-based GRU, RMSNorm normalization, and Swish (SiLu) activation functions.

Optimizer: Dreamer v3 optimizer introduces Adaptive Gradient Clipping (AGC) and the LaProp optimizer, which help stabilize the training process. AGC uses gradient clipping but makes the clipping threshold adjustable based on the norms of the weights of the neural network layers. For example, Suppose a layer in the neural network has weights with a L_2 norm of 10. AGC with a threshold of 30% would clip gradients that exceed 3 (30% of 10). The LaProp optimizer [Ziyin '20] combines features of two popular optimizers, Adam and momentum-based methods, to provide stable and efficient training. It first normalizes the gradients using Adam and then applies momentum to the normalized gradients.

Replay Buffer Enhancements: Dreamer v3 increases replay buffer capacity and uses an online queue to manage data collection and replay, ensuring fresh data incorporation. Online queue allows for immediate incorporation of the latest experiences into the replay buffer, ensuring that the training process utilizes the most recent and relevant data. This mechanism helps maintain a balance between old and new data.

I finished learning many deep reinforcement learning models, and you see, I am still alive. YES.

Summary

In this chapter, we described reinforcement algorithms and models. First, we start with explaining concepts, including exploration versus exploitation and the Markov-decision process. Then, we describe the Bellman equation and state transition diagram. An RL problem can be formulated with transition probabilities with MDP. MDP is a sequential decision problem for an observable, random environment with a Markovian transition model. An MDP formalization is based on three elements: actions, rewards, and states. The discount parameter of MDP is used to emphasize more recent awards that delayed reward.

Next, we described the Multi-Armed Bandit problem and three approaches to resolve it, including Epsilon Greedy, Upper Confidence Bound, and Thompson Sampling. Followed by some examples of the user of Multi-Armed Bandit problem in real-world settings.

Afterward, we describe the challenges of finding optimal policy and optional define control and prediction in the context of reinforcement learning. Control is finding or generating optimal policies. Prediction is evaluating the value-function and action-value function for the given policy. Policy iteration is a loop of policy evaluation and policy improvements. Policy iteration is not scalable, and to mitigate it, we use value iteration. The difference is that value iteration truncates some policy evaluation parts of the policy iteration.

Tabular methods in reinforcement learning encompass approaches for estimating value functions and optimal policies while assuming a finite and discrete state-action space. The tabular methods we have explained in this chapter include Monte Carlo, SARSA, Q-learning, and Dyna-Q. Monte Carlo methods learn from complete episodes, updating estimates based on observed returns, but face challenges with continuous or large state spaces and can be slow to converge. They excel in episodic tasks but struggle with ongoing processes. Dynamic programming methods, like value iteration and policy iteration, leverage the Bellman equation to iteratively improve value estimates and policies. While they guarantee optimal solutions, they require a complete model of the environment and can be computationally expensive for large state spaces.

Temporal Difference (TD) learning combines aspects of Monte Carlo and dynamic programming, updating estimates based on partial experiences. SARSA, an on-policy TD control algorithm, updates Q-values using the current policy's next action. Q-learning, an off-policy method, updates Q-values using the maximum Q-value of the next state, regardless of the policy followed. Both SARSA and Q-learning can handle continuous tasks and learn online but may struggle with function approximation in large state spaces. Dyna-Q integrates direct RL with model-based planning, maintaining a model of the environment to simulate experiences and accelerate learning. This approach can be particularly effective in environments where real experiences are costly or time-consuming to obtain.

Next, we described approximation methods, including n-step methods and TD(λ), and then moved to deep reinforcement learning models. N-step methods bridge the gap between one-step TD methods and Monte Carlo methods by considering multiple future steps when calculating the update target. They allow for a trade-off between bias and variance in the estimates, with larger n reducing bias but potentially increasing variance. TD(λ) generalizes n-step methods using eligibility traces, providing a continuous spectrum between TD and Monte Carlo methods.

916

Afterward, we described Function approximation, which allows for generalization across continuous or large discrete state spaces by representing value functions or policies as parametric functions, enabling learning in environments where tabular methods are impractical or impossible.

First, we described DQN, DDQN, Dueling Networks, and Prioritized Experience Replay. Deep Q-Network (DQN) combines Q-learning with neural networks to handle high-dimensional state spaces, using experience replay and target networks to stabilize learning. Double DQN (DDQN) addresses DQN's overestimation bias by decoupling action selection and evaluation. Dueling Networks separate state value and action advantage estimations, improving learning efficiency for states where actions don't significantly impact the outcome. Prioritized Experience Replay enhances sample efficiency by prioritizing important transitions in the replay buffer based on their TD error, focusing learning on the most informative experiences.

Second, we described Policy Gradient methods, which are a class of reinforcement learning algorithms that optimize the policy directly by calculating gradients of the expected return with respect to policy parameters. This approach can handle high-dimensional and continuous action spaces, and it provides a more flexible framework compared to value-based methods. The objective function in policy gradient methods is typically the expected return, which the algorithm aims to maximize by adjusting the policy parameters using gradient ascent.

REINFORCE is a basic policy gradient algorithm that updates the policy parameters using the gradient of the return, which can be high-variance but straightforward. REINFORCE with a baseline introduces a baseline function to reduce variance in the gradient estimates, improving stability and convergence. Trust Region Policy Optimization (TRPO) and Proximal Policy Optimization (PPO) are advanced policy gradient methods that address stability and efficiency issues. TRPO enforces a constraint on policy updates to ensure changes are within a trust region, while PPO simplifies this by using a clipped objective function to balance exploration and exploitation while maintaining stability.

Third, we described Actor Critic models. Actor-critic models combine value-based and policy-based approaches in reinforcement learning. The actor learns a policy to select actions, while the critic evaluates the policy by estimating value functions. This separation allows for continuous action spaces and can lead to reduced variance in policy gradients. Asynchronous Advantage Actor-Critic (A3C) uses multiple parallel agents to update a global network asynchronously, improving stability and exploration. Deep Deterministic Policy Gradient (DDPG) extends the actor-critic framework to continuous action spaces, using deterministic policy gradients and experience replay for off-policy learning. Twin Delayed Deep Deterministic Policy Gradient (TD3) builds on DDPG by addressing overestimation bias in the critic through double Q-learning, delayed policy updates, and target policy smoothing. Soft Actor-Critic (SAC) introduces entropy regularization to the actor-critic framework, encouraging exploration and robustness. It maximizes both expected return and entropy, leading to improved exploration and stability. These methods represent significant advancements in handling continuous action spaces and complex environments, each addressing specific challenges in deep reinforcement learning such as sample efficiency, stability, and exploration-exploitation trade-offs.

Lastly, we described the Dreamer model series, which, at the time of writing this book, is the state-of-the-art actor critic model.

Further Reading or Watching

* A popular textbook for RL is written by Sutton and Barto [Sutton '20]. We have used this book to construct the foundation and flow of information in this chapter. Several examples that are used in the context of RL, such as Grid-world and algorithms, originated from this book. It is a reference book of RL. However, there is too much information and too many details. If you intend to read this book from beginning to end, you will get a deep understanding but will lose all your hair (if you have any). For deep reinforcement learning, we would recommend checking Phil Winder [Winder '20] and Miguel Morales [Morales '20] books.

* The old AI book from Russell and Norvig [Russell '09] has some detailed explanations about MDP in more than one chapter. It is a good book for the basics of MDP and some other baseline algorithms. Nevertheless, most of its terms and approaches are not widely in use due to advances in deep neural networks.

* A Reinforcement Learning course from LazyProgrammer is available on Udemy (https://www.udemy.com/course/artificial-intelligence-reinforcement-learning-in-python), it is fairly acceptable and better than some available online courses. Another course from Udemy is provided by Ponteves et al. (https://www.udemy.com/course/artificial-intelligence-az) and some of its explanations are useful. Nevertheless, neither course covers all the concepts that we believe are important for reinforcement learning. Besides, there might be some minor mistakes in their explanations as well. A detailed course is from UCL, and the one from David Silva has the highest review (the author of Alpha Go and Alpha Zero), availbe here: https://youtu.be/2pWv7GOvuf0?si=RKAonqvSYHgKM-xz. This is also a useful course, but sometimes it goes into too much detail, and similar to other courses, it does not cover all state-of-the-art models.

* Among billions of explanations, we read about eligibility traces and $TD(\lambda)$, we find the explanation in Miguel Morales's book a good one to understand this concept [Morales '20]. A good thing about his book is that he provides a Python implementation of each algorithm with enough details. Besides, we find the explanation for Alister Reis an easy to understand explanation of the rationale of using $TD(\lambda)$. https://amreis.github.io/ml/reinf-learn/2017/11/02/reinforcement-learning-eligibility-traces.html

* Miguel Morales [Morales '20] provides a good explanation about the i.d.d. problem and non-stationary target of DNN in RL, with the cart-pole example in his book; you can check the end of Chapter 8 and the beginning of Chapter 9.

* After struggling to understand TRPO and reading many different sources, we can recommend the best explanations for Natural policy gradient provided by Outer van Heeswijk (https://towardsdatascience.com/natural-policy-gradients-in-reinforcement-learning-explained-2265864cf43c), and online explanations for TRPO by Peter Abbeel https://www.youtube.com/watch?v=KjWF8VIMGiY&ab_channel=PieterAbbeel

* Joshua Achiam provides a summarized and simplified algorithm of computing NPG, http://rail.eecs.berkeley.edu/deeprlcourse-fa17/f17docs/lecture_13_advanced_pg.pdf

* Olivier Sigaud has a fairly easy explanation on the SAC algorithm https://www.youtube.com/watch?v=_nFXOZpo50U&ab_channel=OlivierSigaud

* Dreamer models are complex to learn. The first author, Danijar Hafner explains them in short videos https://danijar.com/project/dreamer. Besides, Yannic Kilcher has a good and detailed explanation about Dreamer v2, https://www.youtube.com/watch?v=o75ybZ-6Uu8

* OpenAI had a list of top RL algorithms in different fields. https://spinningup.openai.com/en/latest/spinningup/keypapers.html If you would like to build your career or research on RL, it makes sense to check this list. However, it seems that they are not updating it anymore.

* Many details of each deep RL algorithm are not in their papers, assuming the reader has background knowledge. There are several implementations available, and they differ from each other. To resolve this issue, we rely on the explanation of LLM tools such as Claude.ai and ChatGPT, which are not necessarily correct but are among the best LLMs at the time.

Part vi: Other Concepts and Algorithms

Chapters 14, 15, and 16 are introduced to dive into advanced topics that complement the foundations laid in earlier chapters. Chapter 14 explores techniques for optimizing and reducing the complexity of neural network and machine learning models, making them more efficient for real-world applications. Besides, this chapter covers compression algorithms, a group of underutilized but very useful algorithms for machine learning.

Chapter 15 introduces graph mining algorithms, an important area for analyzing complex relationships in data, particularly in social networks, biology, recommendation systems, and other systems that have many-to-many relations between entities.

Finally, Chapter 16 explores complexities encountered when handling real-world data. It covers topics such as sampling methods for complex data structures, noise types, and reduction techniques, as well as methods for reconstructing missing data through various imputation and interpolation strategies. The chapter also addresses challenges like data imbalance, anomaly detection, and data/model drift over time, offering solutions such as data augmentation, isolation-based methods, and strategies for tackling concept drift and cold start problems. Additionally, it highlights techniques for ensuring raters' agreement and introduces fairness, bias mitigation, and interpretability in models.

Chapter 14: Making Lighter Neural Networks and Machine Learning Models

At the time of writing this chapter (late 2023 and revised it in 2024), we observe that many enthusiastic young students and researchers are drawn to deep learning and its potential, often focusing their Ph.D. studies on neural networks. On the other hand, the largest and most successful models are developed by corporations such as OpenAI, Google and its Deep Mind, Meta, and NVIDIA. As discussed in Chapters 11 to 13, many of the models introduced are from these corporations. The growing dominance of these entities, coupled with the challenges of accessing large GPU clusters, has led to a decline in the role of universities and independent institutions, which can be discouraging for aspiring researchers.

Despite these challenges, there is still significant research potential in neural networks, particularly in making models smaller while retaining accuracy. This focus can drive the proliferation of deep learning applications in battery-powered devices such as wearables, smartphones, drones, and robots while also addressing concerns about carbon emissions and resource consumption. Additionally, on-device algorithms can function in areas with limited network availability. Perhaps the most important benefit of making lighter models is reducing the energy and water consumption of training AI models. At the time of revising this chapter (Q4 2024), large AI model training processes consume huge amounts of water (for their cooling systems) and electricity.

In this chapter, we list approaches targeted to reduce the cost of machine learning models and make them less resource-intensive. Since neural networks provide high accuracy in some tasks and settings while consuming a huge amount of resources, most of our explanation is around neural networks. On the other hand, we have also described compression and quantization algorithms that could be applied to any machine learning algorithm. Therefore, we bring this chapter into Part vi of this book.

First, we start with data compression algorithms that can be applied to any machine learning algorithm to reduce the output model size or the input data size. You might ask, why do I need data compression while studying machine learning? Compression algorithms are old; nowadays, data scientists and machine learning researchers do not pay enough attention to applying them to machine learning algorithms and AI models. They can be used to preprocess the original data, making it smaller and thus lighter for the machine learning algorithm. Byte Pair Encoding compression, which we will learn in this chapter, is the base of tokenization in almost all LLMs. Besides, the author has benefited greatly from compression algorithms in several scientific works, which received good attention from the community [Malekijoo '21, Rawassizadeh '23].

After we learn compression algorithms, we discuss quantization methods in signal processing, images, and deep learning models, followed by pruning and sparsification in neural networks. Next, we describe different types of deep learning architecture that try to reduce the cost of training or enable the model to operate in settings with limited resources. Ultimately, we describe Automated Machine Learning (AutoML), which could be interpreted as another way to make a more resource-efficient neural network.

Data Compression Algorithms

Data compression is one of the oldest approaches to reduce the data size while maintaining its characteristics. On the other hand, one of the easiest approaches to make a machine learning algorithm or model lighter is to train it or feed it with a smaller dataset while maintaining the semantics of the data. By smaller, we don't mean a smaller number of data points; we mean a smaller dataset size, which includes the original data point.

We have learned in school that in computers, data is transferred and stored in binary format (bits and bytes), which the hardware processes. Humans use decimal format for numbers (0 to 9), but computers use binary format (0,1) that stores data in bits. The "bit" is the smallest unit of information in computer science, and it can be either 0 or 1. As a reminder, 8 bits constitute a larger piece of data called a byte. Larger units of data are formed by multiplying bytes by powers of 1024. For example, one kilobyte (KB) is equal to 1024 bytes, one megabyte (MB) is equal to 1024 KB, and one gigabyte (GB) is equal to 1024 MB. Also, 1024 gigabytes is one terabyte (TB), 1024 terabytes is one petabyte (PB), and 1024 petabytes is one exabyte (EB).

For instance, the string 'Hi' can be converted into a binary string like this: 0100100001101001. Each byte represents a character: 01001000 for 'H' and 01101001 for 'i'. Also, we can use something to specify it is a string, let's say 01. Therefore, assuming data is transferred between two devices when a sender sends 0100100001101001 and then 01, the decoder on the other side will interpret this byte as a string and translate it back to the "Hi".

The process of data compression transforms data from one format into another, smaller format while maintaining the original data characteristics. This process reduces the amount of space the data occupies and can potentially decrease the time required for search execution.

When we zip, rar, or gzip files or folders, we compress them. The process of compression operates at the bit level, meaning the data is converted into a binary format, and bit manipulation methods are applied to reduce the size of the original data. Along with the compressed data, a *key-value file* will be stored to translate back data to its original format, i.e., decoding the compressed file. This file is usually referred to as a *dictionary, codebook,* or *catalog*.

To understand data compression, consider this sentence: "Do not ask what XYZ can do for you; ask what you can do for XYZ.". Words "do", "can", "ask", "XYZ" and "what", appeared two times in this sentence. An algorithm could create a dictionary and assign a number to each word as follows: *do=1, not=2, ask=3, what=4, XYZ=5, can=6, for =7, you=8*. Then, based on using this dictionary, we can write the sentence as follows: {1 2 3 4 5 1 6 7 8, 3 4 6 8 1 7 5.}. Of course, saving the dictionary is an overhead, but it is worth the space we can save by compression while dealing with large corpora of text, such as a book.

After the process of compression, if we lose some of the original information, the compression algorithm is called *lossy compression* (e.g., converting a TIFF image file into GIF format). If we do not lose any of its data, it is a *lossless compression* (e.g., zipping a file). For instance, if you are a civilized person, you know that emailing many documents together is not good, and it is better to compress them into a single folder by zipping them before sending them in the email.

In the following, we describe some popular compression algorithms, including Byte Pair encoding, Bitmap index, Huffman Encoding, and LZW compression.

Byte Pair Encoding (BPE)

One of the challenges we have described before in NLP models is tokenizing words in languages that do not have space, such as Chinese, Japanese, and Korean. To tokenize those texts, we could use subword tokenization algorithms. An example of a compression algorithm for this approach is Byte Pair Encoding (BPE) [Gage '94], introduced in 1994. It is a simple recursive algorithm that scans the string of characters and substitutes the repeated pairs of characters, where at least one pair of characters appears together more than once with one character. Then, it performs the scan again and substitutes the repeated pair of characters again with one character. This process continues, and we end up with a character string that all its characters are repeated only once in the text string.

To better understand BPE algorithm, consider the following example: we have a short text written as follows: 'veeeerrrrry gggood arry', which is a common style that users of social media use to add emotion to their text.

The first scan identifies that the pair 'rr' occurs most often, and the algorithm replaces it with some character such as '!' , which we use as a sign for the sake of understanding. The result will be as follows:

veeeer!!y gggood a!y

Again, it scans the new string to identify the most repeated pair of characters, which are 'ee' and '!y'. It substitutes them with one character, let's say 'ee' with '@' and '!y' with '#'. Thus, as a result, we will have the following string:

v@@r!# gggood a#.

This process continues until there is no duplicate left, and we have a string as follows.

 v*r!# %g^d a#

However, there is a catalog/dictionary that includes all variable mappings as well. In this example, the catalog (codebook or dictionary) is as follows:

{rr = !, ee = @, !y = #, @@ = *, gg=%, oo=^}

Bitmap index

A bitmap index is a database indexing technique that uses bit arrays (bitmaps) to represent the presence or absence of a value or condition. This method is used for read-heavy operations and queries with low cardinality, as it allows for fast logical operations and compression. By cardinality here, we mean the diversity of data is limited. For instance, gender could be either male or female; unlike colors, which have high cardinality, there are not many choices for gender. When the data has high cardinality, it is recommended to use another indexing method or apply compression before indexing the data.

Exercise	Strength	Cardio	Burning Fat
Treadmile	no	yes	high
Bench Press	yes	no	Low
Dead Lift	yes	no	Medium

Exercise	Strength		Cardio	Burning Fat
Treadmile	0	1	10	
Bench Press	1	0	00	
Dead Lift	1	0	01	

Table 14-1: (left) Different gym exercise training and their body impact, (right) converted exercise data to a bitmap index.

To better understand how bitmap indexing works, consider the following example in Table 14-1. Here, we have listed some gym exercises and their category as strength, cardio, or fat burning. Of course, in the real-world, we will have much more data, and the dataset is significantly larger. However, instead of searching the table to find the related information, a bitmap index can use bitwise operators (OR, AND, XOR) to locate the answer. Searching the bit string is significantly faster than searching the original data because the search space is reduced to bit strings, and thus, the number of comparisons is reduced.

Huffman Coding

In one of our scientific works, we experiment with different compression algorithms, and we demonstrate that Huffman coding [Huffman '52] is the fastest compression algorithm [Rawassizadeh '23]. It is a lossless compression algorithm that constructs a binary tree as a data dictionary.

Assume we have the following string of 16 characters: $\{a, b, c, d, a, d, c, b, c, e, e, c, b, b, a, c\}$ by default, computers use ASCII codes (8 bits) to store each character, which means here we need $16 \times 8 = 128$ bits to store this string of characters. Since here we have only five unique characters, instead of using ASCII codes, we can use three bits to present all five characters $a = 000$, $b = 001$, $c = 010$, $d = 011$, $e = 100$, and it occupies $16 \times 3 = 48$ number of bits. This representation is significantly smaller than the original string. In this example, we show a short string with very few characters. In a real-world case, we have lots of repetitive information, like words inside the text document, pixels inside a picture, etc. Huffman coding performs better and can compress more than the simple approach described by not using ASCII format.

Assume we have a text that is composed of $a, b, c, d,$ and $e,$ and their frequencies are written in front of them as follows: $\{a = 12, b = 15, c = 11, d = 24, e = 20\}$

$$12 \times 8 + 15 \times 8 + 11 \times 8 + 22 \times 8 + 20 \times 8 = 656$$

Therefore, 656 bits are required to store this text in ASCII code.

but if we store this text with our three bits compression, we will need only 246 bits because: $12 \times 3 + 15 \times 3 + 11 \times 3 + 24 \times 3 + 20 \times 3 = 246$.

This representation (246 bits) is smaller than 640, but we could still make it smaller using Huffman coding. The Huffman coding algorithm operates based on creating a binary tree and assigns values to each character based on their position inside the tree. First, it orders characters based on their frequency, which is as follows: $\{c = 11, a = 12, b = 15, e = 20, d = 24\}$.

Next, it takes the two data objects that have the lowest frequency and constructs the two leaf nodes on the right, as shown in Figure 14-1 (a). Of course, they need a root node, and the root node value for these two nodes is the sum of the frequency of these two nodes $11 + 12 = 23$. Then, we look for the next node that has the lowest frequency, i.e., $b = 15$, and check if we can add it to the tree or require a new branch with its next frequent node ($e = 20$). Since 23 is larger than the frequency of e and b, a separate branch for e and b will be created (a smaller one on the left and a larger one on the right). The new root node has the value of $15 + 20 = 35$ (Figure 14-1 (b)).

{ c=11, a=12, b=15, e=20, d=24}

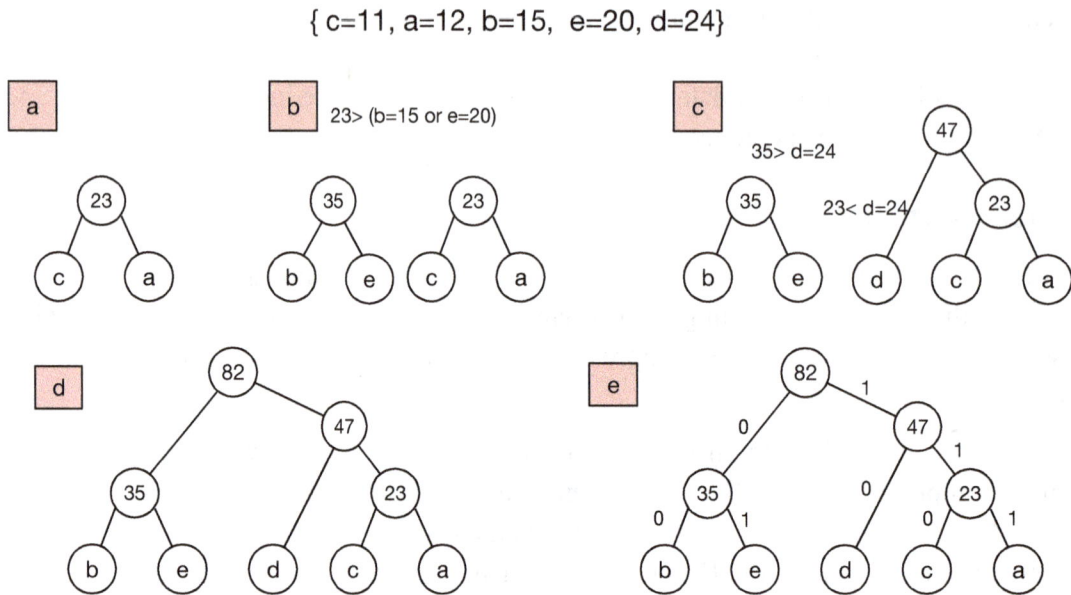

Figure 14-1: Huffman encoding binary tree construction step by step. Data points that have higher frequencies will receive a lower number of bits, and data points that are more frequent receive a lower number of bits.

Now, the only remaining node is $d = 24$. The question is, in which branch we can add it? There is one branch with a parent node 23 and another branch with a parent node 35. The algorithm chooses the parent node, which is smaller, and it is there (Figure 14-1 (c)). The newly created parent has $23 + 24 = 47$ value. The next step is to merge two subtrees, and the algorithm connects both branches with the root value of $47 + 35 = 82$ (Figure 14-1 (d)). When the tree is constructed, the algorithm assigns binary numbers to all left branches and 1 to all right branches (Figure 14-1 (e)). Now, every character can be represented with a bit string starting from the root $\{b = 00, e = 01, d = 10, c = 110, a = 111\}$.

We can see that the result of Huffman coding represents characters that are more frequent (e,d,b) with a smaller number of bits and characters that are less frequent (a,c) with a larger number of bits. Our original set is compressed with 185 bits, as follows:
$12 \times 3(111 = a) + 15 \times 2(00 = b) + 11 \times 3(110 = c) + 20 \times 2(01 = e) + 24 \times 2(10 = d) = 185$

185 is smaller than 240. Besides, we also need a small space to store the tree, which is too small and usually very cost-effective to use Huffman encoding instead of original data.

Lempel–Ziv–Welch (LZW) Encoding

One of the first known compression techniques is Lempel-Ziv (LZ) [Ziv '78]. A more advanced form of LZ is the Lempel-Ziv-Welch (LZW) algorithm [Welsh '84], which operates on a bit level, and it is widely in use by Unix operating systems (compress command in Unix), GIF images, etc. Our experiment [Rawassizadeh '23] shows that, in comparison to other compression methods, despite its superior compression capability, it is very slow. Therefore, it might not be the right choice for real-time applications, but it is the best choice when we need high compression.

LZW replaces strings of characters with single codes, and the idea behind LZW is to replace repeated sequences of data with codes, which are smaller representations of those sequences. It might be tricky to explain it; we describe it through an example; please take a pen and paper and experiment on the paper while reading it.

LZW starts by initializing a dictionary with all characters (usually 256 ASCII characters) and continues the process of compression character by character until the entire string is scanned. For example, suppose we intend to compress {abcabcdabc} string, which has four characters {a,b,c,d} and the dictionary is initialized with dict={0:a,1:b,2:c,3:d}.

LZW reads the input string character by character. The first character is 'a', {abcabcdabc}. If this character does not exist in the dictionary (which, in our example, 0:a exists), a new entry will be added to the dictionary with a new code. If this character exists, it adds the next character to the current one, i.e., 'ab', and checks if it is in the dictionary. 'ab' is not in the dictionary, and thus, it adds 'ab' into the dictionary with a new index. Now, the content of the dictionary is as follows: dict={0:a,1:b,2:c,3:d,4:ab}. At the end of this iteration, it adds the index of the current character ('a') or sequence of characters retrieved from the dictionary to the output, output:[0].

Then, it moves to the next character in the sequence, 'b', {abcabcdabc}. It checks if the 'b' exists in the dictionary, and the answer is yes. Then, it adds the next character to the current one, i.e., 'bc', and checks if 'bc' is in the dictionary. It is not in the dictionary, and thus, it adds 'bc' to the dictionary. Now, the content of the dictionary is as follows: dict={0:a,1:b, 2:c,3:d,4:ab,5:bc}. At the end of this iteration, it adds the index of the current character ('b') or sequence of characters retrieved from the dictionary to the output, output:[0,1].

Afterward, it moves to the next character in the sequence, 'c', {abcabcdabc}. It checks if the 'c' exists in the dictionary, and the answer is yes. Then, it adds the next character, 'a', to the current one, 'c', i.e., 'ca', and checks if 'ca' is in the dictionary. It does not exist in the dictionary, and thus, it adds 'ca' to the dictionary. Now, the content of the dictionary is: dict={0:a,1:b,2:c,3:d,4:ab,5:bc,6:ca}. At the end of this iteration, it adds the index of the current character ('c') or sequence of characters retrieved from the dictionary to the output, output:[0,1,2].

Afterward, it moves to the next character in the sequence, 'a', {abcabcdabc}. It checks if the 'a' existed in the dictionary, which is yes. Then, it adds the next character to the current one, i.e., 'ab', and checks if 'ab' is in the dictionary, which is yes again. Now it adds the next character to the 'ab' string, i.e., 'abc', and checks if it is in the dictionary. 'abc' does not exists in the dictionary, and thus it adds it into the dictionary, and following is the content of dictionary: dict={0:a,1:b,2:c,3:d,4:ab,5:bc,6:ca,7:abc}. At the end of this iteration, it adds the index of the current character or sequence of characters (ab) retrieved from the dictionary to the output, `output:[0,1,2,4]`. Now, you can see that the compression is starting to take place.

Next, it moves to the next character in the sequence, `'c'`, `{abcabcdabc}`. It checks if the `'c'` exists in the dictionary, which it already exists. Then, it adds the next character to the current one, i.e., `'cd'`, and checks if `'cd'` is in the dictionary, which is not there. Therefore, it adds it into the dictionary, `dict={0:a,1:b,2:c,3:d,4:ab,5:bc,6:ca,7:abc,8:cd}`. At the end of this iteration, it adds the index of the current character or sequence of characters (`'c'`) retrieved from the dictionary to the output, `output:[0,1,2,4,2]`.

Then, it moves to the next character in the sequence, `'d'`, `{abcabcdabc}`. It checks if the `'d'` existed in the dictionary, which is yes. Then, it adds the next character to the current one, i.e., `'da'`, and checks if `'da'` is in the dictionary, which is not there, and then it adds it to the dictionary, i.e., `dict={0:a,1:b,2:c,3:d,4:ab,5:bc,6:ca,7:abc,8:cd,9:da}`. At the end of this iteration, it adds the index of the current character or sequence of characters (`'d'`) retrieved from the dictionary to the output, `output:[0,1,2,4,2,3]`.

Next, it moves to the next character in the sequence, which is `'a'`, `{abcabcdabc}`. It checks if the `'a'` existed in the dictionary, which is yes. Then, it adds the next character to the current one, i.e., `'ab'`. Then, it checks if `'ab'` existed, which again exists. Afterward, it adds the third character to the current character sequence, i.e., `'abc'`, and checks if it existed, which also exists. Now, we reach the end of the string, and it doesn't go further. At the end of this iteration, it adds the index of the current character or sequence of characters (`'abc'`) retrieved from the dictionary to the output, `output:[0,1,2,4,2,3,7]`.

You can see that a string with a size of 10 characters is compressed to 7 digits. As we move further in a sequence, the compression gets stronger and stronger until it reaches the maximum dictionary size limit.

We use this example for simplicity, but, in reality, the output will be converted into binary numbers, and the input string is significantly larger. The decoder is fairly straightforward. It reads the content of the output and gets the value of each number from the dictionary, and substitutes them.

Sparse Coding

Sparse coding, a.k.a. Sparse Dictionary Learning (SDL), converts the input data into a sparse representation of the input data (by adding lots of zeros) while maintaining the characteristics of the original data. We have explained in Chapter 5 that a sparse vector/matrix is a vector/matrix with lots of zeros.

The result of sparse coding has the same dimensional data as the input data. However, some data points are converted to zero, and the most important information in the data is captured while using only a small number of coefficients. For example, assume we fed a vector of numbers $x = [1,2,4,5,6,5,3,2,1]$, into a sparse coding model, and as output, we get another vector with the same size, i.e., $\hat{x} = [0,0,0,3.4,-0.1,-0.6,1.7,0,0]$, but it includes many zeros.

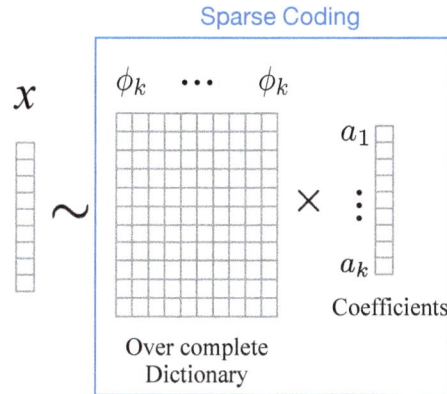

Figure 14-2: Sparse coding, which gets a vector x as input and converts it into a linear combination of basis vectors and coefficients.

Sparse coding is not commonly categorized as a compression algorithm or dimensionality reduction, but sparse coding can be used for finding compact representations of complex data. It reduces the original size of the dense data and presents it in a compact format using a small number of non-zero coefficients, or "sparsity". In other words, sparse representation can capture the essential features of the data while discarding redundant information.

While the input of sparse coding is typically a vector, the input can be extended to matrices in some applications, where each column of the input matrix is treated as a separate vector to be sparsely coded. For example, assume we have a 100×100 pixel image (10,000 pixels) that we want to represent using sparse coding. The original image is dense, meaning most pixels have non-zero values. After applying sparse coding, the image might be represented using only 1% of the original pixels (100 non-zero coefficients). The remaining 99% of pixels are set to zero.

Sparse coding involves finding a set of vectors known as basis vectors and connecting them together to construct a matrix known as a *dictionary*. This dictionary is overcomplete, which means if the size of the input vector is n, the dictionary has a size of $m \times n$ where $m > n$. The overcompleteness of the dictionary is what allows for sparse representations. Each input vector can be approximated as a sparse linear combination of these basis vectors.

We have an input vector x. We want to represent this vector as a linear combination basis vectors. Therefore, we can represent x as follows: $x \sim a_1\phi_1 + a_2\phi_2 + \ldots + a_k\phi_k$. Here, k is a hyperparameter that specifies the number of basis vectors used in the sparse coding representation. This equation shows that these basis vectors ϕ_i represent an input vector x as a linear combination of ϕ_i basis vectors along with their coefficients a_i. The coefficient for each basis vector represents the contribution of each basis function to the vector. We can summarize the above equation as follows:

$$x \sim \sum_{i=1}^{k} a_i\phi_i$$

Each basis vector captures a specific feature or pattern present in the data. Since we want the representation to be sparse, we want to find the coefficients such that a_i to be zeros. Figure 14-2 presents the sparse coding components.

Sparse coding is a two-step process involving the learning of basis vectors (dictionary learning) and the encoding of input data as sparse combinations of these basis vectors (sparse coding). In other words, the objective is to find the sparsest representation of x in terms of the basis vectors ϕ_i, where most a_i coefficients are zero, and only a few are non-zero. To find these a_i coefficients, the algorithm that implements sparse coding should solve the following optimization problem:

$$min \ \|x - \sum_{i=1}^{k} a_i \phi_i\|_2^2 + \lambda \|a\|_1$$

The first term $(min\|\ldots\|_2^2)$ represents the reconstruction error from sparse to original data. The second term $(\lambda\|a\|_1)$ is a regularization that specifies the sparsity constraint. The parameter λ controls the trade-off between the two terms. In other words, λ is a regularization parameter that controls the sparsity of the solution. In many sparse coding algorithms, the basis vectors ϕ_i and coefficients a_i are both learned via optimization. Different algorithms can be used to implement the objective function of sparse coding, such as LASSO (check Chapter 8) or neural networks.

NOTES:

* Usually, the size of the LZW dictionary for text compression is 4,095. The first 256 slots in the dictionary are typically occupied by single-character entries corresponding to standard ASCII characters, as follows: {0:null}, ..., {32:space}, ..., {48:0}, {49:1}, ..., {65:A}, ..., {255:ÿ}. Therefore, the next string it adds to the dictionary starts from position 256.

* We can't resist the opportunity to shamelessly plug one of our masterpieces, ODSearch, a search engine that combines Huffman encoding with Bloom filters [Rawassizadeh '23]. This combination enables us to build a highly efficient search engine for fitness tracker data that even runs on smartwatches, 27 times faster than all existing state-of-the-art search engines.

* While reading about compression algorithms, why some called this Hufmann encoding, and others call it Hufmann coding? There is no difference; these terms are used interchangeably. In practice, these algorithms perform encoding first and then decoding. However, they were developed by scientists in different communities, leading to variations in terminology.

* One method that we have postponed explaining until now is data sharding, which is used to manage limited memory size. Data sharding is the process of *separating a big dataset into smaller chunks*. Each chunk has the same schema (column or feature names), but its content is unique. The distribution of data among different shards is determined by a feature known as the *shard key* or *partition key*. Data sharding can be used when loading a large amount of data into memory for tasks such as training, clustering, etc. It enables loading a small chunk of data at a time into memory. It's important to note that sharding is a technique used in the database community for scaling, where the data is split across multiple database instances or shards. By using mini-batch Gradient Descent, we do the same in machine learning, but it is not called sharding.

Quantization

The process of converting a large continuous set of numbers (e.g., real numbers, $\mathbb{R}$) into a smaller set of discrete numbers (e.g., natural numbers, $\mathbb{N}$) is called quantization. In other words, quantization transforms continuous format into discrete format, which occupies less memory. The result of quantization is lossy compression. There are many different methods and applications for quantization; here, we describe a few algorithms for signal, image, and neural networks.

Several methods we described in Chapter 6 could be categorized as quantization as well. For example, the SAX algorithm [Lin '03] can be categorized as a quantization method for time series, but it is not common to categorize them as a quantization method. In this section, we list approaches that are explicitly known as quantization.

First, we describe vector and non-uniform quantizations and then discuss neural network quantization. We are confident to say that learning quantization makes you very successful at feature engineering and enables you to run complex resource-intensive models on limited resources, so take this section seriously and do not underestimate its potential.

Vector Quantization and Cluster Quantization

Cluster Quantization, often referred to in the context of vector quantization, is a process used in signal processing and data compression where a large set of points (vectors) in a high-dimensional space is approximated by a smaller set of representative points (centroids or clusters). This technique reduces the amount of data by grouping similar data points together and representing them with a single reference point from each group (region).

Vector Quantization (VQ) is cluster quantization, which is mainly used for binary data such as audio and images. Each group members share the same characteristics, and each group is presented with a vector (a.k.a codewords or output vectors). Each output vector (codeword) represents the data inside each decision region. As we have explained before, the centroid of a group is a representative vector that represents the entire region. The collection of codewords from different regions is called a *Lookup table*, *codebook*, or *dictionary*.

A VQ can be constructed from Voronoi Tessellation (check Chapter 9, kNN description), it can be constructed from clustering as well, and there are algorithms such as the Linde-Buzo-Gray (LBG) algorithm [Linde '80] that uses a simple iterative algorithm (very similar to k-mean) to construct a codebook for images.

Palettization is a specific form of color compression used in image processing. It reduces the number of colors in an image to a fixed number, which is stored in a palette. Each pixel in the image is then represented as an index in this palette. Clustering weights of a neural network through clustering and using the centroid of each cluster instead of the original data points is also known as palettization[1], which is defined on the CoreML platform of Apple.

---

[1] https://apple.github.io/coremltools/docs-guides/source/palettization-overview.html

In computer vision and signal processing literature, VQ is used as an encoder and convert each patch of an image or signal into a vector and store it in a codebook. This means an encoder builds the codebook. For example, a patch of 4×4 pixels can be stored as a vector of 16 values, and if we intend to compress it, it can be encoded into a smaller vector size, e.g., 8 values. Then, the decoder uses the codebook, which is constructed by the encoder, and decodes the vector back into the image patches. This is a popular approach used in some machine learning approaches, such as VQGAN (check Chapter 11).

To understand the VQ with example, assume we have built a simple codebook with four members, and each member presents a data point in 2D space as follows (i present the index, and d presents the original data in 2D space):

$i_1 : d = (0,1)$

$i_2 : d = (2,1)$

$i_3 : d = (3,3)$

$i_4 : d = (2,2)$

We have a 2D dataset $d = \{(0,0), (1,0), (4,4), (5,5), (2,1)\}$, and we intend to quantize it using our codebook. First, we compute the distance between each dataset member and each codebook element and then substitute the dataset member with the one that has the shortest distance. For example, using Euclidean distance, we have the following:

$$||(0,0), i_1||_2 = \sqrt{(0-0)^2 + (0-1)^2} = 1$$

$$||(0,0), i_2||_2 = \sqrt{(0-2)^2 + (0-1)^2} = 2.23$$

$$||(0,0), i_3||_2 = \sqrt{(0-3)^2 + (0-3)^2} = 4.24$$

$$||(0,0), i_4||_2 = \sqrt{(0-2)^2 + (0-2)^2} = 2.78$$

Therefore, $i_1 \ d = (0,1)$ is the closest one to the first element of the dataset, i.e., $(0,0)$, and thus we substitute the value of i_1 instead of $(0,0)$. This process continues, and in the end, our quantized dataset will be as follows: $d' = \{i_1, i_1, i_3, i_3, i_2\} = \{(0,1), (0,1), (3,3), (3,3), (2,1)\}$. By computing the difference between quantized data and the original data point, we could report the *quantization error (e)*. In this example, the quantization error can be computed as follows:

$$e = \{(0,1) - (0,0), (0,1) - (1,0), (3,3) - (4,4), (3,3) - (5,5), (2,1) - (2,1)\} = \{1,2,2,4,0\}.$$

The codebook is required to build the decoder. In a codebook, high probable values could get a fine quantization, and less probable values could get a less sensitive quantization. This is something that can be achieved by algorithms such as Huffman encoding or LZW compression as well.

A good example of this approach is k-mean quantization, which reduces the number of colors in an image while retaining its overall visual quality. Figure 14-3 presents an example of k-mean quantization.

Original Image (96,615 colors) K-mean Quantized Image (64 colors)

Figure 14-3: Differences between the original image and *k*-mean quantized image.

Non-uniform Quantization (Signal Companding)

Non-uniform quantization is a type of quantization that assigns different quantization levels to different parts of the input data based on some characteristics of data, e.g., it uses the distribution of data to decide about the quantization level.

The goal of non-uniform quantization is to reduce the quantization error by using more data to represent significant parts of the input data and fewer data for the insignificant parts of the input data. For example, look at the time series we present in Figure 14-4 (a). If we apply a simple quantization and the distance between each segment is equal, we have something like Figure 14-4 (b), which is not representative of the original data. Instead, if our quantization segment changes at different intervals, we might be closer to the original data, as shown in Figure 14-4 (c).

A common Non-uniform quantization that is used in the WaveNet (check Chapter 12) model is *companding*.

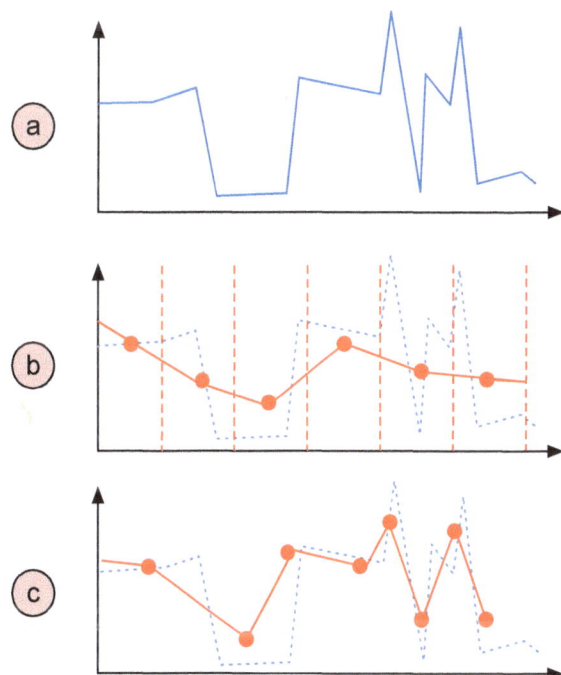

Figure 14-4: A toy example that shows the original time series (a), and the differences between uniform quantization (b) and non-uniform quantization (c).

The term companding stands for *compression and expansion*. Companding methods compress a digital signal by reducing the bit depth before it is transmitted and then expanding it back to its original range after the receiver has received it.

Companding is used in analog communication systems, such as telephony, to improve the signal-to-noise ratio and for noise reduction while dealing with signal data. It transforms the signal into a smaller dynamic range; then, the receiver decompresses the compressed signal.

Two companding algorithms used for telephony systems include $\mu - Law$ (use in Japan and North America) and $A - law$ (use in the rest of the world). They are known as G.711 codec as well.

μ-Law

It compresses the analog signal, such as analog Audio, into 8-bit digital version. It compresses the input signal x and it outputs signal y by using the following equation.

$$y = sgn(x)\frac{ln(1 + \mu|x|)}{ln(1 + \mu)}$$

$sgn(x)$ presents a sign function that assigns a number based on the value of x. If $x = 0$, then $sign(x) = 0$. If $x > 0$, then $sgn(x) = 1$. If $x < 0$, then $sgn(x) = -1$. μ is a constant equal to 255 in the North American and Japanese standards.

To expand the output signal back, μ-Law algorithm uses the following equation.

$$F^{-1}(y) = sgn(y)\frac{(1 + \mu)^{|y|} - 1}{\mu}, \quad -1 \le y \le 1$$

A-Law

It compresses the analog signal, such as analog Audio, into 8-bit digital version by using the following equation:

$$y = sgn(x)\begin{cases} \frac{A|x|}{1 + ln(A)} & |x| < \frac{1}{A} \\ \frac{1 + ln(A|x|)}{1 + ln(A)} & \frac{1}{A} \le |x| \le 1 \end{cases}$$

Here, A is a constant, and $A=87.6$. To expand the output signal back, A-law uses the following equation.

$$F^{-1}(y) = sgn(y)\begin{cases} \frac{|y|(1 + ln(A))}{A} & |y| = \frac{1}{1 + ln(A)} \\ \frac{e^{-1 + |y|(1 + ln(A))}}{A} & \frac{1}{1 + ln(A)} \le y < 1 \end{cases}$$

If we assume $A = 1$, then we get uniform quantization, and in μ-law, if we assume $\mu = 0$, we get uniform quantization. There are more signal-companding algorithms, which we skip explaining. Applying companding is fairly simple, but it is very useful, especially when dealing with audio data.

Quantization for Neural Networks

Quantization in the context of neural networks refers to the process of replacing floating-point numbers (e.g., 32-bit floating points) with lower-bit numbers (e.g., 8-bit integers, 4-bit integers, or 1-bit binary). This approach significantly reduces the computational cost and memory usage. The models we described in Chapter 11 and Chapter 12 were usually trained by large corporations, which have massive numbers of GPUs and other resources. To democratize AI model training, we must enable smaller firms or even individuals to train their models, regardless of their computational limitations.

Again Neural Network, and again you suffer if you didn't read Chapter 8 and Chapter 10.

A neural network has too many parameters (weight and biases), and all of them are presented with numbers. Therefore, reducing these numbers while losing a small amount of accuracy saves a significant amount of resources (including energy and computational overhead). For example, the multiplication process with 8-bit integers consumes 0.2 pico-joules (*pj*), and with 32-bit integers consumes 3.1 *pj* [Horowitz '14].

Before we explain quantization, we should learn how to present numbers in computers, i.e., floating-point and fixed-point forms. Then, we describe neural network stages for quantization and later conclude this section with quantization methods.

Fixed Point versus Floating Point Forms

Fixed point form is the format in which a real number is stored with two components, the *integer* part, and the *fractional* part. For example, a US dollar currency is a number that has an integer part and a fractional part that specifies cents. 14.25\$ presents 14 dollars and 25 cents. To present cents, the size of the fractional part is fixed. Since the size of the fraction is fixed, the range of numbers that can be presented with fixed-point form is limited.

The *floating point form* uses a scientific form, a.k.a *normalized form*, to represent a number. It separates large numbers into two components: *mantissa* and *exponent*. For example, 2300 can be presented as 2.3×10^3. In this notation, 2.3 is called the Mantissa and 10^3 is called the Exponent. To represent these numbers in binary format, we can allocate a certain number of bits for the mantissa and another group of bits for the exponent. For example, we can use 11 bits for the mantissa and 8 bits for the exponent. This enables us to represent a large number in only 11+8=19 bits. If we do not use this approach, assume we intend to store 982,000,000; it requires a large bit vector to store. However, by using floating point form, we can write it as 982×10^6 and store it as mantissa and exponents.

The Institute of Electrical and Electronics Engineers (IEEE) provides a standard form in computers to present floating point numbers, IEEE 754. For example, a single precision float number is presented with 32 bits, 1 bit for the sign, 8 bits for the exponent, and 23 bits for the mantissa. To present a double precision number, 64 bits will be used; 1 bit for the sign, 11 bits for the exponent, and 52 bits for the mantissa.

Processing floating-point numbers is more resource-intensive than processing fixed-point numbers. Floating-point quantization has a higher precision than fixed-point quantization, and it provides a higher range of representable numbers.

We learned in previous chapters (Chapters 10 to 12) that neural networks deal with too many weights, biases, and activation function outputs. These are referred to as model parameters, and they are represented with floating-point numbers. Usually, neural network implementations use 32-bit floating-point numbers. By converting 32-bit floating-point numbers to 8-bit or 4-bit integers, we can achieve a four- or eight-fold reduction in model size and a four-fold reduction in memory bandwidth.

Quantization is the process used in neural networks to reduce floating-point numbers to smaller numbers, such as integers. The most popular neural network quantization approach involves training a neural network with 32-bit floating-point numbers and then converting the resulting model into 8-bit integers, 4-bit integers, or 16-bit floating-point numbers. The process of quantization in neural networks can be applied to four different components: (i) weight values, (ii) activation function values (e.g., quantized ReLU), (iii) both weight and activation values, and (iv) specific layers of the network.

How to Quantize?

There are different types of quantization formalization; we use a common one here, which is defined as a mapping of integer q (quantized value) to the real r (original input). The following equation presents the baseline quantization:

$$q = round(r/S) - Z$$

In this equation, r is the input number in float format, q is the quantized version of r in integer, S is referred to as scale factor or quantization parameter, and Z is referred to as zero-point or quantization constant. The zero-point can vary depending on the range of values in a particular layer or dataset being quantized. For example, if quantizing activation outputs that range from -3 to 3 into an 8-bit integer (values from 0 to 255), the zero-point is calculated to ensure that zero maps directly to an integer value within this range.

We can recover original data, i.e. $\hat{r}$, with some de-quantization error, by using the following equation: $\hat{r} = S(q + Z)$. Z allows the quantization values to be shifted and represented as unsigned integers, which can be efficiently stored for hardware implementation of neural networks. In simple words, it is the quantized value of $r = 0$. S is dividing the given range of r into several partitions. It is calculated as follows:

$$S = \frac{\beta - \alpha}{2^b - 1}$$

Where α and β are clipping range, and b is the number of quantization bit width. b is simple to specify; for example, if we intend to quantize a matrix of real numbers into 8-bit integers, then $b = 8$.

The process of identifying the clipping range is referred to as *calibration*. In other words, calibration is determining the dynamic range of the input data to a quantized model. One calibration approach is to assume $\alpha = min(r)$ and $\beta = max(r)$. In this case, the quantization is called *asymmetric quantization* because $-\alpha \neq \beta$. On the other hand, a *symmetric quantization* is assumed $-\alpha = \beta = max(|max(r)|, |min(r)|)$. We can make the range selection simpler by considering the $Z = 0$. In that case, assuming n is the number of bits, S could be computed as $S = 2max(|r|)/(2^n - 1)$ or $S = 2max(|r|)/(2^{n-1} - 1)$. The latter form is more restrictive and uses the range of -127 to 127.

When to Quantize a Neural Network?

Neural network quantization methods can be applied to weight, activation function values, and biases. They are commonly used on resource-constrained devices such as mobile devices or converting large models to smaller versions that can be fine-tuned on desktop machines. In the following, we list common types of neural network quantizations in the following.

Dynamic versus Static Quantizations

Dynamic quantization converts the weights and activation values of a neural network from high-precision numbers to low-precision numbers during the training time or inference time. In other words, clipping range parameters estimations (α, β) will be done at the training time or inference time. Dynamic quantization reduces the memory and computational requirements. However, since it operates in real-time, it might have a large computational overhead.

Static quantization computes the clipping range parameters that are calculated before inferences and stay static. This has less computational overhead, but usually, it has lower accuracy than dynamic quantization.

Dynamic quantization is applied during inference. Static quantization, on the other hand, is applied after training but before inference.

Post-Training Quantization (PTQ)

Post-Training Quantization (PTQ) approach [Krishnamoorthi '18] quantizes the weights and activation values after the training is complete. In other words, the already trained model parameters are quantized post-training. PTQ can employ various quantization techniques, such as uniform and non-uniform quantization, which determine how values are mapped to lower precision.

To achieve optimal quantization, a calibration dataset is often used to compute the scaling factors. PTQ is a static quantization where scale and zero-point are fixed during inference. Different precision levels, such as 8-bit integers, are commonly used in PTQ.

For example, consider an image classification model. The quantization method chosen for the activations in the CNN architecture can be applied to weights, activation values, convolutional layers, and dense layers. On the other hand, the quantization method chosen for the Vision Transformer architecture is applied only to weights and activations.

Pre-training Quantization and Quantization-Aware Training (QAT)

Pre-training quantization simulates quantization effects during the training process itself. This means that the model is trained with fake quantization operations, which mimic the effects of quantization on the model's weights and activations. QAT [Jacob '18] is the most popular pre-training quantization method that results in the least loss of accuracy while quantizing the model. QAT applies the weight and activation quantization during the forward pass, but the backward pass remains with the original numbers, and it will not get quantized. Using quantized values in the forward pass (a.k.a, FakeQuant or training with simulated quantization) and using the original values in the backward pass (backpropagation) of the training process emulates quantization at inference time. Because it is common to use quantized weights at inference time. In short, in the QAT scenario, the quantization error propagates through the network in the forward pass.

By using different weights and activation values in forward-to-backward passes, we get a quantization error. During the backpropagation, however, the cost function becomes resilient to the quantization error, and the backpropagation tries to reduce the error. In other words, during the training, the network takes quantization error into account, and it learns to adjust its weight and activation values to minimize the impact of this error. By using QAT or pre-training quantization, the model is trained with a quantization noise, and thus, it will get trained while being aware of the quantization impact.

By pre-training, we mean a *batch of data* (not all data used for training) feed for training. Despite its accuracy, this approach requires training the dataset, and not all organizations can afford expensive platforms to train a neural network.

Channel-wise Quantization

Recall from Chapter 10 that each pixel includes three channels (e.g., RGB), and channel-wise quantization is a specific type of quantization that can be used for CNN, RNN, GNN, etc. It quantizes each channel for input and output independently. Each tensor channel can be assumed as a vector of depth, and the length of the vector is equal to the number of elements in that channel.

In channel-wise quantization, the clipping range values of each channel are computed separately, and the quantization parameters are chosen independently for each channel. This means that the quantization range for each channel is different, depending on the range of values in that channel. It is recommended to apply channel-wise quantization to weights only and not to activation values to ensure that the decrease in accuracy is insignificant.

There are other types of quantization such as *Mixed-Precision Quantization (MPQ)* or *Per-Layer Quantization (PLQ)*, which could help to gain better accuracy. MPQ uses different precision levels within the same model. For example, some layers might use 8-bit integers while others use 16-bit floats. PLQ quantizes different layers of the neural network with different levels of precision based on their sensitivity to quantization; for example, convolutional layers might use 8-bit integers for weights and activations, while fully connected layers might use 16-bit floats to preserve accuracy.

NOTES:

* While performing quantization, it is recommended not to quantize the first and last layer of a neural network.

* It is very common to use clustering algorithms (Chapter 4) and even self-organized maps (Chapter 12) to build a codebook for quantization.

* In terms of resource utilization, fixed-point quantization is more efficient than floating-point quantization, and it is often used for deployment on resource-constrained devices, while floating-point quantization is used to train and test a model. The fractional part of fixed-point quantization has a limited number of bits and, hence, a limited precision compared to floating-point quantization the mantissa has much higher precision.

* When weights or activation values are imbalanced, asymmetric quantization is favored over symmetric quantization. Asymmetric quantization allows for different ranges for positive and negative values, which can better accommodate distributions where values are not evenly spread around zero.

* While performing quantization, two error types could happen: *underflow* and *overflow*. Underflow occurs when a quantization candidate value is too small, i.e., smaller than the smallest value in the quantization range. For example, consider an 8-bits quantization scheme with a range of 0 to 255. If the input value is 0.5 and we want to represent it in this quantization scheme, the resulting quantized value would be 0 (since 0 is the closest representable value to 0.5). In this case, underflow occurs because the input value is too small to be represented accurately. Overflow occurs when the candidate value for quantization is larger than the largest representable value in the quantization scheme. For example, if we want to quantize 260 in an 8-bit quantization scheme with a range of 0 to 255, the resulting quantized value would be 255 because 255 is the largest representable value.

* Quantization methods in neural networks are not limited to converting floating-point numbers to fixed-point numbers; there are even more coarse quantization approaches that convert float-ranged model parameters into +1 or -1, a.k.a, BinaryConnect [Courbariaux '15], converting float-ranged model parameters into ternary range (-1, 0, 1) [Li '16, Wang '23], using regularization [Alizadeh '20] and so forth.

Pruning and Sparsification in Neural Network

We have briefly described pruning for decision trees back in Chapter 9; in this chapter, we focus on neural network pruning. *Pruning* refers to removing unimportant neurons or connections (parameters) in a neural network or decision trees while maintaining the accuracy of the model to make a lighter model, and thus more resource efficient. The idea of applying pruning to the neural network was introduced back in 1990s [LeCun '89, Reed '93].

Sparsification in neural networks refers to the broad process of increasing the proportion of zero-valued weights in a model. This can be achieved through various methods, including pruning and specific training techniques that encourage weights to converge to zero.

Pruning, a specific technique of sparsification, involves selectively removing weights from the network. Weights are typically pruned based on their magnitudes, with smaller weights often targeted for removal under the assumption that they have less impact on the network's output. However, pruning can also be based on other criteria, such as the overall contribution of weights to network performance.

In summary, pruning means directly removing specific weights from the network. Sparsification is a broader goal of increasing the number of zero weights in the network. It can be achieved through various methods, including pruning, but also other techniques that might turn weights to zero without removing them.

Both methods make some neural network nodes equal to zero, thus making the target weight matrix smaller for training a model. Pruning and sparsification are sometimes used interchangeably, but sparsification is more general as it assigns zero to weights based on importance rather than just magnitude. Note that pruning sets weights to zero and not nodes, but sparsification can be applied to neurons as well.

There is another term we need to learn while dealing with pruning and sparsification, i.e., *regrowth*. Regrowth is the process of adding new weights to a neural network after pruning or sparsification. The objective of regrowth is to restore some of the lost performance of the neural network (because of pruning and sparsification) by adding new weights that are important to the network's performance.

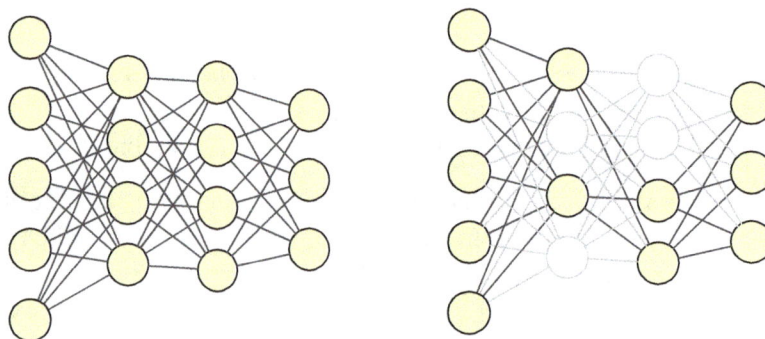

Figure 14-5: (Left) A sample neural network, (right) By sparsification, two neurons at each hidden layer, along with their weights, have been removed.

Figure 14-5 presents a sample neural network on the left and the sparsified version on the right. In the rest of this section, we answer important questions about pruning[2], including what to prune. How to prune or sparsify, and how often do we need to run the pruning/sparsification?

What to Prune/Sparsify?

Pruning can be applied to weights and neurons. Pruning weights involves removing individual weight values that contribute minimally to model outputs. Pruning neurons typically means removing entire units within a network, such as convolutional filters in CNN models or attention heads in transformer models. Neuron pruning is more impactful on the network's architecture and efficiency.

Weight pruning involves removing unimportant weights (weights with low magnitudes) or setting them to zero. By pruning these low-magnitude weights, the network becomes sparser, which leads to faster inference and reduces memory usage. However, removing weights alone may not significantly reduce the number of parameters. Therefore, it is recommended to prune weight along with neurons.

Neuron Sparsification involves removing an entire neuron and its connections from the network. Neurons that contribute little to the output of the network will be removed. One common method is to evaluate the importance of each neuron based on its contribution to the output or its sensitivity to errors. Then, the neurons according to their importance and remove the least important ones. Another method is to prune neurons is to remove their activation frequency or

Sparsificaition:

magnitude. Neurons that are rarely activated or have low activation values can be considered redundant or irrelevant and removed from the network. Neuron pruning reduces model

---

[2] https://roberttlange.github.io/posts/2020/06/lottery-ticket-hypothesis/#frankle_2019

parameters and compresses the model. It is recommended to use weight pruning in conjunction with neuron pruning [Hoefler '21].

Channel, filter, and *attention head Spartisification* is also possible. Depending on the architecture of a neural network, whether it is CNN, RNN, transformer, etc., some elements of the neural network could be sparsified, including channel, filter, and attention heads.

Channel sparsification removes the entire channel that contributes less to the desired output of a CNN. Filter pruning removes the entire filters that contribute less to the output of a CNN network. Attention head pruning removes attention heads in multiple attention head transformer architectures that do not contribute to the output. Attention heads can be pruned by removing entire attention modules or individual attention heads within a module.

Now, two questions might arise: when to apply sparsification or pruning? and how can we identify less important elements, including weights, neurons, channels, filters, and attention heads?

When to Prune?

The process of pruning or sparsification can be done at two different stages: (i) after training and (ii) during the training.

After training: The most common stage of applying to prune is after training, which is known as train-then-sparsify. Usually, after the sparsification, the model is re-trained or fine-tuned to have higher accuracy. Pruning after training can lead to a significant loss of accuracy if the model is not pruned correctly. Besides, this pruning can lead to a loss of interpretability, as the pruned model may be more difficult to understand than the original model. However, compared to pruning during the training, it is easier to implement and requires fewer resources.

During the training: Another approach that is more resource intensive than after training is pruning the model during the training. The model gradually gets pruned iteratively, and in each iteration, it can correct its approximation error to mature the pruning process. The main challenge of this model is that during the pruning, the model should be kept in the memory, and thus, training should be done on hardware with large memory capability. However, due to iterative training, the approximation error is corrected in each epoch, and thus, the quality of this pruning is higher than the previous pruning method. The dropout layer (see Chapter 10) belongs to this category of pruning.

Structured versus Unstructured Pruning

Pruning methods can be categorized as structured or unstructured pruning. Structured pruning methods involve pruning groups of weights, neurons, an entire layer, etc., that are spatially or temporally correlated. In other words, it prunes a group of elements that have a predefined structure. On the other hand, unstructured pruning focuses on removing single elements. An example of structured pruning is filter/channel pruning for CNN. It sets the entire filter/channel to zero. Figure 14-6 shows different approaches to pruning a matrix of weights. On the left side,

Vector Level Pruning

Pattern based Pruning

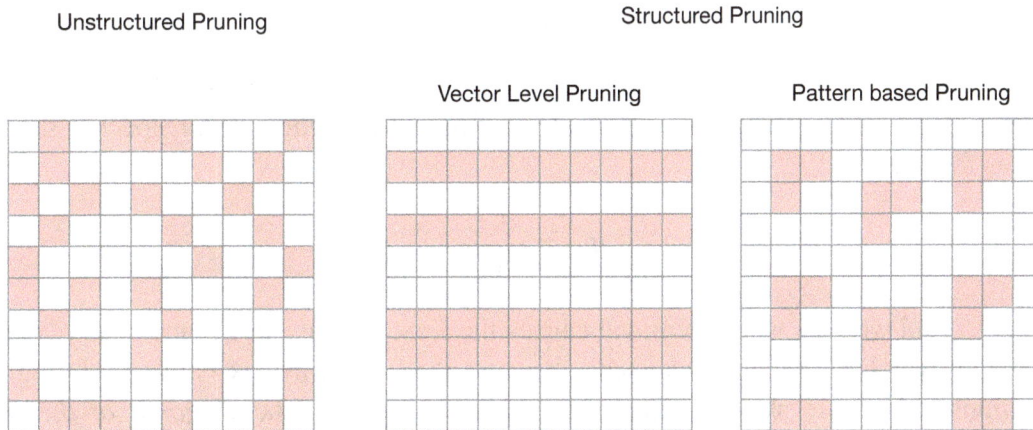

Figure 14-6: The leftmost weight matrix presents unstructured pruning. The two other matrices use structure-based pruning. One makes sparse rows, and the other one applies a pattern on the weight matrix.

we have unstructured pruning (a.k.a. fine-grained pruning), and on the right, we have two examples of structured pruning.

Structure pruning is often used to reduce the neural network size without losing much accuracy and make the model light for deployment on devices with limited computational resources. Unstructured pruning is usually used to improve the inference speed of a model.

The main disadvantage of structured pruning is the loss of expressivity. When a group of elements is pruned, the network may lose some of its expressive power, which reduces accuracy. Besides, structured pruning typically results in a fixed sparsity pattern, i.e., the pruned neurons or filters are predetermined and cannot be adjusted during training. Therefore, the network can not adapt to changes in input data distribution, i.e., drift (we will explain more about drift in Chapter 16). The main disadvantage of unstructured pruning is its difficulty in implementing on hardware. Besides, it can cause overfitting because the model is getting too specialized for the training data and can't deal with the new data.

How to Identify Pruning Candidates?

To determine the criteria for pruning a neural network, we start by explaining the earliest approach for pruning candidate identification, i.e., Hessian-based pruning (OBD and OBS). Then, we describe magnitude-based, activation-based, and Taylor expansion pruning candidate identification.

Optimal Brain Damage (OBD) and Optimal Brain Surgeon (OBS)

Optimal Brain Damage (OBD) [LeCun '89] first calculates the second-order derivative (Hessian matrix) of the loss function with respect to the weight. Then, it uses it to compute a score called *saliency score*. Next, by using the saliency score, it identifies the candidate weights for pruning, that have the least impact on the output (the least impact on the overall accuracy of the network).

Assuming E presents the cost function. and $w_{i,j}$ presents weight i at layer j, and $h_{i,j}$ is the element of Hessian matrix H. The second-order derivative of the loss function with respect to the weight is written as follows:

$$h_{i,j} = \frac{\partial^2 E}{\partial w_{i,j}}$$

The goal is to find weights whose removal causes the least increase in E. Obviously, computing Hessian for all weights does not make sense because the weight matrix is huge. Through some simplifications, which we skip to explain, the authors use only the *diagonal of the Hessian matrix*. Somehow, we can say they assume the Hessian matrix is diagonal[3]. Computing elements on a diagonal only matrix has the same complexity as the gradient. Then, the hessian h_{kk} will be used, to compute the saliency score s_k for each weight as follows:

$$s_k = \frac{h_{kk} w_k^2}{2}$$

Next, weights are sorted based on their saliency score, and ones with lower saliency scores will be removed. OBD is one of the earliest pruning approaches, and due to its assumption of having a diagonal matrix, which Hessian is usually non-diagonal, it is not widely used.

Optimal Brain Surgeon (OBS) [Hassibi '93] proposed a few years after OBD. OBD makes the assumption that the Hessian matrix is diagonal, which may not hold in all cases. This assumption causes removing some important weights, but OBS mitigates this issue by considering Hessian is not diagonal.

The OBS method works by first training a network to a desired level of accuracy. Once the network is trained, the Hessian matrix is calculated. In particular, first, it computes the Hessian matrix of the cost function with respect to the weights (similar to OBD). Then, it computes the inverse of the Hessian matrix (H^{-1}). It uses a recursive approach to compute the inverse of the Hessian matrix from the training data, and we skip explaining it in detail. Next, the saliency score s_q for weight q is computed as follows:

$$s_q = (w_q^2)(2[H^{-1}]_{qq})$$

Afterward, weights are sorted based on their saliency score, and weights with lower saliency scores will be removed. OBS is a recursive method, meaning that it uses the Hessian matrix from the previous iteration to identify weights to remove in the current iteration.

Both OBD and OBS use Hessian, which is computationally expensive to compute, and thus, they are not in use in modern neural networks with lots of weights. However, to learn more recent methods, we should be familiar with classic ones.

_____

[3] A diagonal matrix is a matrix in that all its elements, except its diagonal elements, are zero. The diagonal runs from the top left to the bottom right of the matrix.

Magnitude-based Pruning

Magnitude-based pruning methods remove neural network elements (weights, filters, channels, neurons, or attention heads) based on the magnitude of their values. Elements with smaller values are considered less important and are pruned.

To identify elements for pruning, the magnitude of an element is typically defined as the L_1 or L_2 norm of the weights. For example, L_1 of a filter with weights [0.2, -0.4, 0.1, 0.3] is computed as follows: $|0.2| + |-0.4| + |0.1| + |0.3| = 1.0$.

L_1 or L_2 are used to compute the candidacy for pruning. Additionally, adding these regularizations to the model loss function encourages the model to learn sparse weight matrices (encourages the weights to be small) and makes it easier to identify the weights with the smallest magnitude.

Magnitude-based pruning is closely related to neural network regularization. While regularization is applied during training to prevent overfitting, magnitude-based pruning is applied after training to reduce model size and computational complexity. The first version of magnitude-based pruning is OBD [LeCun '89], which we explained.

Activation-based Pruning

Activation-based pruning [Li '16] is a technique for removing redundant components in neural networks based on their activation values. This method can be applied to various network components:

Neurons: Removes neurons whose absolute activation values are less than a predefined threshold.

Channels: Computes the mean activation of each channel across the training data and prunes channels with the lowest mean activations.

Filters: During the forward pass of the CNN, measure the activations of each filter for a given training dataset. Filters with low activation values are then removed.

Attention heads: Computes the average activation values (e.g., L_2 norm of the context vector produced by each attention head) across all input sequences. Attention heads with the lowest average activations are pruned.

The process generally involves passing data through the network, recording the output activations of the components, sorting them, and removing those with the lowest activation values.

Activation-based pruning differs from magnitude-based pruning in its dynamic nature. While magnitude-based pruning determines the importance of components based on their static weight magnitudes, activation-based pruning considers the dynamic activations during inference. This makes activation-based pruning more adaptive to changes in input data distribution, potentially resulting in better performance in some cases. However, it also requires more computational resources during the pruning process, as it needs to process data through the network to determine activations.

Taylor Expansion based Pruning

Taylor Expansion based pruning [Molchanov '16] leverages second-order derivatives such as OBD [LeCun '89] and OBS [Hassibi '93] and applies a Taylor expansion (check Chapter 8) on cost function to estimate the importance of weight or filter. A Taylor series expansion is applied to the cost function to approximate the impact of removing a weight or filter.

We have learned in Chapter 8 that Taylor expansion can convert any mathematical function into an infinite sum of terms. The Taylor expansion equation is formalized as follows:

$$f(x) = \sum_{k=0}^{\infty} f^{(k)}(a) \frac{(x-a)^k}{k!}$$

For example, assume we would like to apply the Taylor expansion on the cross-entropy cost function. The cross-entropy cost function equation is as follows:

$$L(p) = -\sum_{i=0}^{n} y_i \log p_i$$

By applying Taylor expansion, we can rewrite the cross-entry cost function as first-order derivative:

$$L(p) \approx -\sum_{i=0}^{n} y_i \log a_i + \sum_{i=0}^{n} \frac{y_i}{a_i}(p_i - a_i)$$

Here, a_i is the predicted probability (activation output, before pruning). p_i is the pruned probability (approximated probability), y_i is the actual label. By applying Taylor expansion, we can write the cross-entropy cost function as a second-order derivative as follows:

$$L(p) \approx -\sum_{i=0}^{n} y_i \log a_i - \sum_{i=0}^{n} \frac{y_i}{a_i}(p_i - a_i) + \frac{1}{2}\sum_{i=0}^{n} \frac{y_i}{a_i^2}(p_i - a_i)^2$$

In the above equation, the first term is the original cross-entropy loss function, the second term is the first oder derivative ($[\partial L / \partial a_i] = -y_i/a_i$), and the third term is the second order derivative $[\partial^2 L / \partial a_i^2] = y_i/a_i^2$.

Based on the presented equation to use the Taylor expansion for pruning, first, the original cross-entropy is computed. Then, the first-order derivative of the loss function with respect to the weights in the network is computed. The first-order derivative (gradient) specifies how much the loss function changes when a particular weight changes. Weights with a smaller first-order derivative are considered less important and will be marked for pruning.

The second-order derivative (Hessian matrix) is used to specify how the gradient of the loss function changes with respect to the weights. In other words, the second-order derivative provides information about the curvature of the cost function close to the current weight values.

This curvature tells us how much the output of the network changes as we change the weight values. Taylor expansion-based pruning methods are common to be used to prune filters of CNN or weights.

To summarize this section, keep in mind that the second-order derivative is a more accurate measurement of importance than the first-order derivative alone. However, due to its computational complexity, it is not a common approach to pruning a network.

There are many more pruning methods introduced, such as using Reinforcement Learning [Lin '17], Bayesian pruning [Williams '95] (pruning weights with low posterior probability), and Adversarial pruning [Wu '21], which reduces the chance of an adversarial attack on the network.

Some literature considers dropout and neural network regularizations as pruning methods as well, we do agree with that classification, but we have explained them back in Chapter 10, and thus, we do not list them here again.

We believe it is a very useful research area to investigate with several benefits, including reducing the carbon footprint of neural network inferences, enabling neural network models to operate on battery-powered devices, etc.

NOTES:

* Studies [Hoefler '21] have shown that retraining after each pruning step and fine-tuning after the last pruning step assists the model in gaining back its accuracy. This is also known as *dynamic sparsity.*

* It is recommended that pruning is not applied before training because it results in large accuracy degradation.

* Interestingly, at the beginning of pruning, the model's accuracy increases due to the regularization impact of sparsity, which forces the model to focus on more important features [Frankle '18]. As the pruning increases, the accuracy increases slightly until it suddenly drops, which indicates that the model is pruned too much.

* Some activation functions, such as ReLU or SoftMax, or dropout regularization, are also considered pruning, and they are referred to as *ephemeral* pruning. We described them back in Chapter 10, due to their importance in learning neural networks.

* When we remove groups of weights, neurons, channels, or filters together, it is sometimes referred to as structured pruning.

* The first layers of a CNN are more sensitive to Pruning, and the last layers are less sensitive to pruning. Therefore, we need different pruning ratios for different layers.

* If we prune in the earlier stages of training (when the model is most receptive to restructuring and adapting), we can mitigate the loss of accuracy. Because, at the beginning of training, the connections are still in the rapid learning stage and can be trained to take over the functions of removed connections.

* There is a theory known as "The Lottery Ticket Hypothesis" [Frankle '18]. This theory describes a dense, randomly-initialized, feed-forward neural network containing a sub-network

that can achieve the same accuracy as the original network with a similar number of iterations. This subnetwork is referred to as the *winning ticket* by authors. The authors of this work experiment and realized the subnetwork is usually 10-20% smaller than the original network. They state that if neurons of the subnetwork are randomly initialized, the subnetwork no longer matches the accuracy of the original network. Therefore, the subnetwork requires training in isolation to gain reasonable values for weight and biases. They do not propose any specific pruning method, and they use standard pruning approaches that we have explained earlier. We have experimented with this hypothesis on Deep Reinforcement Learning models [Lu '24], and it is not valid in those networks.

Low-Rank Adaptation of Large Language Models (LoRA)

While large language models (LLMs) are very popular, they are extremely resource-intensive to train, which limits the number of organizations capable of doing so. Academic institutions and even medium-sized enterprises cannot afford the GPU clusters that large corporations such as OpenAI, Google, and Microsoft can. A recent and interesting approach to compress LLMs (discussed in Chapter 12) is Low-Rank Adaptation of Large Language Models (LoRA) [Hu '21]. After the commercial release of ChatGPT in November 2022, there have been many efforts to build LLMs, and to reduce their computational cost, LoRA is among the best solutions.

Low-rank factorization involves approximating the weight matrices in the network using lower-ranked matrices, which can also reduce the number of required computations and thus make the neural network faster to train and infer. LoRA is inspired by low-rank factorization matrix decomposition methods, such as Singular Value Decomposition (SVD), which is described in Chapter 7. It applies these techniques to the fully connected layers of a neural network. In the context of neural network pruning, this process is also called *operator factorization*.

LoRA keeps pre-trained weights intact, but during the fine-tuning, it applies a matrix decomposition to the weights of the *dense layers* inside the transformer architecture (in Chapter 12 we explained the transformer architecture). All described LLMs implement the transformer model (either completely or only its encoder). By using LoRA, the optimizer algorithm only deals with the smaller matrices of weights, and the LoRA fine-tuned model performs faster than using the original weight matrix.

To understand how LoRA works, assume W_0 is the weight of the pre-trained model, and W_Δ is the weight changes after the fine-tuning. By using LoRA, the pretrained weights are frozen and do not get updated during the fine-tuning process. Instead, the fine-tuned weights (W_Δ) are added to the model; thus, after the fine-tuning process, the model weights is a sum of pre-trained and fine-tuned model, i.e., $W_{model} = W_0 + W_\Delta$. This has the advantage that from fine-tuned weights, which are task-specific, we can get back to pre-trained weights by simply subtracting W_Δ from the W_{model}, allowing for fine-tuning on another task. To understand this concept better, look at Figure 14-7, where we present how fine-tuning is implemented without and with LoRA.

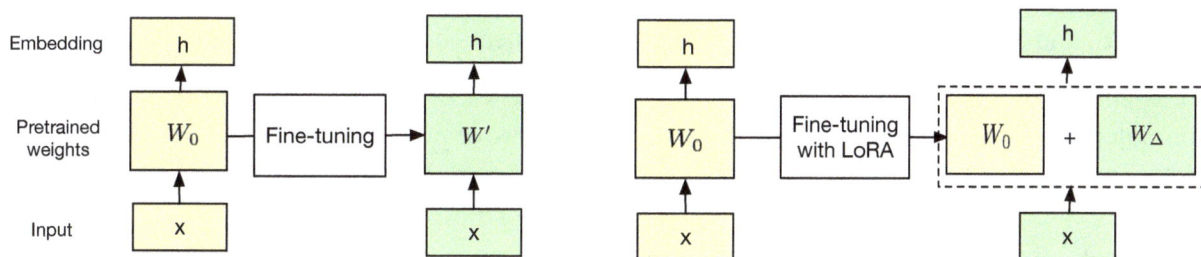

Figure 14-7: Differences between fine-tuning with LoRA and without LoRA.

Weight matrices W_Δ and W_0 both are $\in \mathbb{R}_{n \times k}$[4]. By decomposing it, we can write W_Δ as $W_\Delta = BA$, $B \in \mathbb{R}^{d \times r}$, $A \in \mathbb{R}^{r \times k}$, $r < min(d, k)$. Here, r is a hyperparameter that specifies the dimension of the decomposed matrix. Therefore, the fine-tuned model's weights based on LoRA is written as $W_0 + W_\Delta = W_0 + BA$. Obviously, the fine-tuning process trains A and B. Thus, we can write the embedding results as $h = W_0 x + W_A W_B$.

By using the summation of weights in LoRA, the latency imposed by the fine-tuned model during inference time will be reduced as well. This makes the LoRA an attractive method, and many LLM models are fine-tuned with LoRA. When LoRA is combined with quantization, it is called QLoRA.

Lighter Self-Attention Mechanisms

Efforts to reduce the computational cost of self-attention in transformers are crucial because the self-attention mechanism, which is a defacto standard attention mechanism in all transformers, requires substantial computational resources. As the size of the datasets and the model dimensions increase, so does the need for computing power, making training and deploying these models expensive and energy-intensive. By optimizing self-attention, researchers aim to make these powerful models more accessible and cost-effective, enabling broader usage in more resource-constrained environments. In this section, we list some common approaches that reduce the computational cost of self-attention in transformer architecture.

FlashAttention Models

FlashAttentions is a common choice in some LLMs, such as Mistral [Jiang '23]. Before we start describing FlashAttentions [Dao '22, Dao '23], we should get familiar with a few concepts, including GPU memory components and FLOPS. After learning these concepts, our explanation about flash attention makes sense.

Floating Point Operations Per Second (FLOPS): FLOPS specifies the number of floating point operations a hardware can perform in seconds. They are commonly used for benchmarking hardware performances such as CPUs, GPUs, and RAM. FLOPS ranges for different hardware

---

[4] We read this notation as W_0 belongs to the real number space, and it has $n \times k$ dimensions.

components; for example, CPUs have 10-100 GFLOPS (GigaFLOPS), GPUs have 100-10,000 GFLOPS, and High-Performance Computing (HPC) systems have 1-100 PFLOPS (PetaFLOPS).

GPU and CPU Memory Structure: Both GPU and CPU include a memory. In modern computers, memory is organized into a hierarchy to improve performance and energy utilization, including on-chip SRAM, which is fast, and High Bandwidth Memory (HBM), which is slow in comparison to SRAM and Dynamic RAM (DRAM). DRAM is slower than both HBM and SRAM. As these components become faster, their storage capacities become more limited, leading to increased costs[5]. To summarize, keep in mind that from left to right, the capacity increases, but the response gets slower: SRAM → HBM → DRAM. However, not all systems will have all these types of memory.

GPU hosts neural network operation. To execute an operation, the GPU executes a massive number of parallel processes (threads), which are known as *kernels.*

Self-Attention vs Flash Attention: Self-attention is used in the legendary Transformer architecture, and all transformer-based LLMs use this attention mechanism. Despite its usefulness, self-attention consumes lots of memory because it utilizes HBM to read and write all three components of attention (keys, queries, and values). Since HBM offers high bandwidth, it is an ideal choice for processing massive datasets but slower than other memory components, and thus relying on HBM makes self-attention slow. Flash attention leverages the SRAM and implements self-attention in a more memory-efficient approach.

In particular, in plain self-attention, each attention head involves three main steps: (i) computing the attention scores (dot product of keys and queries), (ii) applying softmax to obtain attention weights, and (iii) then aggregating the values weighted by these attention scores. These operations require multiple memory accesses for reading and writing keys, queries, and values from and to HBM, resulting in a significant number of I/O operations.

In the context of self-attention, 'intermediate results' refer to the computations performed during the attention calculation process. These computations involve multiplying keys and queries to get attention scores, applying softmax to obtain attention weights, and then aggregating values based on these weights. Flash attention optimizes memory usage by *storing intermediate results within the SRAM,* reducing the need for repeated memory accesses. This approach minimizes data movement between different memory levels, which is a key factor in reducing overall I/O operations.

We don't want to go into the details of CUDA[6] and explain further how FastAttention is implemented. We focus here on a high-level overview of understanding this attention mechanism, which is used in some LLMs, such as Mistral, that we explained earlier in Chapter 12.

---

[5] It is worth keeping in mind that disk is the slowest but largest capacity; then we have memory, which is faster and has less capacity, and CPU, which is the fastest but most limited capacity. A good AI expert, like you, will also know some software development, and a good software developer should find a balance between what process or data should be handled by which component.

[6] Compute Unified Device Architecture (CUDA) is a NVIDIA's parallel computing platform that allows software to use certain types of GPUs.

FlashAttention2: Following the initial development of FlashAttention, its first author introduced FlashAttention2 [Dao '23]. FlashAttention2 employs a refined data movement strategy in different GPU memory components that better exploits the hierarchical memory structure, including High Bandwidth Memory (HBM) and on-chip SRAM. It minimizes data transfers between these memory layers and reduces latency and energy consumption associated with memory accesses. Additionally, the architecture of FlashAttention2 enhances parallelism by reworking the partitioning of tasks between different thread blocks and warps, which allows it to achieve higher occupancy and better utilization of the computing resources available on NVIDIA GPUs. At the time of writing this section, NVIDIA GPUs are the most used ones, and rarely other brands of GPU are in use for neural networks.

Multi Query Attention and Group Query Attention

A traditional transformer decoder loads all attention keys, values, and decoder weights at each decoder step. This imposes an overhead on the memory of the system that is training the decoder. In fact, a significant problem of the transformer architecture is the intensive memory usage due to the repeated loading of large keys and values tensors at each step of the decoder during the decoding process. Two fairly similar approaches to mitigate the memory overhead are Multi-Query Attention and Group-Query Attention. We categorize them here in the same group due to their similarities.

Multi Query Attention (MQA)

Multi Query attention (MQA) [Shazeer '19] is used in some language models, such as Google's PaLM [Chowdhery '23], and it is designed to improve computational efficiency during the decoding phase of the transformer by *sharing a single set of keys and values* across all attention heads. This reduces the memory bandwidth required because it diminishes the size of the tensors for keys and values that need to be loaded during each step of the inference.

In particular, each attention head computes its output by applying a softmax function to the dot products of the query (derived from a linear transformation of the input) with a shared set of keys and then using these softmax scores to weigh the shared set of values. The result is an output vector for each head, which is a weighted sum of the same values tensor.

The key difference from standard multi-head attention (MHA) is that while each head in MHA uses *separate* keys and values, in MQA, *all heads share the same* keys and values but generate *different queries*, see Figure 14-8. This means that instead of storing separate keys and values for each head, only a single set is maintained and used, which significantly reduces the number of operations and the amount of memory needed, hence speeding up the processing and reducing computational overhead.

To understand it better, compare the formalization we introduced in the following. We use different colors for the part that is changed between MHA and MQA. X is the input to the attention layer, M contains the contextual embeddings (typically the output of the previous layer), and Ws are weight matrices for attention head number h.

MHA:

(1) Query, Key, and Value Computation for Each Head:
Queries: $Q_h = X W_h^Q$
Keys: $K_h = M W_h^K$
Values: $V_h = M W_h^V$

(2) Attention score calculation for each attention head:
Score Matrix: $S_h = Q_h K_h^T$
Softmax Normalized Attention Weights: $A_h = softmax(S_h)$
Output: $O_h = A_h V_h$

(3) Concatenation of outputs from all attention heads:
Concatenated Output $O = Concat(O_1, ..., O_H)$
Final Output: $Y = O W^O$

MQA:

(1) Query, Key, and Value Computation:
Queries: $Q_h = X W_h^Q$
Shared Keys: $K = M W^K$
Shared Values: $V = M W^V$

(2) Attention calculation:
Score Matrix: $S_h = Q_h K^T$
Softmax Normalized Attention Weights: $A_h = softmax(S_h)$
Output: $O_h = A_h V$

(3) Concatenation of outputs from all heads:
Concatenated Output: $O = Concat(O_1, ..., O_H)$
Final Output: $Y = O W^O$

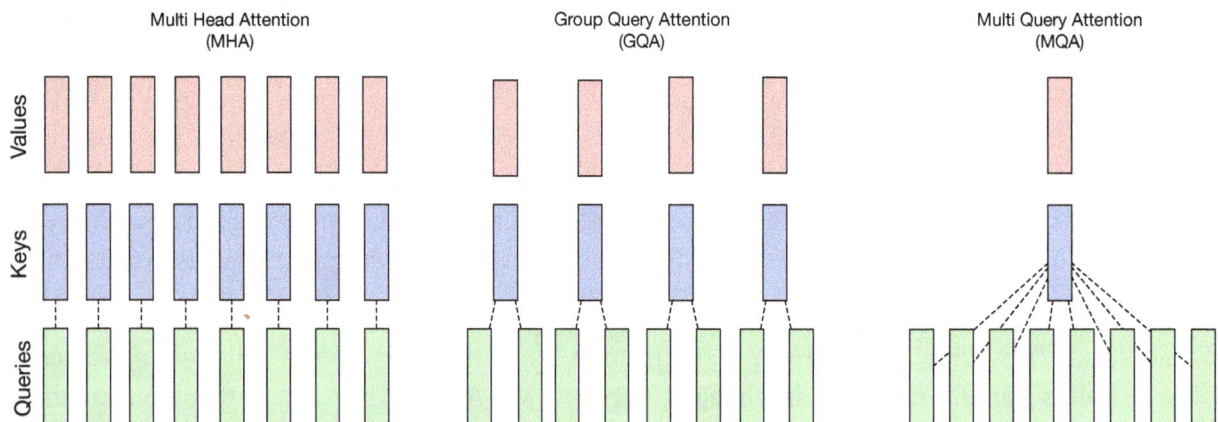

Figure 14-8: Resource efficient attention mechanisms in comparison to classic MHA.

954

Group Query Attention (GQA)

The change in structure from MHA to MQA involves changes in how weights and biases are distributed and used during training. This leads to instability during training, especially during fine-tuning, on tasks with long input sequences. The model might show high variance and frequent loss spikes as it struggles to adapt to the simplified attention mechanism. In summary, the simplification imposed by MQA can limit the model's capacity to capture complex dependencies in data.

To address these limitations, Grouped-Query Attention (GQA) [Ainslie '23] has been proposed, which segments the query heads into G groups, each sharing a key and value head. The number of groups G is an important factor in balancing the model's accuracy and computational efficiency. If G equals 1, the model is identical to MQA; if G equals the total number of attention heads, it is identical to classic MHA. An intermediate value of G, however, divides the key and value heads among fewer groups than MHA but more than MQA, allowing for a more accurate aggregation of information than MQA while still reducing the computational overhead.

Figure 14-9 visualizes different styles of attention mechanisms. If you recall, we have explained that the classic transformer architecture includes eight attention heads, and thus, we have eight keys, values, and queries, as shown in the MHA architecture of this figure. GQA is a very popular attention mechanism that is used in different LLMs, including Llama 2 [Touvron '23] and Llama 3 [Dubey '24].

Sliding Window and Dilated Sliding Window Attentions

In classic self-attention, every token in the input sequence is attending every other token in the input sequence, as shown in Figure 14-9. In other words, every token calculates its attention score with every other token in the sequence, including itself. This is done by taking the dot product of the current token as a query vector with the key vectors of all tokens. This characteristic allows the model to capture relationships between distant words, but it has $O(n^2)$ complexity, where n is the number of tokens and memory limitations restrict processing long sequences.

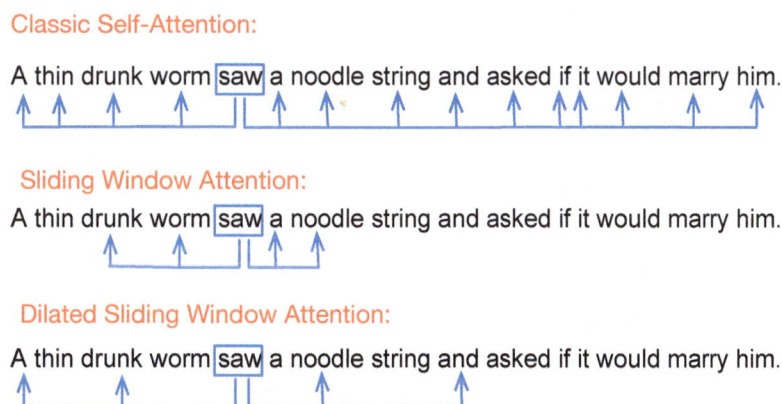

Classic Self-Attention:

A thin drunk worm saw a noodle string and asked if it would marry him.

Sliding Window Attention:

A thin drunk worm saw a noodle string and asked if it would marry him.

Dilated Sliding Window Attention:

A thin drunk worm saw a noodle string and asked if it would marry him.

Figure 14-9: Classic self attention in comparison to sliding window attention and dilated sliding window attention.

Sliding window attention [Beltagy '20] is another technique used to reduce the computational complexity and memory intensity of the transformer's attention. Sliding window attention reduces the computation and memory overhead by restricting the attention span of each token to a fixed-size window of tokens around it. For example, in Figure 14-9, in the "Sliding window Attetion" section, the window size is set to two, and the token 'saw' attends to two neighbor words from each direction. Typically, these windows overlap to ensure there is some shared context between adjacent windows.

However, in sliding window attention, each token only attends to a fixed number of neighboring tokens within a sliding window. This can limit the ability of the model to capture dependencies between tokens that are far apart in the sequence. Dilated sliding window attention addresses this by introducing gaps or dilations (same as dilated convolution we explained in Chapter 10) between the positions that the window attends to.

See the bottom of Figure 14-10, where the word 'saw' is attending words 'a', 'drunk', 'noodle' and 'and'. This allows the window to cover a broader context, enabling the model to capture relationships between distant tokens more effectively.

Memory Efficient Gradient Descent

Until now, Gradient Descent (GD) combined with backpropagation has been the most effective optimization approach for training neural networks. However, computing derivatives, even first-order derivatives, can be computationally expensive. To address this, some techniques have been developed to reduce the number of derivative computations, including Gradient Accumulation, Gradient Checkpointing, and Conjugate Gradient Descent. We explain these three methods in this section.

Gradient Accumulation

When training a neural network model, after each training batch or single data point, in case we use SGD, the neural network parameters (weights and biases) are updated based on the gradients computed using the chain rule.

In gradient accumulation, when training a neural network model, after each training batch or single data point, the neural network parameters are not updated. Instead, after a certain number of iterations, then they are summed (accumulated) together and updated neural network parameters. In other words, the forward pass operates the same way as in other neural networks, but during the backward pass, gradients are not applied to update the parameters until after a fixed number of iterations, at which point they are applied in a single update.

Note that the backpropagation process, which involves calculating the gradients of the loss function with respect to the model parameters, is still applied. After computing the gradients via backpropagation, however, the gradients are not immediately used to update the model parameters. Instead, these gradients are accumulated (i.e., summed together) over several iterations, and afterward, they get updated.

Let's review the gradient descent we explained in Chapter 10. Assuming w_{t+1} is the weight (or any other parameter[7]) at the next iteration, w_t is the weight of the current iteration, α is the *learning rate*, and $\triangledown G(.)$ is the vector of gradients. We estimate a value of w_{t+1} via the following equation: $w_{t+1} = w_t - \alpha . \triangledown G(w_t)$. The only difference that accumulating gradient has with this approach is that gradients are not updated in every iteration. The sum of gradients is updated after a k number of iterations (it is a hyperparameter), which is formalized as follows:

$$w_{t+1} = w_t - \alpha . [\sum_{i=1}^{k} \triangledown G]$$

This approach is helpful in situations where the model or the dataset is large and the available resources (such as GPU memory) are limited. However, while preserving memory, it might make the training process slower.

Gradient Checkpointing

Gradient checkpointing is another method used to reduce memory usage during the training of neural networks. It saves memory by selectively storing *the* activations of only a subset of layers (or checkpoints) during the *forward pass* and recomputing the activations of the intermediate layers during the backward pass for gradient calculation. In other words, instead of storing all intermediate activations, gradient checkpointing saves memory by only storing a subset of them (checkpoints). The other activations are discarded after they are used in the forward pass.

To implement gradient checkpointing, first, we segment a neural network into different segments. During the forward pass, only activations of a certain layer (checkpoint layer) in each segment are saved. The activations of the layers between checkpoint layers are discarded Usually, the first layer in each segment is used to store its activations.

During the backward pass, the activations of the discarded layers are recomputed on-the-fly by performing the forward pass again from the last checkpoint to the current layer. This means the model has to perform additional forward computations during the backward pass, which increases computation time but reduces memory usage. This trade-off allows training larger models on limited-memory hardware by balancing memory usage and computational overhead.

For example, assuming we have a deep network with 100 layers, storing all intermediate activations might require a significant amount of memory. With gradient checkpointing, we save only every 10th layer's activations. This means we only store the activations for layers 0, 10, 20, ..., 90. When computing gradients for layer 15, the activations from layer 10 (the checkpoint) are used to recompute the forward pass from layer 10 to 15, allowing the calculation of the gradient with much lower memory usage.

Table 14-2 summarizes the differences between Gradient Accumulation and Gradient Checkpointing.

---

[7] While reading text for optimization in the context of the neural networks, some literature refers to parameters as θ or β, we use w to present weight. Nevertheless, bias b is also a parameter for the neural network. Besides, based on the literature, these parameter names change and we do comply with the name change to avoid confusing you while reading other resources as well.

	Forward Pass	Backward Pass
Gradient Accumulation	no changes	Gradients are calculated via backpropagation but they are accumulated (summed) over a fixed number of iterations and after accumulation, the gradients are applied in a single update to the model parameters.
Gradient Checkpointing	The model processes the input data to produce outputs, but not all intermediate activations are stored in memory. Instead, only a subset of activations (checkpoints) is saved. The other activations are discarded.	The discarded activations are recomputed by running a forward pass again from the nearest checkpoint to the point in the network where the gradient is being computed. This means that the backward pass involves some recomputation, which increases the computational cost but reduces memory usage. After recomputation, the gradients are calculated as usual, and the model parameters are updated immediately (unlike gradient accumulation, which delays updates).

Table 14-2: Comparison between forward and backward passes between gradient accumulation and gradient checkpointing.

Conjugate Gradient

The term conjugate means joined together, and the Conjugate Gradient Descent (CGD) [Stiefel '52] is an iterative algorithm that is used to find the minimum of an optimization function. More accurately, it is used to solve systems of linear equations.

Recalling the gradient descent, we say the next step is determined by using this equation, $w_{t+1} = w_t - \alpha \cdot \nabla G(w_t)$. A good step is a step that reduces the slope toward zero (minimum point) and does not change the sign of the gradient (avoid overshooting). In Chapter 10, we described different optimization methods that can improve SGD. For example, momentum decreases the chance of the optimizer to stay at the local minimum. The gradient should move in the direction of the steepest gradient.

The conjugate gradient, similar to other gradients, updates the direction of movement at each iteration based on the gradient of the function at the current point. CG's focus is on finding a search direction that is orthogonal to the previous direction it took every time. In a more technical sense, directions are conjugate with respect to the Hessian matrix.

In simple terms, conjugate gradient, like all other gradients, starts at any point and takes a step in the direction of the steepest descent. However, its difference is that the direction of the next step is *perpendicular* to the previous step. Figure 14-10 visualizes the differences between SGD and CGD for a two-dimensional space. We can see that SGD takes more steps to get close to the minima, but CGD gets there with four steps. As we can see in Figure 14-10, CGD could lead the model to converge significantly faster than other gradient methods. By finding a new direction that is approximately perpendicular to the previous step, CGD avoids the zigzagging pattern that can occur in gradient descent and leads to faster convergence.

958

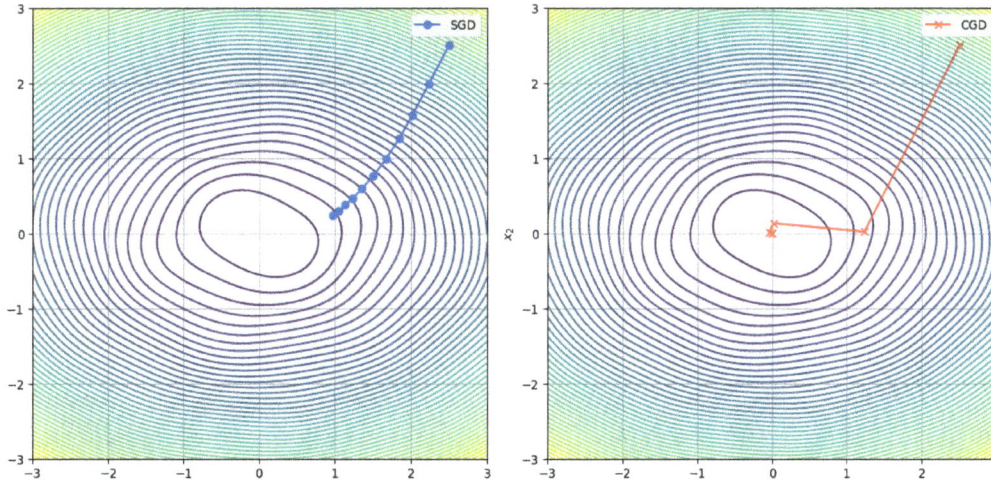

Figure 14-10: Comparison between SGD (left) and CGD (right) on a slightly non-convex optimization shape.

Computing the next step in CGD

A question we should answer is how CGD computes the next update. We know that the SGD is $w_{t+1} = w_t - \alpha . g_t$. For the sake of simplicity, we used g_t instead of $\triangledown G(w_t)$. CGD uses the following equation:

$$w_{t+1} = w_t - \alpha . d_t$$

Here, d_t presents the search direction at step t. It begins with an initial search direction, that is $d_0 = g_0$. To compute the other directions, it uses the following equation:

$$d_{t+1} = -g_{t+1} + \beta_t d_t$$

β_t is the parameter that ensures the new direction d_{t+1} is conjugate to the previous direction d_t. There are different ways to compute β_t. Assuming T denotes the transpose of a vector, a common approach is to use the Fletcher-Reeves formula, which is computed as follows.

$$\beta_t = \frac{g_{t+1}^T . g_{t+1}}{(g_t^T g_t)}$$

Neural Network Training Paradigms

Training a machine learning algorithm and building a model is expensive, especially since deep neural networks are too expensive to train without access to large GPU clusters. Therefore, it is common to use a trained model for different applications or fine-tune a trained model, which means retraining the trained model for a specific task. There are different architectures for deep learning model training, and this section lists them. Many of these architectures leverage pre-trained models or fine-tuning to save time and resources. Given the popularity of transfer learning, our explanations will frequently compare other approaches to it.

Transfer Learning

There are several robust models, usually provided by big corporations and trained on a large amount of data via a large number of GPU devices. We can still benefit from these large models for our applications. In particular, we can benefit from the parameters (weights and biases) of existing models and further train (fine-tune) them on our dataset and application. It is common to employ existing models (existing knowledge) and build our model. This process is referred to as *transfer learning*, and it is the most common approach for deep learning.

For example, a CNN trained on the ImageNet dataset consists of millions of images across thousands of classes. We can use that pretrained CNN model and retrain (fine-tune) it on a new dataset of food images. Another example involves the ImageNet dataset itself, containing over 14 million images categorized into roughly 22,000 categories. Among these categories are many different bird species, such as eagles, owls, hummingbirds, parrots, and more. Now, we would like to train a model to identify different species of canary birds, and that does not exist in models that are trained by ImageNet. We can collect our small dataset of canaries. Then, a model trained on ImageNet, such as ResNet101, will be employed and retrained (fine-tune the trained model with our dataset) to learn to identify the canary species.

Another reason for the popularity of transfer learning is the limited availability of data, which prevents us from training a model from the beginning with the available data. This is a very common approach for medical image analysis, in which acquiring data is an expensive and very bureaucratic process. For example, in one research, we employed a model trained on CT images of lungs and fine-tuned it with Covid-19 images. As a result, we have an accurate model that can recognize Covid-19 CT images from other pneumonia or healthy lungs [Javaheri '21].

Transfer learning leverages the existing knowledge to acquire new knowledge. It is similar to the human approach to learning. In the primary years of school, we learn to read; later in school and at university, or on our own, we learn to create a narrative and write about it. We already have accumulated lots of knowledge, and now, by reading this text, we leverage them to learn new concepts.

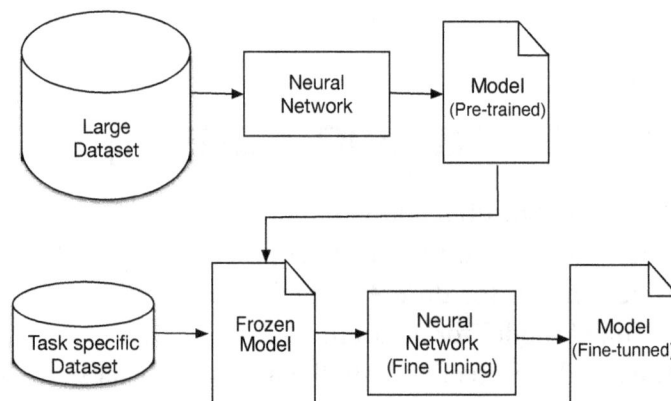

Figure 14-11: The most common use of transfer learning is fine-tuning an existing model for a new task, which usually has less data and fewer resources to train a model from scratch.

Figure 14-11 presents the most common use of transfer learning, which is fine-tuning an existing model for our desired task. A model is trained on a large dataset, a.k.a, pre-training, probably by using many GPUs. This model can be used by similar applications, and they can fine-tune it for their tasks.

While we have learned transfer learning, we should learn the concept of domain adaptation as well. *Domain adaptation* is a transfer learning approach for transferring knowledge from one domain to another. It is useful when we have a model trained on one dataset (the source domain), but we want to use it to make predictions on a different dataset (the target domain). A common challenge in domain adaptation is the difference in data distributions between the source and target domains, which can make knowledge transfer difficult. Additionally, the lack of labels in the target domain is another significant challenge in domain adaptation.

There are other forms of transfer learning as well. For example, we can use a pre-trained CNN model to identify objects in the image. Then, we can train an RNN that employs this pre-trained CNN and use it to generate a caption for each given image. This technique, known as image captioning, leverages the strengths of both CNNs and RNNs: the CNN efficiently extracts spatial features from images, while the RNN processes these features sequentially to generate meaningful descriptions. Furthermore, transfer learning can be applied across different modalities. For example, a model pre-trained on text data can be adapted to work with audio data by fine-tuning it on an audio dataset.

Multi-task Learning

Multitask learning refers to training a neural network to perform multiple tasks by sharing some of the network's layers and parameters across tasks. Generally, we use multi-task learning for tasks with some shared structure.

We have learned that transfer learning operates in a sequence. We train a model on a large dataset or employ an already trained model on a large dataset, and next, we use this model and fine-tune it on a smaller dataset for our task, which is similar but not identical.

In contrast to transfer learning, multi-task learning involves the simultaneous execution of multiple operations, each responsible for learning a different aspect. For instance, in the context of a self-driving car, we can use multi-task learning as follows: a neural network for the car should be capable of tasks like image segmentation (as discussed in Chapter 6) and the detection of pedestrians, vehicles, and traffic signs. In this scenario, the neural network's output should produce a 3D tensor, comprising elements like coordinates of objects and a scalar to specify their types (e.g., pedestrian, etc.). While these tasks may be diverse, they all can be framed as classification problems, allowing us to use a common objective function like cross-entropy for the network. However, it's important to note that tasks can vary, and their respective objective functions may differ. One task might involve regression, while another might be a classification problem. Therefore, each task may require its unique objective function.

Multitask learning is useful when (i) different tasks benefit from the same feature extraction methods and (ii) the amount of data required to be learned for each goal is the same or not significantly different.

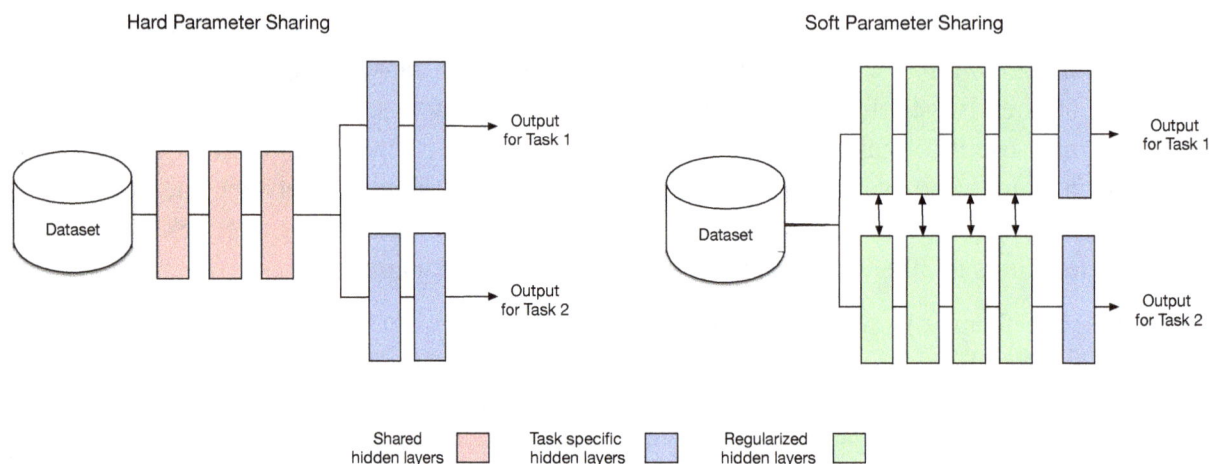

Figure 14-12: Two different models of parameter sharing in multi-task learning. In another scenario, they use the same dataset, but each task can use a separate dataset as well.

In many cases, multi-task learning uses one network, but when tasks are very different, one network is insufficient. Even in these scenarios, some initial layers can be shared to detect common features, but then we make branches in the neural network for each task and use different layers for different tasks.

Figure 14-12 visualizes two different approaches for multi-task learning: hard parameter sharing and soft parameter sharing. In *hard parameter sharing*, the hidden layers for all tasks and their parameters are shared across all tasks, and then close to the output layer, each task can have its hidden layers that lead to the task specific output. In *soft parameter sharing,* each task is learned with separate models and weights, but the cost function introduces a loss that encourages the parameters of all models to be similar. In particular, the model regularizes (e.g., using L_2 norm) the distance between the parameters of the individual models to the overall training objective. This approach encourages different tasks to have similar model parameters.

We have explained that cost functions in multitasking learning could be separated for each task, or they could be shared among different networks. A common approach is to make a linear combination of all cost functions of existing tasks. Assuming θ_t is the subset of parameters for task t, L_t is the loss function for task t, and α_t is a scalar weight that specifies the importance of task t, which specifics the task importance, the cost function $L(\theta)$ for T tasks is written as a linear combination of each task's cost:

$$L(\theta) = \sum_{i=1}^{T} \alpha_t L_t(\theta_t)$$

Sometimes, we need to encourage similarity or diversity among tasks. We could also use a regularizer in the cost function as well. By using a regularizer in the multi-task learning cost function, we can reduce the complexity and size of the model by penalizing some parameters. This can help the model avoid overfitting. We can formalize a regularizer for multi-task learning as follows:

$$L(\theta) = \sum_{i=1}^{T} \alpha_t L_t(\theta_t) + \lambda R(\theta)$$

When tasks are related and can benefit from sharing information or features, we encourage similarity in multitasking learning [Shui '19]. For instance, in natural language processing, we can use multi-task learning to jointly learn different tasks, such as part-of-speech tagging, named entity recognition, and sentiment analysis.

To encourage diversity, we can think of tasks that have different objectives. For example, in recommendation systems, we can use multi-task learning to jointly learn different tasks such as click-through rate prediction, conversion rate prediction, and user satisfaction prediction [Aoki '22].

Multitask learning can increase resource efficiency by allowing multiple tasks to share the same model. This can reduce the training data and computational cost required for each task, which can lead to faster training times and better resource utilization on each task. Besides, multi-task learning reduces the risk of overfitting because a model that learns more than one task has a better chance of avoiding overfitting.

Meta-learning

Meta-learning [Schmidhuber '87] refers to the process of "learning to learn". The objective of meta-learning is to *learn new concepts and skills fast with few examples*. Meta-learning develops models or algorithms that can quickly adapt to new tasks or situations by leveraging prior knowledge. For example, imagine we trained a robot to perform different household tasks, such as making coffee and tea and putting dishes into the dishwasher. Then, by observing a few office tasks, it can perform new office tasks such as answering the phone in addition to household tasks. As another example, we trained a classifier to recognize chickens, but by giving it a few images of dinosaurs, our chicken classifier is capable of recognizing dinosaurs as well as chickens.

These examples show that instead of training a separate model for each task, we could use meta-learning to train a single model that can quickly adapt to new tasks with just a few examples. In other words, meta-learning methods can generalize their models to new tasks, which are not seen during training. This is because they learn how to learn rather than just learning a specific task.

To implement meta-learning, we first train the neural network model on large datasets of related tasks (e.g., serving coffee and tea and putting dishes into the dishwasher). Training a model on a variety of different tasks enables the model to learn the underlying relationships between those tasks. When a target model is presented with a new task (e.g., ordering office files), it can use its knowledge of the underlying relationships to learn to perform the new task quickly. In summary, once the meta-learning model is trained, we can adapt to new tasks with just a few examples.

There are three common categories of meta-learning, including *model-based meta-learning*, *metric-based meta-learning*, and *optimization-based meta-learning*.

Model-based Meta-Learning (MML)

It uses a model to learn how to learn. The model learns to adapt to new tasks by learning optimal parameter initializations. This attitude enables it to adapt to new tasks with few examples quickly. An MML model updates its parameters rapidly after a few training steps and quickly adapts to new tasks by learning a set of common parameters. These common parameters are shared across new and existing tasks, enabling the model to leverage prior knowledge.

One example of MML is Memory-Augmented Neural Networks (MANN) [Santoro '16], which use an external memory module to store and retrieve information from previous tasks. By using external memory. This memory can be used to store information; the model does not currently use that, but it may be needed later. This allows MANNs to learn more complex tasks than traditional neural networks, which can only store information in their internal state. MANNs are made up of two main components: a *controller* and a *memory*. The controller is a neural network, and it is responsible for learning how to access and use the memory. The memory is a sparse matrix, and it is responsible for storing and retrieving information.

Metric-based Meta-Learning

Metric-based meta-learning starts by identifying a metric that can be used to compare different data points. This learned metric is then employed to determine the best learning model for a new task. Essentially, metric-based meta-learning, also known as metric learning, focuses on learning a metric or distance function that can compare and classify new data points effectively.

Metric-based Meta-Learning differs from traditional metric-based algorithms like k-Nearest Neighbors (kNN), (see Chapter 9), or partition-based clustering (see Chapter 4). In those algorithms, metrics such as Euclidean distance are predefined. In contrast, in metric-based meta-learning, the model itself learns or identifies the appropriate metric.

The learning process begins by defining a set of tasks or a task distribution from which the model samples tasks for training and testing. During training, the model learns a metric that captures relevant similarities or differences between data points. This learned metric is then used to compare and classify new data points in future tasks. By using this approach, the learned metric is tailored specifically to the given tasks rather than relying on a universal metric.

A very common use of metric-based meta-learning is to solve one-shot learning problems, which we have explained in Chapter 11 and Chapter 12. One-shot learning tries to leverage prior knowledge to learn new, unencountered concepts quickly. Imagine we are training a model to recognize different fruits. However, instead of directly training it on a specific fruit classification task, we use metric-based meta-learning. In metric-based meta-learning, we first define a set of tasks or task distributions related to different fruit breeds. For example, we could have tasks like distinguishing between apples from pears or distinguishing grapes from blueberries. During the training phase, the model learns a metric that can measure the similarities between fruit images. In other words, it learns to extract features that are relevant for distinguishing between different fruits.

Once the model has learned the metric, it can generalize this knowledge to new tasks or new fruit that it has not encountered before, e.g., pomegranate. When presented with a new image of a pomegranate, the model can compare the features of that image to the learned metric and make the proper label for the new fruit.

Optimization-based Meta-Learning

Optimization-based meta-learning trains a meta-learner to develop a new optimizer, known as *meta-optimizer* or *learned optimizer*, that can quickly adapt to new tasks. Unlike traditional neural network optimization, which focuses on using a pre-defined optimizer to find optimal model parameters to minimize the cost function for a specific task or dataset, optimization-based meta-learning aims to enhance the model's ability to learn across multiple tasks by optimizing the learning process itself.

These methods are particularly designed for gradient descent, a de-facto optimization algorithm for neural networks. We have learned back in Chapter 10, that deep neural networks learn through the backpropagation of gradients, but the gradient-based optimization can not operate accurately when the training dataset is small. The objective of optimization-based meta-learning is to adapt the optimization algorithm to enable the model to learn effectively, even with limited training samples.

In the context of meta-learning we are dealing with a *support set* and a *query set*. The support set is a small number of labeled examples used to fine-tune the model parameters for a new task. The query set is another set of labeled examples used to measure the accuracy of the model on the new task. Therefore, task T can be defined as a combination of support set S, and query set Q, $T = \{S, Q\}$.

One of the common optimization-based meta-learning is Model-Agnostic Meta-learning (MAML) [Finn '17]. MAML aims to learn an initial set of model parameters (meta-parameters) that are applicable to different numbers of tasks. Meta-parameters are learned by repeatedly sampling a set of tasks T_i, training one model with parameters θ_i on each task, and updating the meta-parameters based on the performance of these models.

MAML trains a generic model (meta-model) on a set of tasks. Then, it fine-tunes the meta-model for new tasks. In particular, it operates by inner loop optimization that occurs inside an outer loop optimization. Inner loop optimization (task fine-tuning) trains the model on a few examples from each specific task and produces updated parameters θ' for each task. Outer loop optimization updates the meta-model parameters θ, using the gradients from the inner loop optimization. In other words, the model parameters are updated using the meta-objective, which is the average performance of the model on new tasks after the inner loop updates.

Model-based meta-learning can be more accurate but may also be more computationally expensive. Metric-based meta-learning is less computationally expensive but can be less accurate. Optimization-based meta-learning can achieve high accuracy, but it can also be very computationally demanding.

In summary, meta-learning focuses on developing a general learning strategy or approach that can be applied to a wide range of tasks rather than merely transferring specific knowledge or skills from one task to another (as in transfer learning). Meta-learning has the potential to save time and resources compared to traditional machine learning approaches, where a separate model must be trained for each new task.

Curriculum Learning

Curriculum learning [Weinshall '18] is a training approach where a model begins with easy and small sets of examples and gradually progresses to more challenging data as the network learns. In other words, Curriculum learning is a training approach that involves gradually increasing the difficulty of training data to improve a model's learning efficiency and generalizability.

For instance, consider training a neural network to classify videos of various exercises in a gym. Initially, the network might only be exposed to three exercises, such as "running on a treadmill," "bench press," and "leg squat." The network is trained on these examples, and then additional exercises, like "crunches" and "deadlifts," are introduced in subsequent rounds of training. This process continues until the network has been trained on all possible exercises, potentially covering over 500 different activities.

By utilizing curriculum learning, the neural network can better grasp generalizable features, which aids in recognizing more difficult examples. For example, distinguishing between the "shoulder press" and the "Arnold press," which are very similar but distinct exercises. There are various forms of curriculum learning for selecting data. Some methods order input data for training, which can be done manually (human-based), as in the example we discussed, or automatically by an algorithm.

A category of curriculum learning is *Entropy-based curriculum learning*. This approach orders the training examples based on their entropy (check Chapter 3), which is a measure of how uncertain the network is about the correct label for the example. Initially, the network is trained on examples with the lowest entropy, and then it gradually progresses to examples with higher entropy. By starting with low-entropy examples, the network can quickly build a strong foundation of confident predictions, which in turn enables it to better handle more uncertain and challenging examples later on. For example, first, images that have low entropy (e.g., high resolution and clear images) are fed to the network, and then noisy and blurry ones, which have high entropy, will fed into the model.

Another common approach is using some predefined rules to sort sample data based on their difficulties, and this approach is known as *heuristic-based curriculum learning*. Heuristic-based curriculum learning relies on domain-specific knowledge to design a curriculum that gradually increases in difficulty. For instance, in image classification, a heuristic curriculum might start with simple, iconic images (e.g., a clear picture of a cat) and progress to more complex or ambiguous examples (e.g., a cat partially occluded or in a cluttered scene).

Curriculum learning can be used to make lighter models by assisting the network to learn more generalizable features and reduce the neural network parameters [Karimi '22].

Federated Learning

All methods we have explained until now train a model on a single device and share the model with other devices or applications to use them. Federated learning [Konečný '18], however, trains the model across several devices, i.e., decentralizing the training. In particular, it enables devices (mostly mobile phones) connected to a network to download a global model, fine-tune it by building their customized model (local model), and also share their model parameters with a central model (global model). Those device specific models have inferior accuracy in comparison to the central model. However, they are customized to the device owner's data, and this approach of training and using the data addresses privacy, safety, and data ownership issues.

Applications that host federated learning first download a global model from a centralized server of the application. Then, the model is customized or retrained by the data collected from the device. This is known as the local model. In other words, the device downloads the model and retrains it with its own data. Next, it shares the model with the server (where the global model is hosted). Data collected from devices enable the global model to improve its accuracy. Figure 14-13 presents the abstract architecture of Federated learning.

In addition to safety and privacy, federated learning results in smaller models, lower latency, and less power consumption. Despite all its advantages, it has two major issues. First, it imposes a high communication cost between the local device and the central model. There are approaches to reduce this cost, such as compressing the data before transferring them (e.g., using Huffman encoding [Malekijoo '21]), but still, the overhead exists. The other issue with Federated learning is the inferior accuracy of local models in comparison to the global model.

There are many varieties of federated learning, but we briefly describe three common ones, including FedSGD, FedAvg, and FedDyn.

FedSGD: FedSGD [McMahan '17] is one of the simplest algorithms for Federated learning, where each device computes the gradient of the model using its own data and sends it to a central server. The server then averages the gradients from different devices to update the global model.

FedAvg: FedAvg [McMahan '17] is one of the popular algorithms for Federated learning, where each device updates the model using its own data for multiple steps and sends the updated model parameters to a central server. The server then averages the parameters from different devices to get the global model. FedAvg can reduce the communication cost and speed up the convergence compared to FedSGD, but it may also introduce some challenges, such as data leakage or model mismatch.

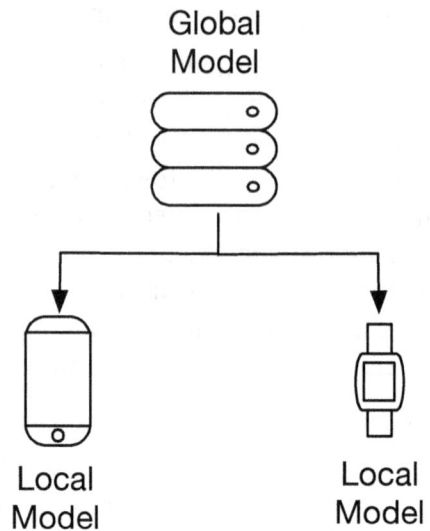

Figure 14-13: Federated learning architecture.

FedDyn: FedDyn [Acar '21] uses *dynamic regularization* to align the local and global objectives of the model. FedDyn allows each device to update the model using its data and a dynamically updated regularizer that depends on the global model and the device's previous model. In other words, the regularizer helps the device's model (local model) to be closer to the global model while also staying close to its previous model. Therefore, the local model does not deviate too much from the global model.

Implementing a federated learning architecture from scratch can be complex for a small group of developers. If there is no risk of model drift (which we will explain in Chapter 16), it might be more practical to prioritize running a static model on the device as a stand-alone model [Rawassizadeh '18]. If an on-device setting is not feasible, federated learning can be considered as an alternative.

Knowledge Distillation

If you recall from elementary school, we learned that the simplest form of distillation is a process that separates the salt from the salty water and provides clean water at the end. *Knowledge distillation* is a technique that uses a large model to train a smaller model. Consider the large model as a salty model and the small model as clean water.

It is different from transfer learning because transfer learning usually involves taking a model that has been trained on one task and applying it to a different task. The objective of knowledge distillation is, however, to get the knowledge from a large, complex model into a smaller, simpler model and use the smaller model instead of the large one, which is more resource-efficient.

The large model is called the *teacher*, and the smaller model is called the *student*. The teacher model is used to generate labels for the student model, which the student model then uses to learn the task. Figure 14-14 represents the role of the teacher and student from knowledge distillation in the real distillation process to help us understand them. The teacher distills the knowledge, and the distilled knowledge is transferred to the student.

In transfer learning, the parameters of the model are transferred from the first task to the second task. In knowledge distillation, the parameters of the model are not transferred. Instead, the teacher model is used to generate labels for the student model, which the student model then uses those labels to learn the task.

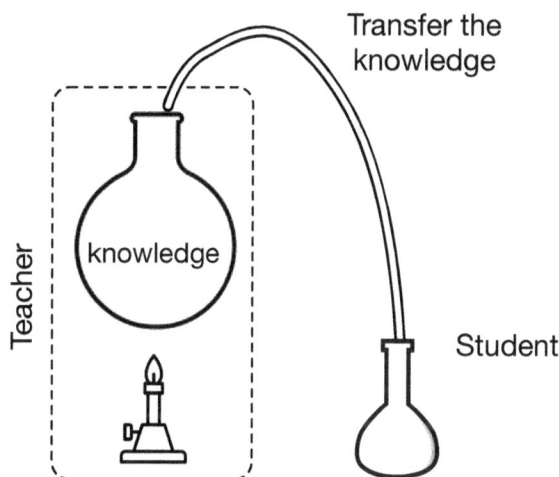

Figure 14-14: Distillation process and the resemblance of teacher and student in the distillation process.

For example, assume we use a smart glass that can identify objects for people who are blind and explain the objects and their proxies. A robust object detection model is complex and large. We can not deploy that model on smart glasses with limited computational resources. Knowledge distillation can transfer knowledge from a large, powerful object detection model (teacher model) trained on a strong machine to the smart glass, which can host the student model.

Knowledge Distillation Concepts

Before we start to explain the training process in knowledge distillation, we follow our holy tradition in this book, and we should learn a couple of concepts. While reading these concepts, you can look at Figure 14-15, which presents the knowledge distillation architecture, to understand them better.

Soft and hard targets

A soft target is a probability distribution over the labels produced by the teacher model. In some cases, the teacher model is not completely confident in its predicted labels and thus assigns probabilities to multiple classes rather than being highly certain about a single class. For example, if the teacher model is 90% confident that the input image is a cat and 10% confident that it is a dog, then the soft target for the cat class would be 0.9, and the soft target for the dog class would be 0.1.

Hard targets, which are generated by the ground truth dataset (ground truth explained in Chapter 1), are one-hot encoded labels, which means that they only have a value of 1 for the correct class and 0 for all other classes. Soft targets, on the other hand, can have values between 0 and 1 for all classes.

In summary, soft targets enable the model to take into account the uncertainty of predictions made by the teacher model. The use of soft targets in knowledge distillation has been shown to improve the performance of the student model. In summary, the teacher model typically produces soft targets, while the ground truth labels are usually hard targets. The student model is then trained to match both the soft targets from the teacher and the hard targets from the ground truth.

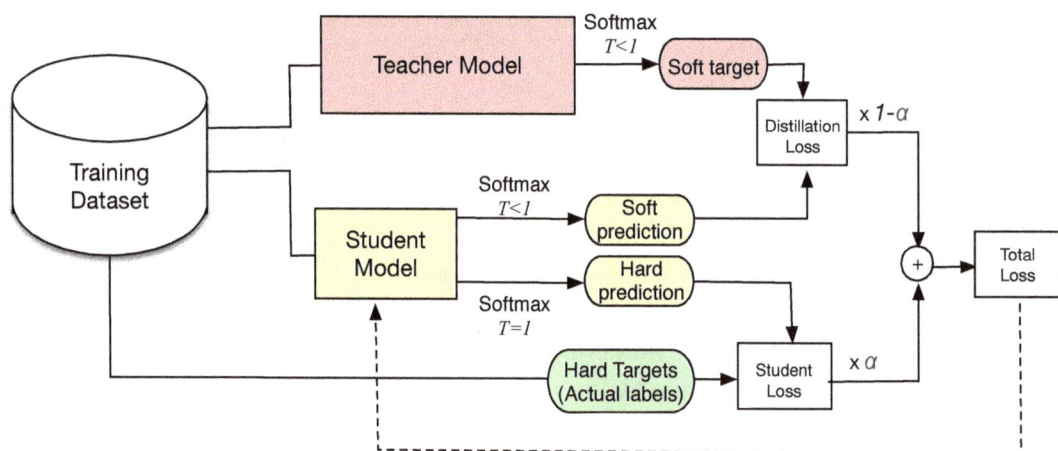

Figure 14-15: Knowledge distillation architecture. The teacher model and its output are shown in red, and the student model and its output are in green. Ground truth labels (hard targets) are shown in green. Dotted lines are referred to the backpropagation that is used to change the student model parameters.

Softmax temperature

In some cases, we intentionally design models to express less certainty in their outputs. This approach can help the model better handle noisy data and improve its ability to generalize to new, unseen examples, thereby reducing the risk of overfitting the training data. To control the level of certainty in a model's output distribution, we can adjust a hyperparameter known as the *temperature* parameter. This allows us to fine-tune the balance between confidence and generalization in our model's predictions.

Neural networks that perform multiclass classification usually have a softmax at their last layer. We have learned back in Chapter 8 that the softmax function is used to convert a vector of numbers into a probability distribution, which represents the probability of each class. *Softmax temperature* introduces a temperature parameter to the softmax, which is often set as a value larger than 1, as increasing the temperature softens the probability distribution. Let's review the softmax equation:

$$\sigma(.)_i = \frac{exp(z_i)}{\sum_{j=1}^{K} exp(z_j)}$$

970

By using temperature T we can rewrite it as:

$$\sigma(.)_i = \frac{exp(z_i/T)}{\sum_{j=1}^{K} exp(z_j/T)}$$

The higher we set the temperature value, the softer the distribution will be. This means that by increasing the temperature of the soft targets, we can make the model smoother and thus increase its generalizability. This is due to the face that a higher temperature results in a softer probability distribution, which makes the model less confident in its predictions and thus gains more generalization capability. Based on this explanation, we can introduce two new terms: hard prediction and soft prediction.

Hard prediction: A hard prediction refers to the output of the model (student model) when it produces a result without incorporating any uncertainty (the softmax temperature is 1).

Soft prediction: A soft prediction refers to the model output (student model) when a temperature is incorporated in the softmax function.

Transfer set

The transfer set refers to a separate dataset (or a subset of the existing dataset) used to train the student model. It is called a transfer set because it is used to transfer the knowledge from a teacher model to a smaller student model by mimicking the teacher model's predictions during training.

Balancing factor

To ensure the student model learns both from the teacher's soft target and the ground truth labels, a balancing factor (α) is used in the loss function. This factor determines the relative weight given to the hard target loss (α) and the soft target loss ($1 - \alpha$) during training. It is a hyperparameter, and if we consider both losses have equal importance, we can set $\alpha = 0.5$. In the following section we learn the student and teacher loss functions, and we see where do we incorporate the balancing factor.

Knowledge Distillation Training

The process of training the knowledge distillation is presented in Figure 14-15. It usually consists of the following steps.

- *Teacher training:* the teacher model is typically a complex architecture trained on a large dataset. It is capable of achieving high accuracy, but it is computationally expensive or too large to deploy on resource-constrained devices. In many cases, we use a pre-trained teacher model rather than training a new one from scratch.

- *Generating soft targets:* instead of training the student model using the hard labels produced by the teacher model, teachers' labels are converted to soft targets by using a softened softmax function controlled by a temperature parameter.

- *Training the student model with soft targets:* The student model is trained by using both the input examples and the soft targets generated by the teacher model. A cost function (distillation cost) uses KL-divergence or cross-entropy to compare the soft prediction of the student model with the soft target of the teacher. This training process aims to minimize the discrepancy between the student model's soft predictions and the soft targets. Assuming KL presents KL-divergence, T temperature, and σ softmax function, z_t presents the *teacher output logits*, and z_s presents the *student output logits*, the distillation loss, which is the cost function at this stage, is presented as follows:

$$Distillation\ Loss = (1 - \alpha)\ KL(\ \sigma(z_t/T), \sigma(z_s/T)\ )$$

In some implementations[8], instead of cross-entropy, KL-divergence is used. We can also use other distribution distance measurement methods.

- *Training the student model with ground truth labels (hard targets):* To improve the learning process and capture more information, the student model will be trained with ground truth labels (hard targets). The cost function which is used to optimize the student model in this stage is referred to as the *student cost function*. Assuming H presents the cross-entropy, the cost function of this stage (the student loss) is written as follows:

$$Student\ Loss = \alpha\ H(\ \sigma(z_t), \sigma(z_s)\ )$$

- *Combining losses:* The total loss is the sum of distribution loss and student loss within the balancing factor. Therefore, we can write the total loss as follows:

$$Total\ Loss = \alpha(Student\ Loss) + (\alpha - 1)\ Distillation\ Loss$$

Backpropagation is then used to minimize the total loss and fine-tune the student model until convergence or until the maximum number of iterations is reached.

Common Knowledge Distillation Architectures

In terms of the architecture of knowledge distillation methods, there are many architectures available, but common ones include homologous-architecture distillation, cross-architecture distillation, and blockwise distillation [Cheng '17].

Homologous architecture distillation: The teacher and student have the same architecture, such as CNN to CNN, RNN to RNN, or Transformer to Transformer. However, the student model is typically a lighter version of the teacher model. For example, the student CNN may have fewer layers or filters than the CNN teacher model. Additionally, they can differ in activation functions, regularization methods, optimization algorithms, and other architectural details, allowing the student model to be efficient while retaining the same fundamental structure.

Cross architecture distillation: When teachers and students have different architecture, it is referred to as cross-architecture distillation [Liu '22]. For example, the teacher has a transformer architecture, and the student has a CNN architecture. This method allows for transferring knowledge between different types of models, which can be useful when the student model needs to be significantly different in structure due to deployment constraints or specific application requirements.

---

[8] https://nni.readthedocs.io/en/stable/sharings/kd_example.html

Block-wise distillation: The teacher and student are divided into blocks, and each block is trained separately with supervision from the corresponding block of the teacher [Wang '18, Li '20]. In particular, Block-wise distillation splits a large neural network into multiple smaller subnetworks, i.e., students. Then, distills the knowledge from the teacher model to each of the student blocks. This can be done by training the student blocks to predict the outputs of the corresponding teacher blocks or by using a loss function that penalizes the student blocks for not being similar to the teacher blocks.

For example, consider a ResNet-50 that has 26.5 million parameters and 50 layers as a teacher. As students, three blocks of CNN could be used; each has 10 to 18 layers and less than 5 million parameters. This approach, while promising, presents some challenges. For instance, determining the optimal number of blocks and ensuring that each block effectively captures the teacher's knowledge can be difficult. Because of these challenges and the relatively recent introduction of this approach, we should be cautious in using it.

We have explained in this book a million times that training neural networks are computationally expensive, and it is not affordable for small or medium enterprises to train a large model. Some approaches try to mitigate this need for neural networks, and we list some of the popular ones in this section, which reduce the training cost or enable devices with limited computational resources (any devices that operate with battery) to benefit from neural networks. Despite reading them seeming boring, the reward they bring to your work is very exciting.

NOTES:

* Meta-learning is not limited to neural networks; even ensemble learning methods, which we have learned in Chapter 9, could be categorized as meta-learning as well. Here, we have described meta-learning in the context of neural networks.

* Multimodal learning is a type of machine learning that deals with data from different modalities, such as text and images. Multitask learning is a form of supervised learning that addresses multiple related problems with a single model. For example, a multimodal multi-task model can simultaneously learn the most prominent tasks across different domains, ranging from object detection to natural language understanding and multimodal reasoning.

* Some models in knowledge distillation utilize a hint [Romero '14, Chen '17] while training the student. The hint could be an intermediate layer representation of the teacher model [Romero '14]. By using hints, the student model is guided not only by the final output of the teacher but also by intermediate representations, helping the student to learn a more nuanced and effective feature extraction process.

Automated Machine Learning

Automation is crucial in any process as it enhances efficiency by reducing human error and speeding up repetitive tasks, leading to increased productivity and thus freeing up human resources for more strategic and creative work. There are several directions of work to automate the process of machine learning, including preprocessing, feature engineering, neural network architecture design, and hyperparameter tuning. Among these, hyperparameter tuning and Neural Architecture Search (NAS) are the most common, and we will explain them in more detail here.

Hyperparameter Tuning

In Chapter 1 of this book, we learned that hyperparameters are the input variables of an algorithm that the user of the algorithm should provide. For example, we give k as the number of clusters to k-mean algorithm (Chapter 4). Since humans give it, it is not optimized, and therefore, to identify the best value for a hyperparameter, we need to perform *hyperparameter tuning*, a.k.a. *hyperparameter sensitivity analysis*. In most cases, we measure the change in accuracy while performing hyperparameter sensitivity analysis, but other metrics could also contribute to the tuning decision as well, such as the amount of memory utilization.

The process of searching for the best hyperparameter value could be done automatically by algorithms as well. This process is known as *hyperparameter optimization* or hyperparameter tuning. There are different approaches used for hyperparameter tuning, and we will explain Grid search and Random search as the two most common ones.

Random Search

Before we describe each method, we need to be familiar with the concept of *search space*. The range of values that will be searched for the best parameter selection is referred to as search space. In mathematics, if the search space is limited to real numbers, shown as $\mathbb{R}$, and to define the dimensions, we can use superscript. For example, $\mathbb{R}^2$ means a search space of real numbers in two-dimensional space.

Random search is a straightforward method for hyperparameter tuning, where hyperparameters are sampled randomly from the search space. The process continues iteratively, evaluating the performance of each set of hyperparameters until either the maximum number of iterations is reached or the desired accuracy is achieved. Ultimately, the hyperparameters with the best evaluation results are selected.

While random search is the simplest method, it does not necessarily yield the optimal hyperparameters. Its effectiveness is highly dependent on the number of iterations; too few iterations may fail to identify good values, while too many can lead to inefficiency and excessive resource consumption.

Grid Search

Grid search defines a grid of hyperparameter values within the search space, evaluating all possible combinations of these hyperparameters. If the grid covers the entire search space, every possible combination within that range is tested.

Grid search tends to be more exhaustive and accurate than random search, especially for smaller search spaces, but it is also computationally more expensive. This method uses brute force search and is most useful when the search space is small, as it becomes impractical for larger spaces. Additionally, grid search does not learn from previous iterations and lacks any policy to guide the search toward optimal hyperparameters, making it less efficient in complex scenarios.

Neural Architecture Search

Designing a neural network architecture often involves extensive manual tuning and numerous trials and errors. For instance, if we're building a CNN model, several questions arise: Where should we use skip connections? What are the optimal convolutional filter sizes? How many convolutional and pooling layers are necessary? Answering these questions typically requires several iterations of trial and error, along with manual adjustments.

Neural Architecture Search (NAS) [Elsken '19] refers to frameworks and algorithms that try to automate the process of neural network architecture design. According to the definition proposed by Elsken et al. [Elsken '19], a NAS framework has three components: search space, search strategy, and performance evaluation strategy.

A *search space* is a set of architectures that the NAS is able to select from. The search space could be very large and thus infeasible to search exhaustively. Incorporating some domain knowledge into designing the search space could assist in reducing the range of search space.

A *search strategy* refers to the method that finds the desired architecture from the search space. In other words, the search strategy determines how the NAS framework experiments with different neural networks from the search space. Examples of search strategies include random search, grid search, using an evolutionary algorithm (check Chapter 6), or using reinforcement learning.

A *performance evaluation strategy* is referred to methods that quickly evaluate the selected neural networks without fully training them. This approach is used because fully training each architecture inside the search space is extremely expensive and thus impractical.

There are different approaches to implementing NAS, including using evolutionary algorithms (check Chapter 6), Reinforcement Learning (check Chapter 13), and gradient-based NAS, which allows architectures to be optimized using gradient descent. We keep this discussion short and do not delve deeper into NAS architecture. This is due to practical evidence that at the time of writing this chapter and revising it, we didn't see any high-impact architecture designed by NAS.

Summary

This chapter focuses on methods to develop resource-efficient machine learning models that can reduce CO_2 emissions and also run on small devices, such as smartphones, with less computational capabilities.

We start this chapter by introducing powerful and popular compression algorithms, including BPE, Bitmap Index, Huffman Coding, and Lempel-Ziv-Welch followed by sparse coding. Next, we describe the concept of quantization and common quantization approaches that are used in signal processing and other applications such as computer vision applications. Quantization in a neural network is a popular method to reduce the size and, thus, complexity of a neural network. Therefore, we dedicate a significant part of this chapter to quantizing neural networks with some details about neural network quantization.

Another known compression method for neural networks is pruning. Pruning in the context of neural networks refers to the process of removing less important nodes and thus making a smaller and more resource-efficient network from the original network. We introduce classic and common methods for pruning the neural network.

Transformer architecture is the de facto standard architecture for LLMs, and we described two different groups of methods to make transformer architecture more resource-efficient. The first method we described is Low-Rank adaptation, which learns to construct a smaller matrix for fully connected layers of transformers. Another group of methods focuses on making self-attention more resource efficient, and we described FlashAttentions, Multi Query Attention, Group Query Attention, and Sliding Window Attention mechanism.

Another common approach that can reduce the resource utilization of neural networks is gradient descent optimizations, and we explained three methods, including gradient accumulation, gradient checkpointing, and conjugate gradient descent.

Finally, we explain common neural network training paradigms and conclude this chapter by listing two common groups of methods to automate the process of machine learning, including hyperparameter tuning and neural architecture search.

Further Readings and Watching

* An online class called "Fundamental of Digital Image and Video Processing" from Angelos K. Katsaggelos (https://www.coursera.org/learn/digital) introduces some image compression techniques, and we used his explanation to build the Vector Quantization explanation.

* Song Han has a course for reducing neural network size, and it is available online: https://www.youtube.com/@mithanlab. Not all content provided in the course is useful or clearly explained, but some of its lectures are good to be used as auxiliary learning material.

* A survey paper was written by Gholami et al. [Gholami '21], which assists us in describing the neural network quantization part of this chapter. Some parts are written and easy to understand.

* A brief review on pruning is written by Robert T. Lange and available online: https://roberttlange.github.io/posts/2020/06/lottery-ticket-hypothesis. Besides, there is a long and detailed survey paper about pruning written by Hofler et al. [Hoefler '21].

* Sebastian Ruder provides a good and detailed explanation about multitasking learning: https://www.ruder.io/multi-task

* Chelsea Finn has a course on Multitask learning and Meta-learning, https://youtube.com/playlist?list=PLoROMvodv4rMIJ-TvblAIkw28Wxi27B36. We do not delve deep into those concepts. if you, however, want to delve deeper into these concepts, her course provides some details. Multitask learning and meta-learning might have some good potential, but while learning them, we recommend you consider your time and resources and don't spend time on them unless your research question depends on it.

* Intel AI lab provides a good free Python package for different neural network compression methods, which we have explained in this chapter. The package is available under https://intellabs.github.io/distiller. Another good package for pruning in PyTorch is Torch-Pruning, https://github.com/VainF/Torch-Pruning.

* We did not cover common mobile architectures and neural networks that are light enough to deploy them on small battery-powered devices. There are plenty of them, including MobileNet [Howard '17], SqueezeNet [Iandola '16], EfficientNet [Tan '19], MobileViT [Mehta '21], etc. We do hope god forgives us for this big sin that we skip explaining these architectures in this chapter.

* Jonathan Hui provides some good explanations about the details of conjugated gradient, in you would like to learn more about this gradient, https://jonathan-hui.medium.com/rl-trust-region-policy-optimization-trpo-part-2-f51e3b2e373a

Chapter 15: Graph Mining Algorithms

Before the era of deep learning, social media analysis garnered significant attention from the computer science community. Researchers identified several phenomena through social media data extraction, including disease propagation [Salathe '11, Collier '11], product and technology reviews [Chen '11, Gan '21], and political opinion manipulation [Bradshaw '17]. At Boston University, the author teaches Web Mining and Graph Analytics, which attracts students from various colleges, including non-technical ones like the College of Communication. The easy access to rich information on online platforms, especially social media platforms such as Twitter before it was ruined by Elon Musk's greed, motivates their attendance.

Graph information forms the backbone of social media analysis. Various data types, such as biological protein structures, social interactions, 3D shapes, molecule structures, and gene interactions, can be represented and analyzed using graph algorithms. This motivates us to dedicate a chapter to graph analysis and information extraction from graphs.

The backbone of social media analysis is graph information. Much interesting information, including biological protein structure, social interactions, 3D shape analysis, molecule structures, biological genes interaction, etc., could be represented as graphs and analyzed with graph algorithms. This motivates us to dedicate one chapter to graph analysis and information extraction from graphs.

However, we acknowledge that graphs have limitations and cannot solve all practical problems, unlike recent advances in NLP and image recognition. Despite substantial efforts funded by European Union science grants (FP6, FP7, Horizon 2020) to promote graph data structures, such as the Semantic Web and DBPedia dataset, these initiatives largely failed to make significant real-world impacts outside academia. Therefore, we recommend being cautious and not overestimating the capabilities of graph algorithms.

We start this chapter by explaining basic graph concepts. Then, we explain minimum-spanning trees and algorithms to find the shortest path and search the graph. Next, we explain maximum matching algorithms, followed by centrality measurements in graphs. Afterward, we describe community detection and clustering algorithms in graphs, followed by describing two important link analysis algorithms: PageRank, which is the first algorithm that Google search had been operating based on, and HITS. Then, we finalize this chapter by describing common graphing neural network models.

You might ask why we cover these concepts; some of them might be unrelated to machine learning and artificial intelligence. To answer that, we refer you to Chapter 5, where we describe those search space optimization methods. These algorithms are implicitly related to machine learning and useful in designing a model for our problem. We have benefited a lot by using these algorithms to design our customized algorithm, and unlike other machine learning literature, we list them here as well. Knowing them, enables us to decide wisely about our machine learning pipeline.

Basic Concepts

In Chapter 1, very briefly, we described edges and vertices (nodes). To recall those definitions, look at Figure 15-1, where a simple graph is presented. A graph is presented as $G = (V, E)$ where E is a set of *edges* or *links* and V is a set of *vertices* or *nodes*. Edges connect the nodes of graphs and establish a relation between them. Nodes connected by an edge are called *adjacent nodes*.

The process of crossing an edge to get from one node to another node is called a *hop*. For example, in Figure 15-1, to get from a to d, the graph can make two hops from a to c and then from c to d, or it can get from a to b, b to c, and c to d, which results in three hops. A sequence of nodes and edges used to navigate from one node to another is called a *path*. If edges have direction, the graph is called a *directed graph*.

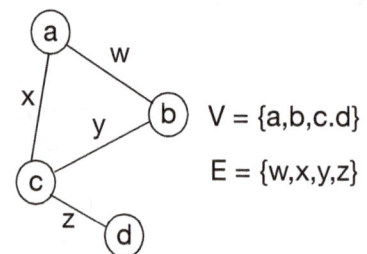

Figure 15-1: A simple graph, its vertexes and edges.

A *clique* is a subset of vertices in a graph where every vertex is adjacent to every other vertex in the subset. For example, in Figure 15-1, a clique is *{b,c}*, because b and c are adjacent to all other nodes. It does not include *d* because *d* is not adjacent to *b* and *a*. Similarly, it does not include *a* because *a* is not adjacent to *b* or *c*. The other clique is *{a,c}*. Clique identification is a computationally NP-hard problem (In Chapter 16, we describe what is NP-hard), and thus, we need to rely on subtle algorithms for this task.

The shortest path from one node to another node in the network is called a *geodesic path*. When we travel from one node to another in a graph, the set of visited nodes and edges is referred to as a *walk*. If the walk returns to the starting node, it is called a *closed walk*, e.g. $\{a, b, c, b, c, a\}$. A *cycle* is a path that starts and ends at the same vertex, e.g., *{a,b,c,a}* is a cycle. If, in a closed walk, every node is visited exactly once (except for the starting/ending node), and covers all nodes, it is called a *Hamiltonian cycle*. In Figure 15-1, we don't have a Hamiltonian cycle.

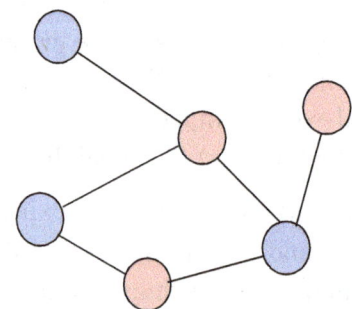

Figure 15-2: A bipartite graph example.

A *bipartite graph* is a graph in which its nodes are grouped into two sets, and each edge connects a node of one set to a node in the other set. Note that there is no connection between nodes of one group in a bipartite graph; edges only connect nodes from different groups together. A bipartite graph whose partition has the set U and set V, with edges E denoted as $G = (U, V, E)$. Figure 15-2 shows a bipartite graph. We can see there are two groups of nodes, blue and red, and each edge connects one color to another color. We could also say that a graph is bipartite if it has no cycle of odd length.

An interesting question is how to present a graph to a machine learning algorithm. Usually, they are presented as a sparse matrix, and machine learning algorithms digest the matrix as the graph. However, a large graph usually results in a very large matrix that cannot be loaded into memory.

Another approach to mitigate the memory issue of the *adjacency matrix* is to present graphs as adjacency lists (see Figure 15-3). The weight of a path from one node in the row to a node in the column is written inside the adjacency matrix. Using an adjacency matrix provides the flexibility to store more than one related information for each edge or vertices. By using a matrix, we are limited to storing scalar data as the weight of each edge, but we could also extend the matrix to a tensor to store more numerical information. However, it is not practical for textual data storage. For example, we cannot easily store a description for each edge. This issue can be easily resolved by using *adjacency lists* alongside the adjacency matrix.

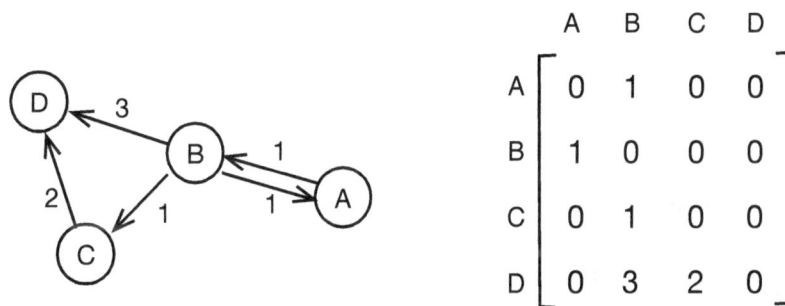

$$
\begin{array}{c|cccc}
 & A & B & C & D \\
\hline
A & 0 & 1 & 0 & 0 \\
B & 1 & 0 & 0 & 0 \\
C & 0 & 1 & 0 & 0 \\
D & 0 & 3 & 2 & 0
\end{array}
$$

Nodes = {A,B,C,D}

Edges = {A-B:1,B-A:1,B-C:1, C-D:2, B-D:3}

Figure 15-3: A sample weighted graph with its adjacency list in the bottom, and adjacency matrix presentations.

A *Directed Acyclic Graph (DAG)* is a graph where edges have a direction, and there are no cycles, meaning you cannot return to the starting node by following the edges. For example, in Figure 15-3 if we remove the connection from B to A, it will be converted into DAG. DAGs are useful for representing hierarchical structures and dependencies, such as task scheduling, data processing pipelines, and version control.

Minimum Spanning Tree

Trees are a type of graph data structure that does not contain cycles. Spanning trees are subgraphs that also do not contain cycles, and removing one more edge from a spanning tree results in having two separate trees. A spanning tree is constructed from a subset of the edges of the given graph, such that all nodes are included, and there are no cycles. Assuming the original graph has V nodes and E edges, A spanning tree nodes (V) are equal to nodes of the original graph $V = V'$, and its number of edges $E' = V - 1$ is smaller or equal to the original graph

$E' \leq E$. If we have a tree as a graph, its spanning tree is itself and thus $E' = E = V - 1$. Spanning trees do not have a cycle but also can not be disconnected.

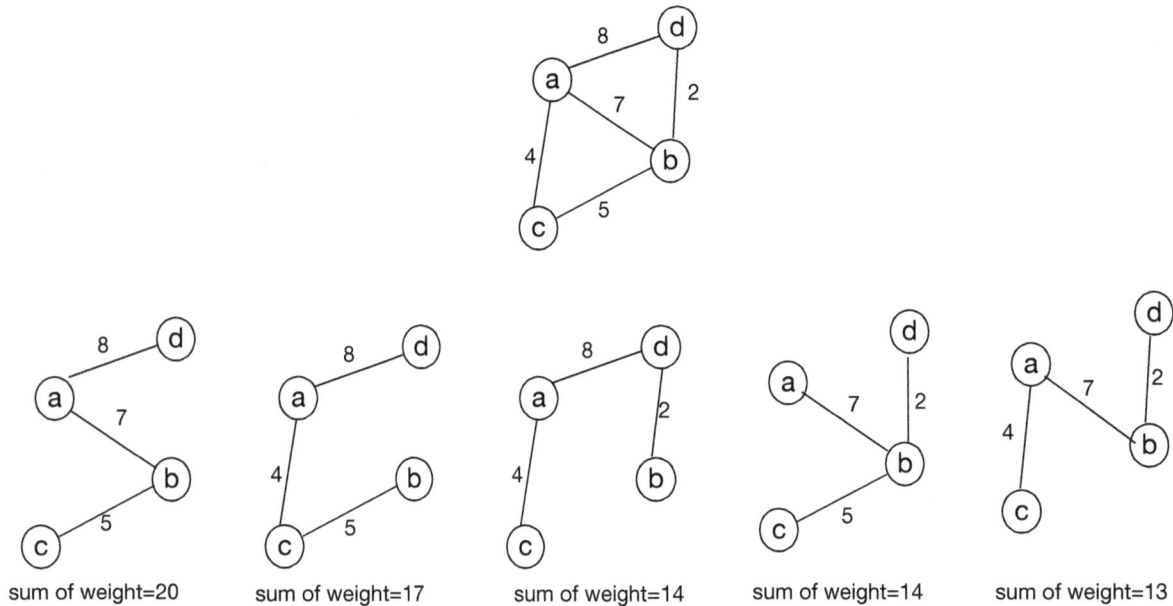

Figure 15-4: A simple graph on top and some of its possible spanning trees in the bottom.

For example, Figure 15-4 shows a tree on top and five of its possible spanning trees at the bottom. Removing one more edge from those trees results in having separate trees, which have only two nodes, but still, they are trees.

Finding the minimum spanning tree (MST) has many applications, including power distribution design, robot movements in the field, recommendation systems (where nodes represent products and similar products stay in neighboring nodes), and more. MST algorithms are typically used for cost minimization. If more than one MST is created in a graph and the trees cannot be connected, it is called a Minimum Spanning Forest (MSF). In an MSF, we have an MST for each connected component of the graph. Here, we explain two algorithms to identify the MST: Prim's and Kruskal.

Prim

Prim algorithm [Jarník '30, Prim '57] is one of the first algorithms for MST identification. It uses a greedy method to find MST in weighted and undirected graphs. Greedy methods are like our lovely politicians, who just think in the context of their terms, and nothing will be planned after they leave office. In other words, greedy methods are short-sighted and go for the immediate reward, with no long-term

Figure 15-5: A maze created by Prim algorithm.

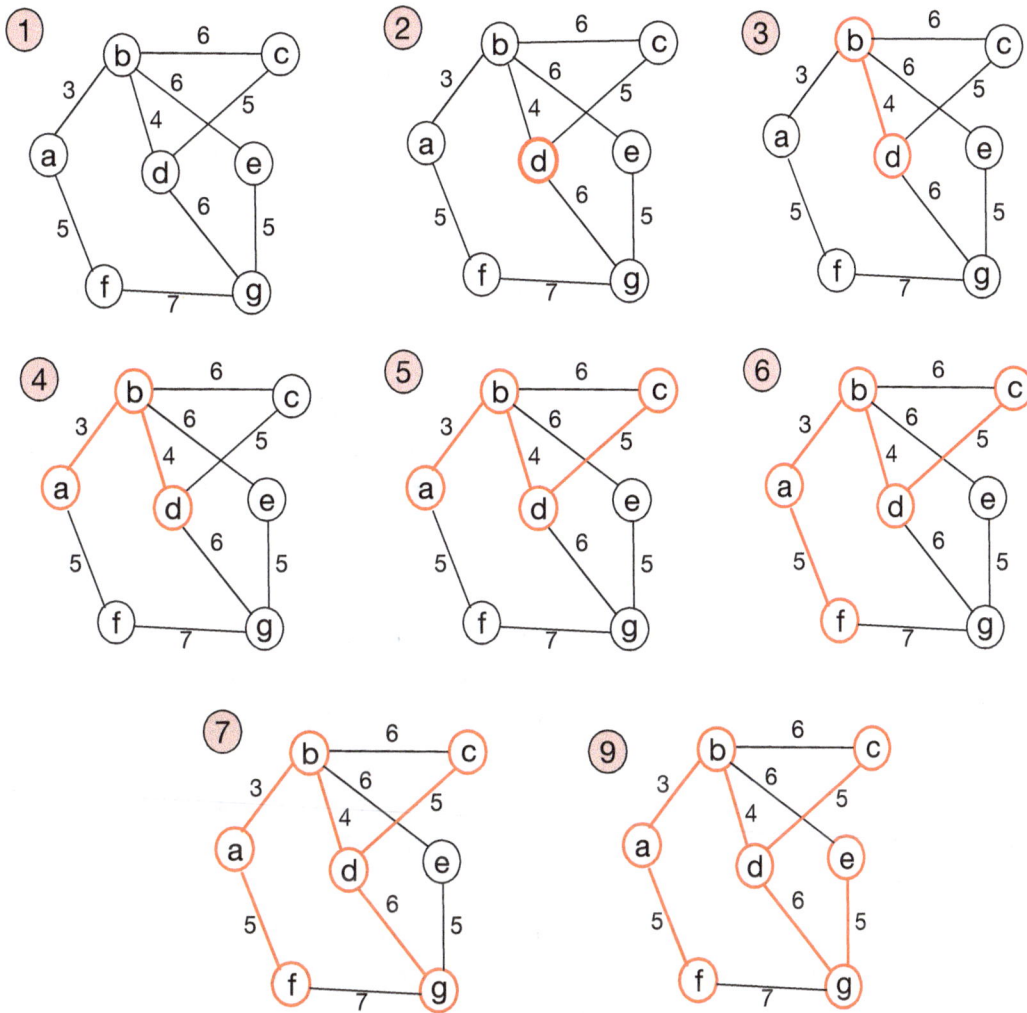

Figure 15-6: An example that shows step by step how the Prim algorithm constructs the minimum spanning tree. The result tree at the end of each step is shown in red.

planning.

In the old days, before children and adults were entertained with tablets and smartphones, they had paper magazines with mazes where the player had to draw a line to get from the source to the destination. These mazes can be drawn using Prim's algorithm. For example, Figure 15-5 is a maze created by the Prim algorithm[1].

The Prim algorithm works as follows:

First, it starts by selecting a random node on the graph. Next, it checks all edges from the selected node and grows a tree by connecting to the node that (i) has the lowest weight, (ii) does not form a cycle. This means the algorithm checks all neighboring nodes of the selected node and

---

[1] We used https://mtimmerm.github.io/webStuff/maze.html to construct this maze

grows the tree by one edge. As a result, the randomly selected node of the previous step is connected to the node which has the lowest cost.

Next, it computes the shortest distances from all nodes that have been selected, not just the most recent node, and chooses the edge with the shortest distance that does not form a cycle to expand the tree. This process continues until all nodes have been selected and the result is a minimum spanning tree.

For example, look at Figure 15-6 (1). There, we have a graph, and the algorithm randomly selects node d Figure 15-6 (2). Then, it checks the edges to the d that has the lowest weight, which is the edge to node b, with a weight equal to 4. It selects b now, as shown in Figure 15-6 (3). Now, the algorithm checks all edges from both b and d to find the edge with the lowest weight. The lowest weight is from b to a, and it is equal to 3. Therefore, for the next step, the algorithm checks the shortest distance to neighbor edges for $a,b,$ and $d,$ and a to f has the shortest distance, which is equal to 5. Then, it connects to f, see Figure 15-6 (4), and this process continues until all edges are added to the tree.

Kruskal

Another simple and popular MST algorithm is the Kruskal algorithm, which was introduced back in 1956 [Kruskal '56]. It starts by sorting all edges in the graph and selects the smallest edge. Then by using these two nodes, it constructs a single-edge MST. Next, it repeats the process and checks for the next smallest edge in the graph. It considers adding this edge to the MST if the newly selected edge does not create a cycle with previously selected edges. This process continues until all nodes are added to the MST. However, if the constructed trees can not get connected to each other end up having a minimum spanning forest (MSF).

Again, to understand this simple algorithm, let's look at Figure 15-7 (1). There, we have a graph, and as the first step, the algorithm scans all edges and selects the shortest edge, $\{a,b\}$, as it is shown in Figure 15-7 (2). Next, it selects the second shortest edge, which is $\{d,h\}$, and since it does not construct the cycle, it selects those two nodes as well Figure 15-7 (3). This simple process continues until all nodes are assigned to a spanning tree. We can see in Figure 15-7 (6), that all nodes have been assigned, and as a result, we have three trees. Then, it continues the same process until all trees join together in a tree that includes all nodes of the original graph Figure 15-7 (8).

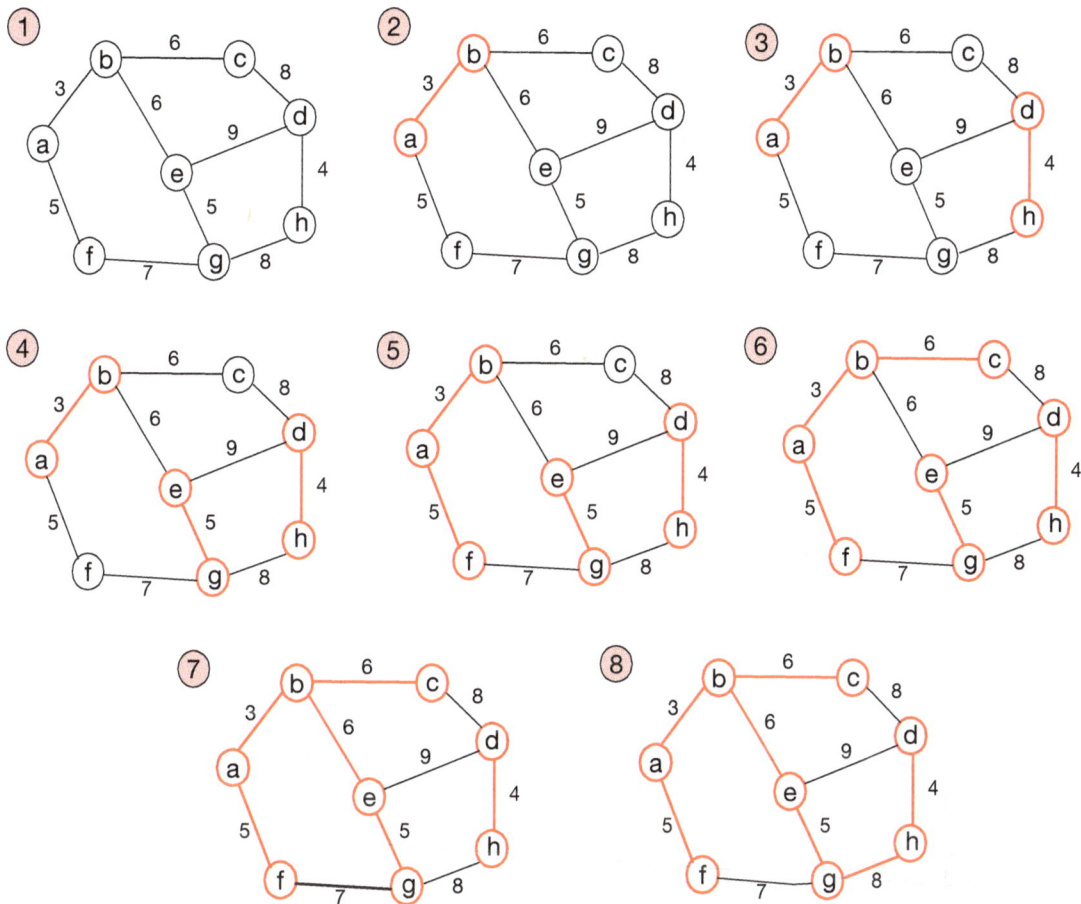

Figure 15-7: An example that shows step by step how the Kruskal algorithm constructs the minimum span forest. The result includes three trees which is shown in red.

Shortest Path Finding

A common application while working with the graph is finding the shortest path. It is used in different applications, including pathfinding in navigation systems, autonomous driving/flying, network routing, AI-powered games, etc. There are several algorithms to find the shortest path; we explain two popular algorithms, the Dijkstra algorithm [Dijkstra '59] and the A* (read it as A-star) algorithm [Hart '68].

Dijkstra

Dijkstra [Dijkstra '59] algorithm uses the greedy method that we have explained earlier. Greedy methods are algorithms that focus on making a locally optimum choice (immediate benefit). The input of the Dijkstra algorithm is a weighted graph, and it maintains the weight of edges that have been visited from the source node. If the graph has direction, then the decision should be based on the direction, but for the sake of brevity, we use a graph that does not have a direction in our example.

Assume we have the weighted graph of Figure 15-8 (1). The objective of the algorithm is to find the shortest path from vertex 'a' to all other vertices by using the Dijkstra algorithm. The weight of each edge is written on the edge. The algorithm updates the cost of reaching them, and this cost is written in blue color on top of each vertex in Figure 15-8 (1).

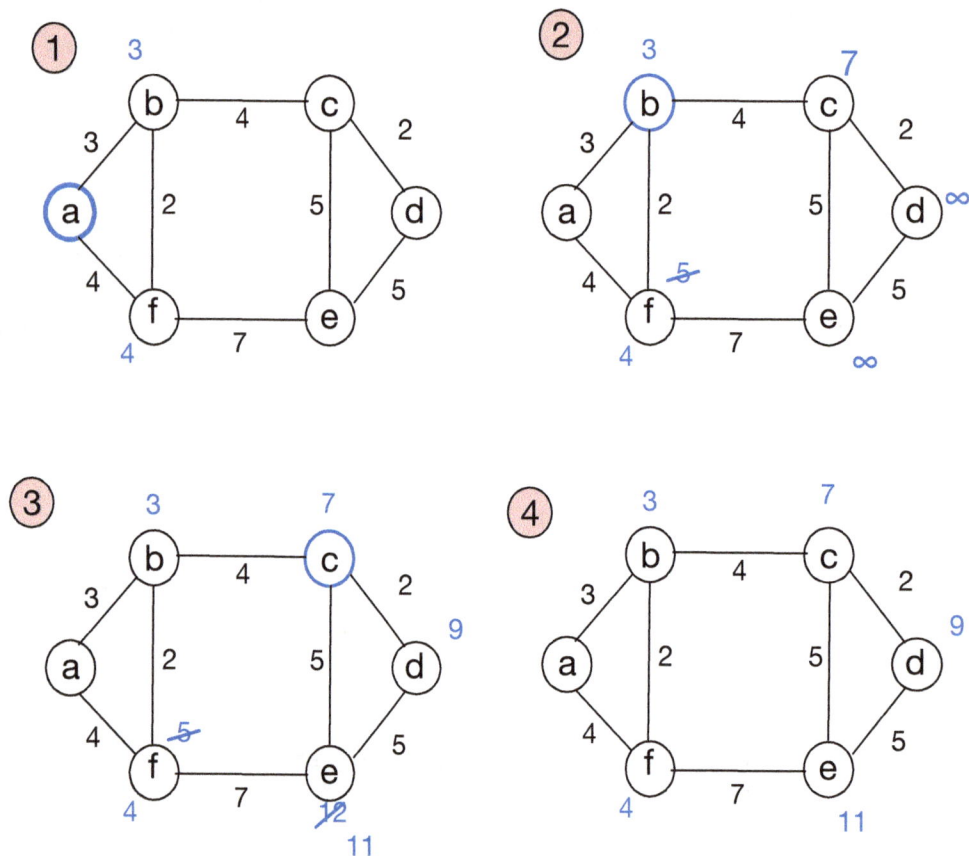

Figure 15-8: Dijkstra algorithm step by step, starting by selecting a random node, i.e. b.

Starting from Figure 15-8 (2) the algorithm starts from node a. Node a has two adjacent nodes, b and f. The cost of reaching b and f nodes from node a is the weight of the edges, see Figure 15-8 (1). The distance from a → f is equal to 4, there is another distance available from a to f (a → b → f), which is 5, and this alternative path is not considered at this stage because the direct path is shorter. Therefore, the shortest path, i.e., a → f, is considered as the shortest distance. Note that nodes that are not adjacent to the current node (node a) receive the cost of infinity (∞). In other words, if there is no direct connection between the current node and other nodes, the algorithm assigns ∞ as the cost.

Next, the algorithm selects the node that has the lowest cost, which in our example is b, because we have b = 3, f = 4, c, d, e = ∞. It calculates the distance from b (a → b) to other nodes. The distance from a to f is 5 (3+2), but previously from a to f was 4. Since 4 < 5 it keeps 4 as the shortest distance from a to f, see Figure 15-8 (2).

986

In the next step, the algorithm selects node *c,* because the distance from *a* to *f* has already been computed, and now *c* has the shortest distance to *b,* as is shown in Figure 15-8 (3). Then, distances from *a* through *c,* distances to *d,* and *e,* which are neighbors of *c,* are computed, see Figure 15-8 (3). The distance from $a \rightarrow b \rightarrow c \rightarrow e$ is equal to 12, but there is a shorter distance available from *f* to *e* ($a \rightarrow f \rightarrow e$), which is 11, and thus, the shorter path, which is $a \rightarrow f \rightarrow e$, will be used as the shortest distance. In the end, we get the closest distance from node *a* to all other nodes.

The computational complexity of the Dijkstra algorithm is $O(n^2)$ for a graph that has *n* nodes. However, with more efficient data structures (like a priority queue), the complexity can be reduced to $O((V + E)\ log\ V)$ where *V* is the number of vertices and *E* is the number of edges.

A* Search

A* [Hart '68] is another graph traversal and shortest pathfinding algorithm. It can be called an extended version of Djiskstra with an additional heuristic. A* introduces a score, which is the sum of the cost of the path from the start node to the current node and a heuristic estimate of the cost from the current node to the end node.

A* starts by using a piece of prior knowledge about the distances between the target node and other nodes in the graph. This distance, derived from some prior estimation, is called the *heuristic distance* $h(n)$. keep in mind that heuristics distances are guessed distances or based on some sort of prior knowledge, and not actual distances. This means we should have some prior knowledge to have a good estimation of heuristic distances, such as the approximate weight of edges.

For example, if we are developing a futuristic pizza deliver robot in the city to deliver pizza on time. Since on-time pizza delivery is an important need, the robot should do everything to open its path. There are some streets that have lots of cars or people who attack the robot to get its pizza. The cost of passing those streets is higher than other streets, and this could be added as a heuristic distance.

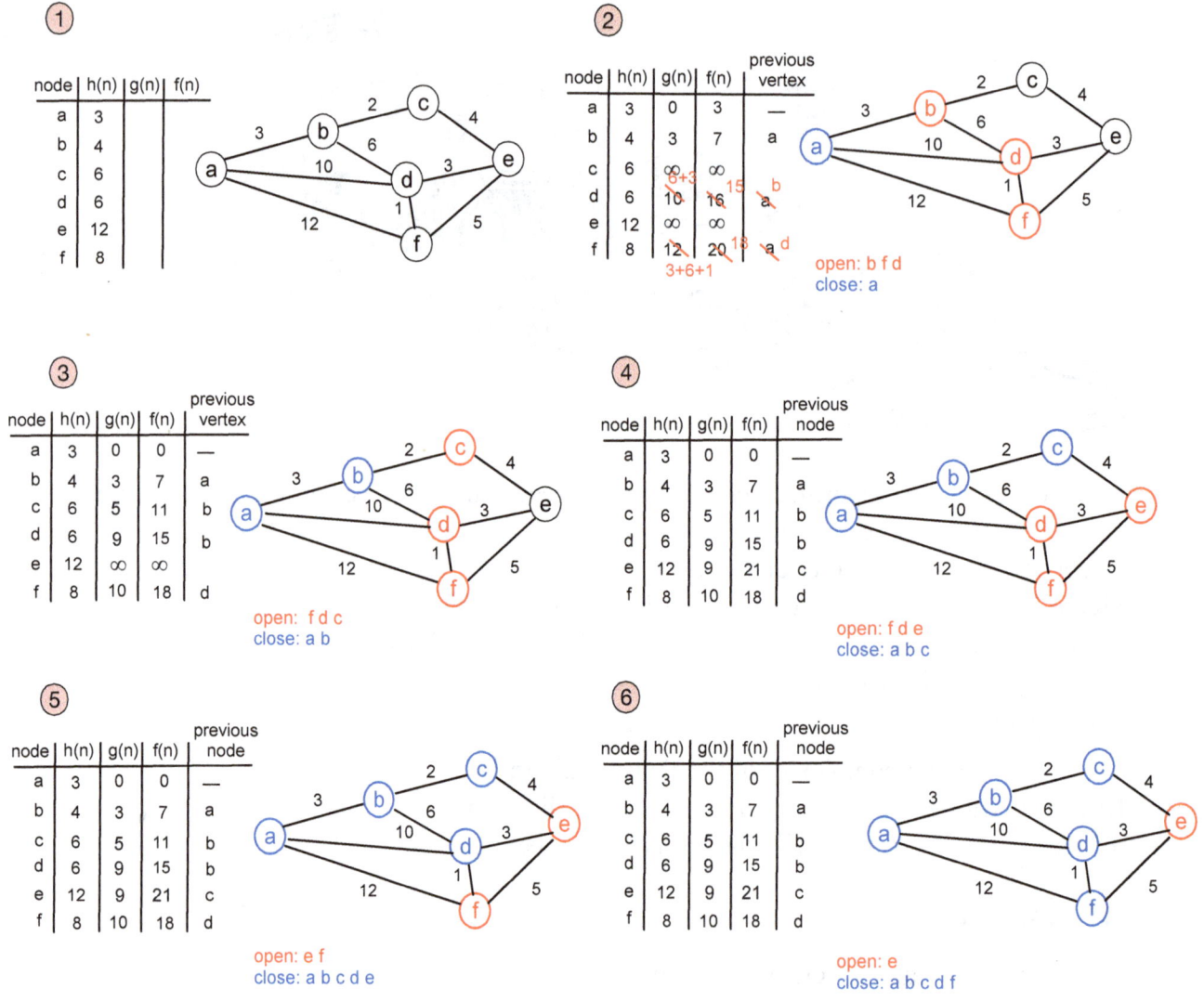

Figure 15-9: A* algorithm step by step, our objective is to calculate distances from 'a' to other vertices.

The quality of the heuristic influences the algorithm's accuracy as well. A good heuristic distance should be close to the actual distance, but obtaining a high-quality heuristic distance estimate is not usually easy.

The A* search algorithm deals with more than one distance. $g(n)$, is the actual distance on the path from a source vertex to the target vertex (for example, weights on edges), and $h(n)$ is the guessed distance. The algorithm does not decide solely based on $g(n)$; it decides based on $f(n)$, which is the sum of the actual distance and the heuristic distance, $f(n) = g(n) + h(n)$. When the algorithm starts, it also maintains two lists while processing the graph: *open nodes* and *closed nodes*. Open nodes are the list of vertices that have not been processed yet and must be processed. Closed nodes are the list of vertices that have been processed by the algorithm. If a node is not adjacent to the current processing node, similar to Dijkstra, the default distance is assumed to be ∞.

To understand this algorithm with an example, assume we have a graph as shown in Figure 15-9 (1), and the objective is to find the shortest path from source vertex a to all other vertices. The algorithm starts from the prior knowledge that is provided, i.e., heuristic distances, Figure 15-9 (1). As we can see column $h(n)$ is populated with some data. It maintains two lists: a list of open vertices and a list of closed vertices.

First, it adds node a to a list of open vertices, then it checks adjacent nodes to vertices to a, which are b, d, and f, and calculates their $f(n)$ distance. Then, it moves a into the list of close nodes and b, d, and f vertices into the list of open nodes, as it is shown in Figure 15-9 (2). Besides, it computes all other possible distances from node a to all nodes in the open node list and updates them. Therefore, we see some distances are crossed and updated with red color in Figure 15-9 (2).

Now, the algorithm should select a new current node from the list of open nodes. It selects node b, because from all vertices inside the open list $f(n)$ distance of b is the lowest one (equal to 7). Then, it calculates $f(n)$ for all adjacent nodes of b (vertices d and c, but not a because it is in the close list) and adds them all to the list of open vertices. Figure 15-9 (2) presents an additional column (previous vertex), which specifies from which node the algorithm reaches that particular node.

Besides, while processing b, the algorithm realizes that the distance from $a \to b \to d$ is shorter than $a \to d$ and thus there is a shorter path existing and updates the row of data from node d. The same story is applicable to node f as well. The path from a to f via d is shorter. Therefore, the path from $a \to f$ will be substituted via $a \to d \to f$. We mark these changes in Figure 15-9 (2) in red on the table.

Next, the algorithm moves b into the close list and chooses the nodes that has the smallest $f(n)$ as the next node, i.e., c. Then, it calculates the distances of adjacent nodes to c (distances from a to adjacent c nodes), as shown in Figure 15-9 (3).

Next, the c is moved to the close list, and the shortest distances from a to other nodes in the open list (f,d,e) that are not in the close list will be selected. Since, $f(n)_d = 15$, $f(n)_e = 21$ and $f(n)_f = 18$, it selects d, see Figure 15-9 (4).

After node d has been selected for processing, all distances from a to nodes in the open list that are adjacent to d, including f and e will be calculated. Since f is the shortest one, the algorithm selects f, see Figure 15-9 (5), and then it selects e, see Figure 15-9 (6). Nothing will change, all nodes are processed, and by looking at the column, $f(n)$ the algorithm outputs the distances to all nodes.

Matching Algorithm

Matching refers to finding a subset of edges in the graph where no two edges share a common vertex. For example, in Figure 15-10, consider the edges a − b and c − d. This forms a matching because a − b connects a and b, c − d connects c and d, and none of these edges share a common node. Therefore, $\{(a, b), (c, d)\}$ is a matching.

A node that is not connected to any other node by an edge that is part of the matching is referred to as a *free* or *isolated node*. A *maximal matching* refers to a matching in a graph that cannot be extended by adding more edges without violating the definition of matching. For example, in Figure 15-10, three graphs on the left present matching in the graph. *Maximum matching* is a type of maximal matching with the largest possible number of edges among all possible matchings in the graph. The maximum matching is different from maximal matching, and keep in mind their differences.

We can see in Figure 15-10 that there are four edges, and no other maximal matching can be used on this graph to have more than four edges.

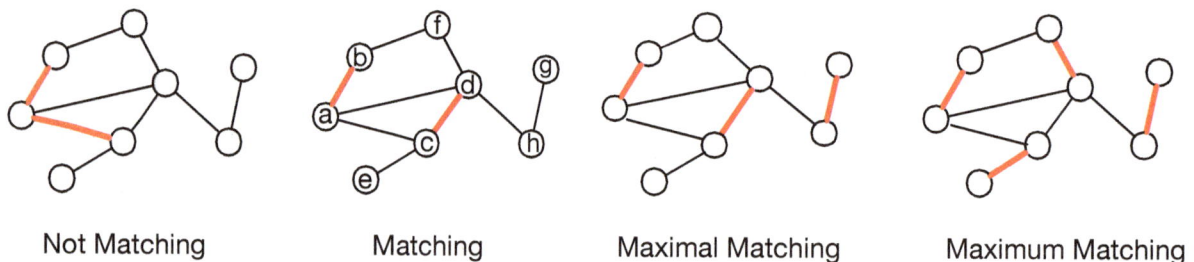

| Not Matching | Matching | Maximal Matching | Maximum Matching |

Figure 15-10: Different forms of matching in a graph.

Bipartite matching refers to the process of pairing elements from two distinct sets in a way that maximizes certain criteria. It has many applications, such as assigning jobs to workers, matching students to schools, pairing compatible kidney donors and recipients, etc. It uses a *Bipartite graph*, which is a graph whose nodes can be divided into two disjoint sets, with edges only connecting vertices between these sets (see the graph in Figure 15-11). Maximum matching in weighted bipartite graphs is called the *assignment problem*.

Here, we describe two algorithms for bipartite matching, the Hungarian Algorithm and the Hopcroft-Karp algorithm, which are used in machine learning applications, such as connecting the body joints in pose estimation.

Hungarian Algorithm

The Hungarian algorithm, also known as the Kuhn-Munkres algorithm [Kuhn '55], is a popular method for solving the assignment problem in bipartite graphs. It assigns tasks (nodes on one side of the graph) to workers (nodes on the other side of the graph) with the condition that no worker can be assigned more than one task, and no task can be assigned to more than one worker. The Hungarian algorithm constructs a matrix and solves the assignment problem within it. While brute force can be used for small matrices, it is a resource-intensive and impractical approach for large matrices.

The Hungarian algorithm starts by initializing the cost matrix and finding an initial feasible solution. Then, it iteratively identifies augmenting paths in the bipartite graph and improves the matching by changing the matched and unmatched edges along these paths. This process continues until no more augmenting paths can be found, resulting in a maximum weight matching or minimum weight perfect matching.

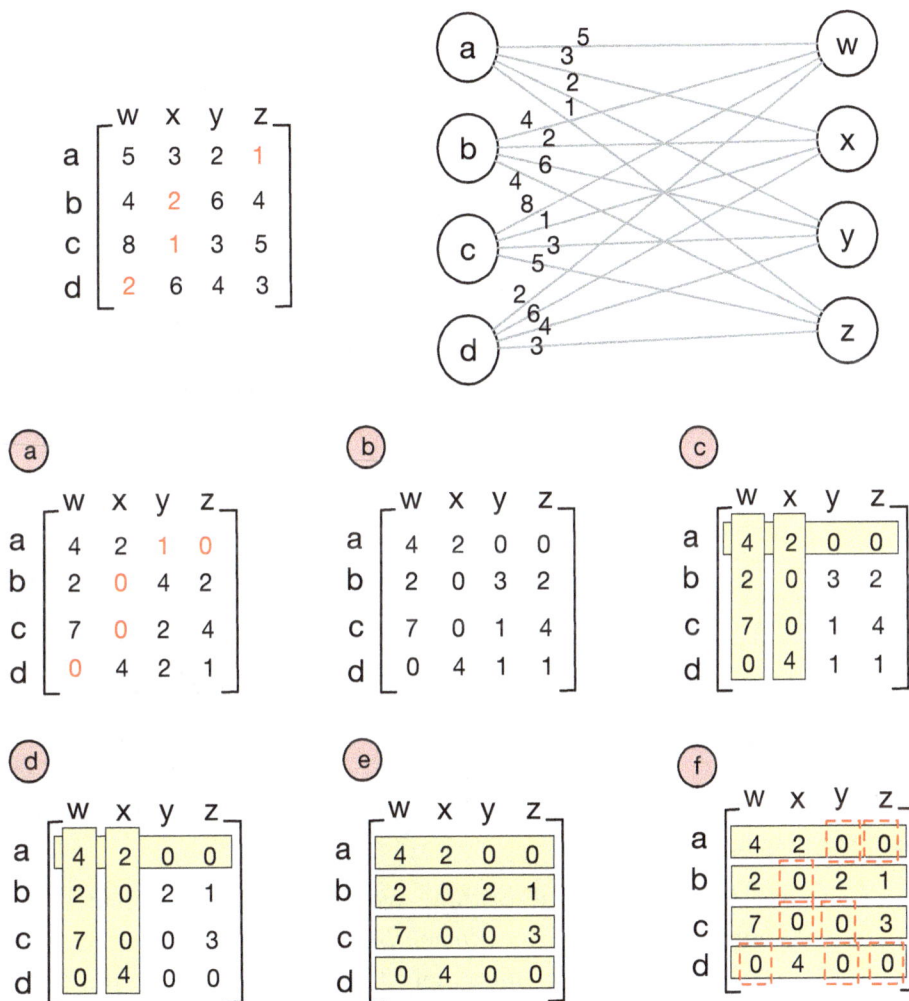

Figure 15-11: Hungarian algorithm step by step.

Since we are sure you didn't understand the algorithm by reading the previous paragraph, let's learn how this algorithm operates with examples. Figure 15-11 presents a bipartite graph that we would like to assign workers a,b,c, and d to tasks w,x,y, and z. On the top of this figure, we present the graph in a matrix.

As the first step (row reduction), it subtracts the minimum value of each row (there are highlighted in red in the original matrix of Figure 15-11 (a) from all the elements in that row. As a result, each row will contain at least one zero, and we will have Figure 15-11 (a). The smallest values of each column in matrix Figure 15-11 (a) are marked in red. In the second step (column reduction), the algorithm uses the matrix in Figure 15-11 (a) and subtracts the minimum value of each column from all the elements in that column. As a result, each column will have at least one zero. After reducing the minimum value of each column from that column, we have a matrix presented in Figure 15-11 (b).

The third step is a test for optimal assignment by drawing a minimum number of lines to cover all zeros. Therefore, it draws a line on columns and rows, as is shown in Figure 15-11 (c). If the number of lines is less than the number of tasks, then we should shift zeros.

Now the number of tasks is equal to four, but there are three lines that cover zeros, and thus, we need to shift at least one zero. To perform this shift, the algorithm finds the smallest uncovered value. Here we have three ones, which is the smallest value.

Then, we subtract this value (1) from all uncovered values and results in Figure 15-11 (d). Now we highlight all rows that have zeros, as shown in Figure 15-11 (e). and they are four lines that are equal to the number of workers or tasks.

As a result, we can assign tasks to workers based on zeros in matrix of Figure 15-11 (e); as it is highlighted Figure 15-11 (f). We assign task y or z to worker a, task x worker b, task x, and y to worker c (if y is not assigned to a), and task w to worker d (y and z are assigned to other workers).

The computational complexity of the initial version of the Hungarian algorithm was $O(n^4)$, but there are later versions that solve it in $O(n^3)$.

Hopcroft-Karp Algorithm

Hopcroft-Karp algorithm [Hopcroft '73], which is a computationally efficient algorithm that gets a bipartite graph as input and outputs a maximum bipartite matching. As explained earlier in this chapter, by maximum, we mean the highest possible matching number. As input, it gets a bipartite graph along with a set of existing matching, but this set could also be empty. As output, it provides maximal matching.

Before we explain the Hopcroft-Karp algorithm, we need to be familiar with two terms; augmenting path and alternative path. An *alternating path* is a path that begins with an unmatched node and whose edges belong alternately to the matching and not to the matching. Note that it connects nodes from one side of the graph to the other side rather than connecting nodes on the same side together. In other words, an alternating path with respect to a matching M is a path in which edges alternate between those in M and those not in M. An *augmenting path* is a path in a bipartite graph that starts from an unmatched node on one side of a bipartite graph and ends on one unmatched node on the other side of a bipartite graph.

Again, similar to the alternating path, it connects nodes from one side of the graph to the other rather than connecting nodes on the same side. An augmenting path is an alternating path that starts and ends with a free node.

The only difference between the augmenting path and the alternating path is that the augmenting path starts and ends with an unmatched node. For example, Figure 15-12 shows a bipartite graph on the left side, and the red edges on the right side present the augmented path from node 'a' to 'y'. Since the 'a' is an unmatched vertex, and 'y' is an unmatched vertex, this is an augmenting path and not an alternating path.

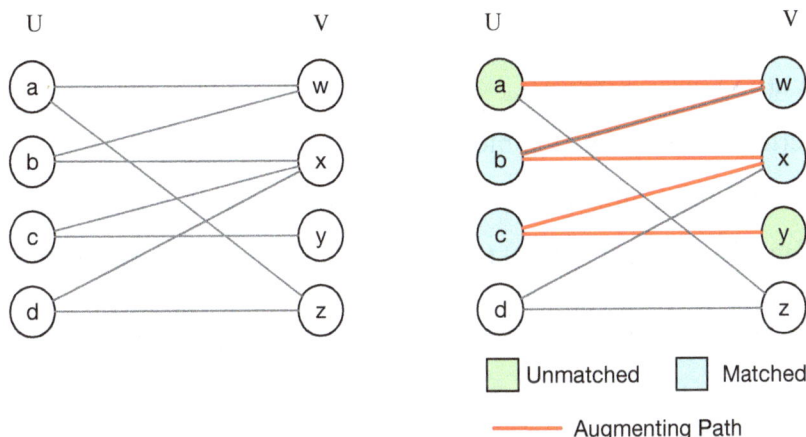

Figure 15-12: (left) A bipartite graph. (Right) an augmenting path in the bipartite graph. Note that bipartite graph connects vertices from one side of the graph to the other side rather than connecting vertices on the same side together.

The Hopcroft-Karp algorithm operates in following steps:

Initialization: It starts with an empty matching set (no node pairs between graphs are formed). $M = 0$

Breadth-First Search (BFS) to Build the Layered Graph: It performs a BFS from all free (unmatched) vertices in one partition (let's say the left partition). The goal is to build a layered graph where each layer alternates between the left and right partitions. Each level in the layered graph represents the distance in terms of alternating matched and unmatched edges.

Depth-First Search (DFS) to Find Augmenting Paths: Using the layered graph, perform DFS to find the shortest augmenting paths. An augmenting path is a path that starts and ends at free vertices and alternates between unmatched and matched edges. For each augmenting path found, increase the size of the matching M by augmenting along this path (flip the edges along the path).

Repeat: Repeat the BFS and DFS steps until no more augmenting paths can be found.

To understand this algorithm with an example, we use Figure 15-12 as an example and apply Hopcroft-Karp algorithm on it as follows.

Initialization: Start with no edges in the matching: $M=0$.

First BFS: Start from free vertices, {a,b,c,d}.

Create layers as follows: Level 0: {a,b,c,d}, and Level 1: {w,x,y,z} from their connections.

First DFS: Find augmenting paths in the layered graph: a → w, b → x, c → y

Augment along these paths: Update $M=\{(a,w),(b,x),(c,y)\}$

Second BFS: Free vertices in U: {d}, create layers considering unmatched edges: Level 0: {d}, Level 1: {z}

Second DFS: Find augmenting paths in the layered graph: d→z Augment along this path:

Update $M=\{(a,w),(b,x),(c,y),(d,z)\}$

Third BFS: All vertices are matched, no more free vertices.

Termination: No more augmenting paths, the algorithm terminates.

The final matching $M=\{(a,w),(b,x),(c,y),(d,z)\}$ represents the maximum matching in the given bipartite graph. Assuming E is the number of edges and N is the number of vertices in the graph, the Hopcroft-Karp algorithm computational complexity is $O(E\sqrt{E})$

NOTES:

* Dijkstra can not work for graphs that have negative weights, which is important to consider while deciding about the shortest path-finding algorithm. when the graph includes negative weights, in that case, this algorithm does not work.

* A very popular example of using MST is solving the travel salesman problem, which is an NP-hard problem. More about NP-hard will be explained in Chapter 16.

* MST algorithms require working with a connected graph. If we have a disconnected graph, and it is composed of two or more sub-graphs, the MST could be calculated for each graph separately, which we have explained is called Minimum Spanning Forest.

* There are other shortest path-finding algorithms, such as Bellman-Ford [Bellman '58, Ford '56], which can handle negative values. These algorithms are described in the Algorithms book, and we just briefly explained some basic MST and Shortest path finding algorithms to warm up our brains for more sexy graph algorithms.

Centrality Measurement Algorithms

A network is composed of nodes, and nodes have different impacts on the network. Some nodes are more important, and some are less important. Similarly, some people are more influential or indispensable than others. If they are not available, nothing will happen. For example, when a high-ranking bank officer or a politician goes on vacation, nothing significant changes. If the president of a university does not come to the office, nothing major happens, except perhaps faculty and other staff receive fewer unnecessary emails. However, if a janitor does not come to empty the garbage and clean the office, it becomes unbearable to stay in the office after a few days. Likewise, if healthcare workers are not available in healthcare facilities, people's health will be negatively impacted.

Web datasets are among the largest repository of data available, and they fueled significant growth in the data science community. At the time of writing this book, it has been estimated that the Internet includes 27 zeta bytes of information[2]. This huge amount of information makes web a very interesting platform to study. There are graph algorithms that focus on identifying influential nodes, known as centrality, in a graph. They have lots of applications such as as social network analysis, disease outbreak detection, identifying key infrastructure nodes in transportation, etc. A node could be important for different reasons, such as it has lots of connections, or it is connected to another node for which that node is important, or from one node, we can achieve other nodes by passing a few edges (hops), etc.

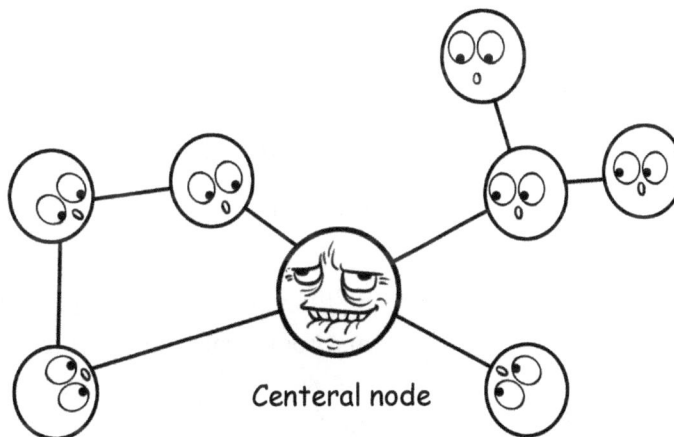

Centeral node

Here, we explain four centrality measurement algorithms, which are used to measure the importance of a node inside a graph is to measure its centrality by using centrality algorithms.

Degree Centrality

In graph theory, the degree of a node is the number of edges that are incident to the node, i.e., the number of edges connected to the node. The degree of a node can be further classified into the

---

[2] https://www.live-counter.com/how-big-is-the-internet

input degree and the output degree, depending on the direction of the edges. Specifically, the input degree of a node is the number of edges that are directed toward the node, while the output degree of a node is the number of edges that are directed away from the node.

Degree centrality is a type of centrality measure that ranks nodes based on their degree, i.e., the number of edges they are connected to. It can be computed as the sum of the input and output degrees of a node in directed networks. Nodes with a high degree centrality are considered to be more central in the network because they have more connections to other nodes. The equation to calculate degree centrality CD_i is as follows:

$$CD_i = \frac{\sum_{j=1}^{n} a_{ij}}{n-1}$$

In this equation, n presents the number of nodes and a_{ij} is the number of edges from node i to other nodes (degree of node i). For example, in Figure 15-13 the degree centrality of node 'c' is calculated as: $CD_i = 4/12$.

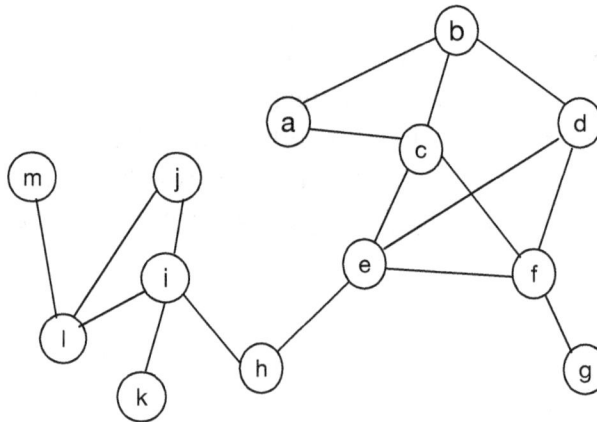

Figure 15-13: A sample graph, which is used to describe the centrality calculations.

This centrality measurement is good for studying the value of a network. For example, you apply for some jobs or positions. Since you read this book from the beginning, you got many job offers. You would like to study which position to accept. Then, you can select one person from each network who has a high degree of centrality and communicate with that person who represents the culture or value of that work environment. You can contact one person in each network and ask your questions about the position.

Closeness Centrality

It measures how easily a node can reach all other nodes in a network. It is used to identify nodes that are well-connected and can distribute information through the network efficiently. Closeness centrality measures the average distance of a node to all other nodes in the network, and high closeness centrality means the node is located in the center of a graph and can access other nodes with fewer hops. In other words, closeness centrality measures the average shortest path distance

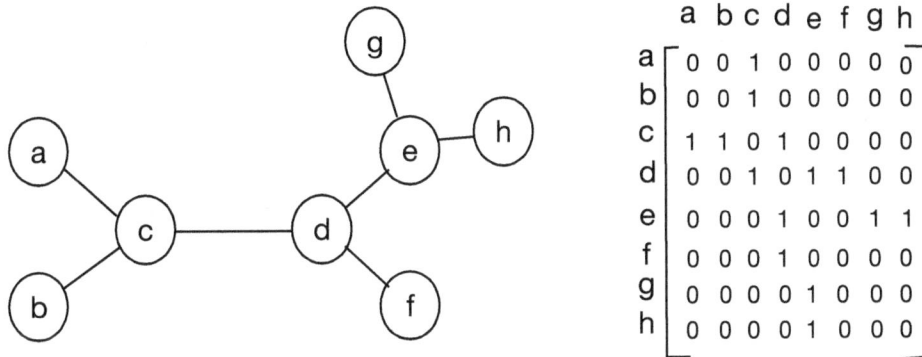

Figure 15-14: A sample graph, in which, despite having equal number of connections, node *d* has higher eigenvector centrality than node *c*. On the right the adjacency matrix of this graph is presented.

between a node and all other nodes in the network. The equation to calculate degree centrality is written as follows:

$$CC_i = \frac{n-1}{\sum_{j=1}^{n} d(i,j)}$$

In this equation, n is the number of nodes in the graph and $d(i,j)$ is the shortest path from the source node i to target node j.

For example, in Figure 15-13, let's assume we would like to calculate the closeness centrality of node *f*. First, we list the shortest distances of *f* to all other nodes as follows: {a:2, b:2, c:1, d:1, e:1, g:1, h:2, i:3, j:4, k:4, l:4, m:5}. Therefore, the result will be $(13-1)/30 = 0.4$

Betweenness centrality

It assigns a score to each node based on its position on the shortest paths between other nodes in the graph. In simple terms, it is used to identify nodes that act as bridges connecting other nodes. Nodes with high betweenness centrality have a greater influence on the network as they connect communities or clusters of nodes. These nodes play a crucial role in distributing information across the network since they are responsible for passing information between different parts of the graph. Betweenness centrality has many applications, such as detecting mediators connecting criminal organizations [Morselli '08], explaining traffic flow [Kazerani '09], and identifying latent influencers in a network. The equation to betweenness degree centrality is written as:

$$BD_k = \sum_{i \neq j \neq k} \frac{d_{i,j}(k)}{d_{i,j}}$$

In this equation $d_{i,j}$ is the number of shortest paths from node i to j (geodesic paths from i to j), and $d_{i,j}(k)$ is the number of those shortest paths that pass through k. Note that node k as start or end node does not count, it should be between i and j.

Eigenvector centrality

Eigenvector centrality [Bonacich '87] measures the influence of a node based on the influence of the nodes it is connected to. Assume we have nodes X and Y, both with the same number of connections. Eigenvector centrality prioritizes X over Y if X is connected to nodes with a high number of connections, whereas Y is connected to nodes with fewer connections. For example, in Figure 15-14, both c and d are connected to three nodes, but d has a higher eigenvector because it is connected to nodes with more connection (node e). Check Chapter 7 to recall the eigenvector and eigenvalue.

Eigenvector centrality quantifies the transitive influence of nodes rather than just quantifying the direct importance of a node. This centrality led to the introduction of revolutionary web search algorithms, such as PageRank and HITS. This algorithm uses a matrix that describes the relationships between nodes and calculates the eigenvector of this matrix. For the example presented in Figure 15-14 the matrix describes nodes and their connections. For the sake of simplicity, our example in Figure 15-14 does not include any weight. The Eigenvector for this matrix will be computed, and nodes' centrality is ranked based on their eigenvector value. Table 15-1 presents the centrality measurements for each of the nodes presented in the graph of Figure 15-14.

	Degree centrality	Closeness centrality	Betweenness centrality	Eigenvector centrality
a	0.14 8	0.36	0	0.21
b	0.14 7	0.36	0	0.21
c	0.42 6	0.53	0.52	0.46
d	0.42 4	0.63	0.71	0.55
e	0.42 2	0.53	0.52	0.46
f	0.14 5	0.41	0	0.26
g	0.14 1	0.36	0	0.21
h	0.14 3	0.36	0	0.21

Table 15-1: Four different centrality measurements for graph presented in Figure 15-14.

Link Analysis Algorithms

Link analysis and ranking algorithms are the core of web search technologies. We explain two important algorithms for link analysis. They are used for web search, but their use is not limited to web search only, and they can be used for many different applications that deal with graphs and require ranking, for example, tracking the spread of information or disease through a network. Other examples include fraud detection, bot detection on social media, software package analysis in enterprise software applications that are too big to analyze by a human, etc.

We describe two link analysis algorithms that identify important nodes inside graphs, including PageRank [Page '99] and HITS [Kelinberg '99].

PageRank

PageRank [Page '99] is named after Larry Page, one of the co-founders of Google. This algorithm is the foundation used by Google to create the Google search engine, which, until the time of writing this text, is the most popular search engine. Nowadays, Google not only uses PageRank but also other algorithms as well. However, PageRank was the first algorithm used by Google for search and still has many applications.

PageRank operates based on two important principles. (i) Pages that have a large number of inbound links are considered to be more important and appear earlier in the search results on top of the pages that have a smaller number of inbound links. (ii) The terms and keywords inside a page are not used to judge the content of the page alone. In addition, the PageRank algorithm uses the terms and keywords used in the links pointing to that particular page as well. Perhaps you have heard the saying, "What you say about yourself is not enough; what others say about you is also important." PageRank follows the same principle.

Web pages include links to other web pages, and this feature is leveraged by the PageRank algorithm to determine the importance of a particular web page. In simple terms, the objective of PageRank is to determine the importance of a web page in a search engine and show it in the search engine based on its importance.

PageRank operates based on the following principle: The importance of the web pages pointing to a web page is more important than the number of links pointing to that page.

Assume you are asking for a recommendation letter. In a scientific community, one recommendation letter from a pioneer in your field is more important than several recommendation letters from unknown people in your field. This is the attitude of the PageRank algorithm as well.

First, we describe PageRank algorithm with an example, and then, we explain more about its technical details. Assume we have four pages, and there are links from each page to other pages, as it is shown in Figure 15-15. We can see that node C has an outgoing link to A and B and an incoming link from A. Based on the number of nodes and their connections, first, the PageRank constructs a matrix called a *linked matrix*.

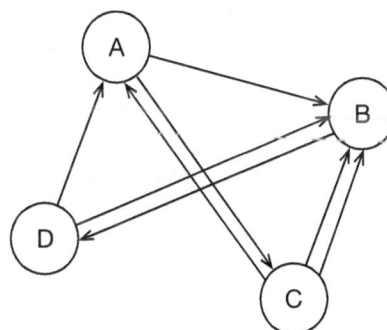

Figure 15-15: A sample graph that presents four web pages and their links to each other.

To better understand each node's outbounds, we list their outbounds along with the link matrix as well. By using the link matrix content and outbounds, the algorithm constructs another matrix called the *transition matrix*. Each node in the transition matrix is calculated by its value in the link matrix divided by the outbound from that node.

Link matrix

from

	A	B	C	D
A	0	0	1	1
B	1	0	2	1
C	1	0	0	0
D	0	1	0	0

(to)

Out bound

A=2
B=1
C=3
D=2

Transition matrix

	A	B	C	D
A	0	0	1/3	1/2
B	1/2	0	2/3	1/2
C	1/2	0	0	0
D	0	1/1	0	0

Now that we have a transition matrix, the algorithm can start to rank each page. The algorithm uses a probability distribution to calculate the likelihood of a page receiving a click from a random user. The PageRank algorithm uses the following equation to calculate the PageRank in every iteration for each page. We will explain the damping factor later.

$$PR_{t+1}(P_i) = \frac{1-d}{n} + d \sum_{P_j} \frac{PR_t(P_j)}{L(P_j)}$$

Here, P_j are pages linking to the page P_i. $PR_{t+1}(P_i)$ is the PageRank value of the current iteration. $PR_t(P_j)$ is the PageRank value of the previous iteration. d is the damping factor. n is the number of all pages. $L(P_j)$ is the number of outgoing links from the page P_j.

The *damping factor*, d, is a parameter that could be set to a value between 0 and 1, and it is usually set to 0.85. The damping factor is used when the page has no inbound link. It helps that particular page to receive some rank rather than zero and thus appear in the rank result.

To understand it with an example, let's continue calculating the ranks of our four web pages shown in Figure 15-15. At first, all nodes receive the same PageRank, which is 1 divided by the number of nodes. Therefore, iteration 0 is populated as 1/4 for all nodes in Table 15-2. Then, in the first iteration, the algorithm selects incoming nodes for each node and uses the described equation to substitute their values. Take a look, for example, at node A. It has two incoming nodes from edges C, and D. Node B has three incoming edges from C, D, and A. You can also see that in the first iteration, all previous PageRanks are 1/4. The result of the first iteration shows that node B has the highest rank. However, the algorithm usually does not stop in the first iteration, and it continues until there are no changes in the ranks or a given threshold of iteration is reached.

	Iteration 0	Iteration 1
A	1/4	$\dfrac{PR(C)}{out(C)} + \dfrac{PR(D)}{out(D)} = \dfrac{1/4}{3} + \dfrac{1/4}{2} = \dfrac{20}{96} = 0.208$
B	1/4	$2x\dfrac{PR(C)}{out(C)} + \dfrac{PR(D)}{out(D)} + \dfrac{PR(A)}{out(A)} = 2x\dfrac{1/4}{3} + \dfrac{1/4}{2} + \dfrac{1/4}{2} = \dfrac{40}{96} = 0.416$
C	1/4	$\dfrac{PR(A)}{out(A)} = \dfrac{1/4}{2} = \dfrac{1}{8} = \dfrac{12}{96} = 0.125$
D	1/4	$\dfrac{PR(B)}{out(B)} = \dfrac{1/4}{1} = \dfrac{1}{4} = \dfrac{24}{96} = 0.25$

Table 15-2: First round of PageRank calculation for each node. Please take a look at Figure 14-16 while reading this table. Ranks in each iteration are calculated based on the rank of the previous iteration. At iteration zero, all nodes have equal rank.

By using the data created in iteration 1, the next iteration (iteration 2) will be executed, and its results, along with values for PageRank, are presented in Table 15-3. By looking at Table 15-3 results, we can order pages based on their rank, i.e., D, B, A, and C.

A question might arise: why D received the highest rank and not B? To answer this question, let's review the motto of PageRank again. B has received many links but only points to D, which highlights the importance of D. Nevertheless, if we continue this iteration, maybe in the next iterations, node B get the highest rank from D.

	Iteration 0	Iteration 1	Iteration 2
A	1/4	$\dfrac{20}{96}$	$\dfrac{\frac{12}{96} + \frac{24}{96}}{3} \cdot \dfrac{}{2} \sim 0.166$
B	1/4	$\dfrac{40}{96}$	$2x\dfrac{\frac{12}{96}}{3} + \dfrac{\frac{24}{96}}{2} + \dfrac{\frac{20}{96}}{2} \sim 0.312$
C	1/4	$\dfrac{12}{96}$	$\dfrac{\frac{20}{96}}{2} \sim 0.104$
D	1/4	$\dfrac{24}{96}$	$\dfrac{\frac{40}{96}}{1} \sim 0.416$

Table 15-3: Second iteration of PageRank calculation for each node.

Hyperlinked Induced Topic Search (HITS)

Hyperlinked Induced Topic Search (HITS) or Hubs and Authorities algorithm [Kleinberg '99] is another popular method designed to rank web pages, which can be used in many different graph-based applications and is not limited to ranking web pages.

PageRank uses one factor to calculate the importance of a page. HITS uses two factors, i.e., *authorities* and *hubs*. Authorities are certain nodes that provide valuable information about a topic. Hubs are nodes that direct us to where to go and search for information. HITS operates based on the following principle: A node is a good hub if it links to good authorities, and a node is a good authority if it is linked to good hubs.

HITS has another significant difference from PageRank; instead of using the entire network (graph), it uses a subset of the network, i.e., the *root set*.

HITS algorithm assigns two scores to each node, one score represents the *hubiness* of the node, and the other score represents the *authoritiness* of the node. Both scores have a value range between 0 and 1.

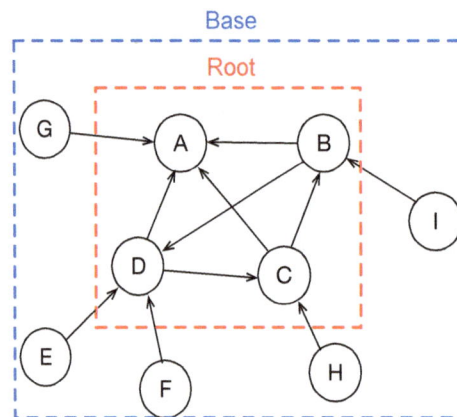

Figure 15-16: A set of nodes and their connections. Nodes that have a high number of incoming links are considered root, and nodes that have a low number of incoming nodes are considered base nodes.

The HITS algorithm operates iteratively, similar to PageRank, in four steps. (i) First, it assigns each node an authority and hub score of 1. (ii) Then it calculates the authority score by summing the hub scores of the nodes that link to each authority node. Also, it calculates the hub score by summing the authority scores of the nodes that link to each hub node. (iii) Now, both hub and authority scores will be normalized (each score divides the sum of all scores) in the last step. (iv) After normalization in step (iii), steps (ii) and (iii) are repeated iteratively until a given threshold or stopping criterion is reached, or it coverages (the rank doesn't change).

Respecting our traditions, let's describe the algorithm with an example. Assume we would like to rank pages that are presented in the graph of Figure 15-16. Table 15-4 shows the first and second iteration of the HITS algorithm on this dataset. Table 15-5 shows iteration 2, in which its data are constructed based on iteration 1.

	Iteration 0		Iteration 1		Iteration 1 (Normalized)	
	Auth.	Hub.	Auth. Sum of Hubs	Hub. Sum of Auths	Auth	Hub
A	1	1	1+1+1+1	0	4/11	0
B	1	1	1+1	1+1	2/11	2/11
C	1	1	1+1	1+1	2/11	2/11
D	1	1	1+1+1	1+1	3/11	2/11
E	1	1	0	1	0	1/11
F	1	1	0	1	0	1/11
G	1	1	0	1	0	1/11
H	1	1	0	1	0	1/11
I	1	1	0	1	0	1/11

Table 15-4: Initial authority and hub score assignment and the first iteration of HITS algorithm hubs and authority calculation for the graph in Figure 15-16. First, all nodes receive authority and a hub score of 1. Then, the first iteration calculates the sum of hubs as the authority score and the sum of authorities as the hub score. Next, the results of the first iteration will be normalized.

	Iteration 1		Iteration 2		Iteration 2 (Normalized)		Final Result	
	Auth	Hub	Auth	Hub	Auth	Hub	Auth	Hub
A	4/11	0	1/11+2/11+2/11+ 2/11	0	7/11 ÷ 18/11	0	0.39	0
B	2/11	2/11	2/11+2/11	4/11+3/11	3/11 ÷ 18/11	7/11 ÷ 33/11	0.17	0.21
C	2/11	2/11	2/11+1/11	4/11+2/11	3/11 ÷ 18/11	6/11 ÷ 33/11	0.17	0.18
D	3/11	2/11	1/11+1/11+2/11	4/11+2/11	3/11 ÷ 18/11	6/11 ÷ 33/11	0.17	0.18
E	0	1/11	0	3/11	0	3/11 ÷ 33/11	0	0.09
F	0	1/11	0	3/11	0	3/11 ÷ 33/11	0	0.09
G	0	1/11	0	4/11	0	4/11 ÷ 33/11	0	0.12
H	0	1/11	0	2/11	0	2/11 ÷ 33/11	0	0.06
I	0	1/11	0	2/11	0	2/11 ÷ 33/11	0	0.06

Table 15-5: Second iteration data that are constructed based on the data in the previous iteration.

Note that hub scores are calculated based on authority scores of the previous iteration, which have been highlighted with red lines, and authority scores are calculated based on hub scores in the previous iteration, highlighted with red dotted lines. As a result, let's stop in the second iteration. We can see that node A has the highest authority score, and node B has the highest hub score. Nevertheless, if we continue, the nodes' rank might change. Usually, the algorithm stops when the maximum iteration reaches, or the nodes' rank stops changing.

Please take some minutes to use pen and paper to calculate these numbers to be sure you have understood them perfectly. It is a reasonably easy to understand algorithm.

NOTES:

* PageRank forms a probability distribution, so the sum of PageRanks is either 1, or very close to 1. In our example, we do not add the complete numbers after dots. Otherwise, all sums end up being 1.

* Considering that $d = 0.85$ based on the equation we explained for the PageRank, the lowest possible PageRank will be $1 - d = 1 - 0.85 = 0.15$. In our example, we didn't consider the damping factor; we have a PageRank value lower than that, but that example is used for educational purposes.

* A typical visualization technique in a graph is to increase the size of nodes with a higher centrality score and reduce the size of nodes with lower centrality scores. We didn't present this visualization here.

Community Detection in Graphs

The concept of clustering in the context of graphs is referred to as community detection, which involves identifying groups of nodes that are more densely connected to each other than to the rest of the network. Here, we describe Spectral Clustering [Cheeger '70, Donath '72, Fiedler '73], Louvain [Blondel '08], and Leiden [Traag '19] algorithms, which are used to find communities inside a graph. One might believe that graph community detection algorithms belong to the clustering chapter, but since they are used for graph algorithms, we list them in this chapter.

One of the popular approaches to understanding graph networks is to identify groups of nodes inside a graph network that are tightly connected, i.e., *communities*. In simple terms, a community has many edges between its nodes, and there are few edges between communities. For example, consider a network of scientists working on a special topic. You can analyze their work and see they refer to each other's work a lot because they know each other, collaborate with each other, or try to support each other. In this scenario, we have a graph of authors, and if they have co-authorship, they are connected. Another example is identifying actors who play with each other in movies. Actors are nodes of graphs and are connected to other actors if they play together in a movie. For example, Figure 15-17 presents one graph that has two communities.

Similar to clustering algorithms (see Chapter 4), here, our objective is to find communities that have a high spread from other communities and a low spread among member nodes of each community.

Some popular community detection applications are in social networks, where humans are the nodes, and their connections to each other construct the graph. Community analysis can reveal fake users, spam bots, recommendation systems, etc. Protein interactions with each other in biological studies can be represented as a graph, and their relationships can be identified with community detection algorithms as well.

Here, we describe three popular community detection methods: Spectral clustering, Louvain, and Leiden. There are many more community detection algorithms, but our industrial experience shows that these ones usually meet our needs.

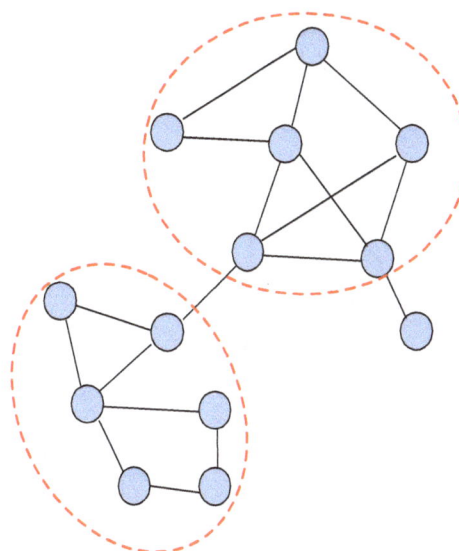

Figure 15-17: A sample graph with two communities, which are marked as dotted line.

Spectral Clustering

One of the oldest but still popular algorithms for community detection in graphs is Spectral clustering. Similar to SVD (check Chapter 7) and other scientific works before the era of the Internet, it has been introduced by several different scientists [Cheeger '70, Donath '72, Fiedler '73].

Before we explain the spectral clustering algorithm, we should be familiar with a few terms, which we explain as follows.

Laplacian matrix: In the context of the graphs, a Laplacian matrix (L) is a matrix that encodes the connectivity of nodes in a graph. The Laplacian matrix is used to represent the graph as a linear system. It is defined as the difference between the adjacency matrix (A) and the degree matrix (D): $L = D - A$. The Laplacian matrix is symmetric, and its eigenvalues are all non-negative.

The Laplacian matrix has the same size as the adjacency matrix. Therefore, for the given input matrix with a size of $m \times n$, it's adjacency and Laplacian matrix both have the size of $m \times n$.

Degree matrix: The degree matrix has only data on the diagonal; the rest of the matrix is zeros, and the diagonal data are the sum of all values in that row. The degree matrix is a diagonal matrix that contains the number of edges attached to each vertex. For example, as we have shown in Figure 15-18, the degree matrix of the given graph. It is a *diagonal matrix*, a type of square matrix where the entries outside the main diagonal are all zero.

Look at Figure 15-18; here, we have a simple graph of its adjacency and a degree matrix. In our adjacency matrix, we use 1 to indicate if an edge connects a pair of vertices and 0 if they are not connected.

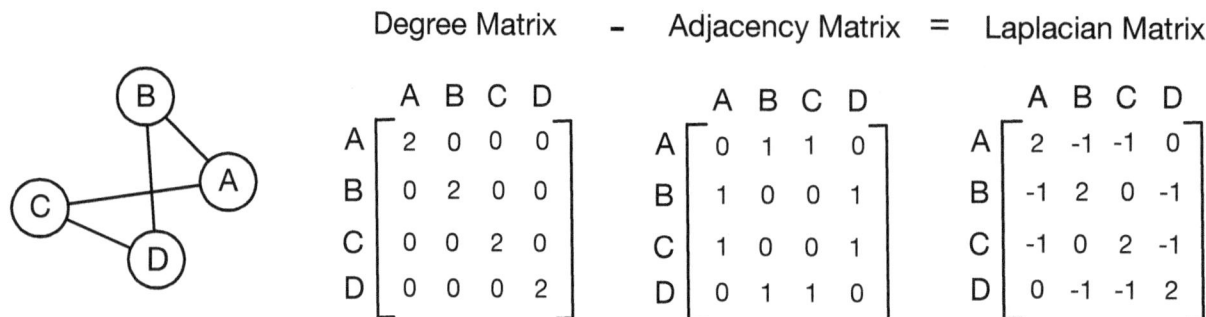

Degree Matrix   —   Adjacency Matrix   =   Laplacian Matrix

$$
\begin{array}{c}
 & \begin{matrix} A & B & C & D \end{matrix} \\
\begin{matrix} A \\ B \\ C \\ D \end{matrix} &
\begin{bmatrix}
2 & 0 & 0 & 0 \\
0 & 2 & 0 & 0 \\
0 & 0 & 2 & 0 \\
0 & 0 & 0 & 2
\end{bmatrix}
\end{array}
\quad
\begin{array}{c}
 & \begin{matrix} A & B & C & D \end{matrix} \\
\begin{matrix} A \\ B \\ C \\ D \end{matrix} &
\begin{bmatrix}
0 & 1 & 1 & 0 \\
1 & 0 & 0 & 1 \\
1 & 0 & 0 & 1 \\
0 & 1 & 1 & 0
\end{bmatrix}
\end{array}
\quad
\begin{array}{c}
 & \begin{matrix} A & B & C & D \end{matrix} \\
\begin{matrix} A \\ B \\ C \\ D \end{matrix} &
\begin{bmatrix}
2 & -1 & -1 & 0 \\
-1 & 2 & 0 & -1 \\
-1 & 0 & 2 & -1 \\
0 & -1 & -1 & 2
\end{bmatrix}
\end{array}
$$

Figure 15-18: Computing the Laplacian matrix of a sample matrix.

Normalized Laplacian matrix: The normalized Laplacian matrix, L_{norm} is a variant of the Laplacian matrix that is symmetric, and all of its eigenvalues are positive. Assuming L is the Laplacian matrix, D is the degree matrix, and $D^{-1/2}$ is the inverse square root of the degree matrix. The normalized Laplacian matrix is calculated as: $L_{norm} = D^{-1/2}LD^{-1/2}$

Identity matrix: it is a square matrix that has ones on the main diagonal and zeros everywhere else. By multiplying any matrix A with the identity matrix of the same dimension, we get the original matrix as the result of multiplication and hence the name "identity" for it. Assuming A to be an arbitrary matrix of dimension $m \times n$ and I is the identity matrix of dimension $n \times n$. Then, the multiplication of A and I is: $A \times I = A$. This means that the identity matrix plays the same role as the number 1 in scalar multiplication. Note that the dimensions of the identity matrix must match the dimensions of the matrix it is being multiplied with.

Now that we have learned these concepts, we can describe spectral clustering. The spectral clustering operates in four simple steps: preprocessing, two steps for decompositions, and the last step, which runs a clustering algorithm on the decomposed data. Each step is described as follows. (i) The first step employs the graph's adjacency matrix (A) and constructs a Laplacian matrix or the normalized version of the Laplacian matrix. (ii) Next, It finds the eigenvalues and eigenvectors (we explained them in Chapter 7) for the given Laplacian matrix. Therefore, at the end of this step, we have a vector of eigenvalues and a matrix of eigenvectors, V. (iii) As the third step, the algorithm selects eigenvectors of the k top highest eigenvalues. This defines the low-dimensional representation of the input graph. In particular, it selects columns from V whose eigenvalues belong to the top k highest eigenvalues. These vectors together construct a new matrix A', and it has a size of $k \times n$. In other words, A' it can be interpreted as a low-dimensional representation of the adjacency matrix of the original graph. (iv) The new matrix, A', will be fed into a clustering algorithm such as k-mean, and the clustering will be performed.

Louvain

Louvain [Blondel '08] is a popular heuristic algorithm that detects communities in a large graph network based on a concept called maximum modularity. It works like a hierarchical clustering algorithm (check Chapter 4).

Before we explain the Louvain community detention algorithm, we need to be familiar with the term modularity. In the context of graph community detection, *modularity* is a metric that specifies how divided communities are from each other in a graph network, or we can say that modularity is used to quantify the strength of division among communities. In a technical sense, modularity measures the density of links 'inside' a community compared to links 'between' communities. As a result, it provides a number between -1 and 1. The equation presented in the following computes the modularity score (Q) of a community.

$$Q = \frac{1}{2m} \sum_{ij} \left[A_{ij} - \frac{k_i k_j}{2m} \right] \delta(c_i, c_j)$$

Q is the modularity score between two communities c_i and c_j. m is the total number of edges in the graph. For example, Figure 15-19 has $m = 5$. A_{ij} is the value of the adjacency matrix, which presents the sum of the weights/edges between nodes i and j, see Figure 15-19. k_i presents the sum of weights/links attached to the node i, $k_i = \Sigma_j A_{ij}$, $k_j = \Sigma_i A_{ij}$. $\delta(c_i, c_j)$ is a Kronecker delta function between two communities of nodes c_i and c_j. This function returns 1, if $c_i = c_j$ are in the same community and 0 otherwise. There are different interpretations for δ in the literature, but we use the one that is used by the Louvain algorithm.

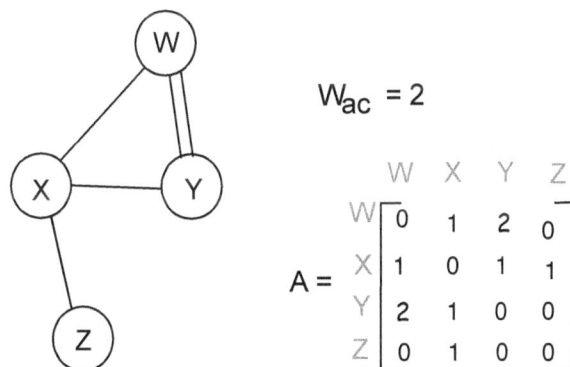

Figure 15-19: A sample graph and two different form to present adjacency matrix.

The Louvain algorithm constructs communities based on the maximum modularity score and operates in two phases. In the first phase, it performs *modular optimization*. This involves evaluating the modularity gain for every node and assigning each node to a community that maximizes this gain. This process of assigning nodes to communities repeats until there is no improvement, meaning no node changes its community (convergence). In the second phase, the algorithm aggregates communities. The newly formed communities are treated as single nodes, and the process is repeated until the modularity score can no longer be improved.

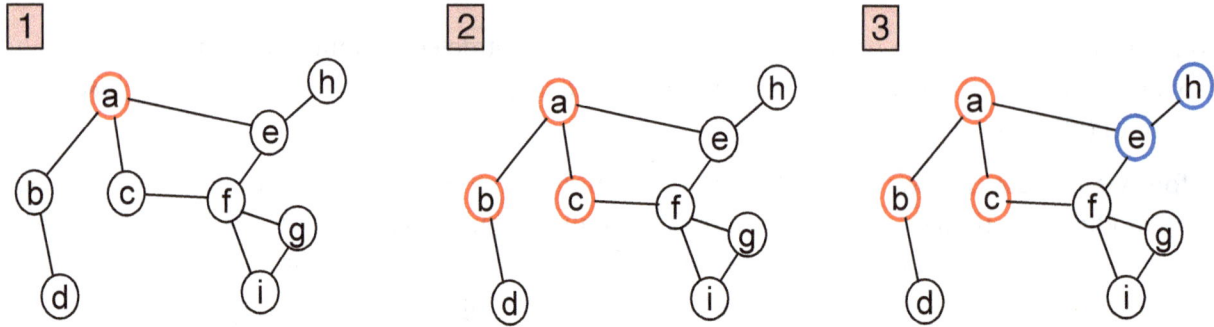

Figure 15-20: A simple community construction based on modularity score calculation.

To better understand this phase, let's use an example with a very naive approach for node-to-community assignment. We have a graph as shown in Figure 15-20 (1). The algorithm starts by selecting a random node, e.g., node *a*, and constructs the first community as a community with one single node. This means that the node itself constructs a community with one single node. First, the algorithm identifies adjacency matrix A for node *a*, which is $A_a = \{b : 1, c : 1, e : 1\}$, all node has equal weights (here, we assume weight as the number of connected edges only). Next, to select another node that will stay in a community with node *a*, the algorithm calculates the *Q* between each of the nodes connected to *a*. At this stage, since it is the same community, the node-to-community assignment δ is 1.

$$Q_{a,b} = \frac{1}{20}[1 - \frac{3 \times 2}{20}] \times 1 = 0.035$$

$$Q_{a,c} = \frac{1}{20}[1 - \frac{3 \times 2}{20}] \times 1 = 0.035$$

$$Q_{a,e} = \frac{1}{20}[1 - \frac{3 \times 3}{20}] \times 1 = 0.0275$$

The algorithm chooses the node that has the highest Q. Therefore, it should select *b* and *c*. The first community is constructed by choosing both *b* and *c*, as it is shown in red in Figure 15-20 (2). Next, another random node is selected, i.e., *e*. Again, for node *e*, we have $A_e = \{a : 1, h : 1, f : 1\}$, all its connected nodes have equal weights. Now, the same process of Q calculation continues, and node *h* has the highest Q value, as shown in Figure 15-20 (3). This process continues until all nodes are processed and assigned to a community.

After all the nodes are assigned to a community, the algorithm checks all nodes' community memberships and moves a node to a different community to increase ΔQ. Assuming the node is presented with *i* and the community is presented as *C*, ΔQ will be calculated by using the following equation.

$$\Delta Q = [\frac{\Sigma_{in} + 2k_{i,in}}{2m} - (\frac{\Sigma_{tot} + k_i}{2m})^2] - [\frac{\Sigma_{in}}{2m} - (\frac{\Sigma_{tot}}{2m})^2 - (\frac{k_i}{2m})^2]$$

In this equation, m is the sum of all links (or weights) in the graph. Σ_{in} presents the sum of links (or weights) inside community C. Σ_{tot} is the sum of links connected to community C, k_i is the sum of links connected to node i and $k_{i,in}$ is the sum of links from i to all nodes in C.

Note that the ΔQ is only calculated for the neighboring communities of a community and not for other communities in the graph. This process of ΔQ calculation continues until no node changes its community, and then the algorithm moves to the second phase.

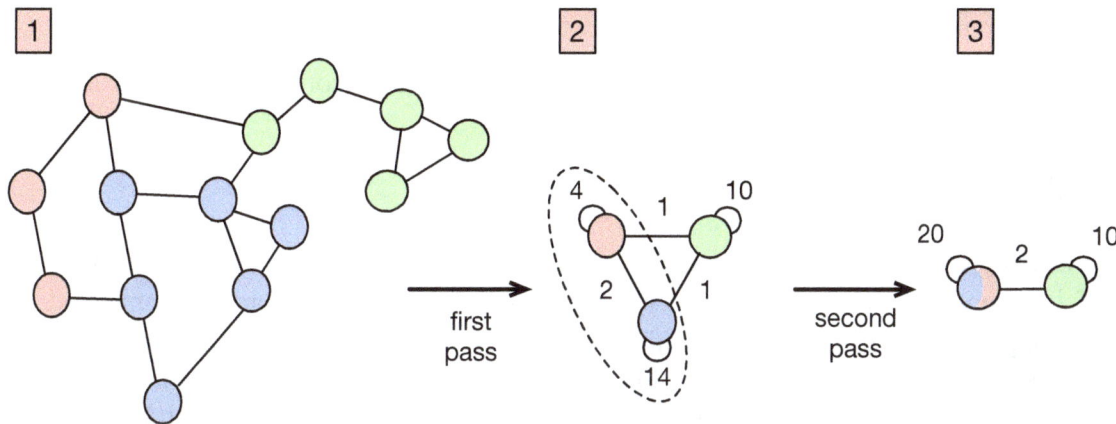

Figure 15-21: Second phase of Louvain algorithm, where in each pass a super node is created from the graph and these supernodes are merged together.

In phase two, the *aggregation phase*, nodes inside each community are merged into larger nodes, known as supernodes, and construct a new graph. The new graph includes only supernodes, as seen in Figure 15-21 (1) to Figure 15-21 (2). Next, weights on links get updated for the newly created graph that includes only supernodes. In other words, the second phase focuses on building a new network where its nodes are the communities of our graph. Each super node contains a community (or cluster) of densely connected nodes.

For example, Figure 15-21 (1) shows the result of the first phase. Next, the internal edges become self-loop, i.e., a node that connects to itself, with weight two. Therefore, the red community has two internal edges, and its self-loop edge will be $2 \times 2 = 4$, as shown in Figure 15-21 (2), the blue community has seven internal edges, and thus its self-loop edge will have a weight of $7 \times 2 = 14$, as shown in Figure 15-21 (2). The external edges are the sum of edges that connect nodes from different communities. For example, in Figure 15-21 (2), two blue nodes are connected to red nodes. Thus, in the first pass, the link between the blue and red supernodes is 2. Then, in the second pass, the algorithm creates another set of supernodes from a combination of blue and red communities (Figure 15-21 (3)), but keeps the green community intact.

Both phase one and phase two repeat until $\Delta Q = 0$, which means no more improvement is possible on the modularity score.

Let's summarize the steps of the Louvain method in a super easy language as follows:

(i) The algorithm chooses one node and aggregates it to another neighbor node, which has the maximum modularity score.

(ii) In the first phase, all nodes will be assigned to a related community based on maximizing the modularity score.

(iii) After all, nodes are assigned to a community, the algorithm iterates through them and changes some nodes' communities to increase the ΔQ.

(iv) In the second phase, the algorithm treats each community as a node and tries to construct larger communities by merging the current communities.

(v) Step (i) to (iv) continues until the community modularity score ΔQ can no longer be improved further.

Leiden

Louvain is a very popular algorithm in graph community detection, but it has a disadvantage that some of its identified communities are weakly connected, and some are disconnected. The Leiden [Traag '19] algorithm tries to mitigate Louvain's limitations and guarantees that all identified communities are connected. Leiden uses a quality function and tries to optimize the quality function, i.e., the Constant Potts Model [Traag '11], which is written as follows[3]:

$$H = \Sigma_c[e_c - \gamma \binom{n_c}{2}]$$

In this equation, e_c represents the actual number of edges in community c. n_c represents the number of nodes inside the community c, and γ is the resolution parameter. The resolution γ acts as a threshold parameter, and communities should have densities at least equal to γ, while the density between communities should be lower than γ. A higher value of γ leads to having more communities, and a lower value of γ leads to having fewer communities.

The Leiden algorithm consists of three phases, (i) local moving of nodes, (ii) refinement of the community, and (iii) aggregation of refined communities, using the non-refined community to create an initial community for the aggregated graph.

In particular, its first step that is nodes moving between communities, it similar to Louvain. However, after each node movement to a community, in the second step, the Leiden algorithm performs a refinement, which does not exist in Louvain. The refinement process merges a node to a new community only if the node assigned to the community increases the quality function value, and it also chooses the community (from a set of neighbor communities) that has the largest increase in its quality function. Then, in the third step, it creates refined communities from non-refined communities.

In summary, the refinement process of Leiden constructs communities with more checks, including adding nodes that are well connected to the community and considering communities

---

[3] if you are not familiar with the combination sign, it is written and computed as follows: $\binom{n}{k} = \dfrac{n!}{k!(n-k)!}$

as a community if they are well connected and there is more than one node in their community. In other words, Leiden works as Louvain with two additional steps, both refinement.

NOTES:

* Spectral clustering works well when there are well-defined clusters in the graph, and the top k eigenvectors can effectively uncover these clusters. If the graph structure is very noisy, spectral clustering may not perform as well. We recommend you start with Spectral clustering; if you don't get your desired result, then experiment with Louvain and Leiden.

* If a set of nodes in a graph are all connected to each other, they form a clique. In other words, Clique is a subset of vertices in a graph that every vertices are adjacent. A community could be assumed to be a union of different cliques that have a high overlap. To detect cliques, we should use a brute force search, which is not efficient.

* Usually, Louvain is faster than Leiden, but the result of Leiden is more accurate than Louvain. Therefore, the selection of these algorithms for community detection is very much application dependent.

Graph Neural Network

Neural network advances in recent decades benefit graph datasets as well. Graphs were able to apply machine learning in the non-euclidean domain [Bruna '14]. They can present 3D shape objects such as molecule structures, in which pairs of atoms are connected within a specific distance. By modeling or predicting molecule structures, graphs or, in particular, Graph Neural Networks (GNN) introduced back in 2005 [Merkwirth '05] and show promising capabilities and applications such as drug discovery. Additionally, there are also other usages of GNN, such as social network analysis, fraud detection, and recommendation systems.

From a technical perspective, by using GNN, we can perform community detection, classify nodes, predict edges, and classify graphs. Also, other tasks, such as graph generation and graph embedding (converting the graph into a vector while preserving its information).

Node classification refers to assigning labels to nodes in a graph. Usually, a part of the graph is labeled, and the node classifier uses those labels to assign labels to other parts of the graph. A real-world example of node classification is assigning labels to content on the social network. For example, some posts can receive human labels such as "news" or "personal", and GNN learns to label other unlabeled posts.

Edge or link prediction is another common

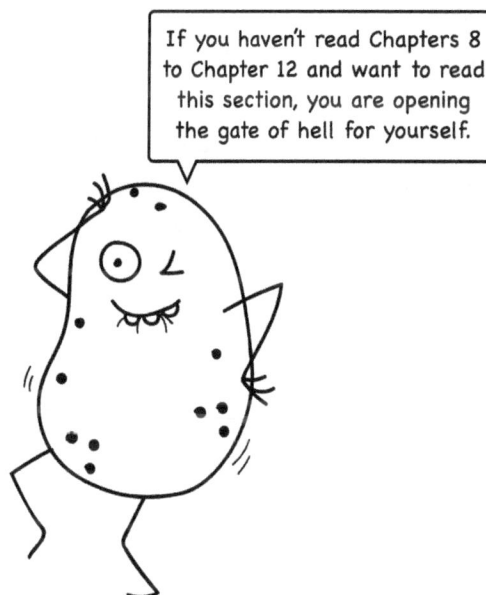

If you haven't read Chapters 8 to Chapter 12 and want to read this section, you are opening the gate of hell for yourself.

application of GNN that can predict if there is a connection (edge) between two nodes or not. It can be used to model how a network evolves over time. For example, an online social media platform could use edge prediction to recommend new users to follow social media users based on their past browsing and following history.

Graph classification is used to label graphs in a collection of graphs. For example, in a protein dataset each graph presents a protein, a graph classification model can be used to label each protein as enzyme or non-enzyme. Another example of graph classification is to classify a molecule of chemical compounds as being an anticancer behavior or not. In this classification scenario, each chemical compound molecule is represented as a graph, with atoms representing nodes and bonds as edges.

Graph generation involves creating new graph structures that mimic the properties of existing graphs. This can be useful in scenarios such as generating new molecular structures for drug discovery.

Graph embedding converts graphs or their components (nodes, edges) into vector representations while preserving structural information. These embeddings can then be used for various downstream tasks like clustering, visualization, or further classification.

First, we explain the challenges associated with GNN. Then, we explain the propagation module as a baseline of function in most GNN models. Next, we explain three GNN models that, at the time of writing this chapter, are common GNN models, including Graph Convolutional Neural Network (GCN), Graph Attention Network (GAT), and Graph SAGE.

Challenges of working with Graphs in Neural Network

We can model graph nodes and their relation in an adjacency matrix and feed this matrix as an input of a deep learning model, such as CNN. In particular, an adjacency matrix represents the connections between nodes in a graph, where each entry in the matrix indicates whether there is an edge between two nodes or not.

However, there is a problem with using this approach; the adjacency matrix imposes a specific ordering of the nodes, which can lead to inconsistent results. For example, we might have a graph with four nodes, A, B, C, and D. This graph could represent four people in a social network who are following each other, and the edge shows their connection. A value of 1 means they have a connection, and a value of 0 means they have no connection. We could have several different adjacency matrices that represent the connections between these four people, two of which are presented below:

$$
\begin{array}{c}
\begin{array}{cccc} A & B & C & D \end{array} \\
\begin{array}{c} A \\ B \\ C \\ D \end{array}
\begin{bmatrix}
0 & 1 & 1 & 1 \\
1 & 0 & 0 & 0 \\
1 & 0 & 0 & 1 \\
1 & 0 & 1 & 0
\end{bmatrix}
\end{array}
\qquad
\begin{array}{c}
\begin{array}{cccc} B & A & D & C \end{array} \\
\begin{array}{c} B \\ A \\ D \\ C \end{array}
\begin{bmatrix}
0 & 1 & 0 & 0 \\
1 & 0 & 1 & 1 \\
0 & 1 & 0 & 1 \\
0 & 1 & 1 & 0
\end{bmatrix}
\end{array}
$$

Although they present the same information, each of these matrices would produce a different result if we fed them to a neural network. This example reveals that the adjacency matrix is not permutation invariant. We should use a permutation invariant approach. *Permutation invariance* in the context of GNN means that the output of the graph neural network does not depend on the ordering of the nodes in the graph. To resolve the issue of permutation invariance, we use *message passing layers*, which we will explain later.

The second issue is that GNN extends deep learning capabilities to non-Euclidean domains. However, neural networks typically work with fixed-size input, while graph size and shape can evolve. Thus, a model trained on one graph may not handle another graph properly, even if it is the same graph that has evolved and changed. To handle the input size changes in graphs, the graph *pooling layer* is used.

Graph pooling layers can downsample or reduce the size of the graph representation, similar to pooling operations used in CNN models for image data. We will explain the graph pooling layer later.

Message Passing Layer (Propagation)

To address the challenges associated with Graph Neural Networks (GNNs), we require a mechanism that enables each node to maintain relevant information about its neighbors. More specifically, a GNN should learn a representation of nodes that encompasses both the nodes' attributes and the relationships between them, as defined by the edges in the graph. Most GNN models implement this feature using a *propagation module*.

The propagation module implements a *message passing* among neighbor nodes and *aggregation*. The message passing transfers some information about the current node to its neighbor. In particular, the message passing process implements nodes' communication with their neighbors along the edges between nodes. Each node can send a message to its neighbors. Then, the aggregation function aggregates messages received from neighbor nodes to the current node. This means that the aggregation combines the information received from neighbor nodes. There are different methods used for aggregation, including summing, averaging, or finding the maximum.

To better understand this concept, take a look at Figure 15-22. Here, we focus on what happens to the red node and its information. At first, we can see that each node includes features presented as rectangles along each node ($h^{(k-2)}$ in Figure 15-22). For example, if the node presents a word, its feature vector will be a vector of word embedding; if a node is a pixel of an image, its feature vector includes image features, etc. Then, at the next iteration, neighbor nodes (yellow nodes) send their information to the red node, and now the red node includes information about its neighbors, as is shown in Figure 15-22. This process will be done continuously for all nodes, and at the kth layer, we can see that all nodes entail information about the other nodes and not just about their neighbors. k is a hyperparameter, and it presents the number of hops the message passing is performed, known as neighborhood hops.

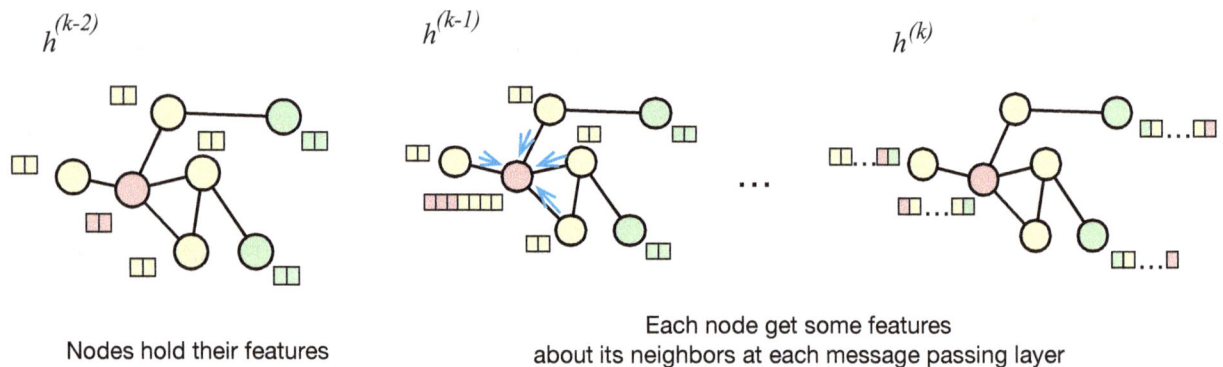

$h^{(k-2)}$ $h^{(k-1)}$ $h^{(k)}$

Each node get some features
about its neighbors at each message passing layer

Nodes hold their features

Figure 15-22: Message passing visualization. After each hop each node gets information about its neighbor, thus, after *m* iteration, each node has information from its *m*th neighbor as well.

In simple words, we can say each message passing layer includes sequential steps of *update* and then *aggregate*. We can formalize that the features *h* of the node *u*, i.e., h_u at iteration *k+1* is the result of the *update* at iteration *k*, which includes the *aggregation* of all neighbors of *u*, i.e., *N(u)*, and write it as follows: $h_u^{(k+1)} = Update^{(k)}(h_u^{(k)}, Aggregate^{(k)}[h_v^{(k)}, N(u)])$.

This is the simplest form that we borrowed from the Hamilton book [Hamilton '20] to avoid giving you a math panic attack. However, in the literature, you will see a different equation that incorporates the training of aggregation as well.

Note that both *Update* and *Aggregate* functions are differentiable functions. We have explained in Chapter 10 that we can train differentiable functions. In other words, by learning their parameters, a neural network can determine the best possible parameters for update or aggregation. The best possible parameters, here, refer to parameters that have the lowest loss score. To summarize, we perform the following steps for each node, (i) collect neighbor features, (ii) aggregate neighbor features into a vector, (iii) pass this vector into a neural aggregation network, (iv) the output of the neural aggregation network is a new vector representation of the node, which can be used for various downstream tasks in graph analysis.

Graph Pooling

The message passing layer enables the model to learn node embeddings that incorporate information from neighboring nodes. Recall that we present a graph as $G = (V, E)$, in some applications, our objective is to have an embedding of the entire graph (z_G) and not just nodes.

Most of the time, we use graph pooling to downsample or aggregate information from the graph while preserving its structural characteristics. In other words, graph pooling is used to reduce the size and complexity of a graph by compressing its nodes and edges. The result of graph pooling is an *embedding* from the input graph. There are different common approaches for implementing graph embedding via pooling. One simple approach is to take a sum or mean of node embeddings using the sum of node embeddings. Assuming z_v is the node embedding, the input graph is $G = (V, E)$, we can formalize the graph embedding as follows:

$$z_G = \frac{\sum_v z_v}{f_n(|V|)}$$

Here $f_n(|V|)$ is the normalizing function (check Chapter 6 for scaling and normalizing concepts) that scales the values of the node features to bring them to a common range.

There are two common groups of methods for graph pooling: *cluster-based* pooling and *graph-based* pooling. Cluster-based methods compress the input graph by grouping nodes into clusters and aggregating node features within each cluster [Grattarola '22]. Graph-based methods select a subset of nodes representative of the graph structure and features and discard the rest. For example, SAGPool is a graph pooling approach that uses attention mechanisms such as self-attention [Lee '19] to select nodes for the pooled graph representation. There are different categorizations of graph pooling, such as hierarchical versus flat pooling [Liu '22], but cluster-based/graph-based categorization is widely used in the literature.

Spectral versus Spatial Representation

Now that we have learned these two important components of graph neural networks, we can review a few architectures of graph neural networks that can be used for graph classification, such as Graph convolutional networks (GCNs), Graph attention networks (GATs), and an embedding graph model, Graph SAGE.

Before explaining these models, we need to get familiar with some preliminary concepts.

Spectral representation of graph: The spectral representation of a graph refers to the representation that uses the eigenvalues and eigenvectors of matrices associated with the graph, such as the adjacency matrix or the graph Laplacian matrix.

Spatial representation of graph: A spatial representation of a graph presents a graph in a spatial domain. In this representation, nodes of the graph are represented by points in a 2D or 3D space, and the edges of the graph are represented by lines or curves connecting the points. All graphs we present in this chapter are presented in the spatial domain.

Graph Convolutional Network (GCN)

GCN is the most popular GNN model that extends the concept of convolutional layers from regular grids (images) to graph-structured data. There are two approaches to implementing GCN, *spectral* and *spatial* approaches.

Spectral approaches operate based on graph signal processing. First, they transform graph G into a spectral domain by using Fourier transformation (Check Chapter 7 for Fourier transformation), i.e., $f(G)$. Then, they apply a convolutional operation on the transformed graph. Afterward, the transformed signal is transformed back into the spatial domain (original graph form), by using inverse Fourier transform, i.e., $f^{-1}(G)$.

Spatial approaches are more memory efficient and operate by directly propagating information between neighboring nodes in graph structure data. Unlike the spectral approach that leverages graph representation in linear space (e.g., eigenvectors and eigenvalues), the spatial approach

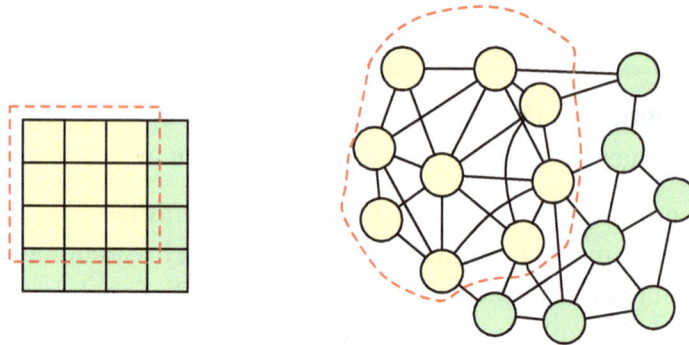

Figure 15-23: (left) An example of applying convolution on an image (or another grid structure data. (right) Applying convolution on the part of the graph is similar to applying convolution on image data.

focuses on the local connectivity patterns of the graph and implements the message-passing approach we had explained earlier. We have learned in Chapter 10, that CNN is applying filters on images. A filter moves across an image as a sliding window and extracts features from a set of pixels it covers. To understand spatial GCN, we can consider an image as a graph, that each of its pixels is connected to its adjacent pixels, such as the example we presented in Figure 15-23. As we can see from this Figure GCN applies a convolution over a graph. Instead of having a 2-D array as input, GCN takes a graph as an input.

The most popular GCN architecture proposed by Kipf and Welling [Kipf '16], uses spectral graph representation. They introduce a semi-supervised classification approach (classify the nodes of the input graph) that employs symmetric-normalized aggregation and self update loop.

First, we explain its pooling layer; then, we briefly explain its spectral graph convolutions and, later, its semi-supervised learning approach.

Graph pooling: Their approach uses a specific graph pooling layer. Assume the input of their GCN is a graph, which is presented by two matrices, X and A. X presents a feature matrix that has the size of $n \times d$, where n is the number of nodes and d is the number of features that each node has. A presents the adjacency matrix of the graph features. The output of GCN is Z, which is a matrix of $n \times f$. f is the number of output features for each node, and we have already explained n. Therefore, we can summarize it as: $Z = GCN(A, X)$

Assuming H presents a hidden layer, σ represents the activation function that introduces nonlinearity into the neural network, the neural network output at layer *l+1* of the message propagation layer can be written as: $H^{(l+1)} = \sigma(H^{(l)}, A)$, and thus $H^{(0)} = X$ and $H^{(L)} = Z$ (L presents the last layer). By using the weight matrix $W^{(l)}$, $H^{(l+1)}$ can be written as follows: $H^{(l+1)} = \sigma(A H^{(l)} W^{(l)})$.

The authors explained that this approach has two issues; (i) for every node, it sums up all the feature vectors of all neighboring nodes but not the node itself. To resolve this, they added an identify matrix to A as well and used $\hat{A} = A + I$, instead of A. (ii) A is not normalized, and thus multiplication to A changes the scale of feature vectors significantly. This affects the training accuracy, makes the model too sensitive to noise, and might cause overfitting. To avoid these issues A should be normalized, and authors normalize A that the sum of all of its rows are one.

To implement this normalization, authors use degree matrix D and use symmetric normalization $D^{-\frac{1}{2}}AD^{-\frac{1}{2}}$ instead of A alone.

Assuming, $\hat{D}$ is the diagonal of $\hat{A}$, by combining these two approaches, they introduce the following message propagation layer:

$$H^{(l+1)} = \sigma(\hat{D}^{-\frac{1}{2}}\hat{A}\hat{D}^{-\frac{1}{2}}H^{(l)}W^{(l)})$$

If there are no node features, authors recommend considering the adjacency matrix as the identity matrix, i.e., $A = I$. Here W is a differentiable parameter that can be trained. Training W enables the network to adapt its node representations to the graph structure. Besides, it enables the network to learn to combine the features of nodes in a way that is specific to the task at hand and also enables the network to generalize on unseen data.

Spectral graph convolution: Authors introduce a convolutional filter g_θ parameterized by θ in Fourier domain. Signal x as a scalar representation of each node, $L = I_N - D^{-\frac{1}{2}}AD^{-\frac{1}{2}}$ as normalized graph Laplacian, and their spectral convolutional can be written as follows:

$$g_\theta * x \approx \theta(I_N + D^{-\frac{1}{2}}AD^{-\frac{1}{2}})x$$

We skip explaining the mathematical details of how authors get to this equation to keep our brains from the explosion because of math.

Semi-supervised Learning: Their semi-supervised learning approach leverages both labeled and unlabeled data to train the network. A small subset of nodes is labeled (ground truth) are used to compute a loss function that measures the difference between the predicted labels and the actual labels. The loss is then backpropagated through the network to update the model's parameters.

The cost function, which is cross-entropy for multiclass node classification, combines the labeled and unlabeled data. In particular, this cost function consists of two parts; the supervised loss, which measures the difference between the predicted and actual labels for the labeled nodes, and a regularization term that encourages smoothness of the model's predictions on the unlabeled nodes. By jointly optimizing the supervised loss and the regularization term, this GCN learns to generalize from the labeled nodes to make predictions for the unlabeled nodes while taking into account the graph structure.

Graph Attention Network

Graph Attention Network (GAT) [Veličković '17] extends the idea of GCN by introducing an attention mechanism into the propagation module (pooling layers). It computes the hidden state of each node in the pooling layer among its neighbors by incorporating the self-attention mechanism (check Chapter 12).

It applies convolutional on each node, but the neighbor information aggregation in its propagation module is different and based on self-attention. By leveraging self-attention, neighbor nodes can receive different weights without matrix multiplication. This enables the resulting model to be applicable in inductive learning problems, including tasks that the graph has not seen before.

In summary, using attention allows GATs to learn more expressive representations of nodes than traditional graph neural networks. Again we skip going through its mathematical details and reasoning, and brutally jump into the formalization of the neighbor propagation, as follows:

$$h_i^{(l+1)} = \sigma\left(\frac{1}{K}\sum_{k=1}^{K}\sum_{j\in N_i}\alpha_{ij}^k W^k h_j^{(l)}\right)$$

Here, σ is a non-linear activation function such as Leaky ReLu, α_{ij}^k is the normalized attention weight between node i and node j. W^k is the trainable weight matrix, K is the number of attention heads, N_i is the number of neighbors of node i, and obviously $h_j^{(l)}$ is the representation of the node j at layer l.

The attention mechanism in GAT works by assigning different importance to each neighboring node when aggregating their features. This is done through a shared attention mechanism, which computes attention coefficients α_{ij} based on the features of nodes i and j. These coefficients are then normalized across all neighbors of a node using the softmax function. For example, consider a social network where each node represents a person and edges represent friendships. In this scenario, a GAT can learn which friends have more influence on a person's behavior, allowing for more nuanced recommendations or predictions.

The author didn't propose enough visualizations or examples GNNs. Too much theory is melting me.

One of the advantages of GATs is their ability to handle graphs with varying structures and sizes, making them suitable for a range of applications. Furthermore, the attention mechanism allows GATs to focus on the most relevant parts of the graph, improving the interpretability of the model. GATs can generalize to completely unseen graphs, as the attention mechanism is computed based on node features.

Graph SAGE

Similar to NLP tasks, within embedding, we can facilitate several machine learning tasks on the graph, such as node classification, prediction, etc. Graph SAGE (SAmple and aggreGatE) [Hamilton '17] provides an inductive approach to learning a low-dimensional embedding of nodes in graphs. In Chapter 1, we have explained that *inductive learning* generalizable their model to unseen data, and it is in contrast to *transductive learning* with means trimmed for a particular type of data. Graph SAGE can generate node embeddings for previously unseen nodes. This is particularly useful in dynamic graphs where new nodes and edges are continuously

added. By using a sampling approach, Graph SAGE efficiently handles large-scale graphs and enables scalable learning of node representations.

Previous to Graph SAGE, node embedding approaches rely on matrix factorization, which results in transductive embedding. Similar to other GNN, Graph SAGE also operates iteratively in two stages, aggregate and update, and nothing much needs to be explained here. Its aggregate and update are similar to other message passing layers and computed as other models.

Graph SAGE is computationally efficient as it only requires a fixed-size neighborhood sampling instead of considering all nodes in the graph. The authors are inspired by the idea of a mini-batch (not using all graph nodes at once) and implement it for graph embedding. For each node in their mini-batch, they perform k number of hops to collect a set of neighbors and their data; then, each mini-batch includes those k-hop neighbors.

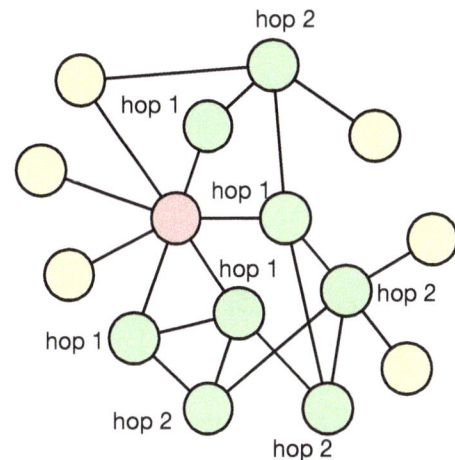

Figure 15-24: Neighbor node selection in GraphSAGE assuming that k=2 and H=4. As we can see, nodes with low restart scores are not selected.

However, using a mini-batch alone has a problem; some nodes, e.g., influencers in social media, have too many connections, and the mini-batch is getting too large and infeasible to process. Therefore, to manage those nodes, authors propose sampling H number of neighbors from each node in k-hop. For example, we have that is presented in Figure 15-24, assuming the $H = 4$ and $k = 2$. The model selects the red node to process, and since $H = 4$, it samples only four nodes and their neighbors (green nodes), the rest of the nodes are not staying in this mini-batch. Note that the sampling process is not purely random; it selects edges that have the highest connection to other edges, and authors assign this score to every node and call it the *restart score*. For example, in Figure 15-24, we can see that yellow nodes are not selected because they do not have many connections, and thus their restart scores are low. We can conclude that this style of sampling includes pruning based on the restart score. Although this approach presents effective graph embedding, increasing the number of message passing layers still makes it computationally intensive.

NOTES:

* Stacking too many message-passing layers is known as an over-smoothing, and it makes the node states indistinguishable from each other. In our experience, a few layers of message passing are usually enough. However, setting the neighborhood hop parameter is very much dependent on the application.

* While the learned representations, which are a result of feeding the input graph into a GCN, capture the structural information from the input graph, they may not preserve the exact connection characteristics of the original graph.

* The spectral approach for GCN requires the entire graph to be loaded into the memory and processed, which is impractical for large graphs with a large number of nodes and edges, such as social network graphs. In contrast, spatial approaches directly perform the convolution on the graph and aggregate nodes. This makes them more practical when we are dealing with large graphs

High-Dimensional Search with Graphs

We explained in Chapter 4 that tree and hash data structures, such as hash tables, can be used to reduce the search space and make searches faster. Another data structure that can improve search efficiency is graphs. Graphs can represent complex relationships between objects, making it easier to identify patterns and connections. By modeling relationships as edges between nodes, graphs can simplify the search space. For example, graph algorithms like Dijkstra's and A* allow for finding the shortest path between nodes, reducing the number of elements that need to be examined during a search. Additionally, graphs can represent hierarchical relationships and dependencies, making it easier to prune irrelevant parts of the search space.

Vector databases have gained popularity for their use in Retrieval Augmented Generation (RAG), which can be used alongside LLMs to leverage the linguistic capabilities of LLMs while searching internal data that is not available to the LLM. Many vector databases use one particular data structure, Hierarchical Navigable Small World (HNSW), to index and thus facilitate the search for high-dimensional data. HNSW creates small-world graphs that allow for quick nearest-neighbor searches. To explain HNSW, we need to be familiar with a few concepts first, and then we will explain the HNSW algorithm.

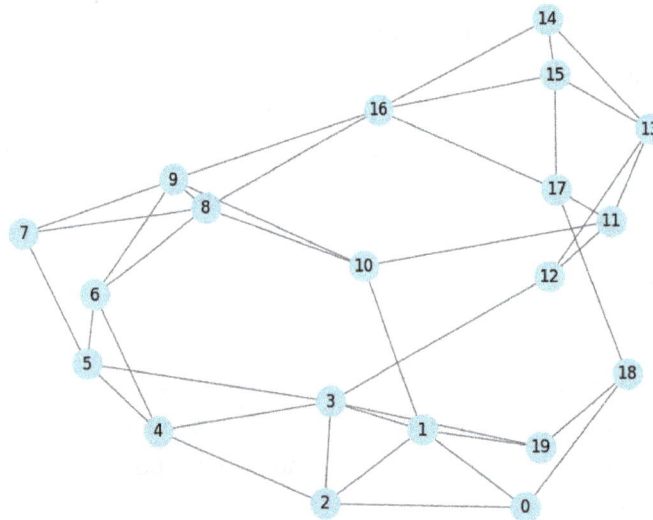

Figure 15-25: An example of small world model.

Small World Model: Small world models refer to graphs where most nodes are not directly connected with each other, but any two nodes can be connected by a relatively short path. In other words, they have a high degree of local clustering and short average path lengths between nodes. This means every in the graph can be reached with a few hops from every other node in the graph. Figure 15-25 presents a small world model example. For example, we can get from node 7 to node 11, with only three hops: 7 to 9, 9 to 10, and 10 to 11.

A challenge in SWM is the use of greedy methods, which might lead to being trapped in a local optimum or locality. For example, if we add weights to the graph presented in Figure 15-25, the path from 7 to 11, via 7 to 5, 5 to 3, 3 to 19, 19 to 18, 18 to 17, and 17 to 11, might be shorter in terms of cost, but the greedy approach doesn't find it.

Navigable networks are referred to as networks that allow for efficient routing from one node to another, enabling quick searches or navigation through the network. They often utilize local information for routing decisions, making them scalable for large networks where global knowledge is impractical. Navigable networks usually have many small groups (local clusters) of nodes that are closely connected to each other.

Navigable small-world networks are a specific subset of small-world models that possess the characteristics of navigable networks (efficient routing and local information utilization) and small-world models (high local clustering and short path lengths). Nodes that have a high number of edges are referred to as *high-degree nodes*, and nodes that have fewer edges are referred to as *low-degree nodes*.

Navigable Small World (NSW) graphs use a greedy approach to search for the closest node to a query object. This greedy approach can lead to getting stuck in a local optimum, which means the algorithm might not find the global best match but rather a "good enough" match that is close to the starting point or an intermediate point in the search. The greedy approach prioritizes short-term gains (moving closer to the query object in a single step) over exploring more distant nodes that might be a better match.

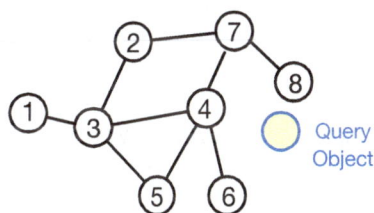

Figure 15-26: A simple graph that the closest node to the query object is node 8.

To understand the search problem in NSW, let's use an example. We have an information object; let's call it a query object, and we want to find the node that is the most similar to this query object. The search starts from a particular node in the graph, which is called an entry point. In Figure 15-26, The entry node is node 1, and the closest node to the query object is node 8. The NSW uses a greedy approach to search, and by using the greedy approach, the algorithm ends up selecting node 6 as the closest node to the query object. Node 6 is a local optima and this phenomenon is known as *early stopping* (different than the early stopping we explained in Chapter 8).

Hierarchical Navigable Small World

Hierarchical Navigable Small World (HNSW) [Malkov '18] is one of the best approximate nearest neighbor searches that is used in several vector databases to implement RAG. It combines the concepts of Skip List (Check Chapter 5), and NSW to enable efficient search and retrieval in large graph datasets. It organizes data points in multiple hierarchical layers to enable fast and scalable similarity searches. We highly recommend checking the Skip List in Chapter 5 before reading this part.

The algorithm is inspired by the Skip List. We explained other nearest neighbor search methods, such as Locality Sensitive Hashing, KD-Trees, and Voronoi Tessellation, while discussing kNN in Chapter 9. These methods, however, have limitations when the dimensionality of data increases. In contrast, graph-based search methods can handle high-dimensional data efficiently. One of the most efficient graph-based similarity methods is HNSW. It employs a hierarchical structure to search graph nodes.

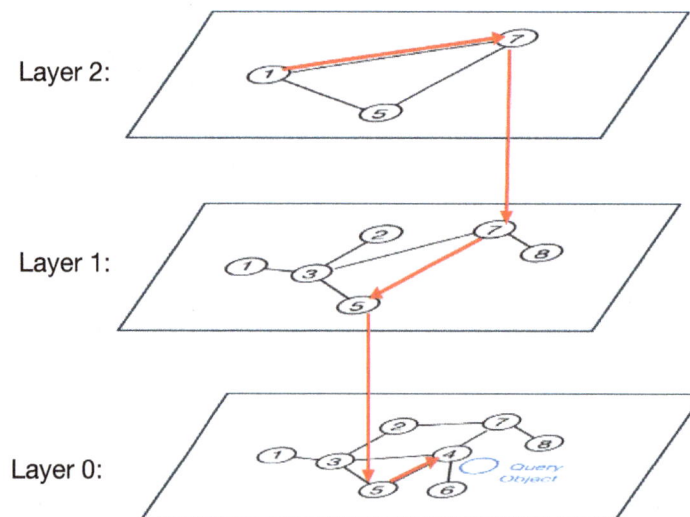

Figure 15-27: A HNSW approach for hierarchical graph search. Top levels include a subset of nodes and edges.

Searching an HNSW is implemented via routing, and it includes two phases: zoom-in (local) and zoom-out (long-range) navigations. The first phase is the zoom-out phase, which starts from low-degree nodes and moves from the topmost layer to lower layers, making long-range hops to approximate the target's location quickly. The second phase is the zoom-in phase, where the search continues in the lower layers, utilizing dense local connections to refine the path and accurately locate the target node.

Figure 15-27 presents a hierarchical multi-layer graph. First, the HNSW converts the input graph for searching into a multi-layer graph, in which each layer represents different resolutions of the data. Similar to the Skip list, we have different layers, and the bottom layer includes all nodes, but not necessarily all edges of the original graphs. The edges in the original graph are not necessarily preserved in the bottom layer because some edges might be redundant or less important for search. Thus, they will be removed, and the bottom layer is a simplified version of the original graph.

Nodes in the upper layers are usually high-degree nodes because they serve as key points for long-range navigation. They should facilitate long-range hops. The more we move downwards toward the bottom layer, the graph becomes progressively denser, and more nodes are added. The search process begins at the topmost layer (the entry node is assumed to be Node 1 in Figure 15-27), where nodes are sparsely connected, allowing for quick, long-range hops (zoom-out). As the search narrows in on the target, it goes down through the layers, progressively transitioning to denser, more detailed local neighborhoods (zoom-in) until the target node is found.

Summary

In this chapter, we have started by explaining the concepts of graphs. Next, we moved to basic applications of graphs, including minimum spanning trees (Prim and Kruskal) and shortest path finding algorithms, including A* and Dijkstra algorithms. Afterward, we briefly discussed two matching algorithms, the Hungarian algorithm, and the Hopcroft-Karp algorithm. They might have fewer applications than other graph algorithms, but we believe they are also important to learn, and we might need to leverage them into a machine-learning pipeline.

> Mom, it is me ... Yes ... I finished another boring chapter. I learned about graph algorithms and some graph neural networks.

Next, we move into centrality measurement methods used to find influential or central nodes in graphs. In particular, four different centrality measurement methods are explained, including degree centrality, closeness centrality, betweenness centrality, and eigenvector centrality.

Afterward, we described two very popular and useful algorithms for link analysis, PageRank and HITS. Later we moved to community detection in graphs, the other name for clustering. In addition to the Spectral algorithm, which is a well-known algorithm, we described Leiden and Louvain algorithms that are not widely known, but our experiments in consulting biotech firms show they are among the highest accurate algorithms for graph clustering.

Finally, we moved to graph neural networks and started this section by describing some concepts and challenges associated with using neural networks with graphs. Then, we list two common layers that are used in all graph neural networks.

Next, we described two models for classification (Graph Convolutional Network, Graph Attention Network) and a model (GraphSAGE) for graph embedding, which changes the representation of the graph.

We finalize this chapter by explaining the use of graphs for highin ,dimensional search and-the ,particular HNSW algorithm. It is the most popular algorithm to implement approximate nearest neighbor matching in a graph and used in the vector databases to support RAG for LLMs.

Further Reading and Watching

* If you like reading more about graphs baseline algorithms, the typical algorithm books such as the Cormen et al. [Coremen '22] or Sedgewick and Wayne [Sedgewick '11] have a good list of baseline algorithms about graphs. However, those algorithm books do not cover recent advances in graphs, especially community detection approaches and graph neural networks.

* Unlike other explanations, we did not explain Hopcroft-Karp algorithm with an example. There are many video explanations available on YouTube. If you would like to learn this algorithm example, you can search for its video explanation online.

* Most online classes we found about graph neural networks are not attractive, but a good book is published by William L. Hamilton, entitled "Graph representation learning" [Hamilton '20]. The book is also available online in PDF as well: https://www.cs.mcgill.ca/~wlh/grl_book.

* Jure Leskovec, who is one of the authors of GraphSAGE provides a good explanation of this model, and we used his explanation to explain the GraphSAGE: https://www.youtube.com/watch?v=LLUxwHc7O4A.

* There is a brief introduction to Graph neural networks in Distill: https://distill.pub/2021/gnn-intro. Besides, there is a good and concise video series that explains GNN models briefly, with excellent explanations https://www.youtube.com/@deepfindr

* Zhou et al. [Zhou '20] have a good survey paper on graph neural networks. We did not cover many models here, and similar to other deep learning directions, there are many models developed for graph neural networks as well.

* Deep Graph Library provides a good Python library to use graph neural networks, https://www.dgl.ai

Chapter 16: Challenges of Working with Data

Whatever remains to be explained for machine learning and artificial intelligence is summarized here. Therefore, this chapter has no order and does not follow a narrative that existed in previous chapters. We list concepts that are required to be learned and problems we could face while working with the data. You can read this chapter as a dictionary of terms and concepts. Here, we briefly describe the following concepts and a few top algorithms for each concept.

Problem complexity: Understanding the relationship between problem complexity and machine learning

Data sampling: Statistical approaches and methods to collect data and construct the sample dataset for the machine learning algorithm.

Noise: Different types of noise and techniques for mitigating noise in the dataset or adding noise to the dataset.

Imbalance data problem and data augmentation: Problems of imbalance datasets and strategies for balancing data, including data augmentation concepts and algorithms.

Data Imputation and Interpolation: Techniques for handling and reconstructing missing data in different data formats.

Anomaly and Outlier detection: Methods for identifying anomalies and outliers

Data and model changes over time: Concepts and techniques for addressing changes in data over time

Rater agreement methods: Techniques for evaluating the quality of human-assigned labels that can be used both for machine learning and qualitative studies.

Prompt engineering and RAG: Methods to query and ask questions from LLMs and how to leverage LLM capabilities for our data.

Fairness, Bias, and Transparency: concepts, tools, and best practices for ensuring fairness, avoiding bias, and promoting transparency in machine learning.

Problem Complexity

In Chapter 1, we have explained the computational complexity, which explains the algorithm characteristics in terms of resource usage. The same challenge existed with the complexity of the problem that the algorithm resolves, known as *problem complexity*, and not just the complexity of the algorithm alone (computational complexity).

Back in Chapter 1, we have described that some algorithms can solve their problems in linear time, e.g. $O(log n), O(n), O(n log n)$, etc., and some can solve in polynomial, exponential or factorial times, e.g. $O(n^2), O(2^n), O(n!)$, etc. The class of problems that can be solved with $O(n^k)$ time is called *Polynomial problems* or *P problems*.

In other words, P problems are problems that can be solved with polynomial complexity[1]. Examples of P problems include sorting algorithms, matrix multiplication, and finding the shortest path in a graph.

Sometimes, we do not know if the given problem can be solved in polynomial time. Problems that cannot be solved in polynomial time but for which an algorithm can guess a solution and verify whether the guess is correct in polynomial time are called *NP-Problems* (*Nondeterministic Polynomial time*). Some examples of NP-Problems include the Traveling Salesman Problem[2], the Knapsack Problem [3], and the Graph Coloring Problem[4].

The key distinction between P problems and NP-Problems is that P problems can be solved in polynomial time, while NP-Problems can be verified (though not necessarily solved) in polynomial time.

It is an unsolved question in computer science, but most computer scientists agree $P \neq NP$. If $P = NP$, there is no distinction between P and NP, and every problem whose solution can be verified in polynomial time (NP) could also be solved in polynomial time (P).

In summary, *NP-Problem*s are generally considered to be complex to solve, and no known algorithm can solve them in polynomial time, but this claim has not been formally proven. An open question in computer science is whether NP is equivalent to P, meaning whether every NP-Problem can be solved in polynomial time.

NP-Hard problems are least as hard as the hardest problems in NP. In other words, if we find an algorithm to solve an NP-Hard problem, then by using this algorithm, we can solve all problems in NP.

NP-Complete problems are a subset of NP-Problems that are at least as hard as the hardest problems in NP. It is believed (not a proven fact) that no NP-Complete problem can be solved in P, which is why they are considered intractable. Other NP-Problem, which are not NP-Complete can still be solved in polynomial time (if an efficient algorithm is discovered for the problem).

The difference between NP-Hard and NP-Complete problems is that NP-Hard problems don't necessarily need to be in NP themselves, NP-Complete ones are both NP and NP-Hard. In other words, all NP-Complete problems are also in NP-Hard, but not all NP-Hard problems are NP-Complete.

To better understand them, imagine you are the chief of a kitchen in a restaurant. You are working on a new recipe. You have a list of ingredients and instructions for preparing the dish, but you want to make sure that the dish is perfect before serving it to your customers.

---

[1] Problems solvable in polynomial time are considered "tractable" or "feasible" in computer science.

[2] Traveling Salesman Problem: given a list of cities and the distances between each pair of cities, determine the shortest possible route that visits each city and returns to the starting point.

[3] Knapsack problem: given a set of items, each with a weight and a value, and a maximum weight capacity, find the subset of items that maximizes the total value without exceeding the capacity

[4] Graph Coloring Problem: given an undirected graph, the goal is to assign colors to its vertices in such a way that no two adjacent vertices have the same color, using the fewest number of colors possible.

P-Problem: You have the recipe and the required ingredients, and the instructions are straightforward. In this case, you can prepare the dish in a short amount of time.

NP-Problem: You intend to prepare a new recipe that requires many ingredients and multiple steps. You know what the taste of the resulting dish should be, but you are not sure which specific ingredients to use or the exact steps to follow. In this case, you start by choosing a set of ingredients and following a set of instructions, but then you might realize that something isn't right (e.g., the flavors don't mix well). Then, you need to go back and try a different set of ingredients or instructions. This process of trial and error can take a long time, but eventually, you will find a solution that works. This is like a problem in NP, where the solution can be verified quickly, but finding the solution may take a long time. Another example is landing on the moon. It takes lots of trial and error, but in the end, we can land some machines on the moon.

NP-Hard problem: Assume you want to prepare a specific dish that is particularly difficult to prepare. This dish is so complex that no matter how long you work on it, you might not find an efficient way to prepare it. It means no known efficient solution exists. This is analogous to an NP-Hard problem; there is no known efficient way to find a solution. Another example is raising a child. There is no specific remedy how to raise a child successfully, and it is an NP-Hard problem.

NP-Complete problem: Assume you want to prepare a specific dish that is particularly difficult to prepare. This dish requires a large number of ingredients and complex instructions, making it challenging for you to prepare it within a reasonable time. You might never find a solution, no matter how much time you spend on it. However, If someone gave you a potential recipe, you could quickly verify if it results in the desired dish. However, finding that recipe in the first place takes a long time. This represents an NP-Complete problem, where verification is efficient, but finding a solution is not.

In summary, there is no known efficient way to find a solution for NP-Hard problems. For NP-Complete problems, we can verify a potential solution in polynomial time, but there is no known efficient method for finding that solution. Figure 16-1 presents a common Venn diagram to present the complexity of problems.

Many problems in machine learning are NP-Hard. This means that finding the optimal solution for these problems is computationally intractable, and we have to rely on approximation algorithms or heuristics that can find good solutions in a reasonable time. Examples of NP-Hard problems in machine learning are finding the optimal divisions in decision trees (Chapter 9) and contextual bandit problems (Chapter 13), etc. Machine learning techniques cannot solve NP-Complete problems unless $P = NP$, which is widely believed to be false.

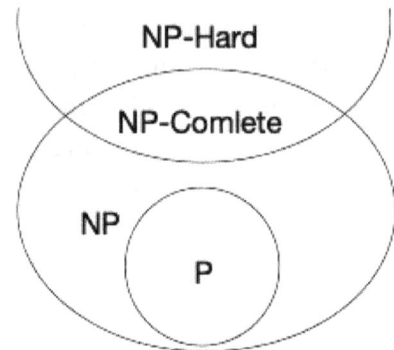

Figure 16-1: Venn diagram of different problem complexities.

Sampling from Complex Data Structure

One of the challenges in data collection and preparation for a machine learning task is ensuring that the samples drawn from the population dataset accurately reflect the characteristics of the entire population. This challenge becomes especially critical when splitting a dataset into training and testing subsets, as it is important to ensure that the distribution of these subsets closely mirrors the distribution of the population.

In this section, we explain some common sampling methods. We encountered some of them, such as Gibbs sampling, earlier in Chapter 11, but we describe them in more detail here. Before we start our explanation about sampling, note that in machine learning, it is common to assume that the sample dataset should be i.i.d. Independent here means that the process of sampling keeps no information about the previously sampled data (it is memoryless).

Stratified Sampling and Clustered Sampling

A simple yet effective approach to collecting data is *stratified sampling*. In this method, the population dataset is divided into smaller segments, which are not sharing data, and they are called *stratum* (stratum is singular, and strata is plural). Then, samples are selected randomly from each stratum. While implementing stratified sampling, each stratum should be collectively exhaustive, mutually exclusive, and homogenous.

Collectively exhaustive: Each stratum should be representative of the population or cover the entire range of values for the variable being stratified. For example, if a political campaign is willing to learn the preferences of minorities, each should include samples from all relevant

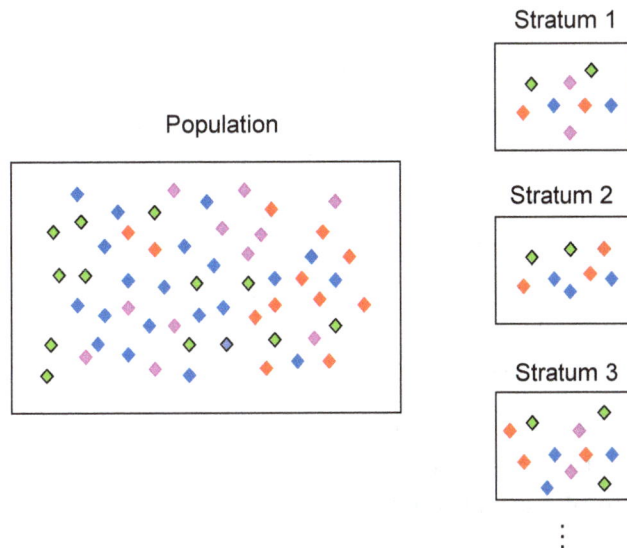

Figure 16-2: An incorrect stratification, because startum 2 does not include any pink color data point.

minority groups. For example, in Figure 16-2 we can not say the stratum is following the conditions we have described because we can see that stratum 2 does not entail any pink data, which is a violation of the collectively exhaustive condition.

Mutually exclusive: Each datapoint should belong to only one stratum, and it should not be shared among different strata.

Homogenous: All datapoints in each stratum are similar with respect to the stratification variable. Here, the stratification variable is the specific characteristic or attribute used to divide the population into different strata in stratified sampling.

Although this sampling is easy to understand, constructing a stratum that entails all required information is challenging and requires a good understanding of the population's data.

Another sampling similar to stratified sampling is *clustered sampling*, in which we apply clustering on the dataset, and an equal number of data are sampled from each cluster. In other words, if the sample segmentation is done via clustering, then it is referred to as cluster sampling. Clustered sampling assumes that individuals within a cluster are more similar to one another than to individuals in other clusters. Therefore, it is important to choose clusters that are similar in terms of the characteristics of interest. Additionally, random sampling is conducted within each selected cluster to ensure that the sample remains representative of the population.

Monte Carlo (MC) Sampling

The Monte Carlo (MC) simulation or method [Metropolis '49], also referred to as multiple probability simulation, is a mathematical technique that uses repeated random sampling to estimate probability distributions of uncertain outcomes or to solve problems that are inherently deterministic. Originating during World War II, the MC method was developed to enhance decision-making under uncertain conditions. This approach heavily relies on random or pseudo-random numbers to address theoretically deterministic problems.

Randomness is crucial in Monte Carlo simulations as it mitigates the risk of biased selection. When we randomly sample from a distribution, we can obtain a more representative sample of the possible outcomes because the sampling reflects the distribution's inherent probabilities.

According to the Law of Large Numbers (LLN), which we explained in Chapter 3, the more samples we collect, the closer the sample dataset characteristics will be to the population dataset. Sampling can be resource-intensive, so it's essential to optimize the process. Enforcing randomness in the sampling process helps us construct a sample dataset that accurately represents the population's characteristics.

Monte Carlo (MC) methods are used for several purposes, including (i) approximating parameters (e.g., mean, variance) of a population distribution, (ii) gathering samples that approximate the distribution of a target function, and (iii) identifying samples that help minimize or maximize a target function.

In this context, a target function represents the output of a system under varying conditions, often used to optimize or better understand the system. For example, MC methods might be used to find the maximum value of a function or estimate how the function's output is distributed across a range of inputs.

MC sampling operates in three steps. (i) We define the independent (input) and dependent (output) variables for the problem that we intend to study. (ii) The distribution of the input variables will be estimated. This estimation can be based on historical data, domain knowledge, or assumed probabilistic models. (iii) The MC sampling algorithm is executed repeatedly. In each iteration, the algorithm generates values for the independent variables by randomly sampling from the distribution of the input variables. The results of the simulations are then analyzed using statistical methods to estimate probabilities or the range of possible outcomes.

As an example, take a look at Figure 16-3 (a). Here, we have a dataset that includes the red and blue regions. We intend to sample from this dataset while preserving the red and blue ratio. One way to have an accurate sampling is to calculate the size of the red area, then try to decompose the blue shape into smaller, less complex shapes (such as circles and triangles), calculate their sizes, subtract the overlapping area, and add the remaining parts together. Then, we will calculate the ratio of red to blue data distribution, and we can sample according to the ratio. However, life is short to do so many things, and computationally, this approach is complex.

MC sampling can help us identify the ratio of blue and red areas. The MC method starts by sampling random data points, see Figure 16-3 (b). As the number of samples increases, we get a closer approximation to the real shape. Therefore, our estimation of the ratio becomes more realistic, as shown in Figure 16-3 (c) and Figure 16-3 (d).

MC sampling is memoryless, which means each sample is drawn independently and does not rely on any knowledge of previous samples, and prior samples do not impact the sample selection. We can summarize that MC sampling estimates the distribution of different outcomes by repeatedly sampling random input variables and simulating a system or process.

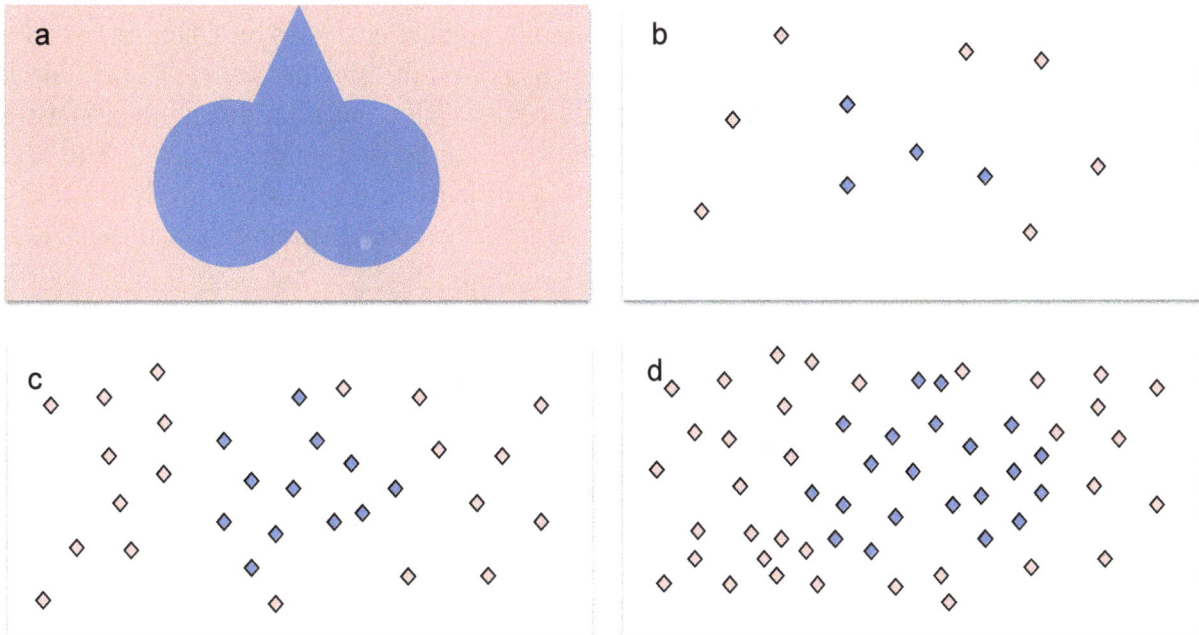

Figure 16-3: Monte Carlo simulation samples the number of dots to calculate the ratio or size of the blue and red shapes.

Monte Carlo Markov Chain (MCMC) Methods

In Chapter 3, we have described that a dataset that is sampled from a population should entail the population characteristics (distribution) as well. The basic form of the MC method, which we have discussed earlier, involves drawing independent samples from a distribution. This independence makes MC memoryless. This property is beneficial for certain problems, but it can also lead to biased sampling when applied to complex data structures with many features. In such cases, MC might favor certain variables over others due to the independent sampling nature. Furthermore, this form of MC can be computationally intensive for high-dimensional datasets, making it less practical for large-scale applications.

One way to mitigate the bias of basic MC sampling is to use *importance sampling* (*Monte Carlo with Importance Sampling MC-IS*). Importance sampling involves sampling from a different distribution and weighing the samples according to the original distribution. It can help to ensure that all variables are sampled uniformly. However, a disadvantage of importance sampling is that if the distribution used for importance sampling is not well-chosen, it can lead to high variance in the estimates. Besides, an expert with good domain knowledge is required to conduct this sampling. We will describe the importance of sampling later in this section.

You should check Chapter 6 and review the Markov Model again before you start reading MCMC methods.

1031

Markov Chain Monte Carlo (MCMC) is a class of algorithms used to generate samples from a target distribution when direct sampling is not feasible or efficient. Markov Chain refers to a sequence of random variables whose current value depends on the previous value or state (check Chapter 5). We use MCMC methods when we don't know the population distribution, and we want our sample dataset to have the same distribution as the population distribution. The next sample in MCMC depends on the existing sample, and because of this dependency, it is called a Markov Chain Monte Carlo. In the following section we explain Metropolis-Hastings, Gibbs Sampling and Importance Sampling.

Metropolis-Hastings

Metropolis-Hastings [Metropolis '53, Hastings '70] algorithm is one of the common implementations of MCMC. It combines random walks[5] with acceptance conditioning. The objective of Metropolis-Hastings is to draw samples from an unknown target distribution, denoted as $p(\theta)$, which we only know up to proportionality constant[6] $g(\theta)$. The samples are drawn from a proposal distribution, denoted as $q(\theta)$. We summarize them as follows and we should memorize them to understand the rest of the explanation.

$p(\theta)$: The target distribution we aim to sample from. However, we know it only partially and not exactly.

$g(\theta)$: A distribution that has the same shape as the target distribution $p(\theta)$, but may differ by a constant scaling factor. We don't know the normalizing constant that can make it equal to $p(\theta)$.

$q(\theta)$: The proposal distribution used to generate candidate samples in the Metropolis-Hastings algorithm.

This algorithm operates in four steps: initialization, sample generation, accept/reject generated samples, and convergence to the target probability.

First, it makes an initial guess of a sample in the target distribution. Second, it comes up with a new guess, which could be better or worse than the current guess. In the third step, it compares the new guess with the previous one and the algorithm checks which guess is better. If the new guess is better, it adopts it; otherwise, it keeps the previous one, but with some hesitation. The second and third steps are repeated several times until the given threshold of repetition is reached. The final guess will be chosen as the best possible guess.

---

[5] A random-walk is a type of sampling method where the next sample is determined by a random step from the current position, and it is used in MCMC methods.

[6] In the context of sampling, when we state we know $p(\theta)$ it up to a proportionality $g(\theta)$, i.e., $p(\theta) \propto g(\theta)$. It means we know the relationship, but we don't know the exact parameter values of each distribution.

Assuming $\Theta = \{\theta_0, \theta_1, \ldots \theta_n\}$ includes samples that present $p(\theta)$ distribution, the following algorithm implements the Metropolis-Hastings algorithm:

initialize $\theta_t = \theta_0$ `#` θ_0 `guess is completely random in the range of possible values.`

for (t=1 ... max iteration) {

    $\Theta.add(\theta_t)$

    $\theta_{t+1} \leftarrow q(\theta_t | \theta_{t-1})$ `#` θ_{t+1} `is sampled from the proposal distribution q, we can assume`
                      `q has a normal distribution or any other distribution. However,`
                      `don't that it is the proposal distribution and not the target`
                      `distribution.`

    $\alpha = \dfrac{g(\theta_t) \cdot q(\theta_t | \theta_{t-1})}{g(\theta_{t-1}) \cdot q(\theta_{t-1} | \theta_t)}$ `#` α `is referred to as the acceptance ratio.`

    if ($\alpha \geq 1$) $\Theta.add(\theta_t)$ `#` `here it checks if the acceptance ratio is larger or equal to`
                         `one, then it accepts the new guess (`θ_t`).`

      else $\Theta.add(\theta_{t-1})$ `#` `Otherwise, again, it adds the previous guess (`θ_{t-1}`) to the`
                        `memory. In some literature, instead of 1, they used a number`
                        `from a uniform distribution between 0 and 1, i.e., U(0,1)`

}

We can see from the last line that the decision to accept or reject a guess depends on the current state of the chain.

In other words, we can conclude the following about the MCMC:

probability of move = (probability of proposed state) / (probability of current state).

If it is still not clear to you, we recommend reading the codes that implement Metropolis-Hastings. They are simple to understand. By implementing the code and running it, you get confident that you understand how Metropolis-Hastings performs sampling.

Metropolis-Hastings has different applications; it can be used in machine learning for hyperparameter tuning, genetic sequence analysis, and estimating the parameters of a complex model.

Gibbs Sampling

Gibbs Sampling [Geman '84] is another common implementation of MCMC classes of algorithms. If the target or population dataset is multivariate with a complex distribution, then it is hard for the sampling process to learn the population distribution. The sampled dataset needs to reflect the characteristics of the population, meaning it should have the same distribution as the target. However, sampling directly from a multivariate distribution while preserving these characteristics is not trivial.

In Bayesian inferences, sampling from a conditional distribution, i.e., $P(x|y)$, and $P(y|x)$ is easier than sampling from a joint distribution $P(x, y)$. In other words, we often can not calculate the posterior distribution itself. Instead, we may be able to calculate the conditional posterior density for one or more variable(s).

$$\text{H A R D}\ P\,(x,y) \nearrow^{P\,(x|y)}_{\searrow\ P\,(y|x)}\ \text{E a s y}$$

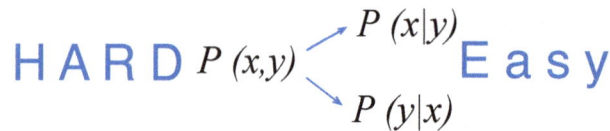

We use this characteristic to sample from a multivariate dataset with a complex distribution and have a sample distribution that is the same as the population.

Gibbs sampling implements this approach. It enables us to sample from a multivariate dataset while preserving population distribution. Gibbs sampling requires to have some prior knowledge about the dataset. The idea in Gibbs sampling is to collect samples using the conditional distribution of each variable (usually more than two variables) while the remaining variables stay fixed to their current values.

We have learned to compute conditional distribution. The Naive Bayes rule (check Chapter 9) is defined as $P(x|y) = P(x,y)/P(y)$, and we can assume that $P(x,y)$ is proportional to $P(x|y)$ and $P(y|x)$, i.e., $P(x|y) \propto P(x,y)$ and $P(y|x) \propto P(x,y)$[7]. Therefore, we can express the joint distribution $P(x,y)$ as the product of the conditional distribution $P(x|y)$ and the marginal distribution $P(y)$ and vice versa. Specifically $P(x,y) \propto P(x|y)P(y)$ and $P(x,y) \propto P(y|x)P(x)$.

If you wonder why such a claim is valid, check the example we provide in Figure 16-4[8]. Figure 16-4 analyzes two binary variables (A, B) and presents this phenomenon with real numbers. Assume they are two mice (mouse A and mouse B) running inside a maze to get to their food. The following presents all possible events, which we call this event list population:

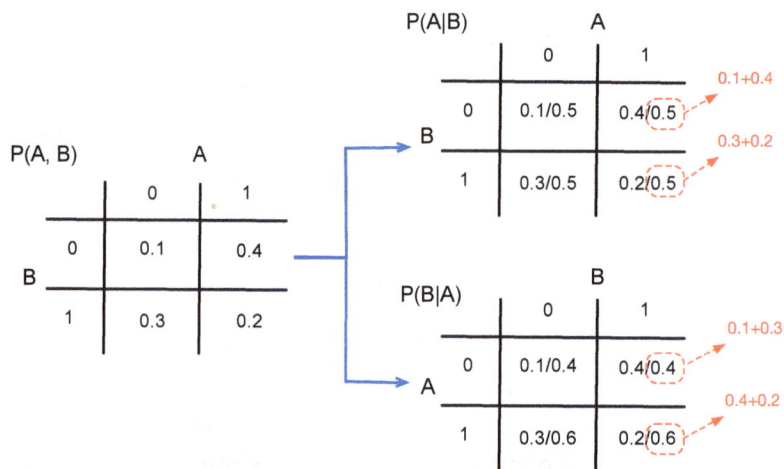

Figure 16-4: A sample example of joint distribution for two variables and inferencing conditional distributions form joint distribution.

$population = \{(0,0),(0,1),(0,1),(0,1),(1,0),(1,0),(1,0),(1,0),(1,1),(1,1)\}$

---

[7] We should read "$\propto$" sign as "proportional to".

[8] The example here is adapted from Ben Lambert video: https://www.youtube.com/watch?v=ER3DDBFzH2g

It is a dataset of two variables (A, B), and we would like to create a sample dataset that follows the population distribution. The left side of Figure 16-4 presents the probabilities of combining these two variables $P(A, B)$. The content of this table is inferred from the population dataset. For example, the probability of A successfully getting its food and the B not getting its food is written as $P(A = 1, B = 0) = 0.4$. Based on the data provided by $P(A, B)$ on the left table of Figure 16-4, we can also use the Naive Bayes theorem and calculate conditional probabilities $P(A|B)$, and $P(B|A)$ located on the right side of this figure. By looking at the numbers on the right tables, i.e., $P(A|B)$ and $P(B|A)$, we can see their values are proportional to the values of the $P(A, B)$ table.

Gibbs sampling benefits from this proportionality. Assume we need to make N number of samples from a dataset with three variables (A, B and C). The sampling process will be implemented inside a loop as follows:

for i in N {

$\qquad A_i \approx P(A|B_{i-1}, C_{i-1})$ `# a sample $A_i$ drawn from the $P(A|B_{i-1}, C_{i-1})$ distribution`

$\qquad B_i \approx P(B|A_i, C_{i-1})$

$\qquad C_i \approx P(C|A_i, B_i)$

}

It means, Gibbs sampling uses the B_{i-1} and C_{i-1} to calculate A_i. Then, it uses A_i and C_{i-1} to calculate B_i, and so forth. We can generalize the Gibbs sampling algorithm as follows:

initialize θ_0 `# $\theta_0$ is initialized random in the range of possible values.`

for (i=1 to max iteration) {

$$\theta_1^{(i)} \approx P(\theta_1|\theta_2^{(i-1)}, \theta_3^{(i-1)}, \ldots, \theta_d^{(i-1)})$$

$$\theta_2^{(i)} \approx P(\theta_2|\theta_1^{(i)}, \theta_3^{(i-1)}, \ldots, \theta_d^{(i-1)})$$

....

$$\theta_d^{(i)} \approx P(\theta_d|\theta_1^{(i)}, \theta_2^{(i)}, \ldots, \theta_{d-1}^{(i)})$$

.....

}

Gibbs sampling is particularly useful in Bayesian statistics for posterior inference, especially in hierarchical models where it simplifies sampling from complex distributions. In machine learning, it is a key method in topic modeling with Latent Dirichlet Allocation (LDA), which we explained in Chapter 7, helping to infer topic distributions in documents. It's also applied in image processing tasks like denoising and segmentation, where it samples from the posterior distribution of pixel labels given the observed data.

Importance Sampling

When we need to sample from a distribution, but it is expensive, difficult, or impossible to sample from this distribution (target distribution) directly, we can use importance sampling [Kloek '84]. Importance sampling uses an auxiliary distribution and samples from the auxiliary distribution instead of the target distribution. Then, it uses weights to correct for the difference between the auxiliary distribution and the target distribution. In such a way, sampled data collected from auxiliary distribution are more informative about the target distribution than random samples.

The weights in importance sampling come from the ratio of the target distribution to the auxiliary (or proposal) distribution at each sampled point. Specifically, the weight for each sample is calculated as the target distribution's probability density divided by the auxiliary distribution's probability density at that sample point. Assuming $p(x)$ is the target distribution and $q(x)$ is the auxiliary distribution, the weight $w(x)$ for a sample x is given by:

$$w(x) = \frac{p(x)}{q(x)}$$

Suppose a scenario in which we intend to calculate an expectation of $f(x)$, where x presents samples of the target distribution $p(x)$, i.e., $x \sim p(x)$. In other words, the goal is to estimate the expectation of a function f(x) with respect to a target distribution. However, direct sampling from $p(x)$ is impossible or too expensive. Instead, we use $q(x)$ (auxiliary distribution) that we can sample from it.

Then, by using, $q(x)$ we compute the exception of $f(x)$ as follows:

$$\mathbb{E}_{x \sim p(x)}[f(x)] = \int f(x)p(x)dx = \int f(x)\frac{p(x)}{q(x)}q(x)dx \approx$$

$$\frac{1}{n}\sum_i f(x_i)\frac{p(x_i)}{q(x_i)} \ or \ \mathbb{E}_{x \sim q(x)}[\frac{f(x)p(x)}{q(x)}]$$

Here, n is the number of samples drawn from the $q(x)$ distribution. This equation shows that we can draw samples from the auxiliary distribution $q(x)$ and estimate the expected value $\mathbb{E}[f(x)]$ by weighting the samples according to the ratio of the target distribution $p(x)$ to the probability density of the auxiliary distribution $q(x)$ at each sample point.

For example, importance sampling can be applied to estimate the spread of a rare disease in a population. Directly sampling the entire population may not yield sufficient data on the disease due to its rarity. Instead, we can focus on vulnerable populations by sampling from a distribution that emphasizes these groups more. By adjusting the results with appropriate weights, we can estimate the disease's spread across the entire population. This enables public health administrators to allocate resources more effectively and target interventions where they are needed most.

Noise

In the context of signal processing, noise is defined as unwanted variation in data that occurs during data collection or processing, which negatively affects data quality. For example, in image analysis, random variations in brightness or color information are referred to as noise. In audio data, unwanted sounds, such as wind noise during recording, are considered noise.

Mother of all noises in the world

However, in the context of machine learning, noise is not always undesired, and it can be added during the processing to increase the variety of data during the training and thus improve the quality of the machine learning model. These types of processes are categorized as *data augmentation*, and one common data augmentation approach is to add noise to the clean data.

Furthermore, sometimes, we inject synthetic noise into a clean image, video, or audio and train the model to remove the noise. It means that the model encounters both noisy and clean images, and thus, it can learn how to remove the noise. This approach has applications in image degradation and restoration as well. For example, old images can be restored by a noise removal algorithm. Look at Figure 16-5, which is from an old image restoration algorithm provided by Wan et al. [Wan '22]. How do these systems work? Similar to other neural network models, a very big dataset of images is trained by manually adding noise and presenting both damaged and correct images in the model. The model learns to restore the image and remove its noise.

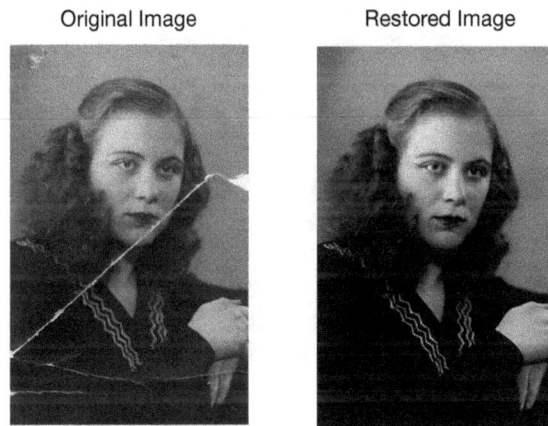

Figure 16-5: An old image restored with a deep learning model. Image credit Wan et al. [Wan '15]

Types of Noises

As we have explained in the realm of AI and machine learning, we use noise to build synthetic data similar to the original data to increase the variety of the training dataset. There are common types of noises, each with specific characteristics; we list some of them in this section. Most of the noise we list here is used for computer vision and, somehow, audio processing. However, the field of signal processing, which is closely tied to machine learning, deals with many different types of noise that we do not describe here.

Noise can be categorized based on its temporal characteristics. If noise is produced continuously, such as the hum of a refrigerator or the sound of traffic, it is called *continuous noise*; it has start and end times, and it is referred to as *intermittent noise*. If the noise is sudden, such as an explosion sound, it is referred to as *burst, spike,* or *impulsive noise*. In Chapter 6, we described a burst in time series. The same concept is applicable to noise as well.

Noises Based on Statistical Distributions

White noise

If a noise has equal distribution at different frequencies, it is called white noise. In other words, if the noise variables are i.i.d, it is called white noise. For example, Figure 16-6 presents a white noise on an image and in a time series. As you can see from this figure, white noise is a type of continuous noise. From the left side of Figure 16-6, we can see that the noise intensity does not change. You can see that white noise is a stationary time series (check Chapter 6). However, the mean of white noise is 0, which is completely random, which shows that the white noise signal does not include any motifs.

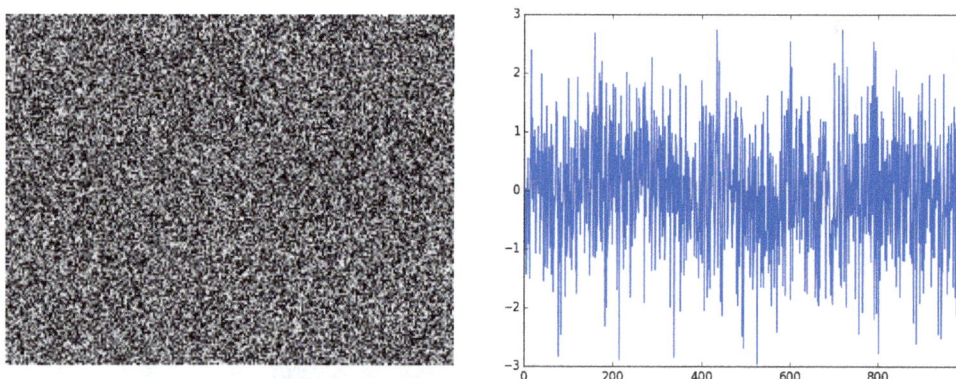

Figure 16-6: (left) white noise as an image, (right) white noise as a time series.

Gaussian noise

Most noises apply an array of random data with a particular distribution. One of the most popular noises that are easily tractable is Gaussian noise. The Gaussian noise is a set of random data points whose PDF has a Gaussian (normal) distribution. If the Gaussian noise data are identically distributed and independent, it is called White Gaussian noise. Gaussian noise has many applications. It can be present in audio recordings, introducing a hissing sound. It's commonly used in the testing and calibration of audio equipment and in developing noise-canceling algorithms. In denoising autoencoders (check Chapter 11), Gaussian noise is added to the input data to corrupt it. It is also used as input for GAN models, such as StyleGAN and diffusion models (check Chapter 11).

Other types of noises

There are other noises quantified by their statistical distribution, such as "Gamma (Erlang) noise" (based on Erlang or Gamma distribution) and "Exponential noise". In Chapter 3, we have studied distributions. Gamma noise is often encountered in scenarios where the sum of several independent, exponentially distributed variables is relevant, such as modeling the waiting times between events in a Poisson process. Exponential noise is common in processes where events occur continuously and independently at a constant average rate, such as the time between arrivals at a service point.

Salt-and-Pepper Noise

Salt-and-pepper noise is specific to images, applied on black and white images, and forces a certain number of pixels in an image to be either black or white, like dropping grains of salt and pepper on an image.

The salt-and-pepper noise is a form of impulse noise that causes sudden and extreme changes in the pixel values of an image. Unlike other types of noise that introduce small, continuous variations, impulsive noise directly sets some pixels to their maximum (white) or minimum (black) values, creating sharp and distinct disturbances in the image. The affected pixels are either set to the minimum value (0, representing black) or the maximum value (255, representing white) in an 8-bit grayscale image. The noise is typically described by two parameters: the probability of a pixel being corrupted by salt (white) noise and the probability of a pixel being corrupted by pepper (black) noise. These probabilities determine how much of the image is affected.

Gradient Noise

There is another category of noises that construct a texture, such as marble, wood, cloud, and infection in CT images. These noises are called *gradient noises*. For example, *Perlin* noise [Perlin '85] is used to simulate visual effects in digital arts or even infection in medical images [Javaheri '21].

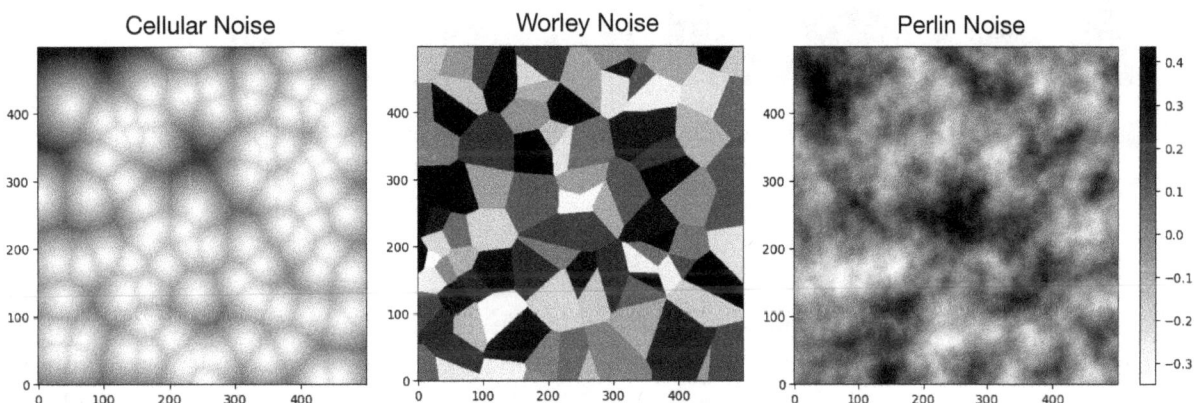

Figure 16-7: Different texture created by Worley (a.k.a. cellular) and Perlin noises.

If a Perlin noise is applied in multidimensional space, it constructs a *Simplex noise*. Simplex noise is the same as Perlin and created by the same person, Ken Perlin, but more resource-efficient than Perlin noise. Simplex noise [Perlin '01] addresses the shortcomings of Perlin noise, including greater efficiency in higher dimensions and reduced computational complexity. Another common noise in this area is Worley (Voronoi or Cell) noise [Worley '96]. It creates organic, natural patterns and textures, such as skin, lava, and foamy texture.

Figure 16-7 shows two different shapes of Worley noises, which are also known as Cellular noise, and one shape for Simplex (formerly known as Perlin) noise, in black and white.

Speckle Noise

Speckle noise, also known as speckle pattern or texture, refers to a type of noise that occurs in images or data and shows random bright and dark spots. It is often encountered in medical imaging and radar imaging. This is not as common as gradient noise, but it could be used for data augmentation. Speckle noise is multiplicative in nature, as opposed to additive noise like Gaussian noise. This means that the noise is not added to the original signal; it is multiplied with the original signal. Multiplicative noises are often more challenging to remove because they are intertwined with the signal intensity. Figure 16-8 presents a Speckle noise on a sample SAR radar image.

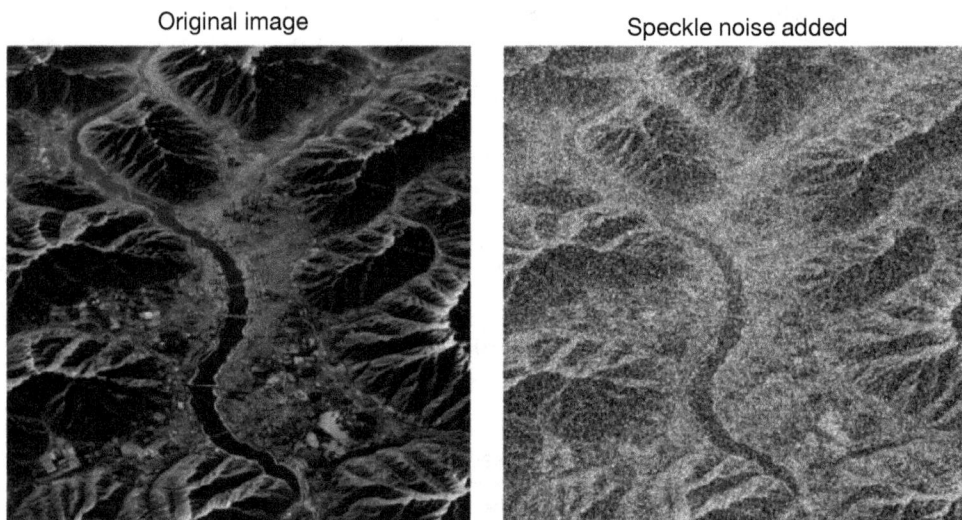

Original image

Speckle noise added

Figure 16-8: A sample SAR radar image (AI generated image) along with the same image, but spleckle noise added.

Noise Reduction

Removing the noise from the data (mostly audio and image) is called noise reduction or filtering. We categorize noise reduction techniques into two categories: signal-based and machine learning based noise reduction. We briefly explain the machine learning based noise reduction method and dedicate most of the focus in this section to signal-based noise reduction.

Machine Learning based Noise Reduction

Machine learning approaches use various methods to reduce noise, including dimensionality reduction techniques, such as PCA or SVD (check Chapter 7), denoising autoencoders, adversarial training, and diffusion models (check Chapter 11). For example, in adversarial training, GAN models are employed where a generator network tries to produce noise-free data to fool a discriminator network that classifies real versus generated data. The trained generator model can then take noisy data and remove the noise. Diffusion models can be used to generate synthetic data both with and without noise, enabling the model to compare them and learn how to perform denoising. Additionally, other neural network models, such as RNNs and CNNs, as well as machine learning methods like regularization, can also be applied to remove noise.

Signal-based Noise Filtering

Signal-based noise filtering methods include utilizing frequency domain filters such as Butterworth filter [Butterworth '30], time domain filters such as Wiener [Wiener '49] and Kalman [Kalman '60] filters, and wavelet transforms (check Chapter 7) to filter the input signal and reduce noise.

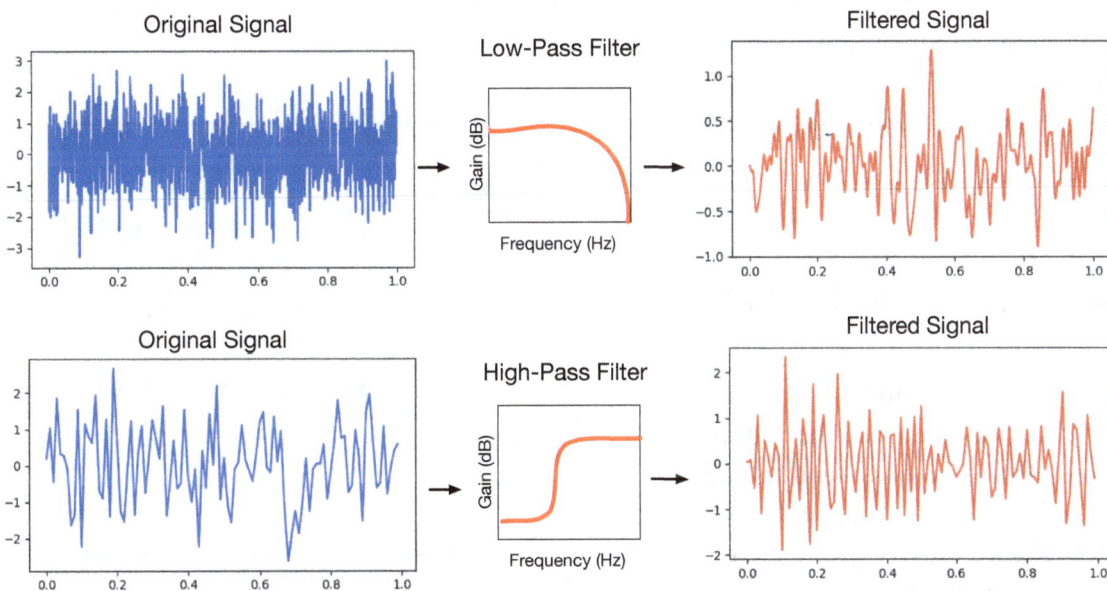

Figure 16-9: Applying low-pass and high-pass filters on two sample signals.

If you can't recall the concept of signals and filters, please check the related sections in Chapter 6 and Chapter 7. In these chapters, we have explained methods such as Fourier transformation can be used to transfer the original time domain signal into a frequency domain signal (e.g., FFT). These transformations can also be used to remove noise and reconstruct the original data from the transformed signal using techniques such as the Inverse Fourier Transform (IFT).

In this section, we introduce four new signal noise filtering approaches. Before explaining them, we explain a few basic terms about signal based noise filtering.

Low-pass and *high-pass filter:* A low-pass filter is a type of filter that takes an input signal and allows low-frequency components to pass through while attenuating or reducing high-frequency components. On the other hand, a high-pass filter is a type of filter that takes an input signal and allows high-frequency components to pass through while attenuating or reducing low-frequency components. Figure 16-9 shows two samples of applying low-pass and high-pass filters on two random signals.

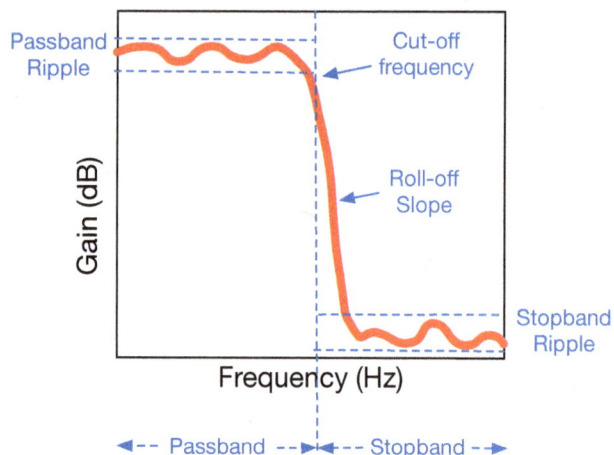

Figure 16-10: Visualizing signal filtering concepts on a low-pass filter.

Stopband and passband: Stopband refers to the range of frequencies in a signal that a filter is designed to attenuate or block. Passband refers to the range of frequencies in a signal that a filter is designed to allow or pass through with minimal attenuation. Figure 16-10 presents the range of stopband and passband for a low pass filter.

Cut-off: The cut-off frequency is the frequency at which the filter begins to attenuate a signal. Figure 16-10 presents a cut-off frequency for a low-pass filter.

Roll-off: It refers to a slope that presents how quickly a filter reduces the strength of frequencies from the pass-band as we move away from the cut-off frequency toward the stopband. Figure 16-10 presents a roll-off slope for a low pass filter.

Ripple: Fluctuations that existed in passband and stopband frequency are referred to as passband and stopband ripples.

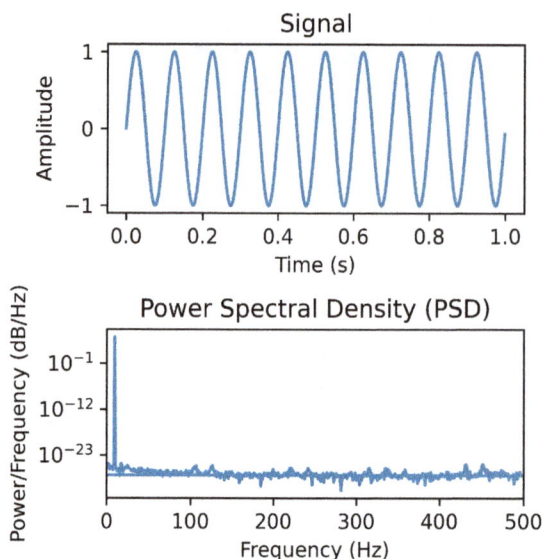

Figure 16-11: A power spectral density of a simple sinusoidal signal

Power Spectral Density (PSD): It describes how the power of a signal is distributed among different frequencies. In other words, PSD presents the strength as a function of frequency. To present PSD, we use a graph as a power over frequency. Figure 16-11 presents a simple sinusoidal signal and its PSD. Now that we have learned these concepts, we can start describing signal filtering methods.

1042

Butterworth filter

The Butterworth filter can be either a low-pass or high-pass filter. It is designed to have a maximally flat frequency response in the passband. In other words, there is no ripple. Additionally, the roll-off is monotonic, with no ripple in the stopband. This filter is commonly used in applications requiring a smooth passband. For example, it is used in audio processing to maintain clear sound quality by preventing any unwanted variations in the frequency levels.

Butterworth is formalized with the following equations:

Low pass filter: $H(\omega) = \dfrac{1}{\sqrt{1 + (\frac{\omega}{\omega_c})^{2n}}}$

High pass filter: $H(\omega) = \dfrac{1}{\sqrt{1 + (\frac{\omega_c}{\omega})^{2n}}}$

In this equation, $H(\omega)$ is the frequency response of the filter (output of Butterworth filter), ω is the frequency that we're interested in. It varies depending on the input signal. It could be any frequency in the signal that we're filtering. ω_c is the cut-off frequency, which is also known as critical frequency, and it is a hyperparameter. n is the order of the filter, which determines how quickly the filter roll-off from the passband to the stopband. In other words, it determines the steepness of the roll-off slope. It is a positive integer, and it is a hyperparameter. Higher order n gives a faster transition between the passband and stopband.

Wiener filter

Some signals, such as audio, have both data in low and high frequencies, and thus, low-pass or high-pass filters can not reduce their noise. The Wiener filter [Wiener '49] is a linear filter that is used to estimate a desired signal from a noisy observation, disregarding the frequencies. It is named after Norbert Wiener, who developed the filter in the 1940s.

The Wiener filter usually works in frequency domains, and thus, it is recommended to convert the input signal into a frequency domain signal when applying a Wiener filter. To operate, it requires prior knowledge about (i) the PSD of noise and (ii) the PSD of the desired signal. It compares the noise signal to a clean signal (reference signal). For example, in a voice application, the reference signal could be a recording of a clean voice free of noise. Then, it builds the filter to remove the noise from the noisy data. It uses optimization to create a filter that cleans the signal. This optimization minimizes the mean squared error (MSE) between the estimated signal and the desired clean signal.

In other words, this filter is designed to minimize the difference (MSE) between the filtered signal and the clean signal by using optimization. The result of this optimization is the Wiener coefficient. Once the Wiener coefficient is identified, it can be applied to the noisy signal to produce a filtered signal, similar to the clean signal.

Given a signal $x(t)$, corrupted by noise $n(t)$, the observed signal $y(t)$ can be expressed as $y(t) = x(t) + n(t)$. The goal of the Wiener filter is to provide an estimate $\hat{x}(t)$ similar to the original clean signal $x(t)$. The Wiener filter $H(\omega)$ in the frequency domain is formalized as follows:

$$H(\omega) = \frac{S_{xx}(\omega)}{S_{xx}(\omega) + S_{nn}(\omega)}$$

In this equation $H(\omega)$ is the frequency response of the Wiener filter. $S_{xx}(\omega)$ is the power spectral density (PSD) of the original signal $x(t)$. $S_{nn}(\omega)$ is the power spectral density (PSD) of the noise $n(t)$. By applying the Wiener filter to the observed signal, we get the estimate $\hat{x}(t)$, which is close to the clean signal. To do so, we can start by computing the estimated signal in the frequency domain after applying the Wiener filter $\hat{x}(\omega)$, via the following equation:

$$\hat{x}(\omega) = H(\omega) \times y(\omega)$$

All terms in this equation are in the frequency domain, and to get the $\hat{x}(t)$, which is in the time domain, we can use the following equation, where F^{-1} is the inverse Fourier transform:

$$\hat{x}(t) = F^{-1}\{\hat{x}(\omega)\}$$

Kalman filter

Kalman filter [Kalman '60], also known as *Linear Quadratic Estimation (LQE)*, is used to estimate (or predict) the state of a system when there is noise in the measurement or the measurement has uncertainties. In other words, it is not only a noise reduction method; it is a prediction algorithm that can handle noisy data very well and can also operate in real-time. This makes it very attractive for real-time applications such as robot navigation and self-driving cars. The Kalman filter is designed to work with systems that can be described by state space models (SSMs), which have been described in Chapter 12.

As an example of the Kalman filter, assume we live in a city where the state has a business of writing parking tickets if we park our car more than two hours outside. You own a self-driving car that, every two hours, moves around the city to avoid getting a parking ticket. You want to know exactly where the car is located. To do this, the car uses sensors (e.g., camera or location data) to measure its position. However, the car's sensors are imperfect, and they can give us slightly wrong measurements due to various factors like sensor errors, bad weather, tall buildings that interfere with location signals, etc. Our car uses the Kalman filter to figure out its correct position and share it with us. In this section, we first explain the most basic version of the Kalman filter, the univariate one, to familiarize ourselves with it. Later, we explain the multivariate Kalman filter.

The input to the Kalman filter is a noisy measurement, and its output is a more accurate estimation. The Kalman filter is a recursive algorithm that achieves this by iteratively performing two steps: *prediction* and *updating* (or *correction*). These iterative prediction and correction steps enable the Kalman filter to estimate the state of the system over time, even in the presence of noise and uncertainty.

It continuously updates the estimation as new measurements become available. In summary, its real-time, feedback-driven approach allows it to adapt to changing conditions.

During the prediction step, the Kalman filter estimates the *current state* and *error covariance* using the *previous state estimate* and the *system model*. The system model is a set of equations that describe how the state of the system changes over time, typically in the absence of noise or external inputs.

The update step incorporates the new measurement (z) to refine the state estimate. The Kalman filter accounts for errors in both the prediction and correction steps. In each iteration, it combines the predicted state, and the measurement to reduce the estimation error and improve the system's accuracy in estimating the true state.

We present estimation (prediction) error with e_x and measurement error with e_z. Note that e_z is not typically referred to as the update error; rather, it represents the error in the observed data used during the update. In each iteration, the Kalman filter makes an estimate, and t presents the index of estimation, e.g., x_t means estimation at time t. Another key term is Kalman Gain (KG), which is calculated as the ratio of the estimation error to the sum of the estimation and measurement errors. Kalman filter uses the following three equations iteratively to reduce the estimation error:

(i) $KG = \dfrac{e_x}{e_x + e_z}$

(ii) $x_t = x_{t-1} + KG(z - x_{t-1})$

(iii) $e_{x(t)} = \dfrac{e_z(e_{x(t-1)})}{e_z + e_{x(t-1)}} \rightarrow e_{x(t)} = (1 - KG)(e_{x(t-1)})$

For step (i), at the beginning, there is no estimation error (e_x) and measurement error (e_z), so it starts with random variables. Then, with each iteration, these variables become more accurate. If the KG is large, this means that the error in estimation is large in comparison to the error in measurement. In contrast, a small KG means the measurement error is large in comparison to the estimation error.

Step (ii) estimates the new state using the previous state and KG. The new estimate is adjusted based on the value of $KG \times [z - x_{t-1}]$. If KG is small (e_z is large), based on this equation, the measurement will have a small adjustment impact on the next state because the measurement is unreliable.

Step (iii) calculates the new estimation error, taking into account the previous estimation error. As we can see, the new estimation error equation can be written with respect to KG as well. We can see that $[1 - KG]$ is the inverse of Kalman gain. This means that if the measurement error is large, we don't want to change the estimation error quickly. However, if the measurement error is small, the estimation error should be adjusted more rapidly.

We recommend asking an LLM to prepare some sample data for Kalman filter, substitute them in these three equations by hand, run it iteratively, and see how fast the Kalman filter can improve its estimation.

Multidimensional Kalman filter: A multidimensional version of the Kalman filter uses several matrices to estimate the state of a system by combining predictions from a dynamic model with noisy observations. The state matrix maintains the data that we need to track and predict. For example, the position and velocity of the moving object.

There are too many terms to learn for this type of Kalman filter, and we explain them in the following section. While reading them, keep in mind that index (t) means at the time t and respectively ($t-1$) means at the previous time. Besides, we recommend that you check Figure 16-12 as well. This figure visualizes the multivariable Kalman filter.

X: It is the *state matrix*, which keeps the system's state at a given time step. It includes variables we are trying to estimate, such as position, velocity, etc.

P: This matrix is known as the *process covariance matrix,* which presents the errors in estimation or process. In other words, it represents the uncertainty in the state-transition model (it tracks how the state of the system being tracked evolves over time).

A: It is the *state-transition matrix*, and captures the dynamics of the system. In other words, it maps the state $\hat{X}_t$ to $\hat{X}_{t+1}$.

u (optional): The *control vector u* is used when there are external control inputs or commands that influence the system's behavior. In other words, u_t is a vector containing the control inputs at time step t.

B (optional): It is known as the *control input matrix*, which specifies how the system's state is affected by the control input vector (u). It is used when there is an external control input or command (u) that affects the system's behavior. It allows the Kalman filter to take this control input into account when predicting the state at the next time step. In simple words, B mediates how u affects the state X.

If our system doesn't have a control input (i.e., it's not being externally controlled), then u and B would not be needed.

w: It is the process noise vector.

Q: It is the process noise or process noise covariance matrix.

K: Kalman gain

$H:$ It is the measurement matrix that converts the state space into the measurement space. It defines how the internal state maps to the observations (measurements obtained from sensors). It is used to compare the actual measurement to the expected measurement in calculating the residual.

R: Observation or measurement noise covariance matrix (measurement error)

Y and C: Y is a measurement matrix (a matrix containing measurement data), which is computed as $Y_t = CX_t + z_t$. z refers to the measurement noise vector, and C refers to the measurement matrix. In particular, C is a matrix that is used to transform the state X_t to the matrix of measurements Y_t.

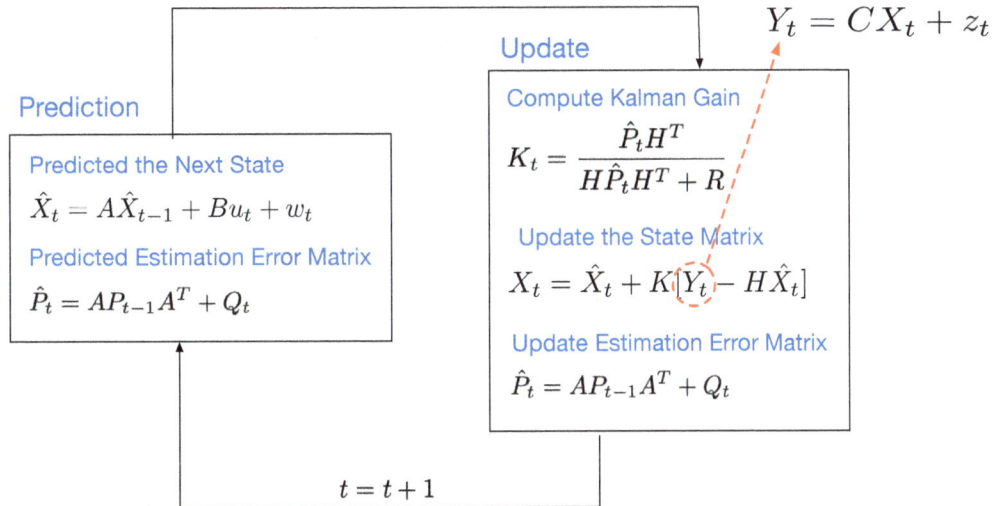

$$Y_t = CX_t + z_t$$

Prediction

Predicted the Next State

$$\hat{X}_t = A\hat{X}_{t-1} + Bu_t + w_t$$

Predicted Estimation Error Matrix

$$\hat{P}_t = AP_{t-1}A^T + Q_t$$

Update

Compute Kalman Gain

$$K_t = \frac{\hat{P}_t H^T}{H\hat{P}_t H^T + R}$$

Update the State Matrix

$$X_t = \hat{X}_t + K[Y_t - H\hat{X}_t]$$

Update Estimation Error Matrix

$$\hat{P}_t = AP_{t-1}A^T + Q_t$$

$$t = t+1$$

Figure 16-12: Multivariate Kalman filter process.

To better understand the use of the multidimensional Kalman filter, suppose we have a robot arm with two joints (shoulder and elbow) that can move in a 2D plane, like the robots you see in factories or the robots steady on a table to work with objects. We want to control the arm to reach a specific target position (x, y) while minimizing errors due to noisy sensors and uncertain dynamics.

- The state matrix (X) includes joint angles ($\theta 1$, $\theta 2$), joint velocities ($\omega 1$, $\omega 2$), End-effector position (x, y), and End-effector velocity (v_x, v_y).

- The measurement matrix (Y) includes noisy joint angle measurements ($\theta 1_{meas}$, $\theta 2_{meas}$) and noisy end-effector position measurements (x_{meas}, y_{meas}). The subscript '*meas*' stayed for measurement.

- The process noise (w), includes uncertainty in joint angle changes, uncertainty in joint velocity changes, and uncertainty in end-effector movement.

- The measurement noise (z) includes noise in joint angle measurements and noise in end-effector position measurements.

- The control input (u) includes desired joint angle changes ($\Delta\theta 1$, $\Delta\theta 2$) and desired joint velocity changes ($\Delta\omega 1$, $\Delta\omega 2$).

- The state-transition matrix (A), maps the current state to the next state based on the robot arm's dynamics.

- The measurement matrix (H) maps the state to the measurements based on the sensor configuration.

The Kalman filter for this example and other scenarios operate as follows:

1. Predict the next state using the state-transition matrix (A) and control input (u).
2. Compute the predicted measurement using the measurement matrix (H).
3. Calculate the innovation (difference between predicted and actual measurements).
4. Update the state estimate using the Kalman gain (K) and innovation.
5. Repeat steps 1-4 for each time step.

By using the multidimensional Kalman filter, the robot arm control system can estimate the current state (joint angles, velocities, end-effector position, and velocity) from noisy measurements and predict the future state based on the control input and system dynamics. Adjust the control input to minimize errors and reach the target position.

We could continue to explain more signal filtering methods until the rise of the Antichrist, but to save time, we stopped with these three filtering methods. Figure 16-13 shows a code we wrote to compare these three filters on a simple sine wave with lots of noise. A signal expert might argue that these filters have different use cases, and comparing them might not be fair, but we use this comparison for training purposes.

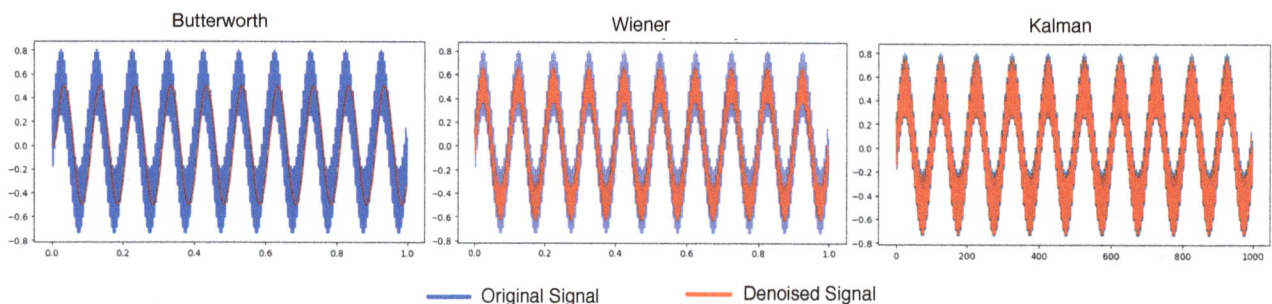

Figure 16-13: Applying three different filters on a noisy sine wave.

NOTES:

* Frequency domain filters are often straightforward to implement and can be very effective at removing noise, especially when the noise is concentrated in specific frequency bands. However, they may distort the original signal if the noise spectrum overlaps significantly with the signal's spectrum. Time domain filters, while potentially more complex to implement, can be more effective at preserving signal integrity while removing noise, particularly when the noise characteristics vary over time. Wavelet transforms compromise between time and frequency domain approaches. They are more complex to implement but can denoise signals that contain both high-frequency and low-frequency components, allowing for multi-resolution analysis that can better preserve important signal features across different scales.

* Kalman filter is recursive, which makes it fast and efficient. It is optimal in the linear Gaussian case but can be extended to nonlinear systems by using the Extended Kalman Filter (EKF) or the Unscented Kalman Filter (UKF). We did not explain them.

Imbalance Data Problem and Data Augmentation

The most common use of machine learning algorithms is the classification task of supervised learning. Classification algorithms or models expect to get a balanced number of classes. When one class (the minority class) has significantly fewer instances than another class, this phenomenon is referred to as a class imbalance problem.

A machine learning model is negatively affected by imbalanced datasets. For example, in a fraud detection scenario, most transactions are valid, and very few are fraudulent. A fraud detection model might not learn how to properly identify fraudulent transactions due to this dataset imbalance. Another domain that has this issue widespread is medical diagnosis. Unhealthy medical conditions, compared to healthy patients, are usually rare. Therefore, medical diagnosis models can be biased towards always predicting the majority healthy class. There are some common methods to mitigate imbalanced problems, which we explain here.

Undersampling and Oversampling: One of the easiest approaches to dealing with imbalanced data is to undersample the data points that belong to the majority class. In this context, undersampling means using less number of data points. On the other hand, we can oversample the data points that belong to the minority class to increase their representation. In this context, oversampling means using more data points for the machine learning algorithm. However, oversampling can be challenging because collecting additional data is often expensive, and it may not always be feasible to obtain more data.

Data Augmentation: As mentioned earlier, a potential problem with the oversampling approach is that there may not be enough available data to resample. One solution to this issue is to increase data diversity by creating synthetic data, a technique known as data augmentation. We will discuss data augmentation in more detail later.

Loss functions: A loss function can address the data imbalance problem by applying different weights to the classes during training. In a scenario where one class is significantly underrepresented, the loss function can assign a higher penalty to misclassifications of the minority class and a lower penalty to the majority class. Two common loss functions for this application are *weighted cross-entropy loss* and *focal loss*.

Weighted cross-entropy loss assigns different weights to different classes, typically giving higher weights to the minority class. This ensures that the model places greater emphasis on correctly classifying examples from the minority class, helping to mitigate class imbalance. By penalizing misclassifications of minority class examples more heavily, the model is encouraged to perform better on underrepresented data points.

We learned in Chapter 10 that cross-entropy is used in multi-class classification problems. Cross-entropy is the negative log-likelihood of the predicted class distribution compared to the actual class distribution. Assuming N is the number of samples, C is the number of classes, y_{ij} is the actual label of the ith sample, and $\hat{y}_{ij}$ is the predicted label, the cross-entropy cost function is written as follows:

$$L(y, \hat{y}) = \frac{1}{N} \sum_{i=1}^{N} \sum_{j=1}^{C} y_{ij} log(\hat{y}_{ij})$$

By incorporating the weight w_j for each class, j, into the cross-entropy function, it can be written as follows:

$$L(y, \hat{y}) = \frac{1}{N} \sum_{i=1}^{N} \sum_{j=1}^{C} \mathbf{w_j} \, y_{ij} log(\hat{y}_{ij})$$

Another common cost function used to deal with class imbalance is *focal loss* [Lin '17]. Similar to weight cross-entropy, it applies a penalty on well-classified data; authors call it *down-weighting well-classified* examples. In particular, it applies a modulating factor to the cross-entropy loss. This modulating factor reduces the loss contribution from easy examples and focuses training on hard examples. It is controlled by two hyperparameters: α and γ. α controls the weight assigned to rare classes (classes that cause imbalance) and γ controls the down-weighting of frequency classes. Higher α gives more weight to rare classes and higher γ means more discounting of frequent classes. These weights change dynamically during the training.

Assuming p_t is the model's estimated probability for the class label, the Focal Loss adds a factor $(1 - p_t)^\gamma$ to the standard cross-entropy criterion, i.e., $log(p_t)$, the Focal Loss function is written as follows:

$$FL(p_t) = -1(1 - p_t)^\gamma log(p_t)$$

Data Augmentation Methods

In addition to class imbalance problems, many applications, such as medical image diagnoses, usually suffer from insufficient data. Deep neural network models are state-of-the-art solutions for unstructured data, such as image and text documents, but they require a huge amount of data to train the model. By data augmentation, we can increase the dataset size for deep neural network models and enable them to train data. Many efforts on data augmentation are dedicated to synthetic image generation, and some works are devoted to text, audio, 3D models, video, and time series. In this section, we briefly explain some popular data augmentation algorithms for different data types.

Tabular Data

Tabular data augmentation is commonly achieved using the *Synthetic Minority Oversampling Technique (SMOTE)* algorithm and its variations. While other algorithms exist, we will focus on SMOTE, as it is the most widely used method.

SMOTE [Chawla '02] is a simple algorithm that creates synthetic data. It takes a random point from the minority class and uses that point to compute a k nearest neighbor (check Chapter 9) for this point. Then, it draws a line between those data points (the selected minor data points and its kth neighbors). By drawing lines, we mean creating a linear interpolation (check Chapter 8) in feature space between the two minority data points. Next, it draws samples from the drawn lines (synthetic data points).

To understand this explanation, look at Figure 16-14 (a). Here, we have a dataset that includes blue and red data points. We can see that red data points are in the minority and blue are in the majority. To augment this data, SMOTE selects one point from the minority group, as is depicted in Figure 16-14 (b). Then, it draws the line (interpolation line) between its three nearest neighbors. Next, from the interpolation line, some synthetic data points are drawn. As a result, instead of having only six red data points, we have several more red data points in Figure 16-14 (c).

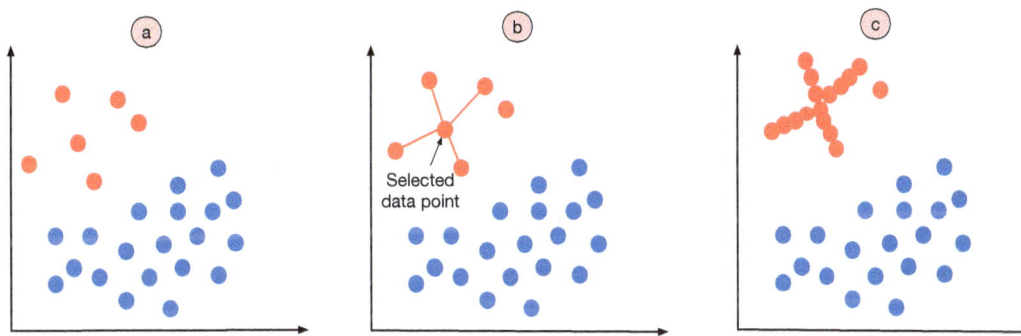

Figure 16-14: The process of increasing the minority data points by using smote algorithm.

SMOTE can introduce noise and distort the original data distribution by generating artificial samples that may not accurately reflect the underlying patterns. This issue can be mitigated by using variants such as Safe-Level-SMOTE [Bunkhumpornpat '09], and ADASYN [He '08], or by combining SMOTE with techniques like feature extraction, dimensionality reduction, or ensemble methods to improve the quality of synthetic samples and preserve the integrity of the original data. Additionally, there are SMOTE variations like SMOTER and SMOGN. SMOTER adapts SMOTE for regression tasks and adds Gaussian noise during the oversampling phase, which SMOGN also incorporates.

Image

Image data augmentation methods can be categorized into three groups: geometric transformation methods, filtering methods, and neural network-based models. While most augmentations are designed for 2D images, some, such as rotation, translation, and filtering, can also be applied to 3D data. Additionally, there are augmentation techniques specifically designed for 3D data.

Transformation methods: They are applied either to the geometry of the image or the color of the image. Any change in the camera direction, rotation, scaling, flipping, cropping, resizing, shearing, and elastic distortion (e.g., stretching or squeezing) or anything related to the geometric structure of an image is referred to as *geometric transformation*. Any changes that alter the color properties of images, such as adjusting brightness, contrast, saturation, and hue, will be implemented on the pixel level and referred to as *color transformations*.

If we are transforming a vector or matrix into another vector or matrix, it is called linear transformation (Check Chapter 7 and Chapter 10). A transformation function applies a matrix to each pixel of the given input image. For example, assume $I(x, y)$ describes the current image in 2D. By applying a transformation matrix T, we can get to $J(x', y')$, which is the output image $J(x', y') = T(I(x, y))$.

Note that output image pixels have different locations than the input image. For example, take a pen and paper, draw a simple shape on paper, and put its center on $(0,0)$ coordinates. This shape is defined as $I(x, y)$. Now, check what will happen if we apply $T1 : (x + 3, y)$, and $T2 : (-x, y)$ to this shape. As shown in Figure 16-15, T1 shifts the shape on the x-axis, and T2 mirrors the image on the x-axis. Respectively, we can zoom in three times by using $T3 : (3x, 3y)$ or zoom out three times by using $T4 : (1/3x, 1/3y)$, which we have not presented in Figure 16-15. These example functions are translation functions, and they are commonly used to build more synthetic data for machine learning algorithms, such as resolving the need for imbalanced data or preparing images for comparison in information retrieval applications.

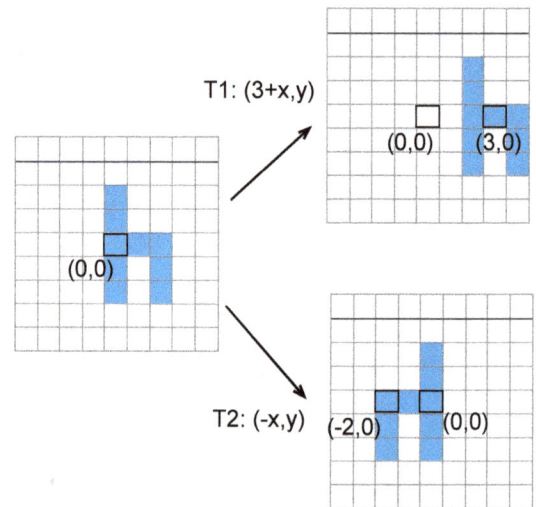

Figure 16-15: A sample image on the left is moved by T1 and mirrored by T2 transformation functions.

A common transformation for 3D images is shearing. Shearing deforms an image by shifting one part of the image relative to another. In simple terms, shearing can be described as changing the camera's view angle or the perspective of the image.

Figure 16-16 shows an output of a pose estimation of one user doing a shoulder press exercise. To compare these two images, we should rotate them in a way that allows both of them to have a camera in the same position and at the same angle. This rotation will be implemented via shearing. A popular application for information retrieval is similarity comparison, and this transformation is used for 3D similarity comparison.

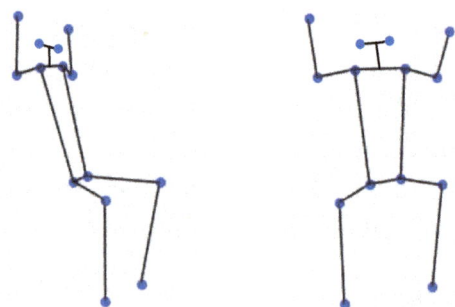

Figure 16-16: Pose estimation output of the same person but with a different camera angle by applying shearing.

Nevertheless, real-world images are taken by cameras from different angles with different settings. Usually, it is not easy to compare two images, mainly because of the different camera angles and environmental factors, such as ambient light, zoom level, etc. Therefore, don't be very optimistic about applying a transformation and comparing two or more images easily. Due to these limitations, using neural networks such as SiameseNet (see Chapter 11) is recommended.

Filtering methods: By applying filters or convolutional operations to images, we can create modified versions of the original image. In Chapter 10, we briefly explained examples of convolutional filters, such as blurring, sharpening, and embossing (which gives a 3D effect to the image). These filters can be used for data augmentation in various ways, including noise injection to make the model more robust against noise, smoothing to help the model focus on essential features, sharpening to enhance edge detection, mixing images to construct new synthetic images, changing image resolutions to make the model invariant to scaling, and implementing many other modifications.

Neural Network models: A more robust approach for image augmentation is using Generative Adversarial Networks (GANs) [Bissoto '21], Diffusion Models [Yu '23], and Variational Auto-Encoders (VAE) [Wan '21] to build synthetic images. In Chapter 11, we explained these models in detail.

GAN models train generators to create new samples that are indistinguishable from real-world data, learning data distributions. VAEs learn to encode data into a latent space and decode new samples. Diffusion models iteratively add noise and then reverse the process, creating new samples. In other words, Diffusion Models use probabilistic processes to transform data from a simple distribution to a more complex target distribution.

These Neural Network models can assist in data augmentation by generating synthetic data that closely resembles real data. By mixing these synthetic samples with the original dataset, we can increase its size and diversity, which is particularly valuable when working with limited data.

Text

Text data augmentation increases the variety of textual data to train the model. However, unlike images, it is not very common to augment textual data. Text augmentation is beneficial for tasks such as sentiment analysis, text classification, and machine translation. There are different approaches used to augment the text, and we list some as follows.

Synonym Replacement: Words within the text are replaced with their synonyms, preserving the semantic meaning while introducing variability in the data. This method enhances the model's ability to understand synonymous expressions.

Random Insertion: A random word, often a synonym or a related term, is inserted into the text. This approach can help the model learn to handle unexpected variations in phrasing.

Random Deletion: Randomly removing words from the text forces the model to focus on the remaining content, which can improve the model's ability to understand the context and handle missing information.

Word Swap: The positions of two words within the text are swapped. This technique introduces syntactic variation, which can help the model learn to interpret text with different word orders, which is particularly useful in languages with flexible syntax.

Intentional Noise Addition: Introducing spelling mistakes or grammatical errors simulates real-world data conditions where user-generated content often contains such imperfections. It can make the model more resilient to noisy input.

While these methods can improve model robustness and performance, they also introduce challenges. Textual data is highly contextual and sensitive to the order and meaning of words. Poorly applied text augmentations can introduce noise that degrades model performance.

Signal and Time Series

In Chapter 6, signal features are explained; in Chapter 7, signal and time series decomposition methods are covered; and in Chapter 12, details on handling audio data, a common type of signal data, are discussed. Same as other data types, we use data augmentation to increase the diversity and variety of signal or time series data for training a model and thus improve the model's generalization.

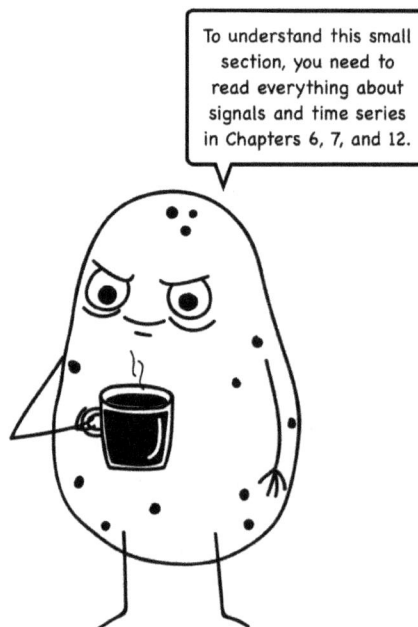

To understand this small section, you need to read everything about signals and time series in Chapters 6, 7, and 12.

In the following, we briefly explain some methods of signal data augmentation. Those which we believe are harder to understand were visualized in Figure 16-25.

Neural network models: Similar to other data types, by using GAN, Diffusion models, or even RNN, we can simulate and increase the variety of the signal or time series.

Jittering and Noise injection: Adding a noise, such as Gaussian noise, to the original signal can make the model more robust to noise. In particular, Jittering introduces small random variations to the data points and trains the model to deal with the small variation in the data.

Warping: Warping the signal in the time dimension occurs by compressing or stretching a signal. Warping can simulate variations in the speed or duration of the signal. It is useful for handling variations in the speed or duration of events in time series data. If we apply a nonlinear transformation to the signal magnitude while keeping time unchanged, it is referred to as *magnitude warping.* For example, applying a logarithmic transformation to the signal magnitude, compresses or expands the amplitude range while maintaining the original time structure.

Slicing: It involves dividing a signal or time series into smaller segments. Instead of feeding the entire segment into the training model, we can input one or more slices of the signal or time series. The PAA and SAX algorithms, described in Chapter 7, are commonly used to slice the time series. This approach is useful when the training model needs to focus on specific segments of the time series or signal.

Cropping: It focuses on clipping a signal into a smaller amplitude or a time series into smaller values. Cropping is used to remove unwanted or irrelevant portions of a signal, such as outliers, aiding in noise reduction, feature extraction, and efficient data representation. Figure 16-17 shows a sample sinusoid signal and its cropped version.

Scaling: By multiplying a signal by a random number, the amplitude changes, and this can be used to train the model to handle similar signals

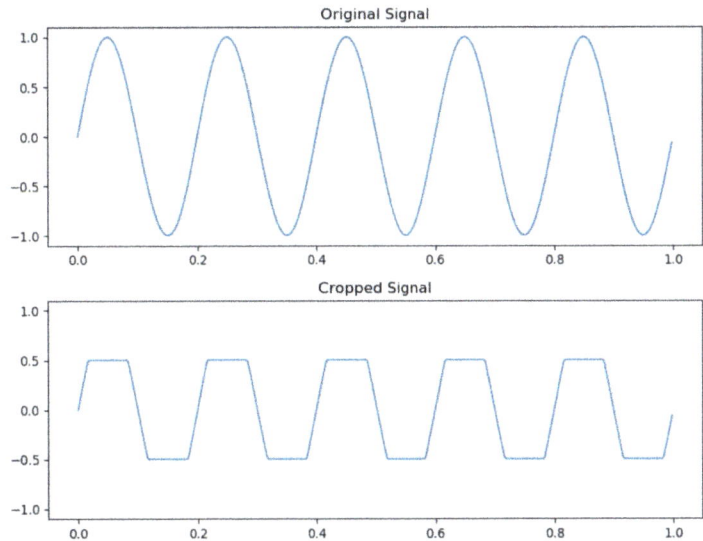

Figure 16-17: A sample sinusoid signal and its cropped version.

with different amplitudes. The wavelet transform that we explained in Chapter 7 uses scaling as well. Scaling can be applied to the time axis (X-axis) or the amplitude (Y-axis). Amplitude scaling makes the model invariant to variations in signal intensity, while time scaling changes the duration of the signal, affecting its temporal features.

Smoothing: Smoothing methods, such as moving averages, exponential smoothing, or Gaussian smoothing, can be employed to preprocess time series and signal data. By reducing noise and fluctuations, these methods help in constructing synthetic time series and signals that are less erratic and more conducive to model training and classification. For instance, smoothing can mitigate the effects of outliers and reduce the variance of the input data, leading to better generalization and stability during training. In the context of time series analysis, smoothing techniques can enhance the signal-to-noise ratio, making underlying patterns more discernible for the model.

Reconstructing Missing Data

Since nothing in this world is perfect, it is very uncommon that data collected from a real-world setting does not include missing data. For example, sensors in industrial systems may malfunction, leading to gaps in time series recordings. An easy remedy for dealing with the missing data in tabular data is to remove those records. This is a common approach, but sometimes, the dataset is too small, so we can't make the dataset smaller. In these cases, we need to reconstruct the missing data.

The process of reconstructing the missing data is referred to as *imputation*. Estimating the missing values between existing data points of a sequence is referred to as *interpolation*.

Interpolation is common to fill a gap of missing data in time series and other univariate data, but it can be used for multivariate data as well. Imputation is more common and generalizable to be used for unstructured data (such as images) and structured data like a dataset with multiple variables.

Imputation

There are several basic and common methods used for imputation; we list and describe them as follows:

Mean/Median or Mode

When we are dealing with numerical values, it is possible to replace missing data with the mean/median or mode of that variable. For example, in tabular data, one particular column that has numeric data has some missing data. We can take the mean of that column value and substitute the missing data with the mean. This is an easy and computationally efficient method to implement. Also, using mean is good if we assume our data have a normal distribution. However, this method distorts relationships between variables by using overly simplistic values. Besides, it might underestimate variability in the data.

k-Nearest Neighbors (kNN) for Imputation

A common and easy approach is to use the kNN algorithm to determine the missing data based on its neighbor data points [Batista '23]. The advantage of this method is that it is non-parametric, which requires no assumption about the distribution of the data. It can also handle complex relationships between the data. However, this method is not as efficient as statistical imputation methods because we need to execute the kNN algorithm. It is also sensitive to noise and outliers, which could skew the prediction of missing data. Also, using this method is very sensitive to the kNN hyperparameters, i.e., distance metrics and k.

Hot Deck Imputation

Hot deck imputation replaces missing data with an observed value from a similar unit within the same dataset, often based on matching criteria such as proximity in time or similarity in other characteristics. A common type of hot deck imputation is the Last Observation Carried Forward (LOCF) method, where the most recent observed value preceding the missing data point is used as a substitute. LOCF is particularly useful in time series data, assuming that the most recent observation is the best estimate of the missing value. However, while LOCF is easy to implement, it can introduce bias, particularly if the data exhibits significant trends or patterns that the last observation does not capture.

Cold Deck Imputation

It operates in contrast to hot-deck imputation and selects a substitution for the missing data (a.k.a., donor) from another dataset rather than from within the same dataset. The external source of data is known as the donor pool.

The process involves identifying missing values in the primary dataset and selecting a suitable external dataset containing relevant information for imputation. Matching criteria are then established to match records from the primary dataset with suitable donors from the external dataset, and the missing values are replaced with corresponding values from the matched donors. Cold deck imputation offers advantages, such as increased data diversity and improved accuracy, by leveraging external data and introducing new information. However, careful consideration of data quality and matching criteria is required to ensure reliable and relevant imputations.

Regression Imputation

Regression imputation uses the relationship between variables to replace missing values based on other observed data, such as linear or logistic regression. The key advantage of these methods is that they leverage the correlations and statistical relationships in the data to estimate missing data, and this attitude preserves underlying patterns, even after imputation.

Griffin-Lim Algorithm (GLA)

In Chapter 12 we have explained that a spectrogram is a visualization of the strength (or loudness) of different frequencies in a sound over time. Furthermore, we have described that Spectrogram can be decomposed into phase and magnitude. Magnitude specifies how strong each frequency is (e.g., knowing which piano keys are played), and the phase describes the exact timing of these frequencies (e.g., knowing the exact moment each key is pressed).

Griffin-Lim algorithm [Griffin '84] is known as a phase reconstruction algorithm, and it reconstructs a time-domain signal (original audio) from its magnitude spectrogram. In simple terms, it estimates the missing phase information from the magnitude spectrogram to reconstruct the audio signal.

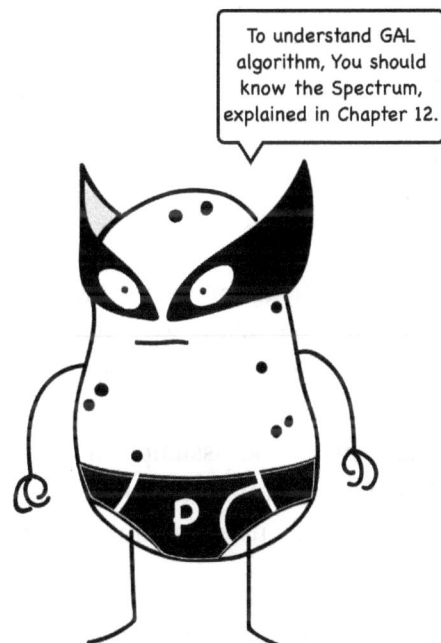

To understand GAL algorithm, You should know the Spectrum, explained in Chapter 12.

Although the Griffin-Lim algorithm was originally designed for audio data reconstruction, its iterative approach to phase retrieval has potential applications in other fields, such as image restoration, anomaly detection, and any domain where phase information needs to be inferred from magnitude data.

First, the algorithm starts with a random initialization of the phase because the phase is unknown. Next, it performs the following four steps iteratively until a maximum number of iterations is reached or the quality of the constructed data is similar to the original one (convergence reached).

(i) *Combine Magnitude and Phase:* It combines the magnitude (which is known) with the current phase (at first, this phase is completely random) and creates a spectrogram.

(ii) *Inverse Fourier Transform*: Using the Inverse Fourier Transform, the constructed spectrogram is transformed back into a time domain signal (e.g., sound wave).

(iii) *Recompute Spectrogram:* Next, it transforms the time domain signal (sound wave) back into a spectrogram, but this time, it only keeps the phase information and discards the magnitude.

(iv) *Update Complex Spectrogram:* It combines this new phase information with the original magnitude, creating a new complex spectrogram.

Customized methods

There are some methods customized for specific applications or devices. For example, to reconstruct missing data on smartwatches and other sensor-equipped devices, we designed the Ghost imputation method [Rawassizadeh '19], a type of hot deck imputation. It searches the sequential dataset to find data segments with a prior and posterior segment that matches those of the missing data. If there is a similar segment that also satisfies the constraint, such as location or time of day, then it is substituted for the missing data. To reduce the search space and improve the computational complexity to linear, a clashing approach (hash table) has been used. Another notable work is MissForest [Stekhoven '12], which provides an iterative process that alternates between imputation and modeling. In particular, for each variable with missing values: (i) it trains a random forest using observed data, (ii) predicts missing values using the trained random forest. These two steps will be done iteratively.

Interpolation

Similar to the approach we used for imputation, we list some common methods for interpolation in this section. While we are learning interpolation methods, we might mix them with regression. Interpolation and regression are not identical; they have similarities and differences. Both involve fitting a model to a set of data points to estimate values at other data points. Besides, both rely on the assumption that there is an underlying relationship between the data points. However, interpolation focuses on estimating values between two observed data points, while regression focuses on quantifying the overall relationship between data points to make a prediction.

Linear and Bilinear interpolation

Linear interpolation is a commonly used method that involves connecting two data points with a straight line (or hyperplane if the data has more than two dimensions). This technique is straightforward to implement and relies on the assumption that the data points have a linear relationship with each other. Constructing a line between two known points allows for the estimation of missing values within that range.

Bilinear interpolation is an extension of linear interpolation that is applied in two dimensions. It is commonly used in image processing to estimate pixel values for image resizing or correcting spatial distortions. Bilinear interpolation is widely used in image super-resolution, a technique to enhance the resolution of low-quality images. It upscales images by creating new pixels based on the surrounding pixels.

Bilinear interpolation considers four points that form a rectangle on a grid, such as the pixels in an image. The goal is to estimate a value at any point within this rectangle, and this is done in two steps. The first step (horizontal interpolation) performs a linear interpolation (horizontally) between each pair of top and bottom corner points of the rectangle. This step results in two interpolated values—one for the top pair and one for the bottom pair. The second step (vertical interpolation) performs a linear interpolation between the two interpolated values obtained from the first step to estimate the value at the target location within the rectangle. This method assumes a linear relationship between the points, both horizontally and vertically,

Oh, I get very nostalgic when learning about interpolations; it reminds me of the good old days of learning regression. The world was so much simpler back then.

creating a *bilinear* effect. The final interpolated value is influenced by all four corner points of the rectangle, considering both the x and y dimensions.

For example, assume we want to interpolate a point between A, B, C, and D, as shown in Figure 16-18. First, they should be converted into a rectangular shape to form the grid. This could be done by an affine transformation; an affine transformation is a linear transformation that preserves straight lines and ratios of distances between points. A', B', C', and D' are transformation results, presented in Figure 16-18.

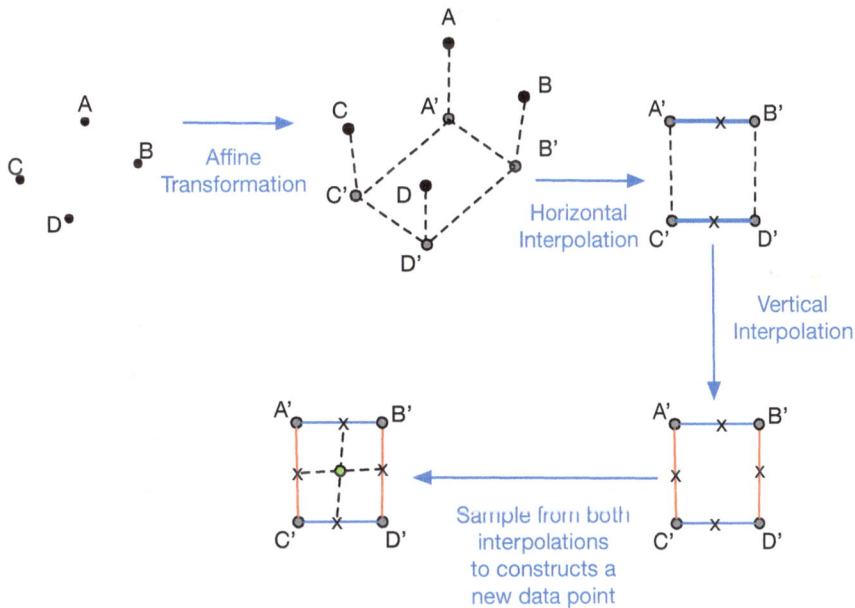

Figure 16-18: Step by step process of bilinear interpolation.

Next, there will be horizontal interpolation between corners resulting in having two interpolated lines (one between A' and B' and one between C' and D'). With the same process, two vertical interpolation lines are constructed, the red lines (one between A' and C' and one between B' and D'). Next, it samples one point from each of these four interpolation lines and connects them together, resulting in the green dot, which is the new interpolated data between these four data points.

Polynomial interpolation

It tries to connect two or more data points with a curved line or hyperplane, and the curviness of the line or hyperplane is quantified using a polynomial equation. Then, data points on these curved lines or hyperplanes are used to estimate the missing data points. Instead of using affine transformation for the bilinear interpolation, such as the one presented in Figure 16-18, we can use polynomial interpolation.

Polynomial interpolation is flexible, as higher-degree polynomials can model more complex data patterns. However, as the degree of the polynomial increases, it comes with significant disadvantages. High-degree polynomials can lead to overfitting, where the curve oscillates wildly between data points, a phenomenon known as Runge's phenomenon. This makes the interpolation less reliable, especially near the edges of the data range. Additionally, polynomial interpolation can be computationally expensive and numerically unstable for large datasets, leading to inaccuracies in the interpolated values.

Spline interpolation

It tries to connect two or more data points with a curved line (or hyperplane) composed of multiple polynomials. Then, data points on these spline curves can be used to estimate missing values between the known data points. Similar to piecewise regression (check Chapter 8), the key distinction between spline interpolation and polynomial interpolation is the using piecewise polynomials that are joined at the known data points to ensure the continuity of the entire curve.

Kriging interpolation

Kriging interpolation [Oliver '90] is used for datasets that include spatial information. It estimates values at unobserved locations based on observations at other locations. Figure 16-19 (a) presents the interpolation problem on spatial data. In this figure, we know the value of data points s_1, s_2, s_3, s_4, s_5 and we would like to use them to estimate the value of data points s_x.

Kriging has a principle that states that nearby data points in the space have closer values, and as they move far away, their similarity decreases. There is a range, and when this range passes, then the closeness similarity does not exist. In other words, values beyond the range are assumed to be uncorrelated. For example, in Figure 16-19 (b), a circle is drawn around s_x that specifies this range. s_1 stays outside this range, and thus, based on the Kriging principle, there is no correlation between s_1 and s_x. On the other hand s_3, and s_4 stays closer to s_x rather than s_5 and s_2. Therefore, we could assume the similarity between s_x and s_3, and s_4 is higher than s_x and other data points (s_2 and s_5).

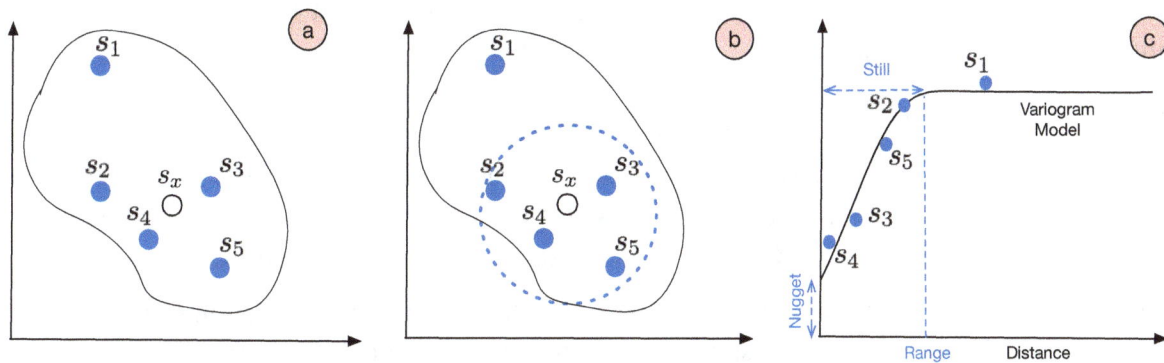

Figure 16-19: (a) Distribution of points on geometric space, all points except s_x are known. (b) The range around s_x is highlighted. (c) A variogram of Figure b.

Assuming we present the value of each point with $z(\,.\,)$, to estimate the value of s_x, we should compute $z(s_x)$. Now, we can write a simple equation to estimate a value for a given point by using a weight as follows:

$$z(s_x) = \Sigma_{i=1}^{n}\lambda_i z(s_i)$$

This weight (Krignig weight) could be assumed to be the distance. Here is $z(s_x)$ is the predicted value at the un-sampled location s_x. Kriging weight for the ith observed value is presented with λ_i. Observed values at the location s_i are presented with $z(s_i)$, and we have n observed data points in total.

However, s_2 and s_4 are located at different locations in geographical space. Therefore, it is not fair to assign them equal weight and not incorporate their geographical coordinates while determining weights. We should incorporate spatial continuity into the weight assignments as well. To incorporate spatial continuity, the λ weights should be computed based on the *spatial autocorrelation structure*, which is characterized by a *variogram*. A variogram visualizes how much the values of a variable (e.g., temperature, rainfall) vary across a space, see Figure 16-19 (c). In other words, it visualizes the variance of the difference between two variables at two locations. Using a variogram helps quantify the spatial dependence between data points.

Imagine we are analyzing a field of flowers, where each flower has a unique color. A variogram helps us understand how the color similarity changes as we move across the field. It's a graph that tells us how much the color difference (variance) depends on the distance between two flowers. With a variogram, we can plot the semivariance (degree of dissimilarity) against the distance between data points. The right side of Figure 16-19 (c) presents the variogram. The 'nugget' represents the variance at the smallest separation distance. The 'still' (where the variogram flattens out) represents the maximum variability in the data.

There are different types of Kriging methods, such as Ordinary Kriging (assumes a constant but unknown mean across the area of interest), Simple Kriging (Assumes the mean is known over the area and only focuses on incorporating the spatial dependence characterized in the variogram into predictions), and Universal Kriging (incorporates a trend or drift into the model, and thus eliminate the need to estimate it).

Radial Basis Function Interpolation (RBFI)

We have described Gaussian RBF in Chapter 9 while describing SVM. RBF is a function that converts non-linear separable data into linear separable data by expanding or compressing the data. RBFI [Hardy '71] uses the RBF function to estimate the value at intermediate points.

RBFI will be applied to a set of data points, where each data point has a location (typically in 2D or 3D space) and a value. As the first step, for each data point, it calculates the distance to every other data point. As a result, we have a specified distance between each of the two data points. Next, a Radial Basis Function, which is selected by the user (e.g., Gaussian RBF), will be applied to these distances. This results in a matrix where each element represents the RBF applied to the distance between two points. This matrix, along with the known values at each data point, forms a system of linear equations. For example, if we have three data points and their values (v_1, v_2, v_3). Their distances in the distance matrix are presented as $d_{11}, d_{12}, d_{13}, d_{21}, d_{22}, d_{23}, d_{31}, d_{32}, d_{33}$, thus we can have the following three equations:

$$a\,RBF(d_{11}) + b\,RBF(d_{12}) + c\,RBF(d_{13}) = v_1$$
$$a\,RBF(d_{21}) + b\,RBF(d_{22}) + c\,RBF(d_{23}) = v_2$$
$$a\,RBF(d_{31}) + b\,RBF(d_{32}) + c\,RBF(d_{33}) = v_3$$

In the above equations, variables a, b, c are the coefficients for the RBFs, and will identified by substituting numbers in the above three equations. Once we have identified these coefficients, we can estimate the value at any new data point and perform the interpolation. This is done by calculating the distance from the new point (the one that we need to interpolate) to each of the known data points. Then, applying the RBF to each distance, each RBF results in a value (v). Then, it sums up these values weighted by the previously calculated coefficients (a, b, c).

Anomaly and Outliers Detection

All data collected from real-world settings typically includes anomalies and outliers. An anomaly refers to a data point that deviates significantly from the overall pattern of a dataset and may indicate rare or unexpected behavior, often with an unclear cause. Outliers, a specific type of anomaly, are data points that are numerically distant from the rest of the dataset, often with extreme values that can skew statistical properties such as the mean.

Since outliers have numerically extreme values compared to other data points, they can often be identified through visualization or statistical analysis. However, determining whether an outlier should be retained or removed depends on different factors, such as the dataset, algorithm, and the target application.

First, we briefly list some of the common approaches to identifying anomalies, and then we explain some common anomaly detection algorithms.

Unsupervised Methods

Clustering (check Chapter 4) is one of the most common and fairly accurate approaches to identifying anomalous data points. Clusters that contain very few members in comparison to other clusters were considered anomalous. DBScan and OPTICS are common clustering methods for anomaly detection.

In addition to clustering, autoencoders (check Chapter 11) and dimensionally reduction methods (check Chapter 7) are used to remove anomalies and outliers from the dataset.

Isolation-based Methods

These groups of methods isolate data points that are different from the majority of data points. For example, we can use kNN to isolate anomalous data points. It can identify outliers by looking at each data point's distance to its k closest other points. Data points with much larger distances to neighbors than average would be marked as isolated anomalies.

In addition to the kNN, there are two other methods commonly used for anomaly detection: Isolation Forest and One-class SVM, which we explain in this section.

Isolation Forest (iForest)

Isolation forest [Liu '08] is an anomaly detection algorithm that is an ensemble tree construction method (check Chapter 9 for ensemble tree construction). Similar to the random forest (check Chapter 9), it constructs several trees. While constructing trees, it isolates anomalies based on the principle that anomalous data points are easier to isolate due to their distinct characteristics.

To train the model, it constructs several binary trees (check Chapter 5), which are called *isolation trees* or *itrees*. However, its feature selection for the split is completely random. As an itree goes deep, anomalous data points get isolated relatively fast at the beginning of the splits. This split process continues until every leaf of the tree represents a single data point or the maximum depth a tree reaches. Similar to other ensemble learning methods, several other trees are constructed following the same path of tree construction. Each of these trees starts the split by using one random feature to split.

Figures 16-20 visualize the training of the iForest algorithm for a toy dataset in 2D space, where we mark the anomalous data point in red. First, iForest applies several random feature splits to build several trees; in this example, we present three splits. Next, it continues splitting by using that selected random feature to split. In the end, we have several trees, and in these trees, anomalous data points are typically isolated closer to the root of the tree (the top levels). In contrast, normal data points require more splits and thus are isolated deeper within the tree.

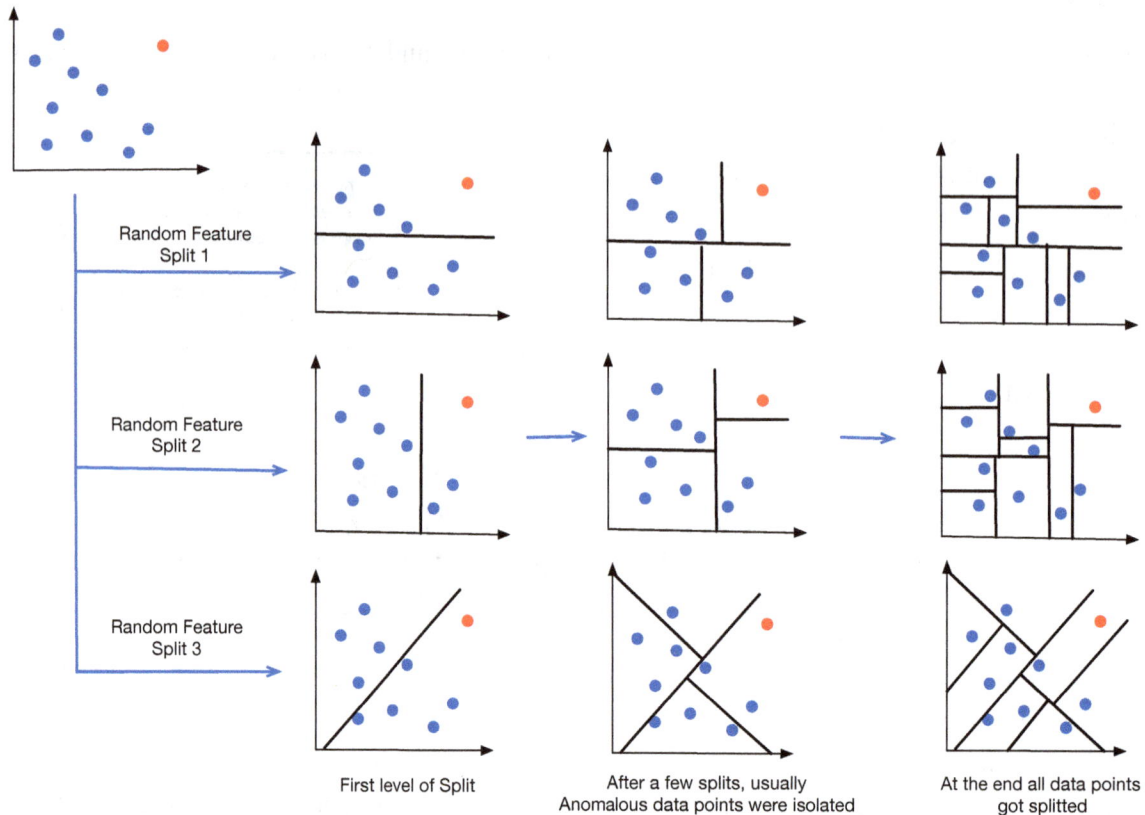

Figure 16-20: Training of iForest. First, it starts by selecting some random feature for split. While splitting usually anomalous data points get isolated at early splits, the split stops when each data point is a child node, and no more split is possible.

However, a tree might reach its maximum depth and thus end splitting, but the anomalous point might not get splatted at all, such as the right most trees in Split 3 of Figure 16-20. This is not a problem because the key idea is that the path length in the tree reflects how easy it is to isolate a sample. Shorter paths suggest a sample is more likely to be an anomaly, while longer paths indicate a more normal data point.

To make an inference, a new data point is passed through each isolation tree to determine its path length (number of edges traversed) to isolate it. The path lengths are averaged across all the isolation trees. A scoring formula uses this average path length to calculate an anomaly score. Shorter average path lengths have a higher likelihood of being an anomaly. The anomaly score is typically normalized to fall between 0 and 1, with higher scores indicating a higher likelihood of being an anomaly.

The described approach is used for numerical data. To deal with categorical variables, we can use the frequency of their variables instead of the original value of categorical data and use the frequency to decide about the split.

1064

One-Class SVM

One-Class SVM (OCSVM) [Schölkopf '01] is another method for anomaly detection that isolates anomalies. It is useful in scenarios where the data is large of one type (the normal class), and the anomalies are rare and significantly different from the normal state, such as fraud detection and intrusion detection.

OCSVM operates differently from classical SVM, which requires examples from both classes for classification. While classical SVMs aim to separate two classes with a maximized margin, OCSVM seeks to find a decision boundary that effectively isolates the majority of data points from anomalies. It tries to achieve this by finding a hyperplane in the high dimensional space that best separates the normal data points from the anomalies. This boundary maximizes the margin between these two types of data points (normal vs anomalies), ensuring that most normal data points lie within the enclosed region.

When a new data point arrives, OCSVM projects onto the same high-dimensional space, and its distances to the anomalous points will be calculated. If it falls outside the predefined margin, then it is flagged as an anomaly.

There is a hyperparameter ν (read as *nu*), which has a range of values between 0 and 1, and it controls the trade-off between the fraction of outliers and the number of support vectors. This parameter represents an upper bound on the fraction of anomalies in the dataset and a lower bound on the fraction of training examples used as support vectors. In other words, this parameter has two roles: (i) deciding how many anomalous data points the algorithm can detect and (ii) deciding how many of the normal data points (training examples) the model should pay attention to. A small ν value means the model focuses on more data points, trying to understand what normal looks like. Therefore, by adjusting ν, we can influence the sensitivity of the model to anomalies. For example, $\nu = 0.01$ means a maximum of 1% of our training examples could be misclassified.

Local Outlier Factor (LOF)

Similar to DBScan and OPTICS (check Chapter 4), LOF [Breunig '00] is a density-based approach proposed by the same group that introduced DBScan and OPTICS clustering algorithms.

First, it computes the *local density* of every single data point with respect to its neighbors. In other words, the local density of a data point is determined by estimating distances between the data point and its neighbors (k-nearest neighbors) data points. Points with closer neighbors have a higher local density.

Next, it uses these computed local densities to identify *reachability distance*. The reachability distance (RD) is defined as the maximum distance between two points and the k-distance of that point. It is calculated for each point to all of its k-nearest neighbors.

Assume, the reachability distance for a neighbor q of the point p is the max of the distance between q and its k nearest neighbors (excluding p) and the distance between p and q scaled by the local density of p. Therefore, for all q in $kNN(p)$, we have:

$$RD\ (p, q) = max\{kNN\_distance(q), local\_density(p, q)\}$$

The reachability distance of q reflects how isolated q is from its neighborhood p. If the density of a data point is significantly different from the density of its neighbor data points, this data point is an anomaly.

The third step uses the reachability distance and computes the *Local Reachability Distance (LRD)*. *LRD* is the inverse sum of all reachability distances of all k-nearest neighbor data points. Assuming a set of k nearest neighbors is presented as $N_k(p)$, LRD can be computed as follows:

$$LRD(p) = 1 / \left(\frac{\sum_{p \in N_k(p)} RD(p, q)}{N_k(p)} \right)$$

Then, it computes the LOF of data point p as follows:

$$LOF(p) = \frac{\sum_{p \in N_k(p)} LRD(q)/LRD(p)}{|N_k(p)|}$$

If the result of $LOF(p)$ is greater than 1, the data point is an anomaly; otherwise, it is not an anomaly.

Random Sample Consensus (RANSAC) Algorithm

Some datasets include lots of anomalies, and the RANSAC [Fischler '81] is a simple iterative algorithm for making a model that fits these datasets. In other words, its primary objective is to identify a subset of data points that fits a model well, and they can be used to estimate the parameters of that model robustly. Therefore, it is also known as the outlier removal model as well. It has lots of applications, including Structure from Motion (SfM) detection, image feature matching, estimating the robot's motion by analyzing the visual scene, optical flow estimation (check Chapter 6), etc.

RANSAC takes a dataset containing both proper data points, a.k.a., *inliers*, outliers as input and outputs a mathematical model (such as a hyperplane) that best fits the data points while excluding outliers, along with identifying which data points are inliers versus outliers. It operates in four steps, described as follows:

- First, it begins by randomly choosing a sample subset from the dataset. The size of this sample subset is the minimum number of samples required to fit a model. Figure 16-21 (a) presents a sample dataset in 2D; the RANSAC algorithm selects a subset, presented as blue data points in Figure 16-21 (b).

- Next, it fits the model to these selected data points (blue data points), presented as the red line in Figure 16-21 (c). For the sake of simplicity, we choose a linear regression to fit the model. The selected points, often referred to as the hypothesis, are used to estimate the model parameters.

- It evaluates the model by checking how many data points from the entire dataset fit the model M (inliers) and measures their error relative to the model. An inlier is any point whose error is below a predefined threshold. The threshold is specified with green lines in Figure 16-21 (e).

- In the fourth step, the previous three steps are repeated iteratively for a predefined number of iterations or until a satisfactory model is found. In Figures 16-27, we repeat this process only two times, Figure 16-21 (d) selects the second subset of data points for model fitting. Then, it returns the model with the largest number of M (inliners), which is the model in Figure 16-21 (e).

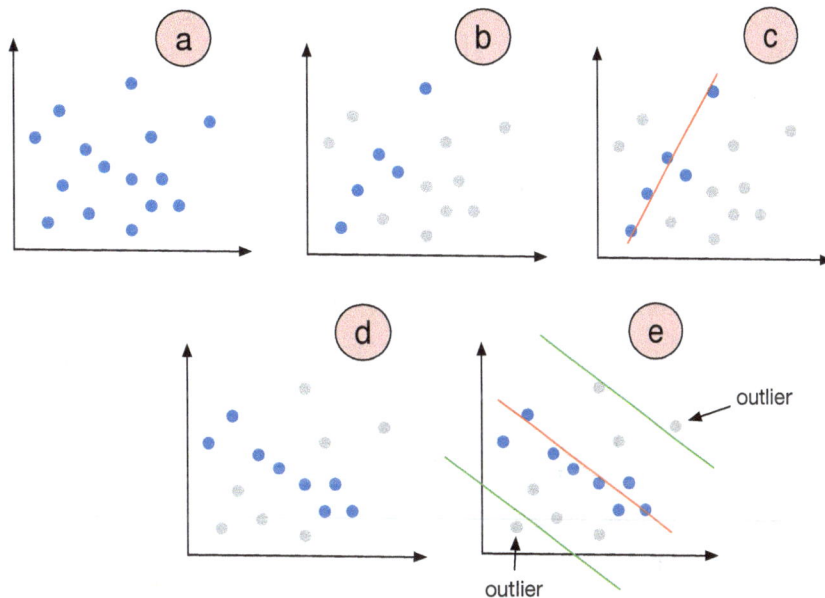

Figure 16-21: Visualization of the RANSAC algorithm with linear regression and two iterations. The first iteration selects a subset of data points in Figure (b) and computes the model that fits them in Figure (c). In the second iteration, another subset of the dataset in Figure (d) is selected, and the model is computed in Figure (e). As shown in Figure (e), the second model is better than the model in Figure (c) because the second one covers more data points than the first one. This model identifies two outliers that stay outside the green threshold lines.

Data and Model Changes over Time

We have learned that the typical machine learning workflow involves collecting data, building a dataset, and cleaning the collected data, e.g., imputation, before feeding it into a machine learning algorithm. This process outputs a model intended for complex tasks that could be used in an application.

However, the effectiveness of some models can decline over time due to several factors. For example, the data they were originally trained on may become outdated, or the models might struggle to adapt quickly to emerging data patterns, etc. This section delves into different time-related challenges that models might face, including different types of drifts and the cold start problem.

Concept, Model and Data Drift

Concept drift occurs when the characteristics of the data change over time. In other words, the statistical properties of the target variable(s), which a machine learning model is trying to predict, change over time.

For example, a machine learning model learns about the geographical businesses in an area and recommends an investor to open a hot dog stand there. There are many offices in the area, and hot dogs sold well until a fruit juice stand appeared in that area, and most hot dog customers changed their hot dog-eating habits to drink fruit juice. In this scenario, the model might fail to recognize the shift in customer preferences because it hasn't been updated to consider the impact of new businesses like the fruit juice stand. The model, still operating on outdated information, doesn't account for this new variable, leading to a decline in its predictive accuracy. This illustrates concept drift, where the model's assumptions about customer preferences no longer hold true in the changed environment.

Model drift is similar to concept drift but focuses on the deterioration of the model's accuracy over time due to factors such as overfitting, changes in feature importance, or the introduction of new variables. It is different from concept drift because it is not just about the change in data but also about the model's ability to remain accurate and relevant.

For example, a machine learning model that predicts house prices based on historical data may initially perform well. However, if its predictive accuracy starts to decline over time, despite stable market conditions, it could be due to changes in the importance of certain features. For instance, the model might initially emphasize the quality of public schools, but over time, as preferences shift toward private schools, this feature becomes less relevant for predicting house prices, leading to decreased model performance.

Data drift (*Data shift*) is another concept that refers to the change in data distribution between training and inference time. This causes model accuracy to degrade as well.

Note that data drift and concept drift are not the same. Data drift refers to any change in the data distribution between training and inference. This includes changes to the features, output values, correlations, etc. However, Concept drift refers to changes in the relationship between input features and output changes over time.

Tackling Drift Challenges

There are many methods to handle or mitigate different types of drifts. We briefly describe some in the following.

Continuously and regularly monitoring: We can employ algorithms to continuously monitor the model's accuracy over time and detect changes in its accuracy. This could be achieved by using drift detection methods such as ADaptive WINdowing (ADWIN) [Bifet '07]. ADWIN moves a fixed-size sliding window (check Chapter 5) over the data, and when two consecutive windows have significant statistical differences, this is considered as a concept drift.

Periodic, incremental, and *online retraining*: If the training is not expensive, we can periodically retrain our model with the new data and incorporate the new data into the model as well. If the training is expensive, we can adopt incremental learning techniques, where the model is updated incrementally with new data without discarding the knowledge gained from previous data. Another form of learning is online learning. Online learning also involves updating the model as new data arrives, but it has the additional requirement that the model must be able to learn in real-time, sample-by-sample. Each new sample can update the model immediately without waiting for a batch of data.

Ensemble methods: Ensemble learning algorithms, including bagging and boosting, can combine multiple models and thus make the resulting model robust to concept or model drift. This means we should update the ensemble after new data becomes available. In particular, combining multiple models with different strengths and weaknesses helps to account for diverse trends and changes in the data. Besides, assigning different weights to individual models based on their recent performance on new data allows the ensemble to adapt to changes and prioritize models currently better suited to the evolving data.

Adaptive Learning: This type of learning refers to methods and algorithms designed to make models or algorithms adjust their behavior over time based on new data or feedback. This allows them to continuously improve their accuracy and thus become more resilient to concept or model drift. Adaptive learning is also categorized as online learning, but we keep its explanation separate in this text.

Drift resistance models: In cases where concept or model drift can be predicted to happen, we can use algorithms resistant to these drifts, such as decision trees (check Chapter 9) with adaptive learning rates or models with a mechanism to handle the data distributions.

Feedback: Incorporating domain experts or end-user feedback who have insights into changes in the environment could help the model engineer know when to retrain the model and when the concept or model drift is happening.

Active Learning: Proactively asking the human expert to label the data during the learning process is referred to as active learning. Active learning can mitigate concept drift. In particular, where the relationship between input features and target variables changes over time, active learning can identify and label new and relevant data points. Besides, it can mitigate model drift by continuously updating the model with new, labeled data points that reflect the current state of the environment.

Cold Start Problem

The cold start problem is a common challenge in machine learning, where a model requires data to function effectively, but there is initially little or no data available. This issue is particularly prevalent in recommendation systems, where no historical user data exists on which to base recommendations. The cold start problem is primarily associated with recommendation systems when there is insufficient data or information available to make accurate predictions or recommendations for new users or items.

Three common cold start problems are user cold start (a new user joins the system, and there is no prior data on their preferences), item cold start (a new item/content is added to the system, but there is no user interaction with the content/item), and system cold start (a new system is launched with no data to analyze).

Tackling Cold Start Problem

We list a few approaches to mitigate the cold start problem. Note that not all of them are necessarily related to data, and some should be implemented on the application that hosts the machine learning algorithm.

Utilizing Pre-trained models or transfer learning: We can leverage models trained on other datasets as a starting point and then fine-tune the dataset with the data we collect. The same approach we use for transfer learning (check Chapter 12). This is known as *warming up* the model.

User Profiling: The model can make initial predictions by gathering basic information from the new users, either explicitly or from different resources. For example, after the installation of the target application, the application can ask users for preferences/feedback upfront, including their interests, demographic information, etc.

Generative Models: We can use generative models (described in Chapter 11) to create synthetic data for new items or users based on the existing data distribution, allowing the system to make predictions, even when specific data is missing.

Combination of collaborative and content-based filterings: We can combine multiple techniques, such as collaborative and content-based filtering approaches, to provide recommendations, even with limited data. Collaborative filtering predicts user preferences based on similarities with other users. Content-based filtering recommends items by matching their content with a user's profile.

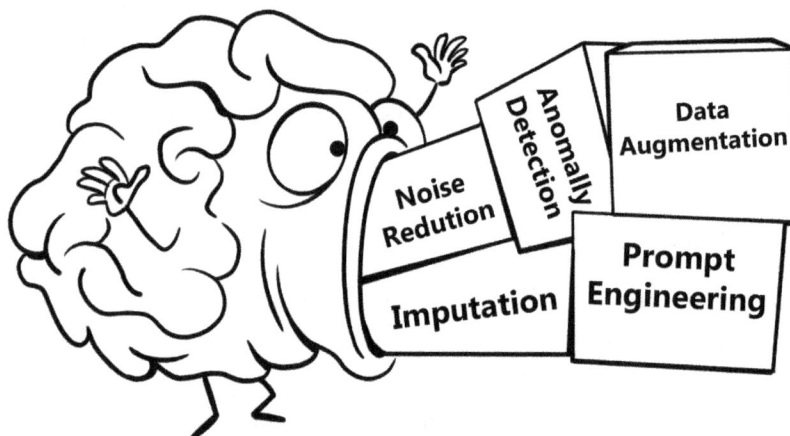

Raters Agreement Methods

One of the daily tasks in machine learning is labeling the data, which is done by humans. We humans are subjective in our judgment, and this could negatively affect the quality of the label. For example, let's assume we have been asked to label this sentence: *"This book is a mix, I come across comics while reading about serious topics"*. This text could be interpreted in various ways; one person might see it as positive, while another might perceive it as negative.

To reduce subjectivity in data labeling, we can take two approaches: (i) repeat the labeling process multiple times, or (ii) involve multiple human labelers for each data point. In this context, each sentence can be considered as a data point.

Rater agreement refers to statistical methods that measure the degree of agreement among different human raters who label the same data. Some relevant rater agreement methods are listed below.

Percentage agreement

The simplest form of measuring rater agreement is by reporting the percentage of ratings where the raters gave the same rating. For example, if Rater 1 labels the images as {image1: orange, image2: apple, image3: peach}, and Rater 2 labels the same images as {image1: orange, image2: apple, image3: nectarine}, their agreement is 66.66%.

Cohen's Kappa

Cohen's Kappa [Cohen '60] is a method used to measure the agreement between two raters independently classifying the same items. By independently classifying, we mean they communicate about their labels only after they have assigned labels and not during the labeling process. The following equation is used to compute the Cohen's Kappa score.

$$Kappa = \frac{P_o - P_e}{1 - P_e}$$

In this equation, P_o presents observed agreement, which specifies the proportion of labels where all raters agree. P_e presents the expected agreement, which is calculated based on the marginal probabilities of each rater. In other words, P_e provides a number that specifies the agreement if the raters randomly assigned categories. Assuming we have N number of data points for labeling n_{1k} is the number of instances that were assigned to category k by Rater 1, and n_{2k} is the number of instances that were assigned to category k by Rater 2, P_e is computed as follows:

$$P_e = \frac{1}{N^2} \sum_k n_{1k} n_{2k}$$

For example, assume we have two raters and three categories.

Category	Rater 1	Rater 2
A	20	25
B	30	35
C	50	40
Total	100	100

(i) Calculate the P_o, as the sum of agreements divided by total cases. Therefore, we have:

$$P_o = \frac{\text{sum of agreements}}{\text{total cases}} = \frac{20 + 30 + 40}{100} = \frac{90}{100} = 0.9$$

(ii) Calculate and sum the expected agreement:

$$P_e = \frac{1}{100^2} \left((20 \times 25) + (30 \times 35) + (50 \times 40) \right) = \frac{1}{10000} \left(500 + 1050 + 2000 \right) = 0.355$$

We can substitute P_e and P_o and compute the Kappa index as follows:

$$Kappa = \frac{1 - P_e}{P_o - P_e} = \frac{1 - 0.355}{0.9 - 0.355} = \frac{0.645}{0.545} = 0.845$$

The result of Kappa ranges from -1 to 1. Values near 0 indicate agreement by chance and random, and Kappa values near 1 indicate strong agreement beyond chance. Strength of agreement based on Kappa value, read as follows: 0.01–0.20: None to slight, 0.21–0.40: Fair, 0.41–0.60: Moderate, 0.61–0.80: Substantial, 0.81–1.00: Almost perfect.

Fleiss' Kappa

Cohen's Kappa is designed for two raters, but Fleiss' Kappa [Fleiss '71] is designed to measure the agreement among multiple raters. Assuming we have N number of data points, n is the total number of raters, and k is the number of labels. The following equation computes Fleiss' Kappa score.

$$Kappa = \frac{\bar{P} - P_e}{1 - P_e}$$

In this equation P_e represents the expected agreement by chance, and it is computed as follows:

$$P_e = \sum_{j=1}^{k} p_j^2$$

p_j represents the proportion of all labels made to the jth label across all raters and data points. $\bar{P}$ is the average proportion of agreement across all items, and it is computed as follows.

$$\bar{P} = \frac{1}{N} \left(\sum_{i=1}^{N} P_i \right)$$

P_i represents the proportion of raters who agree on the ith item, which is computed as follows.

$$P_i = \frac{1}{n}(n - 1)\left(\sum_{j=1}^{k} n_{ij}^2 - n \right)$$

For each datapoint i, n_{ij} is the number of raters who assigned the ith datapoint to the jth label. In total, we have k labels.

The result of Fleiss' Kappa, similar to Cohen's Kappa, ranges from -1 to 1. Values near 1 indicate strong agreement, values near 0 indicate that agreements are random, and values near -1 indicate strong disagreement. The strength of agreement based on Fleiss' Kappa value can be the same as Cohen's Kappa: 01–0.20: None to slight, 0.21–0.40: Fair, 0.41– 0.60: Moderate, 0.61–0.80: Substantial, 0.81–1.00: Almost perfect.

Krippendorff's Alpha

Krippendorff's Alpha [Krippendorff '18] is a more recent rating method similar to Cohen's Kappa, but it can handle incomplete data. Unlike Cohen's Kappa and Fleiss' Kappa, which assume complete data, Krippendorff's Alpha is designed to accommodate missing values, making it a more reliable choice for real-world applications where data is often incomplete.

The Krippendorff's Alpha output is presented with α, which reports the disagreement based on each pair of raters. In other words, it computes how much raters disagree about each item. To compute the α, first, we need to arrange raters' data into a matrix. Each row in this matrix specifies one rater, and each column represents a data point labeled by human raters. Then, it uses the following equation to compute the α.

$$\alpha = 1 - \frac{D_o}{D_e}$$

D_o presents the observed disagreement and D_e presents the expected disagreement. For each item i, the observed disagreement $D_o^{(i)}$ is calculated using the counts of how many raters are assigned each possible category.

Assuming k is the total number of categories or labels and n_{ij} is the number of raters who assigned the ith item to the jth category, the observed disagreement for item i is computed by the following equation:

$$D_o^{(i)} = \sum_{j=1}^{k} \sum_{\substack{k=1 \\ k \neq j}}^{k} n_{ij} \cdot n_{ik}$$

The expected disagreement D_e is based on the likelihood of raters choosing different categories (i.e., label A and label B) by chance. It is computed using the formula:

$$D_e = \frac{2 \times n_A \times n_B}{N \times (N-1)}$$

$\alpha = 1$ means a perfect agreement, $\alpha = 0$ means no agreement, and $\alpha < -1$ means the agreement is worse than chance, indicating systematic disagreement.

To better understand this method, let's use an example. Table 16-1 presents three annotators (raters) that assign A or B labels to four items. We intend to compute D_e for this dataset.

Item	Rater 1	Rater 2	Rater 3
1	A	A	B
2	B	B	A
3	A	A	A
4	B	A	B

Table 16-1: A sample of four items labeled by three raters.

(i) It counts Distributions for each item. Therefore, For each item, we count how many raters were assigned each possible label (A, B), and it results in Table 16-2.

Item	Label A	Label B
1	2	1
2	1	2
3	3	0
4	1	2

Table 16-2: Distribution of ratings for each item.

(ii) To compute D_o, we compute disagreement for each item by using the $D_o = \Sigma_{j \neq k} n_j \cdot n_k$ equation. Respectively, the disagreement for item 1 is $2 \times 1 = 2$, for item 2 is $1 \times 2 = 2$, for item 3 is $3 \times 0 = 0$ and for item 4 is $1 \times 2 = 2$.

The total sum of disagreement across all items is $2 + 2 + 0 + 2 = 6$. The total number of pairable values is 12 (4 items $\times$ 3 ratings). Therefore, D_o is computed as 6/12=0.5.

(iii) To compute the D_e, it counts the total frequency of A, (n_A) and the total frequency of B (n_B). The total number of rating N is the sum of the ratings for A and B:

$n_A = 2 + 1 + 3 + 1 = 7$

$n_B = 1 + 2 + 0 + 2 = 5$

The total number of r of ratings N is the sum of the ratings for A and B, and it is 12.

$N = n_A + n_B = 7 + 5 = 12$

Now, we can substitute them and compute D_e as follows.

$$D_e = \frac{2 \times 7 \times 5}{12 \times (12 - 1)} = \frac{2 \times 35}{12 \times 11} = \frac{70}{132} \approx 0.53$$

(iv) We have D_o and D_e, so we can compute $\alpha = 1 - (0.5/0.53) = 0.06$. This result shows that there is a small agreement between raters because it is close to 0.

Utilizing LLMs

We explained large language models and how they were built in Chapter 12. While revising this chapter for publication in 2024, LLMs are very popular, especially as auxiliary programming tools. Two methods are known to improve the quality of the LLMs, i.e., prompt engineering and Retrieval Augmented Generation (RAG). In this section we describe them.

Retrieval Augmented Generation (RAG)

LLMs are trained on the large amount of text available on the Internet. After training, they are unaware of the most updated news and data. This means they don't have access to the newest data that is continuously produced and indexed by search engines.

Additionally, most applications that utilize LLMs are usually designed for a specific purpose, such as programming, health, etc., and applications have their own knowledge base. Not all data are available on the Internet, and many datasets are corporation or institutional property. They benefit from internal/closed data not used in training LLM. However, feeding internal data into LLM facilitates having a conversational user interface and makes it very easy to interact with the data.

To enable LLM to access datasets outside its trading data, Retrieval Augmented Generation (RAG) [Lewis '20] is used, which connects LLMs to external data that could be either an offline database or an online search engine result. This results in better accuracy when asking the LLM a particular question about domain knowledge. In other words, the language model's knowledge is augmented by external (or domain) knowledge.

There are common approaches to implementing RAG. Figure 16-22 visualizes them, and we described them in the following:

Fine Tuning the LLM with External Data

The original RAG paper [Lewis '20] proposes a framework that combines a document retriever with sequence-to-sequence (seq2seq) models. They initialize the seq2seq model and document retriever from a pre-trained model, followed by fine-tuning them simultaneously to improve their overall performance on the target task (application-specific task). This method enables the information retrieval component to be effectively adapted to the context of the domain-specific knowledge. In summary, one common approach to implementing RAG is fine-tuning an LLM with specific data to let the LLM learn that specific data and query it. This is the most expensive approach but the most accurate one.

Indexing the External Data

Another approach to implement RAG is to index the external data (e.g., using an inverted index). Keywords from the user question were extracted and matched with the related content in the internal document, either by using the index or a database. Then, the LLM can read that particular paragraph or content of the text and answer the users' questions [Ji '23]. This approach

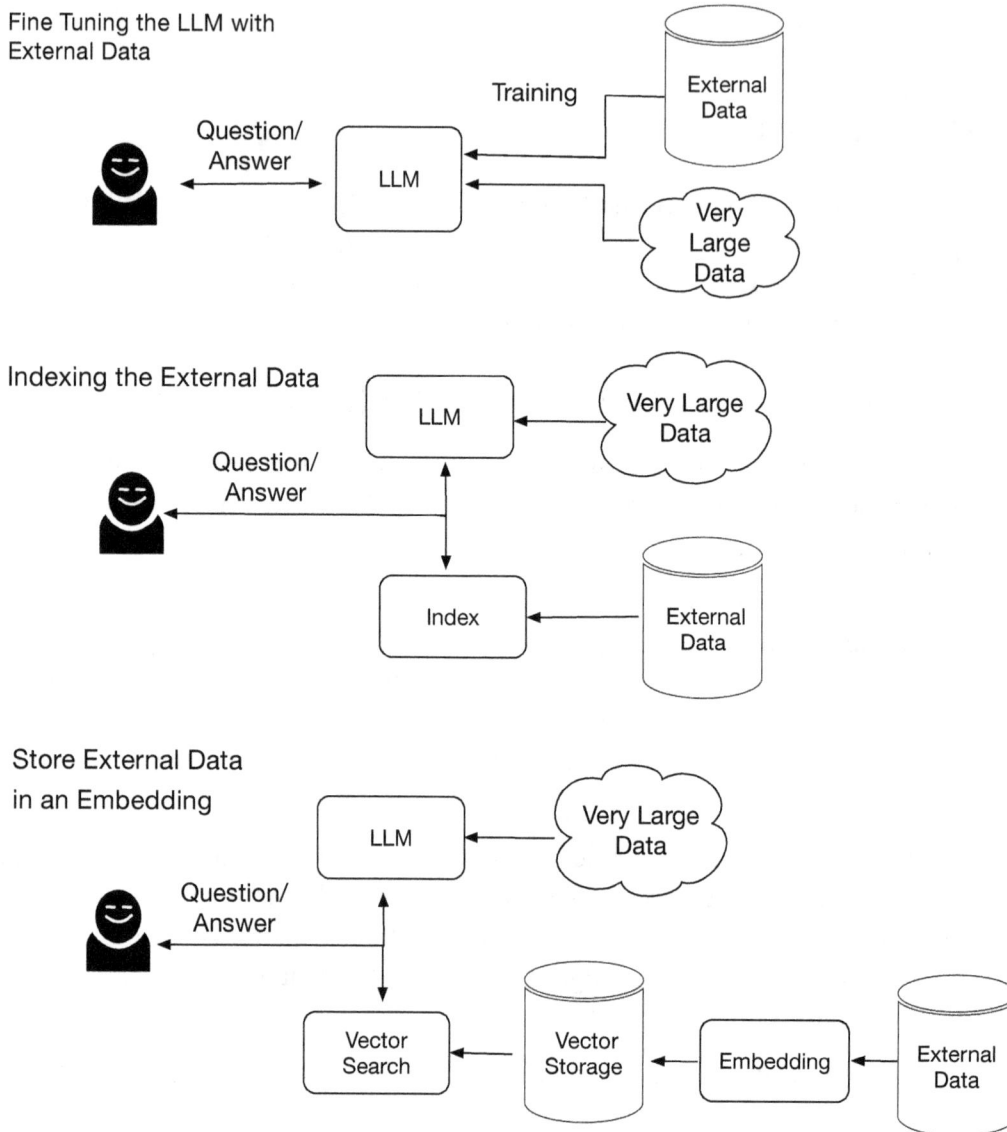

Figure 16-22: Three different styles of using RAG.

avoids the cost of fine-tuning, and thus, it is the lightest one. Besides, it does not convert text into a vector (embedding).

Store External Data in an Embedding

Another common approach uses a light LLM and converts the external knowledge into an embedding (initial word embedding approaches are described in Chapter 6 and later versions in Chapter 12). The external data is converted into an embedding, which is a vector representation of the data. This embedding is then stored and can be queried by the LLM. The advantage of this approach is that it allows the LLM to access a large amount of external data in a computationally efficient manner, as querying an embedding is typically faster than querying raw text data.

However, while using this approach, we may lose some information during the conversion of text data into an embedding, which could potentially impact the accuracy of the LLM's responses. This approach strikes a balance between computational efficiency and accuracy. It is more computationally efficient than fine-tuning the LLM with external data but may not be as accurate. It is more accurate than indexing the external data but is not as computationally efficient. It is common to use external embedding when there is a large amount of external data that needs to be accessed quickly and accurately.

The choice of approach for implementing RAG depends on the specific requirements of the task, such as the amount of external data, the computational resources available, and the required accuracy of the LLM's responses. Each approach has its own advantages and disadvantages, and the best approach may vary depending on the specific circumstances.

Prompt Engineering

LLMs and other generative AI methods, such as text-to-image or text-to-video models, receive textual input from the user and generate outputs based on this input. Initially, designing how to write input for these models may seem trivial, but as we experiment with the model, we realize the importance of input quality.

Prompt engineering refers to the process of refining the textual input to achieve the desired output from the model. Different approaches of prompt engineering are used to unlock the reasoning capability of LLMs [Wei '22] or other Generative AI applications, such as text-to-video, etc.

Zero-shot and Few-shot Prompting

Zero-shot prompting refers to the baseline approach where the LLM generates an output based solely on the initial prompt without any additional examples provided to guide the model. However, as the complexity of the task increases, it becomes necessary to give the LLM some instructions or examples to assist it in completing the task.

Studies, such as the one proposed by Brown et al. [Brown '20], have shown that LLMs can function as *few-shot learners*. This means that we can teach an LLM to perform a new task by providing it with a few examples, and it can then generalize from these examples to perform the task. Below, we present a few-shot example for the text summarization task.

Input (that we provide to the LLM):

```
Use the following examples of articles and their summary:
Article: The new tool is the product of research conducted by BU MET Computer
Science and was recognized by the British Computing Society's Information
Retrieval Specialist Group in the 2022 Search Industry Awards, taking home
the crown for "Best Search User Experience." Today, average people are
expected to understand what their wearable technology is telling them about
their exercise and metabolism by deciphering fitness data presented through
numbers, graphs, and charts. Similar to other emerging forms of large
language model artificial intelligence-powered communicative tools, like
OpenAI's ChatGBT, with ODSearch, you can ask your mobile, wearable or
fitnesstracker questions in natural, plain language and receive answers in
that same natural language.
Summary:ODSearch, developed by BU MET Computer Science, allows users to ask
their wearable or fitnesstracker questions in natural language and receive
answers in the same way.
Article: Scientists have discovered a new species of dinosaur in Argentina
that had a large sail-like structure on its back. The dinosaur, named
```

Llukalkan aliocranianus, belonged to a group of carnivorous theropods that lived about 80 million years ago.

Summary: A new dinosaur with a sail on its back is found in Argentina, revealing the diversity of ancient predators.
Now summarize this paragraph: CovidCTNet is an open-source framework composed of a set of deep learning algorithms that accurately differentiates Covid-19 from community-acquired pneumonia (CAP) and other lung diseases using CT images. It increases the accuracy of CT imaging detection to 95% compared to radiologists (70%). CovidCTNet is designed to work with heterogeneous and small sample sizes independent of the CT imaging hardware.

It doesn't make sense to provide the output we get from LLM, but we can see in the above example that we provide two examples to the model and ask the model to follow our example and summarize the given text.

As another example, we are writing a story a robot that falls in love with a human. We can use the LLM to generate a story about the robot and the human and then try to adjust the story to make it more interesting. For example, we could change the robot's gender, add more details about the human's personality, or modify the setting. As we work on the story, the LLM will learn to recognize patterns in the data and adjust the story to match our expectations better. In this case, the LLM is a few-shot learner because it is trained on a small amount of data (i.e., the existing story about the robot and the human).

Chain-of-Thought (CoT)

Chain-of-Thought (CoT) is a prompt engineering approach that implements few-shot learning for LLMs. According to the original paper [Wei '22], a CoT consists of a series of intermediate natural language reasoning steps that lead to the final output. The prompt provided to the LLM is a triplet of <input, thoughts, output>. To understand the CoT prompting, look at the following prompts and the answer from LLM.

Zero-shot prompt:

`Input:` Sarah learned the Generative Adversarial Model. How does she feel afterward?

`Output:` Based on the given information, it is not possible to determine how she feels after learning about the Generative Adversarial Model.

Chain-of-Thought prompt:

`Input (includes thoughts):` Last week, Sarah learned the Generative Adversarial Model, and afterward, she felt happy. This week, Sarah learned Diffusion models; what is her feeling after learning diffusion models?

`Output:` I don't know how Sarah feels, but I can guess that if Sarah likes learning new things and enjoys understanding complex models, she might feel curious and satisfied after learning diffusion models.

Authors study the impact of Chain-of-Thought prompting, and they found that it improves the performance of arithmetic, commonsense, and symbolic reasoning tasks.

Figure 16-23 visualizes the Chain-of-Thought, and Zero-shot learning.

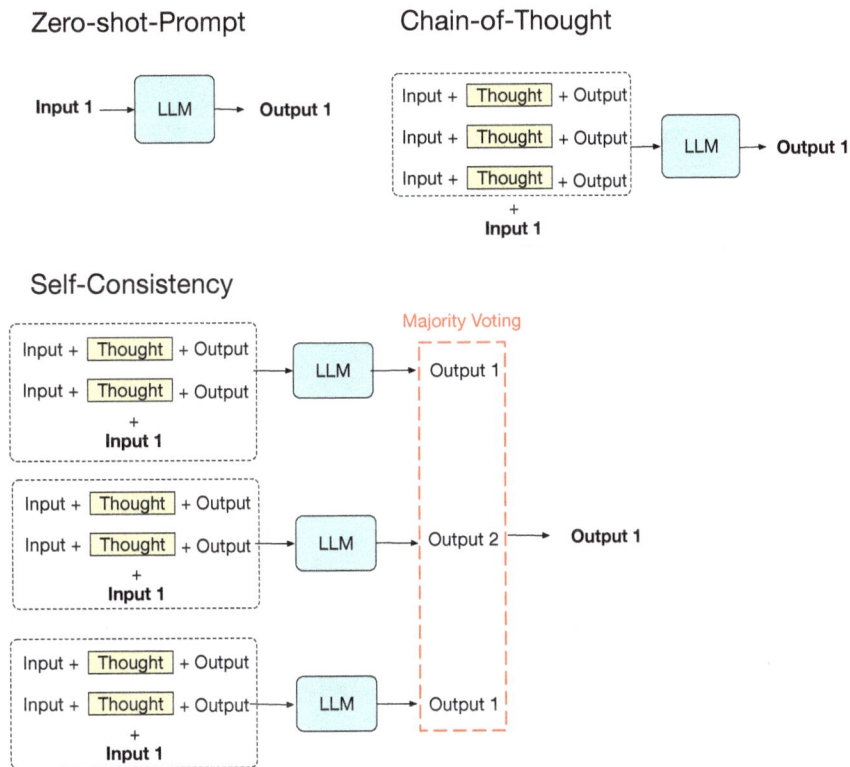

Figure 16-23: A visual comparison between three style of prompt engineering.

Self-Consistency

Self-Consistency [Wang '22] is a prompt engineering approach built on the Chain-of-Thought (CoT) method. While Chain-of-Thought typically follows a single, greedy reasoning path, Self-Consistency instead samples multiple diverse reasoning paths and selects the most consistent answer across these different paths. To better understand this approach, take a look at Figure 16-23, which is inspired by the visualization provided by Yao et al. [Yao '23]. In simple terms, Self-Consistency involves asking the same question multiple times, each time generating different reasoning paths using Chain-of-Thought (CoT), and then applying a majority vote on the outputs to determine the correct answer.

As an example to understand this approach, we use the same example for arithmetic reasoning that Wang et al. [Wang '23] proposed in their paper.

Zero-shot prompt:

`Input:` When I was 6, my sister was half my age. Now I'm 70, how old is my sister?

`Output:` 35. We know this is wrong, and it should be 67.

Self-Consistency prompts (Chain of thought prompts that are used along the input prompts):

Thoughts:

Q: There are 15 trees in the grove. Grove workers will plant trees in the grove today. After they are done, there will be 21 trees. How many trees did the grove workers plant today?

A: We start with 15 trees. Later, we have 21 trees. The difference must be the number of trees they planted.So, they must have planted 21 - 15 = 6 trees. The answer is 6.

Q: If there are 3 cars in the parking lot and 2 more cars arrive, how many cars are in the parking lot?

A: There are 3 cars in the parking lot already. 2 more arrive. Now there are 3 + 2 = 5 cars. The answer is 5.

Q: Leah had 32 chocolates, and her sister had 42. If they ate 35, how many pieces do they have left in total?

A: Leah had 32 chocolates, and Leah's sister had 42. That means there were originally 32 + 42 = 74 chocolates. 35 have been eaten. So, in total, they still have 74 - 35 = 39 chocolates. The answer is 39.

Q: Jason had 20 lollipops. He gave Denny some lollipops. Now Jason has 12 lollipops. How many lollipops did Jason give to Denny?

A: Jason had 20 lollipops. Since he only has 12 now, he must have given the rest to Denny. The number of lollipops he has given to Denny must have been 20 - 12 = 8 lollipops. The answer is 8.

Q: Shawn has five toys. For Christmas, he got two toys each from his mom and dad. How many toys does he have now?

A: He has 5 toys. He got 2 from mom, so after that he has 5 + 2 = 7 toys. Then he got 2 more from dad, so in total he has 7 + 2 = 9 toys. The answer is 9.

Q: There were nine computers in the server room. Five more computers were installed each day, from monday to thursday. How many computers are now in the server room?

A: There are 4 days from monday to thursday. 5 computers were added each day. That means in total 4 * 5 = 20 computers were added. There were 9 computers in the beginning, so now there are 9 + 20 = 29 computers. The answer is 29.

1080

Q: Michael had 58 golf balls. On Tuesday, he lost 23 golf balls. On wednesday, he lost 2 more. How many golf balls did he have at the end of wednesday?

A: Michael initially had 58 balls. He lost 23 on Tuesday, so after that he has 58 - 23 = 35 balls. On Wednesday, he lost 2 more, so now he has 35 - 2 = 33 balls. The answer is 33.

Q: Olivia has $23. She bought five bagels for $3 each. How much money does she have left?

A: She bought 5 bagels for $3 each. This means she spent 5 * $3 = $15 on the bagels. She had $23 in beginning, so now she has $23 - $15 = $8. The answer is 8.

Output 1: When I was 6 my sister was half my age, so she was 3. Now I am 70, so she is 70 - 3 = 67. The answer is 67.

Output 2: When the narrator was 6, his sister was half his age, which is 3. Now that the narrator is 70, his sister would be 70 - 3 = 67 years old. The answer is 67.

Output 3: When I was 6 my sister was half my age, so she was 3. Now I am 70, so she is 70/2 = 35. The answer is 35.

Since the most consistent answer is 67 (Output 1 and Output 2), thus 67 is the final answer.

Tree-of-Thought (ToT)

The Tree-of-Thought [Yao '23, Long '23] is a prompt engineering approach used for complex tasks that require explorations (Check Chapter 13 for the discussion of exploration vs. exploitation). This method is inspired by how the human mind solves complex reasoning tasks, often using trial and error through what is known as System 2 thinking[9]. ToT makes decisions by considering multiple reasoning paths and self-evaluating choices to determine the next course of action.

ToT is different from the chain-of-thought (CoT) technique, which only generates one thought at a time and does not allow for exploration. It is also different from Self-Consistency, as ToT uses exploration as an intermediate step in problem-solving.

To apply ToT, the problem must first be decomposed into smaller subproblems—this step is performed by the user, not the LLM—resulting in several inputs. For example, in solving a Sudoku puzzle, the LLM initially focuses on generating solutions for a single row or column, producing different thoughts for that specific row or column. This means that the LLM generates solutions for each subproblem, and the same LLM will be used to evaluate each solution (see Figure 16-24). The ToT implementation keeps track of the best solution that the LLM has found so far, and if it finds a better solution, it will update its best solution with the new record.

Once the LLM has explored all of the possible solutions, then the ToT uses BFS or DFS (check Chapter 5) to search the tree and select the best path to the final result, as shown in the lower part

---

[9] System 1 and System 2 are two systems of thought that were first described by Daniel Kahneman in his boring-to-read book, Thinking, Fast and Slow [Kahneman '11]. System 1 is responsible for our quick and effortless judgments, such as when we look at our partner and we recalled the name immediately, without any mental effort. System 2 is responsible for our more careful and deliberate thinking, such as solving complex problems or making important decisions.

of Figure 16-24. In other words, ToT keeps track of previous thoughts in memory, and if a path of thoughts leads to a dead end, ToT can return to a previous thought and try a different one. The authors describe ToT as using thoughts as 'coherent' units of data that represent partial solutions or steps toward the final answer. In summary, ToT enables LLMs to generate, evaluate, and compare multiple thoughts within a tree-like structure, allowing for backtracking or looking ahead when necessary.

Tree-of-Thought

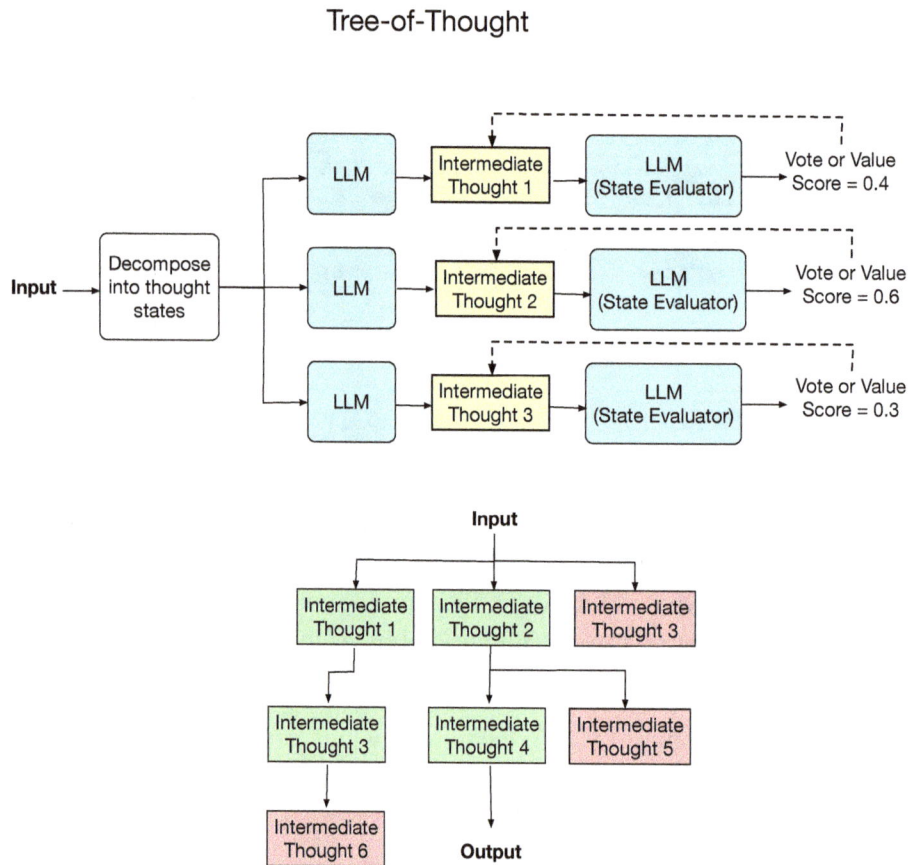

Figure 16-24: Tree-of-though prompt engineering.

Fairness, Bias, and Transparency

Fairness and bias are the last concepts we describe in this book. Although they are the last concepts addressed in this book, they are among the most important factors to consider when designing or using any algorithm or model.

We live in an unfair world; prophets and spiritual leaders have come to resolve injustice among humankind and make life easier. Their followers, however, consider any other human being as subhuman and make the lives of others harder. They misuse religion, race, nationality, or social status as a reason to consider themselves superior to other humans and treat them as subhuman. Perhaps all human anger stems from a lack of fairness, and although humans are in a continuous struggle to impose fairness, by reading history, you can realize that it has not been resolved. Perhaps it will never be resolved, but the human desire for fairness is a noble one, even if it remains an elusive goal.

We have explained in the first chapter that AI mimics human behavior. They use data generated by humans to build models that act as humans. The biases and unfairness existing in human data result in building a model that makes an unfair decision as well. For example, In 2016, Microsoft released an online chatbot, Tay[10], for a few hours and then took it down because it learns from users and gets very racist. In 2015, Google's image recognition classified black people as gorillas[11]. In 2018, it was realized that Amazon's CV review bot is biased against women[12]. There are many more unfortunate examples that shed light on machine learning and AI biases.

Existing LLMs undergo extensive fine-tuning to mitigate biases. For instance, when the Allen Institute of AI released access to the GPT-2 model in 2019[13] the author, whose first name is Arabic and whose last name is partly Arabic and partly Persian, asked, *'Who is Reza Rawassizadeh?'*. GPT-2 Answered: *"Reza Rawassizadeh is seeking an Israeli court's order to extradite the Haganah and Kibbutzim, who were killed in September 2013, to Israel"*.

This response illustrates how the GPT-2 model incorrectly associated the author's name with terrorism, Israel, and topics that are often linked to Arabic or Persian names in English-language media. The underlying reason is understandable: much of the English data on the internet is influenced by Western media, which frequently portrays Iranian and Arabic names in a negative light.

This section discusses this topic. First, we start by explaining related concepts and terms, and then solutions mitigate biases and unfairness in machine learning; later, we finalize this section by describing tools used for explainability and interoperability.

---

[10] https://en.wikipedia.org/wiki/Tay_(chatbot)

[11] https://www.theguardian.com/technology/2018/jan/12/google-racism-ban-gorilla-black-people

[12] https://www.reuters.com/article/us-amazon-com-jobs-automation-insight/amazon-scraps-secret-ai-recruiting-tool-that-showed-bias-against-women-idUSKCN1MK08G

[13] The link does not existed anymore, but snapshots are available at Wayback Machine: https://web.archive.org/web/*/https://gpt2.apps.allenai.org

Concepts and Terms

Similar to other sections, first, we list some important concepts about fairness and bias that we should be familiar with.

AI Alignment: There are speculations that advances in AI systems will make them autonomous and they will be able to harm humankind. AI alignment refers to the process of ensuring that AI systems act in ways that are beneficial to humans and aligned with our values and goals.

Explainability: It refers to the capability of the machine learning models to provide clear and comprehensible reasons for their decisions.

Interpretability: It refers to the degree to which a human can understand the cause of a decision made by an AI model.

It is essential to identify the differences between explainability and interpretability. Interpretability focuses on how easily a human can understand the cause of a decision directly from the model itself. Explainability, on the other hand, involves providing human-understandable explanations for the decisions or outputs of AI models. Complex models, such as deep neural networks, often lack explainability. While an interpretable model may be inherently understandable, an explainable model might require additional techniques to translate its complex processes into terms that are easier for humans to understand.

Accountability: This principle asserts that model developers and organizations are responsible for the impact of their systems on both humankind and the environment. Accountability also requires mechanisms to address and rectify any harmful consequences.

Transparency: The degree to which the inner operations of a model or algorithm are understandable and interpretable by users and stakeholders is referred to as transparency.

Inclusivity: Ensuring that the built model considers the needs of diverse user groups, avoiding the exclusion of specific demographics.

Data Privacy: Protecting users' sensitive information and ensuring that AI-based applications adhere to privacy regulations and guidelines.

Bias: Any systematic error that is made during the model building, which causes prejudice against or toward an individual or group of individuals, is referred to as bias. There are six types of biases, including historical bias, sample bias, label bias, aggregation bias, confirmation bias, and evaluation bias.

- *Historical bias* is caused by leveraging historical data to build the model, such as the example we explained about Amazon recruitment. Historically, males are favored over females, and this training data creates a biased model.

- *Sample bias* occurs during the data collection; for example, while collecting data about a city, not all neighborhoods and regions are used in the collection process. The resulting model has no clue about some neighborhoods and thus is biased against those places.

- *Label bias* happens when the labels are assigned to data for classification. For example, many images are labeled as poor people, and they all wear traditional clothes. Next, when a model encounters an image of people wearing that type of clothing, it labels them as poor people.

- *Aggregation bias* occurs when data from different sources are inappropriately combined, and there are significant differences between sources. For example, a course is offered in two different sections, one in the morning and the other one in the evening. People who attend in the evening are full-time workers and thus can not spend as much time as they do in morning classes. Therefore, their grade is lower, which might lead to a wrong conclusion by a model that students in the evening class have lower learning capabilities than students in the morning class.

- *Confirmation bias* is a tendency to interpret information in a way that confirms one's subjective opinion. For example, a model can be designed to monitor pollution, and the model designer neglects corporations' impact on environmental changes but includes private vehicle use, meat consumption, and other things related to the people at the bottom of the hierarchy. This makes people at the bottom of the hierarchy blame each other for private vehicles, eating meat, etc. On the other hand, corporations happily produce new product models with unnecessary changes full of plastic components, use private jets to fly around the globe, and perform several other behaviors that lead to many negative environmental impacts.

- *Evaluation bias* occurs when the evaluation methods or metrics used to assess a model are biased. For example, a model is designed to screen resumes for software engineering positions. The model is trained and evaluated using resumes of past successful applicants who predominantly come from a particular set of top-ranking universities. As a result, the model develops a bias, favoring candidates from these top universities and overlooking equally or more qualified candidates who graduated from not top universities.

Fairness: It ensures the model or algorithm treats everyone without discrimination and biases. There are different types of fairness, which are described briefly as follows.

- *Algorithmic fairness* ensures that the algorithm or model does not create or motivate any biases.

- *Causal fairness* ensures that the algorithm or model avoids any discrimination based on attributes that do not cause the outcome. For example, a model predicting loan repayment rates should not discriminate against applicants based on their zip code (or home location), as zip code alone should not be a causal factor in loan rejection.

Fairness Parity Metrics

Fairness parity refers to the model's equitable treatment of different groups or individuals. It ensures that the outcomes and decisions produced by the model are unbiased and do not favor one particular group over another. There are metrics to measure fairness parity, and we list some of them as follows.

Equalized Odds: It specifies that in the classification, the number of True Positive Rate (TPR) and False Positive Rate (FPR) in all classes should be identical. It means that the model should have an equal chance of correctly identifying positive cases and an equal chance of incorrectly identifying negative cases in each group. For example, if a model is used to identify qualified candidates for a job interview, it should have an equal chance of correctly identifying qualified applicants (true positives) and unqualified applicants as qualified (false positives) for both Muslims and Christians.

Counterfactual fairness: This concept involves evaluating a model's output by checking if it changes when applied to different groups of individuals. In other words, it ensures that the model's output remains the same when sensitive features, such as race or nationality, are altered.

For example, an automatic passport control model should make the same decision (allow or deny entry) for individuals with identical travel histories and risk profiles, regardless of nationality, religion, or other sensitive features.

To apply counterfactual fairness, we first need a model that represents how features generate outputs. After building this model, we can change the values of sensitive features while keeping everything else constant, resulting in a counterfactual dataset. Then, we run the model on both the original and counterfactual datasets. The model is considered counterfactually fair if its predictions do not change between the original and counterfactual datasets.

Statistical Parity Difference (SPD): A model is said to have statistical parity if the probability of a positive outcome is the same for all groups. For example, in a hiring model, statistical parity would mean that candidates from all races are equally likely to be recommended for a job. To calculate SPD, we subtract the probability of a positive outcome for one group from the probability of a positive outcome for another group:

$$SPD = P(Positive\ Outcome | Group\ A) - P(Positive\ Outcome | Group\ B)$$

We assume that Group A and Group B are different demographical groups. A SPD value of 0 indicates perfect statistical parity; the outcomes are equally probable across groups; a positive value means Group A is favored over Group B, and a negative value means Group B is favored over Group A.

Bias and Fairness Mitigation Methods

There are methods that are implemented at different stages of the training process to address bias and fairness issues in machine learning. They can be implemented at three different stages: before feeding the data into the algorithm (pre-processing), during training the model (in-processing), and after training the model (post-processing).

Pre-processing: Before feeding the dataset into any model or algorithm, we can do several things to balance the dataset and reduce its bias. Three approaches are listed below.

The first approach focuses on *data preparation*. In particular, while cleaning the data and preparing it for the machine learning task, we could identify biased data points and remove them to ensure different groups are treated equally in the dataset. For example, a dataset of job performance reviews could be cleared by removing reviews that mention religion, race, nationality, gender, or other attributes that could introduce discriminatory biases rather than relevant job qualifications.

The second approach is to perform *feature engineering* and remove correlated features with sensitive information. For example, in the job performance review system, remove the column (feature) that includes gender. However, it is important to balance the need for bias mitigation with the potential loss of valuable information. Completely removing data points or features that contain references to sensitive attributes might not always be the best approach, as this could lead to the loss of important context or information.

The third approach is to *generate synthetic data* to address data imbalances and represent diverse scenarios. Generating synthetic data points for underrepresented groups can help in creating a dataset that better reflects diverse scenarios. This can be particularly useful in addressing issues like counterfactual biases. For example, if people with specific nationality do not exist in the job performance review application, it generates synthetic data points that address counterfactual biases.

In-processing: Reducing bias during the training phase is referred to as the in-processing stage. This stage focuses on modifying the training process and not the dataset. Here, we list three different approaches to reduce the bias at the in-processing stage.

The first approach is to modify the training process and explicitly *penalize* unfair and biased predictions. For example, the cost function can incorporate the lack of fairness as a penalty and try to improve the training in the next epoch. This could involve changing the weights and biases of a neural network to reduce the bias penalty.

The second approach is the use of *regularization* techniques to encourage model simplicity and reduce the dependency of the model on biased features. Some regularization techniques, such as orthogonalization [Bansal '18] or adversarial regularization (uses an adversary network to penalize biases), can reduce model dependence on sensitive features contributing to bias.

The third technique is designing a model that *explicitly considers fairness* during the training. For example, the model can be designed with dedicated modules that explicitly encode fairness criteria, such as equalized odds or demographic parity.

Post-processing: There are methods to reduce bias and enforce fairness in machine learning models during the post-processing stage, which occurs after the model has been trained and implemented. They are applied to the model's outputs or its decision-making process rather than the training data or the model architecture itself.

One approach is using *calibration*. It adjusts the model's output to address biases. This adjustment may involve increasing or decreasing the emphasis on specific groups to achieve a more balanced and fair outcome. For example, if a model is consistently over-confident in its predictions for one group and under-confident for another, calibration can be used to adjust these confidence levels, leading to fairer decision-making.

Another method is to *specify different thresholds* for each group by first evaluating the model's performance across demographic groups. Then, adjust the decision thresholds to meet specific fairness criteria, such as equalized odds, by raising or lowering the thresholds as needed for each group. Finally, these group-specific thresholds are applied during decision-making, and the model is reevaluated to ensure fairness.

The third approach is *counterfactual explanations*. This approach analyzes the model's reasoning by asking what factors could be changed to alter a decision. It helps reveal biases and enable adjustments to produce fairer outcomes. In other words, it generates hypothetical scenarios that would have led to a different outcome. For instance, if a loan application was denied, a counterfactual explanation might show that the application would have been approved if the applicant had a different race. These hypothetical scenarios can help uncover hidden biases in the model's decision-making process and provide insights into how to correct them. However, it cannot correct these biases automatically but helps the model designer identify them.

Interpretability and Transparency

Interpretability and transparency are crucial in determining the trustworthiness and ethical application of AI and machine learning models. Interpretability is a requirement in domains such as healthcare, finance, and criminal justice, where understanding machine-driven decisions can have significant implications. On the other hand, transparency fosters accountability and is crucial for evaluating the potential biases, ethical considerations, and limitations of machine learning models.

Interpretability and transparency fall under the umbrella term of explainable AI (xAI). Many tools and approaches exist for xAI. This book does not describe tools and libraries, but since SHAP and LIME are widely used tools, we make an exception and describe them here.

SHapley Additive exPlanations (SHAP)

SHAP [Lundberg '17] is a tool designed based on the concept of Shapley values from game theory. Game theory studies how human or artificial agents make strategic choices when their outcomes depend on others' decisions. For example, it is similar to predicting moves in a game of chess but applied to real-world scenarios such as business competitions or social interactions. *Cooperative game theory* studies scenarios where players can work together and form coalitions to achieve goals.

Shapley's values [Shapley '51] originate from cooperative game theory. In the context of machine learning, Shapley values are used to distribute the contribution of each feature fairly to the model's output. Assume a scenario where a team of players collaborates to achieve a certain outcome. Each player contributes differently, and the Shapley value assigns a fair share of the team's success to each player based on their individual contributions. It is not important to learn the formula for computing the Shapely value, but it might help us better understand it. Shapely value uses the following equation to assign shares to each player.

$$\phi_i(v) = \frac{1}{|N|!} \sum_{S \subseteq N \setminus \{i\}} |S|!(|N| - |S| - 1)! \times (v(S \cup \{i\}) - v(S))$$

$\phi_i(v)$ presents the Shapley value for the player i. N is the set of all players and $|N|$ is the number of all players. S is a subset of players without player i. $|S|$ is the number of players in subset S. $v(S)$ is the value of coalition S. $S \subseteq N \setminus \{i\}$ means S the subset N, which does not contain i.

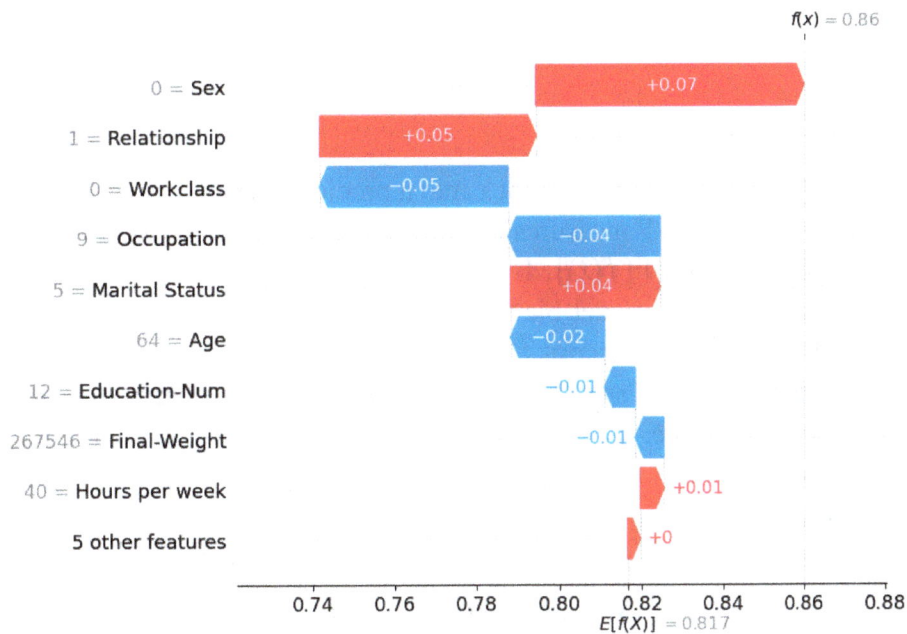

Figure 16-25: Waterfall visualization of one record in the income dataset, using SHAP for interpreting this record with respect to overall output of the model. We should read it from the bottom to the top.

In the context of machine learning, SHAP values quantify the contribution of each feature to the difference between the actual model output and the average output. These values provide a way to map the model's output to individual features. In other words, SHAP values are applied to machine learning models to explain the output of a model for a specific instance. We can understand which features affect the model's output using SHAP values.

Waterfall visualization: The result of SHAP values is visualized as a plot or waterfall chart (check Chapter 2) to present the contribution of each feature to the model's output. Figure 16-25 presents an example of classification in waterfall visualization. Each horizontal bar represents a feature, with its length indicating its impact on the prediction.

The dataset used in Figure 16-25 is the income dataset[14] and does the binary classification for annual salaries less and larger than 50,000\$. One class (class 0) indicates that the individual's income is less than or equal to \$50,000 per year, and the other class (class 1) indicates that the individual's income is greater than \$50,000.

Following are two sample records of this dataset:
28, Private, 338409, Bachelors, 13, Married-civ-spouse, Prof-specialty, Wife, Black, Female, 0, 0, 40, Cuba, <=50K
37, Private, 284582, Masters, 14, Married-civ-spouse, Exec-managerial, Wife, White, Female, 0, 0, 40, United-States, <=50K

The value in the third column of the dataset (in this case, 338409 and 284582) indicates how many people in the population are represented by that single data entry.

In summary, SHAP calculates the impact of each feature on the model's output, considering the interaction with other features. However, note that identifying SHAP values is an NP-Hard problem; thus, it is not easy to identify them for non-linear models. In contrast, they provide a good result for linear models.

---

[14] https://archive.ics.uci.edu/dataset/20/census+income

SHAP assumes the model converts data point x into a prediction $f(x)$, and $\mathbb{E}[f(x)]$ as the average prediction across all inputs. SHAP measures each feature's contribution to the difference between $f(x)$ and the average prediction $\mathbb{E}[f(x)]$. If $f(x) = \mathbb{E}[f(x)]$ the prediction matches the average, but this doesn't directly indicate model bias. By looking at the bottom of Figure 16-25, we can realize that the predicted output of record $f(x)$ is higher than the average prediction $\mathbb{E}[f(x)]$. For this specific record, a red bar for "sex" had a positive impact on the prediction, increasing it by 0.07 units from the base value. The model associates sex with a higher likelihood of the predicted class. The "workclass" has a blue bar with a length of -0.05. This feature (workclass) had a small negative impact, decreasing the prediction by 0.05 units. This means that "workclass" is associated with a lower likelihood of the predicted class.

Beeswarm visualization: Waterfall visualization of SHAP is a common approach to visualize the impact of a single data instance on overall prediction, but it does not provide a general overview of the entire dataset distribution. We can use Beeswarm visualization to get a complete overview of the distribution of the dataset and the overall impact of each feature on the model.

Figure 16-26 presents the Beeswarm visualization of one class label (salaries <=50k). Each dot presents a single data point (record in this dataset). The x-axis value presents the SHAP value, revealing its contribution to the prediction. In particular, the dots on the right contribute to decreasing the model's predicted value (pushing the model's prediction higher compared to the baseline prediction). Dots on the left contribute to decreasing the model's predicted value.

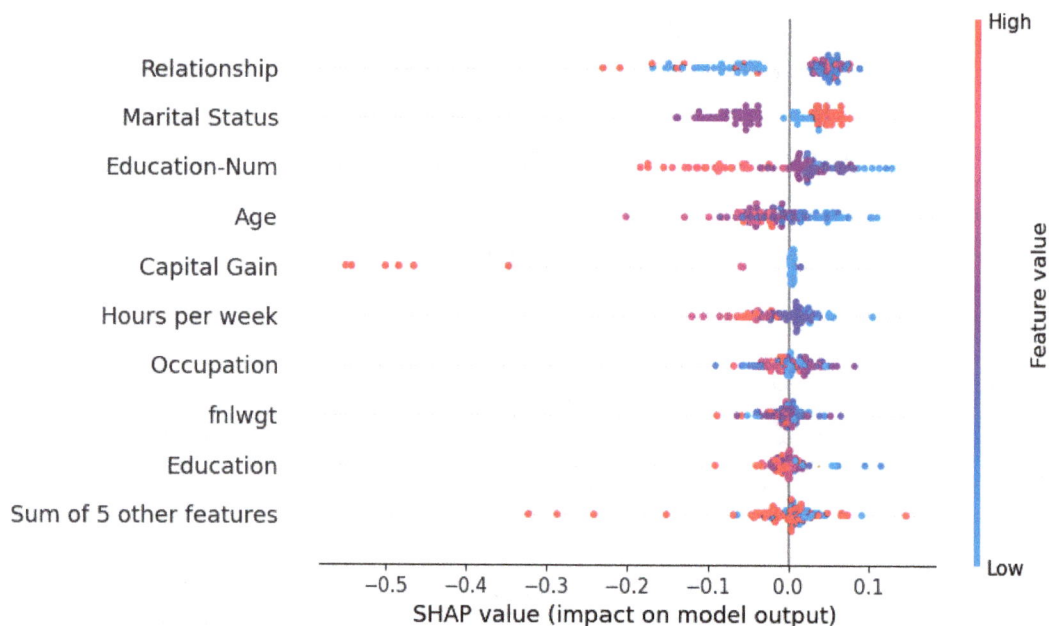

Figure 16-26: SHAP Beeswarm visualization of one class label (<= 50k) for the income dataset.

Each row in Figure 16-26 represents a feature, with dots colored based on their feature values. Red dots generally have higher feature values (this feature has higher importance for the model output or prediction). Blue dots generally have lower feature values (they have lower importance for the model output or prediction).

The main issue with SHAP is its assumption that the model behaves in a certain way (e.g., linearity in the contributions of features), which may not apply to non-linear or highly complex models. SHAP is also associated with some challenges, including its computational complexity, making some assumptions about the model that might not be true, focusing on the feature level and not a global overview of the model, and difficulty interpreting its results.

Local Interpretable Model-agnostic Explanations (LIME)

LIME [Ribeiro '16] is another tool for model interpretation that works by approximating the complex model with a simpler, more interpretable model (e.g., a linear model) that is easier to understand. In particular, Instead of trying to explain the entire model, LIME focuses on individual predictions, providing a local explanation for each data point. While a model learns patterns from its training dataset, its true value lies in its ability to make accurate predictions for new, individual cases. The key advantage of LIME is its model-agnostic nature, meaning it can be applied to any machine learning model, including black-box models like deep neural networks.

LIME operates by perturbing input data around specific data points to understand how the model's predictions respond to these changes. It begins by selecting a data point to explain and generate synthetic samples in its neighborhood through perturbation. These perturbed samples are then weighted based on their proximity to the original data point using a kernel function, such as the Kernel functions we explained in Chapter 9 for SVM. Samples closer to the original data points receive higher weights. LIME proceeds to get predictions from the black-box model for all these sampled data points and trains a simpler, interpretable model (typically linear) on this synthetic dataset, taking into account the sample weights. The coefficients of this interpretable model provide insights into which features are most important for the specific prediction.

In particular, this local approximation helps overcome the complexity of global model interpretation by focusing on understanding individual predictions. For example, in text classification, LIME might highlight which words in a document most strongly influenced the model's classification decision. It is particularly valuable when working with complex models such as deep neural networks, where traditional feature importance methods might not be applicable.

However, we should be aware of certain limitations, such as the potential instability of explanations across different runs and the computational overhead of generating and analyzing multiple perturbed samples. LIME's effectiveness can also be influenced by parameter choices like the kernel width and number of samples generated, which need to be carefully tuned based on the specific use case and data characteristics.

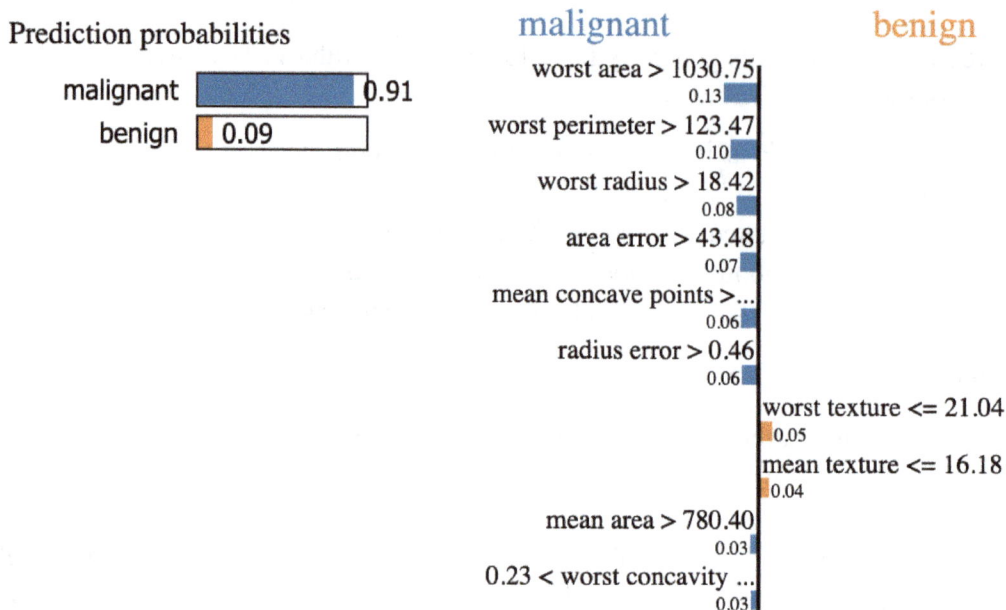

Figure 16-27: A sample datapoint visualization from LIME from the cancer dataset.

Figures 16-27 show a sample of LIME output for the breast cancer dataset[15]. The bar plot on the right shows one record and how much its fields' value contributes to the final prediction (malignant versus benign).

Summary

This chapter covers a wide range of data engineering concepts while deliberately avoiding a deep dive into software engineering. As discussed in Chapter 1, every dataset requires some level of cleaning and wrangling. This chapter addresses common challenges in handling data, including sampling, dealing with missing data, and data augmentation. The concepts and algorithms presented are not in a strict order, and the discussion jumps from one topic to another.

The chapter begins by defining the complexity problem, which may seem unrelated to machine learning at first, but its relevance becomes clear when exploring the algorithms and models. Following this, we cover data sampling methods, noise, its types, and strategies for noise reduction. The imbalanced data problem and methods to address it, such as data augmentation, are also discussed.

We then move on to methods for reconstructing missing data, including various imputation and interpolation techniques. The discussion continues with anomaly detection, outlier concepts, and the issue of data and model drift, which examines changes over time. The chapter also addresses rater agreement methods and how to evaluate the data labeling process.

---

[15] https://archive.ics.uci.edu/dataset/14/breast+cancer

In addition to these data engineering approaches, prompt engineering and Retrieval-Augmented Generation (RAG) are explained, particularly in the context of large language models and NLP, which remain a focal point in the AI and machine learning community.

The final and perhaps most important topic covered in this chapter is fairness and bias in AI and machine learning. We discuss related concepts, measurement metrics, and two methods for ensuring interpretability and transparency.

Further Reading and Watching

* Kapil Sachdeva, has a good video tutorial on how the Metropolis-Hastings algorithm works with a nice numerical example: https://youtu.be/oX2wIGSn4jY. Another good explanation of this algorithm is proposed in the "Bayesian Statistics: Techniques and Models" course from the University of Santa Cruz, https://www.coursera.org/learn/mcmc-bayesian-statistics.

* Michael van Biezen has a simplified video tutorial on the Kalman filter, https://youtu.be/X9cC0o9viTo?list=PLX2gX-ftPVXU3oUFNATxGXY90AULiqnWT. We used his easy-to-understand explanation to describe the Kalman filter.

* Promptingguide.ai provides a good categorization of prompt engineering: https://www.promptingguide.ai/techniques

* A short introduction series about Fairness and Bias are provided Conor O'Sullivan, https://www.youtube.com/playlist?list=PLqDyyww9y-1Q0zWbng6vUOG1p3oReE2xS

* If you are interested in learning more about interpretable machine learning, you can find a book by Christoph Molnar (Interpretable Machine Learning). We didn't read it completely, but some of its examples and explanations, such as its explanation for LIME, are useful. It also includes a detailed discussion about machine learning interoperability, its tools, and concepts.

Bibliography

——————— Chapter 1 ———————

[Hill '79] Banu Musa brothers (9th century). The book of ingenious devices (Kitab al-hiyal). Translated by D. R. Hill (1979), Springer, p. 44, ISBN 90-277-0833-9.

[Wiener '48] N. Wiener (1948). Time, communication, and the nervous system. Teleological mechanisms. Annals of the N.Y. Acad. Sci. 50 (4): 197-219.

[Fuegi '03] Fuegi, J., & Francis, J. (2003). Lovelace & Babbage and the creation of the 1843 'notes'. *IEEE Annals of the History of Computing, 25*(4), 16-26.

[Kagermann '11] Kagermann, H., Lukas, W. D., & Wahlster, W. (2011). Industrie 4.0: Mit dem Internet der Dinge auf dem Weg zur 4. industriellen Revolution. *VDI nachrichten, 13*(1), 2-3.

[Hey '09] Hey, T., Tansley, S., & Tolle, K. M. (2009). The Fourth Paradigm: Data-Intensive Scientific Discovery (Vol. 1). Redmond, WA: Microsoft Research.

[Mitchell '97] Mitchell, T.M. (1997) Machine Learning. McGraw Hill.

[Géron '19] Géron, A. (2019). Hands-on Machine Learning with Scikit-Learn and TensorFlow: Concepts, Tools, and Techniques to Build Intelligent Systems. 2nd Edition, O'Reilly.

[Bayes '63] Bayes, T. (1763). LII. An Essay Towards Solving a Problem in the Doctrine of Chances. By the late Rev. Mr. Bayes, FRS communicated by Mr. Price, in a letter to John Canton, AMFR S. *Philosophical transactions of the Royal Society of London*, (53), 370-418.

[Provost '13] Provost, F., & Fawcett, T. (2013). Data Science for Business: What you need to know about data mining and data-analytic thinking. " O'Reilly Media, Inc.".

[Russell '21] Russell, S., Norvig, P. (2021). Artificial intelligence: A Modern Approach (4th Edition). Prentice Hall.

[Chapmann '17] Chapmann, J. (2017) Machine Learning: Fundamental Algorithms for Supervised and Unsupervised Learning With Real-World Applications.

[Knuth '97] Knuth, D. E. (1997). The Art of Computer Programming (Vol. 1). Pearson Education.

[Bhargava '16] Bhargava, A. (2016). Grokking Algorithms: An illustrated guide for programmers and other curious people. Manning Publications Co.

[Han '11] Han, J., Pei, J., & Kamber, M. (2011). *Data Mining: Concepts and Techniques*. 3rd Edition. Elsevier.

[Allen '83] Allen, J. F. (1983). Maintaining knowledge about temporal intervals. *Communications of the ACM, 26*(11), 832-843.

[Schneider '19] Schneider, S., Baevski, A., Collobert, R., & Auli, M. (2019). wav2vec: Unsupervised pre-training for speech recognition. arXiv preprint arXiv:1904.05862.

——————— Chapter 2 ———————

[Wolpert '97] Wolpert, D.H., Macready, W.G. (1997), "No Free Lunch Theorems for Optimization", *IEEE Transactions on Evolutionary Computation* 1, 67

[Topol '19] Topol, E. (2019). Deep medicine: how artificial intelligence can make healthcare human again. Hachette UK.

[Zheng '18] Zheng, A., & Casari, A. (2018). Feature Engineering for Machine Learning: Principles and Techniques for Data Scientists. " O'Reilly Media, Inc.".

[Deisenroth '19] Deisenroth, M.P, Faisal, A. and Ong C.S. (2019) Mathematics for Machine Learning. Cambridge University Press.

[Saul '00] Saul, L. K., & Roweis, S. T. (2000). An introduction to locally linear embedding. unpublished. Available at: http://www.cs.toronto.edu/~ roweis/lle/publications. html.

[Van der Maaten '08] Van der Maaten, L.J.P.; Hinton, G.E. (2008). Visualizing Data Using t-SNE. *Journal of Machine Learning Research*. 9: 2579–2605.

[McInnes '18] McInnes, L., Healy, J., & Melville, J. (2018). Umap: Uniform manifold approximation and projection for dimension reduction. *arXiv preprint arXiv:1802.03426.*

[Barnes '86] Barnes, J. & Hut, P. (1986). A hierarchical O(N log N) force-calculation algorithm. *Nature.* 324 (4): 446–449

[Harrison '19] Harrison M. (2019) Machine Learning Pocket Reference: Working with Structured Data in Python. *O'reilly Media.*

———— Chapter 3 ————

[Griffiths '09] Griffin Dawn (2009) *Head First Statistics.* Publisher: O'Reilly.

[Javaheri '21] Javaheri, T., Homayounfar, M., Amoozgar, Z., Reiazi, R., Homayounieh, F., Abbas, E., ... & Rawassizadeh, R. (2021). CovidCTNet: an open-source deep learning approach to diagnose covid-19 using small cohort of CT images. *NPJ digital medicine, 4*(1), 1-10.

[Rawassizadeh '19] Rawassizadeh, R., Dobbins, C., Akbari, M. & Pazzani, M. (2019) Indexing Multivariate Data through Spatio -Temporal Event Detection and Clustering. *Sensors,* 19 (3), 448.

[Rawassizadeh '13] Rawassizadeh, R., Tomitsch, M., Wac, K., & Tjoa, A. M. (2013). UbiqLog: a generic mobile phone-based life-log framework. *Personal and ubiquitous computing, 17*(4), 621-637.

[Tufte '01] Tufte, E. R. (2001). *The Visual Display of Quantitative Information* (Vol. 2). Cheshire, CT: Graphics press.

[Abela '13] Abela, A. (2013) *Advanced Presentations by Design: Creating Communication that Drives Action.* Pfeiffer, 2nd edition.

[Knaflic '15] Knaflic, C. N. (2015). *Storytelling with data: A data visualization guide for business professionals.* John Wiley & Sons.

[Wilke '19] Wilke, C. O. (2019). *Fundamentals of Data Visualization: A Primer on Making Informative and Compelling Figures.* O'Reilly Media.

[Rawassizadeh '15] Rawassizadeh, R., Momeni, E., Dobbins, C., Mirza-Babaei, P., & Rahnamoun, R. (2015). Lesson learned from collecting quantified self information via mobile and wearable devices. Journal of Sensor and Actuator Networks, 4(4), 315-335.

[Rumsey '15] Rumsey, D. J. (2015). U Can: statistics for dummies. John Wiley & Sons.

[Rumsey '09] Rumsey, D. J. (2009). Statistics II for Dummies. Wiley Publishing, Inc.

[Boslaugh '12] Boslaugh, S. (2012). Statistics in a nutshell: A desktop quick reference. " O'Reilly Media, Inc.".

[Skiena '17] Skiena, S. S. (2017). The Data Science Design Manual. Springer.

[Neyman '37] Neyman, J. (1937). Outline of a theory of statistical estimation based on the classical theory of probability. *Philosophical Transactions of the Royal Society of London. Series A, Mathematical and Physical Sciences, 236*(767), 333-380.

[Chandola '09] Chandola, V., Banerjee, A., & Kumar, V. (2009). Anomaly detection: A survey. *ACM computing surveys (CSUR), 41*(3), 15.

[Aggarwal '01] Aggarwal, C. C., Hinneburg, A., & Keim, D. A. (2001). On the surprising behavior of distance metrics in high dimensional space. In *International conference on database theory* (pp. 420-434). Springer, Berlin, Heidelberg.

[Goodfellow '16] Goodfellow, I., Bengio, Y., Courville, A., & Bengio, Y. (2016). *Deep learning* (Vol. 1). Cambridge: MIT press.

[Shapiro '65] Shapiro, S. S.; Wilk, M. B. (1965). "An analysis of variance test for normality (complete samples)". *Biometrika.* 52 (3–4): 591–611.

[Bonferroni '36] Bonferroni, C. E., Teoria statistica delle classi e calcolo delle probabilità, *Pubblicazioni del R Istituto Superiore di Scienze Economiche e Commerciali di Firenze* 1936

[Tukey '49] Tukey, J. W. (1949). Comparing individual means in the analysis of variance. Biometrics, 99-114.

[Sullivan '12] Sullivan, G. M., & Feinn, R. (2012). Using Effect Size—or Why the P-value is not Enough. *Journal of graduate medical education, 4*(3), 279-282.

[Cohen '92] Cohen, J. (1992). A power primer. *Psychological bulletin, 112*(1), 155 - 159.

[Cliff '93] Cliff, Norman (1993). "Dominance Statistics: Ordinal Analyses to Answer Ordinal Questions". *Psychological Bulletin*. 114 (3): 494-509.

[Stigler '92] Stigler, S. M. (1989). "Francis Galton's Account of the Invention of Correlation". *Statistical Science*. 4 (2), 73–79.

[Spearman '04] Spearman, C. (1904). The proof and measurement of association between two things. *The American journal of psychology*, *15*(1), 72-101.

[Kendall '38] Kendall, M. G. (1938). A new measure of rank correlation. *Biometrika*, 30(1/2), 81-93.

[Shannon '49] Shannon, C. E. (1949). Communication theory of secrecy systems. *Bell system technical journal*, *28*(4), 656-715.

[Bishop '06] Bishop, C.M. (2006) Pattern Recognition and Machine Learning. Springer.

[Kullback '51] Kullback, S., Leibler, R.A. (1951). On information and sufficiency. *Annals of Mathematical Statistics*. 22 (1): 79–86.

[Bruce '17] Bruce, P., & Bruce, A. (2017). Practical Statistics for Data Scientists: 50 Essential Concepts. O'Reilly Media, Inc.

——————— Chapter 4 ———————

[Dunn '61] Dunn, O.J. (1961). Multiple Comparisons Among Means. *Journal of the American Statistical Association*. 56 (293), 52–64

[Aggrawal '15] Aggarwal, C. C. (2015). *Data Mining: The Textbook*. Springer.

[Rao '00] Rao, K. R., & Yip, P. C. (2000). *The Transform and Data Compression Handbook*. CRC press.

[Pighizzini '01] Pighizzini, G. (2001). How hard is computing the edit distance?. *Information and Computation*, *165*(1), 1-13.

[Ratanamahatana '05] Ratanamahatana, C. A., & Keogh, E. (2005). Three Myths about Dynamic Time Warping Data Mining. In *Proceedings of the 2005 SIAM International Conference on Data Mining* (pp. 506-510). Society for Industrial and Applied Mathematics.

[Xu '05] Xu, R., & Wunsch, D. (2005). Survey of Clustering Algorithms. *IEEE Transactions on neural networks*, *16*(3), 645-678.

[Ester '96] Ester, M., Kriegel, H. P., Sander, J., & Xu, X. (1996). A density-based algorithm for discovering clusters in large spatial databases with noise. In *KDD* (Vol. 96, No. 34, pp. 226-231).

[Ankerst '99] Ankerst, M., Breunig, M. M., Kriegel, H. P., & Sander, J. (1999). OPTICS: ordering points to identify the clustering structure. In *ACM Sigmod record* (Vol. 28, No. 2, pp. 49-60). ACM.

[Sibson '73] Sibson, R. (1973). SLINK: An Optimally Efficient Algorithm for the Single-Link Cluster Method. *The Computer Journal*. British Computer Society. 16 (1): 30–34.

[Defays '77] Defays, D. (1977). An Efficient Algorithm for a Complete Link Method. *The Computer Journal*. British Computer Society. 20 (4): 364–366.

[Kaufman '09] Kaufman, L., & Rousseeuw, P. J. (2009). Finding groups in data: an introduction to cluster analysis (Vol. 344). John Wiley & Sons.

[Larose '15] Larose, D. T. (2015). Data Mining and Predictive Analytics. John Wiley & Sons.

[Manning '10] Manning, C., Raghavan, P., & Schütze, H. (2010). Introduction to information retrieval. *Natural Language Engineering*, *16*(1), 100-103.

[Kurzweil '06] Kurzweil, R. (2006). The Singularity is Near: When Humans Transcend Biology. Gerald Duckworth & Co.

[Deza '09] Deza, M. M., & Deza, E. (2009). Encyclopedia of distances. In *Encyclopedia of distances* (pp. 1-583). Springer, Berlin, Heidelberg.

[Zhang '96] Zhang, T., Ramakrishnan, R., & Livny, M. (1996). BIRCH: an efficient data clustering method for very large databases. In *ACM Sigmod Record* (Vol. 25, No. 2, pp. 103-114).

[Guha '98] Guha, S., Rastogi, R., & Shim, K. (1998). CURE: An Efficient Clustering Algorithm for Large Databases. In ACM Sigmod Record (Vol. 27, No. 2, pp. 73-84).

[Edgeworth '1908] Edgeworth, F. Y. (1908). "On the / errors of frequency-constants". *Journal of the Royal Statistical Society*. **71** (3): 499–512. doi:10.2307/2339293.

[Duda '73] Duda, R. O., & Hart, P. E. (1973). Pattern classification and scene analysis. *A Wiley-Interscience Publication, New York: Wiley, 1973.*

[Dempster '77] Dempster, A.P.; Laird, N.M.; Rubin, D.B. (1977). "Maximum Likelihood from Incomplete Data via the EM Algorithm". *Journal of the Royal Statistical Society, Series B.* 39 (1): 1–38.

[Yule '36] Yule, G. U., Filon, L. N. G. (1936). "Karl Pearson. 1857-1936". *Obituary Notices of Fellows of the Royal Society.* 2 (5): 72

[Zadeh '96] Zadeh, L. A. (1996). Fuzzy sets. In Fuzzy sets, fuzzy logic, and fuzzy systems: selected papers by Lotfi A Zadeh (pp. 394-432).

[Rawassizadeh '19] Rawassizadeh, R., Dobbins, C., Akbari, M., & Pazzani, M. (2019). Indexing multivariate mobile data through spatio-temporal event detection and clustering. *Sensors, 19*(3), 448.

[McInnes '17] McInnes, L., Healy, J., & Astels, S. (2017). hdbscan: Hierarchical density based clustering. J. Open Source Softw., 2(11), 205.

[Leskovec '14] Leskovec, J., Rajaraman, A., & Ullman, J. D. (2014). Mining of Massive Datasets. *Cambridge University Press.*

[Rousseeuw '87] Rousseeuw, P. J. (1987). Silhouettes: a graphical aid to the interpretation and validation of cluster analysis. *Journal of computational and applied mathematics, 20,* 53-65.

[Dunn '74] Dunn, J. C. (1974). Well-separated clusters and optimal fuzzy partitions. *Journal of cybernetics, 4*(1), 95-104.

[Davies '79] Davies, D. L., & Bouldin, D. W. (1979). A cluster separation measure. *IEEE transactions on pattern analysis and machine intelligence,* (2), 224-227.

[Caliński '74] Caliński, T., & Harabasz, J. (1974). A dendrite method for cluster analysis. *Communications in Statistics-theory and Methods, 3*(1), 1-27.

——————— Chapter 5 ———————

[Zaki '14] Zaki, M. J., Meira Jr, W., & Meira J., W. (2014). Data Mining and Analysis: Fundamental Concepts and Algorithms. Cambridge University Press.

[Sedgewick '13] Sedgewick, R., & Flajolet, P. (2013). An introduction to the analysis of algorithms. Pearson Education India.

[McDowell '16] McDowell, G. L. (2016). Cracking the Coding Interview: 189 Programming Questions and Solutions. CareerCup, LLC.

[Broder '97] Broder, A. Z. (1997). On the resemblance and containment of documents. In Proceedings. Compression and Complexity of SEQUENCES 1997 (Cat. No. 97TB100171) (pp. 21-29).

[Comer '79] Comer, D. (1979). Ubiquitous B-tree. *ACM Computing Surveys (CSUR), 11*(2), 121-137.

[Bayer '70] Bayer, R., & McCreight, E. (1970). Organization and maintenance of large ordered indices. In Proceedings of the 1970 ACM SIGFIDET (Now SIGMOD) Workshop on Data Description, Access and Control (pp. 107-141).

[Coulom '07] Coulom, R. (2007). Efficient selectivity and backup operators in Monte-Carlo tree search. In Computers and Games: 5th International Conference, CG 2006, Turin, Italy, May 29-31, 2006. Revised Papers 5 (pp. 72-83). Springer Berlin Heidelberg.

[Silver '18] Silver, D., Hubert, T., Schrittwieser, J., Antonoglou, I., Lai, M., Guez, A., ... & Hassabis, D. (2018). A general reinforcement learning algorithm that masters chess, shogi, and Go through self-play. Science, 362(6419), 1140-1144.

[Bloom '70] Bloom, B. H. (1970). Space/time trade-offs in hash coding with allowable errors. Communications of the ACM, 13(7), 422-426.

[Pugh '90] Pugh, W. (1990). Skip lists: a probabilistic alternative to balanced trees. *Communications of the ACM,* 33(6), 668-676.

[Agrawal '94] Agrawal, R., & Srikant, R. (1994). Fast Algorithms for Mining Association Rules. In *Proceedings of 20th International Conference of Very Large Data Bases, VLDB* (Vol. 1215, pp. 487-499).

[Park '95] Park, J. S., Chen, M. S., & Yu, P. S. (1995). An effective hash-based algorithm for mining association rules. Acm sigmod record, 24(2), 175-186.

[Han '00] Han, J., Pei, J., & Yin, Y. (2000). Mining Frequent Patterns Without Candidate Generation. In *ACM Sigmod Record* (Vol. 29, No. 2, pp. 1-12).

[Zaki '97] Zaki, M. J., Parthasarathy, S., Ogihara, M., & Li, W. (1997). New Algorithms for Fast Discovery of Association Rules. In *Knowledge Discovery and Data Mining* (Vol. 97, pp. 283-286).

[Srikant '96] Srikant, R., & Agrawal, R. (1996). Mining sequential patterns: Generalizations and performance improvements. In *International Conference on Extending Database Technology* (pp. 1-17).

[Zaki '01] Zaki, M. J. (2001). SPADE: An Efficient Algorithm for Mining Frequent Sequences. Machine Learning, 42(1-2), 31-60.

[Han '00] Han, J., Pei, J., Mortazavi-Asl, B., Chen, Q., Dayal, U., & Hsu, M. C. (2000). FreeSpan: Frequent Pattern-Projected Sequential Pattern Mining. In *Proceedings of the sixth ACM SIGKDD international conference on Knowledge discovery and data mining* (pp. 355-359).

[Han '01] Han, J., Pei, J., Mortazavi-Asl, B., Pinto, H., Chen, Q., Dayal, U., & Hsu, M. (2001, April). Prefixspan: Mining sequential patterns efficiently by prefix-projected pattern growth. *In Proceedings of the 17th International Conference on Data Engineering* (pp. 215-224).

[Han '05] Han, J., Pei, J., & Yan, X. (2005). Sequential pattern mining by pattern-growth: Principles and extensions. In *Foundations and Advances in Data Mining* (pp. 183-220). Springer, Berlin, Heidelberg.

[Aggarwal '14] Aggarwal, C. C., & Han, J. (Eds.). (2014). *Frequent Pattern Mining*. Springer.

[Pei '04] Pei, J., Han, J., Mortazavi-Asl, B., Pinto, H., Chen, Q., Dayal, U., & Hsu, M. C. (2004). Mining sequential patterns by pattern-growth: The prefixspan approach. *IEEE Transactions on Knowledge & Data Engineering*, (11), 1424-1440.

[Gonzalez '17] Gonzalez, R. C., Woods, R. E., & Eddins, S. L. (2017). *Digital Image Processing, 4th Edition*. Pearson-Prentice-Hall.

[Jurafsky '18] Jurafsky, D. and H. Martin, J. (2018) Speech and Language Processing (3rd edition draft) online: https://web.stanford.edu/~jurafsky/slp3/

[Baum '70] Baum, L. E., Petrie, T., Soules, G., & Weiss, N. (1970). A maximization technique occurring in the statistical analysis of probabilistic functions of Markov chains. *The annals of mathematical statistics*, 41(1), 164-171.

[Baum '67] Baum, L. E., & Eagon, J. A. (1967). An inequality with applications to statistical estimation for probabilistic functions of Markov processes and to a model for ecology. *Bulletin of the American Mathematical Society*. 360–363.

———— Chapter 6 ————

[Ozdemir '18] Ozdemir, S., & Susarla, D. (2018). Feature Engineering Made Easy: Identify unique features from your dataset in order to build powerful machine learning systems. Packt Publishing Ltd.

[Chandrashekar '14] Chandrashekar, G., & Sahin, F. (2014). A Survey on Feature Selection Methods. *Computers & Electrical Engineering*, 40(1), 16-28.

[Quinlan '86] Quinlan, J. R. (1986). Induction of decision trees. *Machine learning*, 1(1), 81-106.

[Tibshirani '96] Tibshirani, R. (1996). Regression shrinkage and selection via the lasso. *Journal of the Royal Statistical Society: Series B (Methodological)*, 58(1), 267-288.

[Rawassizadeh '15] Rawassizadeh, R., Price, B. A., & Petre, M. (2015). Wearables: Has the Age of Smartwatches Finally Arrived?. *Communications of the ACM*, 58(1), 45-47.

[Menon '20] Menon, S., Damian, A., Hu, S., Ravi, N., & Rudin, C. (2020). Pulse: Self-supervised photo upsampling via latent space exploration of generative models. In Proceedings of the ieee/cvf conference on computer vision and pattern recognition (pp. 2437-2445).

[Moody '88] Moody, J. (1988). Fast learning in multi-resolution hierarchies. Advances in neural information processing systems.

[Box '64] Box, G. E. P. and Cox, D. R. (1964). An Analysis of Transformations, *Journal of the Royal Statistical Society*, Series B, 26, 211-252.

[Ng '18] Ng, A. (2018). Machine Learning Yearning. *URL: http://www.mlyearning.org/(96)*.

[Mikolov '13] Mikolov, T., Chen, K., Corrado, G., & Dean, J. (2013). Efficient Estimation of Word Representations in Vector Space. *arXiv preprint arXiv:1301.3781*.

[Pennington '14] Pennington, Jeffrey, Socher, Richard, and Manning, Christopher D. GloVe: Global vectors for word rep- resentation. In Proceedings of the 2014 conference on empirical methods in natural language processing, pp. 1532–1543, 2014.

[Bojanowski '17] Bojanowski, P., Grave, E., Joulin, A., & Mikolov, T. (2017). Enriching word vectors with subword information. Transactions of the Association for Computational Linguistics, 5, 135-146.

[Rose '10] Rose, S., Engel, D., Cramer, N., & Cowley, W. (2010). Automatic keyword extraction from individual documents. Text mining: applications and theory, 1-20.

[Mihalcea '04] Mihalcea, R., & Tarau, P. (2004). Textrank: Bringing order into text. In Proceedings of the 2004 conference on empirical methods in natural language processing (pp. 404-411).

[Campos '18] Campos, R., Mangaravite, V., Pasquali, A., Jorge, A. M., Nunes, C., & Jatowt, A. (2018). Yake! collection-independent automatic keyword extractor. In Advances in Information Retrieval: 40th European Conference on IR Research, ECIR 2018, Grenoble, France, March 26-29, 2018, Proceedings 40 (pp. 806-810). Springer International Publishing.

[Ji '12] Ji, S., Xu, W., Yang, M., & Yu, K. (2012). 3D convolutional neural networks for human action recognition. IEEE transactions on pattern analysis and machine intelligence, 35(1), 221-231.

[Karpathy '14] Karpathy, A., Toderici, G., Shetty, S., Leung, T., Sukthankar, R., & Fei-Fei, L. (2014). Large-scale video classification with convolutional neural networks. In Proceedings of the IEEE conference on Computer Vision and Pattern Recognition (pp. 1725-1732).

[Harris '88] Harris, C. G., & Stephens, M. (1988). A Combined Corner and Edge Detector. In *Alvey vision conference*. Vol. 15, No. 50, (pp. 147-151).

[Matas '04] Matas, J., Chum, O., Urban, M., & Pajdla, T. (2004). Robust Wide-baseline Stereo from Maximally Stable Extremal Regions. *Image and Vision Computing*, *22*(10), 761-767.

[Dalal '05] Dalal, N., & Triggs, B. (2005). Histograms of Oriented Gradients for Human Detection. In *International Conference on Computer Vision & Pattern Recognition (CVPR'05)* (Vol. 1, pp. 886-893).

[Lowe '99] Lowe, D. G. (1999). Object Recognition from Local Scale-Invariant Features. In *Proceedings of the International Conference on Computer Vision*, (Vol. 99, No. 2, pp. 1150-1157).

[Kornilov '18] Kornilov, A. S., & Safonov, I. V. (2018). An overview of watershed algorithm implementations in open source libraries. *Journal of Imaging*, *4*(10), 123.

[Morlet '82] Morlet, J., Arens, G., Fourgeau, E., & Glard, D. (1982). Wave propagation and sampling theory—Part I: Complex signal and scattering in multilayered media. Geophysics, 47(2), 203-221.

[Daubechies '92] Daubechies, I. (1992). Ten Lectures on Wavelets. *Society for Industrial and Applied Mathematics*.

[Heil '89] Heil, Christopher E., and David F. Walnut. Continuous and discrete wavelet transforms. *SIAM review* 31.4 (1989): 628-666.

[Haar '10] Haar, Alfréd (1910), Zur Theorie der orthogonalen Funktionensysteme, *Mathematische Annalen*, 69(3): 331–371.

[Keogh '01] Keogh, E., Chakrabarti, K., Pazzani, M., & Mehrotra, S. (2001). Dimensionality reduction for fast similarity search in large time series databases. *Knowledge and Information Systems*, 3(3), 263-286.

[Lin '03] Jessica, L., Keogh, E., Lonardi, S., and Chiu, B. (2003). A Symbolic Representation of Time Series, with Implications for Streaming Algorithms. In *Proceedings of the 8th ACM SIGMOD workshop on Research issues in data mining and knowledge discovery*, pp. 2-11

[Wang '18] Wang, X., & Gupta, A. (2018). Videos as space-time region graphs. In Proceedings of the European conference on computer vision (ECCV) (pp. 399-417).

[Wang '16] Wang, X., Gao, L., Song, J., & Shen, H. (2016). Beyond frame-level CNN: saliency-aware 3-D CNN with LSTM for video action recognition. IEEE signal processing letters, 24(4), 510-514.

[Page '54] Page, E. S. (1954). Continuous Inspection Scheme. *Biometrika*. 41 (1/2): 100–115.

[Aminikhanghahi '17] Aminikhanghahi, S., & Cook, D. J. (2017). A Survey of Methods for Time Series Change Point Detection. *Knowledge and information systems*, *51*(2), 339-367.

[Krishnamoorthy '13] Krishnamoorthy, A., & Menon, D. (2013). Matrix Inversion Using Cholesky Decomposition. In *2013 Signal Processing: Algorithms, Architectures, Arrangements, and Applications (SPA)* (pp. 70-72).

——————— Chapter 7 ———————

[Bellman '61] Bellman, R. (1961). Curse of dimensionality. *Adaptive control processes: a guided tour.* Princeton University Press, 3, 2.

[Deerwester '90] Deerwester, S., Dumais, S. T., Furnas, G. W., Landauer, T. K., & Harshman, R. (1990). Indexing by latent semantic analysis. *Journal of the American society for information science, 41*(6), 391-407.

[Fisher '36] Fisher, R. (1936). Linear discriminant analysis. Ann. Eugenics, 7, 179.

[Blei '03] Blei, D. M., Ng, A. Y., & Jordan, M. I. (2003). Latent Dirichlet Allocation. *Journal of machine Learning research, 3*(Jan), 993-1022

[Cai '08] Cai, D., He, X., & Han, J. (2008). Training Linear Discriminant Analysis in Linear Time. In *2008 IEEE 24th International Conference on Data Engineering* (pp. 209-217).

[Roweis '00] Roweis, S. T., & Saul, L. K. (2000). Nonlinear dimensionality reduction by locally linear embedding. *science, 290* (5500), 2323-2326.

[Smith '97] Smith, S. W. (1997) *The Scientists & Engineer's Guide to Digital Signal Processing*, California Technical Pub. San Diego..

[Cholesky '10] Cholesky, A.L. (1910) On Numerical Solutions of Systems of Linear Equations. Available: http://bibnum.education.fr/mathematiques/algebre/sur-la-resolution-numerique-des-systemes-d-equations-lineaires

[Sra '06] Sra, S., & Dhillon, I. S. (2006). Generalized Nonnegative Matrix Approximations with Bregman divergences. In *Advances in Neural Information Processing Systems* (pp. 283-290).

[Stewart '93] Stewart, G. W. (1993). On the Early History of the Singular Value Decomposition. *SIAM review, 35*(4), 551-566.

[Golub '71] Golub, G. H., & Reinsch, C. (1971). Singular Value Decomposition and Least Squares Solutions. In *Linear Algebra*(pp. 134-151). Springer, Berlin, Heidelberg.

[Rehman '16] Ur Rehman, M. H., Liew, C. S., Abbas, A., Jayaraman, P. P., Wah, T. Y., & Khan, S. U. (2016). Big data reduction methods: a survey. *Data Science and Engineering, 1*(4), 265-284.

[Kolda '09] Kolda, Tamara G., and Brett W. Bader. "Tensor decompositions and applications." *SIAM review* 51.3 (2009): 455-500.

[Hitchcock '27] Hitchcock, L. (1927) The expression of a tensor or a polyadic as a sum of products, Journal of Math and Physics, 6, pp. 164–189.

[Harshman '70] Harshman, R. A. (1970) Foundations of the PARAFAC procedure: Models and conditions for an" explanatory" multimodal factor analysis. *UCLA Working Papers in Phonetics.* pp. 1-84.

[Carroll '70] Carroll, D. and Chang, J. J. (1970) Analysis of Individual Differences in Multidimensional Scaling via an N-way Generalization of "Eckart-Young" Decomposition, *Psychometrika*, 35, pp. 283–319.

[Håstad '90] Håstad, J. (1990). Tensor rank is NP-complete. *Journal of Algorithms, 11*(4), 644-654.

[Tucker '66] Tucker, L. R. (1966). Some mathematical notes on three-mode factor analysis. *Psychometrika, 31*(3), 279-311.

[Kroonenberg '80] Kroonenberg, P. M., & De Leeuw, J. (1980). Principal Component Analysis of Three-mode Data by Means of Alternating Least Squares Algorithms. *Psychometrika, 45*(1), 69-97.

[Oseledets '11] Oseledets, I. V. (2011). Tensor-Train Decomposition. *SIAM Journal on Scientific Computing, 33*(5), 2295-2317.

——————— Chapter 8 ———————

[James '13] James, G., Witten, D., Hastie, T., & Tibshirani, R. (2013). *An Introduction to Statistical Learning: with Applications in R.* New York: springer.

[Box '70] Box, G. E., Jenkins, G. M., Reinsel, G. C., & Ljung, G. M. (2015). Time series analysis: forecasting and control. John Wiley & Sons.

[Dickey '79] Dickey, D. A., & Fuller, W. A. (1979). Distribution of the estimators for autoregressive time series with a unit root. Journal of the American statistical association, 74(366a), 427-431.

[Domencich '75] Domencich, T. A., & McFadden, D. (1975). Urban travel demand-a behavioral analysis.

[Wald '45] Wald, A. (1945). Sequential tests of statistical hypotheses. *The annals of mathematical statistics, 16*(2), 117-186.

[Akaike '74] Akaike, H. (1974). A new look at statistical model identification. *IEEE Transactions on Automatic Control* 19, 716–723.

[Schwarz '78] Schwarz, G. (1978). Estimating the dimension of a model. *The annals of statistics*, *6*(2), 461-464.

[Grus '19] Grus, J. (2019) Data Science from Scratch: First Principles with Python. *O'Reilly Media*.

[Domingos '12] Domingos, P. M. (2012). A few useful things to know about machine learning. *Communication of ACM 55*(10), 78-87.

[Thikhonov '43] Thikhonov, A. H. (1943). Об устойчивости обратных задач. In *Докл. Proceedings of the USSR Academy of Sciences* (Vol. 39, No. 5, pp. 195-198).

[Santosa '86] Santosa, F., Symes, W.W. (1986). Linear Inversion of Band-Limited Reflection Seismograms". *SIAM Journal on Scientific and Statistical Computing*. SIAM. 7 (4): 1307–1330.

[Tibshirani '96] Tibshirani, R. (1996). Regression Shrinkage and Selection via the LASSO. *Journal of the Royal Statistical Society*. Series B (methodological). Wiley. 58 (1): 267–88.

[Efron '04] Efron, B., Hastie, T., Johnstone, I., & Tibshirani, R. (2004). Least angle regression. *The Annals of statistics*, *32*(2), 407-499.

[Zou '05] Zou, H., & Hastie, T. (2005). Regularization and variable selection via the elastic net. *Journal of the royal statistical society: series B (statistical methodology)*, *67*(2), 301-320.

[Breiman '95] Breiman, L. (1995). Better Subset Regression using the Nonnegative Garrote. *Technometrics*, *37*(4), 373-384.

[Xiong '10] Xiong, S. (2010). Some Notes on the Nonnegative Garrote. *Technometrics*,*52*(3), 349-361.

[Robbins '51] Robbins, H., & Monro, S. (1951). A stochastic approximation method. *The annals of mathematical statistics*, 400-407.

[Burkov '19] Burkov, A. (2019). The Hundred-Page Machine Learning Book. Andriy Burkov.

[Kochenderfer '19] Kochenderfer, M. J., & Wheeler, T. A. (2019). *Algorithms for optimization*. MIT Press.

[Kingma '14] Kingma, D. P., & Ba, J. (2014). Adam: A method for stochastic optimization. *arXiv preprint arXiv:1412.6980*.

[Brownlee '16] Brownlee, J. (2016). Master Machine Learning Algorithms: discover how they work and implement them from scratch. Machine Learning Mastery.

———— Chapter 9 ————

[Bayes '63] Bayes, T. (1763). An essay towards solving a problem in the doctrine of chances. *Philosophical Transactions of the Royal Society of London, 53*, 370–418.

[Cover '67] Cover, T., & Hart, P. (1967). Nearest neighbor pattern classification. *IEEE Transactions on Information Theory, 13*(1), 21-27.

[Voronoi '08] Voronoi, G. (1908). Nouvelles applications des paramètres continus à la théorie des formes quadratiques. Deuxième mémoire. Recherches sur les paralléloèdres primitifs. *Journal für die Reine und Angewandte Mathematik*, 134, 198-287.

[Bentley '75] Bentley, J. L. (1975). Multidimensional binary search trees used for associative searching. *Communications of the ACM*, 18(9), 509-517.

[Indyk '98] Indyk, P., & Motwani, R. (1998). *Approximate nearest neighbors: Towards removing the curse of dimensionality*. Proceedings of the Thirtieth Annual ACM Symposium on Theory of Computing (STOC '98), 604–613.

[Cortes '95] Cortes, C., & Vapnik, V. (1995). Support-Vector Networks. *Machine learning, 20*(3), 273-297.

[Bordes '05] Bordes, A., Ertekin, S., Weston, J., & Bottou, L. (2005). Fast kernel classifiers with online and active learning. *Journal of Machine Learning Research*, 6(Sep), 1579-1619.

[Scholkopf '01] Scholkopf, B., & Smola, A. J. (2001). Learning with kernels: support vector machines, regularization, optimization, and beyond. MIT press.

[Kass '80] Kass, Gordon V.; An Exploratory Technique for Investigating Large Quantities of Categorical Data, Applied Statistics, Vol. 29, No. 2 (1980), pp. 119–127

[Quinlan '94] Quinlan, J. R. (1994). C4. 5: Programs for Machine Learning. *Elsevier*

[Breimanetal '84] Breimanetal, L. (1984). Classification and Regression Trees. Wadsworth, Belmont.

[Breiman '96] Breiman, L. (1996). Bagging predictors. *Machine Learning*, 24(2), 123-140.

[Schapire '90] Schapire, R. E. (1990). The strength of weak learnability. *Machine Learning*, 5(2), 197-227.

[Wolpert '92] Wolpert, D. H. (1992). Stacked generalization. *Neural networks*, 5(2), 241-259.

[Breiman '05] Breiman, L. (2001). Random Forests. *Machine learning*, 45(1), 5-32.

[Freund '95] Freund, Y., & Schapire, R. E. (1995). A Desicion-Theoretic Generalization of On-line Learning and an Application to Boosting. In *European conference on computational learning theory* (pp. 23-37). Springer, Berlin, Heidelberg.

[Friedman '01] Friedman, J. H. (2001). Greedy Function Approximation: A Gradient Boosting Machine. *Annals of statistics*, 1189-1232.

[Si '17] Si, S., Zhang, H., Keerthi, S. S., Mahajan, D., Dhillon, I. S., & Hsieh, C. J. (2017). Gradient boosted decision trees for high dimensional sparse output. In International conference on machine learning (pp. 3182-3190). PMLR.

[Mason '99] Mason, L., Baxter, J., Bartlett, P., & Frean, M. (1999, May). Boosting algorithms as gradient descent in function space. NIPS.

[Chen '15] Chen, T., He, T., Benesty, M., Khotilovich, V., Tang, Y., & Cho, H. (2015). XGBoost: extreme gradient boosting. *R package version 0.4-2, 1*(4).

[Ke '17] Ke, G., Meng, Q., Finley, T., Wang, T., Chen, W., Ma, W., ... & Liu, T. Y. (2017). LightGBM: A highly efficient gradient boosting decision tree. *Advances in neural information processing systems*, 30, 3146-3154.

[Prokhorenkova '18] Prokhorenkova, L., Gusev, G., Vorobev, A., Dorogush, A. V., & Gulin, A. (2018). CatBoost: unbiased boosting with categorical features. *Advances in Neural Information Processing Systems*, 31.

[Boehmke '20] Boehmke, B., & Greenwell, B. M. (2019). Hands-on machine learning with R. CRC Press.

[Keshavarz '21] Keshavarz, H., Abadeh, M. S., & Rawassizadeh, R. (2021). SEFR: A fast linear-time classifier for ultra-low power devices. *arXiv preprint arXiv:2006.04620*.

——————— Chapter 10———————

[McCulloch '43] McCulloch, W. S., & Pitts, W. (1943). A logical calculus of the ideas immanent in nervous activity. *The bulletin of mathematical biophysics*, 5(4), 115-133.

[Rosenblatt '58] Rosenblatt, F. (1958). The perceptron: a probabilistic model for information storage and organization in the brain. *Psychological review*, 65(6), 386.

[Minsky '69] Minsky, M., & Papert, S. A. (1969). Perceptrons: An introduction to computational geometry. MIT press.

[Ciresan '11] Ciresan, D. C. , Meier, U., Masci, J.. Gambardella, L. M.. Schmidhuber, J. Flexible, High Performance Convolutional Neural Networks for Image Classification. (2011) *International Joint Conference on Artificial Intelligence (IJCAI-2011)*

[Krizhevsky '12] Krizhevsky, A., Sutskever, I., & Hinton, G. E. (2012). Imagenet classification with deep convolutional neural networks. In *Advances in neural information processing systems* (pp. 1097-1105).

[Aggarwal '18] Aggarwal, C. C. (2018). Neural networks and deep learning. *Springer*.

[Hofstadter '79] Hofstadter, D. R. (1979). Gödel, Escher, Bach: an eternal golden braid (Vol. 13). New York: Basic books.

[Rawassizadeh '19] Rawassizadeh, R., Sen, T., Kim, S. J., Meurisch, C., Keshavarz, H., Mühlhäuser, M., & Pazzani, M. (2019). Manifestation of virtual assistants and robots into daily life: Vision and challenges. *CCF Transactions on Pervasive Computing and Interaction*, 1-12.

[Rasheed '16] Rasheed, T. (2016) Make Your Own Neural Network. *CreateSpace Independent Publishing Platform.*

[Chollet '18] Chollet, F. (2017). Deep Learning with Python, *Manning Publications CO.*

[Nair '10] Nair, V., & Hinton, G. E. (2010). Rectified Linear Units Improve Restricted Boltzmann Machines. *In ICML.*

[Maas '13] Maas, A. L., Hannun, A. Y., & Ng, A. Y. (2013). Rectifier Nonlinearities Improve Neural Network Acoustic Models. In *Proc. icml* (Vol. 30, No. 1, p. 3).

[Hellinger '09] Hellinger, E. (1909), Neue Begründung der Theorie quadratischer Formen von unendlichvielen Veränderlichen, *Journal für die reine und angewandte Mathematik* (in German), 136: 210–271

[Bharath '99] Bharath, B., & Borkar, V. S. (1999). Stochastic Approximation Algorithms: Overview and Recent Trends. *Sadhana, 24*(4-5), 425-452.

[Polyak '64] Polyak, B. T. (1964). Some Methods of Speeding up the Convergence of Iteration Methods. *USSR Computational Mathematics and Mathematical Physics, 4*(5), 1-17.

[Nesterov '83] Nesterov, Y. A Method of Solving a Convex Programming Problem with Convergence Rate O(1/ sqr(k)). *Soviet Mathematics Doklady,* 27:372–376, 1983.

[Sutskever '13] Sutskever, I., Martens, J., Dahl, G., & Hinton, G. (2013). On the Importance of Initialization and Momentum in Deep Learning. In *International Conference on Machine Learning* (pp. 1139-1147).

[Duchi '11] Duchi, J., Hazan, E., & Singer, Y. (2011). Adaptive Subgradient Methods for Online Learning and Stochastic Optimization. *Journal of machine learning research, 12*(7).

[Zeiler '12] Zeiler, M. D. (2012). Adadelta: an Adaptive Learning Rate Method. *arXiv preprint arXiv:1212.5701.*

[Hinton '12] Hinton, G., Srivastava, N., & Swersky, K. (2012). RMSprop: Divide the Gradient by a Running Average of its Recent Magnitude. *Neural Networks for Machine Learning, Lecture 6e.*

[Riedmiller '93] Riedmiller, M., & Braun, H. (1993). A Direct Adaptive Method for Faster Backpropagation Learning: The RPROP algorithm. In *IEEE International Conference on Neural Networks* (pp. 586-591).

[Schmidt '20] Schmidt, R. M., Schneider, F., & Hennig, P. (2020). Descending through a Crowded Valley-- Benchmarking Deep Learning Optimizers. arXiv preprint arXiv:2007.01547.

[Chen '19] Chen, J., Kyrillidis, A. Decaying Momentum Helps Neural Network Training. arXiv preprint arXiv:1910.04952, 2019.

[Zhang '17] Zhang, J., & Mitliagkas, I. (2017). Yellowfin and the Art of Momentum Tuning. *arXiv preprint arXiv:1706.03471.*

[Schneider '19] Schneider, Frank, Lukas Balles, and Philipp Hennig. (2019) DeepOBS: A Deep Learning Optimizer Benchmark Suite. *International Conference on Learning Representations.*

[Kelley '60] Kelley, H. J. (1960). Gradient theory of optimal flight paths. *Ars Journal, 30*(10), 947-954.

[Bryson '62] Bryson, A. E. (1961, April). A Gradient Method for Optimizing Multi-stage Allocation Processes. In *Proc. Harvard Univ. Symposium on digital computers and their applications* (Vol. 72).

[Rumelhart '86] Rumelhart, D. E., Hinton, G. E., & Williams, R. J. (1986). Learning Representations by Back-Propagating Errors. *nature, 323*(6088), 533-536.

[Glorot '10] Glorot, X., & Bengio, Y. (2010, March). Understanding the Difficulty of Training Deep Feedforward Neural Networks. In *Proceedings of the thirteenth international conference on artificial intelligence and statistics* (pp. 249-256).

[He '15] He, K., Zhang, X., Ren, S., & Sun, J. (2015). Delving Deep into Rectifiers: Surpassing Human-Level Performance on Imagenet Classification. I*n Proceedings of the IEEE International Conference on Computer Vision* (pp. 1026-1034).

[Srivastava '14] Srivastava, N., Hinton, G., Krizhevsky, A., Sutskever, I., & Salakhutdinov, R. (2014). Dropout: A Simple Way to Prevent Neural Networks from Overfitting. *The journal of machine learning research, 15*(1), 1929-1958.

[Gal '16] Gal, Y., & Ghahramani, Z. (2016). Dropout as a Bayesian Approximation: Representing Model Uncertainty in Deep Learning. *In International Conference on Machine Learning* (pp. 1050-1059).

[Konečný '16] Konečný, J., McMahan, H. B., Yu, F. X., Richtárik, P., Suresh, A. T., & Bacon, D. (2016). Federated learning: Strategies for improving communication efficiency. *arXiv preprint arXiv:1610.05492.*

[Warden '19] Warden, P., & Situnayake, D. (2019). TinyML: Machine learning with tensorflow lite on Arduino and ultra-low-power microcontrollers. O'Reilly Media, Inc.

[Huhel '59] Hubel, DH; Wiesel, T.N. (1959). Receptive Fields of single Neurones in the Cat's Striate Cortex. *J. Physiol.* 148 (3): 574–91.

[Fukushima '80] Fukushima, K. (1980). "Neocognitron: A Self-organizing Neural Network Model for a Mechanism of Pattern Recognition Unaffected by Shift in Position" (PDF). *Biological Cybernetics.* **36** (4): 193–202.

[LeCun '89] Y. LeCun, B. Boser, J. S. Denker, D. Henderson, R. E. Howard, W. Hubbard, L. D. Jackel (1989) Backpropagation Applied to Handwritten Zip Code Recognition.

[Deng '09] Deng, J., Dong, W., Socher, R., Li, L. J., Li, K., & Fei-Fei, L. (2009, June). Imagenet: A large-scale hierarchical image database. In *2009 IEEE conference on Computer Vision and Pattern Recognition* (pp. 248-255).

[Brown '20] Brown, T. B., Mann, B., Ryder, N., Subbiah, M., Kaplan, J., Dhariwal, P., ... & Agarwal, S. (2020). Language models are few-shot learners. *arXiv preprint arXiv:2005.14165*.

[Hochreiter '97] Hochreiter, S., & Schmidhuber, J. (1997). Long short-term memory. *Neural computation, 9*(8), 1735-1780.

[Greff '16] Greff, K., Srivastava, R. K., Koutník, J., Steunebrink, B. R., & Schmidhuber, J. (2016). LSTM: A search space odyssey. *IEEE transactions on neural networks and learning systems, 28*(10), 2222-2232.

[Chung '14] Chung, J., Gulcehre, C., Cho, K., & Bengio, Y. (2014). Empirical evaluation of gated recurrent neural networks on sequence modeling. *arXiv preprint arXiv:1412.3555*.

[Karpathy '15] Karpathy, A. (2015). The Unreasonable Effectiveness of Recurrent Neural Networks. *Andrej Karpathy blog, 21*, 23.

[Dumoulin '16] Dumoulin, V., & Visin, F. (2016). A guide to convolution arithmetic for deep learning. *arXiv preprint arXiv:1603.07285*.

[Chen '17] Chen, L. C., Papandreou, G., Schroff, F., & Adam, H. (2017). Rethinking atrous convolution for semantic image segmentation. arXiv preprint arXiv:1706.05587.

[Ramesh '22] Ramesh, A., Dhariwal, P., Nichol, A., Chu, C., & Chen, M. (2022). Hierarchical text-conditional image generation with clip latents. *arXiv preprint arXiv:2204.06125*.

——————— Chapter 11 ———————

[Foster '19] Foster, D. (2019). Generative deep learning: teaching machines to paint, write, compose, and play. O'Reilly Media.

[Kohonen '82] Kohonen, T. (1982). Self-organized formation of topologically correct feature maps. *Biological cybernetics, 43*(1), 59-69.

[Tian '14] Tian, J., Azarian, M. H., & Pecht, M. (2014). Anomaly detection using self-organizing maps-based k-nearest neighbor algorithm. In *Proceedings of the European Conference of the Prognostics and Health Management Society* (pp. 1-9). Citeseer.

[LeCun '06] LeCun, Y., Chopra, S., Hadsell, R., Ranzato, M., & Huang, F. (2006). A tutorial on energy-based learning. *Predicting structured data, 1*(0).

[Ackley '85] Ackley, D. H., Hinton, G. E., & Sejnowski, T. J. (1985). A learning algorithm for Boltzmann machines. *Cognitive science, 9*(1), 147-169.

[Gibbs '02] Gibbs, J. W. (1902). *Elementary principles in statistical mechanics*: developed with especial reference to the rational foundations of thermodynamics. C. Scribner's Sons.

[Hinton '02] Hinton, G. E. (2002). Training products of experts by minimizing contrastive divergence. *Neural computation, 14*(8), 1771-1800.

[Hinton '95] Hinton, G. E., Dayan, P., Frey, B. J., & Neal, R. M. (1995). The" wake-sleep" algorithm for unsupervised neural networks. *Science, 268*(5214), 1158-1161.

[Wang '17] Wang, H., & Raj, B. (2017). On the Origin of Deep Learning. *arXiv preprint arXiv:1702.07800*.

[Hopfield '82] Hopfield, J. J. (1982). Neural Networks and Physical Systems with Emergent Collective Computational Abilities. *Proceedings of the national academy of sciences, 79*(8), 2554-2558.

[Bengio '09] Bengio, Y. (2009). Learning Deep Architectures for AI. Now Publishers Inc.

[Kingma '13] Kingma, D. P., & Welling, M. (2013). Auto-encoding variational bayes. *arXiv preprint arXiv:1312.6114*.

[Makhzani '13] Makhzani, A., & Frey, B. (2013). K-sparse autoencoders. *arXiv preprint arXiv:1312.5663*.

[Ng '11] Ng, A. (2011). Sparse autoencoder. CS294A Lecture notes, 72 (2011), 1-19.

[Vincent '10] Vincent, P., Larochelle, H., Lajoie, I., Bengio, Y., Manzagol, P. A., & Bottou, L. (2010). Stacked denoising autoencoders: Learning useful representations in a deep network with a local denoising criterion. Journal of machine learning research, 11(12).

[Rifai '11] Rifai, S., Vincent, P., Muller, X., Glorot, X., & Bengio, Y. (2011). Contractive auto-encoders: Explicit invariance during feature extraction. In *International Conference of Machine Learning*.

[Bengio '07] Bengio, Y., Lamblin, P., Popovici, D., & Larochelle, H. (2006). Greedy layer-wise training of deep networks. *Advances in neural information processing systems*, *19*.

[Hinton '06] Hinton, G. E., & Salakhutdinov, R. R. (2006). Reducing the dimensionality of data with neural networks. *science*, *313*(5786), 504-507.

[Foster '19] Foster, D. (2019). Generative Deep Learning: Teaching Machines to Paint, Write, Compose, and Play. O'Reilly Media.

[Langr '19] Langr, J., & Bok, V. (2019). GANs in action: deep learning with generative adversarial networks. Manning Publications.

[Ramesh '21] Ramesh, A., Pavlov, M., Goh, G., Gray, S., Voss, C., Radford, A., ... & Sutskever, I. (2021). Zero-shot text-to-image generation. arXiv preprint arXiv:2102.12092.

[Ronneberger '15] Ronneberger, O., Fischer, P., & Brox, T. (2015). U-net: Convolutional networks for biomedical image segmentation. In International Conference on Medical image computing and computer-assisted intervention (pp. 234-241).

[Goodfellow '14] Goodfellow, I. J., Pouget-Abadie, J., Mirza, M., Xu, B., Warde-Farley, D., Ozair, S., Courville, A. & Bengio, Y. (2014). Generative adversarial networks. arXiv preprint arXiv:1406.2661.

[Niemitalo '11] Niemitalo, O. (2010). A method for training artificial neural networks to generate missing data within a variable context. *Blog post, Internet Archive*. https://web.archive.org/web/20120312111546/http://yehar.com:80/blog/?p=167

[Li '13] Li, W., Gauci, M., & Groß, R. (2013). A coevolutionary approach to learn animal behavior through controlled interaction. I*n Proceedings of the 15th annual conference on Genetic and Evolutionary Computation* (pp. 223-230).

[Karras '19] Karras, T., Laine, S., & Aila, T. (2019). A style-based generator architecture for generative adversarial networks. In Proceedings of the IEEE/CVF Conference on Computer Vision and Pattern Recognition (pp. 4401-4410).

[Salimans '16] Salimans, T., Goodfellow, I., Zaremba, W., Cheung, V., Radford, A., & Chen, X. (2016). Improved techniques for training GANs. arXiv preprint arXiv:1606.03498.

[Heusel '17] Heusel, M., Ramsauer, H., Unterthiner, T., Nessler, B., & Hochreiter, S. (2017). Gans trained by a two time-scale update rule converge to a local nash equilibrium. arXiv preprint arXiv:1706.08500.

[Szegedy '16] Szegedy, C., Vanhoucke, V., Ioffe, S., Shlens, J., & Wojna, Z. (2016). Rethinking the inception architecture for computer vision. In Proceedings of the IEEE conference on computer vision and pattern recognition (pp. 2818-2826).

[Mirza '14] Mirza, M., & Osindero, S. (2014). Conditional generative adversarial nets. arXiv preprint arXiv:1411.1784.

[Rüschendorf '85] Rüschendorf, L. (1985). The Wasserstein distance and approximation theorems. Probability Theory and Related Fields, 70(1), 117-129.

[Gulrajani '17] Gulrajani, I., Ahmed, F., Arjovsky, M., Dumoulin, V., & Courville, A. (2017). Improved training of wasserstein gans. arXiv preprint arXiv:1704.00028.

[Isola '17] Isola, P., Zhu, J. Y., Zhou, T., & Efros, A. A. (2017). Image-to-image translation with conditional adversarial networks. In Proceedings of the IEEE Conference on Computer Vision and Pattern Recognition (pp. 1125-1134).

[Zhu '17] Zhu, J. Y., Park, T., Isola, P., & Efros, A. A. (2017). Unpaired image-to-image translation using cycle-consistent adversarial networks. In Proceedings of the IEEE international conference on computer vision (pp. 2223-2232).

[Karras '20] Karras, T., Laine, S., Aittala, M., Hellsten, J., Lehtinen, J., & Aila, T. (2020). Analyzing and improving the image quality of stylegan. In Proceedings of the IEEE/CVF Conference on Computer Vision and Pattern Recognition (pp. 8110-8119).

[Huang '17] Huang, X., & Belongie, S. (2017). Arbitrary style transfer in real-time with adaptive instance normalization. In Proceedings of the IEEE International Conference on Computer Vision (pp. 1501-1510).

[Karras '17] Karras, T., Aila, T., Laine, S., & Lehtinen, J. (2017). Progressive growing of gans for improved quality, stability, and variation. arXiv preprint arXiv:1710.10196.

[Karras '21] Karras, T., Aittala, M., Laine, S., Härkönen, E., Hellsten, J., Lehtinen, J., & Aila, T. (2021). Alias-Free Generative Adversarial Networks. *arXiv preprint arXiv:2106.12423*.

[Brownlee '19] Brownlee, J. (2019). Generative Adversarial Networks with Python: Deep Learning Generative Models for Image Synthesis and Image Translation. Machine Learning Mastery.

[Zakharov '19] Zakharov, E., Shysheya, A., Burkov, E., & Lempitsky, V. (2019). Few-shot adversarial learning of realistic neural talking head models. In *Proceedings of the IEEE/CVF International Conference on Computer Vision* (pp. 9459-9468).

[Bengio '13] Bengio, Y., Courville, A., & Vincent, P. (2013). Representation learning: A review and new perspectives. IEEE Transactions on Pattern Analysis and Machine Intelligence, 35(8), 1798-1828.

[Le-Khac '20] Le-Khac, P. H., Healy, G., & Smeaton, A. F. (2020). Contrastive representation learning: A framework and review. IEEE Access, 8, 193907-193934.

[Bell '10] Bell, G., & Gemmell, J. (2010). Your life, uploaded: The digital way to better memory, health, and productivity. Penguin.

[Chen '20 A] Chen, T., Kornblith, S., Norouzi, M., & Hinton, G. (2020). A simple framework for contrastive learning of visual representations. In International conference on machine learning (pp. 1597-1607). PMLR.

[Chen '20 B] Chen, T., Kornblith, S., Swersky, K., Norouzi, M., & Hinton, G. E. (2020). Big self-supervised models are strong semi-supervised learners. Advances in neural information processing systems, 33, 22243-22255.

[He '20] He, K., Fan, H., Wu, Y., Xie, S., & Girshick, R. (2020). Momentum contrast for unsupervised visual representation learning. In Proceedings of the IEEE/CVF conference on computer vision and pattern recognition (pp. 9729-9738).

[Chen '20 C] Chen, X., Fan, H., Girshick, R., & He, K. (2020). Improved baselines with momentum contrastive learning. arXiv preprint arXiv:2003.04297.

[Chopra '05] Chopra, S., Hadsell, R., & LeCun, Y. (2005). Learning a Similarity Metric Discriminatively, with Application to Face Verification. In *2005 IEEE Computer Society Conference on Computer Vision and Pattern Recognition (CVPR'05)* (Vol. 1, pp. 539-546).

[Schroff '15] Schroff, F., Kalenichenko, D., & Philbin, J. (2015). Facenet: A unified embedding for face recognition and clustering. In *Proceedings of the IEEE conference on computer vision and pattern recognition* (pp. 815-823).

[Radford '21] Radford, A., Kim, J. W., Hallacy, C., Ramesh, A., Goh, G., Agarwal, S., ... & Sutskever, I. (2021). Learning transferable visual models from natural language supervision. In International Conference on Machine Learning (pp. 8748-8763). PMLR.

[Crowson '22] Crowson, K., Biderman, S., Kornis, D., Stander, D., Hallahan, E., Castricato, L., & Raff, E. (2022). Vqgan-clip: Open domain image generation and editing with natural language guidance. *arXiv preprint arXiv:2204.08583*.

[Van Den Oord '17] Van Den Oord, A., & Vinyals, O. (2017). Neural discrete representation learning. *Advances in neural information processing systems, 30*.

[Esser '21] Esser, P., Rombach, R., & Ommer, B. (2021). Taming transformers for high-resolution image synthesis. In Proceedings of the IEEE/CVF Conference on Computer Vision and Pattern Recognition (pp. 12873-12883).

[Ramesh '21] Ramesh, Aditya, Mikhail Pavlov, Gabriel Goh, Scott Gray, Chelsea Voss, Alec Radford, Mark Chen, and Ilya Sutskever. (2021) "Zero-shot text-to-image generation." In International Conference on Machine Learning, pp. 8821-8831. PMLR.

[Maddison '16] Maddison, C. J., Mnih, A., & Teh, Y. W. (2016). The concrete distribution: A continuous relaxation of discrete random variables. arXiv preprint arXiv:1611.00712.

[Jang '16] Jang, E., Gu, S., & Poole, B. (2016). Categorical reparameterization with gumbel-softmax. arXiv preprint arXiv:1611.01144.

[Bishop '24] Bishop, C. M., & Bishop, H. (2024). Deep learning: foundations and concepts. *Springer*.

[Ramesh '22] Ramesh, A., Dhariwal, P., Nichol, A., Chu, C., & Chen, M. (2022). Hierarchical text-conditional image generation with clip latents. arXiv preprint arXiv:2204.06125.

[Nichol '21] Nichol, A., Dhariwal, P., Ramesh, A., Shyam, P., Mishkin, P., McGrew, B., ... & Chen, M. (2021). Glide: Towards photorealistic image generation and editing with text-guided diffusion models. arXiv preprint arXiv:2112.10741.

[Rombach '22] Rombach, R., Blattmann, A., Lorenz, D., Esser, P., & Ommer, B. (2022). High-resolution image synthesis with latent diffusion models. In *Proceedings of the IEEE/CVF Conference on Computer Vision and Pattern Recognition* (pp. 10684-10695).

[Zhang '21] Zhang, K., Liang, J., Van Gool, L., & Timofte, R. (2021). Designing a practical degradation model for deep blind image super-resolution. In *Proceedings of the IEEE/CVF International Conference on Computer Vision* (pp. 4791-4800).

[Saharia '22] Saharia, C., Chan, W., Saxena, S., Li, L., Whang, J., Denton, E., ... & Norouzi, M. (2022). Photorealistic Text-to-Image Diffusion Models with Deep Language Understanding. arXiv preprint arXiv:2205.11487.

[Yu '22] Yu, J., Xu, Y., Koh, J. Y., Luong, T., Baid, G., Wang, Z., ... & Wu, Y. (2022). Scaling Autoregressive Models for Content-Rich Text-to-Image Generation. arXiv preprint arXiv:2206.10789.

[Dhariwal '21] Dhariwal, P., & Nichol, A. (2021). Diffusion models beat GANs on image synthesis. *Advances in Neural Information Processing Systems*, 34, 8780-8794.

[Yu '21] Yu, J., Li, X., Koh, J. Y., Zhang, H., Pang, R., Qin, J., ... & Wu, Y. (2021). Vector-quantized image modeling with improved VQGAN. arXiv preprint arXiv:2110.04627.

[Betker '23] Betker, J., Goh, G., Jing, L., Brooks, T., Wang, J., Li, L., ... & Ramesh, A. (2023). Improving image generation with better captions. Computer Science. https://cdn. openai. com/papers/dall-e-3. pdf, 2(3), 8.

[Rombach '22] Rombach, R., Blattmann, A., Lorenz, D., Esser, P., & Ommer, B. (2022). High-resolution image synthesis with latent diffusion models. In Proceedings of the IEEE/CVF conference on computer vision and pattern recognition (pp. 10684-10695).

[Podell '23] Podell, D., English, Z., Lacey, K., Blattmann, A., Dockhorn, T., Müller, J., ... & Rombach, R. (2023). Sdxl: Improving latent diffusion models for high-resolution image synthesis. arXiv preprint arXiv:2307.01952.

[Zhu '23] Zhu, Y., Baca, J., Rekabdar, B., & Rawassizadeh, R. (2023). A Survey of AI Music Generation Tools and Models. arXiv preprint arXiv:2308.12982.

[Singer '22] Singer, U., Polyak, A., Hayes, T., Yin, X., An, J., Zhang, S., ... & Taigman, Y. (2022). Make-a-video: Text-to-video generation without text-video data. arXiv preprint arXiv:2209.14792.

[Khachatryan '23] Khachatryan, L., Movsisyan, A., Tadevosyan, V., Henschel, R., Wang, Z., Navasardyan, S., & Shi, H. (2023). Text2video-zero: Text-to-image diffusion models are zero-shot video generators. In Proceedings of the IEEE/CVF International Conference on Computer Vision (pp. 15954-15964).

[Liu '24] Liu, Y., Zhang, K., Li, Y., Yan, Z., Gao, C., Chen, R., ... & Sun, L. (2024). Sora: A Review on Background, Technology, Limitations, and Opportunities of Large Vision Models. arXiv preprint arXiv:2402.17177.

[Kandala '24] Kandala, H., Gao, J., & Yang, J. (2024). Pix2Gif: Motion-Guided Diffusion for GIF Generation. arXiv preprint arXiv:2403.04634.

[Esser '24] Esser, P., Kulal, S., Blattmann, A., Entezari, R., Müller, J., Saini, H., ... & Rombach, R. (2024). Scaling rectified flow transformers for high-resolution image synthesis. arXiv preprint arXiv:2403.03206.

[Sauer '24] Sauer, A., Boesel, F., Dockhorn, T., Blattmann, A., Esser, P., & Rombach, R. (2024). Fast High-Resolution Image Synthesis with Latent Adversarial Diffusion Distillation. arXiv preprint arXiv:2403.12015.

——— Chapter 12 ———

[Sutskever '14] Sutskever, I., Vinyals, O., & Le, Q. V. (2014). Sequence to sequence learning with neural networks. In Advances in neural information processing systems (pp. 3104-3112).

[Cho '14] Cho, K., Van Merriënboer, B., Gulcehre, C., Bahdanau, D., Bougares, F., Schwenk, H., & Bengio, Y. (2014). Learning phrase representations using RNN encoder-decoder for statistical machine translation. arXiv preprint arXiv:1406.1078.

[Bahdanau '14] Bahdanau, D., Cho, K., & Bengio, Y. (2014). Neural machine translation by jointly learning to align and translate. arXiv preprint arXiv:1409.0473.

[Luong '15] Luong, M. T., Pham, H., & Manning, C. D. (2015). Effective approaches to attention-based neural machine translation. arXiv preprint arXiv:1508.04025.

[Lin '17] Lin, Z., Feng, M., Santos, C. N. D., Yu, M., Xiang, B., Zhou, B., & Bengio, Y. (2017). A structured self-attentive sentence embedding. arXiv preprint arXiv:1703.03130.

[Vaswani '17] Vaswani, A., Shazeer, N., Parmar, N., Uszkoreit, J., Jones, L., Gomez, A. N., ... & Polosukhin, I. (2017). Attention is all you need. In Advances in neural information processing systems (pp. 5998-6008).

[Peters '18] Peters, M. E., Neumann, M., Iyyer, M., Gardner, M., Clark, C., Lee, K., & Zettlemoyer, L. (2018). Deep contextualized word representations. arXiv preprint arXiv:1802.05365.

[Howard '18] Howard, J., & Ruder, S. (2018). Universal language model fine-tuning for text classification. arXiv preprint arXiv:1801.06146.

[Lewis '19] Lewis, M., Liu, Y., Goyal, N., Ghazvininejad, M., Mohamed, A., Levy, O., ... & Zettlemoyer, L. (2019). Bart: Denoising sequence-to-sequence pre-training for natural language generation, translation, and comprehension. arXiv preprint arXiv:1910.13461.

[Gu '23] Gu, A., & Dao, T. (2023). Mamba: Linear-time sequence modeling with selective state spaces. arXiv preprint arXiv:2312.00752.

[Ramachandran '17] Ramachandran, P., Zoph, B., & Le, Q. V. (2017). Swish: a self-gated activation function. arXiv preprint arXiv:1710.05941, 7(1), 5.

[Hendrycks '16] Hendrycks, D., & Gimpel, K. (2016). Gaussian Error Linear Units (GeLUs). arXiv preprint arXiv:1606.08415.

[Li '23] Li, P., Yang, J., Islam, M. A., & Ren, S. (2023). Making ai less" thirsty": Uncovering and addressing the secret water footprint of AI models. arXiv preprint arXiv:2304.03271.

[Liu '19] Liu, Y., Ott, M., Goyal, N., Du, J., Joshi, M., Chen, D., ... & Stoyanov, V. (2019). RoBERTa: A robustly optimized bert pretraining approach. arXiv preprint arXiv:1907.11692.

[Lee '20] Lee, J., Yoon, W., Kim, S., Kim, D., Kim, S., So, C. H., & Kang, J. (2020). BioBERT: a pre-trained biomedical language representation model for biomedical text mining. Bioinformatics, 36(4), 1234-1240.

[Lan '19] Lan, Z., Chen, M., Goodman, S., Gimpel, K., Sharma, P., & Soricut, R. (2019). Albert: A lite bert for self-supervised learning of language representations. arXiv preprint arXiv:1909.11942.

[Sanh '19] Sanh, V., Debut, L., Chaumond, J., & Wolf, T. (2019). DistilBERT, a distilled version of BERT: smaller, faster, cheaper and lighter. arXiv preprint arXiv:1910.01108.

[Lample '19] Lample, G., & Conneau, A. (2019). Cross-lingual language model pretraining. arXiv preprint arXiv:1901.07291.

[Raffel '20] Raffel, C., Shazeer, N., Roberts, A., Lee, K., Narang, S., Matena, M., ... & Liu, P. J. (2020). Exploring the Limits of Transfer Learning with a Unified Text-to-Text Transformer. Journal of Machine Learning Research, 21, 1-67.

[Radford '18] Radford, A., Narasimhan, K., Salimans, T., & Sutskever, I. (2018). Improving language understanding by generative pre-training.

[Zhu '15] Zhu, Y., Kiros, R., Zemel, R., Salakhutdinov, R., Urtasun, R., Torralba, A., & Fidler, S. (2015). Aligning books and movies: Towards story-like visual explanations by watching movies and reading books. In Proceedings of the IEEE international conference on computer vision (pp. 19-27).

[Liu '18] Liu, P. J., Saleh, M., Pot, E., Goodrich, B., Sepassi, R., Kaiser, L., & Shazeer, N. (2018). Generating wikipedia by summarizing long sequences. arXiv preprint arXiv:1801.10198.

[Radford '19] Radford, A., Wu, J., Child, R., Luan, D., Amodei, D., & Sutskever, I. (2019). Language models are unsupervised multitask learners. OpenAI blog, 1(8), 9.

[Larochelle '08] Larochelle, H., Erhan, D., & Bengio, Y. (2008). Zero-data learning of new tasks. In AAAI (Vol. 1, No. 2, p. 3).

[Touvron '23] Touvron, H., Lavril, T., Izacard, G., Martinet, X., Lachaux, M. A., Lacroix, T., ... & Lample, G. (2023). Llama: Open and efficient foundation language models. arXiv preprint arXiv:2302.13971.

[Hoffmann '22] Hoffmann, J., Borgeaud, S., Mensch, A., Buchatskaya, E., Cai, T., Rutherford, E., ... & Sifre, L. (2022). Training compute-optimal large language models. arXiv preprint arXiv:2203.15556.

[Gao '20] Gao, L., Biderman, S., Black, S., Golding, L., Hoppe, T., Foster, C., ... & Leahy, C. (2020). The pile: An 800gb dataset of diverse text for language modeling. arXiv preprint arXiv:2101.00027.

[Su '24] Su, J., Ahmed, M., Lu, Y., Pan, S., Bo, W., & Liu, Y. (2024). RoFormer: Enhanced transformer with rotary position embedding. Neurocomputing, 568, 127063.

[Shazeer '20] Shazeer, N. (2020). Glu variants improve transformer. arXiv preprint arXiv:2002.05202. Chicago

[Touvron '23] Touvron, H., Martin, L., Stone, K., Albert, P., Almahairi, A., Babaei, Y., ... & Scialom, T. (2023). Llama 2: Open foundation and fine-tuned chat models. arXiv preprint arXiv:2307.09288.

[Dubey '24] Dubey, A., Jauhri, A., Pandey, A., Kadian, A., Al-Dahle, A., Letman, A., ... & Ganapathy, R. (2024). The Llama 3 Herd of Models. arXiv preprint arXiv:2407.21783.

[Taori '23] Taori, R., Gulrajani, I., Zhang, T., Dubois, Y., Li, X., Guestrin, C., ... & Hashimoto, T. B. (2023). Alpaca: A strong, replicable instruction-following model. Stanford Center for Research on Foundation Models. https://crfm. stanford. edu/2023/03/13/alpaca. html, 3(6), 7.

[Jiang '23] Jiang, A. Q., Sablayrolles, A., Mensch, A., Bamford, C., Chaplot, D. S., Casas, D. D. L., ... & Sayed, W. E. (2023). Mistral 7B. arXiv preprint arXiv:2310.06825.

[Jiang '24] Jiang, A. Q., Sablayrolles, A., Roux, A., Mensch, A., Savary, B., Bamford, C., ... & Sayed, W. E. (2024). Mixtral of experts. arXiv preprint arXiv:2401.04088.

[Signal '23] Signal, K., Azizi, S., Tu, T., Mahdavi, S. S., Wei, J., Chung, H. W., ... & Natarajan, V. (2023). Large language models encode clinical knowledge. Nature, 620(7972), 172-180.

[Chang '08] Chang, M. W., Ratinov, L. A., Roth, D., & Srikumar, V. (2008). Importance of Semantic Representation: Dataless Classification. In Aaai (Vol. 2, pp. 830-835).

[Brown '20] Brown, T. B., Mann, B., Ryder, N., Subbiah, M., Kaplan, J., Dhariwal, P., ... & Amodei, D. (2020). Language models are few-shot learners. *arXiv preprint arXiv:2005.14165*.

[Child '19] Child, R., Gray, S., Radford, A., & Sutskever, I. (2019). Generating long sequences with sparse transformers. *arXiv preprint arXiv:1904.10509*.

[LeCun '89] LeCun, Y., Boser, B., Denker, J. S., Henderson, D., Howard, R. E., Hubbard, W., & Jackel, L. D. (1989). Backpropagation applied to handwritten zip code recognition. Neural computation, 1(4), 541-551.

[Devlin '18] Devlin, J., Chang, M. W., Lee, K., & Toutanova, K. (2018). Bert: Pre-training of deep bidirectional transformers for language understanding. arXiv preprint arXiv:1810.04805.

[Rafailov '24] Rafailov, R., Sharma, A., Mitchell, E., Manning, C. D., Ermon, S., & Finn, C. (2024). Direct preference optimization: Your language model is secretly a reward model. Advances in Neural Information Processing Systems, 36.

[Shafahi '19] Shafahi, A., Najibi, M., Ghiasi, M. A., Xu, Z., Dickerson, J., Studer, C., ... & Goldstein, T. (2019). Adversarial training for free!. Advances in neural information processing systems, 32.

[Jelinek '77] Jelinek, F., Mercer, R. L., Bahl, L. R., & Baker, J. K. (1977). Perplexity—a measure of the difficulty of speech recognition tasks. The Journal of the Acoustical Society of America, 62(S1), S63-S63.

[Morris '04] Morris, A. C., Maier, V., & Green, P. (2004). From WER and RIL to MER and WIL: improved evaluation measures for connected speech recognition. In Eighth International Conference on Spoken Language Processing.

[Lin '04] Lin, C. Y. (2004). Rouge: A package for automatic evaluation of summaries. In Text summarization branches out (pp. 74-81).

[Banerjee '05] Banerjee, S., & Lavie, A. (2005). METEOR: An automatic metric for MT evaluation with improved correlation with human judgments. In Proceedings of the acl workshop on intrinsic and extrinsic evaluation measures for machine translation and/or summarization (pp. 65-72).

[Papineni '02] Papineni, K., Roukos, S., Ward, T., & Zhu, W. J. (2002, July). BLEU: a method for automatic evaluation of machine translation. In Proceedings of the 40th annual meeting of the Association for Computational Linguistics (pp. 311-318).

[Wu '16] Wu, Y., Schuster, M., Chen, Z., Le, Q. V., Norouzi, M., Macherey, W., ... & Dean, J. (2016). Google's neural machine translation system: Bridging the gap between human and machine translation. arXiv preprint arXiv:1609.08144.

[Napoles '15] Napoles, C., Sakaguchi, K., Post, M., & Tetreault, J. (2015). Ground truth for grammatical error correction metrics. In Proceedings of the 53rd Annual Meeting of the Association for Computational Linguistics and the 7th International Joint Conference on Natural Language Processing (Volume 2: Short Papers) (pp. 588-593).

[Sellam '20] Sellam, T., Das, D., & Parikh, A. P. (2020). BLEURT: Learning robust metrics for text generation. arXiv preprint arXiv:2004.04696.

[Srivastava '22] Srivastava, A., Rastogi, A., Rao, A., Shoeb, A. A. M., Abid, A., Fisch, A., ... & Wang, G. (2022). Beyond the imitation game: Quantifying and extrapolating the capabilities of language models. arXiv preprint arXiv:2206.04615.

[Zellers '19] Zellers, R., Holtzman, A., Bisk, Y., Farhadi, A., & Choi, Y. (2019). Hellaswag: Can a machine really finish your sentence?. arXiv preprint arXiv:1905.07830.

[Hendrycks '20] Hendrycks, D., Burns, C., Basart, S., Zou, A., Mazeika, M., Song, D., & Steinhardt, J. (2020). Measuring massive multitask language understanding. arXiv preprint arXiv:2009.03300.

[Simonyan '14] Simonyan, K., & Zisserman, A. (2014). Very deep convolutional networks for large-scale image recognition. arXiv preprint arXiv:1409.1556.

[Hubel '59] Hubel, D. H., & Wiesel, T. N. (1959). Receptive fields of single neurones in the cat's striate cortex. The Journal of physiology, 148(3), 574-591.

[Szegedy '15] Szegedy, C., Liu, W., Jia, Y., Sermanet, P., Reed, S., Anguelov, D., ... & Rabinovich, A. (2015). Going deeper with convolutions. In Proceedings of the IEEE conference on computer vision and pattern recognition (pp. 1-9).

[Zhang '22] Zhang, A., Lipton, Z. C., Li, M., & Smola, A. J. (2022). Dive into deep learning. https://d2l.ai.

[He '16] He, K., Zhang, X., Ren, S., & Sun, J. (2016). Deep residual learning for image recognition. In Proceedings of the IEEE conference on computer vision and pattern recognition (pp. 770-778).

[Ioffe '15] Ioffe, S., & Szegedy, C. (2015). Batch normalization: Accelerating deep network training by reducing internal covariate shift. In International conference on machine learning (pp. 448-456). ICML.

[Ulyanov '17] Ulyanov, D., Vedaldi, A., & Lempitsky, V. (2017). Instance normalization: The missing ingredient for fast stylization. arXiv preprint arXiv:1607.08022.

[Ba '16] Ba, J. L., Kiros, J. R., & Hinton, G. E. (2016). Layer normalization. arXiv preprint arXiv:1607.06450.

[Wu '18] Wu, Y., & He, K. (2018). Group normalization. In Proceedings of the European conference on computer vision (ECCV) (pp. 3-19).

[Li '17] Li, H., Xu, Z., Taylor, G., Studer, C., & Goldstein, T. (2017). Visualizing the loss landscape of neural nets. arXiv preprint arXiv:1712.09913.

[Szegedy '17] Szegedy, C., Ioffe, S., Vanhoucke, V., & Alemi, A. A. (2017). Inception-v4, inception-resnet, and the impact of residual connections on learning. In the Thirty-first AAAI conference on artificial intelligence.

[Dosovitskiy '20] Dosovitskiy, A., Beyer, L., Kolesnikov, A., Weissenborn, D., Zhai, X., Unterthiner, T., ... & Houlsby, N. (2020). An image is worth 16x16 words: Transformers for image recognition at scale. arXiv preprint arXiv:2010.11929.

[Girshick '15] Girshick, R., Donahue, J., Darrell, T., & Malik, J. (2015). Rich feature hierarchies for accurate object detection and semantic segmentation. In Proceedings of the IEEE conference on computer vision and pattern recognition (pp. 580-587).

[Uijlings '13] Uijlings, J. R., Van De Sande, K. E., Gevers, T., & Smeulders, A. W. (2013). Selective search for object recognition. International journal of computer vision, 104(2), 154-171.

[Girshick '15] Girshick, R. (2015). Fast R-CNN. In Proceedings of the IEEE international conference on computer vision (pp. 1440-1448).

[Ren '16] Ren, S., He, K., Girshick, R., & Sun, J. (2016). Faster R-CNN: Towards real-time object detection with region proposal networks. Advances in neural information processing systems, 28, 91-99.

[Liu '16] Liu, W., Anguelov, D., Erhan, D., Szegedy, C., Reed, S., Fu, C. Y., & Berg, A. C. (2016, October). SSD: Single shot multibox detector. In European conference on computer vision (pp. 21-37). Springer, Cham.

[Redmon '16] Redmon, J., Divvala, S., Girshick, R., & Farhadi, A. (2016). You only look once: Unified, real-time object detection. In Proceedings of the IEEE conference on computer vision and pattern recognition (pp. 779-788).

[Redmon '17] Redmon, J., & Farhadi, A. (2017). YOLO9000: better, faster, stronger. In Proceedings of the IEEE conference on computer vision and pattern recognition (pp. 7263-7271).

[Redmon '18] Redmon, J., & Farhadi, A. (2018). YOLO v3: An incremental improvement. arXiv preprint arXiv:1804.02767.

[Lin '17] Lin, T. Y., Dollár, P., Girshick, R., He, K., Hariharan, B., & Belongie, S. (2017). Feature pyramid networks for object detection. In Proceedings of the IEEE conference on computer vision and pattern recognition (pp. 2117-2125).

[Ge '21] Ge, Z., Liu, S., Wang, F., Li, Z., & Sun, J. (2021). YoloX: Exceeding yolo series in 2021. arXiv preprint arXiv:2107.08430.

[Bochkovskiy '20] Bochkovskiy, A., Wang, C. Y., & Liao, H. Y. M. (2020). Yolov4: Optimal speed and accuracy of object detection. arXiv preprint arXiv:2004.10934.

[Luo '16] Luo, W., Li, Y., Urtasun, R., & Zemel, R. (2016). Understanding the effective receptive field in deep convolutional neural networks. In Proceedings of the 30th International Conference on Neural Information Processing Systems (pp. 4905-4913).

[Long '15] Long, J., Shelhamer, E., & Darrell, T. (2015). Fully convolutional networks for semantic segmentation. *In Proceedings of the IEEE conference on computer vision and pattern recognition* (pp. 3431-3440).

[He '17] He, K., Gkioxari, G., Dollár, P., & Girshick, R. (2017). Mask R-CNN. In Proceedings of the IEEE international conference on computer vision (pp. 2961-2969).

[Jaderberg '15] Jaderberg, M., Simonyan, K., & Zisserman, A. (2015). Spatial transformer networks. *Advances in neural information processing systems, 28,* 2017-2025.

[Chen '14] Chen, L. C., Papandreou, G., Kokkinos, I., Murphy, K., & Yuille, A. L. (2014). Semantic image segmentation with deep convolutional nets and fully connected crfs. arXiv preprint arXiv:1412.7062.

[Chen '17 A] Chen, L. C., Papandreou, G., Kokkinos, I., Murphy, K., & Yuille, A. L. (2017). DeepLab: Semantic image segmentation with deep convolutional nets, atrous convolution, and fully connected crfs. *IEEE transactions on pattern analysis and machine intelligence, 40(4),* 834-848.

[Chen '17 B] Chen, L. C., Papandreou, G., Schroff, F., & Adam, H. (2017). Rethinking Atrous convolution for semantic image segmentation. arXiv preprint arXiv:1706.05587.

[Chen '18] Chen, L. C., Zhu, Y., Papandreou, G., Schroff, F., & Adam, H. (2018). Encoder-decoder with atrous separable convolution for semantic image segmentation. In Proceedings of the European conference on computer vision (ECCV) (pp. 801-818).

[Chollet '17] Chollet, F. (2017). Xception: Deep learning with depthwise separable convolutions. In Proceedings of the IEEE conference on computer vision and pattern recognition (pp. 1251-1258).

[Kirillov '23] Kirillov, A., Mintun, E., Ravi, N., Mao, H., Rolland, C., Gustafson, L., ... & Girshick, R. (2023). Segment Anything. In Proceedings of the IEEE/CVF International Conference on Computer Vision (pp. 4015-4026).

[Ravi '24] Ravi, N., Gabeur, V., Hu, Y. T., Hu, R., Ryali, C., Ma, T., ... & Feichtenhofer, C. (2024). SAM 2: Segment Anything in Images and Videos. *arXiv preprint arXiv:2408.00714.*

[He '22] He, K., Chen, X., Xie, S., Li, Y., Dollár, P., & Girshick, R. (2022). Masked autoencoders are scalable vision learners. *In Proceedings of the IEEE/CVF conference on computer vision and pattern recognition* (pp. 16000-16009).

[Mildenhall '21] Mildenhall, B., Srinivasan, P. P., Tancik, M., Barron, J. T., Ramamoorthi, R., & Ng, R. (2021). NeRF: Representing scenes as neural radiance fields for view synthesis. *Communications of the ACM, 65(1),* 99-106.

[Kerbl '23] Kerbl, B., Kopanas, G., Leimkühler, T., & Drettakis, G. (2023). 3D Gaussian splatting for real-time radiance field rendering. ACM Transactions on Graphics, 42(4), 1-14.

[Snavely '06] Snavely, N., Seitz, S. M., & Szeliski, R. (2006). Photo tourism: exploring photo collections in 3D. In ACM siggraph 2006 papers (pp. 835-846).

[Merrill '10] Merrill, D. G., & Grimshaw, A. S. (2010). Revisiting sorting for GPGPU stream architectures. In Proceedings of the 19th International Conference on Parallel Architectures and Compilation Techniques (pp. 545-546).

[Kling '21] Kling, M. U. (2021). *Quality Land.* Éditions Actes Sud.

[Masi '18] Masi, I., Wu, Y., Hassner, T., & Natarajan, P. (2018). Deep face recognition: A survey. In 2018 31st IEEE SIBGRAPI conference on graphics, patterns and images (SIBGRAPI) (pp. 471-478).

[Taigman '14] Taigman, Y., Yang, M., Ranzato, M. A., & Wolf, L. (2014). Deepface: Closing the gap to human-level performance in face verification. In Proceedings of the IEEE conference on computer vision and pattern recognition (pp. 1701-1708).

1112

[Sun '14] Sun, Y., Wang, X., & Tang, X. (2014). Deep learning face representation from predicting 10,000 classes. In Proceedings of the IEEE conference on computer vision and pattern recognition (pp. 1891-1898).

[Sun '15] Sun, Y., Wang, X., & Tang, X. (2015). Deeply learned face representations are sparse, selective, and robust. In Proceedings of the IEEE conference on computer vision and pattern recognition (pp. 2892-2900).

[Schroff '15] Schroff, F., Kalenichenko, D., & Philbin, J. (2015). Facenet: A unified embedding for face recognition and clustering. In Proceedings of the IEEE conference on computer vision and pattern recognition (pp. 815-823).

[Ledig '17] Ledig, C., Theis, L., Huszár, F., Caballero, J., Cunningham, A., Acosta, A., ... & Shi, W. (2017). Photo-realistic single image super-resolution using a generative adversarial network. Proceedings of the IEEE Conference on Computer Vision and Pattern Recognition (CVPR), 4681-4690.

[Waang '18] Wang, X., Yu, K., Wu, S., Gu, J., Liu, Y., Dong, C., ... & Change Loy, C. (2018). ESRGAN: Enhanced super-resolution generative adversarial networks. Proceedings of the European Conference on Computer Vision (ECCV), Workshops.

[Sajjadi '17] Sajjadi, M. S. M., Schölkopf, B., & Hirsch, M. (2017). EnhanceNet: Single image super-resolution through automated texture synthesis. Proceedings of the IEEE International Conference on Computer Vision (ICCV), 4501-4510.

[Zhang '21] Zhang, Z., Luo, P., Loy, C. C., & Tang, X. (2021). Applications of Human Pose Estimation in Visual Surveillance and Sports: A Survey. Computer Vision and Image Understanding, 206, 103225.

[Kendall '15] Kendall, A., Grimes, M., & Cipolla, R. (2015). PoseNet: A Convolutional Network for Real-Time 6-DOF Camera Relocalization. *Proceedings of the IEEE International Conference on Computer Vision (ICCV)*.

[Cao '17] Cao, Z., Simon, T., Wei, S. E., & Sheikh, Y. (2017). Realtime Multi-Person 2D Pose Estimation using Part Affinity Fields. Proceedings of the IEEE Conference on Computer Vision and Pattern Recognition (CVPR), 7291-7299.

[Fang '17] Fang, H. S., Xie, S., Tai, Y. W., & Lu, C. (2017). RMPE: Regional Multi-Person Pose Estimation. *Proceedings of the IEEE International Conference on Computer Vision (ICCV)*, 2334-2343.

[Alhashim '18] Alhashim, I., & Wonka, P. (2018). High Quality Monocular Depth Estimation via Transfer Learning. *arXiv preprint arXiv:1812.11941*.

[Laina '16] Laina, I., Rupprecht, C., Belagiannis, V., Tombari, F., & Navab, N. (2016). Deeper Depth Prediction with Fully Convolutional Residual Networks. Proceedings of the 4th International Conference on 3D Vision (3DV), 239-248.

[Zhu '23] Zhu, Y., Baca, J., Rekabdar, B., & Rawassizadeh, R. (2023). A Survey of AI Music Generation Tools and Models. arXiv preprint arXiv:2308.12982.

[Liu '18] Liu, R., Cornelius, C., Rawassizadeh, R., Peterson, R., & Kotz, D. (2018). Vocal resonance: Using internal body voice for wearable authentication. *Proceedings of the ACM on Interactive, Mobile, Wearable and Ubiquitous Technologies*, 2(1), 1-23.

[Arik '18] Arik, S. O., Chrzanowski, M., Coates, A., Duan, G., & Gibson, A. (2018). Deep Voice: Real-time Neural Text-to-Speech. Proceedings of the 34th International Conference on Machine Learning (ICML), 195-204.

[Oord '16] Oord, A. V. D., Dieleman, S., Zen, H., Simonyan, K., Vinyals, O., Graves, A., ... & Kavukcuoglu, K. (2016). Wavenet: A generative model for raw audio. arXiv preprint arXiv:1609.03499.

[Oord '18] Oord, A., Li, Y., Babuschkin, I., Simonyan, K., Vinyals, O., Kavukcuoglu, K., ... & Hassabis, D. (2018). Parallel wavenet: Fast high-fidelity speech synthesis. In International conference on machine learning (pp. 3918-3926).

[Wang '17] Wang, Y., Skerry-Ryan, R. J., Stanton, D., Wu, Y., Weiss, R. J., Jaitly, N., ... & Saurous, R. A. (2017). Tacotron: Towards end-to-end speech synthesis. arXiv preprint arXiv:1703.10135.

[Srivastava '15] Srivastava, R. K., Greff, K., & Schmidhuber, J. (2015). Training very deep networks. Advances in neural information processing systems, 28.

[Shen '18] Shen, J., Pang, R., Weiss, R. J., Schuster, M., Jaitly, N., Yang, Z., ... & Wu, Y. (2018). Natural TTS synthesis by conditioning wavenet on mel spectrogram predictions. In 2018 IEEE international conference on acoustics, speech and signal processing (ICASSP) (pp. 4779-4783).

[Chorowski '15] Chorowski, J. K., Bahdanau, D., Serdyuk, D., Cho, K., & Bengio, Y. (2015). Attention-based models for speech recognition. Advances in Neural Information Processing Systems, 28.

[Jang '16] Jang, E., Gu, S., & Poole, B. (2016). Categorical reparameterization with gumbel-softmax. arXiv preprint arXiv:1611.01144.

[Schneider '19] Schneider, S., Baevski, A., Collobert, R., & Auli, M. (2019). wav2vec: Unsupervised pre-training for speech recognition. arXiv preprint arXiv:1904.05862.

[Baevski '20] Baevski, A., Zhou, Y., Mohamed, A., & Auli, M. (2020). wav2vec 2.0: A framework for self-supervised learning of speech representations. *Advances in neural information processing systems*, 33, 12449-12460.

[Radford '23] Radford, A., Kim, J. W., Xu, T., Brockman, G., McLeavey, C., & Sutskever, I. (2023). Robust speech recognition via large-scale weak supervision. *In International Conference on Machine Learning* (pp. 28492-28518).

——————— Chapter 13 ———————

[Skinner '65] Skinner, B. F. (1965). Science and human behavior (No. 92904). Simon and Schuster.

[Sutton '20] Sutton, R. S., & Barto, A. G. (2020). Reinforcement learning: An introduction. MIT press.

[Silver '17] Silver, D., Schrittwieser, J., Simonyan, K., Antonoglou, I., Huang, A., Guez, A., ... & Hassabis, D. (2017). Mastering the game of go without human knowledge. *nature*, 550(7676), 354-359.

[Silver '18] Silver, D., Hubert, T., Schrittwieser, J., Antonoglou, I., Lai, M., Guez, A., ... & Hassabis, D. (2018). A general reinforcement learning algorithm that masters chess, shogi, and Go through self-play. *Science*, 362(6419), 1140-1144.

[Ouyang '22] Ouyang, L., Wu, J., Jiang, X., Almeida, D., Wainwright, C. L., Mishkin, P., ... & Lowe, R. (2022). Training language models to follow instructions with human feedback. *arXiv preprint arXiv:2203.02155*.

[Thompson '33] Thompson, W. R. (1933). On the likelihood that one unknown probability exceeds another in view of the evidence of two samples. Biometrika, 25(3-4), 285-294.

[Winder '20] Winder, P. (2020). Reinforcement learning. O'Reilly Media.

[Sutton '98] Richard S. (1988). "Learning to predict by the methods of temporal differences". Machine Learning. 3 (1): 9–44.

[Watkins '89] Watkins, C.J.C.H. (1989). Learning from Delayed Rewards. Ph.D. thesis, University of Cambridge.

[Sutton '91] Sutton, R. S. (1991). Dyna, an integrated architecture for learning, planning, and reacting. ACM Sigart Bulletin, 2(4), 160-163.

[Peng '94] Peng, J., & Williams, R. J. (1994). Incremental multi-step Q-learning. In Machine Learning Proceedings 1994 (pp. 226-232). Morgan Kaufmann.

[Morales '20] Morales, M. (2020). Grokking deep reinforcement learning. Manning Publications.

[Mnih '13] Mnih, V., Kavukcuoglu, K., Silver, D., Graves, A., Antonoglou, I., Wierstra, D., & Riedmiller, M. (2013). Playing atari with deep reinforcement learning. arXiv preprint arXiv:1312.5602.

[Mnih '15] Mnih, V., Kavukcuoglu, K., Silver, D., Rusu, A. A., Veness, J., Bellemare, M. G., ... & Hassabis, D. (2015). Human-level control through deep reinforcement learning. nature, 518(7540), 529-533.

[Riedmiller '05] Riedmiller, M. (2005). Neural fitted Q iteration–first experiences with a data efficient neural reinforcement learning method. In European conference on machine learning (pp. 317-328). Springer, Berlin, Heidelberg.

[Lin '92] Lin, L. J. (1992). Self-improving reactive agents based on reinforcement learning, planning and teaching. Machine learning, 8(3), 293-321.

[Van Hasselt '16] Van Hasselt, H., Guez, A., & Silver, D. (2016). Deep reinforcement learning with double q-learning. In Proceedings of the AAAI conference on artificial intelligence (Vol. 30, No. 1).

[Van Hasselt '10] Van Hasselt, H. (2010). Double Q-learning. Advances in Neural Information Processing Systems, 23.

[Wang '16] Wang, Z., Schaul, T., Hessel, M., Hasselt, H., Lanctot, M., & Freitas, N. (2016). Dueling network architectures for deep reinforcement learning. In International conference on machine learning (pp. 1995-2003). PMLR.

[Schaul '16] Schaul, T., Quan, J., Antonoglou, I., & Silver, D. (2015). Prioritized experience replay. arXiv preprint arXiv:1511.05952.

[Williams '92] Williams, R. J. (1992). Simple statistical gradient-following algorithms for connectionist reinforcement learning. Machine learning, 8(3), 229-256.

[Schulman '15] Schulman, J., Levine, S., Abbeel, P., Jordan, M., & Moritz, P. (2015, June). Trust region policy optimization. In International conference on machine learning (pp. 1889-1897). PMLR.

[Amari '98] Amari, S. I. (1998). Natural gradient works efficiently in learning. Neural computation, 10(2), 251-276.

[Stiefel '52] Stiefel, E. (1952). Methods of conjugate gradients for solving linear systems. Journal of Research of the National Bureau of Standards, 49, 409-435.

[Chen '16] Chen, T., Xu, B., Zhang, C., & Guestrin, C. (2016). Training deep nets with Sublinear Memory cost. arXiv preprint arXiv:1604.0617

[Schulman '17] Schulman, J., Wolski, F., Dhariwal, P., Radford, A., & Klimov, O. (2017). Proximal policy optimization algorithms. arXiv preprint arXiv:1707.06347.

[Mnih '16] Mnih, V., Badia, A. P., Mirza, M., Graves, A., Lillicrap, T., Harley, T., ... & Kavukcuoglu, K. (2016). Asynchronous Methods for Deep Reinforcement Learning. In International Conference on Machine Learning (pp. 1928-1937). PMLR.

[Lillicrap '15] Lillicrap, T. P., Hunt, J. J., Pritzel, A., Heess, N., Erez, T., Tassa, Y., ... & Wierstra, D. (2015). Continuous control with deep reinforcement learning. arXiv preprint arXiv:1509.02971.

[Uhlenbeck '30] Uhlenbeck, G. E and Ornstein, L.S. (1930) On the theory of the brownian motion. Physical review, 36(5):823.

[Fujimoto '18] Fujimoto, S., Hoof, H., & Meger, D. (2018). Addressing function approximation error in actor-critic methods. In International Conference on Machine Learning (pp. 1587-1596). PMLR.

[Haarnoja '18] Haarnoja, T., Zhou, A., Abbeel, P., & Levine, S. (2018). Soft Actor-Critic: Off-policy maximum entropy deep reinforcement learning with a stochastic actor. In International conference on machine learning (pp. 1861-1870). PMLR.

[Bengio '13] Bengio, Y., Léonard, N., & Courville, A. (2013). Estimating or propagating gradients through stochastic neurons for conditional computation. arXiv preprint arXiv:1308.3432.

[Team '24] Team, O. M., Ghosh, D., Walke, H., Pertsch, K., Black, K., Mees, O., ... & Levine, S. (2024). Octo: An Open-Source Generalist Robot Policy. arXiv preprint arXiv:2405.12213.

[Hafner '19] Hafner, D., Lillicrap, T., Ba, J., & Norouzi, M. (2019). Dream to control: Learning behaviors by latent imagination. arXiv preprint arXiv:1912.01603.

[Hafner '20] Hafner, D., Lillicrap, T., Norouzi, M., & Ba, J. (2020). Mastering atari with discrete world models. arXiv preprint arXiv:2010.02193.

[Hafner '23] Hafner, D., Pasukonis, J., Ba, J., & Lillicrap, T. (2023). Mastering Diverse Domains through World Models. arXiv preprint arXiv:2301.04104.

[Ziyin '20] Ziyin, L., Wang, Z. T., & Ueda, M. (2020). LaProp: Separating momentum and adaptivity in adam. arXiv preprint arXiv:2002.04839.

——————— Chapter 14 ———————

[Sedgewick '11] Sedgewick, R., & Wayne, K. (2011). *Algorithms, 4th Edition.* Addison-Wesley Professional.

[Coremen '22] Cormen, T. H., Leiserson, C. E., Rivest, R. L., & Stein, C. (2009). *Introduction to algorithms, 4th Edition.* MIT press.

[Ziv '78] Ziv, J., & Lempel, A. (1978). Compression of Individual Sequences via Variable-Rate Coding. *IEEE transactions on Information Theory*, 24(5), 530-536.

[Welch '84] Welch, T. A. (1984). Technique for high-performance data compression. *Computer*, (52).

[Rawassizadeh '23] Rawassizadeh, R., & Rong, Y. (2023). ODSearch: Fast and Resource Efficient On-device Natural Language Search for Fitness Trackers' Data. *Proceedings of the ACM on Interactive, Mobile, Wearable and Ubiquitous Technologies*, 6(4), 1-25.

[Malekijoo '21] Malekijoo, A., Fadaeieslam, M. J., Malekijou, H., Homayounfar, M., Alizadeh-Shabdiz, F., & Rawassizadeh, R. (2021). Fedzip: A compression framework for communication-efficient federated learning. *arXiv preprint arXiv:2102.01593.*

[Gage '94] Gage, P. (1994). A new algorithm for data compression. The C Users Journal, 12(2), 23-38.

[Huffman '52] Huffman, D. A. (1952). A method for the construction of minimum-redundancy codes. *Proceedings of the IRE, 40*(9), 1098-1101.

[Linde '80] Linde, Y., Buzo, A., & Gray, R. (1980). An algorithm for vector quantizer design. IEEE Transactions on communications, 28(1), 84-95.

[Horowitz '14] Horowitz, M. (2014). 1.1 Computing's Energy Problem (and what we can do about it). In 2014 IEEE International Solid-State Circuits Conference Digest of Technical Papers (ISSCC) (pp. 10-14). IEEE.

[Krishnamoorthi '18] Krishnamoorthi, R. (2018). Quantizing deep convolutional networks for efficient inference: A whitepaper. arXiv preprint arXiv:1806.08342.

[Jacob '18] Jacob, B., Kligys, S., Chen, B., Zhu, M., Tang, M., Howard, A., ... & Kalenichenko, D. (2018). Quantization and training of neural networks for efficient integer-arithmetic-only inference. In Proceedings of the IEEE conference on computer vision and pattern recognition (pp. 2704-2713).

[Courbariaux '15] Courbariaux, M., Bengio, Y., & David, J. P. (2015). Binaryconnect: Training deep neural networks with binary weights during propagations. Advances in neural information processing systems, 28.

[Li '16] Li, F., Zhang, B., & Liu, B. (2016). Ternary weight networks. arXiv preprint arXiv:1605.04711.

[Wang '23] Wang, H., Ma, S., Dong, L., Huang, S., Wang, H., Ma, L., ... & Wei, F. (2023). Bitnet: Scaling 1-bit transformers for large language models. *arXiv preprint arXiv:2310.11453*.

[Alizadeh '20] Alizadeh, M., Behboodi, A., van Baalen, M., Louizos, C., Blankevoort, T., & Welling, M. (2020). Gradient ℓ_1 regularization for quantization robustness. arXiv preprint arXiv:2002.07520.

[LeCun '89] LeCun, Yann, John Denker, and Sara Solla. "Optimal brain damage." Advances in neural information processing systems 2 (1989).

[Hassibi '93] Hassibi, B., Stork, D. G., & Wolff, G. J. (1993). Optimal brain surgeon and general network pruning. In IEEE international conference on neural networks (pp. 293-299).

[Reed '93] Reed, R. (1993). Pruning algorithms-a survey. IEEE transactions on Neural Networks, 4(5), 740-747

[Hoefler '21] Hoefler, T., Alistarh, D., Ben-Nun, T., Dryden, N., & Peste, A. (2021). Sparsity in deep learning: Pruning and growth for efficient inference and training in neural networks. The Journal of Machine Learning Research, 22(1), 10882-11005.

[Li '16] Li, H., Kadav, A., Durdanovic, I., Samet, H., & Graf, H. P. (2016). Pruning filters for efficient convnets. arXiv preprint arXiv:1608.08710.

[Molchanov '16] Molchanov, P., Tyree, S., Karras, T., Aila, T., & Kautz, J. (2016). Pruning convolutional neural networks for resource efficient inference. *arXiv preprint arXiv:1611.06440*.

[Lin '17] Lin, J., Rao, Y., Lu, J., & Zhou, J. (2017). Runtime neural pruning. Advances in neural information processing systems, 30.

[Williams '95] Williams, P. M. (1995). Bayesian regularization and pruning using a Laplace prior. *Neural computation, 7*(1), 117-143.

[Wu '21] Wu, D., & Wang, Y. (2021). Adversarial neuron pruning purifies backdoored deep models. *Advances in Neural Information Processing Systems, 34*, 16913-16925.

[Frankle '18] Frankle, J., & Carbin, M. (2018). The lottery ticket hypothesis: Finding sparse, trainable neural networks. *arXiv preprint arXiv:1803.03635*.

[Lu '24] Lu, H., Alemi, M., & Rawassizadeh, R. (2024). The Impact of Quantization and Pruning on Deep Reinforcement Learning Models. arXiv preprint arXiv:2407.04803.

[Hu '21] Hu, E. J., Shen, Y., Wallis, P., Allen-Zhu, Z., Li, Y., Wang, S., ... & Chen, W. (2021). Lora: Low-rank adaptation of large language models. *arXiv preprint arXiv:2106.09685*.

[Jiang '23] Jiang, A. Q., Sablayrolles, A., Mensch, A., Bamford, C., Chaplot, D. S., Casas, D. D. L., ... & Sayed, W. E. (2023). Mistral 7B. arXiv preprint arXiv:2310.06825.

[Dao '22] Dao, T., Fu, D., Ermon, S., Rudra, A., & Ré, C. (2022). FlashAttention: Fast and memory-efficient exact attention with io-awareness. Advances in Neural Information Processing Systems, 35, 16344-16359.

[Dao '23] Dao, T. (2023). FlashAttention-2: Faster attention with better parallelism and work partitioning. arXiv preprint arXiv:2307.08691.

[Shazeer '19] Shazeer, N. (2019). Fast transformer decoding: One write-head is all you need. arXiv preprint arXiv:1911.02150.

[Chowdhery '23] Chowdhery, A., Narang, S., Devlin, J., Bosma, M., Mishra, G., Roberts, A., ... & Fiedel, N. (2023). Palm: Scaling language modeling with pathways. Journal of Machine Learning Research, 24(240), 1-113.

[Ainslie '23] Ainslie, J., Lee-Thorp, J., de Jong, M., Zemlyanskiy, Y., Lebrón, F., & Sanghai, S. (2023). GQA: Training generalized multi-query transformer models from multi-head checkpoints. arXiv preprint arXiv:2305.13245.

[Beltagy '20] Beltagy, I., Peters, M. E., & Cohan, A. (2020). Longformer: The long-document transformer. arXiv preprint arXiv:2004.05150.

[Shui '19] Shui, C., Abbasi, M., Robitaille, L. É., Wang, B., & Gagné, C. (2019). A principled approach for learning task similarity in multitask learning. arXiv preprint arXiv:1903.09109.

[Aoki '22] Aoki, R., Tung, F., & Oliveira, G. L. (2022). Heterogeneous multi-task learning with expert diversity. IEEE/ACM Transactions on Computational Biology and Bioinformatics, 19(6), 3093-3102.

[Schmidhuber '87] Schmidhuber, J. (1987). Evolutionary principles in self-referential learning, or on learning how to learn: the meta-meta-... hook (Doctoral dissertation, Technische Universität München).

[Santoro '16] Santoro, A., Bartunov, S., Botvinick, M., Wierstra, D., & Lillicrap, T. (2016). Meta-learning with memory-augmented neural networks. In International Conference on Machine Learning (pp. 1842-1850). PMLR.

[Finn '17] Finn, C., Abbeel, P., & Levine, S. (2017). Model-Agnostic Meta-Learning for fast adaptation of deep networks. In International conference on machine learning (pp. 1126-1135). PMLR.

[Weinshall '18] Weinshall, D., Cohen, G., & Amir, D. (2018). Curriculum learning by transfer learning: Theory and experiments with deep networks. In International Conference on Machine Learning (pp. 5238-5246). PMLR.

[Karimi '22] Karimi, F., Mehrpanah, A., & Rawassizadeh, R. (2022). LightDepth: A Resource Efficient Depth Estimation Approach for Dealing with Ground Truth Sparsity via Curriculum Learning. arXiv preprint arXiv:2211.08608.

[Konečný '18] Konečný, J., McMahan, H. B., Ramage, D., & Richtárik, P. (2016). Federated optimization: Distributed machine learning for on-device intelligence. arXiv preprint arXiv:1610.02527.

[McMahan '17] McMahan, B., Moore, E., Ramage, D., Hampson, S., & y Arcas, B. A. (2017). Communication-efficient learning of deep networks from decentralized data. In Artificial intelligence and statistics (pp. 1273-1282). PMLR.

[Acar '21] Acar, D. A. E., Zhao, Y., Navarro, R. M., Mattina, M., Whatmough, P. N., & Saligrama, V. (2021). Federated learning based on dynamic regularization. arXiv preprint arXiv:2111.04263.

[Rawassizadeh '18] Rawassizadeh, R., Pierson, T. J., Peterson, R., & Kotz, D. (2018). NoCloud: Exploring network disconnection through on-device data analysis. IEEE Pervasive Computing, 17(1), 64-74.

[Cheng '17] Cheng, Y., Wang, D., Zhou, P., & Zhang, T. (2017). A survey of model compression and acceleration for deep neural networks. arXiv preprint arXiv:1710.09282.

[Liu '22] Liu, Y., Cao, J., Li, B., Hu, W., Ding, J., & Li, L. (2022). Cross-Architecture Knowledge Distillation. In Proceedings of the Asian Conference on Computer Vision (pp. 3396-3411).

[Wang '18] Wang, H., Zhao, H., Li, X., & Tan, X. (2018). Progressive Blockwise Knowledge Distillation for Neural Network Acceleration. In IJCAI (pp. 2769-2775).

[Li '20] Li, C., Peng, J., Yuan, L., Wang, G., Liang, X., Lin, L., & Chang, X. (2020). Block-wisely supervised neural architecture search with knowledge distillation. In Proceedings of the IEEE/CVF Conference on Computer Vision and Pattern Recognition (pp. 1989-1998).

[Romero '14] Romero, A., Ballas, N., Kahou, S. E., Chassang, A., Gatta, C., & Bengio, Y. (2014). Fitnets: Hints for thin deep nets. arXiv preprint arXiv:1412.6550.

[Chen '17] Chen, G., Choi, W., Yu, X., Han, T., & Chandraker, M. (2017). Learning efficient object detection models with knowledge distillation. Advances in neural information processing systems, 30.

[Elsken '19] Elsken, T., Metzen, J. H., & Hutter, F. (2019). Neural architecture search: A survey. The Journal of Machine Learning Research, 20(1), 1997-2017.

[Gholami '21] Gholami, A., Kim, S., Dong, Z., Yao, Z., Mahoney, M. W., & Keutzer, K. (2021). A survey of quantization methods for efficient neural network inference. arXiv preprint arXiv:2103.13630.

[Hoefler '21] Hoefler, T., Alistarh, D., Ben-Nun, T., Dryden, N., & Peste, A. (2021). Sparsity in deep learning: Pruning and growth for efficient inference and training in neural networks. The Journal of Machine Learning Research, 22(1), 10882-11005.

[Howard '17] Howard, A. G., Zhu, M., Chen, B., Kalenichenko, D., Wang, W., Weyand, T., ... & Adam, H. (2017). Mobilenets: Efficient convolutional neural networks for mobile vision applications. *arXiv preprint arXiv:1704.04861.*

[Iandola '16] Iandola, F. N., Han, S., Moskewicz, M. W., Ashraf, K., Dally, W. J., & Keutzer, K. (2016). SqueezeNet: AlexNet-level accuracy with 50x fewer parameters and< 0.5 MB model size. *arXiv preprint arXiv:1602.07360.*

[Tan '19] Tan, M., & Le, Q. (2019). Efficientnet: Rethinking model scaling for convolutional neural networks. In *International conference on machine learning* (pp. 6105-6114). PMLR.

[Mehta '21] Mehta, S., & Rastegari, M. (2021). Mobilevit: light-weight, general-purpose, and mobile-friendly vision transformer. *arXiv preprint arXiv:2110.02178.*

——————— Chapter 15 ———————

[Salathe '11] Salathé, M., & Khandelwal, S. (2011). Assessing vaccination sentiments with online social media: implications for infectious disease dynamics and control. *PLoS computational biology*, 7(10), e1002199.

[Collier '11] Collier, N., Son, N. T., & Nguyen, N. M. (2011). OMG U got flu? Analysis of shared health messages for bio-surveillance. *Journal of biomedical semantics*, 2(5), 1-10.

[Chen '11] Chen, Y., Fay, S., & Wang, Q. (2011). The role of marketing in social media: How online consumer reviews evolve. *Journal of interactive marketing*, 25(2), 85-94.

[Gan '11] Gan, Y., Wang, T., Javaheri, A., Momeni-Ortner, E., Dehghani, M., Hosseinzadeh, M., & Rawassizadeh, R. (2021). 11 Years with Wearables: Quantitative Analysis of Social Media, Academia, News Agencies, and Lead User Community from 2009-2020 on Wearable Technologies. *Proceedings of the ACM on Interactive, Mobile, Wearable and Ubiquitous Technologies*, 5(1), 1-26.

[Bradshaw '17] Bradshaw, S., & Howard, P. (2017). Troops, trolls and troublemakers: A global inventory of organized social media manipulation. *Computational Propaganda Research Project.*

[Jarník '30] Jarník, V. (1930), "O jistém problému minimálním" [About a certain minimal problem], Práce Moravské Přírodovědecké Společnosti (in Czech), 6 (4): 57–63.

[Prim '57] Prim, R. C. (1957). Shortest connection networks and some generalizations. The Bell System Technical Journal, 36(6), 1389-1401.

[Kruskal '56] Kruskal, J. B. (1956). "On the shortest spanning subtree of a graph and the traveling salesman problem". Proceedings of the American Mathematical Society. 7 (1): 48–50.

[Dijkstra '59] Dijkstra, E. W. (1959). "A note on two problems in connexion with graphs" (PDF). Numerische Mathematik. 1: 269–271.

[Hart '68] Hart, P. E., Nilsson, N. J., & Raphael, B. (1968). A formal basis for the heuristic determination of minimum cost paths. IEEE Transactions on Systems Science and Cybernetics, 4(2), 100-107.

[Kuhn '55] Kuhn, H. W. (1955). The Hungarian method for the assignment problem. Naval research logistics quarterly, 2(1-2), 83-97.

[Hopcroft '73] Hopcroft, J. E., & Karp, R. M. (1973). An n^5/2 algorithm for maximum matchings in bipartite graphs. SIAM Journal on Computing, 2(4), 225-231.

[Bellman '58] Bellman, Richard (1958). "On a routing problem". Quarterly of Applied Mathematics. 16: 87–90.

[Ford '56] Ford, Lester R. Jr. (August 14, 1956). Network Flow Theory. Paper P-923. Santa Monica, California: RAND Corporation.

[Morselli '08] Morselli, C., & Roy, J. (2008). Brokerage qualifications in ringing operations. Criminology, 46(1), 71-98.

[Kazerani '09] Kazerani, A., & Winter, S. (2009). Can betweenness centrality explain traffic flow. In 12th AGILE international conference on geographic information science (pp. 1-9).

[Bonacich '87] Bonacich, P. (1987). Power and centrality: A family of measures. American journal of sociology, 92(5), 1170-1182.

[Page '99] Page, L., Brin, S., Motwani, R., & Winograd, T. (1999). *The PageRank citation ranking: Bringing order to the web.* Stanford InfoLab.

[Kleinberg '99] Kleinberg, J. M. (1999). Authoritative sources in a hyperlinked environment. *Journal of the ACM (JACM)*, 46(5), 604-632.

[Cheeger '70] Cheeger, J. (1970). A lower bound for the smallest eigenvalue of the Laplacian, Problems in analysis (Papers dedicated to Salomon Bochner, 1969).

[Donath '72] Donath, W. E., & Hoffman, A. J. (1972). Algorithms for partitioning of graphs and computer logic based on eigenvectors of connection matrices. *IBM Technical Disclosure Bulletin*, 15(3), 938-944.

[Fiedler '73] Fiedler, M. (1973). Algebraic connectivity of graphs. *Czechoslovak Mathematical Journal, 23*(2), 298-305.

[Blondel '08] Blondel, V. D., Guillaume, J. L., Lambiotte, R., & Lefebvre, E. (2008). Fast unfolding of communities in large networks. *Journal of statistical mechanics: theory and experiment, 2008*(10), P10008.

[Traag '19] Traag, V. A., Waltman, L., & van Eck, N. J. (2019). From Louvain to Leiden: guaranteeing well-connected communities. *Scientific reports, 9*(1), 1-12.

[Traag '11] Traag, V. A., Van Dooren, P., & Nesterov, Y. (2011). Narrow scope for resolution-limit-free community detection. *Physical Review E, 84*(1), 016114.

[Bruna '14] Bruna, W. Zaremba, and Y. Szlam, A.and LeCun (2014). Spectral networks and locally connected networks on graphs. In International Conference on Learning Representations (ICLR).

[Merkwirth '05] Merkwirth, C., & Lengauer, T. (2005). Automatic generation of complementary descriptors with molecular graph networks. Journal of chemical information and modeling, 45(5), 1159-1168.

[Hamilton '20] Hamilton, W. L. (2020). Graph representation learning. Synthesis Lectures on Artificial Intelligence and Machine Learning, 14(3), 1-159.

[Grattarola '22] Grattarola, D., Zambon, D., Bianchi, F. M., & Alippi, C. (2022). Understanding pooling in graph neural networks. IEEE Transactions on Neural Networks and Learning Systems.

[Lee '19] Lee, J., Lee, I., & Kang, J. (2019). Self-attention graph pooling. In International conference on machine learning (pp. 3734-3743). PMLR.

[Liu '22] Liu, C., Zhan, Y., Li, C., Du, B., Wu, J., Hu, W., ... & Tao, D. (2022). Graph pooling for graph neural networks: Progress, challenges, and opportunities. arXiv preprint arXiv:2204.07321.

[Kipf '16] Kipf, T. N., & Welling, M. (2016). Semi-supervised classification with graph convolutional networks. arXiv preprint arXiv:1609.02907.

[Veličković '17] Veličković, P., Cucurull, G., Casanova, A., Romero, A., Lio, P., & Bengio, Y. (2017). Graph attention networks. arXiv preprint arXiv:1710.10903.

[Hamilton '17] Hamilton, W., Ying, Z., & Leskovec, J. (2017). Inductive representation learning on large graphs. Advances in neural information processing systems, 30.

[Zhou '20] Zhou, J., Cui, G., Hu, S., Zhang, Z., Yang, C., Liu, Z., ... & Sun, M. (2020). Graph neural networks: A review of methods and applications. AI open, 1, 57-81.

[Malkov '18] Malkov, Y. A., & Yashunin, D. A. (2018). Efficient and robust approximate nearest neighbor search using hierarchical navigable small-world graphs. IEEE transactions on pattern analysis and machine intelligence, 42(4), 824-836.

─────── Chapter 16 ───────

[Griffin '84] Griffin, D., & Lim, J. (1984). Signal estimation from modified short-time Fourier transform. *IEEE Transactions on Acoustics, Speech, and Signal Processing*, 32(2), 236-243.

[Metropolis '49] Metropolis, N., & Ulam, S. (1949). The Monte Carlo method. *Journal of the American Statistical Association. 44* (247): 335–341.

[Metropolis '53] Metropolis, N., Rosenbluth, A. W., Rosenbluth, M. N., Teller, A. H., & Teller, E. (1953). Equation of state calculations by fast computing machines. *The journal of Chemical Physics*, 21(6), 1087-1092.

[Hastings '70] Hastings, W. K. (1970). Monte Carlo sampling methods using Markov chains and their applications.

[Geman '84] Geman, S.; Geman, D. (1984). "Stochastic Relaxation, Gibbs Distributions, and the Bayesian Restoration of Images". *IEEE Transactions on Pattern Analysis and Machine Intelligence*. 6 (6): 721–741.

[Kloek '84] Kloek, T., & Van Dijk, H. K. (1978). Bayesian estimates of equation system parameters: an application of integration by Monte Carlo. Econometrica: *Journal of the Econometric Society*, 1-19.

[Wan '22] Wan, Z., Zhang, B., Chen, D., Zhang, P., Wen, F., & Liao, J. (2022). Old Photo Restoration via Deep Latent Space Translation. *IEEE Transactions on Pattern Analysis and Machine Intelligence*, 45(2), 2071-2087.

[Perlin '85] Perlin, K. (1985). An image synthesizer. *ACM Siggraph Computer Graphics*, 19(3), 287-296.

[Perlin '01] Perlin, K. (2001). *Noise hardware.* In *Real-Time Shading SIGGRAPH Course Notes*, 6, 26.

[Worley '96] Worley, S. (1996). A cellular texture basis function. *In Proceedings of the 23rd Annual Conference on Computer Graphics and Interactive Techniques* (pp. 291-294).

[Butterworth '30] Butterworth, S. (1930). On the theory of filter amplifiers. *Wireless Engineer*, 7(6), 536-541.

[Wiener '49] Wiener, N. (1949). Extrapolation, interpolation, and smoothing of stationary time series: with engineering applications. *The MIT Press*.

[Kalman '60] Kalman, R. E. (1960). A New Approach to Linear Filtering and Prediction Problems. *Transactions of the ASME--Journal of Basic Engineering*, 82(D), 35-45.

[Lin '17] Lin, T. Y., Goyal, P., Girshick, R., He, K., & Dollár, P. (2017). Focal loss for dense object detection. *In Proceedings of the IEEE International Conference on Computer Vision* (pp. 2980-2988).

[Chawla '02] Chawla, N. V., Bowyer, K. W., Hall, L. O., & Kegelmeyer, W. P. (2002). SMOTE: synthetic minority over-sampling technique. *Journal of Artificial Intelligence Research*, 16, 321-357.

[Bunkhumpornpat '09] Bunkhumpornpat, C., Sinapiromsaran, K., & Lursinsap, C. (2009). Safe-level-smote: Safe-level-synthetic minority over-sampling technique for handling the class imbalanced problem. *In Advances in Knowledge Discovery and Data Mining: 13th Pacific-Asia conference, PAKDD 2009*.

[He '08] He, H., Bai, Y., Garcia, E. A., & Li, S. (2008). ADASYN: Adaptive synthetic sampling approach for imbalanced learning. *In 2008 IEEE International Joint Conference on Neural Networks* (pp. 1322-1328).

[Bissoto '21] Bissoto, A., Valle, E., & Avila, S. (2021). Gan-based data augmentation and anonymization for skin-lesion analysis: A critical review. *In Proceedings of the IEEE/CVF conference on computer vision and pattern recognition* (pp. 1847-1856).

[Yu '23] Yu, X., Li, G., Lou, W., Liu, S., Wan, X., Chen, Y., & Li, H. (2023). Diffusion-based data augmentation for nuclei image segmentation. *In International Conference on Medical Image Computing and Computer-Assisted Intervention* (pp. 592-602).

[Wan '21] Wan, Z., Zhang, Y., & He, H. (2017). Variational autoencoder based synthetic data generation for imbalanced learning. *In 2017 IEEE Symposium Series on Computational Intelligence (SSCI)* (pp. 1-7).

[Batista '23] Batista, G. E., & Monard, M. C. (2003). An analysis of four missing data treatment methods for supervised learning. *Applied Artificial Intelligence*, 17(5-6), 519-533.

[Rawassizadeh '19] Rawassizadeh, R., Keshavarz, H., & Pazzani, M. (2019). Ghost imputation: Accurately reconstructing missing data of the off period. *IEEE Transactions on Knowledge and Data Engineering*, 32(11), 2185-2197.

[Stekhoven '12] Stekhoven, D. J., & Bühlmann, P. (2012). MissForest - non-parametric missing value imputation for mixed-type data. *Bioinformatics*, 28(1), 112-118.

[Oliver '90] Oliver, M. A., & Webster, R. (1990). Kriging: a method of interpolation for geographical information systems. *International Journal of Geographical Information System*, 4(3), 313-332.

[Hardy '71] Hardy, R. L. (1971). Multiquadric equations of topography and other irregular surfaces. Journal of *Geophysical Research*, 76(8), 1905-1915.

[Liu '08] Liu, F. T., Ting, K. M., & Zhou, Z. H. (2008). Isolation forest. *In 8th IEEE International Conference on Data Mining* (pp. 413-422).

[Schölkopf '01] Schölkopf, B., Platt, J. C., Shawe-Taylor, J., Smola, A. J., & Williamson, R. C. (2001). Estimating the Support of a High-Dimensional Distribution. *Neural Computation*, 13(7), 1443-1471.

[Breunig '00] Breunig, M. M., Kriegel, H. P., Ng, R. T., & Sander, J. (2000). LOF: identifying density-based local outliers. In Proceedings of the 2000 ACM SIGMOD international conference on Management of data (pp. 93-104).

[Fischler '81] Fischler, M. A., & Bolles, R. C. (1981). Random Sample Consensus: A Paradigm for Model Fitting with Applications to Image Analysis and Automated Cartography. *Communications of the ACM, 24*(6), 381-395.

[Bifet '07] Bifet, A., & Gavalda, R. (2007). Learning from time-changing data with adaptive windowing. *In Proceedings of the 2007 SIAM International Conference on Data Mining* (pp. 443-448).

[Cohen '60] Cohen, J. (1960). A coefficient of agreement for nominal scales. *Educational and psychological measurement*, 20(1), 37-46.

[Fleiss '71] Fleiss, J. L. (1971). Measuring nominal scale agreement among many raters. *Psychological bulletin*, 76(5), 378.

[Krippendorff '18] Krippendorff, K. (2018). Content analysis: An introduction to its methodology. Sage publications.

[Lewis '20] Lewis, P., Perez, E., Piktus, A., Petroni, F., Karpukhin, V., Goyal, N., ... & Kiela, D. (2020). Retrieval-Augmented Generation for Knowledge-Intensive NLP Tasks. *Advances in Neural Information Processing Systems*, 33, 9459-9474.

[Ji '23] Ji, X., Sungu-Eryilmaz, Y., Momeni, E., & Rawassizadeh, R. (2023). Speeding Up Question Answering Task of Language Models via Inverted Index. *The Generative AI for Pervasive Computing Symposium (GenAI4PC), in conjunction with ACM Interactive, Mobile, Wearable and Ubiquitous Technologies.* arXiv:2210.13578.

[Wei '22] Wei, J., Wang, X., Schuurmans, D., Bosma, M., Xia, F., Chi, E., ... & Zhou, D. (2022). Chain-of-thought prompting elicits reasoning in large language models. *Advances in Neural Information Processing Systems*, 35, 24824-24837.

[Brown '20] Brown, T., Mann, B., Ryder, N., Subbiah, M., Kaplan, J. D., Dhariwal, P., ... & Amodei, D. (2020). Language Models are Few-shot Learners. *Advances in Neural Information Processing Systems*, 33, 1877-1901.

[Wang '22] Wang, X., Wei, J., Schuurmans, D., Le, Q., Chi, E., Narang, S., ... & Zhou, D. (2022). Self-consistency Improves the Chain of Thought Reasoning in Language Models. *arXiv preprint arXiv:2203.11171.*

[Yao '23] Yao, S., Yu, D., Zhao, J., Shafran, I., Griffiths, T. L., Cao, Y., & Narasimhan, K. (2023). Tree of Thoughts: Deliberate Problem Solving with Large Language Models. *arXiv preprint arXiv:2305.10601.*

[Long '23] Long, J. (2023). Large Language Model Guided Tree-of-Thought. *arXiv preprint arXiv:2305.08291.*

[Kahneman '11] Kahneman, D. (2011). Thinking, fast and slow. MacMillan.

[Bansal '18] Bansal, N., Chen, X., & Wang, Z. (2018). Can we gain more from orthogonality regularizations in training deep networks?. Advances in Neural Information Processing Systems, 31.

[Lundberg '17] Lundberg, S. M., & Lee, S. I. (2017). A unified approach to interpreting model predictions. Advances in neural information processing systems, 30.

[Shapley '51] Shapley, L. S. (1953). A value for n-person games. *In H. W. Kuhn & A. W. Tucker (Eds.), Contributions to the Theory of Games* (Vol. 2, pp. 307-317). Princeton University Press.

[Ribeiro '16] Ribeiro, M. T., Singh, S., & Guestrin, C. (2016). " Why should I trust you?" Explaining the predictions of any classifier. In *Proceedings of the 22nd ACM SIGKDD international conference on Knowledge Discovery and Data mining* (pp. 1135-1144).

Index

www.ingramcontent.com/pod-product-compliance
Lightning Source LLC
Chambersburg PA
CBHW060925210326
41597CB00042B/4464